DIGITAL COMMUNICATION SYSTEMS

Simon Haykin
McMaster University

WILEY

ASSOCIATE PUBLISHER	Daniel Sayre
EDITORIAL ASSISTANT	Jessica Knecht
MARKETING MANAGER	Christopher Ruel
PRODUCTION MANAGEMENT SERVICES	Publishing Services
CREATIVE DIRECTOR	Harry Nolan
COVER DESIGNER	Kristine Carney

Cover Image: The figure on the cover, depicting the UMTS-turbo code, is adapted from the doctoral thesis of Dr. Liang Li, Department of Electronics and Computer Science, University of Southampton, United Kingdom, with the permission of Dr. Li, his Supervisor Dr. Robert Maunder, and Professor Lajos Hanzo; the figure also appears on page 654 of the book.

This book was set in Times by Publishing Services and printed and bound by RR Donnelley. The cover was printed by RR Donnelley.

This book is printed on acid free paper. ∞

ISBN: 978-0-471-64735-5

Printed in the United States of America

10 9 8 7 6 5 4 3 2 1

In loving memory of
Vera

Preface

The study of digital communications is an essential element of the undergraduate and postgraduate levels of present-day electrical and computer engineering programs. This book is appropriate for both levels.

A Tour of the Book

The introductory chapter is motivational, beginning with a brief history of digital communications, and continuing with sections on the communication process, digital communications, multiple-access and multiplexing techniques, and the Internet. Four themes organize the remaining nine chapters of the book.

Theme 1 *Mathematics of Digital Communications*

The first theme of the book provides a detailed exposé of the mathematical underpinnings of digital communications, with *continuous mathematics* aimed at the communication channel and interfering signals, and *discrete mathematics* aimed at the transmitter and receiver:

- Chapter 2, *Fourier Analysis of Signals and Systems*, lays down the fundamentals for the representation of signals and linear time-invariant systems, as well as analog modulation theory.
- Chapter 3, *Probability Theory and Bayesian Inference*, presents the underlying mathematics for dealing with uncertainty and the Bayesian paradigm for probabilistic reasoning.
- Chapter 4, *Stochastic Processes*, focuses on weakly or wide-sense stationary processes, their statistical properties, and their roles in formulating models for Poisson, Gaussian, Rayleigh, and Rician distributions.
- Chapter 5, *Information Theory*, presents the notions of entropy and mutual information for discrete as well continuous random variables, leading to Shannon's celebrated theorems on source coding, channel coding, and information capacity, as well as rate-distortion theory.

Theme 2 *From Analog to Digital Communications*

The second theme of the book, covered in Chapter 6, describes how analog waveforms are transformed into coded pulses. It addresses the challenge of performing the transformation with robustness, bandwidth preservation, or minimal computational complexity.

Theme 3 *Signaling Techniques*

Three chapters address the third theme, each focusing on a specific form of *channel impairment*:

- In Chapter 7, *Signaling over Additive White Gaussian Noise (AWGN) Channels,* the impairment is the unavoidable presence of *channel noise*, which is modeled as *additive white Gaussian noise* (AWGN). This model is well-suited for the *signal-space diagram*, which brings insight into the study of phase-shift keying (PSK), quadrature-amplitude modulation (QAM), and frequency-shift keying (FSK) as different ways of accommodating the transmission and reception of binary data.

- In Chapter 8, *Signaling over Band-Limited Channels,* bandwidth limitation assumes center stage, with *intersymbol interference* (ISI) as the source of channel impairment.
- Chapter 9, *Signaling over Fading Channels,* focuses on *fading channels* in wireless communications and the practical challenges they present. The channel impairment here is attributed to the *multipath phenomenon*, so called because the transmitted signal reaches the receiver via a multiplicity of paths.

Theme 4 *Error-control Coding*

Chapter 10 addresses the practical issue of *reliable communications*. To this end, various techniques of the feedforward variety are derived therein, so as to satisfy Shannon's celebrated *coding theorem.*

Two families of error-correcting codes are studied in the chapter:

- *Legacy (classic) codes*, which embody linear block codes, cyclic codes, and convolutional codes. Although different in their structural compositions, they look to algebraic mathematics as the procedure for approaching the *Shannon limit.*
- *Probabilistic compound codes*, which embody turbo codes and low-density parity-check (LDPC) codes. What is remarkable about these two codes is that they both approach the Shannon limit with doable computational complexity in a way that was not feasible until 1993. The trick behind this powerful information-processing capability is the adoption of *random codes*, the origin of which could be traced to Shannon's 1948 classic paper.

Features of the Book

Feature 1 *Analog in Digital Communication*

When we think of digital communications, we must not overlook the fact that such a system is of a *hybrid nature*. The channel across which data are transmitted is analog, exemplified by traditional telephone and wireless channels, and many of the sources responsible for the generation of data (e.g., speech and video) are of an analog kind. Moreover, certain principles of analog modulation theory, namely double sideband-suppressed carrier (DSB-SC) and vestigial sideband (VSB) modulation schemes, include binary phase-shift keying (PSK) and offset QPSK as special cases, respectively.

It is with these points in mind that Chapter 2 includes

- detailed discussion of communication channels as examples of linear systems,
- analog modulation theory, and
- phase and group delays.

Feature 2 *Hilbert Transform*

The Hilbert transform, discussed in Chapter 2, plays a key role in the complex representation of signals and systems, whereby

- a band-pass signal, formulated around a sinusoidal carrier, is transformed into an equivalent complex low-pass signal;

- a band-pass system, be it a linear channel or filter with a midband frequency, is transformed into an equivalent complex low-pass system.

Both transformations are performed without loss of information, and their use changes a difficult task into a much simpler one in mathematical terms, suitable for simulation on a computer. However, one must accommodate the use of complex variables.

The Hilbert transform also plays a key role in Chapter 7. In formulating the *method of orthogonal modulation*, we show that one can derive the well-known formulas for the noncoherent detection of binary frequency-shift keying (FSK) and differential phase-shift keying (DPSK) signals, given unknown phase, in a much simpler manner than following traditional approaches that involve the use of Rician distribution.

Feature 3 *Discrete-time Signal Processing*

In Chapter 2, we briefly review *finite-direction impulse response* (*FIR*) or *tapped-delay line* (*TDL*) *filters*, followed by the *discrete Fourier transform* (*DFT*) and a well-known *fast Fourier transform* (*FFT*) algorithm for its computational implementations. FIR filters and FFT algorithms feature prominently in:

- Modeling of the *raised-cosine spectrum* (*RCS*) and its square-root version (SQRCS), which are used in Chapter 8 to mitigate the ISI in band-limited channels;
- Implementing the *Jakes model* for fast fading channels, demonstrated in Chapter 9;
- Using FIR filtering to simplify the mathematical exposition of the most difficult form of channel fading, namely, the *doubly spread channel* (in Chapter 9).

Another topic of importance in discrete-time signal processing is *linear adaptive filtering*, which appears:

- In Chapter 6, dealing with *differential pulse-code modulation* (*DPCM*), where an adaptive predictor constitutes a key functional block in both the transmitter and receiver. The motivation here is to preserve channel bandwidth at the expense of increased computational complexity. The algorithm described therein is the widely used *least mean-square* (*LMS*) *algorithm*.
- In Chapter 7, dealing with the need for *synchronizing* the receiver to the transmitter, where two algorithms are described, one for recursive estimation of the group delay (essential for timing recovery) and the other for recursive estimation of the unknown carrier phase (essential for carrier recovery). Both algorithms build on the LMS principle so as to maintain linear computational complexity.

Feature 4 *Digital Subscriber Lines*

Digital subscriber lines (DSLs), covered in Chapter 8, have established themselves as an essential tool for transforming a linear wideband channel, exemplified by the twisted-wire pair, into a *discrete multitone* (*DMT*) *channel* that is capable of accommodating data transmission at multiple megabits per second. Moreover, the transformation is afforded practical reality by exploiting the FFT algorithm, with the inverse FFT used in the transmitter and the FFT used in the receiver.

Feature 5 *Diversity Techniques*

As already mentioned, the wireless channel is one of the most challenging media for digital communications. The difficulty of reliable data transmission over a wireless channel is attributed to the multipath phenomenon. Three diversity techniques developed to get around this practical difficulty are covered in Chapter 9:

- *Diversity on receive,* the traditional approach, whereby an array of multiple antennas operating independently is deployed at the receiving end of a wireless channel.
- *Diversity on transmit*, which operates by deploying two or more independent antennas at the transmit end of the wireless channel.
- *Multiple-input multiple-output* (*MIMO*) *channels*, where multiple antennas (again operating independently) are deployed at both ends of the wireless channel.

Among these three forms of diversity, the MIMO channel is naturally the most powerful in information-theoretic terms: an advantage gained at the expense of increased computational complexity.

Feature 6 *Turbo Codes*

Error-control coding has established itself as the most commonly used technique for *reliable* data transmission over a noisy channel. Among the challenging legacies bestowed by Claude Shannon was how to design a code that would closely approach the so-called *Shannon limit*. For over four decades, increasingly more powerful coding algorithms were described in the literature; however it was the *turbo code* that had the honor of closely approaching the Shannon limit, and doing so in a computationally feasible manner.

Turbo codes, together with the associated *maximum a posteriori (MAP) decoding algorithm*, occupy a large portion of Chapter 10, which also includes:

- Detailed derivation of the MAP algorithm and an illustrative example of how it operates;
- The *extrinsic information transfer* (*EXIT*) *chart*, which provides an experimental tool for the design of turbo codes;
- *Turbo equalization*, for demonstrating applicability of the turbo principle beyond error-control coding.

Feature 7 *Placement of Information Theory*

Typically, information theory is placed just before the chapter on error-control coding. In this book, it is introduced early because:

> Information theory is not only of basic importance to error-control coding but also other topics in digital communications.

To elaborate:

- Chapter 6 presents the relevance of *source coding* to pulse-code modulation (PCM), differential pulse-code modulation (DPCM), and delta modulation.
- Comparative evaluation of *M*-ary PSK versus *M*-ary FSK, done in Chapter 7, requires knowledge of *Shannon's information capacity law.*
- Analysis and design of *DSL*, presented in Chapter 8, also builds on Shannon's information capacity law.
- *Channel capacity* in Shannon's coding theorem is important to diversity techniques, particularly of the MIMO kind, discussed in Chapter 9.

Examples, Computer Experiments, and Problems

Except for Chapter 1, each of the remaining nine chapters offers the following:

- Illustrative examples are included to strengthen the understanding of a theorem or topic in as much detail as possible. Some of the examples are in the form of computer experiments.
- An extensive list of end-of-chapter problems are grouped by section to fit the material covered in each chapter. The problems range from relatively easy ones all the way to more challenging ones.
- In addition to the computer-oriented examples, nine computer-oriented experiments are included in the end-of-chapter problems.

The Matlab codes for all of the computer-oriented examples in the text, as well as other calculations performed on the computer, are available at www.wiley.com/college/haykin.

Appendices

Eleven appendices broaden the scope of the theoretical as well as practical material covered in the book:

- Appendix A, *Advanced Probabilistic Models*, covers the chi-square distribution, log-normal distribution, and Nakagami distribution that includes the Rayleigh distribution as a special case and is somewhat similar to the Rician distribution. Moreover, an experiment is included therein that demonstrates, in a step-by-step manner, how the Nakagami distribution evolves into the log-normal distribution in an approximate manner, demonstrating its adaptive capability.
- Appendix B develops tight bounds on the *Q-function.*
- Appendix C discussed the ordinary Bessel function and its modified form.
- Appendix D describes the method of Lagrange multipliers for solving constrained optimization problems.
- Appendix E derives the formula for the *channel capacity of the MIMO channel* under two scenarios: one that assumes no knowledge of the channel by the transmitter, and the other that assumes this knowledge is available to the transmitter via a narrowband feedback link.
- Appendix F discusses the idea of *interleaving*, which is needed for dealing with bursts of interfering signals experienced in wireless communications.
- Appendix G addresses the *peak-to-average power reduction (PAPR) problem*, which arises in the use of orthogonal frequency-division multiplexing (OFDM) for both wireless and DSL applications.
- Appendix H discusses *solid-state nonlinear power amplifiers*, which play a critical role in the limited life of batteries in wireless communications.
- Appendix I presents a short exposé of *Monte Carlo integration:* a theorem that deals with mathematically intractable problems.
- Appendix J studies *maximal-length sequences*, also called *m-sequences*, which are used for implementing linear feedback shift registers (LFSRs). An important application of maximal-length sequences (viewed as pseudo-random noise) is in

designing direct-sequence spread-spectrum communications for code-division multiple access (CDMA).

- Finally, Appendix K provides a useful list of *mathematical formulas and functions*.

Two Noteworthy Symbols

Typically, the square-root of minus one is denoted by the italic symbol *j*, and the differential operator (used in differentiation as well as integration) is denoted by the italic symbol *d*. In reality, however, both of these terms are *operators*, each one in its own way: it is therefore incorrect to use italic symbols for their notations. Furthermore, italic *j* and italic *d* are also frequently used as indices or to represent other matters, thereby raising the potential for confusion. According, throughout the book, *roman* j and *roman* d are used to denote the square root of minus one and the differential operator, respectively.

Concluding Remarks

In writing this book every effort has been made to present the material in the manner easiest to read so as to enhance understanding of the topics covered. Moreover, cross-references within a chapter as well as from chapter to chapter have been included wherever the need calls for it.

Finally, every effort has been made by the author as well as compositor of the book to make it as error-free as humanly possible. In this context, the author would welcome receiving notice of any errors discovered after publication of the book.

Acknowledgements

In writing this book I have benefited enormously from technical input, persistent support, and permissions provided by many.

I am grateful to colleagues around the world for technical inputs that have made a significant difference in the book; in alphabetical order, they are:

- Dr. Daniel Costello, Jr., *University of Notre Dame*, for reading and providing useful comments on the maximum likelihood decoding and maximum a posteriori decoding materials in Chapter 10.
- Dr. Dimitri Bertsekas, *MIT*, for permission to use Table 3.1 on the *Q*-function in Chapter 3, taken from his co-authored book on the theory of probability.
- Dr. Lajos Hanzo, *University of Southampton*, UK, for many useful comments on turbo codes as well as low-density parity-check codes in Chapter 10. I am also indebted to him for putting me in touch with his colleagues at the University of Southampton, Dr. R. G. Maunder and Dr. L. Li, who were extremely helpfully in performing the insightful computer experiments on UMTS-turbo codes and EXIT charts in Chapter 10.
- Dr. Phillip Regalia, *Catholic University*, Washington DC, for contributing a section on serial-concatenated turbo codes in Chapter 10. This section has been edited by myself to follow the book's writing style, and for its inclusion I take full responsibility.
- Dr. Sam Shanmugan, *University of Kansas*, for his insightful inputs on the use of FIR filters and FFT algorithms for modeling the raised-cosine spectrum (RCS) and

its square-root version (SQRCS) in Chapter 8, implementing the Jakes model in Chapter 9, as well as other simulation-oriented issues.

- Dr. Yanbo Xue, *University of Alberta*, Canada, for performing computer-oriented experiments and many other graphical computations throughout the book, using well-developed Matlab codes.
- Dr. Q. T. Zhang, *The City University of Hong Kong*, for reading through an early version of the manuscript and offering many valuable suggestions for improving it. I am also grateful to his student, Jiayi Chen, for performing the graphical computations on the Nakagami distribution in Appendix A.

I'd also like to thank the reviewers who read drafts of the manuscript and provided valuable commentary:

- Ender Ayanoglu, *University of California, Irvine*
- Tolga M. Duman, *Arizona State University*
- Bruce A. Harvey, *Florida State University*
- Bing W. Kwan, *FAMU-FSU College of Engineering*
- Chung-Chieh Lee, *Northwestern University*
- Heung-No Lee, *University of Pittsburgh*
- Michael Rice, *Brigham Young University*
- James Ritcey, *University of Washington*
- Lei Wei, *University of Central Florida*

Production of the book would not have been possible without the following:

- Daniel Sayre, Associate Publisher at John Wiley & Sons, who maintained not only his faith in this book but also provided sustained support for it over the past few years. In am deeply indebted to Dan for what he has done to make this book a reality.
- Cindy Johnson, Publishing Services, Newburyport, MA, for her dedicated commitment to the beautiful layout and composition of the book. I am grateful for her tireless efforts to print the book in as errorless manner as humanly possible.

I salute everyone, and others too many to list, for their individual and collective contributions, without which this book would not have been a reality.

Simon Haykin
Ancaster, Ontario
Canada
December, 2012

Contents

CHAPTER

1

Introduction

1.1 Historical Background

In order to provide a sense of motivation, this introductory treatment of digital communications begins with a historical background of the subject, brief but succinct as it may be. In this first section of the introductory chapter we present some historical notes that identify the pioneering contributors to digital communications specifically, focusing on three important topics: information theory and coding, the Internet, and wireless communications. In their individual ways, these three topics have impacted digital communications in revolutionary ways.

Information Theory and Coding

In 1948, the theoretical foundations of digital communications were laid down by Claude Shannon in a paper entitled "A mathematical theory of communication." Shannon's paper was received with immediate and enthusiastic acclaim. It was perhaps this response that emboldened Shannon to amend the title of his classic paper to "The mathematical theory of communication" when it was reprinted later in a book co-authored with Warren Weaver. It is noteworthy that, prior to the publication of Shannon's 1948 classic paper, it was believed that increasing the rate of transmission over a channel would increase the probability of error; the communication theory community was taken by surprise when Shannon proved that this was not true, provided the transmission rate was below the channel capacity.

Shannon's 1948 paper was followed by three ground-breaking advances in coding theory, which include the following:

1. Development of the first nontrivial error-correcting code by Golay in 1949 and Hamming in 1950.
2. Development of turbo codes by Berrou, Glavieux and Thitimjshima in 1993; turbo codes provide near-optimum error-correcting coding and decoding performance in additive white Gaussian noise.
3. Rediscovery of *low-density parity-check (LDPC) codes*, which were first described by Gallager in 1962; the rediscovery occurred in 1981 when Tanner provided a new interpretation of LDPC codes from a graphical perspective. Most importantly, it was the discovery of turbo codes in 1993 that reignited interest in LDPC codes.

The Internet

From 1950 to 1970, various studies were made on computer networks. However, the most significant of them all in terms of impact on computer communications was the Advanced Research Project Agency Network (ARPANET), which was put into service in 1971. The development of ARPANET was sponsored by the Advanced Research Projects Agency (ARPA) of the United States Department of Defense. The pioneering work in *packet switching* was done on the ARPANET. In 1985, ARPANET was renamed the *Internet.* However, the turning point in the evolution of the Internet occurred in 1990 when Berners-Lee proposed a hypermedia software interface to the Internet, which he named the *World Wide Web.* Thereupon, in the space of only about 2 years, the Web went from nonexistence to worldwide popularity, culminating in its commercialization in 1994. The Internet has dramatically changed the way in which we communicate on a daily basis, using a wirelined network.

Wireless Communications

In 1864, James Clerk Maxwell formulated the *electromagnetic theory of light* and predicted the existence of radio waves; the set of four equations that connect electric and magnetic quantities bears his name. Later on in 1984, Henrich Herz demonstrated the existence of radio waves experimentally.

However, it was on December 12, 1901, that Guglielmo Marconi received a radio signal at Signal Hill in Newfoundland; the radio signal had originated in Cornwall, England, 2100 miles away across the Atlantic. Last but by no means least, in the early days of wireless communications, it was Fessenden, a self-educated academic, who in 1906 made history by conducting the first radio broadcast, transmitting music and voice using a technique that came to be known as *amplitude modulation (AM) radio.*

In 1988, the first digital cellular system was introduced in Europe; it was known as the *Global System for Mobile (GSM) Communications.* Originally, GSM was intended to provide a pan-European standard to replace the myriad of incompatible analog wireless communication systems. The introduction of GSM was soon followed by the North American IS-54 digital standard. As with the Internet, wireless communication has also dramatically changed the way we communicate on a daily basis.

What we have just described under the three headings, namely, information theory and coding, the Internet, and wireless communications, have collectively not only made communications essentially digital, but have also changed the world of communications and made it global.

1.2 The Communication Process

Today, *communication* enters our daily lives in so many different ways that it is very easy to overlook the multitude of its facets. The telephones as well as mobile smart phones and devices at our hands, the radios and televisions in our living rooms, the computer terminals with access to the Internet in our offices and homes, and our newspapers are all capable of providing rapid communications from every corner of the globe. Communication provides the senses for ships on the high seas, aircraft in flight, and rockets and satellites in space. Communication through a wireless telephone keeps a car driver in touch with the office or home miles away, no matter where. Communication provides the means for social networks to engage in different ways (texting, speaking, visualizing), whereby people are brought together around the world. Communication keeps a weather forecaster informed of conditions measured by a multitude of sensors and satellites. Indeed, the list of applications involving the use of communication in one way or another is almost endless.

In the most fundamental sense, communication involves implicitly the transmission of *information* from one point to another through a succession of processes:

1. The generation of a *message signal* – voice, music, picture, or computer data.
2. The description of that message signal with a certain measure of precision, using a set of *symbols* – electrical, aural, or visual.
3. The *encoding* of those symbols in a suitable form for transmission over a physical medium of interest.
4. The *transmission* of the encoded symbols to the desired destination.
5. The *decoding* and *reproduction* of the original symbols.
6. The *re-creation* of the original message signal with some definable degradation in quality, the degradation being caused by unavoidable imperfections in the system.

There are, of course, many other forms of communication that do not directly involve the human mind in real time. For example, in *computer communications* involving communication between two or more computers, human decisions may enter only in setting up the programs or commands for the computer, or in monitoring the results.

Irrespective of the form of communication process being considered, there are three basic elements to every communication system, namely, *transmitter*, *channel*, and *receiver*, as depicted in Figure 1.1. The transmitter is located at one point in space, the receiver is located at some other point separate from the transmitter, and the channel is the physical medium that connects them together as an integrated communication system. The purpose of the transmitter is to convert the *message signal* produced by the *source of information* into a form suitable for transmission over the channel. However, as the transmitted signal propagates along the channel, it is distorted due to channel imperfections. Moreover, noise and interfering signals (originating from other sources) are added to the channel output, with the result that the *received signal* is a corrupted version of the *transmitted signal*. The receiver has the task of operating on the received signal so as to reconstruct a recognizable form of the original message signal for an end user or information sink.

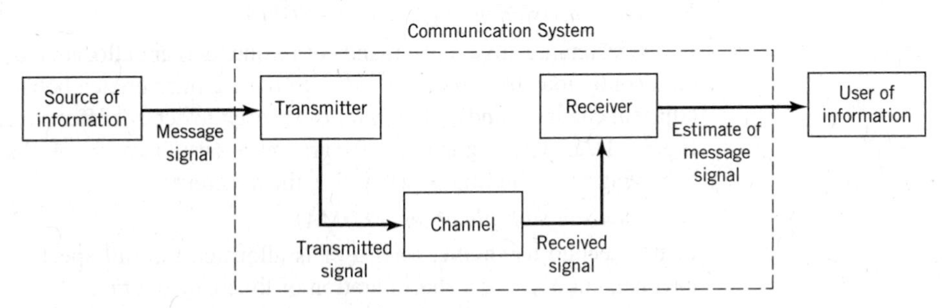

Figure 1.1 Elements of a communication system.

There are two basic modes of communication:

1. *Broadcasting*, which involves the use of a single powerful transmitter and numerous receivers that are relatively inexpensive to build. Here, information-bearing signals flow only in one direction.
2. *Point-to-point communication*, in which the communication process takes place over a link between a single transmitter and a receiver. In this case, there is usually a bidirectional flow of information-bearing signals, which requires the combined use of a transmitter and receiver (i.e., a *transceiver)* at each end of the link.

The underlying communication process in every communication system, irrespective of its kind, is *statistical* in nature. Indeed, it is for this important reason that much of this book is devoted to the statistical underpinnings of digital communication systems. In so doing, we develop a wealth of knowledge on the fundamental issues involved in the study of digital communications.

1.3 Multiple-Access Techniques

Continuing with the communication process, *multiple-access* is a technique whereby many subscribers or local stations can share the use of a communication channel at the same time or nearly so, despite the fact that their individual transmissions may originate from widely different locations. Stated in another way, a multiple-access technique permits the communication resources of the channel to be shared by a large number of users seeking to communicate with each other.

There are subtle differences between multiple access and multiplexing that should be noted:

- Multiple access refers to the remote sharing of a communication channel such as a satellite or radio channel by users in highly dispersed locations. On the other hand, multiplexing refers to the sharing of a channel such as a telephone channel by users confined to a local site.
- In a multiplexed system, user requirements are ordinarily fixed. In contrast, in a multiple-access system user requirements can change dynamically with time, in which case provisions are necessary for dynamic channel allocation.

For obvious reasons it is desirable that in a multiple-access system the sharing of resources of the channel be accomplished without causing serious interference between users of the system. In this context, we may identify four basic types of multiple access:

1. *Frequency-division multiple access (FDMA).*

   In this technique, disjoint subbands of frequencies are allocated to the different users on a continuous-time basis. In order to reduce interference between users allocated adjacent channel bands, *guard bands* are used to act as buffer zones, as illustrated in Figure 1.2a. These guard bands are necessary because of the impossibility of achieving ideal filtering or separating the different users.
2. *Time-division multiple access (TDMA).*

   In this second technique, each user is allocated the full spectral occupancy of the channel, but only for a short duration of time called a *time slot*. As shown in Figure 1.2b, buffer zones in the form of *guard times* are inserted between the assigned time slots. This is done to reduce interference between users by allowing for time uncertainty that arises due to system imperfections, especially in synchronization schemes.

3. *Code-division multiple access (CDMA).*

   In FDMA, the resources of the channel are shared by dividing them along the frequency coordinate into disjoint frequency bands, as illustrated in Figure 1.2a. In TDMA, the resources are shared by dividing them along the time coordinate into disjoint time slots, as illustrated in Figure 1.2b. In Figure 1.2c, we illustrate another technique for sharing the channel resources by using a hybrid combination of FDMA and TDMA, which represents a specific form of code-division multiple access (CDMA). For example, *frequency hopping* may be employed to ensure that during each successive time slot, the frequency bands assigned to the users are reordered in an essentially random manner. To be specific, during time slot 1, user 1 occupies frequency band 1, user 2 occupies frequency band 2, user 3 occupies frequency band 3, and so on. During time slot 2, user 1 hops to frequency band 3, user 2 hops to frequency band 1, user 3 hops to frequency band 2, and so on. Such an arrangement has the appearance of the users playing a game of musical chairs. An important advantage of CDMA over both FDMA and TDMA is that it can provide for *secure* communications. In the type of CDMA illustrated in Figure 1.2c, the frequency hopping mechanism can be implemented through the use of a pseudo-noise (PN) sequence.

4. *Space-division multiple access (SDMA).*

   In this multiple-access technique, resource allocation is achieved by exploiting the spatial separation of the individual users. In particular, *multibeam antennas* are used to separate radio signals by pointing them along different directions. Thus, different users are enabled to access the channel simultaneously on the same frequency or in the same time slot.

These multiple-access techniques share a common feature: allocating the communication resources of the channel through the use of disjointedness (or orthogonality in a loose sense) in time, frequency, or space.

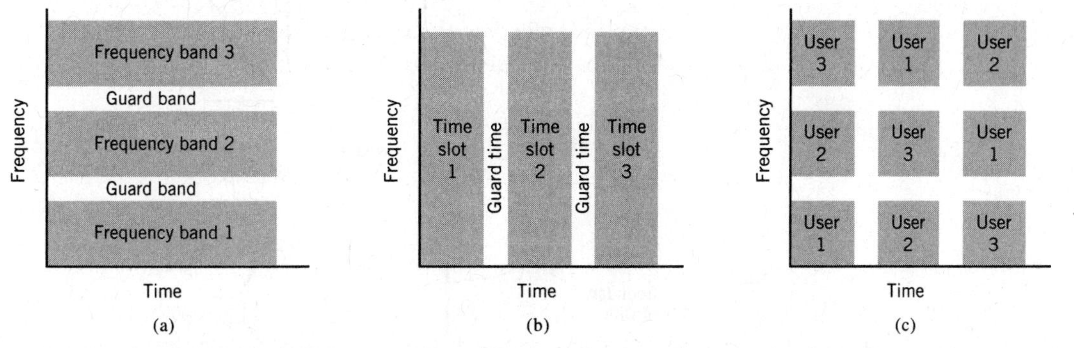

Figure 1.2 Illustrating the ideas behind multiple-access techniques. (a) Frequency-division multiple access. (b) Time-division multiple access. (c) Frequency-hop multiple access.

1.4 Networks

A *communication network* or simply *network*[1], illustrated in Figure 1.3, consists of an interconnection of a number of *nodes* made up of intelligent processors (e.g., microcomputers). The primary purpose of these nodes is to route data through the network. Each node has one or more *stations* attached to it; stations refer to devices wishing to communicate. The network is designed to serve as a shared resource for moving data exchanged between stations in an efficient manner and also to provide a framework to support new applications and services. The traditional telephone network is an example of a communication network in which *circuit switching* is used to provide a dedicated communication path or *circuit* between two stations. The circuit consists of a connected sequence of links from source to destination. The links may consist of time slots in a time-division multiplexed (TDM) system or frequency slots in a frequency-division multiplexed (FDM) system. The circuit, once in place, remains uninterrupted for the entire duration of transmission. Circuit switching is usually controlled by a centralized hierarchical control mechanism with knowledge of the network's organization. To establish a circuit-switched connection, an available path through the network is seized and then dedicated to the exclusive use of the two stations wishing to communicate. In particular, a call-request signal must propagate all the way to the destination, and be acknowledged, before transmission can begin. Then, the network is effectively transparent to the users. This means that, during the connection time, the bandwidth and resources allocated to the circuit are essentially "owned" by the two stations, until the circuit is disconnected. The circuit thus represents an efficient use of resources only to the extent that the allocated bandwidth is properly utilized. Although the telephone network is used to transmit data, voice constitutes the bulk of the network's traffic. Indeed, circuit switching is well suited to the transmission of voice signals, since voice conversations tend to be of long duration (about 2 min on average) compared with the time required for setting up the circuit (about 0.1–0.5 s). Moreover, in most voice conversations, there is information flow for a relatively large percentage of the connection time, which makes circuit switching all the more suitable for voice conversations.

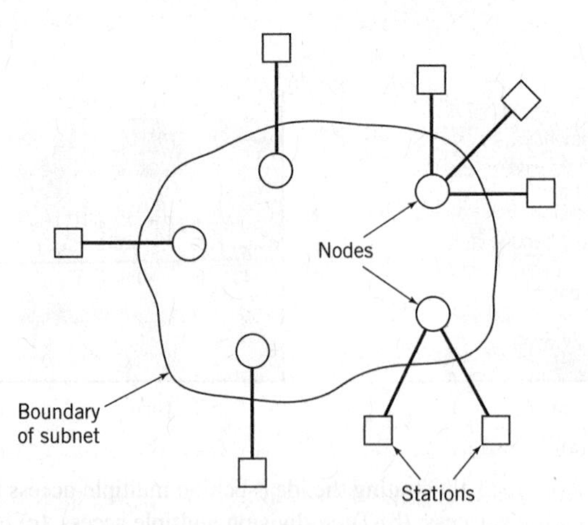

Figure 1.3 Communication network.

In circuit switching, a communication link is shared between the different sessions using that link on a *fixed* allocation basis. In *packet switching*, on the other hand, the sharing is done on a *demand* basis and, therefore, it has an advantage over circuit switching in that when a link has traffic to send, the link may be more fully utilized.

The basic network principle of packet switching is "store and forward." Specifically, in a *packet-switched network*, any message larger than a specified size is subdivided prior to transmission into segments not exceeding the specified size. The segments are commonly referred to as *packets*. The original message is reassembled at the destination on a packet-by-packet basis. The network may be viewed as a distributed pool of *network resources* (i.e., channel bandwidth, buffers, and switching processors) whose capacity is *shared dynamically* by a community of competing users (stations) wishing to communicate. In contrast, in a circuit-switched network, resources are dedicated to a pair of stations for the entire period they are in session. Accordingly, packet switching is far better suited to a computer-communication environment in which "bursts" of data are exchanged between stations on an occasional basis. The use of packet switching, however, requires that careful *control* be exercised on user demands; otherwise, the network may be seriously abused.

The design of a *data network* (i.e., a network in which the stations are all made up of computers and terminals) may proceed in an orderly way by looking at the network in terms of a *layered architecture*, regarded as a hierarchy of nested layers. A *layer* refers to a process or device inside a computer system, designed to perform a specific function. Naturally, the designers of a layer will be intimately familiar with its internal details and operation. At the system level, however, a user views the layer merely as a "black box" that is described in terms of the inputs, the outputs, and the functional relationship between outputs and inputs. In a layered architecture, each layer regards the next lower layer as one or more black boxes with some given functional specification to be used by the given higher layer. Thus, the highly complex communication problem in data networks is resolved as a manageable set of well-defined interlocking functions. It is this line of reasoning that has led to the development of the *open systems interconnection* (OSI)[2] *reference model* by a subcommittee of the International Organization for Standardization. The term "open" refers to the ability of any two systems conforming to the reference model and its associated standards to interconnect.

In the OSI reference model, the communications and related-connection functions are organized as a series of *layers* or *levels* with well-defined *interfaces*, and with each layer built on its predecessor. In particular, each layer performs a related subset of primitive functions, and it relies on the next lower layer to perform additional primitive functions. Moreover, each layer offers certain services to the next higher layer and shields the latter from the implementation details of those services. Between each pair of layers, there is an *interface*. It is the interface that defines the services offered by the lower layer to the upper layer.

The OSI model is composed of seven layers, as illustrated in Figure 1.4; this figure also includes a description of the functions of the individual layers of the model. Layer k on system A, say, communicates with layer k on some other system B in accordance with a set of rules and conventions, collectively constituting the layer k *protocol*, where $k = 1, 2, ..., 7$. (The term "protocol" has been borrowed from common usage, describing conventional social behavior between human beings.) The entities that comprise the corresponding layers on different systems are referred to as *peer processes*. In other words,

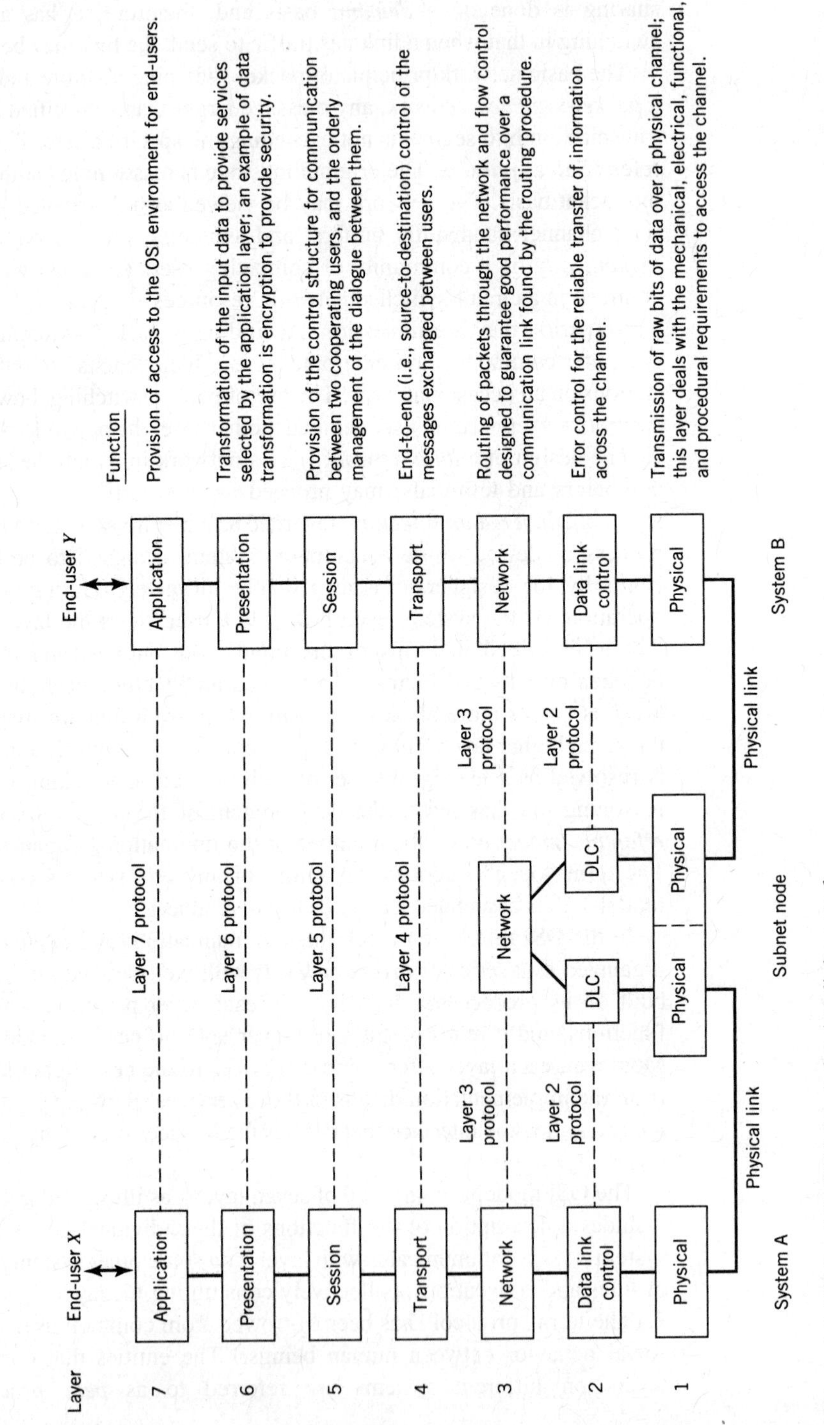
Layer
7
6
5
4
3
2
1
End-user X
Application
Presentation
Session
Transport
Network
Data link control
Physical
System A
Physical link
Layer 7 protocol
Layer 6 protocol
Layer 5 protocol
Layer 4 protocol
Layer 3 protocol
Layer 2 protocol
Network
DLC
DLC
Physical
Physical
Subnet node
Layer 3 protocol
Layer 2 protocol
Physical link
End-user Y
Application
Presentation
Session
Transport
Network
Data link control
Physical
System B
Function
Provision of access to the OSI environment for end-users.
Transformation of the input data to provide services selected by the application layer; an example of data transformation is encryption to provide security.
Provision of the control structure for communication between two cooperating users, and the orderly management of the dialogue between them.
End-to-end (i.e., source-to-destination) control of the messages exchanged between users.
Routing of packets through the network and flow control designed to guarantee good performance over a communication link found by the routing procedure.
Error control for the reliable transfer of information across the channel.
Transmission of raw bits of data over a physical channel; this layer deals with the mechanical, electrical, functional, and procedural requirements to access the channel.

Figure 1.4 OSI model; DLC stands for data link control.

communication is achieved by having the peer processes in two different systems communicate via a protocol, with the protocol itself being defined by a set of rules of procedure. Physical communication between peer processes exits only at layer 1. On the other hand, layers 2 through 7 are in *virtual communication* with their distant peers. However, each of these six layers can exchange data and control information with its neighboring layers (below and above) through layer-to-layer interfaces. In Figure 1.4, physical communication is shown by solid lines and virtual communication by dashed lines. The major principles involved in arriving at seven layers of the OSI reference model are as follows:

1. Each layer performs well-defined functions.
2. A boundary is created at a point where the description of services offered is small and the number of interactions across the boundary is the minimum possible.
3. A layer is created from easily localized functions, so that the architecture of the model may permit modifications to the layer protocol to reflect changes in technology without affecting the other layers.
4. A boundary is created at some point with an eye toward standardization of the associated interface.
5. A layer is created only when a different level of abstraction is needed to handle the data.
6. The number of layers employed should be large enough to assign distinct functions to different layers, yet small enough to maintain a manageable architecture for the model.

Note that the OSI reference model is not a network architecture; rather, it is an international standard for computer communications, which just tells what each layer should do.

1.5 Digital Communications

Today's public communication networks are highly complicated systems. Specifically, public switched telephone networks (collectively referred to as PSTNs), the Internet, and wireless communications (including satellite communications) provide seamless connections between cities, across oceans, and between different countries, languages, and cultures; hence the reference to the world as a "global village."

There are three layers of the OSI model where it can affect the design of digital communication systems, which is the subject of interest of this book:

1. *Physical layer.* This lowest layer of the OSI model embodies the physical mechanism involved in transmitting *bits* (i.e., *bi*nary digi*ts*) between any pair of nodes in the communication network. Communication between the two nodes is accomplished by means of modulation in the transmitter, transmission across the channel, and demodulation in the receiver. The module for performing *mo*dulation and *dem*odulation is often called a *modem.*
2. *Data-link layer.* Communication links are nearly always corrupted by the unavoidable presence of noise and interference. One purpose of the data-link layer, therefore, is to perform *error correction* or *detection*, although this function is also shared with the physical layer. Often, the data-link layer will retransmit packets that are received in error but, for some applications, it discards them. This layer is also

responsible for the way in which different users share the transmission medium. A portion of the data-link layer, called the *medium access control (MAC)* sublayer, is responsible for allowing frames to be sent over the shared transmission media without undue interference with other nodes. This aspect is referred to as *multiple-access* communications.

3. *Network layer.* This layer has several functions, one of which is to determine the *routing* of information, to get it from the source to its ultimate destination. A second function is to determine the *quality of service*. A third function is *flow control*, to ensure that the network does not become congested.

These are three layers of a seven-layer model for the functions that occur in the communications process. Although the three layers occupy a subspace within the OSI model, the functions that they perform are of critical importance to the model.

Block Diagram of Digital Communication System

Typically, in the design of a digital communication system the information source, communication channel, and information sink (end user) are all specified. The challenge is to design the transmitter and the receiver with the following guidelines in mind:

- Encode/modulate the message signal generated by the source of information, transmit it over the channel, and produce an "estimate" of it at the receiver output that satisfies the requirements of the end user.
- Do all of this at an affordable cost.

In a *digital communication* system represented by the block diagram of Figure 1.6, the rationale for which is rooted in information theory, the functional blocks of the transmitter and the receiver starting from the far end of the channel are paired as follows:

- source encoder–decoder;
- channel encoder–decoder;
- modulator–demodulator.

The source encoder removes redundant information from the message signal and is responsible for efficient use of the channel. The resulting sequence of symbols is called the *source codeword*. The data stream is processed next by the channel encoder, which produces a new sequence of symbols called the *channel codeword*. The channel codeword is longer than the source code word by virtue of the *controlled* redundancy built into its construction. Finally, the modulator represents each symbol of the channel codeword by a corresponding analog symbol, appropriately selected from a finite set of possible analog symbols. The sequence of analog symbols produced by the modulator is called a *waveform*, which is suitable for transmission over the channel. At the receiver, the channel output (received signal) is processed in reverse order to that in the transmitter, thereby reconstructing a recognizable version of the original message signal. The reconstructed message signal is finally delivered to the user of information at the destination. From this description it is apparent that the design of a digital communication system is rather complex in conceptual terms but easy to build. Moreover, the system is *robust*, offering greater tolerance of physical effects (e.g., temperature variations, aging, mechanical vibrations) than its analog counterpart; hence the ever-increasing use of digital communications.

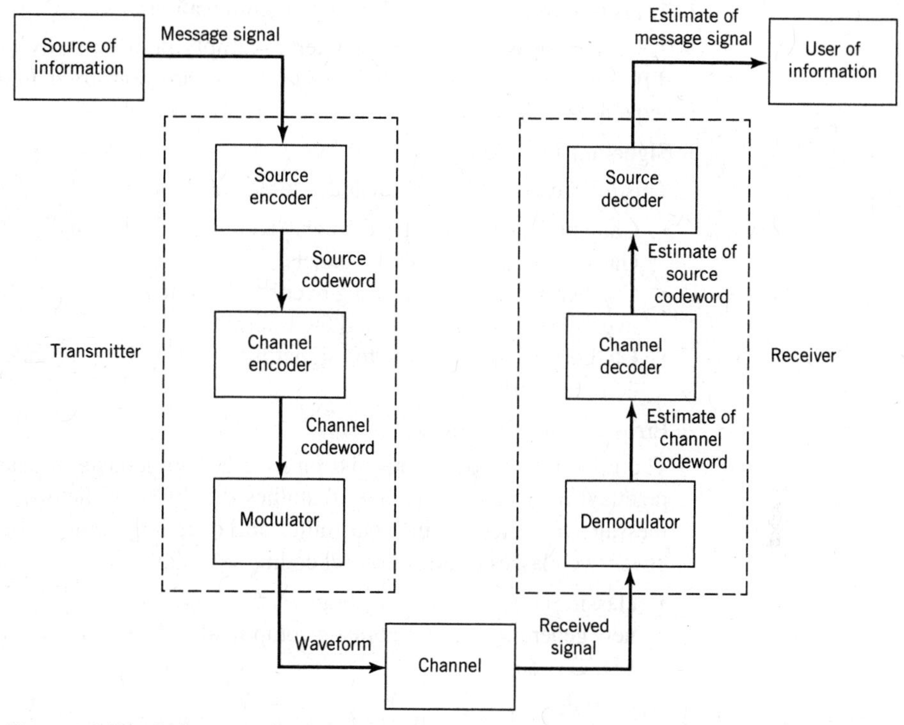

Figure 1.6 Block diagram of a digital communication system.

1.6 Organization of the Book

The main part of the book is organized in ten chapters, which, after this introductory chapter, are organized into five parts of varying sizes as summarized herein.

1. Mathematical Background

   Chapter 2 presents a detailed treatment of the Fourier transform, its properties and algorithmic implementations. This chapter also includes two important related topics:

   - The Hilbert transform, which provides the mathematical basis for transforming real-valued band-pass signals and systems into their low-pass equivalent representations without loss of information.
   - Overview of analog modulation theory, thereby facilitating an insightful link between analog and digital communications.

   Chapter 3 presents a mathematical review of probability theory and Bayesian inference, the understanding of which is essential to the study of digital communications.

   Chapter 4 is devoted to the study of stochastic processes, the theory of which is basic to the characterization of sources of information and communication channels.

   Chapter 5 discusses the fundamental limits of information theory, postulated in terms of source coding, channel capacity, and rate-distortion theory.

2. Transition from Analog to Digital Communications

   This material is covered in Chapter 6. Simply put, the study therein discusses the different ways in which analog waveforms are converted into digitally encoded sequences.

3. Signaling Techniques

   This third part of the book includes three chapters:

   - Chapter 7 discusses the different techniques for signaling over additive white Gaussian noise (AWGN) channels.
   - Chapter 8 discusses signaling over band-limited channels, as in data transmission over telephonic channels and the Internet.
   - Chapter 9 is devoted to signaling over fading channels, as in wireless communications.

4. Error-Control Coding

   The reliability of data transmission over a communication channel is of profound practical importance. Chapter 10 studies the different methods for the encoding of message sequences in the transmitter and decoding them in the receiver. Here, we cover two classes of error-control coding techniques:

   - classic codes rooted in algebraic mathematics, and
   - new generation of probabilistic compound codes, exemplified by turbo codes and LDPC codes.

5. Appendices

   Last but by no means least, the book includes appendices to provide back-up material for different chapters in the book, as they are needed.

Notes

1. For a detailed discussion on communication networks, see the classic book by Tanenbaum, entitled *Computer Networks* (2003).
2. The OSI reference model was developed by a subcommittee of the International Organization for Standardization (ISO) in 1977. For a discussion of the principles involved in arriving at the seven layers of the OSI model and a description of the layers themselves, see Tanenbaum (2003).

CHAPTER
2

Fourier Analysis of Signals and Systems

2.1 Introduction

The study of communication systems involves:

- the processing of a modulated message signal generated at the transmitter output so as to facilitate its transportation across a physical channel and
- subsequent processing of the received signal in the receiver so as to deliver an estimate of the original message signal to a user at the receiver output.

In this study, the *representation of signals and systems* features prominently. More specifically, the *Fourier transform* plays a key role in this representation.

The Fourier transform provides the mathematical link between the time-domain representation (i.e., waveform) of a signal and its frequency-domain description (i.e., spectrum). Most importantly, we can go back and forth between these two descriptions of the signal with no loss of information. Indeed, we may invoke a similar transformation in the representation of linear systems. In this latter case, the time-domain and frequency-domain descriptions of a linear time-invariant system are defined in terms of its impulse response and frequency response, respectively.

In light of this background, it is in order that we begin a mathematical study of communication systems by presenting a review of Fourier analysis. This review, in turn, paves the way for the formulation of simplified representations of band-pass signals and systems to which we resort in subsequent chapters. We begin the study by developing the transition from the Fourier series representation of a periodic signal to the Fourier transform representation of a nonperiodic signal; this we do in the next two sections.

2.2 The Fourier Series

Let $g_{T_0}(t)$ denote a *periodic signal,* where the subscript T_0 denotes the duration of periodicity. By using a *Fourier series expansion* of this signal, we are able to resolve it into an infinite sum of sine and cosine terms, as shown by

$$g_{T_0}(t) = a_0 + 2\sum_{n=1}^{\infty} [a_n \cos(2\pi n f_0 t) + b_n \sin(2\pi n f_0 t)] \tag{2.1}$$

where

$$f_0 = \frac{1}{T_0} \tag{2.2}$$

is the *fundamental frequency.* The coefficients a_n and b_n represent the amplitudes of the cosine and sine terms, respectively. The quantity nf_0 represents the nth harmonic of the fundamental frequency f_0. Each of the terms $\cos(2\pi nf_0 t)$ and $\sin(2\pi nf_0 t)$ is called a *basis function.* These basis functions form an *orthogonal set* over the interval T_0, in that they satisfy three conditions:

$$\int_{-T_0/2}^{T_0/2} \cos(2\pi mf_0 t)\cos(2\pi nf_0 t)\,dt = \begin{cases} T_0/2, & m = n \\ 0, & m \neq n \end{cases} \tag{2.3}$$

$$\int_{-T_0/2}^{T_0/2} \cos(2\pi mf_0 t)\sin(2\pi nf_0 t)\,dt = 0, \quad \text{for all } m \text{ and } n \tag{2.4}$$

$$\int_{-T_0/2}^{T_0/2} \sin(2\pi mf_0 t)\sin(2\pi nf_0 t)\,dt = \begin{cases} T_0/2, & m = n \\ 0, & m \neq n \end{cases} \tag{2.5}$$

To determine the coefficient a_0, we integrate both sides of (2.1) over a complete period. We thus find that a_0 is the *mean value* of the periodic signal $g_{T_0}(t)$ over one period, as shown by the *time average*

$$a_0 = \frac{1}{T_0}\int_{-T_0/2}^{T_0/2} g_{T_0}(t)\,dt \tag{2.6}$$

To determine the coefficient a_n, we multiply both sides of (2.1) by $\cos(2\pi nf_0 t)$ and integrate over the interval $-T_0/2$ to $T_0/2$. Then, using (2.3) and (2.4), we find that

$$a_n = \frac{1}{T_0}\int_{-T_0/2}^{T_0/2} g_{T_0}(t)\cos(2\pi nf_0 t)\,dt, \quad n = 1, 2, \ldots \tag{2.7}$$

Similarly, we find that

$$b_n = \frac{1}{T_0}\int_{-T_0/2}^{T_0/2} g_{T_0}(t)\sin(2\pi nf_0 t)\,dt, \quad n = 1, 2, \ldots \tag{2.8}$$

A basic question that arises at this point is the following:

> Given a periodic signal $g_{T_0}(t)$ of period T_0, how do we know that the Fourier series expansion of (2.1) is *convergent* in that the infinite sum of terms in this expansion is exactly equal to $g_{T_0}(t)$?

To resolve this fundamental issue, we have to show that, for the coefficients a_0, a_n, and b_n calculated in accordance with (2.6) to (2.8), this series will indeed converge to $g_{T_0}(t)$. In general, for a periodic signal $g_{T_0}(t)$ of arbitrary waveform, there is no guarantee that the series of (2.1) will converge to $g_{T_0}(t)$ or that the coefficients a_0, a_n, and b_n will even exist. In a rigorous sense, we may say that a periodic signal $g_{T_0}(t)$ can be expanded in a Fourier series if the signal $g_{T_0}(t)$ satisfies the *Dirichlet conditions*:[1]

1. The function $g_{T_0}(t)$ is single valued within the interval T_0.
2. The function $g_{T_0}(t)$ has at most a finite number of discontinuities in the interval T_0.
3. The function $g_{T_0}(t)$ has a finite number of maxima and minima in the interval T_0.
4. The function $g_{T_0}(t)$ is absolutely integrable; that is,

$$\int_{-T_0/2}^{T_0/2} \left| g_{T_0}(t) \right| \mathrm{d}t < \infty$$

From an engineering perspective, however, it suffices to say that the Dirichlet conditions are satisfied by the periodic signals encountered in communication systems.

Complex Exponential Fourier Series

The Fourier series of (2.1) can be put into a much simpler and more elegant form with the use of complex exponentials. We do this by substituting into (2.1) the exponential forms for the cosine and sine, namely:

$$\cos(2\pi n f_0 t) = \frac{1}{2}[\exp(\mathrm{j}2\pi n f_0 t) + \exp(-\mathrm{j}2\pi n f_0 t)]$$

$$\sin(2\pi n f_0 t) = \frac{1}{2\mathrm{j}}[\exp(\mathrm{j}2\pi n f_0 t) - \exp(-\mathrm{j}2\pi n f_0 t)]$$

where $\mathrm{j} = \sqrt{-1}$. We thus obtain

$$g_{T_0}(t) = a_0 + \sum_{n=1}^{\infty} [(a_n - \mathrm{j}b_n)\exp(\mathrm{j}2\pi n f_0 t) + (a_n + \mathrm{j}b_n)\exp(-\mathrm{j}2\pi n f_0 t)] \tag{2.9}$$

Let c_n denote a complex coefficient related to a_n and b_n by

$$c_n = \begin{cases} a_n - \mathrm{j}b_n, & n > 0 \\ a_0, & n = 0 \\ a_n + \mathrm{j}b_n, & n < 0 \end{cases} \tag{2.10}$$

Then, we may simplify (2.9) into

$$g_{T_0}(t) = \sum_{n=-\infty}^{\infty} c_n \exp(\mathrm{j}2\pi n f_0 t) \tag{2.11}$$

where

$$c_n = \frac{1}{T_0}\int_{-T_0/2}^{T_0/2} g_{T_0}(t)\exp(-\mathrm{j}2\pi n f_0 t)\mathrm{d}t, \qquad n = 0, \pm 1, \pm 2, \ldots \tag{2.12}$$

The series expansion of (2.11) is referred to as the *complex exponential Fourier series*. The c_n themselves are called the *complex Fourier coefficients*.

> Given a periodic signal $g_{T_0}(t)$, (2.12) states that we may determine the complete set of complex Fourier coefficients. On the other hand, (2.11) states that, given this set of coefficients, we may reconstruct the original periodic signal $g_{T_0}(t)$ exactly.

The integral on the right-hand side of (2.12) is said to be an *inner product* of the signal $g_{T_0}(t)$ with the *basis functions* $\exp(-j2\pi n f_0 t)$, by whose linear combination all square integrable functions can be expressed as in (2.11).

According to this representation, a periodic signal contains all frequencies (both positive and negative) that are harmonically related to the fundamental frequency f_0. The presence of negative frequencies is simply a result of the fact that the mathematical model of the signal as described by (2.11) requires the use of negative frequencies. Indeed, this representation also requires the use of complex-valued basis functions, namely $\exp(j2\pi n f_0 t)$, which have no physical meaning either. The reason for using complex-valued basis functions and negative frequency components is merely to provide a compact mathematical description of a periodic signal, which is well-suited for both theoretical and practical work.

2.3 The Fourier Transform

In the previous section, we used the Fourier series to represent a periodic signal. We now wish to develop a similar representation for a signal $g(t)$ that is nonperiodic. In order to do this, we first construct a periodic function $g_{T_0}(t)$ of period T_0 in such a way that $g(t)$ defines exactly one cycle of this periodic function, as illustrated in Figure 2.1. In the limit, we let the period T_0 become infinitely large, so that we may express $g(t)$ as

$$g(t) = \lim_{T_0 \to \infty} g_{T_0}(t) \tag{2.13}$$

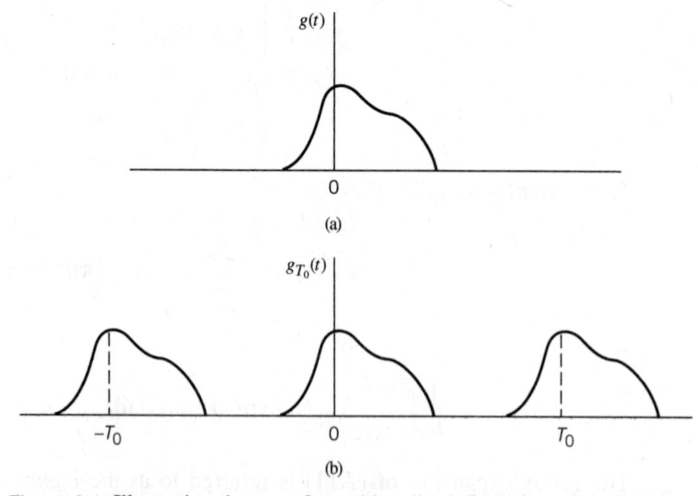

Figure 2.1 Illustrating the use of an arbitrarily defined function of time to construct a periodic waveform. (a) Arbitrarily defined function of time $g(t)$. (b) Periodic waveform $g_{T_0}(t)$ based on $g(t)$.

Representing the periodic function $g_{T_0}(t)$ in terms of the complex exponential form of the Fourier series, we write

$$g_{T_0}(t) = \sum_{n=-\infty}^{\infty} c_n \exp\left(\frac{j2\pi nt}{T_0}\right)$$

where

$$c_n = \frac{1}{T_0}\int_{-T_0/2}^{T_0/2} g_{T_0}(t)\exp\left(-\frac{j2\pi nt}{T_0}\right) dt$$

Here, we have purposely replaced f_0 with $1/T_0$ in the exponents. Define

$$\Delta f = \frac{1}{T_0}$$

$$f_n = \frac{n}{T_0}$$

and

$$G(f_n) = c_n T_0$$

We may then go on to modify the original Fourier series representation of $g_{T_0}(t)$ given in (2.11) into a new form described by

$$g_{T_0}(t) = \sum_{n=-\infty}^{\infty} G(f_n)\exp(j2\pi f_n t)\Delta f \tag{2.14}$$

where

$$G(f_n) = \int_{-T_0/2}^{T_0/2} g_{T_0}(t)\exp(-j2\pi f_n t)\, dt \tag{2.15}$$

Equations (2.14) and (2.15) apply to a periodic signal $g_{T_0}(t)$. What we would like to do next is to go one step further and develop a corresponding pair of formulas that apply to a nonperiodic signal $g(t)$. To do this transition, we use the defining equation (2.13). Specifically, two things happen:

1. The discrete frequency f_n in (2.14) and (2.15) approaches the continuous frequency variable f.
2. The discrete sum of (2.14) becomes an integral defining the area under the function $G(f)\exp(j2\pi ft)$, integrated with respect to time t.

Accordingly, piecing these points together, we may respectively rewrite the limiting forms of (2.15) and (2.14) as

$$G(f) = \int_{-\infty}^{\infty} g(t)\exp(-j2\pi ft)dt \tag{2.16}$$

and

$$g(t) = \int_{-\infty}^{\infty} G(f)\exp(j2\pi ft)df \tag{2.17}$$

In words, we may say:

- the *Fourier transform* of the nonperiodic signal $g(t)$ is defined by (2.16);
- given the Fourier transform $G(f)$, the original signal $g(t)$ is recovered exactly from the inverse *Fourier transform* of (2.17).

Figure 2.2 illustrates the interplay between these two formulas, where we see that the frequency-domain description based on (2.16) plays the role of *analysis* and the time-domain description based on (2.17) plays the role of *synthesis*.

From a notational point of view, note that in (2.16) and (2.17) we have used a lowercase letter to denote the time function and an uppercase letter to denote the corresponding frequency function. Note also that these two equations are of identical mathematical form, except for changes in the algebraic signs of the exponents.

For the Fourier transform of a signal $g(t)$ to exist, it is sufficient but not necessary that the nonperiodic signal $g(t)$ satisfies three *Dirichlet's conditions* of its own:

1. The function $g(t)$ is single valued, with a finite number of maxima and minima in any finite time interval.
2. The function $g(t)$ has a finite number of discontinuities in any finite time interval.
3. The function $g(t)$ is absolutely integrable; that is,

$$\int_{-\infty}^{\infty} |g(t)| \mathrm{d}t < \infty$$

In practice, we may safely ignore the question of the existence of the Fourier transform of a time function $g(t)$ when it is an accurately specified description of a physically realizable signal. In other words, physical realizability is a sufficient condition for the existence of a Fourier transform. Indeed, we may go one step further and state:

All energy signals are Fourier transformable.

A signal $g(t)$ is said to be an *energy signal* if the condition

$$\int_{-\infty}^{\infty} |g(t)|^2 \mathrm{d}t < \infty \tag{2.18}$$

holds.[2]

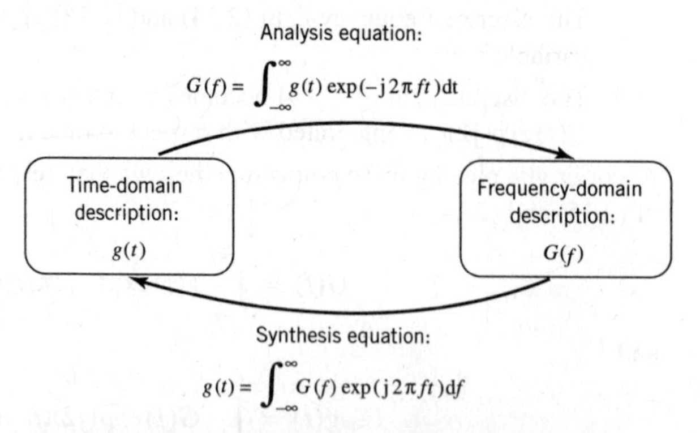

Figure 2.2 Sketch of the interplay between the synthesis and analysis equations embodied in Fourier transformation.

The Fourier transform provides the mathematical tool for measuring the frequency content, or spectrum, of a signal. For this reason, the terms *Fourier transform* and *spectrum* are used interchangeably. Thus, given a signal $g(t)$ with Fourier transform $G(f)$, we may refer to $G(f)$ as the spectrum of the signal $g(t)$. By the same token, we refer to $|G(f)|$ as the *magnitude spectrum* of the signal $g(t)$, and refer to $\arg[G(f)]$ as its *phase spectrum.*

If the signal $g(t)$ is real valued, then the magnitude spectrum of the signal is an even function of frequency f, while the phase spectrum is an odd function of f. In such a case, knowledge of the spectrum of the signal for positive frequencies uniquely defines the spectrum for negative frequencies.

Notations

For convenience of presentation, it is customary to express (2.17) in the short-hand form

$$G(f) = \mathbf{F}[g(t)]$$

where **F** plays the role of an *operator.* In a corresponding way, (2.18) is expressed in the short-hand form

$$g(t) = \mathbf{F}^{-1}[G(f)]$$

where $\mathbf{F}^{-1}$ plays the role of an *inverse operator.*

The time function $g(t)$ and the corresponding frequency function $G(f)$ are said to constitute a *Fourier-transform pair.* To emphasize this point, we write

$$g(t) \rightleftharpoons G(f)$$

where the top arrow indicates the forward transformation from $g(t)$ to $G(f)$ and the bottom arrow indicates the inverse transformation. One other notation: the asterisk is used to denote complex conjugation.

Tables of Fourier Tranformations

To assist the user of this book, two tables of Fourier transformations are included:

1. Table 2.1 on page 23 summarizes the properties of Fourier transforms; proofs of them are presented as end-of-chapter problems.
2. Table 2.2 on page 24 presents a list of Fourier-transform pairs, where the items listed on the left-hand side of the table are time functions and those in the center column are their Fourier transforms.

EXAMPLE 1

Binary Sequence for Energy Calculations

Consider the five-digit binary sequence 10010. This sequence is represented by two different waveforms, one based on the rectangular function $\text{rect}(t)$, and the other based on the sinc function $\text{sinc}(t)$. Despite this difference, both waveforms are denoted by $g(t)$, which implies they both have exactly the same total energy, to be demonstrated next.

Case 1: $\text{rect}(t)$ *as the basis function.*

Let binary symbol 1 be represented by $+\text{rect}(t)$ and binary symbol 0 be represented by $-\text{rect}(t)$. Accordingly, the binary sequence 10010 is represented by the waveform

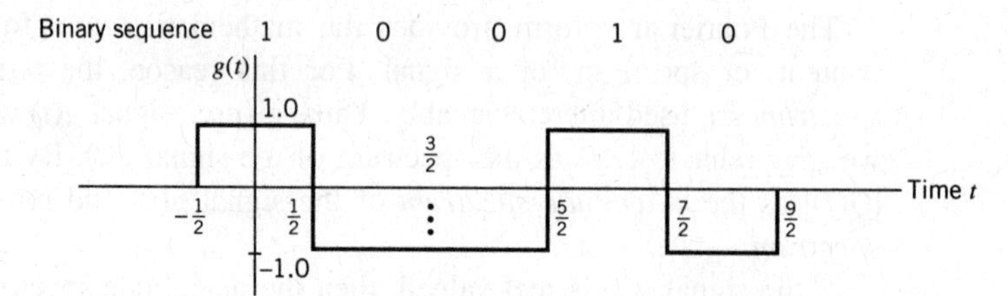

Figure 2.3 Waveform of binary sequence 10010, using rect(t) for symbol 1 and –rect(t) for symbol 0. See Table 2.2 for the definition of rect(t).

shown in Figure 2.3. From this figure, we readily see that, regardless of the representation ±rect(t), each symbol contributes a single unit of energy; hence the total energy for Case 1 is five units.

Case 2: sinc(t) *as the basis function.*

Consider next the representation of symbol 1 by +sinc(t) and the representation of symbol 0 by −sinc(t), which do not interfere with each other in constructing the waveform for the binary sequence 10010. Unfortunately, this time around, it is difficult to calculate the total waveform energy in the time domain. To overcome this difficulty, we do the calculation in the frequency domain.

To this end, in parts a and b of Figure 2.4, we display the waveform of the sinc function in the time domain and its Fourier transform, respectively. On this basis, Figure 2.5 displays the frequency-domain representation of the binary sequence 10010, with part a of the figure displaying the magnitude response $|G(f)|$, and part b displaying the corresponding phrase response $\arg[G(f)]$ expressed in radians. Then, applying Rayleigh's energy theorem, described in Property 14 in Table 2.2, to part a of Figure 2.5, we readily find that the energy of the pulse, ±sinc(t), is equal to one unit, regardless of its amplitude. The total energy of the sinc-based waveform representing the given binary sequence is also exactly five units, confirming what was said at the beginning of this example.

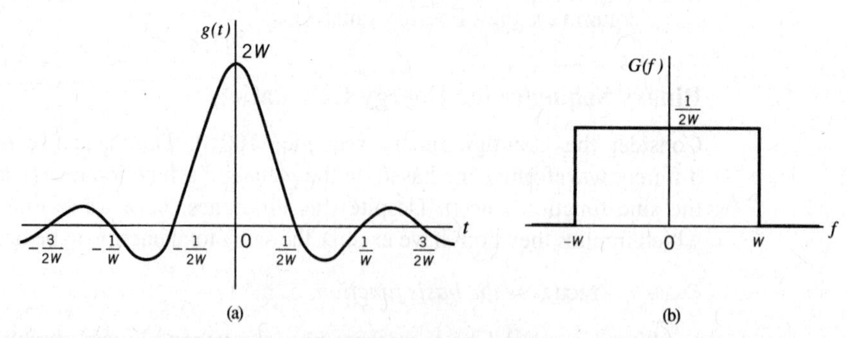

Figure 2.4 (a) Sinc pulse $g(t)$. (b) Fourier transform $G(f)$.

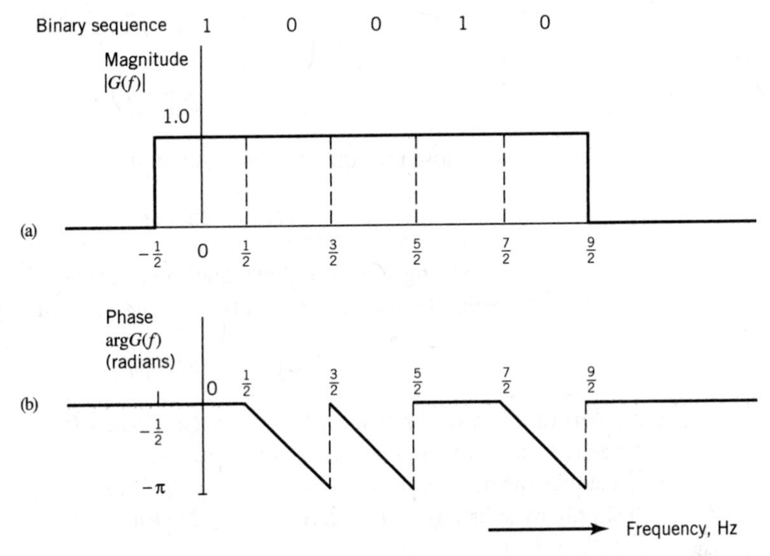

Figure 2.5 (a) Magnitude spectrum of the sequence 10010. (b) Phase spectrum of the sequence.

Observations

1. The dual basis functions, rect(t) and sinc(t), are *dilated* to their simplest forms, each of which has an energy of one unit, hence the equality of the results presented under Cases 1 and 2.
2. Examining the waveform $g(t)$ in Figure 2.3, we clearly see the discrimination between binary symbols 1 and 0. On the other hand, it is the phase response $\arg[G(f)]$ in part b of Figure 2.5 that shows the discrimination between binary symbols 1 and 0.

EXAMPLE 2

Unit Gaussian Pulse

Typically, a pulse signal $g(t)$ and its Fourier transform $G(f)$ have different mathematical forms. This observation is illustrated by the Fourier-transform pair studied in Example 1. In this second example, we consider an exception to this observation. In particular, we use the differentiation property of the Fourier transform to derive the particular form of a *pulse signal that has the same mathematical form as its own Fourier transform.*

Let $g(t)$ denote the pulse signal expressed as a function of time t and $G(f)$ denote its Fourier transform. Differentiating the Fourier transform formula of (2.6) with respect to frequency f yields

$$-\mathrm{j}2\pi t g(t) \rightleftharpoons \frac{\mathrm{d}}{\mathrm{d}f}G(f)$$

or, equivalently,

$$2\pi t g(t) \rightleftharpoons \mathrm{j}\frac{\mathrm{d}}{\mathrm{d}f}G(f) \tag{2.19}$$

Use of the Fourier-transform property on differentiation in the time domain listed in Table 2.1 yields

$$\frac{\mathrm{d}}{\mathrm{d}t}g(t) \rightleftharpoons \mathrm{j}2\pi f G(f) \tag{2.20}$$

Suppose we now impose the equality condition on the left-hand sides of (2.19) and (2.20):

$$\frac{\mathrm{d}}{\mathrm{d}t}g(t) = 2\pi t g(t) \tag{2.21}$$

Then, in a corresponding way, it follows that the right-hand sides of these two equations must (after canceling the common multiplying factor j) satisfy the condition

$$\frac{\mathrm{d}}{\mathrm{d}f}G(f) = 2\pi f G(f) \tag{2.22}$$

Equations (2.21) and (2.22) show that the pulse signal $g(t)$ and its Fourier transform $G(f)$ have exactly the same mathematical form. In other words, provided that the pulse signal $g(t)$ satisfies the differential equation (2.21), then $G(f) = g(f)$, where $g(f)$ is obtained from $g(t)$ simply by substituting f for t. Solving (2.21) for $g(t)$, we obtain

$$g(t) = \exp(-\pi t^2) \tag{2.23}$$

which has a bell-shaped waveform, as illustrated in Figure 2.6. Such a pulse is called a *Gaussian pulse*, the name of which follows from the similarity of the function $g(t)$ to the Gaussian probability density function of probability theory, to be discussed in Chapter 3. By applying the Fourier-transform property on the area under $g(t)$ listed in Table 2.1, we have

$$\int_{-\infty}^{\infty} \exp(-\pi t^2)\,\mathrm{d}t = 1 \tag{2.24}$$

When the central ordinate and the area under the curve of a pulse are both unity, as in (2.23) and (2.24), we say that the Gaussian pulse is a *unit pulse*. Therefore, we may state that the unit Gaussian pulse is its own Fourier transform, as shown by

$$\exp(-\pi t^2) \rightleftharpoons \exp(-\pi f^2) \tag{2.25}$$

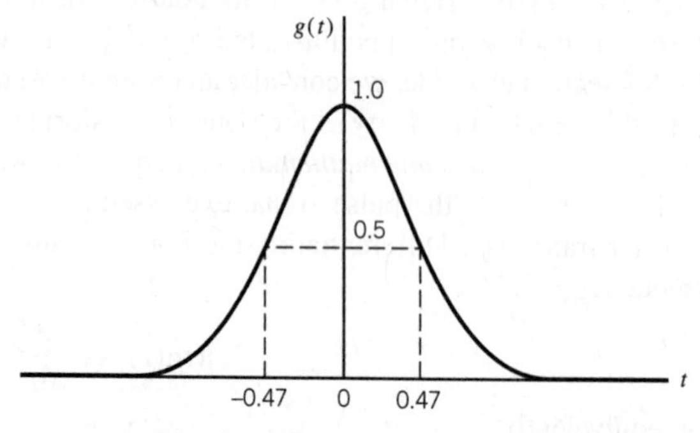

Figure 2.6 Gaussian pulse.

Table 2.1 **Fourier-transform theorems**

Property	Mathematical description
1. Linearity	$ag_1(t) + bg_2(t) \rightleftharpoons aG_1(f) + bG_2(f)$ where a and b are constants
2. Dilation	$g(at) \rightleftharpoons \frac{1}{\lvert a\rvert}G\left(\frac{f}{a}\right)$ where a is a constant
3. Duality	If $g(t) \rightleftharpoons G(f)$, then $G(t) \rightleftharpoons g(-f)$
4. Time shifting	$g(t-t_0) \rightleftharpoons G(f)\exp(-j2\pi f t_0)$
5. Frequency shifting	$g(t)\exp(-j2\pi f_0 t) \rightleftharpoons G(f-f_0)$
6. Area under $g(t)$	$\int_{-\infty}^{\infty} g(t)\mathrm{d}t = G(0)$
7. Area under $G(f)$	$g(0) = \int_{-\infty}^{\infty} G(f)\mathrm{d}f$
8. Differentiation in the time domain	$\frac{\mathrm{d}}{\mathrm{d}t}g(t) \rightleftharpoons j2\pi f G(f)$
9. Integration in the time domain	$\int_{-\infty}^{t} g(\tau)\mathrm{d}\tau \rightleftharpoons \frac{1}{j2\pi f}G(f) + \frac{G(0)}{2}\delta(f)$
10. Conjugate functions	If $g(t) \rightleftharpoons G(f)$, then $g^*(t) \rightleftharpoons G^*(-f)$
11. Multiplication in the time domain	$g_1(t)g_2(t) \rightleftharpoons \int_{-\infty}^{\infty} G_1(\lambda)G_2(f-\lambda)\mathrm{d}\lambda$
12. Convolution in the time domain	$\int_{-\infty}^{t} g_1(\tau)g_2(t-\tau)\mathrm{d}\tau \rightleftharpoons G_1(f)G_2(f)$
13. Correlation theorem	$\int_{-\infty}^{\infty} g_1(t)g_2^*(t-\tau)\mathrm{d}\tau \rightleftharpoons G_1(f)G_2^*(f)$
14. Rayleigh's energy theorem	$\int_{-\infty}^{\infty} \lvert g(t)\rvert^2\mathrm{d}t = \int_{-\infty}^{\infty} \lvert G(f)\rvert^2\mathrm{d}f$
15. Parseval's power theorem for periodic signal of period T_0	$\frac{1}{T_0}\int_{-T_0/2}^{T_0/2} \lvert g(t)\rvert^2\mathrm{d}t = \sum_{n=-\infty}^{\infty} \lvert G(f_n)\rvert^2, \quad f_n = n/T_0$

Table 2.2 **Fourier-transform pairs and commonly used time functions**

Time function	Fourier transform	Definitions
1. $\text{rect}\left(\frac{t}{T}\right)$	$T\,\text{sinc}(fT)$	Unit step function:
2. $\text{sinc}(2Wt)$	$\frac{1}{2W}\text{rect}\left(\frac{f}{2Wf}\right)$	$u(t) = \begin{cases} 1, & t > 0 \\ \frac{1}{2}, & t = 0 \\ 0, & t < 0 \end{cases}$
3. $\exp(-at)u(t), \quad a > 0$	$\frac{1}{a + j2\pi f}$	
4. $\exp(-a\|t\|), \quad a > 0$	$\frac{2a}{a^2 + (2\pi f)^2}$	Dirac delta function: $\delta(t) = 0$ for $t \neq 0$ and
5. $\exp(-\pi t^2)$	$\exp(-\pi f^2)$	$\int_{-\infty}^{\infty} \delta(t)dt = 1$
6. $\begin{cases} 1 - \frac{\|t\|}{T}, & \|t\| < T \\ 0, & \|t\| \geq T \end{cases}$	$T\,\text{sinc}^2(fT)$	Rectangular function: $\text{rect}(t) = \begin{cases} 1, & -\frac{1}{2} < t \leq \frac{1}{2} \\ 0, & \text{otherwise} \end{cases}$
7. $\delta(t)$	1	
8. 1	$\delta(f)$	Signum function:
9. $\delta(t - t_0)$	$\exp(-j2\pi f t_0)$	$\text{sgn}(t) = \begin{cases} +1, & t > 0 \\ 0, & t = 0 \\ -1, & t < 0 \end{cases}$
10. $\exp(j2\pi f_c t)$	$\delta(f - f_c)$	
11. $\cos(2\pi f_c t)$	$\frac{1}{2}[\delta(f - f_c) + \delta(f + f_c)]$	Sinc function:
12. $\sin(2\pi f_c t)$	$\frac{1}{2}[\delta(f - f_c) - \delta(f + f_c)]$	$\text{sinc}(t) = \frac{\sin(\pi t)}{\pi t}$
13. $\text{sgn}(t)$	$\frac{1}{j\pi f}$	Gaussian function: $g(t) = \exp(-\pi t^2)$
14. $\frac{1}{\pi t}$	$-j\,\text{sgn}(f)$	
15. $u(t)$	$\frac{1}{2}\delta(f) + \frac{1}{j2\pi f}$	
16. $\sum_{i=-\infty}^{\infty} \delta(t - iT_0)$	$f_0 \sum_{n=-\infty}^{\infty} \delta(f - nf_0), \quad f_0 = \frac{1}{T_0}$	

2.4 The Inverse Relationship between Time-Domain and Frequency-Domain Representations

The time-domain and frequency-domain descriptions of a signal are *inversely* related. In this context, we may make four important statements:

1. If the time-domain description of a signal is changed, the frequency-domain description of the signal is changed in an *inverse* manner, and vice versa. This inverse relationship prevents arbitrary specifications of a signal in both domains. In other words:

   We may specify an arbitrary function of time or an arbitrary spectrum, but we cannot specify them both together.

2. If a signal is strictly limited in frequency, then the time-domain description of the signal will trail on indefinitely, even though its amplitude may assume a progressively smaller value. To be specific, we say:

   A signal is strictly limited in frequency (i.e., strictly band limited) if its Fourier transform is exactly zero outside a finite band of frequencies.

   Consider, for example, the band-limited sinc pulse defined by

   $$\text{sinc}(t) = \frac{\sin(\pi t)}{\pi t}$$

   whose waveform and spectrum are respectively shown in Figure 2.4: part a shows that the sinc pulse is *asymptotically limited in time* and part b of the figure shows that the sinc pulse is indeed *strictly band limited*, thereby confirming statement 2.

3. In a dual manner to statement 2, we say:

   If a signal is strictly limited in time (i.e., the signal is exactly zero outside a finite time interval), then the spectrum of the signal is infinite in extent, even though the magnitude spectrum may assume a progressively smaller value.

   This third statement is exemplified by a *rectangular pulse*, the waveform and spectrum of which are defined in accordance with item 1 in Table 2.2.

4. In light of the duality described under statements 2 and 3, we now make the final statement:

   A signal cannot be strictly limited in both time and frequency.

The Bandwidth Dilemma

The statements we have just made have an important bearing on the *bandwidth* of a signal, which provides a measure of the *extent of significant spectral content of the signal for positive frequencies*. When the signal is strictly band limited, the bandwidth is well defined. For example, the sinc pulse sinc($2Wt$) has a bandwidth equal to W. However, when the signal is not strictly band limited, as is often the case, we encounter difficulty in defining the bandwidth of the signal. The difficulty arises because the meaning of "significant" attached to the spectral content of the signal is mathematically imprecise. Consequently, there is no universally accepted definition of bandwidth. It is in this sense that we speak of the "bandwidth dilemma."

Nevertheless, there are some commonly used definitions for bandwidth, as discussed next. When the spectrum of a signal is symmetric with a main lobe bounded by well-defined nulls (i.e., frequencies at which the spectrum is zero), we may use the main lobe as the basis for defining the bandwidth of the signal. Specifically:

> If a signal is low-pass (i.e., its spectral content is centered around the origin $f = 0$), the bandwidth is defined as one-half the total width of the main spectral lobe, since only one-half of this lobe lies inside the positive frequency region.

For example, a rectangular pulse of duration T seconds has a main spectral lobe of total width $(2/T)$ hertz centered at the origin. Accordingly, we may define the bandwidth of this rectangular pulse as $(1/T)$ hertz.

If, on the other hand, the signal is *band-pass* with main spectral lobes centered around $\pm f_c$, where f_c is large enough, the bandwidth is defined as the width of the main lobe for positive frequencies. This definition of bandwidth is called the *null-to-null bandwidth*. Consider, for example, a radio-frequency (RF) pulse of duration T seconds and frequency f_c, shown in Figure 2.7. The spectrum of this pulse has main spectral lobes of width $(2/T)$ hertz centered around $\pm f_c$, where it is assumed that f_c is large compared with $(1/T)$. Hence, we define the null-to-null bandwidth of the RF pulse of Figure 2.7 as $(2/T)$ hertz.

On the basis of the definitions presented here, we may state that shifting the spectral content of a low-pass signal by a sufficiently large frequency has the effect of doubling the bandwidth of the signal; this frequency translation is attained by using the process of modulation. Basically, the modulation moves the spectral content of the signal for negative frequencies into the positive frequency region, whereupon the negative frequencies become physically measurable.

Another popular definition of bandwidth is the *3 dB bandwidth*. Specifically, if the signal is low-pass, we say:

> The 3 dB bandwidth of a low-pass signal is defined as the separation between zero frequency, where the magnitude spectrum attains its peak value, and the positive frequency at which the amplitude spectrum drops to $1/\sqrt{2}$ of its peak value.

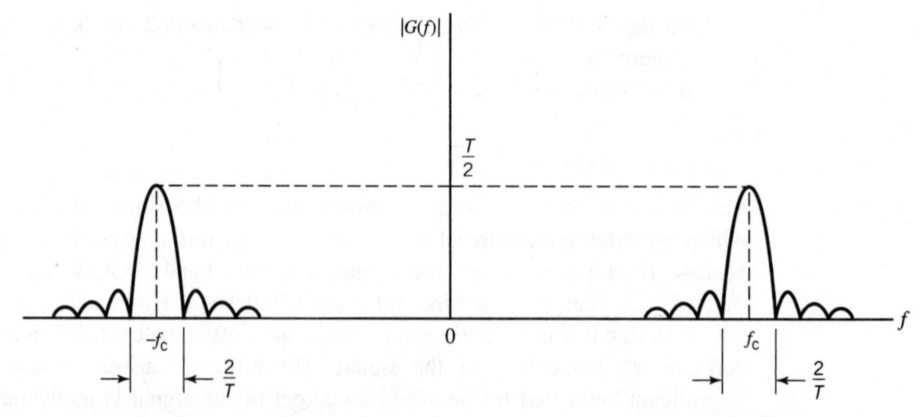

Figure 2.7 Magnitude spectrum of the RF pulse, showing the null-to-null bandwidth to be $2/T$, centered on the mid-band frequency f_c.

For example, the decaying exponential function $\exp(-at)$ has a 3 dB bandwidth of $(a/2\pi)$ hertz.

If, on the other hand, the signal is of a band-pass kind, centered at $\pm f_c$, the 3 dB bandwidth is defined as the separation (along the positive frequency axis) between the two frequencies at which the magnitude spectrum of the signal drops to $1/\sqrt{2}$ of its peak value at f_c.

Regardless of whether we have a low-pass or band-pass signal, the 3 dB bandwidth has the advantage that it can be read directly from a plot of the magnitude spectrum. However, it has the disadvantage that it may be misleading if the magnitude spectrum has slowly decreasing tails.

Time–Bandwidth Product

For any family of pulse signals that differ by a time-scaling factor, the product of the signal's duration and its bandwidth is always a constant, as shown by

$$\text{duration} \times \text{bandwidth} = \text{constant}$$

This product is called the *time–bandwidth product*. The constancy of the time–bandwidth product is another manifestation of the inverse relationship that exists between the time-domain and frequency-domain descriptions of a signal. In particular, if the duration of a pulse signal is decreased by reducing the time scale by a factor a, the frequency scale of the signal's spectrum, and therefore the bandwidth of the signal is increased by the same factor a. This statement follows from the *dilation property* of the Fourier transform (defined in Property 2 of Table 2.1). The time–bandwidth product of the signal is therefore maintained constant. For example, a rectangular pulse of duration T seconds has a bandwidth (defined on the basis of the positive-frequency part of the main lobe) equal to $(1/T)$ hertz; in this example, the time–bandwidth product of the pulse equals unity.

The important point to take from this discussion is that whatever definitions we use for the bandwidth and duration of a signal, the time–bandwidth product remains constant over certain classes of pulse signals; the choice of particular definitions for bandwidth and duration merely change the value of the constant.

Root-Mean-Square Definitions of Bandwidth and Duration

To put matters pertaining to the bandwidth and duration of a signal on a firm mathematical basis, we first introduce the following definition for bandwidth:

> The root-mean-square (rms) bandwidth is defined as the square root of the second moment of a normalized form of the squared magnitude spectrum of the signal about a suitably chosen frequency.

To be specific, we assume that the signal $g(t)$ is of a low-pass kind, in which case the second moment is taken about the origin $f = 0$. The squared magnitude spectrum of the signal is denoted by $|G(f)|^2$. To formulate a nonnegative function, the total area under whose curve is unity, we use the normalizing function

$$\int_{-\infty}^{\infty} |G(f)|^2 df$$

We thus mathematically define the rms bandwidth of a low-pass signal $g(t)$ with Fourier transform $G(f)$ as

$$W_{\mathrm{rms}} = \left(\frac{\int_{-\infty}^{\infty} f^2 |G(f)|^2 df}{\int_{-\infty}^{\infty} |G(f)|^2 df} \right)^{1/2} \tag{2.26}$$

which describes the dispersion of the spectrum $G(f)$ around $f = 0$. An attractive feature of the rms bandwidth W_{rms} is that it lends itself readily to mathematical evaluation. But, it is not as easily measurable in the laboratory.

In a manner corresponding to the rms bandwidth, the *rms duration* of the signal $g(t)$ is mathematically defined by

$$T_{\mathrm{rms}} = \left(\frac{\int_{-\infty}^{\infty} t^2 |g(t)|^2 dt}{\int_{-\infty}^{\infty} |g(t)|^2 dt} \right)^{1/2} \tag{2.27}$$

where it is assumed that the signal $g(t)$ is centered around the origin $t = 0$. In Problem 2.7, it is shown that, using the rms definitions of (2.26) and (2.27), the time–bandwidth product takes the form

$$T_{\mathrm{rms}} W_{\mathrm{rms}} \geq \frac{1}{4\pi} \tag{2.28}$$

In Problem 2.7, it is also shown that the Gaussian pulse $\exp(-\pi t^2)$ satisfies this condition exactly with the equality sign.

2.5 The Dirac Delta Function

Strictly speaking, the theory of the Fourier transform, presented in Section 2.3, is applicable only to time functions that satisfy the Dirichlet conditions. As mentioned previously, such functions naturally include energy signals. However, it would be highly desirable to extend this theory in two ways:

1. To combine the Fourier series and Fourier transform into a unified theory, so that the Fourier series may be treated as a special case of the Fourier transform.
2. To include power signals in the list of signals to which we may apply the Fourier transform. A signal $g(t)$ is said to be a *power signal* if the condition

$$\frac{1}{T}\int_{-T/2}^{T/2} |g(t)|^2 dt < \infty$$

holds, where T is the observation interval.

It turns out that both of these objectives can be met through the "proper use" of the *Dirac delta function*, or *unit impulse*.

The Dirac delta function[3] or just delta function, denoted by $\delta(t)$, is defined as having zero amplitude everywhere except at $t = 0$, where it is infinitely large in such a way that it contains unit area under its curve; that is,

$$\delta(t) = 0, \qquad t \neq 0 \tag{2.29}$$

and

$$\int_{-\infty}^{\infty} \delta(t)\,\mathrm{d}t = 1 \tag{2.30}$$

An implication of this pair of relations is that the delta function $\delta(t)$ is an even function of time t, centered at the origin $t = 0$. Perhaps, the simplest way of describing the Dirac delta function is to view it as the rectangular pulse

$$g(t) = \frac{1}{T}\,\mathrm{rect}\left(\frac{t}{T}\right)$$

whose duration is T and amplitude is $1/T$, as illustrated in Figure 2.8. As T approaches zero, the rectangular pulse $g(t)$ approaches the Dirac delta function $\delta(t)$ in the limit.

For the delta function to have meaning, however, it has to appear as a factor in the integrand of an integral with respect to time, and then, strictly speaking, only when the other factor in the integrand is a continuous function of time. Let $g(t)$ be such a function, and consider the product of $g(t)$ and the time-shifted delta function $\delta(t - t_0)$. In light of the two defining equations (2.29) and (2.30), we may express the integral of this product as

$$\int_{-\infty}^{\infty} g(t)\,\delta(t - t_0)\mathrm{d}t = g(t_0) \tag{2.31}$$

The operation indicated on the left-hand side of this equation sifts out the value $g(t_0)$ of the function $g(t)$ at time $t = t_0$, where $-\infty < t < \infty$. Accordingly, (2.31) is referred to as the *sifting property* of the delta function. This property is sometimes used as the defining equation of a delta function; in effect, it incorporates (2.29) and (2.30) into a single relation.

Noting that the delta function $\delta(t)$ is an even function of t, we may rewrite (2.31) so as to emphasize its resemblance to the convolution integral, as shown by

$$\int_{-\infty}^{\infty} g(\tau)\,\delta(t - \tau)\,\mathrm{d}\tau = g(t) \tag{2.32}$$

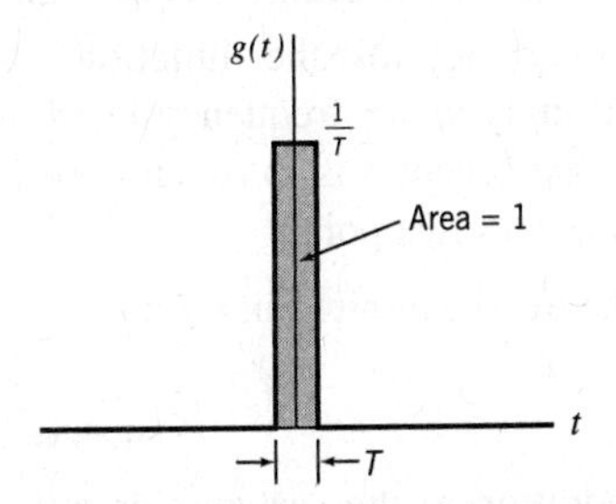

Figure 2.8 Illustrative example of the Dirac delta function as the limiting form of rectangular pulse $\frac{1}{T}\,\mathrm{rect}\left(\frac{t}{T}\right)$ as T approaches zero.

In words, the convolution of any function with the delta function leaves that function unchanged. We refer to this statement as the *replication property* of the delta function.

It is important to realize that no function in the ordinary sense has the two properties of (2.29) and (2.30) or the equivalent sifting property of (2.31). However, we can imagine a sequence of functions that have progressively taller and thinner peaks at $t = 0$, with the area under the curve consistently remaining equal to unity; as this progression is being performed, the value of the function tends to zero at every point except $t = 0$, where it tends to infinity, as illustrated in Figure 2.8, for example. We may therefore say:

> The delta function may be viewed as the limiting form of a pulse of unit area as the duration of the pulse approaches zero.

It is immaterial what sort of pulse shape is used, so long as it is symmetric with respect to the origin; this symmetry is needed to maintain the "even" function property of the delta function.

Two other points are noteworthy:

1. Applicability of the delta function is not confined to the time domain. Rather, it can equally well be applied in the frequency domain; all that we have to do is to replace time t by frequency f in the defining equations (2.29) and (2.30).
2. The area covered by the delta function defines its "strength." As such, the units, in terms of which the strength is measured, are determined by the specifications of the two coordinates that define the delta function.

EXAMPLE 3 The Sinc Function as a Limiting Form of the Delta Function in the Time Domain

As another illustrative example, consider the scaled sinc function $2W\text{sinc}(2Wt)$, whose waveform covers an area equal to unity for all W.

Figure 2.9 displays the evolution of this time function toward the delta function as the parameter W is varied in three stages: $W = 1$, $W = 2$, and $W = 5$. Referring back to Figure 2.4, we may infer that as the parameter W characterizing the sinc pulse is increased, the amplitude of the pulse at time $t = 0$ increases linearly, while at the same time the duration of the main lobe of the pulse decreases inversely. With this objective in mind, as the parameter W is progressively increased, Figure 2.9 teaches us two important things:

1. The scaled sinc function becomes more like a delta function.
2. The constancy of the function's spectrum is maintained at unity across an increasingly wider frequency band, in accordance with the constraint that the area under the function is to remain constant at unity; see Property 6 of Table 2.1 for a validation of this point.

Based on the trend exhibited in Figure 2.9, we may write

$$\delta(t) = \lim_{W \to \infty} 2W\,\text{sinc}(2Wt) \tag{2.33}$$

which, in addition to the rectangular pulse considered in Figure 2.8, is another way of realizing a delta function in the time domain.

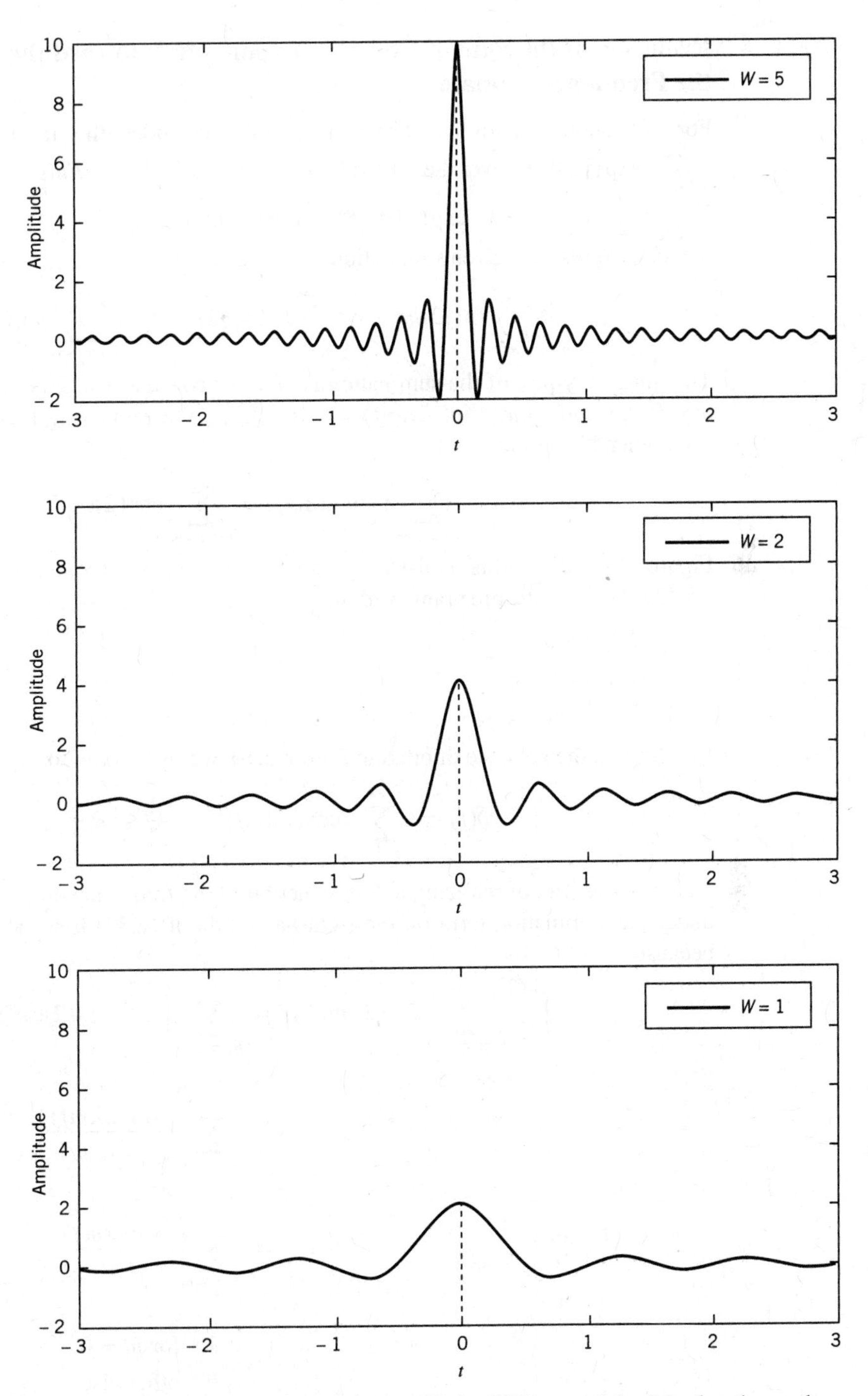

Figure 2.9 Evolution of the sinc function $2W\,\mathrm{sinc}(2Wt)$ toward the delta function as the parameter W progressively increases.

EXAMPLE 4 **Evolution of the Sum of Complex Exponentials toward the Delta Function in the Frequency Domain**

For yet another entirely different example, consider the infinite summation term $\sum_{m=-\infty}^{\infty} \exp(j2\pi mf)$ over the interval $-1/2 \le f < 1/2$. Using *Euler's formula*

$$\exp(j2\pi mf) = \cos(2\pi mf) + j\sin(2\pi mf)$$

we may express the given summation as

$$\sum_{m=-\infty}^{\infty} \exp(j2\pi mf) = \sum_{m=-\infty}^{\infty} \cos(2\pi mf) + j \sum_{m=-\infty}^{\infty} \sin(2\pi mf)$$

The imaginary part of the summation is zero for two reasons. First, $\sin(2\pi mf)$ is zero for $m = 0$. Second, since $\sin(-2\pi mf) = -\sin(2\pi mf)$, the remaining imaginary terms cancel each other. Therefore,

$$\sum_{m=-\infty}^{\infty} \exp(j2\pi mf) = \sum_{m=-\infty}^{\infty} \cos(2\pi mf)$$

Figure 2.10 plots this real-valued summation versus frequency f over the interval $-1/2 \le f < 1/2$ for three ranges of m:

1. $-5 \le m \le 5$
2. $-10 \le m \le 10$
3. $-20 \le m \le 20$

Building on the results exhibited in Figure 2.10, we may go on to say

$$\delta(f) = \sum_{m=-\infty}^{\infty} \cos(2\pi mf), \quad -\frac{1}{2} \le f < \frac{1}{2} \tag{2.34}$$

which is one way of realizing a delta function in the frequency domain. Note that the area under the summation term on the right-hand side of (2.34) is equal to unity; we say so because

$$\begin{aligned}
\int_{-1/2}^{1/2} \sum_{m=-\infty}^{\infty} \cos(2\pi mf)\,df &= \sum_{m=-\infty}^{\infty} \int_{-1/2}^{1/2} \cos(2\pi mf)\,df \\
&= \sum_{m=-\infty}^{\infty} \left[\frac{\sin(2\pi mf)}{2\pi m}\right]_{f=-1/2}^{1/2} \\
&= \sum_{m=-\infty}^{\infty} \left[\frac{\sin(\pi m)}{\pi m}\right] \\
&= \begin{cases} 1 & \text{for } m = 0 \\ 0 & \text{otherwise} \end{cases}
\end{aligned}$$

This result, formulated in the frequency domain, confirms (2.34) as one way of defining the delta function $\delta(f)$.

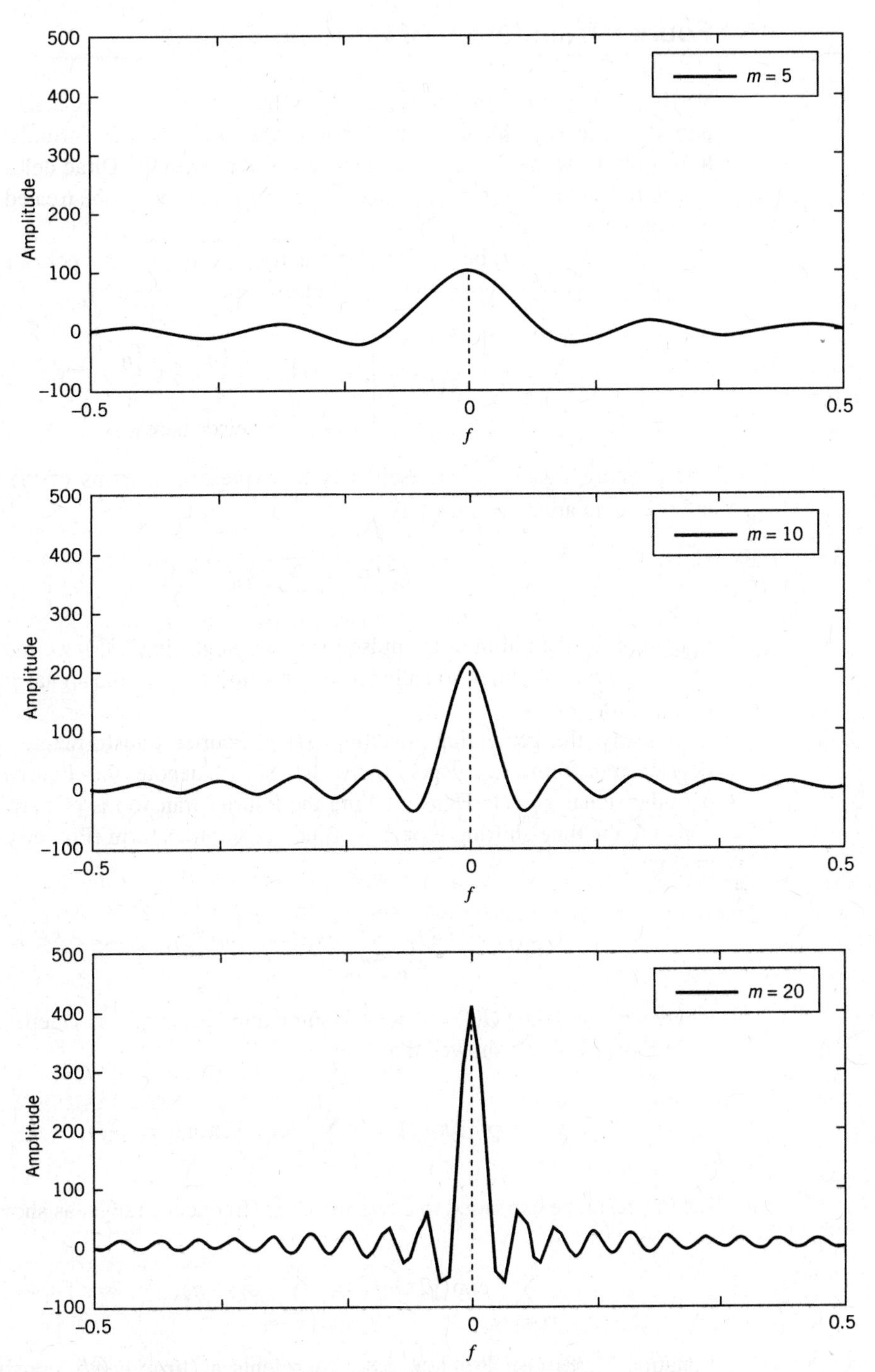

Figure 2.10 Evolution of the sum of m complex exponentials toward a delta function in the frequency domain as m becomes increasingly larger.

2.6 Fourier Transforms of Periodic Signals

We began the study of Fourier analysis by reviewing the Fourier series expansion of periodic signals, which, in turn, paved the way for the formulation of the Fourier transform. Now that we have equipped ourselves with the Dirac delta function, we would like to revisit the Fourier series and show that it can indeed be treated as a special case of the Fourier transform.

To this end, let $g(t)$ be a pulse-like function, which equals a periodic signal $g_{T_0}(t)$ over one period T_0 of the signal and is zero elsewhere, as shown by

$$g(t) = \begin{cases} g_{T_0}(t), & -\dfrac{T_0}{2} < t \leq \dfrac{T_0}{2} \\ 0, & \text{elsewhere} \end{cases} \tag{2.35}$$

The periodic signal $g_{T_0}(t)$ itself may be expressed in terms of the function $g(t)$ as an infinite summation, as shown by

$$g_{T_0}(t) = \sum_{m=-\infty}^{\infty} g(t - mT_0) \tag{2.36}$$

In light of the definition of the pulselike function $g(t)$ in (2.35), we may view this function as a *generating function*, so called as it generates the periodic signal $g_{T_0}(t)$ in accordance with (2.36).

Clearly, the generating function $g(t)$ is Fourier transformable; let $G(f)$ denote its Fourier transform. Correspondingly, let $G_{T_0}(f)$ denote the Fourier transform of the periodic signal $g_{T_0}(t)$. Hence, taking the Fourier transforms of both sides of (2.36) and applying the time-shifting property of the Fourier transform (Property 4 of Table 2.1), we may write

$$G_{T_0}(f) = G(f) \sum_{m=-\infty}^{\infty} \exp(-j2\pi m f T_0), \qquad -\infty < f < \infty \tag{2.37}$$

where we have taken $G(f)$ outside the summation because it is independent of m.

In Example 4, we showed that

$$\sum_{m=-\infty}^{\infty} \exp(j2\pi m f) = \sum_{m=-\infty}^{\infty} \cos(j2\pi m f) = \delta(f), \qquad -\frac{1}{2} \leq f < \frac{1}{2}$$

Let this result be expanded to cover the entire frequency range, as shown by

$$\sum_{m=-\infty}^{\infty} \exp(j2\pi m f) = \sum_{n=-\infty}^{\infty} \delta(f - n), \qquad -\infty < f < \infty \tag{2.38}$$

Equation (2.38) (see Problem 2.8c) represents a *Dirac comb*, consisting of an infinite sequence of uniformly spaced delta functions, as depicted in Figure 2.11.

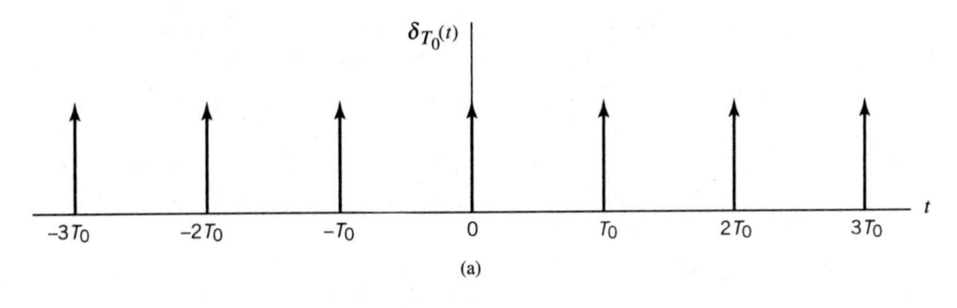

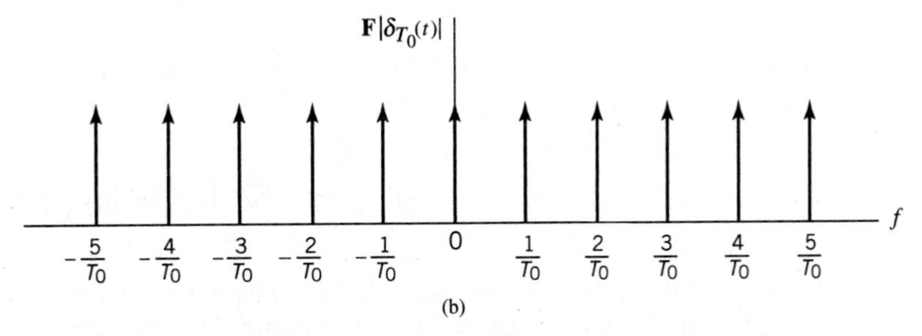

Figure 2.11 (a) Dirac comb. (b) Spectrum of the Dirac comb.

Next, introducing the frequency-scaling factor $f_0 = 1/T_0$ into (2.38), we correspondingly write

$$\sum_{m=-\infty}^{\infty} \exp(\mathrm{j}2\pi m f T_0) = f_0 \sum_{n=-\infty}^{\infty} \delta(f - nf_0), \qquad -\infty < f < \infty \tag{2.39}$$

Hence, substituting (2.39) into the right-hand side of (2.37), we get

$$\begin{aligned} G_{T_0}(f) &= f_0 G(f) \sum_{n=-\infty}^{\infty} \delta(f - nf_0) \\ &= f_0 \sum_{n=-\infty}^{\infty} G(f_n)\delta(f - f_n), \qquad -\infty < f < \infty \end{aligned} \tag{2.40}$$

where $f_n = nf_0$.

What we have to show next is that the inverse Fourier transform of $G_{T_0}(f)$ defined in (2.40) is exactly the same as in the Fourier series formula of (2.14). Specifically, substituting (2.40) into the inverse Fourier transform formula of (2.17), we get

$$g_{T_0}(t) = f_0 \int_{-\infty}^{\infty} \left[\sum_{n=-\infty}^{\infty} G(f_n)\delta(f - f_n) \right] \exp(\mathrm{j}2\pi f t)\, \mathrm{d}f$$

Interchanging the order of summation and integration, and then invoking the sifting property of the Dirac delta function (this time in the frequency domain), we may go on to write

$$g_{T_0}(t) = f_0 \sum_{n=-\infty}^{\infty} \int_{-\infty}^{\infty} G(f_n) \exp(j2\pi ft)\delta(f-f_n)\,df$$

$$= f_0 \sum_{n=-\infty}^{\infty} G(f_n) \exp(j2\pi f_n t)$$

which is an exact rewrite of (2.14) with $f_0 = \Delta f$. Equivalently, in light of (2.36), we may formulate the Fourier transform pair

$$\sum_{m=-\infty}^{\infty} g(t-mT_0) = f_0 \sum_{n=-\infty}^{\infty} G(f_n) \exp(j2\pi f_n t) \tag{2.41}$$

The result derived in (2.41) is one form of *Poisson's sum formula*.

We have thus demonstrated that the Fourier series representation of a periodic signal is embodied in the Fourier transformation of (2.16) and (2.17), provided, of course, we permit the use of the Dirac delta function. In so doing, we have closed the "circle" by going from the Fourier series to the Fourier transform, and then back to the Fourier series.

Consequences of Ideal Sampling

Consider a Fourier transformable pulselike signal $g(t)$ with its Fourier transform denoted by $G(f)$. Setting $f_n = nf_0$ in (2.41) and using (2.38), we may express Poisson's sum formula

$$\sum_{m=-\infty}^{\infty} g(t-mT_0) \rightleftharpoons f_0 \sum_{n=-\infty}^{\infty} G(nf_0)\delta(f-nf_0) \tag{2.42}$$

where $f_0 = 1/T_0$. The summation on the left-hand side of this Fourier-transform pair is a periodic signal with period T_0. The summation on the right-hand side of the pair is a uniformly sampled version of the spectrum $G(f)$. We may therefore make the following statement:

> Uniform sampling of the spectrum $G(f)$ in the frequency domain introduces periodicity of the function $g(t)$ in the time domain.

Applying the duality property of the Fourier transform (Property 3 of Table 2.1) to (2.42), we may also write

$$T_0 \sum_{m=-\infty}^{\infty} g(mT_0)\delta(t-mT_0) \rightleftharpoons \sum_{n=-\infty}^{\infty} G(f-nf_0) \tag{2.43}$$

in light of which we may make the following dual statement:

> Uniform sampling of the Fourier transformable function $g(t)$ in the time domain introduces periodicity of the spectrum $G(f)$ in the frequency domain.

2.7 Transmission of Signals through Linear Time-Invariant Systems

A *system* refers to any physical entity that produces an output signal in response to an input signal. It is customary to refer to the input signal as the *excitation* and to the output signal as the *response*. In a linear system, the *principle of superposition* holds; that is, the response of a linear system to a number of excitations applied simultaneously is equal to the sum of the responses of the system when each excitation is applied individually.

In the time domain, a linear system is usually described in terms of its *impulse response*, which is formally defined as follows:

> The impulse response of a linear system is the response of the system (with zero initial conditions) to a unit impulse or delta function $\delta(t)$ applied to the input of the system at time $t = 0$.

If the system is also *time invariant*, then the shape of the impulse response is the same no matter when the unit impulse is applied to the system. Thus, with the unit impulse or delta function applied to the system at time $t = 0$, the impulse response of a linear time-invariant system is denoted by $h(t)$.

Suppose that a system described by the impulse response $h(t)$ is subjected to an arbitrary excitation $x(t)$, as depicted in Figure 2.12. The resulting response of the system $y(t)$, is defined in terms of the impulse response $h(t)$ by

$$y(t) = \int_{-\infty}^{\infty} x(\tau)h(t-\tau)\,d\tau \tag{2.44}$$

which is called the *convolution integral*. Equivalently, we may write

$$y(t) = \int_{-\infty}^{\infty} h(\tau)x(t-\tau)\,d\tau \tag{2.45}$$

Equations (2.44) and (2.45) state that convolution is *commutative*.

Examining the convolution integral of (2.44), we see that three different time scales are involved: *excitation time* τ, *response time* t, and *system-memory time* $t - \tau$. This relation is the basis of time-domain analysis of linear time-invariant systems. According to (2.44), the present value of the response of a linear time-invariant system is an integral over the past history of the input signal, weighted according to the impulse response of the system. Thus, the impulse response acts as a *memory function* of the system.

Causality and Stability

A linear system with impulse response $h(t)$ is said to be *causal* if its impulse response $h(t)$ satisfies the condition

$$h(t) = 0 \quad \text{for} \quad t < 0$$

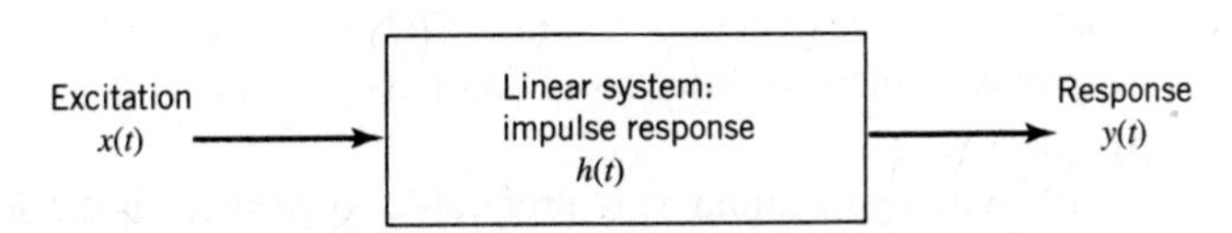

Figure 2.12 Illustrating the roles of excitation $x(t)$, impulse response $h(t)$, and response $y(t)$ in the context of a linear time-invariant system.

The essence of causality is that no response can appear at the output of the system before an excitation is applied to its input. Causality is a necessary requirement for on-line operation of the system. In other words, for a system operating in *real time* to be physically realizable, it has to be causal.

Another important property of a linear system is *stability.* A necessary and sufficient condition for the system to be stable is that its impulse response $h(t)$ must satisfy the inequality

$$\int_{-\infty}^{\infty} |h(t)|\,dt < \infty$$

This requirement follows from the commonly used criterion of *bounded input–bounded output*. Basically, for the system to be stable, its impulse response must be *absolutely integrable*.

Frequency Response

Let $X(f)$, $H(f)$, and $Y(f)$ denote the Fourier transforms of the excitation $x(t)$, impulse response $h(t)$, and response $y(t)$, respectively. Then, applying Property 12 of the Fourier transform in Table 2.1 to the convolution integral, be it written in the form of (2.44) or (2.45), we get

$$Y(f) = H(f)X(f) \tag{2.46}$$

Equivalently, we may write

$$H(f) = \frac{Y(f)}{X(f)} \tag{2.47}$$

The new frequency function $H(f)$ is called the *transfer function* or *frequency response* of the system; these two terms are used interchangeably. Based on (2.47), we may now formally say:

> The frequency response of a linear time-invariant system is defined as the ratio of the Fourier transform of the response of the system to the Fourier transform of the excitation applied to the system.

In general, the frequency response $H(f)$ is a complex quantity, so we may express it in the form

$$H(f) = |H(f)|\exp[j\beta(f)] \tag{2.48}$$

where $|H(f)|$ is called the *magnitude response*, and $\beta(f)$ is the *phase response*, or simply *phase*. When the impulse response of the system is real valued, the frequency response exhibits conjugate symmetry, which means that

$$|H(f)| = |H(-f)|$$

and

$$\beta(f) = -\beta(-f)$$

That is, the magnitude response $|H(f)|$ of a linear system with real-valued impulse response is an even function of frequency, whereas the phase $\beta(f)$ is an odd function of frequency.

In some applications it is preferable to work with the logarithm of $H(f)$ expressed in polar form, rather than with $H(f)$ itself. Using ln to denote the natural logarithm, let

$$\ln H(f) = \alpha(f) + j\beta(f) \tag{2.49}$$

where

$$\alpha(f) = \ln|H(f)| \tag{2.50}$$

The function $\alpha(f)$ is called the *gain* of the system; it is measured in *nepers*. The phase $\beta(f)$ is measured in *radians*. Equation (2.49) indicates that the gain $\alpha(f)$ and phase $\beta(f)$ are, respectively, the real and imaginary parts of the (natural) logarithm of the transfer function $H(f)$. The gain may also be expressed in *decibels* (dB) by using the definition

$$\alpha'(f) = 20\log_{10}|H(f)|$$

The two gain functions $\alpha(f)$ and $\alpha'(f)$ are related by

$$\alpha'(f) = 8.69\,\alpha(f)$$

That is, 1 neper is equal to 8.69 dB.

As a means of specifying the constancy of the magnitude response $|H(f)|$ or gain $\alpha(f)$ of a system, we use the notion of *bandwidth*. In the case of a low-pass system, the bandwidth is customarily defined as the frequency at which the magnitude response $|H(f)|$ is $1/\sqrt{2}$ times its value at zero frequency or, equivalently, the frequency at which the gain $\alpha'(f)$ drops by 3 dB below its value at zero frequency, as illustrated in Figure 2.13a. In the case of a band-pass system, the bandwidth is defined as the range of frequencies over which the magnitude response $|H(f)|$ remains within $1/\sqrt{2}$ times its value at the mid-band frequency, as illustrated in Figure 2.13b.

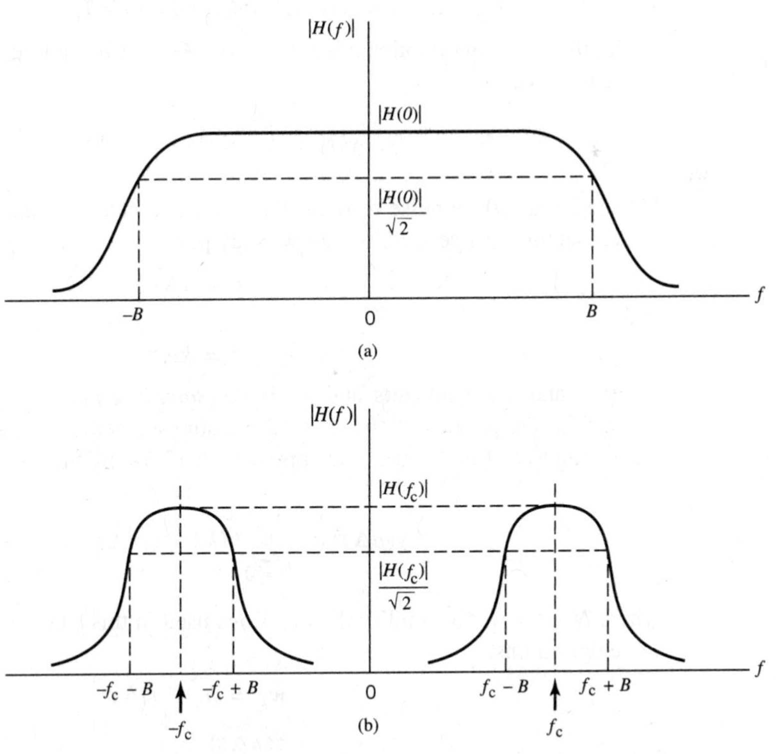

Figure 2.13 Illustrating the definition of system bandwidth. (a) Low-pass system. (b) Band-pass system.

Paley–Wiener Criterion: Another Way of Assessing Causality

A necessary and sufficient condition for a function $\alpha(f)$ to be the gain of a causal filter is the convergence of the integral

$$\int_{-\infty}^{\infty} \frac{|\alpha(f)|}{1+f^2} \mathrm{d}f < \infty \tag{2.51}$$

This condition is known as the *Paley–Wiener criterion.*[4] The criterion states that provided the gain $\alpha(f)$ satisfies the condition of (2.51), then we may associate with this gain a suitable phase $\beta(f)$, such that the resulting filter has a causal impulse response that is zero for negative time. In other words, the Paley–Wiener criterion is the frequency-domain equivalent of the causality requirement. A system with a realizable gain characteristic may have infinite attenuation for a discrete set of frequencies, but it cannot have infinite attenuation over a band of frequencies; otherwise, the Paley–Wiener criterion is violated.

Finite-Duration Impulse Response (FIR) Filters

Consider next a linear time-invariant filter with impulse response $h(t)$. We make two assumptions:

1. *Causality,* which means that the impulse response $h(t)$ is zero for $t < 0$.
2. *Finite support*, which means that the impulse response of the filter is of some finite duration T_f, so that we may write $h(t) = 0$ for $t \geq T_f$.

Under these two assumptions, we may express the filter output $y(t)$ produced in response to the input $x(t)$ as

$$y(t) = \int_0^{T_f} h(\tau)x(t-\tau)\,\mathrm{d}\tau \tag{2.52}$$

Let the input $x(t)$, impulse response $h(t)$, and output $y(t)$ be *uniformly sampled* at the rate $(1/\Delta\tau)$ samples per second, so that we may put

$$t = n\Delta\tau$$

and

$$\tau = k\Delta\tau$$

where k and n are integers and $\Delta\tau$ is the *sampling period.* Assuming that $\Delta\tau$ is small enough for the product $h(\tau)x(t-\tau)$ to remain essentially constant for $k\Delta\tau \leq \tau \leq (k+1)\Delta\tau$ for all values of k and τ, we may approximate (2.52) by the *convolution sum*

$$y(n\Delta\tau) = \sum_{k=0}^{N-1} h(k\Delta\tau)x(n\Delta\tau - k\Delta\tau)\Delta\tau$$

where $N\Delta\tau = T_f$. To simplify the notations used in this summation formula, we introduce three definitions:

$$w_k = h(k\Delta\tau)\Delta\tau$$

$$x(n\Delta\tau) = x_n$$

$$y(n\Delta\tau) = y_n$$

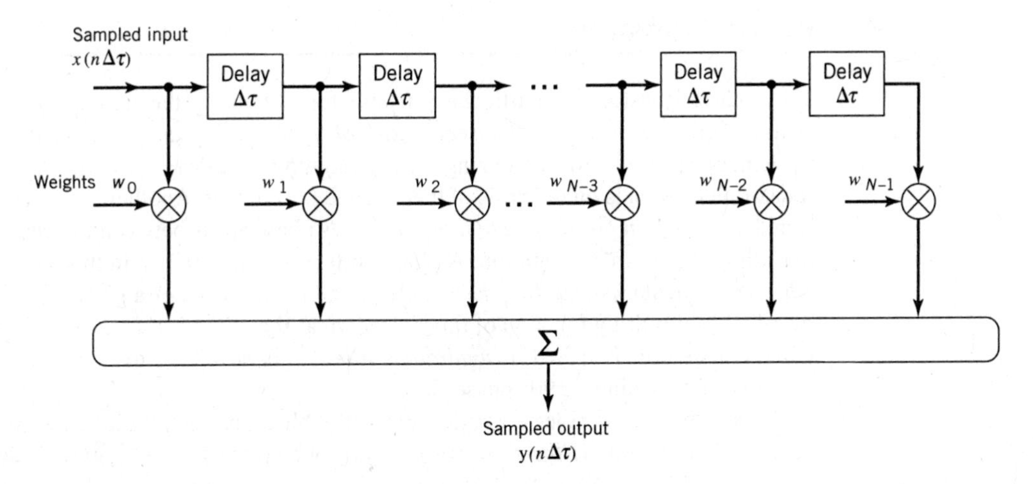

Figure 2.14 Tapped-delay-line (TDL) filter; also referred to as FIR filter.

We may then rewrite the formula for $y(n\Delta\tau)$ in the compact form

$$y_n = \sum_{k=0}^{N-1} w_k x_{n-k}, \qquad n = 0, \pm 1, \pm 2, \ldots \tag{2.53}$$

Equation (2.53) may be realized using the structure shown in Figure 2.14, which consists of a set of *delay elements* (each producing a delay of $\Delta\tau$ seconds), a set of *multipliers* connected to the *delay-line taps*, a corresponding set of *weights* supplied to the multipliers, and a *summer* for adding the multiplier outputs. The sequences x_n and y_n, for integer values of n as described in (2.53), are referred to as the *input* and *output sequences,* respectively.

In the digital signal-processing literature, the structure of Figure 2.14 is known as a *finite-duration impulse response (FIR) filter.* This filter offers some highly desirable practical features:

1. The filter is inherently *stable*, in the sense that a bounded input sequence produces a bounded output sequence.
2. Depending on how the weights $\{w_k\}_{k=0}^{N-1}$ are designated, the filter can perform the function of a low-pass filter or band-pass filter. Moreover, the phase response of the filter can be configured to be a linear function of frequency, which means that there will be no delay distortion.
3. In a digital realization of the filter, the filter assumes a *programmable* form whereby the application of the filter can be changed merely by making appropriate changes to the weights, leaving the structure of the filter completely unchanged; this kind of flexibility is not available with analog filters.

We will have more to say on the FIR filter in subsequent chapters of the book.

2.8 Hilbert Transform

The Fourier transform is particularly useful for evaluating the frequency content of an energy signal or, in a limiting sense, that of a power signal. As such, it provides the mathematical basis for analyzing and designing *frequency-selective filters* for the separation of signals on the basis of their frequency content. Another method of separating signals is based on *phase selectivity*, which uses phase shifts between the pertinent signals to achieve the desired separation. A phase shift of special interest in this context is that of ±90°. In particular, when the phase angles of all components of a given signal are shifted by ±90°, the resulting function of time is known as the *Hilbert transform* of the signal. The Hilbert transform is called a *quadrature filter*; it is so called to emphasize its distinct property of providing a ±90° phase shift.

To be specific, consider a Fourier transformable signal $g(t)$ with its Fourier transform denoted by $G(f)$. The *Hilbert transform* of $g(t)$, which we denote by $\hat{g}(t)$, is defined by[5]

$$\hat{g}(t) = \frac{1}{\pi}\int_{-\infty}^{\infty} \frac{g(\tau)}{t-\tau}\, d\tau \tag{2.54}$$

Table 2.3 **Hilbert-transform pairs***

Time function	Hilbert transform
1. $m(t)\cos(2\pi f_c t)$	$m(t)\sin(2\pi f_c t)$
2. $m(t)\sin(2\pi f_c t)$	$-m(t)\cos(2\pi f_c t)$
3. $\cos(2\pi f_c t)$	$\sin(2\pi f_c t)$
4. $\sin(2\pi f_c t)$	$-\cos(2\pi f_c t)$
5. $\dfrac{\sin t}{t}$	$\dfrac{1-\cos t}{t}$
6. $\mathrm{rect}(t)$	$-\dfrac{1}{\pi}\ln\left\lvert\dfrac{t-1/2}{t+1/2}\right\rvert$
7. $\delta(t)$	$\dfrac{1}{\pi t}$
8. $\dfrac{1}{1+t^2}$	$\dfrac{t}{1+t^2}$
9. $\dfrac{1}{t}$	$-\pi\delta(t)$

Notes: $\delta(t)$ denotes Dirac delta function; rect(t) denotes rectangular function; ln denotes natural logarithm.
* In the first two pairs, it is assumed that $m(t)$ is band limited to the interval $-W \le f \le W$, where $W < f_c$.

Clearly, Hilbert transformation is a linear operation. The *inverse Hilbert transform*, by means of which the original signal $g(t)$ is linearly recovered from $\hat{g}(t)$, is defined by

$$g(t) = -\frac{1}{\pi}\int_{-\infty}^{\infty} \frac{\hat{g}(\tau)}{t-\tau}\, d\tau \tag{2.55}$$

The functions $g(t)$ and $\hat{g}(t)$ are said to constitute a *Hilbert-transform pair.* A short table of Hilbert-transform pairs is given in Table 2.3 on page 42.

The definition of the Hilbert transform $\hat{g}(t)$ given in (2.54) may be interpreted as the convolution of $g(t)$ with the time function $1/(\pi t)$. We know from the convolution theorem listed in Table 2.1 that the convolution of two functions in the time domain is transformed into the multiplication of their Fourier transforms in the frequency domain.

For the time function $1/(\pi t)$, we have the Fourier-transform pair (see Property 14 in Table 2.2)

$$\frac{1}{\pi t} \rightleftharpoons -j\,\mathrm{sgn}(f)$$

where sgn(f) is the *signum function*, defined in the frequency domain as

$$\mathrm{sgn}(f) = \begin{cases} 1, & f>0 \\ 0, & f=0 \\ -1, & f<0 \end{cases} \tag{2.56}$$

It follows, therefore, that the Fourier transform $\hat{G}(f)$ of $\hat{g}(t)$ is given by

$$\hat{G}(f) = -j\,\mathrm{sgn}(f)G(f) \tag{2.57}$$

Equation (2.57) states that given a Fourier transformable signal $g(t)$, we may obtain the Fourier transform of its Hilbert transform $\hat{g}(t)$ by passing $g(t)$ through a linear time-invariant system whose frequency response is equal to $-j$sgn(f). This system may be considered as one that produces a phase shift of $-90°$ for all positive frequencies of the input signal and $+90°$ degrees for all negative frequencies, as in Figure 2.15. The amplitudes of all frequency components in the signal, however, are unaffected by transmission through the device. Such an ideal system is referred to as a *Hilbert transformer*, or *quadrature filter.*

Properties of the Hilbert Transform

The Hilbert transform differs from the Fourier transform in that it operates exclusively in the time domain. It has a number of useful properties of its own, some of which are listed next. The signal $g(t)$ is assumed to be real valued, which is the usual domain of application of the Hilbert transform. For this class of signals, the Hilbert transform has the following properties.

PROPERTY 1 *A signal $g(t)$ and its Hilbert transform $\hat{g}(t)$ have the same magnitude spectrum.*

That is to say,

$$|G(f)| = |\hat{G}(f)|$$

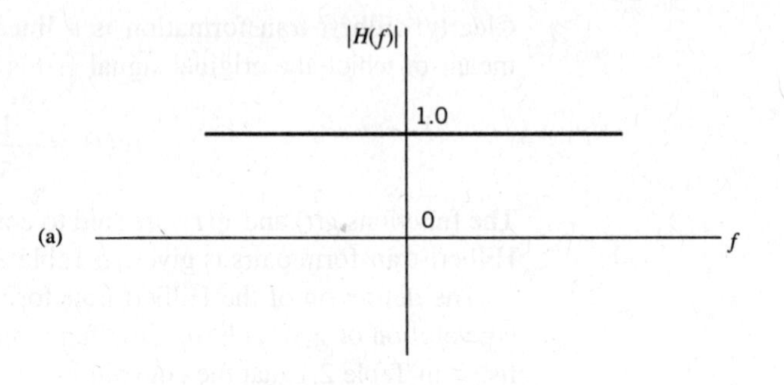

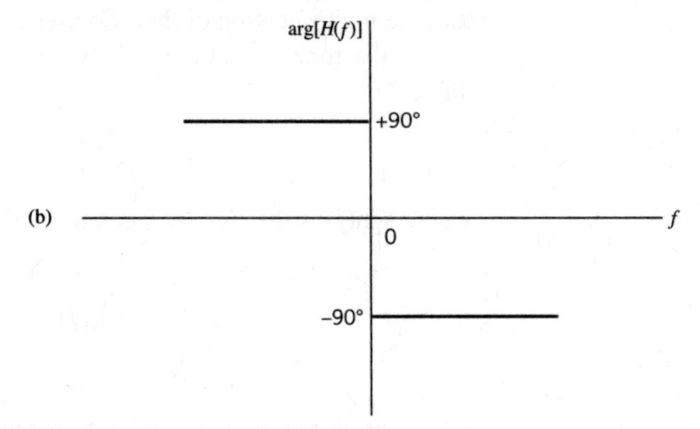

Figure 2.15
(a) Magnitude response and (b) phase response of Hilbert transform.

PROPERTY 2 *If $\hat{g}(t)$ is the Hilbert transform of $g(t)$, then the Hilbert transform of $\hat{g}(t)$ is $-g(t)$.*

Another way of stating this property is to write

$$\arg[G(f)] = -\arg\{\hat{G}(f)\}$$

PROPERTY 3 *A signal $g(t)$ and its Hilbert transform $\hat{g}(t)$ are orthogonal over the entire time interval $(-\infty, \infty)$.*

In mathematical terms, the orthogonality of $g(t)$ and $\hat{g}(t)$ is described by

$$\int_{-\infty}^{\infty} g(t)\hat{g}(t)dt = 0$$

Proofs of these properties follow from (2.54), (2.55), and (2.57).

EXAMPLE 5 **Hilbert Transform of Low-Pass Signal**

Consider Figure 2.16a that depicts the Fourier transform of a low-pass signal $g(t)$, whose frequency content extends from $-W$ to W. Applying the Hilbert transform to this signal yields a new signal $\hat{g}(t)$ whose Fourier transform, $\hat{G}(f)$, is depicted in Figure 2.16b. This figure illustrates that the frequency content of a Fourier transformable signal can be radically changed as a result of Hilbert transformation.

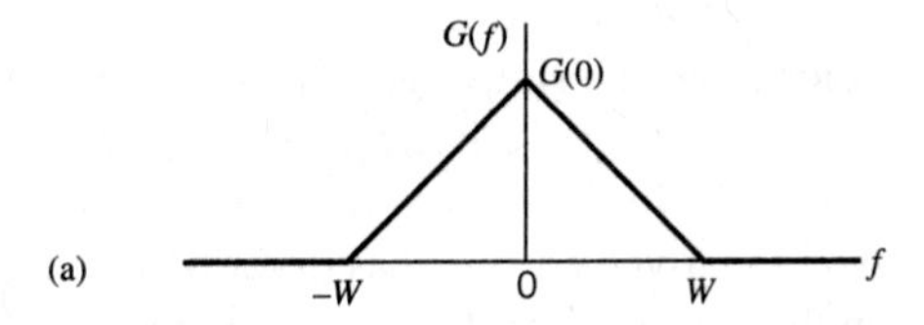

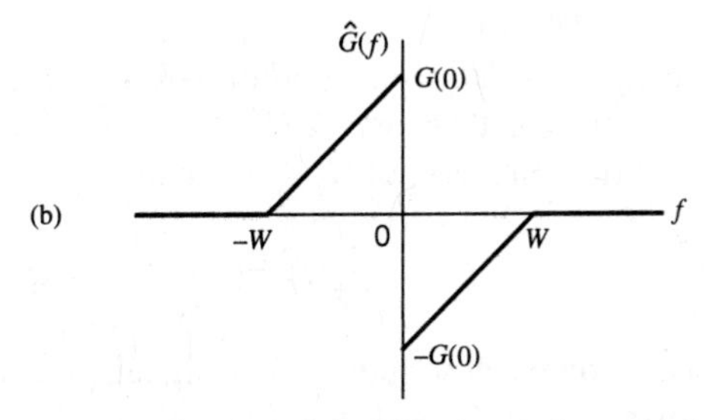

Figure 2.16 Illustrating application of the Hilbert transform to a low-pass signal: (a) Spectrum of the signal $g(t)$; (b) Spectrum of the Hilbert transform $\hat{g}(t)$.

2.9 Pre-envelopes

The Hilbert transform of a signal is defined for both positive and negative frequencies. In light of the spectrum shaping illustrated in Example 5, a question that begs itself is:

> How can we modify the frequency content of a real-valued signal $g(t)$ such that all negative frequency components are completely eliminated?

The answer to this fundamental question lies in the idea of a complex-valued signal called the *pre-envelope*[6] of $g(t)$, formally defined as

$$g_+(t) = g(t) + j\hat{g}(t) \tag{2.58}$$

where $\hat{g}(t)$ is the Hilbert transform of $g(t)$. According to this definition, the given signal $g(t)$ is the real part of the pre-envelope $g_+(t)$, and the Hilbert transform $\hat{g}(t)$ is the imaginary part of the pre-envelope. An important feature of the pre-envelope $g_+(t)$ is the behavior of its Fourier transform. Let $G_+(f)$ denote the Fourier transform of $g_+(t)$. Then, using (2.57) and (2.58) we may write

$$G_+(f) = G(f) + \text{sgn}(f)G(f) \tag{2.59}$$

Next, invoking the definition of the signum function given in (2.56), we may rewrite (2.59) in the equivalent form

$$G_+(f) = \begin{cases} 2G(f), & f>0 \\ G(0), & f=0 \\ 0, & f<0 \end{cases} \tag{2.60}$$

where $G(0)$ is the value of $G(f)$ at the origin $f = 0$. Equation (2.60) clearly shows that the pre-envelope of the signal $g(t)$ has no frequency content (i.e., its Fourier transform vanishes) for all negative frequencies, and the question that was posed earlier has indeed been answered. Note, however, in order to do this, we had to introduce the complex-valued version of a real-valued signal as described in (2.58).

From the foregoing analysis it is apparent that for a given signal $g(t)$ we may determine its pre-envelope $g_+(t)$ in one of two equivalent procedures.

1. *Time-domain procedure*. Given the signal $g(t)$, we use (2.58) to compute the pre-envelope $g_+(t)$.
2. *Frequency-domain procedure*. We first determine the Fourier transform $G(f)$ of the signal $g(t)$, then use (2.60) to determine $G_+(f)$, and finally evaluate the inverse Fourier transform of $G_+(f)$ to obtain

$$g_+(t) = 2\int_0^{\infty} G(f)\exp(j2\pi ft)\,df \tag{2.61}$$

Depending on the description of the signal, procedure 1 may be easier than procedure 2, or vice versa.

Equation (2.58) defines the pre-envelope $g_+(t)$ for positive frequencies. Symmetrically, we may define the pre-envelope for *negative frequencies* as

$$g_-(t) = g(t) - j\hat{g}(t) \tag{2.62}$$

The two pre-envelopes $g_+(t)$ and $g_-(t)$ are simply the complex conjugate of each other, as shown by

$$g_-(t) = g_+^*(t) \tag{2.63}$$

where the asterisk denotes complex conjugation. The spectrum of the pre-envelope $g_+(t)$ is nonzero only for *positive* frequencies; hence the use of a plus sign as the subscript. On the other hand, the use of a minus sign as the subscript is intended to indicate that the spectrum of the other pre-envelope $g_-(t)$ is nonzero only for *negative* frequencies, as shown by the Fourier transform

$$G_-(f) = \begin{cases} 0, & f > 0 \\ G(0), & f = 0 \\ 2G(f), & f < 0 \end{cases} \tag{2.64}$$

Thus, the pre-envelope $g_+(t)$ and $g_-(t)$ constitute a complementary pair of complex-valued signals. Note also that the sum of $g_+(t)$ and $g_-(t)$ is exactly twice the original signal $g(t)$.

Given a real-valued signal, (2.60) teaches us that the pre-envelope $g_+(t)$ is uniquely defined by the spectral content of the signal for positive frequencies. By the same token, (2.64) teaches us that the other pre-envelope $g_-(t)$ is uniquely defined by the spectral content of the signal for negative frequencies. Since $g_-(t)$ is simply the complex conjugate of $g_+(t)$ as indicated in (2.63), we may now make the following statement:

> The spectral content of a Fourier transformable real-valued signal for positive frequencies uniquely defines that signal.

In other words, given the spectral content of such a signal for positive frequencies, we may uniquely define the spectral content of the signal for negative frequencies. Here then is the mathematical justification for basing the bandwidth of a Fourier transformable signal on its spectral content exclusively for positive frequencies, which is exactly what we did in Section 2.4, dealing with bandwidth.

EXAMPLE 6 Pre-envelopes of Low-Pass Signal

Continuing with the low-pass signal $g(t)$ considered in Example 5, Figure 2.17a and b depict the corresponding spectra of the pre-envelope $g_+(t)$ and the second pre-envelope $g_-(t)$, both of which belong to $g(t)$. Whereas the spectrum of $g(t)$ is defined for $-W \leq f \leq W$ as in Figure 2.16a, we clearly see from Figure 2.17 that the spectral content of $g_+(t)$ is confined entirely to $0 \leq f \leq W$, and the spectral content of $g_-(t)$ is confined entirely to $-W \leq f \leq 0$.

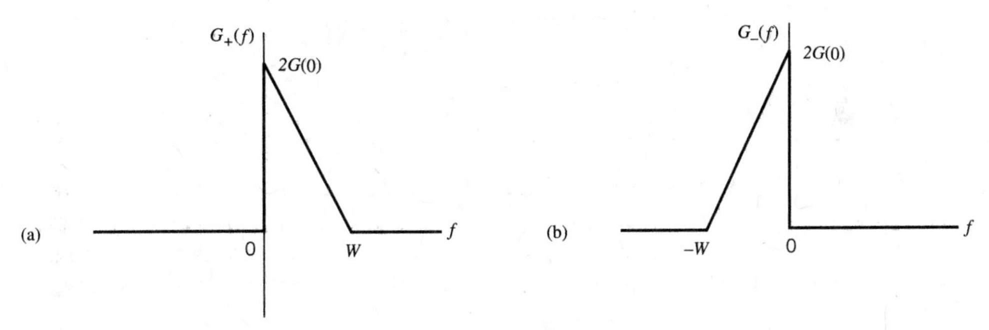

Figure 2.17 Another illustrative application of the Hilbert transform to a low-pass signal: (a) Spectrum of the pre-envelope $g_+(t)$; (b) Spectrum of the other pre-envelope $g_-(t)$.

Practical Importance of the Hilbert Transformation

An astute reader may see an analogy between the use of *phasors* and that of *pre-envelopes*. In particular, just as the use of phasors simplifies the manipulations of alternating currents and voltages in the study of circuit theory, so we find the pre-envelope simplifies the analysis of band-pass signals and band-pass systems in signal theory.

More specifically, by applying the concept of pre-envelope to a band-pass signal, the signal is transformed into an equivalent low-pass representation. In a corresponding way, a band-pass filter is transformed into its own equivalent low-pass representation. Both transformations, rooted in the Hilbert transform, play a key role in the formulation of modulated signals and their demodulation, as demonstrated in what follows in this and subsequent chapters.

2.10 Complex Envelopes of Band-Pass Signals

The idea of pre-envelopes introduced in Section 2.9 applies to any real-valued signal, be it of a low-pass or band-pass kind; the only requirement is that the signal be Fourier transformable. From this point on and for the rest of the chapter, we will restrict attention to band-pass signals. Such signals are exemplified by signals modulated onto a sinusoidal carrier. In a corresponding way, when it comes to systems we restrict attention to band-pass systems. The primary reason for these restrictions is that the material so presented is directly applicable to analog modulation theory, to be covered in Section 2.14, as well as other digital modulation schemes covered in subsequent chapters of the book. With this objective in mind and the desire to make a consistent use of notation with respect to material to be presented in subsequent chapters, henceforth we will use $s(t)$ to denote a modulated signal. When such a signal is applied to the input of a band-pass system, such as a communication channel, we will use $x(t)$ to denote the resulting system (e.g., channel) output. However, as before, we will use $h(t)$ as the impulse response of the system.

To proceed then, let the band-pass signal of interest be denoted by $s(t)$ and its Fourier transform be denoted by $S(f)$. We assume that the Fourier transform $S(f)$ is essentially confined to a band of frequencies of total extent $2W$, centered about some frequency $\pm f_c$, as illustrated in Figure 2.18a. We refer to f_c as the *carrier frequency*; this terminology is

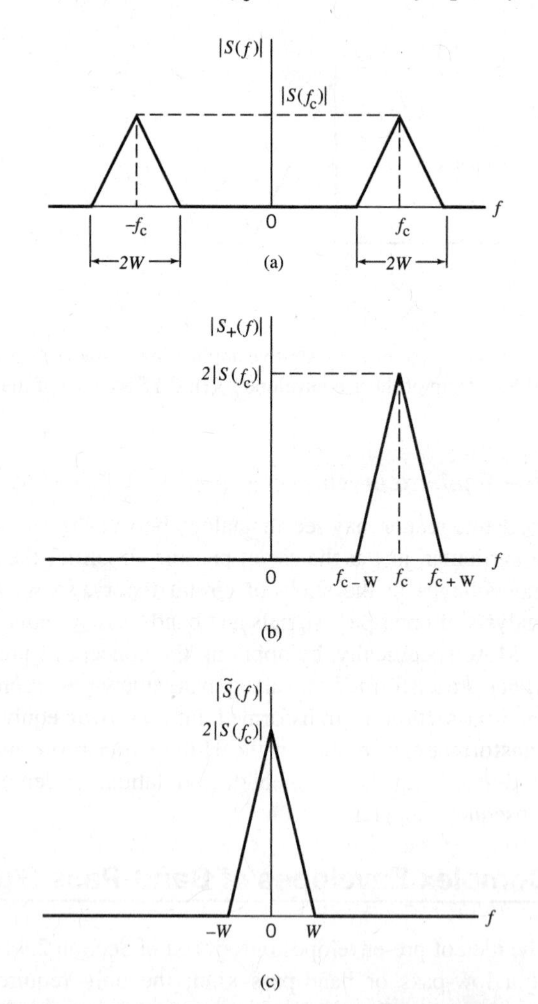

Figure 2.18 (a) Magnitude spectrum of band-pass signal $s(t)$; (b) Magnitude spectrum of pre-envelope $s_+(t)$; (c) Magnitude spectrum of complex envelope $\tilde{s}(t)$.

borrowed from modulation theory. In the majority of communication signals encountered in practice, we find that the bandwidth $2W$ is small compared with f_c, so we may refer to the signal $s(t)$ as a *narrowband signal.* However, a precise statement about how small the bandwidth must be for the signal to be considered narrowband is not necessary for our present discussion. Hereafter, the terms band-pass and narrowband are used interchangeably.

Let the pre-envelope of the narrowband signal $s(t)$ be expressed in the form

$$s_+(t) = \tilde{s}(t)\exp(j2\pi f_c t) \tag{2.65}$$

We refer to $\tilde{s}(t)$ as the *complex envelope* of the band-pass signal $s(t)$. Equation (2.65) may be viewed as the basis of a definition for the complex envelope $\tilde{s}(t)$ in terms of the pre-envelope $s_+(t)$. In light of the narrowband assumption imposed on the spectrum of the band-pass signal $s(t)$, we find that the spectrum of the pre-envelope $s_+(t)$ is limited to the positive frequency band $f_c - W \le f \le f_c + W$, as illustrated in Figure 2.18b. Therefore, applying the frequency-shifting property of the Fourier transform to (2.65), we find that the spectrum of the complex envelope $\tilde{s}(t)$ is correspondingly limited to the band $-W \le f \le W$ and centered at the origin $f = 0$, as illustrated in Figure 2.18c. In other words, the complex envelope $\tilde{s}(t)$ of the band-pass signal $s(t)$ is a *complex low-pass signal.* The essence of the *mapping* from the band-pass signal $s(t)$ to the complex low-pass signal $\tilde{s}(t)$ is summarized in the following threefold statement:

- The information content of a modulated signal $s(t)$ is fully preserved in the complex envelope $\tilde{s}(t)$.
- Analysis of the band-pass signal $s(t)$ is complicated by the presence of the carrier frequency f_c; in contrast, the complex envelope $\tilde{s}(t)$ dispenses with f_c, making its analysis simpler to deal with.
- The use of $\tilde{s}(t)$ requires having to handle complex notations.

2.11 Canonical Representation of Band-Pass Signals

By definition, the real part of the pre-envelope $s_+(t)$ is equal to the original band-pass signal $s(t)$. We may therefore express the band-pass signal $s(t)$ in terms of its corresponding complex envelope $\tilde{s}(t)$ as

$$s(t) = \text{Re}[\tilde{s}(t)\exp(j2\pi f_c t)] \tag{2.66}$$

where the operator Re[.] denotes the real part of the quantity enclosed inside the square brackets. Since, in general, $\tilde{s}(t)$ is a complex-valued quantity, we emphasize this property by expressing it in the Cartesian form

$$\tilde{s}(t) = s_I(t) + js_Q(t) \tag{2.67}$$

where $s_I(t)$ and $s_Q(t)$ are both real-valued low-pass functions; their low-pass property is inherited from the complex envelope $\tilde{s}(t)$. We may therefore use (2.67) in (2.66) to express the original band-pass signal $s(t)$ in the *canonical* or *standard* form

$$s(t) = s_I(t)\cos(2\pi f_c t) - s_Q(t)\sin(2\pi f_c t) \tag{2.68}$$

We refer to $s_I(t)$ as the *in-phase component* of the band-pass signal $s(t)$ and refer to $s_Q(t)$ as the *quadrature-phase component* or simply the *quadrature component* of the signal $s(t)$.

This nomenclature follows from the following observation: if $\cos(2\pi f_c t)$, the multiplying factor of $s_I(t)$, is viewed as the reference sinusoidal carrier, then $\sin(2\pi f_c t)$, the multiplying factor of $s_Q(t)$, is in phase quadrature with respect to $\cos(2\pi f_c t)$.

According to (2.66), the complex envelope $\tilde{s}(t)$ may be pictured as a *time-varying phasor* positioned at the origin of the (s_I, s_Q)-plane, as indicated in Figure 2.19a. With time t varying continuously, the end of the phasor moves about in the plane. Figure 2.19b depicts the phasor representation of the complex exponential $\exp(2\pi f_c t)$. In the definition given in (2.66), the complex envelope $\tilde{s}(t)$ is multiplied by the complex exponential $\exp(j2\pi f_c t)$. The angles of these two phasors, therefore, add and their lengths multiply, as shown in Figure 2.19c. Moreover, in this latter figure, we show the (s_I, s_Q)-phase rotating with an angular velocity equal to $2\pi f_c$ radians per second. Thus, in the picture portrayed in the figure, the phasor representing the complex envelope $\tilde{s}(t)$ moves in the (s_I, s_Q)-plane, while at the very same time the plane itself rotates about the origin. The original band-pass signal $s(t)$ is the projection of this time-varying phasor on a *fixed line* representing the real axis, as indicated in Figure 2.19c.

Since both $s_I(t)$ and $s_Q(t)$ are low-pass signals limited to the band $-W \leq f \leq W$, they may be extracted from the band-pass signal $s(t)$ using the scheme shown in Figure 2.20a. Both low-pass filters in this figure are designed identically, each with a bandwidth equal to W.

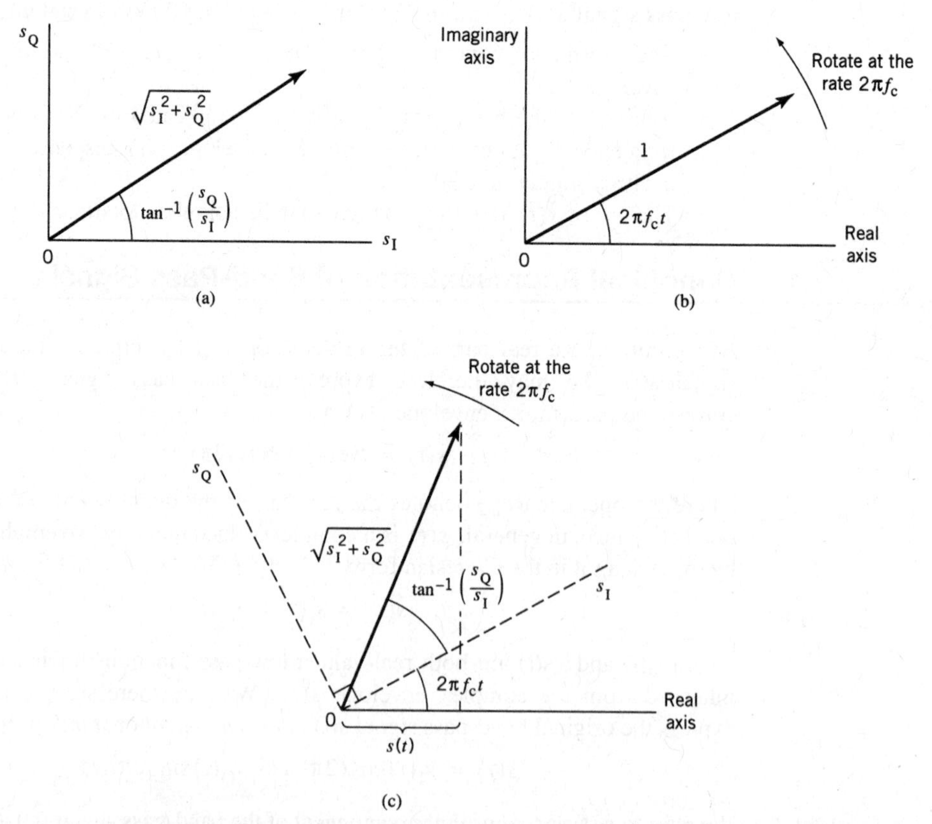

Figure 2.19 Illustrating an interpretation of the complex envelope $\tilde{s}(t)$ and its multiplication by $\exp(j2\pi f_c t)$.

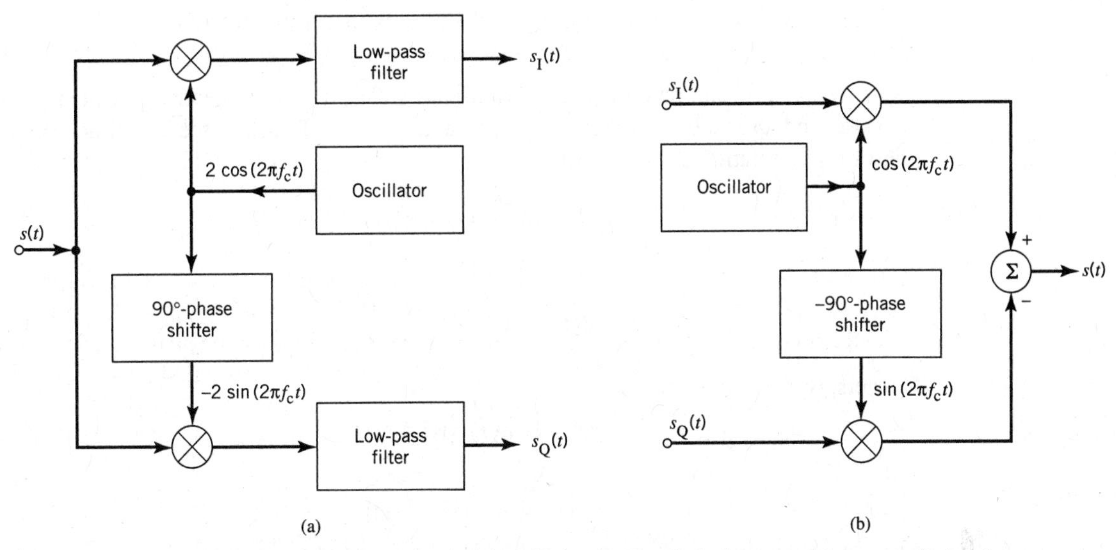

Figure 2.20 (a) Scheme for deriving the in-phase and quadrature components of a band-pass signal $g(t)$. (b) Scheme for reconstructing the band-pass signal from its in-phase and quadrature components.

To reconstruct $s(t)$ from its in-phase and quadrature components, we may use the scheme shown in Figure 2.20b. In light of these statements, we may refer to the scheme in Figure 2.20a as an *analyzer*, in the sense that it extracts the in-phase and quadrature components, $s_I(t)$ and $s_Q(t)$, from the band-pass signal $s(t)$. By the same token, we may refer to the second scheme in Figure 2.20b as a *synthesizer*, in the sense it reconstructs the band-pass signal $s(t)$ from its in-phase and quadrature components, $s_I(t)$ and $s_Q(t)$.

The two schemes shown in Figure 2.20 are basic to the study of *linear modulation schemes, be they of an analog or digital kind.* Multiplication of the low-pass in-phase component $s_I(t)$ by $\cos(2\pi f_c t)$ and multiplication of the quadrature component $s_Q(t)$ by $\sin(2\pi f_c t)$ represent linear forms of modulation. Provided that the carrier frequency f_c is larger than the low-pass bandwidth W, the resulting band-pass function $s(t)$ defined in (2.68) is referred to as a *passband signal waveform.* Correspondingly, the mapping from $s_I(t)$ and $s_Q(t)$ combined into $s(t)$ is known as *passband modulation.*

Polar Representation of Band-Pass Signals

Equation (2.67) is the Cartesian form of defining the complex envelope $\tilde{s}(t)$ of the band-pass signal $s(t)$. Alternatively, we may define $\tilde{s}(t)$ in the *polar form* as

$$\tilde{s}(t) = a(t)\exp[\mathrm{j}\phi(t)] \tag{2.69}$$

where $a(t)$ and $\phi(t)$ are both real-valued low-pass functions. Based on the polar representation of (2.69), the original band-pass signal $s(t)$ is itself defined by

$$\tilde{s}(t) = a(t)\cos[2\pi f_c t + \phi(t)] \tag{2.70}$$

We refer to $a(t)$ as the *natural envelope* or simply the *envelope* of the band-pass signal $s(t)$ and refer to $\phi(t)$ as the *phase* of the signal. We now see why the term "pre-envelope" was used in referring to (2.58), the formulation of which *preceded* that of (2.70).

Relationship Between Cartesian and Polar Representations of Band-Pass Signal

The envelope $a(t)$ and phase $\phi(t)$ of a band-pass signal $s(t)$ are respectively related to the in-phase and quadrature components $s_I(t)$ and $s_Q(t)$ as follows (see the time-varying phasor representation of Figure 2.19a):

$$a(t) = \sqrt{s_I^2(t) + s_Q^2(t)} \tag{2.71}$$

and

$$\phi(t) = \tan^{-1}\left(\frac{s_Q(t)}{s_I(t)}\right) \tag{2.72}$$

Conversely, we may write

$$s_I(t) = a(t)\cos[\phi(t)] \tag{2.73}$$

and

$$s_Q(t) = a(t)\sin[\phi(t)] \tag{2.74}$$

Thus, both the in-phase and quadrature components of a band-pass signal contain amplitude and phase information, both of which are uniquely defined for a prescribed phase $\phi(t)$, modulo 2π.

2.12 Complex Low-Pass Representations of Band-Pass Systems

Now that we know how to handle the complex low-pass representation of band-pass signals, it is logical that we develop a corresponding procedure for handling the representation of linear time-invariant band-pass systems. Specifically, we wish to show that the analysis of band-pass systems is greatly simplified by establishing an *analogy*, more precisely an *isomorphism*, between band-pass and low-pass systems. For example, this analogy would help us to facilitate the computer simulation of a wireless communication channel driven by a sinusoidally modulated signal, which otherwise could be a difficult proposition.

Consider a narrowband signal $s(t)$, with its Fourier transform denoted by $S(f)$. We assume that the spectrum of the signal $s(t)$ is limited to frequencies within $\pm W$ hertz of the carrier frequency f_c. We also assume that $W < f_c$. Let the signal $s(t)$ be applied to a linear time-invariant band-pass system with impulse response $h(t)$ and frequency response $H(f)$. We assume that the frequency response of the system is limited to frequencies within $\pm B$ of the carrier frequency f_c. The *system bandwidth* $2B$ is usually narrower than or equal to the input *signal bandwidth* $2W$. We wish to represent the band-pass impulse response $h(t)$ in terms of two quadrature components, denoted by $h_I(t)$ and $h_Q(t)$. In particular, by analogy to the representation of band-pass signals, we express $h(t)$ in the form

$$h(t) = h_I(t)\cos(2\pi f_c t) - h_Q(t)\sin(2\pi f_c t) \tag{2.75}$$

Correspondingly, we define the *complex impulse response* of the band-pass system as

$$\tilde{h}(t) = h_I(t) + jh_Q(t) \tag{2.76}$$

Hence, following (2.66), we may express $h(t)$ in terms of $\tilde{h}(t)$ as

$$h(t) = \text{Re}[\tilde{h}(t)\exp(\text{j}2\pi f_c t)] \tag{2.77}$$

Note that $h_I(t)$, $h_Q(t)$, and $\tilde{h}(t)$ are all low-pass functions, limited to the frequency band $-B \le f \le B$.

We may determine the complex impulse response $\tilde{h}(t)$ in terms of the in-phase and quadrature components $h_I(t)$ and $h_Q(t)$ of the band-pass impulse response $h(t)$ by building on (2.76). Alternatively, we may determine it from the band-pass frequency response $H(f)$ in the following way. We first use (2.77) to write

$$2h(t) = \tilde{h}(t)\exp(\text{j}2\pi f_c t) + \tilde{h}^*(t)\exp(-\text{j}2\pi f_c t) \tag{2.78}$$

where $\tilde{h}^*(t)$ is the *complex conjugate* of $\tilde{h}(t)$; the rationale for introducing the factor of 2 on the left-hand side of (2.78) follows from the fact that if we add a complex signal and its complex conjugate, the sum adds up to twice the real part and the imaginary parts cancel. Applying the Fourier transform to both sides of (2.78) and using the complex-conjugation property of the Fourier transform, we get

$$2H(f) = \tilde{H}(f-f_c) + \tilde{H}^*(-f-f_c) \tag{2.79}$$

where $H(f) \rightleftharpoons h(t)$ and $\tilde{H}(f) \rightleftharpoons \tilde{h}(t)$. Equation (2.79) satisfies the requirement that $H^*(f) = H(-f)$ for a real-valued impulse response $h(t)$. Since $\tilde{H}(f)$ represents a low-pass frequency response limited to $|f| \le B$ with $B < f_c$, we infer from (2.79) that

$$\tilde{H}(f-f_c) = 2H(f), \qquad f > 0 \tag{2.80}$$

Equation (2.80) states:

> For a specified band-pass frequency response $H(f)$, we may determine the corresponding complex low-pass frequency response $\tilde{H}(f)$ by taking the part of $H(f)$ defined for positive frequencies, shifting it to the origin, and scaling it by the factor 2.

Having determined the complex frequency response $\tilde{H}(f)$, we decompose it into its in-phase and quadrature components, as shown by

$$\tilde{H}(f) = \tilde{H}_I(f) + \text{j}\tilde{H}_Q(f) \tag{2.81}$$

where the *in-phase component* is defined by

$$\tilde{H}_I(f) = \frac{1}{2}[\tilde{H}(f) + \tilde{H}^*(-f)] \tag{2.82}$$

and the *quadrature component* is defined by

$$\tilde{H}_Q(f) = \frac{1}{2\text{j}}[\tilde{H}(f) - \text{j}\tilde{H}^*(-f)] \tag{2.83}$$

Finally, to determine the complex impulse response $\tilde{h}(t)$ of the band-pass system, we take the inverse Fourier transform of $\tilde{H}(f)$, obtaining

$$\tilde{h}(t) = \int_{-\infty}^{\infty} \tilde{H}(f)\exp(\text{j}2\pi ft)\, df \tag{2.84}$$

which is the formula we have been seeking.

2.13 Putting the Complex Representations of Band-Pass Signals and Systems All Together

Examining (2.66) and (2.77), we immediately see that these two equations share a common multiplying factor: the exponential $\exp(\mathrm{j}2\pi f_c t)$. In practical terms, the inclusion of this factor accounts for a sinusoidal carrier of frequency f_c, which facilitates transmission of the modulated (band-pass) signal $s(t)$ across a band-pass channel of midband frequency f_c. In analytic terms, however, the presence of this exponential factor in both (2.66) and (2.77) complicates the analysis of the band-pass system driven by the modulated signal $s(t)$. This analysis can be simplified through the combined use of complex low-pass equivalent representations of both the modulated signal $s(t)$ and the band-pass system characterized by the impulse response $h(t)$. The simplification can be carried out in the time domain or frequency domain, as discussed next.

The Time-Domain Procedure

Equipped with the complex representations of band-pass signals and systems, we are ready to derive an analytically efficient method for determining the output of a band-pass system driven by a corresponding band-pass signal. To proceed with the derivation, assume that $S(f)$, denoting the spectrum of the input signal $s(t)$, and $H(f)$, denoting the frequency response of the system, are both centered around the same frequency f_c. In practice, there is no need to consider a situation in which the carrier frequency of the input signal is not aligned with the midband frequency of the band-pass system, since we have considerable freedom in choosing the carrier or midband frequency. Thus, changing the carrier frequency of the input signal by an amount Δf_c, for example, simply corresponds to absorbing (or removing) the factor $\exp(\pm \mathrm{j}2\pi\Delta f_c t)$ in the complex envelope of the input signal or the complex impulse response of the band-pass system. We are therefore justified in proceeding on the assumption that $S(f)$ and $H(f)$ are both centered around the same carrier frequency f_c.

Let $x(t)$ denote the output signal of the band-pass system produced in response to the incoming band-pass signal $s(t)$. Clearly, $x(t)$ is also a band-pass signal, so we may represent it in terms of its own low-pass complex envelope $\tilde{x}(t)$ as

$$x(t) = \mathrm{Re}[\tilde{x}(t)\exp(\mathrm{j}2\pi f_c t)] \tag{2.85}$$

The output signal $x(t)$ is related to the input signal $s(t)$ and impulse response $h(t)$ of the system in the usual way by the convolution integral

$$x(t) = \int_{-\infty}^{\infty} h(\tau)s(t-\tau)\,\mathrm{d}\tau \tag{2.86}$$

In terms of pre-envelopes, we have $h(t) = \mathrm{Re}[h_+(t)]$ and $s(t) = \mathrm{Re}[s_+(t)]$. We may therefore rewrite (2.86) in terms of the pre-envelopes $s_+(t)$ and $h_+(t)$ as

$$x(t) = \int_{-\infty}^{\infty} \mathrm{Re}[h_+(\tau)]\mathrm{Re}[s_+(t-\tau)]\,\mathrm{d}\tau \tag{2.87}$$

To proceed further, we make use of a basic property of pre-envelopes that is described by the following relation:

$$\int_{-\infty}^{\infty} \mathrm{Re}[h_+(\tau)]\mathrm{Re}[s_+(\tau)]\,\mathrm{d}\tau = \frac{1}{2}\mathrm{Re}\left[\int_{-\infty}^{\infty} h_+(\tau)s_+^*(\tau)\,\mathrm{d}\tau\right] \tag{2.88}$$

where we have used τ as the integration variable to be consistent with that in (2.87); details of (2.88) are presented in Problem 2.20. Next, from Fourier-transform theory we note that using $s(-\tau)$ in place of $s(\tau)$ has the effect of removing the complex conjugation on the right-hand side of (2.88). Hence, bearing in mind the algebraic difference between the argument of $s_+(\tau)$ in (2.88) and that of $s_+(t - \tau)$ in (2.87), and using the relationship between the pre-envelope and complex envelope of a band-pass signal, we may express (2.87) in the equivalent form

$$\begin{aligned} x(t) &= \frac{1}{2}\mathrm{Re}\left[\int_{-\infty}^{\infty} h_+(\tau)s_+(t-\tau)\,\mathrm{d}\tau\right] \\ &= \frac{1}{2}\mathrm{Re}\left\{\int_{-\infty}^{\infty} \tilde{h}(\tau)\exp(\mathrm{j}2\pi f_c\tau)\tilde{s}(t-\tau)\exp[\mathrm{j}2\pi f_c(t-\tau)]\,\mathrm{d}\tau\right\} \\ &= \frac{1}{2}\mathrm{Re}\left[\exp(\mathrm{j}2\pi f_c t)\int_{-\infty}^{\infty} \tilde{h}(\tau)\tilde{s}(t-\tau)\,\mathrm{d}\tau\right] \end{aligned} \tag{2.89}$$

Thus, comparing the right-hand sides of (2.85) and (2.89), we readily find that for a large enough carrier frequency f_c, the complex envelope $\tilde{x}(t)$ of the output signal is simply defined in terms of the complex envelope $\tilde{s}(t)$ of the input signal and the complex impulse response $\tilde{h}(t)$ of the band-pass system as follows:

$$\tilde{x}(t) = \frac{1}{2}\int_{-\infty}^{\infty} \tilde{h}(t)\tilde{s}(t-\tau)\,\mathrm{d}\tau \tag{2.90}$$

This important relationship is the result of the *isomorphism* between a band-pass function and the corresponding complex low-pass function, in light of which we may now make the following summarizing statement:

> Except for the scaling factor 1/2, the complex envelope $\tilde{x}(t)$ of the output signal of a band-pass system is obtained by convolving the complex impulse response $\tilde{h}(t)$ of the system with the complex envelope $\tilde{s}(t)$ of the input band-pass signal.

In computational terms, the significance of this statement is profound. Specifically, in dealing with band-pass signals and systems, we need only concern ourselves with the functions $\tilde{s}(t)$, $\tilde{x}(t)$, and $\tilde{h}(t)$, representing the complex low-pass equivalents of the excitation applied to the input of the system, the response produced at the output of the system, and the impulse response of the system respectively, as illustrated in Figure 2.21. The essence of the filtering process performed in the original system of Figure 2.21a is completely retained in the complex low-pass equivalent representation depicted in Figure 2.21b.

The complex envelope $\tilde{s}(t)$ of the input band-pass signal and the complex impulse response $\tilde{h}(t)$ of the band-pass system are defined in terms of their respective in-phase

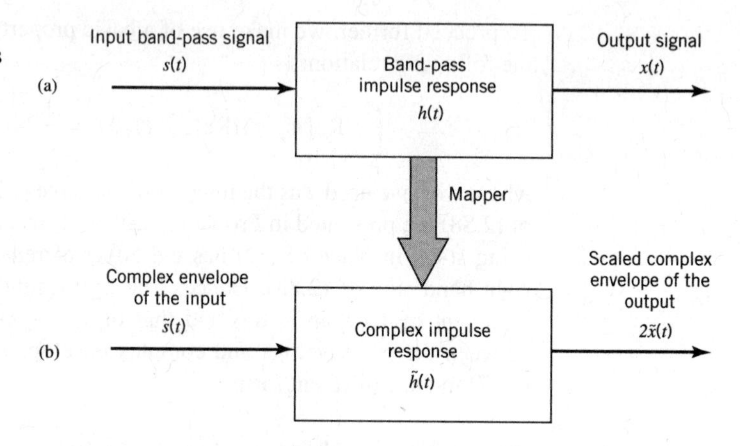

Figure 2.21
(a) Input–output description of a band-pass system; (b) Complex low-pass equivalent model of the band-pass system.

and quadrature components by (2.67) and (2.76), respectively. Substituting these relations into (2.90), we get

$$\begin{aligned} 2\tilde{x}(t) &= \tilde{h}(t)\bigstar\tilde{s}(t) \\ &= [h_I(t)+jh_Q(t)]\bigstar[s_I(t)+js_Q(t)] \end{aligned} \tag{2.91}$$

where the symbol $\bigstar$ denotes convolution. Because convolution is *distributive*, we may rewrite (2.91) in the equivalent form

$$2\tilde{x}(t) = [h_I(t)\bigstar s_I(t) - h_Q(t)\bigstar s_Q(t)] + j[h_Q(t)\bigstar s_I(t) + h_I\bigstar s_Q(t)] \tag{2.92}$$

Let the complex envelope $\tilde{x}(t)$ of the response be defined in terms of its in-phase and quadrature components as

$$\tilde{x}(t) = x_I(t) + jx_Q(t) \tag{2.93}$$

Then, comparing the real and imaginary parts in (2.92) and (2.93), we find that the in-phase component $x_I(t)$ is defined by the relation

$$2x_I(t) = h_I(t)\bigstar s_I(t) - h_Q(t)\bigstar s_Q(t) \tag{2.94}$$

and its quadrature component $x_Q(t)$ is defined by the relation

$$2x_Q(t) = h_Q(t)\bigstar s_I(t) + h_I(t)\bigstar s_Q(t) \tag{2.95}$$

Thus, for the purpose of evaluating the in-phase and quadrature components of the complex envelope $\tilde{x}(t)$ of the system output, we may use the *low-pass equivalent model* shown in Figure 2.22. All the signals and impulse responses shown in this model are real-valued low-pass functions; hence a time-domain procedure for simplifying the analysis of band-pass systems driven by band-pass signals.

The Frequency-Domain Procedure

Alternatively, Fourier-transforming the convolution integral of (2.90) and recognizing that convolution in the time domain is changed into multiplication in the frequency domain, we get

$$\tilde{X}(f) = \frac{1}{2}\tilde{H}(f)\tilde{S}(f) \tag{2.96}$$

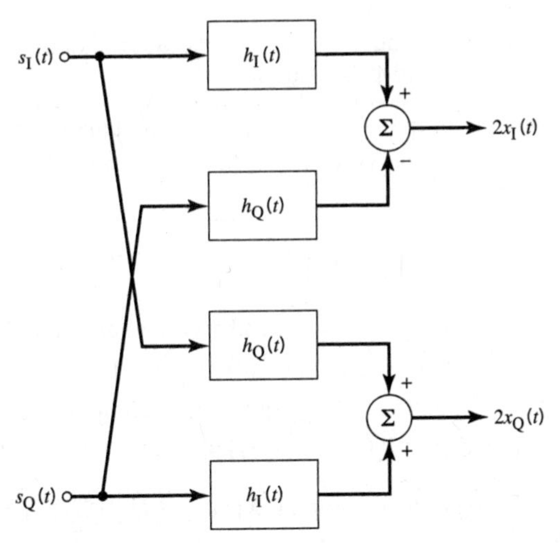

Figure 2.22 Block diagram illustrating the relationship between the in-phase and quadrature components of the response of a band-pass filter and those of the input signal.

where $\tilde{s}(t) \rightleftharpoons \tilde{S}(f)$, $\tilde{h}(t) \rightleftharpoons \tilde{H}(f)$, and $\tilde{x}(t) \rightleftharpoons \tilde{X}(f)$. The $\tilde{H}(f)$ is itself related to the frequency response $H(f)$ of the band-pass system by (2.80). Thus, assuming that $H(f)$ is known, we may use the frequency-domain procedure summarized in Table 2.4 for computing the system output $x(t)$ in response to the system input $s(t)$.

In actual fact, the procedure of Table 2.4 is the frequency-domain representation of the low-pass equivalent to the band-pass system, depicted in Figure 2.21b. In computational terms, this procedure is of profound practical significance. We say so because its use alleviates the analytic and computational difficulty encountered in having to include the carrier frequency f_c in the pertinent calculations.

As discussed earlier in the chapter, the theoretical formulation of the low-pass equivalent in Figure 2.21b is rooted in the Hilbert transformation, the evaluation of which poses a practical problem of its own, because of the wideband 90°-phase shifter involved in its theory. Fortunately, however, we do not need to invoke the Hilbert transform in constructing the low-pass equivalent. This is indeed so, when a message signal modulated onto a sinusoidal carrier is processed by a band-pass filter, as explained here:

1. Typically, the message signal is band limited for all practical purposes. Moreover, the carrier frequency is larger than the highest frequency component of the signal; the modulated signal is therefore a band-pass signal with a well-defined passband. Hence, the in-phase and quadrature components of the modulated signal $s(t)$, represented respectively by $s_I(t)$ and $s_Q(t)$, are readily obtained from the canonical representation of $s(t)$, described in (2.68).
2. Given the well-defined frequency response $H(f)$ of the band-pass system, we may readily evaluate the corresponding complex low-pass frequency response $\tilde{H}(f)$; see (2.80). Hence, we may compute the system output $x(t)$ produced in response to the carrier-modulated input $s(t)$ without invoking the Hilbert transform.

Table 2.4 **Procedure for the computational analysis of a band-pass system driven by a band-pass signal**

Given the frequency response $H(f)$ of a band-pass system, computation of the output signal $x(t)$ of the system in response to an input band-pass signal $s(t)$ is summarized as follows:

1. Use (2.80), namely $\tilde{H}(f-f_c) = 2H(f)$, for $f > 0$ to determine $\tilde{H}(f)$.
2. Expressing the input band-pass signal $s(t)$ in the canonical form of (2.68), evaluate the complex envelope $\tilde{s}(t) = s_I(t) + js_Q(t)$ where $s_I(t)$ is the in-phase component of $s(t)$ and $s_Q(t)$ is its quadrature component. Hence, compute the Fourier transform $\tilde{S}(f) = \mathbf{F}[\tilde{s}(t)]$
3. Using (2.96), compute $\tilde{X}(f) = \frac{1}{2}\tilde{H}(f)\tilde{S}(f)$, which defines the Fourier transform of the complex envelope $\tilde{x}(t)$ of the output signal $x(t)$.
4. Compute the inverse Fourier transform of $\tilde{X}(f)$, yielding $\tilde{x}(t) = \mathbf{F}^{-1}[\tilde{X}(f)]$
5. Use (2.85) to compute the desired output signal $x(t) = \text{Re}[\tilde{x}(t)\exp(j2\pi f_c t)]$

Procedure for Efficient Simulation of Communication Systems

To summarize, the frequency-domain procedure described in Table 2.4 is well suited for the efficient simulation of communication systems on a computer for two reasons:

1. The low-pass equivalents of the incoming band-pass signal and the band-pass system work by eliminating the exponential factor $\exp(j2\pi f_c t)$ from the computation without loss of information.
2. The *fast Fourier transform (FFT) algorithm*, discussed later in the chapter, is used for numerical computation of the Fourier transform. This algorithm is used twice in Table 2.4, once in step 2 to perform Fourier transformation, and then again in step 4 to perform inverse Fourier transformation.

The procedure of this table, rooted largely in the frequency domain, assumes availability of the band-pass system's frequency response $H(f)$. If, however, it is the system's impulse response $h(t)$ that is known, then all we need is an additional step to Fourier transform $h(t)$ into $H(f)$ before initiating the procedure of Table 2.4.

2.14 Linear Modulation Theory

The material presented in Sections 2.8–2.13 on the complex low-pass representation of band-pass signals and systems is of profound importance in the study of communication theory. In particular, we may use the canonical formula of (2.68) as the mathematical basis for a unified treatment of linear modulation theory, which is the subject matter of this section.

We start this treatment with a formal definition:

> Modulation is a process by means of which one or more parameters of a sinusoidal carrier are varied in accordance with a message signal so as to facilitate transmission of that signal over a communication channel.

The message signal (e.g., voice, video, data sequence) is referred to as the *modulating signal*, and the result of the modulation process is referred to as the *modulated signal.* Naturally, in a communication system, modulation is performed in the transmitter. The reverse of modulation, aimed at recovery of the original message signal in the receiver, is called *demodulation.*

Consider the block diagram of Figure 2.23, depicting a modulator, where $m(t)$ is the message signal, $\cos(2\pi f_c t)$ is the carrier, and $s(t)$ is the modulated signal. To apply (2.68) to this modulator, the in-phase component $s_I(t)$ in that equation is treated simply as a scaled version of the message signal denoted by $m(t)$. As for the quadrature component $s_Q(t)$, it is defined by a *spectrally shaped* version of $m(t)$ that is performed linearly. In such a scenario, it follows that a modulated signal $s(t)$ defined by (2.68) is a *linear function* of the message signal $m(t)$; hence the reference to this equation as the mathematical basis of *linear modulation theory.*

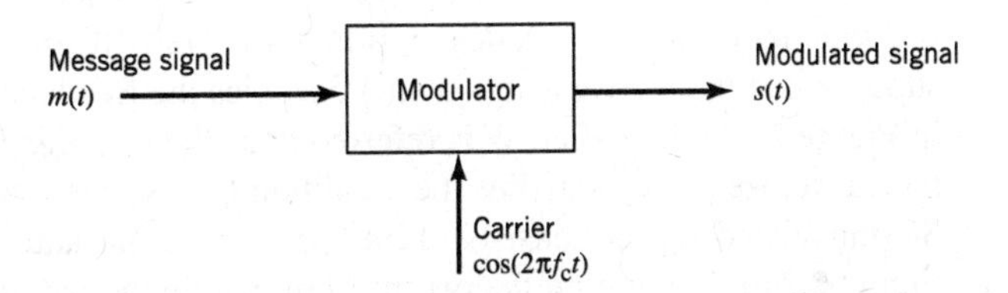

Figure 2.23 Block diagram of a modulator.

To recover the original message signal $m(t)$ from the modulated signal $s(t)$, we may use a demodulator, the block diagram of which is depicted in Figure 2.24. An elegant feature of linear modulation theory is that demodulation of $s(t)$ is also achieved using linear operations. However, for linear demodulation of $s(t)$ to be feasible, the locally generated carrier in the demodulator of Figure 2.24 has to be synchronous with the original sinusoidal carrier used in the modulator of Figure 2.23. Accordingly, we speak of *synchronous demodulation* or *coherent detection.*

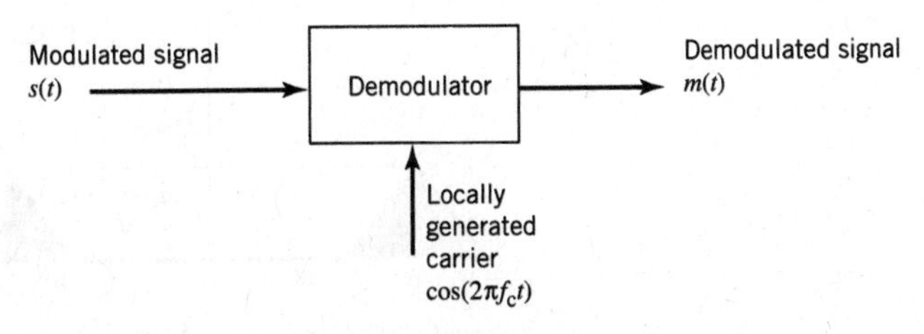

Figure 2.24 Block diagram of a demodulator.

Depending on the spectral composition of the modulated signal, we have three kinds of linear modulation in analog communications:

- double sideband-suppressed carrier (DSB-SC) modulation;
- vestigial sideband (VSB) modulation;
- single sideband (SSB) modulation.

These three methods of modulation are discussed in what follows and in this order.

DSB-SC Modulation

DSB-SC modulation is the simplest form of linear modulation, which is obtained by setting

$$s_{\mathrm{I}}(t) = m(t)$$

and

$$s_{\mathrm{Q}}(t) = 0$$

Accordingly, (2.68) is reduced to

$$s(t) = m(t)\cos(2\pi f_c t) \tag{2.97}$$

the implementation of which simply requires a *product modulator* that multiplies the message signal $m(t)$ by the carrier $\cos(2nf_c t)$, assumed to be of unit amplitude.

For a frequency-domain description of the DSB-SC-modulated signal defined in (2.97), suppose that the message signal $m(t)$ occupies the frequency band $-W \leq f \leq W$, as depicted in Figure 2.25a; hereafter, W is referred to as the *message bandwidth*. Then, provided that the carrier frequency satisfies the condition $f_c > W$, we find that the spectrum of the DSB-SC-modulated signal consists of an *upper sideband* and *lower sideband*, as depicted in Figure 2.25b. Comparing the two parts of this figure, we immediately see that the *channel bandwidth*, *B*, required to support the transmission of the DSB-SC-modulated signal from the transmitter to the receiver is twice the message bandwidth.

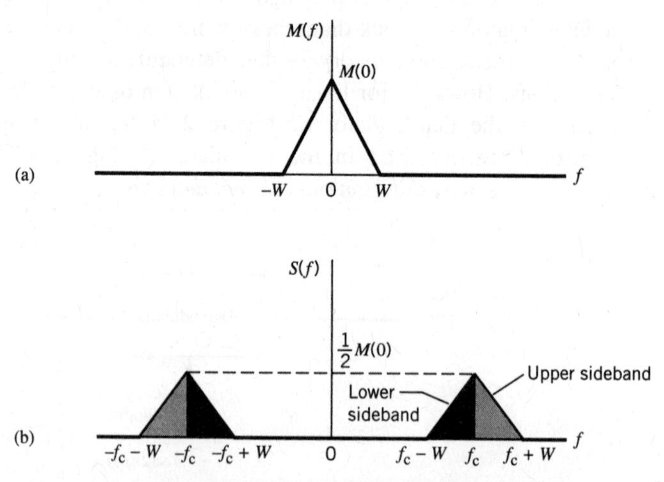

Figure 2.25 (a) Message spectrum. (b) Spectrum of DSB-SC modulated wave $s(t)$, assuming $f_c > W$.

One other interesting point apparent from Figure 2.25b is that the spectrum of the DSB-SC modulated signal is entirely void of delta functions. This statement is further testimony to the fact that the carrier is suppressed from the generation of the modulated signal $s(t)$ of (2.97).

Summarizing the useful features of DSB-SC modulation:

- *suppression of the carrier*, which results in saving of transmitted power;
- *desirable spectral characteristics*, which make it applicable to the modulation of band-limited message signals;
- ease of synchronizing the receiver to the transmitter for coherent detection.

On the downside, DSB-SC modulation is wasteful of channel bandwidth. We say so for the following reason. The two sidebands, constituting the spectral composition of the modulated signal $s(t)$, are actually the *image* of each other with respect to the carrier frequency f_c; hence, the transmission of either sideband is sufficient for transporting $s(t)$ across the channel.

VSB Modulation

In VSB modulation, one sideband is partially suppressed and a *vestige* of the other sideband is configured in such a way to compensate for the partial sideband suppression by exploiting the fact that the two sidebands in DSB-SC modulation are the image of each other. A popular method of achieving this design objective is to use the *frequency discrimination method*. Specifically, a DSB-SC-modulated signal is first generated using a product modulator, followed by a band-pass filter, as shown in Figure 2.26. The desired spectral shaping is thereby realized through the appropriate design of the band-pass filter.

Suppose that a vestige of the lower sideband is to be transmitted. Then, the frequency response of the band-pass filter, $H(f)$, takes the form shown in Figure 2.27; to simplify matters, only the frequency response for positive frequencies is shown in the figure. Examination of this figure reveals two characteristics of the band-pass filter:

1. *Normalization* of the frequency response, which means that

$$H(f) = \begin{cases} 1 & \text{for } f_c + f_v \le |f| < f_c + W \\ \frac{1}{2} & \text{for } |f| = f_c \end{cases} \tag{2.98}$$

   where f_v is the *vestigial bandwidth* and the other parameters are as previously defined.

2. *Odd symmetry of the cutoff portion inside the transition interval* $f_c - f_v \le |f| \le f_c + f_v$, which means that values of the frequency response $H(f)$ at any two frequencies equally spaced above and below the carrier frequency add up to unity.

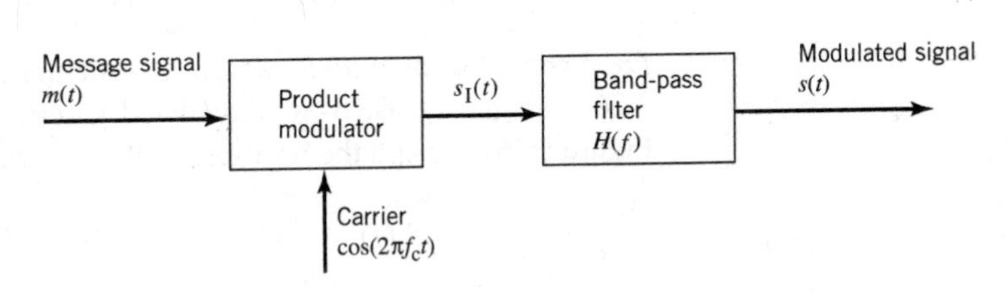

Figure 2.26
Frequency-discrimination method for producing VSB modulation where the intermediate signal $s_I(t)$ is DSB-SC modulated.

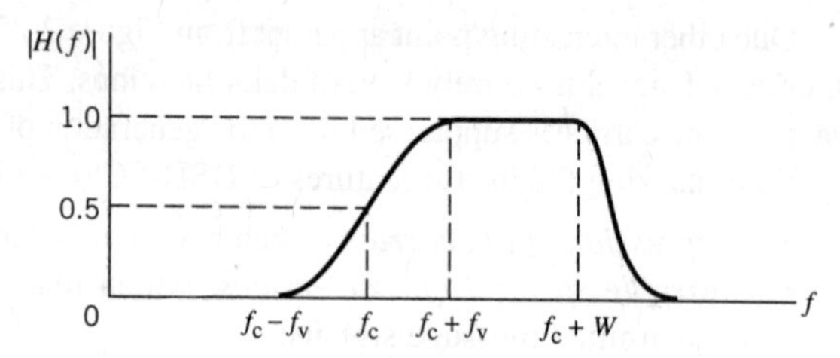

Figure 2.27 Magnitude response of VSB filter; only the positive-frequency portion is shown

Consequently, we find that shifted versions of the frequency response $H(f)$ satisfy the condition

$$H(f - f_c) + H(f + f_c) = 1 \qquad \text{for} - W \le |f| \le W \tag{2.99}$$

Outside the frequency band of interest defined by $|f| \ge f_c + W$, the frequency response $H(f)$ can assume arbitrary values. We may thus express the channel bandwidth required for the transmission of VSB-modulated signals as

$$B = W + f_v \tag{2.100}$$

With this background, we now address the issue of how to specify $H(f)$. We first use the canonical formula of (2.68) to express the VSB-modulated signal $s_1(t)$, containing a vestige of the lower sideband, as

$$s_1(t) = \frac{1}{2}m(t)\cos(2\pi f_c t) - \frac{1}{2}m_Q(t)\sin(2\pi f_c t) \tag{2.101}$$

where $m(t)$ is the message signal, as before, and $m_Q(t)$ is the spectrally shaped version of $m(t)$; the reason for the factor 1/2 will become apparent later. Note that if $m_Q(t)$ is set equal to zero, (2.101) reduces to DSB-SC modulation. It is therefore in the *quadrature signal* $m_Q(t)$ that VSB modulation distinguishes itself from DSB-SC modulation. In particular, the role of $m_Q(t)$ is to interfere with the message signal $m(t)$ in such a way that power in one of the sidebands of the VSB-modulated signal $s(t)$ (e.g., the lower sideband in Figure 2.27) is appropriately reduced.

To determine $m_Q(t)$, we examine two different procedures:

1. *Phase-discrimination,* which is rooted in the time-domain description of (2.101); transforming this equation into the frequency domain, we obtain

$$S_1(f) = \frac{1}{4}[M(f - f_c) + M(f + f_c)] - \frac{1}{4j}[M_Q(f - f_c) - M_Q(f + f_c)] \tag{2.102}$$

   where

$$M(f) = \mathbf{F}[m(t)] \qquad \text{and} \qquad M_Q(f) = \mathbf{F}[m_Q(t)]$$

2. *Frequency-discrimination,* which is structured in the manner described in Figure 2.26; passing the DSB-SC-modulated signal (i.e., the intermediate signal $s_I(t)$ in Figure 2.26) through the band-pass filter, we write

$$S_1(f) = \frac{1}{2}[M(f - f_c) + M(f + f_c)]H(f) \tag{2.103}$$

In both (2.102) and (2.103), the spectrum $S_1(f)$ is defined in the frequency interval

$$f_c - W \le |f| \le f_c + W$$

Equating the right-hand sides of these two equations, we get (after canceling common terms)

$$\begin{aligned}\frac{1}{2}[M(f-f_c)+M(f+f_c)] - \frac{1}{2j}[M_Q(f-f_c) - M_Q(f+f_c)]\\ = [M(f-f_c)+M(f+f_c)]H(f)\end{aligned} \tag{2.104}$$

Shifting both sides of (2.104) to the left by the amount f_c, we get (after canceling common terms)

$$\frac{1}{2}M(f) - \frac{1}{2j}M_Q(f) = M(f)H(f+f_c), \qquad -W \le |f| \le W \tag{2.105}$$

where the terms $M(f+2f_c)$ and $M_Q(f+2f_c)$ are ignored as they both lie outside the interval $-W \le |f| \le W$. Next, shifting both sides of (2.104) by the amount f_c, but this time to the *right*, we get (after canceling common terms)

$$\frac{1}{2}M(f) + \frac{1}{2j}M_Q(f) = M(f)H(f-f_c), \qquad -W \le |f| \le W \tag{2.106}$$

where, this time, the terms $M(f-2f_c)$ and $M_Q(f-2f_c)$ are ignored as they both lie outside the interval $-W \le |f| \le W$.

Given (2.105) and (2.106), all that remains to be done now is to follow two simple steps:

1. Adding these two equations and then factoring out the common term $M(f)$, we get the condition of (2.99) previously imposed on $H(f)$; indeed, it is with this condition in mind that we introduced the scaling factor 1/2 in (2.101).
2. Subtracting (2.105) from (2.106) and rearranging terms, we get the desired relationship between $M_Q(f)$ and $M(f)$:

$$M_Q(f) = j[H(f-f_c) - H(f+f_c)]M(f), \qquad -W \le |f| \le W \tag{2.107}$$

Let $H_Q(f)$ denote the frequency response of a *quadrature filter* that operates on the message spectrum $M(f)$ to produce $M_Q(f)$. In light of (2.107), we may readily define $H_Q(f)$ in terms of $H(f)$ as

$$\begin{aligned}H_Q(f) &= \frac{M_Q(f)}{M(f)}\\ &= j[H(f-f_c) - H(f+f_c)], \qquad -W \le |f| \le W\end{aligned} \tag{2.108}$$

Equation (2.108) provides the frequency-domain basis for the *phase-discrimination method* for generating the VSB-modulated signal $s_1(t)$, where only a vestige of the lower sideband is retained. With this equation at hand, it is instructive to plot the frequency response $H_Q(f)$. For the frequency interval $-W \le f \le W$, the term $H(f-f_c)$ is defined by the response $H(f)$ for negative frequencies shifted to the right by f_c, whereas the term $H(f+f_c)$ is defined by the response $H(f)$ for positive frequencies shifted to the left by f_c. Accordingly, building on the positive frequency response plotted in Figure 2.27, we find that the corresponding plot of $H_Q(f)$ is shaped as shown in Figure 2.28.

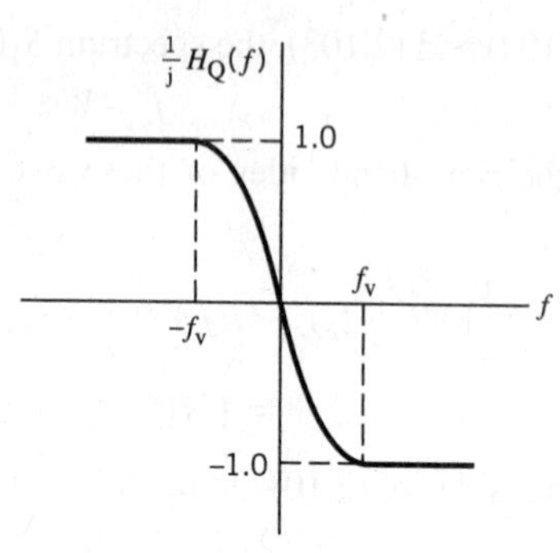

Figure 2.28 Frequency response of the quadrature filter for producing the quadrature component of the VSB wave.

The discussion on VSB modulation has thus far focused on the case where a vestige of the lower sideband is transmitted. For the alternative case when a vestige of the upper sideband is transmitted, we find that the corresponding VSB-modulated wave is described by

$$s_2(t) = \frac{1}{2}m(t)\cos(2\pi f_c t) + \frac{1}{2}m_Q(t)\sin(2\pi f_c t) \tag{2.109}$$

where the quadrature signal $m_Q(t)$ is constructed from the message signal $m(t)$ in exactly the same way as before.

Equations (2.101) and (2.109) are of the same mathematical form, except for an algebraic difference; they may, therefore, be combined into the single formula

$$s(t) = \frac{1}{2}m(t)\cos(2\pi f_c t) \mp \frac{1}{2}m_Q(t)\sin(2\pi f_c t) \tag{2.110}$$

where the minus sign applies to a VSB-modulated signal containing a vestige of the lower sideband and the plus sign applies to the alternative case when the modulated signal contains a vestige of the upper sideband.

The formula of (2.110) for VSB modulation includes DSB-SC modulation as a special case. Specifically, setting $m_Q(t) = 0$, this formula reduces to that of (2.97) for DSB-SC modulation, except for the trivial scaling factor of 1/2.

SSB Modulation

Next, considering *SSB modulation*, we may identify two choices:

1. The carrier and the lower sideband are both suppressed, leaving the upper sideband for transmission in its full spectral content; this first SSB-modulated signal is denoted by $s_{USB}(t)$.
2. The carrier and the upper sideband are both suppressed, leaving the lower sideband for transmission in its full spectral content; this second SSB-modulated signal is denoted by $s_{LSB}(t)$.

The Fourier transforms of these two modulated signals are the *image* of each other with respect to the carrier frequency f_c, which, as mentioned previously, emphasizes that the transmission of either sideband is actually sufficient for transporting the message signal $m(t)$ over the communication channel. In practical terms, both $s_{USB}(t)$ and $s_{LSB}(t)$ require

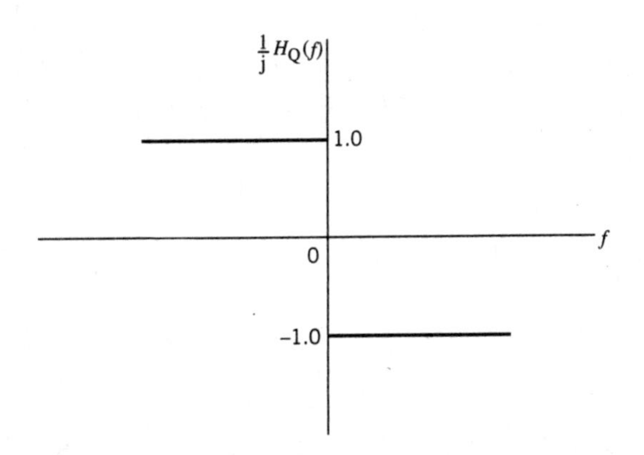

Figure 2.29
Frequency response of the quadrature filter in SSB modulation.

the smallest feasible channel bandwidth, $B = W$, without compromising the perfect recovery of the message signal under noiseless conditions. It is for these reasons that we say SSB modulation is the *optimum form of linear modulation* for analog communications, preserving both the transmitted power and channel bandwidth in the best manner possible.

SSB modulation may be viewed as a special case of VSB modulation. Specifically, setting the vestigial bandwidth $f_V = 0$, we find that the frequency response of the quadrature filter plotted in Figure 2.28 takes the limiting form of the *signum function* shown in Figure 2.29. In light of the material presented in (2.60) on Hilbert transformation, we therefore find that for $f_V = 0$ the quadrature component $m_Q(t)$ becomes the Hilbert transform of the message signal $m(t)$, denoted by $\hat{m}(t)$. Accordingly, using $\hat{m}(t)$ in place of $m_Q(t)$ in (2.110) yields the SSB formula

$$s(t) = \frac{1}{2}m(t)\cos(2\pi f_c t) \mp \frac{1}{2}\hat{m}(t)\sin(2\pi f_c t) \tag{2.111}$$

where the minus sign applies to the SSB-modulated signal $s_{USB}(t)$ and the plus sign applies to the alternative SSB-modulated signal $s_{LSB}(t)$.

Unlike DSB-SC and VSB methods of modulation, SSB modulation is of limited applicability. Specifically, we say:

> For SSB modulation to be feasible in practical terms, the spectral content of the message signal $m(t)$ must have an energy gap centered on the origin.

This requirement, illustrated in Figure 2.30, is imposed on the message signal $m(t)$ so that the band-pass filter in the frequency-discrimination method of Figure 2.26 has a *finite transition* band for the filter to be physically realizable. With the transition band separating the pass-band from the stop-band, it is only when the transition band is finite that the undesired sideband can be suppressed. An example of message signals for which the energy-gap requirement is satisfied is voice signals; for such signals, the energy gap is about 600 Hz, extending from –300 to +300 Hz.

In contrast, the spectral contents of television signals and wideband data extend practically to a few hertz, thereby ruling out the applicability of SSB modulation to this second class of message signals. It is for this reason that VSB modulation is preferred over SSB modulation for the transmission of wideband signals.

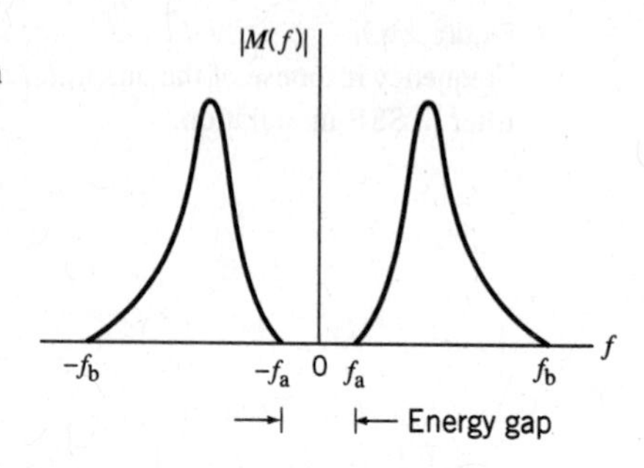

Figure 2.30
Spectrum of a message signal $m(t)$ with an energy gap centered around the origin.

Summary of Linear Modulation Methods

Equation (2.97) for DSB-SC modulation, (2.110) for VSB modulation, and (2.111) for SSB modulation are summarized in Table 2.5 as special cases of the canonical formula of (2.68). Correspondingly, we may treat the time-domain generations of these three linearly modulated signals as special cases of the "synthesizer" depicted in Figure 2.20b.

Table 2.5 **Summary of linear modulation methods viewed as special cases of the canonical formula** $s(t) = s_I(t)\cos(2\pi f_c t) - s_Q(t)\sin(2\pi f_c t)$

Type of modulation	In-phase component, $s_I(t)$	Quadrature component, $s_Q(t)$	Comments
DSB-SC	$m(t)$	zero	$m(t)$ = message signal
VSB	$\frac{1}{2}m(t)$	$\pm\frac{1}{2}m_Q(t)$	Plus sign applies to using vestige of lower sideband and minus sign applies to using vestige of upper sideband
SSB	$\frac{1}{2}m(t)$	$\pm\frac{1}{2}\hat{m}(t)$	Plus sign applies to transmission of upper sideband and minus sign applies to transmission of lower sideband

2.15 Phase and Group Delays

A discussion of signal transmission through linear time-invariant systems is incomplete without considering the phase and group delays involved in the signal transmission process.

Whenever a signal is transmitted through a dispersive system, exemplified by a communication channel (or band-pass filter), some *delay* is introduced into the output signal, the delay being measured with respect to the input signal. In an ideal channel, the phase response varies *linearly* with frequency inside the passband of the channel, in which case the filter introduces a constant delay equal to t_0, where the parameter t_0 controls the slope of the linear phase response of the channel. Now, what if the phase response of the channel is a nonlinear function of frequency, which is frequently the case in practice? The purpose of this section is to address this practical issue.

To begin the discussion, suppose that a steady sinusoidal signal at frequency f_c is transmitted through a dispersive channel that has a phase-shift of $\beta(f_c)$ radians at that frequency. By using two phasors to represent the input signal and the received signal, we see that the received signal phasor lags the input signal phasor by $\beta(f_c)$ radians. The time taken by the received signal phasor to sweep out this phase lag is simply equal to the ratio $\beta(f_c)/(2\pi f_c)$ seconds. This time is called the *phase delay* of the channel.

It is important to realize, however, that the phase delay is not necessarily the true signal delay. This follows from the fact that a steady sinusoidal signal does *not* carry information, so it would be incorrect to deduce from the above reasoning that the phase delay is the true signal delay. To substantiate this statement, suppose that a slowly varying signal, over the interval $-(T/2) \le t \le (T/2)$, is multiplied by the carrier, so that the resulting modulated signal consists of a narrow group of frequencies centered around the carrier frequency; the DSB-SC waveform of Figure 2.31 illustrates such a modulated signal. When this modulated signal is transmitted through a communication channel, we find that there is indeed a delay between the envelope of the input signal and that of the received signal. This delay, called the *envelope* or *group delay* of the channel, represents the true signal delay insofar as the information-bearing signal is concerned.

Assume that the dispersive channel is described by the transfer function

$$H(f) = K\exp[j\beta(f)] \tag{2.112}$$

where the amplitude K is a constant scaling factor and the phase $\beta(f)$ is a nonlinear function of frequency f; it is the nonlinearity of $\beta(f)$ that is responsible for the dispersive

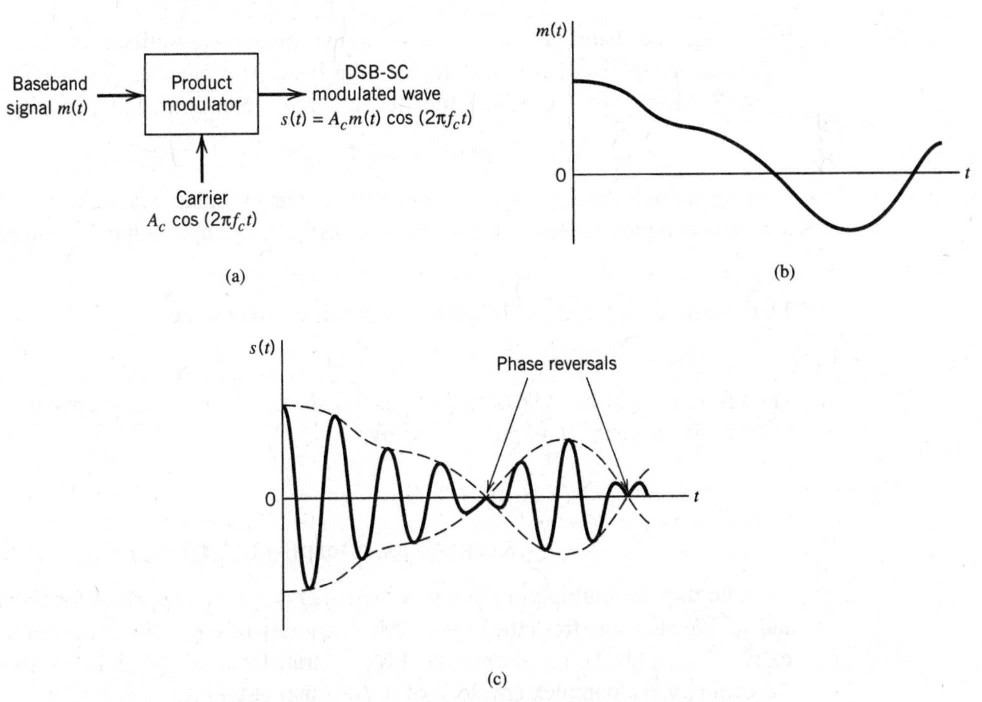

Figure 2.31 (a) Block diagram of product modulator; (b) Baseband signal; (c) DSB-SC modulated wave.

nature of the channel. The input signal $s(t)$ is assumed to be of the kind displayed in Figure 2.31; that is, the DSB-SC-modulated signal

$$s(t) = m(t)\cos(2\pi f_c t) \tag{2.113}$$

where $m(t)$ is the message signal, assumed to be of a low-pass kind and limited to the frequency interval $|f| \le W$. Moreover, we assume that the carrier frequency $f_c > W$. By expanding the phase $\beta(f)$ in a *Taylor series* about the point $f = f_c$ and retaining only the first two terms, we may approximate $\beta(f)$ as

$$\beta(f) \approx \beta(f_c) + (f - f_c)\frac{\partial \beta(f)}{\partial f}\bigg|_{f=f_c} \tag{2.114}$$

Define two new terms:

$$\tau_p = -\frac{\beta(f_c)}{2\pi f_c} \tag{2.115}$$

and

$$\tau_g = -\frac{1}{2\pi}\frac{\partial \beta(f)}{\partial f}\bigg|_{f=f_c} \tag{2.116}$$

Then, we may rewrite (2.114) in the equivalent form

$$\beta(f) \approx -2\pi f_c \tau_p - 2\pi(f - f_c)\tau_g \tag{2.117}$$

Correspondingly, the transfer function of the channel takes the approximate form

$$H(f) \approx K\exp[-j2\pi f_c \tau_p - j2\pi(f - f_c)\tau_g] \tag{2.118}$$

Following the band-pass-to-low-pass transformation described in Section 2.12, in particular using (2.80), we may replace the band-pass channel described by $H(f)$ by an equivalent low-pass filter whose transfer function is approximately given by

$$\tilde{H}(f) \approx 2K\exp(-j2\pi f_c \tau_p - j2\pi f \tau_g), \qquad f > f_c \tag{2.119}$$

Correspondingly, using (2.67) we may replace the modulated signal $s(t)$ of (2.113) by its low-pass complex envelope, which, for the DSB-SC example at hand, is simply defined by

$$\tilde{s}(t) = m(t) \tag{2.120}$$

Transforming $\tilde{s}(t)$ into the frequency domain, we may write

$$\tilde{S}(f) = M(f) \tag{2.121}$$

Therefore, in light of (2.96), the Fourier transform of the complex envelope of the signal received at the channel output is given by

$$\begin{aligned}\tilde{X}(f) &= \frac{1}{2}\tilde{H}(f)\tilde{S}(f) \\ &\approx K\exp(-j2\pi f_c \tau_p)\exp(-j2\pi f_c \tau_g)M(f)\end{aligned} \tag{2.122}$$

We note that the multiplying factor $K\exp(-j2\pi f_c \tau_p)$ is a constant for fixed values of f_c and τ_p. We also note from the time-shifting property of the Fourier transform that the term $\exp(-j2\pi f_c \tau_g)M(f)$ represents the Fourier transform of the delayed signal $m(t - \tau_g)$. Accordingly, the complex envelope of the channel output is

$$\tilde{x}(t) = K\exp(-j2\pi f_c \tau_p)m(t - \tau_g) \tag{2.123}$$

Finally, using (2.66) we find that the actual channel output is itself given by

$$\begin{aligned} x(t) &= \mathrm{Re}[\tilde{x}(t)\exp(\mathrm{j}2\pi f_c t)] \\ &= Km(t-\tau_g)\cos[2\pi f_c(t-\tau_p)] \end{aligned} \tag{2.124}$$

Equation (2.124) reveals that, as a result of transmitting the modulated signal $s(t)$ through the dispersive channel, two different delay effects occur at the channel output:

1. The sinusoidal carrier wave $\cos(2\pi f_c t)$ is delayed by τ_p seconds; hence, τ_p represents the *phase delay*; sometimes τ_p is referred to as the *carrier delay.*
2. The envelope $m(t)$ is delayed by τ_g seconds; hence, τ_g represents the *envelope* or *group delay.*

Note that τ_g is related to the slope of the phase $\beta(f)$, measured at $f = f_c$. Note also that when the phase response $\beta(f)$ varies linearly with frequency f and $\beta(f_c)$ is zero, the phase delay and group delay assume a common value. It is only then that we can think of these two delays being equal.

2.16 Numerical Computation of the Fourier Transform

The material presented in this chapter clearly testifies to the importance of the Fourier transform as a theoretical tool for the representation of deterministic signals and linear time-invariant systems, be they of the low-pass or band-pass kind. The importance of the Fourier transform is further enhanced by the fact that there exists a class of algorithms called FFT algorithms[6] for numerical computation of the Fourier transform in an efficient manner.

The FFT algorithm is derived from the discrete Fourier transform (DFT) in which, as the name implies, both time and frequency are represented in discrete form. The DFT provides an *approximation* to the Fourier transform. In order to properly represent the information content of the original signal, we have to take special care in performing the sampling operations involved in defining the DFT. A detailed treatment of the sampling process is presented in Chapter 6. For the present, it suffices to say that, given a band-limited signal, the sampling rate should be greater than twice the highest frequency component of the input signal. Moreover, if the samples are uniformly spaced by T_s seconds, the spectrum of the signal becomes periodic, repeating every $f_s = (1/T_s)$ hz in accordance with (2.43). Let N denote the number of frequency samples contained in the interval f_s. Hence, the *frequency resolution* involved in numerical computation of the Fourier transform is defined by

$$\Delta f = \frac{f_s}{N} = \frac{1}{NT_s} = \frac{1}{T} \tag{2.125}$$

where T is the total duration of the signal.

Consider then a *finite data sequence* $\{g_0, g_1, \ldots, g_{N-1}\}$. For brevity, we refer to this sequence as g_n, in which the subscript is the *time index* $n = 0, 1, \ldots, N-1$. Such a sequence may represent the result of sampling an analog signal $g(t)$ at times $t = 0, T_s, \ldots, (N-1)T_s$, where T_s is the sampling interval. The ordering of the data sequence defines the sample time in that $g_0, g_1, ..., g_{N-1}$ denote samples of $g(t)$ taken at times $0, T_s, ..., (N-1)T_s$, respectively. Thus we have

$$g_n = g(nT_s) \tag{2.126}$$

We formally define the DFT of g_n as

$$G_k = \sum_{n=0}^{N-1} g_n \exp\left(-\frac{j2\pi}{N}kn\right) \qquad k = 0, 1, \ldots, N-1 \tag{2.127}$$

The sequence $\{G_0, G_1, ..., G_{N-1}\}$ is called the *transform sequence*. For brevity, we refer to this second sequence simply as G_k, in which the subscript is the *frequency index* $k = 0, 1, ..., N-1$.

Correspondingly, we define the *inverse discrete Fourier transform* (IDFT) of G_k as

$$g_n = \frac{1}{N}\sum_{k=0}^{N-1} G_k \exp\left(\frac{j2\pi}{N}kn\right) \qquad n = 0, 1, \ldots, N-1 \tag{2.128}$$

The DFT and the IDFT form a discrete transform pair. Specifically, given a data sequence g_n, we may use the DFT to compute the transform sequence G_k; and given the transform sequence G_k, we may use the IDFT to recover the original data sequence g_n. A distinctive feature of the DFT is that, for the finite summations defined in (2.127) and (2.128), there is no question of convergence.

When discussing the DFT (and algorithms for its computation), the words "sample" and "point" are used interchangeably to refer to a sequence value. Also, it is common practice to refer to a sequence of length N as an *N-point sequence* and to refer to the DFT of a data sequence of length N as an *N-point DFT.*

Interpretation of the DFT and the IDFT

We may visualize the DFT process described in (2.127) as a collection of N *complex heterodyning* and *averaging* operations, as shown in Figure 2.32a. We say that the heterodyning is complex in that samples of the data sequence are multiplied by *complex exponential sequences*. There is a total of N complex exponential sequences to be considered, corresponding to the frequency index $k = 0, 1, ..., N-1$. Their periods have been selected in such a way that each complex exponential sequence has precisely an integer number of cycles in the total interval 0 to $N-1$. The zero-frequency response, corresponding to $k = 0$, is the only exception.

For the interpretation of the IDFT process, described in (2.128), we may use the scheme shown in Figure 2.32b. Here we have a collection of N *complex signal generators*, each of which produces the complex exponential sequence

$$\begin{aligned}\exp\left(\frac{j2\pi}{N}kn\right) &= \cos\left(\frac{2\pi}{N}kn\right) + j\sin\left(\frac{2\pi}{N}kn\right)\\ &= \left\{\cos\left(\frac{2\pi}{N}kn\right), \sin\left(\frac{2\pi}{N}kn\right)\right\}_{k=0}^{N-1}\end{aligned} \tag{2.129}$$

Thus, in reality, each complex signal generator consists of a pair of generators that output a cosinusoidal and a sinusoidal sequence of k cycles per observation interval. The output of each complex signal generator is weighted by the complex Fourier coefficient G_k. At each time index n, an output is formed by summing the weighted complex generator outputs.

It is noteworthy that although the DFT and the IDFT are similar in their mathematical formulations, as described in (2.127) and (2.128), their interpretations as depicted in Figure 2.32a and b are so completely different.

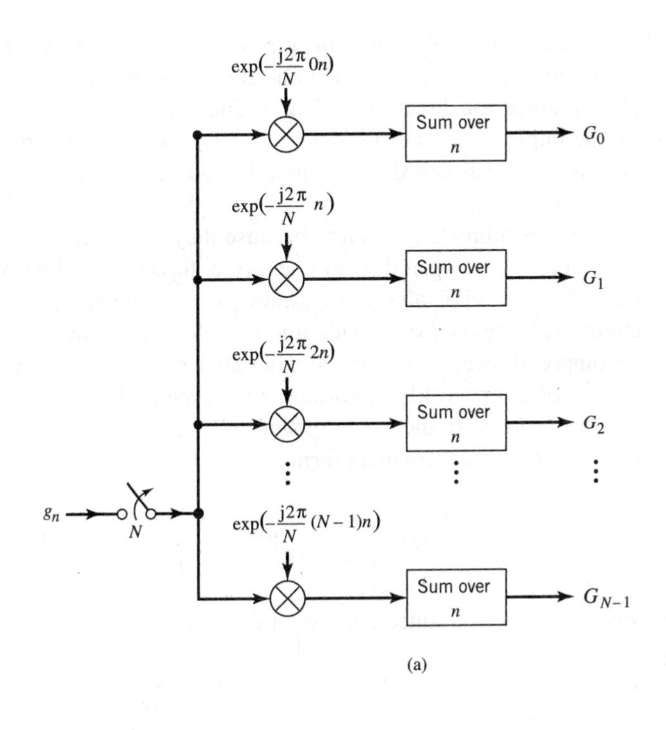

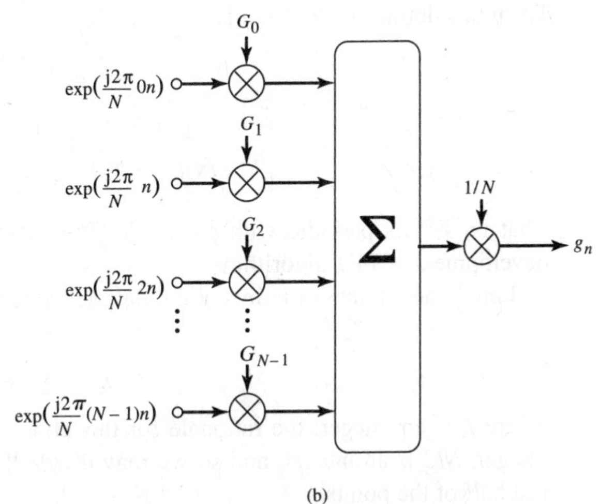

Figure 2.32 Interpretations of (a) the DFT and (b) the IDFT.

Also, the addition of harmonically related periodic signals, involved in these two parts of the figure, suggests that their outputs G_k and g_n must be both periodic. Moreover, the processors shown in Figure 2.32 are linear, suggesting that the DFT and IDFT are both linear operations. This important property is also obvious from the defining equations (2.127) and (2.128).

FFT Algorithms

In the DFT both the input and the output consist of sequences of numbers defined at uniformly spaced points in time and frequency, respectively. This feature makes the DFT ideally suited for direct numerical evaluation on a computer. Moreover, the computation can be implemented most efficiently using a class of algorithms, collectively called *FFT algorithms*. An algorithm refers to a "recipe" that can be written in the form of a computer program.

FFT algorithms are efficient because they use a greatly reduced number of arithmetic operations as compared with the brute force (i.e., direct) computation of the DFT. Basically, an FFT algorithm attains its computational efficiency by following the engineering strategy of "divide and conquer," whereby the original DFT computation is decomposed successively into smaller DFT computations. In this section, we describe one version of a popular FFT algorithm, the development of which is based on such a strategy.

To proceed with the development, we first rewrite (2.127), defining the DFT of g_n, in the convenient mathematical form

$$G_k = \sum_{n=0}^{N-1} g_n W^{kn}, \qquad k = 0, 1, \ldots, N-1 \tag{2.130}$$

where we have introduced the complex parameter

$$W = \exp\left(-\frac{j2\pi}{N}\right) \tag{2.131}$$

From this definition, we readily see that

$$\begin{aligned} W^N &= 1 \\ W^{N/2} &= -1 \\ W^{(l+lN)(n+mN)} &= W^{kn}, \qquad (m, l) = 0, \pm 1, \pm 2, \ldots \end{aligned}$$

That is, W^{kn} is periodic with period N. The periodicity of W^{kn} is a key feature in the development of FFT algorithms.

Let N, the number of points in the data sequence, be an integer power of two, as shown by

$$N = 2^L$$

where L is an integer; the rationale for this choice is explained later. Since N is an even integer, $N/2$ is an integer, and so we may divide the data sequence into the first half and last half of the points.

Thus, we may rewrite (2.130) as

$$\begin{aligned} G_k &= \sum_{n=0}^{(N/2)-1} g_n W^{kn} + \sum_{n=N/2}^{N-1} g_n W^{kn} \\ &= \sum_{n=0}^{(N/2)-1} g_n W^{kn} + \sum_{n=0}^{(N/2)-1} g_{n+N/2} W^{k(n+N/2)} \\ &= \sum_{n=0}^{(N/2)-1} (g_n + g_{n+N/2} W^{kN/2}) W^{kn} \qquad k = 0, 1, \ldots, N-1 \end{aligned} \tag{2.132}$$

Since $W^{N/2} = -1$, we have

$$W^{kN/2} = (-1)^k$$

Accordingly, the factor $W^{kN/2}$ in (2.132) takes on only one of two possible values, namely +1 or –1, depending on whether the frequency index k is even or odd, respectively. These two cases are considered in what follows.

First, let k be even, so that $W^{kN/2} = 1$. Also let

$$k = 2l, \qquad l = 0, 1, \ldots, \frac{N}{2} - 1$$

and define

$$x_n = g_n + g_{n+N/2} \tag{2.133}$$

Then, we may put (2.132) into the new form

$$\begin{aligned} G_{2l} &= \sum_{n=0}^{(N/2)-1} x_n W^{2ln} \\ &= \sum_{n=0}^{(N/2)-1} x_n (W^2)^{ln} \qquad l = 0, 1, \ldots, \frac{N}{2} - 1 \end{aligned} \tag{2.134}$$

From the definition of W given in (2.131), we readily see that

$$\begin{aligned} W^2 &= \exp\left(-\frac{j4\pi}{N}\right) \\ &= \exp\left(-\frac{j2\pi}{N/2}\right) \end{aligned}$$

Hence, we recognize the sum on the right-hand side of (2.134) as the $(N/2)$-point DFT of the sequence x_n.

Next, let k be odd so that $W^{kN/2} = -1$. Also, let

$$k = 2l + 1, \qquad l = 0, 1, \ldots, \frac{N}{2} - 1$$

and define

$$y_n = g_n - g_{n+N/2} \tag{2.135}$$

Then, we may put (2.132) into the corresponding form

$$\begin{aligned} G^{2l+1} &= \sum_{n=0}^{(N/2)-1} y_n W^{(2l+1)n} \\ &= \sum_{n=0}^{(N/2)-1} [y_n W^n](W^2)^{ln} \qquad l = 0, 1, \ldots, \frac{N}{2} - 1 \end{aligned} \tag{2.136}$$

We recognize the sum on the right-hand side of (2.136) as the ($N/2$)-point DFT of the sequence $y_n W^n$. The parameter W^n associated with y_n is called the *twiddle factor.*

Equations (2.134) and (2.136) show that the even- and odd-valued samples of the transform sequence G_k can be obtained from the ($N/2$)-point DFTs of the sequences x_n and $y_n W^n$, respectively. The sequences x_n and y_n are themselves related to the original data sequence g_n by (2.133) and (2.135), respectively. Thus, the problem of computing an N-point DFT is reduced to that of computing two ($N/2$)-point DFTs. The procedure just described is repeated a second time, whereby an ($N/2$)-point DFT is decomposed into two ($N/4$)-point DFTs. The decomposition procedure is continued in this fashion until (after $L = \log_2 N$ stages) we reach the trivial case of N single-point DFTs.

Figure 2.33 illustrates the computations involved in applying the formulas of (2.134) and (2.136) to an eight-point data sequence; that is, $N = 8$. In constructing left-hand portions of the figure, we have used signal-flow graph notation. A *signal-flow graph* consists of an interconnection of *nodes* and *branches*. The *direction* of signal transmission along a branch is indicated by an arrow. A branch multiplies the variable at a node (to which it is connected) by the branch *transmittance*. A node sums the outputs of all incoming branches. The convention used for branch transmittances in Figure 2.33 is as follows. When no coefficient is indicated on a branch, the transmittance of that branch is assumed to be unity. For other branches, the transmittance of a branch is indicated by -1 or an integer power of W, placed alongside the arrow on the branch.

Thus, in Figure 2.33a the computation of an eight-point DFT is reduced to that of two four-point DFTs. The procedure for the eight-point DFT may be mimicked to simplify the computation of the four-point DFT. This is illustrated in Figure 2.33b, where the computation of a four-point DFT is reduced to that of two two-point DFTs. Finally, the computation of a two-point DFT is shown in Figure 2.33c.

Combining the ideas described in Figure 2.33, we obtain the complete signal-flow graph of Figure 2.34 for the computation of the eight-point DFT. A repetitive structure, called the *butterfly* with two inputs and two outputs, can be discerned in the FFT algorithm of Figure 2.34. Examples of butterflies (for the three stages of the algorithm) are shown by the bold-faced lines in Figure 2.34.

For the general case of $N = 2^L$, the algorithm requires $L = \log_2 N$ stages of computation. Each stage requires ($N/2$) butterflies. Each butterfly involves one complex multiplication and two complex additions (to be precise, one addition and one subtraction). Accordingly, the FFT structure described here requires $(N/2)\log_2 N$ complex multiplications and $N \log_2 N$

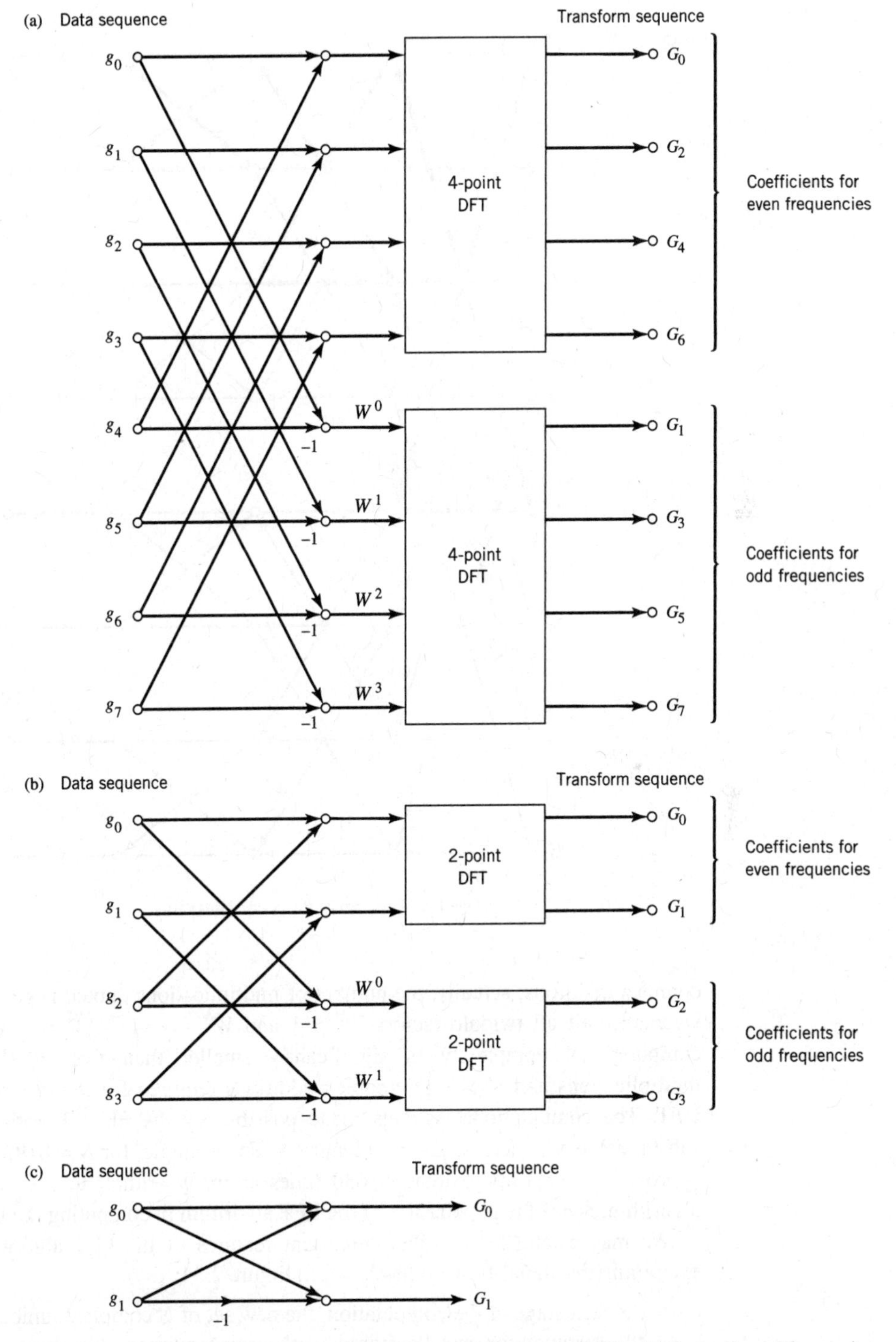

Figure 2.33 (a) Reduction of eight-point DFT into two four-point DFTs. (b) Reduction of four-point DFT into two two-point DFTs. (c) Trivial case of two-point DFT.

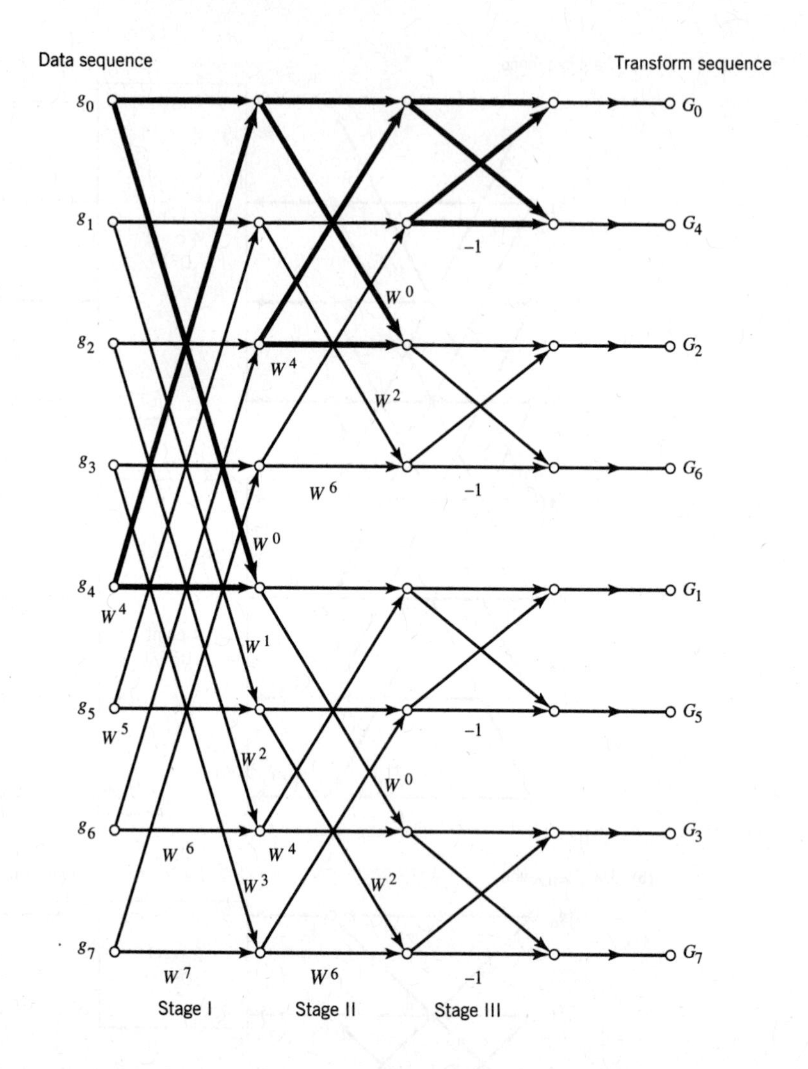

Figure 2.34
Decimation-in-frequency FFT algorithm.

complex additions; actually, the number of multiplications quoted is pessimistic, because we may omit all twiddle factors $W^0 = 1$ and $W^{N/2} = -1$, $W^{N/4} = j$, $W^{3N/4} = -j$. This computational complexity is significantly smaller than that of the N^2 complex multiplications and $N(N - 1)$ complex additions required for *direct* computation of the DFT. The computational savings made possible by the FFT algorithm become more substantial as we increase the data length N. For example, for $N = 8192 = 2^{11}$, the direct approach requires approximately 630 times as many arithmetic operations as the FFT algorithm, hence the popular use of the FFT algorithm in computing the DFT.

We may establish two other important features of the FFT algorithm by carefully examining the signal-flow graph shown in Figure 2.34:

1. At each stage of the computation, the new set of N complex numbers resulting from the computation can be stored in the same memory locations used to store the previous set. This kind of computation is referred to as *in-place computation*.

2. The samples of the transform sequence G_k are stored in a bit-reversed order. To illustrate the meaning of this terminology, consider Table 2.6 constructed for the case of $N = 8$. At the left of the table, we show the eight possible values of the frequency index k (in their natural order) and their 3-bit binary representations. At the right of the table, we show the corresponding bit-reversed binary representations and indices. We observe that the bit-reversed indices in the rightmost column of Table 2.6 appear in the same order as the indices at the output of the FFT algorithm in Figure 2.34.

Table 2.6 **Illustrating bit reversal**

Frequency index, k	Binary representation	Bit-reversed binary representation	Bit-reversed index
0	000	000	0
1	001	100	4
2	010	010	2
3	011	110	6
4	100	001	1
5	101	101	5
6	110	011	3
7	111	111	7

The FFT algorithm depicted in Figure 2.34 is referred to as a *decimation-in-frequency algorithm*, because the transform (frequency) sequence G_k is divided successively into smaller subsequences. In another popular FFT algorithm, called a *decimation-in-time algorithm*, the data (time) sequence g_n is divided successively into smaller subsequences. Both algorithms have the same computational complexity. They differ from each other in two respects. First, for decimation-in-frequency, the input is in natural order, whereas the output is in bit-reversed order; the reverse is true for decimation-in-time. Second, the butterfly for decimation-in-time is slightly different from that for decimation-in-frequency. The reader is invited to derive the details of the decimation-in-time algorithm using the divide-and-conquer strategy that led to the development of the algorithm described in Figure 2.34.

In devising the FFT algorithm presented herein, we placed the factor $1/N$ in the formula for the forward DFT, as shown in (2.128). In some other FFT algorithms, location of the factor $1/N$ is reversed. In yet other formulations, the factor $1/\sqrt{N}$ is placed in the formulas for both the forward and inverse DFTs for the sake of symmetry.

Computation of the IDFT

The IDFT of the transform G_k is defined by (2.128). We may rewrite this equation in terms of the complex parameter W as

$$g_n = \frac{1}{N}\sum_{k=0}^{N-1} G_k W^{-kn}, \qquad n = 0, 1, \ldots, N-1 \tag{2.137}$$

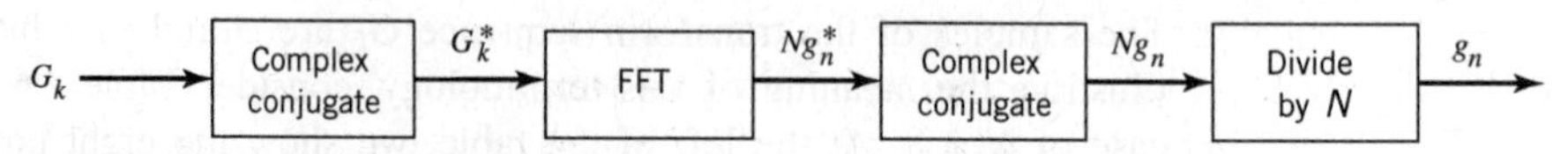

Figure 2.35 Use of the FFT algorithm for computing the IDFT.

Taking the complex conjugate of (2.137) and multiplying by N, we get

$$Ng_n^* = \sum_{k=0}^{N-1} G_k^* W^{-kn}, \qquad n = 0, 1, \ldots, N-1 \tag{2.138}$$

The right-hand side of (2.138) is recognized as the N-point DFT of the complex-conjugated sequence G_k^*. Accordingly, (2.138) suggests that we may compute the desired sequence g_n using the scheme shown in Figure 2.35, based on an N-point FFT algorithm. Thus, the same FFT algorithm can be used to handle the computation of both the IDFT and the DFT.

2.17 Summary and Discussion

In this chapter we have described the Fourier transform as a fundamental tool for relating the time-domain and frequency-domain descriptions of a deterministic signal. The signal of interest may be an energy signal or a power signal. The Fourier transform includes the exponential Fourier series as a special case, provided that we permit the use of the Dirac delta function.

An inverse relationship exists between the time-domain and frequency-domain descriptions of a signal. Whenever an operation is performed on the waveform of a signal in the time domain, a corresponding modification is applied to the spectrum of the signal in the frequency domain. An important consequence of this inverse relationship is the fact that the time–bandwidth product of an energy signal is a constant; the definitions of signal duration and bandwidth merely affect the value of the constant.

An important signal-processing operation frequently encountered in communication systems is that of linear filtering. This operation involves the convolution of the input signal with the impulse response of the filter or, equivalently, the multiplication of the Fourier transform of the input signal by the transfer function (i.e., Fourier transform of the impulse response) of the filter. Low-pass and band-pass filters represent two commonly used types of filters. Band-pass filtering is usually more complicated than low-pass filtering. However, through the combined use of a complex envelope for the representation of an input band-pass signal and the complex impulse response for the representation of a band-pass filter, we may formulate a complex low-pass equivalent for the band-pass filtering problem and thereby replace a difficult problem with a much simpler one. It is also important to note that there is no loss of information in establishing this equivalence. A rigorous treatment of the concepts of complex envelope and complex impulse response as presented in this chapter is rooted in Hilbert transformation.

The material on Fourier analysis, as presented in this chapter, deals with signals whose waveforms can be nonperiodic or periodic, and whose spectra can be continuous or discrete functions of frequency. In this sense, the material has general appeal.

Building on the canonical representation of a band-pass signal involving the in-phase and quadrature components of the signal, we showed that this representation provides an elegant way of describing the three basic forms of linear modulation, namely DSB-SC, VSB, and SSB.

With the Fourier transform playing such a pervasive role in the study of signals and linear systems, we finally described the FFT algorithm as an efficient tool for numerical computation of the DFT that represents the uniformly sampled versions of the forward and inverse forms of the ordinary Fourier transform.

Problems

The Fourier Transform

2.1 Prove the dilation property of the Fourier transform, listed as Property 2 in Table 2.1.

2.2 a. Prove the duality property of the Fourier transform, listed as Property 3 in Table 2.1.

b. Prove the time-shifting property, listed as Property 4; and then use the duality property to prove the frequency-shifting property, listed as Property 5 in the table.

c. Using the frequency-shifting property, determine the Fourier transform of the radio frequency RF pulse

$$g(t) = A\,\mathrm{rect}\left(\frac{t}{T}\right)\cos(2\pi f_c t)$$

assuming that f_c is larger than $(1/T)$.

2.3 a. Prove the multiplication-in-the-time-domain property of the Fourier transform, listed as Property 11 in Table 2.1.

b. Prove the convolution in the time-domain property, listed as Property 12.

c. Using the result obtained in part b, prove the correlation theorem, listed as Property 13.

2.4 Prove Rayleigh's energy theorem listed as Property 14 in Table 2.1.

2.5 The following expression may be viewed as an approximate representation of a pulse with finite rise time:

$$g(t) = \frac{1}{\tau}\int_{t-T}^{t+T} \exp\left(-\frac{\pi u^2}{\tau^2}\right) du$$

where it is assumed that $T >> \tau$. Determine the Fourier transform of $g(t)$. What happens to this transform when we allow τ to become zero? *Hint:* Express $g(t)$ as the superposition of two signals, one corresponding to integration from $t - T$ to 0, and the other from 0 to $t + T$.

2.6 The Fourier transform of a signal $g(t)$ is denoted by $G(f)$. Prove the following properties of the Fourier transform:

a. If a real signal $g(t)$ is an even function of time t, the Fourier transform $G(f)$ is purely real. If a real signal $g(t)$ is an odd function of time t, the Fourier transform $G(f)$ is purely imaginary.

b.

$$t^n g(t) \rightleftharpoons \left(\frac{\mathrm{j}}{2\pi}\right)^n G^{(n)}(f)$$

where $G^{(n)}(f)$ is the nth derivative of $G(f)$ with respect to f.

c.

$$\int_{-\infty}^{\infty} t^n g(t)\,\mathrm{d}t = \left(\frac{\mathrm{j}}{2\pi}\right)^n G^{(n)}(0)$$

d. Assuming that both $g_1(t)$ and $g_2(t)$ are complex signals, show that:

$$g_1(t)g_2^*(t) \rightleftharpoons \int_{-\infty}^{\infty} G_1(\lambda)G_2^*(\lambda - f)\,d\lambda$$

and

$$\int_{-\infty}^{\infty} g_1(t)g_2^*(t)dt = \int_{-\infty}^{\infty} G_1(f)G_2^*(f)\,df$$

2.7 a. The *root mean-square (rms) bandwidth* of a low-pass signal $g(t)$ of finite energy is defined by

$$W_{\text{rms}} = \left[\frac{\int_{-\infty}^{\infty} f^2|G(f)|^2 df}{\int_{-\infty}^{\infty} |G(f)|^2 df}\right]^{1/2}$$

where $|G(f)|^2$ is the energy spectral density of the signal. Correspondingly, the *root mean-square (rms) duration* of the signal is defined by

$$T_{\text{rms}} = \left[\frac{\int_{-\infty}^{\infty} t^2|g(t)|^2 dt}{\int_{-\infty}^{\infty} |g(t)|^2 dt}\right]^{1/2}$$

Using these definitions, show that

$$T_{\text{rms}}W_{\text{rms}} \geq \frac{1}{4\pi}$$

Assume that $|g(t)| \to 0$ faster than $1/\sqrt{|t|}$ as $|t| \to \infty$.

b. Consider a Gaussian pulse defined by

$$g(t) = \exp(-\pi t^2)$$

Show that for this signal the equality

$$T_{\text{rms}}W_{\text{rms}} = \frac{1}{4\pi}$$

is satisfied.

Hint: Use Schwarz's inequality

$$\left(\int_{-\infty}^{\infty} [g_1^*(t)g_2(t) + g_1(t)g_2^*(t)]dt\right)^2 \leq 4\int_{-\infty}^{\infty} |g_1(t)|^2 dt \int_{-\infty}^{\infty} |g_2(t)|^2 dt$$

in which we set

$$g_1(t) = tg(t)$$

and

$$g_2(t) = \frac{dg(t)}{dt}$$

2.8 The *Dirac comb*, formulated in the time domain, is defined by

$$\delta_{T_0}(t) = \sum_{m=-\infty}^{\infty} \delta(t - mT_0)$$

where T_0 is the period.

a. Show that the Dirac comb is its own Fourier transform. That is, the Fourier transform of $\delta_{T_0}(t)$ is also an infinitely long periodic train of delta functions, weighted by the factor $f_0 = (1/T_0)$ and *regularly spaced by* f_0 along the frequency axis.

b. Hence, prove the pair of dual relations:

$$\sum_{m=-\infty}^{\infty} \delta(t - mT_0) = f_0 \sum_{n=-\infty}^{\infty} \exp(\mathrm{j}2\pi n f_0 t)$$

$$T_0 \sum_{m=-\infty}^{\infty} \exp(\mathrm{j}2\pi m f T_0) = \sum_{n=-\infty}^{\infty} \delta(f - nf_0)$$

c. Finally, prove the validity of (2.38).

Signal Transmission through Linear Time-invariant Systems

2.9 The periodic signal

$$x(t) = \sum_{m=-\infty}^{\infty} x(nT_0)\delta(t - nT_0)$$

is applied to a linear system of impulse response $h(t)$. Show that the average power of the signal $y(t)$ produced at the system output is defined by

$$P_{\mathrm{av},y} = \sum_{n=-\infty}^{\infty} |x(nT_0)|^2 |H(nf_0)|^2$$

where $H(f)$ is the frequency response of the system, and $f_0 = 1/T_0$.

2.10 According to the bounded input–bounded output stability criterion, the impulse response $h(t)$ of a linear-invariant system must be absolutely integrable; that is,

$$\int_{-\infty}^{\infty} |h(t)|\,\mathrm{d}t < \infty$$

Prove that this condition is both necessary and sufficient for stability of the system.

Hilbert Transform and Pre-envelopes

2.11 Prove the three properties of the Hilbert transform itemized on pages 43 and 44.

2.12 Let $\hat{g}(t)$ denote the Hilbert transform of $g(t)$. Derive the set of Hilbert-transform pairs listed as items 5 to 8 in Table 2.3.

2.13 Evaluate the inverse Fourier transform $g(t)$ of the one-sided frequency function:

$$G(f) = \begin{cases} \exp(-f), & f > 0 \\ \frac{1}{2}, & f = 0 \\ 0, & f < 0 \end{cases}$$

Show that $g(t)$ is complex, and that its real and imaginary parts constitute a Hilbert-transform pair.

2.14 Let $\hat{g}(t)$ denote the Hilbert transform of a Fourier transformable signal $g(t)$. Show that $\frac{\mathrm{d}}{\mathrm{d}t}\hat{g}(t)$ is equal to the Hilbert transform of $\frac{\mathrm{d}}{\mathrm{d}t}g(t)$.

2.15 In this problem, we revisit Problem 2.14, except that this time we use integration rather than differentiation. Doing so, we find that, in general, the integral $\int_{-\infty}^{\infty} \hat{g}(t)\,dt$ is *not* equal to the Hilbert transform of the integral $\int_{-\infty}^{\infty} g(t)\,dt$.

a. Justify this statement.

b. Find the condition for which exact equality holds.

2.16 Determine the pre-envelope $g_+(t)$ corresponding to each of the following two signals:

a. $g(t) = \text{sinc}(t)$

b. $g(t) = [1 + k\cos(2\pi f_m t)]\cos(2\pi f_c t)$

Complex Envelope

2.17 Show that the complex envelope of the sum of two narrowband signals (with the same carrier frequency) is equal to the sum of their individual complex envelopes.

2.18 The definition of the complex envelope $\tilde{s}(t)$ of a band-pass signal given in (2.65) is based on the pre-envelope $s_+(t)$ for positive frequencies. How is the complex envelope defined in terms of the pre-envelope $s_-(t)$ for negative frequencies? Justify your answer.

2.19 Consider the signal

$$s(t) = c(t)m(t)$$

whose $m(t)$ is a low-pass signal whose Fourier transform $M(f)$ vanishes for $|f| > W$, and $c(t)$ is a high-pass signal whose Fourier transform $C(f)$ vanishes for $|f| < W$. Show that the Hilbert transform of $s(t)$ is $\hat{s}(t) = \hat{c}(t)m(t)$, where $\hat{c}(t)$ is the Hilbert transform of $c(t)$.

2.20 a. Consider two real-valued signals $s_1(t)$ and $s_2(t)$ whose pre-envelopes are denoted by $s_{1+}(t)$ and $s_{2+}(t)$, respectively. Show that

$$\int_{-\infty}^{\infty} \text{Re}[s_{1+}(t)]\text{Re}[s_{2+}(t)]dt = \frac{1}{2}\text{Re}\left[\int_{-\infty}^{\infty} s_{1+}(t)s_{2+}^*(t)\,dt\right]$$

b. Suppose that $s_2(t)$ is replaced by $s_2(-t)$. Show that this modification has the effect of removing the complex conjugation in the right-hand side of the formula given in part a.

c. Assuming that $s(t)$ is a narrowband signal with complex envelope $\tilde{s}(t)$ and carrier frequency f_c, use the result of part a to show that

$$\int_{-\infty}^{\infty} s^2(t)\,dt = \frac{1}{2}\int_{-\infty}^{\infty} |\tilde{s}(t)|^2\,dt$$

2.21 Let a narrow-band signal $s(t)$ be expressed in the form

$$s(t) = s_I(t)\cos(2\pi f_c t) - s_Q(t)\sin(2\pi f_c t)$$

Using $S_+(f)$ to denote the Fourier transform of the pre-envelope of $s_+(t)$, show that the Fourier transforms of the in-phase component $s_I(t)$ and quadrature component $s_Q(t)$ are given by

$$S_I(f) = \frac{1}{2}[S_+(f+f_c) + S_+^*(-f+f_c)]$$

$$S_Q(f) = \frac{1}{2j}[S_+(f+f_c) - S_+^*(-f+f_c)]$$

respectively, where the asterisk denotes complex conjugation.

2.22 The block diagram of Figure 2.20a illustrates a method for extracting the in-phase component $s_I(t)$ and quadrature component $s_Q(t)$ of a narrowband signal $s(t)$. Given that the spectrum of $s(t)$ is limited to the interval $f_c - W \leq |f| f_c + W$, demonstrate the validity of this method. Hence, show that

$$S_I(f) = \begin{cases} S(f-f_c) + S(f+f_c), & -W \le f \le W \\ 0, & \text{elsewhere} \end{cases}$$

and

$$S_Q(f) = \begin{cases} j[S(f-f_c) - S(f+f_c)], & -W \le f \le W \\ 0, & \text{elsewhere} \end{cases}$$

where $S_I(f)$, $S_Q(f)$, and $S(f)$ are the Fourier transforms of $s_I(t)$, $s_Q(t)$, and $s(t)$, respectively.

Low-Pass Equivalent Models of Band-Pass Systems

2.23 Equations (2.82) and (2.83) define the in-phase component $\tilde{H}_I(f)$ and the quadrature component $\tilde{H}_Q(f)$ of the frequency response $\tilde{H}(f)$ of the complex low-pass equivalent model of a band-pass system of impulse response $h(t)$. Prove the validity of these two equations.

2.24 Explain what happens to the low-pass equivalent model of Figure 2.21b when the amplitude response of the corresponding bandpass filter has even symmetry and the phase response has odd symmetry with respect to the mid-band frequency f_c.

2.25 The rectangular RF pulse

$$x(t) = \begin{cases} A\cos(2\pi f_c t), & 0 \le t \le T \\ 0, & \text{elsewhere} \end{cases}$$

is applied to a linear filter with impulse response

$$h(t) = x(T-t)$$

Assume that the frequency f_c equals a large integer multiple of $1/T$. Determine the response of the filter and sketch it.

2.26 Figure P2.26 depicts the frequency response of an idealized band-pass filter in the receiver of a communication system, namely $H(f)$, which is characterized by a bandwidth of $2B$ centered on the carrier frequency f_c. The signal applied to the band-pass filter is described by the modulated sinc function:

$$x(t) = 4A_cB\,\text{sinc}(2Bt)\cos[2\pi(f_c \pm \Delta f)t]$$

where Δf is *frequency misalignment* introduced due to the receiver's imperfections, measured with respect to the carrier $A_c\cos(2\pi f_c t)$.

a. Find the complex low-pass equivalent models of the signal $x(t)$ and the frequency response $H(f)$.

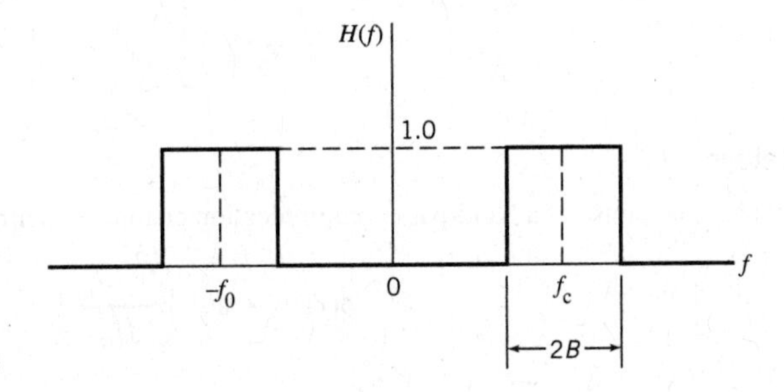

Figure P2.26

b. Then, go on to find the complex low-pass response of the filter output, denoted by $\tilde{y}(t)$, which includes distortion due to $\pm\Delta f$.

c. Building on the formula derived for $\tilde{y}(t)$ obtained in part b, explain how you would mitigate the misalignment distortion in the receiver.

Nonlinear Modulations

2.27 In analog communications, amplitude modulation is defined by

$$s_{\mathrm{AM}}(t) = A_c[1 + k_{\mathrm{a}}m(t)]\cos(2\pi f_c t)$$

where $A_c\cos(2\pi f_c t)$ is the carrier, $m(t)$ is the message signal, and k_{a} is a constant called *amplitude sensitivity* of the modulator. Assume that $|k_{\mathrm{a}}m(t)| < 1$ for all time t.

a. Justify the statement that, in a strict sense, $s_{\mathrm{AM}}(t)$ violates the principle of superposition.

b. Formulate the complex envelope $\tilde{s}_{\mathrm{AM}}(t)$ and its spectrum.

c. Compare the result obtained in part b with the complex envelope of DSB-SC. Hence, comment on the advantages and disadvantages of amplitude modulation.

2.28 Continuing on with analog communications, *frequency modulation* (FM) is defined by

$$s_{\mathrm{FM}}(t) = A_{\mathrm{c}}\left[\cos(2\pi f_{\mathrm{c}}t) + k_{\mathrm{f}}\int_o^t m(\tau)\,d\tau\right]$$

where $A_{\mathrm{c}}\cos(2\pi f_{\mathrm{c}}t)$ is the carrier, $m(t)$ is the message signal, and k_{f} is a constant called the *frequency sensitivity* of the modulator.

a. Show that frequency modulation is nonlinear in that it violates the principle of superposition.

b. Formulate the complex envelope of the FM signal, namely $\tilde{s}_{\mathrm{FM}}(t)$.

c. Consider the message signal to be in the form of a square wave as shown in Figure P2.28. The modulation frequencies used for the positive and negative amplitudes of the square wave, namely f_1 and f_2, are defined as follows:

$$f_1 + f_2 = \frac{2}{T_{\mathrm{b}}}$$

$$f_1 - f_2 = \frac{1}{T_{\mathrm{b}}}$$

where T_{b} is the duration of each positive or negative amplitude in the square wave. Show that under these conditions the complex envelope $\tilde{s}_{\mathrm{FM}}(t)$ maintains *continuity* for all time t, including the switching times between positive and negative amplitudes.

d. Plot the real and imaginary parts of $\tilde{s}_{\mathrm{FM}}(t)$ for the following values:

$$T_b = \frac{1}{3}\text{ s}$$

$$f_1 = 4\frac{1}{2}\text{ Hz}$$

$$f_2 = 1\frac{1}{2}\text{ Hz}$$

Phase and Group Delays

2.29 The phase response of a band-pass communication channel is defined by.

$$\phi(f) = -\tan^{-1}\left(\frac{f^2 - f_{\mathrm{c}}^2}{ff_{\mathrm{c}}}\right)$$

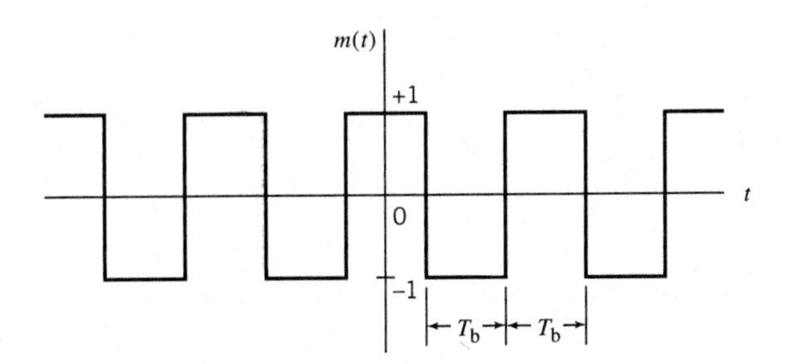

Figure P2.28

A sinusoidally modulated signal defined by

$$s(t) = A_c \cos(2\pi f_m t)\cos(2\pi f_c t)$$

is transmitted through the channel; f_c is the carrier frequency and f_m is the modulation frequency.

a. Determine the phase delay τ_p.

b. Determine the group delay τ_g.

c. Display the waveform produced at the channel output; hence, comment on the results obtained in parts a and b.

Notes

1. For a proof of convergence of the Fourier series, see Kammler (2000).

2. If a time function $g(t)$ is such that the value of the energy $\int_{-\infty}^{\infty} |g(t)|^2 dt$ is defined and finite, then the Fourier transform $G(f)$ of the function $g(t)$ exists and

$$\lim_{A \to \infty}\left[\int_{-\infty}^{\infty}\left|g(t) - \int_{-A}^{A} G(f)\exp(j2\pi f t)\,df\right|^2\right] = 0$$

This result is known as *Plancherel's theorem*. For a proof of this theorem, see Titchmarsh (1950).

3. The notation $\delta(t)$ for a delta function was first introduced into quantum mechanics by Dirac. This notation is now in general use in the signal processing literature. For detailed discussions of the delta function, see Bracewell (1986).

In a rigorous sense, the Dirac delta function is a distribution, not a function; for a rigorous treatment of the subject, see the book by Lighthill (1958).

4. The Paley–Wiener criterion is named in honor of the authors of the paper by Paley and Wiener (1934).

5. The integral in (2.54), defining the Hilbert transform of a signal, is an *improper* integral in that the integrand has a singularity at $\tau = t$. To avoid this singularity, the integration must be carried out in a symmetrical manner about the point $\tau = t$. For this purpose, we use the definition

$$P\int_{-\infty}^{\infty} \frac{g(\tau)}{t-\tau}d\tau = \lim_{g \to 0}\left[\int_{-\infty}^{t = \in} \frac{g(\tau)}{t-\tau}d\tau + \int_{t+\in}^{\infty} \frac{g(\tau)}{t-\tau}d\tau\right]$$

where the symbol P denotes Cauchy's principal value of the integral and $\in$ is incrementally small. For notational simplicity, the symbol P has been omitted from (2.54) and (2.55).

6. The complex representation of an arbitrary signal defined in (2.58) was first described by Gabor (1946). Gabor used the term "analytic signal." The term "pre-envelope" was used in Arens (1957) and Dungundji (1958). For a review of the different envelopes, see the paper by Rice (1982).

7. The FFT is *ubiquitous* in that it is applicable to a great variety of unrelated fields. For a detailed mathematical treatment of this widely used tool and its applications, the reader is referred to Brigham (1988).

CHAPTER

3

Probability Theory and Bayesian Inference

3.1 Introduction

The idea of a *mathematical model* used to describe a physical phenomenon is well established in the physical sciences and engineering. In this context, we may distinguish two classes of mathematical models: deterministic and probabilistic. A model is said to be *deterministic* if there is no uncertainty about its time-dependent behavior at any instant of time; linear time-invariant systems considered in Chapter 2 are examples of a deterministic model. However, in many real-world problems, the use of a deterministic model is inappropriate because the underlying physical phenomenon involves too many unknown factors. In such situations, we resort to a *probabilistic model* that accounts for uncertainty in mathematical terms.

Probabilistic models are needed for the design of systems that are reliable in performance in the face of uncertainty, efficient in computational terms, and cost effective in building them. Consider for example, a digital communication system that is required to provide practically error-free communication across a wireless channel. Unfortunately, the wireless channel is subject to *uncertainties*, the sources of which include:

- *noise*, internally generated due to thermal agitation of electrons in the conductors and electronic devices at the front-end of the receiver;
- *fading* of the channel, due to the multipath phenomenon—an inherent characteristic of wireless channels;
- *interference*, representing spurious electromagnetic waves emitted by other communication systems or microwave devices operating in the vicinity of the receiver.

To account for these uncertainties in the design of a wireless communication system, we need a probabilistic model of the wireless channel.

The objective of this chapter, devoted to probability theory, is twofold:

- the formulation of a logical basis for the mathematical description of probabilistic models and
- the development of probabilistic reasoning procedures for handling uncertainty.

Since the probabilistic models are intended to assign probabilities to the collections (sets) of possible outcomes of random experiments, we begin the study of probability theory with a review of set theory, which we do next.

3.2 Set Theory

Definitions

The objects constituting a set are called the *elements* of the set. Let A be a set and x be an element of the set A. To describe this statement, we write $x \in A$; otherwise, we write $x \notin A$. If the set A is empty (i.e., it has no elements), we denote it by $\varnothing$.

If $x_1, x_2, ..., x_N$ are all elements of the set A, we write

$$A = \{x_1, x_2, ..., x_N\}$$

in which case we say that the set A is countably finite. Otherwise, the set is said to be countably infinite. Consider, for example, an *experiment* involving the throws of a die. In this experiment, there are six possible outcomes: the showing of one, two, three, four, five, and six dots on the upper surface of the die; the set of possible *outcomes* of the experiment is therefore countably finite. On the other hand, the set of all possible odd integers, written as $\{\pm 1, \pm 3, \pm 5, \dots\}$, is countably infinite.

If every element of the set A is also an element of another set B, we say that A is a *subset* of B, which we describe by writing $A \subset B$.

If two sets A and B satisfy the conditions $A \subset B$ and $B \subset A$, then the two sets are said to be *identical* or *equal*, in which case we write $A = B$.

In a discussion of set theory, we also find it expedient to think of a *universal set*, denoted by S. Such a set contains every possible element that could occur in the context of a random experiment.

Boolean Operations on Sets

To illustrate the validity of Boolean operations on sets, the use of *Venn diagrams* can be helpful, as shown in what follows.

Unions and Intersections

The *union* of two sets A and B is defined by the set of elements that belong to A *or* B, or to both. This operation, written as $A \cup B$, is illustrated in the Venn diagram of Figure 3.1. The *intersection* of two sets A and B is defined by the particular set of elements that belong to both A *and* B, for which we write $A \cap B$. The shaded part of the Venn diagram in Figure 3.1 represents this second operation.

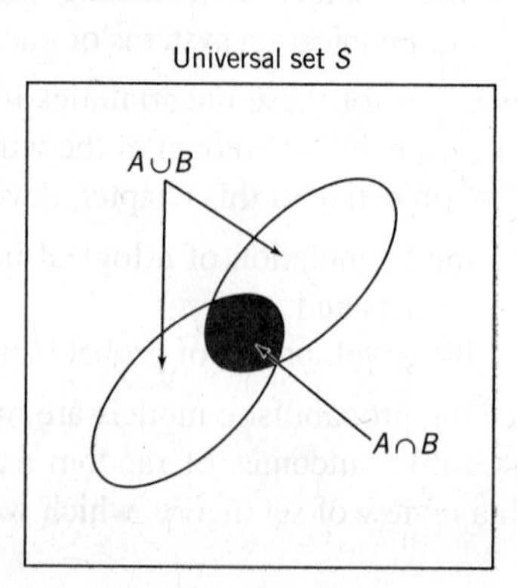

Figure 3.1
Illustrating the union and intersection of two sets, A and B.

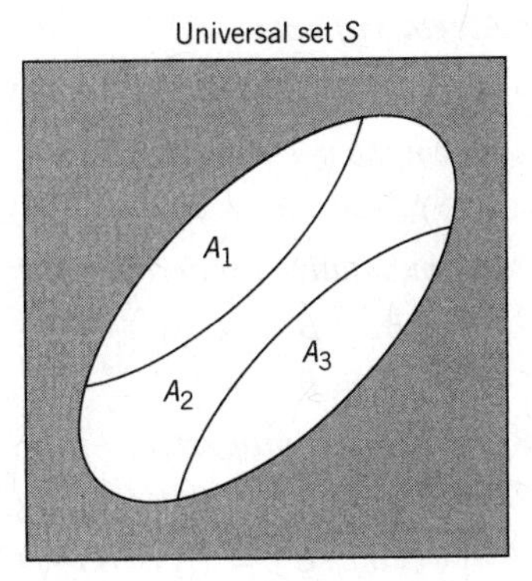

Figure 3.2
Illustrating the partition of set A into three subsets: A_1, A_2, and A_3.

Let x be an element of interest. Mathematically, the operations of union and intersection are respectively described by

$$A \cup B = \{x | x \in A \text{ or } x \in B\}$$

and

$$A \cap B = \{x | x \in A \text{ and } x \in B\}$$

where the symbol | is shorthand for "such that."

Disjoint and Partition Sets

Two sets A and B are said to be *disjoint* if their intersection is empty; that is, they have *no* common elements.

The partition of a set A refers to a collection of disjoint subsets $A_1, A_2, ..., A_N$ of the set A, the union of which equals A; that is,

$$A = A_1 \cup A_2 \ldots \cup A_N$$

The Venn diagram illustrating the partition operation is depicted in Figure 3.2 for the example of $N = 3$.

Complements

The set A^c is said to be the *complement* of the set A, with respect to the universal set S, if it is made up of all the elements of S that do not belong to A, as depicted in Figure 3.3.

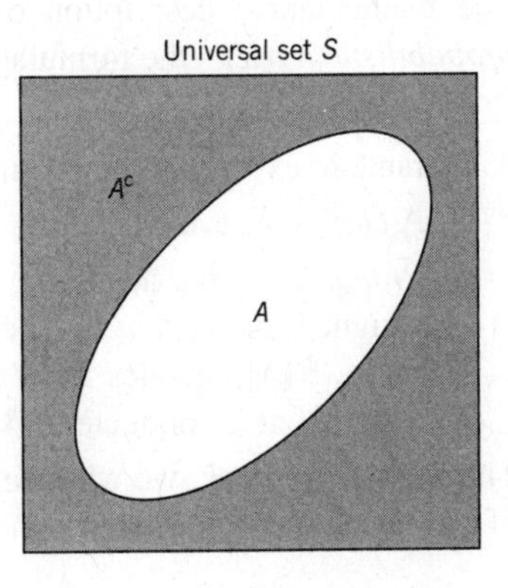

Figure 3.3
Illustrating the complement A^c of set A.

The Algebra of Sets

Boolean operations on sets have several properties, summarized here:

1. *Idempotence property*

   $(A^c)^c = A$

2. *Commutative property*

   $$A \cup B = B \cup A$$
   $$A \cap B = B \cap A$$

3. *Associative property*

   $$A \cup (B \cup C) = (A \cup B) \cup C$$
   $$A \cap (B \cap C) = (A \cap B) \cap C$$

4. *Distributive property*

   $$A \cap (B \cup C) = (A \cap B) \cup (A \cap C)$$
   $$A \cup (B \cap C) = (A \cup B) \cap (A \cup C)$$

   Note that the commutative and associative properties apply to both the union and intersection, whereas the distributive property applies only to the intersection.

5. *De Morgan's laws*

   The complement of the union of two sets A and B is equal to the intersection of their respective complements; that is

   $$(A \cup B)^c = A^c \cap B^c$$

   The complement of the intersection of two sets A and B is equal to the union of their respective complements; that is,

   $$(A \cap B)^c = A^c \cup B^c$$

For illustrations of these five properties and their confirmation, the reader is referred to Problem 3.1.

3.3 Probability Theory

Probabilistic Models

The mathematical description of an experiment with uncertain outcomes is called a *probabilistic model*,[1] the formulation of which rests on three fundamental ingredients:

1. *Sample space* or *universal set* S, which is the set of all conceivable outcomes of a random experiment under study.
2. A *class* E of events that are subsets of S.
3. *Probability law*, according to which a nonnegative measure or number $\mathbb{P}[A]$ is assigned to an *event* A. The measure $\mathbb{P}[A]$ is called the *probability of event* A. In a sense, $\mathbb{P}[A]$ encodes our *belief* in the likelihood of event A occurring when the experiment is conducted.

Throughout the book, we will use the symbol $\mathbb{P}[.]$ to denote the probability of occurrence of the event that appears inside the square brackets.

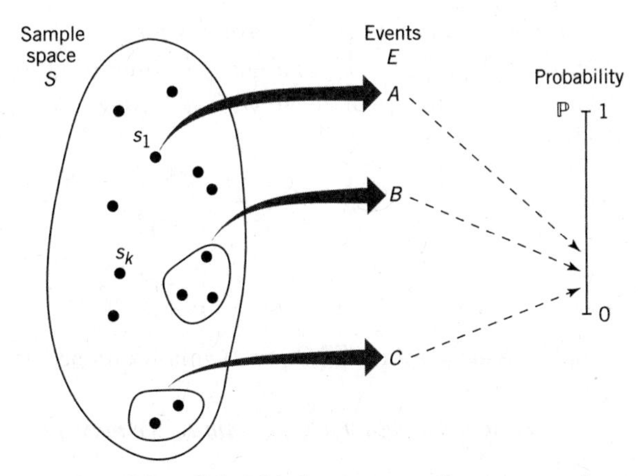

Figure 3.4 Illustration of the relationship between sample space, events, and probability

As illustrated in Figure 3.4, an event may involve a single outcome or a subset of possible outcomes in the sample space *S*. These possibilities are exemplified by the way in which three events, *A, B,* and *C,* are pictured in Figure 3.4. In light of such a reality, we identify two extreme cases:

- *Sure event*, which embodies all the possible outcomes in the sample space *S*.
- *Null* or *impossible event*, which corresponds to the empty set or empty space $\varnothing$.

Axioms of Probability

Fundamentally, the probability measure $\mathbb{P}[A]$, assigned to event *A* in the class *E,* is governed by three axioms:

Axiom I Nonnegativity The first axiom states that *the probability of event A is a nonnegative number bounded by unity,* as shown by

$$O \le \mathbb{P}[A] \le 1 \qquad \text{for any event } A \tag{3.1}$$

Axiom II Additivity The second axiom states that *if A and B are two disjoint events, then the probability of their union satisfies the equality*

$$\mathbb{P}[A \cup B] = \mathbb{P}[A] + \mathbb{P}[B] \tag{3.2}$$

In general, if the sample space has *N* elements and $A_1, A_2, \ldots, A_N$ is a sequence of disjoint events, then the probability of the union of these *N* events satisfies the equality

$$\mathbb{P}[A_1 \cup A_2 \cup \ldots A_N] = \mathbb{P}[A_1] + \mathbb{P}[A_2] + \ldots + \mathbb{P}[A_N]$$

Axiom III Normalization The third and final axiom states that *the probability of the entire sample space S is equal to unity,* as shown by

$$\mathbb{P}[S] = 1 \tag{3.3}$$

These three axioms provide an implicit definition of probability. Indeed, we may use them to develop some other basic properties of probability, as described next.

PROPERTY 1 *The probability of an impossible event is zero.*

To prove this property, we first use the axiom of normalization, then express the sample space S as the union of itself with the empty space $\varnothing$, and then use the axiom of additivity. We thus write

$$\begin{aligned} 1 &= \mathbb{P}[S] \\ &= \mathbb{P}[S \cup \varnothing] \\ &= \mathbb{P}[S] + \mathbb{P}[\varnothing] \\ &= 1 + \mathbb{P}[\varnothing] \end{aligned}$$

from which the property $\mathbb{P}[\varnothing] = 0$ follows immediately.

PROPERTY 2 *Let A^c denote the complement of event A; we may then write*

$$\mathbb{P}[A^c] = 1 - \mathbb{P}[A] \qquad \text{for any event } A \tag{3.4}$$

To prove this property, we first note that the sample space S is the union of the two mutually exclusive events A and A^c. Hence, the use of the additivity and normalization axioms yields

$$\begin{aligned} 1 &= \mathbb{P}[S] \\ &= \mathbb{P}[A \cup A^c] \\ &= \mathbb{P}[A] + \mathbb{P}[A^c] \end{aligned}$$

from which, after rearranging terms, (3.4) follows immediately.

PROPERTY 3 *If event A lies within the subspace of another event B, then*

$$\mathbb{P}[A] \le \mathbb{P}[B] \qquad \text{for } A \subset B \tag{3.5}$$

To prove this third property, consider the *Venn diagram* depicted in Figure 3.5. From this diagram, we observe that event B may be expressed as the union of two disjoint events, one defined by A and the other defined by the intersection of B with the complement of A; that is,

$$B = A \cup (B \cap A^c)$$

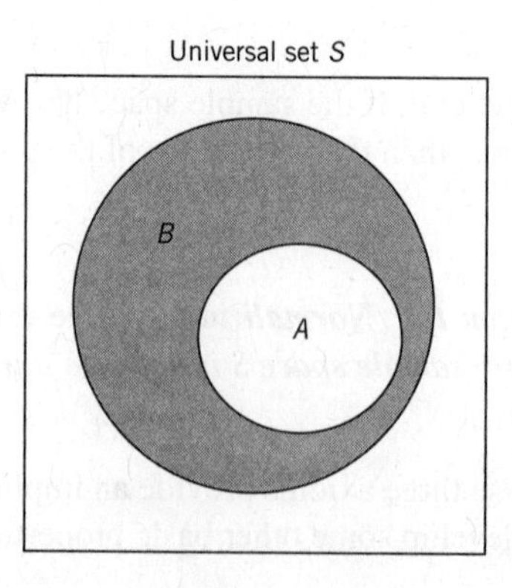

Figure 3.5
The Venn diagram for proving (3.5).

Therefore, applying the additivity axiom to this relation, we get

$$\mathbb{P}[B] = \mathbb{P}[A] + \mathbb{P}[B \cap A^c]$$

Next, invoking the nonnegativity axiom, we immediately find that the probability of event B must be equal to or greater than the probability of event A, as indicated in (3.5).

PROPERTY 4 *Let N disjoint events $A_1, A_2, \ldots, A_N$ satisfy the condition*

$$A_1 \cup A_2 \cup \ldots A_N = S \tag{3.6}$$

then

$$\mathbb{P}[A_1] + \mathbb{P}[A_2] + \ldots + \mathbb{P}[A_N] = 1 \tag{3.7}$$

To prove this fourth property, we first apply the normalization axiom to (3.6) to write

$$\mathbb{P}[A_1 \cup A_2 \cup \ldots \cup A_N] = 1$$

Next, recalling the generalized form of the additivity axiom

$$\mathbb{P}[A_1 \cup A_2 \cup \ldots A_N] = \mathbb{P}[A_1] + \mathbb{P}[A_2] + \ldots + \mathbb{P}[A_N]$$

From these two relations, (3.7) follows immediately.

For the special case of N *equally probable events*, (3.7) reduces to

$$\mathbb{P}[A_i] = \frac{1}{N} \qquad \text{for } i = 1, 2, \ldots, N \tag{3.8}$$

PROPERTY 5 *If two events A and B are not disjoint, then the probability of their union event is defined by*

$$\mathbb{P}[A \cup B] = \mathbb{P}[A] + \mathbb{P}[B] - \mathbb{P}[A \cap B] \qquad \text{for any two events } A \text{ and } B \tag{3.9}$$

where $\mathbb{P}[A \cap B]$ is called the joint probability of A and B.

To prove this last property, consider the Venn diagram of Figure 3.6. From this figure, we first observe that the union of A and B may be expressed as the union of two disjoint events: A itself and $A^c \cap B$, where A^c is the complement of A. We may therefore apply the additivity axiom to write

$$\begin{aligned} \mathbb{P}[A \cup B] &= \mathbb{P}[A \cup (A^c \cap B)] \\ &= \mathbb{P}[A] + \mathbb{P}[A^c \cap B] \end{aligned} \tag{3.10}$$

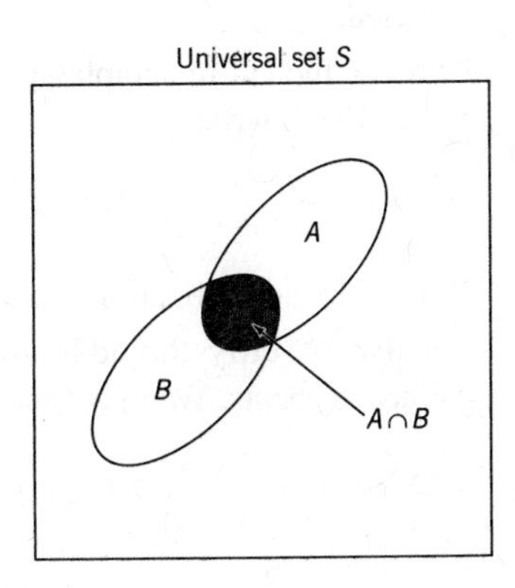

Figure 3.6
The Venn diagram for proving (3.9).

From the Venn diagram of Figure 3.6, we next observe that the event B may be expressed as

$$\begin{aligned} B &= S \cap B \\ &= (A \cup A^c) \cap B \\ &= (A \cap B) \cup (A^c \cap B) \end{aligned}$$

That is, B is the union of two disjoint events: $A \cap B$ and $A^c \cap B$; therefore, applying the additivity axiom to this second relation yields

$$\mathbb{P}[B] = \mathbb{P}[A \cap B] + \mathbb{P}[A^c \cap B] \tag{3.11}$$

Subtracting (3.11) from (3.10), canceling the common term $\mathbb{P}[A^c \cap B]$ and rearranging terms, (3.9) follows and Property 4 is proved.

It is of interest to note that the joint probability $\mathbb{P}[A \cap B]$ accounts for that part of the sample space S where the events A and B coincide. If these two events are disjoint, then the joint probability $\mathbb{P}[A \cap B]$ is zero, in which case (3.9) reduces to the additivity axiom of (3.2).

Conditional Probability

When an experiment is performed and we only obtain *partial information* on the outcome of the experiment, we may *reason* about that particular outcome by invoking the notion of conditional probability. Stated the other way round, we may make the statement:

> Conditional probability provides the premise for probabilistic reasoning.

To be specific, suppose we perform an experiment that involves a pair of events A and B. Let $\mathbb{P}[A|B]$ denote the probability of event A given that event B has occurred. The probability $\mathbb{P}[A|B]$ is called the *conditional probability of A given B*. Assuming that B has nonzero probability, the conditional probability $\mathbb{P}[A|B]$ is formally defined by

$$\mathbb{P}[A|B] = \frac{\mathbb{P}[A \cap B]}{\mathbb{P}[B]} \tag{3.12}$$

where $\mathbb{P}[A \cap B]$ is the joint probability of events A and B, and $\mathbb{P}[B]$ is nonzero.

For a fixed event B, the conditional probability $\mathbb{P}[A|B]$ is a legitimate probability law as it satisfies all three axioms of probability:

1. Since by definition, $\mathbb{P}[A|B]$ is a probability, the nonnegativity axiom is clearly satisfied.
2. Viewing the entire sample space S as event A and noting that $S \cup B = B$, we may use (3.12) to write

$$\mathbb{P}[S|B] = \frac{\mathbb{P}[S|B]}{\mathbb{P}[B]} = \frac{\mathbb{P}[B]}{\mathbb{P}[B]} = 1$$

   Hence, the normalization axiom is also satisfied.
3. Finally, to verify the additivity axiom, assume that A_1 and A_2 are two mutually exclusive events. We may then use (3.12) to write

$$\mathbb{P}[A_1 \cup A_2|B] = \frac{\mathbb{P}[(A_1 \cup A_2) \cap B]}{\mathbb{P}[B]}$$

Applying the distributive property to the numerator on the right-hand side, we have

$$\mathbb{P}[A_1 \cup A_2 | B] = \frac{\mathbb{P}[(A_1 \cup B) \cup (A_2 \cap B)]}{\mathbb{P}[B]}$$

Next, recognizing that the two events $A_1 \cap B$ and $A_2 \cap B$ are actually disjoint, we may apply the additivity axiom to write

$$\begin{aligned} \mathbb{P}[A_1 \cup A_2 | B] &= \frac{\mathbb{P}[A_1 \cap B] + \mathbb{P}[A_2 \cap B]}{\mathbb{P}[B]} \\ &= \frac{\mathbb{P}[A_1 \cap B]}{\mathbb{P}[B]} + \frac{\mathbb{P}[A_2 \cap B]}{\mathbb{P}[B]} \end{aligned} \tag{3.13}$$

which proves that the conditional probability also satisfies the additivity axiom.

We therefore conclude that all three axioms of probability (and therefore all known properties of probability laws) are equally valid for the conditional probability $\mathbb{P}[A|B]$. In a sense, this conditional probability captures the *partial information that the occurrence of event B provides about event A*; we may therefore view the conditional probability $\mathbb{P}[A|B]$ as a probability law concentrated on event B.

Bayes' Rule

Suppose we are confronted with a situation where the conditional probability $\mathbb{P}[A|B]$ and the individual probabilities $\mathbb{P}[A]$ and $\mathbb{P}[B]$ are all easily determined directly, but the conditional probability $\mathbb{P}[B|A]$ is desired. To deal with this situation, we first rewrite (3.12) in the form

$$\mathbb{P}[A \cap B] = \mathbb{P}[A|B]\mathbb{P}[B]$$

Clearly, we may equally write

$$\mathbb{P}[A \cap B] = \mathbb{P}[B|A]\mathbb{P}[A]$$

The left-hand parts of these two relations are identical; we therefore have

$$\mathbb{P}[A|B]\mathbb{P}[B] = \mathbb{P}[B|A]\mathbb{P}[A]$$

Provided that $\mathbb{P}[A]$ is nonzero, we may determine the desired conditional probability $\mathbb{P}[B|A]$ by using the relation

$$\mathbb{P}[B|A] = \frac{\mathbb{P}[A|B]\mathbb{P}[B]}{\mathbb{P}[A]} \tag{3.14}$$

This relation is known as *Bayes' rule*.

As simple as it looks, Bayes' rule provides the correct language for describing *inference*, the formulation of which cannot be done without making assumptions.[2] The following example illustrates an application of Bayes' rule.

EXAMPLE 1 **Radar Detection**

Radar, a remote sensing system, operates by transmitting a sequence of pulses and has its receiver listen to echoes produced by a target (e.g., aircraft) that could be present in its surveillance area.

Let the events A and B be defined as follows:

$A = \{$a target is present in the area under surveillance$\}$
$A^c = \{$there is no target in the area$\}$
$B = \{$the radar receiver detects a target$\}$

In the radar detection problem, there are three probabilities of particular interest:

$\mathbb{P}[A]$ probability that a target is present in the area; this probability is called the *prior probability.*

$\mathbb{P}[B|A]$ probability that the radar receiver detects a target, given that a target is actually present in the area; this second probability is called the *probability of detection.*

$\mathbb{P}[B|A^c]$ probability that the radar receiver detects a target in the area, given that there is no target in the surveillance area; this third probability is called the *probability of false alarm.*

Suppose these three probabilities have the following values:

$$\mathbb{P}[A] = 0.02$$
$$\mathbb{P}[B|A] = 0.99$$
$$\mathbb{P}[B|A^c] = 0.01$$

The problem is to calculate the conditional probability $\mathbb{P}[A|B]$ which defines the probability that a target is present in the surveillance area given that the radar receiver has made a target detection.

Applying Bayes' rule, we write

$$\begin{aligned}\mathbb{P}[A|B] &= \frac{\mathbb{P}[B|A]\mathbb{P}[A]}{\mathbb{P}[B]} \\ &= \frac{\mathbb{P}[B|A]\mathbb{P}[A]}{\mathbb{P}[B|A]\mathbb{P}[A] + \mathbb{P}[B|A^c]\mathbb{P}[A^c]} \\ &= \frac{0.99 \times 0.02}{0.99 \times 0.02 + 0.01 \times 0.98} \\ &= \frac{0.0198}{0.0296} \\ &\approx 0.69\end{aligned}$$

Independence

Suppose that the occurrence of event A provides no information whatsoever about event B; that is,

$$\mathbb{P}[B|A] = \mathbb{P}[B]$$

Then, (3.14) also teaches us that

$$\mathbb{P}[A|B] = \mathbb{P}[A]$$

In this special case, we see that knowledge of the occurrence of either event, A or B, tells us no more about the probability of occurrence of the other event than we knew without that knowledge. Events A and B that satisfy this condition are said to be *independent*.

From the definition of conditional probability given in (3.12), namely,

$$\mathbb{P}[A|B] = \frac{\mathbb{P}[A \cap B]}{\mathbb{P}[B]}$$

we see that the condition $\mathbb{P}[A|B] = \mathbb{P}[A]$ is equivalent to

$$\mathbb{P}[A \cap B] = \mathbb{P}[A]\mathbb{P}[B]$$

We therefore adopt this latter relation as the formal definition of independence. The important point to note here is that the definition still holds even if the probability $\mathbb{P}[B]$ is zero, in which case the conditional probability $\mathbb{P}[A|B]$ is undefined. Moreover, the definition has a *symmetric* property, in light of which we can say the following:

> If an event A is independent of another event B, then B is independent of A, and A and B are therefore independent events.

3.4 Random Variables

It is customary, particularly when using the language of sample space pertaining to an experiment, to describe the outcome of the experiment by using one or more real-valued quantities or measurements that help us think in probabilistic terms. These quantities are called *random variables*, for which we offer the following definition:

> The random variable is a function whose domain is a sample space and whose range is some set of real numbers.

The following two examples illustrate the notion of a random variable embodied in this definition.

Consider, for example, the sample space that represents the integers 1, 2, ..., 6, each one of which is the number of dots that shows uppermost when a die is thrown. Let the sample point k denote the event that k dots show in one throw of the die. The random variable used to describe the probabilistic event k in this experiment is said to be a *discrete random variable*.

For an entirely different experiment, consider the noise being observed at the front end of a communication receiver. In this new situation, the random variable, representing the amplitude of the noise voltage at a particular instant of time, occupies a continuous range of values, both positive and negative. Accordingly, the random variable representing the noise amplitude is said to be a *continuous random variable*.

The concept of a continuous random variable is illustrated in Figure 3.7, which is a modified version of Figure 3.4. Specifically, for the sake of clarity, we have suppressed the events but show subsets of the sample space S being mapped directly to a subset of a real line representing the random variable. The notion of the random variable depicted in Figure 3.7 applies in exactly the same manner as it applies to the underlying events. The benefit of random variables, pictured in Figure 3.7, is that probability analysis can now be developed in terms of real-valued quantities, regardless of the form or shape of the underlying events of the random experiment under study.

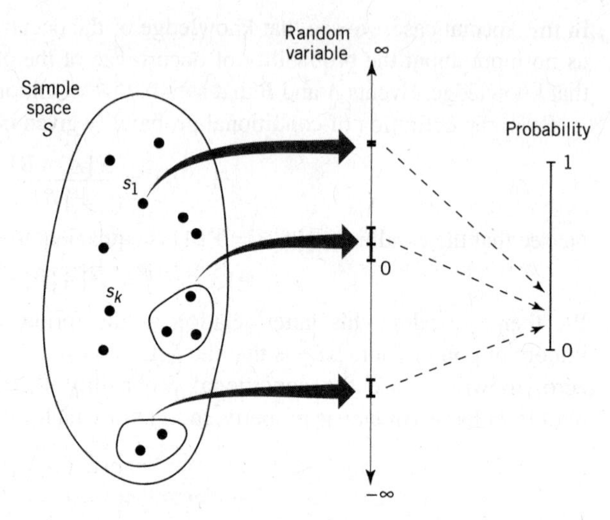

Figure 3.7 Illustration of the relationship between sample space, random variables, and probability.

One last comment is in order before we proceed further. Throughout the whole book, we will be using the following notation:

Uppercase characters denote random variables and lowercase characters denote real values taken by random variables.

3.5 Distribution Functions

To proceed with the probability analysis in mathematical terms, we need a probabilistic description of random variables that works equally well for discrete and continuous random variables. Let us consider the random variable X and the probability of the event $X \le x$. We denote this probability by $\mathbb{P}[X \le x]$. It is apparent that this probability is a function of the *dummy variable* x. To simplify the notation, we write

$$F_X(x) = \mathbb{P}[X \le x] \quad \text{for all } x \tag{3.15}$$

The function $F_X(x)$ is called the *cumulative distribution function* or simply the *distribution function* of the random variable X. Note that $F_X(x)$ is a function of x, not of the random variable X. For any point x in the sample space, the distribution function $F_X(x)$ expresses the probability of an event.

The distribution function $F_X(x)$, applicable to both continuous and discrete random variables, has two fundamental properties:

PROPERTY 1 **Boundedness of the Distribution**

The distribution function $F_X(x)$ is a bounded function of the dummy variable x that lies between zero and one.

Specifically, $F_X(x)$ tends to zero as x tends to $-\infty$, and it tends to one as x tends to ∞.

PROPERTY 2 **Monotonicity of the Distribution**

The distribution function $F_X(x)$ is a monotone nondecreasing function of x.

In mathematical terms, we write

$$F_X(x_1) \le F_X(x_2) \qquad \text{for } x_1 < x_2$$

Both of these properties follow directly from (3.15).

The random variable X is said to be *continuous* if the distribution function $F_X(x)$ is differentiable with respect to the dummy variable x everywhere, as shown by

$$f_X(x) = \frac{\mathrm{d}}{\mathrm{d}x} F_X(x) \qquad \text{for all } x \tag{3.16}$$

The new function $f_X(x)$ is called the *probability density function* of the random variable X. The name, density function, arises from the fact that the probability of the event $x_1 < X \le x_2$ is

$$\begin{aligned} \mathbb{P}[x_1 < X \le x_2] &= \mathbb{P}[X \le x_2] - \mathbb{P}[X \le x_1] \\ &= F_X(x_2) - F_X(x_1) \\ &= \int_{x_1}^{x_2} f_X(x)\,\mathrm{d}x \end{aligned} \tag{3.17}$$

The probability of an interval is therefore the area under the probability density function in that interval. Putting $x_1 = -\infty$ in (3.17) and changing the notation somewhat, we readily see that the distribution function is defined in terms of the probability density function as

$$F_X(x) = \int_{-\infty}^{x} f_X(\xi)\,\mathrm{d}\xi \tag{3.18}$$

where ξ is a dummy variable. Since $F_X(\infty) = 1$, corresponding to the probability of a sure event, and $F_X(-\infty) = 0$, corresponding to the probability of an impossible event, we readily find from (3.17) that

$$\int_{-\infty}^{\infty} f_X(x)\,\mathrm{d}x = 1 \tag{3.19}$$

Earlier we mentioned that a distribution function must always be a monotone nondecreasing function of its argument. It follows, therefore, that the probability density function must always be nonnegative. Accordingly, we may now formally make the statement:

> The probability density function $f_X(x)$ of a continuous random variable X has two defining properties: nonnegativity and normalization.

PROPERTY 3 **Nonnegativity**

The probability density function $f_X(x)$ is a nonnegative function of the sample value x of the random variable X.

PROPERTY 4 **Normalization**

The total area under the graph of the probability density function $f_X(x)$ is equal to unity.

An important point that should be stressed here is that the probability density function $f_X(x)$ contains all the conceivable information needed for statistical characterization of the random variable X.

EXAMPLE 2 **Uniform Distribution**

To illustrate the properties of the distribution function $F_X(x)$ and the probability density function $f_X(x)$ for a continuous random variable, consider a *uniformly distributed random variable*, described by

$$f_X(x) = \begin{cases} 0, & x \le a \\ \dfrac{1}{b-a}, & a < x \le b \\ 0, & x > b \end{cases} \tag{3.20}$$

Integrating $f_X(x)$ with respect to x yields the associated distribution function

$$F_X(x) = \begin{cases} 0, & x \le a \\ \dfrac{x-a}{b-a}, & a < x \le b \\ 0, & x > b \end{cases} \tag{3.21}$$

Plots of these two functions versus the dummy variable x are shown in Figure 3.8.

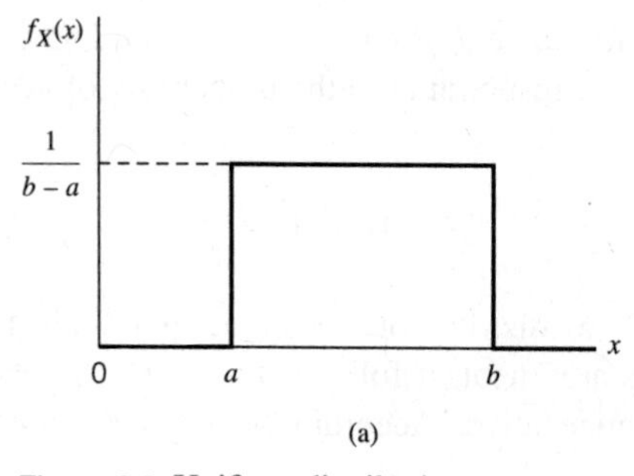

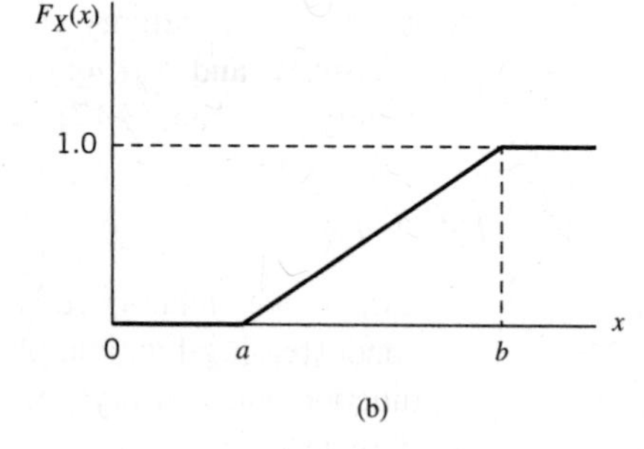

Figure 3.8 Uniform distribution.

Probability Mass Function

Consider next the case of a *discrete* random variable, X, which is a real-valued function of the outcome of a probabilistic experiment that can take a finite or countably infinite number of values. As mentioned previously, the distribution function $F_X(x)$ defined in (3.15) also applies to discrete random variables. However, unlike a continuous random variable, the distribution function of a discrete random variable is not differentiable with respect to its dummy variable x.

To get around this mathematical difficulty, we introduce the notion of the *probability mass function* as another way of characterizing discrete random variables. Let X denote a discrete random variable and let x be any possible value of X taken from a set of real numbers. We may then make the statement:

> The probability mass function of x, denoted by $p_X(x)$, is defined as the probability of the event $X = x$, which consists of all possible outcomes of an experiment that lead to a value of X equal to x.

Stated in mathematical terms, we write

$$p_X(x) = \mathbb{P}[X = x] \tag{3.22}$$

which is illustrated in the next example.

EXAMPLE 3

The Bernoulli Random Variable

Consider a probabilistic experiment involving the discrete random variable X that takes one of two possible values:

- the value 1 with probability p;
- the value 0 with probability $1 - p$.

Such a random variable is called the *Bernoulli random variable*, the probability mass function of which is defined by

$$p_X(x) = \begin{cases} 1-p & x = 0 \\ p, & x = 1 \\ 0, & \text{otherwise} \end{cases} \tag{3.23}$$

This probability mass function is illustrated in Figure 3.9. The two delta functions, each of weight $1/2$, depicted in Figure 3.9 represent the probability mass function at each of the sample points $x = 0$ and $x = 1$.

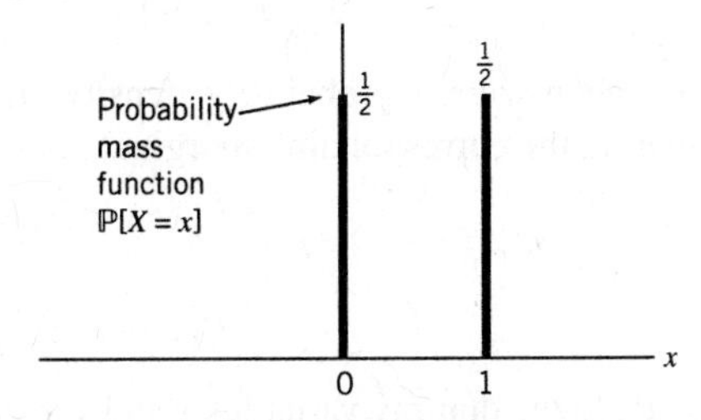

Figure 3.9 Illustrating the probability mass function for a fair coin-tossing experiment.

From here on, we will, largely but not exclusively, focus on the characterization of continuous random variables. A parallel development and similar concepts are possible for discrete random variables as well.[3]

Multiple Random Variables

Thus far we have focused attention on situations involving a single random variable. However, we frequently find that the outcome of an experiment requires several random variables for its description. In what follows, we consider situations involving two random variables. The probabilistic description developed in this way may be readily extended to any number of random variables.

Consider two random variables X and Y. In this new situation, we say:

> The joint distribution function $F_{X,Y}(x,y)$ is the probability that the random variable X is less than or equal to a specified value x, and that the random variable Y is less than or equal to another specified value y.

The variables X and Y may be two separate one-dimensional random variables or the components of a single two-dimensional random vector. In either case, the joint sample space is the xy-plane. The *joint distribution function* $F_{X,Y}(x,y)$ is the probability that the outcome of an experiment will result in a sample point lying inside the quadrant $(-\infty < X \le x, -\infty < Y \le y)$ of the joint sample space. That is,

$$F_{X,Y}(x, y) = \mathbb{P}[X \le x, Y \le y] \tag{3.24}$$

Suppose that the joint distribution function $F_{X,Y}(x,y)$ is continuous everywhere and that the second-order partial derivative

$$f_{X,Y}(x, y) = \frac{\partial^2 F_{X,Y}(x, y)}{\partial x \partial y} \tag{3.25}$$

exists and is continuous everywhere too. We call the new function $f_{X,Y}(x,y)$ the *joint probability density function* of the random variables X and Y. The joint distribution function $F_{X,Y}(x,y)$ is a monotone nondecreasing function of both x and y. Therefore, from (3.25) it follows that the joint probability density function $f_{X,Y}(x,y)$ is always nonnegative. Also, the total volume under the graph of a joint probability density function must be unity, as shown by the double integral

$$\int_{-\infty}^{\infty}\int_{-\infty}^{\infty} f_{X,Y}(x, y)\,\mathrm{d}x\,\mathrm{d}y = 1 \tag{3.26}$$

The so-called *marginal* probability density functions, $f_X(x)$ and $f_Y(y)$, are obtained by differentiating the corresponding marginal distribution functions

$$F_X(x) = f_{X,Y}(x, \infty)$$

and

$$F_Y(y) = f_{X,Y}(\infty, y)$$

with respect to the dummy variables x and y, respectively. We thus write

$$\begin{aligned} f_X(x) &= \frac{\mathrm{d}}{\mathrm{d}x}F_X(x) \\ &= \frac{\mathrm{d}}{\mathrm{d}x}\int_{-\infty}^{x}\left[\int_{-\infty}^{\infty} f_{X,Y}(\xi, y)\,\mathrm{d}y\right]\mathrm{d}\xi \\ &= \int_{-\infty}^{\infty} f_{X,Y}(x, y)\,\mathrm{d}y \end{aligned} \tag{3.27}$$

Similarly, we write

$$f_Y(y) = \int_{-\infty}^{\infty} f_{X,Y}(x,y)\,dx \tag{3.28}$$

In words, the first marginal probability density function $f_X(x)$, defined in (3.27), is obtained from the joint probability density function $f_{X,Y}(x,y)$ by simply integrating it over all possible values of the undesired random variable Y. Similarly, the second marginal probability density function $f_Y(y)$, defined in (3.28), is obtained from $f_{X,Y}(x,y)$ by integrating it over all possible values of the undesired random variable; this time, the undesirable random variable is X. Henceforth, we refer to $f_X(x)$ and $f_Y(y)$, obtained in the manner described herein, as the *marginal densities* of the random variables X and Y, whose joint probability density function is $f_{X,Y}(x,y)$. Here again, we conclude the discussion on a pair of random variables with the following statement:

> The joint probability density function $f_{X,Y}(x,y)$ contains all the conceivable information on the two continuous random variables X and Y that is needed for the probability analysis of joint random variables.

This statement can be generalized to cover the joint probability density function of many random variables.

Conditional Probability Density Function

Suppose that X and Y are two continuous random variables with their joint probability density function $f_{X,Y}(x,y)$. The *conditional probability density function* of Y, such that $X = x$, is defined by

$$f_Y(y|x) = \frac{f_{X,Y}(x,y)}{f_X(x)} \tag{3.29}$$

provided that $f_X(x) > 0$, where $f_X(x)$ is the marginal density of X; $f_Y(y|x)$ is a shortened version of $f_{Y|X}(y|x)$, both of which are used interchangeably. The function $f_Y(y|x)$ may be thought of as a function of the variable Y, with the variable x arbitrary but fixed; accordingly, it satisfies all the requirements of an ordinary probability density function for any x, as shown by

$$f_Y(y|x) \geq 0$$

and

$$\int_{-\infty}^{\infty} f_Y(y|x)\,dy = 1 \tag{3.30}$$

Cross-multiplying terms in (3.29) yields

$$f_{X,Y}(x,y) = f_Y(y|x) f_X(x)$$

which is referred to as the *multiplication rule*.

Suppose that knowledge of the outcome of X can, in no way, affect the distribution of Y. Then, the conditional probability density function $f_Y(y|x)$ reduces to the marginal density $f_Y(y)$, as shown by

$$f_Y(y|x) = f_Y(y)$$

In such a case, we may express the joint probability density function of the random variables X and Y as the product of their respective marginal densities; that is,

$$f_{X,Y}(x, y) = f_X(x)f_Y(y)$$

On the basis of this relation, we may now make the following statement on the *independence* of random variables:

> If the joint probability density function of the random variables X and Y equals the product of their marginal densities, then X and Y are statistically independent.

Sum of Independent Random Variables: Convolution

Let X and Y be two continuous random variables that are statistically independent; their respective probability density functions are denoted by $f_X(x)$ and $f_Y(y)$. Define the sum

$$Z = X + Y$$

The issue of interest is to find the probability density function of the new random variable Z, which is denoted by $f_Z(z)$.

To proceed with this evaluation, we first use probabilistic arguments to write

$$\begin{aligned} \mathbb{P}[Z \le z | X = x] &= \mathbb{P}[X + Y \le z | X = x] \\ &= \mathbb{P}[x + Y \le z | X = x] \end{aligned}$$

where, in the second line, the given value x is used for the random variable X. Since X and Y are statistically independent, we may simplify matters by writing

$$\begin{aligned} \mathbb{P}[Z \le z | X = x] &= \mathbb{P}[x + Y \le z] \\ &= \mathbb{P}[Y \le z - x] \end{aligned}$$

Equivalently, in terms of the pertinent distribution functions, we may write

$$F_Z(z|x) = F_Y(z - x)$$

Hence, differentiating both sides of this equation, we get the corresponding probability density functions

$$f_z(z|x) = f_Y(z - x)$$

Using the multiplication rule described in (3.30), we have

$$f_{Z,X}(z, x) = f_Y(z - x)f_X(x) \tag{3.31}$$

Next, adapting the definition of the marginal density given in (3.27) to the problem at hand, we write

$$f_Z(z) = \int_{-\infty}^{\infty} f_{Z,X}(z, x)\,dx \tag{3.32}$$

Finally, substituting (3.31) into (3.32), we find that the desired $f_Z(z)$ is equal to the convolution of $f_X(x)$ and $f_Y(y)$, as shown by

$$f_Z(z) = \int_{-\infty}^{\infty} f_X(x)f_Y(z - x)\,dx \tag{3.33}$$

In words, we may therefore state:

> The summation of two independent continuous random variables leads to the convolution of their respective probability density functions.

Note, however, that no assumptions were made in arriving at this statement except for the random variables X and Y being continuous random variables.

3.6 The Concept of Expectation

As pointed out earlier, the probability density function $f_X(x)$ provides a complete statistical description of a continuous random variable X. However, in many instances, we find that this description includes more detail than is deemed to be essential for practical applications. In situations of this kind, simple *statistical averages* are usually considered to be adequate for the statistical characterization of the random variable X.

In this section, we focus attention on the *first-order* statistical average, called the expected value or mean of a random variable; *second-order* statistical averages are studied in the next section. The rationale for focusing attention on the mean of a random variable is its practical importance in statistical terms, as explained next.

Mean

The *expected value* or *mean* of a continuous random variable X is formally defined by

$$\mu_X = \mathbb{E}[X] = \int_{-\infty}^{\infty} x f_X(x)\,\mathrm{d}x \tag{3.34}$$

where $\mathbb{E}$ denotes the *expectation* or *averaging operator*. According to this definition, the expectation operator $\mathbb{E}$, applied to a continuous random variable x, produces a single number that is derived uniquely from the probability density function $f_X(x)$.

To describe the meaning of the defining equation (3.34), we may say the following:

> The mean μ_X of a random variable X, defined by the expectation $\mathbb{E}[x]$, locates the center of gravity of the area under the probability density curve of the random variable X.

To elaborate on this statement, we write the integral in (3.34) as the limit of an approximating sum formulated as follows. Let $\{x_k | k = 0, \pm 1, \pm 2, \ldots\}$ denote a set of uniformly spaced points on the real line

$$x_k = \left(k + \frac{1}{2}\right)\Delta, \qquad k = 0, \pm 1, \pm 2, \ldots \tag{3.35}$$

where Δ is the spacing between adjacent points on the line. We may thus rewrite (3.34) in the form of a limit as follows:

$$\begin{aligned}
\mathbb{E}[X] &= \lim_{\Delta \to 0} \sum_{k=-\infty}^{\infty} \int_{k\Delta}^{(k+1)\Delta} x_k f_X(x)\,\mathrm{d}x \\
&= \lim_{\Delta \to 0} \sum_{k=-\infty}^{\infty} x_k \mathbb{P}\left[x_k - \frac{\Delta}{2} < X \le x_k + \frac{\Delta}{2}\right]
\end{aligned}$$

For a physical interpretation of the sum in the second line of the right-hand side of this equation, suppose that we make n independent observations of the random variable X. Let $N_n(k)$ denote the number of times that the random variable X falls inside the kth bin, defined by

$$x_k - \frac{\Delta}{2} < X \le x_k + \frac{\Delta}{2}, \qquad k = 0, \pm 1, \pm 2, \dots$$

Arguing heuristically, we may say that, as the number of observations n is made large, the ratio $N_n(k)/n$ approaches the probability $\mathbb{P}[x_k - \Delta/2 < X \le x_k + \Delta/2]$. Accordingly, we may approximate the expected value of the random variable X as

$$\begin{aligned} \mathbb{E}[X] &\approx \sum_{k=-\infty}^{\infty} x_k \left(\frac{N_n(k)}{n}\right) \\ &= \frac{1}{n} \sum_{k=-\infty}^{\infty} x_k N_n(k), \quad \text{for large } n \end{aligned} \tag{3.36}$$

We now recognize the quantity on the right-hand side of (3.36) simply as the "sample average." The sum is taken over all the values x_k, each of which is weighted by the number of times it occurs; the sum is then divided by the total number of observations to give the sample average. Indeed, (3.36) provides the basis for computing the expectation $\mathbb{E}[X]$.

In a loose sense, we may say that the *discretization*, introduced in (3.35), has changed the expectation of a continuous random variable to the sample averaging over a discrete random variable. Indeed, in light of (3.36), we may formally define the expectation of a discrete random variable X as

$$\mathbb{E}[X] = \sum_{x} x p_X(x) \tag{3.37}$$

where $p_X(x)$ is the probability mass function of X, defined in (3.22), and where the summation extends over all possible discrete values of the dummy variable x. Comparing the summation in (3.37) with that of (3.36), we see that, roughly speaking, the ratio $N_n(x)/n$ plays a role similar to that of the probability mass function $p_X(x)$, which is intuitively satisfying.

Just as in the case of a continuous random variable, here again we see from the defining equation (3.37) that the expectation operator $\mathbb{E}$, applied to a discrete random variable X, produces a single number derived uniquely from the probability mass function $p_X(x)$.

Simply put, the expectation operator $\mathbb{E}$ applies equally well to discrete and continuous random variables.

Properties of the Expectation Operator

The expectation operator $\mathbb{E}$ plays a dominant role in the statistical analysis of random variables (as well as random processes studied in Chapter 4). It is therefore befitting that we study two important properties of this operation in this section; other properties are addressed in the end-of-chapter Problem 3.13.

PROPERTY 1 **Linearity**

Consider a random variable Z, defined by

$$Z = X + Y$$

where X and Y are two continuous random variables whose probability density functions are respectively denoted by $f_X(x)$ and $f_Y(y)$. Extending the definition of expectation introduced in (3.34) to the random variable Z, we write

$$\mathbb{E}[Z] = \int_{-\infty}^{\infty} z f_Z(z)\,dz$$

where $f_Z(z)$ is defined by the convolution integral of (3.33). Accordingly, we may go on to express the expectation $\mathbb{E}[Z]$ as the double integral

$$\begin{aligned}\mathbb{E}[Z] &= \int_{-\infty}^{\infty}\int_{-\infty}^{\infty} z f_X(x) f_Y(z-x)\,dx\,dz \\ &= \int_{-\infty}^{\infty}\int_{-\infty}^{\infty} z f_{X,Y}(x, z-x)\,dx\,dz\end{aligned}$$

where the joint probability density function

$$f_{X,Y}(x, z-x) = f_X(x) f_Y(z-x)$$

Making the one-to-one change of variables

$$y = z - x$$

and

$$x = x$$

we may now express the expectation $\mathbb{E}[Z]$ in the expanded form

$$\begin{aligned}\mathbb{E}[Z] &= \int_{-\infty}^{\infty}\int_{-\infty}^{\infty} (x+y) f_{X,Y}(x, y)\,dx\,dy \\ &= \int_{-\infty}^{\infty}\int_{-\infty}^{\infty} x f_{X,Y}(x, y)\,dx\,dy + \int_{-\infty}^{\infty}\int_{-\infty}^{\infty} y f_{X,Y}(x, y)\,dx\,dy\end{aligned}$$

Next, we recall from (3.27) that the first marginal density of the random variable X is

$$f_X(x) = \int_{-\infty}^{\infty} f_{X,Y}(x, y)\,dy$$

and, similarly, for the second marginal density

$$f_Y(y) = \int_{-\infty}^{\infty} f_{X,Y}(x, y)\,dx$$

The formula for the expectation $\mathbb{E}[Z]$ is therefore simplified as follows:

$$\begin{aligned}\mathbb{E}[Z] &= \int_{-\infty}^{\infty} x f_X(x)\,dx + \int_{-\infty}^{\infty} y f_Y(y)\,dy \\ &= \mathbb{E}[X] + \mathbb{E}[Y]\end{aligned}$$

We may extend this result to the sum of many random variables by *the method of induction* and thus write that, in general,

$$\mathbb{E}\left[\sum_{i=1}^{n} X_i\right] = \sum_{i=1}^{n} \mathbb{E}[X_i]) \tag{3.38}$$

In words, we may therefore state:

> The expectation of a sum of random variables is equal to the sum of the individual expectations.

This statement proves the linearity property of the *expectation operator*, which makes this operator all the more appealing.

PROPERTY 2 **Statistical Independence**

Consider next the random variable Z, defined as the product of two independent random variables X and Y, whose probability density functions are respectively denoted by $f_X(x)$ and $f_Y(y)$. As before, the expectation of Z is defined by

$$\mathbb{E}[Z] = \int_{-\infty}^{\infty} z f_Z(z)\, dz$$

except that, this time, we have

$$\begin{aligned} f_Z(z) &= f_{X,Y}(x, y) \\ &= f_X(x) f_Y(y) \end{aligned}$$

where, in the second line, we used the statistical independence of X and Y. With $Z = XY$, we may therefore recast the expectation $\mathbb{E}[Z]$ as

$$\begin{aligned} \mathbb{E}[XY] &= \int_{-\infty}^{\infty} xy f_X(x) f_Y(y)\, dx\, dy \\ &= \int_{-\infty}^{\infty} x f_X(x)\, dx \int_{-\infty}^{\infty} y f_Y(y)\, dy \\ &= \mathbb{E}[X]\mathbb{E}[Y] \end{aligned} \tag{3.39}$$

In words, we may therefore state:

> The expectation of the product of two statistically independent random variables is equal to the product of their individual expectations.

Here again, by induction, we may extend this statement to the product of many independent random variables.

3.7 Second-Order Statistical Averages

Function of a Random Variable

In the previous section we studied the mean of random variables in some detail. In this section, we expand on the mean by studying different second-order statistical averages.

These statistical averages, together with the mean, complete the *partial characterization* of random variables.

To this end, let X denote a random variable and let $g(X)$ denote a real-valued function of X defined on the real line. The quantity obtained by letting the argument of the function $g(X)$ be a random variable is also a random variable, which we denote as

$$Y = g(X) \tag{3.40}$$

To find the expectation of the random variable Y, we could, of course, find the probability density function $f_Y(y)$ and then apply the standard formula

$$\mathbb{E}[Y] = \int_{-\infty}^{\infty} y f_Y(y)\, dy$$

A simpler procedure, however, is to write

$$\mathbb{E}[g(X)] = \int_{-\infty}^{\infty} g(x) f_X(x)\, dx \tag{3.41}$$

Equation (3.41) is called the *expected value rule*; validity of this rule for a continuous random variable is addressed in Problem 3.14.

EXAMPLE 4

The Cosine Transformation of a Random Variable

Let

$$Y = g(X) = \cos(X)$$

where X is a random variable uniformly distributed in the interval $(-\pi,\pi)$; that is,

$$f_X(x) = \begin{cases} \dfrac{1}{2\pi}, & -\pi \le x \le \pi \\ 0, & \text{otherwise} \end{cases}$$

According to (3.41), the expected value of Y is

$$\begin{aligned} \mathbb{E}[Y] &= \int_{-\pi}^{\pi} (\cos x)\left(\frac{1}{2\pi}\right) dx \\ &= -\frac{1}{2\pi} \sin x \Big|_{x=-\pi}^{\pi} \\ &= 0 \end{aligned}$$

This result is intuitively satisfying in light of what we know about the dependence of a cosine function on its argument.

Second-order Moments

For the special case of $g(X) = X^n$, the application of (3.41) leads to the nth *moment* of the probability distribution of a random variable X; that is,

$$\mathbb{E}[X^n] = \int_{-\infty}^{\infty} x^n f_X(x)\, dx \tag{3.42}$$

From an engineering perspective, however, the most important moments of X are the first two moments. Putting $n = 1$ in (3.42) gives the mean of the random variable, which was discussed in Section 3.6. Putting $n = 2$ gives the *mean-square value* of X, defined by

$$\mathbb{E}[X^2] = \int_{-\infty}^{\infty} x^2 f_X(x)\, dx \tag{3.43}$$

Variance

We may also define *central moments*, which are simply the moments of the difference between a random variable X and its mean μ_X. Thus, the nth central moment of X is

$$\mathbb{E}[(X-\mu_X)^n] = \int_{-\infty}^{\infty} (x-\mu_X)^n f_X(x)\, dx \tag{3.44}$$

For $n = 1$, the central moment is, of course, zero. For $n = 2$, the second central moment is referred to as the *variance* of the random variable X, defined by

$$\begin{aligned} \text{var}[X] &= \mathbb{E}(X-\mu_x)^2 \\ &= \int_{-\infty}^{\infty} (x-\mu_X)^2 f_X(x)\, dx \end{aligned} \tag{3.45}$$

The variance of a random variable X is commonly denoted by σ_X^2. The square root of the variance, namely σ_X, is called the *standard deviation* of the random variable X.

In a sense, the variance σ_X^2 of the random variable X is a measure of the variable's "randomness" or "volatility." By specifying the variance σ_X^2 we essentially constrain the effective width of the probability density function $f_X(x)$ of the random variable X about the mean μ_X. A precise statement of this constraint is contained in the *Chebyshev inequality*, which states that for any positive number ε, we have the probability

$$\mathbb{P}[|X-\mu_X| \geq \varepsilon] \leq \frac{\sigma_X^2}{\varepsilon^2} \tag{3.46}$$

From this inequality we see that the mean and variance of a random variable provide a *weak description* of its probability distribution; hence the practical importance of these two statistical averages.

Using (3.43) and (3.45), we find that the variance σ_X^2 and the mean-square value $\mathbb{E}[X^2]$ are related by

$$\begin{aligned} \sigma_X^2 &= \mathbb{E}[X^2 - 2\mu_X X + \mu_X^2] \\ &= \mathbb{E}[X^2] - 2\mu_X \mathbb{E}[X] + \mu_X^2 \\ &= \mathbb{E}[X^2] - \mu_X^2 \end{aligned} \tag{3.47}$$

where, in the second line, we used the linearity property of the statistical expectation operator $\mathbb{E}$. Equation (3.47) shows that if the mean μ_X is zero, then the variance σ_X^2 and the mean-square value $\mathbb{E}\,[X^2]$ of the random variable X are equal.

Covariance

Thus far, we have considered the characterization of a single random variable. Consider next a pair of random variables X and Y. In this new setting, a set of statistical averages of importance is the *joint moments*, namely the expectation of $X^i Y^k$, where i and k may assume any positive integer values. Specifically, by definition, we have

$$\mathbb{E}[X^i Y^k] = \int_{-\infty}^{\infty} \int_{-\infty}^{\infty} x^i y^k f_{X,Y}(x, y)\, dx\, dy \tag{3.48}$$

A joint moment of particular importance is the *correlation*, defined by $\mathbb{E}[XY]$, which corresponds to $i = k = 1$ in this equation.

More specifically, the correlation of the centered random variables $(X - \mathbb{E}[X])$ and $(Y - \mathbb{E}[Y])$, that is, the joint moment

$$\operatorname{cov}[XY] = \mathbb{E}[(X - \mathbb{E}[X])(Y - \mathbb{E}[Y])] \tag{3.49}$$

is called the *covariance* of X and Y. Let $\mu_X = \mathbb{E}[X]$ and $\mu_Y = \mathbb{E}[Y]$; we may then expand (3.49) to obtain the result

$$\operatorname{cov}[XY] = \mathbb{E}[XY] - \mu_X \mu_Y \tag{3.50}$$

where we have made use of the linearity property of the expectation operator $\mathbb{E}$. Let σ_X^2 and σ_Y^2 denote the variances of X and Y, respectively. Then, the covariance of X and Y, normalized with respect to the product $\sigma_X \sigma_Y$, is called the *correlation coefficient of X and Y*, expressed as

$$\rho(X, Y) = \frac{\operatorname{cov}[XY]}{\sigma_X \sigma_Y} \tag{3.51}$$

The two random variables X and Y are said to be *uncorrelated* if, and only if, their covariance is zero; that is,

$$\operatorname{cov}[XY] = 0$$

They are said to be *orthogonal* if and only if their correlation is zero; that is,

$$\mathbb{E}[XY] = 0$$

In light of (3.50), we may therefore make the following statement:

> If one of the random variables X and Y or both have zero means, and if they are orthogonal, then they are uncorrelated, and vice versa.

3.8 Characteristic Function

In the preceding section we showed that, given a continuous random variable X, we can formulate the probability law defining the expectation of X^n (i.e., nth moment of X) in terms of the probability density function $f_X(x)$, as shown in (3.42). We now introduce another way of formulating this probability law; we do so through the *characteristic function*.

For a formal definition of this new concept, we say:

The characteristic function of a continuous random variable X, denoted by $\Phi_X(\nu)$, is defined as the expectation of the complex exponential function $\exp(j\nu X)$, that is

$$\Phi_X(\nu) = \mathbb{E}[\exp(j\nu X)] = \int_{-\infty}^{\infty} f_X(x)\exp(j\nu x)\,dx \tag{3.52}$$

where ν is real and $j = \sqrt{-1}$.

According to the second expression on the right-hand side of (3.52), we may also view the characteristic function $\Phi_X(\nu)$ of the random variable X as the Fourier transform of the associated probability density function $f_X(x)$, except for a sign change in the exponent. In this interpretation of the characteristic function we have used $\exp(j\nu x)$ rather than $\exp(-j\nu x)$ so as to conform with the convention adopted in probability theory.

Recognizing that ν and x play roles analogous to the variables $2\pi f$ and t respectively in the Fourier-transform theory, we may appeal to the Fourier transform theory of Chapter 2 to recover the probability density function $f_X(x)$ of the random variable X given the characteristic function $\Phi_X(\nu)$. Specifically, we may use the *inversion formula* to write

$$f_X(x) = \frac{1}{2\pi}\int_{-\infty}^{\infty} \Phi_X(\nu)\exp(-j\nu x)\,dx \tag{3.53}$$

Thus, with $f_X(f)$ and $\Phi_X(f)$ forming a Fourier-transform pair, we may obtain the moments of the random variable X from the function $\Phi_X(f)$. To pursue this issue, we differentiate both sides of (3.52) with respect to ν a total of n times, and then set $\nu = 0$; we thus get the result

$$\frac{d^n}{d\nu^n}\Phi_X(\nu)\big|_{\nu=0} = (j)^n \int_{-\infty}^{\infty} x^n f_X(x)\,dx \tag{3.54}$$

The integral on the right-hand side of this relation is recognized as the nth moment of the random variable X. Accordingly, we may recast (3.54) in the equivalent form

$$\mathbb{E}[X^n] = (-j)^n \frac{d^n}{d\nu^n}\Phi_X(\nu)\big|_{\nu=0} \tag{3.55}$$

This equation is a mathematical statement of the so-called *moment theorem*. Indeed, it is because of (3.55) that the characteristic function $\Phi_X(\nu)$ is also referred to as a *moment-generating function*.

EXAMPLE 5 **Exponential Distribution**

The exponential distribution is defined by

$$f_X(x) = \begin{cases} \lambda\exp(-\lambda x), & x \geq 0 \\ 0, & \text{otherwise} \end{cases} \tag{3.56}$$

where λ is the only parameter of the distribution. The characteristic function of the distribution is therefore

$$\begin{aligned}\Phi(\nu) &= \int_0^\infty \lambda \exp(-\lambda x) \exp(j\nu x)\, dx \\ &= \frac{\lambda}{\lambda - j\nu}\end{aligned}$$

We wish to use this result to find the mean of the exponentially distributed random variable X. To do this evaluation, we differentiate the characteristic function $\Phi(\nu)$ with respect to ν once, obtaining

$$\Phi'_X(\nu) = \frac{\lambda j}{(\lambda - j\nu)^2}$$

where the prime in $\Phi'_X(\nu)$ signifies first-order differentiation with respect to the argument ν. Hence, applying the moment theorem of (3.55), we get the desired result

$$\begin{aligned}\mathbb{E}[X] &= -j\Phi'_X(\nu)\big|_{\nu=0} \\ &= \frac{1}{\lambda}\end{aligned} \tag{3.57}$$

3.9 The Gaussian Distribution

Among the many distributions studied in the literature on probability theory, the *Gaussian distribution* stands out, by far, as the most commonly used distribution in the statistical analysis of communications systems, for reasons that will become apparent in Section 3.10. Let X denote a continuous random variable; the variable X is said to be *Gaussian distributed* if its probability density function has the general form

$$f_X(x) = \frac{1}{\sqrt{2\pi}\sigma} \exp\left[-\frac{(x-\mu)^2}{2\sigma^2}\right] \tag{3.58}$$

where μ and σ are two scalar parameters that characterize the distribution. The parameter μ can assume both positive and negative values (including zero), whereas the parameter σ is always positive. Under these two conditions, the $f_X(x)$ of (3.58) satisfies all the properties of a probability density function, including the normalization property; namely,

$$\frac{1}{\sqrt{2\pi}\sigma} \int_{-\infty}^{\infty} \exp\left[-\frac{(x-\mu)^2}{2\sigma^2}\right] dx = 1 \tag{3.59}$$

Properties of the Gaussian Distribution

A Gaussian random variable has many important properties, four of which are summarized on the next two pages.

PROPERTY 1 **Mean and Variance**

In the defining (3.58), the parameter μ is the mean of the Gaussian random variable X and σ^2 is its variance. We may therefore state:

> A Gaussian random variable is uniquely defined by specifying its mean and variance.

PROPERTY 2 **Linear Function of a Gaussian Random Variable**

Let X be a Gaussian random variable with mean μ and variance σ^2. Define a new random variable

$$Y = aX + b$$

where a and b are scalars and $a \neq 0$. Then Y is also Gaussian with mean

$$\mathbb{E}[Y] = a\mu + b$$

and variance

$$\text{var}[Y] = a^2\sigma^2$$

In words, we may state:

> Gaussianity is preserved by a linear transformation.

PROPERTY 3 **Sum of Independent Gaussian Random Variables**

Let X and Y be independent Gaussian random variables with means μ_X and μ_Y, respectively, and variances σ_X^2 and σ_Y^2, respectively. Define a new random variable

$$Z = X + Y$$

The random variable Z is also Gaussian with mean

$$\mathbb{E}[Z] = \mu_X + \mu_Y \tag{3.60}$$

and variance

$$\text{var}[Z] = \sigma_X^2 + \sigma_Y^2 \tag{3.61}$$

In general, we may therefore state:

> The sum of independent Gaussian random variables is also a Gaussian random variable, whose mean and variance are respectively equal to the sum of the means and the sum of the variances of the constituent random variables.

PROPERTY 4 **Jointly Gaussian Random Variables**

Let X and Y be a pair of jointly Gaussian random variables with zero means and variances σ_X^2 and σ_Y^2, respectively. The joint probability density function of X and Y is completely determined by σ_X, σ_Y, and ρ, where ρ is the correlation coefficient defined in (3.51). Specifically, we have

$$f_{X,Y}(x, y) = c\exp(-q(x, y)) \tag{3.62}$$

where the normalization constant c is defined by

$$c = \frac{1}{2\pi\sqrt{1-\rho^2}\,\sigma_X\sigma_Y} \tag{3.63}$$

and the exponential term is defined by

$$q(x, y) = \frac{1}{2(1-\rho^2)}\left(\frac{x^2}{\sigma_X^2} - 2\rho\frac{xy}{\sigma_X\sigma_Y} + \frac{y^2}{\sigma_Y^2}\right) \tag{3.64}$$

In the special case where the correlation coefficient ρ is zero, the joint probability density function of X and Y assumes the simple form

$$\begin{aligned} f_{X,Y}(x, y) &= \frac{1}{2\pi\sigma_X\sigma_Y}\exp\left(-\frac{x^2}{2\sigma_X^2} - \frac{y^2}{2\sigma_Y^2}\right) \\ &= f_X(x)f_Y(y) \end{aligned} \tag{3.65}$$

Accordingly, we may make the statement:

> If the random variables X and Y are both Gaussian with zero mean and if they are also orthogonal (that is, $\mathbb{E}[XY] = 0$), then they are statistically independent.

By virtue of Gaussianity, this statement is stronger than the last statement made at the end of the subsection on covariance.

Commonly Used Notation

In light of Property 1, the notation $\mathcal{N}(\mu, \sigma^2)$ is commonly used as the shorthand description of a Gaussian distribution parameterized in terms of its mean μ and variance σ^2. The symbol $\mathcal{N}$ is used in recognition of the fact that the Gaussian distribution is also referred to as the *normal distribution*, particularly in the mathematics literature.

The Standard Gaussian Distribution

When $\mu = 0$ and $\sigma^2 = 1$, the probability density function of (3.58) reduces to the special form:

$$f_X(x) = \frac{1}{\sqrt{2\pi}}\exp\left(-\frac{x^2}{2}\right) \tag{3.66}$$

A Gaussian random variable X so described is said to be in its *standard form.*[4] Correspondingly, the distribution function of the standard Gaussian random variable is defined by

$$F_X(x) = \frac{1}{\sqrt{2\pi}}\int_{-\infty}^{x}\exp\left(-\frac{t^2}{2}\right)dt \tag{3.67}$$

Owing to the frequent use of integrals of the type described in (3.67), several related functions have been defined and tabulated in the literature. The related function commonly used in the context of communication systems is the *Q-function*, which is formally defined as

$$\begin{aligned} Q(x) &= 1 - F_X(x) \\ &= \frac{1}{\sqrt{2\pi}}\int_{x}^{\infty}\exp\left(-\frac{t^2}{2}\right)dt \end{aligned} \tag{3.68}$$

In words, we may describe the *Q-function* as follows:

> The Q-function, $Q(x)$, is equal to the area covered by the tail of the probability density function of the standard Gaussian random variable X, extending from x to infinity.

Unfortunately, the integral of (3.67) defining the standard Gaussian distribution $F_X(x)$ does not have a closed-form solution. Rather, with accuracy being an issue of importance, $F_X(x)$ is usually presented in the form of a table for varying x. Table 3.1 is one such recording. To utilize this table for calculating the Q-function, we build on two defining equations:

1. For nonnegative values of x, the first line of (3.68) is used.
2. For negative values of x, use is made of the *symmetric property* of the Q-function:

$$Q(-x) = 1 - Q(x) \tag{3.69}$$

Standard Gaussian Graphics

To visualize the graphical formats of the commonly used standard Gaussian functions, $F_X(x), f_X(x)$, and $Q(x)$, three plots are presented at the bottom of this page:

1. Figure 3.10a plots the distribution function, $F_X(x)$, defined in (3.67).
2. Figure 3.10b plots the density function, $f_X(x)$, defined in (3.66).
3. Figure 3.11 plots the Q-function defined in (3.68).

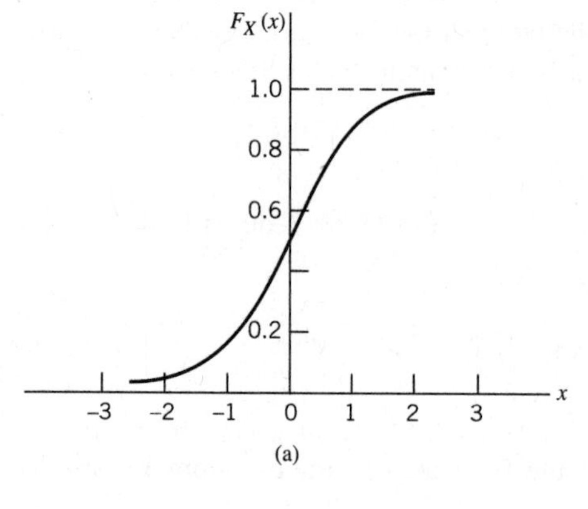

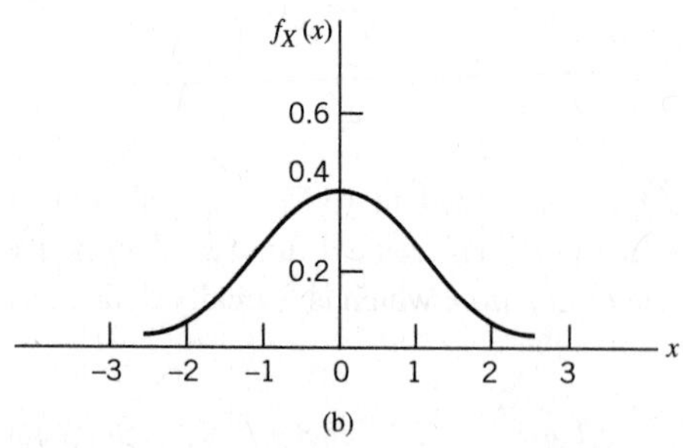

Figure 3.10 The normalized Gaussian (a) distribution function and (b) probability density function.

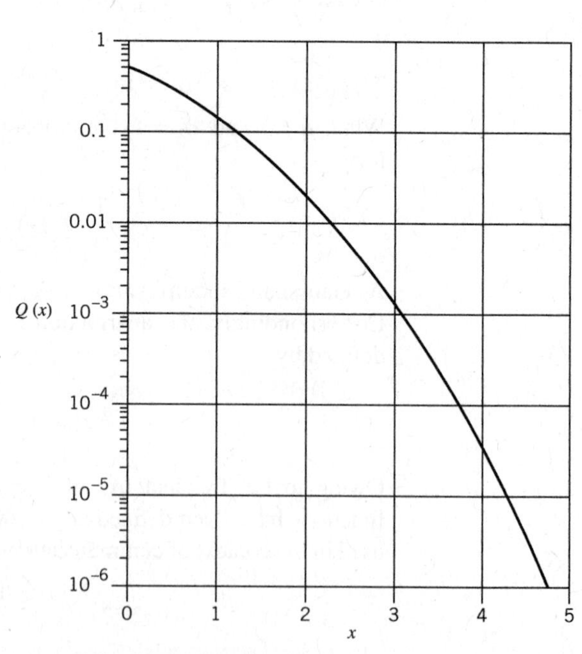

Figure 3.11 The Q-function.

Table 3.1 **The standard Gaussian distribution (Q-function) table**[5]

	.00	.01	.02	.03	.04	.05	.06	.07	.08	.09
0.0	.5000	.5040	.5080	.5120	.5160	.5199	.5239	.5279	.5319	.5359
0.1	.5398	.5438	.5478	.5517	.5557	.5596	.5636	.5675	.5714	.5753
0.2	.5793	.5832	.5871	.5910	.5948	.5987	.6026	.6064	.6103	.6141
0.3	.6179	.6217	.6255	.6293	.6331	.6368	.6406	.6443	.6460	.6517
0.4	.6554	.6591	.6628	.6664	.6700	.6736	.6772	.6808	.6844	.6879
0.5	.6915	.6950	.6985	.7019	.7054	.7088	.7123	.7157	.7190	.7224
0.6	.7257	.7291	.7324	.7357	.7389	.7422	.7454	.7485	.7517	.7549
0.7	.7580	.7611	.7642	.7673	.7704	.7734	.7764	.7794	.7823	.7852
0.8	.7881	.7910	.7939	.7967	.7995	.8023	.8051	.8078	.8106	.8133
0.9	.8159	.8186	.8212	.8238	.8264	.8289	.8315	.8340	.8365	.8389
1.0	.8413	.8438	.8461	.8485	.8508	.8531	.8554	.8577	.8599	.8621
1.1	.8643	.8665	.8686	.8708	.8729	.8749	.8770	.8790	.8810	.8830
1.2	.8849	.8869	.8888	.8907	.8925	.8944	.8962	.8980	.8997	.9015
1.3	.9032	.9049	.9066	.9082	.9099	.9115	.9131	.9149	.9162	.9177
1.4	.9192	.9207	.9222	.9236	.9251	.9265	.9279	.9292	.9306	.9319
1.5	.9332	.9345	.9357	.9370	.9382	.9394	.9406	.9418	.9429	.9441
1.6	.9452	.9463	.9474	.9484	.9495	.9505	.9515	.9525	.9535	.9545
1.7	.9554	.9564	.9573	.9582	.9591	.9599	.9608	.9616	.9625	.9633
1.8	.9641	.9649	.9656	.9664	.9671	.9678	.9686	.9693	.9699	.9706
1.9	.9713	.9719	.9726	.9732	.9738	.9744	.9750	.9756	.9761	.9767
2.0	.9772	.9778	.9783	.9788	.9793	.9798	.9803	.9808	.9812	.9817
2.1	.9821	.9826	.9830	.9834	.9838	.9842	.9846	.9850	.9854	.9857
2.2	.9861	.9864	.9868	.9871	.9875	.9878	.9881	.9884	.9887	.9890
2.3	.9893	.9896	.9898	.9901	.9904	.9906	.9909	.9911	.9913	.9916
2.4	.9918	.9920	.9922	.9925	.9927	.9929	.9931	.9932	.9934	.9936
2.5	.9938	.9940	.9941	.9943	.9945	.9946	.9948	.9949	.9951	.9952
2.6	.9953	.9955	.9956	.9957	.9959	.9960	.9961	.9962	.9963	.9964
2.7	.9965	.9966	.9967	.9968	.9969	.9970	.9971	.9972	.9973	.9974
2.8	.9974	.9975	.9976	.9977	.9977	.9978	.9979	9979	.9980	.9981
2.9	.9981	.9982	.9982	.9983	.9984	.9984	.9985	.9985	.9986	.9986
3.0	.9987	.9987	.9987	.9988	.9988	.9989	.9989	.9989	.9990	.9990
3.1	.9990	.9991	.9991	.9991	.9992	.9992	.9992	.9992	.9993	.9993
3.2	.9993	.9993	.9994	.9994	.9994	.9994	.9994	.9995	.9995	.9995
3.3	.9995	.9995	.9995	.9996	.9996	.9996	.9996	.9996	.9996	.9997
3.4	.9997	.9997	.9997	.9997	.9997	.9997	.9997	.9997	.9997	.9998

1. The entries in this table, x say, occupy the range [0.0, 3.49]; the x is sample value of the random variable X.
2. For each value of x, the table provides the corresponding value of the Q-function:

$$Q(x) = 1 - F_X(x) = \frac{1}{\sqrt{2\pi}}\int_x^\infty \exp(-t^2/2)\,dt$$

3.10 The Central Limit Theorem

The *central limit theorem* occupies an important place in probability theory: it provides the mathematical justification for using the Gaussian distribution as a *model* for an observed random variable that is known to be the result of a large number of random events.

For a formal statement of the central limit theorem, let $X_1, X_2, ..., X_n$ denote a sequence of *independently and identically distributed (iid) random variables* with common mean μ and variance σ^2. Define the related random variable

$$Y_n = \frac{1}{\sigma\sqrt{n}}\left(\sum_{i=1}^{n} X_i - n\mu\right) \tag{3.70}$$

The subtraction of the product term $n\mu$ from the sum $\sum_{i=1}^{n} X_i$ ensures that the random variable Y_n has zero mean; the division by the factor $\sigma\sqrt{n}$ ensures that Y_n has unit variance.

Given the setting described in (3.70), the central limit theorem formally states:

> As the number of random variables n in (3.70) approaches infinity, the normalized random variable Y_n converges to the standard Gaussian random variable with the distribution function
>
> $$F_Y(y) = \frac{1}{\sqrt{2\pi}} \int_{-\infty}^{y} \exp\left(-\frac{x^2}{2}\right) dx$$
>
> in the sense that
>
> $$\lim_{n \to \infty} \mathbb{P}(Y_n \le y) = Q(y) \tag{3.71}$$
>
> where $Q(y)$ is the Q-function.

To appreciate the practical importance of the central limit theorem, suppose that we have a physical phenomenon whose occurrence is attributed to a large number of random events. The theorem, embodying (3.67)–(3.71), permits us to calculate certain probabilities simply by referring to a Q-function table (e.g., Table 3.1). Moreover, to perform the calculation, all that we need to know are means and variances.

However, a word of caution is in order here. The central limit theorem gives only the "limiting" form of the probability distribution of the standardized random variable Y_n as n approaches infinity. When n is finite, it is sometimes found that the Gaussian limit provides a relatively poor approximation for the actual probability distribution of Y_n, even though n may be large.

EXAMPLE 6 **Sum of Uniformly Distributed Random Variables**

Consider the random variable

$$Y_n = \sum_{i=1}^{n} X_i$$

where the X_i are independent and uniformly distributed random variables on the interval from −1 to +1. Suppose that we generate 20000 samples of the random variable Y_n for $n = 10$, and then compute the probability density function of Y_n by forming a histogram of the results. Figure 3.11a compares the computed histogram (scaled for unit area) with the probability density function of a Gaussian random variable with the same mean and variance. The figure clearly illustrates that in this particular example the number of independent distributions n does not have to be large for the sum Y_n to closely approximate a Gaussian distribution. Indeed, the results of this example confirm how powerful the central limit theorem is. Moreover, the results explain why Gaussian models are so ubiquitous in the analysis of random signals not only in the study of communication systems, but also in so many other disciplines.

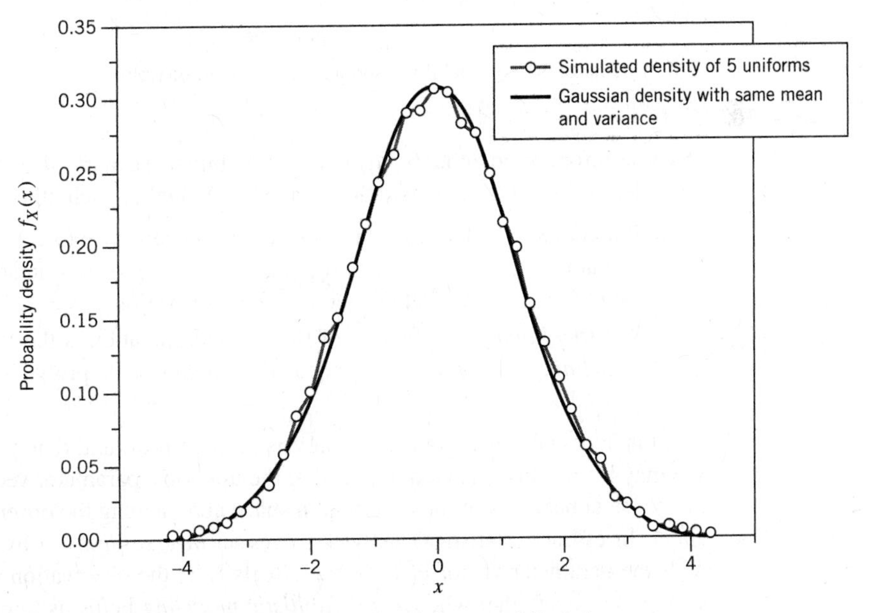

Figure 3.11 Simulation supporting validity of the central limit theorem.

3.11 Bayesian Inference

The material covered up to this point in the chapter has largely addressed issues involved in the mathematical description of probabilistic models. In the remaining part of the chapter we will study the role of probability theory in probabilistic reasoning based on the Bayesian[5] paradigm, which occupies a central place in statistical communication theory.

To proceed with the discussion, consider Figure 3.12, which depicts two finite-dimensional spaces: a *parameter space* and an *observation space*, with the parameter space being hidden from the *observer*. A parameter vector $\boldsymbol{\theta}$, drawn from the parameter space, is mapped probabilistically onto the observation space, producing the observation vector $\mathbf{x}$. The vector $\mathbf{x}$ is the sample value of a random vector $\mathbf{X}$, which provides the

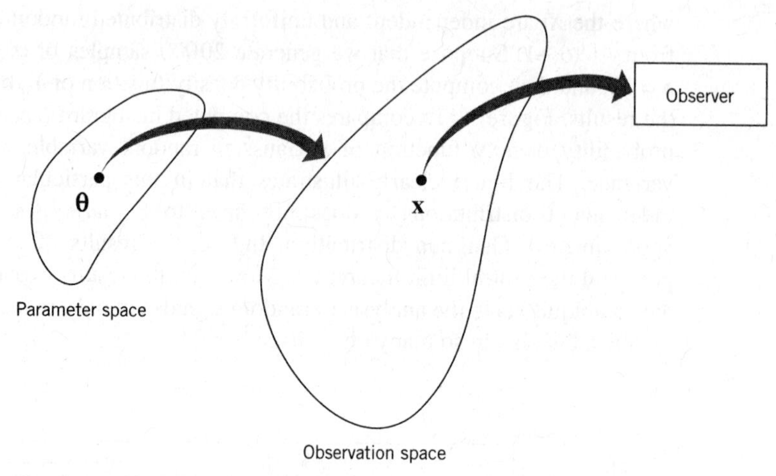

Figure 3.12 Probabilistic model for Bayesian inference.

observer information about $\boldsymbol{\theta}$. Given the probabilistic scenario depicted in Figure 3.12, we may identify two different operations that are the dual of each other.[6]

1. *Probabilistic modeling*. The aim of this operation is to formulate the conditional probability density function $f_{\mathbf{X}|\boldsymbol{\Theta}}(\mathbf{x}|\boldsymbol{\theta})$, which provides an adequate description of the underlying physical behavior of the observation space.
2. *Statistical analysis*. The aim of this second operation is the *inverse of probabilistic modeling*, for which we need the conditional probability density function $f_{\boldsymbol{\Theta}|\mathbf{X}}(\boldsymbol{\theta}|\mathbf{x})$.

In a fundamental sense, statistical analysis is more profound than probabilistic modeling. We may justify this assertion by viewing the unknown parameter vector $\boldsymbol{\theta}$ as the *cause* for the physical behavior of the observation space and viewing the observation vector **x** as the *effect*. In essence, statistical analysis solves an *inverse problem* by retrieving the causes (i.e., the parameter vector $\boldsymbol{\theta}$) from the effects (i.e., the observation vector **x**). Indeed, we may go on to say that whereas *probabilistic modeling* helps us to characterize the *future behavior* of **x** conditional on $\boldsymbol{\theta}$, statistical analysis permits us to make *inference* about $\boldsymbol{\theta}$ given **x**.

To formulate the conditional probability density function of $f_{\mathbf{X}|\boldsymbol{\Theta}}(\mathbf{x}|\boldsymbol{\theta})$, we recast Bayes' theorem of (3.14) in its *continuous version*, as shown by

$$f_{\boldsymbol{\Theta}|\mathbf{X}}(\boldsymbol{\theta}|\mathbf{x}) = \frac{f_{\mathbf{X}|\boldsymbol{\Theta}}(\mathbf{x}|\boldsymbol{\theta})f_{\boldsymbol{\Theta}}(\boldsymbol{\theta})}{f_{\mathbf{X}}(\mathbf{x})} \tag{3.72}$$

The denominator is itself defined in terms of the numerator as

$$\begin{aligned} f_{\mathbf{X}}(\mathbf{x}) &= \int_{\boldsymbol{\Theta}} f_{\mathbf{X}|\boldsymbol{\Theta}}(\mathbf{x}|\boldsymbol{\theta})f_{\boldsymbol{\Theta}}(\boldsymbol{\theta})\,d\boldsymbol{\theta} \\ &= \int_{\boldsymbol{\Theta}} f_{\mathbf{X},\boldsymbol{\Theta}}(\mathbf{x},\boldsymbol{\theta})\,d\boldsymbol{\theta} \end{aligned} \tag{3.73}$$

which is the marginal density of $\mathbf{X}$, obtained by integrating out the dependence of the joint probability density function $f_{\mathbf{X}|\Theta}(\mathbf{x}|\boldsymbol{\theta})$. In words, $f_{\mathbf{X}}(\mathbf{x})$ is a marginal density of the joint probability density function $f_{\mathbf{X},\Theta}(\mathbf{x}, \boldsymbol{\theta})$. The inversion formula of (3.72) is sometimes referred to as the *principle of inverse probability.*

In light of this principle, we may now introduce four notions:

1. *Observation density.* This stands for the conditional probability density function $f_{\mathbf{X}|\Theta}(\mathbf{x}|\boldsymbol{\theta})$, referring to the "observation" vector $\mathbf{x}$ given the parameter vector $\boldsymbol{\theta}$.
2. *Prior.* This stands for the probability density function $f_{\Theta}(\boldsymbol{\theta})$, referring to the parameter vector $\boldsymbol{\theta}$ "prior" to receiving the observation vector $\mathbf{x}$.
3. *Posterior.* This stands for the conditional probability density function $f_{\Theta|\mathbf{X}}(\boldsymbol{\theta}|\mathbf{x})$, referring to the parameter vector $\boldsymbol{\theta}$ "after" receiving the observation vector $\mathbf{x}$.
4. *Evidence.* This stands for the probability density function $f_{\mathbf{X}}(\mathbf{x})$, referring to the "information" contained in the observation vector $\mathbf{X}$ for statistical analysis.

The posterior $f_{\Theta|\mathbf{X}}(\boldsymbol{\theta}|\mathbf{x})$ is central to Bayesian inference. In particular, we may view it as the updating of information available on the parameter vector $\boldsymbol{\theta}$ in light of the information contained in the observation vector $\mathbf{x}$, while the prior $f_{\Theta}(\boldsymbol{\theta})$ is the information available on $\boldsymbol{\theta}$ prior to receiving the observation vector $\mathbf{x}$.

Likelihood

The inversion aspect of statistics manifests itself in the notion of the *likelihood function.*[7] In a formal sense, the likelihood, denoted by $l(\boldsymbol{\theta}|\mathbf{x})$, is just the observation density $f_{\mathbf{X}|\Theta}(\mathbf{x}|\boldsymbol{\theta})$ reformulated in a different order, as shown by

$$l(\boldsymbol{\theta}|\mathbf{x}) = f_{\mathbf{X}|\Theta}(\mathbf{x}|\boldsymbol{\theta}) \tag{3.74}$$

The important point to note here is that the likelihood and the observation density are both governed by exactly the same function that involves the parameter vector $\boldsymbol{\theta}$ and the observation vector $\mathbf{x}$. There is, however, a difference in interpretation: the likelihood function $l(\boldsymbol{\theta}|\mathbf{x})$ is treated as a function of the parameter vector $\boldsymbol{\theta}$ given $\mathbf{x}$, whereas the observation density $f_{\mathbf{X}|\Theta}(\mathbf{x}|\boldsymbol{\theta})$ is treated as a function of the observation vector $\mathbf{x}$ given $\boldsymbol{\theta}$.

Note, however, unlike $f_{\mathbf{X}|\Theta}(\mathbf{x}|\boldsymbol{\theta})$, the likelihood $l(\boldsymbol{\theta}|\mathbf{x})$ is *not* a distribution; rather, it is a function of the parameter vector $\boldsymbol{\theta}$, given $\mathbf{x}$.

In light of the terminologies introduced, namely the posterior, prior, likelihood, and evidence, we may now express Bayes' rule of (3.72) in words as follows:

$$\text{posterior} = \frac{\text{likelihood} \times \text{prior}}{\text{evidence}}$$

The Likelihood Principle

For convenience of presentation, let

$$\pi(\boldsymbol{\theta}) = f_{\Theta}(\boldsymbol{\theta}) \tag{3.75}$$

Then, recognizing that the evidence defined in (3.73) plays merely the role of a normalizing function that is independent of $\boldsymbol{\theta}$, we may now sum up (3.72) on the principle of inverse probability succinctly as follows:

> The Bayesian statistical model is essentially made up of two components: the likelihood function $l(\boldsymbol{\theta}|\mathbf{x})$ and the prior $\pi(\boldsymbol{\theta})$, where $\boldsymbol{\theta}$ is an unknown parameter vector and $\mathbf{x}$ is the observation vector.

To elaborate on the significance of the defining equation (3.74), consider the likelihood functions $l(\boldsymbol{\theta}|\mathbf{x}_1)$ and $l(\boldsymbol{\theta}|\mathbf{x}_2)$ on parameter vector $\boldsymbol{\theta}$. If, for a prescribed prior $\pi(\boldsymbol{\theta})$, these two likelihood functions are scaled versions of each other, then the corresponding posterior densities of $\boldsymbol{\theta}$ are essentially identical, the validity of which is a straightforward consequence of Bayes' theorem. In light of this result we may now formulate the so-called *likelihood principle*[8] as follows:

If $\mathbf{x}_1$ and $\mathbf{x}_2$ are two observation vectors depending on an unknown parameter vector $\boldsymbol{\theta}$, such that

$$l(\boldsymbol{\theta}\,|\,\mathbf{x}_1) = c\, l(\boldsymbol{\theta}\,|\,\mathbf{x}_2) \quad \text{for all } \boldsymbol{\theta}$$

where c is a scaling factor, then these two observation vectors lead to an identical inference on $\boldsymbol{\theta}$ for any prescribed prior $f_{\boldsymbol{\Theta}}(\boldsymbol{\theta})$.

Sufficient Statistic

Consider a model, parameterized by the vector $\boldsymbol{\theta}$ and given the observation vector $\mathbf{x}$. In statistical terms, the model is described by the posterior density $f_{\boldsymbol{\Theta}|\mathbf{X}}(\boldsymbol{\theta}|\mathbf{x})$. In this context, we may now introduce a function $\mathbf{t}(\mathbf{x})$, which is said to be a *sufficient statistic* if the probability density function of the parameter vector $\boldsymbol{\theta}$ given $\mathbf{t}(\mathbf{x})$ satisfies the condition

$$f_{\boldsymbol{\Theta}|\mathbf{X}}(\boldsymbol{\theta}|\mathbf{x}) = f_{\boldsymbol{\Theta}|\mathrm{T}(\mathbf{x})}(\boldsymbol{\theta}|\mathbf{t}(\mathbf{x})) \tag{3.76}$$

This condition imposed on $\mathbf{t}(\mathbf{x})$, for it to be a sufficient statistic, appears intuitively appealing, as evidenced by the following statement:

> The function $\mathbf{t}(\mathbf{x})$ provides a sufficient summary of the whole information about the unknown parameter vector $\boldsymbol{\theta}$, which is contained in the observation vector $\mathbf{x}$.

We may thus view the notion of sufficient statistic as a tool for "data reduction," the use of which results in considerable simplification in analysis.[9] The data reduction power of the sufficient statistic $\mathbf{t}(\mathbf{x})$ is well illustrated in Example 7.

3.12 Parameter Estimation

As pointed out previously, the posterior density $f_{\boldsymbol{\Theta}|\mathbf{X}}(\boldsymbol{\theta}|\mathbf{x})$ is central to the formulation of a Bayesian probabilistic model, where $\boldsymbol{\theta}$ is an unknown parameter vector and $\mathbf{x}$ is the observation vector. It is logical, therefore, that we use this conditional probability density function for parameter estimation.[10] Accordingly, we define the *maximum a posteriori (MAP) estimate* of $\boldsymbol{\theta}$ as

$$\begin{aligned} \hat{\boldsymbol{\theta}}_{\mathrm{MAP}} &= \arg\max_{\boldsymbol{\theta}} f_{\boldsymbol{\Theta}|\mathbf{X}}(\boldsymbol{\theta}|\mathbf{x}) \\ &= \arg\max_{\boldsymbol{\theta}} l(\boldsymbol{\theta}|\mathbf{x})\pi(\boldsymbol{\theta}) \end{aligned} \tag{3.77}$$

where $l(\boldsymbol{\theta}|\mathbf{x})$ is the likelihood function defined in (3.74), and $\pi(\boldsymbol{\theta})$ is the prior defined in (3.75). To compute the estimate $\hat{\boldsymbol{\theta}}_{\mathrm{MAP}}$, we require availability of the prior $\pi(\boldsymbol{\theta})$.

In words, the right-hand side of (3.77) reads as follows:

> Given the observation vector $\mathbf{x}$, the estimate $\hat{\boldsymbol{\theta}}_{\text{MAP}}$ is that particular value of the parameter vector $\boldsymbol{\theta}$ in the argument of the posterior density $f_{\boldsymbol{\Theta}|\mathbf{X}}(\boldsymbol{\theta}|\mathbf{x})$, for which this density attains its maximum value.

Generalizing the statement made at the end of the discussion on multiple random variables in Section 3.5, we may now go on to say that, for the problem at hand, the conditional probability density function $f_{\boldsymbol{\Theta}|\mathbf{X}}(\boldsymbol{\theta}|\mathbf{x})$ *contains all the conceivable information about the multidimensional parameter vector* $\boldsymbol{\theta}$ *given the observation vector* $\mathbf{x}$. The recognition of this fact leads us to make the follow-up important statement, illustrated in Figure 3.13 for the simple case of a one-dimensional parameter vector:

> The maximum a posterior estimate $\hat{\boldsymbol{\theta}}_{\text{MAP}}$ of the unknown parameter vector $\boldsymbol{\theta}$ is the globally optimal solution to the parameter-estimation problem, in the sense that there is no other estimator that can do better.

In referring to $\hat{\boldsymbol{\theta}}_{\text{MAP}}$ as the MAP estimate, we have made a slight change in our terminology: we have, in effect, referred to $f_{\boldsymbol{\Theta}|\mathbf{X}}(\boldsymbol{\theta}|\mathbf{x})$ as the *a posteriori density* rather than the *posterior density* of $\boldsymbol{\theta}$. We have made this minor change so as to conform to the MAP terminology that is well and truly embedded in the literature on statistical communication theory.

In another approach to parameter estimation, known as *maximum likelihood estimation*, the parameter vector $\boldsymbol{\theta}$ is estimated using the formula

$$\hat{\boldsymbol{\theta}}_{\text{ML}} = \arg \sup_{\boldsymbol{\theta}} l(\boldsymbol{\theta}|\mathbf{x}) \tag{3.78}$$

That is, the *maximum likelihood estimate* $\hat{\boldsymbol{\theta}}_{\text{ML}}$ is that value of the parameter vector $\boldsymbol{\theta}$ that maximizes the conditional distribution $f_{\mathbf{X}|\boldsymbol{\Theta}}(\mathbf{x}|\boldsymbol{\theta})$ at the observation vector $\mathbf{x}$. Note that this second estimate ignores the prior $\pi(\boldsymbol{\theta})$ and, therefore, lies at the fringe of the Bayesian paradigm. Nevertheless, maximum likelihood estimation is widely used in the literature on statistical communication theory, largely because in ignoring the prior $\pi(\boldsymbol{\theta})$, it is less demanding than maximum posterior estimation in computational complexity.

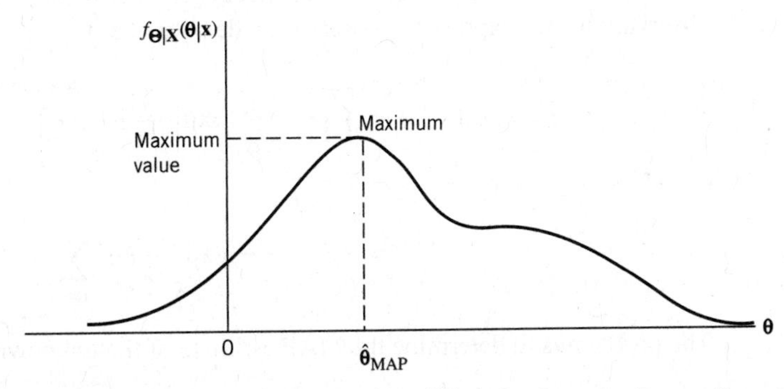

Figure 3.13 Illustrating the a posteriori $f_{\boldsymbol{\Theta}|\mathbf{X}}(\boldsymbol{\theta}|\mathbf{x})$ for the case of a one-dimensional parameter space.

The MAP and ML estimates do share a common possibility, in that the maximizations in (3.77) and (3.78) may lead to more than one global maximum. However, they do differ in one important result: the maximization indicated in (3.78) may *not* always be possible; that is, the procedure used to perform the maximization may *diverge*. To overcome this difficulty, the solution to (3.78) has to be *stabilized* by incorporating prior information on the parameter space, exemplified by the distribution $\pi(\boldsymbol{\theta})$, into the solution, which brings us back to the Bayesian approach and, therefore, (3.77). The most critical part in the Bayesian approach to statistical modeling and parameter estimation is how to choose the prior $\pi(\boldsymbol{\theta})$. There is also the possibility of the Bayesian approach requiring high-dimensional computations. We should not, therefore, underestimate the challenges involved in applying the Bayesian approach, on which note we may say the following:

There is no free lunch: for every gain made, there is a price to be paid.

EXAMPLE 7 **Parameter Estimation in Additive Noise**

Consider a set N of scalar observations, defined by

$$x_i = \theta + n_i, \qquad i = 1, 2, \ldots, N \tag{3.79}$$

where the unknown parameter θ is drawn from the Gaussian distribution $\mathcal{N}(0, \sigma_\theta^2)$; that is,

$$f_\Theta(\theta) = \frac{1}{\sqrt{2\pi}\sigma_\theta} \exp\left(-\frac{\theta^2}{2\sigma_\theta^2}\right) \tag{3.80}$$

Each n_i is drawn from another Gaussian distribution $\mathcal{N}(0, \sigma_n^2)$; that is,

$$f_{N_i}(n_i) = \frac{1}{\sqrt{2\pi}\sigma_n} \exp\left(-\frac{n_i^2}{2\sigma_n^2}\right), \qquad i = 1, 2, \ldots, N$$

It is assumed that the random variables N_i are all independent of each other, and also independent from Θ. The issue of interest is to find the MAP of the parameter θ.

To find the distribution of the random variable X_i, we invoke Property 2 of the Gaussian distribution, described in Section 3.9, in light of which we may say that X_i is also Gaussian with mean θ and variance σ_n^2. Furthermore, since the N_i are independent, by assumption, it follows that the X_i are also independent. Hence, using the vector $\mathbf{x}$ to denote the N observations, we express the observation density of $\mathbf{x}$ as

$$\begin{aligned} f_{\mathbf{X}|\Theta}(\mathbf{x}|\theta) &= \prod_{i=1}^{N} \frac{1}{\sqrt{2\pi}\sigma_n} \exp\left[-\frac{(x_i - \theta)^2}{2\sigma_n^2}\right] \\ &= \frac{1}{(\sqrt{2\pi}\sigma_n)^N} \exp\left[-\frac{1}{2\sigma_n^2} \sum_{i=1}^{N} (x_i - \theta)^2\right] \end{aligned} \tag{3.81}$$

The problem is to determine the MAP estimate of the unknown parameter θ.

To solve this problem, we need to know the posterior density $f_{\Theta|\mathbf{X}}(\theta|\mathbf{x})$. Applying (3.72), we write

$$f_{\Theta|\mathbf{X}}(\theta|\mathbf{x}) = c(\mathbf{x}) \exp\left|-\frac{1}{2}\left(\frac{\theta^2}{\sigma_\theta^2} + \frac{\sum_{i=1}^{N}(x_i - \theta)^2}{\sigma_n^2}\right)\right| \tag{3.82}$$

where

$$c(\mathbf{x}) = \frac{\frac{1}{\sqrt{2\pi}\sigma_\theta} \times \frac{1}{(\sqrt{2\pi}\sigma_n)^N}}{f_{\mathbf{X}}(\mathbf{x})} \tag{3.83}$$

The normalization factor $c(\mathbf{x})$ is independent of the parameter θ and, therefore, has no relevance to the MAP of θ. We therefore need only pay attention to the exponent in (3.82).

Rearranging terms and completing the square in the exponent in (3.82), and introducing a new normalization factor $c'(\mathbf{x})$ that absorbs all the terms involving x_i^2, we get

$$f_{\Theta|\mathbf{X}}(\theta|\mathbf{x}) = c'(\mathbf{x}) \exp\left\{-\frac{1}{2\sigma_p^2}\left(\frac{\sigma_n^2}{\sigma_\theta^2 + (\sigma_n^2/N)}\left(\frac{1}{N}\sum_{i=1}^{N} x_i\right) - \theta\right)^2\right\} \tag{3.84}$$

where

$$\sigma_p^2 = \frac{\sigma_\theta^2 \sigma_n^2}{N\sigma_\theta^2 + \sigma_n^2} \tag{3.85}$$

Equation (3.84) shows that the posterior density of the unknown parameter θ is Gaussian with mean θ and variance σ_p^2. We therefore readily find that the MAP estimate of θ is

$$\hat{\theta}_{\text{MAP}} = \frac{\sigma_n^2}{\sigma_\theta^2 + (\sigma_n^2/N)}\left(\frac{1}{N}\sum_{i=1}^{N} x_i\right) \tag{3.86}$$

which is the desired result.

Examining (3.84), we also see that the N observations enter the posterior density of θ only through the sum of the x_i. It follows, therefore, that

$$t(\mathbf{x}) = \sum_{i=1}^{N} x_i \tag{3.87}$$

is a sufficient statistic for the example at hand. This statement merely confirms that (3.84) and (3.87) satisfy the condition of (3.76) for a sufficient statistic.

3.13 Hypothesis Testing

The Bayesian paradigm discussed in Section 3.11 focused on two basic issues: predictive modeling of the observation space and statistical analysis aimed at parameter estimation. As mentioned previously in that section, these two issues are the dual of each other. In this section we discuss another facet of the Bayesian paradigm, aimed at *hypothesis testing*,[11] which is basic to signal detection in digital communications, and beyond.

Binary Hypotheses

To set the stage for the study of hypothesis testing, consider the model of Figure 3.14. A *source of binary data* emits a sequence of 0s and 1s, which are respectively denoted by hypotheses H_0 and H_1. The source (e.g., digital communication transmitter) is followed by a *probabilistic transition mechanism* (e.g., communication channel). According to some probabilistic law, the transition mechanism generates an *observation vector* **x** that defines a specific point in the observation space.

The mechanism responsible for probabilistic transition is *hidden* from the observer (e.g., digital communication receiver). Given the observation vector **x** and knowledge of the probabilistic law characterizing the transition mechanism, the observer chooses whether hypothesis H_0 or H_1 is true. Assuming that a *decision* must be made, the observer has to have a *decision rule* that works on the observation vector **x**, thereby dividing the observation space Z into two regions: Z_0 corresponding to H_0 being true and Z_1 corresponding to H_1 being true. To simplify matters, the decision rule is not shown in Figure 3.14.

In the context of a digital communication system, for example, the channel plays the role of the probabilistic transition mechanism. The observation space of some finite

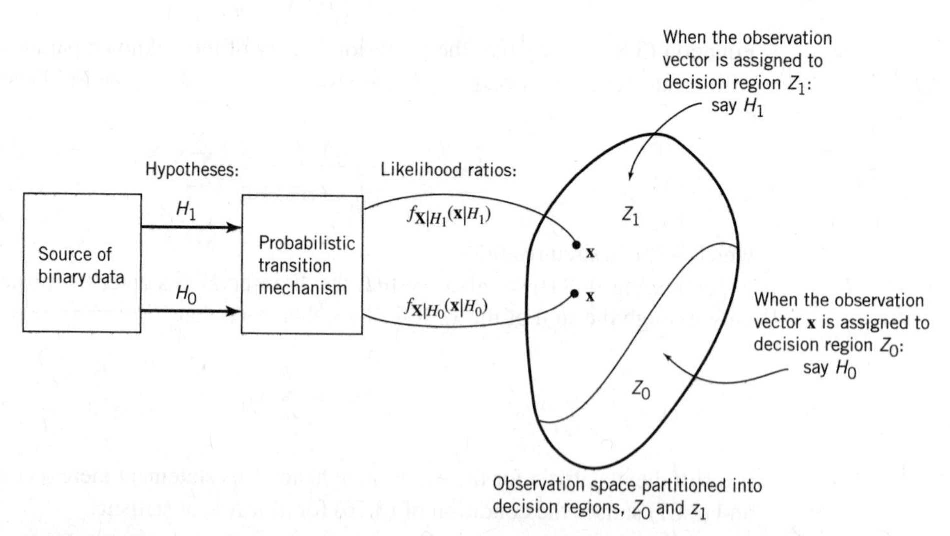

Figure 3.14 Diagram illustrating the binary hypothesis-testing problem. *Note:* according to the likelihood ration test, the bottom observation vector **x** is incorrectly assigned to Z_1.

dimension corresponds to the ensemble of channel outputs. Finally, the receiver performs the decision rule.

Likelihood Receiver

To proceed with the solution to the binary hypothesis-testing problem, we introduce the following notations:

1. $f_{\mathbf{X}|H_0}(\mathbf{x}|H_0)$, which denotes the conditional density of the observation vector $\mathbf{x}$ given that hypothesis H_0 is true.
2. $f_{\mathbf{X}|H_1}(\mathbf{x}|H_1)$, denotes the conditional density of $\mathbf{x}$ given that the other hypothesis H_1 is true.
3. π_0 and π_1 denote the priors of hypotheses H_0 and H_1, respectively.

In the context of hypothesis testing, the two conditional probability density functions $f_{\mathbf{X}|H_0}(\mathbf{x}|H_0)$ and $f_{\mathbf{X}|H_1}(\mathbf{x}|H_1)$ are referred to as *likelihood functions*, or just simply *likelihoods*.

Suppose we perform a measurement on the transition mechanism's output, obtaining the observation vector $\mathbf{x}$. In processing $\mathbf{x}$, there are two kinds of errors that can be made by the decision rule:

1. *Error of the first kind*. This arises when hypothesis H_0 is true but the rule makes a decision in favor of H_1, as illustrated in Figure 3.14.
2. *Error of the second kind*. This arises when hypothesis H_1 is true but the rule makes a decision in favor of H_0.

The conditional probability of an error of the first kind is

$$\int_{Z_1} f_{\mathbf{X}|H_0}(\mathbf{x}|H_0)\,d\mathbf{x}$$

where Z_1 is part of the observation space that corresponds to hypothesis H_1. Similarly, the conditional probability of an error of the second kind is

$$\int_{Z_0} f_{\mathbf{X}|H_1}(\mathbf{x}|H_1)\,d\mathbf{x}$$

By definition, an *optimum* decision rule is one for which a prescribed *cost function* is minimized. A logical choice for the cost function in digital communications is the *average probability of symbol error*, which, in a Bayesian context, is referred to as the *Bayes risk*. Thus, with the probable occurrence of the two kinds of errors identified above, we define the Bayes risk for the binary hypothesis-testing problem as

$$\mathcal{R} = \pi_0 \int_{Z_1} f_{\mathbf{X}|H_0}(\mathbf{x}|H_0)\,d\mathbf{x} + \pi_1 \int_{Z_0} f_{\mathbf{X}|H_1}(\mathbf{x}|H_1)\,d\mathbf{x} \tag{3.88}$$

where we have accounted for the prior probabilities for which hypotheses H_0 and H_1 are known to occur. Using the language of set theory, let the union of the disjoint subspaces Z_0 and Z_1 be

$$Z = Z_0 \cup Z_1 \tag{3.89}$$

Then, recognizing that the subspace Z_1 is the complement of the subspace Z_0 with respect to the total observation space Z, we may rewrite (3.88) in the equivalent form:

$$\begin{aligned}\mathcal{R} &= \pi_0 \int_{Z-Z_0} f_{\mathbf{X}|H_0}(\mathbf{x}|H_0)\,d\mathbf{x} + \pi_1 \int_{Z_0} f_{\mathbf{X}|H_1}(\mathbf{x}|H_1)\,d\mathbf{x} \\ &= \pi_0 \int_{Z} f_{\mathbf{X}|H_0}(\mathbf{x}|H_0)\,d\mathbf{x} + \int_{Z_0} [\pi_1 f_{\mathbf{X}|H_1}(\mathbf{x}|H_1) - \pi_0 f_{\mathbf{X}|H_0}(\mathbf{x}|H_0)]\,d\mathbf{x}\end{aligned} \tag{3.90}$$

The integral $\int_Z f_{\mathbf{X}|H_0}(\mathbf{x}|H_0)\,d\mathbf{x}$ represents the total volume under the conditional density $f_{\mathbf{X}|H_0}(\mathbf{x}|H_0)$, which, by definition, equals unity. Accordingly, we may reduce (3.90) to

$$\mathcal{R} = \pi_0 + \int_{Z_0} [\pi_1 f_{\mathbf{X}|H_1}(\mathbf{x}|H_1)\,d\mathbf{x} - \pi_0 f_{\mathbf{X}|H_0}(\mathbf{x}|H_0)]\,d\mathbf{x} \tag{3.91}$$

The term π_0 on its own on the right-hand side of (3.91) represents a *fixed* cost. The integral term represents the cost controlled by how we assign the observation vector $\mathbf{x}$ to Z_0. Recognizing that the two terms inside the square brackets are both positive, we must therefore insist on the following plan of action for the average risk $\mathcal{R}$ to be minimized:

> Make the integrand in (3.91) negative for the observation vector $\mathbf{x}$ to be assigned to Z_0.

In light of this statement, the optimum decision rule proceeds as follows:

1. If

$$\pi_0 f_{\mathbf{X}|H_0}(\mathbf{x}|H_0) > \pi_1 f_{\mathbf{X}|H_1}(\mathbf{x}|H_1)$$

then the observation vector $\mathbf{x}$ should be assigned to Z_0, because these two terms contribute a negative amount to the integral in (3.91). In this case, we say H_0 is true.

2. If, on the other hand,

$$\pi_0 f_{\mathbf{X}|H_0}(\mathbf{x}|H_0) < \pi_1 f_{\mathbf{X}|H_1}(\mathbf{x}|H_1)$$

then the observation vector $\mathbf{x}$ should be excluded from Z_0 (i.e., assigned to Z_1), because these two terms would contribute a positive amount to the integral in (3.91). In this second case, H_1 is true.

When the two terms are equal, the integral would clearly have no effect on the average risk $\mathcal{R}$; in such a situation, the observation vector $\mathbf{x}$ may be assigned arbitrarily.

Thus, combining points (1) and (2) on the action plan into a *single* decision rule, we may write

$$\frac{f_{\mathbf{X}|H_1}(\mathbf{x}|H_1)}{f_{\mathbf{X}|H_0}(\mathbf{x}|H_0)} \underset{H_0}{\overset{H_1}{\gtrless}} \frac{\pi_0}{\pi_1} \tag{3.92}$$

The observation-dependent quantity on the left-hand side of (3.92) is called the *likelihood ratio*; it is defined by

$$\Lambda(\mathbf{x}) = \frac{f_{\mathbf{X}|H_1}(\mathbf{x}|H_1)}{f_{\mathbf{X}|H_0}(\mathbf{x}|H_0)} \tag{3.93}$$

From this definition, we see that $\Lambda(\mathbf{x})$ is the ratio of two functions of a random variable; therefore, it follows that $\Lambda(\mathbf{x})$ is itself a random variable. Moreover, it is a one-dimensional variable, which holds regardless of the dimensionality of the observation vector $\mathbf{x}$. Most importantly, the likelihood ratio is a sufficient statistic.

The scalar quantity on the right-hand side of (3.92), namely,

$$\eta = \frac{\pi_0}{\pi_1} \tag{3.94}$$

is called the *threshold* of the test. Thus, minimization of the Bayes risk $\mathcal{R}$ leads to the *likelihood ratio test*, described by the combined form of two decisions:

$$\Lambda(\mathbf{x}) \underset{H_0}{\overset{H_1}{\gtrless}} \eta \tag{3.95}$$

Correspondingly, the hypothesis testing structure built on (3.93)–(3.95) is called the *likelihood receiver;* it is shown in the form of a block diagram in Figure 3.15a. An elegant characteristic of this receiver is that all the necessary data processing is confined to computing the likelihood ratio $\Lambda(\mathbf{x})$. This characteristic is of considerable practical importance: adjustments to our knowledge of the priors π_0 and π_1 are made simply through the assignment of an appropriate value to the threshold η.

The natural logarithm is known to be a monotone function of its argument. Moreover, both sides of the likelihood ratio test in (3.95) are positive. Accordingly, we may express the test in its *logarithmic* form, as shown by

$$\ln \Lambda(\mathbf{x}) \underset{H_0}{\overset{H_1}{\gtrless}} \ln \eta \tag{3.96}$$

where ln is the symbol for the natural logarithm. Equation (3.96) leads to the equivalent *log-likelihood ratio receiver*, depicted in Figure 3.15b.

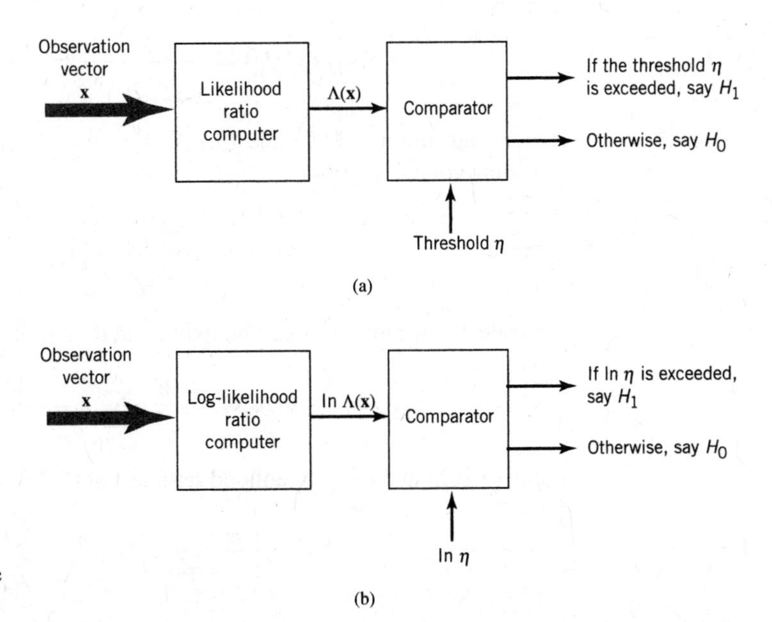

Figure 3.15 Two versions of the likelihood receiver: (a) based on the likelihood ratio $\Lambda(\mathbf{x})$; (b) based on the log-likelihood ratio $\ln \Lambda(\mathbf{x})$.

EXAMPLE 8 **Binary Hypothesis Testing**

Consider a binary hypothesis testing problem, described by the pair of equations:

$$\begin{aligned} &\text{Hypothesis } H_1\text{: } x_i = m + n_i, \quad i = 1, 2, \ldots, N \\ &\text{Hypothesis } H_0\text{: } x_i = n_i, \quad i = 1, 2, \ldots, N \end{aligned} \tag{3.97}$$

The term m is a constant that is nonzero only under hypothesis H_1. As in Example 7, the n_i are independent and Gaussian $\mathcal{N}(0, \sigma_n^2)$. The requirement is to formulate a likelihood ratio test for this example to come up with a decision rule.

Following the discussion presented in Example 7, under hypothesis H_1 we write

$$f_{x_i|H_1}(x_i|H_1) = \frac{1}{\sqrt{2\pi}\sigma_n}\exp\left[-\frac{(x_i - m)^2}{2\sigma_n^2}\right] \tag{3.98}$$

As in Example 7, let the vector $\mathbf{x}$ denote the set of N observations x_i for $i = 1, 2, \ldots, N$. Then, invoking the independence of the n_i, we may express the joint density of the x_i under hypothesis H_1 as

$$\begin{aligned} f_{\mathbf{X}|H_1}(\mathbf{x}|H_1) &= \prod_{i=1}^{N}\frac{1}{\sqrt{2\pi}\sigma_n}\exp\left[-\frac{(x_i - m)^2}{2\sigma_n^2}\right] \\ &= \frac{1}{(\sqrt{2\pi}\sigma_n)^N}\exp\left[-\frac{1}{2\sigma_n^2}\sum_{i=1}^{N}(x_i - m)^2\right] \end{aligned} \tag{3.99}$$

Setting m to zero in (3.99), we get the corresponding joint density of the x_i under hypothesis H_0 as

$$f_{\mathbf{X}|H_0}(\mathbf{x}|H_0) = \frac{1}{(\sqrt{2\pi}\sigma_n)^N}\exp\left(-\frac{1}{2\sigma_n^2}\sum_{i=1}^{N}x_i^2\right) \tag{3.100}$$

Hence, substituting (3.99) and (3.100) into the likelihood ratio of (3.93), we get (after canceling common terms)

$$\Lambda(\mathbf{x}) = \exp\left(\frac{m}{\sigma_n^2}\sum_{i=1}^{N}x_i - \frac{Nm^2}{2\sigma_n^2}\right) \tag{3.101}$$

Equivalently, we may express the likelihood ratio in its logarithmic form

$$\ln\Lambda(\mathbf{x}) = \frac{m}{\sigma_n^2}\sum_{i=1}^{N}x_i - \frac{Nm^2}{2\sigma_n^2} \tag{3.102}$$

Using (3.102) in the log-likelihood ratio test of (3.96), we get

$$\left(\frac{m}{\sigma_n^2}\sum_{i=1}^{N}x_i - \frac{Nm^2}{2\sigma_n^2}\right) \underset{H_0}{\overset{H_1}{\gtrless}} \ln\eta$$

Dividing both sides of this test by (m/σ_n^2) and rearranging terms, we finally write

$$\sum_{i=1}^{N} x_i \underset{H_0}{\overset{H_1}{\gtrless}} \left(\frac{\sigma_n^2}{m} \ln \eta + \frac{Nm^2}{2}\right) \tag{3.103}$$

where the threshold η is itself defined by the ratio of priors, namely π_0/π_1. Equation (3.103) is the desired formula for the decision rule to solve the binary hypothesis-testing problem of (3.97).

One last comment is in order. As with Example 7, the sum of the x_i over the N observations; that is,

$$t(\mathbf{x}) = \sum_{i=1}^{N} x_i$$

is a sufficient statistic for the problem at hand. We say so because the only way in which the observations can enter the likelihood ratio $\Lambda(\mathbf{x})$ is in the sum; see (3.101).

Multiple Hypotheses

Now that we understand binary hypothesis testing, we are ready to consider the more general scenario where we have M possible source outputs to deal with. As before, we assume that a decision must be made as to which one of the M possible source outputs was actually emitted, given an observation vector $\mathbf{x}$.

To develop insight into how to construct a decision rule for testing multiple hypotheses, we consider first the case of $M = 3$ and then generalize the result. Moreover, in formulating the decision rule, we will use *probabilistic reasoning* that builds on the findings of the binary hypothesis-testing procedure. In this context, however, we find it more convenient to work with likelihood functions rather than likelihood ratios.

To proceed then, suppose we make a measurement on the probabilistic transition mechanism's output, obtaining the observation vector $\mathbf{x}$. We use this observation vector and knowledge of the probability law characterizing the transition mechanism to construct three likelihood functions, one for each of the three possible hypotheses. For the sake of illustrating what we have in mind, suppose further that in formulating the three possible probabilistic inequalities, each with its own inference, we get the following three results:

1. $\pi_1 f_{\mathbf{X}|H_1}(\mathbf{x}|H_1) < \pi_0 f_{\mathbf{X}|H_0}(\mathbf{x}|H_0)$
   from which we infer that hypothesis H_0 or H_2 is true.
2. $\pi_2 f_{\mathbf{X}|H_2}(\mathbf{x}|H_2) < \pi_0 f_{\mathbf{X}|H_0}(\mathbf{x}|H_0)$
   from which we infer that hypothesis H_0 or H_1 is true.
3. $\pi_2 f_{\mathbf{X}|H_2}(\mathbf{x}|H_2) < \pi_1 f_{\mathbf{X}|H_1}(\mathbf{x}|H_1)$
   from which we infer that hypothesis H_1 or H_0 is true.

Examining these three possible results for $M = 3$, we immediately see that hypothesis H_0 is the only one that shows up in all three inferences. Accordingly, for the particular scenario we have picked, the decision rule *should* say that hypothesis H_0 is true. Moreover, it is a straightforward matter for us to make similar statements pertaining to hypothesis H_1

or H_2. The rationale just described for arriving at this test is an *example* of what we mean by probabilistic reasoning: the use of multiple inferences to reach a specific decision.

For an equivalent test, let both sides of each inequality under points 1, 2, and 3 be divided by the evidence $f_{\mathbf{X}}(\mathbf{x})$. Let H_i, $i = 1, 2, 3$, denote the three hypotheses. We may then use the definition of joint probability density function to write

$$\begin{aligned}
\frac{\pi_i f_{\mathbf{X}|H_i}(\mathbf{x}|H_i)}{f_{\mathbf{X}}(\mathbf{x})} &= \frac{\mathbb{P}(H_i) f_{\mathbf{X}|H_i}(\mathbf{x}|H_i)}{f_{\mathbf{X}}(\mathbf{x})} \qquad \text{where } \mathbb{P}(H_i) = p_i \\
&= \frac{\mathbb{P}(H_i, \mathbf{x})}{f_{\mathbf{X}}(\mathbf{x})} \\
&= \frac{\mathbb{P}[H_i|\mathbf{x}] f_{\mathbf{X}}(\mathbf{x})}{f_{\mathbf{X}}(\mathbf{x})} \\
&= \mathbb{P}[H_i|\mathbf{x}] \qquad \text{for } i = 0, 1, \ldots, M-1
\end{aligned} \tag{3.104}$$

Hence, recognizing that the conditional probability $\mathbb{P}[H_i|\mathbf{x}]$ is actually the posterior probability of hypothesis H_i *after* receiving the observation vector $\mathbf{x}$, we may now go on to generalize the equivalent test for M possible source outputs as follows:

> Given an observation vector $\mathbf{x}$ in a multiple hypothesis test, the average probability of error is minimized by choosing the hypothesis H_i for which the posterior probability $\mathbb{P}[H_i|\mathbf{x}]$ has the largest value for $i = 0, 1, \ldots, M-1$.

A processor based on this decision rule is frequently referred to as the *MAP probability computer.* It is with this general hypothesis testing rule that earlier we made the supposition embodied under points 1, 2, and 3.

3.14 Composite Hypothesis Testing

Throughout the discussion presented in Section 3.13, the hypotheses considered therein were all *simple*, in that the probability density function for each hypothesis was completely specified. However, in practice, it is common to find that one or more of the probability density functions are *not* simple due to imperfections in the probabilistic transition mechanism. In situations of this kind, the hypotheses are said to be *composite*.

As an illustrative example, let us revisit the binary hypothesis-testing problem considered in Example 8. This time, however, we treat the mean m of the observable x_i under hypothesis H_1 not as a constant, but as a variable inside some interval $[m_a, m_b]$. If, then, we were to use the likelihood ratio test of (3.93) for simple binary hypothesis testing, we would find that the likelihood ratio $\Lambda(x_i)$ involves the unknown mean m. We cannot therefore compute $\Lambda(x_i)$, thereby negating applicability of the simple likelihood ratio test.

The message to take from this illustrative example is that we have to modify the likelihood ratio test to make it applicable to composite hypotheses. To this end, consider the model depicted in Figure 3.16, which is similar to that of Figure 3.14 for the simple case except for one difference: the transition mechanism is now characterized by the conditional probability density function $f_{\mathbf{X}|\boldsymbol{\Theta}, H_i}(\mathbf{x}|\boldsymbol{\theta}, H_i)$, where $\boldsymbol{\theta}$ is a realization of the unknown parameter vector $\boldsymbol{\Theta}$, and the index $i = 0, 1$. It is the conditional dependence on $\boldsymbol{\theta}$ that makes

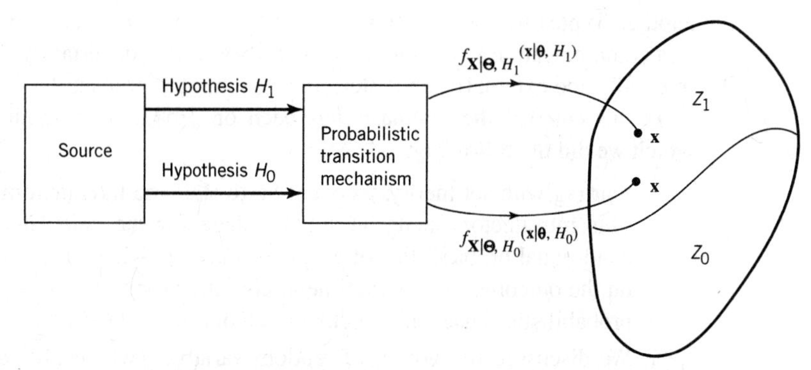

Figure 3.16 Model of composite hypothesis-testing for a binary scenario.

the hypotheses H_0 and H_1 to be of the composite kind. Unlike the simple model of Figure 3.14, we now have two spaces to deal with: an observation space and a parameter space. It is assumed that the conditional probability density function of the unknown parameter vector $\boldsymbol{\Theta}$, that is, $f_{\boldsymbol{\Theta}|H_i}(\boldsymbol{\theta}, H_i)$, is known for $i = 0, 1$.

To formulate the likelihood ratio for the composite hypotheses described in the model of Figure 3.16, we require the likelihood function $f_{\mathbf{X}|H_i}(\mathbf{x}|H_i)$ for $i = 1, 2$. We may satisfy this requirement by reducing the composite hypothesis-testing problem to a simple one by integrating over $\boldsymbol{\theta}$, as shown by

$$f_{\mathbf{X}|H_i}(\mathbf{x}|H_i) = \int_{\boldsymbol{\Theta}} f_{\mathbf{X}|\boldsymbol{\Theta}, H_i}(\mathbf{x}|\boldsymbol{\theta}, H_i) f_{\boldsymbol{\Theta}|H_i}(\boldsymbol{\theta}|H_i)\, d\boldsymbol{\theta} \tag{3.105}$$

the evaluation of which is contingent on knowing the conditional probability density function of $\boldsymbol{\theta}$ given the H_i for $i = 1, 2$. With this specification at hand, we may now formulate the likelihood ratio for composite hypotheses as

$$\Lambda(\mathbf{x}) = \frac{\int_{\boldsymbol{\Theta}} f_{\mathbf{X}|\boldsymbol{\Theta}, H_1}(\mathbf{x}|\boldsymbol{\theta}, H_1) f_{\boldsymbol{\Theta}|H_1}(\boldsymbol{\theta}|H_1)\, d\boldsymbol{\theta}}{\int_{\boldsymbol{\Theta}} f_{\mathbf{X}|\boldsymbol{\Theta}, H_0}(\mathbf{x}|\boldsymbol{\theta}, H_0) f_{\boldsymbol{\Theta}|H_0}(\boldsymbol{\theta}|H_0)\, d\boldsymbol{\theta}} \tag{3.106}$$

Accordingly, we may now extend applicability of the likelihood ratio test described in (3.95) to composite hypotheses.

From this discussion, it is clearly apparent that hypothesis testing for composite hypotheses is computationally more demanding than it is for simple hypotheses. Chapter 7 presents applications of composite hypothesis testing to noncoherent detection, in the course of which the phase information in the received signal is accounted for.

3.15 Summary and Discussion

The material presented in this chapter on probability theory is another mathematical pillar in the study of communication systems. Herein, the emphasis has been on how to deal with *uncertainty*, which is a natural feature of every communication system in one form or another. Typically, uncertainties affect the behavior of channels connecting the transmitter of a communication system to its receiver. Sources of uncertainty include noise, generated internally and externally, and interference from other transmitters.

In this chapter, the emphasis has been on probabilistic modeling, in the context of which we did the following:

1. Starting with set theory, we went on to state the three axioms of probability theory. This introductory material set the stage for the calculation of probabilities and conditional probabilities of events of interest. When partial information is available on the outcome of an experiment, conditional probabilities permit us to reason in a probabilistic sense and thereby enrich our understanding of a random experiment.
2. We discussed the notion of random variables, which provide the natural tools for formulating probabilistic models of random experiments. In particular, we characterized continuous random variables in terms of the cumulative distribution function and probability density function; the latter contains all the conceivable information about a random variable. Through focusing on the mean of a random variable, we studied the expectation or averaging operator, which occupies a dominant role in probability theory. The mean and the variance, considered in that order, provide a weak characterization of a random variable. We also introduced the characteristic function as another way of describing the statistics of a random variable. Although much of the material in the early part of the chapter focused on continuous random variables, we did emphasize important aspects of discrete random variables by describing the concept of the probability mass function (unique to discrete random variables) and the parallel development and similar concepts that embody these two kinds of random variables.
3. Table 3.2 on page 135 summarizes the probabilistic descriptions of some important random variances under two headings: discrete and random. Except for the Rayleigh random variable, these random variables were discussed in the text or are given as end-of-chapter problems; the Rayleigh random variable is discussed in Chapter 4. Appendix A presents advanced probabilistic models that go beyond the contents of Table 3.2.
4. We discussed the characterization of a pair of random variables and introduced the basic concepts of covariance and correlation, and the independence of random variables.
5. We provided a detailed description of the Gaussian distribution and discussed its important properties. Gaussian random variables play a key role in the study of communication systems.

The second part of the chapter focused on the Bayesian paradigm, wherein inference may take one of two forms:

- Probabilistic modeling, the aim of which is to develop a model for describing the physical behavior of an observation space.
- Statistical analysis, the aim of which is the inverse of probabilistic modeling.

In a fundamental sense, statistical analysis is more profound than probabilistic modeling, hence the focused attention on it in the chapter.

Table 3.2 **Some important random variables**

Discrete random variables

1. Bernoulli

$$p_X(x) = \begin{cases} 1-p & \text{if } x = 0 \\ p & \text{if } x = 1 \\ 0 & \text{otherwise} \end{cases}$$

$$\mathbb{E}[X] = p$$

$$\text{var}[X] = p(1-p)$$

2. Poisson

$$p_X(k) = \frac{\lambda^k}{k!}\exp(-\lambda), \qquad k = 0, 1, 2, \ldots, \text{ and } \lambda > 0$$

$$\mathbb{E}[X] = \lambda$$

$$\text{var}[X] = \lambda$$

Continuous random variables

1. Uniform

$$f_X(x) = \frac{1}{b-a}, \qquad a \le x \le b$$

$$\mathbb{E}[X] = \frac{1}{2}(a+b)$$

$$\text{var}[X] = \frac{1}{12}(b-a)^2$$

2. Exponential

$$f_X(x) = \lambda \exp(-\lambda x), \qquad x \ge 0 \text{ and } \lambda > 0$$

$$\mathbb{E}[X] = 1/\lambda$$

$$\text{var}[X^2] = 1/\lambda^2$$

3. Gaussian

$$f_X(x) = \frac{1}{\sqrt{2\pi}\sigma}\exp[-(x-\mu)^2/2\sigma^2], \qquad -\infty < x < \infty$$

$$\mathbb{E}[X] = \mu$$

$$\text{var}[X] = \sigma^2$$

4. Rayleigh

$$f_X(x) = \frac{x}{\sigma^2}\exp(-x^2/2\sigma^2), \qquad x \ge 0 \text{ and } \sigma > 0$$

$$\mathbb{E}[X] = \sigma\sqrt{\pi/2}$$

$$\text{var}[X] = \left(2 - \frac{\pi}{2}\right)\sigma^2$$

5. Laplacian

$$f_X(x) = \frac{\lambda}{2}\exp(-\lambda|x|), \qquad -\infty < x < \infty \text{ and } \lambda > 0$$

$$\mathbb{E}[X] = 0$$

$$\text{var}[X] = 2/\lambda^2$$

Under statistical analysis, viewed from a digital communications perspective, we discussed the following:

1. Parameter estimation, where the requirement is to estimate an unknown parameter given an observation vector; herein we covered:
   - the maximum a posteriori (MAP) rule that requires prior information, and
   - the maximum likelihood procedure that by-passes the need for the prior and therefore sits on the fringe of the Bayesian paradigm.
2. Hypothesis testing, where in a simple but important scenario, we have two hypotheses to deal with, namely H_1 and H_0. In this case, the requirement is to make an optimal decision in favor of hypothesis H_1 or hypothesis H_0 given an observation vector. The likelihood ratio test plays the key role here.

To summarize, the material on probability theory sets the stage for the study of stochastic processes in Chapter 4. On the other hand, the material on Bayesian inference plays a key role in Chapters 7, 8, and 9 in one form or another.

Problems

Set Theory

3.1 Using Venn diagrams, justify the five properties of the algebra of sets, which were stated (without proofs) in Section 3.1:

a. idempotence property

b. commutative property

c. associative property

d. distributive property

e. De Morgan's laws.

3.2 Let A and B denote two different sets. Validate the following three equalities:

a. $A^c = (A^c \cap B) \cup (A^c \cap B^c)$

b. $B^c = (A \cap B^c) \cup (A^c \cap B^c)$

c. $(A \cap B)^c = (A^c \cap B) \cup (A^c \cap B^c) \cup (A \cap B^c)$

Probability Theory

3.3 Using the Bernoulli distribution of Table 3.2, develop an experiment that involves three independent tosses of a fair coin. Irrespective of whether the toss is a head or tail, the probability of every toss is to be conditioned on the results of preceding tosses. Display graphically the sequential evolution of the results.

3.4 Use Bayes' rule to convert the conditioning of event B given event A_i into the conditioning of event A_i given event B for the $i = 1, 2, ..., N$.

3.5 A discrete memoryless channel is used to transmit binary data. The channel is discrete in that it is designed to handle discrete messages and it is memoryless in that at any instant of time the channel output depends on the channel input only at that time. Owing to the unavoidable presence of noise in the channel, errors are made in the received binary data stream. The channel is symmetric in that the probability of receiving symbol 1 when symbol 0 is sent is the same as the probability of receiving symbol 0 when symbol 1 is sent.

The transmitter sends 0s across the channel with probability p_0 and 1s with probability p_1. The receiver occasionally makes random decision errors with probability p; that is, when symbol 0 is sent across the channel, the receiver makes a decision in favor of symbol 1, and vice versa.

Referring to Figure P3.5, determine the following a posteriori probabilities:

a. The conditional probability of sending symbol A_0 given that symbol B_0 was received.

b. The conditional probability of sending symbol A_1 given that symbol B_1 was received.

Hint: Formulate expressions for the probability of receiving event B_0, and likewise for event B_1.

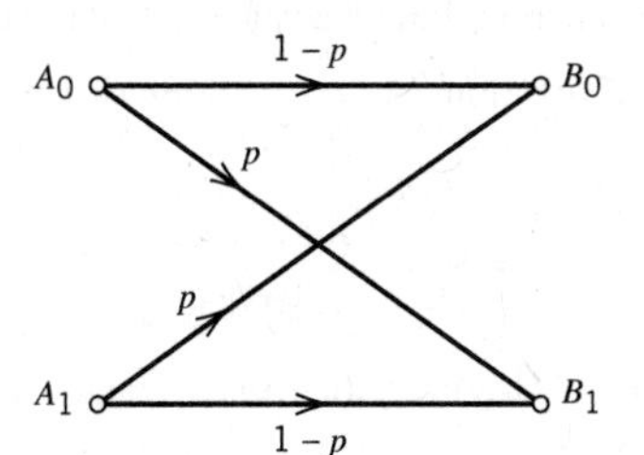

Figure P3.5

3.6 Let $B_1, B_2, \ldots, B_n$ denote a set of joint events whose union equals the sample space S, and assume that $\mathbb{P}[B_i] > 0$ for all i. Let A be any event in the sample space S.

a. Show that

$$A = (A \cap B_1) \cup (A \cap B_2) \cup \ldots \ (A \cap B_n)$$

b. The *total probability theorem* states:

$$\mathbb{P}[A] = \mathbb{P}[A|B_1]\mathbb{P}[B_1] + \mathbb{P}[A|B_2]\mathbb{P}[B_2] + \ldots + \mathbb{P}[A|B_n]\mathbb{P}[B_n]$$

This theorem is useful for finding the probability of event B when the conditional probabilities $\mathbb{P}[A|B_i]$ are known or easy to find for all i. Justify the theorem.

3.7 Figure P3.7 shows the connectivity diagram of a computer network that connects node A to node B along different possible paths. The labeled branches of the diagram display the probabilities for which the links in the network are up; for example, 0.8 is the probability that the link from node A to intermediate node C is up, and so on for the other links. Link failures in the network are assumed to be independent of each other.

a. When all the links in the network are up, find the probability that there is a path connecting node A to node B.

b. What is the probability of complete failure in the network, with no connection from node A to node B?

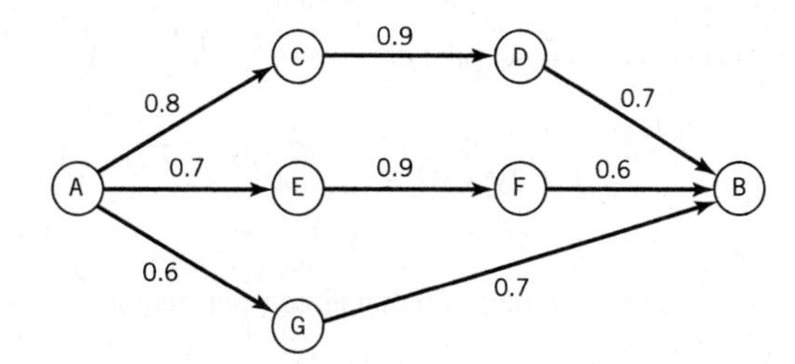

Figure P3.7

Distribution Functions

3.8 The probability density function of a continuous random variable X is defined by

$$f_X(x) = \begin{cases} \frac{c}{\sqrt{x}} & \text{for } 0 \le x \le 1 \\ 0 & \text{otherwise} \end{cases}$$

Despite the fact that this function becomes infinitely large as x approaches zero, it may qualify to be a legitimate probability density function. Find the value of scalar c for which this condition is satisfied.

3.9 The joint probability density function of two random variables X and Y is defined by the *two-dimensional uniform distribution*

$$f_{X,Y}(x, y) = \begin{cases} c & \text{for } a \le x \le b \text{ and } a \le y \le b \\ 0 & \text{otherwise} \end{cases}$$

Find the scalar c for which $f_{X,Y}(x,y)$ satisfies the normalization property of a two-dimensional probability density function.

3.10 In Table 3.2, the probability density function of a Rayleigh random variable is defined by

$$f_X(x) = \frac{x}{\sigma^2}\exp\left(-\frac{x^2}{2\sigma^2}\right) \qquad \text{for } x \ge 0 \text{ and } \sigma > 0$$

a. Show that the mean of X is

$$\mathbb{E}[X] = \sigma\sqrt{\frac{\pi}{2}}$$

b. Using the result of part a, show that the variance of X is

$$\text{var}[X] = \left(2 - \frac{\pi}{2}\right)\sigma^2$$

c. Use the results of a and b to determine the Rayleigh cumulative distribution function.

3.11 The probability density function of an exponentially distributed random variable X is defined by

$$f_X(x) = \begin{cases} \lambda\exp(-\lambda x), & \text{for } 0 \le x \le \infty \\ 0, & \text{otherwise} \end{cases}$$

where λ is a positive parameter.

a. Show that $f_X(x)$ is a legitimate probability density function.

b. Determine the cumulative distribution function of X.

3.12 Consider the one-sided conditional exponential distribution

$$f_X(x|\lambda) = \begin{cases} \frac{\lambda}{Z(\lambda)}\exp(-\lambda x), & 1 \le x \le 20 \\ 0, & \text{otherwise} \end{cases}$$

where $\lambda > 0$ and $Z(\lambda)$ is the normalizing constant required to make the area under $f_X(x|\lambda)$ equal unity.

a. Determine the normalizing constant $Z(\lambda)$.

b. Given N independent values of x, namely $x_1, x_2, \ldots, x_N$, use Bayes' rule to formulate the conditional probability density function of the parameter λ, given this data set.

Expectation Operator

3.13 In Section 3.6 we described two properties of the expectation operator $\mathbb{E}$, one on linearity and the other on statistical independence. In this problem, we address two other important properties of the expectation operator.

a. *Scaling property:* Show that

$$\mathbb{E}(ax) = a\mathbb{E}[X]$$

where a is a constant scaling factor.

b. *Linearity of conditional expectation:* Show that

$$\mathbb{E}[X_1 + X_2|Y] = \mathbb{E}[X_1|Y] + \mathbb{E}[X_2|Y]$$

3.14 Validate the expected value rule of (3.41) by building on two expressions:

a. $g(x) = \max[g(x), 0] - \max[-g(x), 0]$

b. For any $a \geq 0$, $g(x) > a$ provided that $\max[g(x), 0] > a$

3.15 Let X be a discrete random variable with probability mass function $p_X(x)$ and let $g(X)$ be a function of the random variable X. Prove the following rule:

$$\mathbb{E}[g(X)] = \sum_x g(x)p_X(x)$$

where the summation is over all possible discrete values of X.

3.16 Continuing with the Bernoulli random variable X in (3.23), find the mean and variance of X.

3.17 The mass probability function of the *Poisson random variable* X is defined by

$$p_X(k) = \frac{1}{k!}\lambda^k \exp(-\lambda), \qquad k = 0, 1, 2, \ldots, \text{ and } \lambda > 0$$

Find the mean and variance of X.

3.18 Find the mean and variance of the exponentially distributed random variable X in Problem 3.11.

3.19 The probability density function of the *Laplacian random variable* X in Table 3.2 is defined by

$$f_X(x) = \begin{cases} \frac{1}{2}\lambda \exp(-\lambda x) & \text{for } x \geq 0 \\ \frac{1}{2}\lambda \exp(\lambda x) & \text{for } x < 0 \end{cases}$$

for the parameter $\lambda > 0$. Find the mean and variance of X.

3.20 In Example 5 we used the characteristic function $\Phi(j\nu)$ to calculate the mean of an exponentially distributed random variable X. Continuing with that example, calculate the variance of X and check your result against that found in Problem 3.18.

3.21 The characteristic function of a continuous random variable X, denoted by $\Phi(\nu)$, has some important properties of its own:

a. The transformed version of the random variable X, namely, $aX + b$, has the following characteristic function

$$\mathbb{E}[\exp(j\nu(aX + b))] = \exp(jb\nu) \cdot \Phi_X(a\nu)$$

where a and b are constants.

b. The characteristic function $\Phi(\nu)$ is real if, and only if, the distribution function $F_X(x)$, pertaining to the random variable X, is symmetric.

Prove the validity of these two properties, and demonstrate that property b is satisfied by the two-sided exponential distribution described in Problem 3.19.

3.22 Let X and Y be two continuous random variables. One version of the *total expectation theorem* states

$$\mathbb{E}[X] = \int_{-\infty}^{\infty} \mathbb{E}[X|Y = y] f_Y(y)\,dy$$

Justify this theorem.

Inequalities and Theorems

3.23 Let X be a continuous random variable that can only assume nonnegative values. The *Markov inequality* states

$$\mathbb{P}[X \geq a] \leq \frac{1}{a}\mathbb{E}[X], \qquad a > 0$$

Justify this inequality.

3.24 In (3.46) we stated the Chebyshev inequality without proof. Justify this inequality.
Hint: consider the probability $\mathbb{P}[(X-\mu)^2 \geq \varepsilon^2]$ and then apply the Markov inequality, considered in Problem 3.23, with $a = \varepsilon^2$.

3.25 Consider a sequence $X_1, X_2, ..., X_n$ of independent and identically distributed random variables with mean μ and variance σ^2. The sample mean of this sequence is defined by

$$M_n = \frac{1}{n}\sum_{i=1}^{n} X_i$$

The *weak law of large numbers* states

$$\lim_{n \to \infty} \mathbb{P}[|M_n - \mu| < \varepsilon] = 0 \qquad \text{for } \varepsilon > 0$$

Justify this law. *Hint:* use the Chebyshev inequality.

3.26 Let event A denote one of the possible outcomes of a random experiment. Suppose that in n independent trials of the experiment the event A occurs n_A times. The ratio

$$M_n = \frac{n_A}{n}$$

is called the *relative frequency* or *empirical frequency* of the event A. Let $p = \mathbb{P}[A]$ denote the probability of the event A. The experiment is said to exhibit "statistical regularity" if the relative frequency M_n is most likely to be within ε of p for large n. Use the weak law of large numbers, considered in Problem 3.25, to justify this statement.

The Gaussian Distribution

3.27 In the literature on signaling over additive white Gaussian noise (AWGN) channels, formulas are derived for probabilistic error calculations using the *complementary error function*

$$\text{erfc}(x) = 1 - \frac{1}{\sqrt{\pi}}\int_0^x \exp(-t^2)\,dt$$

Show that the erfc(x) is related to the Q-function as follows

a. $Q(x) = \frac{1}{2}\,\text{erfc}\left(\frac{x}{\sqrt{2}}\right)$

b. $\text{erfc}(x) = 2Q(\sqrt{2}\,x)$

3.28 Equation (3.58) defines the probability density function of a Gaussian random variable X. Show that the area under this function is unity, in accordance with the normalization property described in (3.59).

3.29 Continuing with Problem3.28, justify the four properties of the Gaussian distribution stated in Section 3.8 without proofs.

3.30 a. Show that the characteristic function of a Gaussian random variable X of mean μ_X and variance σ_X^2 is

$$\phi_X(\nu) = \exp\left(j\nu\mu_X - \frac{1}{2}\nu^2\sigma_X^2\right)$$

b. Using the result of part a, show that the nth central moment of this Gaussian random variable is as follows:

$$\mathbb{E}[(X-\mu_X)^n] = \begin{cases} 1 \times 3 \times 5 \ldots\ (n-1)\sigma_X^n & \text{for } n \text{ even} \\ 0 & \text{for } n \text{ odd} \end{cases}$$

3.31 A Gaussian-distributed random variable X of zero mean and variance σ_X^2 is transformed by a piecewise-linear rectifier characterized by the input–output relation (see Figure P3.31):

$$Y = \begin{cases} X, & X \geq 0 \\ 0, & X < 0 \end{cases}$$

The probability density function of the new random variable Y is described by

$$f_Y(y) = \begin{cases} 0, & y < 0 \\ k\delta(y) & y = 0 \\ \dfrac{1}{\sqrt{2\pi}\sigma_X}\exp\left(-\dfrac{y^2}{2\sigma_X^2}\right) & y > 0 \end{cases}$$

a. Explain the physical reasons for the functional form of this result.

b. Determine the value of the constant k by which the delta function $\delta(y)$ is weighted.

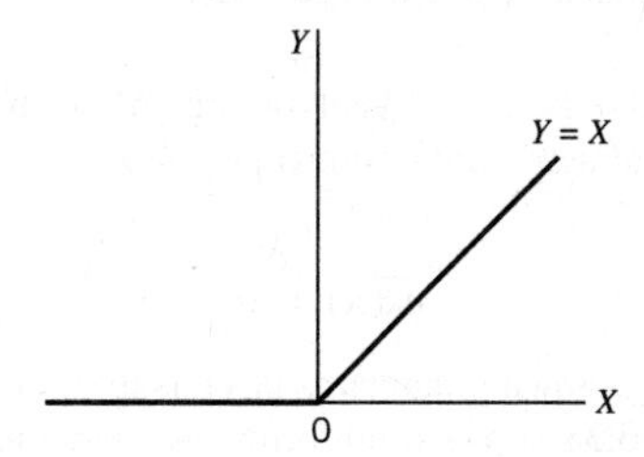

Figure P3.31

3.32 In Section 3.9 we stated the central limit theorem embodied in (3.71) without proof. Justify this theorem.

Bayesian Inference

3.33 Justify the likelihood principle stated (without proof) in Section 3.11.

3.34 In this problem we address a procedure for estimating the mean of the random variable; the procedure was discussed in Section 3.6.

Consider a Gaussian-distributed variable X with unknown mean μ_X and unit variance. The mean μ_X is itself a random variable, uniformly distributed over the interval $[a, b]$. To do the estimation, we are given N independent observations of the random variable X. Justify the estimator of (3.36).

3.35 In this problem, we address the issue of estimating the standard deviation σ of a Gaussian-distributed random variable X of zero mean. The standard deviation itself is uniformly distributed inside the interval $[\sigma_1, \sigma_2]$. For the estimation, we have N independent observations of the random variable X, namely, $x_1, x_2, ..., x_N$.

a. Derive a formula for the estimator $\hat{\sigma}$ using the MAP rule.
b. Repeat the estimation using the maximum likelihood criterion.
c. Comment on the results of parts a and b.

3.36 A binary symbol X is transmitted over a noisy channel. Specifically, symbol $X = 1$ is transmitted with probability p and symbol $X = 0$ is transmitted with probability $(1 - p)$. The received signals at the channel output are defined by

$$Y = X + N$$

The random variable N represents channel noise, modeled as a Gaussian-distributed random variable with zero mean and unit variance. The random variables X and N are independent.

a. Describe how the conditional probability $\mathbb{P}[X = 0|Y = y]$ varies with increasing y, all the way from $-\infty$ to $+\infty$.
b. Repeat the problem for the conditional probability $\mathbb{P}[X = 1|Y = y]$.

3.37 Consider an experiment involving the Poisson distribution, whose parameter λ is unknown. Given that the distribution of λ follows the exponential law

$$f_n(\lambda) = \begin{cases} a\exp(-a\lambda), & \lambda \geq 0 \\ 0, & \text{otherwise} \end{cases}$$

where $a > 0$, show that the MAP estimate of the parameter λ is given by

$$\hat{\lambda}_{\mathrm{MAP}}(k) = \frac{k}{1+a}$$

where k is the number of events used in the observation.

3.38 In this problem we investigate the use of analytic arguments to justify the optimality of the MAP estimate for the simple case of a one-dimensional parameter vector.

Define the estimation error

$$e_\theta(\mathbf{x}) = \theta - \hat{\theta}(\mathbf{x})$$

where θ is the value of an unknown parameter, $\hat{\theta}(\mathbf{x})$ is the estimator to be optimized, and $\mathbf{x}$ is the observation vector. Figure P3.38 shows a uniform cost function, $C(e)$, for this problem, with zero cost being incurred only when the absolute value of the estimation error $e_\theta(\mathbf{x})$ is less than or equal to $\Delta/2$.

a. Formulate the Bayes' risk $\mathcal{R}$ for this parameter estimation problem, accounting for the joint probability density function $f_{\mathbf{A},\mathbf{X}}(\theta, \mathbf{x})$.
b. Hence, determine the MAP estimate $\hat{\theta}_{\mathrm{MAP}}$ by minimizing the risk $\mathcal{R}$ with respect to $\hat{\theta}(\mathbf{x})$. For this minimization, assume that Δ is an arbitrarily small number but nonzero.

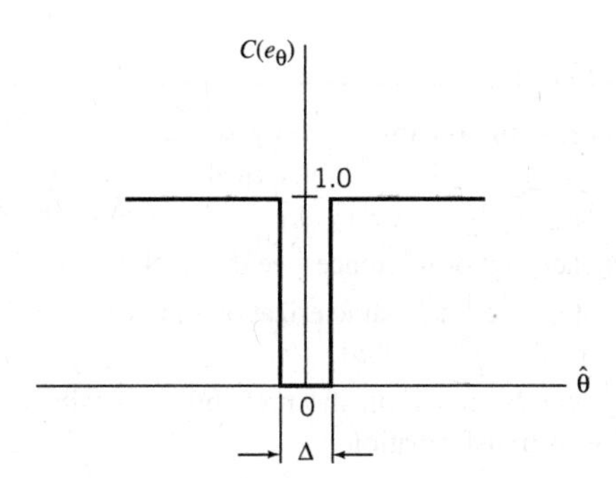

Figure P3.38

3.39 In this problem we generalize the likelihood ratio test for simple binary hypotheses by including costs incurred in the decision-making process. Let C_{ij} denote the cost incurred in deciding in favor of hypothesis H_i when hypothesis H_j is true. Hence, show that the likelihood ratio test of (3.95) still holds, except for the fact that the threshold of the test is now defined by

$$\eta = \frac{\pi_0(C_{10} - C_{00})}{\pi_1(C_{01} - C_{11})}$$

3.40 Consider a binary hypothesis-testing procedure where the two hypotheses H_0 and H_1 are described by different Poisson distributions, characterized by the parameters λ_0 and λ_1, respectively. The observation is simply a number of events k, depending on whether H_0 or H_1 is true. Specifically, for these two hypotheses, the probability mass functions are defined by

$$p_{X_i}(k) = \frac{(\lambda_i)^k}{k!}\exp(-\lambda_i), \qquad k = 0, 1, 2, \ldots,$$

where $i = 0$ for hypothesis H_0 and $i = 1$ for hypothesis H_1. Determine the log-likelihood ratio test for this problem.

3.41 Consider the binary hypothesis-testing problem

$$H_1 : X = M + N$$
$$H_0 : X = N$$

The M and N are independent exponentially distributed random variables, as shown by

$$p_M(m) = \begin{cases} \lambda_m \exp(-\lambda_m), & m \geq 0 \\ 0, & \text{otherwise} \end{cases}$$

$$p_N(n) = \begin{cases} \lambda_n \exp(-\lambda_n), & n \geq 0 \\ 0, & \text{otherwise} \end{cases}$$

Determine the likelihood ratio test for this problem.

3.42 In this problem we revisit Example 8. But this time we assume that the mean m under hypothesis H_1 is Gaussian distributed, as shown by

$$f_{M|H_1}(m|H_1) = \frac{1}{\sqrt{2\pi}\sigma_m}\exp\left(-\frac{m^2}{2\sigma_m^2}\right)$$

a. Derive the likelihood ratio test for the composite hypothesis scenario just described.

b. Compare your result with that derived in Example 8.

Notes

1. For a readable account of probability theory, see Bertsekas and Tsitsiklis (2008). For an advanced treatment of probability theory aimed at electrical engineering, see the book by Fine (2006). For an advanced treatment of probability theory, see the two-volume book by Feller (1968, 1971).

2. For an interesting account of inference, see the book by MacKay (2003).

3. For a detailed treatment of the characterization of discrete random variables, see Chapter 2 of the book by Bertsekas and Tsitsiklis (2008).

4. Indeed, we may readily transform the probability density function of (3.58) into the standard form by using the linear transformation

$$Y = \frac{1}{\sigma}(X - \mu)$$

In so doing, (3.58) is simplified as follows:

$$f_Y(y) = \frac{1}{\sqrt{2\pi}}\exp(-y^2/2)$$

which has exactly the same mathematical form as (3.65), except for the use of y in place of x.

5. Calculations based on Bayes' rule, presented previously as (3.14), are referred to as "Bayesian." In actual fact, Bayes provided a continuous version of the rule; see (3.72). In a historical context, it is also of interest to note that the full generality of (3.72) was not actually perceived by Bayes; rather, the task of generalization was left to Laplace.

6. It is because of this duality that the Bayesian paradigm is referred to as a *principle of duality;* see Robert (2001). Robert's book presents a detailed and readable treatment of the Bayesian paradigm. For a more advanced treatment of the subject, see Bernardo and Smith (1998).

7. In a paper published in 1912, R.A. Fisher moved away from the Bayesian approach. Then, in a classic paper published in 1922, he introduced the likelihood.

8. In Appendix B of their book, Bernardo and Smith (1998) show that many non-Bayesian inference procedures do not lead to identical inferences when applied to such proportional likelihoods.

9. For detailed discussion of the sufficient statistic, see Bernardo and Smith (1998).

10. A more detailed treatment of parameter-estimation theory is presented in the classic book by Van Trees (1968); the notation used by Van Trees is somewhat different from that used in this chapter. See also the book by McDonough and Whalen (1995).

11. For a more detailed treatment and readable account of hypothesis testing, see the classic book by Van Trees (1968). See also the book by McDonough and Whalen (1995).

CHAPTER

4

Stochastic Processes

4.1 Introduction

Stated in simple terms, we may say:

> A stochastic process is a set of random variables indexed in time.

Elaborating on this succinct statement, we find that in many of the real-life phenomena encountered in practice, *time* features prominently in their description. Moreover, their actual behavior has a random appearance. Referring back to the example of wireless communications briefly described in Section 3.1, we find that the received signal at the wireless channel output varies randomly with time. Processes of this kind are said to be *random* or *stochastic*;[1] hereafter, we will use the term "stochastic." Although probability theory does not involve time, the study of stochastic processes naturally builds on probability theory.

The way to think about the relationship between probability theory and stochastic processes is as follows. When we consider the statistical characterization of a stochastic process at a particular instant of time, we are basically dealing with the characterization of a *random variable* sampled (i.e., observed) at that instant of time. When, however, we consider a single realization of the process, we have a *random waveform* that evolves across time. The study of stochastic processes, therefore, embodies two approaches: one based on *ensemble averaging* and the other based on *temporal averaging*. Both approaches and their characterizations are considered in this chapter.

Although it is not possible to predict the exact value of a signal drawn from a stochastic process, it is possible to characterize the process in terms of *statistical parameters* such as average power, correlation functions, and power spectra. This chapter is devoted to the mathematical definitions, properties, and measurements of these functions, and related issues.

4.2 Mathematical Definition of a Stochastic Process

To summarize the introduction: stochastic processes have two properties. First, they are functions of time. Second, they are random in the sense that, before conducting an experiment, it is not possible to define the waveforms that will be observed in the future exactly.

In describing a stochastic process, it is convenient to think in terms of a sample space. Specifically, each realization of the process is associated with a *sample point*. The totality of sample points corresponding to the aggregate of all possible realizations of the stochastic process is called the *sample space*. Unlike the sample space in probability theory, each sample point of the sample space pertaining to a stochastic process is a function of time. We may therefore think of a stochastic process as the sample space or ensemble composed of functions of time. As an integral part of this way of thinking, we assume the existence of a probability distribution defined over an appropriate class of sets in the sample space, so that we may speak with confidence of the probability of various events observed at different points of time.[2]

Consider, then, a stochastic process specified by

a. outcomes s observed from some *sample space S;*

b. events defined on the sample space S; and

c. probabilities of these events.

Suppose that we assign to each sample point s a function of time in accordance with the rule

$$X(t, s), \qquad -T \le t \le T$$

where $2T$ is the *total observation interval.* For a fixed sample point s_j, the graph of the function $X(t, s_j)$ versus time t is called a *realization* or *sample function* of the stochastic process. To simplify the notation, we denote this sample function as

$$x_j(t) = X(t, s_j), \qquad -T \le t \le T \tag{4.1}$$

Figure 4.1 illustrates a set of sample functions $\{x_j(t) | j = 1, 2, \dots, n\}$. From this figure, we see that, for a fixed time t_k inside the observation interval, the set of numbers

$$\{x_1(t_k), x_2(t_k), \dots, x_n(t_k)\} = \{X(t_k, s_1), X(t_k, s_2), \dots, X(t_k, s_n)\}$$

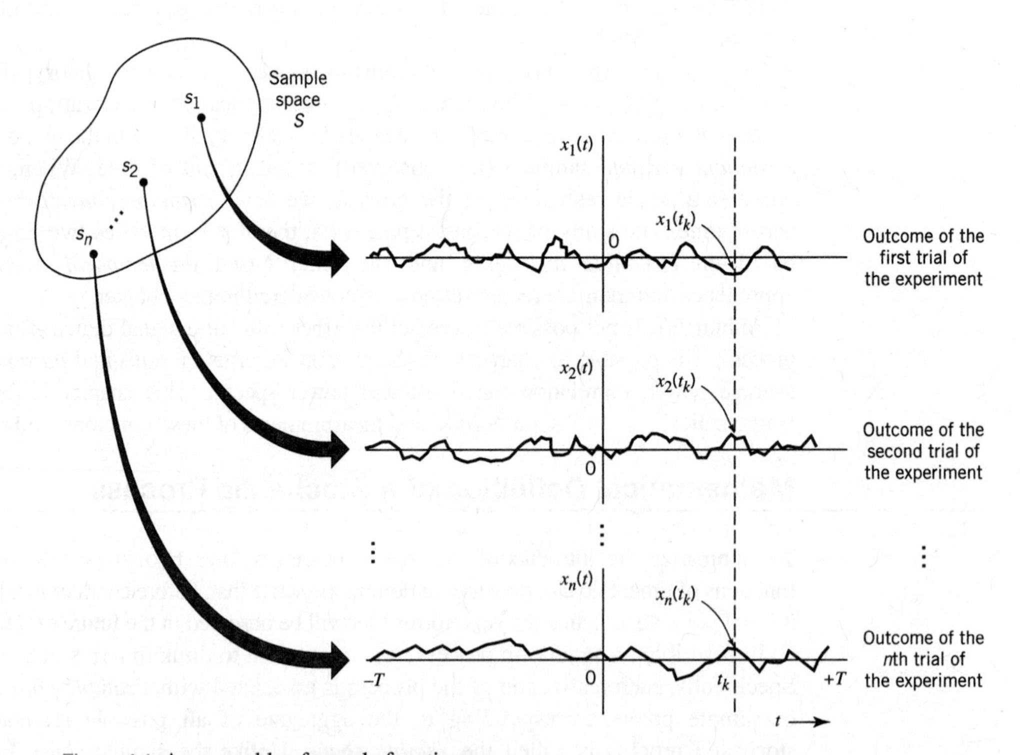

Figure 4.1 An ensemble of sample functions.

constitutes a *random variable*. Thus, a stochastic process $X(t, s)$ is represented by the time-indexed ensemble (family) of random variables $\{X(t, s)\}$. To simplify the notation, the customary practice is to suppress the s and simply use $X(t)$ to denote a stochastic process. We may now formally introduce the definition:

> A stochastic process $X(t)$ is an ensemble of time functions, which, together with a probability rule, assigns a probability to any meaningful event associated with an observation of one of the sample functions of the stochastic process.

Moreover, we may distinguish between a random variable and a random process as follows. For a random variable, the outcome of a stochastic experiment is mapped into a number. On the other hand, for a stochastic process, the outcome of a stochastic experiment is mapped into a waveform that is a function of time.

4.3 Two Classes of Stochastic Processes: Strictly Stationary and Weakly Stationary

In dealing with stochastic processes encountered in the real world, we often find that the statistical characterization of a process is independent of the time at which observation of the process is initiated. That is, if such a process is divided into a number of time intervals, the various sections of the process exhibit essentially the same statistical properties. Such a stochastic process is said to be *stationary*. Otherwise, it is said to be *nonstationary*. Generally speaking, we may say:

> A stationary process arises from a stable phenomenon that has evolved into a steady-state mode of behavior, whereas a nonstationary process arises from an unstable phenomenon.

To be more precise, consider a stochastic process $X(t)$ that is initiated at $t = -\infty$. Let $X(t_1), X(t_2), \ldots, X(t_k)$ denote the random variables obtained by sampling the process $X(t)$ at times $t_1, t_2, \ldots, t_k$, respectively. The joint (cumulative) distribution function of this set of random variables is $F_{X(t_1), \ldots, X(t_k)}(x_1, \ldots, x_k)$. Suppose next we shift all the sampling times by a fixed amount τ denoting the *time shift*, thereby obtaining the new set of random variables: $X(t_1+\tau), X(t_2+\tau), \ldots, X(t_k+\tau)$. The joint distribution function of this latter set of random variables is $F_{X(t_1+\tau), \ldots, X(t_k+\tau)}(x_1, \ldots, x_k)$. The stochastic process $X(t)$ is said to be *stationary in the strict sense*, or *strictly stationary, if the invariance condition*

$$F_{X(t_1+\tau), \ldots, X(t_k+\tau)}(x_1, \ldots, x_k) = F_{X(t_1), \ldots, X(t_k)}(x_1, \ldots, x_k) \tag{4.2}$$

holds for all values of time shift τ, all positive integers k, and any possible choice of sampling times $t_1, \ldots, t_k$. In other words, we may state:

> A stochastic process $X(t)$, initiated at time $t = -\infty$, is strictly stationary if the joint distribution of any set of random variables obtained by observing the process $X(t)$ is invariant with respect to the location of the origin $t = 0$.

Note that the finite-dimensional distributions in (4.2) depend on the relative time separation between random variables, but not on their absolute time. That is, the stochastic process has the same probabilistic behavior throughout the global time t.

Similarly, we may say that two stochastic processes $X(t)$ and $Y(t)$ are *jointly strictly stationary* if the joint finite-dimensional distributions of the two sets of stochastic variables $X(t_1), \ldots, X(t_k)$ and $Y(t'_1), \ldots, Y(t'_j)$ are invariant with respect to the origin $t = 0$ for all positive integers k and j, and all choices of the sampling times $t_1, \ldots, t_k$ and $t'_1, \ldots, t'_j$.

Returning to (4.2), we may identify two important properties:

1. For $k = 1$, we have

$$F_{X(t)}(x) = F_{X(t+\tau)}(x) = F_X(x) \qquad \text{for all } t \text{ and } \tau \tag{4.3}$$

In words, *the first-order distribution function of a strictly stationary stochastic process is independent of time t.*

2. For $k = 2$ and $\tau = -t_2$, we have

$$F_{X(t_1), X(t_2)}(x_1, x_2) = F_{X(0), X(t_1 - t_2)}(x_1, x_2) \qquad \text{for all } t_1 \text{ and } t_2 \tag{4.4}$$

In words, *the second-order distribution function of a strictly stationary stochastic process depends only on the time difference between the sampling instants and not on the particular times at which the stochastic process is sampled.*

These two properties have profound practical implications for the statistical parameterization of a strictly stationary stochastic process, as discussed in Section 4.4.

EXAMPLE 1 Multiple Spatial Windows for Illustrating Strict Stationarity

Consider Figure 4.2, depicting three spatial windows located at times t_1, t_2, t_3. We wish to evaluate the probability of obtaining a sample function $x(t)$ of a stochastic process $X(t)$ that passes through this set of windows; that is, the probability of the joint event

$$\mathbb{P}(A) = F_{X(t_1), X(t_2), X(t_3)}(b_1, b_2, b_3) = F_{X(t_1), X(t_2), X(t_3)}(a_1, a_2, a_3)$$

Suppose now the stochastic process $X(t)$ is known to be strictly stationary. An implication of strict stationarity is that the probability of the set of sample functions of this process passing through the windows of Figure 4.3a is equal to the probability of the set of sample functions passing through the corresponding time-shifted windows of Figure 4.3b. Note, however, that it is not necessary that these two sets consist of the same sample functions.

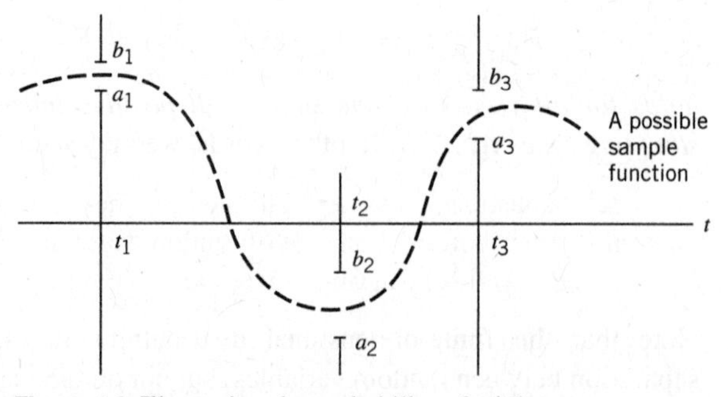

Figure 4.2 Illustrating the probability of a joint event.

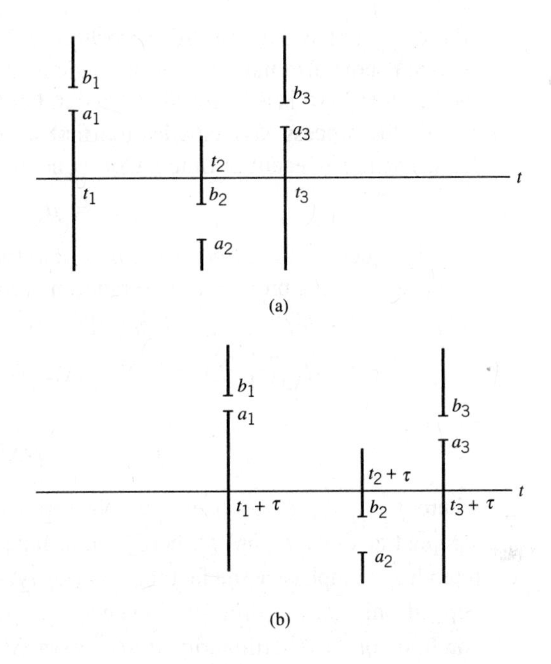

Figure 4.3
Illustrating the concept of stationarity in Example 1.

Another important class of stochastic processes is the so-called *weakly stationary processes.* To be specific, a stochastic process $X(t)$ is said to be weakly stationary if its second-order moments satisfy the following two conditions:

1. *The mean of the process $X(t)$ is constant for all time t.*
2. *The autocorrelation function of the process $X(t)$ depends solely on the difference between any two times at which the process is sampled;* the "auto" in autocorrelation refers to the correlation of the process with itself.

In this book we focus on weakly stationary processes whose second-order statistics satisfy conditions 1 and 2; both of them are easy to measure and considered to be adequate for practical purposes. Such processes are also referred to as *wide-sense stationary processes* in the literature. Henceforth, both terminologies are used interchangeably.

4.4 Mean, Correlation, and Covariance Functions of Weakly Stationary Processes

Consider a real-valued stochastic process $X(t)$. We define the *mean* of the process $X(t)$ as the expectation of the random variable obtained by sampling the process at some time t, as shown by

$$\begin{aligned} \mu_X(t) &= \mathbb{E}[X(t)] \\ &= \int_{-\infty}^{\infty} x f_{X(t)}(x)\,\mathrm{d}x \end{aligned} \tag{4.5}$$

where $f_{X(t)}(x)$ is the first-order probability density function of the process $X(t)$, observed at time t; note also that the use of single X as subscript in $\mu_X(t)$ is intended to emphasize the fact that $\mu_X(t)$ is a first-order moment. For the mean $\mu_X(t)$ to be a constant for all time t so that the process $X(t)$ satisfies the first condition of weak stationarity, we require that $f_{X(t)}(x)$ be independent of time t. Consequently, (4.5) simplifies to

$$\mu_X(t) = \mu_X \quad \text{for all } t \tag{4.6}$$

We next define the *autocorrelation function* of the stochastic process $X(t)$ as the expectation of the product of two random variables, $X(t_1)$ and $X(t_2)$, obtained by sampling the process $X(t)$ at times t_1 and t_2, respectively. Specifically, we write

$$\begin{aligned} M_{XX}(t_1, t_2) &= \mathbb{E}[X(t_1)X(t_2)] \\ &= \int_{-\infty}^{\infty}\int_{-\infty}^{\infty} x_1 x_2 f_{X(t_1), X(t_2)}(x_1, x_2)\, dx_1\, dx_2 \end{aligned} \tag{4.7}$$

where $f_{X(t_1), X(t_2)}(x_1, x_2)$ is the joint probability density function of the process $X(t)$ sampled at times t_1 and t_2; here, again, note that the use of the double X subscripts is intended to emphasize the fact that $M_{XX}(t_1, t_2)$ is a second-order moment. For $M_{XX}(t_1, t_2)$ to depend only on the time difference $t_2 - t_1$ so that the process $X(t)$ satisfies the second condition of weak stationarity, it is necessary for $f_{X(t_1), X(t_2)}(x_1, x_2)$ to depend only on the time difference $t_2 - t_1$. Consequently, (4.7) reduces to

$$\begin{aligned} M_{XX}(t_1, t_2) &= \mathbb{E}[X(t_1)X(t_2)] \\ &= R_{XX}(t_2 - t_1) \quad \text{for all } t_1 \text{ and } t_2 \end{aligned} \tag{4.8}$$

In (4.8) we have purposely used two different symbols for the autocorrelation function: $M_{XX}(t_1, t_2)$ for any stochastic process $X(t)$ and $R_{XX}(t_2 - t_1)$ for a stochastic process that is weakly stationary.

Similarly, the *autocovariance function* of a weakly stationary process $X(t)$ is defined by

$$\begin{aligned} C_{XX}(t_1, t_2) &= \mathbb{E}[(X(t_1) - \mu_X)(X(t_2) - \mu_X)] \\ &= R_{XX}(t_2 - t_1) - \mu_X^2 \end{aligned} \tag{4.9}$$

Equation (4.9) shows that, like the autocorrelation function, the autocovariance function of a weakly stationary process $X(t)$ depends only on the time difference $(t_2 - t_1)$. This equation also shows that if we know the mean and the autocorrelation function of the process $X(t)$, we can uniquely determine the autocovariance function. The mean and autocorrelation function are therefore sufficient to describe the first two moments of the process.

However, two important points should be carefully noted:

1. The mean and autocorrelation function only provide a *weak description* of the distribution of the stochastic process $X(t)$.
2. The conditions involved in defining (4.6) and (4.8) are *not* sufficient to guarantee the stochastic process $X(t)$ to be strictly stationary, which emphasizes a remark that was made in the preceding section.

Nevertheless, practical considerations often dictate that we simply limit ourselves to a weak description of the process given by the mean and autocorrelation function because the computation of higher order moments can be computationally intractable.

Henceforth, the treatment of stochastic processes is confined to weakly stationary processes, for which the definitions of the second-order moments in (4.6), (4.8), and (4.9) hold.

Properties of the Autocorrelation Function

For convenience of notation, we reformulate the definition of the autocorrelation function of a weakly stationary process $X(t)$, presented in (4.8), as

$$R_{XX}(\tau) = \mathbb{E}[X(t+\tau)X(t)] \quad \text{for all } t \tag{4.10}$$

where τ denotes a *time shift*; that is, $t = t_2$ and $\tau = t_1 - t_2$. This autocorrelation function has several important properties.

PROPERTY 1 **Mean-square Value**

The mean-square value of a weakly stationary process $X(t)$ is obtained from $R_{XX}(\tau)$ simply by putting $\tau = 0$ in (4.10), as shown by

$$R_{XX}(0) = \mathbb{E}[X^2(t)] \tag{4.11}$$

PROPERTY 2 **Symmetry**

The autocorrelation function $R_{XX}(\tau)$ of a weakly stationary process $X(t)$ is an even function of the time shift τ; that is,

$$R_{XX}(\tau) = R_{XX}(-\tau) \tag{4.12}$$

This property follows directly from (4.10). Accordingly, we may also define the autocorrelation function $R_{XX}(\tau)$ as

$$R_{XX}(\tau) = \mathbb{E}[X(t)X(t-\tau)]$$

In words, we may say that a graph of the autocorrelation function $R_{XX}(\tau)$, plotted versus τ, is symmetric about the origin.

PROPERTY 3 **Bound on the Autocorrelation Function**

The autocorrelation function $R_{XX}(\tau)$ attains its maximum magnitude at $\tau = 0$; that is,

$$|R_{XX}(\tau)| \le R_{XX}(0) \tag{4.13}$$

To prove this property, consider the nonnegative quantity

$$\mathbb{E}[(X(t+\tau) \pm X(t))^2] \ge 0$$

Expanding terms and taking their individual expectations, we readily find that

$$\mathbb{E}[X^2(t+\tau)] \pm 2\mathbb{E}[X(t+\tau)] + \mathbb{E}[X^2(t)] \ge 0$$

which, in light of (4.11) and (4.12), reduces to

$$2R_{XX}(0) \pm 2R_{XX}(\tau) \ge 0$$

Equivalently, we may write

$$-R_{XX}(0) \le R_{XX}(\tau) \le R_{XX}(0)$$

from which (4.13) follows directly.

PROPERTY 4 **Normalization**

Values of the normalized autocorrelation function

$$\rho_{XX}(\tau) = \frac{R_{XX}(\tau)}{R_{XX}(0)} \tag{4.14}$$

are confined to the range [–1, 1].

This last property follows directly from (4.13).

Physical Significance of the Autocorrelation Function

The autocorrelation function $R_{XX}(\tau)$ is significant because it provides *a means of describing the interdependence of two random variables obtained by sampling the stochastic process X(t) at times τ seconds apart.* It is apparent, therefore, that the more rapidly the stochastic process $X(t)$ changes with time, the more rapidly will the autocorrelation function $R_{XX}(\tau)$ decrease from its maximum $R_{XX}(0)$ as τ increases, as illustrated in Figure 4.4. This behavior of the autocorrelation function may be characterized by a *decorrelation time* τ_{dec}, such that, for $\tau > \tau_{dec}$, the magnitude of the autocorrelation function $R_{XX}(\tau)$ remains below some prescribed value. We may thus introduce the following definition:

> The decorrelation time τ_{dec} of a weakly stationary process $X(t)$ of zero mean is the time taken for the magnitude of the autocorrelation function $R_{XX}(t)$ to decrease, for example, to 1% of its maximum value $R_{XX}(0)$.

For the example used in this definition, the parameter τ_{dec} is referred to as the *one-percent decorrelation time.*

EXAMPLE 2 **Sinusoidal Wave with Random Phase**

Consider a sinusoidal signal with random phase, defined by

$$X(t) = A\cos(2\pi f_c t + \Theta) \tag{4.15}$$

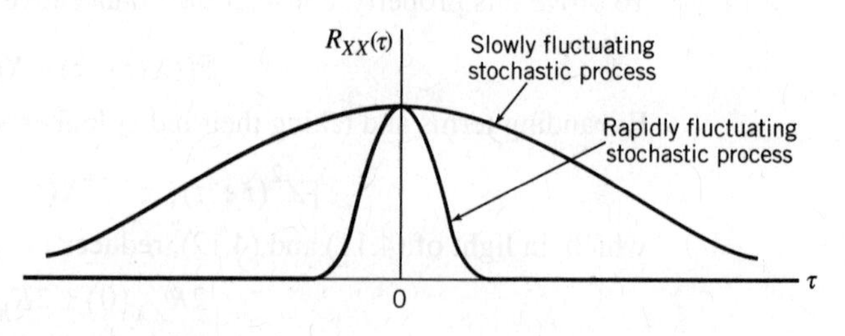

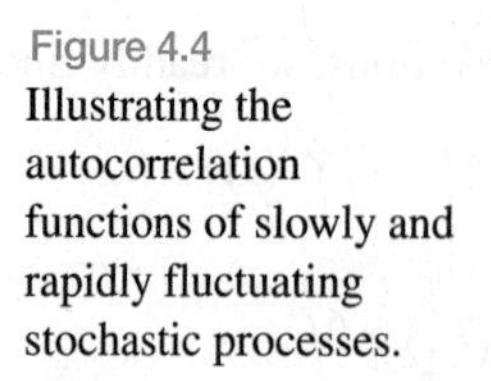
Figure 4.4 Illustrating the autocorrelation functions of slowly and rapidly fluctuating stochastic processes.

where A and f_c are constants and Θ is a random variable that is *uniformly distributed* over the interval $[-\pi, \pi]$; that is,

$$f_\Theta(\theta) = \begin{cases} \dfrac{1}{2\pi}, & -\pi \le \theta \le \pi \\ 0, & \text{elsewhere} \end{cases} \tag{4.16}$$

According to (4.16), the random variable Θ is equally likely to have any value θ in the interval $[-\pi, \pi]$. Each value of θ corresponds to a point in the sample space S of the stochastic process $X(t)$.

The process $X(t)$ defined by (4.15) and (4.16) may represent a locally generated carrier in the receiver of a communication system, which is used in the demodulation of a received signal. In such an application, the random variable Θ in (4.15) accounts for uncertainties experienced in the course of signal transmission across the communication channel.

The autocorrelation function of $X(t)$ is

$$\begin{aligned} R_{XX}(\tau) &= \mathbb{E}[X(t+\tau)X(t)] \\ &= \mathbb{E}[A^2\cos(2\pi f_c t + 2\pi f_c\tau + \Theta)\cos(2\pi f_c t + \Theta)] \\ &= \frac{A^2}{2}\mathbb{E}[\cos(4\pi f_c t + 2\pi f_c \tau + 2\Theta)] + \frac{A^2}{2}\mathbb{E}[\cos(2\pi f_c \tau)] \\ &= \frac{A^2}{2}\int_{-\pi}^{\pi}\cos(4\pi f_c t + 2\pi f_c\tau + 2\theta)\,d\theta + \frac{A^2}{2}\cos(2\pi f_c\tau) \end{aligned}$$

The first term integrates to zero, so we simply have

$$R_{XX}(\tau) = \frac{A^2}{2}\cos(2\pi f_c \tau) \tag{4.17}$$

which is plotted in Figure 4.5. From this figure we see that the autocorrelation function of a sinusoidal wave with random phase is another sinusoid at the same frequency in the "local time domain" denoted by the time shift τ rather than the global time domain denoted by t.

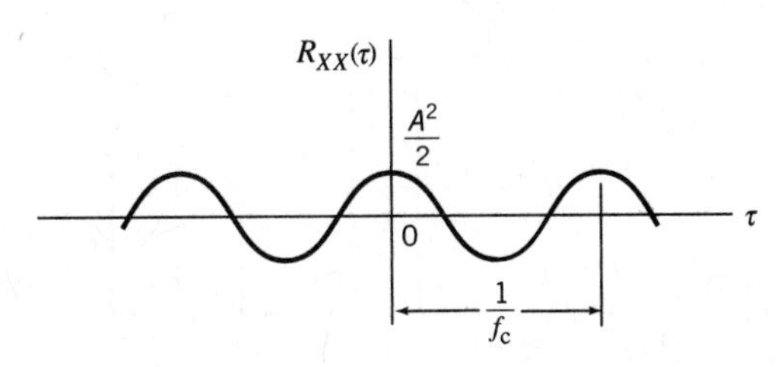

Figure 4.5 Autocorrelation function of a sine wave with random phase.

EXAMPLE 3 **Random Binary Wave**

Figure 4.6 shows the sample function $x(t)$ of a weakly stationary process $X(t)$ consisting of a random sequence of *binary symbols* 1 and 0. Three assumptions are made:

1. The symbols 1 and 0 are represented by pulses of amplitude $+A$ and $-A$ volts respectively and duration T seconds.
2. The pulses are not synchronized, so the starting time t_d of the first complete pulse for positive time is equally likely to lie anywhere between zero and T seconds. That is, t_d is the sample value of a uniformly distributed random variable T_d, whose probability density function is defined by

$$f_{T_d}(t_d) = \begin{cases} \dfrac{1}{T}, & 0 \le t_d \le T \\ 0, & \text{elsewhere} \end{cases}$$

3. During any time interval $(n - 1)T < t - t_d < nT$, where n is a positive integer, the presence of a 1 or a 0 is determined by tossing a fair coin. Specifically, if the outcome is heads, we have a 1; if the outcome is tails, we have a 0. These two symbols are thus equally likely, and the presence of a 1 or 0 in any one interval is independent of all other intervals.

Since the amplitude levels $-A$ and $+A$ occur with equal probability, it follows immediately that $\mathbb{E}[X(t)] = 0$ for all t and the mean of the process is therefore zero.

To find the autocorrelation function $R_{XX}(t_k,t_i)$, we have to evaluate the expectation $\mathbb{E}[X(t_k)X(t_i)]$, where $X(t_k)$ and $X(t_i)$ are random variables obtained by sampling the stochastic process $X(t)$ at times t_k and t_i respectively. To proceed further, we need to consider two distinct conditions:

Condition 1: $|t_k - t_i| > T$

Under this condition, the random variables $X(t_k)$ and $X(t_i)$ occur in different pulse intervals and are therefore independent. We thus have

$$\mathbb{E}[X(t_k)X(t_i)] = \mathbb{E}[X(t_k)]\mathbb{E}[X(t_i)] = 0, \qquad |t_k - t_i| > T$$

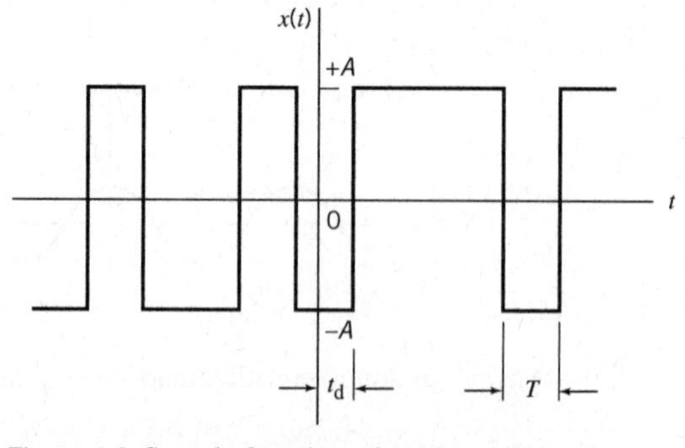

Figure 4.6 Sample function of random binary wave.

Condition 2: $|t_k - t_i| > T$, with $t_k = 0$ and $t_i < t_k$

Under this second condition, we observe from Figure 4.6 that the random variables $X(t_k)$ and $X(t_i)$ occur in the same pulse interval if, and only if, the delay t_d satisfies the condition $t_d < T - |t_k - t_i|$. We thus have the conditional expectation

$$\mathbb{E}[X(t_k)X(t_i)|t_d] = \begin{cases} A^2, & t_d < T - |t_k - t_i| \\ 0, & \text{elsewhere} \end{cases}$$

Averaging this result over all possible values of t_d, we get

$$\begin{aligned} \mathbb{E}[X(t_k)X(t_i)] &= \int_0^{T - |t_k - t_i|} A^2 f_{T_d}(t_d)\, dt_d \\ &= \int_0^{T - |t_k - t_i|} \frac{A^2}{T}\, dt_d \\ &= A^2\left(1 - \frac{|t_k - t_i|}{T}\right), \quad |t_k - t_i| < T \end{aligned}$$

By similar reasoning for any other value of t_k, we conclude that the autocorrelation function of a random binary wave, represented by the sample function shown in Figure 4.6, is only a function of the time difference $\tau = t_k - t_i$, as shown by

$$R_{XX}(\tau) = \begin{cases} A^2\left(1 - \frac{|\tau|}{T}\right), & |\tau| < T \\ 0, & |\tau| \geq T \end{cases} \tag{4.18}$$

This triangular result, described in (4.18), is plotted in Figure 4.7.

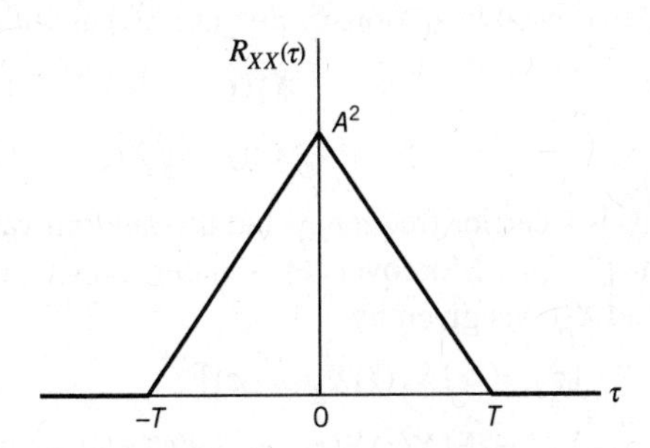

Figure 4.7 Autocorrelation function of random binary wave.

Cross-correlation Functions

Consider next the more general case of two stochastic processes $X(t)$ and $Y(t)$ with autocorrelation functions $M_{XX}(t, u)$ and $M_{YY}(t, u)$ respectively. There are two possible *cross-correlation functions* of $X(t)$ and $Y(t)$ to be considered.

Specifically, we have

$$M_{XY}(t, u) = \mathbb{E}[X(t)Y(u)] \tag{4.19}$$

and

$$M_{YX}(t, u) = \mathbb{E}[Y(t)X(u)] \tag{4.20}$$

where t and u denote two values of the global time at which the processes are observed. All four correlation parameters of the two stochastic processes $X(t)$ and $Y(t)$ may now be displayed conveniently in the form of the two-by-two matrix

$$\mathbf{M}(t, u) = \begin{bmatrix} M_{XX}(t, u) & M_{XY}(t, u) \\ M_{YX}(t, u) & M_{YY}(t, u) \end{bmatrix}$$

which is called the *cross-correlation matrix* of the stochastic processes $X(t)$ and $Y(t)$. If the stochastic processes $X(t)$ and $Y(t)$ are each weakly stationary and, in addition, they are jointly stationary, then the correlation matrix can be expressed by

$$\mathbf{R}(\tau) = \begin{bmatrix} R_{XX}(\tau) & R_{XY}(\tau) \\ R_{YX}(\tau) & R_{YY}(\tau) \end{bmatrix} \tag{4.21}$$

where the time shift $\tau = u - t$.

In general, the cross-correlation function is *not* an even function of the time-shift τ as was true for the autocorrelation function, nor does it have a maximum at the origin. However, it does obey a certain symmetry relationship, described by

$$R_{XY}(\tau) = R_{YX}(-\tau) \tag{4.22}$$

EXAMPLE 4

Quadrature-Modulated Processes

Consider a pair of quadrature-modulated processes $X_1(t)$ and $X_2(t)$ that are respectively related to a weakly stationary process $X(t)$ as follows:

$$X_1(t) = X(t)\cos(2\pi f_c t + \Theta)$$

$$X_2(t) = X(t)\sin(2\pi f_c t + \Theta)$$

where f_c is a carrier frequency and the random variable Θ is uniformly distributed over the interval $[0, 2\pi]$. Moreover, Θ is independent of $X(t)$. One cross-correlation function of $X_1(t)$ and $X_2(t)$ is given by

$$\begin{aligned} R_{12}(\tau) &= \mathbb{E}[X_1(t)X_2(t-\tau)] \\ &= \mathbb{E}[X(t)X(t-\tau)\cos(2\pi f_c t + \Theta)\sin(2\pi f_c t - 2\pi f_c \tau + \Theta)] \\ &= \mathbb{E}[X(t)X(t-\tau)]\mathbb{E}[\cos(2\pi f_c t + \Theta)\sin(2\pi f_c t - 2\pi f_c \tau + \Theta)] \\ &= \frac{1}{2}R_{XX}(\tau)\mathbb{E}[\sin(4\pi f_c \tau - 2\pi f_c t + 2\Theta) - \sin(2\pi f_c \tau)] \\ &= -\frac{1}{2}R_{XX}(\tau)\sin(2\pi f_c \tau) \end{aligned} \tag{4.23}$$

where, in the last line, we have made use of the uniform distribution of the random variable Θ, representing phase. Invoking (4.22), we find that the other cross-correlation function of $X_1(t)$ and $X_2(t)$ is given by

$$\begin{aligned} R_{21}(\tau) &= \frac{1}{2}R_{XX}(-\tau)\sin(2\pi f_c \tau) \\ &= \frac{1}{2}R_{XX}(\tau)\sin(2\pi f_c \tau) \end{aligned}$$

At $\tau = 0$, the factor $\sin(2\pi f_c \tau)$ is zero, in which case we have

$$R_{12}(0) = R_{21}(0) = 0$$

This result shows that the random variables obtained by simultaneously sampling the quadrature-modulated processes $X_1(t)$ and $X_2(t)$ at some fixed value of time t are orthogonal to each other.

4.5 Ergodic Processes

Ergodic processes are subsets of weakly stationary processes. Most importantly, from a practical perspective, the *property of ergodicity* permits us to substitute time averages for ensemble averages.

To elaborate on these two succinct statements, we know that the expectations or ensemble averages of a stochastic process $X(t)$ are averages "across the process." For example, the mean of a stochastic process $X(t)$ at some fixed time t_k is the expectation of the random variable $X(t_k)$ that describes *all possible values* of sample functions of the process $X(t)$ sampled at time $t = t_k$. Naturally, we may also define *long-term sample averages* or *time averages* that are averages "along the process." Whereas in ensemble averaging we consider a set of independent realizations of the process $X(t)$ sampled at some fixed time t_k, in time averaging we focus on a single waveform evolving across time t and representing one waveform realization of the process $X(t)$.

With time averages providing the basis of a practical method for possible *estimation* of ensemble averages of a stochastic process, we would like to explore the conditions under which this estimation is justifiable. To address this important issue, consider the sample function $x(t)$ of a weakly stationary process $X(t)$ observed over the interval $-T \le t \le T$. The time-average value of the sample function $x(t)$ is defined by the definite integral

$$\mu_x(T) = \frac{1}{2T}\int_{-T}^{T} x(t)\,dt \tag{4.24}$$

Clearly, the time average $\mu_x(T)$ is a random variable, as its value depends on the observation interval and which particular sample function of the process $X(t)$ is picked for use in (4.24). Since the process $X(t)$ is assumed to be weakly stationary, the mean of the time average $\mu_x(T)$ is given by (after interchanging the operations of expectation and integration, which is permissible because both operations are linear)

$$\begin{aligned} \mathbb{E}[\mu_x(T)] &= \frac{1}{2T}\int_{-T}^{T} \mathbb{E}[x(t)]\,\mathrm{d}t \\ &= \frac{1}{2T}\int_{-T}^{T} \mu_X\,\mathrm{d}t \\ &= \mu_X \end{aligned} \tag{4.25}$$

where μ_X is the mean of the process $X(t)$. Accordingly, the time average $\mu_x(T)$ represents an *unbiased* estimate of the ensemble-averaged mean μ_X. Most importantly, we say that the process $X(t)$ is *ergodic in the mean* if two conditions are satisfied:

1. The time average $\mu_x(T)$ approaches the ensemble average μ_X in the limit as the observation interval approaches infinity; that is,

$$\lim_{T \to \infty} \mu_x(T) = \mu_X$$

2. The variance of $\mu_x(T)$, treated as a random variable, approaches zero in the limit as the observation interval approaches infinity; that is,

$$\lim_{T \to \infty} \mathrm{var}[\mu_x(T)] = 0$$

The other time average of particular interest is the autocorrelation function $R_{xx}(\tau, T)$, defined in terms of the sample function $x(t)$ observed over the interval $-T \le t \le T$. Following (4.24), we may formally define the *time-averaged autocorrelation function* of $x(t)$ as

$$R_{xx}(\tau, T) = \frac{1}{2T}\int_{-T}^{T} x(t+\tau)x(t)\,\mathrm{d}t \tag{4.26}$$

This second time average should also be viewed as a random variable with a mean and variance of its own. In a manner similar to ergodicity of the mean, we say that the process $x(t)$ is *ergodic in the autocorrelation function* if the following two limiting conditions are satisfied:

$$\lim_{T \to \infty} R_{xx}(\tau, T) = R_{XX}(\tau)$$

$$\lim_{T \to \infty} \mathrm{var}[R_{xx}(\tau, T)] = 0$$

With the property of ergodicity confined to the mean and autocorrelation functions, it follows that ergodic processes are subsets of weakly stationary processes. In other words, all ergodic processes are weakly stationary; however, the converse is not necessarily true.

4.6 Transmission of a Weakly Stationary Process through a Linear Time-invariant Filter

Suppose that a stochastic process $X(t)$ is applied as input to a linear time-invariant filter of impulse response $h(t)$, producing a new stochastic process $Y(t)$ at the filter output, as depicted in Figure 4.8. In general, it is difficult to describe the probability distribution of the output stochastic process $Y(t)$, even when the probability distribution of the input stochastic process $X(t)$ is completely specified for the entire time interval $-\infty < t < \infty$.

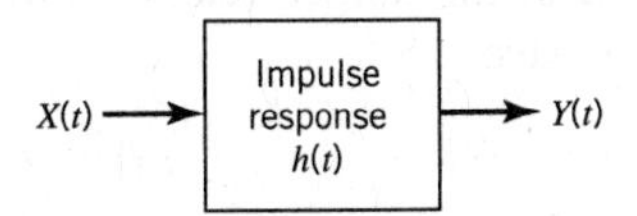

Figure 4.8 Transmission of a stochastic process through a linear time-invariant filter.

For the sake of mathematical tractability, we limit the discussion in this section to the time-domain form of the input–output relations of the filter for defining the mean and autocorrelation functions of the output stochastic process $Y(t)$ in terms of those of the input $X(t)$, assuming that $X(t)$ is a weakly stationary process.

The transmission of a process through a linear time-invariant filter is governed by the *convolution integral*, which was discussed in Chapter 2. For the problem at hand, we may thus express the output stochastic process $Y(t)$ in terms of the input stochastic process $X(t)$ as

$$Y(t) = \int_{-\infty}^{\infty} h(\tau_1)X(t-\tau_1)\,d\tau_1$$

where τ_1 is a local time. Hence, the mean of $Y(t)$ is

$$\begin{aligned}\mu_Y(t) &= \mathbb{E}[Y(t)]\\ &= \mathbb{E}\left[\int_{-\infty}^{\infty} h(\tau_1)X(t-\tau_1)\,d\tau_1\right]\end{aligned} \tag{4.27}$$

Provided that the expectation $\mathbb{E}[X(t)]$ is finite for all t and the filter is stable, we may interchange the order of expectation and integration in (4.27), in which case we obtain

$$\begin{aligned}\mu_Y(t) &= \int_{-\infty}^{\infty} h(\tau_1)\mathbb{E}[X(t-\tau_1)]\,d\tau_1\\ &= \int_{-\infty}^{\infty} h(\tau_1)\mu_X(t-\tau_1)\,d\tau_1\end{aligned} \tag{4.28}$$

When the input stochastic process $X(t)$ is weakly stationary, the mean $\mu_X(t)$ is a constant μ_X; therefore, we may simplify (4.28) as

$$\begin{aligned}\mu_Y &= \mu_X\int_{-\infty}^{\infty} h(\tau_1)\,d\tau_1\\ &= \mu_X H(0)\end{aligned} \tag{4.29}$$

where $H(0)$ is the zero-frequency response of the system. Equation (4.29) states:

> The mean of the stochastic process $Y(t)$ produced at the output of a linear time-invariant filter in response to a weakly stationary process $X(t)$, acting as the input process, is equal to the mean of $X(t)$ multiplied by the zero-frequency response of the filter.

This result is intuitively satisfying.

Consider next the autocorrelation function of the output stochastic process $Y(t)$. By definition, we have

$$M_{YY}(t, u) = \mathbb{E}[Y(t)Y(u)]$$

where t and u denote two values of the time at which the output process $Y(t)$ is sampled. We may therefore apply the convolution integral twice to write

$$M_{YY}(t, u) = \mathbb{E}\left[\int_{-\infty}^{\infty} h(\tau_1)X(t-\tau_1)\, d\tau_1 \int_{-\infty}^{\infty} h(\tau_2)X(u-\tau_2)\, d\tau_2\right] \tag{4.30}$$

Here again, provided that the mean-square value $\mathbb{E}[X^2(t)]$ is finite for all t and the filter is stable, we may interchange the order of the expectation and the integrations with respect to τ_1 and τ_2 in (4.30), obtaining

$$\begin{aligned} M_{YY}(t, u) &= \int_{-\infty}^{\infty}\left[h(\tau_1)\int_{-\infty}^{\infty} d\tau_2\, h(\tau_2)\mathbb{E}[X(t-\tau_1)X(u-\tau_2)]\right] d\tau_1 \\ &= \int_{-\infty}^{\infty}\left[h(\tau_1)\int_{-\infty}^{\infty} d\tau_2\, h(\tau_2)M_{XX}(t-\tau_1, u-\tau_2)\right] d\tau_1 \end{aligned} \tag{4.31}$$

When the input $X(t)$ is a weakly stationary process, the autocorrelation function of $X(t)$ is only a function of the difference between the sampling times $t - \tau_1$ and $u - \tau_2$. Thus, putting $\tau = u - t$ in (4.31), we may go on to write

$$R_{YY}(\tau) = \int_{-\infty}^{\infty}\int_{-\infty}^{\infty} h(\tau_1)h(\tau_2)R_{XX}(\tau + \tau_1 - \tau_2)\, d\tau_1\, d\tau_2 \tag{4.32}$$

which depends only on the time difference τ.

On combining the result of (4.32) with that involving the mean μ_Y in (4.29), we may now make the following statement:

> If the input to a stable linear time-invariant filter is a weakly stationary process, then the output of the filter is also a weakly stationary process.

By definition, we have $R_{YY}(0) = \mathbb{E}[Y^2(t)]$. In light of Property 1 of the autocorrelation function $R_{YY}(\tau)$, it follows, therefore, that the *mean-square value* of the output process $Y(t)$ is obtained by putting $\tau = 0$ in (4.32), as shown by

$$\mathbb{E}[Y^2(t)] = \int_{-\infty}^{\infty}\int_{-\infty}^{\infty} h(\tau_1)h(\tau_2)R_{XX}(\tau_1 - \tau_2)\, d\tau_1\, d\tau_2 \tag{4.33}$$

which, of course, is a constant.

4.7 Power Spectral Density of a Weakly Stationary Process

Thus far we have considered the time-domain characterization of a weakly stationary process applied to a linear filter. We next study the characterization of linearly filtered weakly stationary processes by using frequency-domain ideas. In particular, we wish to derive the frequency-domain equivalent to the result of (4.33), defining the mean-square value of the filter output $Y(t)$. The term "filter" used here should be viewed in a generic sense; for example, it may represent the channel of a communication system.

From Chapter 2, we recall that the impulse response of a linear time-invariant filter is equal to the inverse Fourier transform of the frequency response of the filter. Using $H(f)$ to denote the *frequency response* of the filter, we may thus write

$$h(\tau_1) = \int_{-\infty}^{\infty} H(f) \exp(j2\pi f \tau_1)\, df \tag{4.34}$$

Substituting this expression for $h(\tau_1)$ into (4.33) and then changing the order of integrations, we get the triple integral

$$\begin{aligned} \mathbb{E}[Y^2(t)] &= \int_{-\infty}^{\infty}\int_{-\infty}^{\infty}\left[\int_{-\infty}^{\infty} H(f) \exp(j2\pi f \tau_1)\, df\right] h(\tau_2) R_{XX}(\tau_1 - \tau_2)\, d\tau_1\, d\tau_2 \\ &= \int_{-\infty}^{\infty}\left[H(f)\int_{-\infty}^{\infty} d\tau_2 h(\tau_2) \int_{-\infty}^{\infty} R_{XX}(\tau_1 - \tau_2) \exp(j2\pi f \tau_1)\, d\tau_1\right] df \end{aligned} \tag{4.35}$$

At first, the expression on the right-hand side of (4.35) looks rather overwhelming. However, we may simplify it considerably by first introducing the variable

$$\tau = \tau_1 - \tau_2$$

Then, we may rewrite (4.35) in the new form

$$\mathbb{E}[Y^2(t)] = \int_{-\infty}^{\infty} H(f)\left[\int_{-\infty}^{\infty} h(\tau_2) \exp(j2\pi f \tau_2)\, d\tau_2 \int_{-\infty}^{\infty} R_{XX}(\tau) \exp(-j2\pi f \tau)\, d\tau\right] df \tag{4.36}$$

The middle integral involving the variable τ_2 inside the square brackets on the right-hand side in (4.36) is simply $H^*(f)$, the complex conjugate of the frequency response of the filter. Hence, using $|H(f)|^2 = H(f)H^*(f)$, where $|H(f)|$ is the *magnitude response* of the filter, we may simplify (4.36) as

$$\mathbb{E}[Y^2(t)] = \int_{-\infty}^{\infty} |H(f)|^2 \left[\int_{-\infty}^{\infty} R_{XX}(\tau) \exp(-j2\pi f \tau)\, d\tau\right] df \tag{4.37}$$

We may further simplify (4.37) by recognizing that the integral inside the square brackets in this equation with respect to the variable τ is simply the Fourier transform of the autocorrelation function $R_{XX}(\tau)$ of the input process $X(t)$. In particular, we may now define a new function

$$S_{XX}(f) = \int_{-\infty}^{\infty} R_{XX}(\tau) \exp(-j2\pi f \tau)\, d\tau \tag{4.38}$$

The new function $S_{XX}(f)$ is called the *power spectral density,* or *power spectrum,* of the weakly stationary process $X(t)$. Thus, substituting (4.38) into (4.37), we obtain the simple formula

$$\mathbb{E}[Y^2(t)] = \int_{-\infty}^{\infty} |H(f)|^2 S_{XX}(f)\, df \tag{4.39}$$

which is the desired frequency-domain equivalent to the time-domain relation of (4.33). In words, (4.39) states:

> The mean-square value of the output of a stable linear time-invariant filter in response to a weakly stationary process is equal to the integral over all

frequencies of the power spectral density of the input process multiplied by the squared magnitude response of the filter.

Physical Significance of the Power Spectral Density

To investigate the physical significance of the power spectral density, suppose that the weakly stationary process $X(t)$ is passed through an ideal narrowband filter with a magnitude response $|H(f)|$ centered about the frequency f_c, depicted in Figure 4.9; we may thus write

$$|H(f)| = \begin{cases} 1, & |f \pm f_c| < \frac{1}{2}\Delta f \\ 0, & |f \pm f_c| > \frac{1}{2}\Delta f \end{cases} \tag{4.40}$$

where Δf is the *bandwidth* of the filter. From (4.39) we readily find that if the bandwidth Δf is made sufficiently small compared with the midband frequency f_c of the filter and $S_{XX}(f)$ is a continuous function of the frequency f, then the mean-square value of the filter output is approximately given by

$$\mathbb{E}[Y^2(t)] \approx (2\Delta f) S_{XX}(f) \quad \text{for all } f \tag{4.41}$$

where, for the sake of generality, we have used f in place of f_c. According to (4.41), however, the filter passes only those frequency components of the input random process $X(t)$ that lie inside the narrow frequency band of width Δf. We may, therefore, say that $S_X(f)$ represents the density of the average power in the weakly stationary process $X(t)$, evaluated at the frequency f. The power spectral density is therefore measured in *watts per hertz* (W/Hz).

The Wiener–Khintchine Relations

According to (4.38), the power spectral density $S_{XX}(f)$ of a weakly stationary process $X(t)$ is the Fourier transform of its autocorrelation function $R_{XX}(\tau)$. Building on what we know about Fourier theory from Chapter 2, we may go on to say that the autocorrelation function $R_{XX}(\tau)$ is the inverse Fourier transform of the power spectral density $S_{XX}(f)$.

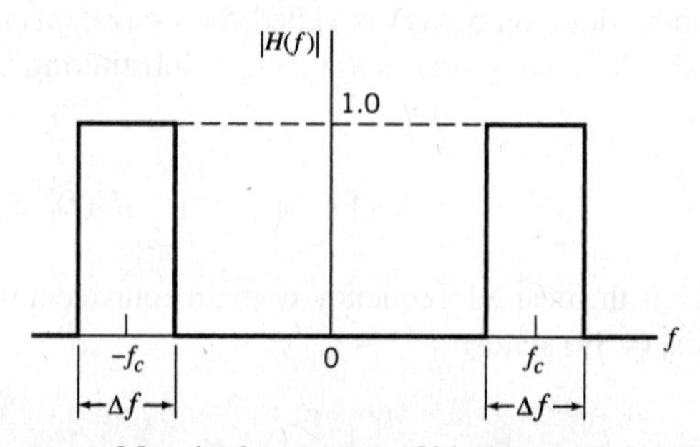

Figure 4.9 Magnitude response of ideal narrowband filter.

Simply put, $R_{XX}(\tau)$ and $S_{XX}(f)$ form a *Fourier-transform pair*, as shown by the following pair of related equations:

$$S_{XX}(f) = \int_{-\infty}^{\infty} R_{XX}(\tau) \exp(-j2\pi f\tau)\, d\tau \tag{4.42}$$

$$R_{XX}(\tau) = \int_{-\infty}^{\infty} S_{XX}(f) \exp(j2\pi f\tau)\, df \tag{4.43}$$

These two equations are known as the *Wiener–Khintchine relations*,[3] which play a fundamental role in the spectral analysis of weakly stationary processes.

The Wiener–Khintchine relations show that if either the autocorrelation function or power spectral density of a weakly stationary process is known, then the other can be found exactly. Naturally, these functions display different aspects of correlation-related information about the process. Nevertheless, it is commonly accepted that, for practical purposes, the power spectral density is the more useful function of the two for reasons that will become apparent as we progress forward in this chapter and the rest of the book.

Properties of the Power Spectral Density

PROPERTY 1 Zero Correlation among Frequency Components

The individual frequency components of the power spectral density $S_{XX}(f)$ of a weakly stationary process $X(t)$ are uncorrelated with each other.

To justify this property, consider Figure 4.10, which shows two adjacent narrow bands of the power spectral density $S_{XX}(f)$, with the width of each band being denoted by Δf. From this figure, we see that there is no overlap, and therefore no correlation, between the contents of these two bands. As Δf approaches zero, the two narrow bands will correspondingly evolve into two adjacent frequency components of $S_{XX}(f)$, remaining uncorrelated with each other. This important property of the power spectral density $S_{XX}(f)$ is attributed to the weak stationarity assumption of the stochastic process $X(t)$.

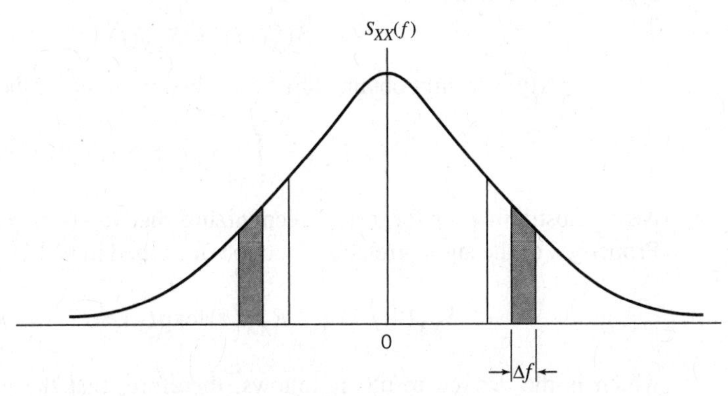

Figure 4.10 Illustration of zero correlation between two adjacent narrow bands of an example power spectral density.

PROPERTY 2 **Zero-frequency Value of Power Spectral Density**

The zero-frequency value of the power spectral density of a weakly stationary process equals the total area under the graph of the autocorrelation function; that is,

$$S_{XX}(0) = \int_{-\infty}^{\infty} R_{XX}(\tau)\,d\tau \tag{4.44}$$

This second property follows directly from (4.42) by putting $f = 0$.

PROPERTY 3 **Mean-square Value of Stationary Process**

The mean-square value of a weakly stationary process X(t) equals the total area under the graph of the power spectral density of the process; that is,

$$\mathbb{E}[X^2(t)] = \int_{-\infty}^{\infty} S_{XX}(f)\,df \tag{4.45}$$

This third property follows directly from (4.43) by putting $\tau = 0$ and using Property 1 of the autocorrelation function described in (4.11) namely $R_X(0) = \mathbb{E}[X^2(t)]$ for all t.

PROPERTY 4 **Nonnegativeness of Power Spectral Density**

The power spectral density of a stationary process X(t) is always nonnegative; that is,

$$S_{XX}(f) \ge 0 \qquad \text{for all } f \tag{4.46}$$

This property is an immediate consequence of the fact that, since the mean-square value $\mathbb{E}[Y^2(t)]$ is always nonnegative in accordance with (4.41), it follows that $S_{XX}(f) \approx \mathbb{E}[Y^2(t)]/(2\Delta f)$ must also be nonnegative.

PROPERTY 5 **Symmetry**

The power spectral density of a real-valued weakly stationary process is an even function of frequency; that is,

$$S_{XX}(-f) = S_{XX}(f) \tag{4.47}$$

This property is readily obtained by first substituting $-f$ for the variable f in (4.42):

$$S_{XX}(-f) = \int_{-\infty}^{\infty} R_{XX}(\tau)\exp(j2\pi f\tau)\,d\tau$$

Next, substituting $-\tau$ for τ, and recognizing that $R_{XX}(-\tau) = R_{XX}(\tau)$ in accordance with Property 2 of the autocorrelation function described in (4.12), we get

$$S_{XX}(-f) = \int_{-\infty}^{\infty} R_{XX}(\tau)\exp(-j2\pi f\tau)\,d\tau = S_{XX}(f)$$

which is the desired result. It follows, therefore, that the graph of the power spectral density $S_{XX}(f)$, plotted versus frequency f, is symmetric about the origin.

PROPERTY 6 **Normalization**

The power spectral density, appropriately normalized, has the properties associated with a probability density function in probability theory.

The normalization we have in mind here is with respect to the total area under the graph of the power spectral density (i.e., the mean-square value of the process). Consider then the function

$$p_{XX}(f) = \frac{S_{XX}(f)}{\int_{-\infty}^{\infty} S_{XX}(f)\,\mathrm{d}f} \tag{4.48}$$

In light of Properties 3 and 4, we note that $p_{XX}(f) \geq 0$ for all f. Moreover, the total area under the function $p_{XX}(f)$ is unity. Hence, the normalized power spectral density, as defined in (4.48), behaves in a manner similar to a probability density function.

Building on Property 6, we may go on to define the *spectral distribution function* of a weakly stationary process $X(t)$ as

$$F_{XX}(f) = \int_{-\infty}^{f} p_{XX}(\nu)\,\mathrm{d}\nu \tag{4.49}$$

which has the following properties:

1. $F_{XX}(-\infty) = 0$
2. $F_{XX}(\infty) = 1$
3. $F_{XX}(f)$ is a nondecreasing function of the frequency f.

Conversely, we may state that every nondecreasing and bounded function $F_{XX}(f)$ is the spectral distribution function of a weakly stationary process.

Just as important, we may also state that the spectral distribution function $F_{XX}(f)$ has all the properties of the cumulative distribution function in probability theory, discussed in Chapter 3.

EXAMPLE 5 **Sinusoidal Wave with Random Phase (continued)**

Consider the stochastic process $X(t) = A\cos(2\pi f_c t + \Theta)$, where Θ is a uniformly distributed random variable over the interval $[-\pi, \pi]$. The autocorrelation function of this stochastic process is given by (4.17), which is reproduced here for convenience:

$$R_{XX}(\tau) = \frac{A^2}{2}\cos(2\pi f_c \tau)$$

Let $\delta(f)$ denote the delta function at $f = 0$. Taking the Fourier transform of both sides of the formula defining $R_{XX}(\tau)$, we find that the power spectral density of the sinusoidal process $X(t)$ is

$$S_{XX}(f) = \frac{A^2}{4}[\delta(f - f_c) + \delta(f + f_c)] \tag{4.50}$$

which consists of a pair of delta functions weighted by the factor $A^2/4$ and located at $\pm f_c$, as illustrated in Figure 4.11. Since the total area under a delta function is one, it follows that the total area under $S_{XX}(f)$ is equal to $A^2/2$, as expected.

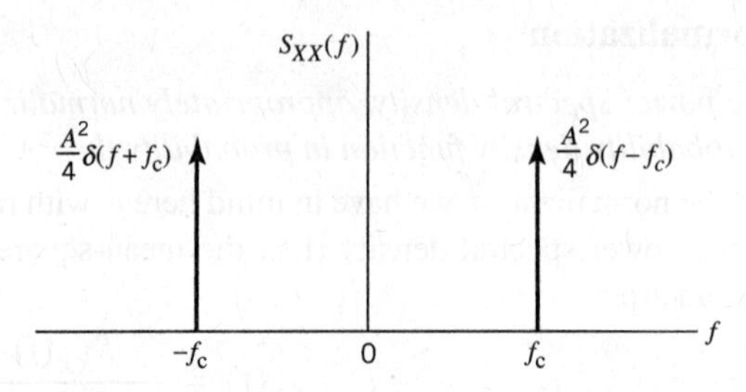

Figure 4.11 Power spectral density of sine wave with random phase; $\delta(f)$ denotes the delta function at $f = 0$.

EXAMPLE 6 **Random Binary Wave (continued)**

Consider again a random binary wave consisting of a sequence of 1s and 0s represented by the values $+A$ and $-A$ respectively. In Example 3 we showed that the autocorrelation function of this random process has the triangular form

$$R_{XX}(\tau) = \begin{cases} A^2\left(1 - \frac{|\tau|}{T}\right), & |\tau| < T \\ 0, & |\tau| \geq T \end{cases}$$

The power spectral density of the process is therefore

$$S_{XX}(f) = \int_{-T}^{T} A^2\left(1 - \frac{|\tau|}{T}\right) \exp(-j2\pi f\tau)\, d\tau$$

Using the Fourier transform of a triangular function (see Table 2.2 of Chapter 2), we obtain

$$S_{XX}(f) = A^2 T \operatorname{sinc}^2(fT) \tag{4.51}$$

which is plotted in Figure 4.12. Here again we see that the power spectral density is non-negative for all f and that it is an even function of f. Noting that $R_{XX}(0) = A^2$ and using

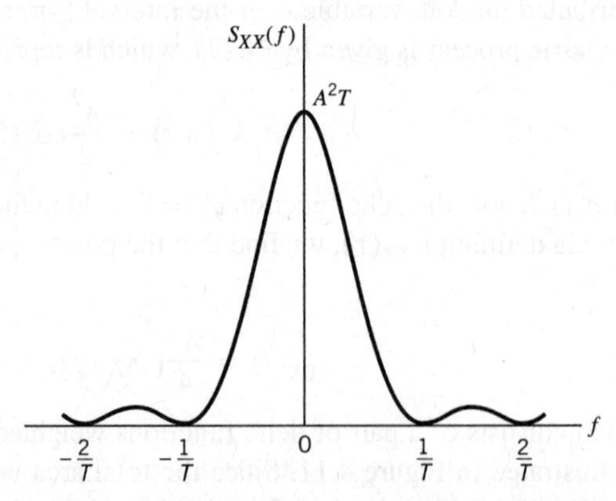

Figure 4.12 Power spectral density of random binary wave.

Property 2 of power spectral density, we find that the total area under $S_{XX}(f)$, or the average power of the random binary wave described here, is A^2, which is intuitively satisfying.

Generalization of Equation (4.51)

It is informative to generalize (4.51) so that it assumes a more broadly applicable form. With this objective in mind, we first note that the energy spectral density (i.e., the squared magnitude of the Fourier transform) of a rectangular pulse $g(t)$ of amplitude A and duration T is given by

$$E_g(f) = A^2T^2 \operatorname{sinc}^2(fT) \tag{4.52}$$

We may therefore express (4.51) in terms of $E_g(f)$ simply as

$$S_{XX}(f) = \frac{E_g(f)}{T} \tag{4.53}$$

In words, (4.53) states:

> For a random binary wave $X(t)$ in which binary symbols 1 and 0 are represented by pulses $g(t)$ and $-g(t)$ respectively, the power spectral density $S_{XX}(f)$ is equal to the energy spectral density $E_g(f)$ of the symbol-shaping pulse $g(t)$ divided by the symbol duration T.

EXAMPLE 7 Mixing of a Random Process with a Sinusoidal Process

A situation that often arises in practice is that of *mixing* (i.e., multiplication) of a weakly stationary process $X(t)$ with a sinusoidal wave $\cos(2\pi f_c t + \Theta)$, where the phase Θ is a random variable that is uniformly distributed over the interval $[0, 2\pi]$. The addition of the random phase Θ in this manner merely recognizes the fact that the time origin is arbitrarily chosen when both $X(t)$ and $\cos(2\pi f_c t + \Theta)$ come from physically independent sources, as is usually the case in practice. We are interested in determining the power spectral density of the stochastic process

$$Y(t) = X(t)\cos(2\pi f_c t + \Theta) \tag{4.54}$$

Using the definition of autocorrelation function of a weakly stationary process and noting that the random variable Θ is independent of $X(t)$, we find that the autocorrelation function of the process $Y(t)$ is given by

$$\begin{aligned} R_{YY}(\tau) &= \mathbb{E}[Y(t+\tau)Y(t)] \\ &= \mathbb{E}[X(t+\tau)\cos(2\pi f_c t + 2\pi f_c \tau + \Theta)X(t)\cos(2\pi f_c t + \Theta)] \\ &= \mathbb{E}[X(t+\tau)x(t)]\mathbb{E}[\cos(2\pi f_c t + 2\pi f_c \tau + \Theta)\cos(2\pi f_c t + \Theta)] \\ &= \frac{1}{2}R_{XX}(\tau)\mathbb{E}[\cos(2\pi f_c t) + \cos(4\pi f_c t + 2\pi f_c \tau + 2\Theta)] \\ &= \frac{1}{2}R_{XX}(\tau)\cos(2\pi f_c t) \end{aligned} \tag{4.55}$$

Since the power spectral density of a weakly stationary process is the Fourier transform of its autocorrelation function, we may go on to express the relationship between the power spectral densities of the processes $X(t)$ and $Y(t)$ as follows:

$$S_{YY}(f) = \frac{1}{4}[S_{XX}(f-f_c) + S_{XX}(f+f_c)] \tag{4.56}$$

Equation (4.56) teaches us that the power spectral density of the stochastic process $Y(t)$ defined in (4.54) can be obtained as follows:

> Shift the given power spectral density $S_{XX}(f)$ of the weakly stationary process $X(t)$ to the right by f_c, shift it to the left by f_c, add the two shifted power spectra, and then divide the result by 4, thereby obtaining the desired power spectral density $S_{YY}(f)$.

Relationship between the Power Spectral Densities of Input and Output Weakly Stationary Processes

Let $S_{YY}(f)$ denote the power spectral density of the output stochastic processes $Y(t)$ obtained by passing the weakly stationary process $X(t)$ through a linear time-invariant filter of frequency response $H(f)$. Then, by definition, recognizing that the power spectral density of a weakly stationary process is equal to the Fourier transform of its autocorrelation function and using (4.32), we obtain

$$\begin{aligned} S_{YY}(f) &= \int_{-\infty}^{\infty} R_{YY}(\tau) \exp(-j2\pi f\tau)\, d\tau \\ &= \int_{-\infty}^{\infty}\int_{-\infty}^{\infty}\int_{-\infty}^{\infty} h(\tau_1)h(\tau_2)R_{XX}(\tau + \tau_1 - \tau_2) \exp(-j2\pi f\tau)\, d\tau_1\, d\tau_2\, d\tau \end{aligned} \tag{4.57}$$

Let $\tau + \tau_1 - \tau_2 = \tau_0$, or equivalently $\tau = \tau_0 - \tau_1 + \tau_2$. By making this substitution into (4.57), we find that $S_{YY}(f)$ may be expressed as the product of three terms:

- the frequency response $H(f)$ of the filter;
- the complex conjugate of $H(f)$; and
- the power spectral density $S_{XX}(f)$ of the input process $X(t)$.

We may thus simplify (4.57) as shown by

$$S_{YY}(f) = H(f)H^*(f)S_{XX}(f) \tag{4.58}$$

Since $|H(f)|^2 = H(f)H^*(f)$, we finally find that the relationship among the power spectral densities of the input and output processes is expressed in the frequency domain by

$$S_{YY}(f) = |H(f)|^2 S_{XX}(f) \tag{4.59}$$

Equation (4.59) states:

> The power spectral density of the output process $Y(t)$ equals the power spectral density of the input process $X(t)$, multiplied by the squared magnitude response of the filter.

By using (4.59), we can therefore determine the effect of passing a weakly stationary process through a stable, linear time-invariant filter. In computational terms, (4.59) is obviously easier to handle than its time-domain counterpart of (4.32) that involves the autocorrelation function.

The Wiener–Khintchine Theorem

At this point in the discussion, a basic question that comes to mind is the following:

> Given a function $\rho_{XX}(\tau)$ whose argument is some time shift τ, how do we know that $\rho_{XX}(\tau)$ is the legitimate normalized autocorrelation function of a weakly stationary process $X(t)$?

The answer to this question is embodied in a theorem that was first proved by Wiener (1930) and at a later date by Khintchine (1934). Formally, the *Wiener–Khintchine theorem*[4] states:

> A necessary and sufficient condition for $\rho_{XX}(\tau)$ to be the normalized autocorrelation function of a weakly stationary process $X(t)$ is that there exists a distribution function $F_{XX}(f)$ such that for all possible values of the time shift τ, the function $\rho_{XX}(\tau)$ may be expressed in terms of the well-known *Fourier–Stieltjes theorem*, defined by

$$\rho_{XX}(\tau) = \int_{-\infty}^{\infty} \exp(j2\pi f\tau)\, dF_{XX}(f) \tag{4.60}$$

The Wiener–Khintchine theorem described in (4.60) is of fundamental importance to a theoretical treatment of weakly stationary processes.

Referring back to the definition of the spectral distribution function $F_{XX}(f)$ given in (4.49), we may express the *integrated spectrum* $dF_{XX}(f)$ as

$$dF_{XX}(f) = p_{XX}(f)\, df \tag{4.61}$$

which may be interpreted as the probability of $X(t)$ contained in the frequency interval $[f, f + df]$. Hence, we may rewrite (4.60) in the equivalent form

$$\rho_{XX}(\tau) = \int_{-\infty}^{\infty} p_{XX}(f) \exp(j2\pi f\tau)\, df \tag{4.62}$$

which expresses $\rho_{XX}(\tau)$ as the inverse Fourier transform of $p_{XX}(f)$. At this point, we proceed by taking three steps:

1. Substitute (4.14) for $\rho_{XX}(\tau)$ on the left-hand side of (4.62).
2. Substitute (4.48) for $p_{XX}(\tau)$ inside the integral on the right-hand side of (4.62).
3. Use Property 3 of power spectral density in Section 4.7.

The end result of these three steps is the reformulation of (4.62) as shown by

$$\frac{R_{XX}(\tau)}{R_{XX}(0)} = \int_{-\infty}^{\infty} \frac{S_{XX}(f)}{R_{XX}(0)} \exp(j2\pi f\tau)\, df$$

Hence, canceling out the common term $R_{XX}(0)$, we obtain

$$R_{XX}(\tau) = \int_{-\infty}^{\infty} S_{XX}(f) \exp(j2\pi f\tau)\, df \tag{4.63}$$

which is a rewrite of (4.43). We may argue, therefore, that basically the two Wiener–Khintchine equations follow from either one of the following two approaches:

1. The definition of the power spectral density as the Fourier transform of the autocorrelation function, which was first derived in (4.38).
2. The Wiener–Khintchine theorem described in (4.60).

4.8 Another Definition of the Power Spectral Density

Equation (4.38) provides one definition of the power spectral density $S_{XX}(f)$ of a weakly stationary process $X(t)$; that is, $S_{XX}(f)$ is the Fourier transform of the autocorrelation function $R_{XX}(\tau)$ of the process $X(t)$. We arrived at this definition by working on the mean-square value (i.e., average power) of the process $Y(t)$ produced at the output of a linear time-invariant filter, driven by a weakly stationary process $X(t)$. In this section, we provide another definition of the power spectral density by working on the process $X(t)$ directly. The definition so developed is not only mathematically satisfying, but it also provides another way of interpreting the power spectral density.

Consider, then, a stochastic process $X(t)$, which is known to be weakly stationary. Let $x(t)$ represent a *sample function* of the process $X(t)$. For the sample function to be Fourier transformable, it must be absolutely integrable; that is,

$$\int_{-\infty}^{\infty} |x(t)|\, dt < \infty$$

This condition can never be satisfied by any sample function $x(t)$ of infinite duration. To get around this problem, we consider a truncated segment of $x(t)$ defined over the observation interval $-T \leq t \leq T$, as illustrated in Figure 4.13, as shown by

$$x_T(t) = \begin{cases} x(t), & -T \leq t \leq T \\ 0, & \text{otherwise} \end{cases} \tag{4.64}$$

Clearly, the truncated signal $x_T(t)$ has finite energy; therefore, it is Fourier transformable. Let $X_T(f)$ denote the Fourier transform of $x_T(t)$, as shown by the transform pair:

$$x_T(t) \rightleftharpoons X_T(f)$$

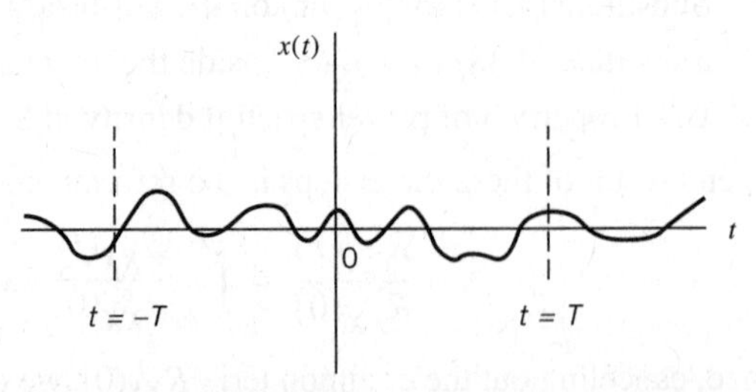

Figure 4.13 Illustration of the truncation of a sample $x(t)$ for Fourier transformability; the actual function $x(t)$ extends beyond the observation interval $(-T, T)$ as shown by the dashed lines.

in light of which we may invoke *Rayleigh's energy theorem* (Property 14 in Table 2.1) to write

$$\int_{-\infty}^{\infty} |x_T(t)|^2 \, dt = \int_{-\infty}^{\infty} |X_T(f)|^2 \, df$$

Since (4.64) implies that

$$\int_{-\infty}^{\infty} |x_T(t)|^2 \, dt = \int_{-T}^{T} |x(t)|^2 \, dt$$

we may also apply Rayleigh's energy theorem to the problem at hand as follows:

$$\int_{-T}^{T} |x(t)|^2 \, dt = \int_{-\infty}^{\infty} |X_T(f)|^2 \, df \tag{4.65}$$

With the two sides of (4.65) based on a single realization of the process $X(t)$, they are both subject to numerical variability (i.e., instability) as we go from one sample function of the process $X(t)$ to another. To mitigate this difficulty, we take the ensemble average of (4.65), and thus write

$$\mathbb{E}\left[\int_{-T}^{T} |x(t)|^2\right] dt = \mathbb{E}\left[\int_{-\infty}^{\infty} |X_T(f)|^2\right] df \tag{4.66}$$

What we have in (4.66) are two energy-based quantities. However, in the weakly stationary process $X(t)$, we have a process with some finite power. To put matters right, we multiply both sides of (4.66) by the scaling factor $1/(2T)$ and take the limiting form of the equation as the observation interval T approaches infinity. In so doing, we obtain

$$\lim_{T \to \infty} \frac{1}{2T} \mathbb{E}\left[\int_{-T}^{T} |x(t)|^2 \, dt\right] = \lim_{T \to \infty} \mathbb{E}\left[\int_{-\infty}^{\infty} \frac{|X_T(f)|^2}{2T} \, df\right] \tag{4.67}$$

The quantity on the left-hand side of (4.67) is now recognized as the average power of the process $X(t)$, denoted by P_{av}, which applies to all possible sample functions of the process $X(t)$. We may therefore recast (4.67) in the equivalent form

$$P_{av} = \lim_{T \to \infty} \mathbb{E}\left[\int_{-\infty}^{\infty} \frac{|X_T(f)|^2}{2T} \, df\right] \tag{4.68}$$

In (4.68), we next recognize that there are two mathematical operations of fundamental interest:

1. Integration with respect to the frequency f.
2. Limiting operation with respect to the total observation interval $2T$ followed by ensemble averaging.

These two operations, viewed in a composite manner, result in a statistically stable quantity defined by P_{av}. Therefore, it is permissible for us to interchange the order of the two operations on the right-hand side of (4.68), recasting this equation in the desired form:

$$P_{av} = \int_{-\infty}^{\infty} \left\{ \lim_{T \to \infty} \mathbb{E}\left[\frac{|X_T(f)|^2}{2T}\right] \right\} df \tag{4.69}$$

With (4.69) at hand, we are now ready to formulate another definition for the power spectral density as[5]

$$S_{XX}(f) = \lim_{T \to \infty} \mathbb{E}\left[\frac{|X_T(f)|^2}{2T}\right] \tag{4.70}$$

This new definition has the following interpretation:

> $S_{XX}(f)\,df$ is the average of the contributions to the total power from components in a weakly stationary process $X(t)$ with frequencies extending from f to $f + df$, and the average is taken over all possible realizations of the process $X(t)$.

This new interpretation of the power spectral density is all the more satisfying when (4.70) is substituted into (4.68), yielding

$$P_{\text{av}} = \int_{-\infty}^{\infty} S_{XX}(f)\,df \tag{4.71}$$

which is immediately recognized as another way of describing Property 3 of the power spectral density (i.e., (4.45). End-of-chapter Problem 4.8 invites the reader to prove other properties of the power spectral density, using the definition of (4.70).

One last comment must be carefully noted: in the definition of the power spectral density given in (4.70), it is *not* permissible to let the observation interval T approach infinity before taking the expectation; in other words, these two operations are *not* commutative.

4.9 Cross-spectral Densities

Just as the power spectral density provides a measure of the frequency distribution of a single weakly stationary process, cross-spectral densities provide measures of the frequency interrelationships between two such processes. To be specific, let $X(t)$ and $Y(t)$ be two jointly weakly stationary processes with their cross-correlation functions denoted by $R_{XY}(\tau)$ and $R_{YX}(\tau)$. We define the corresponding *cross-spectral densities* $S_{XY}(f)$ and $S_{YX}(f)$ of this pair of processes to be the Fourier transforms of their respective cross-correlation functions, as shown by

$$S_{XY}(f) = \int_{-\infty}^{\infty} R_{XY}(\tau) \exp(-j2\pi f\tau)\,d\tau \tag{4.72}$$

and

$$S_{YX}(f) = \int_{-\infty}^{\infty} R_{YX}(\tau) \exp(-j2\pi f\tau)\,d\tau \tag{4.73}$$

The cross-correlation functions and cross-spectral densities form Fourier-transform pairs. Accordingly, using the formula for inverse Fourier transformation, we may also respectively write

$$R_{XY}(\tau) = \int_{-\infty}^{\infty} S_{XY}(f) \exp(j2\pi f\tau)\,df \tag{4.74}$$

and

$$R_{YX}(\tau) = \int_{-\infty}^{\infty} S_{YX}(f) \exp(\mathrm{j}2\pi f\tau)\,\mathrm{d}f \tag{4.75}$$

The cross-spectral densities $S_{XY}(f)$ and $S_{YX}(f)$ are not necessarily real functions of the frequency f. However, substituting the following relationship (i.e., Property 2 of the autocorrelation function)

$$R_{XY}(\tau) = R_{YX}(-\tau)$$

into (4.72) and then using (4.73), we find that $S_{XY}(f)$ and $S_{YX}(f)$ are related as follows:

$$S_{XY}(f) = S_{YX}(-f) = S^*_{YX}(f) \tag{4.76}$$

where the asterisk denotes complex conjugation.

EXAMPLE 8 Sum of Two Weakly Stationary Processes

Suppose that the stochastic processes $X(t)$ and $Y(t)$ have zero mean and let their sum be denoted by

$$Z(t) = X(t) + Y(t)$$

The problem is to determine the power spectral density of the process $Z(t)$.

The autocorrelation function of $Z(t)$ is given by the second-order moment

$$\begin{aligned} M_{ZZ}(t,u) &= \mathbb{E}[Z(t)Z(u)] \\ &= \mathbb{E}[(X(t)+Y(t))(X(u)+Y(u))] \\ &= \mathbb{E}[X(t)X(u)] + \mathbb{E}[X(t)Y(u)] + \mathbb{E}[Y(t)X(u)] + \mathbb{E}[Y(t)Y(u)] \\ &= M_{XX}(t,u) + M_{XY}(t,u) + M_{YX}(t,u) + M_{YY}(t,u) \end{aligned}$$

Defining $\tau = t - u$ and assuming the joint weakly stationarity of the two processes, we may go on to write

$$R_{ZZ}(\tau) = R_{XX}(\tau) + R_{XY}(\tau) + R_{YX}(\tau) + R_{YY}(\tau) \tag{4.77}$$

Accordingly, taking the Fourier transform of both sides of (4.77), we get

$$S_{ZZ}(f) = S_{XX}(f) + S_{XY}(f) + S_{YX}(f) + S_{YY}(f) \tag{4.78}$$

This equation shows that the cross-spectral densities $S_{XY}(f)$ and $S_{YX}(f)$ represent the spectral components that must be added to the individual power spectral densities of a pair of correlated weakly stationary processes in order to obtain the power spectral density of their sum.

When the stationary processes $X(t)$ and $Y(t)$ are uncorrelated, the cross-spectral densities $S_{XY}(f)$ and $S_{YX}(f)$ are zero, in which case (4.78) reduces to

$$S_{ZZ}(f) = S_{XX}(f) + S_{YY}(f) \tag{4.79}$$

We may generalize this latter result by stating:

> When there is a multiplicity of zero-mean weakly stationary processes that are uncorrelated with each other, the power spectral density of their sum is equal to the sum of their individual power spectral densities.

EXAMPLE 9

Filtering of Two Jointly Weakly Stationary Processes

Consider next the problem of passing two jointly weakly stationary processes through a pair of separate, stable, linear time-invariant filters, as shown in Figure 4.14. The stochastic process $X(t)$ is the input to the filter of impulse response $h_1(t)$, and the stochastic process $Y(t)$ is the input to the filter of the impulse response $h_2(t)$. Let $V(t)$ and $Z(t)$ denote the processes at the respective filter outputs. The cross-correlation function of the output processes $V(t)$ and $Z(t)$ is therefore defined by the second-order moment

$$\begin{aligned} M_{VZ}(t, u) &= \mathbb{E}[V(t)Z(u)] \\ &= \mathbb{E}\left[\int_{-\infty}^{\infty} h_1(\tau_1)X(t-\tau_1)\, d\tau_1 \int_{-\infty}^{\infty} h_2(\tau_2)Y(u-\tau_2)\, d\tau_2\right] \\ &= \int_{-\infty}^{\infty}\int_{-\infty}^{\infty} h_1(\tau_1)h_2(\tau_2)\mathbb{E}[X(t-\tau_1)Y(u-\tau_2)]\, d\tau_1\, d\tau_2 \\ &= \int_{-\infty}\int_{-\infty}^{\infty} h_1(\tau_1)h_2(\tau_2)M_{XY}(t-\tau_1, u-\tau_2)\, d\tau_1\, d\tau_2 \end{aligned} \tag{4.80}$$

where $M_{XY}(t, u)$ is the cross-correlation function of $X(t)$ and $Y(t)$. Because the input stochastic processes are jointly weakly stationary, by hypothesis, we may set $\tau = t - u$, and thereby rewrite (4.80) as

$$R_{VZ}(\tau) = \int_{-\infty}^{\infty}\int_{-\infty}^{\infty} h_1(\tau_1)h_2(\tau_2)R_{XY}(\tau - \tau_1 + \tau_2)\, d\tau_1\, d\tau_2 \tag{4.81}$$

Taking the Fourier transform of both sides of (4.81) and using a procedure similar to that which led to the development of (4.39), we finally get

$$S_{VZ}(f) = H_1(f)H_2^*(f)S_{XY}(f) \tag{4.82}$$

where $H_1(f)$ and $H_2(f)$ are the frequency responses of the respective filters in Figure 4.14 and $H_2^*(f)$ is the complex conjugate of $H_2(f)$. This is the desired relationship between the cross-spectral density of the output processes and that of the input processes. Note that (4.82) includes (4.59) as a special case.

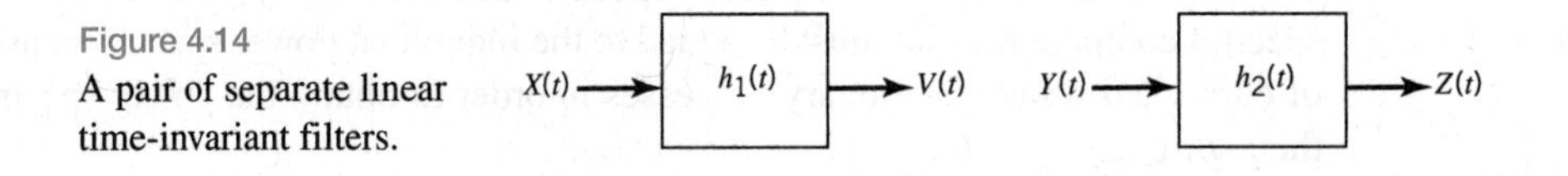

Figure 4.14
A pair of separate linear time-invariant filters.

4.10 The Poisson Process

Having covered the basics of stochastic process theory, we now turn our attention to different kinds of stochastic processes that are commonly encountered in the study of communication systems. We begin the study with the Poisson process,[6] which is the simplest process dealing with the issue of counting the number of occurrences of random events.

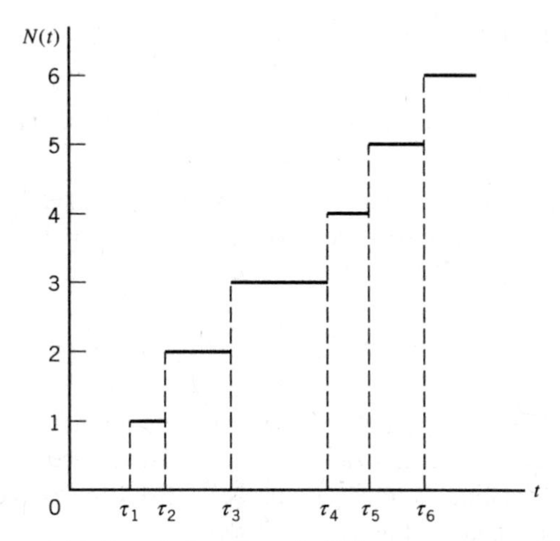

Figure 4.15 Sample function of a Poisson counting process.

Consider, for example, a situation in which events occur at random instants of time, such that the *average rate* of events per second is equal to λ. The sample path of such a random process is illustrated in Figure 4.15, where τ_i denotes the occurrence time of the ith event with $i = 1, 2, \ldots$. Let $N(t)$ be the number of event occurrences in the time interval $[0, t]$. As illustrated in Figure 4.15, we see that $N(t)$ is a nondecreasing, integer-valued, continuous process. Let $p_{k,\tau}$ denote the probability that exactly k events occur during an interval of duration τ; that is,

$$p_{k,\tau} = \mathbb{P}[N(t, t+\tau) = k] \tag{4.83}$$

With this background, we may now formally define the Poisson process:

> A random counting process is said to be a Poisson process with average rate λ if it satisfies the three basic properties listed below.

PROPERTY 1 **Time Homogeneity**

The probability $p_{k,\tau}$ of k event occurrences is the same for all intervals of the same duration τ.

The essence of Property 1 is that the events are equally likely at all times.

PROPERTY 2 **Distribution Function**

The number of event occurrences, $N_{0,t}$ in the interval $[0, t]$ has a distribution function with mean λt, defined by

$$\mathbb{P}[N(t) = k] = \frac{(\lambda t)^k}{k!}\exp(-\lambda t), \qquad k = 0, 1, 2, \ldots \tag{4.84}$$

That is, the time between events is *exponentially distributed.*

From Chapter 3, this distribution function is recognized to be the *Poisson distribution*. It is for this reason that $N(t)$ is called the *Poisson process*.

PROPERTY 3 **Independence**

The numbers of events in nonoverlapping time intervals are statistically independent, regardless of how small or large the intervals happen to be and no matter how close or distant they could be.

Property 3 is the most distinguishing property of the Poisson process. To illustrate the significance of this property, let $[t_i, u_i]$ for $i = 1, 2, \ldots, k$ denote k disjoint intervals on the line $[0, \infty]$. We may then write

$$\mathbb{P}[N(t_1, u_1) = n_1; N(t_2, u_2) = n_2; \ldots; N(t_k, u_k) = t_k] = \prod_{i=1}^{k} \mathbb{P}[N(t_i, u_i) = n_i] \quad (4.85)$$

The important point to take from this discussion is that these three properties provide a complete characterization of the Poisson process.

This kind of stochastic process arises, for example, in the statistical characterization of a special kind of noise called *shot noise* in electronic devices (e.g., diodes and transistors), which arises due to the discrete nature of current flow.

4.11 The Gaussian Process

The second stochastic process of interest is the *Gaussian process*, which builds on the Gaussian distribution discussed in Chapter 3. The Gaussian process is by far the most frequently encountered random process in the study of communication systems. We say so for two reasons: practical applicability and mathematical tractability.[7]

Let us suppose that we observe a stochastic process $X(t)$ for an interval that starts at time $t = 0$ and lasts until $t = T$. Suppose also that we weight the process $X(t)$ by some function $g(t)$ and then integrate the product $g(t)X(t)$ over the observation interval $[0, T]$, thereby obtaining the random variable

$$Y = \int_0^T g(t)X(t)\,\mathrm{d}t \quad (4.86)$$

We refer to Y as a *linear functional* of $X(t)$. The distinction between a function and a functional should be carefully noted. For example, the sum $Y = \Sigma_{i=1}^{N} a_i X_i$, where the a_i are constants and the X_i are random variables, is a linear function of the X_i; for each observed set of values for the random variable X_i, we have a corresponding value for the random variable Y. On the other hand, the value of the random variable Y in (4.86) depends on the course of the *integrand function* $g(t)X(t)$ over the entire observation interval from 0 to T. Thus, a functional is a quantity that depends on the entire course of one or more functions rather than on a number of discrete variables. In other words, the domain of a functional is a space of admissible functions rather than a region of coordinate space.

If, in (4.86), the weighting function $g(t)$ is such that the mean-square value of the random variable Y is finite and if the random variable Y is a *Gaussian-distributed* random variable for every $g(t)$ in this class of functions, then the process $X(t)$ is said to be a *Gaussian process*. In words, we may state:

> A process $X(t)$ is said to be a Gaussian process if every linear functional of $X(t)$ is a Gaussian random variable.

From Chapter 3 we recall that the random variable Y has a *Gaussian distribution* if its probability density function has the form

$$f_Y(y) = \frac{1}{\sqrt{2\pi}\sigma}\exp\left(-\frac{(y-\mu)^2}{2\sigma^2}\right) \tag{4.87}$$

where μ is the mean and σ^2 is the variance of the random variable Y. The distribution of a Gaussian process $X(t)$, sampled at some fixed time t_k, say, satisfies (4.87).

From a theoretical as well as practical perspective, a Gaussian process has two main virtues:

1. The Gaussian process has many properties that make analytic results possible; we will discuss these properties later in the section.
2. The stochastic processes produced by physical phenomena are often such that a Gaussian model is appropriate. Furthermore, the use of a Gaussian model to describe physical phenomena is often confirmed by experiments. Last, but by no means least, the central limit theorem (discussed in Chapter 3) provides mathematical justification for the Gaussian distribution.

Thus, the frequent occurrence of physical phenomena for which a Gaussian model is appropriate and the ease with which a Gaussian process is handled mathematically make the Gaussian process very important in the study of communication systems.

Properties of a Gaussian Process

PROPERTY 1 Linear Filtering

If a Gaussian process $X(t)$ is applied to a stable linear filter, then the stochastic process $Y(t)$ developed at the output of the filter is also Gaussian.

This property is readily derived by using the definition of a Gaussian process based on (4.86). Consider the situation depicted in Figure 4.8, where we have a linear time-invariant filter of impulse response $h(t)$, with the stochastic process $X(t)$ as input and the stochastic process $Y(t)$ as output. We assume that $X(t)$ is a Gaussian process. The process $Y(t)$ is related to $X(t)$ by the convolution integral

$$Y(t) = \int_0^T h(t-\tau)X(\tau)\,d\tau, \qquad 0 \le t < \infty \tag{4.88}$$

We assume that the impulse response $h(t)$ is such that the mean-square value of the output random process $Y(t)$ is finite for all time t in the range $0 \le t < \infty$, for which the process $Y(t)$ is defined. To demonstrate that the output process $Y(t)$ is Gaussian, we must show that any linear functional of it is also a Gaussian random variable. That is, if we define the random variable

$$Z = \int_0^\infty g_Y(t)\left[\int_0^T h(t-\tau)X(\tau)\,d\tau\right]dt \tag{4.89}$$

then Z must be a Gaussian random variable for every function $g_Y(t)$, such that the mean-square value of Z is finite. The two operations performed in the right-hand side of (4.89) are both linear; therefore, it is permissible to interchange the order of integrations, obtaining

$$Z = \int_0^T g(t)X(\tau)\,d\tau \tag{4.90}$$

where the new function

$$g(\tau) = \int_0^T g_Y(t)h(t-\tau)\,d\tau \tag{4.91}$$

Since $X(t)$ is a Gaussian process by hypothesis, it follows from (4.91) that Z must also be a Gaussian random variable. We have thus shown that if the input $X(t)$ to a linear filter is a Gaussian process, then the output $Y(t)$ is also a Gaussian process. Note, however, that although our proof was carried out assuming a time-invariant linear filter, this property is also true for any arbitrary stable linear filter.

PROPERTY 2 **Multivariate Distribution**

Consider the set of random variables $X(t_1)$, $X(t_2)$, ..., $X(t_n)$, obtained by sampling a stochastic process $X(t)$ at times $t_1, t_2, \ldots, t_n$. If the process $X(t)$ is Gaussian, then this set of random variables is jointly Gaussian for any n, with their n-fold joint probability density function being completely determined by specifying the set of means

$$\mu_{X(t_i)} = \mathbb{E}[X(t_i)], \qquad i = 1, 2, \ldots, n \tag{4.92}$$

and the set of covariance functions

$$C_X(t_k, t_i) = \mathbb{E}[(X(t_k) - \mu_{X(t_k)})(X(t_i) - \mu_{X(t_i)})], \qquad k, i = 1, 2, \ldots, n \tag{4.93}$$

Let the n-by-1 vector $\mathbf{X}$ denote the set of random variables $X(t_1)$, $X(t_2)$, ..., $X(t_n)$ derived from the Gaussian process $X(t)$ by sampling it at times $t_1, t_2, \ldots, t_n$. Let the vector $\mathbf{x}$ denote a sample value of $\mathbf{X}$. According to Property 2, the random vector $\mathbf{X}$ has a *multivariate Gaussian distribution,* defined in matrix form as

$$f_{X(t_1), X(t_2), \ldots, X(t_n)}(x_1, x_2, \ldots, x_n) = \frac{1}{(2\pi)^{n/2}\Delta^{1/2}}\exp\left[-\frac{1}{2}(\mathbf{x}-\boldsymbol{\mu})^{\mathrm{T}}\boldsymbol{\Sigma}^{-1}(\mathbf{x}-\boldsymbol{\mu})\right] \tag{4.94}$$

where the superscript T denotes matrix transposition, the mean vector

$$\boldsymbol{\mu} = [\mu_1, \mu_2, \ldots, \mu_n]^{\mathrm{T}}$$

the covariance matrix

$$\boldsymbol{\Sigma} = \{C_X(t_k, t_i)\}_{k, i=1}^{n}$$

$\boldsymbol{\Sigma}^{-1}$ is the inverse of the covariance matrix $\boldsymbol{\Sigma}$, and Δ is the determinant of the covariance matrix $\boldsymbol{\Sigma}$.

Property 2 is frequently used as the definition of a Gaussian process. However, this definition is more difficult to use than that based on (4.86) for evaluating the effects of filtering on a Gaussian process.

Note also that the covariance matrix $\boldsymbol{\Sigma}$ is a symmetric nonnegative definite matrix. For a nondegenerate Gaussian process, $\boldsymbol{\Sigma}$ is positive definite, in which case the covariance matrix is invertible.

PROPERTY 3 **Stationarity**

If a Gaussian process is weakly stationary, then the process is also strictly stationary.

This follows directly from Property 2.

PROPERTY 4 **Independence**

If the random variables $X(t_1)$, $X(t_2)$, ..., $X(t_n)$, obtained by respectively sampling a Gaussian process $X(t)$ at times t_1, t_2, ..., t_n, are uncorrelated, that is

$$\mathbb{E}[(X(t_k) - \mu_{X(t_k)})(X(t_i) - \mu_{X(t_i)})] = 0 \qquad i \neq k \tag{4.95}$$

then these random variables are statistically independent.

The uncorrelatedness of $X(t_1)$, ..., $X(t_n)$ means that the covariance matrix $\mathbf{\Sigma}$ is reduced to a diagonal matrix, as shown by

$$\mathbf{\Sigma} = \begin{bmatrix} \sigma_1^2 & & \mathbf{0} \\ & \ddots & \\ \mathbf{0} & & \sigma_n^2 \end{bmatrix} \tag{4.96}$$

where the $\mathbf{0}$s denote two sets of elements whose values are all zero, and the diagonal terms

$$\sigma_i^2 = \mathbb{E}[X(t_i) - \mathbb{E}[X(t_i)]]^2, \qquad i = 1, 2, \ldots, n \tag{4.97}$$

Under this special condition, the multivariate Gaussian distribution described in (4.94) simplifies to

$$f_{\mathbf{X}}(\mathbf{x}) = \prod_{i=1}^{n} f_{X_i}(x_i) \tag{4.98}$$

where $X_i = X(t_i)$ and

$$f_{X_i}(x_i) = \frac{1}{\sqrt{2\pi}\,\sigma_i}\exp\left[-\frac{(x_i - \mu_{X_i})^2}{2\sigma_i^2}\right], \qquad i = 1, 2, \ldots, n \tag{4.99}$$

In words, if the Gaussian random variables $X(t_1)$, $X(t_2)$, ..., $X(t_n)$ are uncorrelated, then they are statistically independent, which, in turn, means that the joint probability density function of this set of random variables is expressed as the product of the probability density functions of the individual random variables in the set.

4.12 Noise

The term *noise* is used customarily to designate unwanted signals that tend to disturb the transmission and processing of signals in communication systems, and over which we have incomplete control. In practice, we find that there are many potential sources of noise in a communication system. The sources of noise may be external to the system (e.g., atmospheric noise, galactic noise, man-made noise) or internal to the system. The second category includes an important type of noise that arises from the phenomenon of *spontaneous fluctuations* of current flow that is experienced in all electrical circuits. In a physical context, the most common examples of the spontaneous fluctuation phenomenon are *shot noise*, which, as stated in Section 4.10, arises because of the discrete nature of current flow in electronic devices; and *thermal noise*, which is attributed to the random motion of electrons in a conductor.[8] However, insofar as the noise analysis of communication systems is concerned, be they analog or digital, the analysis is customarily based on a source of noise called white-noise, which is discussed next.

White Noise

This source of noise is *idealized*, in that its power spectral density is assumed to be constant and, therefore, independent of the operating frequency. The adjective "white" is used in the sense that white light contains equal amounts of all frequencies within the visible band of electromagnetic radiation. We may thus make the statement:

> White noise, denoted by $W(t)$, is a stationary process whose power spectral density $S_W(f)$ has a constant value across the entire frequency interval $-\infty < f < \infty$.

Clearly, white-noise can only be meaningful as an abstract mathematical concept; we say so because a constant power spectral density corresponds to an unbounded spectral distribution function and, therefore, infinite average power, which is physically nonrealizable. Nevertheless, the utility of white-noise is justified in the study of communication theory by virtue of the fact that it is used to model *channel noise* at the front end of a receiver. Typically, the receiver includes a filter whose frequency response is essentially zero outside a frequency band of some finite value. Consequently, when white-noise is applied to the model of such a receiver, there is no need to describe how the power spectral density $S_{WW}(f)$ falls off outside the usable frequency band of the receiver.[9]

Let

$$S_{WW}(f) = \frac{N_0}{2} \quad \text{for all } f \tag{4.100}$$

as illustrated in Figure 4.16a. Since the autocorrelation function is the inverse Fourier transform of the power spectral density in accordance with the Wiener–Khintchine relations, it follows that for white-noise the autocorrelation function is

$$R_{WW}(\tau) = \frac{N_0}{2}\delta(\tau) \tag{4.101}$$

Hence, the autocorrelation function of white noise consists of a delta function weighted by the factor $N_0/2$ and occurring at the time shift $\tau = 0$, as shown in Figure 4.16b.

Since $R_{WW}(\tau)$ is zero for $\tau \neq 0$, it follows that any two different samples of white noise are uncorrelated no matter how closely together in time those two samples are taken. If the white noise is also Gaussian, then the two samples are statistically independent in accordance with Property 4 of the Gaussian process. In a sense, then, white Gaussian noise represents the ultimate in "randomness."

The utility of a white-noise process in the noise analysis of communication systems is parallel to that of an impulse function or delta function in the analysis of linear systems.

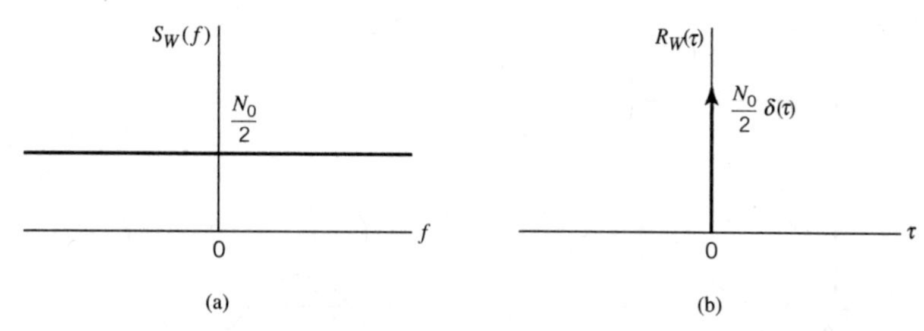

Figure 4.16 Characteristics of white-noise: (a) power spectral density; (b) autocorrelation function.

Just as we may observe the effect of an impulse only after it has been passed through a linear system with a finite bandwidth, so it is with white noise whose effect is observed only after passing through a similar system. We may therefore state:

> As long as the bandwidth of a noise process at the input of a system is appreciably larger than the bandwidth of the system itself, then we may model the noise process as white noise.

EXAMPLE 10 Ideal Low-pass Filtered White Noise

Suppose that a white Gaussian noise of zero mean and power spectral density $N_0/2$ is applied to an ideal low-pass filter of bandwidth B and passband magnitude response of one. The power spectral density of the noise $N(t)$ appearing at the filter output, as shown in Figure 4.17a, is therefore

$$S_{NN}(f) = \begin{cases} \dfrac{N_0}{2}, & -B < f < B \\ 0, & |f| > B \end{cases} \tag{4.102}$$

Since the autocorrelation function is the inverse Fourier transform of the power spectral density, it follows that

$$\begin{aligned} R_{NN}(\tau) &= \int_{-B}^{B} \frac{N_0}{2} \exp(j2\pi f\tau)\, df \\ &= N_0 B \operatorname{sinc}(2B\tau) \end{aligned} \tag{4.103}$$

whose dependence on τ is plotted in Figure 4.17b. From this figure, we see that $R_{NN}(\tau)$ has the maximum value N_0B at the origin and it passes through zero at $\tau = \pm k/(2B)$, where $k = 1, 2, 3, \ldots$.

Since the input noise $W(t)$ is Gaussian (by hypothesis), it follows that the band-limited noise $N(t)$ at the filter output is also Gaussian. Suppose, then, that $N(t)$ is sampled at the rate of $2B$ times per second. From Figure 4.17b, we see that the resulting noise samples are uncorrelated and, being Gaussian, they are statistically independent. Accordingly, the joint probability density function of a set of noise samples obtained in this way is equal to the product of the individual probability density functions. Note that each such noise sample has a mean of zero and variance of N_0B.

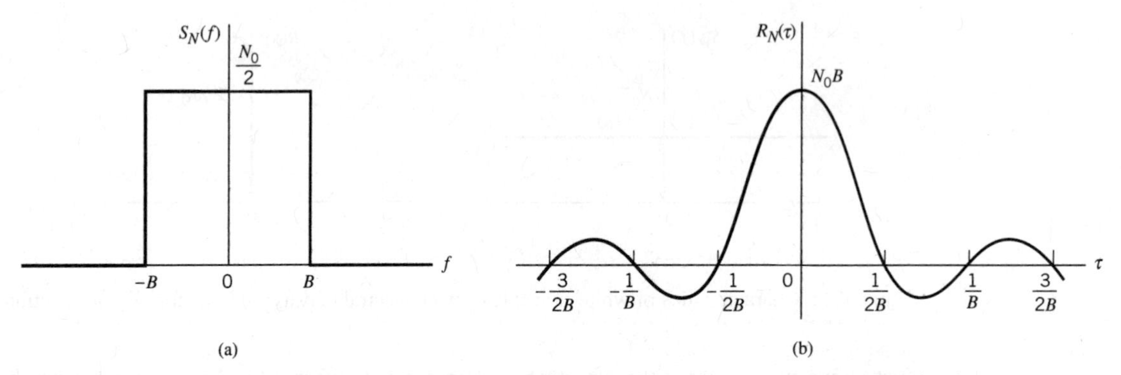

Figure 4.17 Characteristics of low-pass filtered white noise; (a) power spectral density; (b) autocorrelation function.

EXAMPLE 11

Correlation of White Noise with Sinusoidal Wave

Consider the sample function

$$w'(t) = \sqrt{\frac{2}{T}}\int_0^T w(t)\cos(2\pi f_c t)\,dt \tag{4.104}$$

which is the output of a correlator with white Gaussian noise sample function $w(t)$ and sinusoidal wave $\sqrt{2/T}\cos(2\pi f_c t)$ as its two inputs; the scaling factor $\sqrt{2/T}$ is included in (4.104) to make the sinusoidal wave input have unit energy over the interval $0 \le t \le T$. With $w(t)$ having zero mean, it immediately follows that the correlator output $w'(t)$ has zero mean too. The variance of the correlator output is therefore defined by

$$\begin{aligned}
\sigma_{W'}^2 &= \mathbb{E}\left[\frac{2}{T}\int_0^T\int_0^T w(t_1)\cos(2\pi f_c t_1)w(t_2)\cos(2\pi f_c t_2)\,dt_1\,dt_2\right] \\
&= \frac{2}{T}\int_0^T\int_0^T \mathbb{E}[w(t_1)w(t_2)]\cos(2\pi f_c t_1)\cos(2\pi f_c t_2)\,dt_1\,dt_2 \\
&= \frac{2}{T}\int_0^T\int_0^T \frac{N_0}{2}\delta(t_1 - t_2)\cos(2\pi f_c t_1)\cos(2\pi f_c t_2)\,dt_1\,dt_2
\end{aligned} \tag{4.105}$$

where, in the last line, we made use of (4.101). We now invoke the *sifting property* of the delta function, namely

$$\int_{-\infty}^{\infty} g(t)\delta(t)\,dt = g(0) \tag{4.106}$$

where $g(t)$ is a continuous function of time that has the value $g(0)$ at time $t = 0$. Hence, we may further simplify the expression for the noise variance as

$$\begin{aligned}
\sigma_{W'}^2 &= \frac{N_0}{2}\frac{2}{T}\int_{-T}^{T}\cos^2(2\pi f_c t)\,dt \\
&= \frac{N_0}{2T}\int_0^T[1 + \cos(4\pi f_c t)]\,dt \\
&= \frac{N_0}{2}
\end{aligned} \tag{4.107}$$

where, in the last line, it is assumed that the frequency f_c of the sinusoidal wave input is an integer multiple of the reciprocal of T for mathematical convenience.

4.13 Narrowband Noise

The receiver of a communication system usually includes some provision for *preprocessing* the received signal. Typically, the preprocessing takes the form of a *narrowband filter* whose bandwidth is just large enough to pass the modulated component of the received signal essentially undistorted, so as to limit the effect of channel noise passing through the receiver. The noise process appearing at the output of such a filter is called *narrowband noise*. With the spectral components of narrowband noise concentrated about some midband frequency $\pm f_c$ as in Figure 4.18a, we find that a sample function $n(t)$ of such a process appears somewhat similar to a sine wave of frequency f_c. The sample function $n(t)$ may, therefore, undulate slowly in both amplitude and phase, as illustrated in Figure 4.18b.

Consider, then, the $n(t)$ produced at the output of a narrowband filter in response to the sample function $w(t)$ of a white Gaussian noise process of zero mean and unit power spectral density applied to the filter input; $w(t)$ and $n(t)$ are sample functions of the respective processes $W(t)$ and $N(t)$. Let $H(f)$ denote the transfer function of this filter. Accordingly, we may express the power spectral density $S_N(f)$ of the noise $N(t)$ in terms of $H(f)$ as

$$S_{NN}(f) = |H(f)|^2 \tag{4.108}$$

On the basis of this equation, we may now make the following statement:

> Any narrowband noise encountered in practice may be modeled by applying a white-noise to a suitable filter in the manner described in (4.108).

In this section we wish to represent the narrowband noise $n(t)$ in terms of its in-phase and quadrature components in a manner similar to that described for a narrowband signal in Section 2.10. The derivation presented here is based on the idea of pre-envelope and related concepts, which were discussed in Chapter 2 on Fourier analysis of signals and systems.

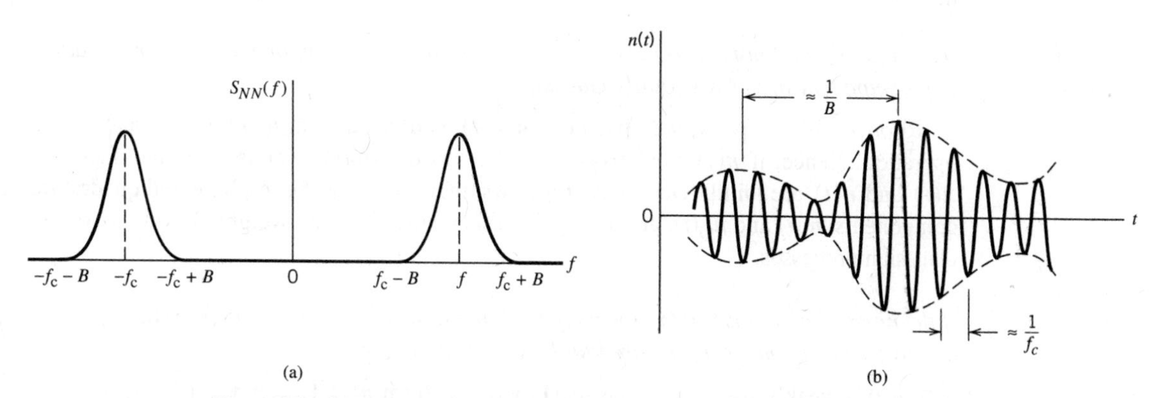

Figure 4.18 (a) Power spectral density of narrowband noise. (b) Sample function of narrowband noise.

Let $n_+(t)$ and $\tilde{n}(t)$, respectively, denote the pre-envelope and complex envelope of the narrowband noise $n(t)$. We assume that the power spectrum of $n(t)$ is centered about the frequency f_c. Then we may write

$$n_+(t) = n(t) + j\hat{n}(t) \tag{4.109}$$

and

$$\tilde{n}(t) = n_+(t)\exp(-j2\pi f_c t) \tag{4.110}$$

where $\hat{n}(t)$ is the Hilbert transform of $n(t)$. The complex envelope $\tilde{n}(t)$ may itself be expressed as

$$\tilde{n}(t) = n_I(t) + jn_Q(t) \tag{4.111}$$

Hence, combining (4.109) through (4.111), we find that the *in-phase component* $n_I(t)$ and the *quadrature component* $n_Q(t)$ of the narrowband noise $n(t)$ are

$$n_I(t) = n(t)\cos(2\pi f_c t) + \hat{n}(t)\sin(j2\pi f_c t) \tag{4.112}$$

and

$$n_Q(t) = \hat{n}(t)\cos(2\pi f_c t) - n(t)\sin(2\pi f_c t) \tag{4.113}$$

respectively. Eliminating $\hat{n}(t)$ between (4.112) and (4.113), we get the desired *canonical form* for representing the narrowband noise $n(t)$, as shown by

$$n(t) = n_I(t)\cos(2\pi f_c t) - n_Q(t)\sin(2\pi f_c t) \tag{4.114}$$

Using (4.112) to (4.114), we may now derive some important properties of the in-phase and quadrature components of a narrowband noise, as described next.

PROPERTY 1 *The in-phase component $n_I(t)$ and quadrature component $n_Q(t)$ of narrowband noise $n(t)$ have zero mean.*

To prove this property, we first observe that the noise $\hat{n}(t)$ is obtained by passing $n(t)$ through a linear filter (i.e., Hilbert transformer). Accordingly, $\hat{n}(t)$ will have zero mean because $n(t)$ has zero mean by virtue of its narrowband nature. Furthermore, from (4.112) and (4.113), we see that $n_I(t)$ and $n_Q(t)$ are weighted sums of $n(t)$ and $\hat{n}(t)$. It follows, therefore, that the in-phase and quadrature components, $n_I(t)$ and $n_Q(t)$, both have zero mean.

PROPERTY 2 *If the narrowband noise $n(t)$ is Gaussian, then its in-phase component $n_I(t)$ and quadrature component $n_Q(t)$ are jointly Gaussian.*

To prove this property, we observe that $\hat{n}(t)$ is derived from $n(t)$ by a linear filtering operation. Hence, if $n(t)$ is Gaussian, the Hilbert transform $\hat{n}(t)$ is also Gaussian, and $n(t)$ and $\hat{n}(t)$ are jointly Gaussian. It follows, therefore, that the in-phase and quadrature components, $n_I(t)$ and $n_Q(t)$, are jointly Gaussian, since they are weighted sums of jointly Gaussian processes.

PROPERTY 3 *If the narrowband noise $n(t)$ is weakly stationary, then its in-phase component $n_I(t)$ and quadrature component $n_Q(t)$ are jointly weakly stationary.*

If $n(t)$ is weakly stationary, so is its Hilbert transform $\hat{n}(t)$. However, since the in-phase and quadrature components, $n_I(t)$ and $n_Q(t)$, are both weighted sums of $n(t)$ and $\hat{n}(t)$

and the weighting functions, $\cos(2\pi f_c t)$ and $\sin(2\pi f_c t)$, vary with time, we cannot directly assert that $n_I(t)$ and $n_Q(t)$ are weakly stationary. To prove Property 3, we have to evaluate their correlation functions.

Using (4.112) and (4.113), we find that the in-phase and quadrature components, $n_I(t)$ and $n_Q(t)$, of a narrowband noise $n(t)$ have the same autocorrelation function, as shown by

$$R_{N_I N_I}(\tau) = R_{N_Q N_Q}(\tau) = R_{NN}(\tau)\cos(2\pi f_c \tau) + \hat{R}_{NN}(\tau)\sin(2\pi f_c \tau) \tag{4.115}$$

and their cross-correlation functions are given by

$$R_{N_I N_Q}(\tau) = -R_{N_Q N_I}(\tau) = R_{NN}(\tau)\sin(2\pi f_c \tau) - \hat{R}_{NN}(\tau)\cos(2\pi f_c \tau) \tag{4.116}$$

where $R_{NN}(\tau)$ is the autocorrelation function of $n(t)$, and $\hat{R}_{NN}(\tau)$ is the Hilbert transform of $R_{NN}(\tau)$. From (4.115) and (4.116), we readily see that the correlation functions $R_{N_I N_I}(\tau)$, $R_{N_Q N_Q}(\tau)$, and $R_{N_I N_Q}(\tau)$ of the in-phase and quadrature components $n_I(t)$ and $n_Q(t)$ depend only on the time shift τ. This dependence, in conjunction with Property 1, proves that $n_I(t)$ and $n_Q(t)$ are weakly stationary if the original narrowband noise $n(t)$ is weakly stationary.

PROPERTY 4 *Both the in-phase noise $n_I(t)$ and quadrature noise $n_Q(t)$ have the same power spectral density, which is related to the power spectral density $S_{NN}(f)$ of the original narrowband noise $n(t)$ as follows:*

$$S_{N_I N_I}(f) = S_{N_Q N_Q}(f) = \begin{cases} S_{NN}(f-f_c) + S_{NN}(f+f_c), & -B \le f \le B \\ 0, & \text{otherwise} \end{cases} \tag{4.117}$$

where it is assumed that $S_{NN}(f)$ occupies the frequency interval $f_c - B \le |f| \le f_c + B$ and $f_c > B$.

To prove this fourth property, we take the Fourier transforms of both sides of (4.115), and use the fact that

$$\begin{aligned} \mathbf{F}[\hat{R}_{NN}(\tau)] &= -j\,\mathrm{sgn}(f)\mathbf{F}[R_{NN}(\tau)] \\ &= -j\,\mathrm{sgn}(f)S_{NN}(f) \end{aligned} \tag{4.118}$$

We thus obtain the result

$$\begin{aligned} S_{N_I N_I}(f) &= S_{N_Q N_Q}(f) \\ &= \frac{1}{2}[S_{NN}(f-f_c) + S_{NN}(f+f_c)] \\ &\quad -\frac{1}{2}[S_{NN}(f-f_c)\,\mathrm{sgn}(f-f_c) - S_{NN}(f+f_c)\,\mathrm{sgn}(f+f_c)] \\ &= \frac{1}{2}S_{NN}(f-f_c)[1-\mathrm{sgn}(f-f_c)] + \frac{1}{2}S_{NN}(f+f_c)[1+\mathrm{sgn}(f+f_c)] \end{aligned} \tag{4.119}$$

Now, with the power spectral density $S_{NN}(f)$ of the original narrowband noise $n(t)$ occupying the frequency interval $f_c - B \le |f| \le f_c + B$, where $f_c > B$, as illustrated in Figure 4.19, we find that the corresponding shapes of $S_{NN}(f - f_c)$ and $S_{NN}(f + f_c)$ are as in Figures 4.19b and 4.19c respectively. Figures 4.19d, 4.19e, and 4.19f show the shapes of sgn(f), sgn($f - f_c$), and sgn($f + f_c$) respectively. Accordingly, we may make the following observation from Figure 4.19:

1. For frequencies defined by $-B \le f \le B$, we have

$$\text{sgn}(f - f_c) = -1$$

and

$$\text{sgn}(f + f_c) = +1$$

Hence, substituting these results into (4.119), we obtain

$$\begin{aligned} S_{N_I N_I}(f) &= S_{N_Q N_Q}(f) \\ &= S_{NN}(f - f_c) + S_{NN}(f + f_c), \qquad -B \le f \le B \end{aligned}$$

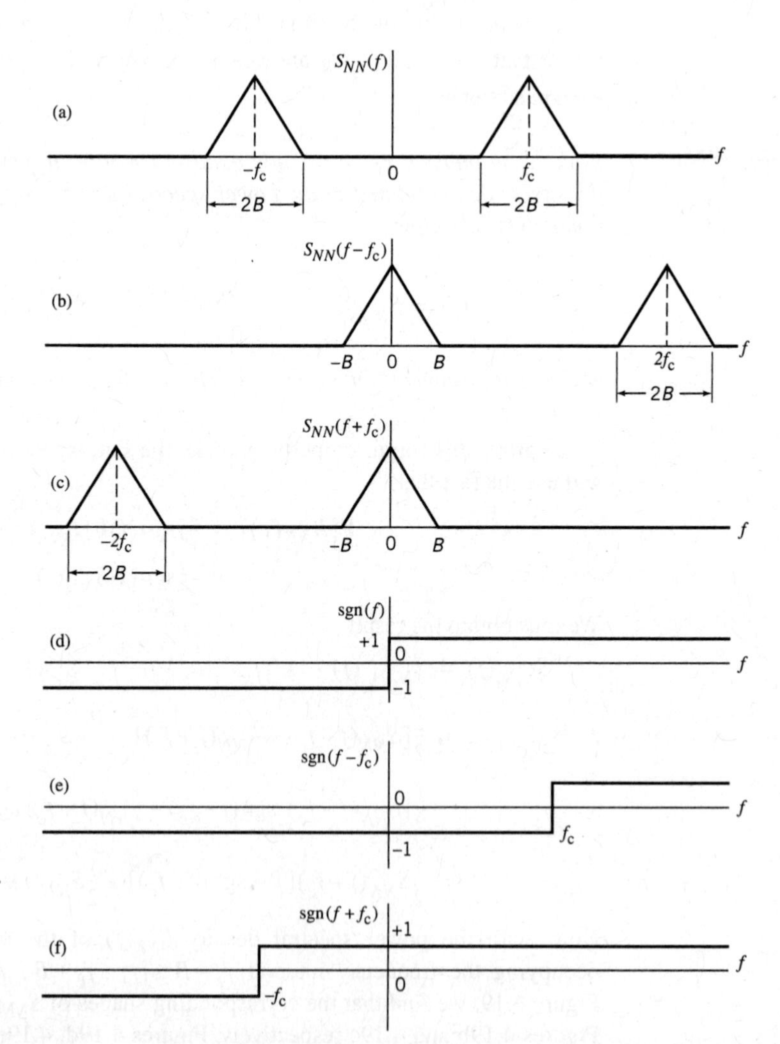

Figure 4.19
(a) Power spectral density $S_{NN}(f)$ pertaining to narrowband noise $n(t)$.
(b), (c) Frequency-shifted versions of $S_{NN}(f)$ in opposite directions.
(d) Signum function sgn(f).
(e), (f) Frequency-shifted versions of sgn(f) in opposite directions.

2. For $2f_c - B \le f \le 2f_c + B$, we have

$$\text{sgn}(f - f_c) = 1$$

and

$$\text{sgn}(f + f_c) = 0$$

with the result that $S_{N_I N_I}(f)$ and $S_{N_Q N_Q}(f)$ are both zero.

3. For $-2f_c - B \le f \le -2f_c + B$, we have

$$\text{sgn}(f - f_c) = 0$$

and

$$\text{sgn}(f + f_c) = -1$$

with the result that, here also, $S_{N_I N_I}(f)$ and $S_{N_Q N_Q}(f)$ are both zero.

4. Outside the frequency intervals defined in points 1, 2, and 3, both $S_{NN}(f - f_c)$ and $S_{NN}(f + f_c)$ are zero, and in a corresponding way, $S_{NN}(f - f_c)$ and $S_{N_Q N_Q}(f)$ are also zero.

Combining these results, we obtain the simple relationship defined in (4.117).

As a consequence of this property, we may extract the in-phase component $n_I(t)$ and quadrature component $n_Q(t)$, except for scaling factors, from the narrowband noise $n(t)$ by using the scheme shown in Figure 4.20a, where both low-pass filters have a cutoff frequency at B. The scheme shown in Figure 4.20a may be viewed as an *analyzer*. Given the in-phase component $n_I(t)$ and the quadrature component $n_Q(t)$, we may generate the narrowband noise $n(t)$ using the scheme shown in Figure 4.20b, which may be viewed as a *synthesizer*.

PROPERTY 5 *The in-phase and quadrature components $n_I(t)$ and $n_Q(t)$ have the same variance as the narrowband noise $n(t)$.*

This property follows directly from (4.117), according to which the total area under the power spectral density curve $n_I(t)$ or $n_Q(t)$ is the same as the total area under the power spectral density curve of $n(t)$. Hence, $n_I(t)$ and $n_Q(t)$ have the same mean-square value as $n(t)$. Earlier we showed that since $n(t)$ has zero mean, then $n_I(t)$ and $n_Q(t)$ have zero mean, too. It follows, therefore, that $n_I(t)$ and $n_Q(t)$ have the same variance as the narrowband noise $n(t)$.

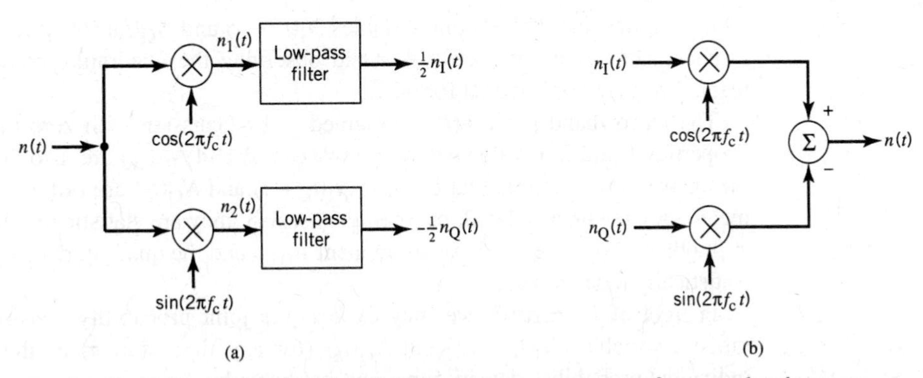

Figure 4.20 (a) Extraction of in-phase and quadrature components of a narrowband process. (b) Generation of a narrowband process from its in-phase and quadrature components.

PROPERTY 6 *The cross-spectral densities of the in-phase and quadrature components of a narrowband noise are purely imaginary, as shown by*

$$S_{N_I N_Q}(f) = -S_{N_Q N_I}(f)$$
$$= \begin{cases} j[S_N(f+f_c) - S_N(f-f_c)], & -B \le f \le B \\ 0, & \text{otherwise} \end{cases} \tag{4.120}$$

To prove this property, we take the Fourier transforms of both sides of (4.116), and use the relation of (4.118), obtaining

$$\begin{aligned} S_{N_I N_Q}(f) &= -S_{N_Q N_I}(f) \\ &= -\frac{j}{2}[S_{NN}(f-f_c) - S_{NN}(f+f_c)] \\ &\quad + \frac{j}{2}[S_{NN}(f-f_c)\operatorname{sgn}(f-f_c) + S_{NN}(f+f_c)\ \operatorname{sgn}(f+f_c)] \\ &= \frac{j}{2}S_{NN}(f+f_c)[1 + \operatorname{sgn}(f+f_c)] - \frac{j}{2}S_{NN}(f-f_c)[1 - \operatorname{sgn}(f-f_c)] \end{aligned} \tag{4.121}$$

Following a procedure similar to that described for proving Property 4, we find that (4.121) reduces to the form shown in (4.120).

PROPERTY 7 *If a narrowband noise n(t) is Gaussian with zero mean and a power spectral density SNN(f) that is locally symmetric about the midband frequency* $\pm f_c$, *then the in-phase noise* $n_I(t)$ *and the quadrature noise* $n_Q(t)$ *are statistically independent.*

To prove this property, we observe that if $S_{NN}(f)$ is locally symmetric about $\pm f_c$, then

$$S_{NN}(f-f_c) = S_{NN}(f+f_c), \qquad -B \le f \le B \tag{4.122}$$

Consequently, we find from (4.120) that the cross-spectral densities of the in-phase and quadrature components, $n_I(t)$ and $n_Q(t)$, are zero for all frequencies. This, in turn, means that the cross-correlation functions $S_{N_I N_Q}(f)$ and $S_{N_Q N_I}(f)$ are zero for all τ, as shown by

$$\mathbb{E}[N_I(t_k+\tau)N_Q(t_k+\tau)] = 0 \tag{4.123}$$

which implies that the random variables $N_I(t_k + \tau)$ and $N_Q(t_k)$ (obtained by observing the in-phase component at time $t_k + \tau$ and observing the quadrature component at time t_k respectively) are orthogonal for all τ.

The narrowband noise $n(t)$ is assumed to be Gaussian with zero mean; hence, from Properties 1 and 2 it follows that both $N_I(t_k + \tau)$ and $N_Q(t_k)$ are also Gaussian with zero mean. We thus conclude that because $N_I(t_k + \tau)$ and $N_Q(t_k)$ are orthogonal and have zero mean, they are uncorrelated, and being Gaussian, they are statistically independent for all τ. In other words, the in-phase component $n_I(t)$ and the quadrature component $n_Q(t)$ are statistically independent.

In light of Property 7, we may express the joint probability density function of the random variables $N_I(t_k + \tau)$ and $N_Q(t_k)$ (for any time shift τ) as the product of their individual probability density functions, as shown by

$$
\begin{aligned}
f_{N_\mathrm{I}(t_k+\tau),\,N_\mathrm{Q}(t_k)}(n_\mathrm{I},n_\mathrm{Q}) &= f_{N_\mathrm{I}(t_k+\tau)}(n_\mathrm{I})f_{N_\mathrm{Q}(t_k)}(n_\mathrm{Q}) \\
&= \frac{1}{\sqrt{2\pi}\,\sigma}\exp\left(-\frac{n_\mathrm{I}^2}{2\sigma^2}\right)\frac{1}{\sqrt{2\pi}\,\sigma}\exp\left(-\frac{n_\mathrm{Q}^2}{2\sigma^2}\right) \\
&= \frac{1}{2\pi\sigma^2}\left(-\frac{n_\mathrm{I}^2+n_\mathrm{Q}^2}{2\sigma^2}\right)
\end{aligned}
\tag{4.124}
$$

where σ^2 is the variance of the original narrowband noise $n(t)$. Equation (4.124) holds if, and only if, the spectral density $S_{NN}(f)$ or $n(t)$ is locally symmetric about $\pm f_\mathrm{c}$. Otherwise, this relation holds only for $\tau = 0$ or those values of τ for which $n_\mathrm{I}(t)$ and $n_\mathrm{Q}(t)$ are uncorrelated.

Summarizing Remarks

To sum up, if the narrowband noise $n(t)$ is zero mean, weakly stationary, and Gaussian, then its in-phase and quadrature components $n_\mathrm{I}(t)$ and $n_\mathrm{Q}(t)$ are both zero mean, jointly stationary, and jointly Gaussian. To evaluate the power spectral density of $n_\mathrm{I}(t)$ or $n_\mathrm{Q}(t)$, we may proceed as follows:

1. Shift the positive frequency portion of the power spectral density $S_{NN}(f)$ of the original narrowband noise $n(t)$ to the left by f_c.
2. Shift the negative frequency portion of $S_{NN}(f)$ to the right by f_c.
3. Add these two shifted spectra to obtain the desired $S_{N_\mathrm{I}N_\mathrm{I}}(f)$ or $S_{N_\mathrm{Q}N_\mathrm{Q}}(f)$.

EXAMPLE 12 Ideal Band-pass Filtered White Noise

Consider a white Gaussian noise of zero mean and power spectral density $N_0/2$, which is passed through an ideal band-pass filter of passband magnitude response equal to one, midband frequency f_c, and bandwidth $2B$. The power spectral density characteristic of the filtered noise $n(t)$ is, therefore, as shown in Figure 4.21a. The problem is to determine the autocorrelation functions of $n(t)$ and those of its in-phase and quadrature components.

The autocorrelation function of $n(t)$ is the inverse Fourier transform of the power spectral density characteristic shown in Figure 4.21a, as shown by

$$
\begin{aligned}
R_{NN}(\tau) &= \int_{-f_\mathrm{c}-B}^{-f_\mathrm{c}+B}\frac{N_0}{2}\exp(\mathrm{j}2\pi f\tau)\,\mathrm{d}f + \int_{f_\mathrm{c}-B}^{f_\mathrm{c}+B}\frac{N_0}{2}\exp(\mathrm{j}2\pi f\tau)\,\mathrm{d}f \\
&= N_0 B\,\mathrm{sinc}(2B\tau)[\exp(-\mathrm{j}2\pi f_\mathrm{c}\tau)+\exp(\mathrm{j}2\pi f_\mathrm{c}\tau)] \\
&= 2N_0 B\,\mathrm{sinc}(2B\tau)\cos(2\pi f_\mathrm{c}\tau)
\end{aligned}
\tag{4.125}
$$

which is plotted in Figure 4.21b.

The spectral density characteristic of Figure 4.21a is symmetric about $\pm f_\mathrm{c}$. The corresponding spectral density characteristics of the in-phase noise component $n_\mathrm{I}(t)$ and the quadrature noise component $n_\mathrm{Q}(t)$ are equal, as shown in Figure 4.21c. Scaling the result of Example 10 by a factor of two in accordance with the spectral characteristics of

Figure 4.21a and 4.21c, we find that the autocorrelation function of $n_{\mathrm{I}}(t)$ or $n_{\mathrm{Q}}(t)$ is given by

$$R_{N_{\mathrm{I}}N_{\mathrm{I}}}(\tau) = R_{N_{\mathrm{Q}}N_{\mathrm{Q}}}(\tau) = 2N_0 B \operatorname{sinc}(2B\tau) \tag{4.126}$$

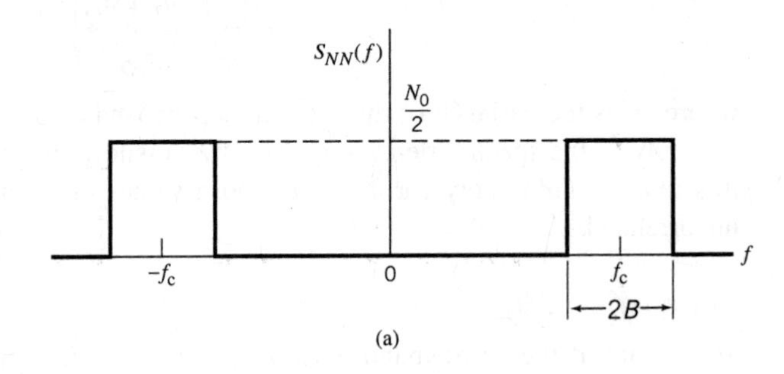

(a)

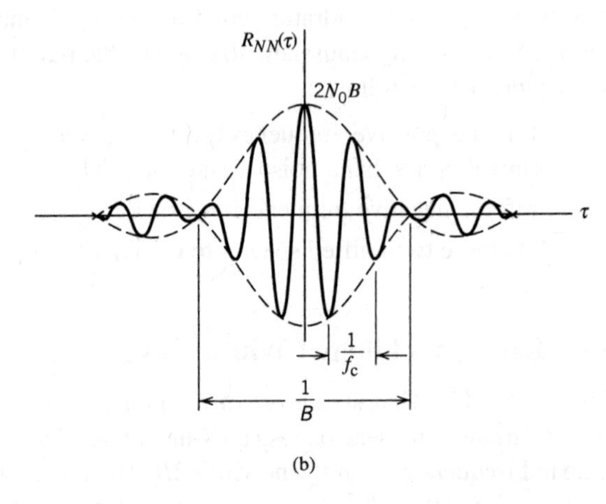

(b)

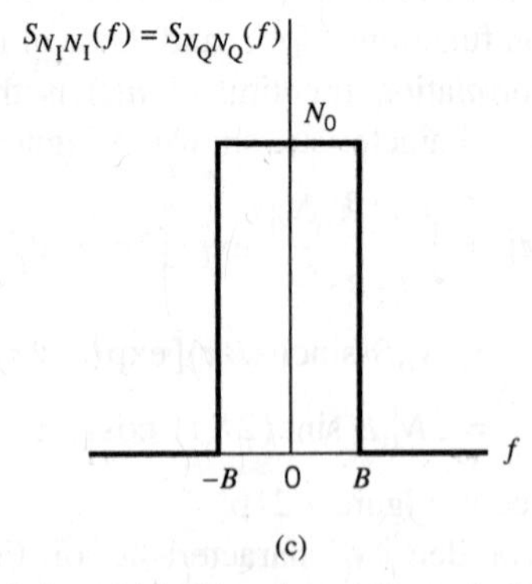

(c)

Figure 4.21 Characteristics of ideal band-pass filtered white noise: (a) power spectral density, (b) autocorrelation function, (c) power spectral density of in-phase and quadrature components.

Representation of Narrowband Noise in Terms of Envelope and Phase Components

In the preceding subsection we used the Cartesian representation of a narrowband noise $n(t)$ in terms of its in-phase and quadrature components. In this subsection we use the polar representation of the noise $n(t)$ in terms of its envelope and phase components, as shown by

$$n(t) = r(t)\cos[2\pi f_c t + \psi(t)] \tag{4.127}$$

where

$$r(t) = [n_I^2(t) + n_Q^2(t)]^{1/2} \tag{4.128}$$

and

$$\psi(t) = \tan^{-1}\left[\frac{n_Q(t)}{n_I(t)}\right] \tag{4.129}$$

The function $r(t)$ is the *envelope* of $n(t)$ and the function $\psi(t)$ is the *phase* of $n(t)$.

The probability density functions of $r(t)$ and $\psi(t)$ may be obtained from those of $n_I(t)$ and $n_Q(t)$ as follows. Let N_I and N_Q denote the random variables obtained by sampling (at some fixed time) the stochastic processes represented by the sample functions $n_I(t)$ and $n_Q(t)$ respectively. We note that N_I and N_Q are independent Gaussian random variables of zero mean and variance σ^2, so we may express their joint probability density function as

$$f_{N_I, N_Q}(n_I, n_Q) = \frac{1}{2\pi\sigma^2}\exp\left(-\frac{n_I^2 + n_Q^2}{2\sigma^2}\right) \tag{4.130}$$

Accordingly, the probability of the joint event that N_I lies between n_I and $n_I + dn_I$ and N_Q lies between $n_Q + dn_Q$ (i.e., the pair of random variables N_I and N_Q lies jointly inside the shaded area of Figure 4.22a) is given by

$$f_{N_I, N_Q}(n_I, n_Q)\, dn_I\, dn_Q = \frac{1}{2\pi\sigma^2}\exp\left(-\frac{n_I^2 + n_Q^2}{2\sigma^2}\right) dn_I dn_Q \tag{4.131}$$

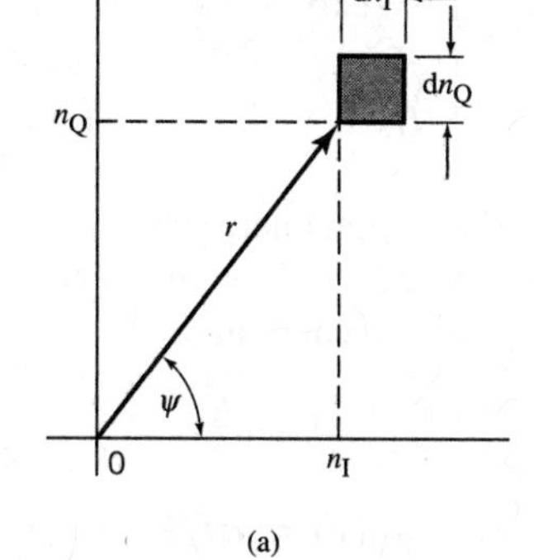

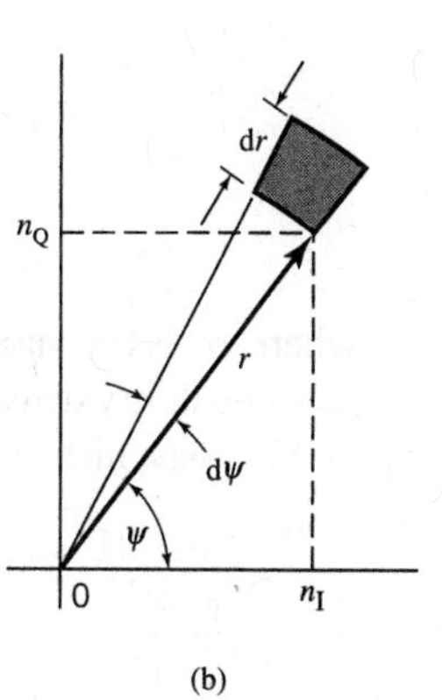

Figure 4.22
Illustrating the coordinate system for representation of narrowband noise: (a) in terms of in-phase and quadrature components; (b) in terms of envelope and phase.

where dn_I and dn_Q are incrementally small. Now, define the transformations (see Figure 4.22b)

$$n_I = r\cos\psi \tag{4.132}$$

$$n_Q = r\sin\psi \tag{4.133}$$

In a limiting sense, we may equate the two incremental areas shown shaded in parts a and b of Figure 4.22 and thus write

$$dn_I dn_Q = r\,dr\,d\psi \tag{4.134}$$

Now, let R and Ψ denote the random variables obtained by observing (at some fixed time t) the stochastic processes represented by the envelope $r(t)$ and phase $\psi(t)$ respectively. Then substituting (4.132)–(4.134) into (4.131), we find that the probability of the random variables R and Ψ lying jointly inside the shaded area of Figure 4.22b is equal to the expression

$$\frac{r}{2\pi\sigma^2}\exp\left(-\frac{r^2}{2\sigma^2}\right)dr\,d\psi$$

That is, the joint probability density function of R and Ψ is given by

$$f_{R,\Psi}(r,\psi) = \frac{r}{2\pi\sigma^2}\exp\left(-\frac{r^2}{2\sigma^2}\right) \tag{4.135}$$

This probability density function is independent of the angle ψ, which means that the random variables R and Ψ are *statistically independent*. We may thus express $f_{R,\Psi}(r,\psi)$ as the product of the two probability density functions: $f_R(r)$ and $f_\Psi(\psi)$. In particular, the random variable Ψ representing the phase is *uniformly distributed* inside the interval $[0, 2\pi]$, as shown by

$$f_\Psi(\psi) = \begin{cases} \frac{1}{2\pi}, & 0 \le \psi \le 2\pi \\ 0, & \text{elsewhere} \end{cases} \tag{4.136}$$

This result leaves the probability density function of the random variable R as

$$f_R(r) = \begin{cases} \frac{r}{\sigma^2}\exp\left(-\frac{r^2}{2\sigma^2}\right), & r \ge 0 \\ 0, & \text{elsewhere} \end{cases} \tag{4.137}$$

where σ^2 is the variance of the original narrowband noise $n(t)$. A random variable having the probability density function of (4.137) is said to be *Rayleigh distributed*.[10]

For convenience of graphical presentation, let

$$v = \frac{r}{\sigma} \tag{4.138}$$

$$f_V(v) = \sigma f_R(r) \tag{4.139}$$

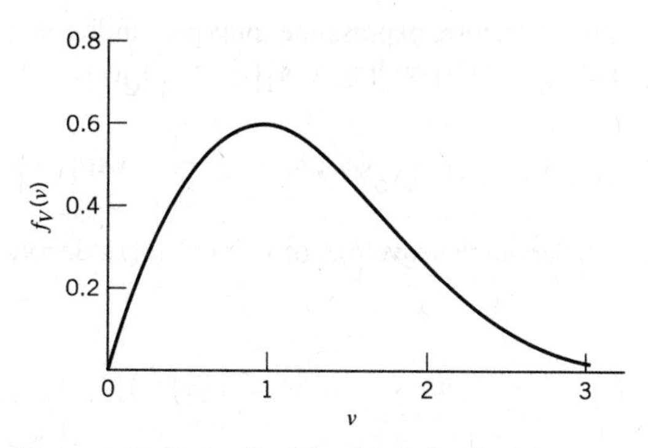

Figure 4.23 Normalized Rayleigh distribution.

Then, we may rewrite the Rayleigh distribution of (4.137) in the *normalized* form

$$f_V(v) = \begin{cases} v \exp\left(-\frac{v^2}{2}\right), & v \ge 0 \\ 0, & \text{elsewhere} \end{cases} \tag{4.140}$$

Equation (4.140) is plotted in Figure 4.23. The peak value of the distribution $f_V(v)$ occurs at $v = 1$ and is equal to 0.607. Note also that, unlike the Gaussian distribution, the Rayleigh distribution is zero for negative values of v, which follows naturally from the fact that the envelope $r(t)$ of the narrowband noise $n(t)$ can only assume nonnegative values.

4.14 Sine Wave Plus Narrowband Noise

Suppose next that we add the sinusoidal wave $A\cos(2\pi f_c t)$ to the narrowband noise $n(t)$, where A and f_c are both constants. We assume that the frequency of the sinusoidal wave is the same as the nominal carrier frequency of the noise. A sample function of the sinusoidal wave plus noise is then expressed by

$$x(t) = A\cos(2\pi f_c t) + n(t) \tag{4.141}$$

Representing the narrowband noise $n(t)$ in terms of its in-phase and quadrature components, we may write

$$x(t) = n_I'(t)\cos(2\pi f_c t) - n_Q(t)\sin(2\pi f_c t) \tag{4.142}$$

where

$$n_I'(t) = A + n_I(t) \tag{4.143}$$

We assume that $n(t)$ is Gaussian with zero mean and variance σ^2. Accordingly, we may state the following:

1. Both $n_I'(t)$ and $n_Q(t)$ are Gaussian and statistically independent.
2. The mean of $n_I'(t)$ is A and that of $n_Q(t)$ is zero.
3. The variance of both $n_I'(t)$ and $n_Q(t)$ is σ^2.

We may, therefore, express the joint probability density function of the random variables N_I' and N_Q, corresponding to $n_I'(t)$ and $n_Q(t)$, as follows:

$$f_{N_I', N_Q}(n_I', n_Q) = \frac{1}{2\pi\sigma^2} \exp\left[-\frac{(n_I' - A)^2 + n_Q^2}{2\sigma^2}\right] \tag{4.144}$$

Let $r(t)$ denote the envelope of $x(t)$ and $\psi(t)$ denote its phase. From (4.142), we thus find that

$$r(t) = \left\{[n_I'(t)]^2 + n_Q^2(t)\right\}^{1/2} \tag{4.145}$$

and

$$\psi(t) = \tan^{-1}\left[\frac{n_Q(t)}{n_I'(t)}\right] \tag{4.146}$$

Following a procedure similar to that described in Section 4.12 for the derivation of the Rayleigh distribution, we find that the joint probability density function of the random variables R and ψ, corresponding to $r(t)$ and $\psi(t)$ for some fixed time t, is given by

$$f_{R,\psi}(r, \psi) = \frac{r}{2\pi\sigma^2} \exp\left(-\frac{r^2 + A^2 - 2Ar\cos\psi}{2\sigma^2}\right) \tag{4.147}$$

We see that in this case, however, we cannot express the joint probability density function $f_{R,\psi}(r, \psi)$ as a product $f_R(r) f_\psi(\psi)$, because we now have a term involving the values of both random variables multiplied together as $r\cos\psi$. Hence, R and ψ are *dependent* random variables for nonzero values of the amplitude A of the sinusoidal component.

We are interested, in particular, in the probability density function of R. To determine this probability density function, we integrate (4.147) over all possible values of ψ, obtaining the desired marginal density

$$\begin{aligned} f_R(r) &= \int_0^{2\pi} f_{R,\psi}(r, \psi)\, d\psi \\ &= \frac{r}{2\pi\sigma^2} \exp\left(-\frac{r^2 + A^2}{2\sigma^2}\right) \int_0^{2\pi} \exp\left(\frac{Ar\cos\psi}{\sigma^2}\right) d\psi \end{aligned} \tag{4.148}$$

An integral similar to that in the right-hand side of (4.148) is referred to in the literature as the *modified Bessel function of the first kind of zero order* (see Appendix C); that is,

$$I_0(x) = \frac{1}{2\pi}\int_0^{2\pi} \exp(x\cos\psi)\, d\psi \tag{4.149}$$

Thus, letting $x = Ar/\sigma^2$, we may rewrite (4.148) in the compact form

$$f_R(r) = \frac{r}{\sigma^2} \exp\left(-\frac{r^2 + A^2}{2\sigma^2}\right) I_0\left(\frac{Ar}{\sigma^2}\right), \qquad r \geq 0 \tag{4.150}$$

This new distribution is called the *Rician distribution*.[11]

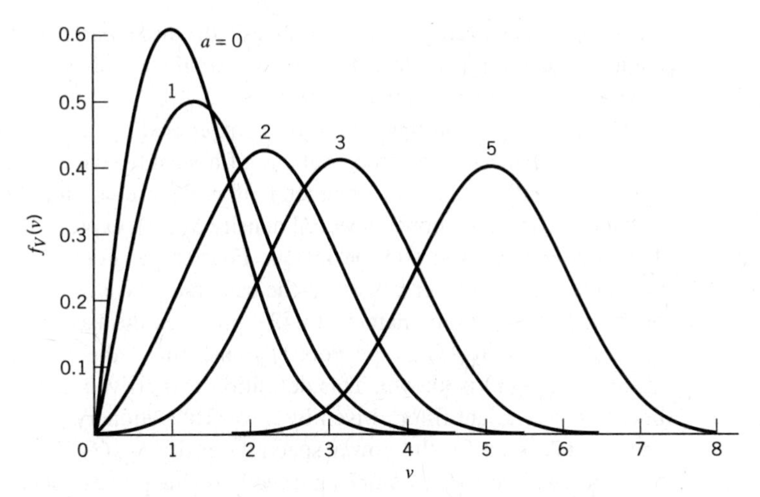

Figure 4.24 Normalized Rician distribution.

As with the Rayleigh distribution, the graphical presentation of the Rician distribution is simplified by putting

$$v = \frac{r}{\sigma} \tag{4.151}$$

$$a = \frac{A}{\sigma} \tag{4.152}$$

$$f_V(v) = \sigma f_R(r) \tag{4.153}$$

Then we may express the Rician distribution of (4.150) in the *normalized* form

$$f_V(v) = v \exp\left(-\frac{v^2 + a^2}{2}\right) I_0(av) \tag{4.154}$$

which is plotted in Figure 4.24 for the values 0, 1, 2, 3, 5, of the parameter a.[12] Based on these curves, we may make two observations:

1. When the parameter $a = 0$, and therefore $I_0(0) = 1$, the Rician distribution reduces to the Rayleigh distribution.
2. The envelope distribution is approximately Gaussian in the vicinity of $v = a$ when a is large; that is, when the sine-wave amplitude A is large compared with σ, the square root of the average power of the noise $n(t)$.

4.15 Summary and Discussion

Much of the material presented in this chapter has dealt with the characterization of a particular class of stochastic processes known to be weakly stationary. The implication of "weak" stationarity is that we may develop a partial description of a stochastic process in terms of two ensemble-averaged parameters: (1) a mean that is independent of time and (2) an autocorrelation function that depends only on the difference between the times at which two samples of the process are drawn. We also discussed ergodicity, which enables us to use time averages as "estimates" of these parameters. The time averages are computed using a sample function (i.e., single waveform realization) of the stochastic process, evolving as a function of time.

The autocorrelation function $R_{XX}(\tau)$, expressed in terms of the time shift τ, is one way of describing the second-order statistic of a weakly (wide-sense) stationary process $X(t)$. Another equally important parameter, if not more so, for describing the second-order statistic of $X(t)$ is the power spectral density $S_{XX}(f)$, expressed in terms of the frequency f. The Fourier transform and the inverse Fourier transform formulas that relate these two parameters to each other constitute the celebrated Wiener–Khintchine equations. The first of these two equations, namely (4.42), provides the basis for a definition of the power spectral density $S_{XX}(f)$ as the Fourier transform of the autocorrelation function $R_{XX}(\tau)$, given that $R_{XX}(\tau)$ is known. This definition was arrived at by working on the output of a linear time-invariant filter, driven by a weakly stationary process $X(t)$. We also described another definition for the power spectral density $S_{XX}(f)$, described in (4.70); this second definition was derived by working directly on the process $X(t)$.

Another celebrated theorem discussed in the chapter is the Wiener–Khintchine theorem, which provides the necessary and sufficient condition for confirming the function $\rho_{XX}(\tau)$ as the normalized autocorrelation function of a weakly stationary process $X(t)$, provided that it satisfies the Fourier–Stieltjes transform, described in (4.60).

The stochastic-process theory described in this chapter also included the topic of cross-power spectral densities $S_{XY}(f)$ and $S_{YX}(f)$, involving a pair of jointly weakly stationary processes $X(t)$ and $Y(t)$, and how these two frequency-dependent parameters are related to the respective cross-correlation functions $R_{XY}(\tau)$ and $R_{YX}(\tau)$.

The remaining part of the chapter was devoted to the statistical characterization of different kinds of stochastic processes:

- The Poisson process, which is well-suited for the characterization of random-counting processes.
- The ubiquitous Gaussian process, which is widely used in the statistical study of communication systems.
- The two kinds of electrical noise, namely shot noise and thermal noise.
- White noise, which plays a fundamental role in the noise analysis of communication systems similar to that of the impulse function in the study of linear systems.
- Narrowband noise, which is produced by passing white noise through a linear band-pass filter. Two different methods for the description of narrowband noise were presented: one in terms of the in-phase and quadrature components and the other in terms of the envelope and phase.
- The Rayleigh distribution, which is described by the envelope of a narrowband noise process.
- The Rician distribution, which is described by the envelope of narrowband noise plus a sinusoidal component, with the midband frequency of the narrowband noise and the frequency of the sinusoidal component being coincident.

We conclude this chapter on stochastic processes by including Table 4.1, where we present a graphical summary of the autocorrelation functions and power spectral densities of important stochastic processes. All the processes described in this table are assumed to have zero mean and unit variance. This table should give the reader a feeling for (1) the interplay between the autocorrelation function and power spectral density of a stochastic process and (2) the role of linear filtering in shaping the autocorrelation function or, equivalently, the power spectral density of a white-noise process.

Table 4.1 **Graphical summary of autocorrelation functions and power spectral densities of random processes of zero mean and unit variance**

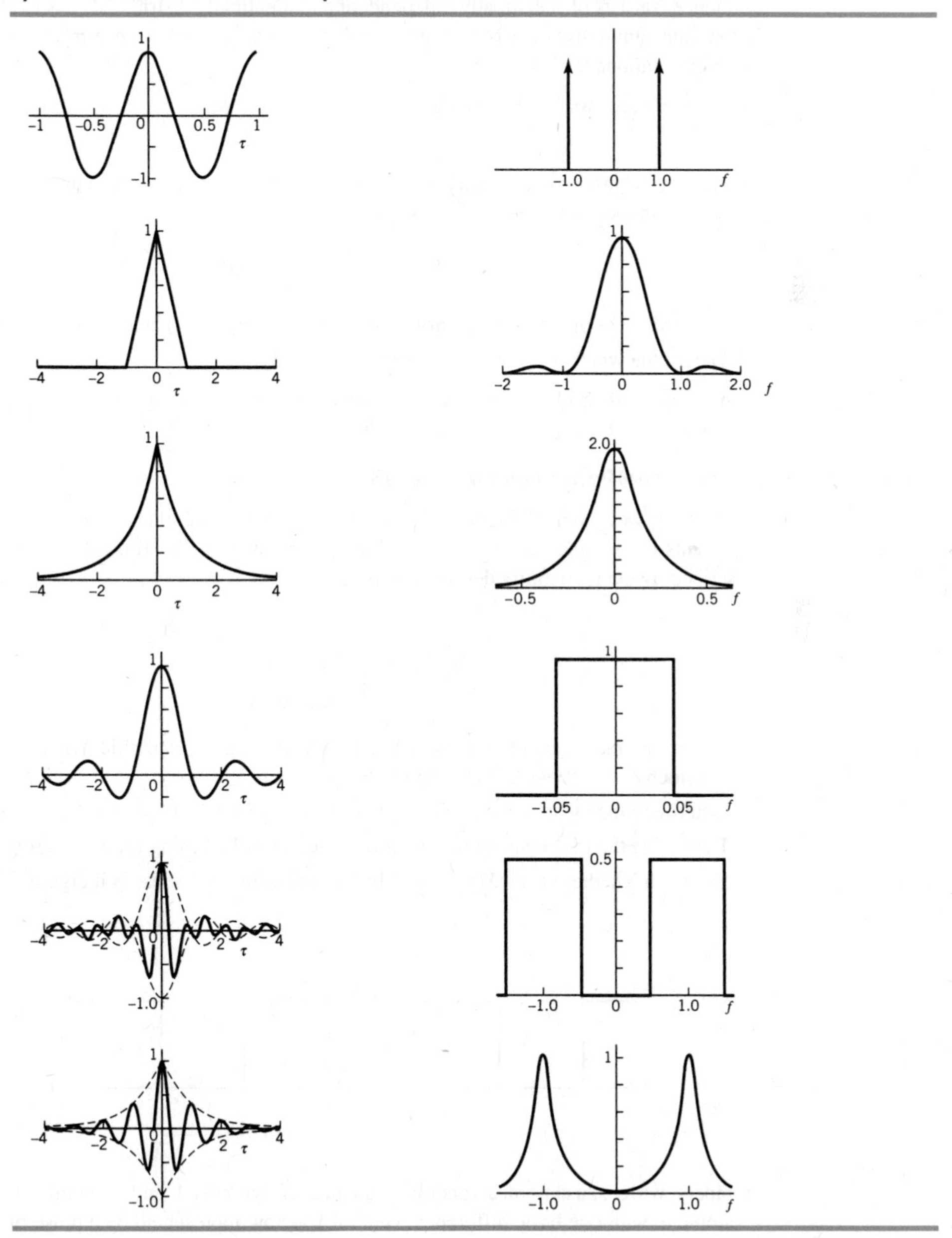

Problems

Stationarity and Ergodicity

4.1 Consider a pair of stochastic processes $X(t)$ and $Y(t)$. In the strictly stationary world of stochastic processes, the statistical independence of $X(t)$ and $Y(t)$ corresponds to their uncorrelatedness in the world of weakly stationary processes. Justify this statement.

4.2 Let $X_1, X_2, \dots, X_k$ denote a sequence obtained by uniformly sampling a stochastic process $X(t)$. The sequence consists of statistically independent and identically distributed (iid) random variables, with a common cumulative distribution function $F_X(x)$, mean μ, and variance σ^2. Show that this sequence is strictly stationary.

4.3 A stochastic process $X(t)$ is defined by

$$X(t) = A\cos(2\pi f_c t)$$

where A is a Gaussian-distributed random variable of zero mean and variance σ_A^2. The process $X(t)$ is applied to an ideal integrator, producing the output

$$Y(t) = \int_0^t X(\tau)\,d\tau$$

a. Determine the probability density function of the output $Y(t)$ at a particular time t_k.

b. Determine whether or not $Y(t)$ is strictly stationary.

4.4 Continuing with Problem 4.3, determine whether or not the integrator output $Y(t)$ produced in response to the input process $X(t)$ is ergodic.

Autocorrelation Function and Power Spectral Density

4.5 The square wave $x(t)$ of Figure P4.5, having constant amplitude A, period T_0, and time shift t_d, represents the sample function of a stochastic process $X(t)$. The time shift t_d is a random variable, described by the probability density function

$$f_{T_d}(t_d) = \begin{cases} \dfrac{1}{T_0}, & -\dfrac{1}{2}T_0 \le t_d \le \dfrac{1}{2}T_0 \\ 0, & \text{otherwise} \end{cases}$$

a. Determine the probability density function of the random variable $X(t_k)$, obtained by sampling the stochastic process $X(t)$ at time t_k.

b. Determine the mean and autocorrelation function of $X(t)$ using ensemble averaging.

c. Determine the mean and autocorrelation function of $X(t)$ using time averaging.

d. Establish whether or not $X(t)$ is weakly stationary. In what sense is it ergodic?

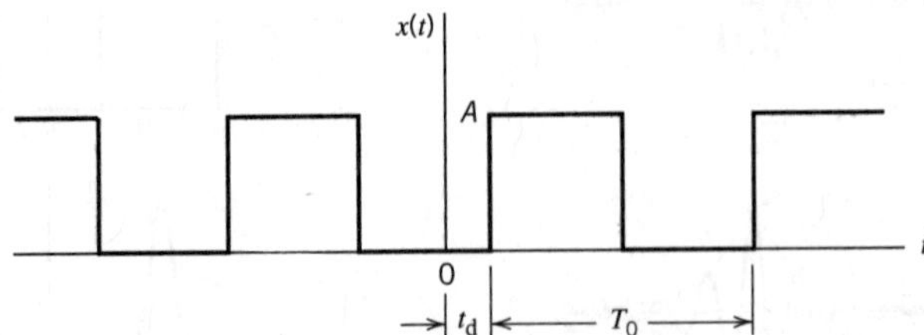

Figure P4.5

4.6 A binary wave consists of a random sequence of symbols 1 and 0, similar to that described in Example 6, with one basic difference: symbol 1 is now represented by a pulse of amplitude A volts, and symbol 0 is represented by zero volts. All other parameters are the same as before. Show that this new random binary wave $X(t)$ is characterized as follows:

a. The autocorrelation function is

$$R_{XX}(\tau) = \begin{cases} \dfrac{A^2}{4} + \dfrac{A^2}{4}\left(1 - \dfrac{|\tau|}{T}\right), & |\tau| < T \\ \dfrac{A^2}{4}, & |\tau| \geq T \end{cases}$$

b. The power spectral density is

$$S_{XX}(f) = \frac{A^2}{4}\delta(f) + \frac{A^2 T}{4}\text{sinc}^2(fT)$$

What is the percentage power contained in the dc component of the binary wave?

4.7 The output of an oscillator is described by

$$X(t) = A\cos(\pi F t + \Theta)$$

where the amplitude A is constant, and F and Θ are independent random variables. The probability density function of Θ is defined by

$$f_{\Theta}(\theta) = \begin{cases} \dfrac{1}{2\pi}, & 0 \leq \theta \leq 2\pi \\ 0, & \text{otherwise} \end{cases}$$

Find the power spectral density of $X(t)$ in terms of the probability density function of the frequency F. What happens to this power spectral density when the randomized frequency F assumes a constant value?

4.8 Equation (4.70) presents the second of two definitions introduced in the chapter for the power spectral density function, $S_{XX}(f)$, pertaining to a weakly stationary process $X(t)$. This definition reconfirms Property 3 of $S_{XX}(f)$, as shown in (4.71).

a. Using (4.70), prove the other properties of $S_{XX}(f)$: zero correlation among frequency components, zero-frequency value, nonnegativity, symmetry, and normalization, which were discussed in Section 4.8.

b. Starting with (4.70), derive (4.43) that defines the autocorrelation function $R_{XX}(\tau)$ of the stationary process $X(t)$ in terms of $S_{XX}(f)$.

4.9 In the definition of (4.70) for the power spectral density of a weakly stationary process $X(t)$, it is not permissible to interchange the order of expectation and limiting operations. Justify the validity of this statement.

The Wiener–Khintchine Theorem

In the next four problems we explore the application of the Wiener–Khintchine theorem of (4.60) to see whether a given function $\rho(\tau)$, expressed in terms of the time shift τ, is a legitimate normalized autocorrelation function or not.

4.10 Consider the Fourier transformable function

$$f(\tau) = \frac{A^2}{2}\sin(2\pi f_c \tau) \qquad \text{for all } \tau$$

By inspection, we see that $f(\tau)$ is an odd function of τ. It cannot, therefore, be a legitimate autocorrelation function as it violates a fundamental property of the autocorrelation function. Apply the Wiener–Khintchine theorem to arrive at this same conclusion.

4.11 Consider the infinite series

$$f(\tau) = \frac{A^2}{2}\left[1 - \frac{1}{2!}(2\pi f_c \tau)^2 + \frac{1}{4!}(2\pi f_c \tau)^4 - \ldots\right] \quad \text{for all } \tau$$

which is an even function of τ, thereby satisfying the symmetry property of the autocorrelation function. Apply the Wiener–Khintchine theorem to confirm that $f(\tau)$ is indeed a legitimate autocorrelation function of a weakly stationary process.

4.12 Consider the Gaussian function

$$f(\tau) = \exp(-\pi\tau^2) \quad \text{for all } \tau$$

which is Fourier transformable. Moreover, it is an even function of τ, thereby satisfying the symmetry property of the autocorrelation function around the origin $\tau = 0$. Apply the Wiener–Khintchine theorem to confirm that $f(\tau)$ is indeed a legitimate normalized autocorrelation function of a weakly stationary process.

4.13 Consider the Fourier transformable function

$$f(\tau) = \begin{cases} \delta\left(\tau - \frac{1}{2}\right), & \tau = \frac{1}{2} \\ -\delta\left(\tau + \frac{1}{2}\right), & \tau = -\frac{1}{2} \\ 0, & \text{otherwise} \end{cases}$$

which is an odd function of τ. It cannot, therefore, be a legitimate autocorrelation function. Apply the Wiener–Khintchine theorem to arrive at this same conclusion.

Cross-correlation Functions and Cross-spectral Densities

4.14 Consider a pair of weakly stationary processes $X(t)$ and $Y(t)$. Show that the cross-correlations $R_{XY}(\tau)$ and $R_{YX}(\tau)$ of these two processes have the following properties:

a. $R_{XY}(\tau) = R_{YX}(-\tau)$

b. $|R_{XY}(\tau)| \le \frac{1}{2}[R_{XX}(0) + R_{YY}(0)]$

where $R_{XX}(\tau)$ and $R_{YY}(\tau)$ are the autocorrelation functions of $X(t)$ and $Y(t)$ respectively.

4.15 A weakly stationary process $X(t)$, with zero mean and autocorrelation function $R_{XX}(\tau)$, is passed through a differentiator, yielding the new process

$$Y(t) = \frac{d}{dt}X(t)$$

a. Determine the autocorrelation function of $Y(t)$.

b. Determine the cross-correlation function between $X(t)$ and $Y(t)$.

4.16 Consider two linear filters connected in cascade as in Figure P4.16. Let $X(t)$ be a weakly stationary process with autocorrelation function $R_{XX}(\tau)$. The weakly stationary process appearing at the first filter output is denoted by $V(t)$ and that at the second filter output is denoted by $Y(t)$.

a. Find the autocorrelation function of $Y(t)$.

b. Find the cross-correlation function $R_{VY}(\tau)$ of $V(t)$ and $Y(t)$.

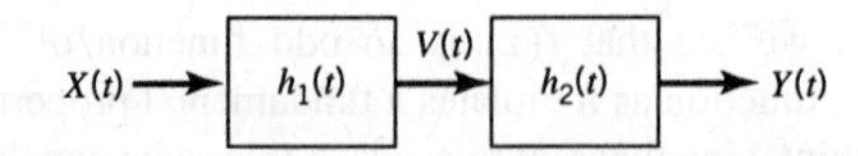

Figure P4.16

4.17 A weakly stationary process $X(t)$ is applied to a linear time-invariant filter of impulse response $h(t)$, producing the output $Y(t)$.

a. Show that the cross-correlation function $R_{YX}(\tau)$ of the output $Y(t)$ and the input $X(t)$ is equal to the impulse response $h(\tau)$ convolved with the autocorrelation function $R_{XX}(\tau)$ of the input, as shown by

$$R_{YX}(\tau) = \int_{-\infty}^{\infty} h(u)R_{XX}(\tau - u)\,du$$

Show that the second cross-correlation function $R_{XY}(\tau)$ is

$$R_{XY}(\tau) = \int_{-\infty}^{\infty} h(-u)R_{XX}(\tau - u)\,du$$

b. Find the cross-spectral densities $S_{YX}(f)$ and $S_{XY}(f)$.

c. Assuming that $X(t)$ is a white-noise process with zero mean and power spectral density $N_0/2$, show that

$$R_{YX}(\tau) = \frac{N_0}{2}h(\tau)$$

Comment on the practical significance of this result.

Poisson Process

4.18 The sample function of a stochastic process $X(t)$ is shown in Figure P4.18a, where we see that the sample function $x(t)$ assumes the values ± 1 in a random manner. It is assumed that at time $t = 0$, the values $X(0) = -1$ and $X(1) = +1$ are equiprobable. From there on, the changes in $X(t)$ occur in accordance with a Poisson process of average rate λ. The process $X(t)$, described herein, is sometimes referred to as a *telegraph signal.*

a. Show that, for any time $t > 0$, the values $X(t) = -1$ and $X(t) = +1$ are equiprobable.

b. Building on the result of part a, show that the mean of $X(t)$ is zero and its variance is unity.

c. Show that the autocorrelation function of $X(t)$ is given by

$$R_{XX}(\tau) = \exp(-2\lambda\tau)$$

d. The process $X(t)$ is applied to the simple low-pass filter of Figure P4.18b. Determine the power spectral density of the process $Y(t)$ produced at the filter output.

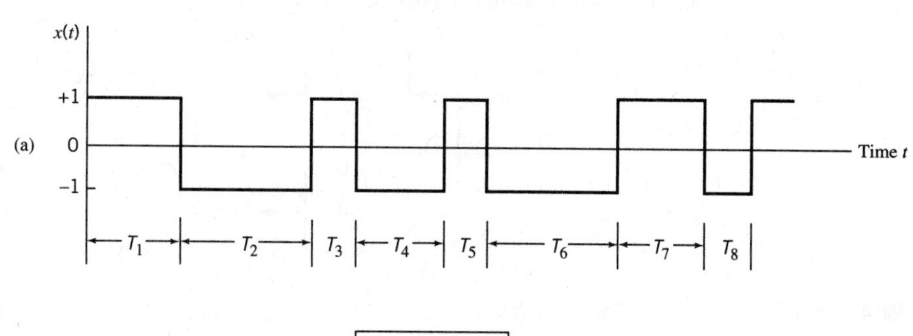

Figure P4.18

Gaussian Process

4.19 Consider the pair of integrals

$$Y_1 = \int_{-\infty}^{\infty} h_1(t)X(t)\,dt$$

and

$$Y_2 = \int_{-\infty}^{\infty} h_2(t)X(t)\,dt$$

where $X(t)$ is a Gaussian process and $h_1(t)$ and $h_2(t)$ are two different weighting functions. Show that the two random variables Y_1 and Y_2, resulting from the integrations, are jointly Gaussian.

4.20 A Gaussian process $X(t)$, with zero mean and variance σ_X^2, is passed through a full-wave rectifier, which is described by the input–output relationship of Figure P4.20. Show that the probability density function of the random variable $Y(t_k)$, obtained by observing the stochastic process $Y(t)$ produced at the rectifier output at time t_k, is one sided, as shown by

$$f_{Y(t_k)}(y) = \begin{cases} \sqrt{\dfrac{2}{\pi}}\dfrac{1}{\sigma_X}\exp\left(-\dfrac{y^2}{2\sigma_X^2}\right), & y \ge 0 \\ 0, & y < 0 \end{cases}$$

Confirm that the total area under the graph of $f_{Y(t_k)}(y)$ is unity.

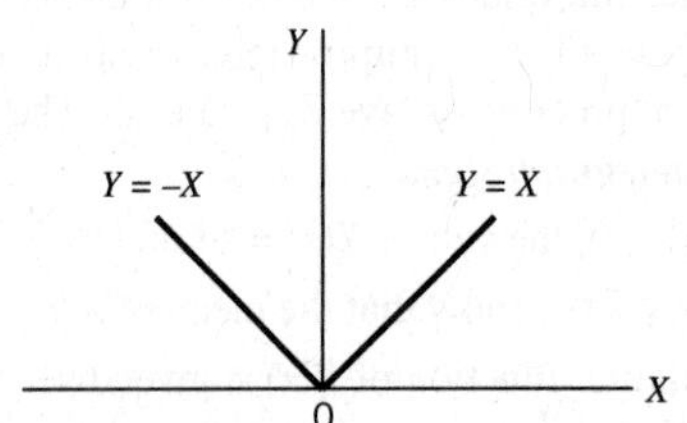

Figure P4.20

4.21 A stationary Gaussian process $X(t)$, with mean μ_X and variance σ_X^2, is passed through two linear filters with impulse responses $h_1(t)$ and $h_2(t)$, yielding the processes $Y(t)$ and $Z(t)$, as shown in Figure P4.21. Determine the necessary and sufficient conditions, for which $Y(t_1)$ and $Z(t_2)$ are statistically independent Gaussian processes.

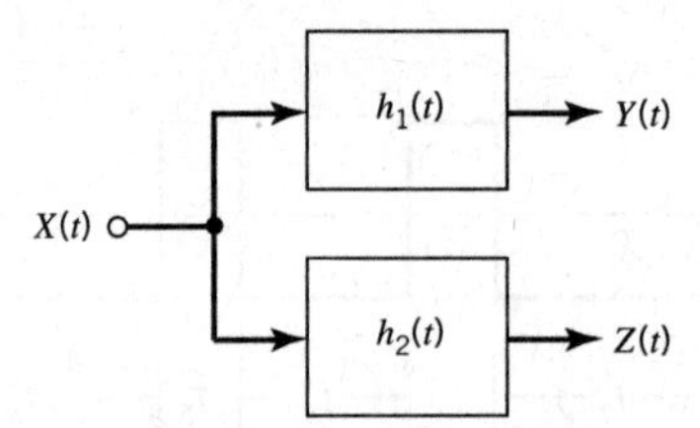

Figure P4.21

White Noise

4.22 Consider the stochastic process

$$X(t) = W(t) + aW(t - t_0)$$

where $W(t)$ is a white-noise process of power spectral density $N_0/2$ and the parameters a and t_0 are constants.

a. Determine the autocorrelation function of the process $X(t)$, and sketch it.

b. Determine the power spectral density of the process $X(t)$, and sketch it.

4.23 The process

$$X(t) = A\cos(2\pi f_0 t + \Theta) + W(t)$$

describes a sinusoidal process that is corrupted by an additive white-noise process $W(t)$ of known power spectral density $N_0/2$. The phase of the sinusoidal process, denoted by Θ, is a uniformly distributed random variable, defined by

$$f_\Theta(\theta) = \begin{cases} \dfrac{1}{2\pi} & \text{for } -\pi \le \theta \le \pi \\ 0 & \text{otherwise} \end{cases}$$

The amplitude A and frequency f_0 are both constant but unknown.

a. Determine the autocorrelation function of the process $X(t)$ and its power spectral density.

b. How would you use the two results of part a to measure the unknown parameters A and f_0?

4.24 A white Gaussian noise process of zero mean and power spectral density $N_0/2$ is applied to the filtering scheme shown in Figure P4.24. The noise at the low-pass filter output is denoted by $n(t)$.

a. Find the power spectral density and the autocorrelation function of $n(t)$.

b. Find the mean and variance of $n(t)$.

c. What is the maximum rate at which $n(t)$ can be sampled so that the resulting samples are essentially uncorrelated?

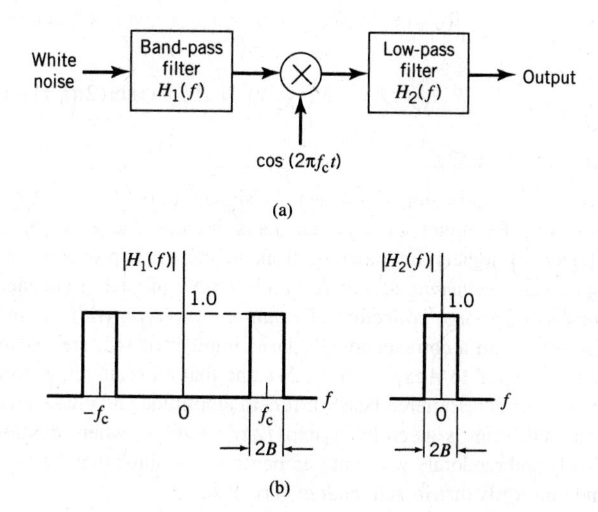

Figure P4.24

4.25 Let $X(t)$ be a weakly stationary process with zero mean, autocorrelation function $R_{XX}(\tau)$, and power spectral density $S_{XX}(f)$. We are required to find a linear filter with impulse response $h(t)$, such that the filter output is $X(t)$ when the input is white-noise of power spectral density $N_0/2$.

a. Determine the condition that the impulse response $h(t)$ must satisfy in order to achieve this requirement.

b. What is the corresponding condition on the transfer function $H(f)$ of the filter?

c. Using the Paley–Wiener criterion discussed in Chapter 2, find the requirement on $S_{XX}(f)$ for the filter to be causal.

Narrowband Noise

4.26 Consider a narrowband noise $n(t)$ with its Hilbert transform denoted by $\hat{n}(t)$.

a. Show that the cross-correlation functions of $n(t)$ and $\hat{n}(t)$ are given by

$$R_{N\hat{N}}(\tau) = -\hat{R}_{NN}(\tau)$$

and

$$R_{\hat{N}N}(\tau) = \hat{R}_{NN}(\tau)$$

where $\hat{R}_{NN}(\tau)$ is the Hilbert transform of the autocorrelation function $R_{NN}(\tau)$ of $n(t)$.

Hint: use the formula

$$\hat{n}(t) = \frac{1}{\pi}\int_{-\infty}^{\infty} \frac{n(\lambda)}{t-\lambda}\,d\lambda$$

b. Show that, for $\tau = 0$, we have $R_{N\hat{N}}(0) = R_{\hat{N}N} = 0$.

4.27 A narrowband noise $n(t)$ has zero mean and autocorrelation function $R_{NN}(\tau)$. Its power spectral density $S_{NN}(f)$ is centered about $\pm f_c$. The in-phase and quadrature components, $n_I(t)$ and $n_Q(t)$, of $n(t)$ are defined by the weighted sums

$$n_I(t) = n(t)\cos(2\pi f_c t) + \hat{n}(t)\sin(2\pi f_c t)$$

and

$$n_Q(t) = \hat{n}(t)\cos(2\pi f_c t) - n(t)\sin(2\pi f_c t)$$

where $\hat{n}(t)$ is the Hilbert transform of the noise $n(t)$. Using the result obtained in part a of Problem 4.26, show that $n_I(t)$ and $n_Q(t)$ have the following autocorrelation functions:

$$R_{N_IN_I}(\tau) = R_{N_QN_Q}(\tau) = R_{NN}(\tau)\cos(2\pi f_c\tau) + \hat{R}_{NN}(\tau)\sin(2\pi f_c\tau)$$

and

$$R_{N_IN_Q}(\tau) = -R_{N_QN_I}(\tau) = R_{NN}(\tau)\sin(2\pi f_c\tau) - \hat{R}_{NN}(\tau)\cos(2\pi f_c\tau)$$

Rayleigh and Rician Distributions

4.28 Consider the problem of propagating signals through so-called *random* or *fading communications channels*. Examples of such channels include the *ionosphere* from which short-wave (high-frequency) signals are reflected back to the earth producing long-range radio transmission, and *underwater communications*. A simple model of such a channel is shown in Figure P4.28, which consists of a large collection of *random scatterers*, with the result that a single incident beam is converted into a correspondingly large number of scattered beams at the receiver. The transmitted signal is equal to $A\exp(j2\pi f_c t)$. Assume that all scattered beams travel at the same mean velocity. However, each scattered beam differs in amplitude and phase from the incident beam, so that the kth scattered beam is given by $A_k\exp(j2\pi f_c t + j\Theta_k)$, where the amplitude A_k and the phase Θ_k vary slowly and randomly with time. In particular, assume that the Θ_k are all independent of one another and uniformly distributed random variables.

a. With the received signal denoted by

$$x(t) = r(t)\exp[j2\pi f_c t + \psi(t)]$$

show that the random variable R, obtained by observing the envelope of the received signal at time t, is Rayleigh-distributed, and that the random variable Ψ, obtained by observing the phase at some fixed time, is uniformly distributed.

b. Assuming that the channel includes a line-of-sight path, so that the received signal contains a sinusoidal component of frequency f_c, show that in this case the envelope of the received signal is Rician distributed.

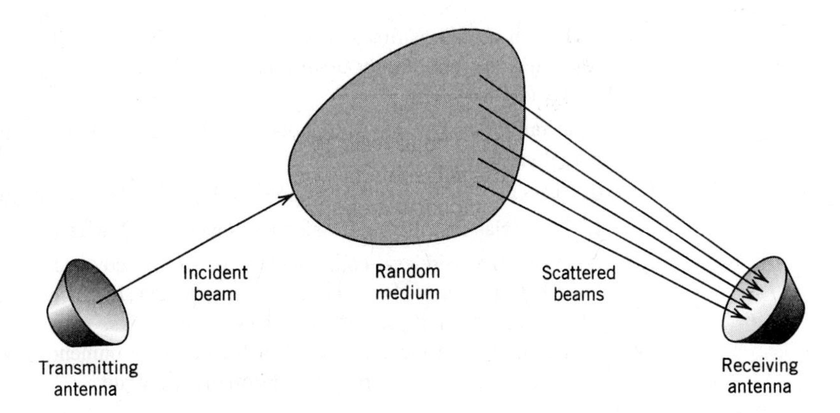

Figure P4.28

4.29 Referring back to the graphical plots of Figure 4.23, describing the Rician envelope distribution for varying parameter a, we see that for the parameter $a = 5$, this distribution is approximately Gaussian. Justify the validity of this statement.

Notes

1. Stochastic is of Greek origin.

2. For rigorous treatment of stochastic processes, see the classic books by Doob (1953), Loève (1963), and Cramér and Leadbetter (1967).

3. Traditionally, (4.42) and (4.43) have been referred to in the literature as the Wiener–Khintchine relations in recognition of pioneering work done by Norbert Wiener and A.I. Khintchine; for their original papers, see Wiener (1930) and Khintchine (1934). The discovery of a forgotten paper by Albert Einstein on time-series analysis (delivered at the Swiss Physical Society's February 1914 meeting in Basel) reveals that Einstein had discussed the autocorrelation function and its relationship to the spectral content of a time series many years before Wiener and Khintchine. An English translation of Einstein's paper is reproduced in the IEEE ASSP Magazine, vol. 4, October 1987. This particular issue also contains articles by W.A. Gardner and A.M. Yaglom, which elaborate on Einstein's original work.

4. For a mathematical proof of the Wiener–Khintchine theorem, see Priestley (1981).

5. Equation (4.70) provides the mathematical basis for estimating the power spectral density of a weakly stationary process. There is a plethora of procedures that have been formulated for performing this estimation. For a detailed treatment of reliable procedures to do the estimation, see the book by Percival and Walden (1993).

6. The Poisson process is named in honor of S.D. Poisson. The distribution bearing his name first appeared in an exposition by Poisson on the role of probability in the administration of justice. The classic book on Poisson processes is Snyder (1975). For an introductory treatment of the subject, see Bertsekas and Tsitsiklis (2008: Chapter 6).

7. The Gaussian distribution and the associated Gaussian process are named after the great mathematician C.F. Gauss. At age 18, Gauss invented *the method of least squares* for finding the best value of a sequence of measurements of some quantity. Gauss later used the method of least squares in fitting orbits of planets to data measurements, a procedure that was published in 1809 in his book entitled *Theory of Motion of the Heavenly Bodies*. In connection with the error of observation, he developed the *Gaussian distribution*.

8. Thermal noise was first studied experimentally by J.B. Johnson in 1928, and for this reason it is sometimes referred to as the *Johnson noise*. Johnson's experiments were confirmed theoretically by Nyquist (1928a).

9. For further insight into white noise, see Appendix I on generalized random processes in the book by Yaglom (1962).

10. The Rayleigh distribution is named in honor of the English physicist J.W. Strutt, Lord Rayleigh.

11. The Rician distribution is named in honor of S.O. Rice (1945).

12. In mobile wireless communications to be covered in Chapter 9, the sinusoidal term $A\cos(2\pi f_c t)$ in (4.141) is viewed as a *line-of-sight (LOS) component* of average power $A^2/2$ and the additive noise term $n(t)$ is viewed as a *Gaussian diffuse component* of average power σ^2, with both being assumed to have zero mean. In such an environment, it is the *Rice factor K* that is used to characterize the Rician distribution. Formally, we write

$$K = \frac{\text{Average power of the LOS component}}{\text{Average power of the diffuse component}}$$

$$= \frac{A^2}{2\sigma^2}$$

In effect, $K = \frac{a^2}{2}$. Thus for the graphical plots of Figure 4.23, the running parameter K would assume the values 0, 0.5, 2, 4.5, 12.5.

CHAPTER
5

Information Theory

5.1 Introduction

As mentioned in Chapter 1 and reiterated along the way, the purpose of a communication system is to facilitate the transmission of signals generated by a source of information over a communication channel. But, in basic terms, what do we mean by the term information? To address this important issue, we need to understand the fundamentals of information theory.[1]

The rationale for studying the fundamentals of information theory at this early stage in the book is threefold:

1. Information theory makes extensive use of probability theory, which we studied in Chapter 3; it is, therefore, a logical follow-up to that chapter.
2. It adds meaning to the term "information" used in previous chapters of the book.
3. Most importantly, information theory paves the way for many important concepts and topics discussed in subsequent chapters.

In the context of communications, information theory deals with mathematical modeling and analysis of a communication system rather than with physical sources and physical channels. In particular, it provides answers to two fundamental questions (among others):

1. What is the irreducible complexity, below which a signal cannot be compressed?
2. What is the ultimate transmission rate for reliable communication over a noisy channel?

The answers to these two questions lie in the entropy of a source and the capacity of a channel, respectively:

1. *Entropy* is defined in terms of the probabilistic behavior of a source of information; it is so named in deference to the parallel use of this concept in thermodynamics.
2. *Capacity* is defined as the intrinsic ability of a channel to convey information; it is naturally related to the noise characteristics of the channel.

A remarkable result that emerges from information theory is that if the entropy of the source is less than the capacity of the channel, then, ideally, error-free communication over the channel can be achieved. It is, therefore, fitting that we begin our study of information theory by discussing the relationships among uncertainty, information, and entropy.

5.2 Entropy

Suppose that a *probabilistic experiment* involves observation of the output emitted by a discrete source during every signaling interval. The source output is modeled as a stochastic process, a sample of which is denoted by the discrete random variable S. This random variable takes on symbols from the fixed finite *alphabet*

$$\mathcal{S} = \{s_0, s_1, \ldots, s_{K-1}\} \tag{5.1}$$

with probabilities

$$\mathbb{P}(S=s_k) = p_k, \qquad k = 0, 1, \ldots, K-1 \tag{5.2}$$

Of course, this set of probabilities must satisfy the normalization property

$$\sum_{k=0}^{K-1} p_k = 1, \qquad p_k \geq 0 \tag{5.3}$$

We assume that the symbols emitted by the source during successive signaling intervals are statistically independent. Given such a scenario, can we find a *measure* of how much information is produced by such a source? To answer this question, we recognize that the idea of information is closely related to that of uncertainty or surprise, as described next.

Consider the event $S = s_k$, describing the emission of symbol s_k by the source with probability p_k, as defined in (5.2). Clearly, if the probability $p_k = 1$ and $p_i = 0$ for all $i \neq k$, then there is no "surprise" and, therefore, no "information" when symbol s_k is emitted, because we know what the message from the source must be. If, on the other hand, the source symbols occur with different probabilities and the probability p_k is low, then there is more surprise and, therefore, information when symbol s_k is emitted by the source than when another symbol s_i, $i \neq k$, with higher probability is emitted. Thus, the words *uncertainty*, *surprise*, and *information* are all related. Before the event $S = s_k$ occurs, there is an amount of uncertainty. When the event $S = s_k$ occurs, there is an amount of surprise. After the occurrence of the event $S = s_k$, there is gain in the amount of information, the essence of which may be viewed as the *resolution of uncertainty*. Most importantly, the amount of information is related to the inverse of the probability of occurrence of the event $S = s_k$.

We define the *amount of information* gained after observing the event $S = s_k$, which occurs with probability p_k, as the logarithmic function[2]

$$I(s_k) = \log\left(\frac{1}{p_k}\right) \tag{5.4}$$

which is often termed "self-information" of the event $S = s_k$. This definition exhibits the following important properties that are intuitively satisfying:

PROPERTY 1

$$I(s_k) = 0 \qquad \text{for } p_k = 1 \tag{5.5}$$

Obviously, if we are absolutely *certain* of the outcome of an event, even before it occurs, there is *no* information gained.

PROPERTY 2

$$I(s_k) \geq 0 \qquad \text{for } 0 \leq p_k \leq 1 \tag{5.6}$$

That is to say, the occurrence of an event $S = s_k$ either provides some or no information, but never brings about a *loss* of information.

PROPERTY 3

$$I(s_k) > I(s_i) \qquad \text{for } p_k < p_i \tag{5.7}$$

That is, the less probable an event is, the more information we gain when it occurs.

PROPERTY 4 $$I(s_k, s_l) = I(s_k) + I(s_l) \text{ if } s_k \text{ and } s_l \text{ are statistically independent}$$

This additive property follows from the logarithmic definition described in (5.4).

The base of the logarithm in (5.4) specifies the units of information measure. Nevertheless, it is standard practice in information theory to use a logarithm to base 2 with binary signaling in mind. The resulting unit of information is called the *bit*, which is a contraction of the words binary digit. We thus write

$$\begin{aligned} I(s_k) &= \log_2\left(\frac{1}{p_k}\right) \\ &= -\log_2 p_k \qquad \text{for } k = 0, 1, \ldots, K-1 \end{aligned} \tag{5.8}$$

When $p_k = 1/2$, we have $I(s_k) = 1$ bit. We may, therefore, state:

> One bit is the amount of information that we gain when one of two possible and equally likely (i.e., equiprobable) events occurs.

Note that the information $I(s_k)$ is positive, because the logarithm of a number less than one, such as a probability, is negative. Note also that if p_k is zero, then the self-information I_{s_k} assumes an unbounded value.

The amount of information $I(s_k)$ produced by the source during an arbitrary signaling interval depends on the symbol s_k emitted by the source at the time. The self-information $I(s_k)$ is a discrete random variable that takes on the values $I(s_0)$, $I(s_1)$, …, $I(s_{K-1})$ with probabilities p_0, p_1, …, p_{K-1} respectively. The *expectation* of $I(s_k)$ over all the probable values taken by the random variable S is given by

$$\begin{aligned} H(S) &= \mathbb{E}[I(s_k)] \\ &= \sum_{k=0}^{K-1} p_k I(s_k) \\ &= \sum_{k=0}^{K-1} p_k \log_2\left(\frac{1}{p_k}\right) \end{aligned} \tag{5.9}$$

The quantity $H(S)$ is called the *entropy*,[3] formally defined as follows:

> The entropy of a discrete random variable, representing the output of a source of information, is a measure of the average information content per source symbol.

Note that the entropy $H(S)$ is independent of the alphabet $\mathcal{S}$; it depends only on the probabilities of the symbols in the alphabet $\mathcal{S}$ of the source.

Properties of Entropy

Building on the definition of entropy given in (5.9), we find that entropy of the discrete random variable S is bounded as follows:

$$0 \le H(S) \le \log_2 K \tag{5.10}$$

where K is the number of symbols in the alphabet $\mathcal{S}$.

Elaborating on the two bounds on entropy in (5.10), we now make two statements:

1. $H(S) = 0$, if, and only if, the probability $p_k = 1$ for some k, and the remaining probabilities in the set are all zero; this lower bound on entropy corresponds to *no uncertainty*.
2. $H(S) = \log K$, if, and only if, $p_k = 1/K$ for all k (i.e., all the symbols in the source alphabet $\mathcal{S}$ are equiprobable); this upper bound on entropy corresponds to maximum uncertainty.

To prove these properties of $H(S)$, we proceed as follows. First, since each probability p_k is less than or equal to unity, it follows that each term $p_k \log_2(1/p_k)$ in (5.9) is always nonnegative, so $H(S) \geq 0$. Next, we note that the product term $p_k \log_2(1/p_k)$ is zero if, and only if, $p_k = 0$ or 1. We therefore deduce that $H(S) = 0$ if, and only if, $p_k = 0$ or 1 for some k and all the rest are zero. This completes the proofs of the lower bound in (5.10) and statement 1.

To prove the upper bound in (5.10) and statement 2, we make use of a property of the natural logarithm:

$$\log_e x \leq x - 1, \qquad x \geq 0 \tag{5.11}$$

where $\log_e$ is another way of describing the *natural logarithm*, commonly denoted by ln; both notations are used interchangeably. This inequality can be readily verified by plotting the functions $\ln x$ and $(x - 1)$ versus x, as shown in Figure 5.1. Here we see that the line $y = x - 1$ always lies above the curve $y = \log_e x$. The equality holds only at the point $x = 1$, where the line is tangential to the curve.

To proceed with the proof, consider first any two different probability distributions denoted by $p_0, p_1, \ldots, p_{K-1}$ and $q_0, q_1, \ldots, q_{K-1}$ on the alphabet $\mathcal{S} = \{s_0, s_1, \ldots, s_{K-1}\}$ of a discrete source. We may then define the *relative entropy* of these two distributions:

$$D(p||q) = \sum_{k=0}^{K-1} p_k \log_2\left(\frac{p_k}{q_k}\right) \tag{5.12}$$

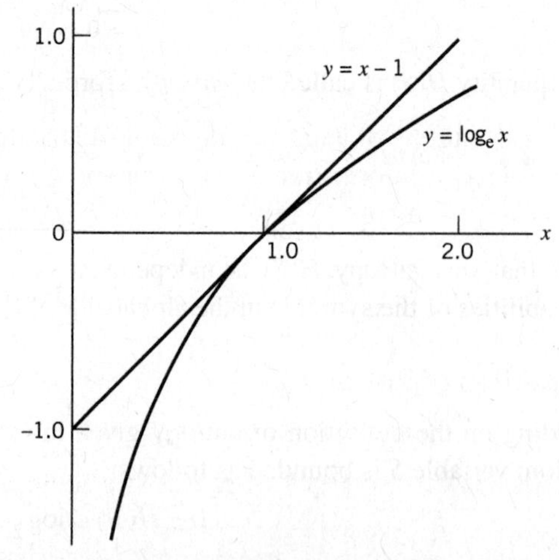

Figure 5.1 Graphs of the functions $x - 1$ and $\log x$ versus x.

Hence, changing to the natural logarithm and using the inequality of (5.11), we may express the summation on the right-hand side of (5.12) as follows:

$$\begin{aligned}\sum_{k=0}^{K-1} p_k \log_2\left(\frac{p_k}{q_k}\right) &= -\sum_{k=0}^{K-1} p_k \log_2\left(\frac{q_k}{p_k}\right)\\ &\geq \frac{1}{\ln 2}\sum_{k=0}^{K-1} p_k\left(\frac{q_k}{p_k}-1\right)\\ &= \frac{1}{\log_e 2}\sum_{k=0}^{K-1}(q_k - p_k)\\ &= 0\end{aligned}$$

where, in the third line of the equation, it is noted that the sums over p_k and q_k are both equal to unity in accordance with (5.3). We thus have the *fundamental property* of probability theory:

$$D(p||q) \geq 0 \tag{5.13}$$

In words, (5.13) states:

> The relative entropy of a pair of different discrete distributions is always nonnegative; it is zero only when the two distributions are identical.

Suppose we next put

$$q_k = \frac{1}{K}, \qquad k = 0, 1, \ldots, K-1$$

which corresponds to a source alphabet $\mathcal{S}$ with *equiprobable* symbols. Using this distribution in (5.12) yields

$$\begin{aligned}D(p||q) &= \sum_{k=0}^{K-1} p_k \log_2 p_k + \log_2 K \sum_{k=0}^{K-1} p_k\\ &= -H(S) + \log_2 K\end{aligned}$$

where we have made use of (5.3) and (5.9). Hence, invoking the fundamental inequality of (5.13), we may finally write

$$H(S) \leq \log_2 K \tag{5.14}$$

Thus, $H(S)$ is always less than or equal to $\log_2 K$. The equality holds if, and only if, the symbols in the alphabet $\mathcal{S}$ are equiprobable. This completes the proof of (5.10) and with it the accompanying statements 1 and 2.

EXAMPLE 1 Entropy of Bernoulli Random Variable

To illustrate the properties of $H(S)$ summed up in (5.10), consider the Bernoulli random variable for which symbol 0 occurs with probability p_0 and symbol 1 with probability $p_1 = 1 - p_0$.

The entropy of this random variable is

$$\begin{aligned} H(S) &= -p_0 \log_2 p_0 - p_1 \log_2 p_1 \\ &= -p_0 \log_2 p_0 - (1-p_0) \log_2 (1-p_0) \text{ bits} \end{aligned} \tag{5.15}$$

from which we observe the following:

1. When $p_0 = 0$, the entropy $H(S) = 0$; this follows from the fact that $x \log_e x \to 0$ as $x \to 0$.
2. When $p_0 = 1$, the entropy $H(S) = 0$.
3. The entropy $H(S)$ attains its maximum value $H_{max} = 1$ bit when $p_1 = p_0 = 1/2$; that is, when symbols 1 and 0 are equally probable.

In other words, $H(S)$ is symmetric about $p_0 = 1/2$.

The function of p_0 given on the right-hand side of (5.15) is frequently encountered in information-theoretic problems. It is customary, therefore, to assign a special symbol to this function. Specifically, we define

$$H(p_0) = -p_0 \log_2 p_0 - (1-p_0) \log_2 (1-p_0) \tag{5.16}$$

We refer to $H(p_0)$ as the *entropy function*. The distinction between (5.15) and (5.16) should be carefully noted. The $H(S)$ of (5.15) gives the entropy of the Bernoulli random variable S. The $H(p_0)$ of (5.16), on the other hand, is a function of the prior probability p_0 defined on the interval [0, 1]. Accordingly, we may plot the entropy function $H(p_0)$ versus p_0, defined on the interval [0, 1], as shown in Figure 5.2. The curve in Figure 5.2 highlights the observations made under points 1, 2, and 3.

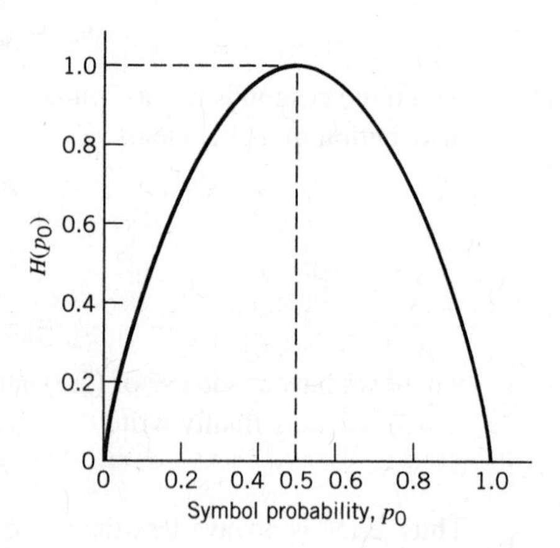

Figure 5.2
Entropy function $H(p_0)$.

Extension of a Discrete Memoryless Source

To add specificity to the discrete source of symbols that has been the focus of attention up until now, we now assume it to be *memoryless* in the sense that the symbol emitted by the source at any time is independent of previous and future emissions.

In this context, we often find it useful to consider *blocks* rather than individual symbols, with each block consisting of n successive source symbols. We may view each such block as being produced by an *extended source* with a source alphabet described by the Cartesian product of a set S^n that has K^n distinct blocks, where K is the number of distinct symbols in the source alphabet S of the original source. With the source symbols being statistically independent, it follows that the probability of a source symbol in S^n is equal to the product of the probabilities of the n source symbols in S that constitute a particular source symbol of S^n. We may thus intuitively expect that $H(S^n)$, the entropy of the extended source, is equal to n times $H(S)$, the entropy of the original source. That is, we may write

$$H(S^{(n)}) = nH(S) \tag{5.17}$$

We illustrate the validity of this relationship by way of an example.

EXAMPLE 2 **Entropy of Extended Source**

Consider a discrete memoryless source with source alphabet $\mathcal{S} = \{s_0, s_1, s_2\}$, whose three distinct symbols have the following probabilities:

$$p_0 = \frac{1}{4}$$

$$p_1 = \frac{1}{4}$$

$$p_2 = \frac{1}{2}$$

Hence, the use of (5.9) yields the entropy of the discrete random variable S representing the source as

$$\begin{aligned} H(S) &= p_0 \log_2\left(\frac{1}{p_0}\right) + p_1 \log_2\left(\frac{1}{p_1}\right) + p_2 \log_2\left(\frac{1}{p_2}\right) \\ &= \frac{1}{4}\log_2(4) + \frac{1}{4}\log_2(4) + \frac{1}{2}\log_2(2) \\ &= \frac{3}{2} \text{ bits} \end{aligned}$$

Consider next the second-order extension of the source. With the source alphabet $\mathcal{S}$ consisting of three symbols, it follows that the source alphabet of the extended source $S^{(2)}$ has nine symbols. The first row of Table 5.1 presents the nine symbols of $S^{(2)}$, denoted by $\sigma_0, \sigma_1, \ldots, \sigma_8$. The second row of the table presents the composition of these nine symbols in terms of the corresponding sequences of source symbols s_0, s_1, and s_2, taken two at a

Table 5.1 **Alphabets of second-order extension of a discrete memoryless source**

Symbols of $S^{(2)}$	σ_0	σ_1	σ_2	σ_3	σ_4	σ_5	σ_6	σ_7	σ_8
Corresponding sequences of symbols of S	s_0s_0	s_0s_1	s_0s_2	s_1s_0	s_1s_1	s_1s_2	s_2s_0	s_2s_1	s_2s_2
Probability $\mathbb{P}(\sigma_i)$, $i = 0, 1, \ldots, 8$	$\frac{1}{16}$	$\frac{1}{16}$	$\frac{1}{8}$	$\frac{1}{16}$	$\frac{1}{16}$	$\frac{1}{8}$	$\frac{1}{8}$	$\frac{1}{8}$	$\frac{1}{4}$

time. The probabilities of the nine source symbols of the extended source are presented in the last row of the table. Accordingly, the use of (5.9) yields the entropy of the extended source as

$$\begin{aligned} H(S^{(2)}) &= \sum_{i=0}^{8} p(\sigma_i) \log_2\left(\frac{1}{p(\sigma_i)}\right) \\ &= \frac{1}{16}\log_2(16) + \frac{1}{16}\log_2(16) + \frac{1}{8}\log_2(8) + \frac{1}{16}\log_2(16) \\ &+ \frac{1}{16}\log_2(16) + \frac{1}{8}\log_2(8) + \frac{1}{8}\log_2(8) + \frac{1}{8}\log_2(8) + \frac{1}{4}\log_2(4) \\ &= 3 \text{ bits} \end{aligned}$$

We thus see that $H(S^{(2)}) = 2H(S)$ in accordance with (5.17).

5.3 Source-coding Theorem

Now that we understand the meaning of entropy of a random variable, we are equipped to address an important issue in communication theory: the representation of data generated by a discrete source of information.

The process by which this representation is accomplished is called *source encoding*. The device that performs the representation is called a *source encoder*. For reasons to be described, it may be desirable to know the statistics of the source. In particular, if some source symbols are known to be more probable than others, then we may exploit this feature in the generation of a *source code* by assigning *short* codewords to *frequent* source symbols, and *long* codewords to *rare* source symbols. We refer to such a source code as a *variable-length code*. The *Morse code*, used in telegraphy in the past, is an example of a variable-length code. Our primary interest is in the formulation of a source encoder that satisfies two requirements:

1. The codewords produced by the encoder are in *binary* form.
2. The source code is *uniquely decodable*, so that the original source sequence can be reconstructed perfectly from the encoded binary sequence.

The second requirement is particularly important: it constitutes the basis for a *perfect source code*.

Consider then the scheme shown in Figure 5.3 that depicts a discrete memoryless source whose output s_k is converted by the source encoder into a sequence of 0s and 1s, denoted by b_k. We assume that the source has an alphabet with K different symbols and that the kth symbol s_k occurs with probability p_k, $k = 0, 1, \ldots, K-1$. Let the binary

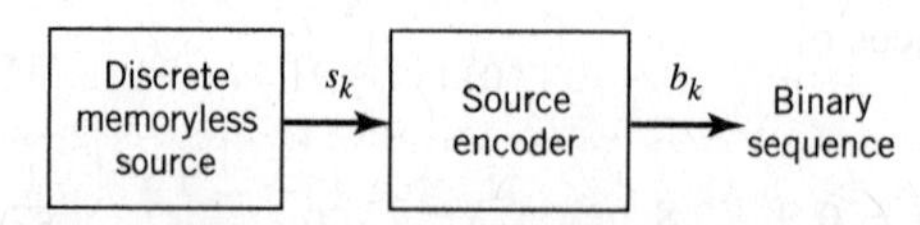

Figure 5.3 Source encoding.

codeword assigned to symbol s_k by the encoder have length l_k, measured in bits. We define the *average codeword length* $\bar{L}$ of the source encoder as

$$\bar{L} = \sum_{k=0}^{K-1} p_k l_k \tag{5.18}$$

In physical terms, the parameter $\bar{L}$ represents the *average number of bits per source symbol* used in the source encoding process. Let $L_{\min}$ denote the *minimum* possible value of $\bar{L}$. We then define the *coding efficiency* of the source encoder as

$$\eta = \frac{L_{\min}}{\bar{L}} \tag{5.19}$$

With $\bar{L} \geq L_{\min}$, we clearly have $\eta \leq 1$. The source encoder is said to be *efficient* when η approaches unity.

But how is the minimum value $L_{\min}$ determined? The answer to this fundamental question is embodied in Shannon's first theorem: the *source-coding theorem*,[4] which may be stated as follows:

> Given a discrete memoryless source whose output is denoted by the random variable S, the entropy $H(S)$ imposes the following bound on the average codeword length $\bar{L}$ for any source encoding scheme:
>
> $$\bar{L} \geq H(S) \tag{5.20}$$

According to this theorem, the entropy $H(S)$ represents a *fundamental limit* on the average number of bits per source symbol necessary to represent a discrete memoryless source, in that it can be made as small as but no smaller than the entropy $H(S)$. Thus, setting $L_{\min} = H(S)$, we may rewrite (5.19), defining the efficiency of a source encoder in terms of the entropy $H(S)$ as shown by

$$\eta = \frac{H(S)}{\bar{L}} \tag{5.21}$$

where as before we have $\eta \leq 1$.

5.4 Lossless Data Compression Algorithms

A common characteristic of signals generated by physical sources is that, in their natural form, they contain a significant amount of *redundant* information, the transmission of which is therefore wasteful of primary communication resources. For example, the output of a computer used for business transactions constitutes a redundant sequence in the sense that any two adjacent symbols are typically correlated with each other.

For efficient signal transmission, the redundant information should, therefore, be removed from the signal prior to transmission. This operation, with no loss of information, is ordinarily performed on a signal in digital form, in which case we refer to the operation as *lossless data compression.* The code resulting from such an operation provides a representation of the source output that is not only efficient in terms of the average number of bits per symbol, but also exact in the sense that the original data can be reconstructed with no loss of information. The entropy of the source establishes the fundamental limit on the removal of redundancy from the data. Basically, lossless data compression is achieved by assigning short descriptions to the most frequent outcomes of the source output and longer descriptions to the less frequent ones.

In this section we discuss some source-coding schemes for lossless data compression. We begin the discussion by describing a type of source code known as a prefix code, which not only is uniquely decodable, but also offers the possibility of realizing an average codeword length that can be made arbitrarily close to the source entropy.

Prefix Coding

Consider a discrete memoryless source of alphabet $\{s_0, s_1, ..., s_{K-1}\}$ and respective probabilities $\{p_0, p_1, ..., p_{K-1}\}$. For a source code representing the output of this source to be of practical use, the code has to be uniquely decodable. This restriction ensures that, for each finite sequence of symbols emitted by the source, the corresponding sequence of codewords is different from the sequence of codewords corresponding to any other source sequence. We are specifically interested in a special class of codes satisfying a restriction known as the *prefix condition*. To define the prefix condition, let the codeword assigned to source symbol s_k be denoted by $(m_{k_1}, m_{k_2}, ..., m_{k_n})$, where the individual elements $m_{k_1}, ..., m_{k_n}$ are 0s and 1s and n is the codeword length. The initial part of the codeword is represented by the elements $m_{k_1}, ..., m_{k_i}$ for some $i \leq n$. Any sequence made up of the initial part of the codeword is called a *prefix* of the codeword. We thus say:

> A *prefix code* is defined as a code in which no codeword is the prefix of any other codeword.

Prefix codes are distinguished from other uniquely decodable codes by the fact that the end of a codeword is always recognizable. Hence, the decoding of a prefix can be accomplished as soon as the binary sequence representing a source symbol is fully received. For this reason, prefix codes are also referred to as *instantaneous codes.*

EXAMPLE 3 **Illustrative Example of Prefix Coding**

To illustrate the meaning of a prefix code, consider the three source codes described in Table 5.2. Code I is not a prefix code because the bit 0, the codeword for s_0, is a prefix of 00, the codeword for s_2. Likewise, the bit 1, the codeword for s_1, is a prefix of 11, the codeword for s_3. Similarly, we may show that code III is not a prefix code but code II is.

Table 5.2 **Illustrating the definition of a prefix code**

Symbol source	Probability of occurrence	Code I	Code II	Code III
s_0	0.5	0	0	0
s_1	0.25	1	10	01
s_2	0.125	00	110	011
s_3	0.125	11	111	0111

Decoding of Prefix Code

To decode a sequence of codewords generated from a prefix source code, the *source decoder* simply starts at the beginning of the sequence and decodes one codeword at a time. Specifically, it sets up what is equivalent to a *decision tree*, which is a graphical portrayal of the codewords in the particular source code. For example, Figure 5.4 depicts the decision tree corresponding to code II in Table 5.2. The tree has an *initial state* and four *terminal states* corresponding to source symbols s_0, s_1, s_2, and s_3. The decoder always starts at the initial state. The first received bit moves the decoder to the terminal state s_0 if it is 0 or else to a second decision point if it is 1. In the latter case, the second bit moves the decoder one step further down the tree, either to terminal state s_1 if it is 0 or else to a third decision point if it is 1, and so on. Once each terminal state emits its symbol, the decoder is reset to its initial state. Note also that each bit in the received encoded sequence is examined only once. Consider, for example, the following encoded sequence:

$$\underbrace{10}_{s_1}\ \underbrace{111}_{s_3}\ \underbrace{110}_{s_2}\ \underbrace{0}_{s_0}\ \underbrace{0}_{s_0}\ \ldots$$

This sequence is readily decoded as the source sequence $s_1 s_3 s_2 s_0 s_0 \ldots$ The reader is invited to carry out this decoding.

As mentioned previously, a prefix code has the important property that it is instantaneously decodable. But the converse is not necessarily true. For example, code III in Table 5.2 does not satisfy the prefix condition, yet it is uniquely decodable because the bit 0 indicates the beginning of each codeword in the code.

To probe more deeply into prefix codes, exemplified by that in Table 5.2, we resort to an inequality, which is considered next.

Kraft Inequality

Consider a discrete memoryless source with source alphabet $\{s_0, s_1, \ldots, s_{K-1}\}$ and source probabilities $\{p_0, p_1, \ldots, p_{K-1}\}$, with the codeword of symbol s_k having length l_k, $k = 0, 1,$

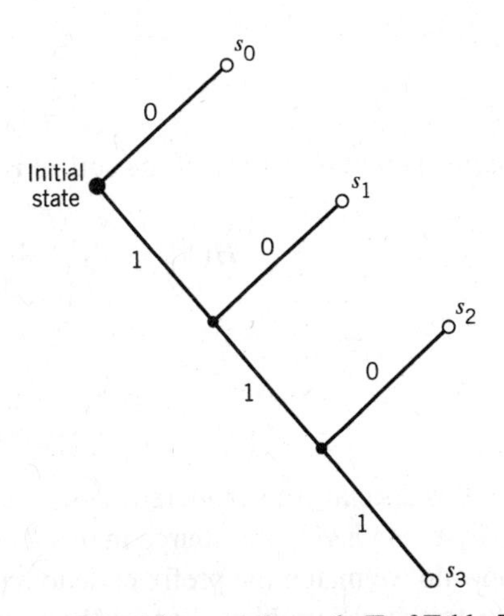

Figure 5.4 Decision tree for code II of Table 5.2.

..., $K-1$. Then, according to the *Kraft inequality*,[5] the codeword lengths always satisfy the following inequality:

$$\sum_{k=0}^{K-1} 2^{-l_k} \le 1 \tag{5.22}$$

where the factor 2 refers to the number of symbols in the binary alphabet. The Kraft inequality is a necessary but not sufficient condition for a source code to be a prefix code. In other words, the inequality of (5.22) is merely a condition on the codeword lengths of a prefix code and not on the codewords themselves. For example, referring to the three codes listed in Table 5.2, we see:

- Code I violates the Kraft inequality; it cannot, therefore, be a prefix code.
- The Kraft inequality is satisfied by both codes II and III, but only code II is a prefix code.

Given a discrete memoryless source of entropy $H(S)$, a prefix code can be constructed with an average codeword length $\bar{L}$, which is bounded as follows:

$$H(S) \le \bar{L} < H(S) + 1 \tag{5.23}$$

The left-hand bound of (5.23) is satisfied with equality under the condition that symbol s_k is emitted by the source with probability

$$p_k = 2^{-l_k} \tag{5.24}$$

where l_k is the length of the codeword assigned to source symbol s_k. A distribution governed by (5.24) is said to be a *dyadic distribution*. For this distribution, we naturally have

$$\sum_{k=0}^{K-1} 2^{-l_k} = \sum_{k=0}^{K-1} p_k = 1$$

Under this condition, the Kraft inequality of (5.22) confirms that we can construct a prefix code, such that the length of the codeword assigned to source symbol s_k is $-\log_2 p_k$. For such a code, the average codeword length is

$$\bar{L} = \sum_{k=0}^{K-1} \frac{l_k}{2^{l_k}} \tag{5.25}$$

and the corresponding entropy of the source is

$$\begin{aligned} H(S) &= \sum_{k=0}^{K-1} \left(\frac{1}{2^{l_k}}\right) \log_2(2^{l_k}) \\ &= \sum_{k=0}^{K-1} \left(\frac{l_k}{2^{l_k}}\right) \end{aligned} \tag{5.26}$$

Hence, in this special (rather meretricious) case, we find from (5.25) and (5.26) that the prefix code is *matched* to the source in that $\bar{L} = H(S)$.

But how do we match the prefix code to an arbitrary discrete memoryless source? The answer to this basic problem lies in the use of an *extended code*. Let $\bar{L}_n$ denote the average codeword length of the extended prefix code. For a uniquely decodable code, $\bar{L}_n$ is the smallest possible. From (5.23), we find that

$$nH(S) \le \bar{L}_n < nH(S) + 1 \tag{5.27}$$

or, equivalently,

$$H(S) \le \frac{\bar{L}_n}{n} < H(S) + \frac{1}{n} \tag{5.28}$$

In the limit, as n approaches infinity, the lower and upper bounds in (5.28) converge as shown by

$$\lim_{n \to \infty} \frac{1}{n}\bar{L}_n = H(S) \tag{5.29}$$

We may, therefore, make the statement:

> By making the order n of an extended prefix source encoder large enough, we can make the code faithfully represent the discrete memoryless source S as closely as desired.

In other words, the average codeword length of an extended prefix code can be made as small as the entropy of the source, provided that the extended code has a high enough order in accordance with the source-coding theorem. However, the price we have to pay for decreasing the average codeword length is increased decoding complexity, which is brought about by the high order of the extended prefix code.

Huffman Coding

We next describe an important class of prefix codes known as Huffman codes. The basic idea behind *Huffman coding*[6] is the construction of a simple algorithm that computes an *optimal* prefix code for a given distribution, optimal in the sense that the code has the *shortest expected length*. The end result is a source code whose average codeword length approaches the fundamental limit set by the entropy of a discrete memoryless source, namely $H(S)$. The essence of the *algorithm* used to synthesize the Huffman code is to replace the prescribed set of source statistics of a discrete memoryless source with a simpler one. This *reduction* process is continued in a step-by-step manner until we are left with a final set of only two source statistics (symbols), for which (0, 1) is an optimal code. Starting from this trivial code, we then work backward and thereby construct the Huffman code for the given source.

To be specific, the Huffman *encoding algorithm* proceeds as follows:

1. The source symbols are listed in order of decreasing probability. The two source symbols of lowest probability are assigned 0 and 1. This part of the step is referred to as the *splitting stage*.
2. These two source symbols are then *combined* into a new source symbol with probability equal to the sum of the two original probabilities. (The list of source symbols, and, therefore, source statistics, is thereby *reduced* in size by one.) The probability of the new symbol is placed in the list in accordance with its value.
3. The procedure is repeated until we are left with a final list of source statistics (symbols) of only two for which the symbols 0 and 1 are assigned.

The code for each (original) source is found by working backward and tracing the sequence of 0s and 1s assigned to that symbol as well as its successors.

EXAMPLE 4 Huffman Tree

To illustrate the construction of a Huffman code, consider the five symbols of the alphabet of a discrete memoryless source and their probabilities, which are shown in the two leftmost columns of Figure 5.5b. Following through the Huffman algorithm, we reach the end of the computation in four steps, resulting in a *Huffman tree* similar to that shown in Figure 5.5; the Huffman tree is not to be confused with the decision tree discussed previously in Figure 5.4. The codewords of the Huffman code for the source are tabulated in Figure 5.5a. The average codeword length is, therefore,

$$\begin{aligned}\bar{L} &= 0.4(2) + 0.2(2) + 0.2(2) + 0.1(3) + 0.1(3)\\ &= 2.2 \text{ binary symbols}\end{aligned}$$

The entropy of the specified discrete memoryless source is calculated as follows (see (5.9)):

$$\begin{aligned}H(S) &= 0.4 \log_2\left(\frac{1}{0.4}\right) + 0.2 \log_2\left(\frac{1}{0.2}\right) + 0.2 \log_2\left(\frac{1}{0.2}\right) + 0.1 \log_2\left(\frac{1}{0.1}\right) + 0.1 \log_2\left(\frac{1}{0.1}\right)\\ &= 0.529 + 0.464 + 0.464 + 0.332 + 0.332\\ &= 2.121 \text{ bits}\end{aligned}$$

For this example, we may make two observations:

1. The average codeword length $\bar{L}$ exceeds the entropy $H(S)$ by only 3.67%.
2. The average codeword length $\bar{L}$ does indeed satisfy (5.23).

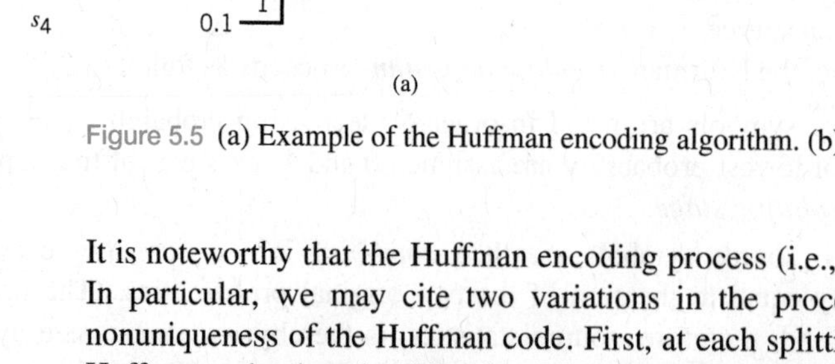

Symbol	Stage I	Stage II	Stage III	Stage IV
s_0	0.4	0.4	0.4	0.6 (0)
s_1	0.2	0.2	0.4 (0)	0.4 (1)
s_2	0.2	0.2 (0)	0.2 (1)	
s_3	0.1 (0)	0.2 (1)		
s_4	0.1 (1)			

(a)

Symbol	Probability	Code word
s_0	0.4	00
s_1	0.2	10
s_2	0.2	11
s_3	0.1	010
s_4	0.1	011

(b)

Figure 5.5 (a) Example of the Huffman encoding algorithm. (b) Source code.

It is noteworthy that the Huffman encoding process (i.e., the Huffman tree) is not unique. In particular, we may cite two variations in the process that are responsible for the nonuniqueness of the Huffman code. First, at each splitting stage in the construction of a Huffman code, there is arbitrariness in the way the symbols 0 and 1 are assigned to the last two source symbols. Whichever way the assignments are made, however, the resulting differences are trivial. Second, ambiguity arises when the probability of a *combined* symbol (obtained by adding the last two probabilities pertinent to a particular step) is found to equal another probability in the list. We may proceed by placing the probability of the new symbol as *high* as possible, as in Example 4. Alternatively, we may place it as *low* as possible. (It is presumed that whichever way the placement is made, high or low, it is consistently adhered to throughout the encoding process.) By this time, noticeable differences arise in that the codewords in the resulting source code can have different lengths. Nevertheless, the average codeword length remains the same.

As a measure of the variability in codeword lengths of a source code, we define the *variance* of the average codeword length $\bar{L}$ over the ensemble of source symbols as

$$\sigma^2 = \sum_{k=0}^{K-1} p_k (l_k - \bar{L})^2 \tag{5.30}$$

where $p_0, p_1, \ldots, p_{K-1}$ are the source statistics and l_k is the length of the codeword assigned to source symbol s_k. It is usually found that when a combined symbol is moved as high as possible, the resulting Huffman code has a significantly smaller variance σ^2 than when it is moved as low as possible. On this basis, it is reasonable to choose the former Huffman code over the latter.

Lempel–Ziv Coding

A drawback of the Huffman code is that it requires knowledge of a probabilistic model of the source; unfortunately, in practice, source statistics are not always known a priori. Moreover, in the modeling of text we find that storage requirements prevent the Huffman code from capturing the higher-order relationships between words and phrases because the codebook grows exponentially fast in the size of each super-symbol of letters (i.e., grouping of letters); the efficiency of the code is therefore compromised. To overcome these practical limitations of Huffman codes, we may use the *Lempel–Ziv algorithm*,[7] which is intrinsically *adaptive* and simpler to implement than Huffman coding.

Basically, the idea behind encoding in the Lempel–Ziv algorithm is described as follows:

> The source data stream is parsed into segments that are the shortest subsequences not encountered previously.

To illustrate this simple yet elegant idea, consider the example of the binary sequence

000101110010100101 ...

It is assumed that the binary symbols 0 and 1 are already stored in that order in the code book. We thus write

Subsequences stored:	0, 1
Data to be parsed:	000101110010100101 ...

The encoding process begins at the left. With symbols 0 and 1 already stored, the *shortest subsequence* of the data stream encountered for the first time and not seen before is 00; so we write

Subsequences stored:	0, 1, 00
Data to be parsed:	0101110010100101 ...

The second shortest subsequence not seen before is 01; accordingly, we go on to write

Subsequences stored: 0, 0, 00, 01
Data to be parsed: 01110010100101 ...

The next shortest subsequence not encountered previously is 011; hence, we write

Subsequences stored: 0, 1, 00, 01, 011
Data to be parsed: 10010100101 ...

We continue in the manner described here until the given data stream has been completely parsed. Thus, for the example at hand, we get the *code book* of binary subsequences shown in the second row of Figure 5.6.[8]

The first row shown in this figure merely indicates the numerical positions of the individual subsequences in the code book. We now recognize that the first subsequence of the data stream, 00, is made up of the concatenation of the *first* code book entry, 0, with itself; it is, therefore, represented by the number 11. The second subsequence of the data stream, 01, consists of the *first* code book entry, 0, concatenated with the *second* code book entry, 1; it is, therefore, represented by the number 12. The remaining subsequences are treated in a similar fashion. The complete set of numerical representations for the various subsequences in the code book is shown in the third row of Figure 5.6. As a further example illustrating the composition of this row, we note that the subsequence 010 consists of the concatenation of the subsequence 01 in position 4 and symbol 0 in position 1; hence, the numerical representation is 41. The last row shown in Figure 5.6 is the binary encoded representation of the different subsequences of the data stream.

The last symbol of each subsequence in the code book (i.e., the second row of Figure 5.6) is an *innovation symbol*, which is so called in recognition of the fact that its appendage to a particular subsequence distinguishes it from all previous subsequences stored in the code book. Correspondingly, the last bit of each uniform block of bits in the binary encoded representation of the data stream (i.e., the fourth row in Figure 5.6) represents the innovation symbol for the particular subsequence under consideration. The remaining bits provide the equivalent binary representation of the "pointer" to the *root subsequence* that matches the one in question, except for the innovation symbol.

The *Lempel–Ziv decoder* is just as simple as the encoder. Specifically, it uses the pointer to identify the root subsequence and then appends the innovation symbol. Consider, for example, the binary encoded block 1101 in position 9. The last bit, 1, is the innovation symbol. The remaining bits, 110, point to the root subsequence 10 in position 6. Hence, the block 1101 is decoded into 101, which is correct.

From the example described here, we note that, in contrast to Huffman coding, the Lempel–Ziv algorithm uses fixed-length codes to represent a variable number of source symbols; this feature makes the Lempel–Ziv code suitable for synchronous transmission.

Numerical positions	1	2	3	4	5	6	7	8	9
Subsequences	0	1	00	01	011	10	010	100	101
Numerical representations			11	12	42	21	41	61	62
Binary encoded blocks			0010	0011	1001	0100	1000	1100	1101

Figure 5.6 Illustrating the encoding process performed by the Lempel–Ziv algorithm on the binary sequence 000101110010100101 ...

In practice, fixed blocks of 12 bits long are used, which implies a code book of $2^{12} = 4096$ entries.

For a long time, Huffman coding was unchallenged as the algorithm of choice for lossless data compression; Huffman coding is still optimal, but in practice it is hard to implement. It is on account of practical implementation that the Lempel–Ziv algorithm has taken over almost completely from the Huffman algorithm. The Lempel–Ziv algorithm is now the standard algorithm for file compression.

5.5 Discrete Memoryless Channels

Up to this point in the chapter we have been preoccupied with discrete memoryless sources responsible for *information generation*. We next consider the related issue of *information transmission*. To this end, we start the discussion by considering a discrete memoryless channel, the counterpart of a discrete memoryless source.

A *discrete memoryless channel* is a statistical model with an input X and an output Y that is a *noisy* version of X; both X and Y are random variables. Every unit of time, the channel accepts an input symbol X selected from an alphabet $\mathscr{X}$ and, in response, it emits an output symbol Y from an alphabet $\mathscr{Y}$. The channel is said to be "discrete" when both of the alphabets $\mathscr{X}$ and $\mathscr{Y}$ have *finite* sizes. It is said to be "memoryless" when the current output symbol depends *only* on the current input symbol and *not* any previous or future symbol.

Figure 5.7a shows a view of a discrete memoryless channel. The channel is described in terms of an *input alphabet*

$$\mathscr{X} = \{x_0, x_1, \ldots, x_{J-1}\} \tag{5.31}$$

and an *output alphabet*

$$\mathscr{Y} = \{y_0, y_1, \ldots, y_{K-1}\} \tag{5.32}$$

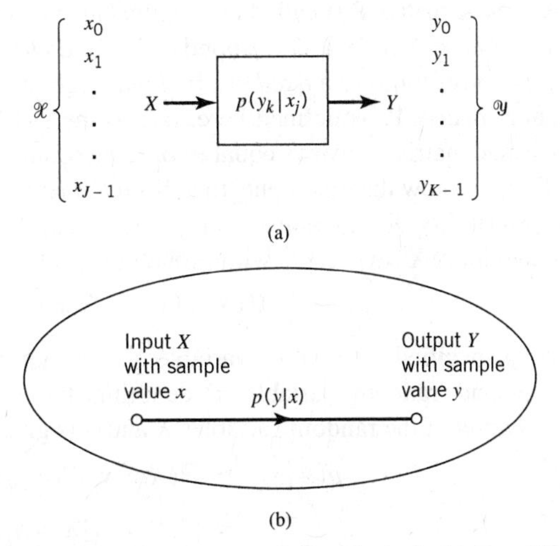

Figure 5.7 (a) Discrete memoryless channel; (b) Simplified graphical representation of the channel.

The *cardinality* of the alphabets $\mathscr{X}$ and $\mathscr{Y}$, or any other alphabet for that matter, is defined as the number of elements in the alphabet. Moreover, the channel is characterized by a set of *transition probabilities*

$$p(y_k|x_j) = \mathbb{P}(Y = y_k|X = x_j) \quad \text{for all } j \text{ and } k \tag{5.33}$$

for which, according to probability theory, we naturally have

$$0 \le p(y_k|x_j) \le 1 \quad \text{for all } j \text{ and } k \tag{5.34}$$

and

$$\sum_k p(y_k|x_j) = 1 \quad \text{for fixed } j \tag{5.35}$$

When the number of input symbols J and the number of output symbols K are not large, we may depict the discrete memoryless channel graphically in another way, as shown in Figure 5.7b. In this latter depiction, each input–output symbol pair (x, y), characterized by the transition probability $p(y|x) > 0$, is joined together by a line labeled with the number $p(y|x)$.

Also, the input alphabet $\mathscr{X}$ and output alphabet $\mathscr{Y}$ need not have the same size; hence the use of J for the size of $\mathscr{X}$ and K for the size of $\mathscr{Y}$. For example, in channel coding, the size K of the output alphabet $\mathscr{Y}$ may be larger than the size J of the input alphabet $\mathscr{X}$; thus, $K \ge J$. On the other hand, we may have a situation in which the channel emits the same symbol when either one of two input symbols is sent, in which case we have $K \le J$.

A convenient way of describing a discrete memoryless channel is to arrange the various transition probabilities of the channel in the form of a *matrix*

$$\mathbf{P} = \begin{bmatrix} p(y_0|x_0) & p(y_1|x_0) & \cdots & p(y_{K-1}|x_0) \\ p(y_0|x_1) & p(y_1|x_1) & \cdots & p(y_{K-1}|x_1) \\ \vdots & \vdots & & \vdots \\ p(y_0|x_{J-1}) & p(y_1|x_{J-1}) & \cdots & p(y_{K-1}|x_{J-1}) \end{bmatrix} \tag{5.36}$$

The J-by-K matrix $\mathbf{P}$ is called the *channel matrix*, or *stochastic matrix*. Note that each *row* of the channel matrix $\mathbf{P}$ corresponds to a *fixed channel input*, whereas each column of the matrix corresponds to a *fixed channel output*. Note also that a fundamental property of the channel matrix $\mathbf{P}$, as defined here, is that the sum of the elements along any row of the stochastic matrix is always equal to one, according to (5.35).

Suppose now that the inputs to a discrete memoryless channel are selected according to the probability distribution $\{p(x_j), j = 0, 1, \ldots, J-1\}$. In other words, the event that the channel input $X = x_j$ occurs with probability

$$p(x_j) = \mathbb{P}(X = x_j) \quad \text{for } j = 0, 1, \ldots, J-1 \tag{5.37}$$

Having specified the random variable X denoting the channel input, we may now specify the second random variable Y denoting the channel output. The *joint probability distribution* of the random variables X and Y is given by

$$\begin{aligned} p(x_j, y_k) &= \mathbb{P}(X = x_j, Y = y_k) \\ &= \mathbb{P}(Y = y_k|X = x_j)\mathbb{P}(X = x_j) \\ &= p(y_k|x_j)p(x_j) \end{aligned} \tag{5.38}$$

The *marginal probability distribution* of the output random variable Y is obtained by averaging out the dependence of $p(x_j, y_k)$ on x_j, obtaining

$$\begin{aligned} p(y_k) &= \mathbb{P}(Y = y_k) \\ &= \sum_{j=0}^{J-1} \mathbb{P}(Y = y_k | X = x_j)\mathbb{P}(X = x_j) \\ &= \sum_{j=0}^{J-1} p(y_k | x_j)p(x_j) \qquad \text{for } k = 0, 1, \ldots, K-1 \end{aligned} \tag{5.39}$$

The probabilities $p(x_j)$ for $j = 0, 1, \ldots, J-1$, are known as the *prior probabilities* of the various input symbols. Equation (5.39) states:

> If we are given the input prior probabilities $p(x_j)$ and the stochastic matrix (i.e., the matrix of transition probabilities $p(y_k|x_j)$), then we may calculate the probabilities of the various output symbols, the $p(y_k)$.

EXAMPLE 5

Binary Symmetric Channel

The *binary symmetric channel* is of theoretical interest and practical importance. It is a special case of the discrete memoryless channel with $J = K = 2$. The channel has two input symbols ($x_0 = 0, x_1 = 1$) and two output symbols ($y_0 = 0, y_1 = 1$). The channel is symmetric because the probability of receiving 1 if 0 is sent is the same as the probability of receiving 0 if 1 is sent. This conditional probability of error is denoted by p (i.e., the probability of a bit flipping). The *transition probability diagram* of a binary symmetric channel is as shown in Figure 5.8. Correspondingly, we may express the stochastic matrix as

$$\mathbf{P} = \begin{bmatrix} 1-p & p \\ p & 1-p \end{bmatrix}$$

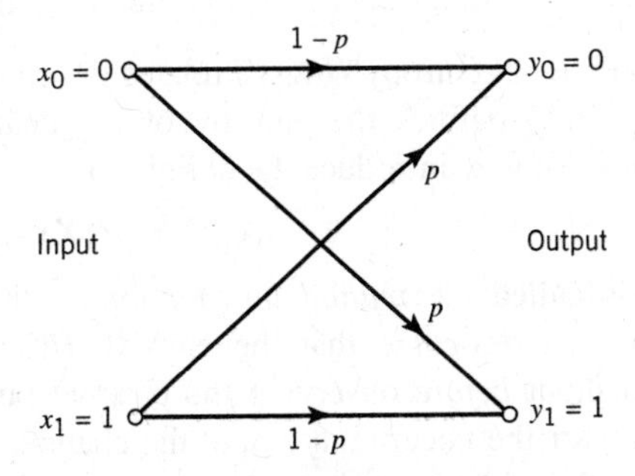

Figure 5.8 Transition probability diagram of binary symmetric channel.

5.6 Mutual Information

Given that we think of the channel output Y (selected from alphabet $\mathcal{Y}$) as a noisy version of the channel input X (selected from alphabet $\mathcal{X}$) and that the entropy $H(X)$ is a measure of the prior uncertainty about X, how can we measure the uncertainty about X after observing Y? To answer this basic question, we extend the ideas developed in Section 5.2 by defining the *conditional entropy* of X selected from alphabet $\mathcal{X}$, given $Y = y_k$. Specifically, we write

$$H(X|Y = y_k) = \sum_{j=0}^{J-1} p(x_j|y_k)\log_2\left(\frac{1}{p(x_j|y_k)}\right) \tag{5.40}$$

This quantity is itself a random variable that takes on the values $H(X|Y = y_0)$, ..., $H(X|Y = y_{K-1})$ with probabilities $p(y_0)$, ..., $p(y_{K-1})$, respectively. The expectation of entropy $H(X|Y = y_k)$ over the output alphabet $\mathcal{Y}$ is therefore given by

$$\begin{aligned} H(X|Y) &= \sum_{k=0}^{K-1} H(X|Y = y_k)p(y_k) \\ &= \sum_{k=0}^{K-1}\sum_{j=0}^{J-1} p(x_j|y_k)p(y_k)\log_2\left(\frac{1}{p(x_j|y_k)}\right) \\ &= \sum_{k=0}^{K-1}\sum_{j=0}^{J-1} p(x_j, y_k)\log_2\left(\frac{1}{p(x_j|y_k)}\right) \end{aligned} \tag{5.41}$$

where, in the last line, we used the definition of the probability of the joint event ($X = x_j$, $Y = y_k$) as shown by

$$p(x_j, y_k) = p(x_j|y_k)p(y_k) \tag{5.42}$$

The quantity $H(X|Y)$ in (5.41) is called the *conditional entropy,* formally defined as follows:

> The conditional entropy, $H(X|Y)$, is the average amount of uncertainty remaining about the channel input after the channel output has been observed.

The conditional entropy $H(X|Y)$ relates the channel output Y to the channel input X. The entropy $H(X)$ defines the entropy of the channel input X by itself. Given these two entropies, we now introduce the definition

$$I(X;Y) = H(X) - H(X|Y) \tag{5.43}$$

which is called the *mutual information* of the channel. To add meaning to this new concept, we recognize that the entropy $H(X)$ accounts for the uncertainty about the channel input *before* observing the channel output and the conditional entropy $H(X|Y)$ accounts for the uncertainty about the channel input *after* observing the channel output. We may, therefore, go on to make the statement:

> The mutual information $I(X;Y)$ is a measure of the uncertainty about the channel input, which is resolved by observing the channel output.

Equation (5.43) is not the only way of defining the mutual information of a channel. Rather, we may define it in another way, as shown by

$$I(Y;X) = H(Y) - H(Y|X) \tag{5.44}$$

on the basis of which we may make the next statement:

> The mutual information $I(Y;X)$ is a measure of the uncertainty about the channel output that is resolved by *sending* the channel input.

On first sight, the two definitions of (5.43) and (5.44) look different. In reality, however, they embody equivalent statements on the mutual information of the channel that are worded differently. More specifically, they could be used interchangeably, as demonstrated next.

Properties of Mutual Information

PROPERTY 1 **Symmetry**

The mutual information of a channel is symmetric in the sense that

$$I(X;Y) = I(Y;X) \tag{5.45}$$

To prove this property, we first use the formula for entropy and then use (5.35) and (5.38), in that order, obtaining

$$\begin{aligned} H(X) &= \sum_{j=0}^{J-1} p(x_j) \log_2\left(\frac{1}{p(x_j)}\right) \\ &= \sum_{j=0}^{J-1} p(x_j) \log_2\left(\frac{1}{p(x_j)}\right) \sum_{k=0}^{K-1} p(y_k|x_j) \\ &= \sum_{j=0}^{J-1} \sum_{k=0}^{K-1} p(y_k|x_j) p(x_j) \log_2\left(\frac{1}{p(x_j)}\right) \\ &= \sum_{j=0}^{J-1} \sum_{k=0}^{K-1} p(x_j, y_k) \log_2\left(\frac{1}{p(x_j)}\right) \end{aligned} \tag{5.46}$$

where, in going from the third to the final line, we made use of the definition of a joint probability. Hence, substituting (5.41) and (5.46) into (5.43) and then combining terms, we obtain

$$I(X;Y) = \sum_{j=0}^{J-1} \sum_{k=0}^{K-1} p(x_j, y_k) \log_2\left(\frac{p(x_j|y_k)}{p(x_j)}\right) \tag{5.47}$$

Note that the double summation on the right-hand side of (5.47) is *invariant* with respect to swapping the x and y. In other words, the symmetry of the mutual information $I(X;Y)$ is already evident from (5.47).

To further confirm this property, we may use *Bayes' rule* for conditional probabilities, previously discussed in Chapter 3, to write

$$\frac{p(x_j|y_k)}{p(x_j)} = \frac{p(y_k|x_j)}{p(y_k)} \tag{5.48}$$

Hence, substituting (5.48) into (5.47) and interchanging the order of summation, we get

$$\begin{aligned} I(X;Y) &= \sum_{k=0}^{K-1} \sum_{j=0}^{J-1} p(x_j, y_k) \log_2\left(\frac{p(y_k|x_j)}{p(y_k)}\right) \\ &= I(Y;X) \end{aligned} \tag{5.49}$$

which proves Property 1.

PROPERTY 2 **Nonnegativity**

The mutual information is always nonnegative; that is;

$$I(X;Y) \geq 0 \tag{5.50}$$

To prove this property, we first note from (5.42) that

$$p(x_j|y_k) = \frac{p(x_j, y_k)}{p(y_k)} \tag{5.51}$$

Hence, substituting (5.51) into (5.47), we may express the mutual information of the channel as

$$I(X;Y) = \sum_{j=0}^{J-1} \sum_{k=0}^{K-1} p(x_j, y_k) \log_2\left(\frac{p(x_j, y_k)}{p(x_j)p(y_k)}\right) \tag{5.52}$$

Next, a direct application of the fundamental inequality of (5.12) on relative entropy confirms (5.50), with equality if, and only if,

$$p(x_j, y_k) = p(x_j)p(y_k) \qquad \text{for all } j \text{ and } k \tag{5.53}$$

In words, Property 2 states the following:

> We cannot lose information, on the average, by observing the output of a channel.

Moreover, the mutual information is zero if, and only if, the input and output symbols of the channel are statistically independent; that is, when (5.53) is satisfied.

PROPERTY 3 **Expansion of the Mutual Information**

The mutual information of a channel is related to the joint entropy of the channel input and channel output by

$$I(X;Y) = H(X) + H(Y) - H(X, Y) \tag{5.54}$$

where the joint entropy $H(X, Y)$ is defined by

$$H(X, Y) = \sum_{j=0}^{J-1} \sum_{k=0}^{K-1} p(x_j, y_k) \log_2\left(\frac{1}{p(x_j, y_k)}\right) \tag{5.55}$$

To prove (5.54), we first rewrite the joint entropy in the equivalent form

$$H(X,Y) = \sum_{j=0}^{J-1}\sum_{k=0}^{K-1} p(x_j, y_k)\log_2\left(\frac{p(x_j)p(y_k)}{p(x_j, y_k)}\right) + \sum_{j=0}^{J-1}\sum_{k=0}^{K-1} p(x_j, y_k)\log_2\left(\frac{1}{p(x_j)p(y_k)}\right) \tag{5.56}$$

The first double summation term on the right-hand side of (5.56) is recognized as the negative of the mutual information of the channel, $I(X;Y)$, previously given in (5.52). As for the second summation term, we manipulate it as follows:

$$\begin{aligned}\sum_{j=0}^{J-1}\sum_{k=0}^{K-1} p(x_j, y_k)\log_2\left(\frac{1}{p(x_j)p(y_k)}\right) &= \sum_{j=0}^{J-1}\log_2\left(\frac{1}{p(x_j)}\right)\sum_{k=0}^{K-1} p(x_j, y_k) \\ &\quad + \sum_{k=0}^{K-1}\log_2\left(\frac{1}{p(y_k)}\right)\sum_{j=0}^{J-1} p(x_j, y_k) \\ &= \sum_{j=0}^{J-1} p(x_j)\log_2\left(\frac{1}{p(x_j)}\right) + \sum_{k=0}^{K-1} p(y_k)\log_2\left(\frac{1}{p(y_k)}\right) \\ &= H(X) + H(Y)\end{aligned} \tag{5.57}$$

where, in the first line, we made use of the following relationship from probability theory:

$$\sum_{k=0}^{K-1} p(x_j, y_k) = p(y_k)$$

and a similar relationship holds for the second line of the equation.

Accordingly, using (5.52) and (5.57) in (5.56), we get the result

$$H(X,Y) = -I(X;Y) + H(X) + H(Y) \tag{5.58}$$

which, on rearrangement, proves Property 3.

We conclude our discussion of the mutual information of a channel by providing a diagramatic interpretation in Figure 5.9 of (5.43), (5.44), and (5.54).

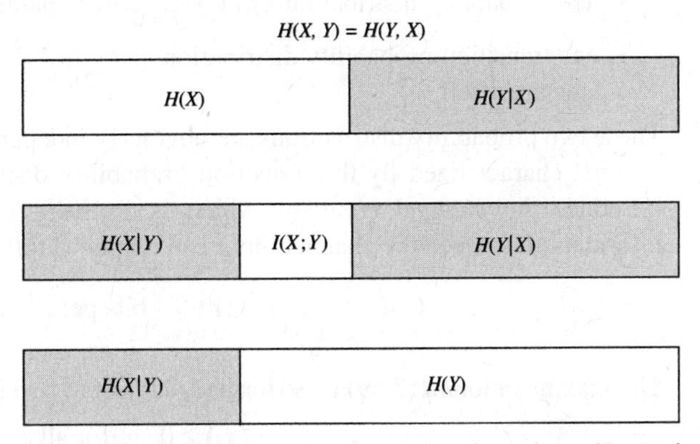

Figure 5.9 Illustrating the relations among various channel entropies.

5.7 Channel Capacity

The concept of entropy introduced in Section 5.2 prepared us for formulating Shannon's first theorem: the source-coding theorem. To set the stage for formulating Shannon's second theorem, namely the channel-coding theorem, this section introduces the concept of *capacity*, which, as mentioned previously, defines the intrinsic ability of a communication channel to convey information.

To proceed, consider a discrete memoryless channel with input alphabet $\mathcal{X}$, output alphabet $\mathcal{Y}$, and transition probabilities $p(y_k|x_j)$, where $j = 0, 1, \ldots, J-1$ and $k = 0, 1, \ldots, K-1$. The mutual information of the channel is defined by the first line of (5.49), which is reproduced here for convenience:

$$I(X;Y) = \sum_{k=0}^{K-1} \sum_{j=0}^{J-1} p(x_j, y_k) \log_2\left(\frac{p(y_k|x_j)}{p(y_k)}\right)$$

where, according to (5.38),

$$p(x_j, y_k) = p(y_k|x_j)p(x_j)$$

Also, from (5.39), we have

$$p(y_k) = \sum_{j=0}^{J-1} p(y_k|x_j)p(x_j)$$

Putting these three equations into a single equation, we write

$$I(X;Y) = \sum_{k=0}^{K-1} \sum_{j=0}^{J-1} p(y_k, x_j)p(x_j) \log_2\left(\frac{p(y_k|x_j)}{\sum_{j=0}^{J-1} p(y_k|x_j)p(x_j)}\right)$$

Careful examination of the double summation in this equation reveals two different probabilities, on which the essence of mutual information $I(X;Y)$ depends:

- the probability distribution $\{p(x_j)\}_{j=0}^{J-1}$ that characterizes the channel input and
- the transition probability distribution $\{p(y_k|x_j)\}_{j=0,\, k=0}^{j=J-1,\, K-1}$ that characterizes the channel itself.

These two probability distributions are obviously independent of each other. Thus, given a channel characterized by the transition probability distribution $\{p(y_k|x_j\}$, we may now introduce the *channel capacity*, which is formally defined in terms of the mutual information between the channel input and output as follows:

$$C = \max_{\{p(x_j)\}} I(X;Y) \quad \text{bits per channel use} \tag{5.59}$$

The maximization in (5.59) is performed, subject to two input probabilistic constraints:

$$p(x_j) \geq 0 \quad \text{for all } j$$

and

$$\sum_{j=0}^{J-1} p(x_j) = 1$$

Accordingly, we make the following statement:

> The channel capacity of a discrete memoryless channel, commonly denoted by C, is defined as the maximum mutual information $I(X;Y)$ in any single use of the channel (i.e., signaling interval), where the maximization is over all possible input probability distributions $\{p(x_j)\}$ on X.

The channel capacity is clearly an intrinsic property of the channel.

EXAMPLE 6 **Binary Symmetric Channel (Revisited)**

Consider again the *binary symmetric channel*, which is described by the *transition probability diagram* of Figure 5.8. This diagram is uniquely defined by the conditional probability of error p.

From Example 1 we recall that the entropy $H(X)$ is maximized when the channel input probability $p(x_0) = p(x_1) = 1/2$, where x_0 and x_1 are each 0 or 1. Hence, invoking the defining equation (5.59), we find that the mutual information $I(X;Y)$ is similarly maximized and thus write

$$C = I(X;Y)|_{p(x_0) = p(x_1) = 1/2}$$

From Figure 5.8 we have

$$p(y_0|x_1) = p(y_1|x_0) = p$$

and

$$p(y_0|x_0) = p(y_1|x_1) = 1 - p$$

Therefore, substituting these channel transition probabilities into (5.49) with $J = K = 2$ and then setting the input probability $p(x_0) = p(x_1) = 1/2$ in (5.59), we find that the capacity of the binary symmetric channel is

$$C = 1 + p \log_2 p + (1 - p) \log_2(1 - p) \tag{5.60}$$

Moreover, using the definition of the entropy function introduced in (5.16), we may reduce (5.60) to

$$C = 1 - H(p)$$

The channel capacity C varies with the probability of error (i.e., transition probability) p in a convex manner as shown in Figure 5.10, which is symmetric about $p = 1/2$. Comparing the curve in this figure with that in Figure 5.2, we make two observations:

1. When the channel is *noise free*, permitting us to set $p = 0$, the channel capacity C attains its maximum value of one bit per channel use, which is exactly the information in each channel input. At this value of p, the entropy function $H(p)$ attains its minimum value of zero.
2. When the conditional probability of error $p = 1/2$ due to channel noise, the channel capacity C attains its minimum value of zero, whereas the entropy function $H(p)$

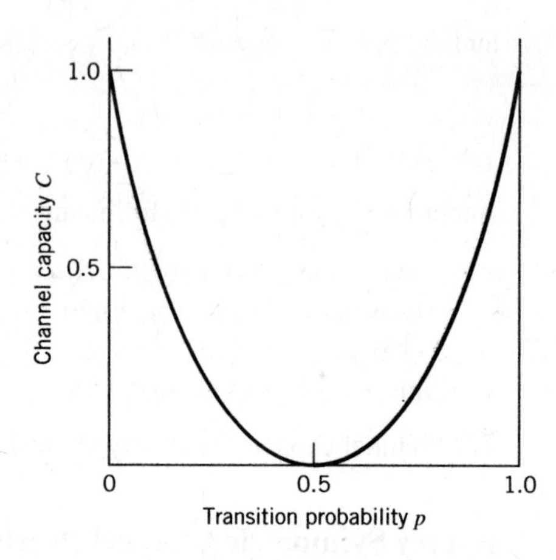

Figure 5.10
Variation of channel capacity of a binary symmetric channel with transition probability p.

attains its maximum value of unity; in such a case, the channel is said to be *useless* in the sense that the channel input and output assume statistically independent structures.

5.8 Channel-coding Theorem

With the entropy of a discrete memoryless source and the corresponding capacity of a discrete memoryless channel at hand, we are now equipped with the concepts needed for formulating Shannon's second theorem: the channel-coding theorem.

To this end, we first recognize that the inevitable presence of *noise* in a channel causes discrepancies (errors) between the output and input data sequences of a digital communication system. For a relatively noisy channel (e.g., wireless communication channel), the probability of error may reach a value as high as 10^{-1}, which means that (on the average) only 9 out of 10 transmitted bits are received correctly. For many applications, this *level of reliability* is utterly unacceptable. Indeed, a probability of error equal to 10^{-6} or even lower is often a necessary practical requirement. To achieve such a high level of performance, we resort to the use of channel coding.

The design goal of channel coding is to increase the resistance of a digital communication system to channel noise. Specifically, *channel coding* consists of *mapping* the incoming data sequence into a channel input sequence and *inverse mapping* the channel output sequence into an output data sequence in such a way that the overall effect of channel noise on the system is minimized. The first mapping operation is performed in the transmitter by a *channel encoder*, whereas the inverse mapping operation is performed in the receiver by a *channel decoder*, as shown in the block diagram of Figure 5.11; to simplify the exposition, we have not included source encoding (before channel encoding) and source decoding (after channel decoding) in this figure.[9]

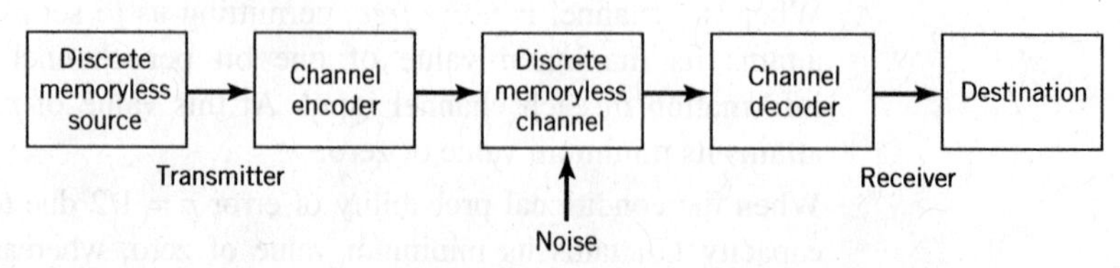

Figure 5.11
Block diagram of digital communication system.

The channel encoder and channel decoder in Figure 5.11 are both under the designer's control and should be designed to optimize the overall reliability of the communication system. The approach taken is to introduce *redundancy* in the channel encoder in a controlled manner, so as to reconstruct the original source sequence as accurately as possible. In a rather loose sense, we may thus view channel coding as the *dual* of source coding, in that the former introduces controlled redundancy to improve reliability whereas the latter reduces redundancy to improve efficiency.

Treatment of the channel-coding techniques is deferred to Chapter 10. For the purpose of our present discussion, it suffices to confine our attention to *block codes*. In this class of codes, the message sequence is subdivided into sequential blocks each k bits long, and each k-bit block is *mapped* into an n-bit block, where $n > k$. The number of redundant bits added by the encoder to each transmitted block is $n - k$ bits. The ratio k/n is called the *code rate*. Using r to denote the code rate, we write

$$r = \frac{k}{n} \tag{5.61}$$

where, of course, r is less than unity. For a prescribed k, the code rate r (and, therefore, the system's coding efficiency) approaches zero as the block length n approaches infinity.

The accurate reconstruction of the original source sequence at the destination requires that the *average probability of symbol error* be arbitrarily low. This raises the following important question:

> Does a channel-coding scheme exist such that the probability that a message bit will be in error is less than any positive number ε (i.e., as small as we want it), and yet the channel-coding scheme is efficient in that the code rate need not be too small?

The answer to this fundamental question is an emphatic "yes." Indeed, the answer to the question is provided by Shannon's second theorem in terms of the channel capacity C, as described in what follows.

Up until this point, *time* has not played an important role in our discussion of channel capacity. Suppose then the discrete memoryless source in Figure 5.11 has the source alphabet $\mathcal{S}$ and entropy $H(S)$ bits per source symbol. We assume that the source emits symbols once every T_s seconds. Hence, the *average information rate* of the source is $H(S)/T_s$ bits per second. The decoder delivers decoded symbols to the destination from the source alphabet S and at the same source rate of one symbol every T_s seconds. The discrete memoryless channel has a channel capacity equal to C bits per use of the channel. We assume that the channel is capable of being used once every T_c seconds. Hence, the *channel capacity per unit time* is C/T_c bits per second, which represents the maximum rate of information transfer over the channel. With this background, we are now ready to state Shannon's second theorem, the *channel-coding theorem*,[10] in two parts as follows:

1. Let a discrete memoryless source with an alphabet $\mathcal{S}$ have entropy $H(S)$ for random variable S and produce symbols once every T_s seconds. Let a discrete memoryless channel have capacity C and be used once every T_c seconds, Then, if

$$\frac{H(S)}{T_s} \le \frac{C}{T_c} \tag{5.62}$$

there exists a coding scheme for which the source output can be transmitted over the channel and be reconstructed with an arbitrarily small probability of error. The parameter C/T_c is called the *critical rate*; when (5.62) is satisfied with the equality sign, the system is said to be signaling at the critical rate.

2. Conversely, if

$$\frac{H(S)}{T_s} > \frac{C}{T_c}$$

it is not possible to transmit information over the channel and reconstruct it with an arbitrarily small probability of error.

The channel-coding theorem is the single most important result of information theory. The theorem specifies the channel capacity C as a *fundamental limit* on the rate at which the transmission of reliable error-free messages can take place over a discrete memoryless channel. However, it is important to note two limitations of the theorem:

1. The channel-coding theorem does not show us how to construct a good code. Rather, the theorem should be viewed as an *existence proof* in the sense that it tells us that if the condition of (5.62) is satisfied, then good codes do exist. Later, in Chapter 10, we describe good codes for discrete memoryless channels.
2. The theorem does not have a precise result for the probability of symbol error after decoding the channel output. Rather, it tells us that the probability of symbol error tends to zero as the length of the code increases, again provided that the condition of (5.62) is satisfied.

Application of the Channel-coding Theorem to Binary Symmetric Channels

Consider a discrete memoryless source that emits equally likely binary symbols (0s and 1s) once every T_s seconds. With the source entropy equal to one bit per source symbol (see Example 1), the information rate of the source is $(1/T_s)$ bits per second. The source sequence is applied to a channel encoder with code rate r. The channel encoder produces a symbol once every T_c seconds. Hence, the encoded symbol transmission rate is $(1/T_c)$ symbols per second. The channel encoder engages a binary symmetric channel once every T_c seconds. Hence, the channel capacity per unit time is (C/T_c) bits per second, where C is determined by the prescribed channel transition probability p in accordance with (5.60). Accordingly, part (1) of the channel-coding theorem implies that if

$$\frac{1}{T_s} \le \frac{C}{T_c} \tag{5.63}$$

then the probability of error can be made arbitrarily low by the use of a suitable channel-encoding scheme. But the ratio T_c/T_s equals the code rate of the channel encoder:

$$r = \frac{T_c}{T_s} \tag{5.64}$$

Hence, we may restate the condition of (5.63) simply as

$$r \le C$$

That is, for $r \le C$, there exists a code (with code rate less than or equal to channel capacity C) capable of achieving an arbitrarily low probability of error.

EXAMPLE 7

Repetition Code

In this example we present a graphical interpretation of the channel-coding theorem. We also bring out a surprising aspect of the theorem by taking a look at a simple coding scheme.

Consider first a binary symmetric channel with transition probability $p = 10^{-2}$. For this value of p, we find from (5.60) that the channel capacity $C = 0.9192$. Hence, from the channel-coding theorem, we may state that, for any $\varepsilon > 0$ and $r \le 0.9192$, there exists a code of large enough length n, code rate r, and an appropriate decoding algorithm such that, when the coded bit stream is sent over the given channel, the average probability of channel decoding error is less than ε. This result is depicted in Figure 5.12 for the limiting value $\varepsilon = 10^{-8}$.

To put the significance of this result in perspective, consider next a simple coding scheme that involves the use of a *repetition code*, in which each bit of the message is repeated several times. Let each bit (0 or 1) be repeated n times, where $n = 2m + 1$ is an odd integer. For example, for $n = 3$, we transmit 0 and 1 as 000 and 111, respectively.

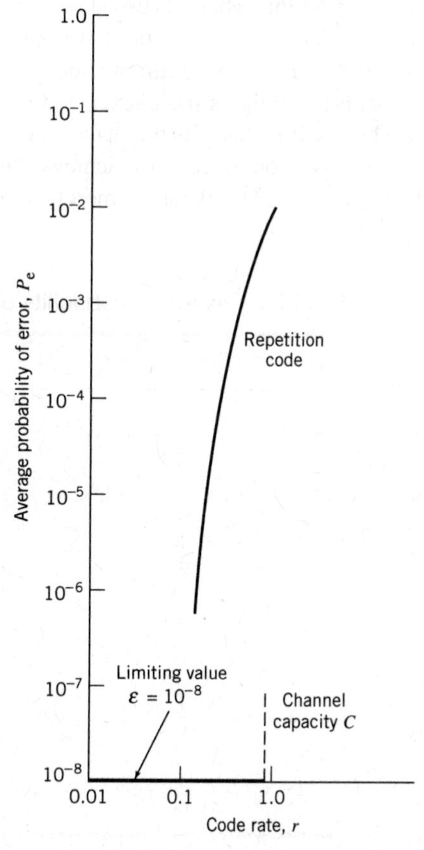

Figure 5.12 Illustrating the significance of the channel-coding theorem.

Intuitively, it would seem logical to use a *majority rule* for decoding, which operates as follows:

> If in a block of n *repeated* bits (representing one bit of the message) the number of 0s exceeds the number of 1s, the decoder decides in favor of a 0; otherwise, it decides in favor of a 1.

Hence, an error occurs when $m + 1$ or more bits out of $n = 2m + 1$ bits are received incorrectly. Because of the assumed symmetric nature of the channel, the *average probability of error,* denoted by P_e, is independent of the *prior* probabilities of 0 and 1. Accordingly, we find that P_e is given by

$$P_e = \sum_{i=m+1}^{n} \binom{n}{i} p^i (1-p)^{n-i} \tag{5.65}$$

where p is the transition probability of the channel.

Table 5.3 gives the average probability of error P_e for a repetition code that is calculated by using (5.65) for different values of the code rate r. The values given here assume the use of a binary symmetric channel with transition probability $p = 10^{-2}$. The improvement in reliability displayed in Table 5.3 is achieved at the cost of decreasing code rate. The results of this table are also shown plotted as the curve labeled "repetition code" in Figure 5.12. This curve illustrates the *exchange of code rate for message reliability,* which is a characteristic of repetition codes.

This example highlights the unexpected result presented to us by the channel-coding theorem. The result is that it is not necessary to have the code rate r approach zero (as in the case of repetition codes) to achieve more and more reliable operation of the communication link. The theorem merely requires that the code rate be less than the channel capacity C.

Table 5.3 Average probability of error for repetition code

Code rate, $r = 1/n$	Average probability of error, P_e
1	10^{-2}
$\frac{1}{3}$	3×10^{-4}
$\frac{1}{5}$	10^{-6}
$\frac{1}{7}$	4×10^{-7}
$\frac{1}{9}$	10^{-8}
$\frac{1}{11}$	5×10^{-10}

5.9 Differential Entropy and Mutual Information for Continuous Random Ensembles

The sources and channels considered in our discussion of information-theoretic concepts thus far have involved ensembles of random variables that are *discrete* in amplitude. In this section, we extend these concepts to *continuous* random variables. The motivation for doing so is to pave the way for the description of another fundamental limit in information theory, which we take up in Section 5.10.

Consider a continuous random variable X with the *probability density function* $f_X(x)$. By analogy with the entropy of a discrete random variable, we introduce the following definition:

$$h(X) = \int_{-\infty}^{\infty} f_X(x) \log_2\left[\frac{1}{f_X(x)}\right] dx \tag{5.66}$$

We refer to the new term $h(X)$ as the *differential entropy* of X to distinguish it from the ordinary or absolute entropy. We do so in recognition of the fact that, although $h(X)$ is a useful mathematical quantity to know, it is *not* in any sense a measure of the randomness of X. Nevertheless, we justify the use of (5.66) in what follows. We begin by viewing the continuous random variable X as the limiting form of a discrete random variable that assumes the value $x_k = k\Delta x$, where $k = 0, \pm 1, \pm 2, \ldots$, and Δx approaches zero. By definition, the continuous random variable X assumes a value in the interval $[x_k, x_k + \Delta x]$ with probability $f_X(x_k)\Delta x$. Hence, permitting Δx to approach zero, the ordinary entropy of the continuous random variable X takes the limiting form

$$\begin{aligned}
H(X) &= \lim_{\Delta x \to 0} \sum_{k=-\infty}^{\infty} f_X(x_k)\Delta x \log_2\left(\frac{1}{f_X(x_k)\Delta x}\right) \\
&= \lim_{\Delta x \to 0} \left(\sum_{k=-\infty}^{\infty} f_X(x_k) \log_2\left(\frac{1}{f_X(x_k)}\right)\Delta x - \log_2 \Delta x \sum_{k=-\infty}^{\infty} f_X(x_k)\Delta x \right) \\
&= \int_{-\infty}^{\infty} f_X(x) \log_2\left(\frac{1}{f_X(x)}\right) dx - \lim_{\Delta x \to 0}\left(\log_2 \Delta x \int_{-\infty}^{\infty} f_X(x_k)\, dx\right) \\
&= h(X) - \lim_{\Delta x \to 0} \log_2 \Delta x
\end{aligned} \tag{5.67}$$

In the last line of (5.67), use has been made of (5.66) and the fact that the total area under the curve of the probability density function $f_X(x)$ is unity. In the limit as Δx approaches zero, the term $-\log_2 \Delta x$ approaches infinity. This means that the entropy of a continuous random variable is infinitely large. Intuitively, we would expect this to be true because a continuous random variable may assume a value anywhere in the interval $(-\infty, \infty)$; we may, therefore, encounter uncountable infinite numbers of probable outcomes. To avoid the problem associated with the term $\log_2 \Delta x$, we adopt $h(X)$ as a *differential entropy*, with the term $-\log_2 \Delta x$ serving merely as a reference. Moreover, since the information transmitted over a channel is actually the difference between two entropy terms that have a common reference, the information will be the same as the difference between the corresponding differential entropy terms. We are, therefore, perfectly justified in using the term $h(X)$, defined in (5.66), as the differential entropy of the continuous random variable X.

When we have a continuous random vector $\mathbf{X}$ consisting of n random variables $X_1, X_2, \dots, X_n$, we define the differential entropy of $\mathbf{X}$ as the *n-fold integral*

$$h(\mathbf{X}) = \int_{-\infty}^{\infty} f_{\mathbf{X}}(\mathbf{x}) \log_2\left[\frac{1}{f_{\mathbf{X}}(\mathbf{x})}\right] d\mathbf{x} \tag{5.68}$$

where $f_{\mathbf{X}}(\mathbf{x})$ is the joint probability density function of $\mathbf{X}$.

EXAMPLE 8 **Uniform Distribution**

To illustrate the notion of differential entropy, consider a random variable X uniformly distributed over the interval $(0, a)$. The probability density function of X is

$$f_X(x) = \begin{cases} \dfrac{1}{a}, & 0 < x < a \\ 0, & \text{otherwise} \end{cases}$$

Applying (5.66) to this distribution, we get

$$\begin{aligned} h(X) &= \int_0^a \frac{1}{a} \log(a)\, dx \\ &= \log a \end{aligned} \tag{5.69}$$

Note that $\log a < 0$ for $a < 1$. Thus, this example shows that, unlike a discrete random variable, the differential entropy of a continuous random variable can assume a negative value.

Relative Entropy of Continuous Distributions

In (5.12) we defined the relative entropy of a pair of different discrete distributions. To extend that definition to a pair of continuous distributions, consider the continuous random variables X and Y whose respective probability density functions are denoted by $f_X(x)$ and $f_Y(x)$ for the same sample value (argument) x. The *relative entropy*[11] of the random variables X and Y is defined by

$$D(f_Y \| f_X) = \int_{-\infty}^{\infty} f_Y(x) \log_2\left(\frac{f_Y(x)}{f_X(x)}\right) dx \tag{5.70}$$

where $f_X(x)$ is viewed as the "reference" distribution. In a corresponding way to the fundamental property of (5.13), we have

$$D(f_Y \| f_X) \geq 0 \tag{5.71}$$

Combining (5.70) and (5.71) into a single inequality, we may thus write

$$\int_{-\infty}^{\infty} f_Y(x) \log_2\left(\frac{1}{f_Y(x)}\right) dx \leq \int_{-\infty}^{\infty} f_Y(x) \log_2\left(\frac{1}{f_X(x)}\right) dx$$

The expression on the left-hand side of this inequality is recognized as the differential entropy of the random variable *Y*, namely $h(Y)$. Accordingly,

$$h(Y) \le \int_{-\infty}^{\infty} f_Y(x) \log_2\left(\frac{1}{f_Y(x)}\right) \mathrm{d}x \tag{5.72}$$

The next example illustrates an insightful application of (5.72).

EXAMPLE 9

Gaussian Distribution

Suppose two random variables, *X* and *Y*, are described as follows:

- the random variables *X* and *Y* have the common mean μ and variance σ^2;
- the random variable *X* is *Gaussian distributed* (see Section 3.9) as shown by

$$f_X(x) = \frac{1}{\sqrt{2\pi}\sigma}\exp\left[-\frac{(x-\mu)^2}{2\sigma^2}\right] \tag{5.73}$$

Hence, substituting (5.73) into (5.72) and changing the base of the logarithm from 2 to e = 2.7183, we get

$$h(Y) \le -\log_2 \mathrm{e}\int_{-\infty}^{\infty} f_Y(x)\left[-\frac{(x-\mu)^2}{2\sigma^2} - \log(\sqrt{2\pi}\sigma)\right]\mathrm{d}x \tag{5.74}$$

where e is the base of the natural algorithm. We now recognize the following characterizations of the random variable *Y* (given that its mean is μ and its variance is σ^2):

$$\int_{-\infty}^{\infty} f_Y(x)\,\mathrm{d}x = 1$$

$$\int_{-\infty}^{\infty} (x-\mu)^2 f_Y(x)\,\mathrm{d}x = \sigma^2$$

We may, therefore, simplify (5.74) as

$$h(Y) \le \frac{1}{2}\log_2(2\pi\mathrm{e}\sigma^2) \tag{5.75}$$

The quantity on the right-hand side of (5.75) is, in fact, the differential entropy of the Gaussian random variable *X*:

$$h(X) = \frac{1}{2}\log_2(2\pi\mathrm{e}\sigma^2) \tag{5.76}$$

Finally, combining (5.75) and (5.76), we may write

$$h(Y) \le h(X), \qquad \begin{cases} X\text{: Gaussian random variable} \\ Y\text{: nonGaussian random variable} \end{cases} \tag{5.77}$$

where equality holds if, and only if, $Y = X$.

We may now summarize the results of this important example by describing two entropic properties of a random variable:

PROPERTY 1 *For any finite variance, a Gaussian random variable has the largest differential entropy attainable by any other random variable.*

PROPERTY 2 *The entropy of a Gaussian random variable is uniquely determined by its variance (i.e., the entropy is independent of the mean).*

Indeed, it is because of Property 1 that the Gaussian channel model is so widely used as a conservative model in the study of digital communication systems.

Mutual Information

Continuing with the information-theoretic characterization of continuous random variables, we may use analogy with (5.47) to define the *mutual information* between the pair of continuous random variables X and Y as follows:

$$I(X;Y) = \int_{-\infty}^{\infty}\int_{-\infty}^{\infty} f_{X,Y}(x,y)\log_2\left[\frac{f_X(x|y)}{f_X(x)}\right]\,dx\,dy \tag{5.78}$$

where $f_{X,Y}(x,y)$ is the joint probability density function of X and Y and $f_X(x|y)$ is the conditional probability density function of X given $Y = y$. Also, by analogy with (5.45), (5.50), (5.43), and (5.44), we find that the mutual information between the pair of Gausian random variables has the following properties:

$$I(X;Y) = I(Y;X) \tag{5.79}$$

$$I(X;Y) \geq 0 \tag{5.80}$$

$$\begin{aligned} I(X;Y) &= h(X) - h(X|Y) \\ &= h(Y) - h(Y|X) \end{aligned} \tag{5.81}$$

The parameter $h(X)$ is the differential entropy of X; likewise for $h(Y)$. The parameter $h(X|Y)$ is the *conditional differential entropy* of X given Y; it is defined by the double integral (see (5.41))

$$h(X|Y) = \int_{-\infty}^{\infty}\int_{-\infty}^{\infty} f_{X,Y}(x,y)\log_2\left[\frac{1}{f_X(x|y)}\right]\,dx\,dy \tag{5.82}$$

The parameter $h(Y|X)$ is the conditional differential entropy of Y given X; it is defined in a manner similar to $h(X|Y)$.

5.10 Information Capacity Law

In this section we use our knowledge of probability theory to expand Shannon's channel-coding theorem, so as to formulate the information capacity for a *band-limited, power-limited Gaussian channel*, depicted in Figure 5.13. To be specific, consider a zero-mean stationary process $X(t)$ that is band-limited to B hertz. Let X_k, $k = 1, 2, \ldots, K$, denote the continuous random variables obtained by uniform sampling of the process $X(t)$ at a rate of $2B$ samples per second. The rate $2B$ samples per second is the smallest permissible rate for a bandwidth B that would not result in a loss of information in accordance with the sampling theorem; this is discussed in Chapter 6. Suppose that these samples are

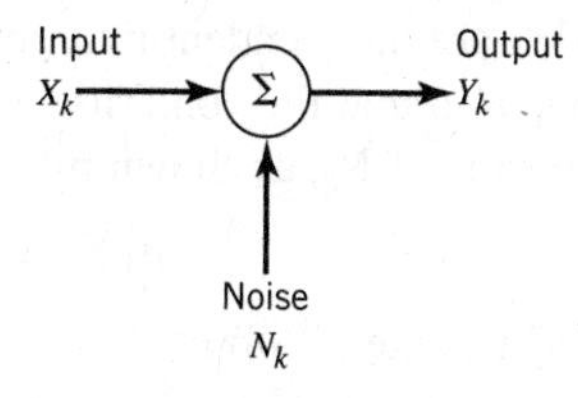

Figure 5.13 Model of discrete-time, memoryless Gaussian channel.

transmitted in T seconds over a noisy channel, also band-limited to B hertz. Hence, the total number of samples K is given by

$$K = 2BT \tag{5.83}$$

We refer to X_k as a sample of the *transmitted signal*. The channel output is perturbed by *additive white Gaussian noise* (AWGN) of zero mean and power spectral density $N_0/2$. The noise is band-limited to B hertz. Let the continuous random variables Y_k, $k = 1, 2, \dots, K$, denote the corresponding samples of the channel output, as shown by

$$Y_k = X_k + N_k, \quad k = 1, 2, \dots, K \tag{5.84}$$

The noise sample N_k in (5.84) is Gaussian with zero mean and variance

$$\sigma^2 = N_0 B \tag{5.85}$$

We assume that the samples Y_k, $k = 1, 2, \dots, K$, are statistically independent.

A channel for which the noise and the received signal are as described in (5.84) and (5.85) is called a *discrete-time, memoryless Gaussian channel*, modeled as shown in Figure 5.13. To make meaningful statements about the channel, however, we have to assign a *cost* to each channel input. Typically, the transmitter is *power limited*; therefore, it is reasonable to define the cost as

$$\mathbb{E}[X_k^2] \le P, \qquad k = 1, 2, \dots, K \tag{5.86}$$

where P is the *average transmitted power*. The *power-limited Gaussian channel* described herein is not only of theoretical importance but also of practical importance, in that it models many communication channels, including line-of-sight radio and satellite links.

The *information capacity* of the channel is defined as the maximum of the mutual information between the channel input X_k and the channel output Y_k over all distributions of the input X_k that satisfy the power constraint of (5.86). Let $I(X_k;Y_k)$ denote the mutual information between X_k and Y_k. We may then define the *information capacity* of the channel as

$$C = \max_{f_{X_k}(x)} I(X_k;Y_k), \qquad \text{subject to the constraint } \mathbb{E}[X_k^2] = P \qquad \text{for all } k \tag{5.87}$$

In words, maximization of the mutual information $I(X_k;Y_k)$ is done with respect to all probability distributions of the channel input X_k, satisfying the power constraint $\mathbb{E}[X_k^2] = P$.

The mutual information $I(X_k;Y_k)$ can be expressed in one of the two equivalent forms shown in (5.81). For the purpose at hand, we use the second line of this equation to write

$$I(X_k;Y_k) = h(Y_k) - h(Y_k|X_k) \tag{5.88}$$

Since X_k and N_k are independent random variables and their sum equals Y_k in accordance with (5.84), we find that the conditional differential entropy of Y_k given X_k is equal to the differential entropy of N_k, as shown by

$$h(Y_k|X_k) = h(N_k) \tag{5.89}$$

Hence, we may rewrite (5.88) as

$$I(X_k;Y_k) = h(Y_k) - h(N_k) \tag{5.90}$$

With $h(N_k)$ being independent of the distribution of X_k, it follows that maximizing $I(X_k;Y_k)$ in accordance with (5.87) requires maximizing the differential entropy $h(Y_k)$. For $h(Y_k)$ to be maximum, Y_k has to be a Gaussian random variable. That is to say, samples of the channel output represent a noiselike process. Next, we observe that since N_k is Gaussian by assumption, the sample X_k of the channel input must be Gaussian too. We may therefore state that the maximization specified in (5.87) is attained by choosing samples of the channel input from a noiselike Gaussian-distributed process of average power P. Correspondingly, we may reformulate (5.87) as

$$C = I(X_k;Y_k)\text{: for Gaussian } X_k \text{ and } \mathbb{E}[X_k^2] = P \qquad \text{for all } k \tag{5.91}$$

where the mutual information $I(X_k;Y_k)$ is defined in accordance with (5.90).

For evaluation of the information capacity C, we now proceed in three stages:

1. The variance of sample Y_k of the channel output equals $P + \sigma^2$, which is a consequence of the fact that the random variables X and N are statistically independent; hence, the use of (5.76) yields the differential entropy

$$h(Y_k) = \frac{1}{2}\log_2[2\pi e(P + \sigma^2)] \tag{5.92}$$

2. The variance of the noisy sample N_k equals σ^2; hence, the use of (5.76) yields the differential entropy

$$h(N_k) = \frac{1}{2}\log_2[2\pi e\sigma^2] \tag{5.93}$$

3. Substituting (5.92) and (5.93) into (5.90), and recognizing the definition of information capacity given in (5.91), we get the formula:

$$C = \frac{1}{2}\log_2\left(1 + \frac{P}{\sigma^2}\right) \text{ bits per channel use} \tag{5.94}$$

With the channel used K times for the transmission of K samples of the process $X(t)$ in T seconds, we find that the information capacity per unit time is (K/T) times the result given in (5.94). The number K equals $2BT$, as in (5.83). Accordingly, we may express the information capacity of the channel in the following equivalent form:

$$C = B\log_2\left(1 + \frac{P}{N_0B}\right) \text{ bits per second} \tag{5.95}$$

where N_0B is the total noise power at the channel output, defined in accordance with (5.85).

Based on the formula of (5.95), we may now make the following statement

> The information capacity of a continuous channel of bandwidth B hertz, perturbed by AWGN of power spectral density $N_0/2$ and limited in bandwidth

to B, is given by the formula

$$C = B \log_2\left(1 + \frac{P}{N_0 B}\right) \text{ bits per second}$$

where P is the average transmitted power.

The *information capacity law*[12] of (5.95) is one of the most remarkable results of Shannon's information theory. In a single formula, it highlights most vividly the interplay among three key system parameters: channel bandwidth, average transmitted power, and power spectral density of channel noise. Note, however, that the dependence of information capacity C on channel bandwidth B is *linear*, whereas its dependence on signal-to-noise ratio $P/(N_0B)$ is *logarithmic*. Accordingly, we may make another insightful statement:

> It is easier to increase the information capacity of a continuous communication channel by expanding its bandwidth than by increasing the transmitted power for a prescribed noise variance.

The information capacity formula implies that, for given average transmitted power P and channel bandwidth B, we can transmit information at the rate of C bits per second, as defined in (5.95), with arbitrarily small probability of error by employing a sufficiently complex encoding system. It is not possible to transmit at a rate higher than C bits per second by any encoding system without a definite probability of error. Hence, the channel capacity law defines the *fundamental limit* on the permissible rate of error-free transmission for a power-limited, band-limited Gaussian channel. To approach this limit, however, the transmitted signal must have statistical properties approximating those of white Gaussian noise.

Sphere Packing

To provide a plausible argument supporting the information capacity law, suppose that we use an encoding scheme that yields K codewords, one for each sample of the transmitted signal. Let n denote the length (i.e., the number of bits) of each codeword. It is presumed that the coding scheme is designed to produce an acceptably low probability of symbol error. Furthermore, the codewords satisfy the power constraint; that is, the average power contained in the transmission of each codeword with n bits is nP, where P is the average power per bit.

Suppose that any codeword in the code is transmitted. The received vector of n bits is Gaussian distributed with a mean equal to the transmitted codeword and a variance equal to $n\sigma^2$, where σ^2 is the noise variance. With a high probability, we may say that the received signal vector at the channel output lies inside a sphere of radius $\sqrt{n\sigma^2}$; that is, centered on the transmitted codeword. This sphere is itself contained in a larger sphere of radius $\sqrt{n(P + \sigma^2)}$, where $n(P + \sigma^2)$ is the average power of the received signal vector.

We may thus visualize the sphere packing[13] as portrayed in Figure 5.14. With everything inside a small sphere of radius $\sqrt{n\sigma^2}$ assigned to the codeword on which it is

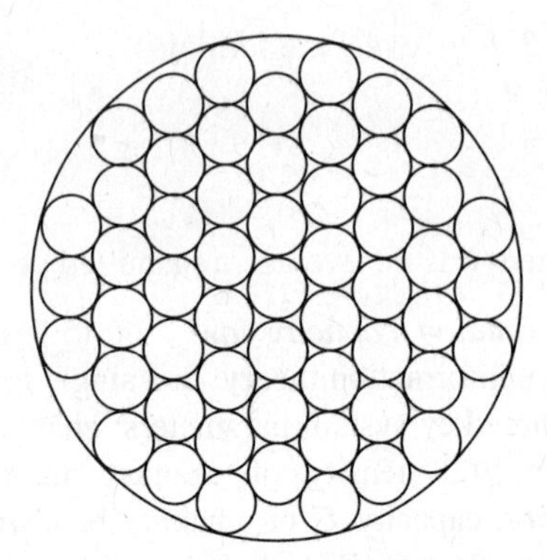

Figure 5.14 The sphere-packing problem.

centered. It is therefore reasonable to say that, when this particular codeword is transmitted, the probability that the received signal vector will lie inside the correct "decoding" sphere is high. The key question is:

> How many decoding spheres can be packed inside the larger sphere of received signal vectors? In other words, how many codewords can we in fact choose?

To answer this question, we want to eliminate the overlap between the decoding spheres as depicted in Figure 5.14. Moreover, expressing the volume of an n-dimensional sphere of radius r as $A_n r^n$, where A_n is a scaling factor, we may go on to make two statements:

1. The volume of the sphere of received signal vectors is $A_n[n(P + \sigma^2)]^{n/2}$.
2. The volume of the decoding sphere is $A_n(n\sigma^2)^{n/2}$.

Accordingly, it follows that the maximum number of *nonintersecting* decoding spheres that can be packed inside the sphere of possible received signal vectors is given by

$$\begin{aligned} \frac{A_n[n(P+\sigma^2)]^{n/2}}{A_n(n\sigma^2)^{n/2}} &= \left(1 + \frac{P}{\sigma^2}\right)^{n/2} \\ &= 2^{(n/2)\log_2(1 + P/\sigma^2)} \end{aligned} \tag{5.96}$$

Taking the logarithm of this result to base 2, we readily see that the maximum number of bits per transmission for a low probability of error is indeed as defined previously in (5.94).

A final comment is in order: (5.94) is an idealized manifestation of Shannon's channel-coding theorem, in that it provides an upper bound on the physically realizable information capacity of a communication channel.

5.11 Implications of the Information Capacity Law

Now that we have a good understanding of the information capacity law, we may go on to discuss its implications in the context of a Gaussian channel that is limited in both power and bandwidth. For the discussion to be useful, however, we need an ideal framework against which the performance of a practical communication system can be assessed. To this end, we introduce the notion of an *ideal system*, defined as a system that transmits data at a bit rate R_b equal to the information capacity C. We may then express the average transmitted power as

$$P = E_b C \tag{5.97}$$

where E_b is the *transmitted energy per bit*. Accordingly, the ideal system is defined by the equation

$$\frac{C}{B} = \log_2\left(1 + \frac{E_b C}{N_0 B}\right) \tag{5.98}$$

Rearranging this formula, we may define the *signal energy-per-bit to noise power spectral density ratio*, E_b/N_0, in terms of the ratio C/B for the ideal system as follows:

$$\frac{E_b}{N_0} = \frac{2^{C/B} - 1}{C/B} \tag{5.99}$$

A plot of the bandwidth efficiency R_b/B versus E_b/N_0 is called the *bandwidth-efficiency diagram*. A generic form of this diagram is displayed in Figure 5.15, where the curve labeled "capacity boundary" corresponds to the ideal system for which $R_b = C$.

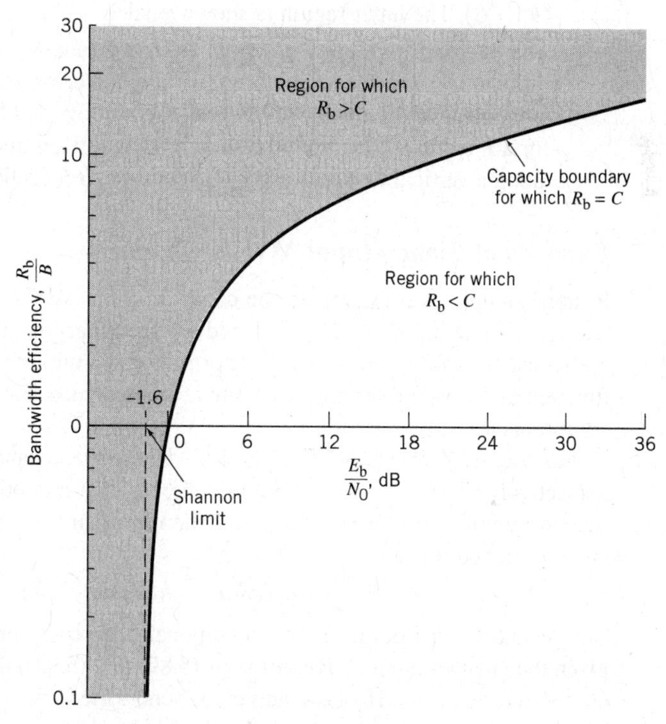

Figure 5.15 Bandwidth-efficiency diagram.

Based on Figure 5.15, we can make three observations:

1. For *infinite channel bandwidth*, the ratio E_b/N_0 approaches the limiting value

$$\left(\frac{E_b}{N_0}\right)_\infty = \lim_{B\to\infty}\left(\frac{E_b}{N_0}\right) = \log_e 2 = 0.693 \tag{5.100}$$

where $\log_e$ stands for the natural logarithm ln. The value defined in (5.100) is called the *Shannon limit* for an AWGN channel, assuming a code rate of zero. Expressed in decibels, the Shannon limit equals –1.6 dB. The corresponding limiting value of the channel capacity is obtained by letting the channel bandwidth B in (5.95) approach infinity, in which case we obtain

$$C_\infty = \lim_{B\to\infty} C = \left(\frac{P}{N_0}\right)\log_2 e \tag{5.101}$$

2. The *capacity boundary* is defined by the curve for the critical bit rate $R_b = C$. For any point on this boundary, we may flip a fair coin (with probability of 1/2) whether we have error-free transmission or not. As such, the boundary separates combinations of system parameters that have the potential for supporting error-free transmission ($R_b < C$) from those for which error-free transmission is not possible ($R_b > C$). The latter region is shown shaded in Figure 5.15.
3. The diagram highlights potential *trade-offs* among three quantities: the E_b/N_0, the ratio R_b/B, and the probability of symbol error P_e. In particular, we may view movement of the operating point along a horizontal line as trading P_e versus E_b/N_0 for a fixed R_b/B. On the other hand, we may view movement of the operating point along a vertical line as trading P_e versus R_b/B for a fixed E_b/N_0.

EXAMPLE 10 Capacity of Binary-Input AWGN Channel

In this example, we investigate the capacity of an AWGN channel using *encoded* binary antipodal signaling (i.e., levels –1 and +1 for binary symbols 0 and 1, respectively). In particular, we address the issue of determining the minimum achievable bit error rate as a function of E_b/N_0 for varying code rate r. It is assumed that the binary symbols 0 and 1 are equiprobable.

Let the random variables X and Y denote the channel input and channel output respectively; X is a discrete variable, whereas Y is a continuous variable. In light of the second line of (5.81), we may express the mutual information between the channel input and channel output as

$$I(X;Y) = h(Y) - h(Y|X)$$

The second term, $h(Y|X)$, is the conditional differential entropy of the channel output Y, given the channel input X. By virtue of (5.89) and (5.93), this term is just the entropy of a Gaussian distribution. Hence, using σ^2 to denote the variance of the channel noise, we write

$$h(Y|X) = \frac{1}{2}\log_2(2\pi e\sigma^2)$$

Next, the first term, $h(Y)$, is the differential entropy of the channel output Y. With the use of binary antipodal signaling, the probability density function of Y, given $X = x$, is a mixture of two Gaussian distributions with common variance σ^2 and mean values −1 and +1, as shown by

$$f_Y(y_i|x) = \frac{1}{2}\left\{\frac{\exp[-(y_i+1)^2/2\sigma^2]}{\sqrt{2\pi}\sigma} + \frac{\exp[-(y_i-1)^2/2\sigma^2]}{\sqrt{2\pi}\sigma}\right\} \tag{5.102}$$

Hence, we may determine the differential entropy of Y using the formula

$$h(Y) = -\int_{-\infty}^{\infty} f_Y(y_i|x)\log_2[f_Y(y_i|x)]\,dy_i$$

where $f_Y(y_i\,|\,x)$ is defined by (5.102). From the formulas of $h(Y|X)$ and $h(Y)$, it is clear that the mutual information is solely a function of the noise variance σ^2. Using $M(\sigma^2)$ to denote this functional dependence, we may thus write

$$I(X;Y) = M(\sigma^2)$$

Unfortunately, there is no closed formula that we can derive for $M(\sigma^2)$ because of the difficulty of determining $h(Y)$. Nevertheless, the differential entropy $h(Y)$ can be well approximated using *Monte Carlo integration*; see Appendix E for details.

Because symbols 0 and 1 are equiprobable, it follows that the channel capacity C is equal to the mutual information between X and Y. Hence, for error-free data transmission over the AWGN channel, the code rate r must satisfy the condition

$$r < M(\sigma^2) \tag{5.103}$$

A robust measure of the ratio E_b/N_0, is

$$\frac{E_b}{N_0} = \frac{P}{N_0 r} = \frac{P}{2\sigma^2 r}$$

where P is the average transmitted power and $N_0/2$ is the two-sided power spectral density of the channel noise. Without loss of generality, we may set $P = 1$. We may then express the noise variance as

$$\sigma^2 = \frac{N_0}{2E_b r} \tag{5.104}$$

Substituting Equation (5.104) into (5.103) and rearranging terms, we get the desired relation:

$$\frac{E_b}{N_0} = \frac{1}{2rM^{-1}(r)} \tag{5.105}$$

where $M^{-1}(r)$ is the *inverse* of the mutual information between the channel input and putput, expressed as a function of the code rate r.

Using the Monte Carlo method to estimate the differential entropy $h(Y)$ and therefore $M^{-1}(r)$, the plots of Figure 5.16 are computed.[14] Figure 5.16a plots the minimum E_b/N_0 versus the code rate r for error-free transmission. Figure 5.16b plots the minimum

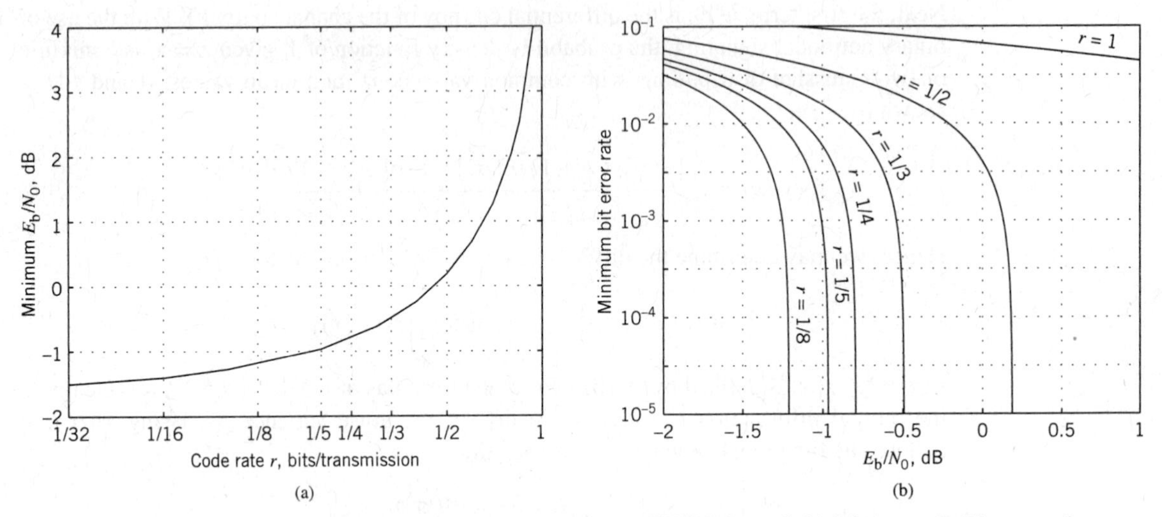

Figure 5.16 Binary antipodal signaling over an AWGN channel. (a) Minimum E_b/N_0 versus the code rate r. (b) Minimum bit error rate versus E_b/N_0 for varying code rate r.

achievable bit error rate versus E_b/N_0 with the code rate r as a running parameter. From Figure 5.16 we may draw the following conclusions:

- For uncoded binary signaling (i.e., $r = 1$), an infinite E_b/N_0 is required for error-free communication, which agrees with what we know about uncoded data transmission over an AWGN channel.
- The minimum E_b/N_0, decreases with decreasing code rate r, which is intuitively satisfying. For example, for $r = 1/2$, the minimum value of E_b/N_0 is slightly less than 0.2 dB.
- As r approaches zero, the minimum E_b/N_0 approaches the limiting value of –1.6 dB, which agrees with the Shannon limit derived earlier; see (5.100).

5.12 Information Capacity of Colored Noisy Channel

The information capacity theorem as formulated in (5.95) applies to a band-limited white noise channel. In this section we extend Shannon's information capacity law to the more general case of a *nonwhite*, or *colored, noisy channel*.[15] To be specific, consider the channel model shown in Figure 5.17a where the transfer function of the channel is denoted by $H(f)$. The channel noise $n(t)$, which appears additively at the channel output, is modeled as the sample function of a stationary Gaussian process of zero mean and power spectral density $S_N(f)$. The requirement is twofold:

1. Find the input ensemble, described by the power spectral density $S_{xx}(f)$, that maximizes the mutual information between the channel output $y(t)$ and the channel input $x(t)$, subject to the constraint that the average power of $x(t)$ is fixed at a constant value P.
2. Hence, determine the optimum information capacity of the channel.

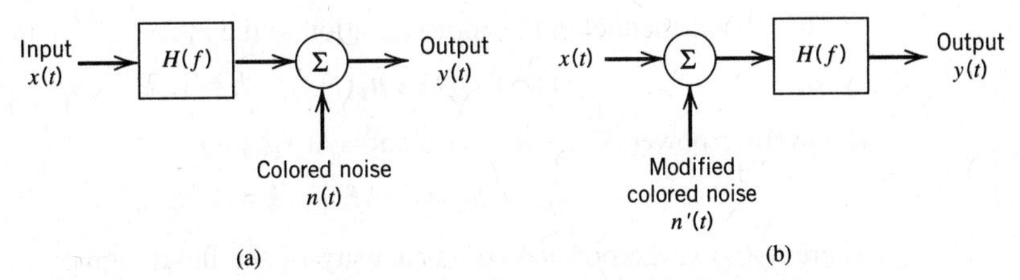

Figure 5.17 (a) Model of band-limited, power-limited noisy channel. (b) Equivalent model of the channel.

This problem is a constrained optimization problem. To solve it, we proceed as follows:

- Because the channel is linear, we may replace the model of Figure 5.17a with the equivalent model shown in Figure 5.17b. From the viewpoint of the spectral characteristics of the signal plus noise measured at the channel output, the two models of Figure 5.17 are equivalent, provided that the power spectral density of the noise $n'(t)$ in Figure 5.17b is defined in terms of the power spectral density of the noise $n(t)$ in Figure 5.17a as

$$S_{N'N'}(f) = \frac{S_{NN}(f)}{|H(f)|^2} \tag{5.106}$$

  where $|H(f)|$ is the magnitude response of the channel.
- To simplify the analysis, we use the "principle of divide and conquer" to approximate the continuous $|H(f)|$ described as a function of frequency f in the form of a staircase, as illustrated in Figure 5.18. Specifically, the channel is divided into a large number of adjoining frequency slots. The smaller we make the incremental frequency interval Δf of each subchannel, the better this approximation is.

The net result of these two points is that the original model of Figure 5.17a is replaced by the parallel combination of a finite number of subchannels, N, each of which is corrupted essentially by "band-limited white Gaussian noise."

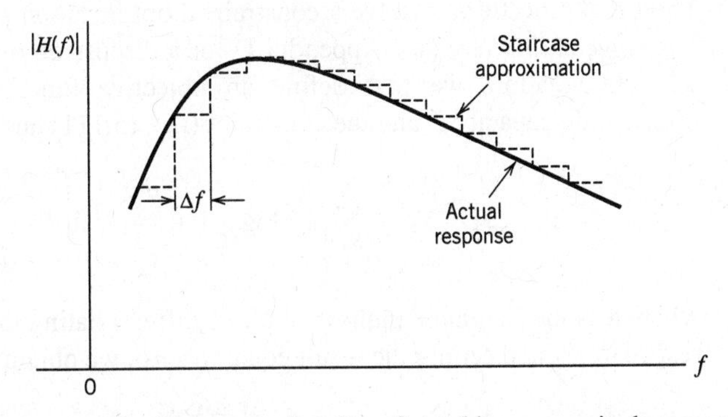

Figure 5.18 Staricase approximation of an arbitrary magnitude response $|H(f)|$; only the positive frequency portion of the response is shown.

The kth subchannel in the approximation to the model of Figure 5.17b is described by

$$y_k(t) = x_k(t) + n_k(t), \qquad k = 1, 2, \ldots, N \tag{5.107}$$

The average power of the signal component $x_k(t)$ is

$$P_k = S_{XX}(f_k)\Delta f, \qquad k = 1, 2, \ldots, N \tag{5.108}$$

where $S_X(f_k)$ is the power spectral density of the input signal evaluated at the frequency $f = f_k$. The variance of the noise component $n_k(t)$ is

$$\sigma_k^2 = \frac{S_{NN}(f_k)}{|H(f_k)|^2}\Delta f, \qquad k = 1, 2, \ldots, N \tag{5.109}$$

where $S_N(f_k)$ and $|H(f_k)|$ are the noise spectral density and the channel's magnitude response evaluated at the frequency f_k, respectively. The information capacity of the kth subchannel is

$$C_k = \frac{1}{2}\Delta f \log_2\left(1 + \frac{P_k}{\sigma_k^2}\right), \qquad k = 1, 2, \ldots, N \tag{5.110}$$

where the factor 1/2 accounts for the fact that Δf applies to both positive and negative frequencies. All the N subchannels are independent of one another. Hence, the total capacity of the overall channel is approximately given by the summation

$$\begin{aligned} C &\approx \sum_{k=1}^{N} C_k \\ &= \frac{1}{2}\sum_{k=1}^{N} \Delta f \log_2\left(1 + \frac{P_k}{\sigma_k^2}\right) \end{aligned} \tag{5.111}$$

The problem we have to address is to maximize the overall information capacity C subject to the constraint

$$\sum_{k=1}^{N} P_k = P = \text{constant} \tag{5.112}$$

The usual procedure to solve a constrained optimization problem is to use the *method of Lagrange multipliers* (see Appendix D for a discussion of this method). To proceed with this optimization, we first define an objective function that incorporates both the information capacity C and the constraint (i.e., (5.111) and (5.112)), as shown by

$$J(P_k) = \frac{1}{2}\sum_{k=1}^{N} \Delta f \log_2\left(1 + \frac{P_k}{\sigma_k^2}\right) + \lambda\left(P - \sum_{k=1}^{N} P_k\right) \tag{5.113}$$

where λ is the Lagrange multiplier. Next, differentiating the objective function $J(P_k)$ with respect to P_k and setting the result equal to zero, we obtain

$$\frac{\Delta f \log_2 e}{P_k + \sigma_k^2} - \lambda = 0$$

To satisfy this optimizing solution, we impose the following requirement:

$$P_k + \sigma_k^2 = K\Delta f \qquad \text{for } k = 1, 2, \ldots, N \tag{5.114}$$

where K is a constant that is the same for all k. The constant K is chosen to satisfy the average power constraint.

Inserting the defining values of (5.108) and (5.109) in the optimizing condition of (5.114), simplifying, and rearranging terms we get

$$S_{XX}(f_k) = K - \frac{S_{NN}(f_k)}{|H(f_k)|^2}, \qquad k = 1, 2, \ldots, N \tag{5.115}$$

Let $\mathcal{F}_A$ denote the frequency range for which the constant K satisfies the condition

$$K \geq \frac{S_{NN}(f_k)}{|H(f_k)|^2}$$

Then, as the incremental frequency interval Δf is allowed to approach zero and the number of subchannels N goes to infinity, we may use (5.115) to formally state that the power spectral density of the input ensemble that achieves the optimum information capacity is a nonnegative quantity defined by

$$S_{XX}(f) = \begin{cases} K - \dfrac{S_{NN}(f)}{|H(f)|^2} & f \in \mathcal{F}_A \\ 0, & \text{otherwise} \end{cases} \tag{5.116}$$

Because the average power of a random process is the total area under the curve of the power spectral density of the process, we may express the average power of the channel input $x(t)$ as

$$P = \int_{f \in \mathcal{F}_A} \left(K - \frac{S_{NN}(f)}{|H(f)|^2} \right) df \tag{5.117}$$

For a prescribed P and specified $S_N(f)$ and $H(f)$, the constant K is the solution to (5.117).

The only thing that remains for us to do is to find the optimum information capacity. Substituting the optimizing solution of (5.114) into (5.111) and then using the defining values of (5.108) and (5.109), we obtain

$$C \approx \frac{1}{2} \sum_{k=1}^{N} \Delta f \log_2 \left(K \frac{|H(f_k)|^2}{S_{NN}(f_k)} \right)$$

When the incremental frequency interval Δf is allowed to approach zero, this equation takes the limiting form

$$C = \frac{1}{2} \int_{-\infty}^{\infty} \log_2 \left(K \frac{|H(f)|^2}{S_{NN}(f)} \right) df \tag{5.118}$$

where the constant K is chosen as the solution to (5.117) for a prescribed input signal power P.

Water-filling Interpretation of the Information Capacity Law

Equations (5.116) and (5.117) suggest the picture portrayed in Figure 5.19. Specifically, we make the following observations:

- The appropriate input power spectral density $S_X(f)$ is described as the bottom regions of the function $S_N(f)/|H(f)|^2$ that lie below the constant level K, which are shown shaded.
- The input power P is defined by the total area of these shaded regions.

The spectral-domain picture portrayed here is called the *water-filling (pouring) interpretation*, in the sense that the process by which the input power is distributed across the function $S_N(f)/|H(f)|^2$ is identical to the way in which water distributes itself in a vessel.

Consider now the idealized case of a band-limited signal in AWGN channel of power spectral density $N(f) = N_0/2$. The transfer function $H(f)$ is that of an ideal band-pass filter defined by

$$H(f) = \begin{cases} 1, & 0 \le f_c - \frac{B}{2} \le |f| \le f_c + \frac{B}{2} \\ 0, & \text{otherwise} \end{cases}$$

where f_c is the midband frequency and B is the channel bandwidth. For this special case, (5.117) and (5.118) reduce respectively to

$$P = 2B\left(K - \frac{N_0}{2}\right)$$

and

$$C = B \log_2\left(\frac{2K}{N_0}\right)$$

Hence, eliminating K between these two equations, we get the standard form of Shannon's capacity theorem, defined by (5.95).

EXAMPLE 11 **Capacity of NEXT-Dominated Channel**

Digital subscriber lines (DSLs) refer to a family of different technologies that operate over a closed transmission loop; they will be discussed in Chapter 8, Section 8.11. For the

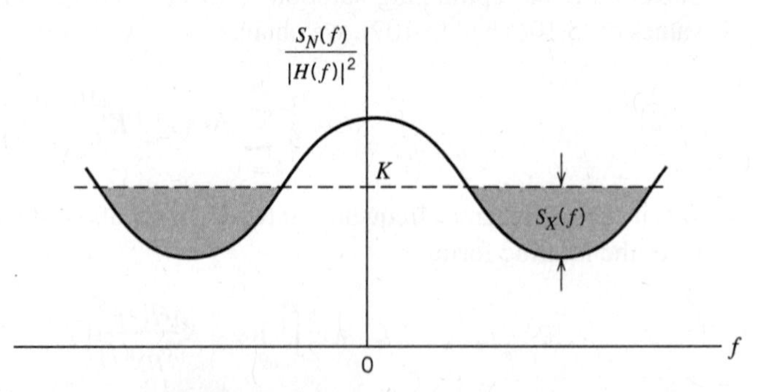

Figure 5.19 Water-filling interpretation of information-capacity theorem for a colored noisy channel.

present, it suffices to say that a DSL is designed to provide for data transmission between a user terminal (e.g., computer) and the central office of a telephone company. A major channel impairment that arises in the deployment of a DSL is the near-end cross-talk (NEXT). The power spectral density of this crosstalk may be taken as

$$S_N(f) = |H_{\text{NEXT}}(f)|^2 S_X(f) \tag{5.119}$$

where $S_X(f)$ is the power spectral density of the transmitted signal and $H_{\text{NEXT}}(f)$ is the transfer function that couples adjacent twisted pairs. The only constraint we have to satisfy in this example is that the power spectral density function $S_X(f)$ be *nonnegative for all f.* Substituting (5.119) into (5.116), we readily find that this condition is satisfied by solving for K as

$$K = \left(1 + \frac{|H_{\text{NEXT}}(f)|^2}{|H(f)|^2}\right) S_X(f)$$

Finally, using this result in (5.118), we find that the capacity of the NEXT-dominated digital subscriber channel is given by

$$C = \frac{1}{2}\int_{\mathcal{F}_A} \log_2\left(1 + \frac{|H(f)|^2}{|H_{\text{NEXT}}(f)|^2}\right) df$$

where $\mathcal{F}_A$ is the set of positive and negative frequencies for which $S_X(f) > 0$.

5.13 Rate Distortion Theory

In Section 5.3 we introduced the source-coding theorem for a discrete memoryless source, according to which the average codeword length must be at least as large as the source entropy for perfect coding (i.e., perfect representation of the source). However, in many practical situations there are constraints that force the coding to be imperfect, thereby resulting in unavoidable *distortion.* For example, constraints imposed by a communication channel may place an upper limit on the permissible code rate and, therefore, on average codeword length assigned to the information source. As another example, the information source may have a continuous amplitude as in the case of speech, and the requirement is to quantize the amplitude of each sample generated by the source to permit its representation by a codeword of finite length as in pulse-code modulation to be discussed in Chapter 6. In such cases, the problem is referred to as *source coding with a fidelity criterion*, and the branch of information theory that deals with it is called *rate distortion theory.*[16] Rate distortion theory finds applications in two types of situations:

- Source coding where the permitted coding alphabet cannot exactly represent the information source, in which case we are forced to do lossy *data compression.*
- Information transmission at a rate greater than channel capacity.

Accordingly, rate distortion theory may be viewed as a natural extension of Shannon's coding theorem.

Rate Distortion Function

Consider a discrete memoryless source defined by an M-ary alphabet $\mathscr{X}$: $\{x_i | i = 1, 2, \ldots, M\}$, which consists of a set of statistically independent symbols together with the associated symbol probabilities $\{p_i | i = 1, 2, \ldots, M\}$. Let R be the average code rate in bits per codeword. The representation codewords are taken from another alphabet $\mathscr{Y}$:$\{y_j | j = 1, 2, \ldots, N\}$. The source-coding theorem states that this second alphabet provides a perfect representation of the source provided that $R > H$, where H is the source entropy. But if we are forced to have $R < H$, then there is unavoidable distortion and, therefore, loss of information.

Let $p(x_i, y_j)$ denote the joint probability of occurrence of source symbol x_i and representation symbol y_j. From probability theory, we have

$$p(x_i, y_j) = p(y_j|x_i)p(x_i) \tag{5.120}$$

where $p(y_j|x_i)$ is a transition probability. Let $d(x_i, y_j)$ denote a measure of the cost incurred in representing the source symbol x_i by the symbol y_j; the quantity $d(x_i, y_j)$ is referred to as a *single-letter distortion measure*. The statistical average of $d(x_i, y_j)$ over all possible source symbols and representation symbols is given by

$$\bar{d} = \sum_{i=1}^{M} \sum_{j=1}^{N} p(x_i)p(y_j|x_i)d(x_i|y_j) \tag{5.121}$$

Note that the average distortion $\bar{d}$ is a nonnegative continuous function of the transition probabilities $p(y_j|x_i)$ that are determined by the source encoder–decoder pair.

A conditional probability assignment $p(y_j|x_i)$ is said to be *D-admissible* if, and only if, the average distortion $\bar{d}$ is less than or equal to some acceptable value D. The set of all D-admissible conditional probability assignments is denoted by

$$\mathscr{P}_D = \{p(y_j|x_i): \bar{d} \le D\} \tag{5.122}$$

For each set of transition probabilities, we have a mutual information

$$I(X;Y) = \sum_{i=1}^{M} \sum_{j=1}^{N} p(x_i)p(y_j|x_i) \log\left(\frac{p(y_j|x_i)}{p(y_j)}\right) \tag{5.123}$$

A *rate distortion function R(D)* is defined as *the smallest coding rate possible for which the average distortion is guaranteed not to exceed D*. Let $\mathscr{P}_D$ denote the set to which the conditional probability $p(y_j|x_i)$ belongs for a prescribed D. Then, for a fixed D we write[17]

$$R(D) = \min_{p(y_j|x_i) \in \mathscr{P}_D} I(X;Y) \tag{5.124}$$

subject to the constraint

$$\sum_{j=1}^{N} p(y_j|x_i) = 1 \quad \text{for } i = 1, 2, \ldots, M \tag{5.125}$$

The rate distortion function $R(D)$ is measured in units of bits if the base-2 logarithm is used in (5.123). Intuitively, we expect the distortion D to decrease as the rate distortion function $R(D)$ is increased. We may say conversely that tolerating a large distortion D permits the use of a smaller rate for coding and/or transmission of information.

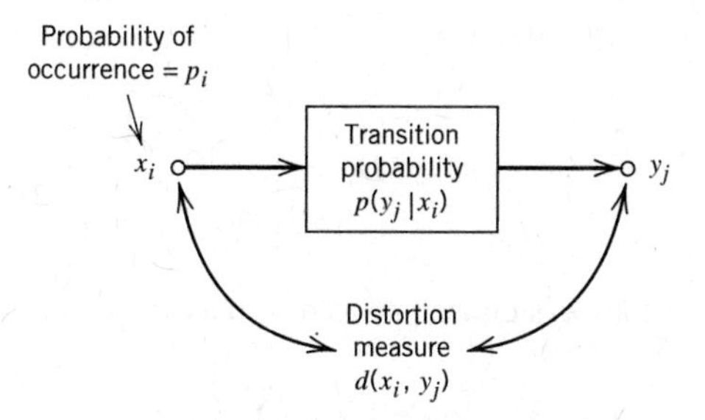

Figure 5.20
Summary of rate distortion theory.

Figure 5.20 summarizes the main parameters of rate distortion theory. In particular, given the source symbols $\{x_i\}$ and their probabilities $\{p_i\}$, and given a definition of the single-letter distortion measure $d(x_i,y_j)$, the calculation of the rate distortion function $R(D)$ involves finding the conditional probability assignment $p(y_j | x_i)$ subject to certain constraints imposed on $p(y_j | x_i)$. This is a variational problem, the solution of which is unfortunately not straightforward in general.

EXAMPLE 12 Gaussian Source

Consider a discrete-time, memoryless Gaussian source with zero mean and variance σ^2. Let x denote the value of a sample generated by such a source. Let y denote a quantized version of x that permits a finite representation of it. The *square-error distortion*

$$d(x, y) = (x - y)^2$$

provides a distortion measure that is widely used for continuous alphabets. The rate distortion function for the Gaussian source with square-error distortion, as described herein, is given by

$$R(D) = \begin{cases} \frac{1}{2}\log\left(\frac{\sigma^2}{D}\right), & 0 \le D \le \sigma^2 \\ 0, & D > \sigma^2 \end{cases} \tag{5.126}$$

In this case, we see that $R(D) \to \infty$ as $D \to 0$, and $R(D) = 0$ for $D = \sigma^2$.

EXAMPLE 13 Set of Parallel Gaussian Sources

Consider next a set of N independent Gaussian random variables $\{X_i\}_{i=1}^{N}$, where X_i has zero mean and variance σ_i^2.Using the distortion measure

$$d = \sum_{i=1}^{N} (x_i - \hat{x}_i)^2, \qquad \hat{x}_i = \text{estimate of } x_i$$

and building on the result of Example 12, we may express the rate distortion function for the set of parallel Gaussian sources described here as

$$R(D) = \sum_{i=1}^{N} \frac{1}{2}\log\left(\frac{\sigma_i^2}{D_i}\right) \tag{5.127}$$

where D_i is itself defined by

$$D_i = \begin{cases} \lambda, & \lambda < \sigma_i^2 \\ \sigma_i^2, & \lambda \geq \sigma_i^2 \end{cases} \tag{5.128}$$

and the constant λ is chosen so as to satisfy the condition

$$\sum_{i=1}^{N} D_i = D \tag{5.129}$$

Compared to Figure 5.19, (5.128) and (5.129) may be interpreted as a kind of "water-filling in reverse," as illustrated in Figure 5.21. First, we choose a constant λ and only the subset of random variables whose variances exceed the constant λ. No bits are used to describe the remaining subset of random variables whose variances are less than the constant λ.

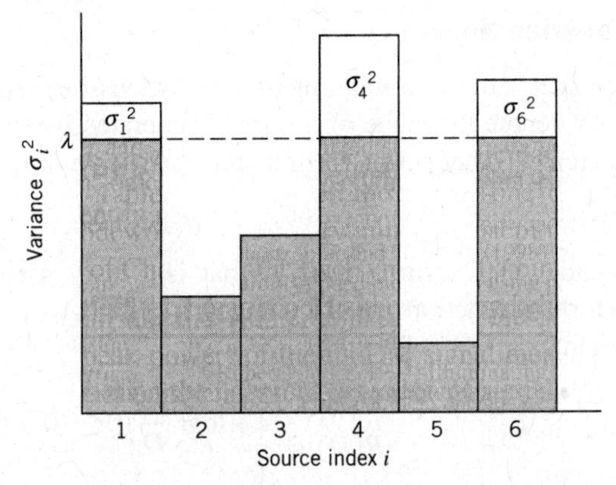

Figure 5.21 Reverse water-filling picture for a set of parallel Gaussian processes.

5.14 Summary and Discussion

In this chapter we established two fundamental limits on different aspects of a communication system, which are embodied in the source-coding theorem and the channel-coding theorem.

The *source-coding theorem*, Shannon's first theorem, provides the mathematical tool for assessing *data compaction;* that is, *lossless compression* of data generated by a discrete memoryless source. The theorem teaches us that we can make the average number of binary code elements (bits) per source symbol as small as, but no smaller than, the entropy of the source measured in bits. The *entropy* of a source is a function of the probabilities of the source symbols that constitute the alphabet of the source. Since entropy is a measure of uncertainty, the entropy is maximum when the associated probability distribution generates maximum uncertainty.

The *channel-coding theorem*, Shannon's second theorem, is both the most surprising and the single most important result of information theory. For a *binary symmetric channel*, the channel-coding theorem teaches us that, for any *code rate* r less than or equal to the *channel capacity* C, codes do exist such that the average probability of error is as small as we want it. A binary symmetric channel is the simplest form of a discrete memoryless channel. It is symmetric, because the probability of receiving symbol 1 if symbol 0 is sent is the same as the probability of receiving symbol 0 if symbol 1 is sent. This probability, the probability that an error will occur, is termed a *transition probability*. The transition probability p is determined not only by the additive noise at the channel output, but also by the kind of receiver used. The value of p uniquely defines the channel capacity C.

The *information capacity law*, an application of the channel-coding theorem, teaches us that there is an upper limit to the rate at which any communication system can operate reliably (i.e., free of errors) when the system is constrained in power. This maximum rate, called the *information capacity*, is measured in bits per second. When the system operates at a rate greater than the information capacity, it is condemned to a high probability of error, regardless of the choice of signal set used for transmission or the receiver used for processing the channel output.

When the output of a source of information is compressed in a lossless manner, the resulting data stream usually contains redundant bits. These redundant bits can be removed by using a lossless algorithm such as Huffman coding or the Lempel–Ziv algorithm for data compaction. We may thus speak of data compression followed by data compaction as two constituents of the *dissection of source coding*, which is so called because it refers exclusively to the sources of information.

We conclude this chapter on Shannon's information theory by pointing out that, in many practical situations, there are constraints that force source coding to be imperfect, thereby resulting in unavoidable *distortion*. For example, constraints imposed by a communication channel may place an upper limit on the permissible code rate and, therefore, average codeword length assigned to the information source. As another example, the information source may have a continuous amplitude, as in the case of speech, and the requirement is to *quantize* the amplitude of each sample generated by the source to permit its representation by a codeword of finite length, as in pulse-code modulation discussed in Chapter 6. In such cases, the information-theoretic problem is referred to as *source coding with a fidelity criterion*, and the branch of information theory that deals with it is called *rate distortion theory*, which may be viewed as a natural extension of Shannon's coding theorem.

Problems

Entropy

5.1 Let p denote the probability of some event. Plot the amount of information gained by the occurrence of this event for $0 \leq p \leq 1$.

5.2 A source emits one of four possible symbols during each signaling interval. The symbols occur with the probabilities

$$p_0 = 0.4$$
$$p_1 = 0.3$$
$$p_2 = 0.2$$
$$p_3 = 0.1$$

which sum to unity as they should. Find the amount of information gained by observing the source emitting each of these symbols.

5.3 A source emits one of four symbols s_0, s_1, s_2, and s_3 with probabilities 1/3, 1/6, 1/4 and 1/4, respectively. The successive symbols emitted by the source are statistically independent. Calculate the entropy of the source.

5.4 Let X represent the outcome of a single roll of a fair die. What is the entropy of X?

5.5 The sample function of a Gaussian process of zero mean and unit variance is uniformly sampled and then applied to a uniform quantizer having the input–output amplitude characteristic shown in Figure P5.5. Calculate the entropy of the quantizer output.

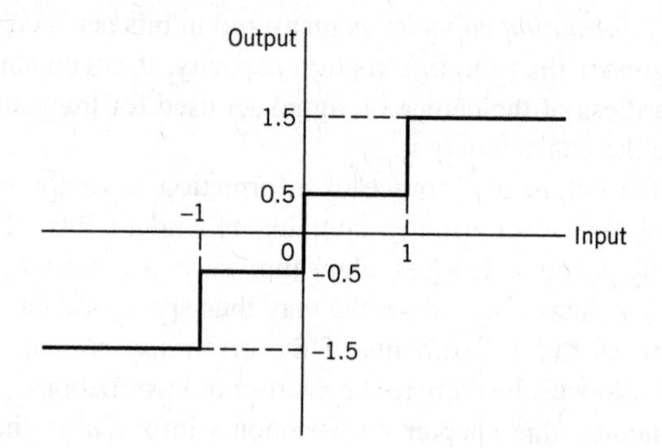

Figure P5.5

5.6 Consider a discrete memoryless source with source alphabet $S = \{s_0, s_1, \ldots, s_{K-1}\}$ and source statistics $\{p_0, p_1, \ldots, p_{K-1}\}$. The nth extension of this source is another discrete memoryless source with source alphabet $S^{(n)} = \{\sigma_0, \sigma_1, \ldots, \sigma_{M-1}\}$, where $M = K^n$. Let $P(\sigma_i)$ denote the probability of σ_i.

a. Show that, as expected,

$$\sum_{i=0}^{M-1} P(\sigma_i) = 1$$

b. Show that

$$\sum_{i=0}^{M-1} P(\sigma_i) \log_2\left(\frac{1}{p_{i_k}}\right) = H(S), \qquad k = 1, 2, \ldots, n$$

where p_{i_k} is the probability of symbol s_{i_k} and $H(S)$ is the entropy of the original source.

c. Hence, show that

$$H(S^{(n)}) = \sum_{i=0}^{M-1} P(\sigma_i) \log_2\left(\frac{1}{P(\sigma_i)}\right)$$
$$= nH(S)$$

5.7 Consider a discrete memoryless source with source alphabet $S = \{s_0, s_1, s_2\}$ and source statistics $\{0.7, 0.15, 0.15\}$.

a. Calculate the entropy of the source.

b. Calculate the entropy of the second-order extension of the source.

5.8 It may come as a surprise, but the number of bits needed to store text is much less than that required to store its spoken equivalent. Can you explain the reason for this statement?

5.9 Let a discrete random variable X assume values in the set $\{x_1, x_2, \ldots, x_n\}$. Show that the entropy of X satisfies the inequality

$$H(X) \leq \log n$$

and with equality if, and only if, the probability $p_i = 1/n$ for all i.

Lossless Data Compression

5.10 Consider a discrete memoryless source whose alphabet consists of K equiprobable symbols.

a. Explain why the use of a fixed-length code for the representation of such a source is about as efficient as any code can be.

b. What conditions have to be satisfied by K and the codeword length for the coding efficiency to be 100%?

5.11 Consider the four codes listed below:

Symbol	Code I	Code II	Code III	Code IV
s_0	0	0	0	00
s_1	10	01	01	01
s_2	110	001	011	10
s_3	1110	0010	110	110
s_4	1111	0011	111	111

a. Two of these four codes are prefix codes. Identify them and construct their individual decision trees.

b. Apply the Kraft inequality to codes I, II, III, and IV. Discuss your results in light of those obtained in part a.

5.12 Consider a sequence of letters of the English alphabet with their probabilities of occurrence

Letter	a	i	l	m	n	o	p	y
Probability	0.1	0.1	0.2	0.1	0.1	0.2	0.1	0.1

Compute two different Huffman codes for this alphabet. In one case, move a combined symbol in the coding procedure as high as possible; in the second case, move it as low as possible. Hence, for each of the two codes, find the average codeword length and the variance of the average codeword length over the ensemble of letters. Comment on your results.

5.13 A discrete memoryless source has an alphabet of seven symbols whose probabilities of occurrence are as described here:

Symbol	s_0	s_1	s_2	s_3	s_4	s_5	s_6
Probability	0.25	0.25	0.125	0.125	0.125	0.0625	0.0625

Compute the Huffman code for this source, moving a "combined" symbol as high as possible. Explain why the computed source code has an efficiency of 100%.

5.14 Consider a discrete memoryless source with alphabet $\{s_0, s_1, s_2\}$ and statistics $\{0.7, 0.15, 0.15\}$ for its output.

a. Apply the Huffman algorithm to this source. Hence, show that the average codeword length of the Huffman code equals 1.3 bits/symbol.

b. Let the source be extended to order two. Apply the Huffman algorithm to the resulting extended source and show that the average codeword length of the new code equals 1.1975 bits/symbol.

c. Extend the order of the extended source to three and reapply the Huffman algorithm; hence, calculate the average codeword length.

d. Compare the average codeword length calculated in parts b and c with the entropy of the original source.

5.15 Figure P5.15 shows a Huffman tree. What is the codeword for each of the symbols A, B, C, D, E, F, and G represented by this Huffman tree? What are their individual codeword lengths?

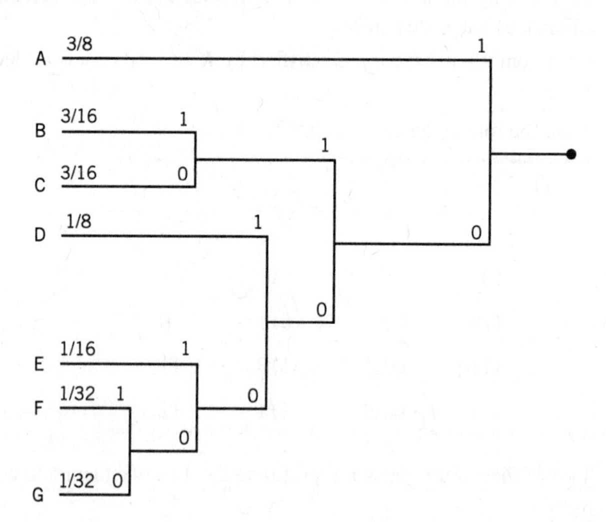

Figure P5.15

5.16 A computer executes four instructions that are designated by the codewords (00, 01, 10, 11). Assuming that the instructions are used independently with probabilities (1/2, 1/8, 1/8, 1/4), calculate the percentage by which the number of bits used for the instructions may be reduced by the use of an optimum source code. Construct a Huffman code to realize the reduction.

5.17 Consider the following binary sequence

11101001100010110100 ...

Use the Lempel–Ziv algorithm to encode this sequence, assuming that the binary symbols 0 and 1 are already in the cookbook.

Binary Symmetric Channel

5.18 Consider the transition probability diagram of a binary symmetric channel shown in Figure 5.8. The input binary symbols 0 and 1 occur with equal probability. Find the probabilities of the binary symbols 0 and 1 appearing at the channel output.

5.19 Repeat the calculation in Problem 5.18, assuming that the input binary symbols 0 and 1 occur with probabilities 1/4 and 3/4, respectively.

Mutual Information and Channel Capacity

5.20 Consider a binary symmetric channel characterized by the transition probability p. Plot the mutual information of the channel as a function of p_1, the a priori probability of symbol 1 at the channel input. Do your calculations for the transition probability $p = 0, 0.1, 0.2, 0.3, 0.5$.

5.21 Revisiting (5.12), express the mutual information $I(X;Y)$ in terms of the relative entropy

$$D(p(x,y)\|p(x)p(y))$$

5.22 Figure 5.10 depicts the variation of the channel capacity of a binary symmetric channel with the transition probability p. Use the results of Problem 5.19 to explain this variation.

5.23 Consider the binary symmetric channel described in Figure 5.8. Let p_0 denote the probability of sending binary symbol $x_0 = 0$ and let $p_1 = 1 - p_0$ denote the probability of sending binary symbol $x_1 = 1$. Let p denote the transition probability of the channel.

a. Show that the mutual information between the channel input and channel output is given by

$$I(X;Y) = H(z) - H(p)$$

where the two entropy functions

$$H(z) = z\log_2\left(\frac{1}{z}\right) + (1-z)\log_2\left(\frac{1}{1-z}\right)$$

$$z = p_0 p + (1-p_0)(1-p)$$

and

$$H(p) = p\log_2\left(\frac{1}{p}\right) + (1-p)\log_2\left(\frac{1}{1-p}\right)$$

b. Show that the value of p_0 that maximizes $I(X;Y)$ is equal to 1/2.

c. Hence, show that the channel capacity equals

$$C = 1 - H(p)$$

5.24 Two binary symmetric channels are connected in cascade as shown in Figure P5.24. Find the overall channel capacity of the cascaded connection, assuming that both channels have the same transition probability diagram of Figure 5.8.

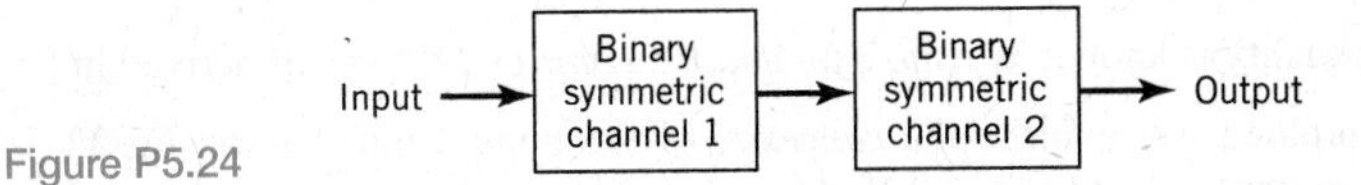

Figure P5.24

5.25 The *binary erasure channel* has two inputs and three outputs as described in Figure P5.25. The inputs are labeled 0 and 1 and the outputs are labeled 0, 1, and e. A fraction α of the incoming bits is erased by the channel. Find the capacity of the channel.

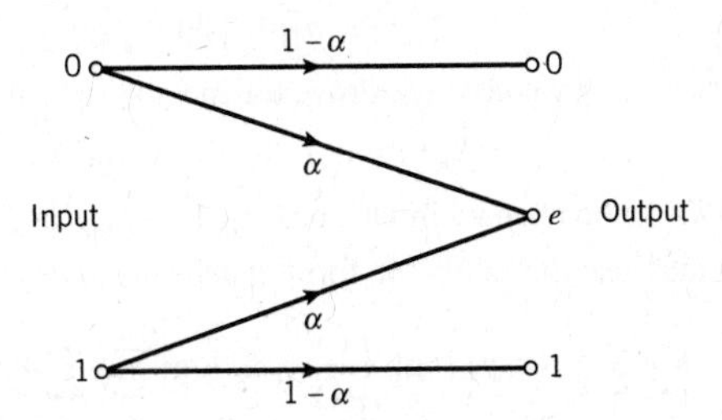

Figure P5.25

5.26 Consider a digital communication system that uses a *repetition code* for the channel encoding/decoding. In particular, each transmission is repeated n times, where $n = 2m + 1$ is an odd integer. The decoder operates as follows. If in a block of n received bits the number of 0s exceeds the number of 1s, then the decoder decides in favor of a 0; otherwise, it decides in favor of a 1. An error occurs when $m + 1$ or more transmissions out of $n = 2m + 1$ are incorrect. Assume a binary symmetric channel.

a. For $n = 3$, show that the average probability of error is given by

$$P_e = 3p^2(1-p) + p^3$$

where p is the transition probability of the channel.

b. For $n = 5$, show that the average probability of error is given by

$$P_e = 10p^3(1-p)^2 + 5p^4(1-p) + p^5$$

c. Hence, for the general case, deduce that the average probability of error is given by

$$P_e = \sum_{i=m+1}^{n} \binom{n}{i} p^i (1-p)^{n-i}$$

5.27 Let X, Y, and Z be three discrete random variables. For each value of the random variable Z, represented by sample z, define

$$A(z) = \sum_x \sum_y p(y) p(z|x, y)$$

Show that the conditional entropy $H(X \mid Y)$ satisfies the inequality

$$H(X|Y) \le H(z) + \mathbb{E}[\log A]$$

where $\mathbb{E}$ is the expectation operator.

5.28 Consider two correlated discrete random variables X and Y, each of which takes a value in the set $\{x_i\}_{i=1}^{n}$. Suppose that the value taken by Y is known. The requirement is to guess the value of X. Let P_e denote the probability of error, defined by

$$P_e = \mathbb{P}[X \ne Y]$$

Show that P_e is related to the conditional entropy of X given Y by the inequality

$$H(X|Y) \le H(P_e) + P_e \log(n-1)$$

This inequality is known as *Fano's inequality. Hint:* Use the result derived in Problem 5.27.

5.29 In this problem we explore the *convexity* of the mutual information $I(X;Y)$, involving the pair of discrete random variables X and Y.

Consider a discrete memoryless channel, for which the transition probability $p(y|x)$ is fixed for all x and y. Let X_1 and X_2 be two input random variables, whose input probability distributions are respectively denoted by $p(x_1)$ and $p(x_2)$. The corresponding probability distribution of X is defined by the convex combination

$$p(x) = a_1 p(x_1) + a_2 p(x_2)$$

where a_1 and a_2 are arbitrary constants. Prove the inequality

$$I(X;Y) \ge a_1 I(X_1;Y_1) + a_2 I(X_2;Y_2)$$

where X_1, X_2, and X are the channel inputs, and Y_1, Y_2, and Y are the corresponding channel outputs. For the proof, you may use the following form of *Jensen's inequality*:

$$\sum_x y \sum_y p_1(x, y) \log\left(\frac{p(y)}{p_1(y)}\right) \le \log\left[\sum_x y \sum_y p_1(x, y)\left(\frac{p(y)}{p_1(y)}\right)\right]$$

Differential Entropy

5.30 The differential entropy of a continuous random variable X is defined by the integral of (5.66). Similarly, the differential entropy of a continuous random vector $\mathbf{X}$ is defined by the integral of (5.68). These two integrals may not exist. Justify this statement.

5.31 Show that the differential entropy of a continuous random variable X is invariant to translation; that is,

$$h(X + c) = h(X)$$

for some constant c.

5.32 Let $X_1, X_2, \ldots, X_n$ denote the elements of a Gaussian vector $\mathbf{X}$. The X_i are independent with mean m_i and variance σ_i^2, $i = 1, 2, \ldots, n$. Show that the differential entropy of the vector $\mathbf{X}$ is given by

$$h(\mathbf{X}) = \frac{n}{2}\log_2[2\pi e(\sigma_1^2\sigma_2^2\ldots\sigma_n^2)^{1/n}]$$

where e is the base of the natural logarithm. What does $h(\mathbf{X})$ reduce to if the variances are all equal?

5.33 A continuous random variable X is constrained to a peak magnitude M; that is,

$$-M < X < M$$

a. Show that the differential entropy of X is maximum when it is uniformly distributed, as shown by

$$f_X(x) = \begin{cases} 1/(2M), & -M < x \le M \\ 0, & \text{otherwise} \end{cases}$$

b. Determine the maximum differential entropy of X.

5.34 Referring to (5.75), do the following:

a. Verify that the differential entropy of a Gaussian random variable of mean μ and variance σ^2 is given by $1/2 \log_2(2\pi e\sigma^2)$, where e is the base of the natural algorithm.

b. Hence, confirm the inequality of (5.75).

5.35 Demonstrate the properties of symmetry, nonnegativity, and expansion of the mutual information $I(X;Y)$ described in Section 5.6.

5.36 Consider the continuous random variable Y, defined by

$$Y = X + N$$

where the random variables X and N are statistically independent. Show that the conditional differential entropy of Y, given X, equals

$$h(Y \mid X) = h(N)$$

where $h(N)$ is the differential entropy of N.

Information Capacity Law

5.37 A voice-grade channel of the telephone network has a bandwidth of 3.4 kHz.

a. Calculate the information capacity of the telephone channel for a signal-to-noise ratio of 30 dB.

b. Calculate the minimum signal-to-noise ratio required to support information transmission through the telephone channel at the rate of 9600 bits/s.

5.38 Alphanumeric data are entered into a computer from a remote terminal through a voice-grade telephone channel. The channel has a bandwidth of 3.4 kHz and output signal-to-noise ratio of 20 dB. The terminal has a total of 128 symbols. Assume that the symbols are equiprobable and the successive transmissions are statistically independent.

a. Calculate the information capacity of the channel.

b. Calculate the maximum symbol rate for which error-free transmission over the channel is possible.

5.39 A black-and-white television picture may be viewed as consisting of approximately 3×10^5 elements, each of which may occupy one of 10 distinct brightness levels with equal probability. Assume that (1) the rate of transmission is 30 picture frames per second and (2) the signal-to-noise ratio is 30 dB.

Using the information capacity law, calculate the minimum bandwidth required to support the transmission of the resulting video signal.

5.40 In Section 5.10 we made the statement that it is easier to increase the information capacity of a communication channel by expanding its bandwidth B than increasing the transmitted power for a prescribed noise variance N_0B. This statement assumes that the noise spectral density N_0 varies inversely with B. Why is this inverse relationship the case?

5.41 In this problem, we revisit Example 5.10, which deals with coded binary antipodal signaling over an additive white Gaussian noise (AWGN) channel. Starting with (5.105) and the underlying theory, develop a software package for computing the minimum E_b/N_0 required for a given bit error rate, where E_b is the signal energy per bit, and $N_0/2$ is the noise spectral density. Hence, compute the results plotted in parts *a* and *b* of Figure 5.16.

As mentioned in Example 5.10, the computation of the mutual information between the channel input and channel output is well approximated using Monte Carlo integration. To explain how this method works, consider a function $g(y)$ that is difficult to sample randomly, which is indeed the case for the problem at hand. (For this problem, the function $g(y)$ represents the complicated integrand in the formula for the differential entropy of the channel output.) For the computation, proceed as follows:

- Find an area A that includes the region of interest and that is easily sampled.
- Choose N points, uniformly randomly inside the area A.

Then the *Monte Carlo integration theorem* states that the integral of the function $g(y)$ with respect to y is approximately equal to the area A multiplied by the fraction of points that reside below the curve of g, as illustrated in Figure P5.41. The accuracy of the approximation improves with increasing N.

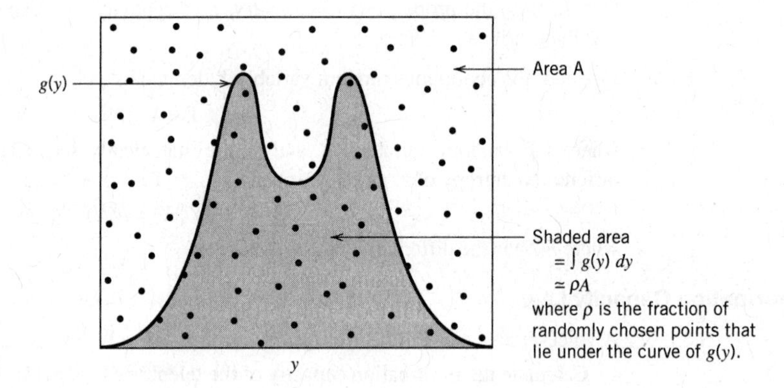

Figure P5.41

Notes

1. According to Lucky (1989), the first mention of the term *information theory* by Shannon occurred in a 1945 memorandum entitled "A mathematical theory of cryptography." It is rather curious that the term was never used in Shannon's (1948) classic paper, which laid down the foundations of information theory. For an introductory treatment of information theory, see Part 1 of the book by McEliece (2004), Chapters 1–6. For an advanced treatment of this subject, viewed in a rather broad context and treated with rigor, and clarity of presentation, see Cover and Thomas (2006).

For a collection of papers on the development of information theory (including the 1948 classic paper by Shannon), see Slepian (1974). For a collection of the original papers published by Shannon, see Sloane and Wyner (1993).

2. The use of a logarithmic measure of information was first suggested by Hartley (1928); however, Hartley used logarithms to base 10.

3. In statistical physics, the entropy of a physical system is defined by (Rief, 1965: 147)

$$L = k_{\mathrm{B}} \ln \Omega$$

where k_{B} is *Boltzmann's constant*, Ω is the number of states accessible to the system, and ln denotes the natural logarithm. This entropy has the dimensions of energy, because its definition involves the constant k_{B}. In particular, it provides a *quantitative measure of the degree of randomness of the system*. Comparing the entropy of statistical physics with that of information theory, we see that they have a similar form.

4. For the original proof of the source coding theorem, see Shannon (1948). A general proof of the source coding theorem is also given in Cover and Thomas (2006). The source coding theorem is also referred to in the literature as the *noiseless coding theorem*, noiseless in the sense that it establishes the condition for error-free encoding to be possible.

5. For proof of the Kraft inequality, see Cover and Thomas (2006). The Kraft inequality is also referred to as the Kraft–McMillan inequality in the literature.

6. The Huffman code is named after its inventor D.A. Huffman (1952). For a detailed account of Huffman coding and its use in data compaction, see Cover and Thomas (2006).

7. The original papers on the Lempel–Ziv algorithm are Ziv and Lempel (1977, 1978). For detailed treatment of the algorithm, see Cover and Thomas (2006).

8. It is also of interest to note that once a "parent" subsequence is joined by its two children, that parent subsequence can be replaced in constructing the Lempel–Ziv algorithm. To illustrate this nice feature of the algorithm, suppose we have the following example sequence:

$$01, 010, 011, \ldots$$

where 01 plays the role of a parent and 010 and 011 play the roles of the parent's children. In this example, the algorithm removes the 01, thereby reducing the length of the table through the use of a pointer.

9. In Cover and Thomas (2006), it is proved that the two-stage method, where the source coding and channel coding are considered separately as depicted in Figure 5.11, is as good as any other method of transmitting information across a noisy channel. This result has practical implications, in that the design of a communication system may be approached in two separate parts: source coding followed by channel coding. Specifically, we may proceed as follows:

- Design a source code for the most efficient representation of data generated by a discrete memoryless source of information.
- Separately and independently, design a channel code that is appropriate for a discrete memoryless channel.

The combination of source coding and channel coding designed in this manner will be as efficient as anything that could be designed by considering the two coding problems jointly.

10. To prove the channel-coding theorem, Shannon used several ideas that were new at the time; however, it was some time later when the proof was made rigorous (Cover and Thomas, 2006: 199). Perhaps the most thoroughly rigorous proof of this basic theorem of information theory is presented in Chapter 7 of the book by Cover and Thomas (2006). Our statement of the theorem, though slightly different from that presented by Cover and Thomas, in essence is the same.

11. In the literature, the relative entropy is also referred to as the *Kullback–Leibler divergence* (KLD).

12. Equation (5.95) is also referred to in the literature as the *Shannon–Hartley law* in recognition of the early work by Hartley on information transmission (Hartley, 1928). In particular, Hartley showed that the amount of information that can be transmitted over a given channel is proportional to the product of the channel bandwidth and the time of operation.

13. A lucid exposition of sphere packing is presented in Cover and Thomas (2006); see also Wozencraft and Jacobs (1965).

14. Parts a and b of Figure 5.16 follow the corresponding parts of Figure 6.2 in the book by Frey (1998).

15. For a rigorous treatment of information capacity of a colored noisy channel, see Gallager (1968). The idea of replacing the channel model of Figure 5.17a with that of Figure 5.17b is discussed in Gitlin, Hayes, and Weinstein (1992)

16. For a complete treatment of rate distortion theory, see the classic book by Berger (1971); this subject is also treated in somewhat less detail in Cover and Thomas (1991), McEliece (1977), and Gallager (1968).

17. For the derivation of (5.124), see Cover and Thomas (2006). An algorithm for computation of the rate distortion function $R(D)$ defined in (5.124) is described in Blahut (1987) and Cover and Thomas (2006).

CHAPTER 6

Conversion of Analog Waveforms into Coded Pulses

6.1 Introduction

In *continuous-wave (CW) modulation*, which was studied briefly in Chapter 2, some parameter of a sinusoidal carrier wave is varied continuously in accordance with the message signal. This is in direct contrast to *pulse modulation*, which we study in this chapter. In pulse modulation, some parameter of a pulse train is varied in accordance with the message signal. On this basis, we may distinguish two families of pulse modulation:

1. *Analog pulse modulation*, in which a periodic pulse train is used as the carrier wave and some characteristic feature of each pulse (e.g., amplitude, duration, or position) is varied in a continuous manner in accordance with the corresponding *sample* value of the message signal. Thus, in analog pulse modulation, information is transmitted basically in analog form but the transmission takes place at discrete times.
2. *Digital pulse modulation*, in which the message signal is represented in a form that is discrete in both time and amplitude, thereby permitting transmission of the message in digital form as a sequence of *coded pulses*; this form of signal transmission has *no* CW counterpart.

The use of coded pulses for the transmission of analog information-bearing signals represents a basic ingredient in digital communications. In this chapter, we focus attention on digital pulse modulation, which, in basic terms, is described as the *conversion of analog waveforms into coded pulses*. As such, the conversion may be viewed as the transition from analog to digital communications.

Three different kinds of digital pulse modulation are studied in the chapter:

1. *Pulse-code modulation* (PCM), which has emerged as the most favored scheme for the digital transmission of analog information-bearing signals (e.g., voice and video signals). The important advantages of PCM are summarized thus:
   - *robustness* to channel noise and interference;
   - efficient *regeneration* of the coded signal along the transmission path;
   - efficient *exchange* of increased channel bandwidth for improved signal-to-quantization noise ratio, obeying an exponential law;
   - a *uniform format* for the transmission of different kinds of baseband signals, hence their integration with other forms of digital data in a common network;

- comparative *ease* with which message sources may be dropped or reinserted in a multiplex system;
- *secure* communication through the use of special modulation schemes or encryption.

These advantages, however, are attained at the cost of increased system complexity and increased transmission bandwidth. Simply stated:

There is no free lunch.

For every gain we make, there is a price to pay.

2. *Differential pulse-code modulation* (DPCM), which exploits the use of *lossy data compression* to remove the redundancy inherent in a message signal, such as voice or video, so as to reduce the bit rate of the transmitted data without serious degradation in overall system response. In effect, increased system complexity is traded off for reduced bit rate, therefore reducing the bandwidth requirement of PCM.
3. *Delta modulation* (DM), which addresses another practical limitation of PCM: the need for simplicity of implementation when it is a necessary requirement. DM satisfies this requirement by intentionally "oversampling" the message signal. In effect, increased transmission bandwidth is traded off for reduced system complexity. DM may therefore be viewed as the dual of DPCM.

Although, indeed, these three methods of analog-to-digital conversion are quite different, they do share two basic signal-processing operations, namely sampling and quantization:

- the process of sampling, followed by
- pulse-amplitude modulation (PAM) and finally
- amplitude quantization

are studied in what follows in this order.

6.2 Sampling Theory

The *sampling process* is usually described in the time domain. As such, it is an operation that is basic to digital signal processing and digital communications. Through use of the sampling process, an analog signal is converted into a corresponding sequence of samples that are usually spaced uniformly in time. Clearly, for such a procedure to have practical utility, it is necessary that we choose the sampling rate properly in relation to the bandwidth of the message signal, so that the sequence of samples uniquely defines the original analog signal. This is the essence of the sampling theorem, which is derived in what follows.

Frequency-Domain Description of Sampling

Consider an arbitrary signal $g(t)$ of finite energy, which is specified for all time t. A segment of the signal $g(t)$ is shown in Figure 6.1a. Suppose that we sample the signal $g(t)$ instantaneously and at a uniform rate, once every T_s seconds. Consequently, we obtain an infinite sequence of samples spaced T_s seconds apart and denoted by $\{g(nT_s)\}$, where n takes on all possible integer values, positive as well as negative. We refer to T_s as the *sampling period*, and to its reciprocal $f_s = 1/T_s$ as the *sampling rate*. For obvious reasons, this ideal form of sampling is called *instantaneous sampling*.

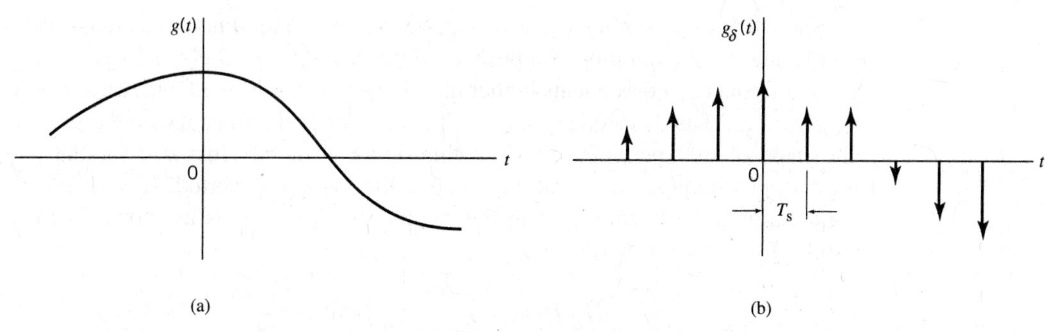

Figure 6.1 The sampling process. (a) Analog signal. (b) Instantaneously sampled version of the analog signal.

Let $g_\delta(t)$ denote the signal obtained by individually weighting the elements of a periodic sequence of delta functions spaced T_s seconds apart by the sequence of numbers $\{g(nT_s)\}$, as shown by (see Figure 6.1b):

$$g_\delta(t) = \sum_{n=-\infty}^{\infty} g(nT_s)\delta(t-nT_s) \tag{6.1}$$

We refer to $g_\delta(t)$ as the *ideal sampled signal.* The term $\delta(t-nT_s)$ represents a delta function positioned at time $t = nT_s$. From the definition of the delta function, we recall from Chapter 2 that such an idealized function has unit area. We may therefore view the multiplying factor $g(nT_s)$ in (6.1) as a "mass" assigned to the delta function $\delta(t-nT_s)$. A delta function weighted in this manner is closely approximated by a rectangular pulse of duration Δt and amplitude $g(nT_s)/\Delta t$; *the smaller we make Δt the better the approximation will be.*

Referring to the table of Fourier-transform pairs in Table 2.2, we have

$$g_\delta(t) \rightleftharpoons f_s \sum_{m=-\infty}^{\infty} G(f-mf_s) \tag{6.2}$$

where $G(f)$ is the Fourier transform of the original signal $g(t)$ and f_s is the sampling rate. Equation (6.2) states:

> The process of uniformly sampling a continuous-time signal of finite energy results in a periodic spectrum with a frequency equal to the sampling rate.

Another useful expression for the Fourier transform of the ideal sampled signal $g_\delta(t)$ may be obtained by taking the Fourier transform of both sides of (6.1) and noting that the Fourier transform of the delta function $\delta(t-nT_s)$ is equal to $\exp(-j2\pi nfT_s)$. Letting $G_\delta(f)$ denote the Fourier transform of $g_\delta(t)$, we may write

$$G_\delta(f) = \sum_{n=-\infty}^{\infty} g(nT_s)\exp(-j2\pi nfT_s) \tag{6.3}$$

Equation (6.3) describes the *discrete-time Fourier transform.* It may be viewed as a complex Fourier series representation of the periodic frequency function $G_\delta(f)$, with the sequence of samples $\{g(nT_s)\}$ defining the coefficients of the expansion.

The discussion presented thus far applies to any continuous-time signal $g(t)$ of finite energy and infinite duration. Suppose, however, that the signal $g(t)$ is *strictly band limited*, with no frequency components higher than W hertz. That is, the Fourier transform $G(f)$ of the signal $g(t)$ has the property that $G(f)$ is zero for $|f| \geq W$, as illustrated in Figure 6.2a; the shape of the spectrum shown in this figure is merely intended for the purpose of illustration. Suppose also that we choose the sampling period $T_s = 1/2W$. Then the corresponding spectrum $G_\delta(f)$ of the sampled signal $g_\delta(t)$ is as shown in Figure 6.2b. Putting $T_s = 1/2W$ in (6.3) yields

$$G_\delta(f) = \sum_{n=-\infty}^{\infty} g\left(\frac{n}{2W}\right) \exp\left(-\frac{j\pi nf}{W}\right) \tag{6.4}$$

Isolating the term on the right-hand side of (6.2), corresponding to $m = 0$, we readily see that the Fourier transform of $g_\delta(t)$ may also be expressed as

$$G_\delta(f) = f_s G(f) + f_s \sum_{\substack{m=-\infty \\ m \neq 0}}^{\infty} G(f - mf_s) \tag{6.5}$$

Suppose, now, we impose the following two conditions:

1. $G(f) = 0$ for $|f| \geq W$.
2. $f_s = 2W$.

We may then reduce (6.5) to

$$G(f) = \frac{1}{2W} G_\delta(f), \qquad -W < f < W \tag{6.6}$$

Substituting (6.4) into (6.6), we may also write

$$G(f) = \frac{1}{2W} \sum_{n=-\infty}^{\infty} g\left(\frac{n}{2W}\right) \exp\left(-\frac{j\pi nf}{W}\right), \qquad -W < f < W \tag{6.7}$$

Equation (6.7) is the desired formula for the frequency-domain description of sampling. This formula reveals that if the sample values $g(n/2W)$ of the signal $g(t)$ are specified for all n, then the Fourier transform $G(f)$ of that signal is uniquely determined. Because $g(t)$ is related to $G(f)$ by the inverse Fourier transform, it follows, therefore, that $g(t)$ is itself uniquely determined by the sample values $g(n/2W)$ for $-\infty < n < \infty$. In other words, the sequence $\{g(n/2W)\}$ has all the information contained in the original signal $g(t)$.

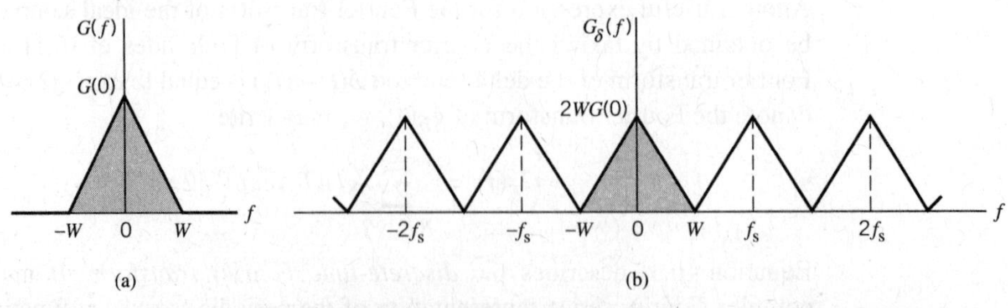

Figure 6.2 (a) Spectrum of a strictly band-limited signal $g(t)$. (b) Spectrum of the sampled version of $g(t)$ for a sampling period $T_s = 1/2W$.

Consider next the problem of reconstructing the signal $g(t)$ from the sequence of sample values $\{g(n/2W)\}$. Substituting (6.7) in the formula for the inverse Fourier transform

$$g(t) = \int_{-\infty}^{\infty} G(f)\exp(j2\pi ft)\,df$$

and interchanging the order of summation and integration, which is permissible because both operations are linear, we may go on to write

$$g(t) = \sum_{n=-\infty}^{\infty} g\left(\frac{n}{2W}\right)\frac{1}{2W}\int_{-W}^{W} \exp\left[j2\pi f\left(t-\frac{n}{2W}\right)\right] df \qquad (6.8)$$

The definite integral in (6.8), including the multiplying factor $1/2W$, is readily evaluated in terms of the sinc function, as shown by

$$\begin{aligned}\frac{1}{2W}\int_{-W}^{W} \exp\left[j2\pi f\left(t-\frac{n}{2W}\right)\right] df &= \frac{\sin(2\pi Wt-n\pi)}{2\pi Wt-n\pi}\\ &= \operatorname{sinc}(2Wt-n)\end{aligned}$$

Accordingly, (6.8) reduces to the infinite-series expansion

$$g(t) = \sum_{n=-\infty}^{\infty} g\left(\frac{n}{2W}\right)\operatorname{sinc}(2Wt-n), \qquad -\infty < t < \infty \qquad (6.9)$$

Equation (6.9) is the desired *reconstruction formula*. This formula provides the basis for reconstructing the original signal $g(t)$ from the sequence of sample values $\{g(n/2W)\}$, with the sinc function $\operatorname{sinc}(2Wt)$ playing the role of a *basis function* of the expansion. Each sample, $g(n/2W)$, is multiplied by a delayed version of the *basis function*, $\operatorname{sinc}(2Wt - n)$, and all the resulting individual waveforms in the expansion are added to reconstruct the original signal $g(t)$.

The Sampling Theorem

Equipped with the frequency-domain description of sampling given in (6.7) and the reconstruction formula of (6.9), we may now state the *sampling theorem* for strictly band-limited signals of finite energy in two equivalent parts:

1. A band-limited signal of finite energy that has no frequency components higher than W hertz is completely described by specifying the values of the signal instants of time separated by $1/2W$ seconds.
2. A band-limited signal of finite energy that has no frequency components higher than W hertz is completely recovered from a knowledge of its samples taken at the rate of $2W$ samples per second.

Part 1 of the theorem, following from (6.7), is performed in the transmitter. Part 2 of the theorem, following from (6.9), is performed in the receiver. For a signal bandwidth of W hertz, the sampling rate of $2W$ samples per second, for a signal bandwidth of W hertz, is called the *Nyquist rate*; its reciprocal $1/2W$ (measured in seconds) is called the *Nyquist interval*; see the classic paper (Nyquist, 1928b).

Aliasing Phenomenon

Derivation of the sampling theorem just described is based on the assumption that the signal $g(t)$ is strictly band limited. In practice, however, a message signal is *not* strictly band limited, with the result that some degree of undersampling is encountered, as a consequence of which *aliasing* is produced by the sampling process. Aliasing refers to the phenomenon of a high-frequency component in the spectrum of the signal seemingly taking on the identity of a lower frequency in the spectrum of its sampled version, as illustrated in Figure 6.3. The aliased spectrum, shown by the solid curve in Figure 6.3b, pertains to the undersampled version of the message signal represented by the spectrum of Figure 6.3a.

To combat the effects of aliasing in practice, we may use two corrective measures:

1. Prior to sampling, a low-pass *anti-aliasing filter* is used to attenuate those high-frequency components of the signal that are not essential to the information being conveyed by the message signal $g(t)$.
2. The filtered signal is sampled at a rate slightly higher than the Nyquist rate.

The use of a sampling rate higher than the Nyquist rate also has the beneficial effect of easing the design of the *reconstruction filter* used to recover the original signal from its sampled version. Consider the example of a message signal that has been anti-alias (low-pass) filtered, resulting in the spectrum shown in Figure 6.4a. The corresponding spectrum of the instantaneously sampled version of the signal is shown in Figure 6.4b, assuming a sampling rate higher than the Nyquist rate. According to Figure 6.4b, we readily see that design of the reconstruction filter may be specified as follows:

- The reconstruction filter is low-pass with a passband extending from $-W$ to W, which is itself determined by the anti-aliasing filter.
- The reconstruction filter has a transition band extending (for positive frequencies) from W to $(f_s - W)$, where f_s is the sampling rate.

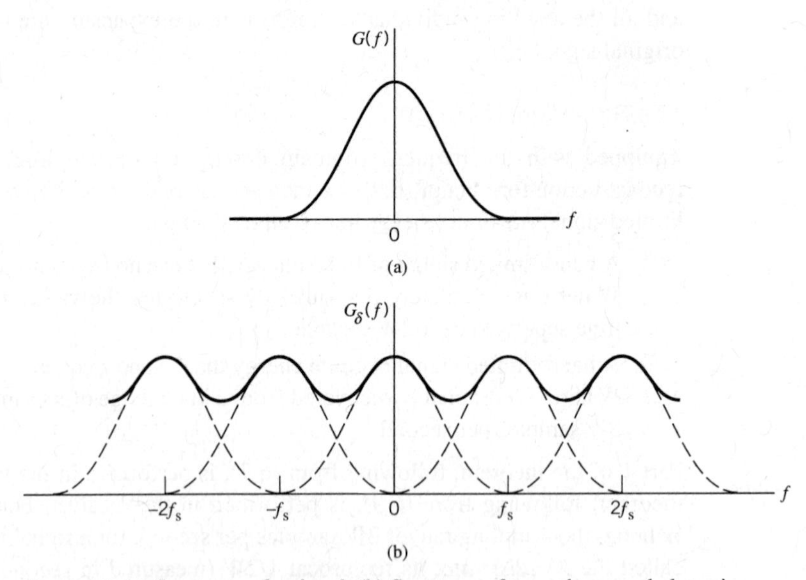

Figure 6.3 (a) Spectrum of a signal. (b) Spectrum of an under-sampled version of the signal exhibiting the aliasing phenomenon.

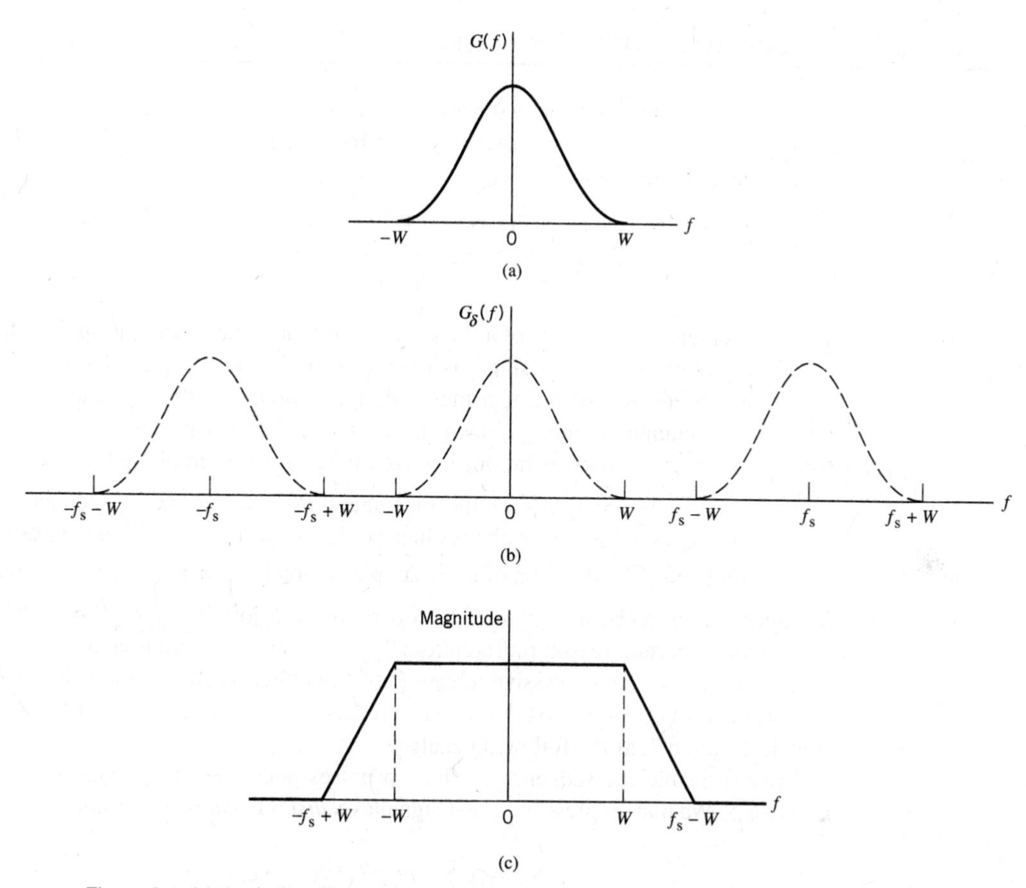

Figure 6.4 (a) Anti-alias filtered spectrum of an information-bearing signal. (b) Spectrum of instantaneously sampled version of the signal, assuming the use of a sampling rate greater than the Nyquist rate. (c) Magnitude response of reconstruction filter.

EXAMPLE 1

Sampling of Voice Signals

As an illustrative example, consider the sampling of voice signals for waveform coding. Typically, the frequency band, extending from 100 Hz to 3.1 kHz, is considered to be adequate for telephonic communication. This limited frequency band is accomplished by passing the voice signal through a low-pass filter with its cutoff frequency set at 3.1 kHz; such a filter may be viewed as an anti-aliasing filter. With such a cutoff frequency, the Nyquist rate is $f_s = 2 \times 3.1 = 6.2$ kHz. The standard sampling rate for the waveform coding of voice signals is 8 kHz. Putting these numbers together, design specifications for the reconstruction (low-pass) filter in the receiver are as follows:

Cutoff frequency	3.1 kHz
Transition band	6.2 to 8 kHz
Transition-band width	1.8 kHz.

6.3 Pulse-Amplitude Modulation

Now that we understand the essence of the sampling process, we are ready to formally define PAM, which is the simplest and most basic form of analog pulse modulation. It is formally defined as follows:

> PAM is a linear modulation process where the amplitudes of regularly spaced pulses are varied in proportion to the corresponding sample values of a continuous message signal.

The pulses themselves can be of a rectangular form or some other appropriate shape.

The waveform of a PAM signal is illustrated in Figure 6.5. The dashed curve in this figure depicts the waveform of a message signal $m(t)$, and the sequence of amplitude-modulated rectangular pulses shown as solid lines represents the corresponding PAM signal $s(t)$. There are two operations involved in the generation of the PAM signal:

1. *Instantaneous sampling* of the message signal $m(t)$ every T_s seconds, where the sampling rate $f_s = 1/T_s$ is chosen in accordance with the sampling theorem.
2. *Lengthening* the duration of each sample so obtained to some constant value T.

In digital circuit technology, these two operations are jointly referred to as "sample and hold." One important reason for intentionally lengthening the duration of each sample is to avoid the use of an excessive channel bandwidth, because bandwidth is inversely proportional to pulse duration. However, care has to be exercised in how long we make the sample duration T, as the following analysis reveals.

Let $s(t)$ denote the sequence of flat-top pulses generated in the manner described in Figure 6.5. We may express the PAM signal as a *discrete convolution sum*:

$$s(t) = \sum_{n=-\infty}^{\infty} m(nT_s)h(t - nT_s) \tag{6.10}$$

where T_s is the *sampling period* and $m(nT_s)$ is the sample value of $m(t)$ obtained at time $t = nT_s$. The $h(t)$ is a Fourier-transformal pulse. With spectral analysis of $s(t)$ in mind, we would like to recast (6.10) in the form of a convolution integral. To this end, we begin by invoking the sifting property of a delta function (discussed in Chapter 2) to express the delayed version of the pulse shape $h(t)$ in (6.10) as

$$h(t - nT_s) = \int_{-\infty}^{\infty} h(t - \tau)\delta(t - nT_s)\, d\tau \tag{6.11}$$

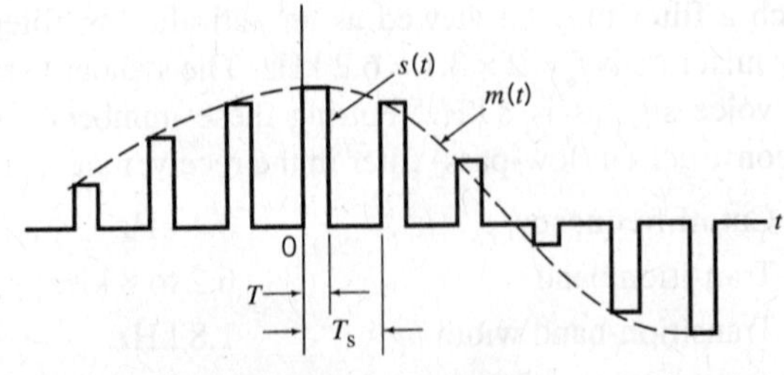

Figure 6.5 Flat-top samples, representing an analog signal.

Hence, substituting (6.11) into (6.10), and interchanging the order of summation and integration, we get

$$s(t) = \int_{-\infty}^{\infty} \left[\sum_{n=-\infty}^{\infty} m(nT_s)\delta(t - nT_s) \right] h(t - \tau)\, d\tau \tag{6.12}$$

Referring to (6.1), we recognize that the expression inside the brackets in (6.12) is simply the instantaneously sampled version of the message signal $m(t)$, as shown by

$$m_\delta(t) = \sum_{n=-\infty}^{\infty} m(nT_s)\delta(t - nT_s) \tag{6.13}$$

Accordingly, substituting (6.13) into (6.12), we may reformulate the PAM signal $s(t)$ in the desired form

$$\begin{aligned} s(t) &= \int_{-\infty}^{\infty} m_\delta(t)h(t - \tau)\, d\tau \\ &= m_\delta(t) \star h(t) \end{aligned} \tag{6.14}$$

which is the convolution of the two time functions; $m_\delta(t)$ and $h(t)$.

The stage is now set for taking the Fourier transform of both sides of (6.14) and recognizing that the convolution of two time functions is transformed into the multiplication of their respective Fourier transforms; we get the simple result

$$S(f) = M_\delta(f)H(f) \tag{6.15}$$

where $S(f) = \mathbf{F}[s(t)]$, $M_\delta(f) = \mathbf{F}[m_\delta(t)]$, and $H(f) = \mathbf{F}[h(t)]$. Adapting (6.2) to the problem at hand, we note that the Fourier transform $M_\delta(f)$ is related to the Fourier transform $M(f)$ of the original message signal $m(t)$ as follows:

$$M_\delta(f) = f_s \sum_{k=-\infty}^{\infty} M(f - kf_s) \tag{6.16}$$

where f_s is the sampling rate. Therefore, the substitution of (6.16) into (6.15) yields the desired formula for the Fourier transform of the PAM signal $s(t)$, as shown by

$$S(f) = f_s \sum_{k=-\infty}^{\infty} M(f - kf_s)H(f) \tag{6.17}$$

Given this formula, how do we recover the original message signal $m(t)$? As a first step in this reconstruction, we may pass $s(t)$ through a low-pass filter whose frequency response is defined in Figure 6.4c; here, it is assumed that the message signal is limited to bandwidth W and the sampling rate f_s is larger than the Nyquist rate $2W$. Then, from (6.17) we find that the spectrum of the resulting filter output is equal to $M(f)H(f)$. This output is equivalent to passing the original message signal $m(t)$ through another low-pass filter of frequency response $H(f)$.

Equation (6.17) applies to any Fourier-transformable pulse shape $h(t)$.

Consider now the special case of a rectangular pulse of unit amplitude and duration T, as shown in Figure 6.6a; specifically:

$$h(t) = \begin{cases} 1, & 0 < t < T \\ \frac{1}{2}, & t = 0, t = T \\ 0, & \text{otherwise} \end{cases} \tag{6.18}$$

Correspondingly, the Fourier transform of $h(t)$ is given by

$$H(f) = T\text{sinc}(fT)\exp(-j\pi fT) \tag{6.19}$$

which is plotted in Figure 6.6b. We therefore find from (6.17) that by using flat-top samples to generate a PAM signal we have introduced *amplitude distortion* as well as a *delay* of $T/2$. This effect is rather similar to the variation in transmission with frequency that is caused by the finite size of the scanning aperture in television. Accordingly, the distortion caused by the use of PAM to transmit an analog information-bearing signal is referred to as the *aperture effect.*

To correct for this distortion, we connect an *equalizer* in cascade with the low-pass reconstruction filter, as shown in Figure 6.7. The equalizer has the effect of decreasing the in-band loss of the reconstruction filter as the frequency increases in such a manner as to

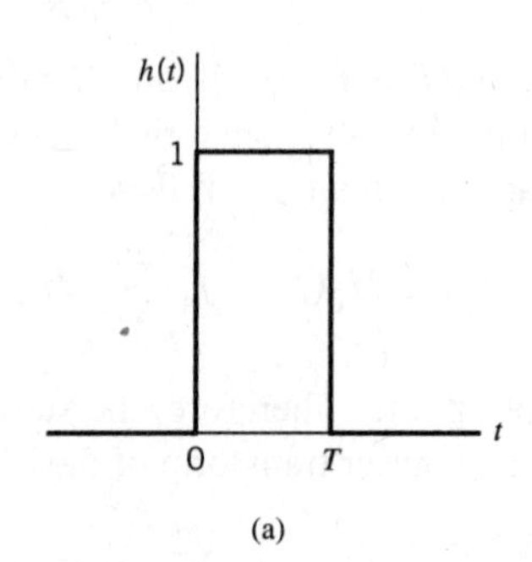

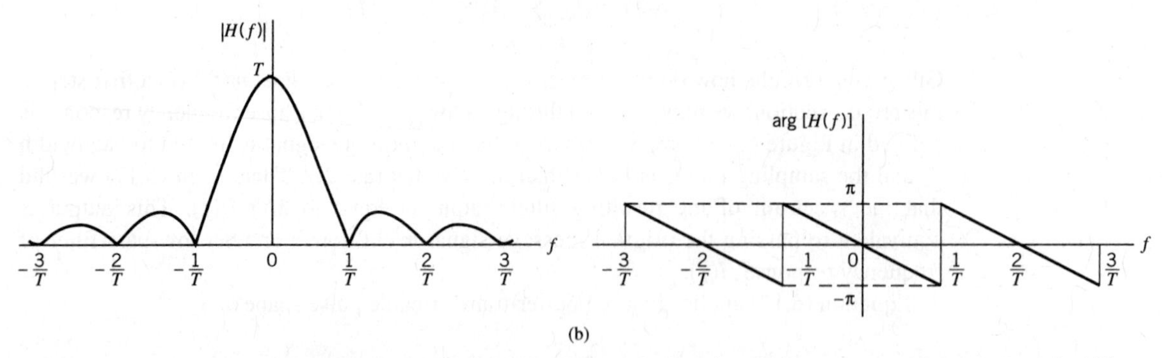

Figure 6.6 (a) Rectangular pulse $h(t)$. (b) Transfer function $H(f)$, made up of the magnitude $|H(f)|$ and phase $\arg[H(f)]$.

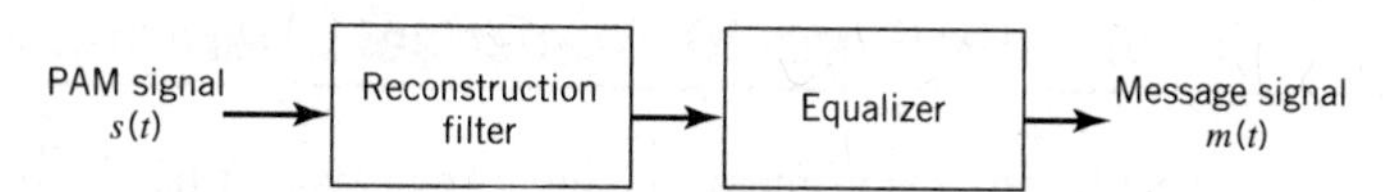

Figure 6.7 System for recovering message signal $m(t)$ from PAM signal $s(t)$.

compensate for the aperture effect. In light of (6.19), the magnitude response of the equalizer should ideally be

$$\frac{1}{|H(f)|} = \frac{1}{T\operatorname{sinc}(fT)} = \frac{\pi f}{\sin(\pi f T)}$$

The amount of equalization needed in practice is usually small. Indeed, for a duty cycle defined by the ratio $T/T_s \le 0.1$, the amplitude distortion is less than 0.5%. In such a situation, the need for equalization may be omitted altogether.

Practical Considerations

The transmission of a PAM signal imposes rather stringent requirements on the frequency response of the channel, because of the relatively short duration of the transmitted pulses. One other point that should be noted: relying on amplitude as the parameter subject to modulation, the noise performance of a PAM system can never be better than baseband-signal transmission. Accordingly, in practice, we find that for transmission over a communication channel PAM is used only as the preliminary means of message processing, whereafter the PAM signal is changed to some other more appropriate form of pulse modulation.

With analog-to-digital conversion as the aim, what would be the appropriate form of modulation to build on PAM? Basically, there are three potential candidates, each with its own advantages and disadvantages, as summarized here:

1. *PCM*, which, as remarked previously in Section 6.1, is robust but demanding in both transmission bandwidth and computational requirements. Indeed, PCM has established itself as the standard method for the conversion of speech and video signals into digital form.
2. *DPCM*, which provides a method for the reduction in transmission bandwidth but at the expense of increased computational complexity.
3. *DM*, which is relatively simple to implement but requires a significant increase in transmission bandwidth.

Before we go on, a comment on terminology is in order. The term "modulation" used herein is a *misnomer*. In reality, PCM, DM, and DPCM are different forms of source coding, with source coding being understood in the sense described in Chapter 5 on information theory. Nevertheless, the terminologies used to describe them have become embedded in the digital communications literature, so much so that we just have to live with them.

Despite their basic differences, PCM, DPCM and DM do share an important feature: the message signal is represented in discrete form in both time and amplitude. PAM takes care of the discrete-time representation. As for the discrete-amplitude representation, we resort to a process known as quantization, which is discussed next.

6.4 Quantization and its Statistical Characterization

Typically, an analog message signal (e.g., voice) has a continuous range of amplitudes and, therefore, its samples have a continuous amplitude range. In other words, within the finite amplitude range of the signal, we find an infinite number of amplitude levels. In actual fact, however, it is not necessary to transmit the exact amplitudes of the samples for the following reason: any human sense (the ear or the eye) as ultimate receiver can detect only finite intensity differences. This means that the message signal may be *approximated* by a signal constructed of discrete amplitudes selected on a minimum error basis from an available set. The existence of a finite number of discrete amplitude levels is a basic condition of waveform coding exemplified by PCM. Clearly, if we assign the discrete amplitude levels with sufficiently close spacing, then we may make the approximated signal practically indistinguishable from the original message signal. For a formal definition of *amplitude quantization*, or just *quantization* for short, we say:

> Quantization is the process of transforming the sample amplitude $m(nT_s)$ of a message signal $m(t)$ at time $t = nT_s$ into a discrete amplitude $v(nT_s)$ taken from a finite set of possible amplitudes.

This definition assumes that the *quantizer* (i.e., the device performing the quantization process) is *memoryless and instantaneous*, which means that the transformation at time $t = nT_s$ is not affected by earlier or later samples of the message signal $m(t)$. This simple form of scalar quantization, though not optimum, is commonly used in practice.

When dealing with a memoryless quantizer, we may simplify the notation by dropping the time index. Henceforth, the symbol m_k is used in place of $m(kT_s)$, as indicated in the block diagram of a quantizer shown in Figure 6.8a. Then, as shown in Figure 6.8b, the signal amplitude m is specified by the index k if it lies inside the *partition cell*

$$J_k: \{m_k < m \le m_{k+1}\}, \qquad k = 1, 2, \ldots, L \tag{6.20}$$

where

$$m_k = m(kT_s) \tag{6.21}$$

and L is the total number of amplitude levels used in the quantizer. The discrete amplitudes m_k, $k = 1, 2, \ldots, L$, at the quantizer input are called *decision levels* or *decision thresholds*. At the quantizer output, the index k is transformed into an amplitude v_k that represents all amplitudes of the cell J_k; the discrete amplitudes v_k, $k = 1, 2, \ldots, L$, are called *representation levels* or *reconstruction levels*. The spacing between two adjacent representation levels is called a *quantum* or *step-size*. Thus, given a quantizer denoted by $g(\cdot)$, the quantized output v equals v_k if the input sample m belongs to the interval J_k. In effect, the mapping (see Figure 6.8a)

$$v = g(m) \tag{6.22}$$

defines the *quantizer characteristic*, described by a staircase function.

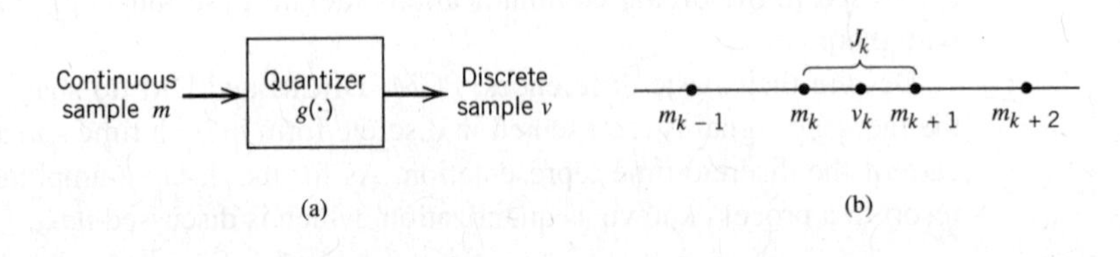

Figure 6.8
Description of a memoryless quantizer.

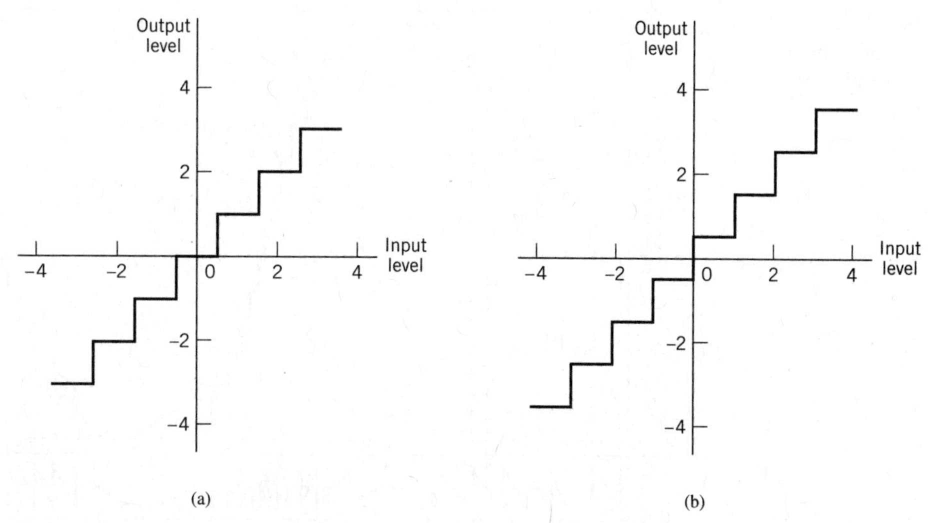

Figure 6.9 Two types of quantization: (a) midtread and (b) midrise.

Quantizers can be of a *uniform* or *nonuniform* type. In a uniform quantizer, the representation levels are uniformly spaced; otherwise, the quantizer is nonuniform. In this section, we consider only uniform quantizers; nonuniform quantizers are considered in Section 6.5. The quantizer characteristic can also be of *midtread* or *midrise type*. Figure 6.9a shows the input–output characteristic of a uniform quantizer of the midtread type, which is so called because the origin lies in the middle of a tread of the staircaselike graph. Figure 6.9b shows the corresponding input–output characteristic of a uniform quantizer of the midrise type, in which the origin lies in the middle of a rising part of the staircaselike graph. Despite their different appearances, both the midtread and midrise types of uniform quantizers illustrated in Figure 6.9 are *symmetric* about the origin.

Quantization Noise

Inevitably, the use of quantization introduces an error defined as the difference between the continuous input sample m and the quantized output sample v. The error is called *quantization noise.*[1] Figure 6.10 illustrates a typical variation of quantization noise as a function of time, assuming the use of a uniform quantizer of the midtread type.

Let the quantizer input m be the sample value of a zero-mean random variable M. (If the input has a nonzero mean, we can always remove it by subtracting the mean from the input and then adding it back after quantization.) A quantizer, denoted by $g(\cdot)$, maps the input random variable M of continuous amplitude into a discrete random variable V; their respective sample values m and v are related by the nonlinear function $g(\cdot)$ in (6.22). Let the quantization error be denoted by the random variable Q of sample value q. We may thus write

$$q = m - v \tag{6.23}$$

or, correspondingly,

$$Q = M - V \tag{6.24}$$

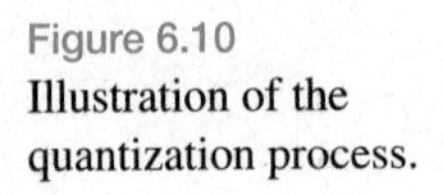
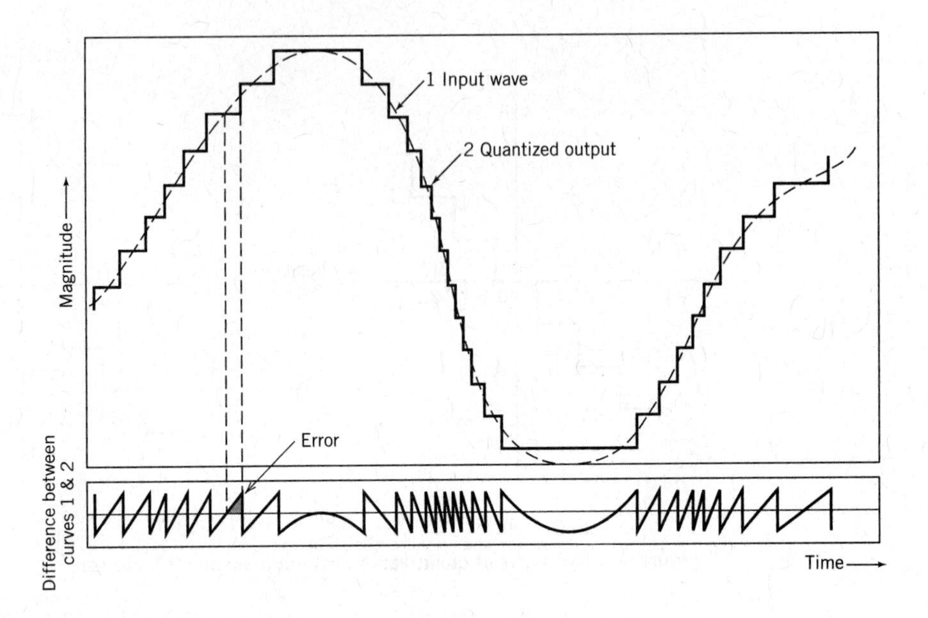

Figure 6.10
Illustration of the quantization process.

With the input M having zero mean and the quantizer assumed to be symmetric as in Figure 6.9, it follows that the quantizer output V and, therefore, the quantization error Q will also have zero mean. Thus, for a partial statistical characterization of the quantizer in terms of output signal-to-(quantization) noise ratio, we need only find the mean-square value of the quantization error Q.

Consider, then, an input m of continuous amplitude, which, symmetrically, occupies the range $[-m_{\max}, m_{\max}]$. Assuming a uniform quantizer of the midrise type illustrated in Figure 6.9b, we find that the step size of the quantizer is given by

$$\Delta = \frac{2m_{\max}}{L} \tag{6.25}$$

where L is the total number of representation levels. For a uniform quantizer, the quantization error Q will have its sample values bounded by $-\Delta/2 \le q \le \Delta/2$. If the step size Δ is sufficiently small (i.e., the number of representation levels L is sufficiently large), it is reasonable to assume that the quantization error Q is a *uniformly distributed* random variable and the interfering effect of the quantization error on the quantizer input is similar to that of thermal noise, hence the reference to quantization error as *quantization noise.* We may thus express the probability density function of the quantization noise as

$$f_Q(q) = \begin{cases} \dfrac{1}{\Delta}, & -\dfrac{\Delta}{2} < q \le \dfrac{\Delta}{2} \\ 0, & \text{otherwise} \end{cases} \tag{6.26}$$

For this to be true, however, we must ensure that the incoming continuous sample does *not* overload the quantizer. Then, with the mean of the quantization noise being zero, its variance σ_Q^2 is the same as the mean-square value; that is,

$$\sigma_Q^2 = \mathbb{E}[Q^2]$$
$$= \int_{-\Delta/2}^{\Delta/2} q^2 f_Q(q)\, dq \tag{6.27}$$

Substituting (6.26) into (6.27), we get

$$\sigma_Q^2 = \frac{1}{\Delta}\int_{-\Delta/2}^{\Delta/2} q^2\, dq$$
$$= \frac{\Delta^2}{12} \tag{6.28}$$

Typically, the L-ary number k, denoting the kth representation level of the quantizer, is transmitted to the receiver in binary form. Let R denote the *number of bits per sample* used in the construction of the binary code. We may then write

$$L = 2^R \tag{6.29}$$

or, equivalently,

$$R = \log_2 L \tag{6.30}$$

Hence, substituting (6.29) into (6.25), we get the step size

$$\Delta = \frac{2m_{\max}}{2^R} \tag{6.31}$$

Thus, the use of (6.31) in (6.28) yields

$$\sigma_Q^2 = \frac{1}{3} m_{\max}^2 2^{-2R} \tag{6.32}$$

Let P denote the average power of the original message signal $m(t)$. We may then express the *output signal-to-noise ratio* of a uniform quantizer as

$$(\mathrm{SNR})_{\mathrm{O}} = \frac{P}{\sigma_Q^2}$$
$$= \left(\frac{3P}{m_{\max}^2}\right) 2^{2R} \tag{6.33}$$

Equation (6.33) shows that the output signal-to-noise ratio of a uniform quantizer $(\mathrm{SNR})_{\mathrm{O}}$ increases *exponentially* with increasing number of bits per sample R, which is intuitively satisfying.

EXAMPLE 2 Sinusoidal Modulating Signal

Consider the special case of a full-load sinusoidal modulating signal of amplitude A_{m}, which utilizes all the representation levels provided. The average signal power is (assuming a load of 1 Ω)

$$P = \frac{A_{\mathrm{m}}^2}{2}$$

The total range of the quantizer input is $2A_m$, because the modulating signal swings between $-A_m$ and A_m. We may, therefore, set $m_{max} = A_m$, in which case the use of (6.32) yields the average power (variance) of the quantization noise as

$$\sigma_Q^2 = \frac{1}{3}A_m^2 2^{-2R}$$

Thus, the output signal-to-noise of a uniform quantizer, for a full-load test tone, is

$$(\text{SNR})_O = \frac{A_m^2/2}{A_m^2 2^{-2R}/3} = \frac{3}{2}(2^{2R}) \tag{6.34}$$

Expressing the signal-to-noise (SNR) in decibels, we get

$$10 \log_{10}(\text{SNR})_O = 1.8 + 6R \tag{6.35}$$

The corresponding values of signal-to-noise ratio for various values of L and R, are given in Table 6.1. For sinusoidal modulation, this table provides a basis for making a quick estimate of the number of bits per sample required for a desired output signal-to-noise ratio.

Table 6.1 Signal-to-(quantization) noise ratio for varying number of representation levels for sinusoidal modulation

No. of representation levels L	No. of bits per sample R	SNR (dB)
32	5	31.8
64	6	37.8
128	7	43.8
256	8	49.8

Conditions of Optimality of Scalar Quantizers

In designing a scalar quantizer, the challenge is how to select the representation levels and surrounding partition cells so as to minimize the average quantization power for a fixed number of representation levels.

To state the problem in mathematical terms: consider a message signal $m(t)$ drawn from a stationary process and whose dynamic range, denoted by $-A \leq m \leq A$, is partitioned into a set of L cells, as depicted in Figure 6.11. The boundaries of the partition cells are defined by a set of real numbers $m_1, m_2, \ldots, m_{L-1}$ that satisfy the following three conditions:

$$m_1 = -A$$

$$m_{L-1} = A$$

$$m_k \leq m_{k-1} \text{ for } k = 1, 2, \ldots, L$$

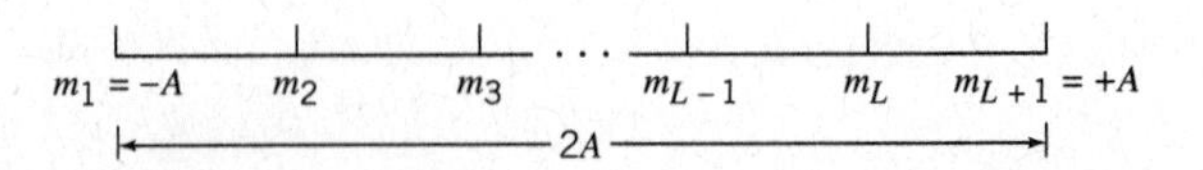

Figure 6.11
Illustrating the partitioning of the dynamic range $-A \leq m \leq A$ of a message signal $m(t)$ into a set of L cells.

The kth partition cell is defined by (6.20), reproduced here for convenience:

$$J_k : m_k < m < m_{k-1} \text{ for } k = 1, 2, \ldots, L \tag{6.36}$$

Let the representation levels (i.e., quantization values) be denoted by v_k, $k = 1, 2, \ldots, L$. Then, assuming that $d(m,v_k)$ denotes a *distortion measure* for using v_k to represent all those values of the input m that lie inside the partition cell J_k, the goal is to find the two sets $\{v_k\}_{k=1}^{L}$ and $\{J_k\}_{k=1}^{L}$ that minimize the *average distortion*

$$D = \sum_{k=1}^{L} \int_{m \in v_k} d(m,v_k) f_M(m)\, dm \tag{6.37}$$

where $f_M(m)$ is the probability density function of the random variable M with sample value m.

A commonly used distortion measure is defined by

$$d(m,v_k) = (m - v_k)^2 \tag{6.38}$$

in which case we speak of the *mean-square distortion*. In any event, the optimization problem stated herein is nonlinear, defying an explicit, closed-form solution. To get around this difficulty, we resort to an *algorithmic approach* for solving the problem in an *iterative manner.*

Structurally speaking, the quantizer consists of two components with interrelated design parameters:

- An encoder characterized by the set of partition cells $\{J_k\}_{k=1}^{L}$; this is located in the transmitter.
- A decoder characterized by the set of representation levels $\{v_k\}_{k=1}^{L}$; this is located in the receiver.

Accordingly, we may identify two critically important conditions that provide the mathematical basis for all algorithmic solutions to the optimum quantization problem. One condition assumes that we are given a decoder and the problem is to find the optimum encoder in the transmitter. The other condition assumes that we are given an encoder and the problem is to find the optimum decoder in the receiver. Henceforth, these two conditions are referred to as condition I and II, respectively.

Condition I: **Optimality of the Encoder for a Given Decoder**

The availability of a decoder means that we have a certain *codebook* in mind. Let the codebook be defined by

$$\mathscr{C} : \{v_k\}_{k=1}^{L} \tag{6.39}$$

Given the codebook $\mathscr{C}$, the problem is to find the set of partition cells $\{J_k\}_{k=1}^{L}$ that minimizes the mean-square distortion D. That is, we wish to find the encoder defined by the nonlinear mapping

$$g(m) = v_k, \qquad k = 1, 2, \ldots, L \tag{6.40}$$

such that we have

$$D = \int_{-A}^{A} d(m, g(m)) f_M(m)\, dM \geq \sum_{k=1}^{L} \int_{m \in J_k} [\min d(m, v_k)] f_M(m)\, dm \tag{6.41}$$

For the lower bound specified in (6.41) to be attained, we require that the nonlinear mapping of (6.40) be satisfied only if the condition

$$d(m, v_k) \le d(m, v_j) \quad \text{holds for all } j \ne k \tag{6.42}$$

The necessary condition described in (6.42) for optimality of the encoder for a specified codebook $\mathscr{C}$ is recognized as the *nearest-neighbor condition.* In words, the nearest neighbor condition requires that the partition cell J_k should embody all those values of the input m that are closer to v_k than any other element of the codebook $\mathscr{C}$. This optimality condition is indeed intuitively satisfying.

Condition II: **Optimality of the Decoder for a Given Encoder**

Consider next the reverse situation to that described under condition I, which may be stated as follows: optimize the codebook $\mathscr{C} = \{v_k\}L_{k=1}$ for the decoder, given that the set of partition cells $\{J_k\}_{k=1}^{L}$ characterizing the encoder is fixed. The criterion for optimization is the average (mean-square) distortion:

$$D = \sum_{k=1}^{L} \int_{m \in J_k} (m - v_k)^2 f_M(m)\, dm \tag{6.43}$$

The probability density function $f_M(m)$ is clearly independent of the codebook $\mathscr{C}$. Hence, differentiating D with respect to the representation level v_k, we readily obtain

$$\frac{\partial D}{\partial v_k} = -2 \sum_{k=1}^{L} \int_{m \in J_k} (m - v_k) f_M(m)\, dm \tag{6.44}$$

Setting $\partial D / \partial v_k$ equal to zero and then solving for v_k, we obtain the optimum value

$$v_{k,\text{opt}} = \frac{\int_{m \in J_k} m f_M(m)\, dm}{\int_{m \in J_k} f_M(m)\, dm} \tag{6.45}$$

The denominator in (6.45) is just the probability p_k that the random variable M with sample value m lies in the partition cell J_k, as shown by

$$\begin{aligned} p_k &= \mathbb{P}(m_k < M \le m_k + 1) \\ &= \int_{m \in J_k} f_M(m)\, dm \end{aligned} \tag{6.46}$$

Accordingly, we may interpret the optimality condition of (6.45) as choosing the representation level v_k to equal the *conditional mean* of the random variable M, given that M lies in the partition cell J_k. We can thus formally state that the condition for optimality of the decoder for a given encoder as follows:

$$v_{k,\text{opt}} = \mathbb{E}[M | m_k < M \le m_{k+1}] \tag{6.47}$$

where $\mathbb{E}$ is the expectation operator. Equation (6.47) is also intuitively satisfying.

Note that the nearest neighbor condition (I) for optimality of the encoder for a given decoder was proved for a generic average distortion. However, the conditional mean requirement (condition II) for optimality of the decoder for a given encoder was proved for the special case of a mean-square distortion. In any event, these two conditions are necessary for optimality of a scalar quantizer. Basically, the algorithm for designing the quantizer consists of alternately optimizing the encoder in accordance with condition I, then optimizing the decoder in accordance with condition II, and continuing in this manner until the average distortion D reaches a minimum. The optimum quantizer designed in this manner is called the *Lloyd–Max quantizer.*[2]

6.5 Pulse-Code Modulation

With the material on sampling, PAM, and quantization presented in the preceding sections, the stage is set for describing PCM, for which we offer the following definition:

> PCM is a discrete-time, discrete-amplitude waveform-coding process, by means of which an analog signal is directly represented by a sequence of coded pulses.

Specifically, the transmitter consists of two components: a *pulse-amplitude modulator* followed by an *analog-to-digital (A/D) converter.* The latter component itself embodies a *quantizer* followed by an *encoder.* The receiver performs the inverse of these two operations: *digital-to-analog (D/A) conversion* followed by *pulse-amplitude demodulation.* The communication channel is responsible for transporting the encoded pulses from the transmitter to the receiver.

Figure 6.12, a block diagram of the PCM, shows the transmitter, the transmission path from the transmitter output to the receiver input, and the receiver.

It is important to realize, however, that once distortion in the form of quantization noise is introduced into the encoded pulses, there is absolutely nothing that can be done at the receiver to compensate for that distortion. The only design precaution that can be taken is to choose a number of representation levels in the receiver that is large enough to ensure that the quantization noise is imperceptible for human use at the receiver output.

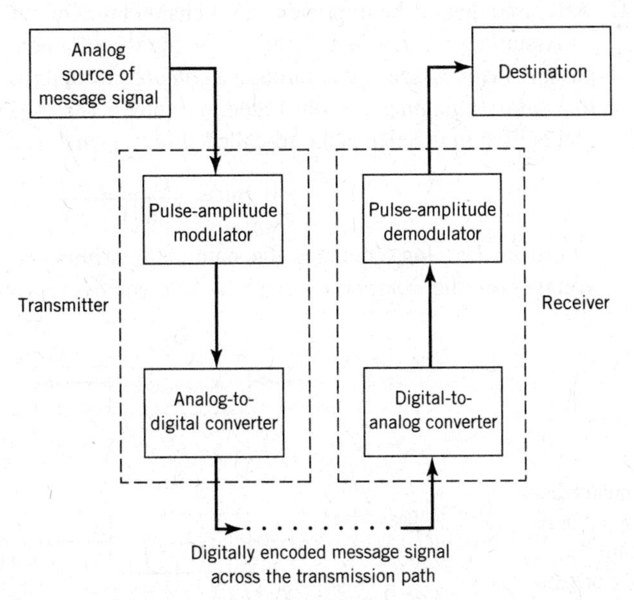

Figure 6.12 Block diagram of PCM system.

Sampling in the Transmitter

The incoming message signal is sampled with a train of rectangular pulses short enough to closely approximate the instantaneous sampling process. To ensure perfect reconstruction of the message signal at the receiver, the sampling rate must be greater than twice the highest frequency component W of the message signal in accordance with the sampling theorem. In practice, a low-pass anti-aliasing filter is used at the front end of the pulse-amplitude modulator to exclude frequencies greater than W before sampling and which are of negligible practical importance. Thus, the application of sampling permits the reduction of the continuously varying message signal to a limited number of discrete values per second.

Quantization in the Transmitter

The PAM representation of the message signal is then quantized in the analog-to-digital converter, thereby providing a new representation of the signal that is discrete in both time and amplitude. The quantization process may follow a uniform law as described in Section 6.4. In telephonic communication, however, it is preferable to use a variable separation between the representation levels for efficient utilization of the communication channel. Consider, for example, the quantization of voice signals. Typically, we find that the range of voltages covered by voice signals, from the peaks of loud talk to the weak passages of weak talk, is on the order of 1000 to 1. By using a *nonuniform quantizer* with the feature that the step size increases as the separation from the origin of the input–output amplitude characteristic of the quantizer is increased, the large end-steps of the quantizer can take care of possible excursions of the voice signal into the large amplitude ranges that occur relatively infrequently. In other words, the weak passages needing more protection are favored at the expense of the loud passages. In this way, a nearly uniform percentage precision is achieved throughout the greater part of the amplitude range of the input signal. The end result is that fewer steps are needed than would be the case if a uniform quantizer were used; hence the improvement in channel utilization.

Assuming memoryless quantization, the use of a nonuniform quantizer is equivalent to passing the message signal through a *compressor* and then applying the compressed signal to a *uniform quantizer*, as illustrated in Figure 6.13a. A particular form of *compression law* that is used in practice is the so-called *μ-law*,[3] which is defined by

$$|v| = \frac{\ln(1 + \mu|m|)}{\ln(1 + \mu)} \tag{6.48}$$

where ln, i.e., $\log_e$, denotes the natural logarithm, m and v are the input and output voltages of the *compressor*, and μ is a positive constant. It is assumed that m and,

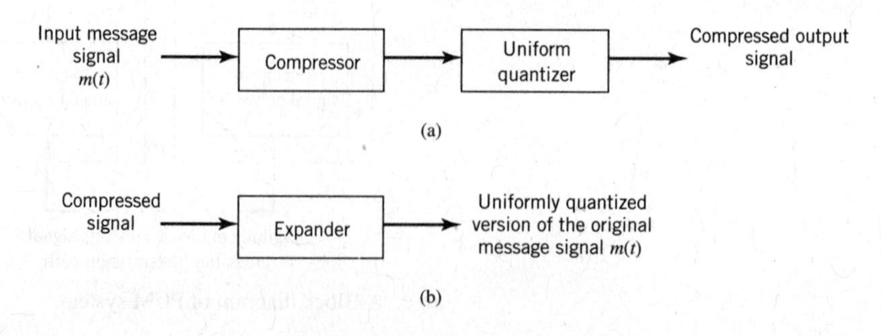

Figure 6.13
(a) Nonuniform quantization of the message signal in the transmitter. (b) Uniform quantization of the original message signal in the receiver.

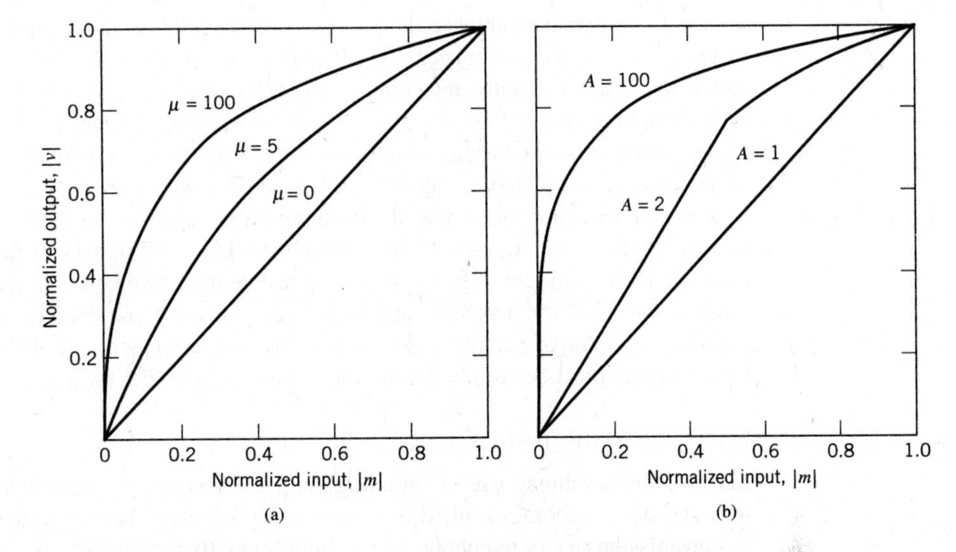

Figure 6.14
Compression laws:
(a) μ-law;
(b) A-law.

therefore, v are scaled so that they both lie inside the interval [–1, 1]. The μ-law is plotted for three different values of μ in Figure 6.14a. The case of uniform quantization corresponds to $\mu = 0$. For a given value of μ, the reciprocal slope of the compression curve that defines the quantum steps is given by the derivative of the absolute value $|m|$ with respect to the corresponding absolute value $|v|$; that is,

$$\frac{\mathrm{d}|m|}{\mathrm{d}|v|} = \frac{\ln(1+\mu)}{\mu}(1+\mu|m|) \tag{6.49}$$

From (6.49) it is apparent that the μ-law is neither strictly linear nor strictly logarithmic. Rather, it is approximately linear at low input levels corresponding to $\mu|m| \ll 1$ and approximately logarithmic at high input levels corresponding to $\mu|m| \gg 1$.

Another compression law that is used in practice is the so-called *A-law*, defined by

$$|v| = \begin{cases} \dfrac{A|m|}{1+\ln A}, & 0 \le |m| \le \dfrac{1}{A} \\[2ex] \dfrac{1+\ln(A|m|)}{1+\ln A}, & \dfrac{1}{A} \le |m| \le 1 \end{cases} \tag{6.50}$$

where A is another positive constant. Equation (6.50) is plotted in Figure 6.14b for varying A. The case of uniform quantization corresponds to $A = 1$. The reciprocal slope of this second compression curve is given by the derivative of $|m|$ with respect to $|v|$, as shown by

$$\frac{\mathrm{d}|m|}{\mathrm{d}|v|} = \begin{cases} \dfrac{1+\ln A}{A}, & 0 \le |m| \le \dfrac{1}{A} \\[2ex] (1+\ln A)|m|, & \dfrac{1}{A} \le |m| \le 1 \end{cases} \tag{6.51}$$

To restore the signal samples to their correct relative level, we must, of course, use a device in the receiver with a characteristic complementary to the compressor. Such a device is called an *expander.* Ideally, the compression and expansion laws are exactly the inverse of each other. With this provision in place, we find that, except for the effect of quantization, the expander output is equal to the compressor input. The cascade combination of a *compressor* and an *expander,* depicted in Figure 6.13, is called a *compander.*

For both the μ-law and A-law, the dynamic range capability of the compander improves with increasing μ and A, respectively. The SNR for low-level signals increases at the expense of the SNR for high-level signals. To accommodate these two conflicting requirements (i.e., a reasonable SNR for both low- and high-level signals), a compromise is usually made in choosing the value of parameter μ for the μ-law and parameter A for the A-law. The typical values used in practice are $\mu = 255$ for the μ–law and $A = 87.6$ for the A-law.[4]

Encoding in the Transmitter

Through the combined use of sampling and quantization, the specification of an analog message signal becomes limited to a discrete set of values, but not in the form best suited to transmission over a telephone line or radio link. To exploit the advantages of sampling and quantizing for the purpose of making the transmitted signal more robust to noise, interference, and other channel impairments, we require the use of an *encoding process* to translate the discrete set of sample values to a more appropriate form of signal. Any plan for representing each of this discrete set of values as a particular arrangement of discrete events constitutes a *code.* Table 6.2 describes the one-to-one correspondence between representation levels and codewords for a binary number system for $R = 4$ bits per sample. Following the terminology of Chapter 5, the two symbols of a binary code are customarily denoted as 0 and 1. In practice, the binary code is the preferred choice for encoding for the following reason:

> The maximum advantage over the effects of noise encountered in a communication system is obtained by using a binary code because a binary symbol withstands a relatively high level of noise and, furthermore, it is easy to regenerate.

The last signal-processing operation in the transmitter is that of *line coding,* the purpose of which is to represent each binary codeword by a sequence of pulses; for example, symbol 1 is represented by the presence of a pulse and symbol 0 is represented by absence of the pulse. Line codes are discussed in Section 6.10. Suppose that, in a binary code, each codeword consists of R bits. Then, using such a code, we may represent a total of 2^R distinct numbers. For example, a sample quantized into one of 256 levels may be represented by an 8-bit codeword.

Inverse Operations in the PCM Receiver

The first operation in the receiver of a PCM system is to *regenerate* (i.e., reshape and clean up) the received pulses. These clean pulses are then regrouped into codewords and decoded (i.e., mapped back) into a quantized pulse-amplitude modulated signal. The *decoding* process involves generating a pulse the amplitude of which is the linear sum of all the pulses in the codeword. Each pulse is weighted by its place value ($2^0, 2^1, 2^2, \ldots, 2^{R-1}$) in the code, where R is the number of bits per sample. Note, however, that whereas the analog-to-digital

Table 6.2 **Binary number system for $T = 4$ bits/sample**

Ordinal number of representation level	Level number expressed as sum of powers of 2	Binary number
0		0000
1	2^0	0001
2	2^1	0010
3	$2^1 + 2^0$	0011
4	2^2	0100
5	$2^2 + 2^0$	0101
6	$2^2 + 2^1$	0110
7	$2^2 + 2^1 + 2^0$	0111
8	2^3	1000
9	$2^3 + 2^0$	1001
10	$2^3 + 2^1$	1010
11	$2^3 + 2^1 + 2^0$	1011
12	$2^3 + 2^2$	1100
13	$2^3 + 2^2 + 2^0$	1101
14	$2^3 + 2^2 + 2^1$	1110
15	$2^3 + 2^2 + 2^1 + 2^0$	1111

converter in the transmitter involves both quantization and encoding, the digital-to-analog converter in the receiver involves decoding only, as illustrated in Figure 6.12.

The final operation in the receiver is that of *signal reconstruction*. Specifically, an estimate of the original message signal is produced by passing the decoder output through a *low-pass reconstruction filter* whose cutoff frequency is equal to the message bandwidth W. Assuming that the transmission link (connecting the receiver to the transmitter) is error free, the reconstructed message signal includes no noise with the exception of the initial distortion introduced by the quantization process.

PCM Regeneration along the Transmission Path

The most important feature of a PCM systems is its ability to control the effects of distortion and noise produced by transmitting a PCM signal through the channel, connecting the receiver to the transmitter. This capability is accomplished by reconstructing the PCM signal through a chain of *regenerative repeaters*, located at sufficiently close spacing along the transmission path.

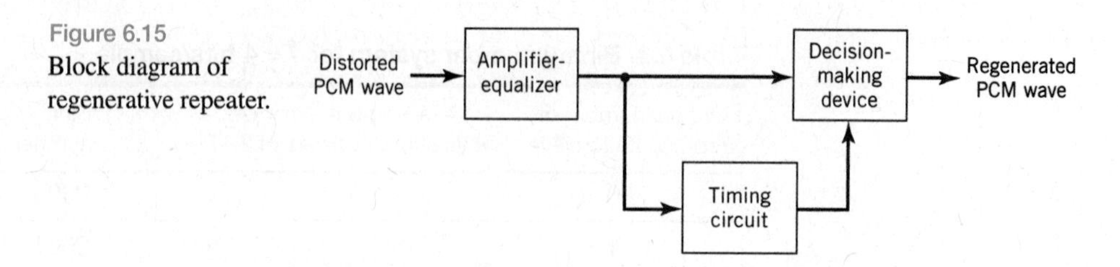

Figure 6.15
Block diagram of regenerative repeater.

As illustrated in Figure 6.15, three basic functions are performed in a regenerative repeater: *equalization*, *timing*, and *decision making*. The equalizer shapes the received pulses so as to compensate for the effects of amplitude and phase distortions produced by the non-ideal transmission characteristics of the channel. The timing circuitry provides a periodic pulse train, derived from the received pulses, for sampling the equalized pulses at the instants of time where the SNR ratio is a maximum. Each sample so extracted is compared with a predetermined *threshold* in the decision-making device. In each bit interval, a decision is then made on whether the received symbol is 1 or 0 by observing whether the threshold is exceeded or not. If the threshold is exceeded, a clean new pulse representing symbol 1 is transmitted to the next repeater; otherwise, another clean new pulse representing symbol 0 is transmitted. In this way, it is possible for the accumulation of distortion and noise in a repeater span to be almost completely removed, provided that the disturbance is not too large to cause an error in the decision-making process. Ideally, except for delay, the regenerated signal is exactly the same as the signal originally transmitted. In practice, however, the regenerated signal departs from the original signal for two main reasons:

1. The unavoidable presence of channel noise and interference causes the repeater to make wrong decisions occasionally, thereby introducing *bit errors* into the regenerated signal.
2. If the spacing between received pulses deviates from its assigned value, a *jitter* is introduced into the regenerated pulse position, thereby causing distortion.

The important point to take from this subsection on PCM is the fact that regeneration along the transmission path is provided across the spacing between individual regenerative repeaters (including the last stage of regeneration at the receiver input) provided that the spacing is short enough. If the transmitted SNR ratio is high enough, then the regenerated PCM data stream is the same as the transmitted PCM data stream, except for a practically negligibly small *bit error rate* (BER). In other words, under these operating conditions, performance degradation in the PCM system is essentially confined to quantization noise in the transmitter.

6.6 Noise Considerations in PCM Systems

The performance of a PCM system is influenced by two major sources of noise:

1. *Channel noise*, which is introduced anywhere between the transmitter output and the receiver input; channel noise is always present, once the equipment is switched on.
2. *Quantization noise*, which is introduced in the transmitter and is carried all the way along to the receiver output; unlike channel noise, quantization noise is *signal dependent*, in the sense that it disappears when the message signal is switched off.

Naturally, these two sources of noise appear simultaneously once the PCM system is in operation. However, the traditional practice is to consider them separately, so that we may develop insight into their individual effects on the system performance.

The main effect of channel noise is to introduce *bit errors* into the received signal. In the case of a binary PCM system, the presence of a bit error causes symbol 1 to be mistaken for symbol 0, or vice versa. Clearly, the more frequently bit errors occur, the more dissimilar the receiver output becomes compared with the original message signal. The fidelity of information transmission by PCM in the presence of channel noise may be measured in terms of the *average probability of symbol error*, which is defined as the probability that the reconstructed symbol at the receiver output differs from the transmitted binary symbol on the average. The average probability of symbol error, also referred to as the BER, assumes that all the bits in the original binary wave are of equal importance. When, however, there is more interest in restructuring the analog waveform of the original message signal, different symbol errors may be *weighted* differently; for example, an error in the most significant bit in a codeword (representing a quantized sample of the message signal) is more harmful than an error in the least significant bit.

To optimize system performance in the presence of channel noise, we need to minimize the average probability of symbol error. For this evaluation, it is customary to model the channel noise as an ideal *additive white Gaussian noise* (AWGN) channel. The effect of channel noise can be made practically negligible by using an adequate signal energy-to-noise density ratio through the provision of short-enough spacing between the regenerative repeaters in the PCM system. In such a situation, the performance of the PCM system is essentially limited by quantization noise acting alone.

From the discussion of quantization noise presented in Section 6.4, we recognize that quantization noise is essentially under the designer's control. It can be made negligibly small through the use of an adequate number of representation levels in the quantizer and the selection of a companding strategy matched to the characteristics of the type of message signal being transmitted. We thus find that the use of PCM offers the possibility of building a communication system that is *rugged* with respect to channel noise on a scale that is beyond the capability of any analog communication system; hence its use as a *standard* against which other waveform coders (e.g., DPCM and DM) are compared.

Error Threshold

The underlying theory of BER calculation in a PCM system is deferred to Chapter 8. For the present, it suffices to say that the average probability of symbol error in a binary encoded PCM receiver due to AWGN depends solely on E_b/N_0, which is defined as *the ratio of the transmitted signal energy per bit* E_b, *to the noise spectral density* N_0. Note that the ratio E_b/N_0 is dimensionless even though the quantities E_b and N_0 have different physical meaning. In Table 6.3, we present a summary of this dependence for the case of a binary PCM system, in which symbols 1 and 0 are represented by rectangular pulses of equal but opposite amplitudes. The results presented in the last column of the table assume a bit rate of 10^5 bits/s.

From Table 6.3 it is clear that there is an *error threshold* (at about 11 dB). For E_b/N_0 below the error threshold the receiver performance involves significant numbers of errors, and above it the effect of channel noise is practically negligible. In other words, provided that the ratio E_b/N_0 exceeds the error threshold, channel noise has virtually no effect on

Table 6.3 Influence of E_b/N_0 on the probability of error

E_b/N_0 (dB)	Probability of error P_e	For a bit rate of 10^5 bits/s, this is about one error every
4.3	10^{-2}	10^{-3} s
8.4	10^{-4}	10^{-1} s
10.6	10^{-6}	10 s
12.0	10^{-8}	20 min
13.0	10^{-10}	1 day
14.0	10^{-12}	3 months

the receiver performance, which is precisely the goal of PCM. When, however, E_b/N_0 drops below the error threshold, there is a sharp increase in the rate at which errors occur in the receiver. Because decision errors result in the construction of incorrect codewords, we find that when the errors are frequent, the reconstructed message at the receiver output bears little resemblance to the original message signal.

An important characteristic of a PCM system is its *ruggedness to interference*, caused by impulsive noise or cross-channel interference. The combined presence of channel noise and interference causes the error threshold necessary for satisfactory operation of the PCM system to increase. If, however, an adequate margin over the error threshold is provided in the first place, the system can withstand the presence of relatively large amounts of interference. In other words, a PCM system is *robust* with respect to channel noise and interference, providing further confirmation to the point made in the previous section that performance degradation in PCM is essentially confined to quantization noise in the transmitter.

PCM Noise Performance Viewed in Light of the Information Capacity Law

Consider now a PCM system that is known to operate above the error threshold, in which case we would be justified to ignore the effect of channel noise. In other words, the noise performance of the PCM system is essentially determined by quantization noise acting alone. Given such a scenario, how does the PCM system fare compared with the information capacity law, derived in Chapter 5?

To address this question of practical importance, suppose that the system uses a codeword consisting of n symbols with each symbol representing one of M possible discrete amplitude levels; hence the reference to the system as an "M-ary" PCM system. For this system to operate above the error threshold, there must be provision for a large enough noise margin.

For the PCM system to operate above the error threshold as proposed, the requirement for a noise margin that is sufficiently large to maintain a negligible error rate due to channel noise. This, in turn, means there must be a certain separation between the M discrete amplitude levels. Call this separation $c\sigma$, where c is a constant and $\sigma^2 = N_0B$ is the noise variance measured in a channel bandwidth B. The number of amplitude levels M is usually an integer power of 2. The average transmitted power will be least if the amplitude range is symmetrical about zero. Then, the discrete amplitude levels, normalized with respect to the separation $c\sigma$, will have the values $\pm 1/2, \pm 3/2, \ldots, \pm(M-1)/2$. We assume that these M different amplitude levels are equally likely. Accordingly, we find that the average transmitted power is given by

$$\begin{aligned} P &= \frac{2}{M}\left[\left(\frac{1}{2}\right)^2 + \left(\frac{3}{2}\right)^2 + \cdots + \left(\frac{M-1}{2}\right)^2\right](c\sigma)^2 \\ &= c^2\sigma^2\left(\frac{M^2-1}{12}\right) \end{aligned} \tag{6.52}$$

Suppose that the M-ary PCM system described herein is used to transmit a message signal with its highest frequency component equal to W hertz. The signal is sampled at the Nyquist rate of $2W$ samples per second. We assume that the system uses a quantizer of the midrise type, with L equally likely representation levels. Hence, the probability of occurrence of any one of the L representation levels is $1/L$. Correspondingly, the amount of information carried by a single sample of the signal is $\log_2 L$ bits. With a maximum sampling rate of $2W$ samples per second, the maximum rate of information transmission of the PCM system measured in bits per second is given by

$$R_b = 2W \log_2 L \text{ bits/s} \tag{6.53}$$

Since the PCM system uses a codeword consisting of n code elements with each one having M possible discrete amplitude values, we have M^n different possible codewords. For a unique encoding process, therefore, we require

$$L = M^n \tag{6.54}$$

Clearly, the rate of information transmission in the system is unaffected by the use of an encoding process. We may, therefore, eliminate L between (6.53) and (6.54) to obtain

$$R_b = 2Wn \log_2 M \text{ bits/s} \tag{6.55}$$

Equation (6.52) defines the average transmitted power required to maintain an M-ary PCM system operating above the error threshold. Hence, solving this equation for the number of discrete amplitude levels, we may express the number M in terms of the average transmitted power P and channel noise variance $\sigma^2 = N_0 B$ as follows:

$$M = \left(1 + \frac{12P}{c^2 N_0 B}\right)^{1/2} \tag{6.56}$$

Therefore, substituting (6.56) into (6.55), we obtain

$$R_b = Wn \log_2\left(1 + \frac{12P}{c^2 N_0 B}\right) \tag{6.57}$$

The channel bandwidth B required to transmit a rectangular pulse of duration $1/(2nW)$, representing a symbol in the codeword, is given by

$$B = \kappa n W \tag{6.58}$$

where κ is a constant with a value lying between 1 and 2. Using the minimum possible value $\kappa = 1$, we find that the channel bandwidth $B = nW$. We may thus rewrite (6.57) as

$$R_b = B \log_2\left(1 + \frac{12P}{c^2 N_0 B}\right) \text{ bits/s} \tag{6.59}$$

which defines the upper bound on the information capacity realizable by an M-ary PCM system.

From Chapter 5 we recall that, in accordance with Shannon's information capacity law, the *ideal transmission system* is described by the formula

$$C = B \log_2\left(1 + \frac{P}{N_0 B}\right) \text{ bits/s} \tag{6.60}$$

The most interesting point derived from the comparison of (6.59) with (6.60) is the fact that (6.59) is of the right mathematical form in an information-theoretic context. To be more specific, we make the following statement:

> Power and bandwidth in a PCM system are exchanged on a logarithmic basis, and the information capacity of the system is proportional to the channel bandwidth B.

As a corollary, we may go on to state:

> When the SNR ratio is high, the bandwidth-noise trade-off follows an exponential law in PCM.

From the study of noise in analog modulation systems,[5] it is known that the use of frequency modulation provides the best improvement in SNR ratio. To be specific, when the carrier-to-noise ratio is high enough, the bandwidth-noise trade-off follows a *square law* in frequency modulation (FM). Accordingly, in comparing the noise performance of FM with that of PCM we make the concluding statement:

> PCM is more efficient than FM in trading off an increase in bandwidth for improved noise performance.

Indeed, this statement is further testimony for the PCM being viewed as a standard for waveform coding.

6.7 Prediction-Error Filtering for Redundancy Reduction

When a voice or video signal is sampled at a rate slightly higher than the Nyquist rate, as usually done in PCM, the resulting sampled signal is found to exhibit a high degree of *correlation* between adjacent samples. The meaning of this high correlation is that, in an average sense, the signal does not change rapidly from one sample to the next. As a result, the difference between adjacent samples has a variance that is smaller than the variance of the original signal. When these highly correlated samples are encoded, as in the standard PCM system, the resulting encoded signal contains *redundant information*. This kind of signal structure means that symbols that are not absolutely essential to the transmission of information are generated as a result of the conventional encoding process described in Section 6.5. By reducing this redundancy before encoding, we obtain a *more efficient* coded signal, which is the basic idea behind DPCM. Discussion of this latter form of waveform coding is deferred to the next section. In this section we discuss prediction-error filtering, which provides a method for reduction and, therefore, improved waveform coding.

Theoretical Considerations

To elaborate, consider the block diagram of Figure 6.16a, which includes:

- a direct forward path from the input to the output;
- a predictor in the forward direction as well; and
- a comparator for computing the difference between the input signal and the predictor output.

The difference signal, so computed, is called the *prediction error.* Correspondingly, a filter that operates on the message signal to produce the prediction error, illustrated in Figure 6.16a, is called a *prediction-error filter.*

To simplify the presentation, let

$$m_n = m(nT_s) \tag{6.61}$$

denote a sample of the message signal $m(t)$ taken at time $t = nT_s$. Then, with $\hat{m}_n$ denoting the corresponding predictor output, the prediction error is defined by

$$e_n = m_n - \hat{m}_n \tag{6.62}$$

where e_n is the amount by which the predictor fails to predict the input sample m_n exactly. In any case, the objective is to design the predictor so as to *minimize the variance* of the prediction error e_n. In so doing, we effectively end up using a smaller number of bits to represent e_n than the original message sample m_n; hence, the need for a smaller transmission bandwidth.

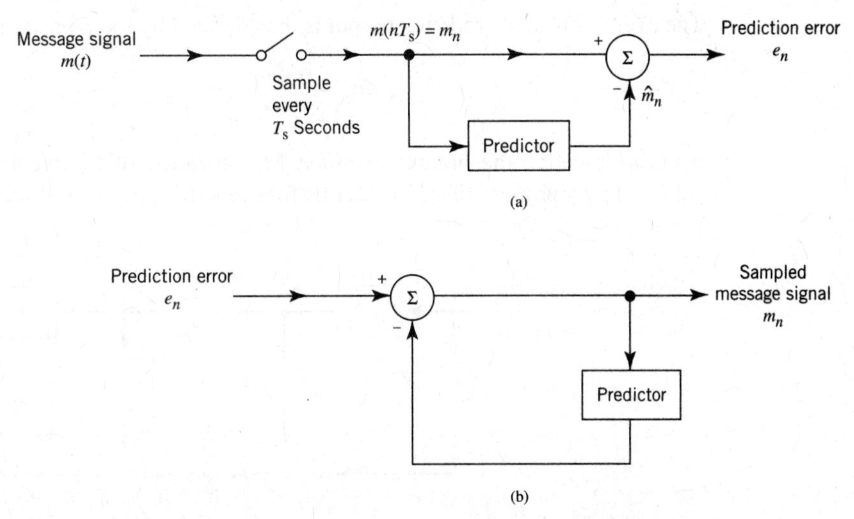

Figure 6.16 Block diagram of (a) prediction-error filter and (b) its inverse.

The prediction-error filter operates on the message signal on a sample-by-sample basis to produce the prediction error. With such an operation performed in the transmitter, how do we recover the original message signal from the prediction error at the receiver? To address this fundamental question in a simple-minded and yet practical way, we invoke the use of *linerarity.* Let the *operator* **L** denote the action of the predictor, as shown by

$$\hat{m}_n = \mathbf{L}[m_n] \tag{6.63}$$

Accordingly, we may rewrite (6.62) in operator form as follows:

$$\begin{aligned} e_n &= m_n - \mathbf{L}[m_n] \\ &= (1 - \mathbf{L})[m_n] \end{aligned} \tag{6.64}$$

Under the assumption of linearity, we may invert (6.64) to recover the message sample from the prediction error, as shown by

$$m_n = \left(\frac{1}{1 - \mathbf{L}}\right)[e_n] \tag{6.65}$$

Equation (6.65) is immediately recognized as the equation of a *feedback system*, as illustrated in Figure 6.16b. Most importantly, in functional terms, this feedback system may be viewed as the *inverse of prediction-error filtering*.

Discrete-Time Structure for Prediction

To simplify the design of the linear predictor in Figure 6.16, we propose to use a discrete-time structure in the form of a *finite-duration impulse response (FIR) filter*, which is well known in the digital signal-processing literature. The FIR filter was briefly discussed in Chapter 2.

Figure 6.17 depicts an FIR filter, consisting of two functional components:

- a set of *p unit-delay elements*, each of which is represented by z^{-1}; and
- a corresponding set of *adders* used to sum the scaled versions of the delayed inputs,

$$m_{n-1}, m_{n-2}, \ldots, m_{n-p}.$$

The overall linearly predicted output is thus defined by the *convolution sum*

$$\hat{m}_n = \sum_{k=1}^{p} w_k m_{n-k} \tag{6.66}$$

where p is called the *prediction order.* Minimization of the prediction-error variance is achieved by a proper choice of the FIR filter-coefficients as described next.

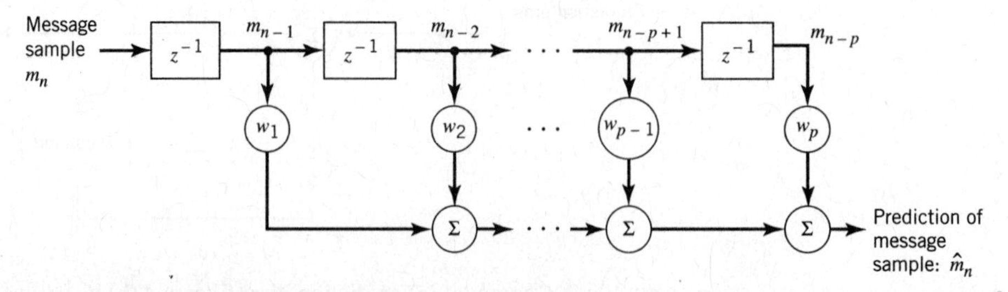

Figure 6.17 Block diagram of an FIR filter of order p.

First, however, we make the following assumption:

> The message signal $m(t)$ is drawn from a stationary stochastic processor $M(t)$ with zero mean.

This assumption may be satisfied by processing the message signal on a block-by-block basis, with each block being just long enough to satisfy the assumption in a *pseudo-stationary manner.* For example, a block duration of 40 ms is considered to be adequate for voice signals.

With the random variable M_n assumed to have zero mean, it follows that the variance of the prediction error e_n is the same as its mean-square value. We may thus define

$$J = \mathbb{E}[e^2(n)] \tag{6.67}$$

as the *index of performance*. Substituting (6.65) and (6.66) into (6.67) and then expanding terms, the index of performance is expressed as follows:

$$J(\mathbf{w}) = \mathbb{E}[m_n^2] - 2\sum_{k=1}^{p} w_k \mathbb{E}[m_n m_{n-k}] + \sum_{j=1}^{p}\sum_{k=1}^{p} w_j w_k \mathbb{E}[m_{n-j} m_{n-k}] \tag{6.68}$$

Moreover, under the above assumption of pseudo-stationarity, we may go on to introduce the following second-order statistical parameters for m_n treated as a sample of the stochastic process $M(t)$ at $t = nT_s$:

1. *Variance*

$$\begin{aligned} \sigma_M^2 &= \mathbb{E}[(m_n - \mathbb{E}[m_n])^2] \\ &= \mathbb{E}[m_n^2] \text{ for } \mathbb{E}[m_n] = 0 \end{aligned} \tag{6.69}$$

2. *Autocorrelation function*

$$R_{M,k-j} = \mathbb{E}[m_{n-j} m_{n-k}] \tag{6.70}$$

Note that to simplify the notation in (6.67) to (6.70), we have applied the expectation operator $\mathbb{E}$ to samples rather than the corresponding random variables.

In any event, using (6.69) and (6.70), we may reformulate the index of performance of (6.68) in the new form involving statistical parameters:

$$J(\mathbf{w}) = \sigma_M^2 - 2\sum_{k=1}^{p} w_k R_{M,k} + \sum_{j=1}^{p}\sum_{k=1}^{p} w_j w_k R_{M,k-j} \tag{6.71}$$

Differentiating this index of performance with respect to the filter coefficients, setting the resulting expression equal to zero, and then rearranging terms, we obtain the following system of simultaneous equations:

$$\sum_{j=1}^{p} w_{o,j} R_{M,k-j} = R_{M,k}, \qquad k = 1, 2, \ldots, p \tag{6.72}$$

where $w_{o,j}$ is the optimal value of the jth filter coefficient w_j. This optimal set of equations is the discrete-time version of the celebrated *Wiener–Hopf equations* for linear prediction.

With compactness of mathematical exposition in mind, we find it convenient to formulate the Wiener–Hopf equations in matrix form, as shown by

$$\mathbf{R}_M \mathbf{w}_o = \mathbf{r}_M \tag{6.73}$$

where

$$\mathbf{w}_o = [w_{o,1}, w_{o,2}, w_{o,p}]^T \tag{6.74}$$

is the p-by-1 *optimum coefficient vector* of the FIR predictor,

$$\mathbf{r}_M = [R_{M,1}, R_{M,2}, \ldots, R_{M,p}]^T \tag{6.75}$$

is the p-by-1 *autocorrelation vector* of the original message signal, excluding the mean-square value represented by $R_{M,0}$, and

$$\mathbf{R}_M = \begin{bmatrix} R_{M,0} & R_{M,1} & \cdots & R_{M,p-1} \\ R_{M,1} & R_{M,0} & \cdots & R_{M,p-2} \\ \cdots & \cdots & \cdots & \cdots \\ R_{M,p-1} & R_{M,p-2} & \cdots & R_{M,0} \end{bmatrix} \tag{6.76}$$

is the p-by-y *correlation matrix* of the original message signal, including $R_{M,0}$.[6]

Careful examination of (6.76) reveals the *Toeplitz property* of the autocorrelation matrix $\mathbf{R}_M$, which embodies two distinctive characteristics:

1. All the elements on the main diagonal of the matrix $\mathbf{R}_M$ are equal to the mean-square value or, equivalently under the zero-mean assumption, the variance of the message sample m_n, as shown by

$$R_M(0) = \sigma_M^2$$

2. The matrix is *symmetric* about the main diagonal.

This Toeplitz property is a direct consequence of the assumption that message signal $m(t)$ is the sample function of a stationary stochastic process. From a practical perspective, the Toeplitz property of the autocorrelation matrix $\mathbf{R}_M$ is important in that all of its elements are uniquely defined by the *autocorrelation sequence* $\{R_{M,k}\}_{k=0}^{p-1}$. Moreover, from the defining equation (6.75), it is clear that the autocorrelation vector $\mathbf{r}_M$ is uniquely defined by the autocorrelation sequence $\{R_{M,k}\}_{k=1}^{p}$. We may therefore make the following statement:

> The p filter coefficients of the optimized linear predictor, configured in the form of an FIR filter, are uniquely defined by the variance $\sigma_M^2 = R_M(0)$ and the autocorrelation sequence $\{R_{M,k}\}_{k=0}^{p-1}$, which pertain to the message signal $m(t)$ drawn from a weakly stationary process.

Typically, we have

$$|R_{M,k}| < R_M(0) \qquad \text{for } k = 1, 2, \ldots, p$$

Under this condition, we find that the autocorrelation matrix $\mathbf{R}_M$ is also invertible; that is, the inverse matrix $\mathbf{R}_M^{-1}$ exists. We may therefore solve (6.73) for the unknown value of the optimal coefficient vector $\mathbf{w}_o$ using the formula[7]

$$\mathbf{w}_o = \mathbf{R}_M^{-1}\mathbf{r}_M \tag{6.77}$$

Thus, given the variance σ_M^2 and autocorrelation sequence $\{R_{M,k}\}_{k=1}^{p}$, we may uniquely determine the optimized coefficient vector of the linear predictor, $\mathbf{w}_o$, defining an FIR filter of order p; and with it our design objective is satisfied.

To complete the linear prediction theory presented herein, we need to find the minimum mean-square value of prediction error, resulting from the use of the optimized predictor. We do this by first reformulating (6.71) in the matrix form:

$$J(\mathbf{w}_o) = \sigma_M^2 - 2\mathbf{w}_o^T\mathbf{r}_M + \mathbf{w}_o^T\mathbf{R}_M\mathbf{w}_o \tag{6.78}$$

where the superscript T denotes *matrix transposition*, $\mathbf{w}_o^T\mathbf{r}_M$ is the *inner product* of the p-by-1 vectors $\mathbf{w}_o$ and $\mathbf{r}_M$, and the matrix product $\mathbf{w}_o^T\mathbf{R}_M\mathbf{w}_o$ is a *quadratic form*. Then, substituting the optimum formula of (6.77) into (6.78), we find that the *minimum mean-square value of prediction error* is given by

$$\begin{aligned} J_{\min} &= \sigma_M^2 - 2(\mathbf{R}_M^{-1}\mathbf{r}_M)^T\mathbf{r}_M + (\mathbf{R}_M^{-1}\mathbf{r}_M)^T\mathbf{R}_M(\mathbf{R}_M^{-1}r_M) \\ &= \sigma_M^2 - 2\mathbf{r}_M^T\mathbf{R}_M^{-1}\mathbf{r}_M + \mathbf{r}_M^T\mathbf{R}_M^{-1}r_M \\ &= \sigma_M^2 - \mathbf{r}_M^T\mathbf{R}_M^{-1}\mathbf{r}_M \end{aligned} \tag{6.79}$$

where we have used the property that the autocorrelation matrix of a weakly stationary process is *symmetric*; that is,

$$\mathbf{R}_M^T = \mathbf{R}_M \tag{6.80}$$

By definition, the quadratic form $\mathbf{r}_M^T\mathbf{R}_M^{-1}\mathbf{r}_M$ is always positive. Accordingly, from (6.79) it follows that the minimum value of the mean-square prediction error $J_{\min}$ is always smaller than the variance σ_M^2 of the zero-mean message sample m_n that is being predicted. Through the use of linear prediction as described herein, we have thus satisfied the objective:

> To design a prediction-error filter the output of which has a smaller variance than the variance of the message sample applied to its input, we need to follow the optimum formula of (6.77).

This statement provides the rationale for going on to describe how the bandwidth requirement of the standard PCM can be reduced through redundancy reduction. However, before proceeding to do so, it is instructive that we consider an adaptive implementation of the linear predictor.

Linear Adaptive Prediction

The use of (6.77) for calculating the optimum weight vector of a linear predictor requires knowledge of the autocorrelation function $R_{m,k}$ of the message signal sequence $\{m_k\}_{k=0}^{p}$ where p is the prediction order. What if knowledge of this sequence is not available? In situations of this kind, which occur frequently in practice, we may resort to the use of an *adaptive predictor.*

The predictor is said to be adaptive in the following sense:

- Computation of the tap weights w_k, $k = 1, 2, \ldots, p$, proceeds in an iterative manner, starting from some arbitrary initial values of the tap weights.
- The algorithm used to adjust the tap weights (from one iteration to the next) is "self-designed," operating solely on the basis of available data.

The aim of the algorithm is to find the minimum point of the *bowl-shaped error surface* that describes the dependence of the cost function J on the tap weights. It is, therefore, intuitively reasonable that successive adjustments to the tap weights of the predictor be made in the direction of the steepest descent of the error surface; that is, in a direction opposite to the *gradient vector* whose elements are defined by

$$g_k = \frac{\partial J}{\partial w_k}, \qquad k = 1, 2, \ldots, p \tag{6.81}$$

This is indeed the idea behind the *method of deepest descent.* Let $w_{k,n}$ denote the value of the kth tap weight at iteration n. Then, the updated value of this weight at iteration $n + 1$ is defined by

$$w_{k,n+1} = w_{k,n} - \frac{1}{2}\mu g_k, \qquad k = 1, 2, \ldots, p \tag{6.82}$$

where μ is a *step-size parameter* that controls the speed of adaptation and the factor 1/2 is included for convenience of presentation. Differentiating the cost function J of (6.68) with respect to w_k, we readily find that

$$g_k = -2\mathbb{E}[m_n m_{n-k}] + \sum_{j=1}^{p} w_j \mathbb{E}[m_{n-j} m_{n-k}] \tag{6.83}$$

From a practical perspective, the formula for the gradient g_k in (6.83) could do with further simplification that ignores the expectation operator. In effect, *instantaneous values are used as estimates of autocorrelation functions*. The motivation for this simplification is to permit the adaptive process to proceed forward on a step-by-step basis in a self-organized manner. Clearly, by ignoring the expectation operator in (6.83), the gradient g_k takes on a time-dependent value, denoted by $g_{k,n}$. We may thus write

$$g_{k,n} = -2m_n m_{n-k} + 2m_{n-k}\sum_{j=1}^{p} \hat{w}_{j,n} m_{n,j}, \qquad k = 1, 2, \ldots, p \tag{6.84}$$

where $\hat{w}_{j,n}$ is an estimate of the filter coefficient $w_{j,n}$ at time n.

The stage is now set for substituting (6.84) into (6.82), where in the latter equation $\hat{w}_{k,n}$ is substituted for $w_{k,n}$; this change is made to account for dispensing with the expectation operator:

$$
\begin{aligned}
\hat{w}_{k,\,n+1} &= \hat{w}_{k,\,n} - \frac{1}{2}\mu g_{k,\,n} \\
&= \hat{w}_{k,\,n} + \mu\left(m_n m_{n-k} - \sum_{j=1}^{p} \hat{w}_{j,\,n} m_{n-j} m_{n-k}\right) \\
&= \hat{w}_{k,\,n} + \mu m_{n-k}\left(m_n - \sum_{j=1}^{p} \hat{w}_{j,\,n} m_{n-j}\right) \\
&= \hat{w}_{k,\,n} + \mu m_{n-k} e_n
\end{aligned}
\tag{6.85}
$$

where e_n is the *new prediction error* defined by

$$
e_n = m_n - \sum_{j=1}^{p} \hat{w}_{j,\,n} m_{n-j}
\tag{6.86}
$$

Note that the current value of the message signal, m_n, plays a role as the *desired response for predicting* the value of m_n given the past values of the message signal: $m_{n-1}, m_{n-2}, \ldots, m_{n-p}$.

In words, we may express the adaptive filtering algorithm of (6.85) as follows:

$$
\begin{pmatrix}\text{Updated value of the } k\text{th} \\ \text{filter coefficient at time } n+1\end{pmatrix} = \begin{pmatrix}\text{Old value of the same} \\ \text{filter coefficient at time } n\end{pmatrix} + \begin{pmatrix}\text{Step-size} \\ \text{parameter}\end{pmatrix} \times \begin{pmatrix}\text{Message signal } m_n \\ \text{delayed by } k \text{ time steps}\end{pmatrix}\begin{pmatrix}\text{Prediction error} \\ \text{computed at time } n\end{pmatrix}
$$

The algorithm just described is the popular *least-mean-square (LMS) algorithm*, formulated for the purpose of linear prediction. The reason for popularity of this adaptive filtering algorithm is the simplicity of its implementation. In particular, the computational complexity of the algorithm, measured in terms of the number of additions and multiplications, is *linear* in the prediction order p. Moreover, the algorithm is not only *computationally efficient* but it is also *effective in performance*.

The LMS algorithm is a *stochastic* adaptive filtering algorithm, stochastic in the sense that, starting from the *initial condition* defined by $\{w_{k,\,0}\}_{k=1}^{p}$, it seeks to find the minimum point of the error surface by following a zig-zag path. However, it never finds this minimum point exactly. Rather, it continues to execute a random motion around the minimum point of the error surface (Haykin, 2013).

6.8 Differential Pulse-Code Modulation

DPCM, the scheme to be considered for *channel-bandwidth conservation*, exploits the idea of linear prediction theory with a practical difference:

> In the transmitter, the linear prediction is performed on a quantized version of the message sample instead of the message sample itself, as illustrated in Figure 6.18.

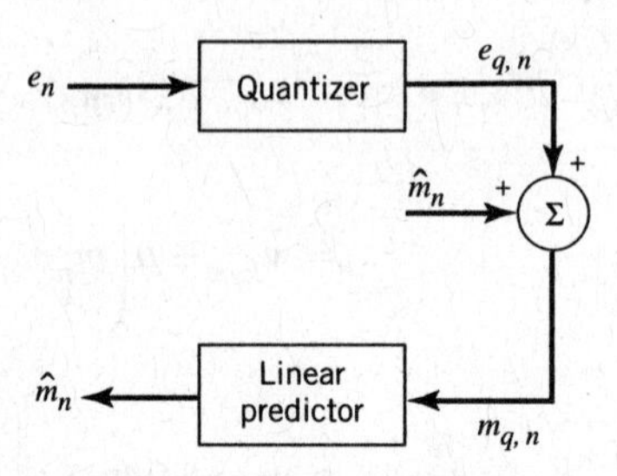

Figure 6.18 Block diagram of a differential quantizer.

The resulting process is referred to as *differential quantization*. The motivation behind the use of differential quantization follows from two practical considerations:

1. Waveform encoding in the transmitter requires the use of quantization.
2. Waveform decoding in the receiver, therefore, has to process a quantized signal.

In order to cater to both requirements in such a way that the *same structure* is used for predictors in both the transmitter and the receiver, the transmitter has to perform prediction-error filtering on the quantized version of the message signal rather than the signal itself, as shown in Figure 6.19a. Then, assuming a noise-free channel, the predictors in the transmitter and receiver operate on exactly the same sequence of quantized message samples.

To demonstrate this highly desirable and distinctive characteristic of differential PCM, we see from Figure 6.19a that

$$e_{q,n} = e_n + q_n \tag{6.87}$$

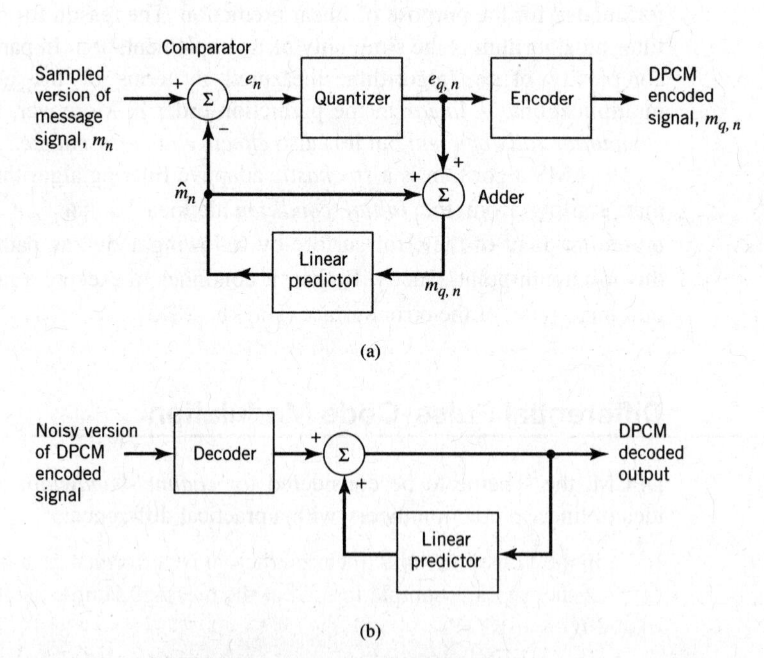

Figure 6.19 DPCM system: (a) transmitter; (b) receiver.

where q_n is the quantization noise produced by the quantizer operating on the prediction error e_n. Moreover, from Figure 6.19a, we readily see that

$$m_{q,n} = \hat{m}_n + e_{q,n} \tag{6.88}$$

where $\hat{m}_n$ is the predicted value of the original message sample m_n; thus, (6.88) is in perfect agreement with Figure 6.18. Hence, the use of (6.87) in (6.88) yields

$$m_{q,n} = \hat{m}_n + e_n + q_n \tag{6.89}$$

We may now invoke (6.88) of linear prediction theory to rewrite (6.89) in the equivalent form:

$$m_{q,n} = m_n + q_n \tag{6.90}$$

which describes a quantized version of the original message sample m_n.

With the differential quantization scheme of Figure 6.19a at hand, we may now expand on the structures of the transmitter and receiver of DPCM.

DPCM Transmitter

Operation of the DPCM transmitter proceeds as follows:

1. Given the predicted message sample $\hat{m}_n$, the comparator at the transmitter input computes the prediction error e_n, which is quantized to produce the quantized version of e_n in accordance with (6.87).
2. With $\hat{m}_n$ and $e_{q,n}$ at hand, the adder in the transmitter produces the quantized version of the original message sample m_n, namely $m_{q,n}$, in accordance with (6.88).
3. The required one-step prediction $\hat{m}_n$ is produced by applying the sequence of quantized samples $\{m_{q,k}\}_{k=1}^{p}$ to a linear FIR predictor of order p.

This multistage operation is clearly *cyclic*, encompassing three steps that are repeated at each time step n. Moreover, at each time step, the encoder operates on the quantized prediction error $e_{q,n}$ to produce the DPCM-encoded version of the original message sample m_n. The DPCM code so produced is a *lossy-compressed* version of the PCM code; it is "lossy" because of the prediction error.

DPCM Receiver

The structure of the receiver is much simpler than that of the transmitter, as depicted in Figure 6.19b. Specifically, first, the decoder reconstructs the quantized version of the prediction error, namely $e_{q,n}$. An estimate of the original message sample m_n is then computed by applying the decoder output to the same predictor used in the transmitter of Figure 6.19a. In the absence of channel noise, the encoded signal at the receiver input is identical to the encoded signal at the transmitter output. Under this ideal condition, we find that the corresponding receiver output is equal to $m_{q,n}$, which differs from the original signal sample m_n only by the quantization error q_n incurred as a result of quantizing the prediction error e_n.

From the foregoing analysis, we thus observe that, in a noise-free environment, the linear predictors in the transmitter and receiver of DPCM operate on the same sequence of samples, $m_{q,n}$. It is with this point in mind that a feedback path is appended to the quantizer in the transmitter of Figure 6.19a.

Processing Gain

The output SNR of the DPCM system, shown in Figure 6.19, is, by definition,

$$(\mathrm{SNR})_\mathrm{O} = \frac{\sigma_\mathrm{M}^2}{\sigma_\mathrm{Q}^2} \tag{6.91}$$

where σ_M^2 is the variance of the original signal sample m_n, assumed to be of zero mean, and σ_Q^2 is the variance of the quantization error q_n, also of zero mean. We may rewrite (6.91) as the product of two factors, as shown by

$$\begin{aligned}(\mathrm{SNR})_\mathrm{O} &= \left(\frac{\sigma_\mathrm{M}^2}{\sigma_\mathrm{E}^2}\right)\left(\frac{\sigma_\mathrm{E}^2}{\sigma_\mathrm{Q}^2}\right) \\ &= G_\mathrm{p}(\mathrm{SNR})_\mathrm{Q}\end{aligned} \tag{6.92}$$

where, in the first line, σ_E^2 is the variance of the prediction error e_n. The factor $(\mathrm{SNR})_\mathrm{Q}$ introduced in the second line is the *signal-to-quantization noise ratio*, which is itself defined by

$$(\mathrm{SNR})_\mathrm{Q} = \frac{\sigma_\mathrm{E}^2}{\sigma_\mathrm{Q}^2} \tag{6.93}$$

The other factor G_p is the *processing gain* produced by the differential quantization scheme; it is formally defined by

$$G_\mathrm{p} = \frac{\sigma_\mathrm{M}^2}{\sigma_\mathrm{E}^2} \tag{6.94}$$

The quantity G_p, when it is greater than unity, represents a *gain in signal-to-noise ratio*, which is due to the differential quantization scheme of Figure 6.19. Now, for a given message signal, the variance σ_M^2 is fixed, so that G_p is maximized by minimizing the variance σ_M^2 of the prediction error e_n. Accordingly, the objective in implementing the DPCM should be to design the prediction filter so as to minimize the prediction-error variance, σ_E^2.

In the case of voice signals, it is found that the optimum signal-to-quantization noise advantage of the DPCM over the standard PCM is in the neighborhood of 4–11 dB. Based on experimental studies, it appears that the greatest improvement occurs in going from no prediction to first-order prediction, with some additional gain resulting from increasing the order p of the prediction filter up to 4 or 5, after which little additional gain is obtained. Since 6 dB of quantization noise is equivalent to 1 bit per sample by virtue of the results presented in Table 6.1 for sinusoidal modulation, the advantage of DPCM may also be expressed in terms of bit rate. For a constant signal-to-quantization noise ratio, and assuming a sampling rate of 8 kHz, the use of DPCM may provide a saving of about 8–16 kHz (i.e., 1 to 2 bits per sample) compared with the standard PCM.

6.9 Delta Modulation

In choosing DPCM for waveform coding, we are, in effect, economizing on transmission bandwidth by increasing system complexity, compared with standard PCM. In other words, DPCM exploits the *complexity–bandwidth tradeoff*. However, in practice, the need may arise for reduced system complexity compared with the standard PCM. To achieve this other objective, transmission bandwidth is traded off for reduced system complexity, which is precisely the motivation behind DM. Thus, whereas DPCM exploits the *complexity–bandwidth tradeoff*, DM exploits the *bandwidth–complexity tradeoff*. We may, therefore, differentiate between the standard PCM, the DPCM, and the DM along the lines described in Figure 6.20. With the bandwidth–complexity tradeoff being at the heart of DM, the incoming message signal $m(t)$ is *oversampled*, which requires the use of a sampling rate higher than the Nyquist rate. Accordingly, the correlation between adjacent samples of the message signal is purposely increased so as to permit the use of a *simple* quantizing strategy for constructing the encoded signal.

DM Transmitter

In the DM transmitter, system complexity is reduced to the minimum possible by using the combination of two strategies:

1. *Single-bit quantizer*, which is the simplest quantizing strategy; as depicted in Figure 6.21, the quantizer acts as a hard limiter with only two decision levels, namely, $\pm\Delta$.
2. *Single unit-delay element*, which is the most primitive form of a predictor; in other words, the only component retained in the FIR predictor of Figure 6.17 is the front-end block labeled z^{-1}, which acts as an *accumulator*.

Thus, replacing the multilevel quantizer and the FIR predictor in the DPCM transmitter of Figure 6.19a in the manner described under points 1 and 2, respectively, we obtain the block diagram of Figure 6.21a for the DM transmitter.

From this figure, we may express the equations underlying the operation of the DM transmitter by the following set of equations (6.95)–(6.97):

$$\begin{aligned} e_n &= m_n - \hat{m}_n \\ &= m_n - m_{q,\,n-1} \end{aligned} \tag{6.95}$$

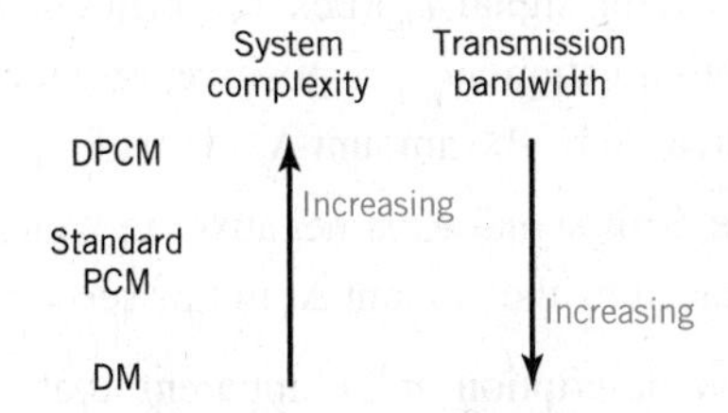

Figure 6.20 Illustrating the tradeoffs between standard PCM, DPCM, and DM.

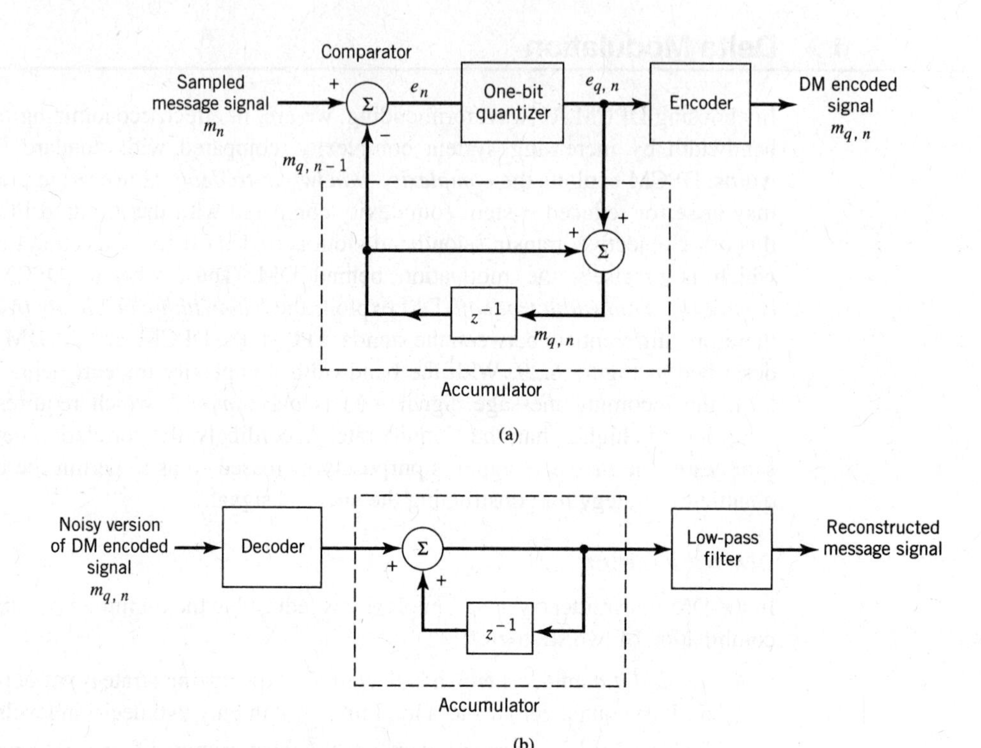

Figure 6.21 DM system: (a) transmitter; (b) receiver.

$$e_{q,n} = \Delta \operatorname{sgn}[e_n]$$

$$= \begin{cases} +\Delta \text{ if } e_n > 0 \\ -\Delta \text{ if } e_n < 0 \end{cases} \tag{6.96}$$

$$m_{q,n} = m_{q,n-1} + e_{q,n} \tag{6.97}$$

According to (6.95) and (6.96), two possibilities may naturally occur:

1. The error signal e_n (i.e., the difference between the message sample m_n and its approximation $\hat{m}_n$) is positive, in which case the approximation $\hat{m}_n = m_{q,n-1}$ is increased by the amount Δ; in this first case, the encoder sends out symbol 1.
2. The error signal e_n is negative, in which case the approximation $\hat{m}_n = m_{q,n-1}$ is reduced by the amount Δ; in this second case, the encoder sends out symbol 0.

From this description it is apparent that the delta modulator produces a staircase approximation to the message signal, as illustrated in Figure 6.22a. Moreover, the rate of data transmission in DM is equal to the sampling rate $f_s = 1/T_s$, as illustrated in the binary sequence of Figure 6.22b.

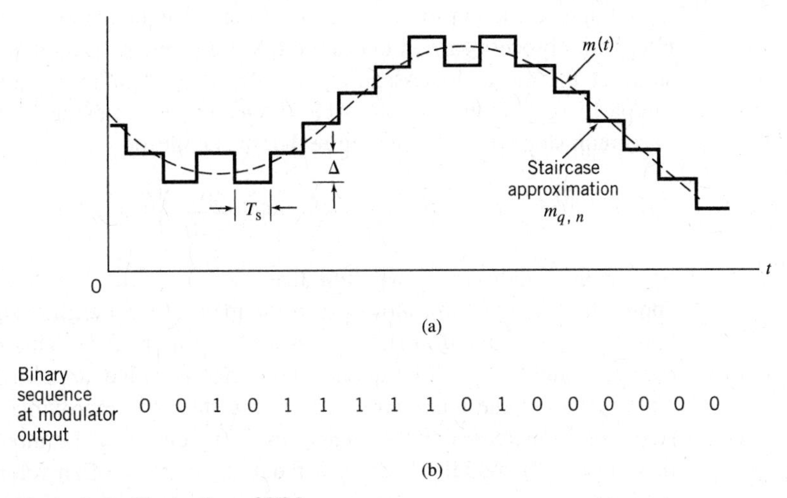

(a)

(b)

Figure 6.22 Illustration of DM.

DM Receiver

Following a procedure similar to the way in which we constructed the DM transmitter of Figure 6.21a, we may construct the DM receiver of Figure 6.21b as a special case of the DPCM receiver of Figure 6.19b. Working through the operation of the DM receiver, we find that reconstruction of the staircase approximation to the original message signal is achieved by passing the sequence of positive and negative pulses (representing symbols 1 and 0, respectively) through the block labeled "accumulator."

Under the assumption that the channel is distortionless, the accumulated output is the desired $m_{q,n}$ given that the decoded channel output is $e_{q,n}$. The out-of-band quantization noise in the high-frequency staircase waveform in the accumulator output is suppressed by passing it through a low-pass filter with a cutoff frequency equal to the message bandwidth.

Quantization Errors in DM

DM is subject to two types of quantization error: slope overload distortion and granular noise. We will discuss the case of slope overload distortion first.

Starting with (6.97), we observe that this equation is the *digital equivalent of integration*, in the sense that it represents the *accumulation* of positive and negative increments of magnitude Δ. Moreover, denoting the quantization error applied to the message sample m_n by q_n, we may express the quantized message sample as

$$m_{q,n} = m_n + q_n \tag{6.98}$$

With this expression for $m_{q,n}$ at hand, we find from (6.98) that the quantizer input is

$$e_n = m_n - (m_{n-1} + q_{n-1}) \tag{6.99}$$

Thus, except for the delayed quantization error q_{n-1}, the quantizer input is a *first backward difference* of the original message sample. This difference may be viewed as a digital approximation to the quantizer input or, equivalently, as the *inverse* of the digital integration process carried out in the DM transmitter. If, then, we consider the maximum slope of the original message signal $m(t)$, it is clear that in order for the sequence of samples $\{m_{q,n}\}$ to increase as fast as the sequence of message samples $\{m_n\}$ in a region of maximum slope of $m(t)$, we require that the condition

$$\frac{\Delta}{T_s} \geq \max\left|\frac{dm(t)}{dt}\right| \tag{6.100}$$

be satisfied. Otherwise, we find that the step-size Δ is too small for the staircase approximation $m_q(t)$ to follow a steep segment of the message signal $m(t)$, with the result that $m_q(t)$ falls behind $m(t)$, as illustrated in Figure 6.23. This condition is called *slope overload*, and the resulting quantization error is called *slope-overload distortion* (*noise*). Note that since the maximum slope of the staircase approximation $m_q(t)$ is fixed by the step size Δ, increases and decreases in $m_q(t)$ tend to occur along straight lines. For this reason, a delta modulator using a fixed step size is often referred to as a *linear delta modulator*.

In contrast to slope-overload distortion, *granular noise* occurs when the step size Δ is too large relative to the local slope characteristics of the message signal $m(t)$, thereby causing the staircase approximation $m_q(t)$ to hunt around a relatively flat segment of $m(t)$; this phenomenon is also illustrated in the tail end of Figure 6.23. Granular noise is analogous to quantization noise in a PCM system.

Adaptive DM

From the discussion just presented, it is appropriate that we need to have a large step size to accommodate a wide dynamic range, whereas a small step size is required for the accurate representation of relatively low-level signals. It is clear, therefore, that the choice of the optimum step size that minimizes the mean-square value of the quantization error in a linear delta modulator will be the result of a compromise between slope-overload distortion and granular noise. To satisfy such a requirement, we need to make the delta modulator "adaptive," in the sense that the step size is made to vary in accordance with the input signal. The step size is thereby made variable, such that it is enlarged during intervals when the slope-overload distortion is dominant and reduced in value when the granular (quantization) noise is dominant.

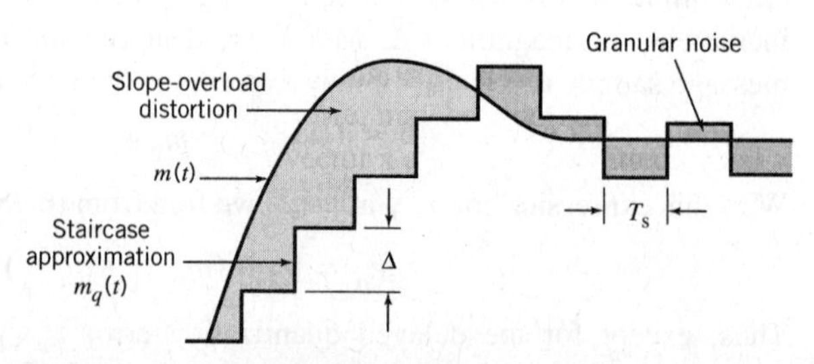

Figure 6.23 Illustration of the two different forms of quantization error in DM.

6.10 Line Codes

In this chapter, we have described three basic waveform-coding schemes: PCM, DPCM, and DM. Naturally, they differ from each other in several ways: transmission–bandwidth requirement, transmitter–receiver structural composition and complexity, and quantization noise. Nevertheless, all three of them have a common need: *line codes* for electrical representation of the encoded binary streams produced by their individual transmitters, so as to facilitate transmission of the binary streams across the communication channel.

Figure 6.24 displays the waveforms of five important line codes for the example data stream 01101001. Figure 6.25 displays their individual power spectra (for positive frequencies) for randomly generated binary data, assuming that first, symbols 0 and 1 are equiprobable, second, the average power is normalized to unity, and third, the frequency f is normalized with respect to the bit rate $1/T_b$. In what follows, we describe the five line codes involved in generating the coded waveforms of Figure 6.24.

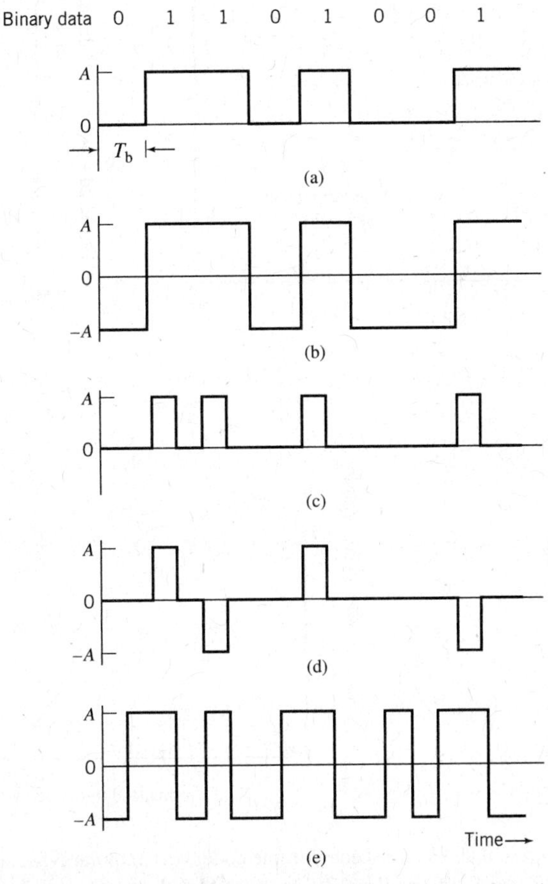

Figure 6.24 Line codes for the electrical representations of binary data: (a) unipolar nonreturn-to-zero (NRZ) signaling; (b) polar NRZ signaling; (c) unipolar return-to-zero (RZ) signaling; (d) bipolar RZ signaling; (e) split-phase or Manchester code.

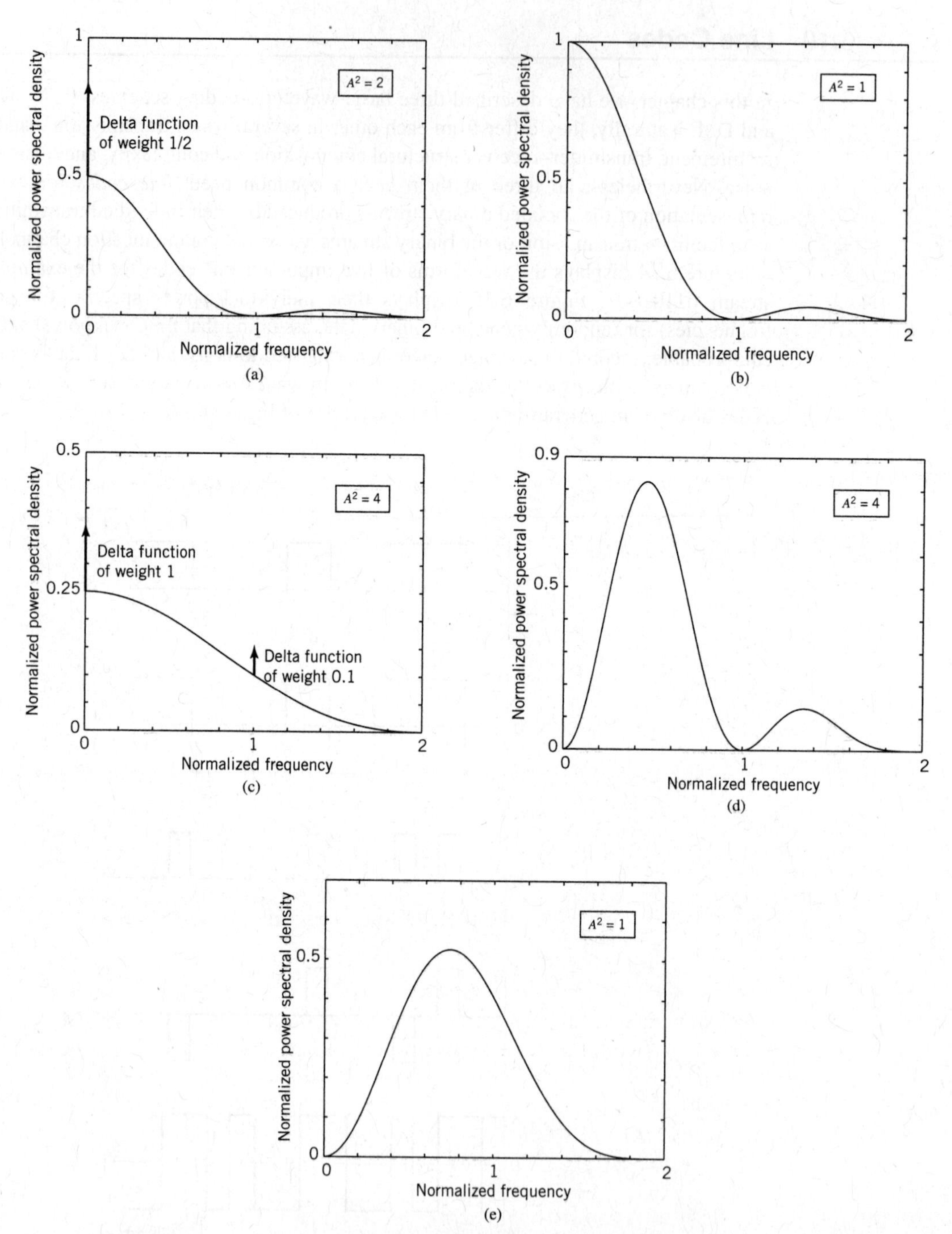

Figure 6.25 Power spectra of line codes: (a) unipolar NRZ signal; (b) polar NRZ signal; (c) unipolar RZ signal; (d) bipolar RZ signal; (e) Manchester-encoded signal. The frequency is normalized with respect to the bit rate $1/T_b$, and the average power is normalized to unity.

Unipolar NRZ Signaling

In this line code, symbol 1 is represented by transmitting a pulse of amplitude A for the duration of the symbol, and symbol 0 is represented by switching off the pulse, as in Figure 6.24a. The unipolar NRZ line code is also referred to as *on–off signaling*. Disadvantages of on–off signaling are the waste of power due to the transmitted DC level and the fact that the power spectrum of the transmitted signal does not approach zero at zero frequency.

Polar NRZ Signaling

In this second line code, symbols 1 and 0 are represented by transmitting pulses of amplitudes $+A$ and $-A$, respectively, as illustrated in Figure 6.24b. The polar NRZ line code is relatively easy to generate, but its disadvantage is that the power spectrum of the signal is large near zero frequency.

Unipolar RZ Signaling

In this third line code, symbol 1 is represented by a rectangular pulse of amplitude A and half-symbol width and symbol 0 is represented by transmitting *no* pulse, as illustrated in Figure 6.24c. An attractive feature of the unipolar RZ line code is the presence of delta functions at $f = 0, \pm 1/T_b$ in the power spectrum of the transmitted signal; the delta functions can be used for *bit-timing recovery* at the receiver. However, its disadvantage is that it requires 3 dB more power than polar RZ signaling for the same probability of symbol error.

Bipolar RZ Signaling

This line code uses three amplitude levels, as indicated in Figure 6.24(d). Specifically, positive and negative pulses of equal amplitude (i.e., $+A$ and $-A$) are used alternately for symbol 1, with each pulse having a half-symbol width; no pulse is always used for symbol 0. A useful property of the bipolar RZ signaling is that the power spectrum of the transmitted signal has no DC component and relatively insignificant low-frequency components for the case when symbols 1 and 0 occur with equal probability. The bipolar RZ line code is also called *alternate mark inversion* (AMI) signaling.

Split-Phase (Manchester Code)

In this final method of signaling, illustrated in Figure 6.24e, symbol 1 is represented by a positive pulse of amplitude A followed by a negative pulse of amplitude $-A$, with both pulses being half-symbol wide. For symbol 0, the polarities of these two pulses are reversed. A unique property of the Manchester code is that it suppresses the DC component and has relatively insignificant low-frequency components, *regardless of the signal statistics*. This property is essential in some applications.

6.11 Summary and Discussion

In this chapter we introduced two fundamental and complementary processes:

- *Sampling*, which operates in the time domain; the sampling process is the link between an analog waveform and its discrete-time representation.
- *Quantization*, which operates in the amplitude domain; the quantization process is the link between an analog waveform and its discrete-amplitude representation.

The sampling process builds on the *sampling theorem*, which states that a strictly band-limited signal with no frequency components higher than W Hz is represented uniquely by a sequence of samples taken at a uniform rate equal to or greater than the Nyquist rate of $2W$ samples per second. The quantization process exploits the fact that any human sense, as ultimate receiver, can only detect finite intensity differences.

The sampling process is basic to the operation of all pulse modulation systems, which may be classified into analog pulse modulation and digital pulse modulation. The distinguishing feature between them is that analog pulse modulation systems maintain a continuous amplitude representation of the message signal, whereas digital pulse modulation systems also employ quantization to provide a representation of the message signal that is discrete in both time and amplitude.

Analog pulse modulation results from varying some parameter of the transmitted pulses, such as amplitude, duration, or position, in which case we speak of PAM, pulse-duration modulation, or pulse-position modulation, respectively. In this chapter we focused on PAM, as it is used in all forms of digital pulse modulation.

Digital pulse modulation systems transmit analog message signals as a sequence of coded pulses, which is made possible through the combined use of sampling and quantization. PCM is an important form of digital pulse modulation that is endowed with some unique system advantages, which, in turn, have made it the standard method of modulation for the transmission of such analog signals as voice and video signals. The advantages of PCM include robustness to noise and interference, efficient regeneration of the coded pulses along the transmission path, and a uniform format for different kinds of baseband signals.

Indeed, it is because of this list of advantages unique to PCM that it has become the method of choice for the construction of public switched telephone networks (PSTNs). In this context, the reader should carefully note that the telephone channel viewed from the PSTN by an Internet service provider, for example, is *nonlinear* due to the use of companding and, most importantly, it is *entirely digital*. This observation has a significant impact on the design of high-speed modems for communications between a computer user and server, which will be discussed in Chapter 8.

DM and DPCM are two other useful forms of digital pulse modulation. The principal advantage of DM is the simplicity of its circuitry, which is achieved at the expense of increased transmission bandwidth. In contrast, DPCM employs increased circuit complexity to reduce channel bandwidth. The improvement is achieved by using the idea of prediction to reduce redundant symbols from an incoming data stream. A further improvement in the operation of DPCM can be made through the use of adaptivity to account for statistical variations in the input data. By so doing, bandwidth requirement may be reduced significantly without serious degradation in system performance.[8]

Problems

Sampling Process

6.1 In natural sampling, an analog signal $g(t)$ is multiplied by a periodic train of rectangular pulses $c(t)$, each of unit area. Given that the pulse repetition frequency of this periodic train is f_s and the duration of each rectangular pulse is T (with $f_s T \ll 1$), do the following:

a. Find the spectrum of the signal $s(t)$ that results from the use of natural sampling; you may assume that time $t = 0$ corresponds to the midpoint of a rectangular pulse in $c(t)$.

b. Show that the original signal $g(t)$ may be recovered exactly from its naturally sampled version, provided that the conditions embodied in the sampling theorem are satisfied.

6.2 Specify the Nyquist rate and the Nyquist interval for each of the following signals:

a. $g(t) = \text{sinc}(200t)$.

b. $g(t) = \text{sinc}^2(200t)$.

c. $g(t) = \text{sinc}(200t) + \text{sinc}^2(200t)$.

6.3 Discussion of the sampling theorem presented in Section 6.2 was confined to the time domain. Describe how the sampling theorem can be applied in the frequency domain.

Pulse-Amplitude Modulation

6.4 Figure P6.4 shows the idealized spectrum of a message signal $m(t)$. The signal is sampled at a rate equal to 1 kHz using flat-top pulses, with each pulse being of unit amplitude and duration 0.1 ms. Determine and sketch the spectrum of the resulting PAM signal.

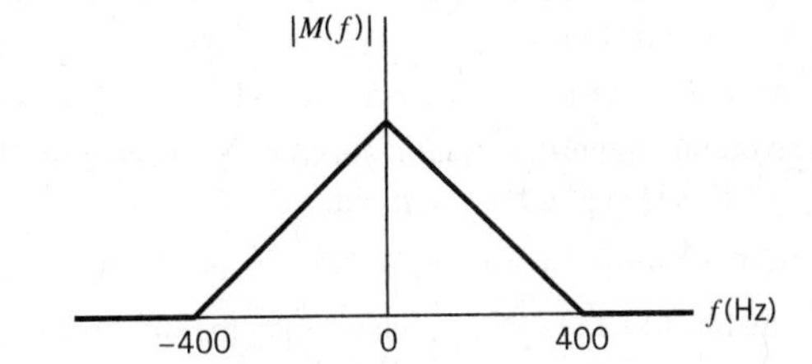

Figure P6.4

6.5 In this problem, we evaluate the equalization needed for the aperture effect in a PAM system. The operating frequency $f = f_s/2$, which corresponds to the highest frequency component of the message signal for a sampling rate equal to the Nyquist rate. Plot $1/\text{sinc}(0.5T/T_s)$ versus T/T_s, and hence find the equalization needed when $T/T_s = 0.1$.

6.6 Consider a PAM wave transmitted through a channel with white Gaussian noise and minimum bandwidth $B_T = 1/2T_s$, where T_s is the sampling period. The noise is of zero mean and power spectral density $N_0/2$. The PAM signal uses a standard pulse $g(t)$ with its Fourier transform defined by

$$G(f) = \begin{cases} \dfrac{1}{2B_T}, & |f| < B_T \\ 0, & |f| > B_T \end{cases}$$

By considering a full-load sinusoidal modulating wave, show that PAM and baseband-signal transmission have equal SNRs for the same average transmitted power.

6.7 Twenty-four voice signals are sampled uniformly and then time-division multiplexed (TDM). The sampling operation uses flat-top samples with 1 μs duration. The multiplexing operation includes provision for synchronization by adding an extra pulse of sufficient amplitude and also 1 μs duration. The highest frequency component of each voice signal is 3.4 kHz.

a. Assuming a sampling rate of 8 kHz, calculate the spacing between successive pulses of the multiplexed signal.

b. Repeat your calculation assuming the use of Nyquist rate sampling.

6.8 Twelve different message signals, each with a bandwidth of 10 kHz, are to be multiplexed and transmitted. Determine the minimum bandwidth required if the multiplexing/modulation method used is time-division multiplexing (TDM), which was discussed in Chapter 1.

Pulse-Code Modulation

6.9 A speech signal has a total duration of 10 s. It is sampled at the rate of 8 kHz and then encoded. The signal-to-(quantization) noise ratio is required to be 40 dB. Calculate the minimum storage capacity needed to accommodate this digitized speech signal.

6.10 Consider a uniform quantizer characterized by the input-output relation illustrated in Figure 6.9a. Assume that a Gaussian-distributed random variable with zero mean and unit variance is applied to this quantizer input.

a. What is the probability that the amplitude of the input lies outside the range –4 to +4?

b. Using the result of part a, show that the output SNR of the quantizer is given by

$$(\mathrm{SNR})_{\mathrm{O}} = 6R - 7.2 \text{ dB}$$

where R is the number of bits per sample. Specifically, you may assume that the quantizer input extends from –4 to +4. Compare the result of part b with that obtained in Example 2.

6.11 A PCM system uses a uniform quantizer followed by a 7-bit binary encoder. The bit rate of the system is equal to 50×10^6 bits/s.

a. What is the maximum message bandwidth for which the system operates satisfactorily?

b. Determine the output signal-to-(quantization) noise when a full-load sinusoidal modulating wave of frequency 1 MHz is applied to the input.

6.12 Show that with a nonuniform quantizer the mean-square value of the quantization error is approximately equal to $(1/12)\Sigma_i \Delta_i^2 p_i$, where Δ_i is the ith step size and p_i is the probability that the input signal amplitude lies within the ith interval. Assume that the step size Δ_i is small compared with the excursion of the input signal.

6.13 a. A sinusoidal signal with an amplitude of 3.25 V is applied to a uniform quantizer of the midtread type whose output takes on the values 0, ±1, ±2, ±3 V. Sketch the waveform of the resulting quantizer output for one complete cycle of the input.

b. Repeat this evaluation for the case when the quantizer is of the midrise type whose output takes on the values 0.5, ±1.5, ±2.5, ±3.5 V.

6.14 The signal

$$m(t) \text{ (volts)} = 6\sin(2\pi t)$$

is transmitted using a 40-bit binary PCM system. The quantizer is of the midrise type, with a step size of 1 V. Sketch the resulting PCM wave for one complete cycle of the input. Assume a sampling rate of four samples per second, with samples taken at t(s) = ±1/8, ±3/8, ±5/8, ...

6.15 Figure P6.15 shows a PCM signal in which the amplitude levels of +1 V and –1 V are used to represent binary symbols 1 and 0, respectively. The codeword used consists of three bits. Find the sampled version of an analog signal from which this PCM signal is derived.

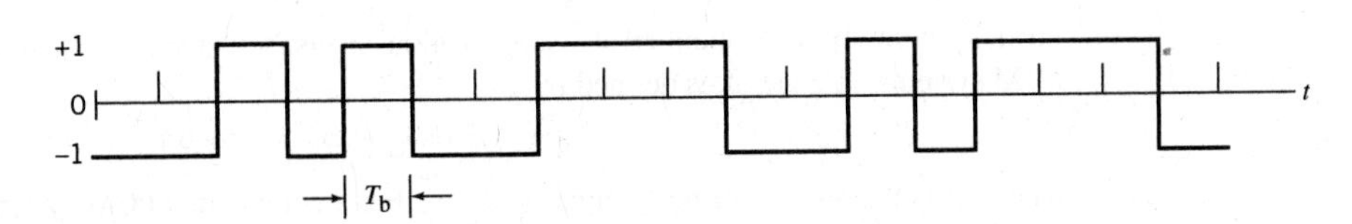

Figure P6.15

6.16 Consider a chain of $(n - 1)$ regenerative repeaters, with a total of n sequential decisions made on a binary PCM wave, including the final decision made at the receiver. Assume that any binary symbol transmitted through the system has an independent probability p_1 of being inverted by any repeater. Let p_n represent the probability that a binary symbol is in error after transmission through the complete system.

a. Show that

$$p_n = \frac{1}{2}[1-(1-2p_1)^n]$$

b. If p_1 is very small and n is not too large, what is the corresponding value of p_n?

6.17 Discuss the basic issues involved in the design of a regenerative repeater for PCM.

Linear Prediction

6.18 A one-step linear predictor operates on the sampled version of a sinusoidal signal. The sampling rate is equal to $10f_0$, where f_0 is the frequency of the sinusoid. The predictor has a single coefficient denoted by w_1.

a. Determine the optimum value of w_1 required to minimize the prediction-error variance.

b. Determine the minimum value of the prediction error variance.

6.19 A stationary process $X(t)$ has the following values for its autocorrelation function:

$$R_X(0) = 1$$

$$R_X(0) = 0.8$$

$$R_X(0) = 0.6$$

$$R_X(0) = 0.4$$

a. Calculate the coefficients of an optimum linear predictor involving the use of three unit-time delays.

b. Calculate the variance of the resulting prediction error.

6.20 Repeat the calculations of Problem 6.19, but this time use a linear predictor with two unit-time delays. Compare the performance of this second optimum linear predictor with that considered in Problem 6.19.

Differential Pulse-Code Modulation

6.21 A DPCM system uses a linear predictor with a single tap. The normalized autocorrelation function of the input signal for a lag of one sampling interval is 0.75. The predictor is designed to minimize the prediction-error variance. Determine the processing gain attained by the use of this predictor.

6.22 Calculate the improvement in processing gain of a DPCM system using the optimized three-tap linear predictor. For this calculation, use the autocorrelation function values of the input signal specified in Problem 6.19.

6.23 In this problem, we compare the performance of a DPCM system with that of an ordinary PCM system using companding.

For a sufficiently large number of representation levels, the signal-to-(quantization) noise ratio of PCM systems, in general, is defined by

$$10 \log_{10}(\text{SNR})_O \text{ (dB)} = \alpha + 6n$$

where 2^n is the number of representation levels. For a companded PCM system using the μ-law, the constant α is itself defined by

$$\alpha(\text{dB}) \approx 4.77 - 20 \log_{10} \log(1 + \mu)$$

For a DPCM system, on the other hand, the constant α lies in the range $-3 < \alpha < 15$ dBs. The formulas quoted herein apply to telephone-quality speech signals.

Compare the performance of the DPCM system against that of the μ-companded PCM system with $\mu = 255$ for each of the following scenarios:

a. The improvement in $(\text{SNR})_O$ realized by DPCM over companded PCM for the same number of bits per sample.

b. The reduction in the number of bits per sample required by DPCM, compared with the companded PCM for the same $(\text{SNR})_O$.

6.24 In the DPCM system depicted in Figure P6.24, show that in the absence of channel noise, the transmitting and receiving prediction filters operate on slightly different input signals.

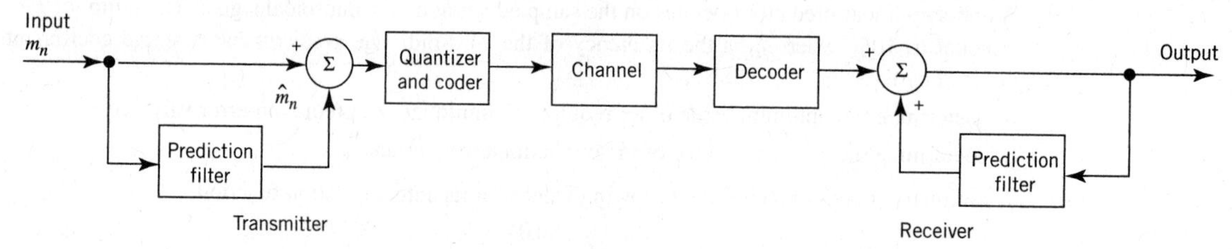

Figure P6.24

6.25 Figure P6.25 depicts the block diagram of adaptive quantization for DPCM. The quantization is of a backward estimation kind because samples of the quantization output and prediction errors are used to continuously derive backward estimates of the variance of the message signal. This estimate computed at time n is denoted by $\hat{\sigma}^2_{m,n}$. Given this estimate, the step size is varied so as to match the actual variance of the message sample m_n, as shown by

$$\Delta_n = \phi \sigma_{m,n}$$

where $\hat{\sigma}^2_{m,n}$ is the estimate of the standard deviation and ϕ is a constant. An attractive feature of the adaptive scheme in Figure P6.25 is that samples of the quantization output and the prediction error are used to compute the predictor's coefficients.

Modify the block diagram of the DPCM transmitter in Figure 6.19a so as to accommodate *adaptive prediction with backward estimation.*

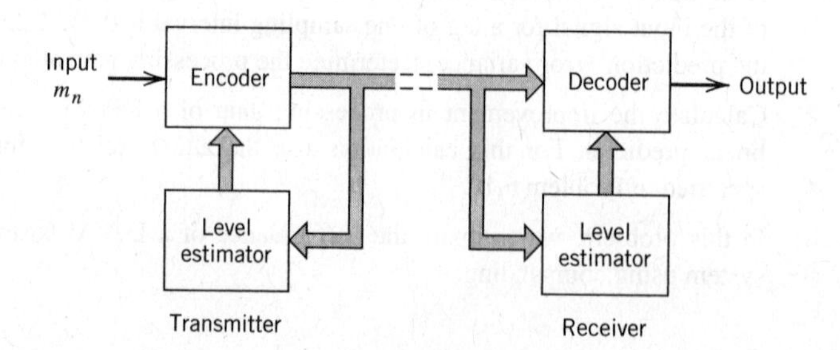

Figure P6.25

Delta Modulation

6.26 Consider a test signal $m(t)$ defined by a hyperbolic tangent function:

$$m(t) = A\tanh(\beta t)$$

where A and β are constants. Determine the minimum step size Δ for DM of this signal, which is required to avoid slope-overload distortion.

6.27 Consider a sine wave of frequency f_m and amplitude A_m, which is applied to a delta modulator of step size Δ. Show that slope-overload distortion will occur if

$$A_m > \frac{\Delta}{2\pi f_m T_s}$$

where T_s is the sampling period. What is the maximum power that may be transmitted without slope-overload distortion?

6.28 A linear delta modulator is designed to operate on speech signals limited to 3.4 kHz. The specifications of the modulator are as follows:

- Sampling rate = $10f_{\text{Nyquist}}$, where f_{Nyquist} is the Nyquist rate of the speech signal.
- Step size Δ = 100 mV.

The modulator is tested with a 1 kHz sinusoidal signal. Determine the maximum amplitude of this test signal required to avoid slope-overload distortion.

6.29 In this problem, we derive an empirical formula for the average signal-to-(quantization) noise ratio of a DM system with a sinusoidal signal of amplitude A and frequency f_m as the test signal. Assume that the power spectral density of the granular noise generated by the system is governed by the formula

$$S_N(f) = \frac{\Delta^2}{6f_s}$$

where f_s is the sampling rate and Δ is the step size. (Note that this formula is basically the same as that for the power spectral density of quantization noise in a PCM system with $\Delta/2$ for PCM being replaced by Δ for DM.) The DM system is designed to handle analog message signals limited to bandwidth W.

a. Show that the average quantization noise power produced by the system is

$$N = \frac{4\pi^2 A^2 f_m^2 W}{3f_s^3}$$

where it is assumed that the step size Δ has been chosen in accordance with the formula used in Problem 6.28 so as to avoid slope-overload distortion.

b. Hence, determine the signal-to-(quantization) noise ratio of the DM system for a sinusoidal input.

6.30 Consider a DM system designed to accommodate analog message signals limited to bandwidth W = 5 kHz. A sinusoidal test signal of amplitude A = 1V and frequency f_m = 1 kHz is applied to the system. The sampling rate of the system is 50 kHz.

a. Calculate the step size Δ required to minimize slope overload distortion.

b. Calculate the signal-to-(quantization) noise ratio of the system for the specified sinusoidal test signal.

For these calculations, use the formula derived in Problem 6.29.

6.31 Consider a low-pass signal with a bandwidth of 3 kHz. A linear DM system with step size Δ = 0.1V is used to process this signal at a sampling rate 10 times the Nyquist rate.

a. Evaluate the maximum amplitude of a test sinusoidal signal of frequency 1 kHz, which can be processed by the system without slope-overload distortion.

b. For the specifications given in part a, evaluate the output SNR under (i) prefiltered and (ii) postfiltered conditions.

6.32 In the conventional form of DM, the quantizer input may be viewed as an approximate to the *derivative* of the incoming message signal $m(t)$. This behavior leads to a drawback of DM: transmission disturbances (e.g., noise) result in an accumulation error in the demodulated signal. This drawback can be overcome by *integrating* the message signal $m(t)$ prior to DM, resulting in three beneficial effects:

a. Low frequency content of $m(t)$ is pre-emphasized.

b. Correlation between adjacent samples of $m(t)$ is increased, tending to improve overall system performance by reducing the variance of the error signal at the quantizer input.

c. Design of the receiver is simplified.

Such a DM scheme is called *delta–sigma modulation*.

Construct a block diagram of the delta–sigma modulation system in such a way that it provides an interpretation of the system as a "smoothed" version of 1-bit PCM in the following composite sense:

- smoothness implies that the comparator output is integrated prior to quantization, and
- 1-bit modulation merely restates that the quantizer consists of a hard limiter with only two representation levels.

Explain how the receiver of the delta–sigma modulation system is simplified, compared with conventional DM.

Line Codes

6.33 In this problem, we derive the formulas used to compute the power spectra of Figure 6.25 for the five line codes described in Section 6.10. In the case of each line code, the bit duration is T_b and the pulse amplitude A is conditioned to normalize the average power of the line code to unity as indicated in Figure 6.25. Assume that the data stream is randomly generated and symbols 0 and 1 are equally likely.

Derive the power spectral densities of these line codes as summarized here:

a. Unipolar NRZ signals:

$$S(f) = \frac{A^2 T_b}{4}\,\text{sinc}^2(fT_b)\left(1 + \frac{1}{T_b}\delta(f)\right)$$

b. Polar NRZ signals:

$$S(f) = A^2 T_b\,\text{sinc}^2(fT_b)$$

c. Unipolar RZ signals:

$$S(f) = \frac{A^2 T_b}{16}\,\text{sinc}^2\left(\frac{fT_b}{2}\right)\left[1 + \frac{1}{T_b}\sum_{n=-\infty}^{\infty}\delta\left(f - \frac{n}{T_b}\right)\right]$$

d. Bipolar RZ signals:

$$S(f) = \frac{A^2 T_b}{4}\,\text{sinc}^2\left(\frac{fT_b}{2}\right)\sin^2(\pi f T_b)$$

e. Manchester-encoded signals:

$$S(f) = \frac{A^2 T_b}{4}\,\text{sinc}^2\left(\frac{fT_b}{2}\right)\sin^2\left(\frac{\pi f T_b}{2}\right)$$

Hence, confirm the spectral plots displayed in Figure 6.25.

6.34 A randomly generated data stream consists of equiprobable binary symbols 0 and 1. It is encoded into a polar NRZ waveform with each binary symbol being defined as follows:

$$s(t) = \begin{cases} \cos\left(\frac{\pi t}{T_b}\right), & -\frac{T_b}{2} < t \le \frac{T_b}{2} \\ 0, & \text{otherwise} \end{cases}$$

a. Sketch the waveform so generated, assuming that the data stream is 00101110.

b. Derive an expression for the power spectral density of this signal and sketch it.

c. Compare the power spectral density of this random waveform with that defined in part b of Problem 6.33.

6.35 Given the data stream 1110010100, sketch the transmitted sequence of pulses for each of the following line codes:

a. unipolar NRZ

b. polar NRZ

c. unipolar RZ

d. bipolar RZ

e. Manchester code.

Computer Experiments

**6.36 A sinusoidal signal of frequency $f_0 = 10^4/2\pi$ Hz is sampled at the rate of 8 kHz and then applied to a sample-and-hold circuit to produce a flat-topped PAM signal $s(t)$ with pulse duration $T = 500$ μs.

a. Compute the waveform of the PAM signal $s(t)$.

b. Compute $|S(f)|$, denoting the magnitude spectrum of the PAM signal $s(t)$.

c. Compute the envelope of $|S(f)|$. Hence confirm that the frequency at which this envelope goes through zero for the first time is equal to $(1/T) = 20$ kHz.

**6.37 In this problem, we use computer simulation to compare the performance of a companded PCM system using the μ-law against that of the corresponding system using a uniform quantizer. The simulation is to be performed for a sinusoidal input signal of varying amplitude.

With a companded PCM system in mind, Table 6.4 describes the 15-segment *pseudo-linear* characteristic that consists of 15 linear segments configured to approximate the logarithmic μ-law

Table 6.4 The 15-segment companding characteristic (μ = 255)

Linear segment number	Step-size	Projections of segment end points onto the horizontal axis
0	2	±31
1*a*, 1*b*	4	±95
2*a*, 2*b*	8	±223
3*a*, 3*b*	16	±479
4*a*, 4*b*	32	±991
5*a*, 5*b*	64	±2015
6*a*, 6*b*	128	±4063
7*a*, 7*b*	256	±8159

of (6.48), with $\mu = 255$. This approximation is constructed in such a way that the segment endpoints in Table 6.4 lie on the compression curve computed from (6.48).

a. Using the μ-law described in Table 6.4, plot the output signal-to-noise ratio as a function of the input signal-to-noise ratio, both ratios being expressed in decibels.

b. Compare the results of your computation in part (a) with a uniform quantizer having 256 representation levels.

**6.38 In this experiment we study the linear adaptive prediction of a signal x_n governed by the following recursion:

$$x_n = 0.8x_{n-1} - 0.1x_{n-2} + 0.1v_n$$

where v_n is drawn from a discrete–time white noise process of zero mean and unit variance. (A process generated in this manner is referred to as an *autoregressive process of order two*.) Specifically, the adaptive prediction is performed using the *normalized LMS algorithm* defined by

$$\hat{x}_n = \sum_{k=1}^{p} w_{k,n} x_{n-k}$$

$$e_n = x_n - \hat{x}_n$$

$$w_{k,n+1} = w_{k,n} + \mu \Big/ \left(\sum_{k=1}^{p} x_{n-k}^2 \right) x_{n-k} e_n \qquad k = 1, 2, \ldots, p$$

where p is the prediction order and μ is the normalized step-size parameter. The important point to note here is that μ is dimensionless and stability of the algorithm is assured by choosing it in accordance with the formula

$$0 < \mu < 2$$

The algorithm is initiated by setting

$$w_{k,0} = 0 \qquad \text{for all } k$$

The *learning curve* of the algorithm is defined as a plot of the mean-square error versus the number of iterations n for specified parameter values, which is obtained by averaging the plot of e_n^2 versus n over a large number of different realizations of the algorithm.

a. Plot the learning curves for the adaptive prediction of x_n for a fixed prediction order $p = 5$ and three different values of step-size parameter: $\mu = 0.0075$, 0.05, and 0.5.

b. What observations can you make from the learning curves of part a?

**6.39 In this problem, we study adaptive delta modulation, the underlying principle of which is two-fold:

1. If successive errors are of opposite polarity, then the delta modulator is operating in the granular mode, in which case the step size Δ is reduced.
2. If, on the other hand, the successive errors are of the same polarity, then the delta modulator is operating in the slope-overload mode, in which case the step size Δ is increased.

Parts a and b of Figure P6.39 depict the block diagrams of the transmitter and receiver of the adaptive delta modulator, respectively, in which the step size, Δ, is increased or decreased by a factor of 50% at each iteration of the adaptive process, as shown by:

$$\Delta_n = \begin{cases} \dfrac{\Delta_{n-1}}{m_{q,n}}(m_{q,n} + 0.5m_{q,n-1}) & \text{if } \Delta_{n-1} \geq \Delta_{\min} \\ \Delta_{\min} & \text{if } \Delta_{n-1} < \Delta_{\min} \end{cases}$$

where Δ_n is the step size at iteration (time step) n of the adaptation algorithm, and $m_{q,n}$ is the 1-bit quantizer output that equals ± 1.

Specifications: The input signal applied to the transmitter is sinusoidal as shown by

$$m_t = A \sin(2\pi f_{\text{m}} t)$$

where $A = 10$ and $f_{\text{m}} = f_{\text{s}}/100$ where f_{s} is the sampling frequency; the step size $\Delta_n = 1$ for all n; $\Delta_{\min} = 1/8$.

a. Using the above-described adaptation algorithm, use a computer to plot the resulting waveform for one complete cycle of the sinusoidal modulating signal, and also display the coded modulator output in the transmitter.

b. For the same specifications, repeat the computation using linear modulation.

c. Comment on the results obtained in parts a and b of the problem.

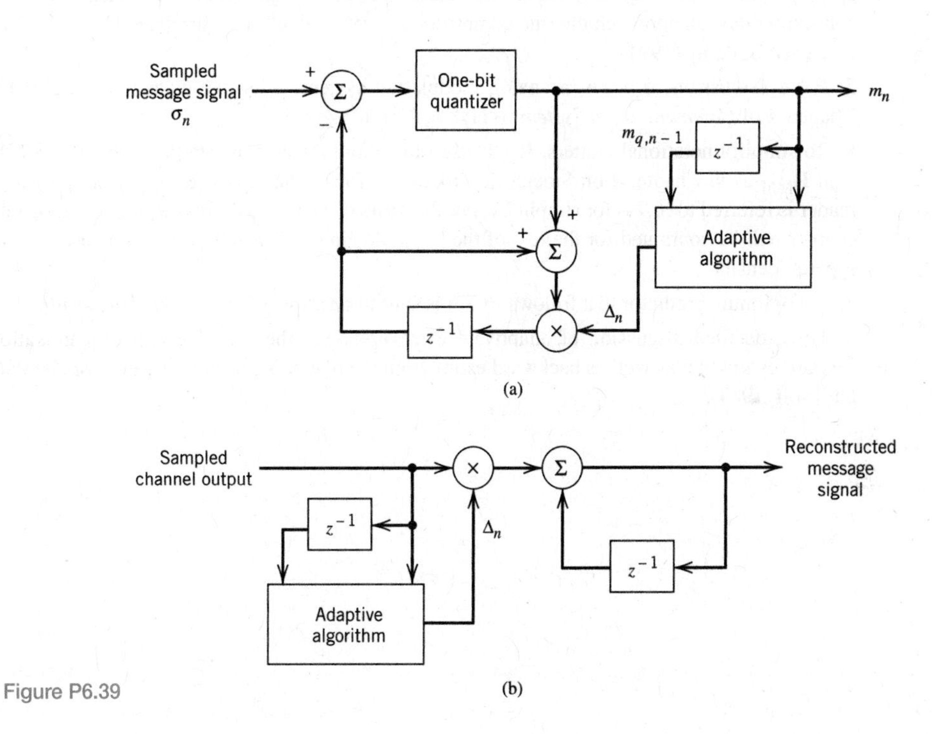

Figure P6.39

Notes

1. For an exhaustive study of quantization noise in signal processing and communications, see Widrow and Kollar (2008).

2. The two necessary conditions of (3.42) and (3.47) for optimality of a scalar quantizer were reported independently by Lloyd (1957) and Max (1960), hence the name "Lloyd–Max quantizer." The derivation of these two optimality conditions presented in this chapter follows the book by Gersho and Gray (1992).

3. The μ-law is used in the USA, Canada, and Japan. On the other hand, in Europe, the A-law is used for signal compression.

4. In actual PCM systems, the companding circuitry does not produce an exact replica of the nonlinear compression curves shown in Figure 6.14. Rather, it provides a *piecewise linear* approximation to the desired curve. By using a large enough number of linear segments, the approximation can approach the true compression curve very closely; for detailed discussion of this issue, see Bellamy (1991).

5. For a discussion of noise in analog modulation systems with particular reference to FM, see Chapter 4 of *Communication Systems* (Haykin, 2001).

6. To simplify notational matters, $\mathbf{R}_{\mathrm{M}}$ is used to denote the autocorrelation matrix in (6.70) rather than $\mathbf{R}_{\mathrm{MM}}$ as in Chapter 4 on Stochastic Processes. To see the rationale for this simplification, the reader is referred to (6.79) for simplicity. For the same reason, henceforth the practice adopted in this chapter will be continued for the rest of the book, dealing with autocorrelation matrices and power spectral density.

7. An optimum predictor that follows (6.77) is said to be a special case of the *Wiener filter.*

8. For a detailed discussion of adaptive DPCM involving the use of adaptive quantization with forward estimation as well as backward estimation, the reader is referred to the classic book (Jayant and Noll, 1984).

CHAPTER

7

Signaling over AWGN Channels

7.1 Introduction

Chapter 6 on the conversion of analog waveforms into coded pulses represents the transition from analog communications to digital communications. This transition has been empowered by several factors:

1. *Ever-increasing advancement of digital silicon chips, digital signal processing, and computers*, which, in turn, has prompted further enhancement in digital silicon chips, thereby repeating the cycle of improvement.
2. *Improved reliability*, which is afforded by digital communications to a much greater extent than is possible with analog communications.
3. *Broadened range of multiplexing of users*, which is enabled by the use of digital modulation techniques.
4. *Communication networks*, for which, in one form or another, the use of digital communications is the preferred choice.

In light of these compelling factors, we may justifiably say that we live in a "digital communications world." For an illustrative example, consider the remote connection of two digital computers, with one computer acting as the information source by calculating digital outputs based on observations and inputs fed into it; the other computer acts as the recipient of the information. The source output consists of a sequence of 1s and 0s, with each *binary symbol* being emitted every T_b seconds. The transmitting part of the digital communication system takes the 1s and 0s emitted by the source computer and encodes them into distinct signals denoted by $s_1(t)$ and $s_2(t)$, respectively, which are suitable for transmission over the analog channel. Both $s_1(t)$ and $s_2(t)$ are *real-valued energy signals*, as shown by

$$E_i = \int_0^{T_b} s_i^2(t)\,dt, \qquad i = 1, 2 \tag{7.1}$$

With the analog channel represented by an AWGN model, depicted in Figure 7.1, the *received signal* is defined by

$$x(t) = s_i(t) + w(t), \qquad \begin{cases} 0 \le t \le T_b \\ i = 1, 2 \end{cases} \tag{7.2}$$

where $w(t)$ is the *channel noise*. The receiver has the task of observing the received signal $x(t)$ for a duration of T_b seconds and then making an *estimate* of the transmitted signal

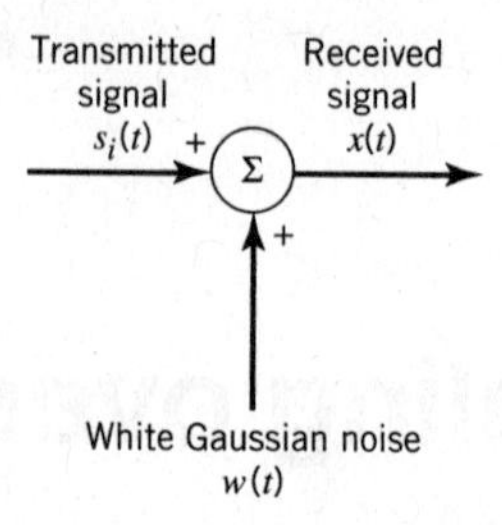

Figure 7.1 AWGN model of a channel.

$s_i(t)$, or equivalently the ith symbol, $i = 1, 2$. However, owing to the presence of channel noise, the receiver will inevitably make occasional *errors*. The requirement, therefore, is to design the receiver so as to *minimize the average probability of symbol error,* defined as

$$P_e = \pi_1 \mathbb{P}(\hat{m} = 0|1 \text{ sent}) + \pi_2 \mathbb{P}(\hat{m} = 1|0 \text{ sent}) \tag{7.3}$$

where π_1 and π_2 are the *prior probabilities* of transmitting symbols 1 and 0, respectively, and $\hat{m}$ is the estimate of the symbol 1 or 0 sent by the source, which is computed by the receiver. The $\mathbb{P}(\hat{m} = 0|1 \text{ sent})$ and $\mathbb{P}(\hat{m} = 1|0 \text{ sent})$ are conditional probabilities.

In minimizing the average probability of symbol error between the receiver output and the symbol emitted by the source, the motivation is to make the digital communication system as *reliable* as possible. To achieve this important design objective in a generic setting that involves an *M-ary alphabet* whose symbols are denoted by $m_1, m_2, \ldots, m_M$, we have to understand two basic issues:

1. How to optimize the design of the receiver so as to minimize the average probability of symbol error.
2. How to choose the set of signals $s_1(t), s_2(t), \ldots, s_M(t)$ for representing the symbols $m_1, m_2, \ldots, m_M$, respectively, since this choice affects the average probability of symbol error.

The key question is how to develop this understanding in a principled as well as insightful manner. The answer to this fundamental question is found in the *geometric representation of signals.*

7.2 Geometric Representation of Signals

The essence of *geometric representation of signals*[1] is to represent any set of M energy signals $\{s_i(t)\}$ as linear combinations of N *orthonormal basis functions*, where $N \le M$. That is to say, given a set of real-valued energy signals, $s_1(t), s_2(t), \ldots, s_M(t)$, each of duration T seconds, we write

$$s_i(t) = \sum_{j=1}^{N} s_{ij}\phi_j(t), \qquad \begin{cases} 0 \le t \le T \\ i = 1, 2, \ldots, M \end{cases} \tag{7.4}$$

where the coefficients of the expansion are defined by

$$s_{ij} = \int_0^T s_i(t)\phi_j(t)\,dt, \qquad \begin{cases} i = 1, 2, \ldots, M \\ j = 1, 2, \ldots, N \end{cases} \tag{7.5}$$

The real-valued basis functions $\phi_1(t)$, $\phi_2(t)$, ..., $\phi_N(t)$ form an *orthonormal set*, by which we mean

$$\int_0^T \phi_i(t)\phi_j(t)\,\mathrm{d}t = \delta_{ij} = \begin{cases} 1 & \text{if } i = j \\ 0 & \text{if } i \neq j \end{cases} \tag{7.6}$$

where δ_{ij} is the *Kronecker delta*. The first condition of (7.6) states that each basis function is *normalized* to have unit energy. The second condition states that the basis functions $\phi_1(t)$, $\phi_2(t)$, ..., $\phi_N(t)$ are *orthogonal* with respect to each other over the interval $0 \le t \le T$.

For prescribed i, the set of coefficients $\{s_{ij}\}_{j=1}^{N}$ may be viewed as an *N-dimensional signal vector*, denoted by $\mathbf{s}_i$. The important point to note here is that the vector $\mathbf{s}_i$ bears a *one-to-one* relationship with the transmitted signal $s_i(t)$:

- Given the N elements of the vector $\mathbf{s}_i$ operating as input, we may use the scheme shown in Figure 7.2a to generate the signal $s_i(t)$, which follows directly from (7.4). This figure consists of a bank of N multipliers with each multiplier having its own basis function followed by a summer. The scheme of Figure 7.2a may be viewed as a *synthesizer.*
- Conversely, given the signals $s_i(t)$, $i = 1, 2, \ldots, M$, operating as input, we may use the scheme shown in Figure 7.2b to calculate the coefficients $s_{i1}, s_{i2}, \ldots, s_{iN}$ which follows directly from (7.5). This second scheme consists of a bank of N *product-integrators* or *correlators* with a common input, and with each one of them supplied with its own basis function. The scheme of Figure 7.2b may be viewed as an *analyzer.*

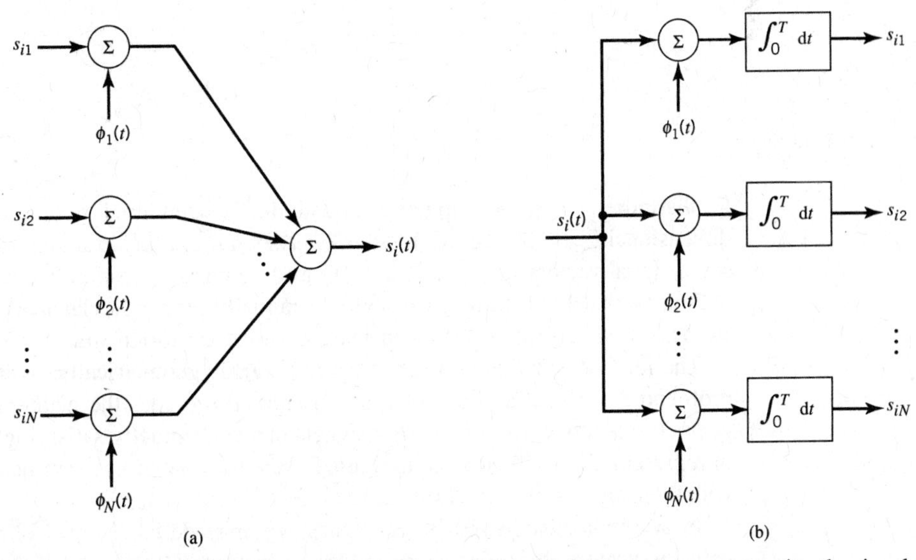

Figure 7.2 (a) Synthesizer for generating the signal $s_i(t)$. (b) Analyzer for reconstructing the signal vector $\{\mathbf{s}_i\}$.

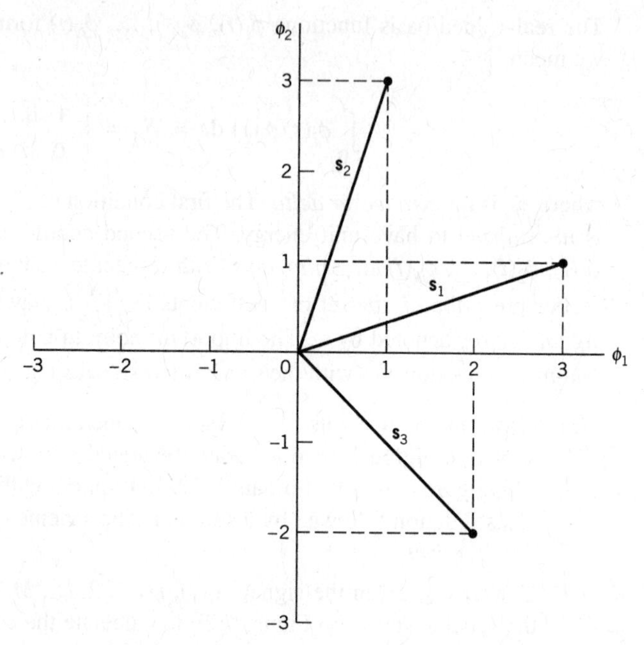

Figure 7.3 Illustrating the geometric representation of signals for the case when $N = 2$ and $M = 3$.

Accordingly, we may state that each signal in the set $\{s_i(t)\}$ is completely determined by the signal vector

$$\mathbf{s}_i = \begin{bmatrix} s_{i1} \\ s_{i2} \\ \vdots \\ s_{iN} \end{bmatrix}, \qquad i = 1, 2, \ldots, M \tag{7.7}$$

Furthermore, if we conceptually extend our conventional notion of two- and three-dimensional Euclidean spaces to an *N-dimensional Euclidean space*, we may visualize the set of signal vectors $\{\mathbf{s}_i | i = 1, 2, \ldots, M\}$ as defining a corresponding set of M points in an N-dimensional Euclidean space, with N mutually perpendicular axes labeled $\phi_1, \phi_2, \ldots, \phi_N$. This N-dimensional Euclidean space is called the *signal space*.

The idea of visualizing a set of energy signals geometrically, as just described, is of profound theoretical and practical importance. It provides the mathematical basis for the geometric representation of energy signals in a conceptually satisfying manner. This form of representation is illustrated in Figure 7.3 for the case of a two-dimensional signal space with three signals; that is, $N = 2$ and $M = 3$.

In an N-dimensional Euclidean space, we may define *lengths* of vectors and *angles* between vectors. It is customary to denote the length (also called the *absolute value* or *norm*) of a signal vector $\mathbf{s}_i$ by the symbol $\|\mathbf{s}_i\|$. The squared length of any signal vector $\mathbf{s}_i$ is defined to be the *inner product* or *dot product* of $\mathbf{s}_i$ with itself, as shown by

$$\|\mathbf{s}_i\|^2 = \mathbf{s}_i^{\mathrm{T}}\mathbf{s}_i$$
$$= \sum_{j=1}^{N} s_{ij}^2, \qquad i = 1, 2, \ldots, M \tag{7.8}$$

where s_{ij} is the jth element of $\mathbf{s}_i$ and the superscript T denotes matrix transposition.

There is an interesting relationship between the energy content of a signal and its representation as a vector. By definition, the energy of a signal $s_i(t)$ of duration T seconds is

$$E_i = \int_0^T s_i^2(t)\,\mathrm{d}t, \qquad i = 1, 2, \ldots, M \tag{7.9}$$

Therefore, substituting (7.4) into (7.9), we get

$$E_i = \int_0^T \left[\sum_{j=1}^{N} s_{ij}\phi_j(t)\right]\left[\sum_{k=1}^{N} s_{ik}\phi_k(t)\right]\mathrm{d}t$$

Interchanging the order of summation and integration, which we can do because they are both linear operations, and then rearranging terms we get

$$E_i = \sum_{j=1}^{N}\sum_{k=1}^{N} s_{ij}s_{ik}\int_0^T \phi_j(t)\phi_k(t)\,\mathrm{d}t \tag{7.10}$$

Since, by definition, the $\phi_j(t)$ form an orthonormal set in accordance with the two conditions of (7.6), we find that (7.10) reduces simply to

$$E_i = \sum_{j=1}^{N} s_{ij}^2$$
$$= \|\mathbf{s}_i\|^2 \tag{7.11}$$

Thus, (7.8) and (7.11) show that the energy of an energy signal $s_i(t)$ is equal to the squared length of the corresponding signal vector $\mathbf{s}_i(t)$.

In the case of a pair of signals $s_i(t)$ and $s_k(t)$ represented by the signal vectors $\mathbf{s}_i$ and $\mathbf{s}_k$, respectively, we may also show that

$$\int_0^T s_i(t)s_k(t)\,\mathrm{d}t = \mathbf{s}_i^{\mathrm{T}}\mathbf{s}_k \tag{7.12}$$

Equation (7.12) states:

> The *inner product* of the energy signals $s_i(t)$ and $s_k(t)$ over the interval $[0,T]$ is equal to the inner product of their respective vector representations $\mathbf{s}_i$ and $\mathbf{s}_k$.

Note that the inner product $\mathbf{s}_i^{\mathrm{T}}\mathbf{s}_k$ is *invariant* to the choice of basis functions $\{\phi_j(t)\}_{j=1}^{N}$, in that it only depends on the components of the signals $s_i(t)$ and $s_k(t)$ projected onto each of the basis functions.

Yet another useful relation involving the vector representations of the energy signals $s_i(t)$ and $s_k(t)$ is described by

$$\begin{aligned} \|\mathbf{s}_i - \mathbf{s}_k\|^2 &= \sum_{j=1}^{N} (s_{ij} - s_{kj})^2 \\ &= \int_0^T (s_i(t) - s_k(t))^2 \, dt \end{aligned} \tag{7.13}$$

where $\|\mathbf{s}_i - \mathbf{s}_k\|$ is the *Euclidean distance* d_{ik} between the points represented by the signal vectors $\mathbf{s}_i$ and $\mathbf{s}_k$.

To complete the geometric representation of energy signals, we need to have a representation for the angle θ_{ik} subtended between two signal vectors $\mathbf{s}_i$ and $\mathbf{s}_k$. By definition, the *cosine of the angle* θ_{ik} is equal to the inner product of these two vectors divided by the product of their individual norms, as shown by

$$\cos(\theta_{ik}) = \frac{\mathbf{s}_i^T \mathbf{s}_k}{\|\mathbf{s}_i\| \|\mathbf{s}_k\|} \tag{7.14}$$

The two vectors $\mathbf{s}_i$ and $\mathbf{s}_k$ are thus orthogonal or perpendicular to each other if their inner product $\mathbf{s}_i^T \mathbf{s}_k$ is zero, in which case $\theta_{ik} = 90°$; this condition is intuitively satisfying.

EXAMPLE 1 The Schwarz Inequality

Consider any pair of energy signals $s_1(t)$ and $s_2(t)$. The *Schwarz inequality* states

$$\left(\int_{-\infty}^{\infty} s_1(t) s_2(t) \, dt\right)^2 \le \left(\int_{-\infty}^{\infty} s_1^2(t) \, dt\right)\left(\int_{-\infty}^{\infty} s_2^2(t) \, dt\right) \tag{7.15}$$

The equality holds if, and only if, $s_2(t) = c s_1(t)$, where c is any constant.

To prove this important inequality, let $s_1(t)$ and $s_2(t)$ be expressed in terms of the pair of orthonormal basis functions $\phi_1(t)$ and $\phi_2(t)$ as follows:

$$s_1(t) = s_{11}\phi_1(t) + s_{12}\phi_2(t)$$

$$s_2(t) = s_{21}\phi_1(t) + s_{22}\phi_2(t)$$

where $\phi_1(t)$ and $\phi_2(t)$ satisfy the orthonormality conditions over the time interval $(-\infty, \infty)$:

$$\int_{-\infty}^{\infty} \phi_i(t)\phi_j(t) \, dt = \delta_{ij} = \begin{cases} 1 & \text{for } j = i \\ 0 & \text{otherwise} \end{cases}$$

On this basis, we may represent the signals $s_1(t)$ and $s_2(t)$ by the following respective pair of vectors, as illustrated in Figure 7.4:

$$\mathbf{s}_1 = \begin{bmatrix} s_{11} \\ s_{12} \end{bmatrix}$$

$$\mathbf{s}_2 = \begin{bmatrix} s_{21} \\ s_{22} \end{bmatrix}$$

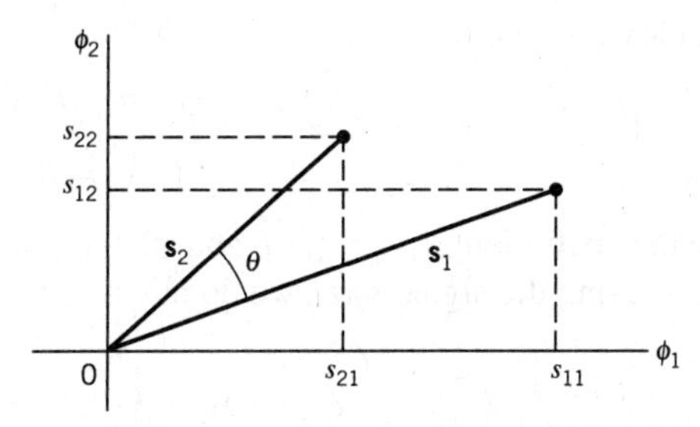

Figure 7.4 Vector representations of signals $s_1(t)$ and $s_2(t)$, providing the background picture for proving the Schwarz inequality.

From Figure 7.4 we readily see that the cosine of angle θ subtended between the vectors $\mathbf{s}_1$ and $\mathbf{s}_2$ is

$$\begin{aligned}\cos\theta &= \frac{\mathbf{s}_1^{\mathrm{T}}\mathbf{s}_2}{\|\mathbf{s}_1\|\|\mathbf{s}_2\|} \\ &= \frac{\int_{-\infty}^{\infty} s_1(t)s_2(t)\,\mathrm{d}t}{\left(\int_{-\infty}^{\infty} s_1^2(t)\,\mathrm{d}t\right)^{1/2}\left(\int_{-\infty}^{\infty} s_2^2(t)\,\mathrm{d}t\right)^{1/2}}\end{aligned} \tag{7.16}$$

where we have made use of (7.14) and (7.12). Recognizing that $|\cos\theta| \le 1$, the Schwarz inequality of (7.15) immediately follows from (7.16). Moreover, from the first line of (7.16) we note that $|\cos\theta| = 1$ if, and only if, $\mathbf{s}_2 = c\mathbf{s}_1$; that is, $s_2(t) = cs_1(t)$, where c is an arbitrary constant.

Proof of the Schwarz inequality, as presented here, applies to real-valued signals. It may be readily extended to complex-valued signals, in which case (7.15) is reformulated as

$$\left|\int_{-\infty}^{\infty} s_1(t)s_2^*(t)\,\mathrm{d}t\right| \le \left(\int_{-\infty}^{\infty} |s_1(t)|^2\,\mathrm{d}t\right)^{1/2}\left(\int_{-\infty}^{\infty} |s_2(t)|^2\,\mathrm{d}t\right)^{1/2} \tag{7.17}$$

where the asterisk denotes complex conjugation and the equality holds if, and only if, $s_2(t) = cs_1(t)$, where c is a constant.

Gram–Schmidt Orthogonalization Procedure

Having demonstrated the elegance of the geometric representation of energy signals with an example, how do we justify it in mathematical terms? The answer to this question lies in the *Gram–Schmidt orthogonalization procedure*, for which we need a *complete orthonormal set of basis functions*. To proceed with the formulation of this procedure, suppose we have a set of M energy signals denoted by $s_1(t), s_2(t), \ldots, s_M(t)$. Starting with $s_1(t)$ chosen from this set arbitrarily, the first basis function is defined by

$$\phi_1(t) = \frac{s_1(t)}{\sqrt{E_1}} \tag{7.18}$$

where E_1 is the energy of the signal $s_1(t)$.

Then, clearly, we have

$$\begin{aligned} s_1(t) &= \sqrt{E_1}\phi_1(t) \\ &= s_{11}(t)\phi_1(t) \end{aligned}$$

where the coefficient $s_{11} = \sqrt{E_1}$ and $\phi_1(t)$ has unit energy as required.

Next, using the signal $s_2(t)$, we define the coefficient s_{21} as

$$s_{21} = \int_0^T s_2(t)\phi_1(t)\,\mathrm{d}t$$

We may thus introduce a new intermediate function

$$g_2(t) = s_2(t) - s_{21}\phi_1(t) \tag{7.19}$$

which is orthogonal to $\phi_1(t)$ over the interval $0 \le t \le T$ by virtue of the definition of s_{21} and the fact that the basis function $\phi_1(t)$ has unit energy. Now, we are ready to define the second basis function as

$$\phi_2(t) = \frac{g_2(t)}{\sqrt{\int_0^T g_2^2(t)\,\mathrm{d}t}} \tag{7.20}$$

Substituting (7.19) into (7.20) and simplifying, we get the desired result

$$\phi_2(t) = \frac{s_2(t) - s_{21}\phi_1(t)}{\sqrt{E_2 - s_{21}^2}} \tag{7.21}$$

where E_2 is the energy of the signal $s_2(t)$. From (7.20) we readily see that

$$\int_0^T \phi_2^2(t)\,\mathrm{d}t = 1$$

in which case (7.21) yields

$$\int_0^T \phi_1(t)\phi_2(t)\,\mathrm{d}t = 0$$

That is to say, $\phi_1(t)$ and $\phi_2(t)$ form an orthonormal pair as required.

Continuing the procedure in this fashion, we may, in general, define

$$g_i(t) = s_i(t) - \sum_{j=1}^{i-1} s_{ij}\phi_j(t) \tag{7.22}$$

where the coefficients s_{ij} are themselves defined by

$$s_{ij} = \int_0^T s_i(t)\phi_j(t)\,\mathrm{d}t, \qquad j = 1, 2, \ldots, i-1$$

For $i = 1$, the function $g_i(t)$ reduces to $s_i(t)$.

Given the $g_i(t)$, we may now define the set of basis functions

$$\phi_i(t) = \frac{g_i(t)}{\sqrt{\int_0^T g_i^2(t)\,\mathrm{d}t}}, \qquad j = 1, 2, \ldots, N \tag{7.23}$$

which form an orthonormal set. The dimension N is less than or equal to the number of given signals, M, depending on one of two possibilities:

- The signals $s_1(t)$, $s_2(t)$, …, $s_M(t)$ form a linearly independent set, in which case $N = M$.
- The signals $s_1(t)$, $s_2(t)$, …, $s_M(t)$ are not linearly independent, in which case $N < M$ and the intermediate function $g_i(t)$ is zero for $i > N$.

Note that the conventional Fourier series expansion of a periodic signal, discussed in Chapter 2, may be viewed as a special case of the Gram–Schmidt orthogonalization procedure. Moreover, the representation of a band-limited signal in terms of its samples taken at the Nyquist rate, discussed in Chapter 6, may be viewed as another special case. However, in saying what we have here, two important distinctions should be made:

1. The form of the basis functions $\phi_1(t)$, $\phi_2(t)$, …, $\phi_N(t)$ has not been specified. That is to say, unlike the Fourier series expansion of a periodic signal or the sampled representation of a band-limited signal, we have not restricted the Gram–Schmidt orthogonalization procedure to be in terms of sinusoidal functions (as in the Fourier series) or sinc functions of time (as in the sampling process).
2. The expansion of the signal $s_i(t)$ in terms of a finite number of terms is not an approximation wherein only the first N terms are significant; rather, it is an exact expression, where N and only N terms are significant.

EXAMPLE 2

2B1Q Code

The 2B1Q code is the North American line code for a special class of modems called digital subscriber lines. This code represents a quaternary PAM signal as shown in the Gray-encoded alphabet of Table 7.1. The four possible signals $s_1(t)$, $s_2(t)$, $s_3(t)$, and $s_4(t)$ are amplitude-scaled versions of a Nyquist pulse. Each signal represents a *dibit* (i.e., pair of bits). The issue of interest is to find the vector representation of the 2B1Q code.

This example is simple enough for us to solve it by inspection. Let $\phi_1(t)$ denote a pulse normalized to have unit energy. The $\phi_1(t)$ so defined is the only basis function for the vector representation of the 2B1Q code. Accordingly, the signal-space representation of this code is as shown in Figure 7.5. It consists of four signal vectors $\mathbf{s}_1$, $\mathbf{s}_2$, $\mathbf{s}_3$, and $\mathbf{s}_4$, which are located on the ϕ_1-axis in a symmetric manner about the origin. In this example, we have $M = 4$ and $N = 1$.

Table 7.1 **Amplitude levels of the 2B1Q code**

Signal	Amplitude	Gray code
$s_1(t)$	−3	00
$s_2(t)$	−1	01
$s_3(t)$	+1	11
$s_4(t)$	+3	10

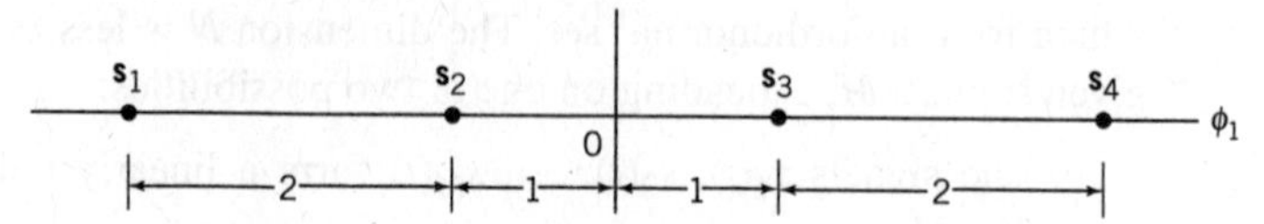

Figure 7.5 Signal-space representation of the 2B1Q code.

We may generalize the result depicted in Figure 7.5 for the 2B1Q code as follows: the signal-space diagram of an M-ary PAM signal, in general, is one-dimensional with M signal points uniformly positioned on the only axis of the diagram.

7.3 Conversion of the Continuous AWGN Channel into a Vector Channel

Suppose that the input to the bank of N product integrators or correlators in Figure 7.2b is not the transmitted signal $s_i(t)$ but rather the received signal $x(t)$ defined in accordance with the AWGN channel of Figure 7.1. That is to say,

$$x(t) = s_i(t) + w(t), \qquad \begin{cases} 0 \le t \le T \\ i = 1, 2, \ldots, M \end{cases} \tag{7.24}$$

where $w(t)$ is a sample function of the white Gaussian noise process $W(t)$ of zero mean and power spectral density $N_0/2$. Correspondingly, we find that the output of correlator j, say, is the sample value of a random variable X_j, whose sample value is defined by

$$\begin{aligned} x_j &= \int_0^T x(t)\phi_j(t)\, dt \\ &= s_{ij} + w_j, \qquad j = 1, 2, \ldots, N \end{aligned} \tag{7.25}$$

The first component, s_{ij}, is the deterministic component of x_j due to the transmitted signal $s_i(t)$, as shown by

$$s_{ij} = \int_0^T s_i(t)\phi_j(t)\, dt \tag{7.26}$$

The second component, w_j, is the sample value of a random variable W_j due to the channel noise $w(t)$, as shown by

$$w_j = \int_0^T w(t)\phi_j(t)\, dt \tag{7.27}$$

Consider next a new stochastic process $X'(t)$ whose sample function $x'(t)$ is related to the received signal $x(t)$ as follows:

$$x'(t) = x(t) - \sum_{j=1}^{N} x_j\phi_j(t) \tag{7.28}$$

Substituting (7.24) and (7.25) into (7.28), and then using the expansion of (7.4), we get

$$\begin{aligned} x'(t) &= s_i(t) + w(t) - \sum_{j=1}^{N} (s_{ij} + w_j)\phi_j(t) \\ &= w(t) - \sum_{j=1}^{N} w_j\phi_j(t) \\ &= w'(t) \end{aligned} \tag{7.29}$$

The sample function $x'(t)$, therefore, depends solely on the channel noise $w(t)$. On the basis of (7.28) and (7.29), we may thus express the received signal as

$$\begin{aligned} x(t) &= \sum_{j=1}^{N} x_j\phi_j(t) + x'(t) \\ &= \sum_{j=1}^{N} x_j\phi_j(t) + w'(t) \end{aligned} \tag{7.30}$$

Accordingly, we may view $w'(t)$ as a *remainder* term that must be included on the right-hand side of (7.30) to preserve equality. It is informative to contrast the expansion of the received signal $x(t)$ given in (7.30) with the corresponding expansion of the transmitted signal $s_i(t)$ given in (7.4): the expansion of (7.4), pertaining to the transmitter, is entirely deterministic; on the other hand, the expansion of (7.30) is random (stochastic) due to the channel noise at the receiver input.

Statistical Characterization of the Correlator Outputs

We now wish to develop a statistical characterization of the set of N correlator outputs. Let $X(t)$ denote the stochastic process, a sample function of which is represented by the received signal $x(t)$. Correspondingly, let X_j denote the random variable whose sample value is represented by the correlator output x_j, $j = 1, 2, \ldots, N$. According to the AWGN model of Figure 7.1, the stochastic process $X(t)$ is a Gaussian process. It follows, therefore, that X_j is a Gaussian random variable for all j in accordance with Property 1 of a Gaussian process (Chapter 4). Hence, X_j is characterized completely by its mean and variance, which are determined next.

Let W_j denote the random variable represented by the sample value w_j produced by the jth correlator in response to the white Gaussian noise component $w(t)$. The random variable W_j has zero mean because the channel noise process $W(t)$ represented by $w(t)$ in the AWGN model of Figure 7.1 has zero mean by definition. Consequently, the mean of X_j depends only on s_{ij}, as shown by

$$\begin{aligned} \mu_{X_j} &= \mathbb{E}[X_j] \\ &= \mathbb{E}[s_{ij} + W_j] \\ &= s_{ij} + \mathbb{E}[W_j] \\ &= s_{ij} \end{aligned} \tag{7.31}$$

To find the variance of X_j, we start with the definition

$$\begin{aligned}\sigma_{X_j}^2 &= \mathrm{var}[X_j] \\ &= \mathbb{E}[(X_j - s_{ij})^2] \\ &= \mathbb{E}[W_j^2]\end{aligned} \tag{7.32}$$

where the last line follows from (7.25) with x_j and w_j replaced by X_j and W_j, respectively. According to (7.27), the random variable W_j is defined by

$$W_j = \int_0^T W(t)\phi_j(t)\, dt$$

We may therefore expand (7.32) as

$$\begin{aligned}\sigma_{X_j}^2 &= \mathbb{E}\int_0^T W(t)\phi_j(t)\, dt \int_0^T W(u)\phi_j(u)\, du \\ &= \mathbb{E}\left[\int_0^T\int_0^T \phi_j(t)\phi_j(u)W(t)W(u)\, dt\, du\right]\end{aligned} \tag{7.33}$$

Interchanging the order of integration and expectation, which we can do because they are both linear operations, we obtain

$$\begin{aligned}\sigma_{X_j}^2 &= \int_0^T\int_0^T \phi_j(t)\phi_j(u)\mathbb{E}[W(t)W(u)]\, dt\, du \\ &= \int_0^T\int_0^T \phi_j(t)\phi_j(u)R_W(t, u)\, dt\, du\end{aligned} \tag{7.34}$$

where $R_W(t,u)$ is the autocorrelation function of the noise process $W(t)$. Since this noise is stationary, $R_W(t,u)$ depends only on the time difference $t - u$. Furthermore, since $W(t)$ is white with a constant power spectral density $N_0/2$, we may express $R_W(t,u)$ as

$$R_W(t, u) = \left(\frac{N_0}{2}\right)\delta(t-u) \tag{7.35}$$

Therefore, substituting (7.35) into (7.34) and then using the sifting property of the delta function $\delta(t)$, we get

$$\begin{aligned}\sigma_{X_j}^2 &= \frac{N_0}{2}\int_0^T\int_0^T \phi_j(t)\phi_j(u)\delta(t-u)\, dt\, du \\ &= \frac{N_0}{2}\int_0^T \phi_j^2(t)\, dt\end{aligned}$$

Since the $\phi_j(t)$ have unit energy, by definition, the expression for noise variance $\sigma_{x,j}^2$ reduces to

$$\sigma_{X_j}^2 = \frac{N_0}{2}, \qquad \text{for all } j \tag{7.36}$$

This important result shows that all the correlator outputs, denoted by X_j with $j = 1, 2, \ldots, N$, have a variance equal to the power spectral density $N_0/2$ of the noise process $W(t)$.

Moreover, since the basic functions $\phi_j(t)$ form an orthonormal set, X_j and X_k are mutually uncorrelated, as shown by

$$\begin{aligned}
\text{cov}[X_j X_k] &= \mathbb{E}[(X_j - \mu_{X_j})(X_k - \mu_{X_k})] \\
&= \mathbb{E}[(X_j - s_{ij})(X_k - s_{ik})] \\
&= \mathbb{E}[W_j W_k] \\
&= \mathbb{E}\left[\int_0^T W(t)\phi_j(t)\,dt \int_0^T W(u)\phi_k(u)\,du\right] \\
&= \int_0^T \int_0^T \phi_j(t)\phi_k(u) R_W(t,u)\,dt\,du \\
&= \frac{N_0}{2}\int_0^T \int_0^T \phi_j(t)\phi_k(u)\delta(t-u)\,dt\,du \\
&= \frac{N_0}{2}\int_0^T \phi_j(t)\phi_k(u)\,dt \\
&= 0, \qquad j \neq k
\end{aligned} \tag{7.37}$$

Since the X_j are Gaussian random variables, (7.37) implies that they are also statistically independent in accordance with Property 4 of a Gaussian process (Chapter 4).

Define the vector of N random variables

$$\mathbf{X} = \begin{bmatrix} X_1 \\ X_2 \\ \vdots \\ X_N \end{bmatrix} \tag{7.38}$$

whose elements are independent Gaussian random variables with mean values equal to s_{ij} and variances equal to $N_0/2$. Since the elements of the vector $\mathbf{X}$ are statistically independent, we may express the conditional probability density function of the vector $\mathbf{X}$, given that the signal $s_i(t)$ or the corresponding symbol m_i was sent, as the product of the conditional probability density functions of its individual elements; that is,

$$f_{\mathbf{X}}(\mathbf{x}|m_i) = \prod_{j=1}^{N} f_{X_j}(x_j|m_i), \qquad i = 1, 2, \ldots, M \tag{7.39}$$

where the vector $\mathbf{x}$ and scalar x_j are sample values of the random vector $\mathbf{X}$ and random variable X_j, respectively. The vector $\mathbf{x}$ is called the *observation vector*; correspondingly, x_j is called an *element* of the observation vector. A channel that satisfies (7.39) is said to be a *memoryless channel.*

Since each X_j is a Gaussian random variable with mean s_{ij} and variance $N_0/2$, we have

$$f_{X_j}(x_j|m_i) = \frac{1}{\sqrt{\pi N_0}} \exp\left[-\frac{1}{N_0}(x_j - s_{ij})^2\right], \qquad \begin{cases} j = 1, 2, \ldots, N \\ i = 1, 2, \ldots, M \end{cases} \tag{7.40}$$

Therefore, substituting (7.40) into (7.39) yields

$$f_{\mathbf{X}}(\mathbf{x}|m_i) = (\pi N_0)^{-N/2} \exp\left[-\frac{1}{N_0}\sum_{j=1}^{N}(x_j - s_{ij})^2\right], \qquad i = 1, 2, \ldots, M \tag{7.41}$$

which completely characterizes the first term of (7.30).

However, there remains the noise term $w'(t)$ in (7.30) to be accounted for. Since the noise process $W(t)$ represented by $w(t)$ is Gaussian with zero mean, it follows that the noise process $W'(t)$ represented by the sample function $w'(t)$ is also a zero-mean Gaussian process. Finally, we note that any random variable $W'(t_k)$, say, derived from the noise process $W'(t)$ by sampling it at time t_k, is in fact statistically independent of the random variable X_j; that is to say:

$$\mathbb{E}[X_j W'(t_k)] = 0, \qquad \begin{cases} j = 1, 2, \ldots, N \\ 0 \le t_k \le T \end{cases} \tag{7.42}$$

Since any random variable based on the remainder noise process $W'(t)$ is independent of the set of random variables $\{X_j\}$ as well as the set of transmitted signals $\{s_i(t)\}$, (7.42) states that the random variable $W'(t_k)$ is irrelevant to the decision as to which particular signal was actually transmitted. In other words, the correlator outputs determined by the received signal $x(t)$ are the only data that are useful for the decision-making process; therefore, they represent *sufficient statistics* for the problem at hand. By definition, sufficient statistics summarize the whole of the relevant information supplied by an observation vector.

We may now summarize the results presented in this section by formulating the *theorem of irrelevance*:

> Insofar as signal detection in AWGN is concerned, only the projections of the noise onto the basis functions of the signal set $\{s_i(t)\}_{i=1}^{M}$ affect the sufficient statistics of the detection problem; the remainder of the noise is irrelevant.

Putting this theorem into a mathematical context, we may say that the AWGN channel model of Figure 7.1a is equivalent to an *N-dimensional vector channel* described by the equation

$$\mathbf{x} = \mathbf{s}_i + \mathbf{w}, \qquad i = 1, 2, \ldots, M \tag{7.43}$$

where the dimension N is the number of basis functions involved in formulating the signal vector $\mathbf{s}_i$ for all i. The individual components of the signal vector $\mathbf{s}_i$ and the additive Gaussian noise vector $\mathbf{w}$ are defined by (7.5) and (7.27), respectively. The theorem of irrelevance and its mathematical description given in (7.43) are indeed basic to the understanding of the signal-detection problem as described next. Just as importantly, (7.43) may be viewed as the *baseband version* of the time-dependent received signal of (7.24).

Likelihood Function

The conditional probability density functions $f_X(\mathbf{x}|m_i)$, $i = 1, 2, \ldots, M$, provide the very characterization of an AWGN channel. Their derivation leads to a functional dependence on the observation vector $\mathbf{x}$ given the transmitted message symbol m_i. However, at the receiver we have the exact opposite situation: we are given the observation vector $\mathbf{x}$ and the requirement is to estimate the message symbol m_i that is responsible for generating $\mathbf{x}$. To emphasize this latter viewpoint, we follow Chapter 3 by introducing the idea of a *likelihood function*, denoted by $l(m_i)$ and defined by

$$l(m_i) = f_X(\mathbf{x}|m_i), \qquad i = 1, 2, \ldots, M \tag{7.44}$$

However, tt is important to recall from Chapter 3 that although $l(m_i)$ and $f_X(\mathbf{x}|m_i)$ have exactly the same mathematical form, their individual meanings are quite different.

In practice, we find it more convenient to work with the *log-likelihood function*, denoted by $L(m_i)$ and defined by

$$L(m_i) = \ln l(m_i), \qquad i = 1, 2, \ldots, M \tag{7.45}$$

where ln denotes the natural logarithm. The log-likelihood function bears a one-to-one relationship to the likelihood function for two reasons:

1. By definition, a probability density function is always nonnegative. It follows, therefore, that the likelihood function is likewise a nonnegative quantity.
2. The logarithmic function is a monotonically increasing function of its argument.

The use of (7.41) in (7.45) yields the log-likelihood function for an AWGN channel as

$$L(m_i) = -\frac{1}{N_0}\sum_{j=1}^{N}(x_j - s_{ij})^2, \qquad i = 1, 2, \ldots, M \tag{7.46}$$

where we have ignored the constant term $-(N/2)\ln(\pi N_0)$ since it bears no relation whatsoever to the message symbol m_i. Recall that the s_{ij}, $j = 1, 2, \ldots, N$, are the elements of the signal vector $\mathbf{s}_i$ representing the message symbol m_i. With (7.46) at our disposal, we are now ready to address the basic receiver design problem.

7.4 Optimum Receivers Using Coherent Detection

Maximum Likelihood Decoding

Suppose that, in each time slot of duration T seconds, one of the M possible signals $s_1(t)$, $s_2(t)$, $\ldots$, $s_M(t)$ is transmitted with equal probability, $1/M$. For geometric signal representation, the signal $s_i(t)$, $i = 1, 2, \ldots, M$, is applied to a bank of correlators with a common input and supplied with an appropriate set of N orthonormal basis functions, as depicted in Figure 7.2b. The resulting correlator outputs define the signal vector $\mathbf{s}_i$. Since knowledge of the signal vector $\mathbf{s}_i$ is as good as knowing the transmitted signal $s_i(t)$ itself, and vice versa, we may represent $s_i(t)$ by a point in a Euclidean space of dimension $N \le M$. We refer to this point as the *transmitted signal point*, or *message point* for short. The set of message points corresponding to the set of transmitted signals $\{s_i(t)\}_{i=1}^{M}$ is called a *message constellation*.

However, representation of the received signal $x(t)$ is complicated by the presence of additive noise $w(t)$. We note that when the received signal $x(t)$ is applied to the bank of N correlators, the correlator outputs define the observation vector $\mathbf{x}$. According to (7.43), the vector $\mathbf{x}$ differs from the signal vector $\mathbf{s}_i$ by the noise vector $\mathbf{w}$, whose orientation is completely random, as it should be.

The noise vector **w** is completely characterized by the channel noise $w(t)$; the converse of this statement, however, is not true, as explained previously. The noise vector **w** represents that portion of the noise $w(t)$ that will interfere with the detection process; the remaining portion of this noise, denoted by $w'(t)$, is tuned out by the bank of correlators and, therefore, irrelevant.

Based on the observation vector **x**, we may represent the received signal $x(t)$ by a point in the same Euclidean space used to represent the transmitted signal. We refer to this second point as the *received signal point*. Owing to the presence of noise, the received signal point wanders about the message point in a completely random fashion, in the sense that it may lie anywhere inside a Gaussian-distributed "cloud" centered on the message point. This is illustrated in Figure 7.6a for the case of a three-dimensional signal space. For a particular realization of the noise vector **w** (i.e., a particular point inside the random cloud of Figure 7.6a) the relationship between the observation vector **x** and the signal vector $\mathbf{s}_i$ is as illustrated in Figure 7.6b.

We are now ready to state the signal-detection problem:

> Given the observation vector **x**, perform a mapping from **x** to an estimate $\hat{m}$ of the transmitted symbol, m_i, in a way that would minimize the probability of error in the decision-making process.

Given the observation vector **x**, suppose that we make the decision $\hat{m} = m_i$. The probability of error in this decision, which we denote by $P_e(m_i|\mathbf{x})$, is simply

$$P_e(m_i|\mathbf{x}) = 1 - \mathbb{P}(m_i \text{ sent}|\mathbf{x}) \tag{7.47}$$

The requirement is to minimize the average probability of error in mapping each given observation vector **x** into a decision. On the basis of (7.47), we may, therefore, state the *optimum decision rule*:

> Set $\hat{m} = m_i$ if
>
> $$\mathbb{P}(m_i|\text{sent}|\mathbf{x}) \geq \mathbb{P}(m_k|\text{sent}|\mathbf{x}) \quad \text{for all } k \neq i \text{ and } k = 1, 2, \ldots, M. \tag{7.48}$$

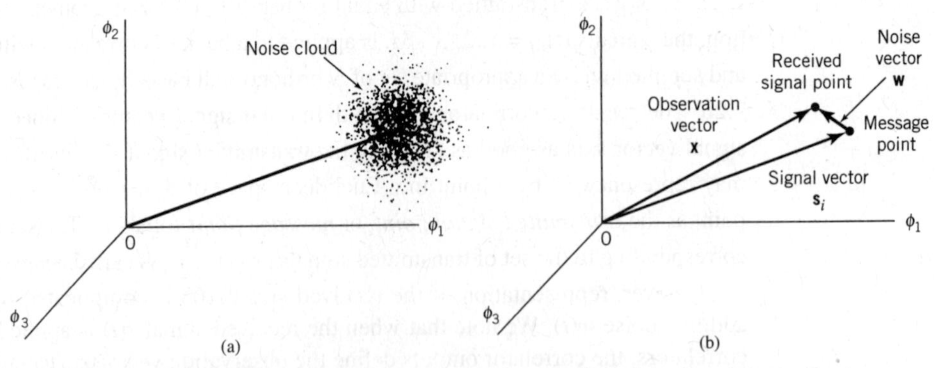

Figure 7.6 Illustrating the effect of (a) noise perturbation on (b) the location of the received signal point.

The decision rule described in (7.48) is referred to as the *maximum a posteriori probability (MAP) rule*. Correspondingly, the system used to implement this rule is called a *maximum a posteriori decoder*.

The requirement of (7.48) may be expressed more explicitly in terms of the *prior* probabilities of the transmitted signals and the likelihood functions, using Bayes' rule discussed in Chapter 3. For the moment, ignoring possible ties in the decision-making process, we may restate the MAP rule as follows:

Set $\hat{m} = m_i$ if

$$\frac{\pi_k f_{\mathbf{X}}(\mathbf{x}|m_i)}{f_{\mathbf{X}}(\mathbf{x})} \quad \text{is maximum for } k = i \tag{7.49}$$

where π_k is the prior probability of transmitting symbol m_k, $f_{\mathbf{X}}(\mathbf{x}|m_i)$ is the conditional probability density function of the random observation vector $\mathbf{X}$ given the transmission of symbol m_k, and $f_{\mathbf{X}}(\mathbf{x})$ is the unconditional probability density function of $\mathbf{X}$.

In (7.49), we now note the following points:

- the denominator term $f_{\mathbf{X}}(\mathbf{x})$ is independent of the transmitted symbol;
- the prior probability $\pi_k = \pi_i$ when all the source symbols are transmitted with equal probability; and
- the conditional probability density function $f_{\mathbf{X}}(\mathbf{x}|m_k)$ bears a one-to-one relationship to the log-likelihood function $L(m_k)$.

Accordingly, we may simply restate the decision rule of (7.49) in terms of $L(m_k)$ as follows:

$$\text{Set } \hat{m} = m_i \text{ if } L(m_k) \text{ is maximum for } k = i. \tag{7.50}$$

The decision rule of (7.50) is known as the *maximum likelihood rule*, discussed previously in Chapter 3; the system used for its implementation is correspondingly referred to as the *maximum likelihood decoder*. According to this decision rule, a maximum likelihood decoder computes the log-likelihood functions as metrics for all the M possible message symbols, compares them, and then decides in favor of the maximum. Thus, the maximum likelihood decoder is a simplified version of the maximum a posteriori decoder, in that the M message symbols are assumed to be equally likely.

It is useful to have a graphical interpretation of the maximum likelihood decision rule. Let Z denote the N-dimensional space of all possible observation vectors $\mathbf{x}$. We refer to this space as the *observation space*. Because we have assumed that the decision rule must say $\hat{m} = m_i$, where $i = 1, 2, \ldots, M$, the total observation space Z is correspondingly partitioned into *M-decision regions*, denoted by $Z_1, Z_2, \ldots, Z_M$. Accordingly, we may restate the decision rule of (7.50) as

$$\text{Observation vector } \mathbf{x} \text{ lies in region } Z_i \text{ if } L(m_k) \text{ is maximum for } k = i. \tag{7.51}$$

Aside from the boundaries between the decision regions $Z_1, Z_2, \ldots, Z_M$, it is clear that this set of regions covers the entire observation space. We now adopt the convention that all ties are resolved at random; that is, the receiver simply makes a random guess. Specifically, if the observation vector $\mathbf{x}$ falls on the boundary between any two decision regions, Z_i and Z_k, say, the choice between the two possible decisions $\hat{m} = m_i$ and $\hat{m} = m_k$ is resolved *a priori* by the flip of a fair coin. Clearly, the outcome of such an event does not affect the ultimate value of the probability of error since, on this boundary, the condition of (7.48) is satisfied with the equality sign.

The maximum likelihood decision rule of (7.50) or its geometric counterpart described in (7.51) assumes that the channel noise $w(t)$ is additive. We next specialize this rule for the case when $w(t)$ is both white and Gaussian.

From the log-likelihood function defined in (7.46) for an AWGN channel, we note that $L(m_k)$ attains its maximum value when the summation term $\sum_{j=1}^{N}(x_j - s_{kj})^2$ is minimized by the choice $k = i$. Accordingly, we may formulate the maximum likelihood decision rule for an AWGN channel as

$$\text{Observation vector } \mathbf{x} \text{ lies in region } Z_i \text{ if } \sum_{j=1}^{N}(x_j - s_{kj})^2 \text{ is minimum for } k = i. \tag{7.52}$$

Note we have used "minimum" as the optimizing condition in (7.52) because the minus sign in (7.46) has been ignored. Next, we note from the discussion presented in Section 7.2 that

$$\sum_{j=1}^{N}(x_j - s_{kj})^2 = \|\mathbf{x} - \mathbf{s}_k\|^2 \tag{7.53}$$

where $\|\mathbf{x} - \mathbf{s}_k\|$ is the Euclidean distance between the observation vector $\mathbf{x}$ at the receiver input and the transmitted signal vector $\mathbf{s}_k$. Accordingly, we may restate the decision rule of (7.53) as

$$\begin{aligned}&\text{Observation vector } \mathbf{x} \text{ lies in region } Z_i \text{ if Euclidean distance } \|\mathbf{x} - \mathbf{s}_k\| \\ &\text{is minimum for } k = i\end{aligned} \tag{7.54}$$

In words, (7.54) states that *the maximum likelihood decision rule is simply to choose the message point closest to the received signal point,* which is intuitively satisfying.

In practice, the decision rule of (7.54) is simplified by expanding the summation on the left-hand side of (7.53) as

$$\sum_{j=1}^{N}(x_j - s_{kj})^2 = \sum_{j=1}^{N}x_j^2 - 2\sum_{j=1}^{N}x_j s_{kj} + \sum_{j=1}^{N}s_{kj}^2 \tag{7.55}$$

The first summation term of this expansion is independent of the index k pertaining to the transmitted signal vector s_k and, therefore, may be ignored. The second summation term is the inner product of the observation vector $\mathbf{x}$ and the transmitted signal vector $\mathbf{s}_k$. The third summation term is the transmitted signal energy

$$E_k = \sum_{j=1}^{N}s_{kj}^2 \tag{7.56}$$

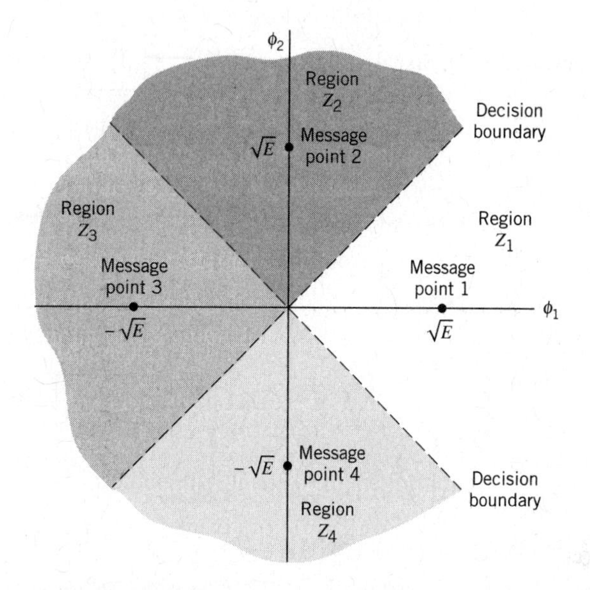

Figure 7.7
Illustrating the partitioning of the observation space into decision regions for the case when $N = 2$ and $M = 4$; it is assumed that the M transmitted symbols are equally likely.

Accordingly, we may reformulate the maximum-likelihood decision rule one last time:

$$\text{Observation vector } \mathbf{x} \text{ lies in region } Z_i \text{ if } \left(\sum_{j=1}^{N} x_j s_{kj} - \frac{1}{2} E_k \right) \text{ is maximim for } k = i, \text{ where } E_k \text{ is transmitted energy.} \tag{7.57}$$

From (7.57) we infer that, for an AWGN channel, the M decision regions are bounded by linear hyperplane boundaries. The example in Figure 7.7 illustrates this statement for $M = 4$ signals and $N = 2$ dimensions, assuming that the signals are transmitted with equal energy E and equal probability.

Correlation Receiver

In light of the material just presented, the optimum receiver for an AWGN channel and for the case when the transmitted signals $s_1(t)$, $s_2(t)$, ..., $s_M(t)$ are equally likely is called a *correlation receiver*; it consists of two subsystems, which are detailed in Figure 7.8:

1. *Detector* (Figure 7.8a), which consists of M *correlators* supplied with a set of orthonormal basis functions $\phi_1(t)$, $\phi_2(t)$, ..., $\phi_N(t)$ that are generated locally; this bank of correlators operates on the received signal $x(t)$, $0 \le t \le T$, to produce the observation vector $\mathbf{x}$.
2. *Maximum-likelihood decoder* (Figure 7.8b), which operates on the observation vector $\mathbf{x}$ to produce an estimate $\hat{m}$ of the transmitted symbol m_i, $i = 1, 2, \ldots, M$, in such a way that the average probability of symbol error is minimized.

In accordance with the maximum likelihood decision rule of (7.57), the decoder multiplies the N elements of the observation vector $\mathbf{x}$ by the corresponding N elements of each of the M signal vectors $\mathbf{s}_1$, $\mathbf{s}_2$, ..., $\mathbf{s}_M$. Then, the resulting products are successively summed in *accumulators* to form the corresponding set of inner products $\{\mathbf{x}^T\mathbf{s}_k | k = 1, 2, \ldots, M\}$.

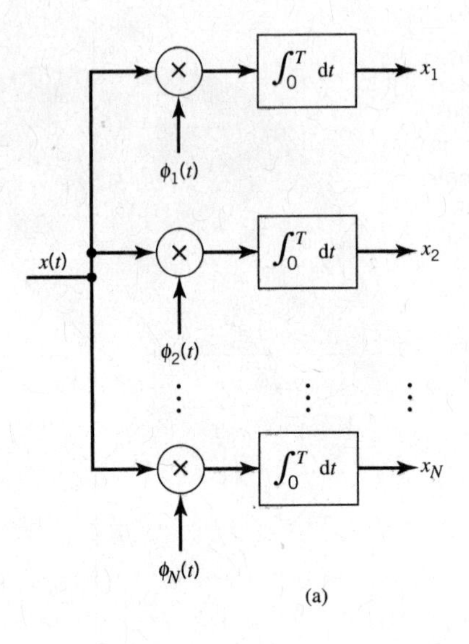

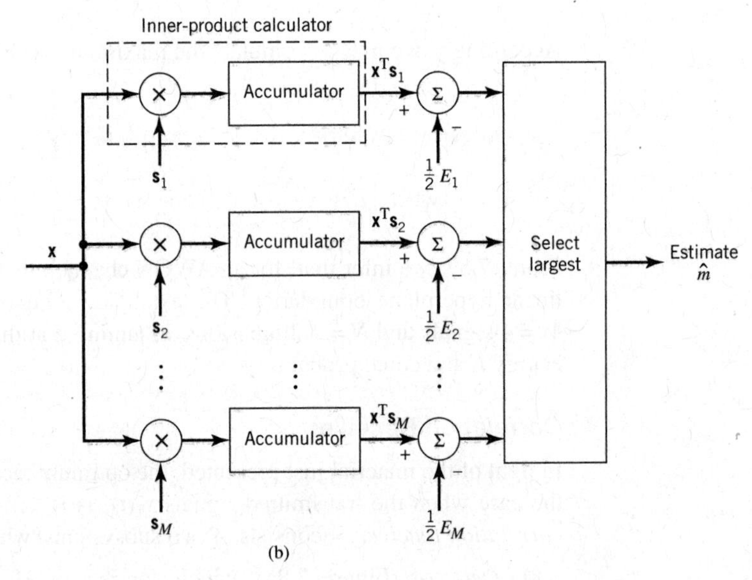

Figure 7.8
(a) Detector or demodulator. (b) Signal transmission decoder.

Next, the inner products are corrected for the fact that the transmitted signal energies may be unequal. Finally, the largest one in the resulting set of numbers is selected, and an appropriate decision on the transmitted message is thereby made.

Matched Filter Receiver

The detector shown in Figure 7.8a involves a set of correlators. Alternatively, we may use a different but equivalent structure in place of the correlators. To explore this alternative method of implementing the optimum receiver, consider a linear time-invariant filter with impulse response $h_j(t)$. With the received signal $x(t)$ operating as input, the resulting filter output is defined by the convolution integral

$$y_j(t) = \int_{-\infty}^{\infty} x(\tau)h_j(t-\tau)\,d\tau$$

To proceed further, we evaluate this integral over the duration of a transmitted symbol, namely $0 \le t \le T$. With time t restricted in this manner, we may replace the variable τ with t and go on to write

$$y_j(T) = \int_0^T x(t)h_j(T-t)\,dt \tag{7.58}$$

Consider next a detector based on a bank of correlators. The output of the jth correlator is defined by the first line of (7.25), reproduced here for convenience of representation:

$$x_j = \int_0^T x(t)\phi_j(t)\,dt \tag{7.59}$$

For $y_j(T)$ to equal x_j, we find from (7.58) and (7.59) that this condition is satisfied provided that we choose

$$h_j(T-t) = \phi_j(t) \qquad \text{for } 0 \le t \le T \text{ and } j = 1, 2, \ldots, M$$

Equivalently, we may express the condition imposed on the desired impulse response of the filter as

$$h_j(t) = \phi_j(T-t), \qquad \text{for } 0 \le t \le T \text{ and } j = 1, 2, \ldots, M \tag{7.60}$$

We may now generalize the condition described in (7.60) by stating:

> Given a pulse signal $\phi(t)$ occupying the interval $0 \le t \le T$, a linear time-invariant filter is said to be matched to the signal $\phi(t)$ if its impulse response $h(t)$ satisfies the condition
>
> $$h(t) = \phi(T-t) \qquad \text{for } 0 \le t \le T \tag{7.61}$$

A time-invariant filter defined in this way is called a *matched filter*. Correspondingly, an optimum receiver using matched filters in place of correlators is called a *matched-filter receiver*. Such a receiver is depicted in Figure 7.9, shown below.

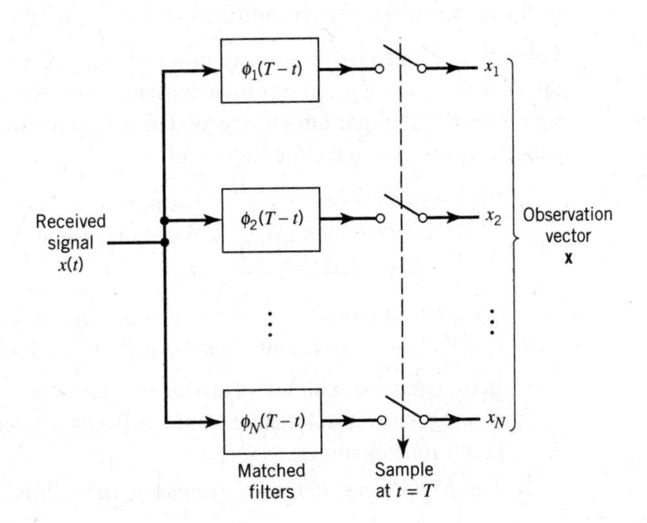

Figure 7.9 Detector part of matched filter receiver; the signal transmission decoder is as shown in Figure 7.8(b).

7.5 Probability of Error

To complete the statistical characterization of the correlation receiver of Figure 7.8a or its equivalent, the matched filter receiver of Figure 7.9, we need to evaluate its performance in the presence of AWGN. To do so, suppose that the observation space Z is partitioned into a set of regions, $\{Z_i\}_{i=1}^{M}$, in accordance with the maximum likelihood decision rule. Suppose also that symbol m_i (or, equivalently, signal vector $\mathbf{s}_i$) is transmitted and an observation vector $\mathbf{x}$ is received. Then, an error occurs whenever the received signal point represented by $\mathbf{x}$ does not fall inside region Z_i associated with the message point $\mathbf{s}_i$. Averaging over all possible transmitted symbols assumed to be equiprobable, we see that the *average probability of symbol error* is

$$\begin{aligned} P_{\mathrm{e}} &= \sum_{i=1}^{M} \pi_i \mathbb{P}(\mathbf{x} \text{ does not lie in } Z_i | m_i \text{ sent}) \\ &= \frac{1}{M} \sum_{i=1}^{M} \mathbb{P}(\mathbf{x} \text{ does not lie in } Z_i | m_i \text{ sent}), \; \pi_i = 1/M \\ &= 1 - \frac{1}{M} \sum_{i=1}^{M} \mathbb{P}(\mathbf{x} \text{ lies in } Z_i | m_i \text{ sent}) \end{aligned} \tag{7.62}$$

where we have used the standard notation to denote the conditional probability of an event. Since $\mathbf{x}$ is the sample value of random vector $\mathbf{X}$, we may rewrite (7.62) in terms of the likelihood function as follows, given that the message symbol m_i is sent:

$$P_{\mathrm{e}} = 1 - \frac{1}{M} \sum_{i=1}^{M} \int_{Z_i} f_{\mathbf{X}}(\mathbf{x}|m_i) \, d\mathbf{x} \tag{7.63}$$

For an N-dimensional observation vector, the integral in (7.63) is likewise N-dimensional.

Invariance of the Probability of Error to Rotation

There is a uniqueness to the way in which the observation space Z is partitioned into the set of regions $Z_1, Z_2, \ldots, Z_M$ in accordance with the maximum likelihood detection of a signal in AWGN; that uniqueness is defined by the message constellation under study. In particular, we may make the statement:

> Changes in the orientation of the message constellation with respect to both the coordinate axes and origin of the signal space do *not* affect the probability of symbol error P_{e} defined in (7.63).

This statement embodies the *invariance property of the average probability of symbol error P_{e} with respect to notation and translation*, which is the result of two facts:

1. In maximum likelihood detection, the probability of symbol error P_{e} depends solely on the relative Euclidean distance between a received signal point and message point in the constellation.
2. The AWGN is *spherically symmetric* in all directions in the signal space.

To elaborate, consider first the invariance of P_e with respect to *rotation*. The effect of a rotation applied to all the message points in a constellation is equivalent to multiplying the N-dimensional signal vector $\mathbf{s}_i$ by an N-by-N *orthonormal matrix* denoted by $\mathbf{Q}$ for all i. By definition, the matrix $\mathbf{Q}$ satisfies the condition

$$\mathbf{Q}\mathbf{Q}^{\mathrm{T}} = \mathbf{I} \tag{7.64}$$

where the superscript T denotes matrix transposition and $\mathbf{I}$ is the *identity matrix* whose diagonal elements are all unity and its off-diagonal elements are all zero. According to (7.64), the inverse of the real-valued orthonormal matrix $\mathbf{Q}$ is equal to its own transpose. Thus, in dealing with rotation, the message vector $\mathbf{s}_i$ is replaced by its rotated version

$$\mathbf{s}_{i,\,\text{rotate}} = \mathbf{Q}\mathbf{s}_i, \qquad i = 1, 2, \ldots, M \tag{7.65}$$

Correspondingly, the N-by-1 noise vector $\mathbf{w}$ is replaced by its rotated version

$$\mathbf{w}_{\text{rotate}} = \mathbf{Q}\mathbf{w} \tag{7.66}$$

However, the statistical characteristics of the noise vector are unaffected by this rotation for three reasons:

1. From Chapter 4 we recall that a linear combination of Gaussian random variables is also Gaussian. Since the noise vector $\mathbf{w}$ is Gaussian, by assumption, then it follows that the rotated noise vector $\mathbf{w}_{\text{rotate}}$ is also Gaussian.
2. Since the noise vector $\mathbf{w}$ has zero mean, the rotated noise vector $\mathbf{w}_{\text{rotate}}$ also has zero mean, as shown by

$$\begin{aligned}\mathbb{E}[\mathbf{w}_{\text{rotate}}] &= \mathbb{E}[\mathbf{Q}\mathbf{w}] \\ &= \mathbf{Q}\mathbb{E}[\mathbf{w}] \\ &= \mathbf{0}\end{aligned} \tag{7.67}$$

3. The covariance matrix of the noise vector $\mathbf{w}$ is equal to $(N_0/2)\mathbf{I}$, where $N_0/2$ is the power spectral density of the AWGN $w(t)$ and $\mathbf{I}$ is the identity matrix; that is

$$\mathbb{E}[\mathbf{w}\mathbf{w}^{\mathrm{T}}] = \frac{N_0}{2}\mathbf{I} \tag{7.68}$$

Hence, the covariance matrix of the rotated noise vector is

$$\begin{aligned}\mathbb{E}[\mathbf{w}_{\text{rotate}}\mathbf{w}_{\text{rotate}}^{\mathrm{T}}] &= \mathbb{E}[\mathbf{Q}\mathbf{w}(\mathbf{Q}\mathbf{w})^{\mathrm{T}}] \\ &= \mathbb{E}[\mathbf{Q}\mathbf{w}\mathbf{w}^{\mathrm{T}}\mathbf{Q}^{\mathrm{T}}] \\ &= \mathbf{Q}\mathbb{E}[\mathbf{w}\mathbf{w}^{\mathrm{T}}]\mathbf{Q}^{\mathrm{T}} \\ &= \frac{N_0}{2}\mathbf{Q}\mathbf{Q}^{\mathrm{T}} \\ &= \frac{N_0}{2}\mathbf{I}\end{aligned} \tag{7.69}$$

where, in the last two lines, we have made use of (7.68) and (7.64).

In light of these three reasons, we may, therefore, express the observation vector in the rotated message constellation as

$$\mathbf{x}_{\text{rotate}} = \mathbf{Q}\mathbf{s}_i + \mathbf{w}, \qquad i = 1, 2, \ldots, M \tag{7.70}$$

Using (7.65) and (7.70), we may now express the Euclidean distance between the rotated vectors $\mathbf{x}_{\text{rotate}}$ and $\mathbf{s}_{\text{rotate}}$ as

$$\begin{aligned}\|\mathbf{x}_{\text{rotate}} - \mathbf{s}_{i,\,\text{rotate}}\| &= \|\mathbf{Q}\mathbf{s}_i + \mathbf{w} - \mathbf{Q}\mathbf{s}_i\| \\ &= \|\mathbf{w}\| \\ &= \|\mathbf{x} - \mathbf{s}_i\|, \qquad i = 1, 2, \dots, M\end{aligned} \tag{7.71}$$

where, in the last line, we made use of (7.43).

We may, therefore, formally state the *principle of rotational invariance*:

> If a message constellation is rotated by the transformation
>
> $$\mathbf{s}_{i,\,\text{rotate}} = \mathbf{Q}\mathbf{s}_i, \qquad i = 1, 2, \dots, M$$
>
> where $\mathbf{Q}$ is an orthonormal matrix, then the probability of symbol error P_e incurred in maximum likelihood signal-detection over an AWGN channel is completely unchanged.

EXAMPLE 3 **Illustration of Rotational Invariance**

To illustrate the principle of rotational invariance, consider the signal constellation shown in Figure 7.10a. The constellation is the same as that of Figure 7.10b, except for the fact that it has been rotated through 45°. Although these two constellations do indeed look different in a geometric sense, the principle of rotational invariance teaches us immediately that the P_e is the same for both of them.

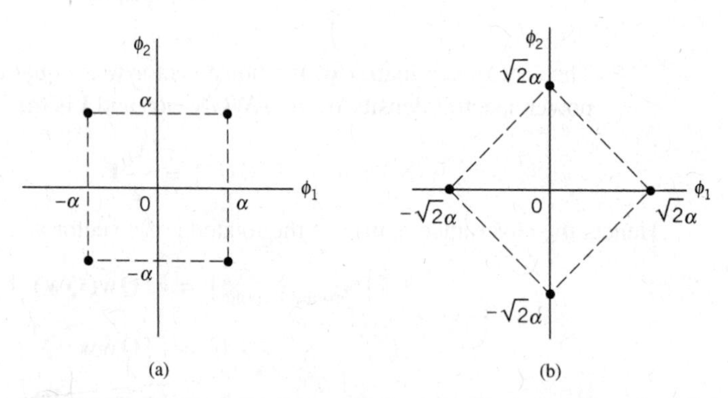

Figure 7.10 A pair of signal constellations for illustrating the principle of rotational invariance.

Invariance of the Probability to Translation

Consider next the invariance of P_e to translation. Suppose all the message points in a signal constellation are translated by a constant vector amount $\mathbf{a}$, as shown by

$$\mathbf{s}_{i,\,\text{translate}} = \mathbf{s}_i - \mathbf{a}, \qquad i = 1, 2, \dots, M \tag{7.72}$$

The observation vector is correspondingly translated by the same vector amount, as shown by

$$\mathbf{x}_{\text{translate}} = \mathbf{x} - \mathbf{a} \tag{7.73}$$

From (7.72) and (7.73) we see that the translation **a** is common to both the translated signal vector $\mathbf{s}_i$ and translated observation vector **x**. We, therefore, immediately deduce that

$$\|\mathbf{x}_{\text{translate}} - \mathbf{s}_{i,\text{ translate}}\| = \|\mathbf{x} - \mathbf{s}_i\|, \qquad \text{for } i = 1, 2, \ldots, M \tag{7.74}$$

and thus formulate the *principle of translational invariance*:

> If a signal constellation is translated by a constant vector amount, then the probability of symbol error P_e incurred in maximum likelihood signal detection over an AWGN channel is completely unchanged.

EXAMPLE 4

Translation of Signal Constellation

As an example, consider the two signal constellations shown in Figure 7.11, which pertain to a pair of different four-level PAM signals. The constellation of Figure 7.11b is the same as that of Figure 7.11a, except for a translation $3\alpha/2$ to the right along the ϕ_1-axis. The principle of translational invariance teaches us that the P_e is the same for both of these signal constellations.

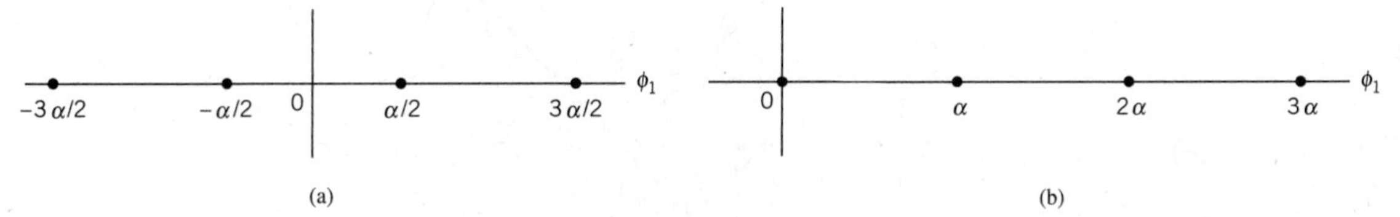

Figure 7.11 A pair of signal constellations for illustrating the principle of translational invariance.

Union Bound on the Probability of Error

For AWGN channels, the formulation of the average probability of symbol error[2] P_e is conceptually straightforward, in that we simply substitute (7.41) into (7.63). Unfortunately, however, numerical computation of the integral so obtained is impractical, except in a few simple (nevertheless, important) cases. To overcome this computational difficulty, we may resort to the use of *bounds*, which are usually adequate to predict the SNR (within a decibel or so) required to maintain a prescribed error rate. The approximation to the integral defining P_e is made by simplifying the integral or simplifying the region of integration. In the following, we use the latter procedure to develop a simple yet useful upper bound, called the *union bound*, as an approximation to the average probability of symbol error for a set of M equally likely signals (symbols) in an AWGN channel.

Let A_{ik}, with $(i,k) = 1, 2, \ldots, M$, denote the event that the observation vector **x** is closer to the signal vector $\mathbf{s}_k$ than to $\mathbf{s}_i$, when the symbol m_i (message vector $\mathbf{s}_i$) is sent. The conditional probability of symbol error when symbol m_i is sent, $P_e(m_i)$, is equal to the probability of the union of events, defined by the set $\{A_{ik}\}_{\substack{k=1 \\ k \neq i}}^{M}$. Probability theory teaches us that the *probability of a finite union of events is overbounded by the sum of the probabilities of the constituent events*. We may, therefore, write

$$P_e(m_i) \leq \sum_{\substack{k=1 \\ k \neq i}}^{M} \mathbb{P}(A_{ik}), \qquad i = 1, 2, \dots, M \tag{7.75}$$

EXAMPLE 5

Constellation of Four Message Points

To illustrate applicability of the union bound, consider Figure 7.12 for the case of $M = 4$. Figure 7.12a shows the four message points and associated decision regions, with the point $\mathbf{s}_1$ assumed to represent a transmitted symbol. Figure 7.12b shows the three constituent signal-space descriptions where, in each case, the transmitted message point $\mathbf{s}_1$ and one other message point are retained. According to Figure 7.12a the conditional probability of symbol error, $P_e(m_i)$, is equal to the probability that the observation vector $\mathbf{x}$

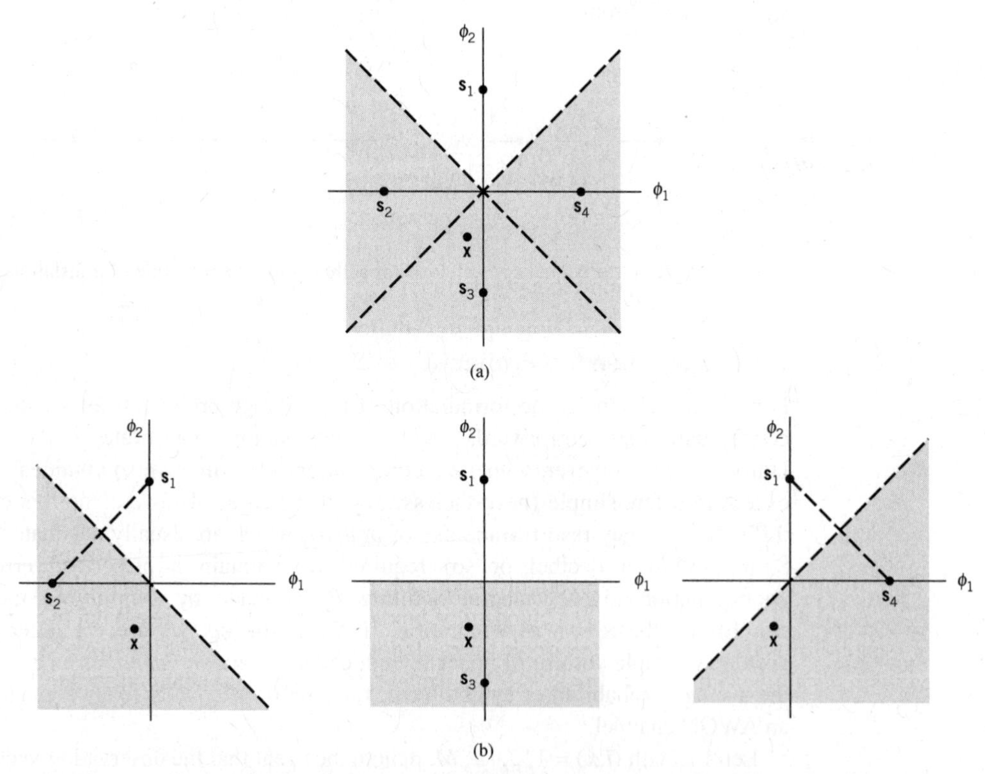

Figure 7.12 Illustrating the union bound. (a) Constellation of four message points. (b) Three constellations with a common message point and one other message point **x** retained from the original constellation.

lies in the shaded region of the two-dimensional signal-space diagram. Clearly, this probability is less than the sum of the probabilities of the three individual events that **x** lies in the shaded regions of the three constituent signal spaces depicted in Figure 7.12b.

Pairwise Error Probability

It is important to note that, in general, the probability $\mathbb{P}(A_{ik})$ is different from the probability $\mathbb{P}(\hat{m} = m_k | m_i)$, which is the probability that the observation vector **x** is closer to the signal vector $\mathbf{s}_k$ (i.e., symbol m_k) than every other when the vector $\mathbf{s}_i$ (i.e., symbol m_i) is sent. On the other hand, the probability $\mathbb{P}(A_{ik})$ depends on only two signal vectors, $\mathbf{s}_i$ and $\mathbf{s}_k$. To emphasize this difference, we rewrite (7.75) by adopting p_{ik} in place of $\mathbb{P}(A_{ik})$. We thus write

$$P_e(m_i) \le \sum_{\substack{k=1 \\ k \ne i}}^{M} p_{ik}, \quad i = 1, 2, \ldots, M \tag{7.76}$$

The probability p_{ik} is called the *pairwise error probability,* in that if a digital communication system uses only a pair of signals, $\mathbf{s}_i$ and $\mathbf{s}_k$, then p_{ik} is the probability of the receiver mistaking $\mathbf{s}_k$ for $\mathbf{s}_i$.

Consider then a simplified digital communication system that involves the use of two equally likely messages represented by the vectors $\mathbf{s}_i$ and $\mathbf{s}_k$. Since white Gaussian noise is identically distributed along any set of orthogonal axes, we may temporarily choose the first axis in such a set as one that passes through the points $\mathbf{s}_i$ and $\mathbf{s}_k$; for three illustrative examples, see Figure 7.12b. The corresponding decision boundary is represented by the bisector that is perpendicular to the line joining the points $\mathbf{s}_i$ and $\mathbf{s}_k$. Accordingly, when the vector $\mathbf{s}_i$ (i.e., symbol m_i) is sent, and if the observation vector **x** lies on the side of the bisector where $\mathbf{s}_k$ lies, an error is made. The probability of this event is given by

$$\begin{aligned} p_{ik} &= \mathbb{P}(\mathbf{x} \text{ is closer to } \mathbf{s}_k \text{ than } \mathbf{s}_i, \text{ when } \mathbf{s}_i \text{ is sent}) \\ &= \int_{d_{ik}/2}^{\infty} \frac{1}{\sqrt{\pi N_0}} \exp\left(-\frac{v^2}{N_0}\right) dv \end{aligned} \tag{7.77}$$

where d_{ik} in the lower limit of the integral is the Euclidean distance between signal vectors $\mathbf{s}_i$ and $\mathbf{s}_k$; that is,

$$d_{ik} = \|\mathbf{s}_i - \mathbf{s}_k\| \tag{7.78}$$

To change the integral of (7.77) into a standard form, define a new integration variable

$$z = \sqrt{\frac{2}{N_0}}\, v \tag{7.79}$$

Equation (7.77) is then rewritten in the desired form

$$p_{ik} = \frac{1}{\sqrt{2\pi}} \int_{d_{ik}/\sqrt{2N_0}}^{\infty} \exp\left(-\frac{z^2}{2}\right) dz \tag{7.80}$$

The integral in (7.80) is the Q-function of (3.68) that was introduced in Chapter 3. In terms of the Q-function, we may now express the probability p_{ik} in the compact form

$$p_{ik} = Q\left(\frac{d_{ik}}{\sqrt{2N_0}}\right) \tag{7.81}$$

Correspondingly, substituting (7.81) into (7.76), we write

$$P_e(m_i) \le \sum_{\substack{k=1 \\ k \ne i}}^{M} Q\left(\frac{d_{ik}}{\sqrt{2N_0}}\right), \quad i = 1, 2, \ldots, M \tag{7.82}$$

The probability of symbol error, averaged over all the M symbols, is, therefore, over-bounded as follows:

$$\begin{aligned} P_e &= \sum_{i=1}^{M} \pi_i P_e(m_i) \\ &\le \sum_{i=1}^{M} \sum_{\substack{k=1 \\ k \ne i}}^{M} \pi_i Q\left(\frac{d_{ik}}{\sqrt{2N_0}}\right) \end{aligned} \tag{7.83}$$

where π_i is the probability of sending symbol m_i.

There are two special forms of (7.83) that are noteworthy:

1. Suppose that the signal constellation is *circularly symmetric about the origin*. Then, the conditional probability of error $P_e(m_i)$ is the same for all i, in which case (7.83) reduces to

$$P_e \le \sum_{\substack{k=1 \\ k \ne i}}^{M} Q\left(\frac{d_{ik}}{\sqrt{2N_0}}\right) \quad \text{for all } i \tag{7.84}$$

Figure 7.10 illustrates two examples of circularly symmetric signal constellations.

2. Define the *minimum distance* of a signal constellation $d_{\min}$ as the smallest Euclidean distance between any two transmitted signal points in the constellation, as shown by

$$d_{\min} = \min_{k \ne i} d_{ik} \quad \text{for all } i \text{ and } k \tag{7.85}$$

Then, recognizing that the Q-function is a monotonically decreasing function of its argument, we have

$$Q\left(\frac{d_{ik}}{\sqrt{2N_0}}\right) \le Q\left(\frac{d_{\min}}{\sqrt{2N_0}}\right) \quad \text{for all } i \text{ and } k \tag{7.86}$$

Therefore, in general, we may simplify the bound on the average probability of symbol error in (7.83) as

$$P_e \le (M-1)Q\left(\frac{d_{\min}}{\sqrt{2N_0}}\right) \tag{7.87}$$

The Q-function in (7.87) is itself upper bounded as[3]

$$Q\left(\frac{d_{\min}}{\sqrt{2N_0}}\right) \le \frac{1}{\sqrt{2\pi}}\exp\left(-\frac{d_{\min}^2}{4N_0}\right) \tag{7.88}$$

Accordingly, we may further simplify the bound on P_e in (7.87) as

$$P_e < \left(\frac{M-1}{\sqrt{2\pi}}\right)\exp\left(-\frac{d_{\min}^2}{4N_0}\right) \tag{7.89}$$

In words, (7.89) states the following:

In an AWGN channel, the average probability of symbol error P_e decreases exponentially as the squared minimum distance, $d_{\min}^2$.

Bit Versus Symbol Error Probabilities

Thus far, the only figure of merit we have used to assess the noise performance of a digital communication system in AWGN has been the average probability of symbol (word) error. This figure of merit is the natural choice when messages of length $m = \log_2 M$ are transmitted, such as alphanumeric symbols. However, when the requirement is to transmit binary data such as digital computer data, it is often more meaningful to use another figure of merit called the BER. Although, in general, there are no unique relationships between these two figures of merit, it is fortunate that such relationships can be derived for two cases of practical interest, as discussed next.

Case 1: *M-tuples Differing in Only a Single Bit*

Suppose that it is possible to perform the mapping from binary to M-ary symbols in such a way that the two binary M-tuples corresponding to any pair of adjacent symbols in the M-ary modulation scheme differ in only one bit position. This mapping constraint is satisfied by using a *Gray code*. When the probability of symbol error P_e is acceptably small, we find that the probability of mistaking one symbol for either one of the two "nearest" symbols is greater than any other kind of symbol error. Moreover, given a symbol error, the most probable number of bit errors is one, subject to the aforementioned mapping constraint. Since there are $\log_2 M$ bits per symbol, it follows that the average probability of symbol error is related to the BER as follows:

$$\begin{aligned} P_e &= \mathbb{P}\left(\bigcup_{i=1}^{\log_2 M}\{i\text{th bit is in error}\}\right) \\ &\le \sum_{i=1}^{\log_2 M}\mathbb{P}(i\text{th bit is in error}) \\ &= \log_2 M \cdot (\text{BER}) \end{aligned} \tag{7.90}$$

where, in the first line, $\cup$ is the symbol for "union" as used in set theory. We also note that

$$P_e \ge \mathbb{P}(i\text{th bit is in error}) = \text{BER} \tag{7.91}$$

It follows, therefore, that the BER is bounded as follows:

$$\frac{P_e}{\log_2 M} \le \text{BER} \le P_e \tag{7.92}$$

Case 2: *Number of Symbols Equal to Integer Power of 2*

Suppose next $M = 2^K$, where K is an integer. We assume that all symbol errors are equally likely and occur with probability

$$\frac{P_e}{M-1} = \frac{P_e}{2^K - 1}$$

where P_e is the average probability of symbol error. To find the probability that the ith bit in a symbol is in error, we note that there are 2^{K-1} cases of symbol error in which this particular bit is changed and there are 2^{K-1} cases in which it is not. Hence, the BER is

$$\text{BER} = \left(\frac{2^{K-1}}{2^K - 1}\right) P_e \tag{7.93}$$

or, equivalently,

$$\text{BER} = \left(\frac{M/2}{M-1}\right) P_e \tag{7.94}$$

Note that, for large M, the BER approaches the limiting value of $P_e/2$. Note also that the bit errors are not independent in general.

7.6 Phase-Shift Keying Techniques Using Coherent Detection

With the background material on the coherent detection of signals in AWGN presented in Sections 7.2–7.4 at our disposal, we are now ready to study specific passband data-transmission systems. In this section, we focus on the family of phase-shift keying (PSK) techniques, starting with the simplest member of the family discussed next.

Binary Phase-Shift Keying

In a *binary PSK system*, the pair of signals $s_1(t)$ and $s_2(t)$ used to represent binary symbols 1 and 0, respectively, is defined by

$$s_1(t) = \sqrt{\frac{2E_b}{T_b}} \cos(2\pi f_c t), \qquad 0 \le t \le T_b \tag{7.95}$$

$$s_2(t) = \sqrt{\frac{2E_b}{T_b}} \cos(2\pi f_c t + \pi) = -\sqrt{\frac{2E_b}{T_b}} \cos(2\pi f_c t), \qquad 0 \le t \le T_b \tag{7.96}$$

where T_b is the *bit duration* and E_b is the *transmitted signal energy per bit*. We find it convenient, although not necessary, to assume that each transmitted bit contains an integral number of cycles of the carrier wave; that is, the carrier frequency f_c is chosen equal to n_c/T_b for some fixed integer n_c. A pair of sinusoidal waves that differ only in a relative phase-shift of 180°, defined in (7.95) and (7.96), is referred to as an *antipodal signal*.

Signal-Space Diagram of Binary PSK Signals

From this pair of equations it is clear that, in the case of binary PSK, there is only one basis function of unit energy:

$$\phi_1(t) = \sqrt{\frac{2}{T_b}}\cos(2\pi f_c t), \qquad 0 \le t \le T_b \tag{7.97}$$

Then, we may respectively express the transmitted signals $s_1(t)$ and $s_2(t)$ in terms of $\phi_1(t)$ as

$$s_1(t) = \sqrt{E_b}\,\phi_1(t), \qquad 0 \le t \le T_b \tag{7.98}$$

$$s_2(t) = -\sqrt{E_b}\,\phi_1(t), \qquad 0 \le t \le T_b \tag{7.99}$$

A binary PSK system is, therefore, characterized by having a signal space that is one-dimensional (i.e., $N = 1$), with a signal constellation consisting of two message points (i.e., $M = 2$). The respective coordinates of the two message points are

$$\begin{aligned} s_{11} &= \int_0^{T_b} s_1(t)\phi_1(t)\,dt \\ &= +\sqrt{E_b} \end{aligned} \tag{7.100}$$

$$\begin{aligned} s_{21} &= \int_0^{T_b} s_2(t)\phi_1(t)\,dt \\ &= -\sqrt{E_b} \end{aligned} \tag{7.101}$$

In words, the message point corresponding to $s_1(t)$ is located at $s_{11} = +\sqrt{E_b}$ and the message point corresponding to $s_2(t)$ is located at $s_{21} = -\sqrt{E_b}$. Figure 7.13a displays the

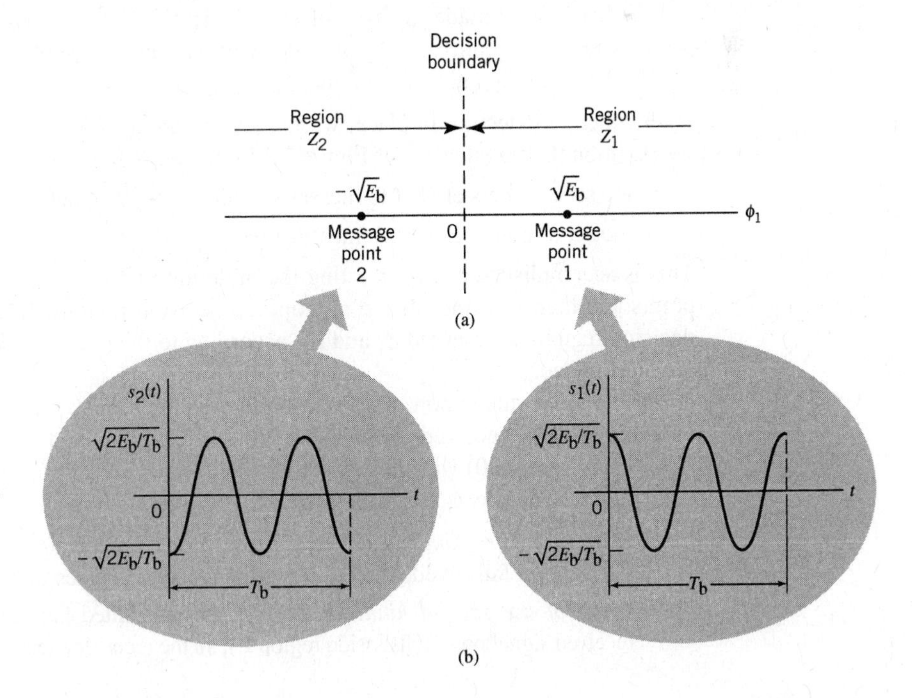

Figure 7.13
(a) Signal-space diagram for coherent binary PSK system. (b) The waveforms depicting the transmitted signals $s_1(t)$ and $s_2(t)$, assuming $n_c = 2$.

signal-space diagram for binary PSK and Figure 7.13b shows example waveforms of antipodal signals representing $s_1(t)$ and $s_2(t)$. Note that the binary constellation of Figure 7.13 has *minimum average energy.*

Generation of a binary PSK signal follows readily from (7.97) to (7.99). Specifically, as shown in the block diagram of Figure 7.14a, the generator (transmitter) consists of two components:

1. *Polar NRZ-level encoder*, which represents symbols 1 and 0 of the incoming binary sequence by amplitude levels $+\sqrt{E_b}$ and $-\sqrt{E_b}$, respectively.
2. *Product modulator*, which multiplies the output of the polar NRZ encoder by the basis function $\phi_1(t)$; in effect, the sinusoidal $\phi_1(t)$ acts as the "carrier" of the binary PSK signal.

Accordingly, binary PSK may be viewed as a special form of DSB-SC modulation that was studied in Section 2.14.

Error Probability of Binary PSK Using Coherent Detection

To make an optimum decision on the received signal $x(t)$ in favor of symbol 1 or symbol 0 (i.e., estimate the original binary sequence at the transmitter input), we assume that the receiver has access to a *locally generated replica of the basis function* $\phi_1(t)$. In other words, the receiver is *synchronized* with the transmitter, as shown in the block diagram of Figure 7.14b. We may identify two basic components in the binary PSK receiver:

1. *Correlator*, which correlates the received signal $x(t)$ with the basis function $\phi_1(t)$ on a bit-by-bit basis.
2. *Decision device*, which compares the correlator output against a zero-threshold, assuming that binary symbols 1 and 0 are equiprobable. If the threshold is exceeded, a decision is made in favor of symbol 1; if not, the decision is made in favor of symbol 0. Equality of the correlator with the zero-threshold is decided by the toss of a fair coin (i.e., in a random manner).

With coherent detection in place, we may apply the decision rule of (7.54). Specifically, we partition the signal space of Figure 7.13 into two regions:

- the set of points closest to message point 1 at $+\sqrt{E_b}$; and
- the set of points closest to message point 2 at $-\sqrt{E_b}$.

This is accomplished by constructing the midpoint of the line joining these two message points and then marking off the appropriate decision regions. In Figure 7.13, these two decision regions are marked Z_1 and Z_2, according to the message point around which they are constructed.

The decision rule is now simply to decide that signal $s_1(t)$ (i.e., binary symbol 1) was transmitted if the received signal point falls in region Z_1 and to decide that signal $s_2(t)$ (i.e., binary symbol 0) was transmitted if the received signal point falls in region Z_2. Two kinds of erroneous decisions may, however, be made:

1. *Error of the first kind.* Signal $s_2(t)$ is transmitted but the noise is such that the received signal point falls inside region Z_1; so the receiver decides in favor of signal $s_1(t)$.
2. *Error of the second kind.* Signal $s_1(t)$ is transmitted but the noise is such that the received signal point falls inside region Z_2; so the receiver decides in favor of signal $s_2(t)$.

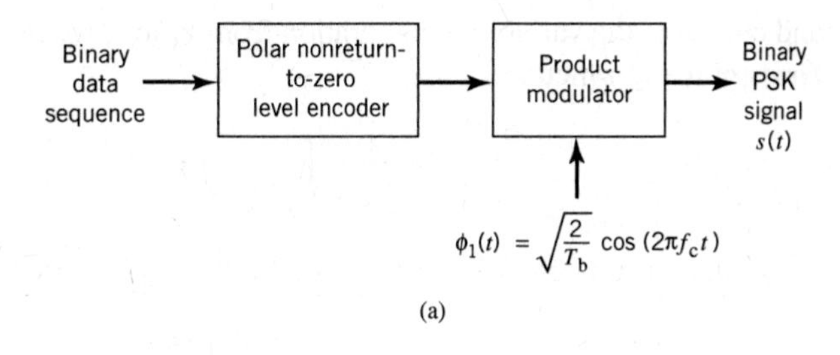

(a)

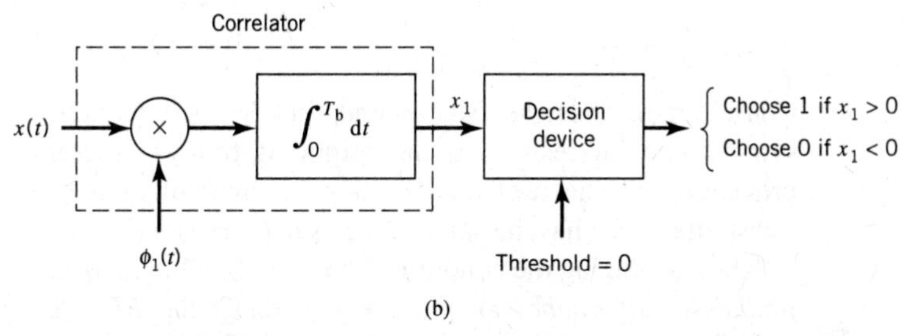

(b)

Figure 7.14 Block diagrams for (a) binary PSK transmitter and (b) coherent binary PSK receiver.

To calculate the probability of making an error of the first kind, we note from Figure 7.13a that the decision region associated with symbol 1 or signal $s_1(t)$ is described by

$$Z_1 : 0 < x_1 < \infty$$

where the observable element x_1 is related to the received signal $x(t)$ by

$$x_1 = \int_0^{T_b} x(t)\phi_1(t)\,dt \tag{7.102}$$

The conditional probability density function of random variable X_1, given that symbol 0 (i.e., signal $s_2(t)$) was transmitted, is defined by

$$f_{X_1}(x_1|0) = \frac{1}{\sqrt{\pi N_0}} \exp\left[-\frac{1}{N_0}(x_1 - s_{21})^2\right] \tag{7.103}$$

Using (7.101) in this equation yields

$$f_{X_1}(x_1|0) = \frac{1}{\sqrt{\pi N_0}} \exp\left[-\frac{1}{N_0}(x_1 + \sqrt{E_b})^2\right] \tag{7.104}$$

The conditional probability of the receiver deciding in favor of symbol 1, given that symbol 0 was transmitted, is therefore

$$p_{10} = \frac{1}{\sqrt{\pi N_0}} \int_0^{\infty} \exp\left[-\frac{1}{N_0}(x_1 + \sqrt{E_b})^2\right] dx_1 \tag{7.105}$$

Putting

$$z = \sqrt{\frac{2}{N_0}}(x_1 + \sqrt{E_b}) \tag{7.106}$$

and changing the variable of integration from x_1 to z, we may compactly rewrite (7.105) in terms of the Q-function:

$$p_{10} = \frac{1}{\sqrt{2\pi}} \int_{\sqrt{2E_b/N_0}}^{\infty} \exp\left(-\frac{z^2}{2}\right) dz \tag{7.107}$$

Using the formula of (3.68) in Chapter 3 for the Q-function in (7.107) we get

$$p_{10} = Q\left(\sqrt{\frac{2E_b}{N_0}}\right) \tag{7.108}$$

Consider next an error of the second kind. We note that the signal space of Figure 7.13a is symmetric with respect to the origin. It follows, therefore, that p_{01}, the conditional probability of the receiver deciding in favor of symbol 0, given that symbol 1 was transmitted, also has the same value as in (7.108).

Thus, averaging the conditional error probabilities p_{10} and p_{01}, we find that the *average probability of symbol error* or, equivalently, the *BER for binary PSK* using coherent detection and assuming equiprobable symbols is given by

$$P_e = Q\left(\sqrt{\frac{2E_b}{N_0}}\right) \tag{7.109}$$

As we increase the transmitted signal energy per bit E_b for a specified noise spectral density $N_0/2$, the message points corresponding to symbols 1 and 0 move further apart and the average probability of error P_e is correspondingly reduced in accordance with (7.109), which is intuitively satisfying.

Power Spectra of Binary PSK Signals

Examining (7.97) and (7.98), we see that a binary PSK wave is an example of DSB-SC modulation that was discussed in Section 2.14. More specifically, it consists of an in-phase component only. Let $g(t)$ denote the underlying *pulse-shaping function* defined by

$$g(t) = \begin{cases} \sqrt{\frac{2E_b}{T_b}}, & 0 \le t \le T_b \\ 0, & \text{otherwise} \end{cases} \tag{7.110}$$

Depending on whether the transmitter input is binary symbol 1 or 0, the corresponding transmitter output is $+g(t)$ or $-g(t)$, respectively. It is assumed that the incoming binary sequence is random, with symbols 1 and 0 being equally likely and the symbols transmitted during the different time slots being statistically independent.

In Example 6 of Chapter 4, it was shown that the power spectral density of a random binary wave so described is equal to the energy spectral density of the symbol shaping function divided by the symbol duration. The energy spectral density of a Fourier-transformable signal $g(t)$ is defined as the squared magnitude of the signal's Fourier transform. For the binary PSK signal at hand, the baseband power spectral density is, therefore, defined by

$$S_B(f) = \frac{2E_b \sin^2(\pi T_b f)}{(\pi T_b f)^2} = 2E_b \operatorname{sinc}^2(T_b f) \tag{7.111}$$

Examining (7.111), we may make the following observations on binary PSK:

1. The power spectral density $S_B(f)$ is *symmetric* about the vertical axis, as expected.
2. $S_B(f)$ goes through zero at multiples of the bit rate; that is, $f = \pm 1/T_b, \pm 2/T_b, \ldots$
3. With $\sin^2(\pi T_b f)$ limited to a maximum value of unity, $S_B(f)$ *falls off as the inverse square of the frequency, f.*

These three observations are all embodied in the plot of $S_B(f)$ versus f, presented in Figure 7.15.

Figure 7.15 also includes a plot of the baseband power spectral density of a binary frequency-shift keying (FSK) signal, details of which are presented in Section 7.8. Comparison of these two spectra is deferred to that section.

Quadriphase-Shift Keying

The provision of reliable performance, exemplified by a very low probability of error, is one important goal in the design of a digital communication system. Another important goal is the efficient utilization of channel bandwidth. In this subsection we study a *bandwidth-conserving modulation scheme* known as *quadriphase-shift keying* (QPSK), using coherent detection.

As with binary PSK, information about the message symbols in QPSK is contained in the carrier phase. In particular, the phase of the carrier takes on one of four equally spaced

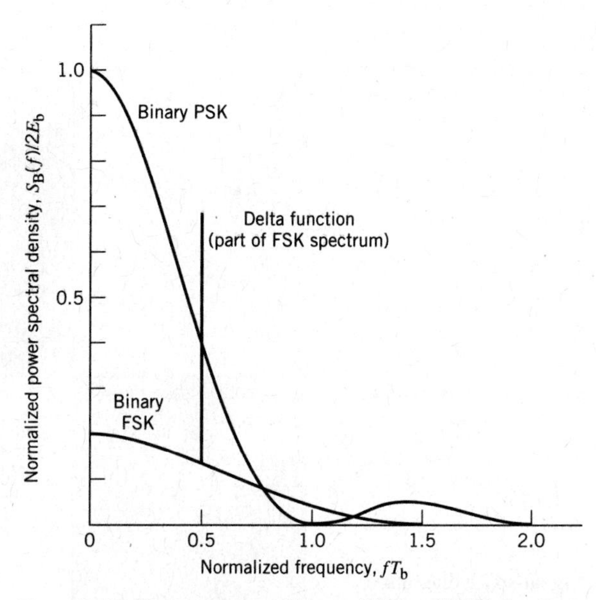

Figure 7.15 Power spectra of binary PSK and FSK signals.

values, such as $\pi/4$, $3\pi/4$, $5\pi/4$, and $7\pi/4$. For this set of values, we may define the transmitted signal as

$$s_i(t) = \begin{cases} \sqrt{\dfrac{2E}{T}}\cos\left[2\pi f_c t + (2i-1)\dfrac{\pi}{4}\right], & \begin{cases} 0 \le t \le T \\ i = 1, 2, 3, 4 \end{cases} \\ 0, & \text{elsewhere} \end{cases} \tag{7.112}$$

where E is the *transmitted signal energy per symbol* and T is the *symbol duration*. The carrier frequency f_c equals n_c/T for some fixed integer n_c. Each possible value of the phase corresponds to a unique *dibit* (i.e., pair of bits). Thus, for example, we may choose the foregoing set of phase values to represent the *Gray-encoded* set of dibits, 10, 00, 01, and 11, where only a single bit is changed from one dibit to the next.

Signal-Space Diagram of QPSK Signals

Using a well-known trigonometric identity, we may expand (7.112) to redefine the transmitted signal in the canonical form:

$$s_i(t) = \sqrt{\frac{2E}{T}}\cos\left[(2i-1)\frac{\pi}{4}\right]\cos(2\pi f_c t) - \sqrt{\frac{2E}{T}}\sin\left[(2i-1)\frac{\pi}{4}\right]\sin(2\pi f_c t) \tag{7.113}$$

where $i = 1, 2, 3, 4$. Based on this representation, we make two observations:

1. There are two orthonormal basis functions, defined by a pair of *quadrature carriers*:

$$\phi_1(t) = \sqrt{\frac{2}{T}}\cos(2\pi f_c t), \qquad 0 \le t \le T \tag{7.114}$$

$$\phi_2(t) = \sqrt{\frac{2}{T}}\sin(2\pi f_c t), \qquad 0 \le t \le T \tag{7.115}$$

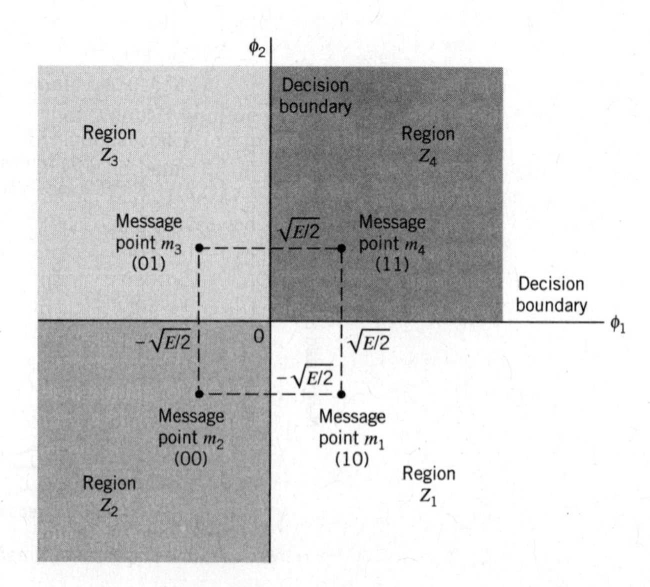

Figure 7.16
Signal-space diagram of QPSK system.

Table 7.2 **Signal-space characterization of QPSK**

Gray-encoded input dibit	Phase of QPSK signal (radians)	Coordinates of message points	
		s_{i1}	s_{i2}
11	$\pi/4$	$+\sqrt{E/2}$	$+\sqrt{E/2}$
01	$3\pi/4$	$-\sqrt{E/2}$	$+\sqrt{E/2}$
00	$5\pi/4$	$-\sqrt{E/2}$	$-\sqrt{E/2}$
10	$7\pi/4$	$+\sqrt{E/2}$	$-\sqrt{E/2}$

2. There are four message points, defined by the two-dimensional signal vector

$$\mathbf{s}_i = \begin{bmatrix} \sqrt{E}\cos\left((2i-1)\frac{\pi}{4}\right) \\ -\sqrt{E}\sin\left((2i-1)\frac{\pi}{4}\right) \end{bmatrix}, \quad i = 1, 2, 3, 4 \tag{7.116}$$

Elements of the signal vectors, namely s_{i1} and s_{i2}, have their values summarized in Table 7.2; the first two columns give the associated dibit and phase of the QPSK signal.

Accordingly, a QPSK signal has a two-dimensional signal constellation (i.e., $N = 2$) and four message points (i.e., $M = 4$) whose phase angles increase in a counterclockwise direction, as illustrated in Figure 7.16. As with binary PSK, the QPSK signal has *minimum average energy*.

EXAMPLE 6 **QPSK Waveforms**

Figure 7.17 illustrates the sequences and waveforms involved in the generation of a QPSK signal. The input binary sequence 01101000 is shown in Figure 7.17a. This sequence is divided into two other sequences, consisting of odd- and even-numbered bits of the input sequence. These two sequences are shown in the top lines of Figure 7.17b and c. The waveforms representing the two components of the QPSK signal, namely $s_{i1}\phi_1(t)$ and $s_{i2}\phi_2(t)$ are also shown in Figure 7.17b and c, respectively. These two waveforms may individually be viewed as examples of a binary PSK signal. Adding them, we get the QPSK waveform shown in Figure 7.17d.

To define the decision rule for the coherent detection of the transmitted data sequence, we partition the signal space into four regions, in accordance with Table 7.2. The individual regions are defined by the set of symbols closest to the message point represented by message vectors $\mathbf{s}_1$, $\mathbf{s}_2$, $\mathbf{s}_3$, and $\mathbf{s}_4$. This is readily accomplished by constructing the perpendicular bisectors of the square formed by joining the four message points and then marking off the appropriate regions. We thus find that the decision regions

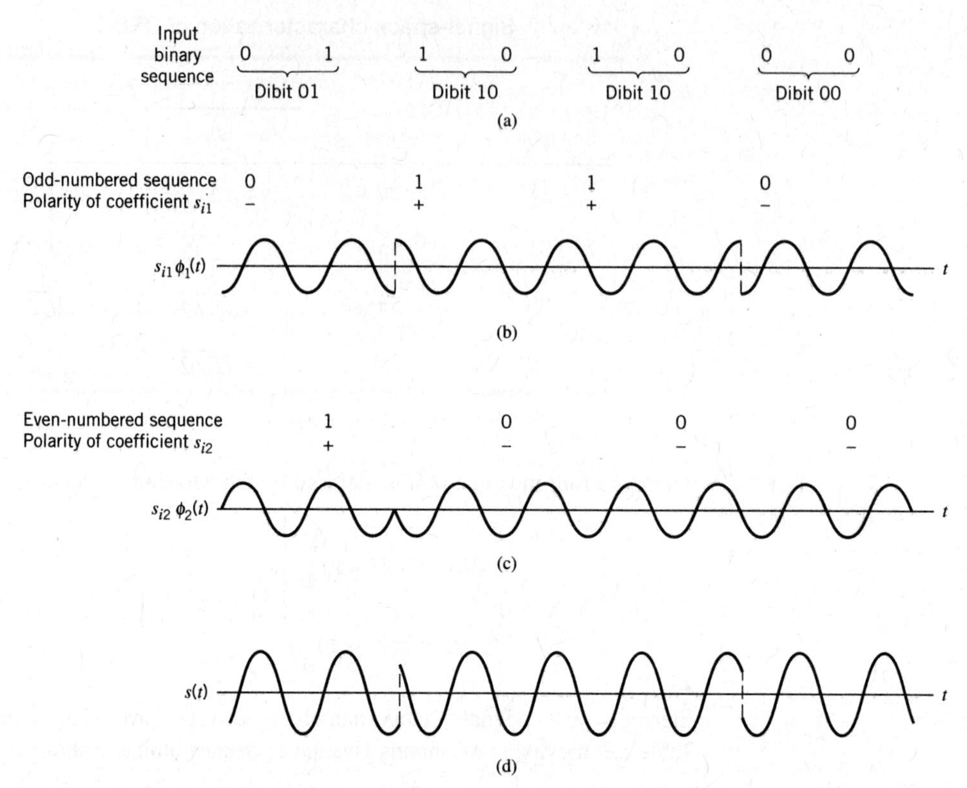

Figure 7.17 (a) Input binary sequence. (b) Odd-numbered dibits of input sequence and associated binary PSK signal. (c) Even-numbered dibits of input sequence and associated binary PSK signal. (d) QPSK waveform defined as $s(t) = s_{i1}\,\phi_1(t) + s_{i2}\,\phi_2(t)$.

are quadrants whose vertices coincide with the origin. These regions are marked Z_1, Z_2, Z_3, and Z_4 in Figure 7.17, according to the message point around which they are constructed.

Generation and Coherent Detection of QPSK Signals

Expanding on the binary PSK transmitter of Figure 7.14a, we may build on (7.113) to (7.115) to construct the QPSK transmitter shown in Figure 7.18a. A distinguishing feature of the QPSK transmitter is the block labeled *demultiplexer.* The function of the demultiplexer is to divide the binary wave produced by the polar NRZ-level encoder into two separate binary waves, one of which represents the odd-numbered dibits in the incoming binary sequence and the other represents the even-numbered dibits. Accordingly, we may make the following statement:

> The QPSK transmitter may be viewed as two binary PSK generators that work in parallel, each at a bit rate equal to one-half the bit rate of the original binary sequence at the QPSK transmitter input.

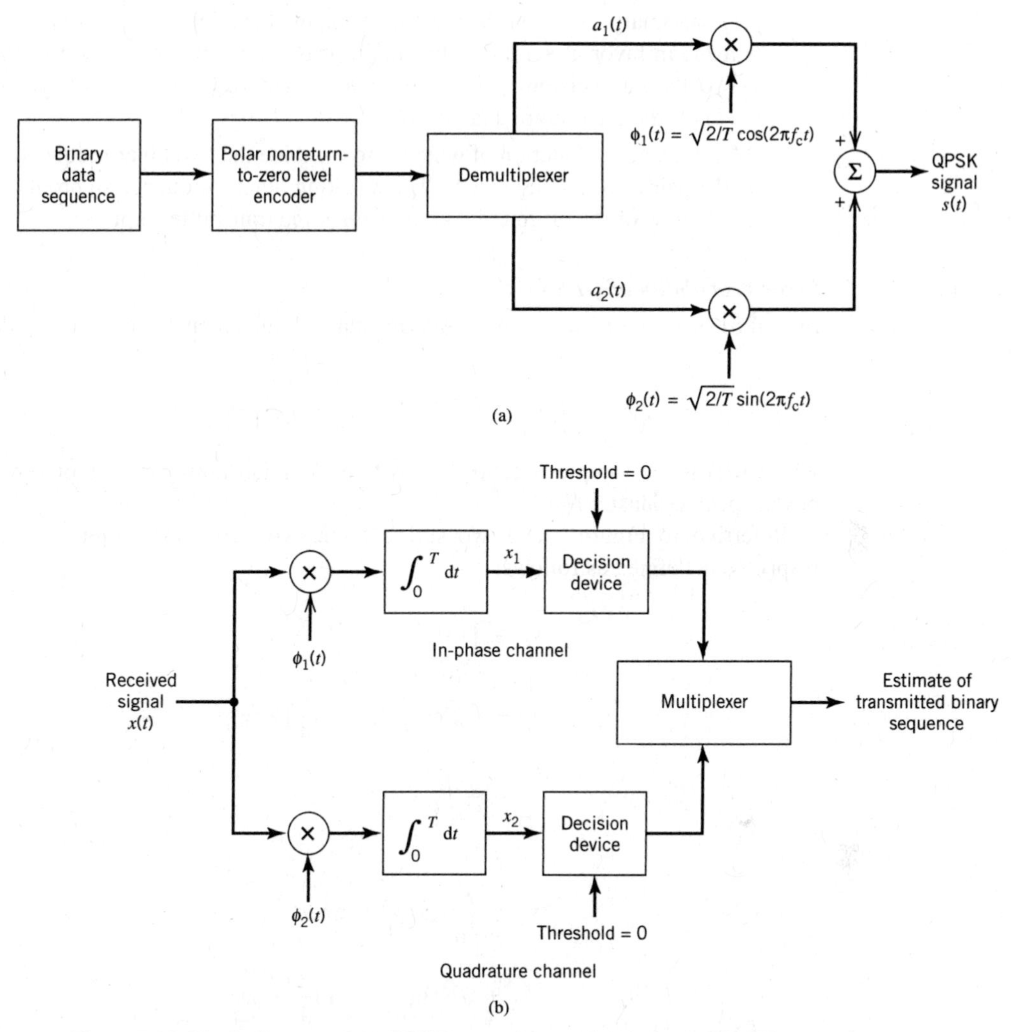

Figure 7.18 Block diagram of (a) QPSK transmitter and (b) coherent QPSK receiver.

Expanding on the binary PSK receiver of Figure 7.14b, we find that the QPSK receiver is structured in the form of an *in-phase path* and a *quadrature path*, working in parallel as depicted in Figure 7.18b. The functional composition of the QPSK receiver is as follows:

1. *Pair of correlators*, which have a common input $x(t)$. The two correlators are supplied with a pair of *locally generated orthonormal basis functions* $\phi_1(t)$ and $\phi_2(t)$, which means that the receiver is synchronized with the transmitter. The correlator outputs, produced in response to the received signal $x(t)$, are denoted by x_1 and x_2, respectively.
2. *Pair of decision devices*, which act on the correlator outputs x_1 and x_2 by comparing each one with a zero-threshold; here, it is assumed that the symbols 1 and 0 in the

original binary stream at the transmitter input are equally likely. If $x_1 > 0$, a decision is made in favor of symbol 1 for the in-phase channel output; on the other hand, if $x_1 < 0$, then a decision is made in favor of symbol 0. Similar binary decisions are made for the quadrature channel.

3. *Multiplexer*, the function of which is to combine the two binary sequences produced by the pair of decision devices. The resulting binary sequence so produced provides an *estimate* of the original binary stream at the transmitter input.

Error Probability of QPSK

In a QPSK system operating on an AWGN channel, the received signal $x(t)$ is defined by

$$x(t) = s_i(t) + w(t), \qquad \begin{cases} 0 \le t \le T \\ i = 1, 2, 3, 4 \end{cases} \tag{7.117}$$

where $w(t)$ is the sample function of a white Gaussian noise process of zero mean and power spectral density $N_0/2$.

Referring to Figure 7.18a, we see that the two correlator outputs, x_1 and x_2, are respectively defined as follows:

$$\begin{aligned} x_1 &= \int_0^T x(t)\phi_1(t)\,dt \\ &= \sqrt{E}\cos\left[(2i-1)\frac{\pi}{4}\right] + w_1 \\ &= \pm\sqrt{\frac{E}{2}} + w_1 \end{aligned} \tag{7.118}$$

and

$$\begin{aligned} x_2 &= \int_0^T x(t)\phi_2(t)\,dt \\ &= \sqrt{E}\sin\left[(2i-1)\frac{\pi}{4}\right] + w_2 \\ &= \mp\sqrt{\frac{E}{2}} + w_2 \end{aligned} \tag{7.119}$$

Thus, the observable elements x_1 and x_2 are sample values of independent Gaussian random variables with mean values equal to $\pm\sqrt{E/2}$ and $\mp\sqrt{E/2}$, respectively, and with a common variance equal to $N_0/2$.

The decision rule is now simply to say that $s_1(t)$ was transmitted if the received signal point associated with the observation vector $\mathbf{x}$ falls inside region Z_1; say that $s_2(t)$ was transmitted if the received signal point falls inside region Z_2, and so on for the other two regions Z_3 and Z_4. An erroneous decision will be made if, for example, signal $s_4(t)$ is transmitted but the noise $w(t)$ is such that the received signal point falls *outside* region Z_4.

To calculate the average probability of symbol error, recall that a QPSK receiver is in fact equivalent to two binary PSK receivers working in parallel and using two carriers that are in phase quadrature. The in-phase channel x_1 and the quadrature channel output x_2

(i.e., the two elements of the observation vector **x**) may be viewed as the individual outputs of two binary PSK receivers. Thus, according to (7.118) and (7.119), these two binary PSK receivers are characterized as follows:

- signal energy per bit equal to $E/2$, and
- noise spectral density equal to $N_0/2$.

Hence, using (7.109) for the average probability of bit error of a coherent binary PSK receiver, we may express the average probability of bit error in the in-phase and quadrature paths of the coherent QPSK receiver as

$$P' = Q\left(\sqrt{\frac{E}{N_0}}\right) \tag{7.120}$$

where E is written in place of $2E_b$. Another important point to note is that the bit errors in the in-phase and quadrature paths of the QPSK receiver are statistically independent. The decision device in the in-phase path accounts for one of the two bits constituting a symbol (dibit) of the QPSK signal, and the decision device in the quadrature path takes care of the other dibit. Accordingly, the *average probability of a correct detection* resulting from the combined action of the two channels (paths) working together is

$$\begin{aligned} P_c &= (1-P')^2 \\ &= \left[1 - Q\left(\sqrt{\frac{E}{N_0}}\right)\right]^2 \\ &= 1 - 2Q\left(\sqrt{\frac{E}{N_0}}\right) + Q^2\left(\sqrt{\frac{E}{N_0}}\right) \end{aligned} \tag{7.121}$$

The average probability of symbol error for QPSK is therefore

$$\begin{aligned} P_e &= 1 - P_c \\ &= 2Q\left(\sqrt{\frac{E}{N_0}}\right) - Q^2\left(\sqrt{\frac{E}{N_0}}\right) \end{aligned} \tag{7.122}$$

In the region where $(E/N_0) \gg 1$, we may ignore the quadratic term on the right-hand side of (7.122), so the average probability of symbol error for the QPSK receiver is approximated as

$$P_e \approx 2Q\left(\sqrt{\frac{E}{N_0}}\right) \tag{7.123}$$

Equation (7.123) may also be derived in another insightful way, using the signal-space diagram of Figure 7.16. Since the four message points of this diagram are circularly symmetric with respect to the origin, we may apply the approximate formula of (7.85) based on the union bound. Consider, for example, message point m_1 (corresponding to dibit 10) chosen as the transmitted message point. The message points m_2 and m_4 (corresponding to dibits 00 and 11) are the closest to m_1. From Figure 7.16 we readily find that m_1 is equidistant from m_2 and m_4 in a Euclidean sense, as shown by

$$d_{12} = d_{14} = \sqrt{2E}$$

Assuming that E/N_0 is large enough to ignore the contribution of the most distant message point m_3 (corresponding to dibit 01) relative to m_1, we find that the use of (7.85) with the equality sign yields an approximate expression for P_e that is the same as that of (7.123). Note that in mistaking either m_2 or m_4 for m_1, a single bit error is made; on the other hand, in mistaking m_3 for m_1, two bit errors are made. For a high enough E/N_0, the likelihood of both bits of a symbol being in error is much less than a single bit, which is a further justification for ignoring m_3 in calculating P_e when m_1 is sent.

In a QPSK system, we note that since there are two bits per symbol, the transmitted signal energy per symbol is twice the signal energy per bit, as shown by

$$E = 2E_b \tag{7.124}$$

Thus, expressing the average probability of symbol error in terms of the ratio E_b/N_0, we may write

$$P_e \approx 2Q\left(\sqrt{\frac{2E_b}{N_0}}\right) \tag{7.125}$$

With Gray encoding used for the incoming symbols, we find from (7.120) and (7.124) that the BER of QPSK is exactly

$$\text{BER} = Q\left(\sqrt{\frac{2E_b}{N_0}}\right) \tag{7.126}$$

We may, therefore, state that a QPSK system achieves the same average probability of bit error as a binary PSK system for the same bit rate and the same E_b/N_0, but uses only half the channel bandwidth. Stated in another way:

> For the same E_b/N_0 and, therefore, the same average probability of bit error, a QPSK system transmits information at twice the bit rate of a binary PSK system for the same channel bandwidth.

For a prescribed performance, QPSK uses channel bandwidth better than binary PSK, which explains the preferred use of QPSK over binary PSK in practice.

Earlier we stated that the binary PSK may be viewed as a special case of DSB-SC modulation. In a corresponding way, we may view the QPSK as a special case of the *quadrature amplitude modulation* (QAM) in analog modulation theory.

Power Spectra of QPSK Signals

Assume that the binary wave at the modulator input is random with symbols 1 and 0 being equally likely, and with the symbols transmitted during adjacent time slots being statistically independent. We then make the following observations pertaining to the in-phase and quadrature components of a QPSK signal:

1. Depending on the dibit sent during the signaling interval $-T_b \le t \le T_b$, the in-phase component equals $+g(t)$ or $-g(t)$, and similarly for the quadrature component. The $g(t)$ denotes the symbol-shaping function defined by

$$g(t) = \begin{cases} \sqrt{\dfrac{E}{T}}, & 0 \le t \le T \\ 0, & \text{otherwise} \end{cases} \tag{7.127}$$

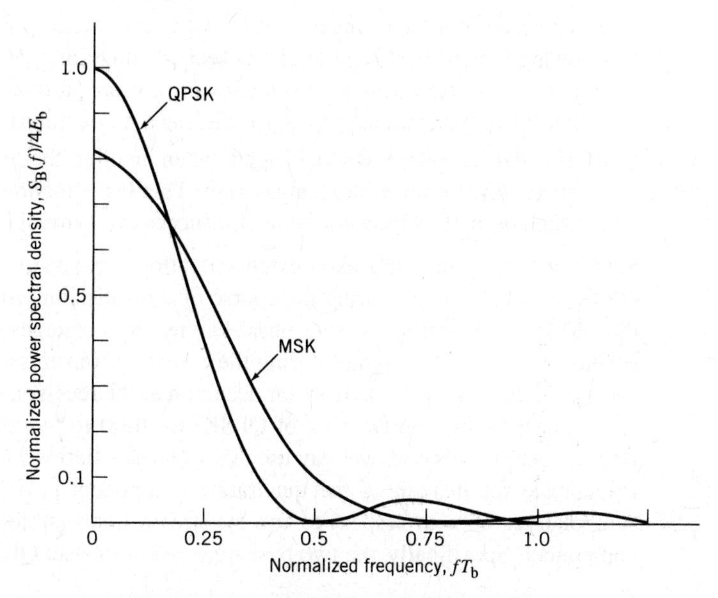

Figure 7.19 Power spectra of QPSK and MSK signals.

Hence, the in-phase and quadrature components have a common power spectral density, namely, $E\,\text{sinc}^2(Tf)$.

2. The in-phase and quadrature components are statistically independent. Accordingly, the baseband power spectral density of the QPSK signal equals the sum of the individual power spectral densities of the in-phase and quadrature components, so we may write

$$\begin{aligned} S_B(f) &= 2E\,\text{sinc}^2(Tf) \\ &= 4E_b\,\text{sinc}^2(2T_b f) \end{aligned} \tag{7.128}$$

Figure 7.19 plots $S_B(f)$, normalized with respect to $4E_b$, versus the *normalized frequency* $T_b f$. This figure also includes a plot of the baseband power spectral density of a certain form of binary FSK called minimum shift keying, the evaluation of which is presented in Section 7.8. Comparison of these two spectra is deferred to that section.

Offset QPSK

For a variation of the QPSK, consider the signal-space diagram of Figure 7.20a that embodies all the possible phase transitions that can arise in the generation of a QPSK signal. More specifically, examining the QPSK waveform illustrated in Figure 7.17 for Example 6, we may make three observations:

1. The carrier phase changes by ±180° whenever both the in-phase and quadrature components of the QPSK signal change sign. An example of this situation is illustrated in Figure 7.17 when the input binary sequence switches from dibit 01 to dibit 10.

2. The carrier phase changes by ±90° whenever the in-phase or quadrature component changes sign. An example of this second situation is illustrated in Figure 7.17 when the input binary sequence switches from dibit 10 to dibit 00, during which the in-phase component changes sign, whereas the quadrature component is unchanged.
3. The carrier phase is unchanged when neither the in-phase component nor the quadrature component changes sign. This last situation is illustrated in Figure 7.17 when dibit 10 is transmitted in two successive symbol intervals.

Situation 1 and, to a much lesser extent, situation 2 can be of a particular concern when the QPSK signal is filtered during the course of transmission, prior to detection. Specifically, the 180° and 90° shifts in carrier phase can result in changes in the carrier amplitude (i.e., envelope of the QPSK signal) during the course of transmission over the channel, thereby causing additional symbol errors on detection at the receiver.

To mitigate this shortcoming of QPSK, we need to reduce the extent of its amplitude fluctuations. To this end, we may use *offset QPSK*.[4] In this variant of QPSK, the bit stream responsible for generating the quadrature component is delayed (i.e., offset) by half a symbol interval with respect to the bit stream responsible for generating the in-phase component. Specifically, the two basis functions of offset QPSK are defined by

$$\phi_1(t) = \sqrt{\frac{2}{T}}\cos(2\pi f_c t), \qquad 0 \le t \le T \tag{7.129}$$

and

$$\phi_2(t) = \sqrt{\frac{2}{T}}\sin(2\pi f_c t), \qquad \frac{T}{2} \le t \le \frac{3T}{2} \tag{7.130}$$

The $\phi_1(t)$ of (7.129) is exactly the same as that of (7.114) for QPSK, but the $\phi_2(t)$ of (7.130) is different from that of (7.115) for QPSK. Accordingly, unlike QPSK, the phase transitions likely to occur in offset QPSK are confined to ±90°, as indicated in the signal-space diagram of Figure 7.20b. However, ±90° phase transitions in offset QPSK occur twice as frequently but with half the intensity encountered in QPSK. Since, in addition to ±90° phase transitions, ±180° phase transitions also occur in QPSK, we find that amplitude fluctuations in offset QPSK due to filtering have a smaller amplitude than in the case of QPSK.

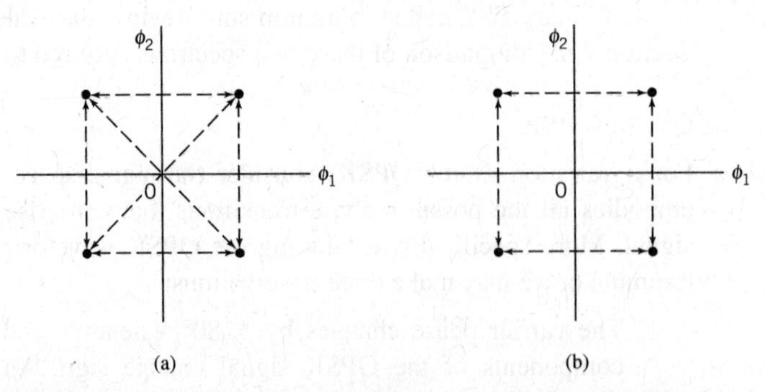

Figure 7.20 Possible paths for switching between the message points in (a) QPSK and (b) offset QPSK.

Despite the delay $T/2$ applied to the basis function $\phi_2(t)$ in (7.130) compared with that in (7.115) for QPSK, the offset QPSK has exactly the same probability of symbol error in an AWGN channel as QPSK. The equivalence in noise performance between these PSK schemes assumes the use of coherent detection at the receiver. The reason for the equivalence is that the statistical independence of the in-phase and quadrature components applies to both QPSK and offset QPSK. We may, therefore, say that Equation (7.123) for the average probability of symbol error applies equally well to the offset QPSK.

M-ary PSK

QPSK is a special case of the generic form of PSK commonly referred to as *M-ary PSK*, where the phase of the carrier takes on one of M possible values: $\theta_i = 2(i-1)\pi/M$, where $i = 1, 2, \ldots, M$. Accordingly, during each signaling interval of duration T, one of the M possible signals

$$s_i(t) = \sqrt{\frac{2E}{T}}\cos\left[2\pi f_c t + \frac{2\pi}{M}(i-1)\right], \qquad i = 1, 2, \ldots, M \tag{7.131}$$

is sent, where E is the signal energy per symbol. The carrier frequency $f_c = n_c/T$ for some fixed integer n_c.

Each $s_i(t)$ may be expanded in terms of the same two basis functions $\phi_1(t)$ and $\phi_2(t)$; the signal constellation of M-ary PSK is, therefore, *two-dimensional*. The M message points are equally spaced on a circle of radius $\sqrt{E}$ and center at the origin, as illustrated in Figure 7.21a for the case of *octaphase-shift-keying* (i.e., $M = 8$).

From Figure 7.21a we see that the signal-space diagram is circularly symmetric. We may, therefore, apply (7.85), based on the union bound, to develop an approximate formula for the average probability of symbol error for M-ary PSK. Suppose that the transmitted signal corresponds to the message point m_1, whose coordinates along the ϕ_1- and ϕ_2-axes are $+\sqrt{E}$ and 0, respectively. Suppose that the ratio E/N_0 is large enough to consider the nearest two message points, one on either side of m_1, as potential candidates for being mistaken for m_1 due to channel noise. This is illustrated in Figure 7.21b for the case of $M = 8$. The Euclidean distance for each of these two points from m_1 is (for $M = 8$)

$$d_{12} = d_{18} = 2\sqrt{E}\sin\left(\frac{\pi}{M}\right)$$

Hence, the use of (7.85) yields the average probability of symbol error for coherent M-ary PSK as

$$P_e \approx 2Q\left[\sqrt{\frac{2E}{N_0}}\sin\left(\frac{\pi}{M}\right)\right] \tag{7.132}$$

where it is assumed that $M \geq 4$. The approximation becomes extremely tight for fixed M, as E/N_0 is increased. For $M = 4$, (7.132) reduces to the same form given in (7.123) for QPSK.

Power Spectra of M-ary PSK Signals

The symbol duration of M-ary PSK is defined by

$$T = T_b \log_2 M \tag{7.133}$$

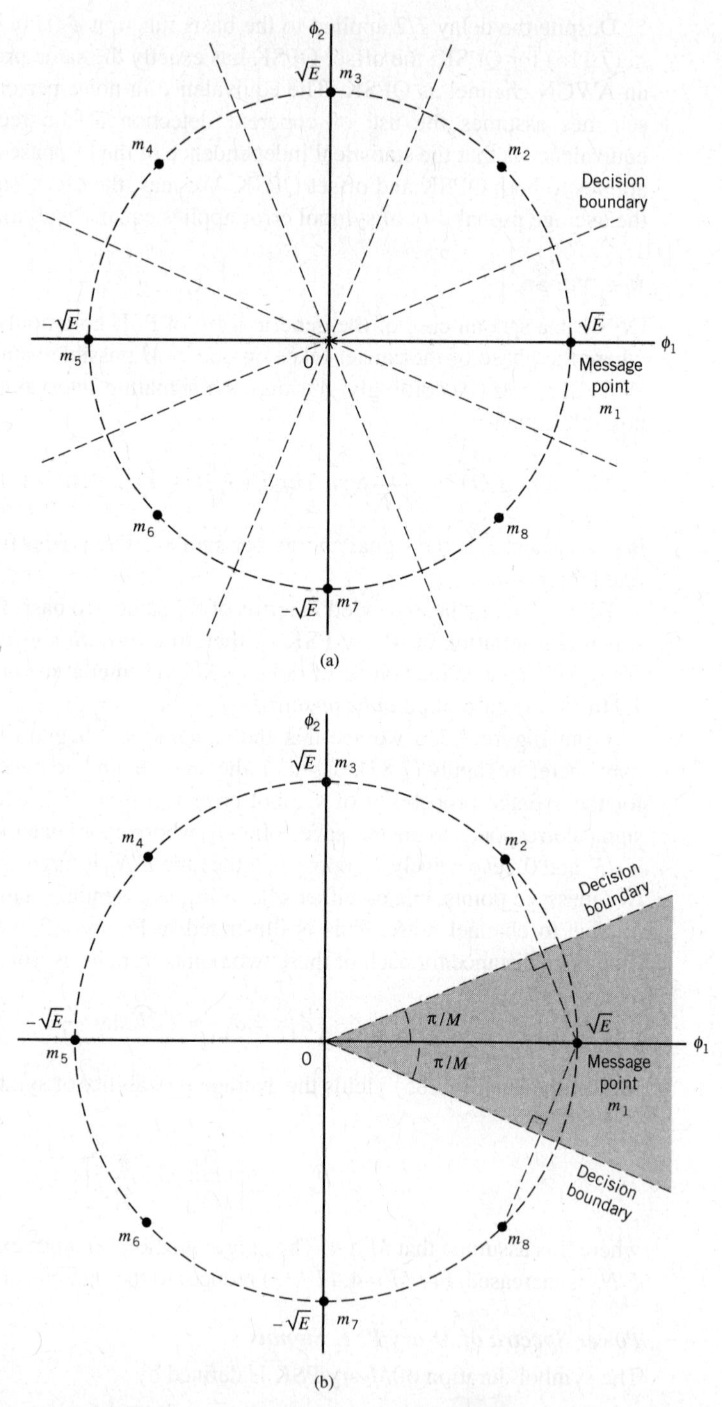

Figure 7.21 (a) Signal-space diagram for octaphase-shift keying (i.e., $M = 8$). The decision boundaries are shown as dashed lines. (b) Signal-space diagram illustrating the application of the union bound for octaphase-shift keying.

where T_b is the bit duration. Proceeding in a manner similar to that described for a QPSK signal, we may show that the baseband power spectral density of an M-ary PSK signal is given by

$$\begin{aligned} S_B(f) &= 2E \operatorname{sinc}^2(Tf) \\ &= 2E_b(\log_2 M)[\operatorname{sinc}^2(T_b f \log_2 M)] \end{aligned} \tag{7.134}$$

Figure 7.22 is a plot of the normalized power spectral density $S_B(f)/2E_b$ versus the normalized frequency $T_b f$ for three different values of M, namely $M = 2, 4, 8$. Equation (7.134) includes (7.111) for $M = 2$ and (7.128) for $M = 4$ as two special cases.

The baseband power spectra of M-ary PSK signals plotted in Figure 7.22 possess a main lobe bounded by well-defined *spectral nulls* (i.e., frequencies at which the power spectral density is zero). In light of the discussion on the bandwidth of signals presented in Chapter 2, we may use the main lobe as a basis for bandwidth assessment. Accordingly, invoking the notion of *null-to-null bandwidth*, we may say that the spectral width of the main lobe provides a simple, yet informative, measure for the bandwidth of M-ary PSK signals. Most importantly, a large fraction of the average signal power is contained inside the main lobe. On this basis, we may define the channel bandwidth required to pass M-ary PSK signals through an analog channel as

$$B = \frac{2}{T} \tag{7.135}$$

where T is the symbol duration. But the symbol duration T is related to the bit duration T_b by (7.133). Moreover, the bit rate $R_b = 1/T_b$. Hence, we may redefine the channel bandwidth of (7.135) in terms of the bit rate as

$$B = \frac{2R_b}{\log_2 M} \tag{7.136}$$

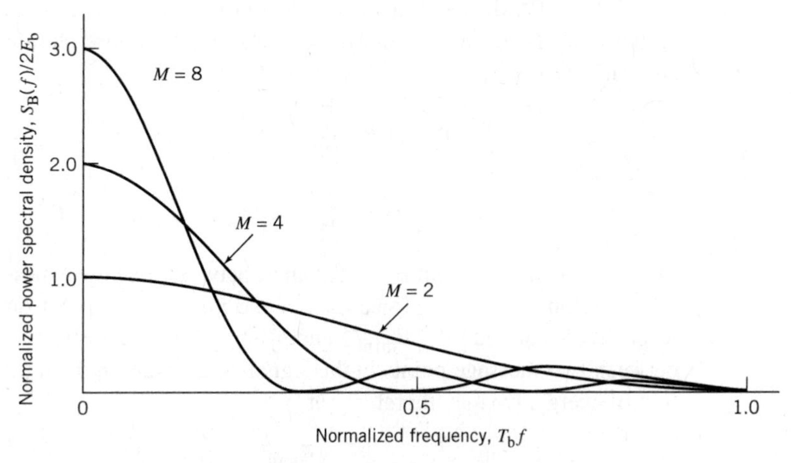

Figure 7.22 Power spectra of M-ary PSK signals for M = 2, 4, 8.

Table 7.3 **Bandwidth efficiency of *M*-ary PSK signals**

M	2	4	8	16	32	64
ρ (bit/(s/Hz))	0.5	1	1.5	2	2.5	3

Based on this formula, the *bandwidth efficiency* of M-ary PSK signals is given by

$$\begin{aligned}\rho &= \frac{R_b}{B} \\ &= \frac{\log_2 M}{2}\end{aligned} \tag{7.137}$$

Table 7.3 gives the values of ρ calculated from (7.137) for varying M. In light of (7.132) and Table 7.3, we now make the statement:

> As the number of states in M-ary PSK is increased, the bandwidth efficiency is improved at the expense of error performance.

However, note that if we are to ensure that there is no degradation in error performance, we have to increase E_b/N_0 to compensate for the increase in M.

7.7 *M*-ary Quadrature Amplitude Modulation

In an M-ary PSK system, the in-phase and quadrature components of the modulated signal are interrelated in such a way that the *envelope is constrained to remain constant*. This constraint manifests itself in a circular constellation for the message points, as illustrated in Figure 7.21a. However, if this constraint is removed so as to permit the in-phase and quadrature components to be independent, we get a new modulation scheme called *M-ary* QAM. The QAM is a *hybrid* form of modulation, in that the carrier experiences amplitude as well as phase-modulation.

In M-ary PAM, the signal-space diagram is one-dimensional. M-ary QAM is a two-dimensional generalization of M-ary PAM, in that its formulation involves two orthogonal passband basis functions:

$$\begin{aligned}\phi_1(t) &= \sqrt{\frac{2}{T}}\cos(2\pi f_c t), \qquad 0 \le t \le T \\ \phi_2(t) &= \sqrt{\frac{2}{T}}\sin(2\pi f_c t), \qquad 0 \le t \le T\end{aligned} \tag{7.138}$$

Let $d_{\min}$ denote the minimum distance between any two message points in the QAM constellation. Then, the projections of the ith message point on the ϕ_1- and ϕ_2-axes are respectively defined by $a_i d_{\min}/2$ and $b_i d_{\min}/2$, where $i = 1, 2, \ldots, M$. With the separation between two message points in the signal-space diagram being proportional to the square root of energy, we may therefore set

$$\frac{d_{\min}}{2} = \sqrt{E_0} \tag{7.139}$$

where E_0 is the *energy of the message signal with the lowest amplitude*. The transmitted M-ary QAM signal for symbol k can now be defined in terms of E_0:

$$s_k(t) = \sqrt{\frac{2E_0}{T}}a_k\cos(2\pi f_c t) - \sqrt{\frac{2E_0}{T}}b_k\sin(2\pi f_c t), \qquad \begin{cases} 0 \le t \le T \\ k = 0, \pm 1, \pm 2, \ldots \end{cases} \tag{7.140}$$

The signal $s_k(t)$ involves two phase-quadrature carriers, each one of which is modulated by a set of discrete amplitudes; hence the terminology "quadrature amplitude modulation."

In M-ary QAM, the constellation of message points depends on the number of possible symbols, M. In what follows, we consider the case of *square constellations*, for which the number of bits per symbol is even.

QAM Square Constellations

With an *even* number of bits per symbol, we write

$$L = \sqrt{M}, \qquad L\text{: positive integer} \tag{7.141}$$

Under this condition, an M-ary QAM square constellation can always be viewed as the *Cartesian product of a one-dimensional L-ary PAM constellation with itself.* By definition, the Cartesian product of two sets of coordinates (representing a pair of one-dimensional constellations) is made up of the set of all possible ordered pairs of coordinates with the first coordinate in each such pair being taken from the first set involved in the product and the second coordinate taken from the second set in the product.

Thus, the ordered pairs of coordinates naturally form a square matrix, as shown by

$$\{a_i, b_i\} = \begin{bmatrix} (-L+1, L-1) & (-L+3, L-1) & \cdots & (L-1, L-1) \\ (-L+1, L-3) & (-L+3, L-3) & \cdots & (L-1, L-3) \\ \vdots & \vdots & & \vdots \\ (-L+1, -L+1) & (-L+3, -L+1) & \cdots & (L-1, -L+1) \end{bmatrix} \tag{7.142}$$

To calculate the probability of symbol error for this M-ary QAM, we exploit the following property:

> A QAM square constellation can be factored into the product of the corresponding L-ary PAM constellation with itself.

To exploit this statement, we may proceed in one of two ways:

Approach 1: We start with a signal constellation of the M-ary PAM for a prescribed M, and then build on it to construct the corresponding signal constellation of the M-ary QAM.

Approach 2: We start with a signal constellation of the M-ary QAM, and then use it to construct the corresponding orthogonal M-ary PAMS.

In the example to follow, we present a systematic procedure based on Approach 1.

EXAMPLE 7 **M-ary QAM for $M = 4$**

In Figure 7.23, we have constructed two signal constellations for the 4-ary PAM, one vertically oriented along the ϕ_1-axis in part a of the figure, and the other horizontally oriented along the ϕ_2-axis in part b of the figure. These two parts are *spatially orthogonal* to each other, accounting for the two-dimensional structure of the M-ary QAM. In developing this structure, the following points should be born in mind:

- The same binary sequence is used for both 4-ary PAM constellations.
- The Gray encoding rule is applied, which means that as we move from one codeword to an adjacent one, only a single bit is changed.
- In constructing the 4-ary QAM constellation, we move from one quadrant to the next in a counterclockwise direction.

With four quadrants constituting the 4-ary QAM, we proceed in four stages as follows:

Stage 1: *First-quadrant constellation.* Referring to Figure 7.23, we use the codewords along the positive parts of the ϕ_2 and ϕ_1-axes, respectively, to write

$$\begin{bmatrix} 11 \\ 10 \end{bmatrix} \begin{bmatrix} 10 & 11 \end{bmatrix} \rightarrow \begin{bmatrix} 1110 & 1111 \\ 1010 & 1011 \end{bmatrix}$$

Top to bottom   Left to right   First quadrant

Stage 2: *Second-quadrant constellation.* Following the same procedure as in Stage 1, we write

$$\begin{bmatrix} 11 \\ 10 \end{bmatrix} \begin{bmatrix} 01 & 00 \end{bmatrix} \rightarrow \begin{bmatrix} 1101 & 1100 \\ 1001 & 1000 \end{bmatrix}$$

Top to bottom   Left to right   Second quadrant

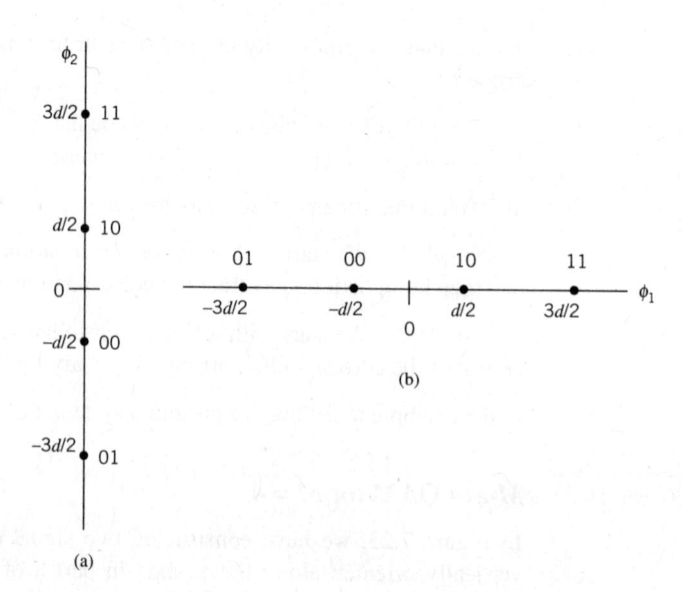

Figure 7.23
The two orthogonal constellations of the 4-ary PAM. (a) Vertically oriented constellation. (b) Horizontally oriented constellation. As mentioned in the text, we move top-down along the ϕ_2-axis and from left to right along the ϕ_1-axis.

Stage 3: *Third-quadrant constellation.* Again, following the same procedure as before, we next write

$$\begin{bmatrix} 00 \\ 01 \end{bmatrix} \begin{bmatrix} 01 & 00 \end{bmatrix} \rightarrow \begin{bmatrix} 0001 & 0000 \\ 0101 & 0100 \end{bmatrix}$$

Top to bottom　Left to right　Third quadrant

Stage 4: *Fourth-quadrant constellation.* Finally, we write

$$\begin{bmatrix} 00 \\ 01 \end{bmatrix} \begin{bmatrix} 10 & 11 \end{bmatrix} \rightarrow \begin{bmatrix} 0010 & 0011 \\ 0110 & 0111 \end{bmatrix}$$

Top to bottom　Left to right　Fourth quadrant

The final step is to piece together these four constituent 4-ary PAM constellations to construct the 4-ary QAM constellations as described in Figure 7.24. The important point to note here is that all the codewords in Figure 7.24 obey the Gray encoding rule, not only within each quadrant but also as we move from one quadrant to the next.

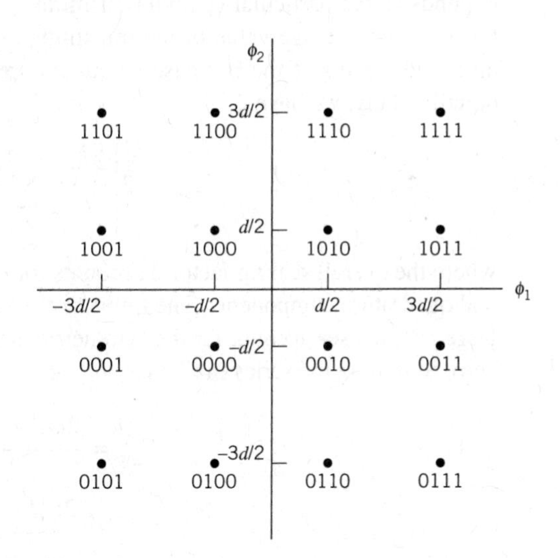

Figure 7.24
(a) Signal-space diagram of M-ary QAM for $M = 16$; the message points in each quadrant are identified with Gray-encoded quadbits.

Average Probability of Error

In light of the equivalence established between the M-ary QAM and M-ary PAM, we may formulate the average probability of error of the M-ary QAM by proceeding as follows:

1. The *probability of correct detection* for M-ary QAM is written as

$$P_c = (1 - P'_e)^2 \tag{7.143}$$

where P'_e is the probability of symbol error for the L-ary PAM.

2. With $L = \sqrt{M}$, the probability of symbol error P'_e is itself defined by

$$P'_e = 2\left(1 - \frac{1}{\sqrt{M}}\right) Q\left(\sqrt{\frac{2E_0}{N_0}}\right) \tag{7.144}$$

3. The probability of symbol error for M-ary QAM is given by

$$\begin{aligned} P_e &= 1 - P_c \\ &= 1 - (1 - P'_e)^2 \\ &\approx 2P'_e \end{aligned} \tag{7.145}$$

where it is assumed that P'_e is small enough compared with unity to justify ignoring the quadratic term.

Hence, using (7.143) and (7.144) in (7.145), we find that the probability of symbol error for M-ary QAM is approximately given by

$$P_e \approx 4\left(1 - \frac{1}{\sqrt{M}}\right) Q\left(\sqrt{\frac{2E_0}{N_0}}\right) \tag{7.146}$$

The transmitted energy in M-ary QAM is variable, in that its instantaneous value naturally depends on the particular symbol transmitted. Therefore, it is more logical to express P_e in terms of the average value of the transmitted energy rather than E_0. Assuming that the L amplitude levels of the in-phase or quadrature component of the M-ary QAM signal are equally likely, we have

$$E_{av} = 2\left[\frac{2E_0}{L} \sum_{i=1}^{L/2} (2i-1)^2\right] \tag{7.147}$$

where the overall scaling factor 2 accounts for the equal contributions made by the in-phase and quadrature components. The limits of the summation and the scaling factor 2 inside the large parentheses account for the symmetric nature of the pertinent amplitude levels around zero. Summing the series in (7.147), we get

$$E_{av} = \frac{2(L^2 - 1)E_0}{3} \tag{7.148}$$

$$= \frac{2(M-1)E_0}{3} \tag{7.149}$$

Accordingly, we may rewrite (7.146) in terms of E_{av} as

$$P_e \approx 4\left(1 - \frac{1}{\sqrt{M}}\right) Q\left[\sqrt{\frac{3E_{av}}{(M-1)N_0}}\right] \tag{7.150}$$

which is the desired result.

The case of $M = 4$ is of special interest. The signal constellation for this particular value of M is the same as that for QPSK. Indeed, putting $M = 4$ in (7.150) and noting that, for this special case, E_{av} equals E, where E is the energy per symbol, we find that the resulting formula for the probability of symbol error becomes identical to that in (7.123) for QPSK; and so it should.

7.8 Frequency-Shift Keying Techniques Using Coherent Detection

M-ary PSK and *M*-ary QAM share a common property: both of them are examples of *linear modulation.* In this section, we study a *nonlinear* method of modulation known as FSK using coherent detection. We begin the study by considering the simple case of binary FSK, for which $M = 2$.

Binary FSK

In *binary FSK,* symbols 1 and 0 are distinguished from each other by transmitting one of two sinusoidal waves that differ in frequency by a fixed amount. A typical pair of sinusoidal waves is described by

$$s_i(t) = \begin{cases} \sqrt{\dfrac{2E_\mathrm{b}}{T_\mathrm{b}}}\cos(2\pi f_i t), & 0 \le t \le T_\mathrm{b} \\ 0, & \text{elsewhere} \end{cases} \tag{7.151}$$

where $i = 1, 2$ and E_b is the transmitted signal energy per bit; the transmitted frequency is set at

$$f_i = \frac{n_\mathrm{c} + 1}{T_\mathrm{b}} \quad \text{for some fixed integer } n_\mathrm{c} \text{ and } i = 1, 2 \tag{7.152}$$

Symbol 1 is represented by $s_1(t)$ and symbol 0 by $s_2(t)$. The FSK signal described here is known as *Sunde's FSK.* It is a *continuous-phase signal,* in the sense that phase continuity is always maintained, including the inter-bit switching times.

From (7.151) and (7.152), we observe directly that the signals $s_1(t)$ and $s_2(t)$ are orthogonal, but not normalized to have unit energy. The most useful form for the set of orthonormal basis functions is described by

$$\phi_i(t) = \begin{cases} \sqrt{\dfrac{2}{T_\mathrm{b}}}\cos(2\pi f_i t), & 0 \le t \le T_\mathrm{b} \\ 0, & \text{elsewhere} \end{cases} \tag{7.153}$$

where $i = 1, 2$. Correspondingly, the coefficient s_{ij} for where $i = 1, 2$ and $j = 1, 2$ is defined by

$$\begin{aligned} s_{ij} &= \int_0^{T_\mathrm{b}} s_i(t)\phi_j(t)\,\mathrm{d}t \\ &= \int_0^{T_\mathrm{b}} \sqrt{\frac{2E_\mathrm{b}}{T_\mathrm{b}}}\cos(2\pi f_i t)\sqrt{\frac{2}{T_\mathrm{b}}}\cos(2\pi f_j t)\,\mathrm{d}t \end{aligned} \tag{7.154}$$

Carrying out the integration in (7.154), the formula for s_{ij} simplifies to

$$s_{ij} = \begin{cases} \sqrt{E_b}, & i = j \\ 0, & i \neq j \end{cases} \tag{7.155}$$

Thus, unlike binary PSK, binary FSK is characterized by having a signal-space diagram that is two-dimensional (i.e., $N = 2$) with two message points (i.e., $M = 2$), as shown in Figure 7.25. The two message points are defined by the vectors

$$\mathbf{s}_1 = \begin{bmatrix} \sqrt{E_b} \\ 0 \end{bmatrix} \tag{7.156}$$

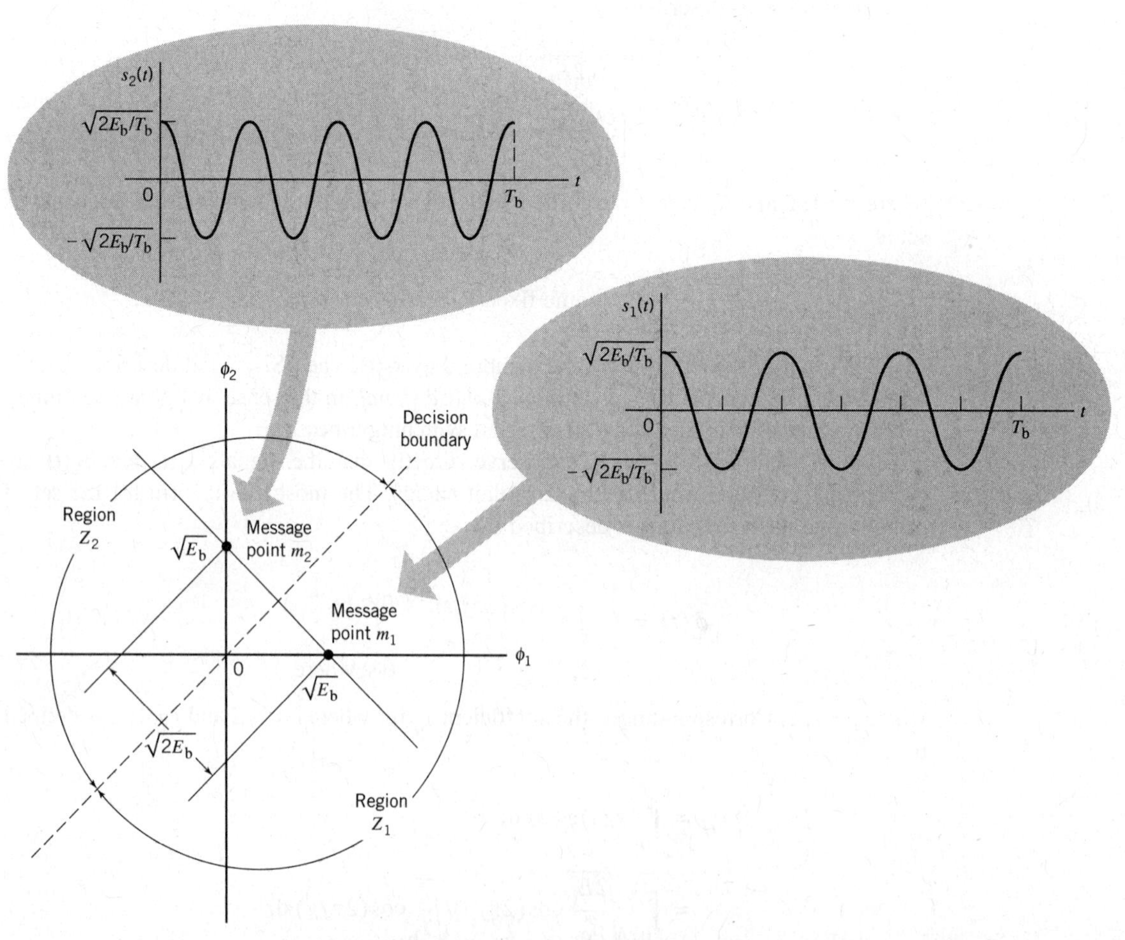

Figure 7.25 Signal-space diagram for binary FSK system. The diagram also includes example waveforms of the two modulated signals $s_1(t)$ and $s_2(t)$.

and

$$\mathbf{s}_2 = \begin{bmatrix} 0 \\ \sqrt{E_b} \end{bmatrix} \tag{7.157}$$

The Euclidean distance $\|\mathbf{s}_1 - \mathbf{s}_2\|$ is equal to $\sqrt{2E_b}$. Figure 7.25 also includes a couple of waveforms representative of signals $s_1(t)$ and $s_2(t)$.

Generation and Coherent Detection of Binary FSK Signals

The block diagram of Figure 7.26a describes a scheme for generating the binary FSK signal; it consists of two components:

1. *On–off level encoder*, the output of which is a constant amplitude of $\sqrt{E_b}$ in response to input symbol 1 and zero in response to input symbol 0.
2. *Pair of oscillators*, whose frequencies f_1 and f_2 differ by an integer multiple of the bit rate $1/T_b$ in accordance with (7.152). The lower oscillator with frequency f_2 is preceded by an inverter. When in a signaling interval, the input symbol is 1, the upper oscillator with frequency f_1 is switched on and signal $s_1(t)$ is transmitted, while the lower oscillator is switched off. On the other hand, when the input symbol is 0, the upper oscillator is switched off, while the lower oscillator is switched on

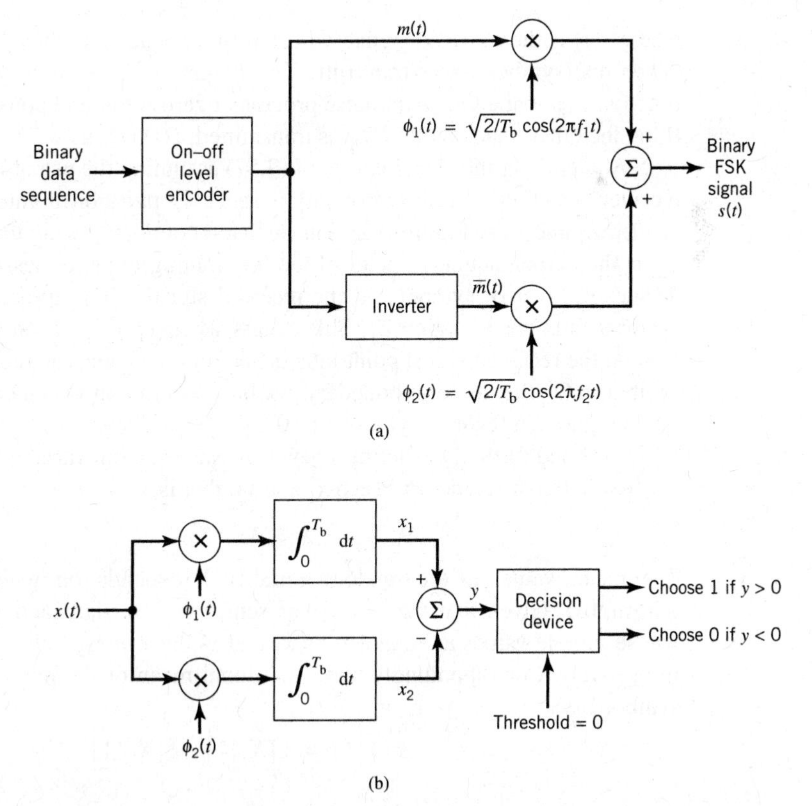

Figure 7.26 Block diagram for (a) binary FSK transmitter and (b) coherent binary FSK receiver.

and signal $s_2(t)$ with frequency f_2 is transmitted. With phase continuity as a requirement, the two oscillators are *synchronized* with each other. Alternatively, we may use a voltage-controlled oscillator, in which case phase continuity is automatically satisfied.

To coherently detect the original binary sequence given the noisy received signal $x(t)$, we may use the receiver shown in Figure 7.26b. It consists of two correlators with a common input, which are supplied with locally generated coherent reference signals $\phi_1(t)$ and $\phi_2(t)$. The correlator outputs are then subtracted, one from the other; the resulting difference y is then compared with a threshold of zero. If $y > 0$, the receiver decides in favor of 1. On the other hand, if $y < 0$, it decides in favor of 0. If y is exactly zero, the receiver makes a random guess (i.e., flip of a fair coin) in favor of 1 or 0.

Error Probability of Binary FSK

The observation vector $\mathbf{x}$ has two elements x_1 and x_2 that are defined by, respectively,

$$x_1 = \int_0^{T_b} x(t)\phi_1(t)\,dt \tag{7.158}$$

and

$$x_2 = \int_0^{T_b} x(t)\phi_2(t)\,dt \tag{7.159}$$

where $x(t)$ is the received signal, whose form depends on which symbol was transmitted. Given that symbol 1 was transmitted, $x(t)$ equals $s_1(t) + w(t)$, where $w(t)$ is the sample function of a white Gaussian noise process of zero mean and power spectral density $N_0/2$. If, on the other hand, symbol 0 was transmitted, $x(t)$ equals $s_2(t) + w(t)$.

Now, applying the decision rule of (7.57) assuming the use of coherent detection at the receiver, we find that the observation space is partitioned into two decision regions, labeled Z_1 and Z_2 in Figure 7.25. The decision boundary, separating region Z_1 from region Z_2, is the perpendicular bisector of the line joining the two message points. The receiver decides in favor of symbol 1 if the received signal point represented by the observation vector $\mathbf{x}$ falls inside region Z_1. This occurs when $x_1 > x_2$. If, on the other hand, we have $x_1 < x_2$, the received signal point falls inside region Z_2 and the receiver decides in favor of symbol 0. On the decision boundary, we have $x_1 = x_2$, in which case the receiver makes a random guess in favor of symbol 1 or 0.

To proceed further, we define a new Gaussian random variable Y whose sample value y is equal to the difference between x_1 and x_2; that is,

$$y = x_1 - x_2 \tag{7.160}$$

The mean value of the random variable Y depends on which binary symbol was transmitted. Given that symbol 1 was sent, the Gaussian random variables X_1 and X_2, whose sample values are denoted by x_1 and x_2, have mean values equal to $\sqrt{E_b}$ and zero, respectively. Correspondingly, the conditional mean of the random variable Y given that symbol 1 was sent is

$$\begin{aligned} \mathbb{E}[Y|1] &= \mathbb{E}[X_1|1] - \mathbb{E}[X_2|1] \\ &= +\sqrt{E_b} \end{aligned} \tag{7.161}$$

On the other hand, given that symbol 0 was sent, the random variables X_1 and X_2 have mean values equal to zero and $\sqrt{E_b}$, respectively. Correspondingly, the conditional mean of the random variable Y given that symbol 0 was sent is

$$\begin{aligned}\mathbb{E}[Y|0] &= \mathbb{E}[X_1|0] - \mathbb{E}[X_2|0] \\ &= -\sqrt{E_b}\end{aligned} \tag{7.162}$$

The variance of the random variable Y is independent of which binary symbol was sent. Since the random variables X_1 and X_2 are statistically independent, each with a variance equal to $N_0/2$, it follows that

$$\begin{aligned}\text{var}[Y] &= \text{var}[X_1] + \text{var}[X_2] \\ &= N_0\end{aligned} \tag{7.163}$$

Suppose we know that symbol 0 was sent. The conditional probability density function of the random variable Y is then given by

$$f_Y(y|0) = \frac{1}{\sqrt{2\pi N_0}} \exp\left[-\frac{(y+\sqrt{E_b})^2}{2N_0}\right] \tag{7.164}$$

Since the condition $x_1 > x_2$ or, equivalently, $y > 0$ corresponds to the receiver making a decision in favor of symbol 1, we deduce that the conditional probability of error given that symbol 0 was sent is

$$\begin{aligned}p_{10} &= \mathbb{P}(y > 0|\text{symbol 0 was sent}) \\ &= \int_0^{\infty} f_Y(y|0)\, dy \\ &= \frac{1}{\sqrt{2\pi N_0}} \int_0^{\infty} \exp\left[-\frac{(y+\sqrt{E_b})^2}{2N_0}\right] dy\end{aligned} \tag{7.165}$$

To put the integral in (7.165) in a standard form involving the Q-function, we set

$$\frac{y+\sqrt{E_b}}{\sqrt{N_0}} = z \tag{7.166}$$

Then, changing the variable of integration from y to z, we may rewrite (7.165) as

$$\begin{aligned}p_{10} &= \frac{1}{\sqrt{2\pi}} \int_{\sqrt{E_b/N_0}}^{\infty} \exp\left(-\frac{z^2}{2}\right) dz \\ &= Q\left(\sqrt{\frac{E_b}{N_0}}\right)\end{aligned} \tag{7.167}$$

Similarly, we may show the p_{01}, the conditional probability of error given that symbol 1 was sent, has the same value as in (7.167). Accordingly, averaging p_{10} and p_{01} and assuming equiprobable symbols, we find that the *average probability of bit error* or, equivalently, the *BER for binary FSK using coherent detection* is

$$P_e = Q\left(\sqrt{\frac{E_b}{N_0}}\right) \tag{7.168}$$

Comparing (7.108) and (7.168), we see that for a binary FSK receiver to maintain the same BER as in a binary PSK receiver, the bit energy-to-noise density ratio, E_b/N_0, has to be doubled. This result is in perfect accord with the signal-space diagrams of Figures 7.13 and 7.25, where we see that in a binary PSK system the Euclidean distance between the two message points is equal to $2\sqrt{E_b}$, whereas in a binary FSK system the corresponding distance is $\sqrt{2E_b}$. For a prescribed E_b, the minimum distance d_{min} in binary PSK is, therefore, $\sqrt{2}$ times that in binary FSK. Recall from (7.89) that the probability of error decreases exponentially as d_{min}^2; hence the difference between (7.108) and (7.168).

Power Spectra of Binary FSK Signals

Consider the case of Sunde's FSK, for which the two transmitted frequencies f_1 and f_2 differ by an amount equal to the bit rate $1/T_b$, and their arithmetic mean equals the nominal carrier frequency f_c; as mentioned previously, phase continuity is always maintained, including inter-bit switching times. We may express this special binary FSK signal as a frequency-modulated signal, defined by

$$s(t) = \sqrt{\frac{2E_b}{T_b}}\cos\left(2\pi f_c t \pm \frac{\pi t}{T_b}\right), \qquad 0 \le t \le T_b \tag{7.169}$$

Using a well-known trigonometric identity, we may reformulate $s(t)$ in the expanded form

$$\begin{aligned} s(t) &= \sqrt{\frac{2E_b}{T_b}}\cos\left(\pm\frac{\pi t}{T_b}\right)\cos(2\pi f_c t) - \sqrt{\frac{2E_b}{T_b}}\sin\left(\pm\frac{\pi t}{T_b}\right)\sin(2\pi f_c t) \\ &= \sqrt{\frac{2E_b}{T_b}}\cos\left(\frac{\pi t}{T_b}\right)\cos(2\pi f_c t) \mp \sqrt{\frac{2E_b}{T_b}}\sin\left(\frac{\pi t}{T_b}\right)\sin(2\pi f_c t) \end{aligned} \tag{7.170}$$

In the last line of (7.170), the plus sign corresponds to transmitting symbol 0 and the minus sign corresponds to transmitting symbol 1. As before, we assume that the symbols 1 and 0 in the binary sequence applied to the modulator input are equally likely, and that the symbols transmitted in adjacent time slots are statistically independent. Then, based on the representation of (7.170), we may make two observations pertaining to the in-phase and quadrature components of a binary FSK signal with continuous phase:

1. The in-phase component is completely independent of the input binary wave. It equals $\sqrt{2E_b/T_b}\cos(\pi t/T_b)$ for all time t. The power spectral density of this component, therefore, consists of two delta functions at $t = \pm 1/2\mathrm{T}_b$ and weighted by the factor $E_b/2T_b$, and occurring at $f = \pm 1/2T_b$.
2. The quadrature component is directly related to the input binary sequence. During the signaling interval $0 \le t \le T_b$, it equals $-g(t)$ when we have symbol 1 and $+g(t)$ when we have symbol 0, with $g(t)$ denoting a symbol-shaping function defined by

$$g(t) = \begin{cases} \sqrt{\dfrac{2E_b}{T_b}}\sin\left(\dfrac{\pi t}{T}\right), & 0 \le t \le T \\ 0, & \text{elsewhere} \end{cases} \tag{7.171}$$

The energy spectral density of $g(t)$ is defined by

$$\Psi_g(f) = \frac{8E_b T_b \cos^2(\pi T_b f)}{\pi^2 (4T_b^2 f^2 - 1)^2} \tag{7.172}$$

The power spectral density of the quadrature component equals $\Psi_g(f)/T_b$. It is also apparent that the in-phase and quadrature components of the binary FSK signal are independent of each other. Accordingly, the baseband power spectral density of Sunde's FSK signal equals the sum of the power spectral densities of these two components, as shown by

$$S_B(f) = \frac{E_b}{2T_b}\left[\delta\left(f - \frac{1}{2T_b}\right) + \delta\left(f + \frac{1}{2T_b}\right)\right] + \frac{8E_b T_b \cos^2(\pi T_b f)}{\pi^2 (4T_b^2 f^2 - 1)^2} \tag{7.173}$$

From Chapter 4, we recall the following relationship between baseband modulated power spectra:

$$S_S(f) = \frac{1}{4}[S_B(f - f_c) + S_B(f + f_c)] \tag{7.174}$$

where f_c is the carrier frequency. Therefore, substituting (7.173) into (7.174), we find that the power spectrum of the binary FSK signal contains two discrete frequency components, one located at $(f_c + 1/2T_b) = f_1$ and the other located at $(f_c - 1/2T_b) = f_2$, with their average powers adding up to one-half the total power of the binary FSK signal. The presence of these two discrete frequency components serves a useful purpose: it provides a practical basis for *synchronizing* the receiver with the transmitter.

Examining (7.173), we may make the following statement:

> The baseband power spectral density of a binary FSK signal with *continuous phase* ultimately falls off as the inverse fourth power of frequency.

In Figure 7.15, we plotted the baseband power spectra of (7.111) and (7.173). (To simplify matters, we have only plotted the results for positive frequencies.) In both cases, $S_B(f)$ is shown normalized with respect to $2E_b$, and the frequency is normalized with respect to the bit rate $R_b = 1/T_b$. The difference in the falloff rates of these spectra can be explained on the basis of the pulse shape $g(t)$. The smoother the pulse, the faster the drop of spectral tails to zero. Thus, since binary FSK with continuous phase has a smoother pulse shape, it has lower sidelobes than binary PSK does.

Suppose, next, the FSK signal exhibits *phase discontinuity* at the inter-bit switching instants, which arises when the two oscillators supplying the basis functions with frequencies f_1 and f_2 operate independently of each other. In this discontinuous scenario, we find that power spectral density ultimately falls off as the inverse square of frequency. Accordingly, we may state:

> A binary FSK signal with continuous phase does not produce as much interference outside the signal band of interest as a corresponding FSK signal with discontinuous phase does.

The important point to take from this statement is summed up as follows: when interference is an issue of practical concern, continuous FSK is preferred over its discontinuous counterpart. However, this advantage of continuous FSK is gained at the expense of increased system complexity.

Minimum Shift Keying

In the coherent detection of binary FSK signal, the phase information contained in the received signal is not fully exploited, other than to provide for synchronization of the receiver to the transmitter. We now show that by proper use of the continuous-phase property when performing detection it is possible to improve the noise performance of the receiver significantly. Here again, this improvement is achieved at the expense of increased system complexity.

Consider a *continuous-phase frequency-shift keying (CPFSK) signal*, which is defined for the signaling interval $0 \le t \le T_b$ as follows:

$$s(t) = \begin{cases} \sqrt{\dfrac{2E_b}{T_b}}\cos(2\pi f_1 t + \theta(0)) & \text{for symbol 1} \\ \sqrt{\dfrac{2E_b}{T_b}}\cos(2\pi f_2 t + \theta(0)) & \text{for symbol 0} \end{cases} \tag{7.175}$$

where E_b is the transmitted signal energy per bit and T_b is the bit duration. The defining equation (7.175) distinguishes itself from that of (7.151) in using the phase $\theta(0)$. This new term, denoting the value of the phase at time $t = 0$, sums up the past history of the FM process up to time $t = 0$. The frequencies f_1 and f_2 are sent in response to binary symbols 1 and 0, respectively, applied to the modulator input.

Another useful way of representing the CPFSK signal $s(t)$ is to express it as a conventional angle-modulated signal:

$$s(t) = \sqrt{\frac{2E_b}{T_b}}\cos[2\pi f_c t + \theta(t)] \tag{7.176}$$

where $\theta(t)$ is the phase of $s(t)$ at time t. When the phase $\theta(t)$ is a continuous function of time, we find that the modulated signal $s(t)$ is itself also continuous at all times, including the inter-bit switching times. The phase $\theta(t)$ of a CPFSK signal increases or decreases linearly with time during each bit duration of T_b seconds, as shown by

$$\theta(t) = \theta(0) \pm \left(\frac{\pi h}{T_b}\right)t, \qquad 0 \le t \le T_b \tag{7.177}$$

where the plus sign corresponds to sending symbol 1 and the minus sign corresponds to sending symbol 0; the dimensionless parameter h is to be defined. Substituting (7.177) into (7.176), and then comparing the angle of the cosine function with that of (7.175), we deduce the following pair of relations:

$$f_c + \frac{h}{2T_b} = f_1 \tag{7.178}$$

$$f_c - \frac{h}{2T_b} = f_2 \tag{7.179}$$

Solving this pair of equations for f_c and h, we get

$$f_c = \frac{1}{2}(f_1 + f_2) \tag{7.180}$$

and

$$h = T_b(f_1 - f_2) \tag{7.181}$$

The nominal carrier frequency f_c is, therefore, the arithmetic mean of the transmitted frequencies f_1 and f_2. The difference between the frequencies f_1 and f_2, normalized with respect to the bit rate $1/T_b$, defines the dimensionless parameter h, which is referred to as the *deviation ratio*.

Phase Trellis

From (7.177) we find that, at time $t = T_b$,

$$\theta(T_b) - \theta(0) = \begin{cases} \pi h & \text{for symbol 1} \\ -\pi h & \text{for symbol 0} \end{cases} \tag{7.182}$$

That is to say, sending symbol 1 increases the phase of a CPFSK signal $s(t)$ by πh radians, whereas sending symbol 0 reduces it by an equal amount.

The variation of phase $\theta(t)$ with time t follows a path consisting of a sequence of straight lines, the slopes of which represent frequency changes. Figure 7.27 depicts possible paths starting from $t = 0$. A plot like that shown in this figure is called a *phase tree*. The tree makes clear the transitions of phase across successive signaling intervals. Moreover, it is evident from the figure that the phase of a CPFSK signal is an odd or even multiple of πh radians at odd or even multiples of the bit duration T_b, respectively.

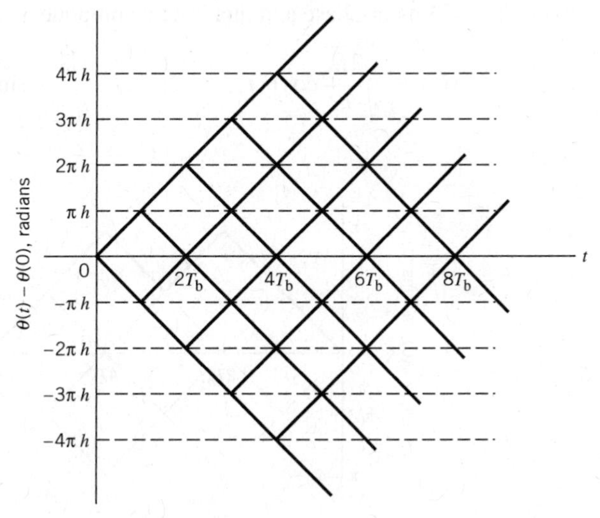

Figure 7.27 Phase tree.

The phase tree described in Figure 7.27 is a manifestation of phase continuity, which is an inherent characteristic of a CPFSK signal. To appreciate the notion of phase continuity, let us go back for a moment to Sunde's FSK, which is also a CPFSK signal as previously described. In this case, the deviation ratio h is exactly unity. Hence, according to Figure 7.27, the phase change over one bit interval is $\pm\pi$ radians. But, a change of $+\pi$ radians is exactly the same as a change of $-\pi$ radians, modulo 2π. It follows, therefore, that in the case of Sunde's FSK there is *no memory*; that is, knowing which particular change occurred in the *previous* signaling interval provides no help in the *current* signaling interval.

In contrast, we have a completely different situation when the deviation ratio h is assigned the special value of 1/2. We now find that the phase can take on only the two values $\pm\pi/2$ at odd multiples of T_b, and only the two values 0 and π at even multiples of T_b, as in Figure 7.28. This second graph is called a *phase trellis*, since a "trellis" is a treelike structure with re-emerging branches. Each path from left to right through the trellis of Figure 7.28 corresponds to a specific binary sequence at the transmitter input. For example, the path shown in boldface in Figure 7.28 corresponds to the binary sequence 1101000 with $\theta(0) = 0$. Henceforth, we focus on $h = 1/2$.

With $h = 1/2$, we find from (7.181) that the frequency deviation (i.e., the difference between the two signaling frequencies f_1 and f_2) equals half the bit rate; hence the following statement:

> The frequency deviation $h = 1/2$ is the minimum frequency spacing that allows the two FSK signals representing symbols 1 and 0 to be *coherently orthogonal.*

In other words, symbols 1 and 0 do not interfere with one another in the process of detection. It is for this reason that a CPFSK signal with a deviation ratio of one-half is commonly referred to as *minimum shift-keying* (MSK).[5]

Signal-Space Diagram of MSK

Using a well-known trigonometric identity in (7.176), we may expand the CPFSK signal $s(t)$ in terms of its in-phase and quadrature components as

$$s(t) = \sqrt{\frac{2E_b}{T_b}}\cos\theta(t)\cos(2\pi f_c t) - \sqrt{\frac{2E_b}{T_b}}\sin\theta(t)\sin(2\pi f_c t) \tag{7.183}$$

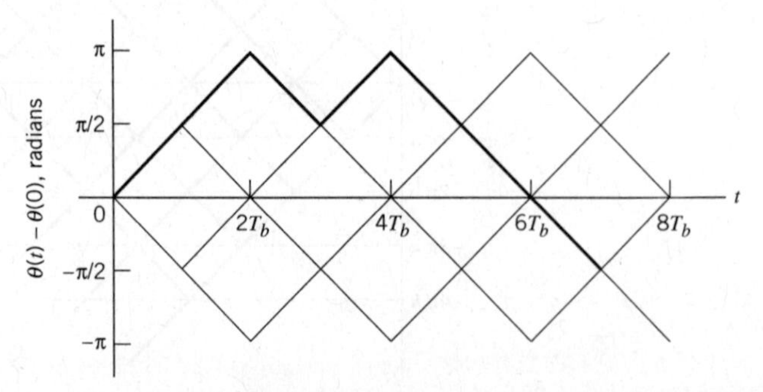

Figure 7.28 Phase trellis; boldfaced path represents the sequence 1101000.

Consider, first, the in-phase component $\sqrt{2E_{\text{b}}/T_{\text{b}}}\cos\theta(t)$. With the deviation ratio $h = 1/2$, we have from (7.177) that

$$\theta(t) = \theta(0) \pm \frac{\pi}{2T_{\text{b}}}, \qquad 0 \le t \le T_{\text{b}} \tag{7.184}$$

where the plus sign corresponds to symbol 1 and the minus sign corresponds to symbol 0. A similar result holds for $\theta(t)$ in the interval $-T_{\text{b}} \le t \le 0$, except that the algebraic sign is not necessarily the same in both intervals. Since the phase $\theta(0)$ is 0 or π depending on the past history of the modulation process, we find that in the interval $-T_{\text{b}} \le t \le T_{\text{b}}$, the polarity of $\cos\theta(t)$ depends only on $\theta(0)$, regardless of the sequence of 1s and 0s transmitted before or after $t = 0$. Thus, for this time interval, the in-phase component consists of the *half-cycle cosine pulse*:

$$\begin{aligned} s_{\text{I}}(t) &= \sqrt{\frac{2E_{\text{b}}}{T_{\text{b}}}}\cos\theta(t) \\ &= \sqrt{\frac{2E_{\text{b}}}{T_{\text{b}}}}\cos\theta(0)\cos\left(\frac{\pi}{2T_{\text{b}}}t\right) \\ &= \pm\sqrt{\frac{2E_{\text{b}}}{T_{\text{b}}}}\cos\left(\frac{\pi}{2T_{\text{b}}}t\right), \qquad -T_{\text{b}} \le t \le T_{\text{b}} \end{aligned} \tag{7.185}$$

where the plus sign corresponds to $\theta(0) = 0$ and the minus sign corresponds to $\theta(0) = \pi$. In a similar way, we may show that, in the interval $0 \le t \le 2T_{\text{b}}$, the quadrature component of $s(t)$ consists of the *half-cycle sine pulse*:

$$\begin{aligned} s_{\text{Q}}(t) &= \sqrt{\frac{2E_{\text{b}}}{T_{\text{b}}}}\sin\theta(t) \\ &= \sqrt{\frac{2E_{\text{b}}}{T_{\text{b}}}}\sin\theta(T_{\text{b}})\sin\left(\frac{\pi}{2T_{\text{b}}}t\right) \\ &= \pm\sqrt{\frac{2E_{\text{b}}}{T_{\text{b}}}}\sin\left(\frac{\pi}{2T_{\text{b}}}t\right), \qquad 0 \le t \le 2T_{\text{b}} \end{aligned} \tag{7.186}$$

where the plus sign corresponds to $\theta(T_{\text{b}}) = \pi/2$ and the minus sign corresponds to $\theta(T_{\text{b}}) = -\pi/2$. From the discussion just presented, we see that the in-phase and quadrature components of the MSK signal differ from each other in two important respects:

- they are in phase quadrature with respect to each other and
- the polarity of the in-phase component $s_{\text{I}}(t)$ depends on $\theta(0)$, whereas the polarity of the quadrature component $s_{\text{Q}}(t)$ depends on $\theta(T_{\text{b}})$.

Moreover, since the phase states $\theta(0)$ and $\theta(T_{\text{b}})$ can each assume only one of two possible values, any one of the following four possibilities can arise:

1. $\theta(0) = 0$ and $\theta(T_{\text{b}}) = \pi/2$, which occur when sending symbol 1.
2. $\theta(0) = \pi$ and $\theta(T_{\text{b}}) = \pi/2$, which occur when sending symbol 0.

3. $\theta(0) = \pi$ and $\theta(T_b) = -\pi/2$ (or, equivalently, $3\pi/2$ modulo 2π), which occur when sending symbol 1.
4. $\theta(0) = 0$ and $\theta(T_b) = -\pi/2$, which occur when sending symbol 0.

This fourfold scenario, in turn, means that the MSK signal itself can assume one of four possible forms, depending on the values of the phase-state pair: $\theta(0)$ and $\theta(T_b)$.

Signal-Space Diagram

Examining the expansion of (7.183), we see that there are two orthonormal basis functions $\phi_1(t)$ and $\phi_2(t)$ characterizing the generation of MSK; they are defined by the following *pair of sinusoidally modulated quadrature carriers*:

$$\phi_1(t) = \sqrt{\frac{2}{T_b}}\cos\left(\frac{\pi}{2T_b}t\right)\cos(2\pi f_c t), \qquad 0 \le t \le T_b \tag{7.187}$$

$$\phi_2(t) = \sqrt{\frac{2}{T_b}}\sin\left(\frac{\pi}{2T_b}t\right)\sin(2\pi f_c t), \qquad 0 \le t \le T_b \tag{7.188}$$

With the formulation of a signal-space diagram in mind, we rewrite (7.183) in the compact form

$$s(t) = s_1\phi_1(t) + s_2\phi_2(t), \qquad 0 \le t \le T_b \tag{7.189}$$

where the coefficients s_1 and s_2 are related to the phase states $\theta(0)$ and $\theta(T_b)$, respectively. To evaluate s_1, we integrate the product $s(t)\phi_1(t)$ with respect to time t between the limits $-T_b$ and T_b, obtaining

$$\begin{aligned} s_1 &= \int_{-T_b}^{T_b} s(t)\phi_1(t)\,\mathrm{d}t \\ &= \sqrt{E_b}\cos[\theta(0)], \qquad -T_b \le t \le T_b \end{aligned} \tag{7.190}$$

Similarly, to evaluate s_2 we integrate the product $s(t)\phi_2(t)$ with respect to time t between the limits 0 and $2T_b$, obtaining

$$\begin{aligned} s_2 &= \int_{0}^{2T_b} s(t)\phi_2(t)\,\mathrm{d}t \\ &= \sqrt{E_b}\sin[\theta(T_b)], \qquad 0 \le t \le T_b \end{aligned} \tag{7.191}$$

Examining (7.190) and (7.191), we now make three observations:

1. Both integrals are evaluated for a time interval equal to twice the bit duration.
2. The lower and upper limits of the integral in (7.190) used to evaluate s_1 are shifted by the bit duration T_b with respect to those used to evaluate s_2.
3. The time interval $0 \le t \le T_b$, for which the phase states $\theta(0)$ and $\theta(T_b)$ are defined, is common to both integrals.

It follows, therefore, that the signal constellation for an MSK signal is two-dimensional (i.e., $N = 2$), with four possible message points (i.e., $M = 4$), as illustrated in the signal-space diagram of Figure 7.29. Moving in a counterclockwise direction, the coordinates of the message points are as follows:

$$(+\sqrt{E_b}, +\sqrt{E_b}),\ (-\sqrt{E_b}, +\sqrt{E_b}),\ (-\sqrt{E_b}, -\sqrt{E_b}), \text{ and } (+\sqrt{E_b}, -\sqrt{E_b}).$$

The possible values of $\theta(0)$ and $\theta(T_b)$, corresponding to these four message points, are also included in Figure 7.29. The signal-space diagram of MSK is thus similar to that of QPSK in that both of them have four message points in a two-dimensional space. However, they differ in a subtle way that should be carefully noted:

- QPSK, moving from one message point to an adjacent one, is produced by sending a two-bit symbol (i.e., dibit).
- MSK, on the other hand, moving from one message point to an adjacent one, is produced by sending a binary symbol, 0 or 1. However, each symbol shows up in two opposite quadrants, depending on the value of the phase-pair: $\theta(0)$ and $\theta(T_b)$.

Table 7.4 presents a summary of the values of $\theta(0)$ and $\theta(T_b)$, as well as the corresponding values of s_1 and s_2 that are calculated for the time intervals $-T_b \le t \le T_b$ and $0 \le t \le 2T_b$, respectively. The first column of this table indicates whether symbol 1 or symbol 0 was sent in the interval $0 \le t \le T_b$. Note that the coordinates of the message points, s_1 and s_2, have opposite signs when symbol 1 is sent in this interval, but the same sign when symbol 0 is sent. Accordingly, for a given input data sequence, we may use the entries of Table 7.4 to derive on a bit-by-bit basis the two sequences of coefficients required to scale $\phi_1(t)$ and $\phi_2(t)$, and thereby determine the MSK signal $s(t)$.

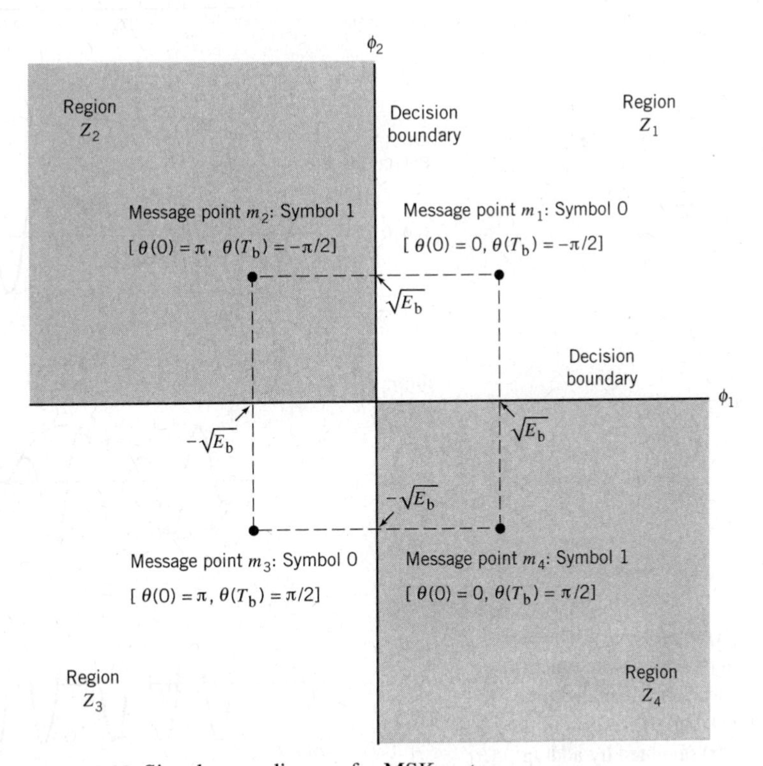

Figure 7.29 Signal-space diagram for MSK system.

Table 7.4 **Signal-space characterization of MSK**

Transmitted binary symbol, $0 \leq t \leq T_b$	Phase states (rad) $\theta(0)$	$\theta(T_b)$	Coordinates of message points s_1	s_2
0	0	$-\pi/2$	$+\sqrt{E_b}$	$+\sqrt{E_b}$
1	π	$-\pi/2$	$-\sqrt{E_b}$	$+\sqrt{E_b}$
0	π	$+\pi/2$	$-\sqrt{E_b}$	$-\sqrt{E_b}$
1	0	$+\pi/2$	$+\sqrt{E_b}$	$-\sqrt{E_b}$

EXAMPLE 8 **MSK Waveforms**

Figure 7.30 shows the sequences and waveforms involved in the generation of an MSK signal for the binary sequence 1101000. The input binary sequence is shown in Figure 7.30a. The two modulation frequencies are $f_1 = 5/4T_b$ and $f_2 = 3/4T_b$. Assuming that at time $t = 0$

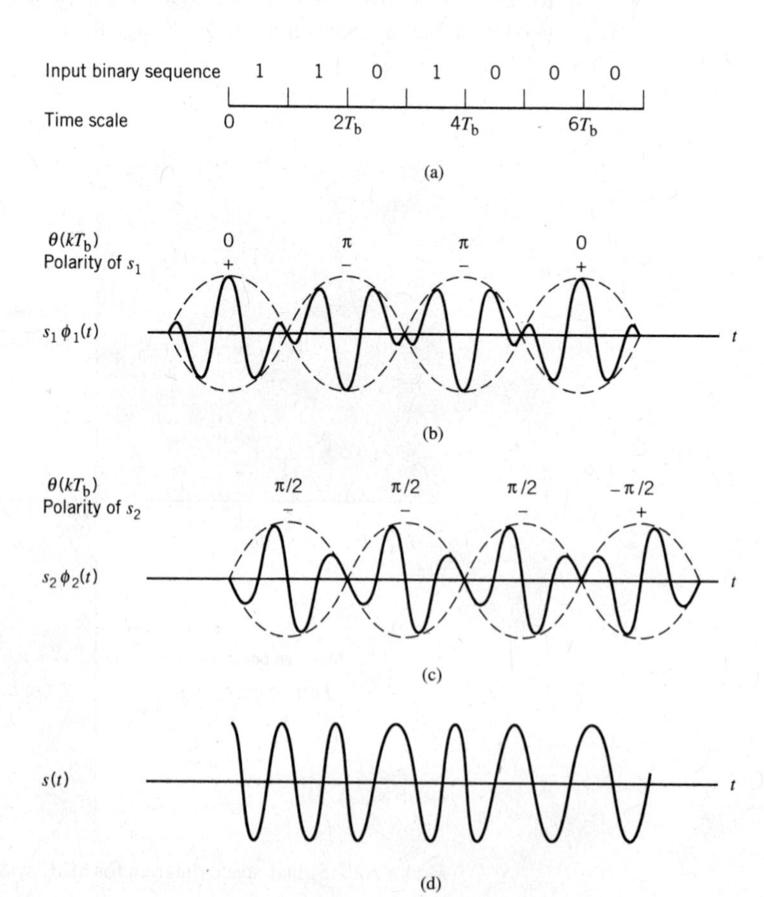

Figure 7.30 (a) Input binary sequence. (b) Waveform of scaled time function $s_1\phi_1(t)$. (c) Waveform of scaled time function $s_2\phi_2(t)$. (d) Waveform of the MSK signal $s(t)$ obtained by adding $s_1\phi_1(t)$ and $s_2\phi_2(t)$ on a bit-by-bit basis.

the phase $\theta(0)$ is zero, the sequence of phase states is as shown in Figure 7.30, modulo 2π. The polarities of the two sequences of factors used to scale the time functions $\phi_1(t)$ and $\phi_2(t)$ are shown in the top lines of Figure 7.30b and c. These two sequences are offset relative to each other by an interval equal to the bit duration T_b. The waveforms of the resulting two components of $s(t)$, namely, $s_1\phi_1(t)$ and $s_2\phi_2(t)$, are shown in Figure 7.30b and c. Adding these two modulated waveforms, we get the desired MSK signal $s(t)$ shown in Figure 7.30d.

Generation and Coherent Detection of MSK Signals

With $h = 1/2$, we may use the block diagram of Figure 7.31a to generate the MSK signal. The advantage of this method of generating MSK signals is that the signal coherence and deviation ratio are largely unaffected by variations in the input data rate. Two input sinusoidal waves, one of frequency $f_c = n_c/4T_b$ for some fixed integer n_c and the other of frequency $1/4T_b$, are first applied to a product modulator. This modulator produces two phase-coherent sinusoidal waves at frequencies f_1 and f_2, which are related to the carrier

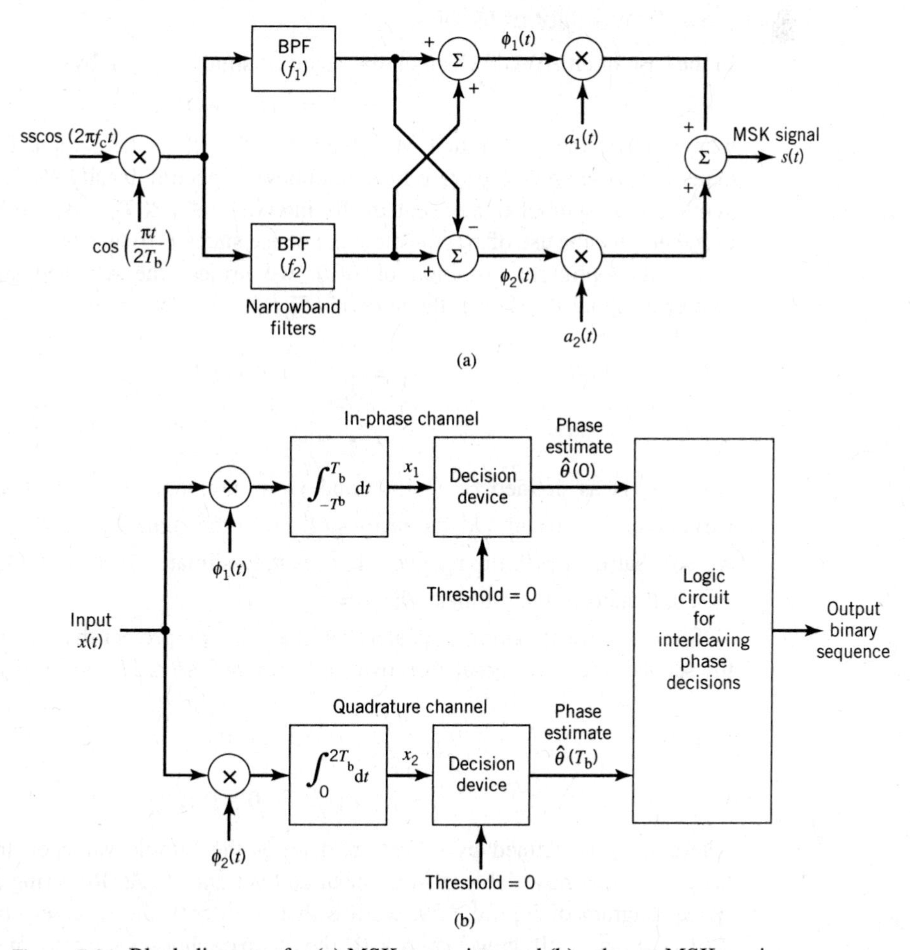

Figure 7.31 Block diagrams for (a) MSK transmitter and (b) coherent MSK receiver.

frequency f_c and the bit rate $1/T_b$ in accordance with (7.178) and (7.179) for deviation ratio $h = 1/2$. These two sinusoidal waves are separated from each other by two narrowband filters, one centered at f_1 and the other at f_2. The resulting filter outputs are next linearly combined to produce the pair of quadrature carriers or orthonormal basis functions $\phi_1(t)$ and $\phi_2(t)$. Finally, $\phi_1(t)$ and $\phi_2(t)$ are multiplied with two binary waves $a_1(t)$ and $a_2(t)$, both of which have a bit rate equal to $1/(2T_b)$. These two binary waves are extracted from the incoming binary sequence in the manner described in Example 7.

Figure 7.31b shows the block diagram of the coherent MSK receiver. The received signal $x(t)$ is correlated with $\phi_1(t)$ and $\phi_2(t)$. In both cases, the integration interval is $2T_b$ seconds, and the integration in the quadrature channel is delayed by T_b seconds with respect to that in the in-phase channel. The resulting in-phase and quadrature channel correlator outputs, x_1 and x_2, are each compared with a threshold of zero; estimates of the phase $\theta(0)$ and $\theta(T_b)$ are then derived in the manner described previously. Finally, these phase decisions are interleaved so as to estimate the original binary sequence at the transmitter input with the minimum average probability of symbol error in an AWGN channel.

Error Probability of MSK

In the case of an AWGN channel, the received signal is given by

$$x(t) = s(t) + w(t)$$

where $s(t)$ is the transmitted MSK signal and $w(t)$ is the sample function of a white Gaussian noise process of zero mean and power spectral density $N_0/2$. To decide whether symbol 1 or symbol 0 was sent in the interval $0 \le t \le T_b$, say, we have to establish a procedure for the use of $x(t)$ to detect the phase states $\theta(0)$ and $\theta(T_b)$.

For the optimum detection of $\theta(0)$, we project the received signal $x(t)$ onto the reference signal $\phi_1(t)$ over the interval $-T_b \le t \le T_b$, obtaining

$$\begin{aligned} x_1 &= \int_{-T_b}^{T_b} x(t)\phi_1(t)\,dt \\ &= s_1 + w_1 \end{aligned} \tag{7.192}$$

where s_1 is as defined by (7.190) and w_1 is the sample value of a Gaussian random variable of zero mean and variance $N_0/2$. From the signal-space diagram of Figure 7.29, we see that if $x_1 > 0$, the receiver chooses the estimate $\hat{\theta}(0) = 0$. On the other hand, if $x_1 < 0$, it chooses the estimate $\hat{\theta}(0) = \pi$.

Similarly, for the optimum detection of $\theta(T_b)$, we project the received signal $x(t)$ onto the second reference signal $\phi_2(t)$ over the interval $0 \le t \le 2T_b$, obtaining

$$\begin{aligned} x_2 &= \int_{0}^{2T_b} x(t)\phi_2(t)\,dt \\ &= s_2 + w_2, \qquad 0 \le t \le 2T_b \end{aligned} \tag{7.193}$$

where s_2 is as defined by (7.191) and w_2 is the sample value of another independent Gaussian random variable of zero mean and variance $N_0/2$. Referring again to the signal-space diagram of Figure 7.29, we see that if $x_2 > 0$, the receiver chooses the estimate $\theta(T_b) = -\pi/2$. If, however, $x_2 < 0$, the receiver chooses the estimate $\hat{\theta}(T_b) = \pi/2$.

To reconstruct the original binary sequence, we interleave the above two sets of phase estimates in accordance with Table 7.4, by proceeding as follows:

- If estimates $\hat{\theta}(0) = 0$ and $\hat{\theta}(0) = (T_\mathrm{b}) = -\pi/2$, or alternatively if $\hat{\theta}(0) = \pi$ and $\hat{\theta}(T_\mathrm{b}) = -\pi/2$, then the receiver decides in favor of symbol 0.
- If, on the other hand, the estimates $\hat{\theta}(0) = \pi$ and $\hat{\theta}(T_\mathrm{b}) = -\pi/2$, or alternatively if $\hat{\theta}(0) = 0$ and $\hat{\theta}(T_\mathrm{b}) = \pi/2$, then the receiver decides in favor of symbol 1.

Most importantly, examining the signal-space diagram of Figure 7.29, we see that the coordinates of the four message points characterizing the MSK signal are identical to those of the QPSK signal in Figure 7.16. Moreover, the zero-mean noise variables in (7.192) and (7.193) have exactly the same variance as those for the QPSK signal in (7.118) and (7.119). It follows, therefore, that the BER for the coherent detection of MSK signals is given by

$$P_\mathrm{e} = Q\left(\sqrt{\frac{2E_\mathrm{b}}{N_0}}\right) \tag{7.194}$$

which is the same as that of QPSK in (7.126). In both MSK and QPSK, this good performance is the result of coherent detection being performed in the receiver on the basis of observations over $2T_\mathrm{b}$ seconds.

Power Spectra of MSK Signals

As with the binary FSK signal, we assume that the input binary wave is random, with symbols 1 and 0 being equally likely and the symbols sent during adjacent time slots being statistically independent. Under these assumptions, we make three observations:

1. Depending on the value of phase state $\theta(0)$, the in-phase component equals $+g(t)$ or $-g(t)$, where the pulse-shaping function

$$g(t) = \begin{cases} \sqrt{\frac{2E_\mathrm{b}}{T_\mathrm{b}}}\cos\left(\frac{\pi t}{2T_\mathrm{b}}\right), & -T_\mathrm{b} \le t \le T_\mathrm{b} \\ 0, & \text{otherwise} \end{cases} \tag{7.195}$$

The energy spectral density of $g(t)$ is

$$\psi_g(f) = \frac{32E_\mathrm{b}T_\mathrm{b}}{\pi^2}\left[\frac{\cos(2\pi T_\mathrm{b}f)}{16T_\mathrm{b}^2 f^2 - 1}\right]^2 \tag{7.196}$$

The power spectral density of the in-phase component equals $\psi_g(f)/2T_\mathrm{b}$.

2. Depending on the value of the phase state $\theta(T_\mathrm{b})$, the quadrature component equals $+g(t)$ or $-g(t)$, where we now have

$$g(t) = \begin{cases} \sqrt{\frac{2E_\mathrm{b}}{T_\mathrm{b}}}\sin\left(\frac{\pi t}{2T_\mathrm{b}}\right), & -0 \le t \le 2T_\mathrm{b} \\ 0, & \text{otherwise} \end{cases} \tag{7.197}$$

Despite the difference in which the time interval over two adjacent time slots is defined in (7.195) and (7.197), we get the same energy spectral density as in (7.196).

Hence, the in-phase and quadrature components have the same power spectral density.

3. The in-phase and quadrature components of the MSK signal are statistically independent; it follows that the baseband power spectral density of $s(t)$ is given by

$$\begin{aligned} S_B(f) &= 2\left(\frac{\Psi_g(f)}{2T_b}\right) \\ &= \frac{32E_b}{\pi^2}\left[\frac{\cos(2\pi T_b f)}{16T_b^2 f^2 - 1}\right]^2 \end{aligned} \quad (7.198)$$

A plot of the baseband power spectrum of (7.198) is included in Figure 7.19, where the power spectrum is normalized with respect to $4E_b$ and the frequency f is normalized with respect to the bit rate $1/T_b$. Figure 7.19 also includes the corresponding plot of (7.128) for the QPSK signal. As stated previously, for $f \gg 1/T_b$ the baseband power spectral density of the MSK signal falls off as the inverse fourth power of frequency, whereas in the case of the QPSK signal it falls off as the inverse square of frequency. Accordingly, MSK does *not* produce as much interference outside the signal band of interest as QPSK does. This is a desirable characteristic of MSK, especially when the digital communication system operates with a bandwidth limitation in an interfering environment.

Gaussian-Filtered MSK

From the detailed study of MSK just presented, we may summarize its desirable properties:

- modulated signal with constant envelope;
- relatively narrow-bandwidth occupancy;
- coherent detection performance equivalent to that of QPSK.

However, the out-of-band spectral characteristics of MSK signals, as good as they are, still do not satisfy the stringent requirements of certain applications such as wireless communications. To illustrate this limitation, we find from (7.198) that, at $T_b f = 0.5$, the baseband power spectral density of the MSK signal drops by only $10 \log_{10} 9 = 9.54$ dB below its midband value. Hence, when the MSK signal is assigned a transmission bandwidth of $1/T_b$, the adjacent channel interference of a wireless-communication system using MSK is not low enough to satisfy the practical requirements of a multiuser-communications environment.

Recognizing that the MSK signal can be generated by direct FM of a voltage-controlled oscillator, we may overcome this practical limitation of MSK by modifying its power spectrum into a more compact form while maintaining the constant-envelope property of the MSK signal. This modification can be achieved through the use of a premodulation low-pass filter, hereafter referred to as a baseband *pulse-shaping filter*. Desirably, the pulse-shaping filter should satisfy the following three conditions:

- frequency response with narrow bandwidth and sharp cutoff characteristics;
- impulse response with relatively low overshoot; and
- evolution of a phase trellis with the carrier phase of the modulated signal assuming the two values $\pm\pi/2$ at odd multiples of the bit duration T_b and the two values 0 and π at even multiples of T_b as in MSK.

The frequency-response condition is needed to suppress the high-frequency components of the modified frequency-modulated signal. The impulse-response condition avoids excessive deviations in the instantaneous frequency of the modified frequency-modulated signal. Finally, the condition imposed on phase-trellis evolution ensures that the modified frequency-modulated signal can be coherently detected in the same way as the MSK signal, or it can be noncoherently detected as a simple binary FSK signal if so desired.

These three conditions can be satisfied by passing an NRZ-level-encoded binary data stream through a baseband pulse-shaping filter whose impulse response (and, likewise, its frequency response) is defined by a *Gaussian function.* The resulting method of binary FM is naturally referred to as *Gaussian-filtered minimum-shift keying* (GMSK).[6]

Let W denote the *3 dB baseband bandwidth* of the pulse-shaping filter. We may then define the transfer function $H(f)$ and impulse response $h(t)$ of the pulse-shaping filter as:

$$H(f) = \exp\left[-\frac{\ln 2}{2}\left(\frac{f}{W}\right)^2\right] \tag{7.199}$$

and

$$h(t) = \sqrt{\frac{2\pi}{\ln 2}}W\exp\left(-\frac{2\pi^2}{\ln 2}W^2t^2\right) \tag{7.200}$$

where ln denotes the natural algorithm. The response of this Gaussian filter to a rectangular pulse of unit amplitude and duration T_b, centered on the origin, is given by

$$\begin{aligned} g(t) &= \int_{-T_b/2}^{T_b/2} h(t-\tau)\,d\tau \\ &= \sqrt{\frac{2\pi}{\ln 2}}W\int_{-T_b/2}^{T_b/2} \exp\left[-\frac{2\pi^2}{\ln 2}W^2(t-\tau)^2\right]d\tau \end{aligned} \tag{7.201}$$

The pulse response $g(t)$ in (7.201) provides the basis for building the GMSK modulator, with the dimensionless *time–bandwidth product* WT_b playing the role of a design parameter.

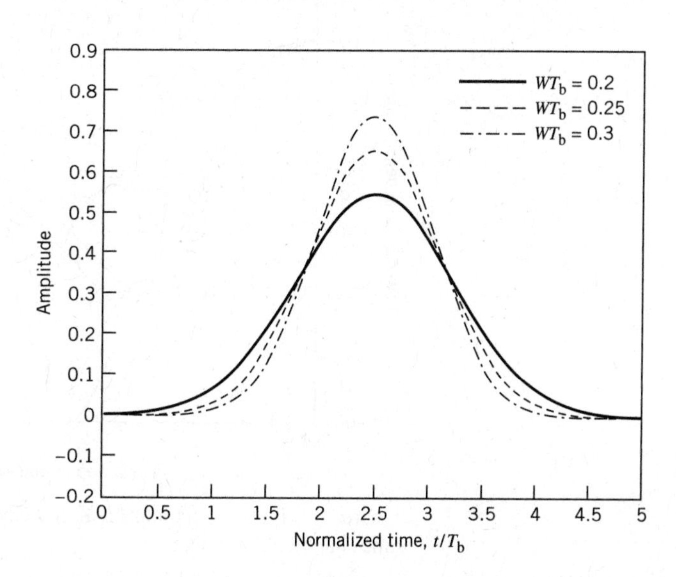

Figure 7.32
Frequency-shaping pulse $g(t)$ of (7.201) shifted in time by $2.5T_b$ and truncated at $\pm 2.5T_b$ for varying time–bandwidth product WT_b.

Unfortunately, the pulse response $g(t)$ is *noncausal* and, therefore, not physically realizable for real-time operation. Specifically, $g(t)$ is nonzero for $t < -T_b/2$, where $t = -T_b/2$ is the time at which the input rectangular pulse (symmetrically positioned around the origin) is applied to the Gaussian filter. For a causal response, $g(t)$ must be truncated and shifted in time. Figure 7.32 presents plots of $g(t)$, which has been truncated at $t = \pm 2.5T_b$ and then shifted in time by $2.5T_b$. The plots shown here are for three different settings: $WT_b = 0.2$, 0.25, and 0.3. Note that as WT_b is reduced, the time spread of the frequency-shaping pulse is correspondingly increased.

Figure 7.33 shows the machine-computed power spectra of MSK signals (expressed in decibels) versus the normalized frequency difference $(f - f_c)T_b$, where f_c is the mid-band frequency and T_b is the bit duration.[7] The results plotted in Figure 7.33 are for varying values of the time–bandwidth product WT_b. From this figure we may make the following observations:

- The curve for the limiting condition $WT_b = \infty$ corresponds to the case of ordinary MSK.
- When WT_b is less than unity, increasingly more of the transmit power is concentrated inside the passband of the GMSK signal.

An undesirable feature of GMSK is that the processing of NRZ binary data by a Gaussian filter generates a modulating signal that is no longer confined to a single bit interval as in ordinary MSK, which is readily apparent from Figure 7.33. Stated in another way, the tails of the Gaussian impulse response of the pulse-shaping filter cause the modulating signal to spread out to adjust symbol intervals. The net result is the generation of *intersymbol interference*, the extent of which increases with decreasing WT_b. In light of this discussion and the various plots presented in Figure 7.33, we find that the value assigned to the time–bandwidth product WT_b offers a tradeoff between spectral compactness and system-performance loss.

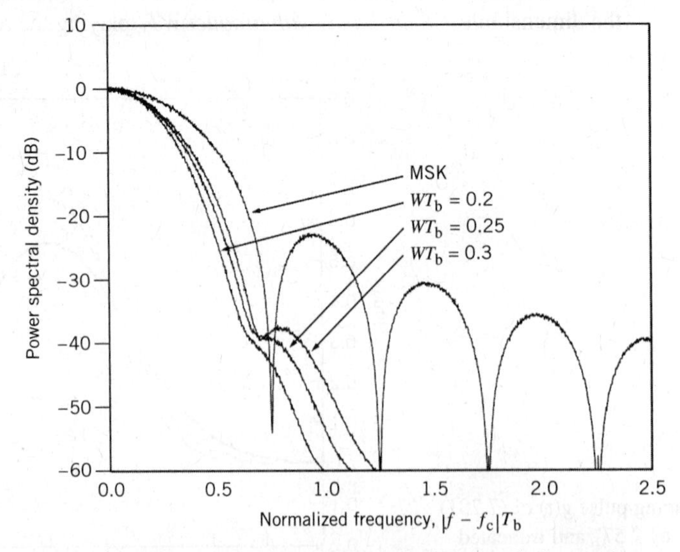

Figure 7.33 Power spectra of MSK and GMSK signals for varying time–bandwidth product.

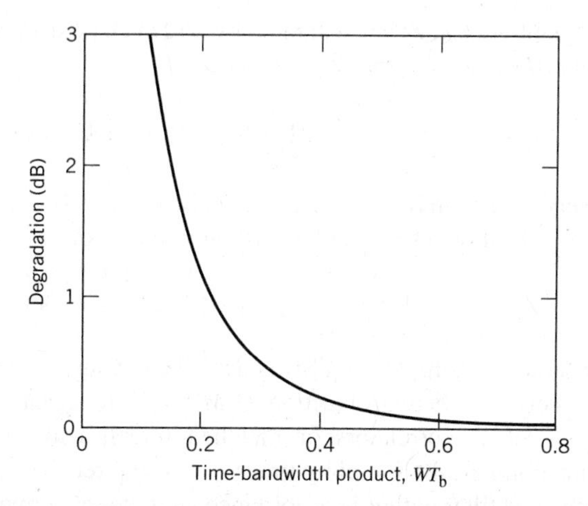

Figure 7.34 Theoretical E_b/N_0 degradation of GMSK for varying time–bandwidth product.

To explore the issue of performance degradation resulting from the use of GMSK compared with MSK, consider the coherent detection in the presence of AWGN. Recognizing that GMSK is a special kind of binary FM, we may express its average probability of symbol error P_e by the empirical formula

$$P_e = Q\left(\sqrt{\frac{\alpha E_b}{N_0}}\right) \tag{7.202}$$

where, as before, E_b is the signal energy per bit and $N_0/2$ is the noise spectral density. The factor α is a constant whose value depends on the time–bandwidth product WT_b. Comparing (7.202) for GMSK with (7.194) for ordinary MSK, we may view $10\log_{10}(\alpha/2)$, expressed in decibels, as a measure of performance degradation of GMSK compared with ordinary MSK. Figure 7.34 shows the machine-computed value of $10\log_{10}(\alpha/2)$ versus WT_b. For ordinary MSK we have $WT_b = \infty$, in which case (7.202) with $\alpha = 2$ assumes exactly the same form as (7.194) and there is no degradation in performance, which is confirmed by Figure 7.34. For GMSK with $WT_b = 0.3$ we find from Figure 7.34 that there is a degradation in performance of about 0.46dB, which corresponds to $\alpha/2 = 0.9$. This degradation in performance is a small price to pay for the highly desirable spectral compactness of the GMSK signal.

M-ary FSK

Consider next the *M*-ary version of FSK, for which the transmitted signals are defined by

$$s_i(t) = \sqrt{\frac{2E}{T}}\cos\left[\frac{\pi}{T}(n_c + i)t\right], \qquad 0 \le t \le T \tag{7.203}$$

where $i = 1, 2, \ldots, M$, and the carrier frequency $f_c = n_c/(2T)$ for some fixed integer n_c. The transmitted symbols are of equal duration T and have equal energy E. Since the individual signal frequencies are separated by $1/(2T)$ Hz, the M-ary FSK signals in (7.203) constitute an *orthogonal set*; that is,

$$\int_0^T s_i(t)s_j(t)\,dt = 0, \qquad i \neq j \tag{7.204}$$

Hence, we may use the transmitted signals $s_i(t)$ themselves, except for energy normalization, as a complete orthonormal set of basis functions, as shown by

$$\phi_i(t) = \frac{1}{\sqrt{E}}s_i(t), \quad \text{for } 0 \le t \le T \text{ and } i = 1, 2, \ldots, M \tag{7.205}$$

Accordingly, the M-ary FSK is described by an M-dimensional signal-space diagram.

For the coherent detection of M-ary FSK signals, the optimum receiver consists of a bank of M correlators or matched filters, with $\phi_i(t)$ of (7.205) providing the basis functions. At the sampling times $t = kT$, the receiver makes decisions based on the largest matched filter output in accordance with the maximum likelihood decoding rule. An exact formula for the probability of symbol error is, however, difficult to derive for a coherent M-ary FSK system. Nevertheless, we may use the union bound of (7.88) to place an upper bound on the average probability of symbol error for M-ary FSK. Specifically, since the minimum distance d_{min} in M-ary FSK is $\sqrt{2E}$, using (7.87) we get (assuming equiprobable symbols)

$$P_e \le (M-1)Q\left(\sqrt{\frac{E}{N_0}}\right) \tag{7.206}$$

For fixed M, this bound becomes increasingly tight as the ratio E/N_0 is increased. Indeed, it becomes a good approximation to P_e for values of $P_e \le 10^{-3}$. Moreover, for $M = 2$ (i.e., binary FSK), the bound of (7.202) becomes an equality; see (7.168).

Power Spectra of M-ary FSK Signals

The spectral analysis of M-ary FSK signals[8] is much more complicated than that of M-ary PSK signals. A case of particular interest occurs when the frequencies assigned to the multilevels make the frequency spacing uniform and the frequency deviation $h = 1/2$. That is, the M signal frequencies are separated by $1/2T$, where T is the symbol duration. For $h = 1/2$, the baseband power spectral density of M-ary FSK signals is plotted in Figure 7.35 for $M = 2, 4, 8$.

Bandwidth Efficiency of M-ary FSK Signals

When the orthogonal signals of an M-ary FSK signal are detected coherently, the adjacent signals need only be separated from each other by a frequency difference $1/2T$ so as to maintain orthogonality. Hence, we may define the channel bandwidth required to transmit M-ary FSK signals as

$$B = \frac{M}{2T} \tag{7.207}$$

For multilevels with frequency assignments that make the frequency spacing uniform and equal to $1/2T$, the bandwidth B of (7.207) contains a large fraction of the signal power.

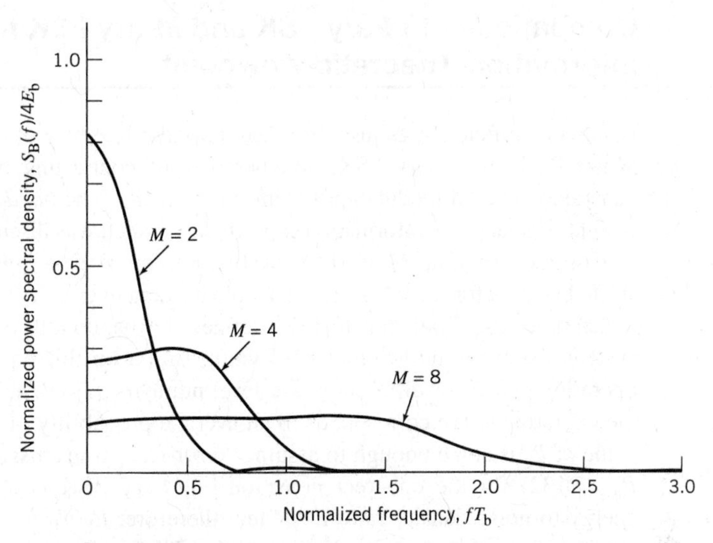

Figure 7.35 Power spectra of M-ary PSK signals for $M = 2, 4, 8$.

This is readily confirmed by looking at the baseband power spectral plots shown in Figure 7.36. From (7.133) we recall that the symbol period T is equal to $T_b \log_2 M$. Hence, using $R_b = 1/T_b$, we may redefine the channel bandwidth B for M-ary FSK signals as

$$B = \frac{R_b M}{2 \log_2 M} \tag{7.208}$$

The bandwidth efficiency of M-ary signals is therefore

$$\begin{aligned}\rho &= \frac{R_b}{B} \\ &= \frac{2 \log_2 M}{M}\end{aligned} \tag{7.209}$$

Table 7.5 gives the values of ρ calculated from (7.207) for varying M.

Comparing Tables 7.3 and 7.5, we see that increasing the number of levels M tends to increase the bandwidth efficiency of M-ary PSK signals, but it also tends to decrease the bandwidth efficiency of M-ary FSK signals. In other words, M-ary PSK signals are spectrally efficient, whereas M-ary FSK signals are spectrally inefficient.

Table 7.5 **Bandwidth efficiency of M-ary FSK signals**

M	2	4	8	16	32	64
ρ (bits/(sHz))	1	1	0.75	0.5	0.3125	0.1875

7.9 Comparison of *M*-ary PSK and *M*-ary FSK from an Information-Theoretic Viewpoint

Bandwidth efficiency, as just discussed, provides one way of contrasting the capabilities of *M*-ary PSK and *M*-ary FSK. Another way of contrasting the capabilities of these two generalized digital modulation schemes is to look at the *bandwidth–power tradeoff* viewed in light of Shannon's information capacity law, which was discussed previously in Chapter 5.

Consider, first, an *M*-ary PSK system that employs a *nonorthogonal* set of *M* phase-shifted signals for the transmission of binary data over an AWGN channel. Referring back to Section 7.6, recall that (7.137) defines the bandwidth efficiency of the *M*-ary PSK system, using the null-to-null bandwidth. Based on this equation, Figure 7.36 plots the operating points for different phase-level numbers $M = 2, 4, 8, 16, 32, 64$. Each point on the operating curve corresponds to an average probability of symbol error $P_e = 10^{-5}$; this value of P_e is small enough to assume "error-free" transmission. Given this fixed value of P_e, (7.132) for the coherent detection of *M*-ary PSK is used to calculate the symbol energy-to-noise density ratio E/N_0 and, therefore, E_b/N_0 for a prescribed *M*; Figure 7.36 also includes the capacity boundary for the ideal transmission system, computed in accordance with (5.99). Figure 7.36 teaches us the following:

> In *M*-ary PSK using coherent detection, increasing *M* improves the bandwidth efficiency, but the E_b/N_0 required for the idealized condition of "error-free" transmission moves away from the Shannon limit as *M* is increased.

Consider next an *M*-ary FSK system that uses an *orthogonal* set of *M* frequency-shifted signals for the transmission of binary data over an AWGN channel. As discussed in Section 7.8, the separation between adjacent signal frequencies in the set is $1/2T$, where T is the symbol period. The bandwidth efficiency of *M*-ary FSK is defined in (7.209), the formulation of which also invokes the null-to-null bandwidth. Using this equation, Figure 7.37 plots the operating points for different frequency-level numbers $M = 2, 4, 8, 16, 32, 64$ for the same average probability of symbol error, namely $P_e = 10^{-5}$. Given this fixed value of P_e, (7.206) is used to calculate the E/N_0 and, therefore, E_b/N_0 required for a prescribed value of *M*. As in Figure 7.36 for *M*-ary PSK, Figure 7.37 for *M*-ary FSK also includes the capacity boundary for the ideal condition of error-free transmission. Figure 7.37 shows that increasing *M* in *M*-ary FSK has the opposite effect to that in *M*-ary PSK. In more specific terms, we may state the following:

> In *M*-ary FSK, as the number of frequency-shift levels *M* is increased—which is equivalent to increased channel-bandwidth requirement—the operating point moves closer to the Shannon limit.

In other words, in an information-theoretic context, *M*-ary FSK behaves better than *M*-ary PSK.

In the final analysis, the choice of *M*-ary PSK or *M*-ary FSK for binary data transmission over an AWGN channel is determined by the design criterion of interest: bandwidth efficiency or the E_b/N_0 needed for reliable data transmission.

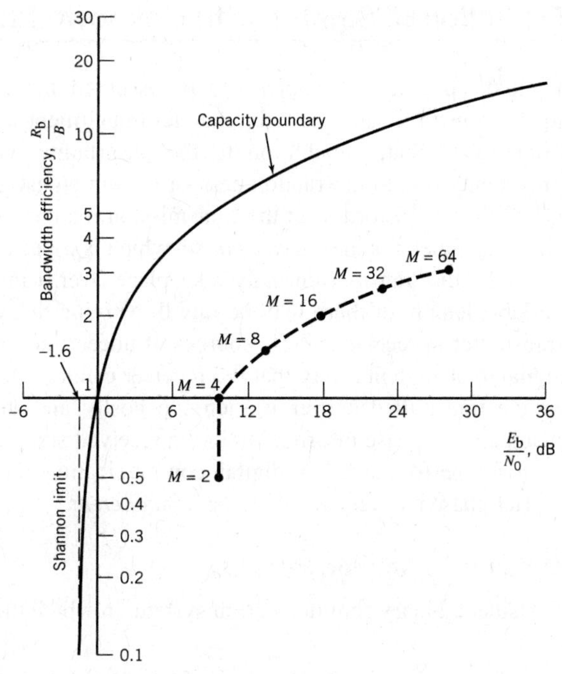

Figure 7.36 Comparison of M-ary PSK with the ideal system for $P_e = 10^{-5}$.

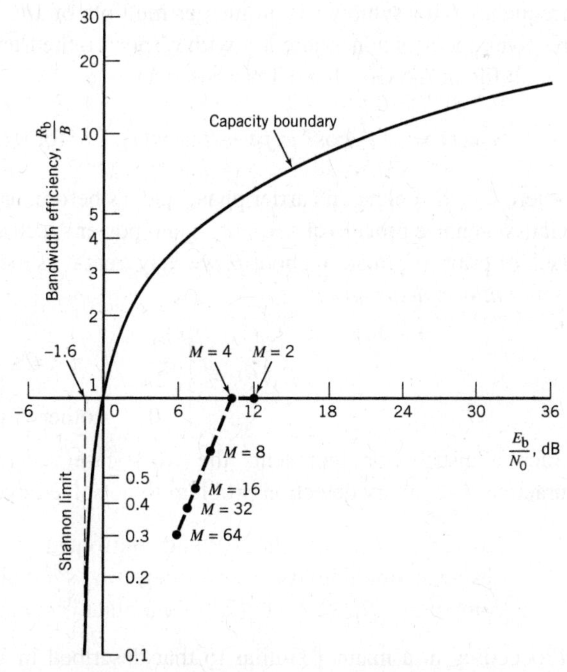

Figure 7.37 Comparison of M-ary FSK with the ideal system for $P_e = 10^{-5}$.

7.10 Detection of Signals with Unknown Phase

Up to this point in the chapter we have assumed that the receiver is perfectly synchronized to the transmitter and the only channel impairment is AWGN. In practice, however, it is often found that, in addition to the uncertainty due to channel noise, there is also uncertainty due to the randomness of certain signal parameters. The usual cause of this uncertainty is distortion in the transmission medium. Perhaps the most common random signal parameter is the *carrier phase*, which is especially true for narrowband signals. For example, the transmission may take place over a multiplicity of paths of different and variable length, or there may be rapidly varying delays in the propagating medium from transmitter to receiver. These sources of uncertainty may cause the phase of the received signal to change in a way that the receiver cannot follow. Synchronization with the phase of the transmitted carrier is then too costly and the designer may simply choose to disregard the phase information in the received signal at the expense of some degradation in noise performance. A digital communication receiver with no provision made for carrier phase recovery is said to be *noncoherent*.

Optimum Quadratic Receiver

Consider a binary communication system, in which the transmitted signal is defined by

$$s_i(t) = \sqrt{\frac{2E}{T}}\cos(2\pi f_i t), \qquad \begin{cases} 0 \le t \le T \\ i = 1, 2 \end{cases} \tag{7.210}$$

where E is the signal energy, T is the duration of the signaling interval, and the carrier frequency f_i for symbol i is an integer multiple of $1/(2T)$. For reasons just mentioned, the receiver operates noncoherently with respect to the transmitter, in which case the received signal for an AWGN channel is written as

$$x(t) = \sqrt{\frac{2E}{T}}\cos(2\pi f_i t + \theta) + w(t), \qquad \text{for } 0 \le t \le T \text{ and } i = 1, 2 \tag{7.211}$$

where θ is the unknown carrier phase and, as before, $w(t)$ is the sample function of a white Gaussian noise process of zero mean and power spectral density $N_0/2$. Assuming complete lack of prior information about θ, we may treat it as the sample value of a random variable with *uniform distribution*:

$$f_\Theta(\theta) = \begin{cases} \dfrac{1}{2\pi}, & -\pi < \theta \le \pi \\ 0, & \text{otherwise} \end{cases} \tag{7.212}$$

Such a distribution represents the worst-case scenario that could be encountered in practice. The binary detection problem to be solved may now be stated as follows:

> Given the received signal $x(t)$ and confronted with the unknown carrier phase θ, design an optimum receiver for detecting symbol s_i represented by the signal component $\sqrt{E/(2T)}\cos(2\pi f_i t + \theta)$ that is contained in $x(t)$.

Proceeding in a manner similar to that described in Section 7.4, we may formulate the likelihood function of symbol s_i given the carrier phase θ as

$$l(s_i(\theta)) = \exp\left[\sqrt{\frac{E}{N_0 T}} \int_0^T x(t) \cos(2\pi f_i t + \theta) \, dt\right] \tag{7.213}$$

To proceed further, we have to remove dependence of $l(s_i(\theta))$ on phase θ, which is achieved by integrating it over all possible values of θ, as shown by

$$\begin{aligned} l(s_i) &= \int_{-\pi}^{\pi} l(s_i(\theta)) f_\Theta(\theta) \, d\theta \\ &= \frac{1}{2\pi}\int_{-\pi}^{\pi} \exp\left[\sqrt{\frac{E}{N_0 T}} \int_0^T x(t) \cos(2\pi f_i t + \theta) \, dt\right] d\theta \end{aligned} \tag{7.214}$$

Using a well-known trigonometric formula, we may expand the cosine term in (7.214) as

$$\cos(2\pi f_i t + \theta) = \cos(2\pi f_i t) \cos\theta - \sin(2\pi f_i t) \sin\theta$$

Correspondingly, we may rewrite the integral in the exponent of (7.214) as

$$\int_0^T x(t) \cos(2\pi f_i t + \theta) \, dt = \cos\theta \int_0^T x(t) \cos(2\pi f_i t) \, dt - \sin\theta \int_0^T x(t) \sin(2\pi f_i t) \, dt \tag{7.215}$$

Define two new terms:

$$\alpha_i = \left\{\left[\int_0^T x(t) \cos(2\pi f_i t) \, dt\right]^2 + \left[\int_0^T x(t) \sin(2\pi f_i t) \, dt\right]^2\right\}^{1/2} \tag{7.216}$$

$$\beta_i = \tan^{-1}\left[\frac{\int_0^T x(t) \sin(2\pi f_i t) \, dt}{\int_0^T x(t) \cos(2\pi f_i t) \, dt}\right] \tag{7.217}$$

Then, we may go one step further and simplify the inner integral in (7.214) to

$$\begin{aligned} \int_0^T x(t) \cos(2\pi f_i t + \theta) \, dt &= \alpha_i(\cos\theta \cos\beta_i - \sin\theta \sin\beta_i) \\ &= \alpha_i \cos(\theta + \beta_i) \end{aligned} \tag{7.218}$$

Accordingly, using (7.218) in (7.214), we obtain

$$\begin{aligned} l(s_i) &= \frac{1}{2\pi} \int_{-\pi}^{\pi} \exp\left[\sqrt{\frac{E}{N_0 T}} \alpha_i \cos(\theta + \beta_i)\right] d\theta \\ &= \frac{1}{2\pi} \int_{-\pi+\beta_i}^{\pi+\beta_i} \exp\left(\sqrt{\frac{E}{N_0 T}} \alpha_i \cos\theta\right) d\theta \\ &= \frac{1}{2\pi} \int_{-\pi}^{\pi} \exp\left(\sqrt{\frac{E}{N_0 T}} \alpha_i \cos\theta\right) d\theta \end{aligned} \tag{7.219}$$

where, in the last line, we have used the fact that the definite integral is unaffected by the phase β_i.

From Appendix C on Bessel functions, we recognize the integral of (7.219) as the *modified Bessel function of zero order*, written in the compact form

$$I_0\left(\sqrt{\frac{E}{N_0T}}\alpha_i\right) = \frac{1}{2\pi}\int_{-\pi}^{\pi}\exp\left(\sqrt{\frac{E}{N_0T}}\alpha_i\cos\theta\right)\mathrm{d}\theta \tag{7.220}$$

Using this formula, we may correspondingly express the likelihood function for the signal-detection problem described herein in the compact form

$$l(s_i) = I_0\left(\sqrt{\frac{E}{N_0T}}\alpha_i\right) \tag{7.221}$$

With binary transmission as the issue of interest, there are two hypotheses to be considered: hypothesis H_1, that signal $s_1(t)$ was sent, and hypothesis H_2, that signal s_2 was sent. In light of (7.221), the binary-hypothesis test may now be formulated as follows:

$$I_0\left(\sqrt{\frac{E}{N_0T}}\alpha_1\right) \underset{H_2}{\overset{H_1}{\gtrless}} \left(I_0\sqrt{\frac{E}{N_0T}}\alpha_2\right)$$

The modified Bessel function $I(\cdot)$ is a monotonically increasing function of its argument. Hence, we may simplify the hypothesis test by focusing on α_i for given E/N_0T. For convenience of implementation, however, the simplified hypothesis test is carried out in terms of α_i^2 rather than α_i; that is to say:

$$\alpha_1^2 \underset{H_2}{\overset{H_1}{\gtrless}} \alpha_2^2 \tag{7.222}$$

For obvious reasons, a receiver based on (7.222) is known as the *quadratic receiver*. In light of the definition of α_i given in (7.216), the receiver structure for computing α_i is as shown in Figure 7.38a. Since the test described in (7.222) is independent of the symbol energy E, this hypothesis test is said to be *uniformly most powerful* with respect to E.

Two Equivalent Forms of the Quadratic Receiver

We next derive two equivalent forms of the quadrature receiver shown in Figure 7.38a. The first form is obtained by replacing each correlator in this receiver with a corresponding equivalent matched filter. We thus obtain the alternative form of quadrature receiver shown in Figure 7.38b. In one branch of this receiver, we have a filter matched to the signal $\cos(2\pi f_i t)$ and in the other branch we have a filter matched to $\sin(2\pi f_i t)$, both of which are defined for the signaling interval $0 \le t \le T$. At time $t = T$, the filter outputs are sampled, squared, and then added together.

To obtain the second equivalent form of the quadrature receiver, suppose we have a filter that is matched to $s(t) = \cos(2\pi f_i t + \theta)$ for $0 \le t \le T$. The envelope of the matched filter output is obviously unaffected by the value of phase θ. Therefore, we may simply choose a matched filter with impulse response $\cos[2\pi f_i(T-t)]$, corresponding to $\theta = 0$. The output of such a filter in response to the received signal $x(t)$ is given by

$$\begin{aligned} y(t) &= \int_0^T x(\tau)\cos[2\pi f_i(T-t+\tau)]\,\mathrm{d}\tau \\ &= \cos[2\pi f_i(T-t)]\int_0^T x(\tau)\cos(2\pi f_i\tau)\,\mathrm{d}\tau - \sin[2\pi f_i(T-t)]\int_0^T x(\tau)\sin(2\pi f_i\tau)\,\mathrm{d}\tau \end{aligned} \tag{7.223}$$

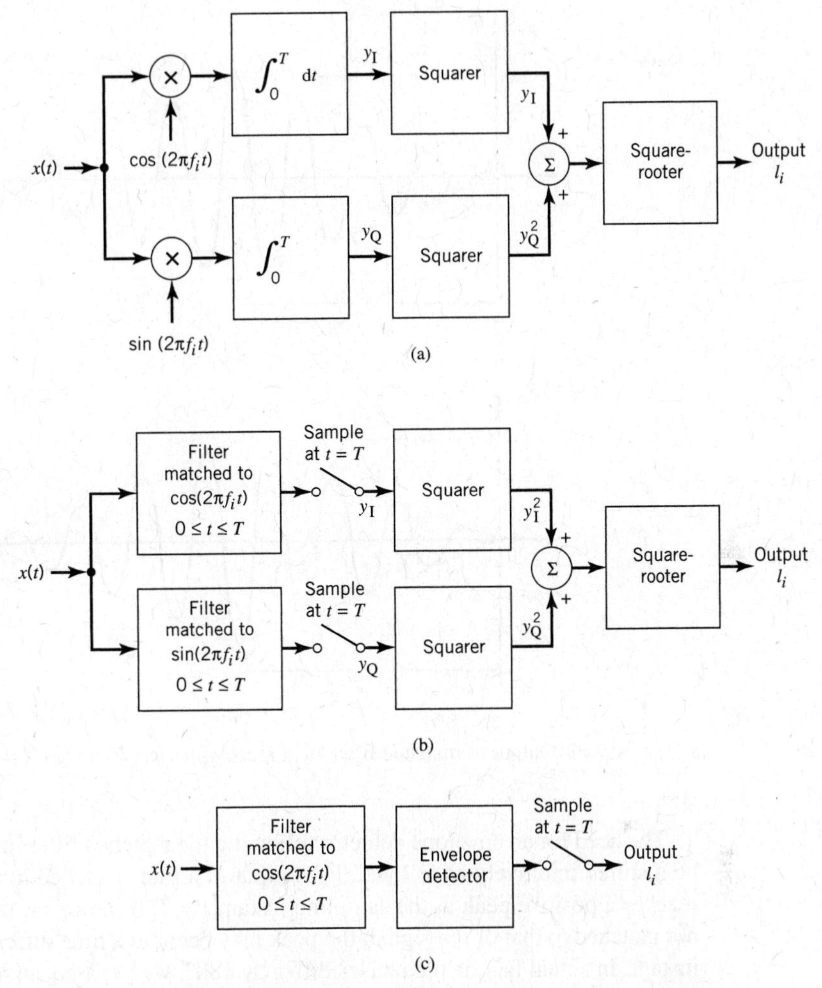

Figure 7.38 Noncoherent receivers: (a) quadrature receiver using correlators; (b) quadrature receiver using matched fiters; (c) noncoherent matched filter.

The envelope of the matched filter output is proportional to the square root of the sum of the squares of the two definite integrals in (7.223). This envelope, evaluated at time $t = T$, is, therefore, given by the following square root:

$$\left\{\left[\int_0^T x(\tau)\cos(2\pi f_i \tau)\,d\tau\right]^2 + \left[\int_0^T x(\tau)\sin(2\pi f_i \tau)\,d\tau\right]^2\right\}^{1/2}$$

But this is just a repeat of the output of the quadrature receiver defined earlier. Therefore, the output (at time T) of a filter matched to the signal $\cos(2\pi f_i t + \theta)$ of arbitrary phase θ, followed by an envelope detector, is the same as the quadrature receiver's output l_i. This form of receiver is shown in Figure 7.38c. The combination of matched filter and envelope detector shown in Figure 7.38c is called a *noncoherent matched filter*.

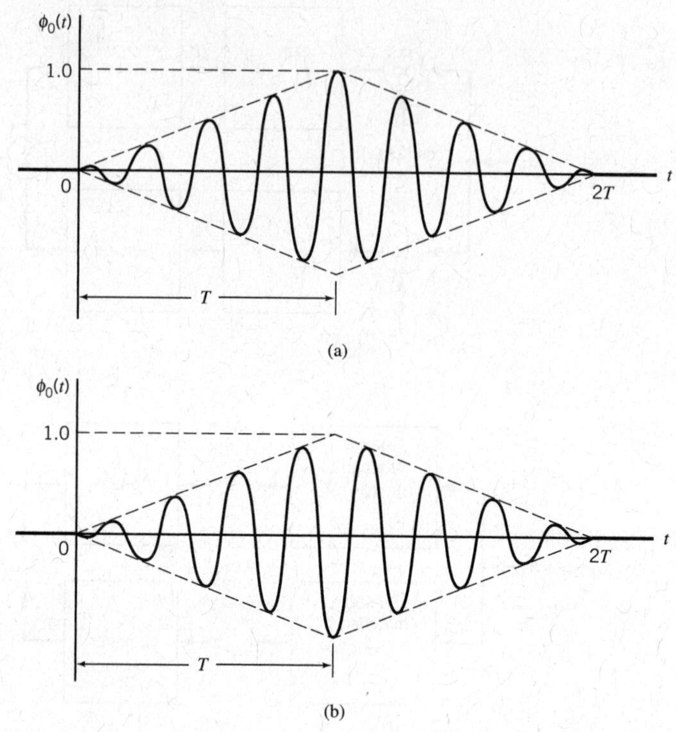

Figure 7.39 Output of matched filter for a rectangular RF wave: (a) $\theta = 0$; (b) $\theta = 180°$.

The need for an envelope detector following the matched filter in Figure 7.38c may also be justified intuitively as follows. The output of a filter matched to a rectangular RF wave reaches a positive peak at the sampling instant $t = T$. If, however, the phase of the filter is not matched to that of the signal, the peak may occur at a time different from the sampling instant. In actual fact, if the phases differ by 180°, we get a negative peak at the sampling instant. Figure 7.39 illustrates the matched filter output for the two limiting conditions: $\theta = 0$ and $\theta = 180°$ for which the respective waveforms of the matched filter output are displayed in parts a and b of the figure. To avoid poor sampling that arises in the absence of prior information about the phase, it is reasonable to retain only the envelope of the matched filter output, since it is completely independent of the phase mismatch θ.

7.11 Noncoherent Orthogonal Modulation Techniques

With the noncoherent receiver structures of Figure 7.38 at our disposal, we may now proceed to study the noise performance of *noncoherent orthogonal modulation* that includes two noncoherent receivers as special cases: noncoherent binary FSK; and differential PSK (called DPSK), which may be viewed as the noncoherent version of binary PSK.

Consider a binary signaling scheme that involves the use of two orthogonal signals $s_1(t)$ and $s_2(t)$, which have equal energy. During the signaling interval $0 \le t \le T$, where T may be different from the bit duration T_b, one of these two signals is sent over an imperfect channel that shifts the carrier phase by an unknown amount. Let $g_1(t)$ and $g_2(t)$ denote the phase-shifted versions of $s_1(t)$ and $s_2(t)$ that result from this transmission, respectively. It is assumed that the signals $g_1(t)$ and $g_2(t)$ remain orthogonal and have the same energy E, regardless of the unknown carrier phase. We refer to such a signaling scheme as *noncoherent orthogonal modulation*, hence the title of the section.

In addition to carrier-phase uncertainty, the channel also introduces AWGN $w(t)$ of zero mean and power spectral density $N_0/2$, resulting in the received signal

$$x(t) = \begin{cases} g_1(t) + w(t), & s_1(t) \text{ sent for } 0 \le t \le T \\ g_2(t) + w(t), & s_2(t) \text{ sent for } 0 \le t \le T \end{cases} \tag{7.224}$$

To tackle the signal detection problem given $x(t)$, we employ the generalized receiver shown in Figure 7.39a, which consists of a pair of filters matched to the transmitted signals $s_1(t)$ and $s_2(t)$. Because the carrier phase is unknown, the receiver relies on amplitude as the only possible discriminant. Accordingly, the matched-filter outputs are envelope-detected, sampled, and then compared with each other. If the upper path in Figure 7.38a has an output amplitude l_1 greater than the output amplitude l_2 of the lower path, the receiver decides in favor of $s_1(t)$; the l_1 and l_2 used here should not be confused with the symbol l denoting the likelihood function in the preceding section. If the converse is true, the receiver decides in favor of $s_2(t)$. When they are equal, the decision may be made by flipping a fair coin (i.e., randomly). In any event, a decision error occurs when the matched filter that rejects the signal component of the received signal $x(t)$ has a larger output amplitude (due to noise alone) than the matched filter that passes it.

From the discussion presented in Section 7.10 we note that a noncoherent matched filter (constituting the upper or lower path in the receiver of Figure 7.40a), may be viewed as being equivalent to a quadrature receiver. The quadrature receiver itself has two channels. One version of the quadrature receiver is shown in Figure 7.40b. In the upper path, called the *in-phase path*, the received signal $x(t)$ is correlated with the function $\psi_i(t)$, which represents a scaled version of the transmitted signal $s_1(t)$ or $s_2(t)$ with zero carrier phase. In the lower path, called the *quadrature path*, on the other hand, $x(t)$ is correlated with another function $\hat{\psi}_i(t)$, which represents the version of $\psi_i(t)$ that results from shifting the carrier phase by –90°. The signals $\psi_i(t)$ and $\hat{\psi}_i(t)$ are orthogonal to each other.

In actual fact, the signal $\hat{\psi}_i(t)$ is the *Hilbert transform* of $\psi_i(t)$; the Hilbert transform was discussed in Chapter 2. To illustrate the nature of this relationship, let

$$\psi_i(t) = m(t)\cos(2\pi f_i t) \tag{7.225}$$

where $m(t)$ is a band-limited message signal. Typically, the carrier frequency f_i is greater than the highest frequency component of $m(t)$. Then the Hilbert transform $\psi_i(t)$ is defined by

$$\hat{\psi}_i(t) = m(t)\sin(2\pi f_i t) \tag{7.226}$$

for which reference should be made in Table 2.3 of Chapter 2. Since

$$\cos\left(2\pi f_i t - \frac{\pi}{2}\right) = \sin(2\pi f_i t)$$

we see that $\hat{\psi}_i(t)$ is indeed obtained from $\psi_i(t)$ by shifting the carrier $\cos(2\pi f_i t)$ by $-90°$. An important property of Hilbert transformation is that a signal and its Hilbert transform are orthogonal to each other. Thus, $\psi_i(t)$ and $\hat{\psi}_i(t)$ are indeed orthogonal to each other, as already stated.

The average probability of error for the noncoherent receiver of Figure 7.40a is given by the simple formula

$$P_e = \frac{1}{2}\exp\left(-\frac{E}{2N_0}\right) \tag{7.227}$$

where E is the signal energy per symbol and $N_0/2$ is the noise spectral density.

Derivation of Equation (7.227)

To derive Equation (7.227)[9] we make use of the equivalence depicted in Figure 7.40. In particular, we observe that, since the carrier phase is unknown, noise at the output of each

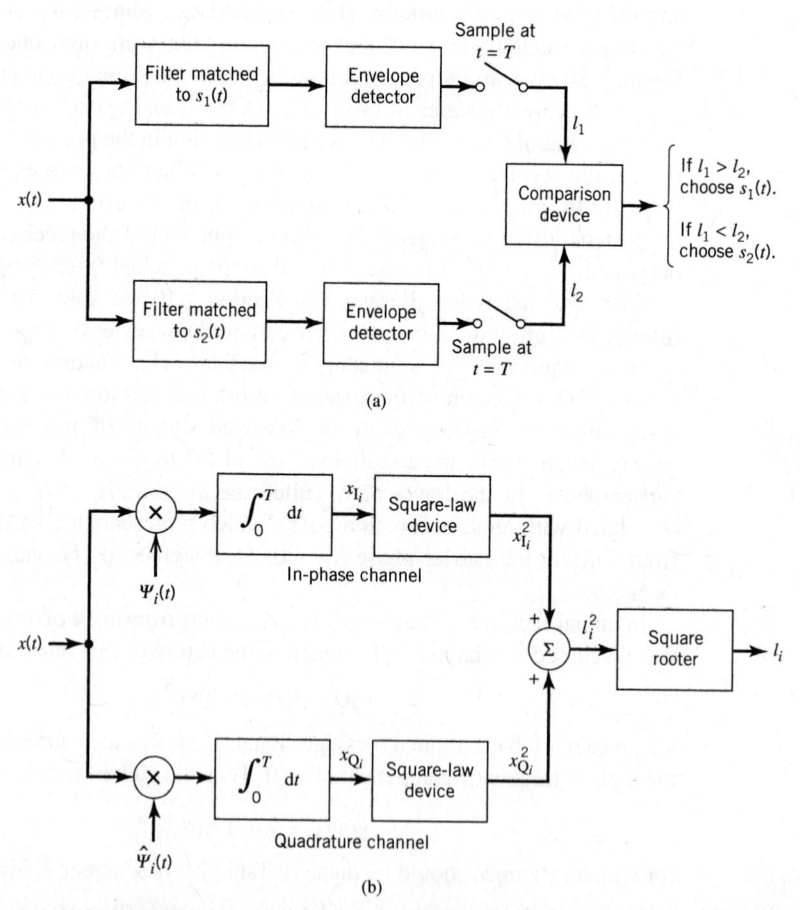

Figure 7.40 (a) Generalized binary receiver for noncoherent orthogonal modulation. (b) Quadrature receiver equivalent to either one of the two matched filters in (a); the index $i = 1, 2$.

matched filter in Figure 7.40a has *two degrees of freedom*: in-phase and quadrature. Accordingly, the noncoherent receiver of Figure 7.40a has a total of four noisy parameters that are *conditionally independent* given the phase θ, and also *identically distributed*. These four noisy parameters have sample values denoted by x_{I1}, x_{Q1}, and x_{I2}, and x_{Q2}; the first two account for degrees of freedom associated with the upper path of Figure 7.40a, and the latter two account for degrees of freedom associated with the lower path of the figure.

The receiver of Figure 7.40a has a *symmetric* structure, meaning that the probability of choosing $s_2(t)$ given that $s_1(t)$ was transmitted is the same as the probability of choosing $s_1(t)$ given that $s_2(t)$ was transmitted. In other words, the average probability of error may be obtained by transmitting $s_1(t)$ and calculating the probability of choosing $s_2(t)$, or vice versa; it is assumed that the original binary symbols and therefore $s_1(t)$ and $s_2(t)$ are equiprobable.

Suppose that signal $s_1(t)$ is transmitted for the interval $0 \le t \le T$. An error occurs if the channel noise $w(t)$ is such that the output l_2 of the lower path in Figure 7.40a is greater than the output l_1 of the upper path. Then, the receiver decides in favor of $s_2(t)$ rather than $s_1(t)$. To calculate the probability of error so made, we must have the probability density function of the random variable L_2 (represented by sample value l_2). Since the filter in the lower path is matched to $s_2(t)$ and $s_2(t)$ is orthogonal to the transmitted signal $s_1(t)$, it follows that the output of this matched filter is due to *noise alone*. Let x_{I2} and x_{Q2} denote the in-phase and quadrature components of the matched filter output in the lower path of Figure 7.40a. Then, from the equivalent structure depicted in this figure, we see that (for $i = 2$)

$$l_2 = \sqrt{x_{I2}^2 + x_{Q2}^2} \tag{7.228}$$

Figure 7.41a shows a geometric interpretation of this relation. The channel noise $w(t)$ is both white (with power spectral density $N_0/2$) and Gaussian (with zero mean). Correspondingly, we find that the random variables X_{I2} and X_{Q2} (represented by sample values x_{I2} and x_{Q2}) are both Gaussian distributed with zero mean and variance $N_0/2$, given the phase θ. Hence, we may write

$$f_{X_{I2}}(x_{I2}) = \frac{1}{\sqrt{\pi N_0}} \exp\left(-\frac{x_{I2}^2}{N_0}\right) \tag{7.229}$$

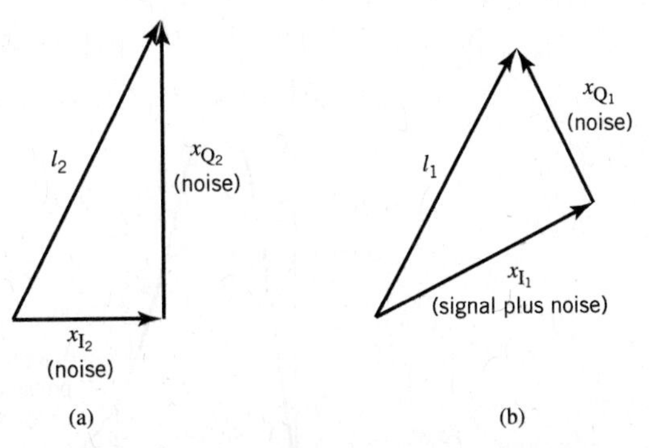

Figure 7.41 Geometric interpretations of the two path outputs l_1 and l_2 in the generalized non-coherent receiver.

and

$$f_{X_{Q2}}(x_{Q2}) = \frac{1}{\sqrt{\pi N_0}} \exp\left(-\frac{x_{Q2}^2}{N_0}\right) \tag{7.230}$$

Next, we use the well-known property presented in Chapter 4 on stochastic processes: the envelope of a Gaussian process represented in polar form is *Rayleigh distributed* and independent of the phase θ. For the situation at hand, therefore, we may state that the random variable L_2 whose sample value l_2 is related to x_{I2} and x_{Q2} by (7.228) has the following probability density function:

$$f_{L_2}(l_2) = \begin{cases} \dfrac{2l_2}{N_0} \exp\left(-\dfrac{l_2^2}{N_0}\right), & l_2 \geq 0 \\ 0, & \text{elsewhere} \end{cases} \tag{7.231}$$

Figure 7.42 shows a plot of this probability density function, where the shaded area defines the conditional probability that $l_2 > l_1$. Hence, we have

$$\mathbb{P}(l_2 > l_1 | l_1) = \int_{l_1}^{\infty} f_{L_2}(l_2)\, \mathrm{d}l_2 \tag{7.232}$$

Substituting (7.231) into (7.232) and integrating, we get

$$\mathbb{P}(l_2 > l_1 | l_1) = \exp\left(-\frac{l_1^2}{N_0}\right) \tag{7.233}$$

Consider next the output amplitude l_1, pertaining to the upper path in Figure 7.40a. Since the filter in this path is matched to $s_1(t)$ and it is assumed that $s_1(t)$ is transmitted, it follows that l_1 is due to signal plus noise. Let x_{I1} and x_{Q1} denote the components at the output of the matched filter in the upper path of Figure 7.39a that are in phase and in quadrature with respect to the received signal, respectively. Then, from the equivalent structure depicted in Figure 7.40b, we see that, for $i = 1$,

$$l_1 = \sqrt{x_{I1}^2 + x_{Q1}^2} \tag{7.234}$$

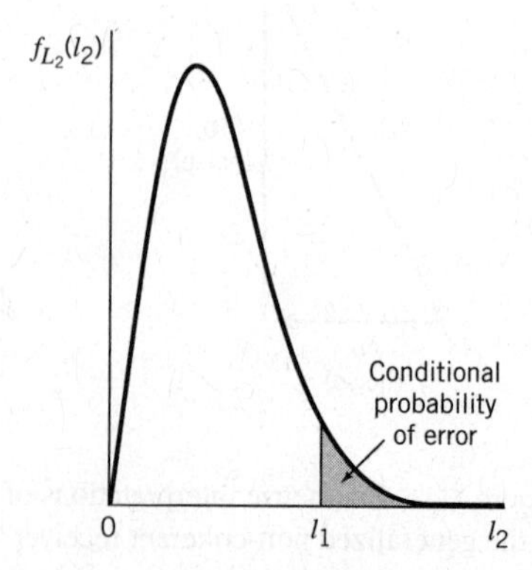

Figure 7.42
Calculation of the conditional probability that $l_2 > l_1$, given l_1.

A geometric interpretation of l_i is presented in Figure 7.41b. Since a Fourier-transformable signal and its Hilbert transform form an orthogonal pair, it follows that x_{I1} is due to signal plus noise, whereas x_{Q1} is due to noise alone. This statement has two implications:

- The random variable X_{I1} represented by the sample value x_{I1} is Gaussian distributed with mean $\sqrt{E}$ and variance $N_0/2$, where E is the signal energy per symbol.
- The random variable X_{Q1} represented by the sample value x_{Q1} is Gaussian distributed with zero mean and variance $N_0/2$.

Hence, we may express the probability density functions of these two independent random variables as

$$f_{X_{I1}}(x_{I1}) = \frac{1}{\sqrt{\pi N_0}} \exp\left[-\frac{(x_{I1} - \sqrt{E})^2}{N_0}\right] \tag{7.235}$$

and

$$f_{X_{Q1}}(x_{Q1}) = \frac{1}{\sqrt{\pi N_0}} \exp\left(-\frac{x_{Q1}^2}{N_0}\right) \tag{7.236}$$

respectively. Since the two random variables X_{I1} and X_{Q1} are statistically independent, their joint probability density function is simply the product of the probability density functions given in (7.235) and (7.236).

To find the average probability of error, we have to average the conditional probability of error given in (7.233) over all possible values of l_1. Naturally, this calculation requires knowledge of the probability density function of random variables L_1 represented by sample value l_1. The standard method is now to combine (7.235) and (7.236) to find the probability density function of L_1 due to signal plus noise. However, this leads to rather complicated calculations involving the use of Bessel functions. This analytic difficulty may be circumvented by the following approach. Given x_{I1} and x_{Q1}, an error occurs when, in Figure 7.40a, the lower path's output amplitude l_2 due to noise alone exceeds l_1 due to signal plus noise; squaring both sides of (7.234), we write

$$l_1^2 = x_{I1}^2 + x_{Q1}^2 \tag{7.237}$$

The probability of the occurrence just described is obtained by substituting (7.237) into (7.233):

$$\mathbb{P}(\text{error}|x_{I1}, x_{Q1}) = \exp\left(-\frac{x_{I1}^2 + x_{Q1}^2}{N_0}\right) \tag{7.238}$$

which is a probability of error conditioned on the output of the matched filter in the upper path of Figure 7.40a taking on the sample values x_{I1} and x_{Q1}. This conditional probability multiplied by the joint probability density function of the random variables X_{I1} and X_{Q1} is the *error-density given x_{I1} and x_{Q1}*. Since X_{I1} and X_{Q1} are statistically independent, their joint probability density function equals the product of their individual probability density functions. The resulting error-density is a complicated expression in x_{I1} and x_{Q1}. However, the average probability of error, which is the issue of interest, may be obtained in a relatively simple manner. We first use (7.234), (7.235), and (7.236) to evaluate the desired error-density as

$$\mathbb{P}(\text{error}|x_{I1}, x_{Q1}) f_{X_{I1}}(x_{I1}) f_{X_{Q1}}(x_{Q1}) = \frac{1}{\pi N_0} \exp\left\{-\frac{1}{N_0}[x_{I1}^2 + x_{Q1}^2 + (x_{I1} - \sqrt{E})^2 + x_{Q1}^2]\right\} \tag{7.239}$$

Completing the square in the exponent of (7.239) without the scaling factor $-1/N_0$, we may rewrite it as follows:

$$x_{I1}^2 + x_{Q1}^2 + (x_{I1} - \sqrt{E})^2 + x_{Q1}^2 = 2\left(x_{I1} - \frac{\sqrt{E}}{2}\right)^2 + 2x_{Q1}^2 + \frac{E}{2} \tag{7.240}$$

Next, we substitute (7.240) into (7.239) and integrate the error-density over all possible values of x_{I1} and x_{Q1}, thereby obtaining the average probability of error:

$$\begin{aligned} P_e &= \int_{-\infty}^{\infty}\int_{-\infty}^{\infty} \mathbb{P}(\text{error}|x_{I1}, x_{Q1}) f_{X_{I1}}(x_{I1}) f_{X_{Q1}}(x_{Q1})\, dx_{I1}\, dx_{Q1} \\ &= \frac{1}{\pi N_0}\exp\left(-\frac{E}{2N_0}\right)\int_{-\infty}^{\infty}\exp\left[-\frac{2}{N_0}\left(x_{I1} - \frac{\sqrt{E}}{2}\right)^2\right] dx_{I1} \int_{-\infty}^{\infty}\exp\left(-\frac{2x_{Q1}^2}{N_0}\right) dx_{Q1} \end{aligned} \tag{7.241}$$

We now use the following two identities:

$$\int_{-\infty}^{\infty}\exp\left[-\frac{2}{N_0}\left(x_{I1} - \frac{\sqrt{E}}{2}\right)^2\right] dx_{I1} = \sqrt{\frac{N_0\pi}{2}} \tag{7.242}$$

and

$$\int_{-\infty}^{\infty}\exp-\left(\frac{2x_{Q1}^2}{N_0}\right) dx_{Q1} = \sqrt{\frac{N_0\pi}{2}} \tag{7.243}$$

The identity of (7.242) is obtained by considering a Gaussian-distributed variable with mean $\sqrt{E/2}$ and variance $N_0/4$ and recognizing the fact that the total area under the curve of a random variable's probability density function is unity. The identity of (7.243) follows as a special case of (7.242). Thus, in light of these two identities, (7.241) reduces to

$$P_e = \frac{1}{2}\exp\left(-\frac{E}{2N_0}\right)$$

which is the desired result presented previously as (7.227). With this formula at our disposal, we are ready to consider noncoherent binary FSK and DPSK as special cases, which we do next in that order.[10]

7.12 Binary Frequency-Shift Keying Using Noncoherent Detection

In binary FSK, the transmitted signal is defined in (7.151) and repeated here for convenience of presentation:

$$s_i(t) = \begin{cases} \sqrt{\dfrac{2E_b}{T_b}}\cos(2\pi f_i t), & 0 \le t \le T_b \\ 0, & \text{elsewhere} \end{cases} \tag{7.244}$$

where T_b is the bit duration and the carrier frequency f_i equals one of two possible values f_1 and f_2; to ensure that the signals representing these two frequencies are orthogonal, we choose $f_i = n_i/T_b$, where n_i is an integer. The transmission of frequency f_1 represents symbol 1 and the transmission of frequency f_2 represents symbol 0. For the noncoherent

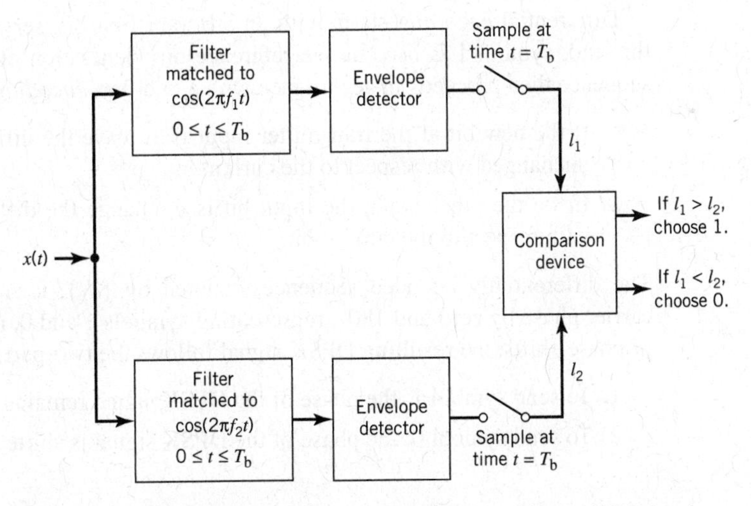

Figure 7.43 Noncoherent receiver for the detection of binary FSK signals.

detection of this frequency-modulated signal, the receiver consists of a pair of matched filters followed by envelope detectors, as in Figure 7.43. The filter in the upper path of the receiver is matched to $\cos(2\pi f_1 t)$ and the filter in the lower path is matched to $\cos(2\pi f_2 t)$ for the signaling interval $0 \le t \le T_b$. The resulting envelope detector outputs are sampled at $t = T_b$ and their values are compared. The envelope samples of the upper and lower paths in Figure 7.43 are shown as l_1 and l_2. The receiver decides in favor of symbol 1 if $l_1 > l_2$ and in favor of symbol 0 if $l_1 < l_2$. If $l_1 = l_2$, the receiver simply guesses randomly in favor of symbol 1 or 0.

The noncoherent binary FSK described herein is a special case of noncoherent orthogonal modulation with $T = T_b$ and $E = E_b$, where E_b is the signal energy per bit. Hence, the *BER for noncoherent binary FSK* is

$$P_e = \frac{1}{2}\exp\left(-\frac{E_b}{2N_0}\right) \tag{7.245}$$

which follows directly from (7.227) as a special case of noncoherent orthogonal modulation.

7.13 Differential Phase-Shift Keying

As remarked at the beginning of Section 7.9, we may view DPSK as the "noncoherent" version of binary PSK. The distinguishing feature of DPSK is that it eliminates the need for synchronizing the receiver to the transmitter by combining two basic operations at the transmitter:

- *differential encoding* of the input binary sequence and
- *PSK* of the encoded sequence,

from which the name of this new binary signaling scheme follows.

Differential encoding starts with an arbitrary first bit, serving as the *reference bit*; to this end, symbol 1 is used as the reference bit. Generation of the differentially encoded sequence then proceeds in accordance with a two-part *encoding rule* as follows:

1. If the new bit at the transmitter input is 1, leave the differentially encoded symbol unchanged with respect to the current bit.
2. If, on the other hand, the input bit is 0, change the differentially encoded symbol with respect to the current bit.

The differentially encoded sequence, denoted by $\{d_k\}$, is used to shift the sinusoidal carrier phase by zero and 180°, representing symbols 1 and 0, respectively. Thus, in terms of phase-shifts, the resulting DPSK signal follows the two-part rule:

1. To send symbol 1, the phase of the DPSK signal remains unchanged.
2. To send symbol 0, the phase of the DPSK signal is shifted by 180°.

EXAMPLE 9 **Illustration of DPSK**

Consider the input binary sequence, denoted $\{b_k\}$, to be 10010011, which is used to derive the generation of a DPSK signal. The differentially encoded process starts with the reference bit 1. Let $\{d_k\}$ denote the differentially encoded sequence starting in this manner and $\{d_{k-1}\}$ denote its delayed version by one bit. The complement of the modulo-2 sum of $\{b_k\}$ and $\{d_{k-1}\}$ defines the desired $\{d_k\}$, as illustrated in the top three lines of Table 7.6. In the last line of this table, binary symbols 1 and 0 are represented by phase-shifts of 1 and π radians.

Table 7.6 **Illustrating the generation of DPSK signal**

	reference								
$\{b_k\}$		1	0	0	1	0	0	1	1
$\{d_{k-1}\}$		1	1	0	1	1	0	1	1
Differentially encoded sequence $\{d_k\}$	1	1	0	1	1	0	1	1	1
Transmitted phase (radians)	0	0	π	0	0	π	0	0	0

Error Probability of DPSK

Basically, the DPSK is also an example of noncoherent orthogonal modulation when its behavior is considered over successive two-bit intervals; that is, $0 \le t \le 2T_b$. To elaborate, let the transmitted DPSK signal be $\sqrt{2E_b/T_b}\cos(2\pi f_c t)$ for the first-bit interval $0 \le t \le T_b$, which corresponds to symbol 1. Suppose, then, the input symbol for the second-bit interval $T_b \le t \le 2T_b$ is also symbol 1. According to part 1 of the DPSK encoding rule, the carrier phase remains unchanged, thereby yielding the DPSK signal

$$s_1(t) = \begin{cases} \sqrt{\dfrac{2E_b}{T_b}}\cos(2\pi f_c t), & \text{symbol 1 for } 0 \le t \le T_b \\ \sqrt{\dfrac{2E_b}{T_b}}\cos(2\pi f_c t), & \text{symbol 0 for } T_b \le t \le 2T_b \end{cases} \tag{7.246}$$

Suppose, next, the signaling over the two-bit interval changes such that the symbol at the transmitter input for the second-bit interval $T_b \le t \le 2T_b$ is 0. Then, according to part 2 of the DPSK encoding rule, the carrier phase is shifted by π radians (i.e., 180°), thereby yielding the new DPSK signal

$$s_2(t) = \begin{cases} \sqrt{\dfrac{2E_b}{T_b}}\cos(2\pi f_c t), & \text{symbol 1 for } 0 \le t \le T_b \\ \sqrt{\dfrac{2E_b}{T_b}}\cos(2\pi f_c t + \pi), & \text{symbol 1 for } T_b \le t \le 2T_b \end{cases} \tag{7.247}$$

We now readily see from (7.246) and (7.247) that $s_1(t)$ and $s_2(t)$ are indeed orthogonal over the two-bit interval $0 \le t \le 2T_b$, which confirms that DPSK is indeed a special form of noncoherent orthogonal modulation with one difference compared with the case of binary FSK: for DPSK, we have $T = 2T_b$ and $E = 2E_b$. Hence, using (7.227), we find that the *BER for DPSK* is given by

$$P_e = \frac{1}{2}\exp\left(-\frac{E_b}{N_0}\right) \tag{7.248}$$

According to this formula, DPSK provides a gain of 3 dB over binary FSK using noncoherent detection for the same E_b/N_0.

Generation of DPSK Signal

Figure 7.44 shows the block diagram of the DPSK transmitter. To be specific, the transmitter consists of two functional blocks:

- *Logic network and one-bit delay (storage) element*, which are interconnected so as to convert the raw input binary sequence $\{b_k\}$ into the differentially encoded sequence $\{d_k\}$.
- *Binary PSK modulator*, the output of which is the desired DPSK signal.

Optimum Receiver for the Detection of DPSK

In the use of DPSK, the carrier phase θ is unknown, which complicates the received signal $x(t)$. To deal with the unknown phase θ in the differentially coherent detection of the DPSK signal in $x(t)$, we equip the receiver with an in-phase and a quadrature path. We thus have a signal-space diagram where the received signal points over the two-bit interval

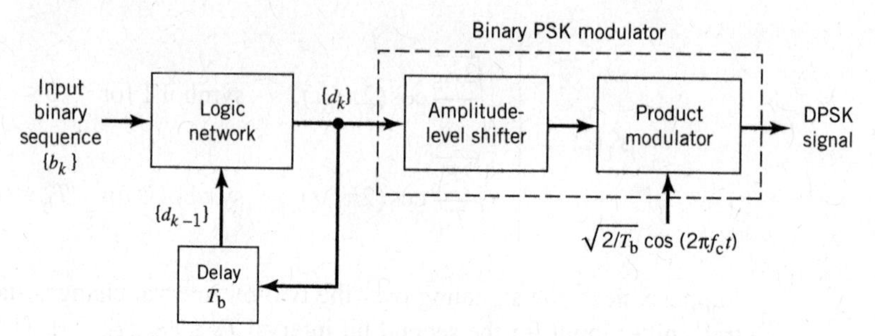

Figure 7.44 Block diagram of a DPSK transmitter.

$0 \leq t \leq 2T_b$ are defined by $(A\cos\theta, A\sin\theta)$ and $(-A\cos\theta, -A\sin\theta)$, where A denotes the carrier amplitude.

This geometry of possible signals is illustrated in Figure 7.45. For the two-bit interval $0 \leq t \leq 2T_b$, the receiver measures the coordinates x_{I_0}, x_{Q_0}, first, at time $t = T_b$ and then measures x_{I_1}, x_{Q_1} at time $t = 2T_b$. The issue to be resolved is whether these two points map to the same signal point or different ones. Recognizing that the vectors $\mathbf{x}_0$ and $\mathbf{x}_1$, with end points x_{I_0}, x_{Q_0} and x_{I_1}, x_{Q_1}, respectively, are points roughly in the same direction if their inner product is positive, we may formulate the binary-hypothesis test with a question:

Is the inner product $\mathbf{x}_0^T\mathbf{x}_1$ positive or negative?

Expressing this statement in analytic terms, we may write

$$x_{I_0}x_{I_1} + x_{Q_0}x_{Q_1} \underset{\text{say } 0}{\overset{\text{say } 1}{\gtrless}} 0 \tag{7.249}$$

where the threshold is zero for equiprobable symbols.

We now note the following identity:

$$x_{I_0}x_{I_1} + x_{Q_0}x_{Q_1} = \frac{1}{4}\left((x_{I_0} + x_{I_1})^2 - (x_{I_0} - x_{I_1})^2 + (x_{Q_0} + x_{Q_1})^2 - (x_{Q_0} - x_{Q_1})^2\right)$$

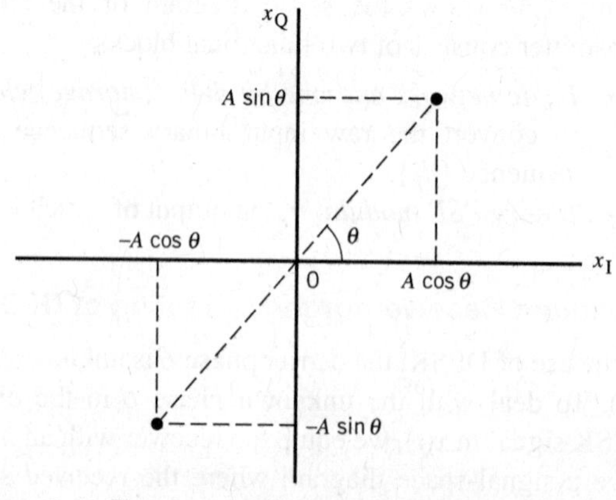

Figure 7.45 Signal-space diagram of received DPSK signal.

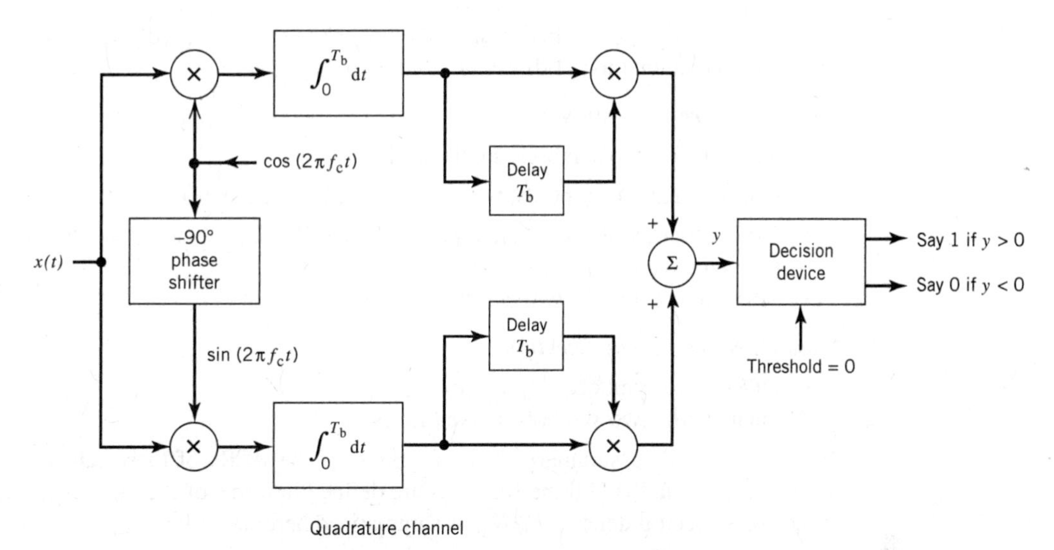

Figure 7.46 Block diagram of a DPSK receiver.

Hence, substituting this identity into (7.249), we get the equivalent test:

$$(x_{I_0} + x_{I_1})^2 + (x_{Q_0} + x_{Q_1})^2 - (x_{I_0} - x_{I_1})^2 - (x_{Q_0} - x_{Q_1})^2 \underset{\text{say } 0}{\overset{\text{say } 1}{\gtrless}} 0 \tag{7.250}$$

where the scaling factor 1/4 is ignored. In light of this equation, the question on the binary hypothesis test for the detection of DPSK may now be restated as follows:

> Given the current signal point (x_{I_0}, x_{Q_0}) received in the time interval $0 < t < 2T_b$, is this point closer to the signal point (x_{I_1}, x_{Q_1}) or its image $(-x_{I_1}, -x_{Q_1})$ received in the next time interval $T_b < t < 2T_b$?

Thus, the *optimum receiver*[11] for the detection of binary DPSK is as shown in Figure 7.46, the formulation of which follows directly from the binary hypothesis test of (7.250). This implementation is simple, in that it merely requires that *sample* values be stored.

The receiver of Figure 7.46 is said to be optimum for two reasons:

1. In structural terms, the receiver avoids the use of fancy delay lines that could be needed otherwise.
2. In operational terms, the receiver makes the decoding analysis straightforward to handle, in that the two signals to be considered are orthogonal over the interval $[0, 2T_b]$ in accordance with the formula of (7.227).

7.14 BER Comparison of Signaling Schemes over AWGN Channels

Much of the material covered in this chapter has been devoted to digital modulation schemes operating over AWGN channels. In this section, we present a summary of the

BERs of some popular digital modulation schemes, classified into two categories, depending on the method of detection used in the receiver:

Class I: Coherent detection

- binary PSK: two symbols, single carrier
- binary FSK: two symbols, two carriers one for each symbol
- QPSK: four symbols, single carrier—the QPSK also includes the QAM, employing four symbols as a special case
- MSK: four symbols, two carriers.

Class II: Noncoherent detection

- DPSK: two symbols, single carrier
- binary FSK: two symbols, two carriers.

Table 7.7 presents a summary of the formulas of the BERs of these schemes separated under Classes I and II. All the formulas are defined in terms of the ratio of energy per bit to the noise spectral density, E_b/N_0, as summarized herein:

1. Under Class I, the formulas are expressed in terms of the *Q-function*. This function is defined as the area under the tail end of the standard Gaussian distribution with zero mean and unit variance; the lower limit in the integral defining the *Q*-function is dependent solely on E_b/N_0, scaled by the factor 2 for binary PSK, QPSK, and MSK. Naturally, as this SNR ratio is increased, the area under the *Q*-function is reduced and with it the BER is correspondingly reduced.
2. Under Class II, the formulas are expressed in terms of an exponential function, where the negative exponent depends on the E_b/N_0 ratio for DPSK and its scaled version by the factor 1/2 for binary FSK. Here again, as the E_b/N_0 is increased, the BER is correspondingly reduced.

The performance curves of the digital modulation schemes listed in Table 7.7 are shown in Figure 7.47 where the BER is plotted versus E_b/N_0. As expected, the BERs for all the

Table 7.7 **Formulas for the BER of digital modulation schemes employing two or four symbols**

	Signaling Scheme	BER
I. Coherent detection	Binary PSK QPSK MSK	$Q\sqrt{2E_b/N_0}$
	Binary FSK	$Q\sqrt{E_b/N_0}$
II. Noncoherent detection	DPSK	$\frac{1}{2}\exp(-E_b/N_0)$
	Binary FSK	$\frac{1}{2}\exp(-E_b/2N_0)$

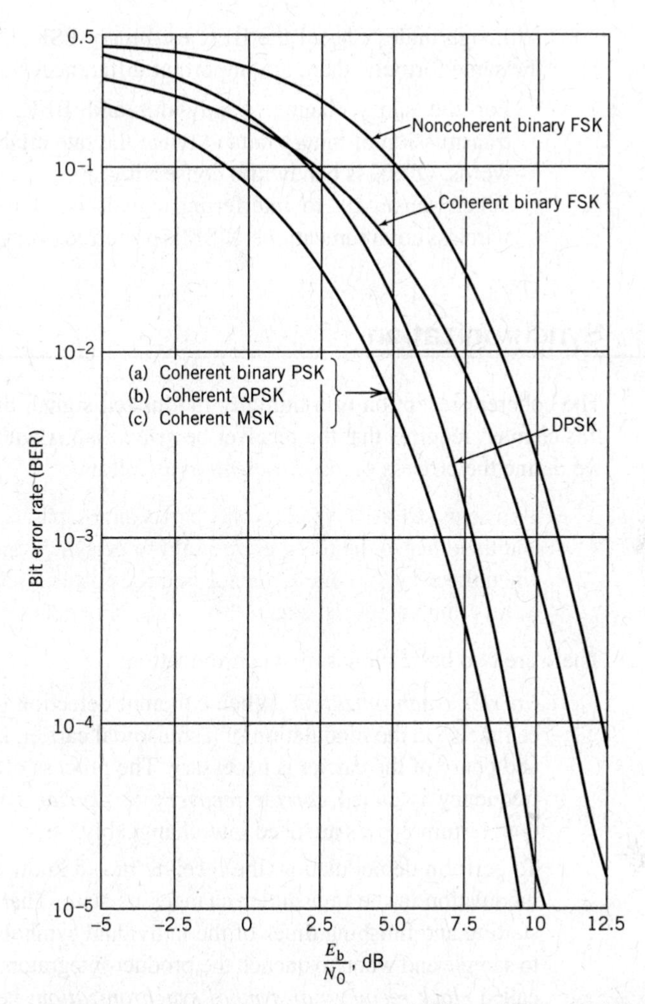

Figure 7.47 Comparison of the noise performance of different PSK and FSK schemes.

schemes decrease *monotonically* with increasing E_b/N_0, with all the graphs having a similar shape in the form of a *waterfall*. Moreover, we can make the following observations from Figure 7.47:

1. For any value of E_b/N_0, the schemes using coherent detection produce a smaller BER than those using noncoherent detection, which is intuitively satisfying.
2. PSK schemes employing two symbols, namely binary PSK with coherent detection and DPSK with noncoherent detection, require an E_b/N_0 that is 3 dB less than their FSK counterpart to realize the same BER.
3. At high values of E_b/N_0, DPSK and binary FSK using noncoherent detection perform almost as well, to within about 1 dB of their respective counterparts using coherent detection for the same BER.

4. Although under Class I the BER for binary PSK, QPSK, and MSK is governed by the same formula, there are important differences between them:
   - For the same channel bandwidth and BER, the QPSK accommodates the transmission of binary data at twice the rate attainable with binary PSK; in other words, QPSK is bandwidth conserving.
   - When sensitivity to interfering signals is an issue of practical concern, as in wireless communications, MSK is preferred over QPSK.

7.15 Synchronization

The coherent reception of a digitally modulated signal, discussed in previous sections of this chapter, requires that the receiver be *synchronous* with the transmitter. In this context, we define the process of *synchronization* as follows:

> Two sequences of related events performed separately, one in the transmitter and the other in the receiver, are said to be synchronous relative to each other when the events in one sequence and the corresponding events in the other occur simultaneously, except for some finite delay.

There are two basic modes of synchronization:

1. *Carrier synchronization.* When coherent detection is used in signaling over AWGN channels via the modulation of a sinusoidal carrier, knowledge of both the frequency and phase of the carrier is necessary. The process of estimating the carrier phase and frequency is called *carrier recovery* or *carrier synchronization*; in what follows, both terminologies are used interchangeably.
2. To perform demodulation, the receiver has to know the instants of time at which the modulation in the transmitter changes its state. That is, the receiver has to know the starting and finishing times of the individual symbols, so that it may determine when to sample and when to quench the product-integrators. The estimation of these times is called *clock recovery or symbol synchronization*; here again, both terminologies are used interchangeably.

We may classify synchronization schemes as follows, depending on whether some form of aiding is used or not:

1. *Data-aided synchronization.* In data-aided synchronization schemes, a *preamble* is transmitted along with the data-bearing signal in a time-multiplexed manner on a periodic basis. The preamble contains information about the symbol timing, which is extracted by appropriate processing of the channel output at the receiver. Such an approach is commonly used in digital satellite and wireless communications, where the motivation is to minimize the time required to synchronize the receiver to the transmitter. Limitations of data-aided synchronization are twofold:
   - reduced data-throughput efficiency, which is incurred by assigning a certain portion of each transmitted frame to the preamble, and
   - reduced power efficiency, which results from the allocation of a certain fraction of the transmitted power to the transmission of the preamble.

2. *Nondata-aided synchronization*. In this second approach, the use of a preamble is avoided and the receiver has the task of establishing synchronization by extracting the necessary information from the noisy distorted modulated signal at the channel output. Both throughput and power efficiency are thereby improved, but at the expense of an increase in the time taken to establish synchronization.

In this section, the discussion is focused on nondata-aided forms of carrier and clock recovery schemes. To be more specific, we adopt an *algorithmic approach*,[12] which is so-called on account of the fact that implementation of the sychronizer enables the receiver to estimate the carrier phase and symbol timing in a *recursive manner* from one time instant to another. The processing is performed on the *baseband* version of the received signal, using discrete-time (digital) signal-processing algorithms.

Algorithmic Approach to Synchronization

Maximum likelihood decoding played a key role in much of the material on signaling techniques in AWGN channels presented in Sections 7.4 through 7.13. *Maximum likelihood parameter estimation* plays a key role of its own in the algorithmic approach to synchronization. Both of these methods were discussed previously in Chapter 3 on probability theory and Bayesian inference. In this context, it may therefore be said that a sense of continuity is being maintained throughout this chapter.

Given the received signal, the maximum likelihood method is used to estimate two parameters: carrier phase and symbol timing, both of which are, of course, unknown. Here, we are assuming that knowledge of the carrier frequency is available at the receiver.

Moreover, in the algorithmic approach, the symbol-timing recovery is performed before phase recovery. The rationale for proceeding in this way is that once we know the envelope delay incurred by signal transmission through a dispersive channel, then one sample per symbol at the matched filter output may be sufficient for estimating the unknown carrier phase. Moreover, computational complexity of the receiver is minimized by using synchronization algorithms that operate at the symbol rate $1/T$.

In light of the remarks just made, we will develop the algorithmic approach to synchronization by proceeding as follows:

1. Through processing the received signal corrupted by channel noise and channel dispersion, the likelihood function is formulated.
2. The likelihood function is maximized to recover the clock.
3. With clock recovery achieved, the next step is to maximize the likelihood function to recover the carrier.

The derivations presented in this chapter focus on the QPSK signal. The resulting formulas may be readily extended to binary PSK symbols as a special case and generalized for M-ary PSK signals.

7.16 Recursive Maximum Likelihood Estimation for Synchronization

In the previous section, we remarked that, in algorithmic synchronization, estimation of the two unknown parameters, namely carrier phase and symbol timing, is performed in a recursive manner from one time instant to another.

In other words:

> Discrete time is an essential dimension of recursive parameter estimation.

Moreover, the estimation is performed at time $t = nT$, where n is an integer and T is the symbol duration. Equivalently, we may say that $n = t/T$ denotes the *normalized (dimensionless) discrete time.*

One other important point to note: recursive estimation of the unknown parameter, be that the carrier phase or symbol time, plays a key role in the synchronization process. Specifically, it proceeds across discrete time in accordance with the following rule:

$$\begin{pmatrix}\text{Updated estimate}\\ \text{of the parameter}\end{pmatrix} = \begin{pmatrix}\text{Old estimate}\\ \text{of the parameter}\end{pmatrix} + \begin{pmatrix}\text{Step-size}\\ \text{parameter}\end{pmatrix} \times \begin{pmatrix}\text{Error}\\ \text{signal}\end{pmatrix} \tag{7.251}$$

In other words, the recursive parameter estimation takes on the structure of an *adaptive filtering algorithm*, in which the product of the step-size parameter and error signal assumes the role of an *algorithmic adjustment*.

In what follows, we derive adaptive filtering algorithms for estimating the unknown synchronization parameters with the error signal being derived from the likelihood function.

Likelihood Functions

The idea of *maximum likelihood parameter estimation* based on continuous-time waveforms was discussed in Chapter 3. To briefly review the material described therein, consider a baseband signal defined by

$$x(t) = s(t, \lambda) + w(t)$$

where λ is an unknown parameter and $w(t)$ denotes an AWGN. Given a sample of the signal $x(t)$, the requirement is to estimate the parameter λ; so, we say:

> The most likely value of the estimate $\hat{\lambda}$ is the particular λ for which the *likelihood* function $l(\lambda)$ is a maximum.

Note that we say "a maximum" rather than "the maximum" because it is possible for the graph of $l(\lambda)$ plotted versus λ to have multiple maxima. In any event, the likelihood function given x, namely $l(\lambda)$, is defined as the probability density function $f(x|\lambda)$ with the roles of x and λ interchanged, as shown by

$$l(\lambda) = f(x|\lambda)$$

where, for convenience of presentation, we have omitted the conditional dependence of λ on x in $l(\lambda)$.

In the algorithmic synchronization procedures derived in this section, we will be concerned only with cases in which the parameter λ is a *scalar.* Such cases are referred to as independent estimation. However, when we are confronted with the synchronization of a digital communication receiver to its transmitter operating over a *dispersive channel*, we have two unknown channel-related parameters to deal with: the phase (carrier) delay τ_c, and the group (envelope) delay τ_g, both of which were discussed in Chapter 2. In the context of these two parameters, when we speak of independent estimation for synchronization, we mean that the two parameters τ_c and τ_g are considered *individually* rather than *jointly.* Intuitively speaking, independent estimation is much easier to tackle and visualize than joint estimation, and it may yield more robust estimates in general.

Let the transmitted signal for symbol i in the QPSK signal be defined by

$$s_i(t) = \sqrt{\frac{2E}{T}}\cos(2\pi f_c t + \alpha_2), \qquad 0 \le t \le T \tag{7.252}$$

where E is the signal energy per symbol, T is the symbol period, and α_i is the carrier phase used for transmitting symbol i. For example, for the QPSK we have

$$\alpha_i = \frac{\pi}{4}(2i-1), \qquad i = 1, 2, 3, 4$$

Equivalently, we may write

$$s_i(t) = \sqrt{\frac{2E}{T}}\cos(2\pi f_c t + \alpha_i)g(t) \tag{7.253}$$

where $g(t)$ is the *shaping pulse*, namely a rectangular pulse of unit amplitude and duration T. By definition, τ_c affects the carrier and τ_g affects the envelope. Accordingly, the received signal at the channel output is given by

$$\begin{aligned} x(t) &= \sqrt{\frac{2E}{T}}\cos(2\pi f_c(t-\tau_c) + \alpha_i)g(t-\tau_g) + w(t) \\ &= \sqrt{\frac{2E}{T}}\cos(2\pi f_c t + \theta + \alpha_i)g(t-\tau_g) + w(t) \end{aligned} \tag{7.254}$$

where $w(t)$ is the channel noise. The new term θ introduced in (7.254) is an additive carrier phase attributed to the phase delay τ_c produced by the dispersive channel; it is defined by

$$\theta = -2\pi f_c \tau_c \tag{7.255}$$

The minus sign is included in the right-hand side of (7.255) to be consistent with previous notation used in dealing with signal detection.

Both the carrier phase θ and group delay τ_g are unknown. However, it is assumed that they remain essentially constant over the observation interval $0 \le t \le T_0$ or through the transmission of a sequence made up of $L_0 = T_0/T$ symbols.

With θ used to account for the carrier delay τ_c, we may simplify matters by using τ in place of τ_g for the group delay; that is, (7.254) is rewritten as

$$x(t) = \sqrt{\frac{2E}{T}}\cos(2\pi f_c t + \theta + \alpha_i)g(t-\tau) + w(t), \qquad \begin{array}{l}\tau \le t \le T+\tau \\ i = 1, 2, 3, 4\end{array} \tag{7.256}$$

At the receiver, the orthogonal pair of basis functions for QPSK signals is defined by

$$\phi_1(t) = \sqrt{\frac{2}{T}}\cos(2\pi f_c t), \qquad \tau \le t \le T+\tau \tag{7.257}$$

$$\phi_2(t) = \sqrt{\frac{2}{T}}\sin(2\pi f_c t), \qquad \tau \le t \le T+\tau \tag{7.258}$$

Here, it is assumed that the receiver has perfect knowledge of the carrier frequency f_c, which is a reasonable assumption; otherwise, a carrier-frequency offset has to be included that will complicate the analysis.

Accordingly, we may represent the received signal $x(t)$ by the baseband vector

$$\mathbf{x}(\tau) = \begin{bmatrix} x_1(\tau) \\ x_2(\tau) \end{bmatrix} \tag{7.259}$$

where

$$x_k(\tau) = \int_{\tau}^{T+\tau} x(t)\phi_k(t)\,\mathrm{d}t, \qquad k = 1, 2 \tag{7.260}$$

In a corresponding fashion, we may express the signal component of $\mathbf{x}(\tau)$ by the vector

$$\mathbf{s}(\alpha_i, \theta, \tau) = \begin{bmatrix} s_1(\alpha_i, \theta, \tau) \\ s_2(\alpha_i, \theta, \tau) \end{bmatrix} \tag{7.261}$$

where

$$s_k(\alpha_i, \theta, \tau) = \int_{\tau}^{T+\tau} \sqrt{\frac{2E}{T}}\cos(2\pi f_c t + \theta + \alpha_i)\phi_k(t)\,\mathrm{d}t, \qquad k = 1, 2 \quad i = 1, 2, 3, 4 \tag{7.262}$$

Assuming that f_c is an integer multiple of the symbol rate $1/T$, evaluation of the integral in (7.262) shows that dependence of s_1 and s_2 on the group delay τ is eliminated, as shown by

$$s_1(\alpha_i, \theta) = \sqrt{E}\cos(\theta + \alpha_i) \tag{7.263}$$

$$s_2(\alpha_i, \theta) = -\sqrt{E}\sin(\theta + \alpha_i) \tag{7.264}$$

We may thus expand on (7.259) to write

$$\mathbf{x}(\tau) = \mathbf{s}(\alpha_i, \theta) + \mathbf{w}(\tau), \qquad i = 1, 2, 3, 4 \tag{7.265}$$

where

$$\mathbf{w}(\tau) = \begin{bmatrix} w_1(\tau) \\ w_2(\tau) \end{bmatrix} \tag{7.266}$$

The two elements of the noise vector $\mathbf{w}$ are themselves defined by

$$w_k = \int_{\tau}^{T+\tau} w(t)\phi_k(t)\,\mathrm{d}t, \qquad k = 1, 2 \tag{7.267}$$

The w_k in (7.267) is the sample value of a Gaussian random variable W of zero mean and variance $N_0/2$, where $N_0/2$ is the power spectral density of the channel noise $w(t)$. Dependence of the baseband signal vector $\mathbf{x}$ on delay τ is inherited from (7.265).

The conditional probability density function of the random vector $\mathbf{X}$, represented by the sample $\mathbf{x}$ at the receiver input given transmission of the ith symbol, and occurrence of the carrier phase θ and group delay τ resulting from the dispersive channel, is defined by

$$f_{\mathbf{X}}(\mathbf{x}|\alpha_i, \theta, \tau) = \frac{1}{\pi N_0}\exp\left(-\frac{1}{N_0}\|\mathbf{x}(\tau) - \mathbf{s}(\alpha_i, \theta)\|^2\right) \tag{7.268}$$

Setting $\mathbf{s}(\alpha_i, \sigma)$ equal to zero, (7.268) reduces to

$$f_{\mathbf{X}}(\mathbf{x}|\mathbf{s} = \mathbf{0}) = \frac{1}{\pi N_0}\exp\left(-\frac{1}{N_0}\|\mathbf{x}(\tau)\|^2\right) \tag{7.269}$$

Equation (7.268) defines the probability density function of the random vector $\mathbf{X}$ in the combined presence of signal and channel noise, whereas (7.269) defines the probability density function of $\mathbf{x}$ in the presence of channel noise acting alone. Accordingly, we may define the *likelihood function* for QPSK as the ratio of these two probability density functions, as shown by

$$\begin{aligned} l(\alpha_i, \theta, \tau) &= \frac{f_{\mathbf{X}}(\mathbf{x}|\alpha_i, \theta, \tau)}{f_{\mathbf{X}}(\mathbf{x}|\mathbf{s} = \mathbf{0})} \\ &= \exp\left(\frac{2}{N_0}\mathbf{x}^{\mathrm{T}}(\tau)\mathbf{s}(\alpha_i, \theta) - \frac{1}{N_0}\|\mathbf{s}(\alpha_i, \theta)\|^2\right) \end{aligned} \tag{7.270}$$

In QPSK, we have

$$\|\mathbf{s}(\alpha_i, \theta)\| = \text{constant}$$

because all four message points lie on a circle of radius $\sqrt{E}$. Hence, ignoring the second term in the exponent in (7.270), we may reduce the likelihood function to

$$l(\alpha_i, \theta, \tau) = \exp\left(\frac{2}{N_0}\mathbf{x}^{\mathrm{T}}(\tau)\mathbf{s}(\alpha_i, \theta)\right) \tag{7.271}$$

Complex Terminology for Algorithmic Synchronization

Before proceeding with the derivations of adaptive filtering algorithms for recovery of the clock and carrier, we find it instructive to reformulate the likelihood function of (7.271) using complex terminology. Such a step is apropos given the fact that the received signal vector as well as its contituent signal and noise vectors in (7.265) are all in their respective baseband forms.

Specifically, the two-dimensional vector $\mathbf{x}(\tau)$ is represented by the *complex envelope of the received signal*

$$\tilde{x}(\tau) = x_1 + \mathrm{j}x_2 \tag{7.272}$$

where $\mathrm{j} = \sqrt{-1}$.

Correspondingly, the signal vector $\mathbf{s}(\alpha_i, \theta)$, comprising the pair of signal components $s_1(\alpha_i, \theta)$ and $s_2(\alpha_i, \theta)$, is represented by the *complex envelope of the transmitter signal corrupted by carrier phase* θ:

$$\begin{aligned} \tilde{s}(\alpha_i, \theta) &= s_1(\alpha_i, \theta) + \mathrm{j}s_2(\alpha_i, \theta) \\ &= \sqrt{E}[\cos(\alpha_i, \theta) + \mathrm{j}\sin(\alpha_i, \theta)] \\ &= \sqrt{E}\tilde{\alpha}_i e^{\mathrm{j}\theta}, \qquad i = 1, 2, 3, 4 \end{aligned} \tag{7.273}$$

The new complex parameter $\tilde{\alpha}_i$ in (7.273) is a *symbol indicator* in the message constellation of the QPSK; it is defined by

$$\begin{aligned} \tilde{\alpha}_i &= e^{\mathrm{j}\alpha_i} \\ &= \cos\alpha_i + \mathrm{j}\sin\alpha_i \end{aligned} \tag{7.274}$$

Correspondingly, the complex experimental factor embodying the carrier phase θ is defined by

$$e^{j\theta} = \cos\theta + j\sin\theta \tag{7.275}$$

Both (7.274) and (7.275) follow from *Euler's formula.*

With the complex representations of (7.272) to (7.275) at hand, we may now reformulate the exponent of the likelihood function in (7.271) in the equivalent complex form:

$$\begin{aligned} \frac{2}{N_0}\mathbf{x}^{\mathrm{T}}\mathbf{s}(\alpha_i, \theta) &= \frac{2\sqrt{E}}{N_0}\operatorname{Re}[\tilde{x}_i(\tau)\tilde{s}^*(\alpha_i, \theta)] \\ &= \frac{2\sqrt{E}}{N_0}\operatorname{Re}[\tilde{x}(\tau)\tilde{\alpha}_i^* e^{-j\theta}] \end{aligned} \tag{7.276}$$

where Re[.] denotes the *real part* of the complex expression inside the square brackets. Hence, we may make the following statement:

> The inner product of the two complex vectors $\mathbf{x}(\tau)$ and $\mathbf{s}(\alpha_i, \theta)$ in (7.276) is replaced by $\sqrt{E}$ times the real part of the inner product of two complex variables: $\tilde{x}(\tau)$ and $\tilde{\alpha}_i e^{j\theta}$.

Two points are noteworthy here:

1. The complex envelope of the received signal is dependent on the group delay τ, hence $\tilde{x}(\tau)$. The product $\tilde{\alpha}_i e^{j\theta}$ is made up of the complex symbol indicator $\tilde{\alpha}_i$ attributed to the QPSK signal generated in the transmitter and the exponential term $e^{j\theta}$ attributed to phase distortion in the channel.
2. In complex variable theory, given a pair of complex terms $\tilde{x}(\tau)$ and $\tilde{\alpha}_i e^{j\theta}$, their inner product could be defined as $\tilde{x}(\tau)(\tilde{\alpha}_i e^{j\theta})^* = \tilde{x}(\tau)\tilde{\alpha}_i^* e^{-j\theta}$, as shown in (7.276).

The complex representation on the right-hand side of (7.276), expressed in *Cartesian form,* is well suited for estimating the unknown phase θ. On the other hand, for estimating the unknown group delay τ, we find it more convenient to use a *polar representation* for the inner product of the two vectors $\mathbf{x}(\tau)$ and $\mathbf{s}(\alpha_i, \theta)$, as shown by

$$\frac{2}{N_0}\mathbf{x}^{\mathrm{T}}(\tau)\mathbf{s}(\alpha_i, \theta) = \frac{2\sqrt{E}}{N_0}\left|\tilde{\alpha}_i\tilde{x}(\tau)\right|\cos(\arg[\tilde{x}(\tau)] - \arg[\tilde{\alpha}_i] - \theta) \tag{7.277}$$

Indeed, it is a straightforward matter to show that the two complex representations on the right-hand side of (7.276) and (7.277) are indeed equivalent. The reasons for why these two representations befit the estimation of carrier phase θ and group delay τ, respectively, will become apparent in the next two subsections.

Moreover, in light of what was said previously, estimation of the group delay should precede that of the carrier phase. Accordingly, the next subsection is devoted to group-delay estimation, followed by the sub-section devoted to carrier-phase estimation.

Recursive Estimation of the Group Delay

To begin the task of estimating the unknown group delay, first of all we have to remove dependence of the likelihood function $l(\alpha_i, \sigma, \tau)$ on the unknown carrier phase θ in

(7.271). To do this, we will *average* the likelihood function over all possible values of θ inside the range $[0, 2\pi]$. To this end, θ is assumed to be uniformly distributed inside this range, as shown by

$$f_{\Theta}(\theta) = \begin{cases} \dfrac{1}{2\pi}, & 0 \le \theta \le 2\pi \\ 0, & \text{otherwise} \end{cases} \tag{7.278}$$

which is the worst possible situation that can arise in practice. Under this assumption, we may thus express the average likelihood function as

$$\begin{aligned} l_{\mathrm{av}}(\tilde{\alpha}_i, \tau) &= \int_0^{2\pi} l(\alpha_i, \theta, \tau) f_{\Theta}(\theta)\, \mathrm{d}\theta \\ &= \frac{1}{2\pi}\int_0^{2\pi} l(\alpha_i, \theta, \tau)\, \mathrm{d}\theta \\ &= \frac{1}{2\pi}\int_0^{2\pi} \exp\left(\frac{2}{N_0}\mathbf{x}^{\mathrm{T}}(\tau)\mathbf{s}(\alpha_i, \theta)\right) \mathrm{d}\theta \end{aligned} \tag{7.279}$$

where, in the last line, we used (7.271).

Examining the two alternative complex representations of the likelihood function's exponent given in (7.276) and (7.277), it is the latter that best suits solving the integration in (7.279). Specifically, we may write

$$\begin{aligned} l_{\mathrm{av}}(\tilde{\alpha}_i, \tau) &= \frac{1}{2\pi}\int_0^{2\pi} \exp\left[\frac{2\sqrt{E}}{N_0}\left|\tilde{\alpha}_i\tilde{x}(\tau)\right| \cos(\arg[\tilde{x}(\tau)] - \arg[\tilde{\alpha}] - \theta)\right] \mathrm{d}\theta \\ &= \frac{1}{2\pi}\int_{-\arg[\tilde{x}(\tau)] + \arg[\tilde{l}]}^{2\pi - \arg[\tilde{x}(\tau)] + \arg[\alpha_i]} \exp\left(\frac{2\sqrt{E}}{N_0}\left|\alpha_i\tilde{x}(\tau)\right| \cos(\varphi) \mathrm{d}\varphi\right) \end{aligned} \tag{7.280}$$

where, in the last line, we have made the substitution

$$\varphi = \arg[\tilde{x}(\tau)] - \arg[\alpha_i] - \theta$$

We now invoke the definition of the *modified Bessel function of zero order*, as shown by (see Appendix C)

$$I_0(x) = \frac{1}{2\pi}\int_0^{2\pi} \mathrm{e}^{x\cos\varphi}\, \mathrm{d}\varphi \tag{7.281}$$

Using this formula, we may, therefore, express the average likelihood function $l_{\mathrm{av}}(\tilde{\alpha}_i, \tau)$ in (7.280) as follows:

$$l_{\mathrm{av}}(\tilde{\alpha}_i, \tau) = I_0\left(\frac{2\sqrt{E}}{N_0}\left|\tilde{\alpha}_i\tilde{x}_i(\tau)\right|\right) \tag{7.282}$$

where $\tilde{x}_i(\tau)$ is the complex envelope of the matched filter output in the receiver. By definition, for QPSK we have

$$|\tilde{\alpha}_i| = 1, \qquad \text{for all } i$$

It follows, therefore, that (7.282) reduces to

$$l_{\mathrm{av}}(\tau) = I_0\left(\frac{2\sqrt{E}}{N_0}|\tilde{x}(\tau)|\right) \tag{7.283}$$

Here, it is important to note that, as a result of averaging the likelihood function over the carrier phase θ, we have also removed dependence on the transmitted symbol $\tilde{\alpha}_i$ for QPSK; this result is intuitively satisfying.

In any event, taking the natural logarithm of $l_{\mathrm{av}}(\tau)$ in (7.283) to obtain the *log-likelihood function* of τ, we write

$$\begin{aligned} L_{\mathrm{av}}(\tau) &= \ln l_{\mathrm{av}}(\tau) \\ &= \ln I_0\left(\frac{2\sqrt{E}}{N_0}|\tilde{x}(\tau)|\right) \end{aligned} \tag{7.284}$$

where ln denotes the natural logarithm. To proceed further, we need to find a good approximation for $L_{\mathrm{av}}(\tau)$. To this end, we first note that the modified Bessel function $I_0(x)$ may itself be expanded in a power series (see Appendix C):

$$I_0(x) = \sum_{m=0}^{\infty} \frac{\left(\frac{1}{2}x\right)^{2m}}{(m!)^2}$$

where x stands for the product term $2\sqrt{E}/(N_0)|\tilde{x}(\tau)|$. For small values of x, we may thus approximate $I_0(x)$ as shown by

$$I_0(x) \approx 1 + \frac{x^2}{4}$$

We may further simplify matters by using the approximation

$$\begin{aligned} \ln I_0(x) &\approx \ln\left(1 + \frac{x^2}{4}\right) \\ &\approx \frac{x^2}{4} \quad \text{for small } x \end{aligned} \tag{7.285}$$

For the problem at hand, small x corresponds to small SNR. Under this condition, we may now approximate the log-likelihood function of (7.284) as follows:

$$L_{\mathrm{av}}(\tau) \approx \frac{E}{N_0^2}|\tilde{x}(\tau)|^2 \tag{7.286}$$

With maximization of $L_{\mathrm{av}}(\tau)$ as the objective, we differentiate it with respect to the envelope delay τ, obtaining

$$\begin{aligned} \frac{\partial L_{\mathrm{av}}(\tau)}{\partial \tau} &= \frac{E}{N_0^2}\frac{\partial}{\partial \tau}|\tilde{x}_i(\tau)|^2 \\ &= \frac{2E}{N_0^2}\mathrm{Re}[\tilde{x}^*(\tau)\tilde{x}'(\tau)] \end{aligned} \tag{7.287}$$

where $\tilde{x}_i^*(\tau)$ is the complex conjugate of $\tilde{x}(\tau)$ and $\tilde{x}'(\tau)$ is its derivative with respect to τ.

The formula in (7.287) is the result of operating on the received signal at the channel output, $x(t)$, defined in (7.254) for a particular symbol of the QPSK signal defined in the interval $[\tau, T+\tau]$. In the course of finding the baseband vector representation of the received signal, namely $\tilde{x}(\tau)$, dependence on time t disappeared in (7.287). Notwithstanding this point, the fact of the matter is the log-likelihood ratio $L_{\text{av}}(\tau)$ in (7.287) pertains to some point in discrete time $n = t/T$, and it changes with n. To go forward with recursive estimation of the group delay τ, we must therefore bring discrete time n into the procedure. To this end, n is assigned as a subscript to both $\tilde{x}^*(\tau)$ and $\tilde{x}'(\tau)$ in (7.287). Thus, with the recursive estimation of τ following the format described in words in (7.251), we may define the *error signal* needed for the recursive estimation of τ (i.e., symbol-timing recovery) as follows:

$$e_n = \text{Re}[\tilde{x}_n^*(\tau)\tilde{x}_n'(\tau)] \tag{7.288}$$

Let $\hat{\tau}_n$ denote the estimate of the unknown group delay τ at discrete time n. Correspondingly, we may introduce two definitions

$$\tilde{x}_n(\tau) = \tilde{x}(nT+\hat{\tau}_n) \tag{7.289}$$

and

$$\tilde{x}'_n(\tau) = \tilde{x}'(nT+\hat{\tau}_n) \tag{7.290}$$

Accordingly, we may reformulate the error signal e_n in (7.288) as follows:

$$e_n = \text{Re}[\tilde{x}^*(nT+\hat{\tau}_n)\tilde{x}'(nT+\hat{\tau}_n)] \tag{7.291}$$

Computation of the error signal e_n, therefore, requires the use of two filters:

1. *Complex matched filter*, which is used for generating $\tilde{x}_n(\tau)$.
2. *Complex derivative matched filter*, which is used for generating $\tilde{x}_n'(\tau)$.

By design, the receiver is already equipped with the first filter. The second one is new. In practice, the additional computational complexity due to the derivative matched filter is found to be an undesireable requirement. To dispense with the need for it, we propose to approximate the derivative using a *finite difference*, as shown by

$$\tilde{x}'(nT+\hat{\tau}_n) \approx \frac{1}{T}\left[\tilde{x}\left(nT+\frac{T}{2}+\hat{\tau}_{n+1/2}\right)-\tilde{x}\left(nT-\frac{T}{2}+\hat{\tau}_{n-1/2}\right)\right] \tag{7.292}$$

Note, however, that in using the finite-difference approximation of (7.292) we have simplified computation of the derivative matched filter by doubling the symbol rate. It is desirable to make one further modification to account for the fact that timing estimates are updated at multiples of the symbol period T and the only available quantities are $\hat{\tau}_n$. Consequently, we replace $\hat{\tau}_{n+1/2}$ by the current (updated estimate) $\hat{\tau}_n$ and replace $\hat{\tau}_{n-1/2}$ by the old estimate $\hat{\tau}_{n-1}$. We may thus rewrite (7.292) as follows:

$$\tilde{x}'(nT+\hat{\tau}_n) \approx \frac{1}{T}\left[\tilde{x}\left(nT+\frac{T}{2}+\hat{\tau}_n\right)-\tilde{x}\left(nT-\frac{T}{2}+\hat{\tau}_{n-1}\right)\right] \tag{7.293}$$

So, we finally redefine the error signal as follows:

$$e_n = \text{Re}\left\{\tilde{x}^*(nT+\hat{\tau}_n)\left[\tilde{x}\left(nT+\frac{T}{2}+\hat{\tau}_n\right)-\tilde{x}\left(nT-\frac{T}{2}+\hat{\tau}_{n-1}\right)\right]\right\} \tag{7.294}$$

where the scaling factor $1/T$ is accounted for in what follows.

Finally, building on the format of the recursive estimation procedure described in (7.251), we may formulate the *adaptive filtering algorithm* for symbol timing recovery:

$$c_{n+1} = c_n + \gamma e_n, \qquad n = 0, 1, 2, 3, \ldots \tag{7.295}$$

where we have the following:

- The γ in (7.295) is the step-size parameter, in which the two scaling factors $2E/N_0^2$ and $1/T$ are absorbed; the factor $2E/N_0^2$ was ignored in moving from (7.287) to (7.288) and the factor $1/T$ was ignored from (7.293) to (7.294).
- The error signal e_n is defined by (7.294).
- The c_n is a real number employed as control for the frequency of an oscillator, referred to as a *number-controlled oscillator* (NCO).

The *closed-loop feedback system* for implementing the timing-recovery algorithm of (7.295) is shown in Figure 7.48. From a historical perspective, the scheme shown in this figure is analogous to the continuous-time version of the traditional early–late gate synchronizer widely used for timing recovery. In light of this analogy, the scheme of Figure 7.48 is referred to as a *recursive early–late delay (NDA-ELD) synchronizer.* At every recursion (i.e., time step), the synchronizer works on three successive samples of the matched filter output, namely:

$$\tilde{x}\left(nT + \frac{T}{2} + \hat{\tau}_n\right),\ \tilde{x}(nT + \hat{\tau}_n), \text{ and } \tilde{x}\left(nT + \frac{T}{2} - \hat{\tau}_{n-1}\right)$$

The first sample is *early* and the last one is *late*, both defined with respect to the middle one.

Recursive Estimation of the Carrier Phase

With estimation of the symbol time τ taken care of, the next step is to estimate the carrier phase θ. This estimation is also based on the likelihood function defined in (7.270), but

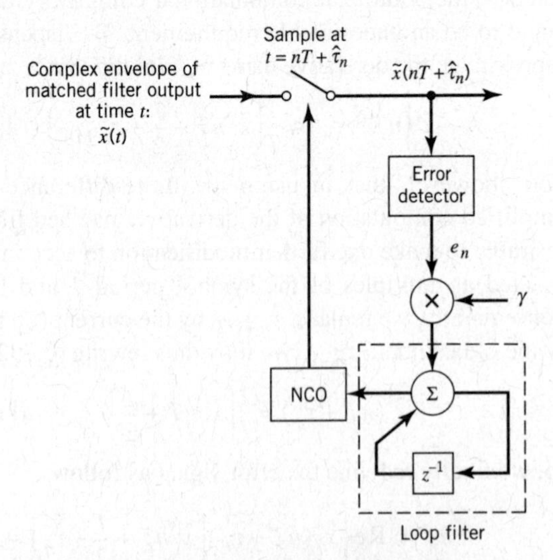

Figure 7.48 Nondata-aided early–late delay synchronizer for estimating the group delay.

with a difference: this time we use the complex representation on the right-hand side of (7.276) for the likelihood function's exponent. Thus, the *likelihood function* of θ is now expressed as follows:

$$l(\theta) = \exp\left(\frac{2\sqrt{E}}{N_0}\mathrm{Re}[\tilde{x}(\tau)\tilde{\alpha}_i^* \mathrm{e}^{-\mathrm{j}\theta}]\right) \tag{7.296}$$

Taking the natural logorithm of both sides of (7.296), the *log-likelihood function* of θ is, therefore, given by

$$L(\theta) = \frac{2\sqrt{E}}{N_0}\mathrm{Re}[\tilde{x}(\tau)\tilde{\alpha}_i^* \mathrm{e}^{-\mathrm{j}\theta}] \tag{7.297}$$

Here again, maximizing the estimate of the carrier phase θ as the issue of interest, we differentiate $L(\theta)$ with respect to θ, obtaining

$$\frac{\partial L(\theta)}{\partial\theta} = \frac{2\sqrt{E}}{N_0}\frac{\partial}{\partial\theta}\mathrm{Re}[\tilde{x}\tilde{\alpha}_i^* \mathrm{e}^{-\mathrm{j}\theta}]$$

The real-part operator Re[·] is linear; therefore, we may interchange this operation with the differentiation. Moreover, we have

$$\frac{\partial}{\partial\theta}\mathrm{e}^{-\mathrm{j}\theta} = -\mathrm{j}\,\mathrm{e}^{-\mathrm{j}\theta}$$

As a result of the differentiation, the argument $\tilde{x}(\tau)\alpha_i^* \mathrm{e}^{-\mathrm{j}\theta}$ in (7.297) is multiplied by –j, which, in turn, has the effect of replacing the real-part operator Re[.] by the corresponding *imaginary-part operator* Im[.] Accordingly, we may express derivative of the log-likelihood function in (7.297) with respect to θ as follows:

$$\frac{\partial L(\theta)}{\partial\theta} = \frac{2\sqrt{E}}{N_0}\mathrm{Im}[\tilde{x}(\tau)\tilde{\alpha}_i^* \mathrm{e}^{-\mathrm{j}\theta}] \tag{7.298}$$

With this equation at hand, we are now ready to formulate the adaptive filtering algorithm for estimating the unknown carrier phase θ. To this end, we incorporate discrete-time n into the recursive estimation procedure for clock recovery in a manner similar to what we did for the group delay; specifically:

1. With the argument of the imaginary-part operator in (7.298) playing the role of error signal, we write:

$$e_n = \mathrm{Im}[\tilde{x}_n(\tau)\tilde{\alpha}_n^* \mathrm{e}^{-\mathrm{j}\theta_n}] \tag{7.299}$$

   where n denotes the normalized discrete-time.
2. The scaling factor $2\sqrt{E}/N_0$ is absorbed in the new step-size parameter μ.
3. With $\hat{\theta}_n$ denoting the *old estimate* of the carrier phase θ and $\hat{\theta}_{n+1}$ denoting its *updated value*, the update rule for the estimation is defined as follows:

$$\hat{\theta}_{n+1} = \hat{\theta}_n + \mu e_n, \qquad n = 0, 1, 2, 3, \ldots \tag{7.300}$$

Equations (7.299) and (7.300) not only define the adaptive filtering algorithm for carrier-phase estimation, but also they provide the basis for implementing the algorithm, as shown in Figure 7.49. This figure may be viewed as a generalization of the well-known *Costas loop* for the analog synchronization of linear quadrature-amplitude modulation schemes that

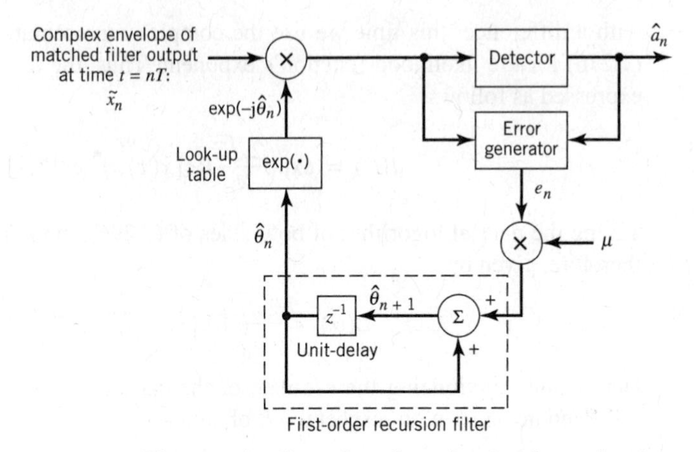

Figure 7.49 The recursive Costas loop for estimating the carrier phase.

involve the combined use of in-phase and quadrature components, of which the QPSK is a special example. As such, we may refer to the closed-loop synchronization scheme of Figure 7.49 as the *recursive Costas loop* for phase synchronization.

The following points should be noted in Figure 7.49:

- The detector supplies an estimate of the symbol indicator $\hat{\alpha}_n$ and, therefore, the transmitted symbol, given the matched filter output.
- For the input $\hat{\theta}_n$, the look-up table in the figure supplies the value of the exponential

$$\exp(-j\hat{\theta}_n) = \cos\hat{\theta}_n - j\sin\hat{\theta}_n$$

- The output of the error generator is the error signal e_n, defined in (7.299).
- The block labeled z^{-1} represents a *unit-time delay.*

The recursive Costas loop of Figure 7.49 uses a *first-order digital filter.* To improve the tracking performance of this synchronization system, we may use a second-order digital filter. Figure 7.50 shows an example of a *second-order recursive filter* made up of a cascade of two first-order sections, with ρ as an adjustable loop parameter. An important property of a second-order recursive filter used in the Costas loop for phase recovery is that it will eventually lock onto the incoming carrier with no static error, provided that the frequency error between the receiver and transmitter is initially small.

Convergence Considerations

The adaptive behavior of the filtering schemes in Figures 7.48 and 7.49 for group-delay and carrier-phase estimation, respectively, is governed by how the step-size parameters

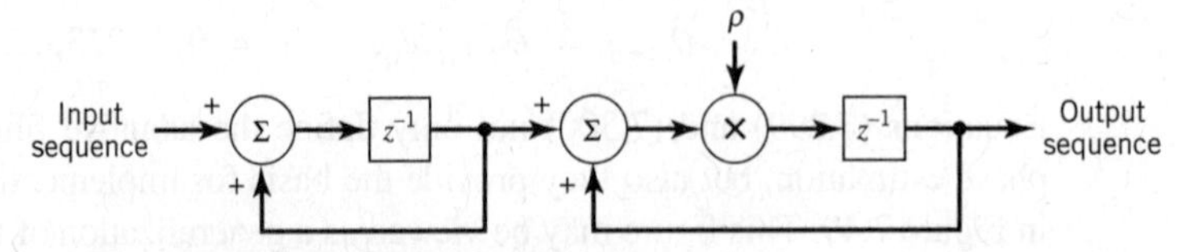

Figure 7.50 Second-order recursive filter.

γ and μ are selected. The smaller we make γ and, likewise, μ, the more *refined* will be the trajectories resulting from application of the algorithms. However, this benefit is attained at the cost of the number of recursions required for convergence of the algorithms. On the other hand, if the step-size parameter γ and μ is assigned a large value, then the trajectories may follow a zig-zag sort of path. Indeed, if γ and μ exceeds a certain critical value of its own, it is quite possible for the algorithm to diverge, which means that the synchronization schemes of Figures 7.48 and 7.49 may become *unstable*. So, from a design perspective, the compromise choice between accuracy of estimation and speed of convergence may require a detailed attention, both theoretical and experimental.

7.17 Summary and Discussion

The primary goal of the material presented in this chapter is the formulation of a systematic procedure for the analysis and design of a digital communication receiver in the presence of AWGN. The procedure, known as *maximum likelihood detection*, decides which particular transmitted symbol is the most likely cause of the noisy signal observed at the channel output. The approach that led to the formulation of the maximum likelihood detector (receiver) is called *signal-space analysis*. The basic idea of the approach is to represent each member of a set of transmitted signals by an N-dimensional vector, where N is the number of orthonormal basis functions needed for a unique geometric representation of the transmitted signals. The set of signal vectors so formed defines a *signal constellation* in an N-dimensional *signal space*.

For a given signal constellation, the (average) probability of symbol error, P_e, incurred in maximum likelihood signal detection over an AWGN channel is invariant to rotation of the signal constellation as well as its translation. However, except for a few simple (but important) cases, the numerical calculation of P_e is an impractical proposition. To overcome this difficulty, the customary practice is to resort to the use of bounds that lend themselves to computation in a straightforward manner. In this context, we described the *union bound* that follows directly from the signal-space diagram. The union bound is based on an intuitively satisfying idea:

> The probability of symbol error P_e is dominated by the nearest neighbors to the transmitted signal in the signal-space diagram.

The results obtained using the union bound are usually fairly accurate, particularly when the SNR is high.

With the basic background theory on optimum receivers covered in the early part of Chapter 7 at our disposal, formulas were derived for, or bounds on, the BER for some important digital modulation techniques in an AWGN channel:

1. PSK, using coherent detection; it is represented by
   - binary PSK;
   - QPSK and its variants, namely, such as the offset QPSK;
   - coherent M-ary PSK, which includes binary PSK and QPSK as special cases with $M = 2$ and $M = 4$, respectively.

   The DPSK may be viewed as the pseudo-noncoherent form of PSK.

2. M-ary QAM, using coherent detection; this modulation scheme is a hybrid form of modulation that combines amplitude and phase-shift keying. For $M = 4$, it includes QPSK as a special case.
3. FSK, using coherent detection; it is represented by
   - binary FSK;
   - MSK and its Gaussian variant known as GMSK;
   - M-ary FSK.
4. Noncoherent detection schemes, involving the use of binary FSK and DPSK.

Irrespective of the digital modulation system of interest, synchronization of the receiver to the transmitter is essential to the operation of the system. Symbol timing recovery is required whether the receiver is coherent or not. If the receiver is coherent, we also require provision for carrier recovery. In the latter part of the chapter we discussed nondata-aided synchronizers to cater to these two requirements with emphasis on M-ary PSK, exemplified by QPSK signals, in which the carrier is suppressed. The presentation focused on recursive synchronization techniques that are naturally suited for the use of discrete-time signal processing algorithms.

We conclude the discussion with some additional notes on the two adaptive filtering algorithms described in Section 7.16 on estimating the unknown parameters: carrier phase and group delay. In a computational context, these two algorithms are in the same class as the celebrated least-mean-square (LMS) algorithm described by Widrow and Hoff over 50 years ago. The LMS algorithm is known for its computational efficiency, effectiveness in performance, and robustness with respect to the nonstationary character of the environment in which it is embedded. The two algorithmic phase and delay synchronizers share the first two properties of the LMS algorithm; for a conjecture, it may well be they are also robust when operating in a nonstationary communication environment.

Problems

Representation of Signals

7.1 In Chapter 6 we described line codes for pulse-code modulation. Referring to the material presented therein, formulate the signal constellations for the following line codes:

a. unipolar nonreturn-to-zero code

b. polar nonreturn-to-zero code

c. unipolar return-to-zero code

d. manchester code.

7.2 An 8-level PAM signal is defined by

$$s_i(t) = A_i \operatorname{rect}\left(\frac{t}{T} - \frac{1}{2}\right)$$

where $A_i = \pm1, \pm3, \pm5, \pm7$. Formulate the signal constellation of $\{s_i(t)\}_{i=1}^{8}$.

7.3 Figure P7.3 displays the waveforms of four signals $s_1(t)$, $s_2(t)$, $s_3(t)$, and $s_4(t)$.

a. Using the Gram–Schmidt orthogonalization procedure, find an orthonormal basis for this set of signals.

b. Construct the corresponding signal-space diagram.

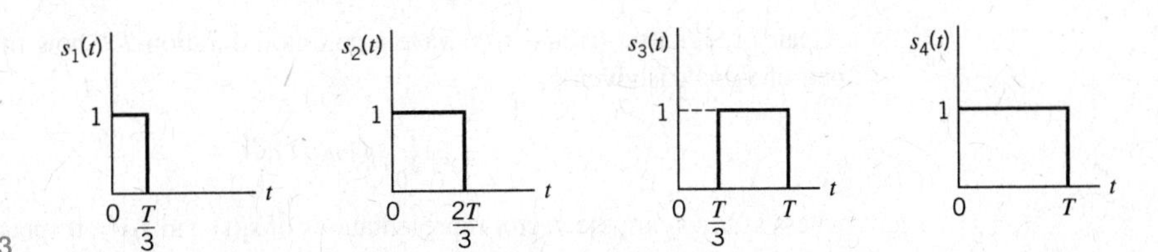

Figure P7.3

7.4 a. Using the Gram–Schmidt orthogonalization procedure, find a set of orthonormal basis functions to represent the three signals $s_1(t)$, $s_2(t)$, and $s_3(t)$ shown in Figure P7.4.

b. Express each of these signals in terms of the set of basis functions found in part a.

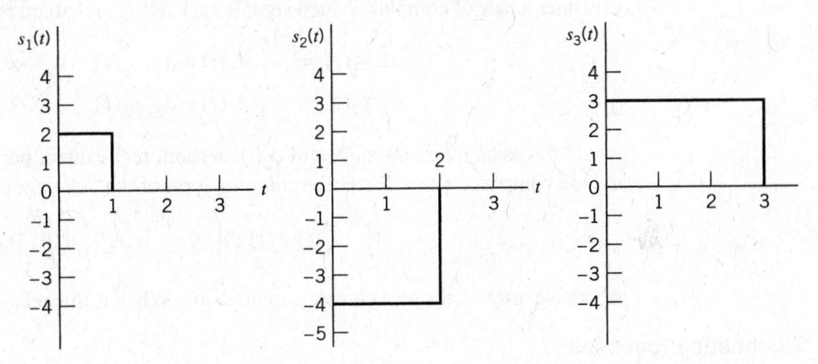

Figure P7.4

7.5 An orthogonal set of signals is characterized by the property that the inner product of any pair of signals in the set is zero. Figure P7.5 shows a pair of signals $s_1(t)$ and $s_2(t)$ that satisfy this definition. Construct the signal constellation for this pair of signals.

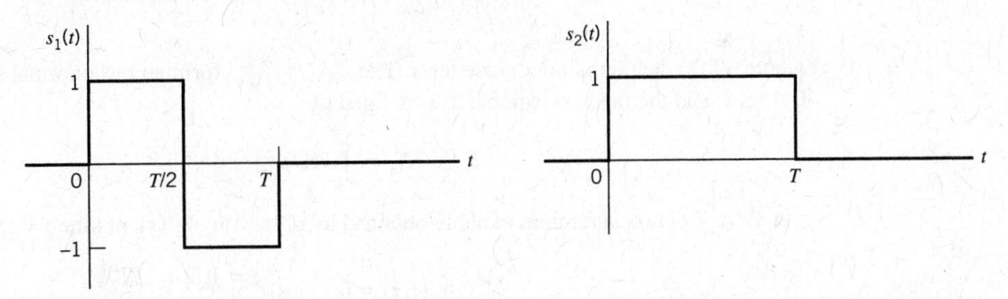

Figure P7.5

7.6 A source of information emits a set of symbols denoted by $\{m_i\}_{i=1}^{M}$. Two candidate modulation schemes, namely pulse-duration modulation (PDM) and pulse-position modulation (PPM), are considered for the electrical representation of this set of symbols. In PDM, the ith symbol is represented by a pulse of unit amplitude and duration $(i/M)T$. On the other hand, in PPM, the ith symbol is represented by a short pulse of unit amplitude and fixed duration, which is transmitted at time $t = (i/M)T$. Show that PPM is the only one of the two that can produce an orthogonal set of signals over the interval $0 \leq t \leq T$.

7.7 A set of $2M$ *biorthogonal signals* is obtained from a set of M ordinary orthogonal signals by augmenting it with the negative of each signal in the set.

a. The extension of orthogonal to biorthogonal signals leaves the dimensionality of the signal space unchanged. Explain how.

b. Construct the signal constellation for the biorthogonal signals corresponding to the pair of orthogonal signals shown in Figure P7.5.

7.8 a. A pair of signals $s_i(t)$ and $s_k(t)$ have a common duration T. Show that the inner product of this pair of signals is given by

$$\int_0^T s_i(t)s_k(t)\,dt = \mathbf{s}_i^{\mathrm{T}}\mathbf{s}_k$$

where $\mathbf{s}_i$ and $\mathbf{s}_k$ are the vector representations of $s_i(t)$ and $s_k(t)$, respectively.

b. As a follow-up to part a of the problem, show that

$$\int_0^T (s_i(t)-s_k(t))^2\,dt = \|\mathbf{s}_i-\mathbf{s}_k\|^2$$

7.9 Consider a pair of complex-valued signals $s_i(t)$ and $s_k(t)$ that are respectively represented by

$$\begin{aligned} s_1(t) &= a_{11}\phi_1(t)+a_{12}\phi_2(t), \qquad -\infty < t < \infty \\ s_2(t) &= a_{21}\phi_1(t)+a_{22}\phi_2(t), \qquad -\infty < t < \infty \end{aligned}$$

where the basis functions $\phi_1(t)$ and $\phi_2(t)$ are both real valued, but the coefficients a_{11}, a_{12}, a_{21}, and a_{22} are complex valued. Prove the complex form of the Schwarz inequality:

$$\left|\int_{-\infty}^{\infty} s_1(t)s_2^*(t)\,dt\right|^2 \le \int_{-\infty}^{\infty}|s_1(t)|^2\,dt\int_{-\infty}^{\infty}|s_2(t)|^2\,dt$$

where the asterisk denotes complex conjugation. When is this relation satisfied with the equality sign?

Stochastic Processes

7.10 Consider a stochastic process $X(t)$ expanded in the form

$$X(t) = \sum_{i=1}^{N} X_i\phi_i(t) + W'(t), \qquad 0 \le t \le T$$

where $W'(t)$ is a remainder noise term. The $\{\phi_i(t)\}_{i=1}^{N}$ form an orthonormal set over the interval $0 \le t \le T$, and the random variable X_i is defined by

$$X_i = \int_0^T X(t)\phi_i(t)\,dt$$

Let $W'(t_k)$ denote a random variable obtained by observing $W'(t)$ at time $t = t_k$. Show that

$$\mathbb{E}[X_jW'(t_k)] = 0, \qquad \begin{cases} j = 1, 2, \ldots, N \\ 0 \le t_k \le T \end{cases}$$

7.11 Consider the optimum detection of the sinusoidal signal in AWGN:

$$s(t) = \sin\left(\frac{8\pi t}{T}\right), \qquad 0 \le t \le T$$

a. Determine the correlator output assuming a noiseless input.

b. Determine the corresponding matched filter output, assuming that the filter includes a delay T to make it causal.

c. Hence, show that these two outputs are exactly the same only at the time instant $t = T$.

Probability of Error

7.12 Figure P7.12 shows a pair of signals $s_1(t)$ and $s_2(t)$ that are orthogonal to each other over the observation interval $0 \le t \le 3T$. The received signal is defined by

$$x(t) = s_k(t) + w(t) \quad \begin{cases} 0 \le t \le 3T \\ k = 1, 2 \end{cases}$$

where $w(t)$ is white Gaussian noise of zero mean and power spectral density $N_0/2$.

a. Design a receiver that decides in favor of signals $s_1(t)$ or $s_2(t)$, assuming that these two signals are equiprobable.

b. Calculate the average probability of symbol error incurred by this receiver for $E/N_0 = 4$, where E is the signal energy.

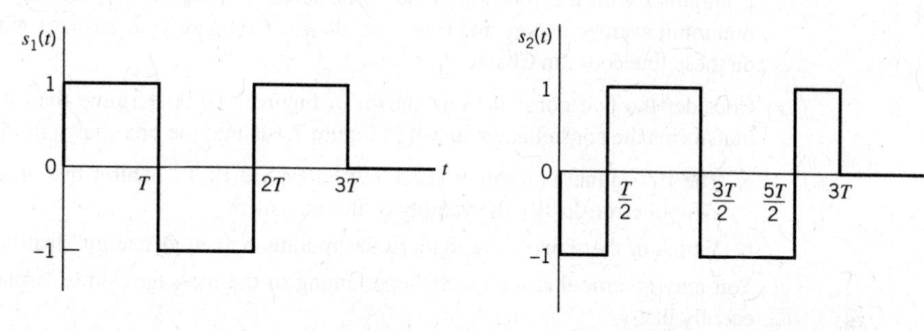

Figure P7.12

7.13 In the Manchester code discussed in Chapter 6, binary symbol 1 is represented by the doublet pulse $s(t)$ shown in Figure P7.13, and binary symbol 0 is represented by the negative of this pulse. Derive the formula for the probability of error incurred by the maximum likelihood detection procedure applied to this form of signaling over an AWGN channel.

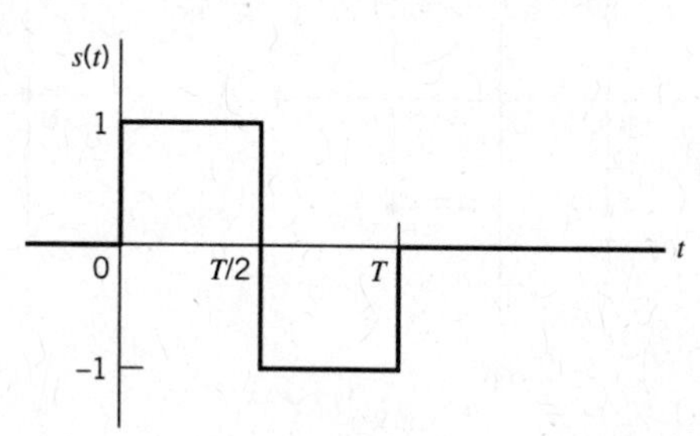

Figure P7.13

7.14 In the *Bayes' test*, applied to a binary hypothesis-testing problem where we have to choose one of two possible hypotheses H_0 and H_1, we minimize the *risk* $\mathcal{R}$ defined by

$$\mathcal{R} = C_{00}p_0(\text{say } H_0|H_0 \text{ is true}) + C_{10}p_0(\text{say } H_1|H_0 \text{ is true}) + C_{11}p_1(\text{say } H_1|H_1 \text{ is true}) + C_{01}p_1(\text{say } H_0|H_1 \text{ is true})$$

The parameters C_{00}, C_{10}, C_{11}, and C_{01} denote the costs assigned to the four possible outcomes of the experiment: the first subscript indicates the hypothesis chosen and the second the hypothesis that is true. Assume that $C_{10} > C_{00}$ and $C_{01} > C_{11}$. The p_0 and p_1 denote the a priori probabilities of hypotheses H_0 and H_1, respectively.

a. Given the observation vector $\mathbf{x}$, show that the partitioning of the observation space so as to minimize the risk $\mathcal{R}$ leads to the *likelihood ratio test*:

$$\text{say } H_0 \text{ if } \Lambda(\mathbf{x}) < \lambda$$

$$\text{say } H_1 \text{ if } \Lambda(\mathbf{x}) > \lambda$$

where $\Lambda(\mathbf{x})$ is the *likelihood ratio* defined by

$$\Lambda(\mathbf{x}) = \frac{f_{\mathbf{X}}(\mathbf{x}|H_1)}{f_{\mathbf{X}}(\mathbf{x}|H_0)}$$

and λ is the *threshold* of the test defined by

$$\lambda = \frac{p_0(C_{10} - C_{00})}{p_1(C_{01} - C_{11})}$$

b. What are the cost values for which the Bayes' criterion reduces to the minimum probability of error criterion?

Principles of Rotational and Translational Invariance

7.15 Continuing with the four line codes considered in Problem 7.1, identify the line codes that have minimum average energy and those that do not. Compare your answers with the observations made on these line codes in Chapter 6.

7.16 Consider the two constellations shown in Figure 7.10. Determine the orthonormal matrix **Q** that transforms the constellation shown in Figure 7.10a into the one shown in Figure 7.10b.

7.17 a. The two signal constellations shown in Figure P7.17 exhibit the same average probability of symbol error. Justify the validity of this statement.

b. Which of these two constellations has minimum average energy? Justify your answer.

You may assume that the symbols pertaining to the message points displayed in Figure P7.17 are equally likely.

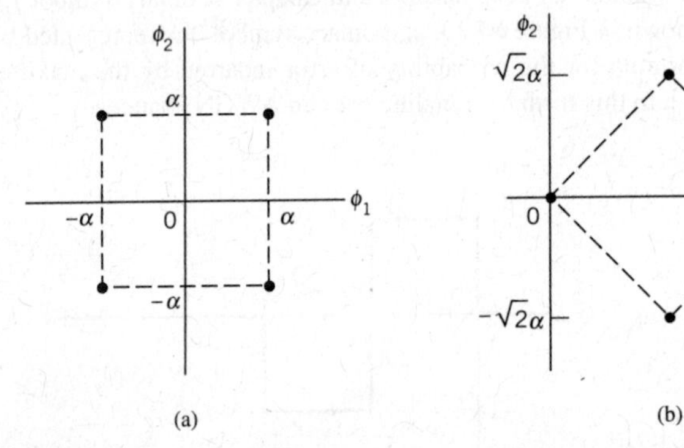

Figure P7.17

7.18 *Simplex (transorthogonal) signals* are equally likely highly-correlated signals with the most negative correlation that can be achieved with a set of M orthogonal signals. That is, the correlation coefficient between any pair of signals in the set is defined by

$$\rho_{ij} = \begin{cases} 1 & \text{for } i = j \\ -1/(M-1) & \text{for } i \neq j \end{cases}$$

One method of constructing simplex signals is to start with a set of M orthogonal signals each with energy E and then apply the minimum energy translate.

Consider a set of three equally likely symbols whose signal constellation consists of the vertices of an equilateral triangle. Show that these three symbols constitute a simplex code.

Amplitude-Shift Keying

7.19 In the *on–off keying* version of an ASK system, symbol 1 is represented by transmitting a sinusoidal carrier of amplitude $\sqrt{2E_b/T_b}$, where E_b is the signal energy per bit and T_b is the bit duration. Symbol 0 is represented by switching off the carrier. Assume that symbols 1 and 0 occur with equal probability.

For an AWGN channel, determine the average probability of error for this ASK system under the following scenarios:

a. Coherent detection.

b. Noncoherent detection, operating with a large value of bit energy-to-noise spectral density ratio E_b/N_0.

Note: when x is large, the modified Bessel function of the first kind of zero order may be approximated as follows (see Appendix C):

$$I_0(x) \approx \frac{\exp(x)}{\sqrt{2\pi x}}$$

Phase-Shift Keying

7.20 The PSK signal is applied to a correlator supplied with a phase reference that lies within φ radians of the exact carrier phase. Determine the effect of the phase error φ on the average probability of error of the system.

7.21 The signal component of a PSK system scheme using coherent detection is defined by

$$s(t) = A_c k \sin(2\pi f_c t) \pm A_c \sqrt{1-k^2} \cos(2\pi f_c t)$$

where $0 \le t \le T_b$, the plus sign corresponds to symbol 1, and the minus sign corresponds to symbol 0; the parameter k lies in the range $0 < k < 1$. The first term of $s(t)$ represents a carrier component included for the purpose of synchronizing the receiver to the transmitter.

a. Draw a signal-space diagram for the scheme described here. What observations can you make about this diagram?

b. Show that, in the presence of AWGN of zero mean and power spectral density $N_0/2$, the average probability of error is

$$P_e = Q\left[\sqrt{\frac{2E_b}{N_0}(1-k^2)}\right]$$

where

$$E_b = \frac{1}{2}A_c^2 T_b$$

c. Suppose that 10% of the transmitted signal power is allocated to the carrier component. Determine the E_b/N_0 required to realize $P_e = 10^{-4}$.

d. Compare this value of E_b/N_0 with that required for a binary PSK scheme using coherent detection, with the same probability of error.

7.22 a. Given the input binary sequence 1100100010, sketch the waveforms of the in-phase and quadrature components of a modulated wave obtained using the QPSK based on the signal set of Figure 7.16.

b. Sketch the QPSK waveform itself for the input binary sequence specified in part a.

7.23 Let P_{eI} and P_{eQ} denote the probabilities of symbol error for the in-phase and quadrature channels, respectively, of a narrowband digital communication system. Show that the average probability of symbol error for the overall system is given by

$$P_e = P_{eI} + P_{eQ} - P_{eI}P_{eQ}$$

7.24 Equation (7.132) is an approximate formula for the average probability of symbol error for M-ary PSK using coherent detection. This formula was derived using the union bound in light of the signal-space diagram of Figure 7.22b. Given that message point m_1 was transmitted, show that the approximation of (7.132) may be derived directly from Figure 7.22b.

7.25 Find the power spectral density of an offset QPSK signal produced by a random binary sequence in which symbols 1 and 0 (represented by ±1) are equally likely and the symbols in different time slots are statistically independent and identically distributed.

7.26 Vestigial sideband modulation (VSB), discussed in Chapter 2, offers another possible modulation method for signaling over an AWGN channel.

a. In particular, a digital VSB transmission system may be viewed as a time-varying one-dimensional system operating at a rate of $2/T$ dimensions per second, where T is the symbol period. Justify the validity of this statement.

b. Show that digital VSB is indeed equivalent in performance to the offset QPSK.

Quadrature Amplitude Modulation

7.27 Referring back to Example 7, develop a systematic procedure for constructing M-ary QAM constellations given the M-ary QAM constellation of Figure 7.24 for $M = 16$. In effect, this problem addresses the opposite approach to that described in Example 7.

7.28 Figure P7.28 describes the block diagram of a *generalized* M-ary QAM modulator. Basically, the modulator includes a *mapper* that produces a complex amplitude a_m input for $m = 0, 1, \ldots, M-1$, The real and imaginary parts of a_m input the basis functions $\phi_1(t)$ and $\phi_2(t)$, respectively. The modulator is generalized in that it embodies M-ary PSK and M-ary PAM as special cases.

a. Formulate the underlying mathematics of the modulator described in Figure P7.28.

b. Hence, show that M-ary PSK and M-ary PAM are indeed special cases of the M-ary QPSK generated by the block diagram of Figure P7.28.

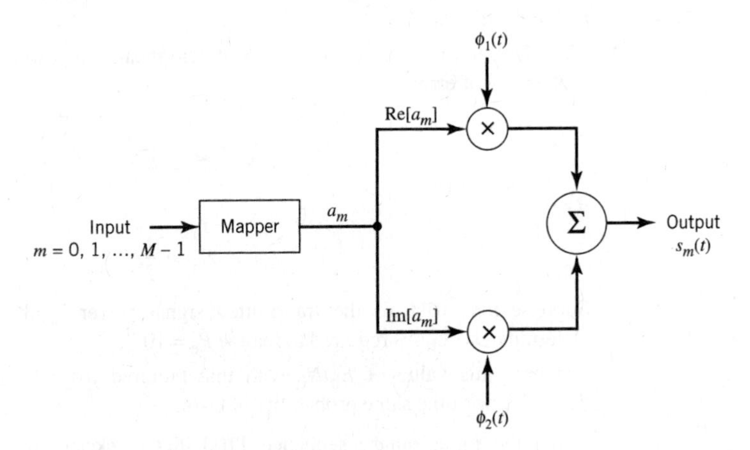

Figure P7.28

Frequency-Shift Keying

7.29 The signal vectors $\mathbf{s}_1$ and $\mathbf{s}_2$ are used to represent binary symbols 1 and 0, respectively, in a binary FSK system using coherent detection. The receiver decides in favor of symbol 1 when

$$\mathbf{x}^T\mathbf{s}_1 > \mathbf{x}^T\mathbf{s}_2$$

where $\mathbf{x}^T\mathbf{s}_i$ is the inner product of the observation vector $\mathbf{x}$ and the signal vector $\mathbf{s}_i$, $i = 1, 2$. Show that this decision rule is equivalent to the condition $x_1 > x_2$, where x_1 and x_2 are the two elements of the observation vector $\mathbf{x}$. Assume that the signal vectors $\mathbf{s}_1$ and $\mathbf{s}_2$ have equal energy.

7.30 An FSK system transmits binary data at the rate of 2.5×10^6 bits/s. During the course of transmission, white Gaussian noise of zero mean and power spectral density 10^{-20} W/Hz is added to the signal. In the absence of noise, the amplitude of the received sinusoidal wave for digit 1 or 0 is 1 mV. Determine the average probability of symbol error for the following system configurations:

a. binary FSK using coherent detection;

b. MSK using coherent detection;

c. binary FSK using noncoherent detection.

7.31 In an FSK system using coherent detection, the signals $s_1(t)$ and $s_2(t)$ representing binary symbols 1 and 0, respectively, are defined by

$$s_1(t), s_2(t) = A_c \cos\left[2\pi\left(f_c \pm \frac{\Delta f}{2}\right)t\right], \qquad 0 \le t \le T_b$$

Assuming that $f_c > \Delta f$, show that the correlation coefficient of the signals $s_1(t)$ and $s_2(t)$ is approximately given by

$$\rho = \frac{\int_0^{T_b} s_1(t)s_2(t)\,dt}{\int_0^{T_b} s_1^2(t)\,dt} \approx \text{sinc}(2\Delta f.T_b)$$

a. What is the minimum value of frequency shift Δf for which the signals $s_1(t)$ and $s_2(t)$ are orthogonal?

b. What is the value of Δf that minimizes the average probability of symbol error?

c. For the value of Δf obtained in part c, determine the increase in E_b/N_0 required so that this FSK scheme has the same noise performance as a binary PSK scheme system, also using coherent detection.

7.32 A binary FSK signal with *discontinuous phase* is defined by

$$s(t) = \begin{cases} \sqrt{\dfrac{2E_b}{T_b}} \cos\left[2\pi\left(f_c + \dfrac{\Delta f}{2}\right)t + \theta_1\right] & \text{for symbol 1} \\ \sqrt{\dfrac{2E_b}{T_b}} \cos\left[2\pi\left(f_c - \dfrac{\Delta f}{2}\right)t + \theta_2\right] & \text{for symbol 0} \end{cases}$$

where E_b is the signal energy per bit, T_b is the bit duration, and θ_1 and θ_2 are sample values of uniformly distributed random variables over the interval 0 to 2π. In effect, the two oscillators supplying the transmitted frequencies $f_c \pm \Delta f/2$ operate independently of each other. Assume that $f_c \gg \Delta f$.

a. Evaluate the power spectral density of the FSK signal.

b. Show that, for frequencies far removed from the carrier frequency f_c, the power spectral density falls off as the inverse square of frequency. How does this result compare with a binary FSK signal with continuous phase?

7.33 Set up a block diagram for the generation of Sunde's FSK signal $s(t)$ with continuous phase by using the representation given in (7.170), which is reproduced here

$$s(t) = \sqrt{\frac{2E_b}{T_b}} \cos\left(\frac{\pi t}{T_b}\right) \cos(2\pi f_c t) \mp \sqrt{\frac{2E_b}{T_b}} \sin\left(\frac{\pi t}{T_b}\right) \sin(2\pi f_c t)$$

7.34 Discuss the similarities between MSK and offset QPSK, and the features that distinguish them.

7.35 There are two ways of detecting an MSK signal. One way is to use a coherent receiver to take full advantage of the phase information content of the MSK signal. Another way is to use a noncoherent receiver and disregard the phase information. The second method offers the advantage of simplicity of implementation at the expense of a degraded noise performance. By how many decibels do we have to increase the bit energy-to-noise density ratio E_b/N_0 in the second method so as to realize the same average probability of symbol error equal to 10^{-5}?

7.36 a. Sketch the waveforms of the in-phase and quadrature components of the MSK signal in response to the input binary sequence 1100100010.

b. Sketch the MSK waveform itself for the binary sequence specified in part a.

7.37 An NRZ data stream of amplitude levels ± 1 is passed through a low-pass filter whose impulse response is defined by the Gaussian function

$$h(t) = \frac{\sqrt{\pi}}{\alpha}\exp\left(-\frac{\pi^2 t^2}{\alpha^2}\right)$$

where α is a design parameter defined in terms of the filter's 3dB bandwidth by

$$\alpha = \sqrt{\frac{\ln 2}{2}}\frac{1}{W}$$

a. Show that the transfer function of the filter is defined by

$$H(f) = \exp(-\alpha^2 f^2)$$

Hence, demonstrate that the 3dB bandwidth of the filter is indeed equal to W. You may use the list of Fourier-transform pairs in Table 2.1.

b. Determine the response of the filter to a rectangular pulse of unit amplitude and duration T centered on the origin.

7.38 Summarize the similarities and differences between the standard MSK and Gaussian filtered MSK signals.

7.39 Summarize the basic similarities and differences between the standard MSK and QPSK.

Noncoherent Receivers

7.40 In Section 7.12 we derived the formula for the BER of binary FSK using noncoherent detection as a special case of noncoherent orthogonal modulation. In this problem we revisit this issue. As before, we assume that symbol 1 is represented by signal $s_1(t)$ and symbol 0 is represented by signal $s_2(t)$. According to the material presented in Section 7.12, we note the following:

- The random variable L_2 represented by the sample value l_2 is Rayleigh distributed.
- The random variable L_1 represented by the sample value l_1 is Rician distributed.

The Rayleigh and Rician distributions were discussed in Chapter 4. Using the probability distributions defined in that chapter, derive (7.245) for the BER of binary FSK, using noncoherent detection.

7.41 Figure P7.41a shows a noncoherent receiver using a matched filter for the detection of a sinusoidal signal of known frequency but random phase and under the assumption of AWGN. An alternative implementation of this receiver is its mechanization in the frequency domain as a *spectrum analyzer receiver*, as in Figure P7.41b, where the correlator computes the finite-time autocorrelation function defined by

$$R_x(\tau) = \int_0^{T-\tau} x(t)x(t+\tau), \qquad 0 \le \tau \le T$$

Show that the square-law envelope detector output sampled at time $t = T$ in Figure P7.41a is twice the spectral output of the Fourier transform sampled at frequency $f = f_c$ in Figure P7.41b.

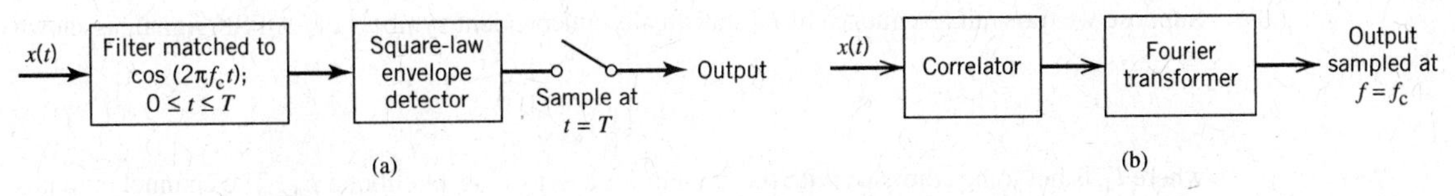

Figure P7.41

7.42 The binary sequence 1100100010 is applied to the DPSK transmitter of Figure 7.44.

a. Sketch the resulting waveform at the transmitter output.

b. Applying this waveform to the DPSK receiver of Figure 7.46, show that in the absence of noise the original binary sequence is reconstructed at the receiver output.

Comparison of Digital Modulation Schemes Using a Single Carrier

7.43 Binary data are transmitted over a microwave link at the rate of 10^6 bits/s and the power spectral density of the noise at the receiver input is 10^{-10} W/Hz. Find the average carrier power required to maintain an average probability of error $P_e \leq 10^{-4}$ for the following schemes:

a. Binary PSK using coherent detection;

b. DPSK.

7.44 The values of E_b/N_0 required to realize an average probability of symbol error $P_e = 10^{-4}$ for binary PSK and binary FSK schemes are equal to 7.2 and 13.5, respectively. Using the approximation

$$Q(u) \approx \frac{1}{\sqrt{2\pi}u}\exp(-2u^2)$$

determine the separation in the values of E_b/N_0 for $P_e = 10^{-4}$, using:

a. binary PSK using coherent detection and DPSK;

b. binary PSK and QPSK, both using coherent detection;

c. binary FSK using (i) coherent detection and (ii) noncoherent detection;

d. binary FSK and MSK, both using coherent detection.

7.45 In Section 7.14 we compared the noise performances of various digital modulation schemes under the two classes of coherent and noncoherent detection; therein, we used the BER as the basis of comparison. In this problem we take a different viewpoint and use the average probability of symbol error P_e, to do the comparison. Plot P_e versus E_b/N_0 for each of these schemes and comment on your results.

Synchronization

7.46 Demonstrate the equivalence of the two complex representations given in (7.276) and (7.277), which pertain to the likelihood function.

7.47 a. In the recursive algorithm of (7.295) for symbol timing recovery, the control signals c_n and c_{n+1} are both dimensionless. Discuss the units in which the error signal e_n and step-size parameter μ are measured.

b. In the recursive algorithm of (7.300) for phase recovery, the old estimate $\hat{\theta}_n$ and the updated estimate $\hat{\theta}_{n+1}$ of the carrier phase θ are both measured in radians. Discuss the units in which the error signal e_n and step-size parameter μ are measured.

7.48 The binary PSK is a special case of QPSK. Using the adaptive filtering algorithms derived in Section 7.16 for estimating the group delay τ and carrier phase θ, find the corresponding adaptive filtering algorithms for binary PSK.

7.49 Repeat Problem 7.48, but this time find the adaptive filtering algorithms for M-ary PSK.

7.50 Suppose we transmit a sequence of L_0 statistically independent symbols of a QPSK signal, as shown by

$$\mathbf{s} = \{s_i\}_{i=0}^{L_0-1}$$

where L_0 is not to be confused with the symbol for average log-likelihood L_{av}. The channel output is corrupted by AWGN of zero mean and power spectral density $N_0/2$, carrier phase θ, and unknown group delay τ.

a. Determine the likelihood function with respect to the group delay τ, assuming that θ is uniformly distributed.

b. Hence, formulate the maximum likelihood estimate of the group delay τ.

c. Compare this feedforward scheme of group-delay estimation with that provided by the NDA-ELD synchronizer of Figure 7.48.

7.51 Repeat Problem 7.50, but this time do the following:

a. Determine the likelihood function with respect to the carrier phase θ, assuming that the group delay τ is known.

b. Hence, formulate the maximum likelihood estimate of the carrier phase θ.

c. Compare this feedforward scheme of a carrier-phase estimation with the recursive Costas loop of Figure 7.49.

7.52 In Section 7.16 we studied a nondata-aided scheme for carrier phase recovery, based on the log-likelihood function of (7.296). In this problem we explore the use of this equation for *data-aided carrier phase recovery*.

a. Consider a receiver designed for a linear modulation system. Given that the receiver has knowledge of a preamble of length L_0, show that the maximum likelihood estimate of the carrier phase is defined by

$$\hat{\theta} = \arg\left\{\sum_{n=0}^{L_0-1} \tilde{a}_n^* \tilde{x}_n\right\}$$

where the preamble $\{\tilde{a}_n\}_{n=0}^{L_0-1}$ is a known sequence of complex symbols and $\{\tilde{x}_n\}_{n=0}^{L_0-1}$ is the complex envelope of the corresponding received signal.

b. Using the result derived in part a, construct a block diagram for the maximum likelihood phase estimator.

7.53 Figure P7.53 shows the block diagram of a phase-synchronization system. Determine the phase estimate $\hat{\theta}$ of the unknown carrier phase in the received signal $x(t)$.

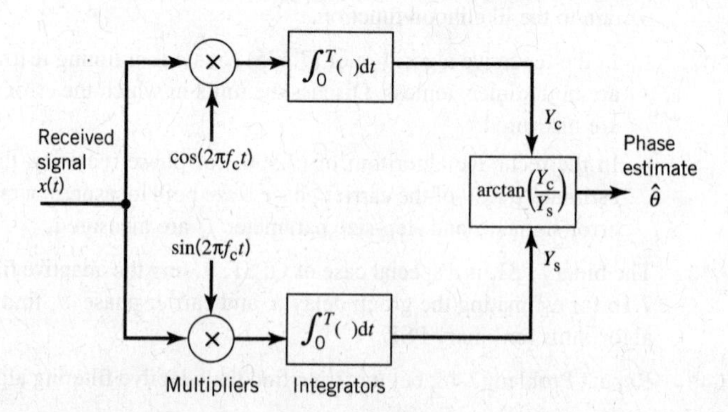

Figure P7.53

Computer Experiments

**7.54 In this *computer-oriented problem*, we study the operation of the NDA-ELD synchronizer for symbol timing recovery by considering a coherent QPSK system with the following specifications:

- The channel response is described by a raised cosine pulse with rolloff factor $\alpha = 0.5$.
- The recursive filter is a first-order digital filter with transfer function

$$H(z) = \frac{z^{-1}}{1-(1-\gamma A)z^{-1}}$$

where z^{-1} denotes unit delay, γ is the step-size parameter, and A is a parameter, to be defined.

- The loop bandwidth B_L is 2% of the symbol rate $1/T$, that is, $B_L T = 0.02$.

With symbol timing recovery as the objective, a logical way to proceed is to plot the *S-curve* for the NDA-ELD under the following conditions:

a. $E_b/N_0 = 10$ dB

b. $E_b/N_0 = \infty$ (i.e., noiseless channel).

For NDA-ELD, the scheme shown in Figure P7.54 is responsible for generating the S-curve that plots the *timing offset* versus the discrete time $n = t/T$.

Using this scheme, plot the S-curves, and comment on the results obtained for parts a and b.

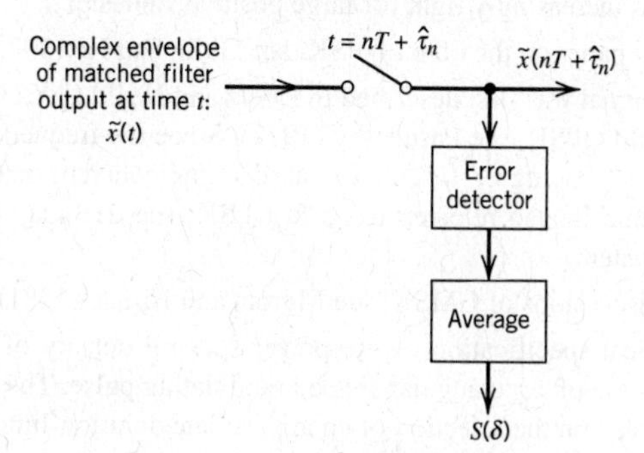

Figure P7.54

7.55 In this follow-up to the *computer-oriented* Problem 7.54, we study the recursive Costas loop for phase recovery using the same system specifications described in Problem 7.54. This time, however, we use the scheme of Figure P7.54 for measuring the *S*-curve to plot the phase error versus discrete-time $n = t/T$.

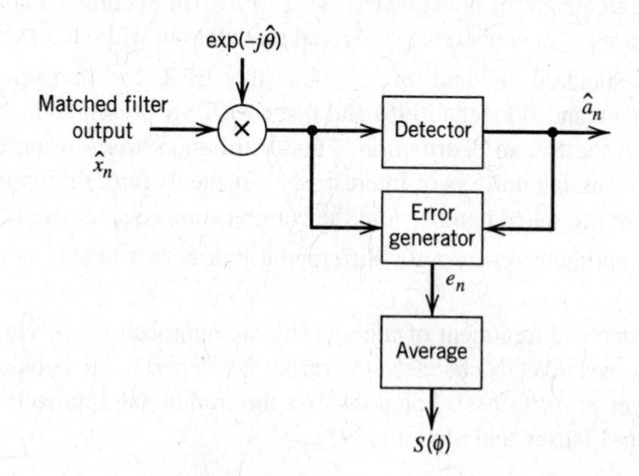

Figure P7.55

The plot is to be carried out under the following conditions:

a. $E_b/N_0 = 5$ dB

b. $E_b/N_0 = 10$ dB

c. $E_b/N_0 = 30$ dB (i.e., practically noiseless channel)

Comment on the results obtained for these three conditions.

Notes

1. The geometric representation of signals was first developed by Kotel'nikov (1947) which is a translation of the original doctoral dissertation presented in January 1947 before the Academic Council of the Molotov Energy Institute in Moscow. In particular, see Part II of the book. This method was subsequently brought to fuller fruition in the classic book by Wozencraft and Jacobs (1965).

2. The classic reference for the union bound is Wozencraft and Jacobs (1965).

3. Appendix C addresses the derivation of simple bounds on the Q-function. In (7.88), we have used the following bound:

$$Q(x) \leq \frac{1}{\sqrt{2\pi}} \exp\left(-\frac{x^2}{2}\right)$$

which becomes increasingly tight for large positive values of x.

4. For an early paper on the offset QPSK, see Gitlin and Ho (1975).

5. The MSK signal was first described in Doelz and Heald (1961). For a tutorial review of MSK and comparison with QPSK, see Pasupathy (1979). Since the frequency spacing is only half as much as the conventional spacing of $1/T_b$ that is used in the coherent detection of binary FSK signals, this signaling scheme is also referred to as fast FSK; see deBuda (1972), who was not aware of the Doelz–Heald patent.

6. For early discussions of GMSK, see Murota and Hirade (1981) and Ishizuke and Hirade (1980).

7. The analytical specification of the power spectral density of digital FM is difficult to handle, except for the case of a rectangular shaped modulating pulse. The paper by Garrison (1975) presents a procedure based on the selection of an appropriate duration-limited/level-quantized approximation for the modulating pulse. The equations developed therein are particularly suitable for machine computation of the power spectra of digital FM signals; see the book by Stüber (1996).

8. A detailed analysis of the spectra of M-ary FSK for an arbitrary value of frequency deviation is presented in the paper by Anderson and Salz (1965).

9. Readers who are not interested in the formal derivation of (7.227) may at this point wish to move on to the treatment of noncoherent binary FSK (in Section 7.12) and DPSK (in Section 7.13), two special cases of noncoherent orthogonal modulation, without loss of continuity.

10. The standard method of deriving the BER for noncoherent binary FSK, presented in McDonough and Whalen (1995) and that for DPSK presented in Arthurs and Dym (1962), involves the use of the Rician distribution. This distribution arises when the envelope of a sine wave plus additive Gaussian noise is of interest; see Chapter 4 for a discussion of the Rician distribution. The derivations presented herein avoid the complications encountered in the standard method.

11. The optimum receiver for differential phase-shift keying is discussed in Simon and Divsalar (1992).

12. For detailed treatment of the algorithmic approach for solving the synchronization problem in signaling over AWGN channels, the reader is referred to the books by Mengali and D'Andrea (1997) and Meyer *et al.* (1998). For books on the traditional approach to synchronization, the reader is referred to Lindsey and Simon (1973).

CHAPTER

8

Signaling over Band-Limited Channels

8.1 Introduction

In Chapter 7 we focused attention on signaling over a channel that is assumed to be distortionless except for the AWGN at the channel output. In other words, there was *no* limitation imposed on the channel bandwidth, with the energy per bit to noise spectral density ratio E_b/N_0 being the only factor to affect the performance of the receiver. In reality, however, every physical channel is not only noisy, but also limited to some finite bandwidth. Hence the title of this chapter: signaling over band-limited channels.

The important point to note here is that if, for example, a rectangular pulse, representing one bit of information, is applied to the channel input, the shape of the pulse will be distorted at the channel output. Typically, the distorted pulse may consist of a main lobe representing the original bit of information surrounded by a long sequence of sidelobes on each side of the main lobe. The sidelobes represent a new source of channel distortion, referred to as *intersymbol interference*, so called because of its degrading influence on the adjacent bits of information.

There is a fundamental difference between intersymbol interference and channel noise that could be summarized as follows:

- Channel noise is independent of the transmitted signal; its effect on data transmission over the band-limited channel shows up at the receiver input, once the data transmission system is switched on.
- Intersymbol interference, on the other hand, is *signal dependent*; it disappears only when the transmitted signal is switched off.

In Chapter 7, channel noise was considered all by itself so as to develop a basic understanding of how its presence affects receiver performance. It is logical, therefore, that in the sequel to that chapter, we initially focus on intersymbol interference acting alone. In practical terms, we may justify a noise-free condition by assuming that the SNR is high enough to ignore the effect of channel noise. The study of signaling over a band-limited channel, under the condition that the channel is effectively "noiseless," occupies the first part of the chapter. The objective here is that of *signal design*, whereby the effect of symbol interference is reduced to zero.

The second part of the chapter focuses on a noisy wideband channel. In this case, data transmission over the channel is tackled by dividing it into a number of subchannels, with each subchannel being narrowband enough to permit the application of Shannon's information capacity law that was considered in Chapter 5. The objective here is that of *system design*, whereby the rate of data transmission through the system is maximized to the highest level physically possible.

8.2 Error Rate Due to Channel Noise in a Matched-Filter Receiver

We begin the study of signaling over band-limited channels by determining the operating conditions that would permit us to view the channel to be effectively "noiseless." To this end, consider the block diagram of Figure 8.1, which depicts the following data-transmission scenario: a binary data stream is applied to a noisy channel where the additive channel noise $w(t)$ is modeled as white and Gaussian with zero mean and power spectral density $N_0/2$. The data stream is based on *polar NRZ signaling*, in which symbols 1 and 0 are represented by positive and negative rectangular pulses of amplitude A and duration T_b. In the signaling interval $0 \le t \le T_b$, the received signal is defined by

$$x(t) = \begin{cases} +A + w(t), & \text{symbol 1 was sent} \\ -A + w(t), & \text{symbol 0 was sent} \end{cases} \tag{8.1}$$

The receiver operates *synchronously* with the transmitter, which means that the *matched filter* at the front end of the receiver has knowledge of the starting and ending times of each transmitted pulse. The matched filter is followed by a *sampler*, and then finally a *decision device*. To simplify matters, it is assumed that the symbols 1 and 0 are equally likely; the threshold in the decision device, namely λ, may then be set equal to zero. If this threshold is exceeded, the receiver decides in favor of symbol 1; if not, it decides in favor of symbol 0. A random choice is made in the case of a tie.

Following the geometric signal-space theory presented in Section 7.6 on binary PSK, the transmitted signal constellation consists of a pair of message points located at $+\sqrt{E_b}$ and $-\sqrt{E_b}$. The energy per bit is defined by

$$E_b = A^2 T_b$$

The only basis function of the signal-space diagram is a rectangular pulse defined as follows:

$$\phi(t) = \begin{cases} \sqrt{E_b/T_b}, & \text{for } 0 \le t \le T_b \\ 0, & \text{otherwise} \end{cases} \tag{8.2}$$

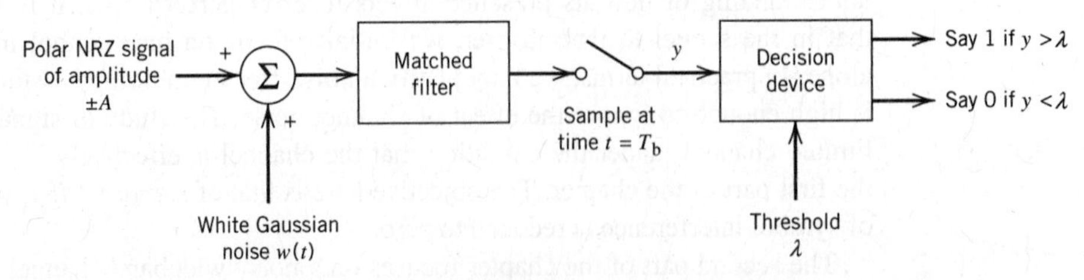

Figure 8.1 Receiver for baseband transmission of binary-encoded data stream using polar NRZ signaling.

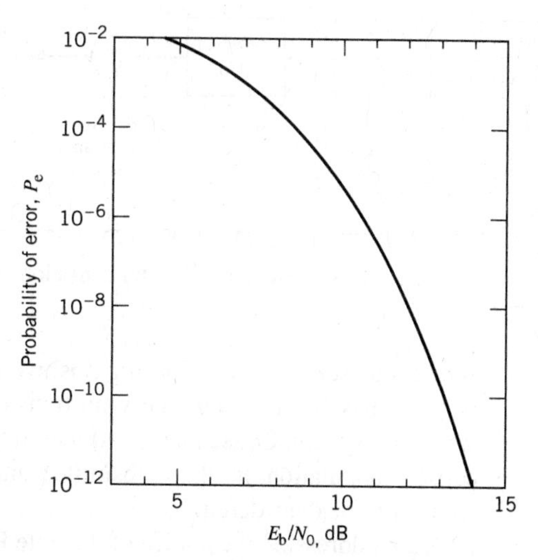

Figure 8.2
Probability of error in the signaling scheme of Figure 8.1.

In mathematical terms, the form of signaling embodied in Figure 8.1 is equivalent to that of binary PSK. Following (7.109), the average probability of symbol error incurred by the matched-filter receiver in Figure 8.1 is therefore defined by the Q-function

$$P_e = Q\left(\sqrt{\frac{2E_b}{N_0}}\right) \tag{8.3}$$

Although this result for NRZ-signaling over an AWGN channel may seem to be special, (8.3) holds for a binary data transmission system where symbol 1 is represented by a generic pulse $g(t)$ and symbol 0 is represented by $-g(t)$ under the assumption that the energy contained in $g(t)$ is equal to E_b. This statement follows from matched-filter theory presented in Chapter 7.

Figure 8.2 plots P_e versus the dimensionless SNR, E_b/N_0. The important message to take from this figure is summed up as follows:

> The matched-filter receiver of Figure 8.1 exhibits an exponential improvement in the average probability of symbol error P_e with the increase in E_b/N_0.

For example, expressing E_b/N_0 in decibels we see from Figure 8.2 that P_e is on the order of 10^{-6} when $E_b/N_0 = 10$ dB. Such a value of P_e is small enough to say that the effect of the channel noise is ignorable.

Henceforth, in the first part of the chapter dealing with signaling over band-limited channels, we assume that the SNR, E_b/N_0, is large enough to leave intersymbol interference as the only source of interference.

8.3 Intersymbol Interference

To proceed with a mathematical study of intersymbol interference, consider a *baseband binary PAM system*, a generic form of which is depicted in Figure 8.3. The term "baseband" refers to an information-bearing signal whose spectrum extends from (or near)

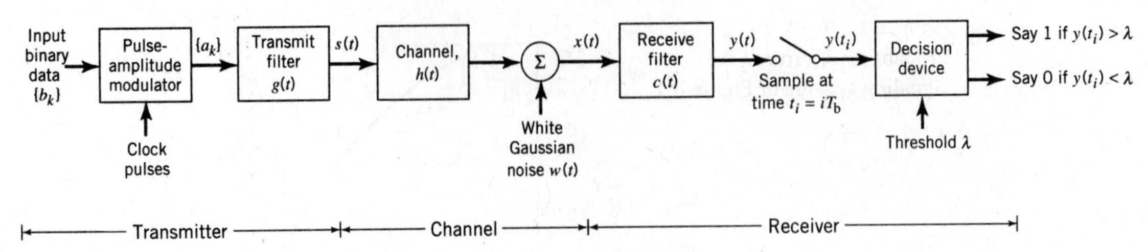

Figure 8.3 Baseband binary data transmission system.

zero up to some finite value for positive frequencies. Thus, with the input data stream being a baseband signal, the data-transmission system of Figure 8.3 is said to be a *baseband system.* Consequently, unlike the subject matter studied in Chapter 7, there is no carrier modulation in the transmitter and, therefore, no carrier demodulation in the receiver to be considered.

Next, addressing the choice of discrete PAM, we say that this form of pulse modulation is one of the most efficient schemes for data transmission over a baseband channel when the utilization of both *transmit power* and *channel bandwidth* is of particular concern. In this section, we consider the simple case of binary PAM.

Referring back to Figure 8.3, the *pulse-amplitude modulator* changes the input binary data stream $\{b_k\}$ into a new sequence of short pulses, short enough to approximate impulses. More specifically, the pulse amplitude a_k is represented in the polar form:

$$a_k = \begin{cases} +1 \text{ if } b_k \text{ is symbol } 1 \\ -1 \text{ if } b_k \text{ is symbol } 0 \end{cases} \tag{8.4}$$

The sequence of short pulses so produced is applied to a *transmit filter* whose impulse response is denoted by $g(t)$. The transmitted signal is thus defined by the sequence

$$s(t) = \sum_k a_k g(t - kT_b) \tag{8.5}$$

Equation (8.5) is a form of *linear modulation*, which may be stated in words as follows:

> A binary data stream represented by the sequence $\{a_k\}$, where $a_k = +1$ for symbol 1 and $a_k = -1$ for symbol 0, *modulates* the basis pulse $g(t)$ and superposes *linearly* to form the transmitted signal $s(t)$.

The signal $s(t)$ is naturally modified as a result of transmission through the *channel* whose impulse response is denoted by $h(t)$. The noisy received signal $x(t)$ is passed through a *receive filter* of impulse response $c(t)$. The resulting filter output $y(t)$ is sampled *synchronously* with the transmitter, with the sampling instants being determined by a *clock* or *timing signal* that is usually extracted from the receive-filter output. Finally, the sequence of samples thus obtained is used to reconstruct the original data sequence by means of a *decision device*. Specifically, the amplitude of each sample is compared with a *zero threshold*, assuming that the symbols 1 and 0 are equiprobable. If the zero threshold is exceeded, a decision is made in favor of symbol 1; otherwise a decision is made in favor of symbol 0. If the sample amplitude equals the zero threshold exactly, the receiver simply makes a random guess.

Except for a trivial scaling factor, we may now express the receive filter output as

$$y(t) = \sum_{k} a_k p(t - kT_b) \tag{8.6}$$

where the pulse $p(t)$ is to be defined. To be precise, an arbitrary time delay t_0 should be included in the argument of the pulse $p(t - kT_b)$ in (8.6) to represent the effect of transmission delay through the system. To simplify the exposition, we have put this delay equal to zero in (8.6) without loss of generality; moreover, the channel noise is ignored.

The scaled pulse $p(t)$ is obtained by a double convolution involving the impulse response $g(t)$ of the transmit filter, the impulse response $h(t)$ of the channel, and the impulse response $c(t)$ of the receive filter, as shown by

$$p(t) = g(t) \star h(t) \star c(t) \tag{8.7}$$

where, as usual, the star denotes convolution. We assume that the pulse $p(t)$ is *normalized* by setting

$$p(0) = 1 \tag{8.8}$$

which justifies the use of a scaling factor to account for amplitude changes incurred in the course of signal transmission through the system.

Since convolution in the time domain is transformed into multiplication in the frequency domain, we may use the Fourier transform to change (8.7) into the equivalent form

$$P(f) = G(f)H(f)C(f) \tag{8.9}$$

where $P(f)$, $G(f)$, $H(f)$, and $C(f)$ are the Fourier transforms of $p(t)$, $g(t)$, $h(t)$, and $c(t)$, respectively.

The receive filter output $y(t)$ is sampled at time $t_i = iT_b$, where i takes on integer values; hence, we may use (8.6) to write

$$\begin{aligned} y(t_i) &= \sum_{k=-\infty}^{\infty} a_k p[(i-k)T_b] \\ &= a_i + \sum_{\substack{k=-\infty \\ k \neq i}}^{\infty} a_k p[(i-k)T_b] \end{aligned} \tag{8.10}$$

In (8.10), the first term a_i represents the contribution of the ith transmitted bit. The second term represents the residual effect of all other transmitted bits on the decoding of the ith bit. This residual effect due to the occurrence of pulses before and after the sampling instant t_i is called *intersymbol interference* (ISI).

In the absence of ISI—and, of course, channel noise—we observe from (8.10) that the summation term is zero, thereby reducing the equation to

$$y(t_i) = a_i$$

which shows that, under these ideal conditions, the ith transmitted bit is decoded correctly.

8.4 Signal Design for Zero ISI

The primary objective of this chapter is to formulate an overall pulse shape $p(t)$ so as to mitigate the ISI problem, given the impulse response of the channel $h(t)$. With this objective in mind, we may now state the problem at hand:

> Construct the overall pulse shape $p(t)$ produced by the entire binary data-transmission system of Figure 8.3, such that the receiver is enabled to reconstruct the original data stream applied to the transmitter input exactly.

In effect, signaling over the band-limited channel becomes distortionless; hence, we may refer to the pulse-shaping requirement as a *signal-design problem.*

In the next section we describe a signal-design procedure, whereby overlapping pulses in the binary data-transmission system of Figure 8.3 are configured in such a way that at the receiver output they *do not interfere with each other at the sampling times* $t_i = iT_b$. So long as the reconstruction of the original binary data stream is accomplished, the behavior of the overlapping pulses outside these sampling times is clearly of no practical consequence. Such a design procedure is rooted in the *criterion for distortionless transmission*, which was formulated by Nyquist (1928b) on telegraph transmission theory, a theory that is as valid then as it is today.

Referring to (8.10), we see that the weighted pulse contribution, $a_k p(iT_b - kT_b)$, must be zero for all k except for $k = 1$ for binary data transmission across the band-limited channel to be *ISI free*. In other words, the overall pulse-shape $p(t)$ must be designed to satisfy the requirement

$$p(iT_b - kT_b) = \begin{cases} 1 & \text{for } i = k \\ 0 & \text{for } i \neq k \end{cases} \tag{8.11}$$

where $p(0)$ is set equal to unity in accordance with the normalization condition of (8.8). A pulse $p(t)$ that satisfies the two-part condition of (8.11) is called a *Nyquist pulse*, and the condition itself is referred to as *Nyquist's criterion for distortionless binary baseband data transmission*. However, there is no unique Nyquist pulse; rather, there are many pulse shapes that satisfy the Nyquist criterion of (8.11). In the next section we describe two kinds of Nyquist pulses, each with its own attributes.

8.5 Ideal Nyquist Pulse for Distortionless Baseband Data Transmission

From a design point of view, it is informative to transform the two-part condition of (8.11) into the frequency domain. Consider then the sequence of samples $\{p(nT_b)\}$, where $n = 0, \pm1, \pm2, \ldots$. From the discussion presented in Chapter 6 on the sampling process, we recall that sampling in the time domain produces periodicity in the frequency domain. In particular, we may write

$$P_\delta(f) = R_b \sum_{n=-\infty}^{\infty} P(f - nR_b) \tag{8.12}$$

where $R_b = 1/T_b$ is the *bit rate* in bits per second; $P_\delta(f)$ on the left-hand side of (8.12) is the Fourier transform of an infinite periodic sequence of delta functions of period T_b whose individual areas are weighted by the respective sample values of $p(t)$. That is, $P_\delta(f)$ is given by

$$P_\delta(f) = \int_{-\infty}^{\infty} \sum_{m=-\infty}^{\infty} [p(mT_b)\delta(t - mT_b)]\exp(-j2\pi ft)\,dt \tag{8.13}$$

Let the integer $m = i - k$. Then, $i = k$ corresponds to $m = 0$ and, likewise, $i \neq k$ corresponds to $m \neq 0$. Accordingly, imposing the conditions of (8.11) on the sample values of $p(t)$ in the integral in (8.13), we get

$$\begin{aligned} P_\delta(f) &= p(0)\int_{-\infty}^{\infty} \delta(t)\exp(-j2\pi ft)\,dt \\ &= p(0) \end{aligned} \tag{8.14}$$

where we have made use of the sifting property of the delta function. Since from (8.8) we have $p(0) = 1$, it follows from (8.12) and (8.14) that the frequency-domain condition for zero ISI is satisfied, provided that

$$\sum_{n=-\infty}^{\infty} P(f - nR_b) = T_b \tag{8.15}$$

where $T_b = 1/R_b$. We may now make the following statement on the *Nyquist criterion*[1] *for distortionless baseband transmission* in the frequency domain:

> The frequency function $P(f)$ eliminates intersymbol interference for samples taken at intervals T_b provided that it satisfies (8.15).

Note that $P(f)$ refers to the overall system, incorporating the transmit filter, the channel, and the receive filter in accordance with (8.9).

Ideal Nyquist Pulse

The simplest way of satisfying (8.15) is to specify the frequency function $P(f)$ to be in the form of a *rectangular function*, as shown by

$$\begin{aligned} P(f) &= \begin{cases} \dfrac{1}{2W}, & -W < f < W \\ 0, & |f| > W \end{cases} \\ &= \frac{1}{2W}\text{rect}\left(\frac{f}{2W}\right) \end{aligned} \tag{8.16}$$

where rect(f) stands for a rectangular function of unit amplitude and unit support centered on $f = 0$ and the overall baseband system bandwidth W is defined by

$$W = \frac{R_b}{2} = \frac{1}{2T_b} \tag{8.17}$$

According to the solution in (8.16), no frequencies of absolute value exceeding half the bit rate are needed. Hence, from Fourier-transform pair 1 of Table 2.2 in Chapter 2, we find that a signal waveform that produces zero ISI is defined by the *sinc function*:

$$p(t) = \frac{\sin(2\pi Wt)}{2\pi Wt} = \text{sinc}(2Wt) \tag{8.18}$$

The special value of the bit rate $R_b = 2W$ is called the *Nyquist rate* and W is itself called the *Nyquist bandwidth*. Correspondingly, the baseband pulse $p(t)$ for distortionless transmission described in (8.18) is called the *ideal Nyquist pulse*, ideal in the sense that the bandwidth requirement is one half the bit rate.

Figure 8.4 shows plots of $P(f)$ and $p(t)$. In part a of the figure, the normalized form of the frequency function $P(f)$ is plotted for positive and negative frequencies. In part b of the figure, we have also included the signaling intervals and the corresponding centered sampling instants. The function $p(t)$ can be regarded as the impulse response of an ideal low-pass filter with passband magnitude response $1/2W$ and bandwidth W. The function $p(t)$ has its peak value at the origin and goes through zero at integer multiples of the bit duration T_b. It is apparent, therefore, that if the received waveform $y(t)$ is sampled at the instants of time $t = 0, \pm T_b, \pm 2T_b, \ldots$, then the pulses defined by $a_i p(t - iT_b)$ with amplitude a_i and index $i = 0, \pm 1, \pm 2, \ldots$ will not interfere with each other. This condition is illustrated in Figure 8.5 for the binary sequence 1011010.

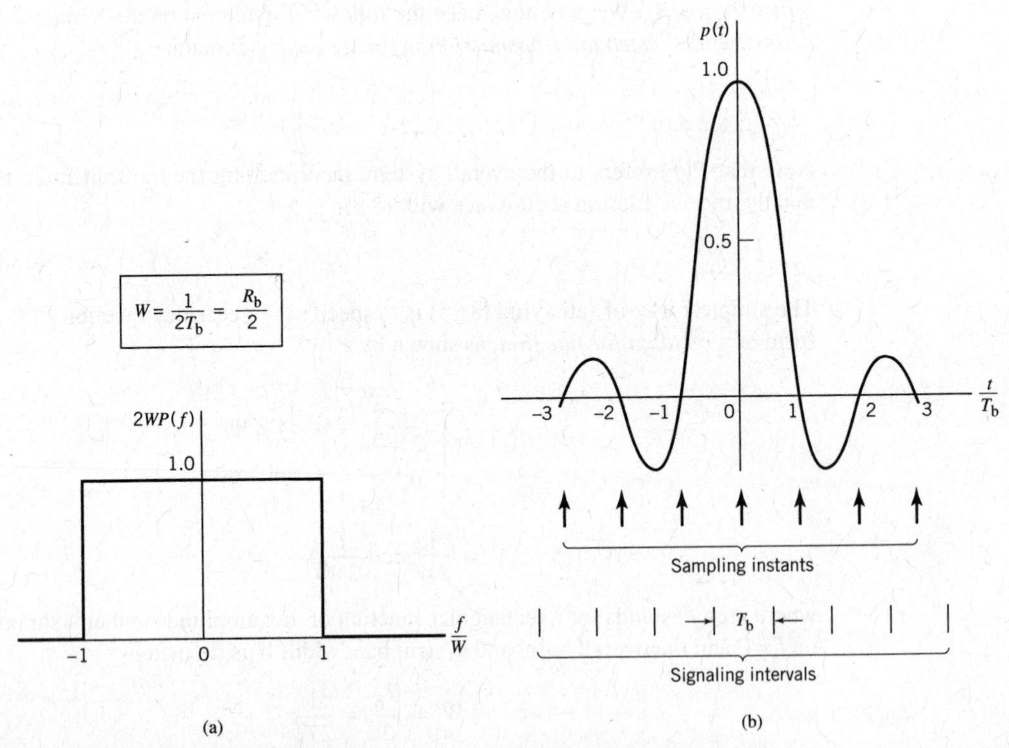

Figure 8.4 (a) Ideal magnitude response. (b) Ideal basic pulse shape.

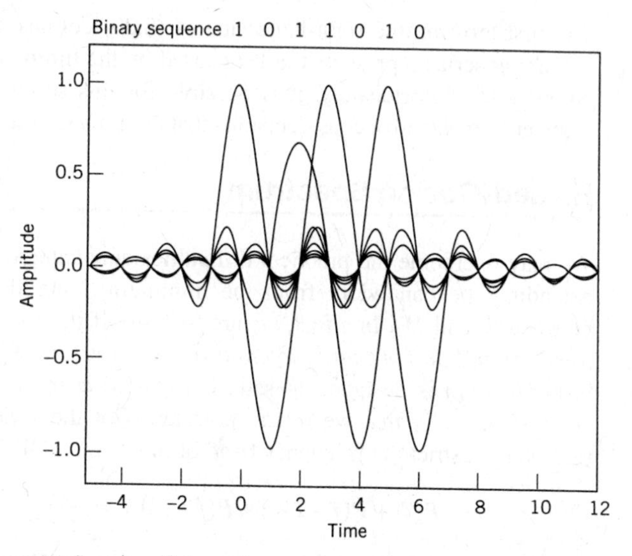

Figure 8.5 A series of sinc pulses corresponding to the sequence 1011010.

Although the use of the ideal Nyquist pulse does indeed achieve economy in bandwidth, in that it solves the problem of zero ISI with the minimum bandwidth possible, there are two practical difficulties that make it an undesirable objective for signal design:

1. It requires that the magnitude characteristic of $P(f)$ be flat from $-W$ to $+W$, and zero elsewhere. This is physically unrealizable because of the abrupt transitions at the band edges $\pm W$, in that the Paley–Wiener criterion discussed in Chapter 2 is violated.
2. The pulse function $p(t)$ decreases as $1/|t|$ for large $|t|$, resulting in a slow rate of decay. This is also caused by the discontinuity of $P(f)$ at $\pm W$. Accordingly, there is practically no margin of error in sampling times in the receiver.

To evaluate the effect of the timing error alluded to under point 2, consider the sample of $y(t)$ at $t = \Delta t$, where Δt is the timing error. To simplify the exposition, we may put the correct sampling time t_i equal to zero. In the absence of noise, we thus have from the first line of (8.10):

$$\begin{aligned} y(\Delta t) &= \sum_{k=-\infty}^{\infty} a_k p(\Delta t - kT_b) \\ &= \sum_{k=-\infty}^{\infty} a_k \left\{ \frac{\sin[2\pi W(\Delta t - kT_b)]}{2\pi W(\Delta t - kT_b)} \right\} \end{aligned} \tag{8.19}$$

Since $2WT_b = 1$, by definition, we may reduce (8.19) to

$$y(\Delta t) = a_0 \operatorname{sinc}(2W\Delta t) + \frac{\sin(2\pi W\Delta t)}{\pi} \sum_{\substack{k=-\infty \\ k \neq 0}}^{\infty} \frac{(-1)^k a_k}{2W\Delta t - k} \tag{8.20}$$

The first term on the right-hand side of (8.20) defines the desired symbol, whereas the remaining series represents the ISI caused by the timing error Δt in sampling the receiver output $y(t)$. Unfortunately, it is possible for this series to diverge, thereby causing the receiver to make erroneous decisions that are undesirable.

8.6 Raised-Cosine Spectrum

We may overcome the practical difficulties encountered with the ideal Nyquist pulse by extending the bandwidth from the minimum value $W = R_b/2$ to an adjustable value between W and $2W$. In effect, we are trading off increased channel bandwidth for a more *robust signal design that is tolerant of timing errors*. Specifically, the overall frequency response $P(f)$ is designed to satisfy a condition more stringent than that for the ideal Nyquist pulse, in that we retain three terms of the summation on the left-hand side of (8.15) and restrict the frequency band of interest to $[-W, W]$, as shown by

$$P(f) + P(f-2W) + P(f+2W) = \frac{1}{2W}, \qquad -W \le f \le W \tag{8.21}$$

where, on the right-hand side, we have set $R_b = 1/2W$ in accordance with (8.17). We may now devise several band-limited functions that satisfy (8.21). A particular form of $P(f)$ that embodies many desirable features is provided by a *raised-cosine (RC) spectrum*. This frequency response consists of a *flat* portion and a *roll-off* portion that has a sinusoidal form, as shown by:

$$P(f) = \begin{cases} \dfrac{1}{2W}, & 0 \le |f| < f_1 \\[2ex] \dfrac{1}{4W}\left\{1 + \cos\left[\dfrac{\pi}{2W\alpha}(|f| - f_1)\right]\right\}, & f_1 \le |f| < 2W - f_1 \\[2ex] 0, & |f| \ge 2W - f_1 \end{cases} \tag{8.22}$$

In (8.22), we have introduced a new frequency f_1 and a dimensionless parameter α, which are related by

$$\alpha = 1 - \frac{f_1}{W} \tag{8.23}$$

The parameter α is commonly called the *roll-off factor*; it indicates the *excess bandwidth* over the ideal solution, W. Specifically, the new *transmission bandwidth* is defined by

$$\begin{aligned} B_T &= 2W - f_1 \\ &= W(1+\alpha) \end{aligned} \tag{8.24}$$

The frequency response $P(f)$, normalized by multiplying it by the factor $2W$, is plotted in Figure 8.6a for $\alpha = 0$, 0.5, and 1. We see that for $\alpha = 0.5$ or 1, the frequency response $P(f)$ rolls off gradually compared with the ideal Nyquist pulse (i.e., $\alpha = 0$) and it is therefore easier to implement in practice. This roll-off is cosine-like in shape, hence the terminology "RC spectrum." Just as importantly, the $P(f)$ exhibits odd symmetry with respect to the Nyquist bandwidth W, which makes it possible to satisfy the frequency-domain condition of (8.15).

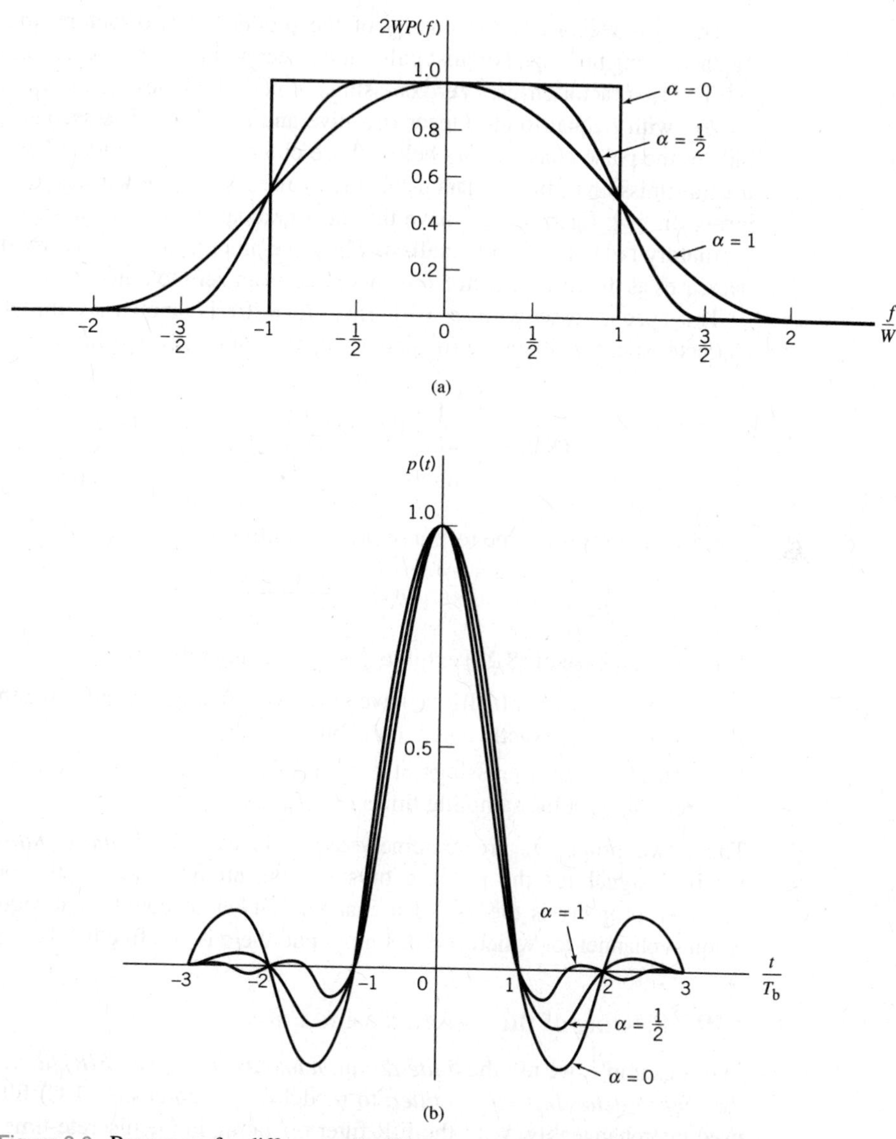

Figure 8.6 Responses for different roll-off factors: (a) frequency response; (b) time response.

The time response $p(t)$ is naturally the inverse Fourier transform of the frequency response $P(f)$. Hence, transforming the $P(f)$ defined in (8.22) into the time domain, we obtain

$$p(t) = \operatorname{sinc}(2Wt)\frac{\cos(2\pi\alpha Wt)}{1-16\alpha^2W^2t^2} \tag{8.25}$$

which is plotted in Figure 8.6b for $\alpha = 0$, 0.5, and 1.

The time response $p(t)$ consists of the product of two factors: the factor sinc$(2Wt)$ characterizing the ideal Nyquist pulse and a second factor that decreases as $1/|t|^2$ for large $|t|$. The first factor ensures zero crossings of $p(t)$ at the desired sampling instants of time $t = iT_b$, with i equal to an integer (positive and negative). The second factor reduces the tails of the pulse considerably below those obtained from the ideal Nyquist pulse, so that the transmission of binary data using such pulses is relatively insensitive to sampling time errors. In fact, for $\alpha = 1$ we have the most gradual roll-off, in that the amplitudes of the oscillatory tails of $p(t)$ are smallest. Thus, the amount of ISI resulting from timing error decreases as the roll-off factor α is increased from zero to unity.

The special case with $\alpha = 1$ (i.e., $f_1 = 0$) is known as the *full-cosine roll-off* characteristic, for which the frequency response of (8.22) simplifies to

$$P(f) = \begin{cases} \dfrac{1}{4W}\left[1 + \cos\left(\dfrac{\pi f}{2W}\right)\right], & 0 < |f| < 2W \\ 0, & |f| \geq 2W \end{cases} \tag{8.26}$$

Correspondingly, the time response $p(t)$ simplifies to

$$p(t) = \frac{\text{sinc}(4Wt)}{1 - 16W^2t^2} \tag{8.27}$$

The time response of (8.27) exhibits two interesting properties:

1. At $t = +T_b/2 = \pm 1/4W$, we have $p(t) = 0.5$; that is, the pulse width measured at half amplitude is exactly equal to the bit duration T_b.
2. There are zero crossings at $t = \pm 3T_b/2, \pm 5T_b/2, \ldots$ in addition to the usual zero crossings at the sampling times $t = \pm T_b, \pm 2T_b, \ldots$.

These two properties are extremely useful in extracting *timing information* from the received signal for the purpose of synchronization. However, the price paid for this desirable property is the use of a channel bandwidth double that required for the ideal Nyquist channel for which $\alpha = 0$: simply put, there is "no free lunch."

EXAMPLE 1 FIR Modeling of the Raised-Cosine Pulse

In this example, we use the *finite-duration impulse response (FIR) filter*, also referred to as the *tapped-delay-line (TDL) filter*, to model the raised-cosine (RC) filter; both terms are used interchangeably. With the FIR filter operating in the discrete-time domain, there are two time-scales to be considered:

1. Discretization of the input signal $a(t)$ applied to the FIR model, for which we write

$$n = \frac{t}{T} \tag{8.28}$$

where T is the sampling period in the FIR model shown in Figure 8.7. The tap inputs in this model are denoted by $a_n, a_{n-1}, \ldots, a_{n-l}, \ldots, a_{n-2l+1}, a_{n-2l}$, which, for some integer l, occupies the duration $2lT$. Note that the FIR model in Figure 8.7 is symmetric about the midpoint, a_{n-l}, which satisfies the *symmetric* structure of the RC pulse.

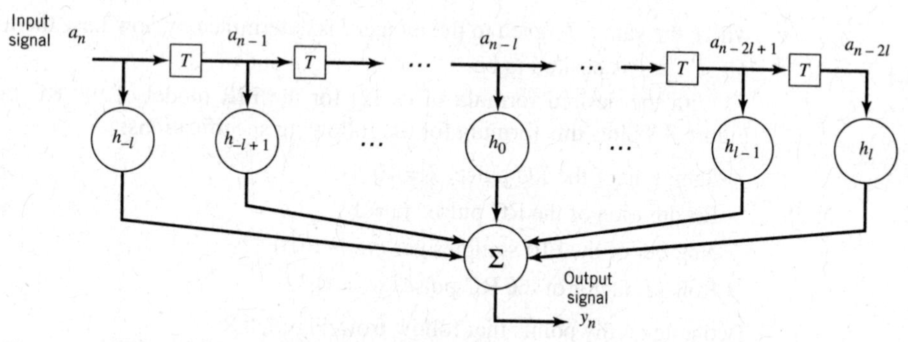

Figure 8.7 TDL model of linear time-invariant system.

2. Discretization of the RC pulse $p(t)$ for which we have

$$m = \frac{T_b}{T} \tag{8.29}$$

where T_b is the bit duration.

To model the RC pulse properly, the sampling rate of the model, $1/T$, must be higher than the bit rate, $1/T_b$. It follows therefore that the integer m defined in (8.29) must be larger than one. In assigning a suitable value to m, we must keep in mind the *tradeoff* between modeling accuracy (requiring large m) and computational complexity (preferring small m).

In any event, using (8.17), (8.28), and (8.29), obtaining the product

$$Wt = \frac{n}{2m} \tag{8.30}$$

and then substituting this result into (8.25), we get the discretized version of RC pulse as shown by

$$p_n = \text{sinc}(n/m)\left[\frac{\cos(\pi\alpha n/m)}{1 - 4\alpha^2(n/m)^2}\right], \qquad n = 0, \pm 1, \pm 2, \ldots \tag{8.31}$$

There are two computational difficulties encountered in the way in which the discretized RC pulse, p_n, is defined in (8.31):

1. The pulse p_n goes on indefinitely with increasing n.
2. The pulse is also noncausal in that the output signal y_n in Figure 8.7 is produced before the input a_n is applied to the FIR model.

To overcome difficulty 1, we truncate the sequence p_n such that it occupies a finite duration $2lT$ for some prescribed integer l, which is indeed what has been done in Figure 8.8. To mitigate the non-causality problem 2, with $T > T_b$, the ratio n/m must be replaced by $(n/m) - l$. In so doing, the truncated causal RC pulse assumes the following modified form:

$$p_n = \begin{cases} \text{sinc}\left(\dfrac{n}{m} - l\right)\left\{\dfrac{\cos\left[\pi\alpha\left(\dfrac{n}{m} - l\right)\right]}{1 - 4\alpha^2\left(\dfrac{n}{m} - l\right)^2}\right\}, & -l \le n \le l \\ 0, & \text{otherwise} \end{cases} \tag{8.32}$$

where the value assigned to the integer l is determined by how long the truncated sequence $\{p_n\}_{n=-l}^{l}$ is desired to be.

With the desired formula of (8.32) for the FIR model of the RC pulse $p(t)$ at hand, Figure 8.8 plots this formula for the following specifications:[2]

Sampling of the RC pulse, $T = 10$
Bit duration of the RC pulse, $T_b = 1$
Number of the FIR samples per bit, $m = 10$
Roll-off factor of the RC pulse, $\alpha = 0.32$

Two noteworthy points that follow from Figure 8.8:

1. The truncated causal RC pulse p_n of length $2l - 10$ is symmetric about the midpoint, $n = 5$.
2. The p_n is exactly zero at integer multiples of the bit duration T_b.

Both points reaffirm exactly what we know and therefore expect about the RC pulse $p(t)$ plotted in Figure 8.6b.

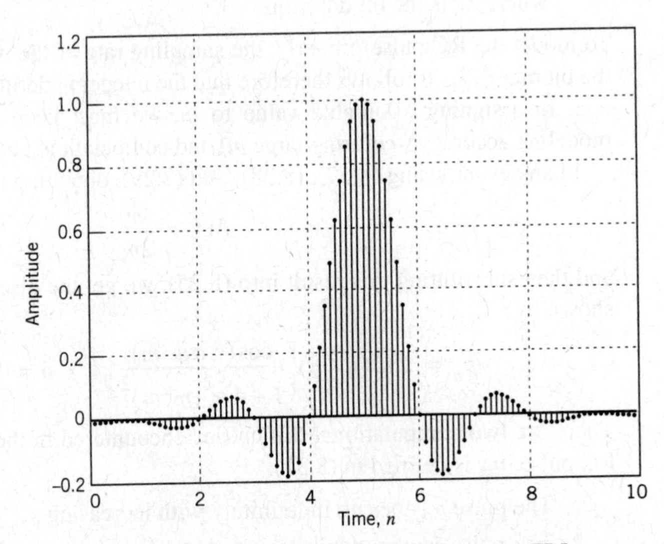

Figure 8.8 Discretized RC pulse, computed using the TDL.

8.7 Square-Root Raised-Cosine Spectrum

A more sophisticated form of pulse shaping uses the square-root raised-cosine (SRRC) spectrum[3] rather than the conventional RC spectrum of (8.22). Specifically, the spectrum of the basic pulse is now defined by the square root of the right-hand side of this equation. Thus, using the trigonometric identity

$$\cos^2\theta = \frac{1}{2}(1 + \cos 2\theta)$$

where, for the problem at hand, the angle

$$\theta = \frac{\pi}{2W\alpha}(|f| - f_1)$$
$$= \frac{\pi}{2W\alpha}[|f| - W(1-\alpha)]$$

To avoid confusion, we use $G(f)$ as the symbol for the SRRC spectrum, and so we may write

$$G(f) = \begin{cases} \dfrac{1}{\sqrt{2W}}, & 0 \le |f| \le f_1 \\ \dfrac{1}{\sqrt{2W}} \cos\left\{\dfrac{\pi}{4W\alpha}[|f| - W(1-\alpha)]\right\}, & f_1 \le |f| < 2W - f_1 \\ 0, & |f| \ge 2W - f_1 \end{cases} \tag{8.33}$$

where, as before, the roll-off factor α is defined in terms of the frequency parameter f_1 and the bandwidth W as in (8.23).

If, now, the transmitter includes a *pre-modulation filter* with the transfer function defined in (8.33) and the receiver includes an identical *post-modulation* filter, then under ideal conditions the overall pulse waveform will experience the squared spectrum $G^2(f)$, which is the regular RC spectrum. In effect, by adopting the SRRC spectrum $G(f)$ of (8.33) for pulse shaping, we would be working with $G^2(f) = P(f)$ in an overall transmitter–receiver sense. On this basis, we find that in wireless communications, for example, if the channel is affected by both fading and AWGN and the pulse-shape filtering is partitioned equally between the transmitter and the receiver in the manner described herein, then effectively the receiver would maximize the output SNR at the sampling instants.

The inverse Fourier transform of (8.33) defines the *SRRC shaping pulse*:

$$g(t) = \frac{\sqrt{2W}}{1-(8\alpha Wt)^2}\left\{\frac{\sin[2\pi W(1-\alpha)t]}{2\pi Wt} + \frac{4\alpha}{\pi}\cos[2\pi W(1+\alpha)t]\right\} \tag{8.34}$$

The important point to note here is the fact that the SRRC shaping pulse $g(t)$ of (8.34) is radically different from the conventional RC shaping pulse of (8.25). In particular, the new shaping pulse has the distinct property of satisfying the *orthogonality constraint under T-shifts*, described by

$$\int_{-\infty}^{\infty} g(t)g(t-nT)\,dt = 0 \qquad \text{for } n = \pm1, \pm2, \ldots \tag{8.35}$$

where T is the symbol duration. Yet, the new pulse $g(t)$ has exactly the same excess bandwidth as the conventional RC pulse.

It is also important to note, however, that despite the added property of orthogonality, the SRRC shaping pulse of (8.34) lacks the zero-crossing property of the conventional RC shaping pulse defined in (8.25).

Figure 8.9a plots the SRRC spectrum $G(f)$ for the roll-off factor $\alpha = 0, 0.5, 1$; the corresponding time-domain plots are shown in Figure 8.9b. These plots are naturally different from those of Figure 8.6 for nonzero α. The following example contrasts the waveform of a specific binary sequence using the SRRC shaping pulse with the corresponding waveform using the regular RC shaping pulse.

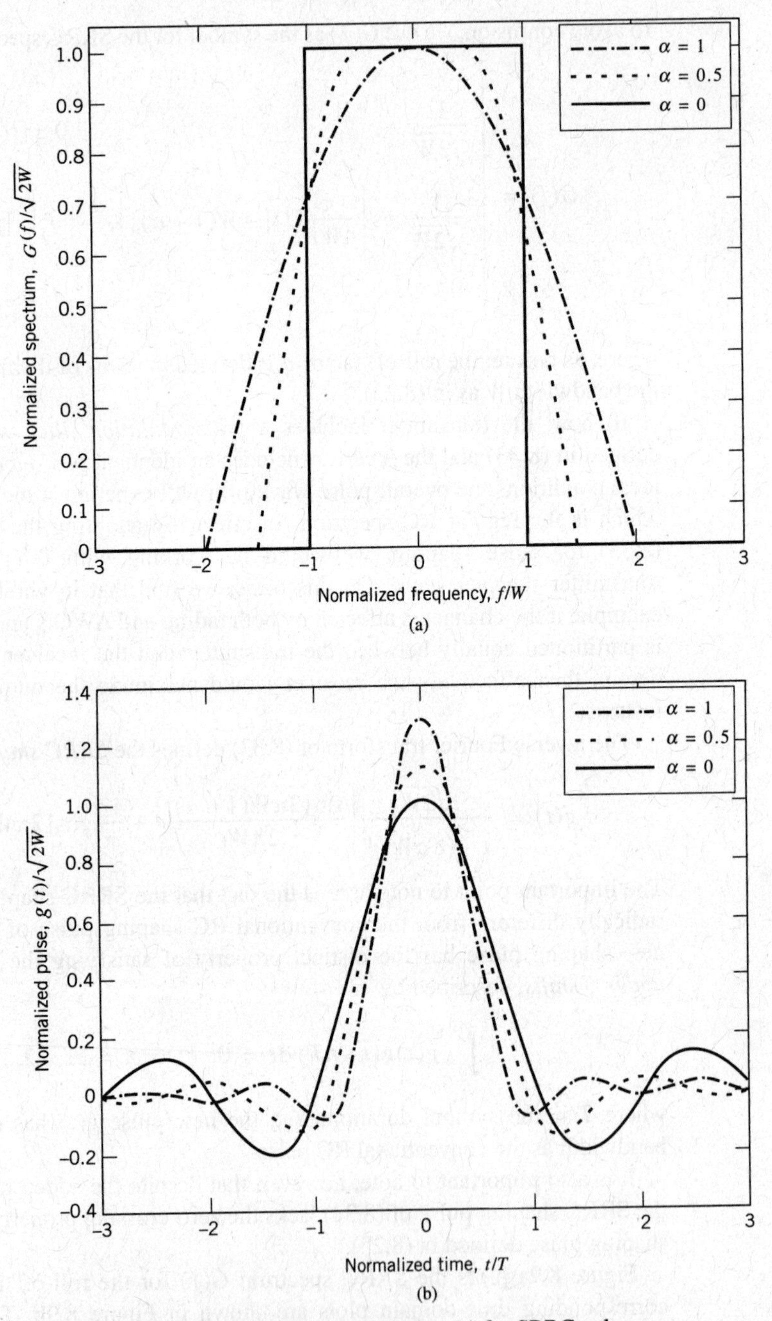

Figure 8.9 (a) $G(f)$ for SRRC spectrum. (b) $g(t)$ for SRRC pulse.

EXAMPLE 2

Pulse Shaping Comparison Between SRRC and RC

Using the SRRC shaping pulse $g(t)$ of (8.34) with roll-off factor $\alpha = 0.5$, the requirement is to plot the waveform for the binary sequence 01100 and compare it with the corresponding waveform obtained by using the conventional RC shaping pulse $p(t)$ of (8.25) with the same roll-off factor.

Using the SRRC pulse $g(t)$ of (8.34) with a multiplying plus sign for binary symbol 1 and multiplying minus sign for binary symbol 0, we get the dashed pulse train shown in Figure 8.10 for the sequence 01100. The solid pulse train shown in the figure corresponds to the use of the conventional RC pulse $p(t)$ of (8.25). The figure clearly shows that the SRRC waveform occupies a larger dynamic range than the conventional RC waveform: a feature that distinguishes one from the other.

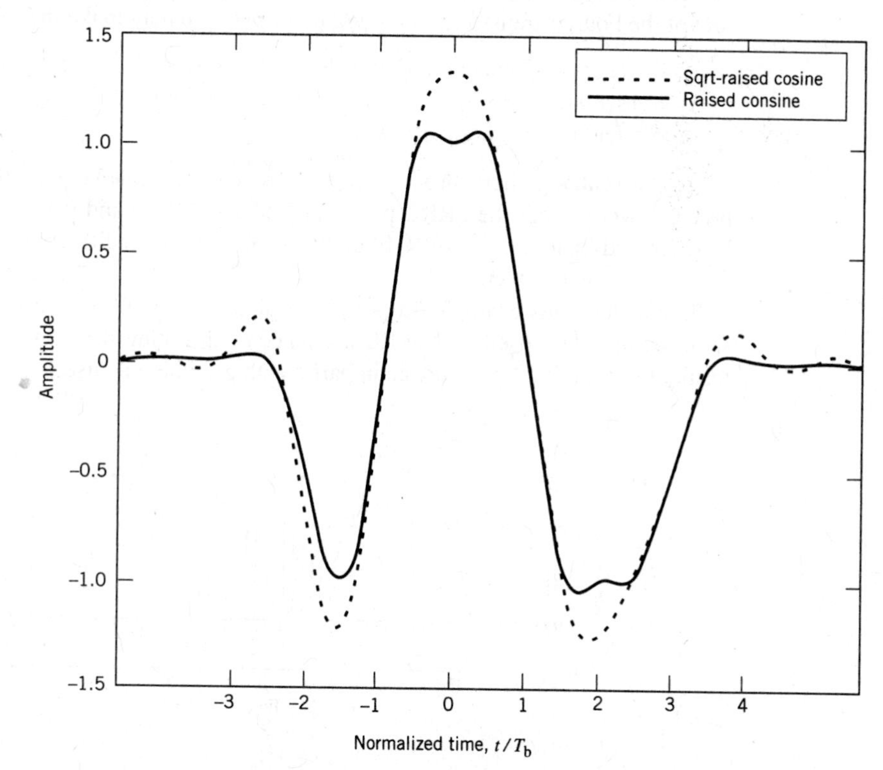

Figure 8.10 Two pulse trains for the sequence 01100, one using regular RC pulse (solid line), and the other using an SRRC pulse (dashed line).

EXAMPLE 3

FIR Modeling of the Square-Root-Raised-Cosine Pulse

In this example, we study FIR modeling of the SRRC pulse described in (8.34). To be specific, we follow a procedure similar to that used for the RC pulse $g(t)$ in Example 1, taking care of the issues of truncation and noncausality. This is done by discretizing the

SRRC pulse, $g(t)$, and substituting the dimensionless parameter, $(n/m) - l$, for Wt in (8.34). In so doing we obtain the following sequence

$$g_n = \begin{cases} \dfrac{4\alpha}{\pi\sqrt{T_b}} \dfrac{\left[\dfrac{\sin\left[\pi(1-\alpha)\left(\dfrac{n}{m}-l\right)\right]}{4\alpha\left(\dfrac{n}{m}-l\right)} + \cos\left[\pi(1+\alpha)\left(\dfrac{n}{m}-l\right)\right]\right]}{1-16\alpha^2\left(\dfrac{n}{m}-l\right)^2}, & -l \le n \le n \\ 0, & \text{otherwise} \end{cases} \tag{8.36}$$

Since, by definition, the Fourier transform of the SRRC pulse, $g(t)$, is equal to the square root of the Fourier transform of the RC pulse $p(t)$, we may make the following statement:

> The cascade connection of two identical FIR filters, each one defined by (8.36), is essentially equivalent to a TDL filter that exhibits zero intersymbol interference in accordance with (8.25).

We say "essentially" here on account of the truncation applied to both (8.32) and (8.36). In practice, when using the SRRC pulse for "ISI-free" baseband data transmission across a band-limited channel, one FIR filter would be placed in the transmitter and the other would be in the receiver.

To conclude this example, Figure 8.11a plots the SRRC sequence g_n of (8.36) for the same set of values used for the RC sequence p_n in Figure 8.8. Figure 8.11b displays the result of convolving the sequence in part a with g_n, which is, itself.

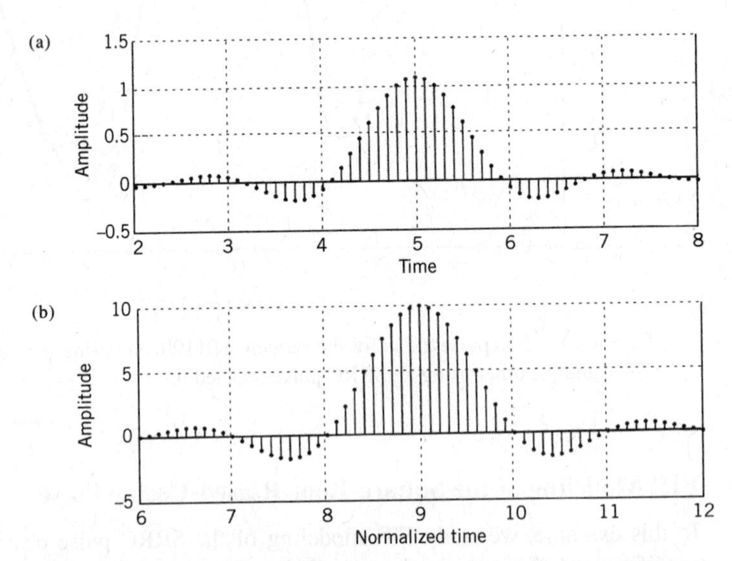

Figure 8.11 (a) Discretized SRRC pulse, computed using FIR modeling. (b) Discretized pulse resulting from the convolution of the pulse in part a with itself.

Two points are noteworthy from Figure 8.11:

1. The zero-crossings of the SRRC sequence g_n do not occur at integer multiples of the bit duration T_b, which is to be expected.
2. The sequence plotted in Figure 8.11b is essentially equivalent to the RC sequence p_n, the zero-crossings of which do occur at integer multiples of the bit duration, and so they should.

8.8 Post-Processing Techniques: The Eye Pattern

The study of signaling over band-limited channels would be incomplete without discussing the idea of *post-processing*, the essence of which is to manipulate a given set of data so as to provide a *visual interpretation* of the data rather than just numerical listing of the data. For an illustrative example, consider the formulas for the BER of digital modulation schemes operating over an AWGN channel, which were summarized in Table 7.7 of Chapter 7. The graphical plots of the schemes, shown in Figure 7.47, provide an immediate comparison on how these different modulation schemes compete with each other in terms of performance measured on the basis of their respective BERs for varying E_b/N_0. In other words, there is much to be gained from graphical plots that are most conveniently made possible by computation.

What we have in mind in this section, however, is the description of a commonly used *post-processor*, namely eye patterns, which are particularly suited for the experimental study of digital communication systems.

The *eye pattern*, also referred to as the *eye diagram*, is produced by the *synchronized superposition of (as many as possible) successive symbol intervals of the distorted waveform appearing at the output of the receive filter prior to thresholding*. As an illustrative example, consider the distorted, but noise-free, waveform shown in part a of Figure 8.12. Part b of the figure displays the corresponding synchronized superposition of the waveform's eight binary symbol intervals. The resulting display is called an "eye pattern" because of its resemblance to a human eye. By the same token, the interior of the eye pattern is called the *eye opening*.

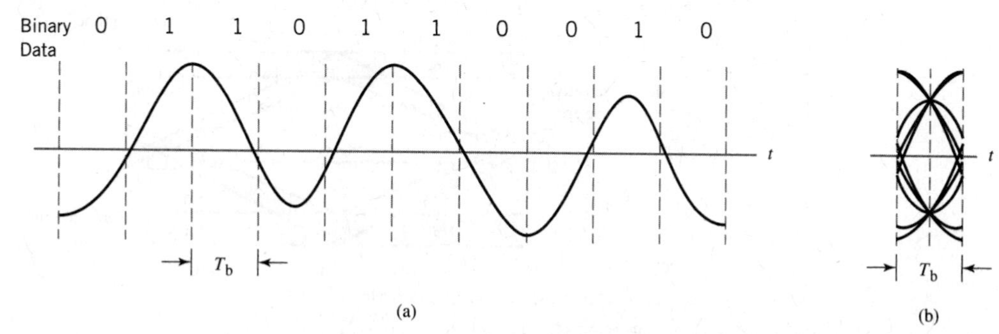

Figure 8.12 (a) Binary data sequence and its waveform. (b) Corresponding eye pattern.

As long as the additive channel noise is not large, then the eye pattern is well defined and may, therefore, be studied experimentally on an oscilloscope. The waveform under study is applied to the deflection plates of the oscilloscope with its time-base circuit operating in a synchronized condition. From an experimental perspective, the eye pattern offers two compelling virtues:

- The simplicity of eye-pattern generation.
- The provision of a great deal of insightful information about the characteristics of the data transmission system. Hence, the wide use of eye patterns as a visual indicator of how well or poorly a data transmission system performs the task of transporting a data sequence across a physical channel.

Timing Features

Figure 8.13 shows a generic eye pattern for distorted but noise-free binary data. The horizontal axis, representing time, spans the symbol interval from $-T_b/2$ to $T_b/2$, where T_b is the bit duration. From this diagram, we may infer three timing features pertaining to a binary data transmission system, exemplified by a PAM system:

1. *Optimum sampling time.* The width of the eye opening defines the time interval over which the distorted binary waveform appearing at the output of the receive filter in the PAM system can be uniformly sampled without decision errors. Clearly, the *optimum sampling time* is the time at which the eye opening is at its widest.
2. *Zero-crossing jitter.* In practice, the timing signal (for synchronizing the receiver to the transmitter) is extracted from the *zero-crossings* of the waveform that appears at the receive-filter output. In such a form of synchronization, there will always be irregularities in the zero-crossings, which, in turn, give rise to *jitter* and, therefore, nonoptimum sampling times.
3. *Timing sensitivity.* Another timing-related feature is the sensitivity of the PAM system to *timing errors*. This sensitivity is determined by the rate at which the eye pattern is closed as the sampling time is varied.

Figure 8.13 indicates how these three timing features of the system (and other insightful attributes) can be measured from the eye pattern.

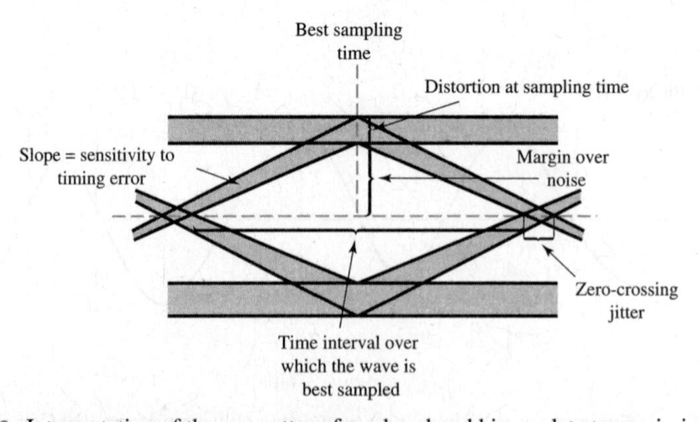

Figure 8.13 Interpretation of the eye pattern for a baseband binary data transmission system.

The Peak Distortion for Intersymbol Interference

Hereafter, we assume that the ideal signal amplitude is scaled to occupy the range from –1 to +1. We then find that, in the absence of channel noise, the eye opening assumes two extreme values:

1. *An eye opening of unity,*[4] which corresponds to zero ISI.
2. *An eye opening of zero*, which corresponds to a completely closed eye pattern; this second extreme case occurs when the effect of intersymbol interference is severe enough for some upper traces in the eye pattern to cross with its lower traces.

It is indeed possible for the receiver to make decision errors even when the channel is noise free. Typically, *an eye opening of 0.5 or better is considered to yield reliable data transmission.*

In a noisy environment, the extent of eye opening at the optimum sampling time provides a measure of the operating margin over additive channel noise. This measure, as illustrated in Figure 8.13, is referred to as the *noise margin.*

From this discussion, it is apparent that the eye opening plays an important role in assessing system performance; hence the need for a formal definition of the eye opening. To this end, we offer the following definition:

$$\text{Eye opening} = 1 - D_{\text{peak}} \tag{8.37}$$

where D_{peak} denotes a new criterion called the *peak distortion.* The point to note here is that peak distortion is a *worst-case* criterion for assessing the effect of ISI on the performance (i.e., error rate) of a data transmission system. The relationship between the eye opening and peak distortion is illustrated in Figure 8.14. With the eye opening being dimensionless, the peak distortion is dimensionless too. To emphasize this statement, the two extreme values of the eye opening translate as follows:

1. *Zero peak distortion*, which occurs when the eye opening is unity.
2. *Unity peak distortion*, which occurs when the eye pattern is completely closed.

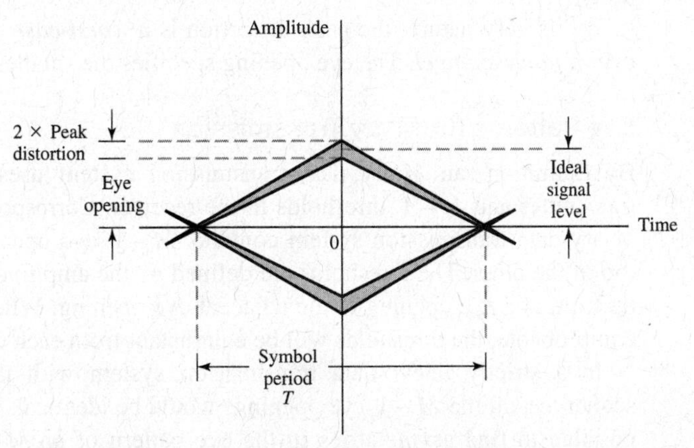

Figure 8.14 Illustrating the relationship between peak distortion and eye opening. *Note:* the ideal signal level is scaled to lie inside the range –1 to +1.

With this background, we offer the following definition:

> The peak distortion is the maximum value assumed by the intersymbol interference over all possible transmitted sequences, with this maximum value divided by a normalization factor equal to the absolute value of the corresponding signal level idealized for zero intersymbol interference.

Referring to (8.10), the two components embodied in this definition are themselves defined as follows:

1. The idealized signal component of the receive filter output is defined by the first term in (8.10), namely a_i, where a_i is the ith encoded symbol and unit transmitted signal energy per bit.
2. The intersymbol interference is defined by the second term, namely

$$\sum_{\substack{k=-\infty \\ k \neq i}}^{\infty} a_k p_{i-k} \tag{8.38}$$

where p_{i-k} stands for the term $p[(i-k)T_b]$. The maximum value of this summation occurs when each encoded symbol a_k has the same algebraic sign as $p_i - k$. Therefore,

$$\text{Maximum ISI} = \sum_{\substack{k=-\infty \\ k \neq i}}^{\infty} |p_{i-k}| \tag{8.39}$$

Hence, invoking the definition of peak distortion, we get the desired formula:

$$D_{\text{peak}} = \sum_{\substack{k=-\infty \\ k \neq i}}^{\infty} |p_{i-k}| \tag{8.40}$$

where $p_0 = 1$ for all $i = k$. Note that, by involving the assumption of a signal amplitude from -1 to $+1$, we have scaled the transmitted signal energy for a binary symbol to be unity.

By its very nature, the peak distortion is a *worst-case criterion for data transmission over a noisy channel*. The eye opening specifies the smallest possible noise margin.

Eye Patterns for *M*-ary Transmission

By definition, an M-ary data transmission system uses M encoded symbols in the transmitter and $M-1$ thresholds in the receiver. Correspondingly, the eye pattern for an M-ary data transmission system contains $M-1$ eye openings stacked vertically one on top of the other. The thresholds are defined by the amplitude-transition levels as we move up from one eye opening to the adjacent eye opening. When the encoded symbols are all equiprobable, the thresholds will be equidistant from each other.

In a strictly linear data transmission system with truly transmitted random data sequences, all the $M-1$ eye openings would be identical. In practice, however, it is often possible to find asymmetries in the eye pattern of an M-ary data transmission system, which are caused by nonlinearities in the communication channel or other distortion-sensitive parts of the system.

EXAMPLE 4

Eye Patterns for Binary and Quaternary Systems

Figure 8.15a and b depict the eye patterns for a baseband PAM transmission system using $M = 2$ and $M = 4$, respectively. The channel has no bandwidth limitation and the source symbols used are obtained from a random number generator. An RC pulse is used in both cases. The system parameters used for the generation of these eye patterns are a bit rate of 1 Hz and roll-off factor $\alpha = 0.5$. For the binary case of $M = 2$ in Figure 8.15a,

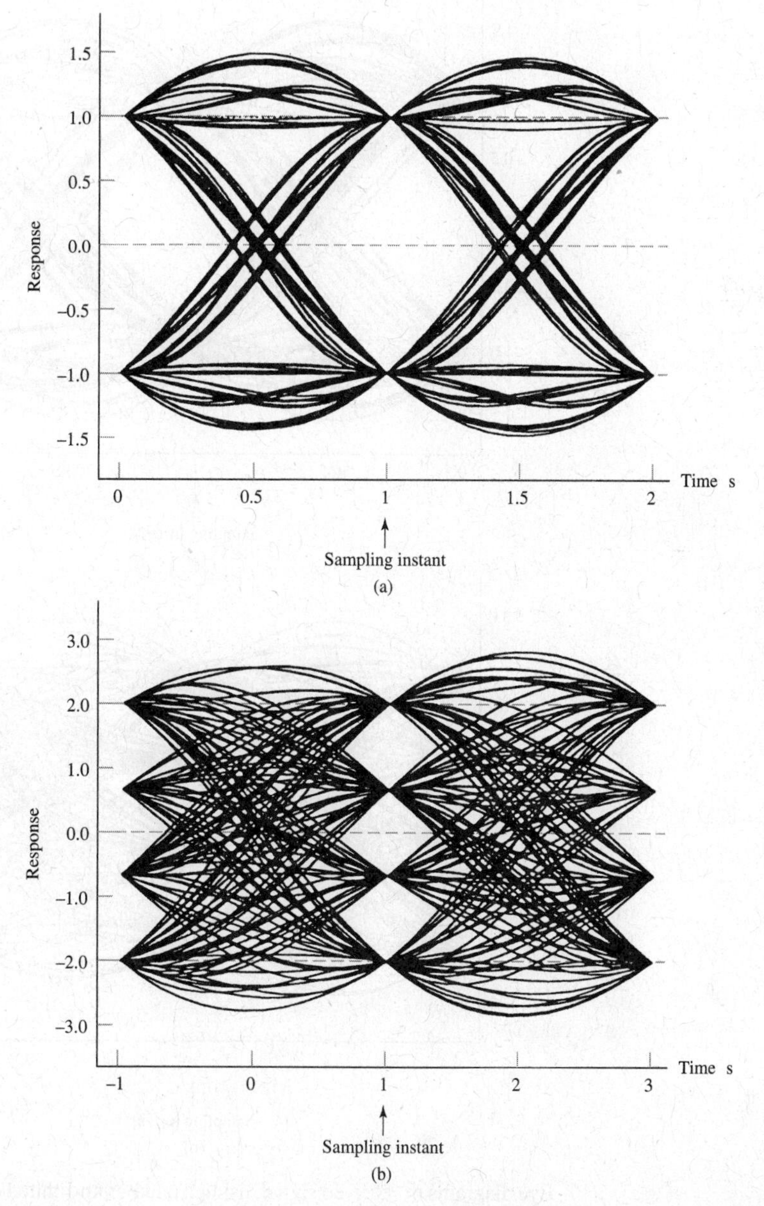

Figure 8.15 Eye diagrams of received signal with no bandwidth limitation: (a) $M = 2$; (b) $M = 4$.

the symbol duration T and the bit duration T_b are the same, with $T_b = 1$ s. For the case of $M = 4$ in Figure 8.15b we have $T = T_b \log_2 M = 2T_b$. In both cases we see that the eyes are open, indicating perfectly reliable operation of the system, perfect in the sense that the ISI is zero.

Figure 8.16a and b show the eye patterns for these two baseband-pulse transmission systems using the same system parameters as before, but this time under a bandwidth-

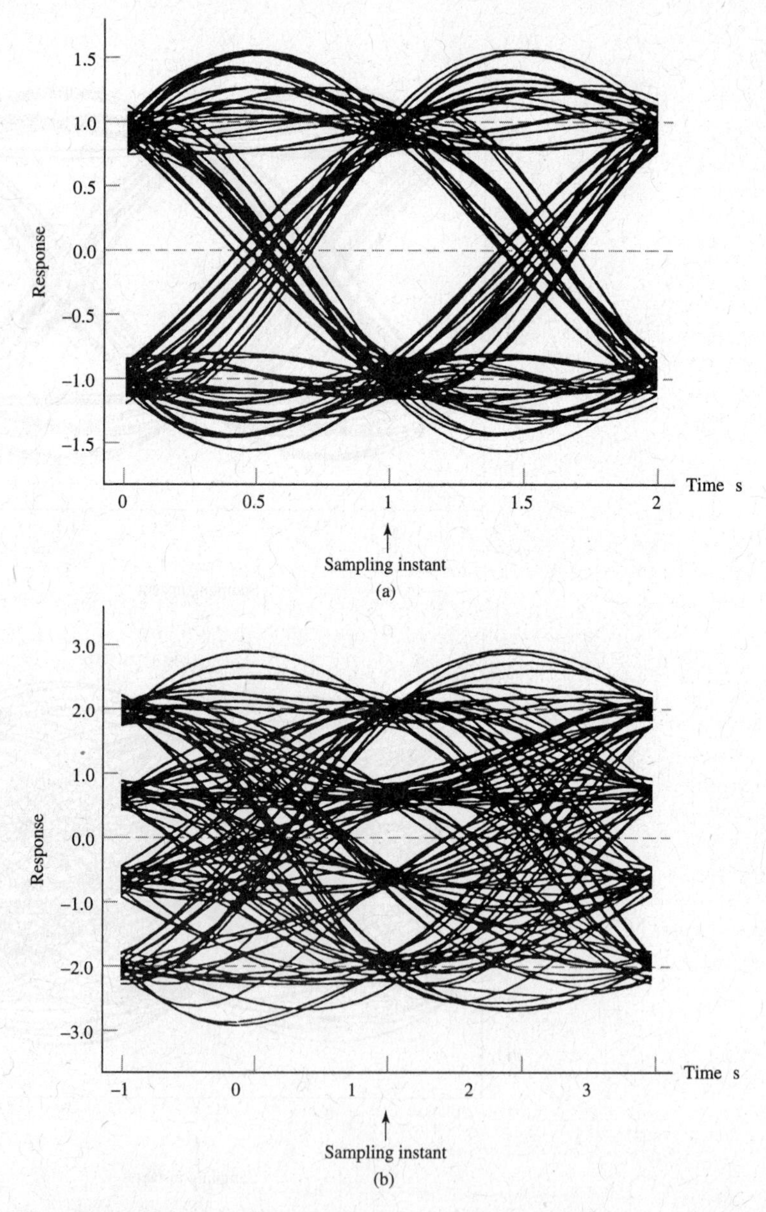

Figure 8.16 Eye diagrams of received signal, using a bandwidth-limited channel: (a) $M = 2$; (b) $M = 4$.

limited condition. Specifically, the channel is now modeled by a low-pass *Butterworth filter*, whose frequency response is defined by

$$|H(f)| = \frac{1}{1 + (f/f_0)^{2N}}$$

where N is the order of the filter, and f_0 is the 3-dB cutoff frequency of the filter. For the results displayed in Figure 8.16, the following filter parameter values were used:

$$N = 3, \text{ and } f_0 = 0.6 \text{ Hz for binary PAM}$$
$$N = 3, \text{ and } f_0 = 0.3 \text{ Hz for 4-PAM}$$

With the roll-off factor $\alpha = 0.5$ and Nyquist bandwidth $W = 0.5$ Hz, for binary PAM, the use of (8.24) defines the transmission bandwidth of the PAM transmission system to be

$$B_T = 0.5(1 + 0.5) = 0.75 \text{ Hz}$$

Although the channel bandwidth cutoff frequency is greater than absolutely necessary, its effect on the passband is observed in a decrease in the size of the eye opening. Instead of the distinct values at time $t = 1$ s, shown in Figure 8.15a and b, now there is a blurred region. If the channel bandwidth were to be reduced further, the eye would close even more until finally no distinct eye opening would be recognizable.

8.9 Adaptive Equalization

In this section we develop a simple and yet effective algorithm for the adaptive equalization of a linear channel of unknown characteristics. Figure 8.17 shows the structure of an adaptive synchronous equalizer, which incorporates the matched filtering action. The algorithm used to adjust the equalizer coefficients assumes the availability of a desired response. One's first reaction to the availability of a replica of the transmitted signal is: If such a signal is available at the receiver, why do we need adaptive equalization? To answer this question, we first note that a typical telephone channel changes little during an average data call. Accordingly, prior to data transmission, the equalizer is adjusted under the

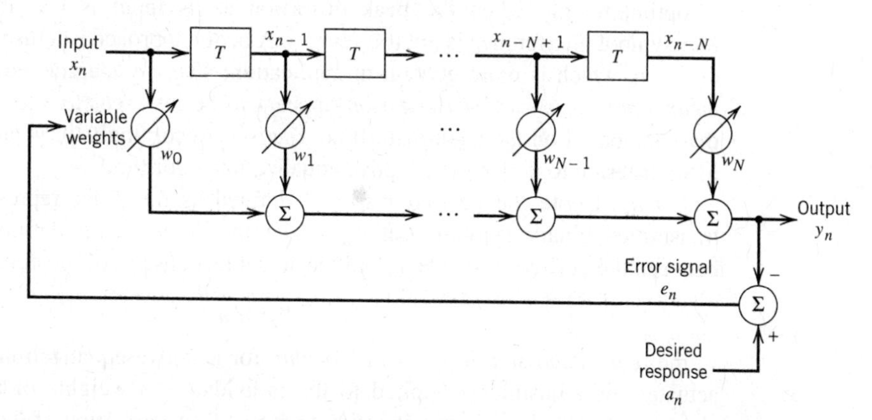

Figure 8.17 Block diagram of adaptive equalizer using an adjustable TDL filter.

guidance of a *training sequence* transmitted through the channel. A synchronized version of this training sequence is generated at the receiver, where (after a time shift equal to the transmission delay through the channel) it is applied to the equalizer as the desired response. A training sequence commonly used in practice is the *pseudonoise* (*PN*) *sequence*, which consists of a deterministic periodic sequence with noise-like characteristics. Two identical PN sequence generators are used, one at the transmitter and the other at the receiver. When the training process is completed, the PN sequence generator is switched off and the adaptive equalizer is ready for normal data transmission. A detailed description of PN sequence generators is presented in Appendix J.

Least-Mean-Square Algorithm (Revisited)

To simplify notational matters, we let

$$x_n = x(nT)$$

$$y_n = y(nT)$$

Then, the output y_n of the tapped-delay-line (TDL) equalizer in response to the input sequence $\{x_n\}$ is defined by the discrete convolution sum (see Figure 8.17)

$$y_n = \sum_{k=0}^{N} w_k x_{n-k} \tag{8.41}$$

where w_k is the weight at the kth tap and $N + 1$ is the total number of taps. The tap weights constitute the adaptive equalizer coefficients. We assume that the input sequence x_n has finite energy. We have used a notation for the equalizer weights in Figure 8.17 that is different from the corresponding notation in Figure 6.17 to emphasize the fact that the equalizer in Figure 8.17 also incorporates matched filtering.

The adaptation may be achieved by observing the error between the desired pulse shape and the actual pulse shape at the equalizer output, measured at the sampling instants, and then using this error to estimate the direction in which the tap weights of the equalizer should be changed so as to approach an optimum set of values. For the adaptation, we may use a criterion based on minimizing the *peak distortion*, defined as the worst-case intersymbol interference at the output of the equalizer. However, the equalizer so designed is optimum only when the peak distortion at its input is less than 100% (i.e., the intersymbol interference is not too severe). A better approach is to use a mean-square error criterion, which is more general in application; also, an adaptive equalizer based on the *mean-square error (MSE) criterion* appears to be less sensitive to timing perturbations than one based on the peak-distortion criterion. Accordingly, in what follows we use the MSE criterion to derive the adaptive equalization algorithm.

Let a_n denote the *desired response* defined as the polar representation of the nth transmitted binary symbol. Let e_n denote the *error signal* defined as the difference between the desired response a_n and the actual response y_n of the equalizer, as shown by

$$e_n = a_n - y_n \tag{8.42}$$

In the *least-mean-square* (*LMS*) *algorithm* for adaptive equalization, the error signal e_n actuates the adjustments applied to the individual tap weights of the equalizer as the algorithm proceeds from one iteration to the next. A derivation of the LMS algorithm for adaptive prediction was presented in Section 6.7 of Chapter 6. Recasting (6.85) into its most general form, we may restate the formula for the LMS algorithm in words as follows:

$$\begin{pmatrix}\text{Updated value}\\\text{of }k\text{th tap weight}\end{pmatrix} = \begin{pmatrix}\text{Old value of}\\k\text{th tap weight}\end{pmatrix} + \begin{pmatrix}\text{Step-size}\\\text{parameter}\end{pmatrix}\begin{pmatrix}\text{Input signal applied}\\\text{to }k\text{th tap weight}\end{pmatrix}\begin{pmatrix}\text{Error}\\\text{signal}\end{pmatrix} \tag{8.43}$$

Let μ denote the step-size parameter. From Figure 8.17 we see that the input signal applied to the kth tap weight at time step n is x_{n-k}. Hence, using $\hat{w}_k(n)$ as the old value of the kth tap weight at time step n, the updated value of this tap weight at time step $n + 1$ is, in light of (8.43), defined by

$$\hat{w}_{k,n+1} = \hat{w}_{k,n} + \mu x_{n-k} e_n, \qquad k = 0, 1, \dots, N \tag{8.44}$$

where

$$e_n = a_n - \sum_{k=0}^{N} \hat{w}_{k,n} x_{n-k} \tag{8.45}$$

These two equations constitute the *LMS algorithm for adaptive equalization*.

We may simplify the formulation of the LMS algorithm using matrix notation. Let the $(N + 1)$-by-1 vector $\mathbf{x}_n$ denote the tap inputs of the equalizer:

$$\mathbf{x}_n = [x_n, \dots, x_{n-N+1}, x_{n-N}]^{\mathrm{T}} \tag{8.46}$$

where the superscript T denotes matrix transposition. Correspondingly, let the $(N + 1)$-by-1 vector $\hat{\mathbf{w}}_n$ denote the tap weights of the equalizer:

$$\hat{\mathbf{w}}_n = [\hat{w}_{0,n}, \hat{w}_{1,n}, \dots, \hat{w}_{N,n}]^{\mathrm{T}} \tag{8.47}$$

We may then use matrix notation to recast the discrete convolution sum of (8.41) in the compact form

$$y_n = \mathbf{x}_n^{\mathrm{T}} \hat{\mathbf{w}}_n \tag{8.48}$$

where $\mathbf{x}_n^{\mathrm{T}} \hat{\mathbf{w}}_n$ is referred to as the *inner product* of the vectors $\mathbf{x}_n$ and $\hat{\mathbf{w}}_n$. We may now summarize the LMS algorithm for adaptive equalization as follows:

1. Initialize the algorithm by setting $\hat{\mathbf{w}}_1 = \mathbf{0}$ (i.e., set all the tap weights of the equalizer to zero at $n = 1$, which corresponds to time $t = T$.
2. For $n = 1, 2, \dots,$ compute

$$y_n = \mathbf{x}_n^{\mathrm{T}} \hat{\mathbf{w}}_n$$
$$e_n = a_n - y_n$$
$$\hat{\mathbf{w}}_{n+1} = \hat{\mathbf{w}}_n + \mu e_n \mathbf{x}_n$$

   where μ is the step-size parameter.
3. Continue the iterative computation until the equalizer reaches a "steady state," by which we mean that the actual mean-square error of the equalizer essentially reaches a constant value.

The LMS algorithm is an example of a feedback system, as illustrated in the block diagram of Figure 8.18, which pertains to the kth filter coefficient. It is therefore possible for the algorithm to diverge (i.e., for the adaptive equalizer to become unstable).

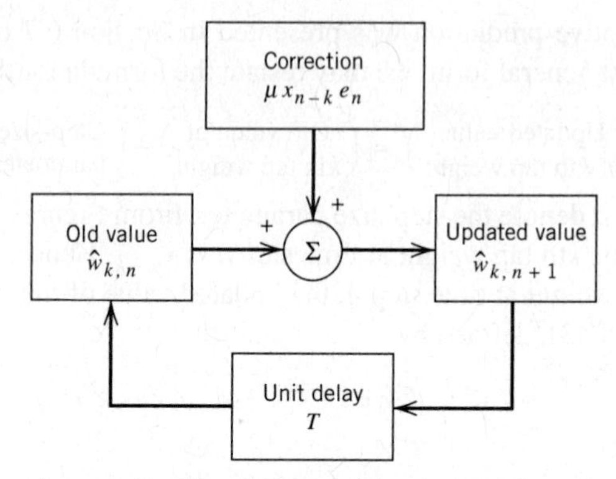

Figure 8.18 Signal-flow graph representation of the LMS algorithm involving the kth tap weight.

Unfortunately, the convergence behavior of the LMS algorithm is difficult to analyze. Nevertheless, provided that the step-size parameter μ is assigned a small value, we find that after a large number of iterations the behavior of the LMS algorithm is roughly similar to that of the *steepest-descent algorithm* (discussed in Chapter 6), which uses the actual gradient rather than a noisy estimate for the computation of the tap weights.

Operation of the Equalizer

There are two modes of operation for an adaptive equalizer, namely the training mode and decision-directed mode, as shown in Figure 8.19. During the *training mode*, a known PN sequence is transmitted and a synchronized version of it is generated in the receiver, where (after a time shift equal to the transmission delay) it is applied to the adaptive equalizer as the desired response; the tap weights of the equalizer are thereby adjusted in accordance with the LMS algorithm.

When the training process is completed, the adaptive equalizer is switched to its second mode of operation: the *decision-directed mode*. In this mode of operation, the error signal is defined by

$$e_n = \hat{a}_n - y_n \tag{8.49}$$

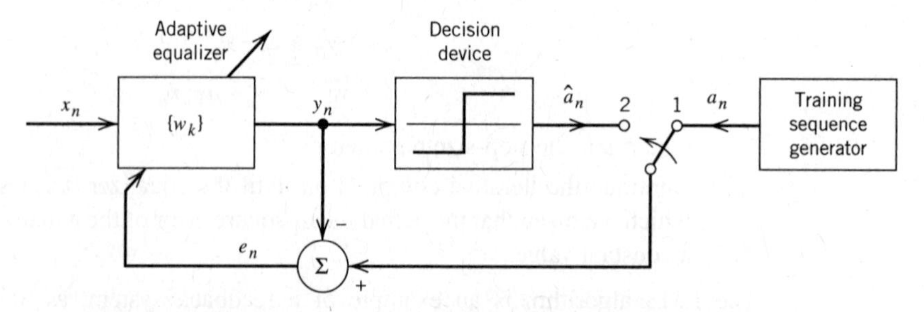

Figure 8.19 Illustrating the two operating modes of an adaptive equalizer: for the training mode, the switch is in position 1; for the tracking mode, it is moved to position 2.

where y_n is the equalizer output at time $t = nT$ and $\hat{a}_n$ is the final (not necessarily) correct estimate of the transmitted symbol a_n. Now, in normal operation the decisions made by the receiver are correct with high probability. This means that the error estimates are correct most of the time, thereby permitting the adaptive equalizer to operate satisfactorily. Furthermore, an adaptive equalizer operating in a decision-directed mode is able to track relatively slow variations in channel characteristics.

It turns out that the larger the step-size parameter μ is, the faster the tracking capability of the adaptive equalizer. However, a large step-size parameter μ may result in an unacceptably high excess *mean-square error*, defined as that part of the mean-square value of the error signal in excess of the minimum attainable value, which results when the tap weights are at their optimum settings. We therefore find that, in practice, the choice of a suitable value for the step-size parameter μ involves making a compromise between fast tracking and reducing the excess mean-square error.

Decision-Feedback Equalization[5]

To develop further insight into adaptive equalization, consider a baseband channel with impulse response denoted in its sampled form by the sequence $\{h_n\}$, where $h_n = h(nT)$. The response of this channel to an input sequence $\{x_n\}$, in the absence of noise, is given by the discrete convolution sum

$$\begin{aligned} y_n &= \sum_k h_k x_{n-k} \\ &= h_0 x_n + \sum_{k<0} h_k x_{n-k} + \sum_{k>0} h_k x_{n-k} \end{aligned} \tag{8.50}$$

The first term of (8.50) represents the desired data symbol. The second term is due to the *precursors* of the channel impulse response that occur before the main sample h_0 associated with the desired data symbol. The third term is due to the *postcursors* of the channel impulse response that occur after the main sample h_0. The precursors and postcursors of a channel impulse response are illustrated in Figure 8.20. The idea of *decision-feedback equalization* is to use data decisions made on the basis of precursors of

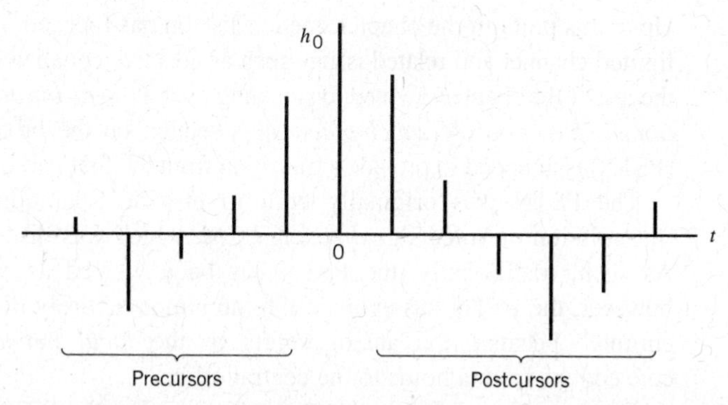

Figure 8.20 Impulse response of a discrete-time channel, depicting the precursors and postcursors.

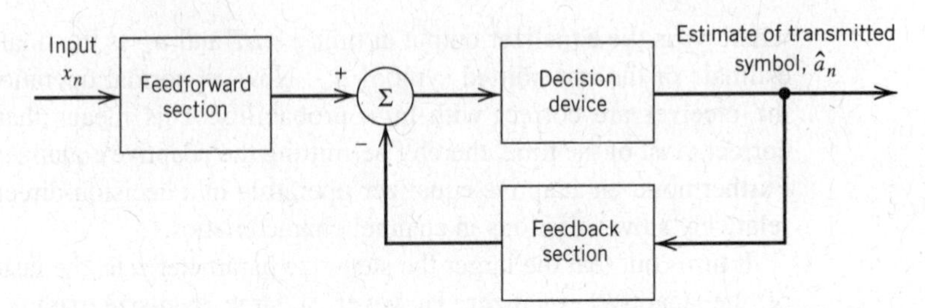

Figure 8.21 Block diagram of decision-feedback equalizer.

the channel impulse response to take care of the postcursors; for the idea to work, however, the decisions would obviously have to be correct for the DFE to function properly most of the time.

A DFE consists of a feedforward section, a feedback section, and a decision device connected together as shown in Figure 8.21. The feedforward section consists of a TDL filter whose taps are spaced at the reciprocal of the signaling rate. The data sequence to be equalized is applied to this section. The feedback section consists of another TDL filter whose taps are also spaced at the reciprocal of the signaling rate. The input applied to the feedback section consists of the decisions made on previously detected symbols of the input sequence. The function of the feedback section is to subtract out that portion of the intersymbol interference produced by previously detected symbols from the estimates of future samples.

Note that the inclusion of the decision device in the feedback loop makes the equalizer intrinsically *nonlinear* and, therefore, more difficult to analyze than an ordinary LMS equalizer. Nevertheless, the mean-square error criterion can be used to obtain a mathematically tractable optimization of a DFE. Indeed, the LMS algorithm can be used to jointly adapt both the feedforward tap weights and the feedback tap weights based on a *common* error signal.

8.10 Broadband Backbone Data Network: Signaling over Multiple Baseband Channels

Up to this point in the chapter, the discussion has focused on signaling over a *single* band-limited channel and related issues such as adaptive equalization. In order to set the stage for the rest of the chapter devoted to signaling over a *linear broadband channel purposely partitioned into a set of subchannels*, this section on the broadband backbone data network (PSTN) is intended to provide a transition from the first part of the chapter to the second part.

The PSTN was originally built to provide a ubiquitous structure for the digital transmission of voice signals using PCM, which was discussed previously in Chapter 6. As such, traditionally, the PSTN has been viewed as an analog network. In reality, however, the PSTN has evolved into an almost entirely digital network. We say "almost entirely" because the analog refers to the *local network*, which stands for short-connections from a home to the central office.

For many decades past, data transmission over the PSTN relied on the use of *modems*; the term "modem" is a contraction of modulator–demodulator. Despite the enormous effort that was put into the design of modems, they could not cope with the ever-increasing rate of data transmission. This situation prevailed until the advent of *digital subscriber line (DSL) technology* in the 1990s. The inquisitive reader may well ask the question: How was it that the modem theorists and designers got it wrong while the DSL theorists and designers got it right? Unfortunately, in the development of modems, the telephone channel was treated as one whole entity. On the other hand, the development of DSL abandoned the traditional approach by viewing the telephone channel as a conglomeration of subchannels extending over a wide frequency band and operating in parallel, and with each subchannel treated as a narrowband channel, thereby exploiting Shannon's information capacity law in a much more effective manner.

It is therefore not surprising that the DSL technology has converted an ordinary telephone line into a broadband communication link, so much so that we may now view the PSTN effectively as a *broadband backbone data network*, which is being widely used all over the world. The data consist of digital signals generated by computers or Internet service providers (ISPs). Most importantly, the deployment of DSL technology has literally made it possible to increase the rate of data transmission across a telephone channel by orders of magnitude compared with the old modems. This transition from modem to DSL technology is indeed an impressive engineering accomplishment, which resulted from "thinking outside the box."

With this brief historical account, it is apropos that we devote the rest of the chapter to the underlying theory of the widely used DSL technology.

8.11 Digital Subscriber Lines

The term *DSL* is commonly used to refer to a family of different technologies that operate over a *local loop* less than 1.5 km to provide for digital signal transmission between a user terminal (e.g., computer) and the central office (CO) of a telephone company. Through the CO, the user is connected directly to the so-called broadband backbone data network, whereby transmission is maintained in the digital domain. In the course of transmission, the digital signal is switched and routed at regular intervals. Figure 8.22 is a schematic diagram illustrating that typically the data rate *upstream* (i.e., in the direction of the ISP) is lower than the data rate *downstream* (i.e., in the direction of the user). It is for this reason that the DSL is said to be *asymmetric*;[6] hence the acronym ADSL.

The twisted wire-pair used in the local loop, the only analog part of the data transmission system as remarked earlier, is *inductively loaded*. Specifically, extra inductance is purposely supplied by local coils, which are inserted at regular intervals across the wire-pair. This addition is made in order to produce a fairly flat frequency

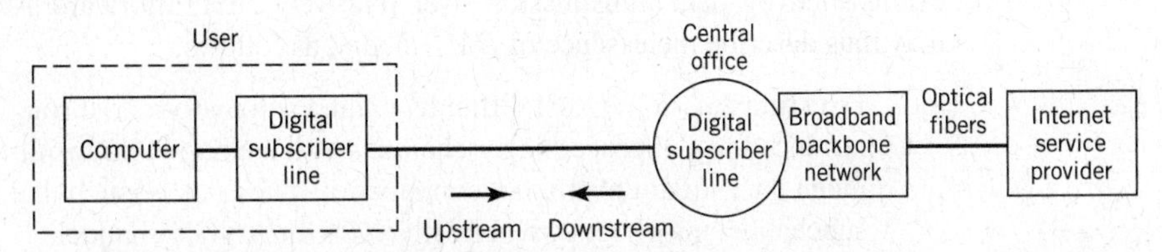

Figure 8.22 Block diagram depicting the operational environment of DSL.

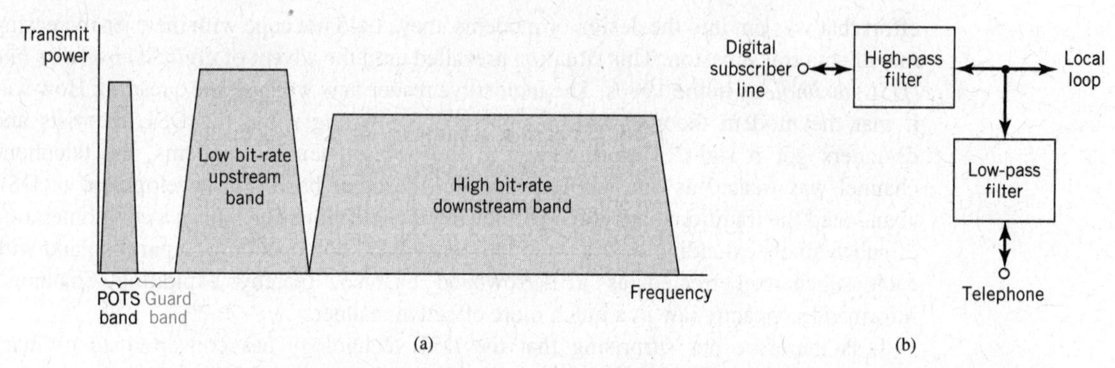

Figure 8.23 (a) Illustrating the different band allocations for an FDM-based ADSL system. (b) Block diagram of splitter performing the function of a multiplexer or demultiplexer. *Note:* both filters in the splitter are bidirectional filters.

response across the effective voice band. However, the improvement so gained for the transmission of voice signals is attained at the expense of continually increasing attenuation at frequencies higher than 3.4 kHz. Figure 8.23a illustrates the two different frequency bands allocated to a frequency-division multiplexive (FDM)-based ADSL; the way in which two filters, one high-pass and the other low-pass, are used to connect the DSL to the local loop is shown in Figure 8.23b.

With access to the wide band represented by frequencies higher than 3.6 kHz, the DSL uses discrete multicarrier transmission (DMT) techniques to convert the twisted wire-pair in the local loop into a broadband communication link; the two terms "multichannel" and "multicarrier" are used interchangeably. The net result is that data rates of 1.5 to 9.0 Mbps downstream in a bandwidth of up to 1 MHz and over a distance of 2.7 to 5.5 km. *Very high-bit-rate digital subscriber lines*[7] (VDSLs) do even better, supporting data rates of 13 to 52 Mbps downstream in a bandwidth of up to 30 MHz and over a distance of 0.3 to 1.5 km. These numbers indicate that the data rates attainable by DSL technology depend on both bandwidth and distance, and the technology continues to improve.

The basic idea behind DMT is rooted in a commonly used engineering paradigm:

> Divide and conquer.

According to this paradigm, a difficult problem is solved by dividing it into a number of simpler problems and then combining the solutions to those simple problems. In the context of our present discussion, the difficult problem is that of data transmission over a wideband channel with severe intersymbol interference, and the simpler problems are exemplified by data transmission over relatively straightforward AWGN channels. We may thus describe the essence of *DMT theory*, as follows:

> Data transmission over a difficult channel is transformed through the use of advanced signal processing techniques into the parallel transmission of the given data stream over a large number of subchannels, such that each subchannel may be viewed effectively as an AWGN channel.

Naturally, the overall data rate is the sum of the individual data rates over the subchannels designed to operate in parallel: this new way of thinking on signaling over wideband channels is entirely different from the approach described in the first part of the chapter, in that it builds on ideas described in Chapter 5 on Shannon's information theory and in Chapter 7 on signaling over AWGN channels.

8.12 Capacity of AWGN Channel Revisited

At the heart of discrete multichannel data transmission theory is *Shannon's information capacity law,* discussed in Chapter 5 on information theory. According to this law, the capacity of an AWGN channel (free from ISI) is defined by

$$C = B \log_2(1 + \mathrm{SNR}) \text{ bits/s} \tag{8.51}$$

where B is the channel bandwidth in hertz and SNR is measured at the channel output. Equation (8.51) teaches us that, for a given SNR, we can transmit data over an AWGN channel of bandwidth B at the maximum rate of B bit/s with arbitrarily small probability of error, provided that we employ an encoding system of sufficiently high complexity. Equivalently, we may express the capacity C in bits per transmission of channel use as

$$C = \frac{1}{2}\log_2(1 + \mathrm{SNR}) \quad \text{bits per transmission} \tag{8.52}$$

In practice, we usually find that a physically realizable encoding system must transmit data at a rate R less than the maximum possible rate C for it to be *reliable*. For an implementable system operating at low enough probability of symbol error, we thus need to introduce an *SNR gap* or just *gap*, denoted by Γ. The gap is a function of the permissible probability of symbol error P_e and the encoding system of interest. It provides a measure of the "efficiency" of an encoding system with respect to the ideal transmission system of (8.52). With C denoting the capacity of the ideal encoding system and R denoting the capacity of the corresponding implementable encoding system, the gap is defined by

$$\begin{aligned}\Gamma &= \frac{2^{2C} - 1}{2^{2R} - 1} \\ &= \frac{\mathrm{SNR}}{2^{2R} - 1}\end{aligned} \tag{8.53}$$

Rearranging (8.53) with R as the focus of interest, we may write

$$R = \frac{1}{2}\log_2\left(1 + \frac{\mathrm{SNR}}{\Gamma}\right) \quad \text{bits per transmission} \tag{8.54}$$

For an encoded PAM or QAM operating at $P_e = 10^{-6}$, for example, the gap Γ is constant at 8.8 dB. Through the use of codes (e.g., trellis codes to be discussed in Chapter 10), the gap Γ may be reduced to as low as 1 dB.

Let P denote the transmitted signal power and σ^2 denote the channel noise variance measured over the bandwidth B. The SNR is therefore

$$\mathrm{SNR} = \frac{P}{\sigma^2}$$

where

$$\sigma^2 = N_0 B$$

We may thus finally define the attainable data rate as

$$R = \frac{1}{2}\log_2\left(1 + \frac{P}{\Gamma\sigma^2}\right) \quad \text{bits per transmission} \tag{8.55}$$

With this modified version of Shannon's information capacity law at hand, we are ready to describe discrete multichannel modulation in quantitative terms.

8.13 Partitioning Continuous-Time Channel into a Set of Subchannels

To be specific in practical terms, consider a linear wideband channel (e.g., twisted wire-pair) with an arbitrary frequency response $H(f)$. Let the magnitude response of the channel, denoted by $|H(f)|$, be approximated by a staircase function as illustrated in Figure 8.24, with Δf denoting the width of each frequency step (i.e., subchannel). In the limit, as the frequency increment Δf approaches zero, the staircase approximation of the channel approaches the actual $H(f)$. Along each step of the approximation, the channel may be assumed to operate as an AWGN channel free from intersymbol interference. The problem of transmitting a single wideband signal is thereby transformed into the transmission of a set of narrowband orthogonal signals. Each orthogonal narrowband signal, with its own carrier, is generated using a spectrally efficient modulation technique such as M-ary QAM, with AWGN being essentially the only primary source of transmission impairment. This scenario, in turn, means that data transmission over each subchannel of bandwidth Δf can be optimized by invoking a modified form of Shannon's information capacity law, with the optimization of each subchannel being performed independently of all the others. Thus, in practical signal-processing terms, we may make the following statement:

> The need for complicated equalization of a wideband channel is replaced by the need for multiplexing and demultiplexing the transmission of an incoming data stream over a large number of narrowband subchannels that are continuous and disjoint.

Although the resulting complexity of a DMT system so described is indeed high for a large number of subchannels, implementation of the entire system can be accomplished in a cost-effective manner through the combined use of efficient digital signal-processing algorithms and very-large-scale integration technology.

Figure 8.25 shows a block diagram of the *DMT* system in its most basic form. The system configured here uses QAM, whose choice is justified by virtue of its spectral efficiency. The incoming binary data stream is first applied to a demultiplexer (not shown in the figure), thereby producing a set of N substreams. Each substream represents a sequence of two-element subsymbols, which, for the symbol interval $0 \leq t \leq T$, is denoted by

$$(a_n, b_n), \qquad n = 1, 2, \ldots, N$$

where a_n and b_n are element values along the two coordinates of subchannel n.

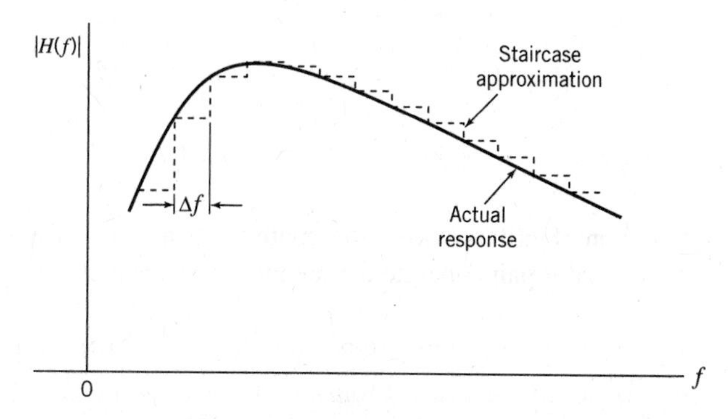

Figure 8.24 Staircase approximation of an arbitrary magnitude response of a channel, |H(f)|; only positive-frequency portion of the response is shown.

Correspondingly, the *passband basis functions* of the quadrature-amplitude modulators are defined by the following function pairs:

$$\{\phi(t)\cos(2\pi f_n t),\ \phi(t)\sin(2\pi f_n t)\},\qquad n = 1, 2, \ldots, N \tag{8.56}$$

The carrier frequency f_n of the nth modulator described in (8.56) is an integer multiple of the symbol rate $1/T$, as shown by

$$f_n = \frac{n}{T},\qquad n = 1, 2, \ldots, N$$

and the low-pass function $\phi(t)$, common to all the subchannels, is the sinc function

$$\phi(t) = \sqrt{\frac{2}{T}}\operatorname{sinc}\left(\frac{t}{T}\right),\qquad -\infty < t < \infty \tag{8.57}$$

The passband basis functions defined here have the following desirable properties, whose proofs are presented as an end-of-chapter problem.

PROPERTY 1 *For each n, the two quadrature-modulated sinc functions form an orthogonal pair, as shown by*

$$\int_{-\infty}^{\infty}[\phi(t)\cos(2\pi f_n t)][\phi(t)\sin(2\pi f_n t)]\,dt = 0\qquad \text{for all } n \tag{8.58}$$

This orthogonal relationship provides the basis for formulating the signal constellation for each of the N modulators in the form of a squared lattice.

PROPERTY 2 *Recognizing that*

$$\exp(j2\pi f_n t) = \cos(2\pi f_n t) + j\sin(2\pi f_n t)$$

we may completely redefine the passband basis functions in the complex form

$$\left\{\frac{1}{\sqrt{2}}\phi(t)\exp(j2\pi f_n t)\right\},\qquad n = 1, 2, \ldots, N \tag{8.59}$$

where the factor $1/\sqrt{2}$ *has been introduced to ensure that the scaled function* $\phi(t)/\sqrt{2}$ *has unit energy. Hence, these passband basis functions form an orthonormal set, as shown by*

$$\int_{-\infty}^{\infty}\left[\frac{1}{\sqrt{2}}\phi(t)\exp(\mathrm{j}2\pi f_n t)\right]\left[\frac{1}{\sqrt{2}}\phi(t)\exp(\mathrm{j}2\pi f_k t)\right]^* dt = \begin{cases} 1, & k = n \\ 0, & k \neq n \end{cases} \tag{8.60}$$

The asterisk assigned to the second factor on the left-hand side denotes complex conjugation.

Equation (8.60) provides the mathematical basis for ensuring that the N modulator-demodulator pairs operate independently of each other.

PROPERTY 3 *The set of channel-output functions $\{h(t) \star \phi(t)\}$ remains orthogonal for a linear channel with arbitrary impulse response $h(t)$, where $\star$ denotes convolution.*

Thus, in light of these three properties, the original wideband channel is partitioned into an ideal setting of independent subchannels operating in continuous time.

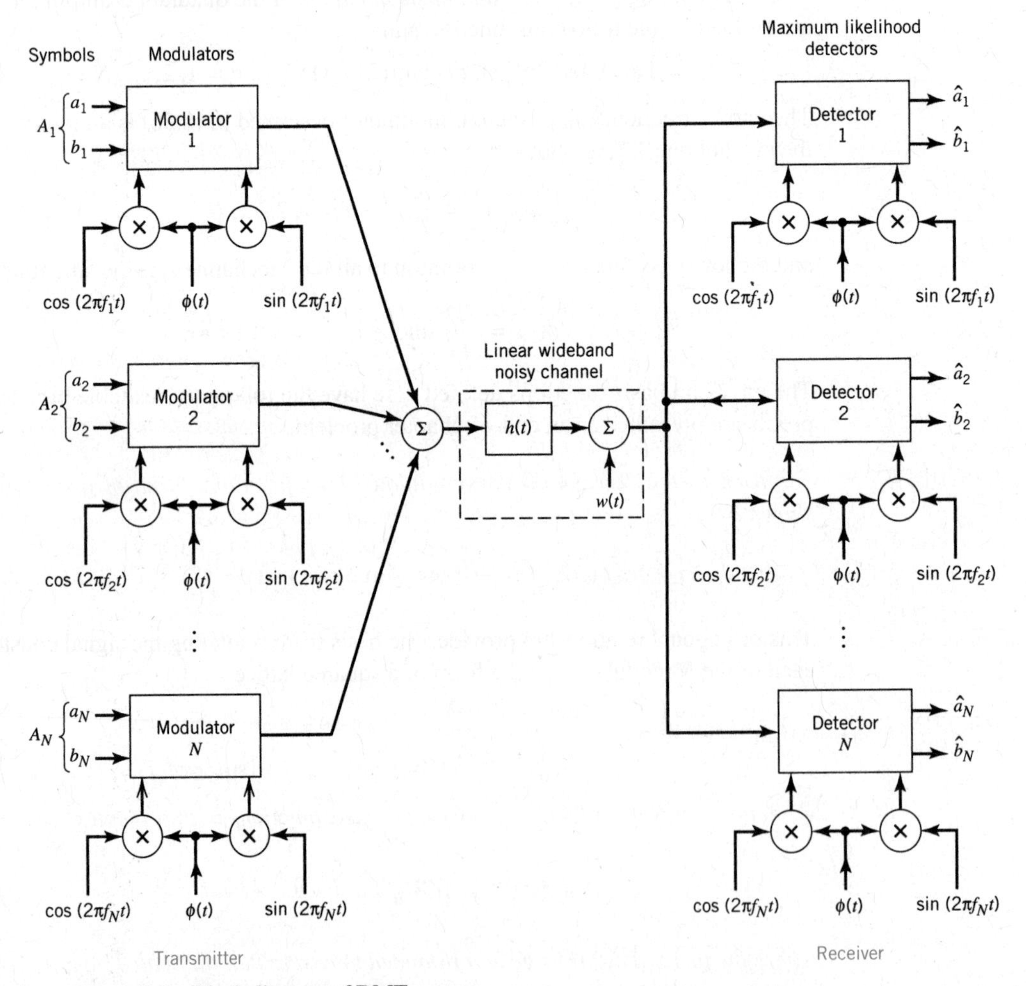

Figure 8.25 Block diagram of DMT system.

Figure 8.25 also includes the corresponding structure of the receiver. It consists of a bank of N coherent detectors, with the channel output being simultaneously applied to the detector inputs, operating in parallel. Each detector is supplied with a locally generated pair of quadrature-modulated sinc functions operating in synchrony with the pair of passband basis function applied to the corresponding modulator in the transmitter.

It is possible for each subchannel to have some residual ISI. However, as the number of subchannels N approaches infinity, the ISI disappears for all practical purposes. From a theoretical perspective, we find that, for a sufficiently large N, the bank of coherent detectors in Figure 8.25 operates as *maximum likelihood detectors*, operating independently of each other and on a subsymbol-by-subsymbol basis. (Maximum likelihood detection was discussed in Chapter 7.)

To define the detector outputs in response to the input subsymbols, we find it convenient to use complex notation. Let A_n denote the subsymbol applied to the nth modulator during the symbol interval $0 \le t \le T$, as shown by

$$A_n = a_n + jb_n, \qquad n = 1, 2, \ldots, N \tag{8.61}$$

The corresponding detector output is expressed as follows:

$$Y_n = H_n A_n + W_n, \qquad n = 1, 2, \ldots, N \tag{8.62}$$

where H_n is the complex-valued frequency response of the channel evaluated at the subchannel carrier frequency $f = f_n$, that is,

$$H_n = H(f_n), \qquad n = 1, 2, \ldots, N \tag{8.63}$$

The W_n in (8.62) is a complex-valued random variable produced by the channel noise $w(t)$; the real and imaginary parts of W_n have zero mean and variance $N_0/2$. With knowledge of the measured frequency response $H(f)$ available, we may therefore use (8.62) to compute a maximum likelihood estimate of the transmitted subsymbol A_n. The estimates $\hat{A}_1, \hat{A}_2, \ldots, \hat{A}_N$ so obtained are finally multiplexed to produce the overall estimate of the original binary data transmitted during the interval $0 \le t \le T$.

To summarize, for a sufficiently large N, we may implement the receiver as an *optimum maximum likelihood detector* that operates as N subsymbol-by-subsymbol detectors. The rationale for building a maximum likelihood receiver in such a simple way is motivated by the following property:

PROPERTY 4 *The passband basis functions constitute an orthonormal set and their orthogonality is maintained for any channel impulse response $h(t)$.*

Geometric SNR

In the DMT system of Figure 8.25, each subchannel is characterized by an SNR of its own. It would be highly desirable, therefore, to derive a single measure for the performance of the entire system in Figure 8.25.

To simplify the derivation of such a measure, we assume that all of the subchannels in Figure 8.25 are represented by one-dimensional constellations. Then, using the modified Shannon information capacity law of (8.55), the channel capacity of the entire system is successively expressed as follows:

$$\begin{aligned}R &= \frac{1}{N}\sum_{n=1}^{N} R_n \\ &= \frac{1}{2N}\sum_{n=1}^{N} \log_2\left(1 + \frac{P_n}{\Gamma\sigma_n^2}\right) \\ &= \frac{1}{2N}\log_2\left[\prod_{n=1}^{N}\left(1 + \frac{P_n}{\Gamma\sigma_n^2}\right)\right] \\ &= \frac{1}{2}\log_2\left[\prod_{n=1}^{N}\left(1 + \frac{P_n}{\Gamma\sigma_n^2}\right)\right]^{1/N} \quad \text{bits per transmission}\end{aligned} \tag{8.64}$$

Let $(\text{SNR})_{\text{overall}}$ denote the overall SNR of the entire DMT system. Then, in light of (8.54), we may express the rate R as

$$R = \frac{1}{2}\log_2\left(1 + \frac{(\text{SNR})_{\text{overall}}}{\Gamma}\right) \quad \text{bits per transmission} \tag{8.65}$$

Accordingly, comparing (8.65) with (8.64) and rearranging terms, we may write

$$(\text{SNR})_{\text{overall}} = \Gamma\left[\prod_{n=1}^{N}\left(1 + \frac{P_n}{\Gamma\sigma_n^2}\right)^{1/N} - 1\right] \tag{8.66}$$

Assuming that the SNR, namely $P_n/(\Gamma\sigma_n^2)$, is large enough to ignore the two unity terms on the right-hand side of (8.66), we may approximate the overall SNR simply as follows:

$$(\text{SNR})_{\text{overall}} \approx \prod_{n=1}^{N}\left(\frac{P_n}{\sigma_n^2}\right)^{1/N} \tag{8.67}$$

which is independent of the gap Γ. We may thus characterize the overall system by an SNR that is the *geometric mean* of the SNRs of the individual subchannels.

The geometric form of the SNR of (8.67) can be improved considerably by distributing the available transmit power among the N subchannels on a nonuniform basis. This objective is attained through the use of loading, which is discussed next.

Loading of the DMT System

Equation (8.64) for the bit rate of the entire DMT system ignores the effect of the channel on system performance. To account for this effect, define

$$g_n = |H(f_n)|, \quad n = 1, 2, \ldots, N \tag{8.68}$$

Then, assuming that the number of subchannels N is large enough, we may treat g_n as a constant over the entire bandwidth Δf assigned to subchannel n for all n. In such a case, we may modify the second line of (8.64) for the overall SNR of the system into

$$R = \frac{1}{2N}\sum_{n=1}^{N}\log_2\left(1+\frac{g_n^2 P_n}{\Gamma\sigma_n^2}\right) \tag{8.69}$$

where the g_n^2 and Γ are usually fixed. The noise variance σ_n^2 is $\Delta f N_0$ for all n, where Δf is the bandwidth of each subchannel and $N_0/2$ is the noise power spectral density of the subchannel. We may therefore optimize the overall bit rate R through a proper allocation of the total transmit power among the various subchannels. However, for this optimization to be of practical value, we must maintain the total transmit power at some constant value denoted by P, as shown by

$$\sum_{n=1}^{N} P_n = P \tag{8.70}$$

The optimization we therefore have to deal with is a *constrained optimization problem*, stated as follows:

> Maximize the bit rate R for the DMT system through an optimal sharing of the total transmit power P between the N subchannels, subject to the constraint that the total transmit power P is maintained constant.

To solve this optimization problem, we first use the *method of Lagrange multipliers*[8] to set up an objective function (i.e., the *Lagrangian*) that incorporates (8.69) and the constraint of (8.70) as shown by

$$\begin{aligned} J &= \frac{1}{2N}\sum_{n=1}^{N}\log_2\left(1+\frac{g_n^2 P_n}{\Gamma\sigma_n^2}\right) + \lambda\left(P - \sum_{n=1}^{N} P_n\right) \\ &= \frac{1}{2N}\log_2 e\sum_{n=1}^{N}\log_e\left(1+\frac{g_n^2 P_n}{\Gamma\sigma_n^2}\right) + \lambda\left(P - \sum_{n=1}^{N} P_n\right) \end{aligned} \tag{8.71}$$

where λ is the *Lagrange multiplier*; in the second line of (8.71) the logarithm to base 2 has been changed to the natural logarithm written as $\log_2 e$. Hence, differentiating the Lagrangian J with respect to P_n, then setting the result equal to zero and finally rearranging terms, we get

$$\frac{\frac{1}{2N}\log_2 e}{P_n + \frac{\Gamma\sigma_n^2}{g_n^2}} = \lambda \tag{8.72}$$

The result of (8.72) indicates that the solution to our constrained optimization problem is to have

$$P_n + \frac{\Gamma\sigma_n^2}{g_n^2} = K \qquad \text{for } n = 1, 2, \ldots, N \tag{8.73}$$

where K is a prescribed constant under the designer's control. That is, the sum of the transmit power and the noise variance (power) scaled by the ratio Γ/g_n^2 must be maintained constant for each subchannel. The process of allocating the transmit power P to the individual subchannels so as to maximize the bit rate of the entire multichannel transmission system is called *loading*; this term is not to be confused with loading coils used in twisted wire-pairs.

8.14 Water-Filling Interpretation of the Constrained Optimization Problem

In solving the constrained optimization problem just described, the two conditions of (8.70) and (8.73) must both be satisfied. The optimum solution so defined has an interesting interpretation, as illustrated in Figure 8.26 for $N = 6$, assuming that the gap Γ is maintained constant over all the subchannels. To simplify the illustration in Figure 8.26, we have set $\sigma_n^2 = N_0 \Delta f = 1$; that is, the average noise power is unity for all N subchannels. Referring to this figure, we may now make three observations:

1. With $\sigma_n^2 = 1$, the sum of power P_n allocated to subchannel n and the scaled noise power Γ/g_n^2 satisfies the constraint of (8.73) for four of the subchannels for a prescribed value of the constant K.
2. The sum of power allocations to these four subchannels consumes all the available transmit power, maintained at the constant value P.
3. The remaining two subchannels have been eliminated from consideration because they would each require negative power to satisfy (8.73) for the prescribed value of the constant K; from a physical perspective, this condition is clearly unacceptable.

The interpretation illustrated in Figure 8.26 prompts us to refer to the optimum solution of (8.73), subject to the constraint of (8.70), as the *water-filling solution*; the principle of water-filling was discussed under Shannon's information theory in Chapter 5. This terminology

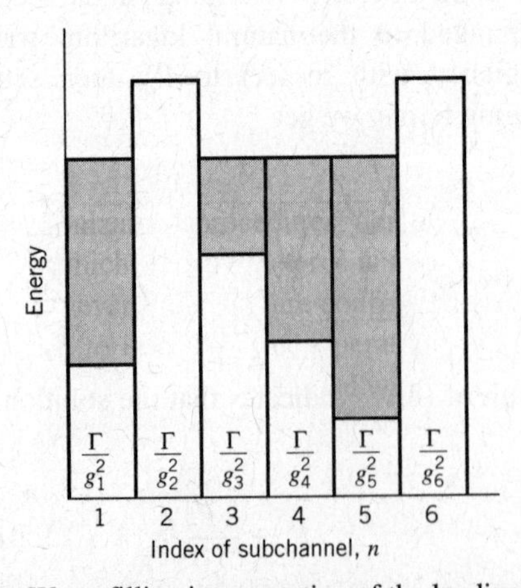

Figure 8.26 Water-filling interpretation of the loading problem.

follows from analogy of our optimization problem with a fixed amount of water—standing for transmit power—being poured into a container with a number of connected regions, each having a different depth—standing for noise power. In such a scenario, the water distributes itself in such a way that a constant water level is attained across the whole container, hence the term "water filling."

Returning to the task of how to allocate the fixed transmit power P among the various subchannels of a multichannel data transmission system so as to optimize the bit rate of the entire system, we may proceed along the following pair of steps:

1. Let the total transmit power be fixed at the constant value P as in (8.70).
2. Let K denote the constant value prescribed for the sum, $P_n + \Gamma\sigma_n^2/g_n^2$, for all n as in (8.73).

On the basis of these two steps, we may then set up the following system of simultaneous equations:

$$\begin{aligned} P_1 + P_2 + \cdots\ P_N &= P \\ P_1 - K &= -\Gamma\sigma^2/g_1^2 \\ P_2 - K &= -\Gamma\sigma^2/g_2^2 \\ &\vdots \\ P_N - K &= -\Gamma\sigma^2/g_N^2 \end{aligned} \tag{8.74}$$

where we have a total of $(N + 1)$ unknowns and $(N + 1)$ equations to solve for them. Using matrix notation, we may rewrite this system of $N + K$ simultaneous equations in the compact form

$$\begin{bmatrix} 1 & 1 & \cdots & 1 & 0 \\ 1 & 0 & \cdots & 0 & -1 \\ 0 & 1 & \cdots & 0 & -1 \\ \vdots & \vdots & & \vdots & \vdots \\ 0 & 0 & \cdots & 1 & -1 \end{bmatrix} \begin{bmatrix} P_1 \\ P_2 \\ P_3 \\ \vdots \\ K \end{bmatrix} = \begin{bmatrix} P \\ -\Gamma\sigma^2/g_1^2 \\ -\Gamma\sigma^2/g_2^2 \\ \vdots \\ -\Gamma\sigma^2/g_N^2 \end{bmatrix} \tag{8.75}$$

Premultiplying both sides of (8.75) by the inverse of the $(N + 1)$-by-$(N + 1)$ matrix on the left-hand side of the equation, we obtain solutions for the unknowns $P_1, P_2, \ldots, P_N$, and K. We should always find that K is positive, but it is possible for some of the P values to be negative. In such a situation, the negative P values are discarded as power cannot be negative for physical reasons.

EXAMPLE 5 **Linear Channel with Squared Magnitude Response**

Consider a linear channel whose squared magnitude response $|H(f)|^2$ has the piecewise-linear form shown in Figure 8.27. To simplify the example, we have set the gap $\Gamma = 1$ and the noise variance $\sigma^2 = 1$.

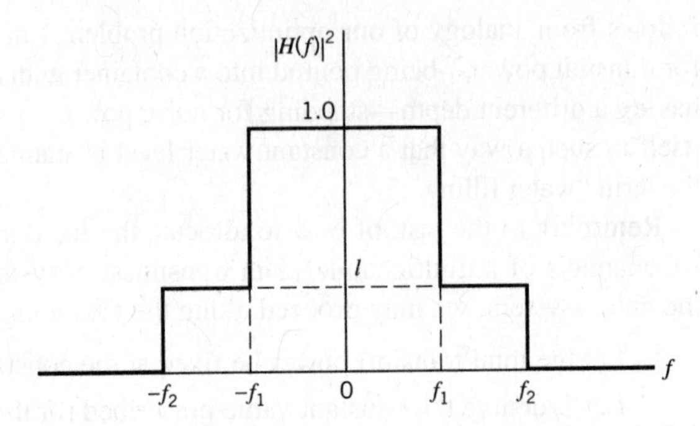

Figure 8.27 Squared magnitude response for Example 5.

Under this set of values, the application of (8.74) yields

$$P_1 + P_2 = P$$
$$P_1 - K = -1$$
$$P_2 - K = -1/l$$

where the new parameter $0 < l < 1$ has been introduced to distinguish the third equation from the second one. Solving these three simultaneous equations for P_1, P_2, and K, we get

$$P_1 = \frac{1}{2}\left(P - 1 + \frac{1}{l}\right)$$

$$P_2 = \frac{1}{2}\left(P + 1 - \frac{1}{l}\right)$$

$$K = \frac{1}{2}\left(P + 1 + \frac{1}{l}\right)$$

Since $0 < l < 1$, it follows that $P_1 > 0$, but it is possible for P_2 to be negative. This latter condition can arise if

$$l < \frac{1}{P+1}$$

But then P_1 exceeds the prescribed value of transmit power P. Therefore, it follows that, in this example, the only acceptable solution is to have $1/(P + 1) < l < 1$. Suppose then we have $P = 10$ and $l = 0.1$; under these two conditions the desired solution is

$$K = 10.5$$
$$P_1 = 9.5$$
$$P_2 = 0.5$$

The corresponding water-filling picture for the problem at hand is portrayed in Figure 8.28.

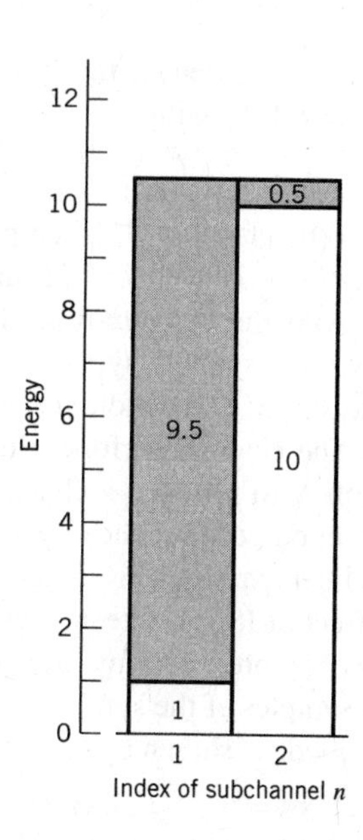

Figure 8.28
Water-filling profile for Example 5.

8.15 DMT System using Discrete Fourier Transform

The material presented in Sections 8.13 and 8.14 provides an insightful introduction to the notion of *multicarrier modulation* in a DMT system. In particular, the continuous-time channel partitioning induced by the passband (modulated) basis functions of (8.56), or equivalently (8.59) in complex terms, exhibits a highly desirable property described as follows:

> Orthogonality of the basis functions, and therefore the channel partitioning, is preserved despite their individual convolutions with the impulse response of the channel.

However, the DSL system so described has two practical shortcomings:

1. The passband basis functions use a sinc function that is nonzero for an infinite time interval, whereas practical considerations favor a finite observation interval.
2. For a finite number of subchannels N the system is suboptimal; optimality of the system is assured only when N approaches infinity.

We may overcome both shortcomings by using DMT, the basic idea of which is to transform a noisy wideband channel into a set of N subchannels operating in parallel. What makes DMT distinctive is the fact that the transformation is performed in discrete time as well as discrete frequency, paving the way for exploiting digital signal processing. Specifically, the transmitter's input–output behavior of the entire communication system admits a linear matrix representation, which lends itself to implementation using the DFT. In the following we know from Chapter 2 on Fourier analysis of signals and systems that the DFT is the result of discretizing the Fourier transform both in time and frequency.

To exploit this new approach, we first recognize that in a realistic situation the channel has its nonzero impulse response $h(t)$ essentially confined to a finite interval $[0, T_b]$. So, let the sequence $h_0, h_1, \ldots, h_\nu$ denote the baseband equivalent impulse response of the channel sampled at the rate $1/T_s$, with

$$T_b = (1 + \nu)T_s \tag{8.76}$$

where the role of ν is to be clarified. The sampling rate $1/T_s$ is chosen to be greater than twice the highest frequency component of interest in accordance with the sampling theorem. To continue with the discrete-time description of the system, let $s_n = s(nT_s)$ denote a sample of the transmitted symbol $s(t)$, $w_n = w(nT_s)$ denote a sample of the channel noise $w(t)$, and $x_n = x(nT_s)$ denote the corresponding sample of the channel output (i.e., received signal). The channel performs linear convolution on the incoming symbol sequence $\{s_n\}$ of length N to produce a channel output sequence $\{x_n\}$ of length $N + \nu$. Extension of the channel output sequence by ν samples compared with the channel input sequence is due to the intersymbol interference produced by the channel.

To overcome the effect of ISI, we create a cyclically extended *guard interval*, whereby each symbol sequence is preceded by a periodic extension of the sequence itself. Specifically, the last ν samples of the symbol sequence are repeated at the beginning of the sequence being transmitted, as shown by

$$s_k = s_{N-K} \qquad \text{for } K = 1, 2, \ldots, \nu \tag{8.77}$$

The condition described in (8.77) is called a *cyclic prefix*. The *excess bandwidth factor* due to the inclusion of the cyclic prefix is therefore ν/N, where N is the number of transmitted samples after the guard interval.

With the cyclic prefix in place, the matrix description of the channel now takes the new form

$$\begin{bmatrix} x_{N-1} \\ x_{N-2} \\ \vdots \\ x_{N-\nu-1} \\ x_{N-\nu-2} \\ \vdots \\ x_0 \end{bmatrix} = \begin{bmatrix} h_0 & h_1 & h_2 & & h_{\nu-1} & h_\nu & 0 & \cdots & 0 \\ 0 & h_0 & h_1 & & h_{\nu-2} & h_{\nu-1} & h_\nu & \cdots & 0 \\ \vdots & \vdots & \vdots & & \vdots & \vdots & \vdots & & \vdots \\ 0 & 0 & 0 & \cdots & 0 & h_0 & h_1 & \cdots & h_\nu \\ h_\nu & 0 & 0 & \cdots & 0 & 0 & h_0 & \cdots & h_{\nu-1} \\ \vdots & \vdots & \vdots & & \vdots & \vdots & \vdots & & \vdots \\ h_1 & h_2 & h_3 & \cdots & h_\nu & 0 & 0 & \cdots & h_0 \end{bmatrix} \begin{bmatrix} s_{N-1} \\ s_{N-2} \\ \vdots \\ s_{N-\nu-1} \\ s_{N-\nu-2} \\ \vdots \\ s_0 \end{bmatrix} + \begin{bmatrix} w_{N-1} \\ w_{N-2} \\ \vdots \\ w_{N-\nu-1} \\ w_{N-\nu-2} \\ \vdots \\ w_0 \end{bmatrix} \tag{8.78}$$

In a compact way, we may describe the discrete-time representation of the channel in the matrix form

$$\mathbf{x} = \mathbf{H}\mathbf{s} + \mathbf{w} \tag{8.79}$$

where the transmitted symbol vector $\mathbf{s}$, the channel noise vector $\mathbf{w}$, and the received signal vector $\mathbf{x}$ are all N-by-1 vectors that are respectively defined as follows:

$$\mathbf{s} = [s_{N-1}, s_{N-2}, \ldots, s_0]^T \tag{8.80}$$

$$\mathbf{w} = [w_{N-1}, w_{N-2}, \ldots, w_0]^T \tag{8.81}$$

$$\mathbf{x} = [x_{N-1}, x_{N-2}, \ldots, x_0]^T \tag{8.82}$$

Figure 8.29
Discrete-time representation of multichannel data transmission system.

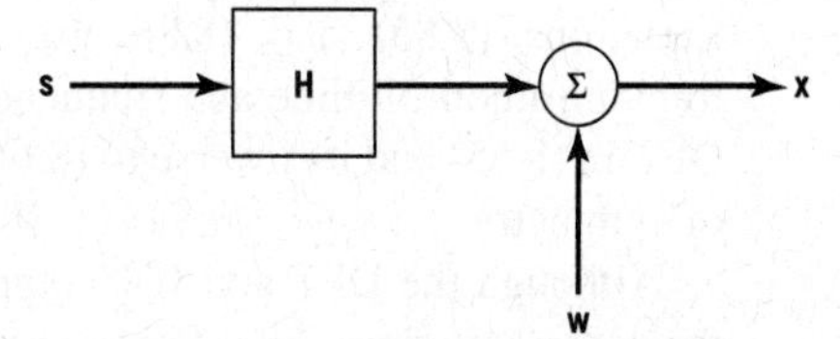

We may thus simply depict the discrete-time representation of the channel as in Figure 8.29. The N-by-N channel matrix $\mathbf{H}$ is itself defined by

$$\mathbf{H} = \begin{bmatrix} h_0 & h_1 & h_2 & \cdots & h_{\nu-1} & h_\nu & 0 & \cdots & 0 \\ 0 & h_0 & h_1 & \cdots & h_{\nu-2} & h_{\nu-1} & h_\nu & \cdots & 0 \\ \vdots & \vdots & \vdots & & \vdots & \vdots & \vdots & & \vdots \\ 0 & 0 & 0 & \cdots & 0 & h_0 & h_1 & \cdots & h_\nu \\ h_\nu & 0 & 0 & \cdots & 0 & 0 & h_0 & \cdots & h_{\nu-1} \\ \vdots & \vdots & \vdots & & \vdots & \vdots & \vdots & & \vdots \\ h_1 & h_2 & h_3 & \cdots & h_\nu & 0 & 0 & \cdots & h_0 \end{bmatrix} \tag{8.83}$$

From the definition in (8.83), we readily see that the matrix $\mathbf{H}$ has the following structural composition:

> Every row of the matrix is obtained by cyclically applying a right-shift to the previous row by one position, with the added proviso that the rightmost element of the previous row spills over in the shifting process to be "circulated" back to the leftmost element of the next row.

Accordingly, the matrix $\mathbf{H}$ is referred to as a *circulant matrix*.

Before proceeding further, it is befitting that we briefly review the DFT and its role in the spectral decomposition of the *circulant matrix* $\mathbf{H}$.

Discrete Fourier Transform

Consider the N-by-1 vector $\mathbf{x}$ of (8.79). Let the DFT of the vector $\mathbf{x}$ be denoted by the N-by-1 vector

$$\mathbf{X} = [X_{N-1}, X_{N-2}, \ldots, X_0]^{\mathrm{T}} \tag{8.84}$$

whose kth element is defined by

$$X_k = \frac{1}{\sqrt{N}} \sum_{n=0}^{N-1} x_n \exp\left(-\mathrm{j}\frac{2\pi}{N}kn\right), \qquad k = 0, 1, \ldots, N-1 \tag{8.85}$$

The exponential term $\exp(-\mathrm{j}2\pi kn/N)$ is the *kernel* of the DFT. Correspondingly, the IDFT (i.e., inverse DFT) of the N-by-1 vector $\mathbf{X}$ is defined by

$$x_n = \frac{1}{\sqrt{N}} \sum_{n=0}^{N-1} X_k \exp\left(\mathrm{j}\frac{2\pi}{N}kn\right), \qquad n = 0, 1, \ldots, N-1 \tag{8.86}$$

Equations (8.85) and (8.86) follow from discretizing the continuous-time Fourier transform both in time and frequency, as discussed in Chapter 2 with one difference: the DFT in (8.65) and its inverse in (8.66) have the same scaling factor, $1/\sqrt{2}$ for the purpose of symmetry.

Although the DFT and IDFT appear to be similar in their mathematical formulations, their interpretations are different, as discussed previously in Chapter 2. As a reminder, we may interpret the DFT process described in (8.85) as a system of *N complex heterodyning and averaging operations*, as shown in Figure 2.32a. In the picture depicted in this part of the figure, *heterodyning* refers to the multiplication of the data sequence x_n by one of N complex exponentials, $\exp(-j2\pi kn/N)$. As such, (8.85) may be viewed as the *analysis* equation. For the interpretation of (8.86), we may view it as the *synthesis equation*: specifically, the complex Fourier coefficient X_k is weighted by one of N complex exponentials $\exp(-j2\pi kn/N)$. At time n, the output x_n is formed by summing the weighted complex Fourier coefficients, as shown in Figure 2.32b.

An important property of a circulant matrix, exemplified by the channel matrix $\mathbf{H}$ of (8.83), is that it permits the *spectral decomposition* defined by

$$\mathbf{H} = \mathbf{Q}^{\dagger}\mathbf{\Lambda}\mathbf{Q} \tag{8.87}$$

where the superscript † denotes *Hermitian transposition* (i.e., the combination of complex conjugation and ordinary matrix transposition). Descriptions of the matrices $\mathbf{Q}$ and $\mathbf{\Lambda}$ are presented in the following in that order. The matrix $\mathbf{Q}$ is a square matrix defined in terms of the kernel of the N-point DFT as shown by

$$\mathbf{Q} = \frac{1}{\sqrt{N}}\begin{bmatrix} \exp\left[-j\frac{2\pi}{N}(N-1)(N-1)\right] & \cdots & \exp\left[-j\frac{2\pi}{N}2(N-1)\right] & \exp\left[-j\frac{2\pi}{N}(N-1)\right] & 1 \\ \exp\left[-j\frac{2\pi}{N}(N-1)(N-2)\right] & \cdots & \exp\left[-j\frac{2\pi}{N}2(N-2)\right] & \exp\left[-j\frac{2\pi}{N}(N-2)\right] & 1 \\ \vdots & & \vdots & \vdots & \\ \exp\left[-j\frac{2\pi}{N}(N-1)\right] & \cdots & \exp\left(-j\frac{2\pi}{N}2\right) & \exp\left(-j\frac{2\pi}{N}\right) & 1 \\ 1 & \cdots & 1 & 1 & 1 \end{bmatrix} \tag{8.88}$$

From this definition, we readily see that the klth element of the N-by-N matrix, $\mathbf{Q}$, starting from the *bottom right* at $k = 0$ and $l = 0$ and counting up step-by-step, is

$$q_{kl} = \frac{1}{\sqrt{N}}\exp\left(-j\frac{2\pi}{N}kl\right), \qquad k, l = 0, 1, \ldots, N-1 \tag{8.89}$$

The matrix $\mathbf{Q}$ is an *orthonormal matrix* or *unitary matrix,* in the sense that it satisfies the condition

$$\mathbf{Q}^{\dagger}\mathbf{Q} = \mathbf{I} \tag{8.90}$$

where $\mathbf{I}$ is the identity matrix. That is, the inverse matrix of $\mathbf{Q}$ is equal to the *Hermitian transpose* of $\mathbf{Q}$.

The matrix Λ in (8.87) is a *diagonal matrix* that contains the N DFT values of the sequence $h_0, h_1, \ldots, h_\nu$ that characterize the channel. Denoting these transform values by $\lambda_{N-1}, \ldots, \lambda_1, \lambda_0$, respectively, we may express Λ as

$$\Lambda = \begin{bmatrix} \lambda_{N-1} & 0 & \cdots & 0 \\ 0 & \lambda_{N-2} & \cdots & 0 \\ \vdots & \vdots & & \vdots \\ 0 & 0 & \cdots & \lambda_0 \end{bmatrix} \tag{8.91}$$

Note that λ used here are not to be confused with the Lagrange multipliers in Section 8.13.

From a system design objective, the DFT has established itself as one of the principal tools of digital signal processing by virtue of its efficient computation using the *FFT algorithm*, which was also described in Chapter 2. Computationally speaking, the FFT algorithm requires on the order of $N\log_2 N$ operations rather than the N^2 operations for direct computation of the DFT. For efficient implementation of the FFT algorithm, we should choose the block length N to be an integer power of 2. The computational savings obtained by using the FFT algorithm are made possible by exploiting the special structure of the DFT defined in (8.85). Moreover, these savings become more substantial as we increase the data length N.

Frequency–Domain Description of the Channel

With this brief review of the DFT and its FFT implementations at hand, we are ready to resume our discussion of the DMT system. First, we define

$$\mathbf{s} = \mathbf{Q}^\dagger \mathbf{S} \tag{8.92}$$

where $\mathbf{S}$ is the frequency-domain vector representation of the transmitter output. Each element of the N-by-1 vector $\mathbf{S}$ may be viewed as a complex-valued point in a two-dimensional QAM signal constellation. Given the channel output vector $\mathbf{x}$, we define its corresponding frequency-domain representation as

$$\mathbf{X} = \mathbf{Q}\mathbf{x} \tag{8.93}$$

Using (8.87), (8.92), and (8.93), we may rewrite (8.79) in the equivalent form

$$\mathbf{X} = \mathbf{Q}(\mathbf{Q}^\dagger \Lambda \mathbf{Q}\mathbf{Q}^\dagger \mathbf{S} + \mathbf{W}) \tag{8.94}$$

Hence, using the equality of (8.90) in (8.94) we may reduce the vector $\mathbf{X}$ to the simple form

$$\mathbf{X} = \Lambda \mathbf{S} + \mathbf{W} \tag{8.95}$$

where

$$\mathbf{W} = \mathbf{Q}\mathbf{w} \tag{8.96}$$

In expanded (scalar) form, the matrix equation (8.95) reads as follows:

$$X_k = \lambda_k S_k + W_k, \qquad k = 0, 1, \ldots, N-1 \tag{8.97}$$

where the set of frequency-domain values $\{\lambda_k\}_{k=0}^{N-1}$ is known for a prescribed channel. Note that X_k is a random variable and w_k is a random variable sampled from a white Gaussian noise process.

For a channel with additive white noise, (8.97) leads us to make the following important statement:

> The receiver of a DMT-based DSL is composed of a set of independent processors operating in parallel.

With the λ_k being all known, we may thus use the block of frequency-domain values $\{X_k\}_{k=0}^{N-1}$ to compute estimates of the corresponding transmitted block of frequency-domain values $\{S_k\}_{k=0}^{N-1}$.

DFT-Based DMT System

Equations (8.95), (8.85), (8.86), and (8.97) provide the mathematical basis for the implementation of DMT using the DFT. Figure 8.30 illustrates the block diagram of the system derived from these equations, setting the stage for their practical roles:

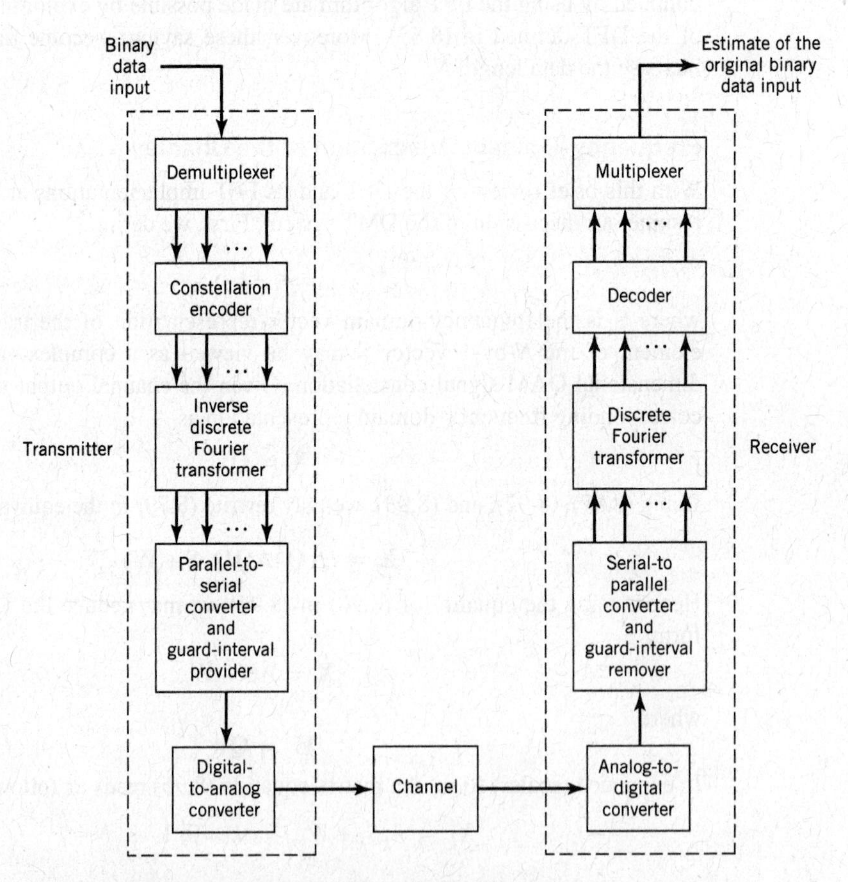

Figure 8.30 Block diagram of the DFT-based DMT system.

1. The transmitter consists of the following functional blocks:
   - *Demultiplexer*, which converts the incoming serial data stream into parallel form.
   - *Constellation encoder*, which maps the parallel data into $N/2$ multibit subchannels with each subchannel being represented by a QAM signal constellation. Bit allocation among the subchannels is also performed here in accordance with a loading algorithm.
   - *IDFT*, which transforms the frequency-domain parallel data at the constellation encoder output into parallel time-domain data. For efficient implementation of the IDFT using the FFT algorithm, we need to choose $N = 2^k$, where k is a positive integer.
   - *Parallel-to-serial converter*, which converts the parallel time-domain data into serial form. Guard intervals stuffed with cyclic prefixes are inserted into the serial data on a periodic basis before conversion into analog form.
   - *Digital-to-analog converter (DAC)*, which converts the digital data into analog form ready for transmission over the channel.

   Typically, the DAC includes a transmit filter. Accordingly, the time function $h(t)$ in Figure 8.25 should be redefined as the combined impulse response of the cascade connection of the transmit filter and the channel.
2. The receiver performs the inverse operations of the transmitter, as described here:
   - *Analog-to-digital converter (ADC)*, which converts the analog channel output into digital form.
   - *Serial-to-parallel converter*, which converts the resulting bit stream into parallel form. Before this conversion takes place, the guard intervals (cyclic prefixes) are removed.
   - *DFT*, which transforms the time-domain parallel data into frequency-domain parallel data; as with the IDFT, the FFT algorithm is used to implement the DFT.
   - *Decoder*, which uses the DFT output to compute estimates of the original multibit subchannel data supplied to the transmitter.
   - *Multiplexer*, which combines the estimates so computed to produce a reconstruction of the transmitted serial data stream.

To sum up:

> Thanks to the computationally efficient FFT algorithm, the DMT has established itself as the standard core technology for the design of asymmetric and very high bit-rate versions of the DSL by virtue of two important operational attributes: effective performance and efficient implementation.

Practical Applications of DMT-based DSL

An important application of DMT is in the transmission of data over two-way channels. Indeed, DMT has been standardized for use on ADSLs using twisted wire-pairs. In ADSL, for example, the DMT provides for the transmission of data downstream (i.e., from an ISP to a subscriber) at the rate of 1.544 Mbits/s and the simultaneous transmission of data upstream (i.e., from the subscriber to the ISP) at 160 kbits/s. This kind of data transmission capability is well suited for handling data-intensive applications such as video-on-demand.

DMT is also a core technology in implementing the asymmetric VDSLs, which differs from all other DSL transmission techniques because of its ability to deliver extremely high data rates. For example, a VDSL can provide data rates of 13 to 26 Mbits/s downstream and 2 to 3 MB/s upstream over twisted wire-pairs that emanate from an optical network unit and connect to the subscriber over distances of less than about 1 km. These high data rates allow the delivery of digital TV, super-fast Web surfing and file transfer, and virtual offices at home.

From a practical perspective, the use of DMT for implementing ADSL and VDSL provides a number of advantages:

- *The ability of DMT to maximize the transmitted bit rate*, which is provided by tailoring the distribution of information-bearing signals across the channel according to channel attenuation and noise conditions.
- *Adaptivity to changing line conditions*, which is realized by virtue of the fact that the channel is partitioned into a number of subchannels.
- *Reduced sensitivity to impulse noise*, which is achieved by spreading its energy over the many subchannels of the receiver. As the name implies, *impulse noise* is characterized by long, quiet intervals followed by narrow pulses of randomly varying amplitude. In an ADSL or VDSL environment, impulse noise arises due to switching transients coupled to twisted wire-pairs in the central office and to various electrical devices on the user's premises.
- Effectively, employment of the DMT system eliminates the need for adaptive channel equalization.

8.16 Summary and Discussion

In this chapter devoted to data transmission over band-limited channels, two important aspects of this practical problem were discussed.

In the first part of the chapter, we assumed that the SNR at the channel input is large enough for the effect of channel noise to be ignored. Under this assumption, the issue of dealing with intersymbol interference was viewed as a signal design problem. That is, the overall pulse shape $p(t)$ is configured in such a way that $p(t)$ is zero at the sampling times nT_b, where T_b is the reciprocal of the bit rate R_b. In so doing, the intersymbol interference is reduced to zero. Finding the pulse shape that satisfies this requirement is best handled in the frequency domain. The ideal solution is a "brick-wall" spectrum that is constant over the interval $-W \leq f \leq W$ where $W = 1/2T_b$. T_b is the bit duration and W is called the Nyquist bandwidth. Unfortunately, this ideal pulse shape is impractical on two accounts: noncausal behavior and sensitivity to timing errors. To overcome these two practical difficulties, we proposed the use of an RC spectrum that rolls off gradually from a constant value over a prescribed band toward zero in a half-cosine-like manner on either side of the band. We finished this first part of the chapter by introducing the SRRC spectrum, where the overall pulse shaping is split equally between the transmitter and receiver; this latter form of signal design finds application in wireless communication.

Turning next to the second part of the chapter, we discussed another way of tackling data transmission over a wideband channel by applying the engineering principle of "divide and conquer." Specifically, a telephone channel, using a twisted wire-pair, is partitioned into a large number of narrowband subchannels, such that each noisy subchannel can be handled by applying Shannon's information-capacity law. Then, through a series of clever mathematical steps, the treatment of a difficult "discrete multicarrier transmission system" is modified into a new "DMT system." Most importantly, by exploiting the computational efficiency of the FFT algorithm, practical implementation of the DMT assumes a well-structured *transceiver* (i.e., pair of transmitter and receiver) that is effective in performance and efficient in computational terms. Indeed, the DMT has established itself as the *standard core technology* for designing the asymmetric and very high bit-rate members of the *digital subscriber line family.* Moreover, the world-wide deployment of DSL technology has converted an ordinary telephone line into a broadband communication link, so much so that we may now view the PSTN as a *broadband backbone data network*. Most importantly, this analog-to-digital network conversion has made it possible to transmit data at rates in the megabits per second region, which is a truly a remarkable engineering achievement.

Problems

Nyquist's Criterion

8.1 The NRZ pulse of Figure P8.1 may be viewed as a very crude form of a Nyquist pulse. Justify this statement by comparing the spectral characteristics of these two pulses.

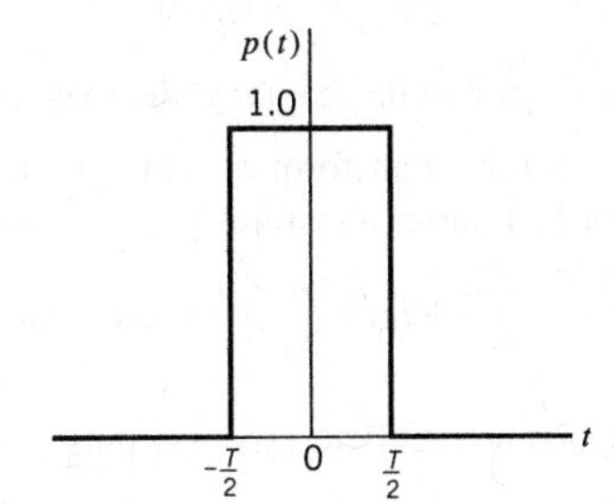

Figure P8.1

8.2 A binary PAM signal is to be transmitted over a baseband channel with an absolute maximum bandwidth of 75 kHz. The bit duration is 10 μs. Find an RC spectrum that satisfies these requirements.

8.3 An analog signal is sampled, quantized, and encoded into a binary PCM. Specifications of the PCM signal include the following:

- Sampling rate, 8 kHz
- Number of representation levels, 64.

The PCM signal is transmitted over a baseband channel using discrete PAM. Determine the minimum bandwidth required for transmitting the PCM signal if each pulse is allowed to take on the following number of amplitude levels: 2, 4, or 8.

8.4 Consider a baseband binary PAM system that is designed to have an RC spectrum $P(f)$. The resulting pulse $p(t)$ is defined in (8.25). How would this pulse be modified if the system is designed to have a linear phase response?

8.5 Determine the Nyquist pulse whose inverse Fourier transform is defined by the frequency function $P(f)$ defined in (8.26).

8.6 Continuing with the defining condition in Problem 8.5, namely

$$\sum_{n=-\infty}^{\infty} P\left(f+\frac{n}{T_b}\right) = T_b \qquad T_b > 0$$

demonstrate that the Nyquist pulse $p(t)$ with the narrowest bandwidth $\frac{1}{2T_b}$ is described by the sinc function:

$$p(t) = \text{sinc}\left(\frac{t}{T_b}\right)$$

8.7 A pulse $p(t)$ is said to be orthogonal under T-shifts if it satisfies the condition

$$\int_{-\infty}^{\infty} p(t)p(t-nT_b)\,dt = 0 \qquad \text{for } n = \pm 1, \pm 2, \ldots$$

where T_b is the bit duration. In other words, the pulse $p(t)$ is uncorrelated with itself when it is shifted by any integer multiple of T_b. Show that this condition is satisfied by a Nyquist pulse.

8.8 Let $P(f)$ be an integrable function, the inverse Fourier transform of which is given by

$$p(t) = \int_{-\infty}^{\infty} P(f)\exp(j2\pi ft)\,df$$

and let T_b be given. The pulse $p(t)$ so defined is a Nyquist pulse of bit duration T_b if, and only if, the Fourier transform $P(f)$ satisfies the condition

$$\sum_{n=-\infty}^{\infty} P\left(f+\frac{n}{T_b}\right) = T_b$$

Using the Poisson sum formula described in Chapter 2, demonstrate the validity of this statement.

8.9 Let $g(t)$ denote a function, the Fourier transform of which is denoted by $G(f)$. The pulse $g(t)$ is *orthogonal under T-shifts* in that its Fourier transform $G(f)$ satisfies the condition

$$\sum_{n=-\infty}^{\infty} \left|G\left(f+\frac{n}{T_b}\right)\right|^2 = \text{constant}$$

Show that this condition is satisfied by the SRRC shaping pulse.

Partial Response Signaling

8.10 The sinc pulse is the optimum Nyquist pulse, optimum in the sense it produces zero intersymbol interference occupying the minimum bandwidth possible $W = 1/2T_b$, where T_b is the bit duration. However, as discussed in Section 8.5, the sinc pulse is prone to timing errors; hence the preference for the RC spectrum that requires twice the minimum bandwidth, $2W$.

In this problem, we explore a new pulse that achieves the minimum possible bandwidth $W = 1/2T_b$ as the sinc pulse, but at the expense of a deterministic (i.e., controlled) intersymbol interference; being controllable, appropriate measures can be taken at the receiver to account for it.

This new pulse is denoted by $g_1(t)$, the Fourier transform of which is denoted by

$$G_1(f) = \begin{cases} 2\cos(\pi f T_b)\exp(-j\pi f T_b), & |f| \le (1/2T_b) \\ 0, & \text{otherwise} \end{cases}$$

a. Plot the magnitude and phase spectrum of $G_1(f)$.

b. Show that the pulse $g_1(t)$ is defined by

$$g_1(t) = \frac{T_b^2 \sin(\pi t/T_b)}{\pi t(T_b - t)}$$

and therefore justify the statement that the tails of $g_1(t)$ decay as $1/|t|^2$, which is faster than the rate of decay $1/|t|$ that characterizes the sinc pulse. Comment on this advantage of $g_1(t)$ over the sinc pulse.

c. Plot the waveform of $g_1(t)$ to demonstrate that $g_1(t)$ has only two distinguishable values at the sampling instants; hence the reference to $g_1(t)$ as a *duobinary code*.

d. Signaling over a band-limited channel with the use of a duobinary code is referred to *partial-response signaling*. Explain why.

8.11 In this problem, we explore another form of partial-response signaling based on the *modified duobinary code*. Let this second code be represented by the pulse $g_2(t)$ whose Fourier transform is defined by

$$G_2(f) = \begin{cases} 2\mathrm{j}\sin(2\pi f T_\mathrm{b})\exp(-\mathrm{j}2\pi f T_\mathrm{b}), & |f| \le 1/2T_\mathrm{b} \\ 0, & \text{otherwise} \end{cases}$$

a. Plot the magnitude and phase spectra of $G_2(f)$.

b. Show that the modified duobinary pulse is itself defined by

$$g_2(t) = \frac{2T_\mathrm{b}^2 \sin(\pi t/T_\mathrm{b})}{\pi t(2T_\mathrm{b} - t)}$$

and therefore demonstrate that it has three distinguishable levels at the sampling instants.

c. What is a practical advantage of the modified duobinary code over the duobinary code in terms of transmission over a band-limited channel?

Multichannel Line Codes

8.12 Consider the passband basis functions defined in (8.56), where $\phi(t)$ is itself defined by (8.57). Demonstrate the validity of Properties 1, 2, and 3 of these passband basis functions.

8.13 The water-filling solution for the loading problem is defined by (8.73) subject to the constraint of (8.70). Using this pair of relations, formulate a recursive algorithm for computing the allocation of the transmit power P among the N subchannels. The algorithm should start with an initial total or sum *noise-to-signal ratio* $\mathrm{NSR}_{(i)} = 0$ for iteration $i = 0$, and the subchannels sorted in terms of those with the smallest power allocation to the largest.

8.14 The squared magnitude response of a linear channel, denoted by $|H(f)|^2$, is shown in Figure P8.14. Assuming that the gap $\Gamma = 1$ and the noise variance $\sigma_n^2 = 1$ for all subchannels, do the following:

a. Derive the formulas for the optimum powers P_1, P_2, and P_3, allocated to the three subchannels of frequency bands $(0, W_1)$, (W_1, W_2), and (W_2, W).

b. Given that the total transmit power $P = 10$, $l_1 = 2/3$, and $l_2 = 1/3$, calculate the corresponding values of P_1, P_2, and P_3.

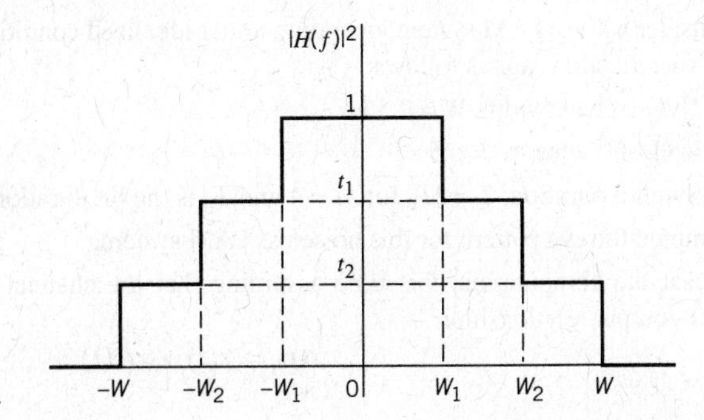

Figure P8.14

8.15 In this problem we explore the use of singular value decomposition (SVD) as an alternative to the DFT for vector coding. This approach avoids the need for a cyclic prefix, with the channel matrix being formulated as

$$\mathbf{H} = \begin{bmatrix} h_0 & h_1 & h_2 & \cdots & h_\nu & 0 & \cdots & 0 \\ 0 & h_0 & h_1 & \cdots & h_{\nu-1} & h_\nu & \cdots & 0 \\ \vdots & \vdots & \vdots & & \vdots & \vdots & & \vdots \\ 0 & 0 & 0 & \cdots & h_0 & h_1 & \cdots & h_\nu \end{bmatrix}$$

where the sequence $h_0, h_1, \ldots, h_\nu$ denotes the sampled impulse response of the channel. The SVD of the matrix $\mathbf{H}$ is defined by

$$\mathbf{H} = \mathbf{U}[\boldsymbol{\Lambda}:\mathbf{0}_{N,\nu}]\mathbf{V}^\dagger$$

where $\mathbf{U}$ is an N-by-N unitary matrix and $\mathbf{V}$ is an $(N+\nu)$-by-$(N+\nu)$ unitary matrix; that is,

$$\mathbf{U}\mathbf{U}^\dagger = \mathbf{I}$$
$$\mathbf{V}\mathbf{V}^\dagger = \mathbf{I}$$

where $\mathbf{I}$ is the identity matrix and the superscript $\dagger$ denotes Hermitian transposition. The $\boldsymbol{\Lambda}$ is an N-by-N diagonal matrix with singular values λ_n, $n = 1, 2, \ldots, N$. The $\mathbf{0}_{N,\nu}$ is an N-by-ν matrix of zeros.

a. Using this decomposition, show that the N subchannels resulting from the use of vector coding are mathematically described by

$$X_n = \lambda_n A_n + W_n$$

The X_n is an element of the matrix product $\mathbf{U}^\dagger\mathbf{x}$, where $\mathbf{x}$ is the received signal (channel output) vector. A_n is the nth symbol $a_n + jb_n$ and W_n is a random variable due to channel noise.

b. Show that the SNR for vector coding as described herein is given by

$$(\text{SNR})_{\text{vector coding}} = \Gamma\left[\prod_{n=1}^{N^*}\left(1 + \frac{(\text{SNR})_n}{\Gamma}\right)\right]^{1/(N+\nu)} - \Gamma$$

where N^* is the number of channels for each of which the allocated transmit power is nonnegative, $(\text{SNR})_n$ is the SNR of subchannel n, and Γ is a prescribed gap.

c. As the block length N approaches infinity, the singular values approach the magnitudes of the channel Fourier transform. Using this result, comment on the relationship between vector coding and discrete multitone.

Computer Experiments

**8.16 In this computer-oriented problem, consisting of two parts, we demonstrate the effect of nonlinearity on eye patterns.

a. Consider a 4-ary PAM system, operating under idealized conditions: no channel noise and no ISI. the specifications are as follows:

Nyquist bandwidth, $W = 0.5$ Hz

Roll-off factor, $\alpha = 0.5$

Symbol duration, $T = 2T_b$ for $M = 4$ and T_b is the bit duration.

Compute the eye pattern for this noiseless PAM system.

b. Repeat the computation, this time assuming that the channel is nonlinear with the following input–output relationship:

$$x(t) = s(t) + as^2(t)$$

where $s(t)$ is the channel input and $x(t)$ is the channel output (i.e., received signal); the a is a constant. Compute the eye pattern for the following three nonlinear conditions:

$$a = 0.05, 0.1, 0.2$$

Hence, discuss how varying the constant a affects the shape of the eye pattern for the 4-ary PAM system.

Notes

1. The criterion described in (8.11) or (8.15) was first formulated by Nyquist in the study of telegraph transmission theory; the Nyquist (1928b) paper is a classic. In the literature, this criterion is referred to as *Nyquist's first criterion*. In the 1928b paper, Nyquist described another method, referred to in the literature as *Nyquist's second criterion*. The second method makes use of the instants of transition between unlike symbols in the received signal rather than centered samples. A discussion of the first and second criteria is presented in Bennett (1970: 78–92) and in the paper by Gibby and Smith (1965). A third criterion attributed to Nyquist is discussed in Sunde (1969); see also the papers by Pasupathy (1974) and Sayar and Pasupathy (1987).
2. The specifications described in Example 1 follow the book by Tranter *et al.* (2004).
3. The SRRC pulse shaping is discussed in Chennakeshu and Saulnier (1993) in the context of $\pi/4$-shifted differential QPSK for digital cellular radio. It is also discussed in Anderson (2005: 27–29).
4. In a strict sense, an eye pattern that is completely open occupies the range from –1 to +1. On this basis, zero intersymbol interference would correspond to an ideal eye opening of 2. However, for two reasons, convenience of presentation and consistency with the literature, we have chosen an eye opening of unity to refer to the ideal condition of zero intersymbol interference.
5. For a detailed treatment of decision feedback equalizers, see the fifth edition of the classic book on Digital Communications by Proakis and Salehi (2008).
6. The idea of an ADSL is attributed to Lechleider (1989) in having had the insight that such an arrangement offers the possibility of more than doubling the information capacity of a symmetric arrangement.
7. For a detailed discussion of VDSL, see Chapter 7 of the book by Starr *et al.* (2003); see also the paper by Cioffi *et al.* (1999).
8. The method of Lagrange multipliers is discussed in Appendix D.

CHAPTER 9

Signaling over Fading Channels

9.1 Introduction

In Chapters 7 and 8 we studied signaling over AWGN and band-limited channels, respectively. In this chapter we go on to study a more complicated communications environment, namely a *fading channel*, which is at the very core of ever-expanding wireless communications. Fading refers to the fact that even though the distance separating a *mobile* receiver from the transmitter is essentially constant, a relatively small movement of the receiver away from the transmitter could result in a significant change in the received power. The physical phenomenon responsible for fading is *multipath*, which means that the transmitted signal reaches the mobile receiver via multiple paths with varying *spatio-temporal characteristics*, hence the challenging nature of the wireless channel for reliable communication.

This chapter consists of three related parts:

First we study signaling over a fading channel by characterizing its statistical behavior in temporal as well as spacial terms. This statistical characterization is carried out from three different perspectives: physical, mathematical, and computational, each of which enriches our understanding of the multipath phenomenon in its own way. This first part of the chapter finishes with:

- BER comparison of different modulation schemes for AWGN and Rayleigh fading channels.
- Graphical display of how different fading channels compare to a corresponding AWGN channel using binary PSK.

This evaluation then prompts the issue of how to combat the degrading effect of multipath and thereby realize reliable communication over a fading channel. Indeed, the second part of the chapter is devoted to this important practical issue. Specifically, we study the use of *space diversity*, which can be one of three kinds:

1. *Diversity-on-receive*, which involves the use of a single transmitter and multiple receivers, with each receiver having its own antenna.
2. *Diversity-on-transmit*, which involves the use of multiple transmitting antennas and a single receiver.
3. *Multiple-input, multiple-output* (MIMO) antenna system, which includes diversity on receive and diversity on transmit in a combined manner.

The use of diversity-on-receive techniques is of long standing in the study of radio communications. On the other hand, diversity-on-transmit and MIMO antenna systems are of recent origin. The study of diversity is closely related to that of information capacity, the evaluation of which is also given special attention in the latter part of the chapter.

For the third and final part of the chapter, we study spread-spectrum signals, which provide the basis of another novel way of thinking about how to mitigate the degrading effects of the multipath phenomenon. In more specific terms, the use of spread-spectrum signaling leads to the formulation of code-division multiple access, a topic that was covered briefly in the introductory Chapter 1.

9.2 Propagation Effects

The major propagation problems[1] encountered in the use of mobile radio in built-up areas are due to the fact that the antenna of a mobile unit may lie well below the surrounding buildings. Simply put, there is no "line-of-sight" path to the base station. Instead, radio propagation takes place mainly by way of scattering from the surfaces of the surrounding buildings and by diffraction over and/or around them, as illustrated in Figure 9.1. The important point to note from Figure 9.1 is that energy reaches the receiving antenna via more than one path. Accordingly, we speak of a *multipath phenomenon,* in that the various incoming radio waves reach their destination from different directions and with different time delays.

To understand the nature of the multipath phenomenon, consider first a "static" multipath environment involving a stationary receiver and a transmitted signal that consists of a narrowband signal (e.g., unmodulated sinusoidal carrier). Let it be assumed that two attenuated versions of the transmitted signal arrive sequentially at the receiver. The effect of the differential time delay is to introduce a relative phase shift between any two components of the received signal. We may then identify one of two extreme cases that can arise:

- The relative phase shift is zero, in which case the two components add *constructively*, as illustrated in Figure 9.2a.
- The relative phase shift is 180°, in which case the two components add *destructively*, as illustrated in Figure 9.2b.

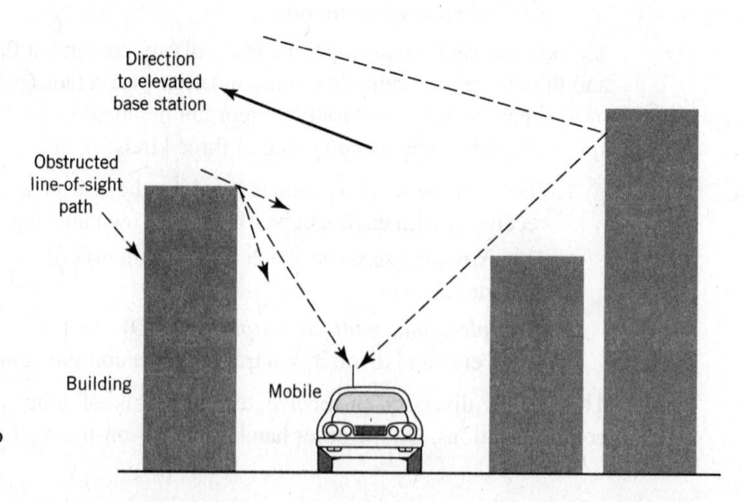

Figure 9.1
Illustrating the mechanism of radio propagation in urban areas.

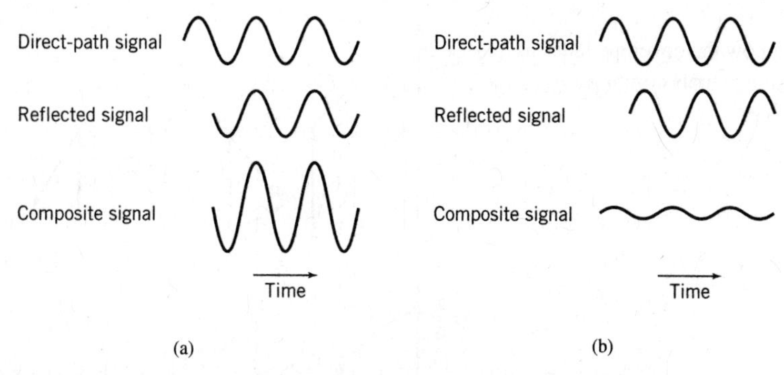

Figure 9.2 (a) Constructive and (b) destructive forms of the multipath phenomenon for sinusoidal signals.

We may also use *phasors* to demonstrate the constructive and destructive effects of multipath, as shown in Figures 9.3a and 9.3b, respectively. Note that, in the static multipath environment described herein, the amplitude of the received signal does not vary with time.

Consider next a “dynamic” multipath environment in which the receiver is in motion and two versions of the transmitted narrowband signal reach the receiver via paths of different

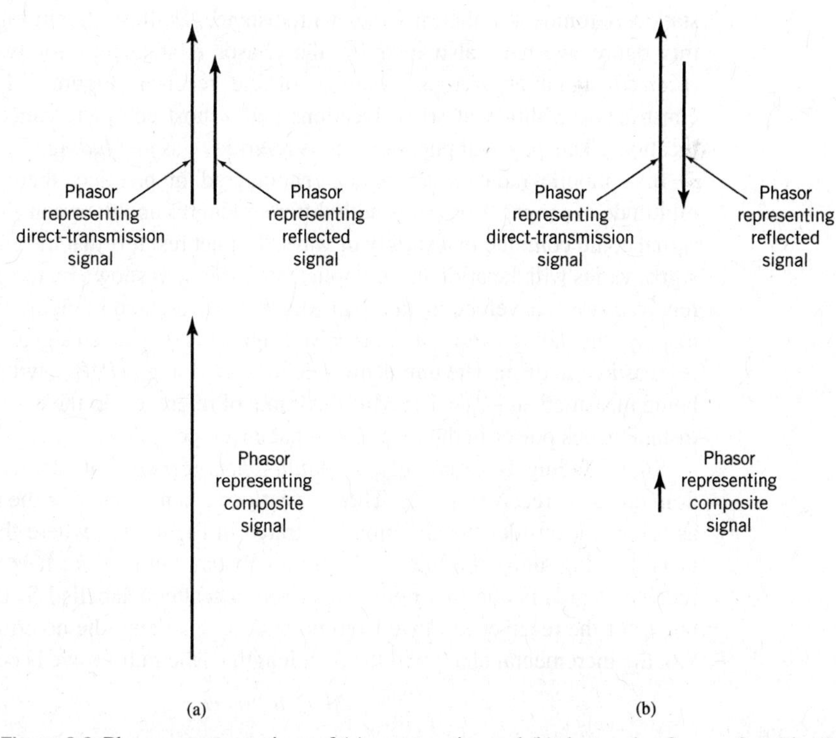

Figure 9.3 Phasor representations of (a) constructive and (b) destructive forms of multipath.

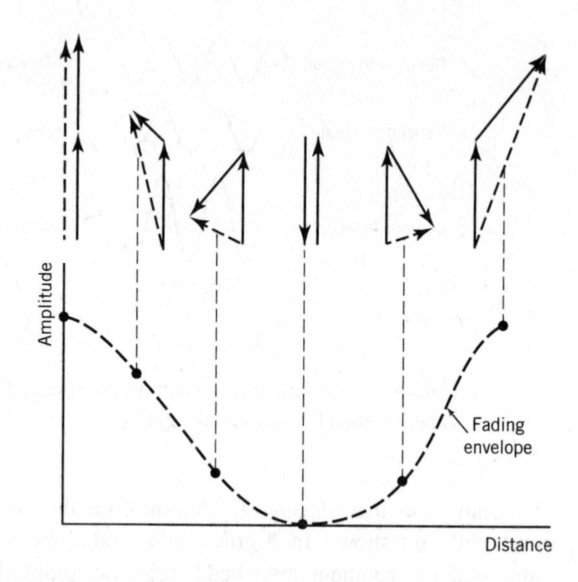

Figure 9.4
Illustrating how the envelope fades as two incoming signals combine with different phases.

lengths. Owing to motion of the receiver, there is a continuous change in the length of each propagation path. Hence, the relative phase shift between the two components of the received signal is a function of spatial location of the receiver. As the receiver moves, we now find that the received amplitude (envelope) is no longer constant, as was the case in a static environment; rather, it varies with distance, as illustrated in Figure 9.4. At the top of this figure, we have also included the phasor relationships for two components of the received signal at various locations of the receiver. Figure 9.4 shows that there is constructive addition at some locations and almost complete cancellation at some other locations. This physical phenomenon is referred to as *fast fading*.

In a mobile radio environment encountered in practice, there may of course be a multitude of propagation paths with different lengths and their contributions to the received signal could combine in a variety of ways. The net result is that the envelope of the received signal varies with location in a complicated fashion, as shown by the experimental record of received signal envelope in an urban area that is presented in Figure 9.5. This figure clearly displays the fading nature of the received signal. The received signal envelope in Figure 9.5 is measured in dBm. The unit dBm is defined as $10 \log_{10}(P/P_0)$, with P denoting the power being measured and $P_0 = 1$ mW as the frame of reference. In the case of Figure 9.5, P is the instantaneous power in the received signal envelope.

Signal fading is essentially a *spatial phenomenon* that manifests itself in the time domain as the receiver moves. These variations can be related to the motion of the receiver as follows. Consider the situation illustrated in Figure 9.6, where the receiver is assumed to be moving along the line AA′ with a constant velocity v. It is also assumed that the received signal is due to a radio wave from a scatterer labelled S. Let Δt denote the time taken for the receiver to move from point A to A′. Using the notation described in Figure 9.6, the incremental change in the path length of the radio wave is deduced to be

$$\begin{aligned} \Delta l &= d \cos \psi \\ &= -v\Delta t \cos \psi \end{aligned} \tag{9.1}$$

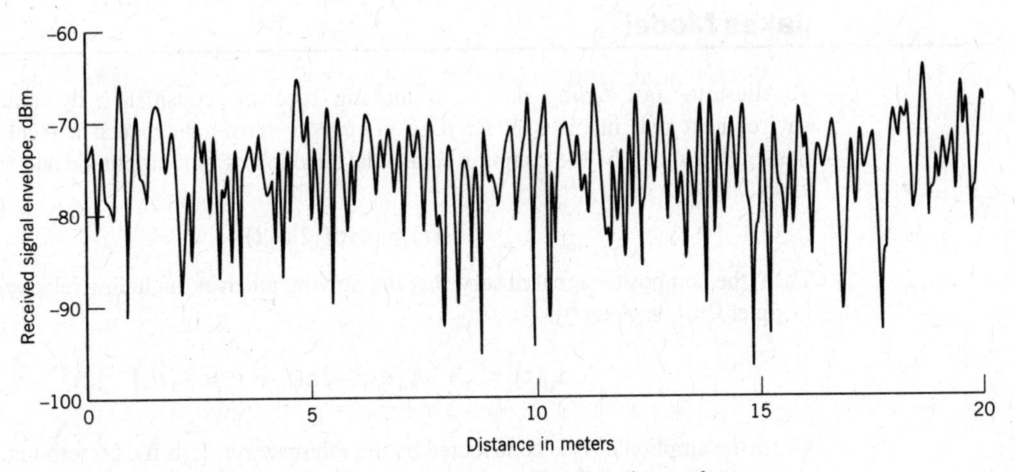

Figure 9.5 Experimental record of received signal envelope in an urban area.

where ψ is the spatial angle subtended between the incoming radio wave and the direction of motion of the receiver. Correspondingly, the change in the phase angle of the received signal at point A′ with respect to that at point A is given by

$$\Delta\phi = \frac{2\pi}{\lambda}\Delta l$$

$$= -\frac{2\pi v \Delta t}{\lambda}\cos\psi$$

where λ is the radio wavelength. The apparent change in frequency, or the *Doppler shift*, is therefore defined by

$$\nu = -\frac{1}{2\pi}\frac{\Delta\phi}{\Delta t}$$
$$= \frac{v}{\lambda}\cos\psi \qquad (9.2)$$

The Doppler shift ν is positive (resulting in an increase in frequency) when the radio waves arrive from ahead of the mobile unit and it is negative when the radio waves arrive from behind the mobile unit.

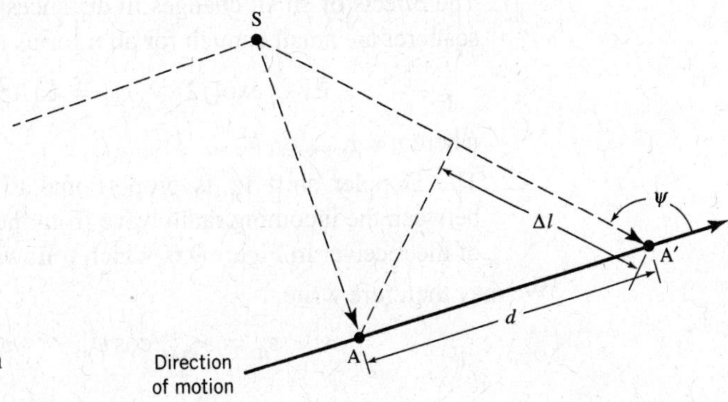

Figure 9.6
Illustrating the calculation of Doppler shift.

9.3 Jakes Model

To illustrate fast fading due to a moving receiver, consider a dynamic multipath environment that involves N *iid fixed scatterers* surrounding such a receiver. Let the transmitted signal be the complex sinusoidal function of unit amplitude and frequency f_c, as shown by

$$s(t) = \exp(j2\pi f_c t)$$

Then, the composite signal observed at the moving receiver, including relative effects of a Doppler shift, is given by

$$x_0(t) = \sum_{n=1}^{N} A_n \exp[j2\pi(f_c + \nu_n)t + j\theta_n]$$

where the amplitude A_n is contributed by the nth scatterer, ν_n is the corresponding Doppler shift, and θ_n is some random phase. The *complex envelope* of the received signal is time varying, as shown by

$$\tilde{x}_0(t) = \sum_{n=1}^{N} A_n \exp[j2\pi\nu_n t + j\theta_n] \tag{9.3}$$

Correspondingly, the autocorrelation function of the complex envelope $\tilde{x}_0(t)$ is defined by

$$R_{\tilde{x}_0}(t) = \mathbb{E}[\tilde{x}_0^*(t)\tilde{x}_0(t+\tau)] \tag{9.4}$$

where $\mathbb{E}$ is the expectation operator with respect to time t and the asterisk in $\tilde{x}_0^*(t)$ denotes complex conjugation. Inserting (9.3) in (9.4) leads to a double summation, one indexed by n and the other indexed by m. Then, simplifying the result under the iid assumption, the autocorrelation function $R_{\tilde{x}_0}(\tau)$ reduces to

$$R_{\tilde{x}_0}(\tau) = \begin{cases} \sum_{n=1}^{N} \mathbb{E}[A_n^2 \exp(j2\pi\nu_n\tau)], & \text{if } m = n \\ 0, & \text{if } m \neq n \end{cases} \tag{9.5}$$

At this point in the discussion, we make two observations:

1. The effects of small changes in distances between the moving receiver and the nth scatterer are small enough for all n for us to write

$$\mathbb{E}[A_n^2 \exp(j2\pi\nu_n\tau)] = \mathbb{E}[A_n^2]\mathbb{E}[\exp(j2\pi\nu_n\tau)] \tag{9.6}$$

where $n = 1, 2, \ldots, N$.

2. The Doppler shift ν_n is proportional to the cosine of the angle ψ_n subtended between the incoming radio wave from the nth scatterer and the direction of motion of the receiver in Figure 9.6, which follows from (9.2).

We may therefore write

$$\nu_n = \nu_{\max} \cos\psi_n, \qquad n = 1, 2, \ldots, N \tag{9.7}$$

where ν_{max} is the *maximum Doppler shift* that occurs when the incoming radio waves propogate in the same direction as the motion of the receiver. Accordingly, using (9.6) and (9.7) in (9.5), we may write

$$R_{\tilde{x}_0}(\tau) = \begin{cases} P_0 \sum_{n=1}^{N} \mathbb{E}[\exp(j2\pi\nu_{max}\tau\cos(\psi_n)], & \text{for } m = n \\ 0, & \text{for } m \neq n \end{cases} \tag{9.8}$$

where the multiplying factor

$$P_0 = \sum_{n=1}^{N} A_n^2 \tag{9.9}$$

is the *average signal power* at the receiver input.

We now make two final assumptions:

1. All the radio waves arrive at the receiver from a *horizontal direction* (Clarke, 1968).
2. The multipath is *uniformly distributed* over the range $[-\pi, \pi]$, as shown by the probability density function (Jakes, 1974):

$$f_\Psi(\psi) = \begin{cases} \frac{1}{2\pi}, & -\pi \le \psi \le \pi \\ 0, & \text{otherwise} \end{cases} \tag{9.10}$$

Under these two assumptions, the remaining expectation in (9.8) becomes independent of n and with it, that equation simplifies further as follows:

$$\begin{aligned} R_{\tilde{x}_0}(\tau) &= P_0\mathbb{E}[\exp(j2\pi\nu_{max}\tau\cos\psi)], \qquad -\pi \le \psi \le \pi \\ &= P_0\int_{-\pi}^{\pi} f_\Psi(\psi)\exp(j2\pi\nu_{max}\tau\cos\psi)\,d\psi \\ &= P_0\left[\frac{1}{2\pi}\int_{-\pi}^{\pi}\exp(j2\pi\nu_{max}\tau\cos\psi)\,d\psi\right] \end{aligned}$$

The definite integral inside the brackets of this equation is recognized as the *Bessel function of the first kind of order zero*,[2] see Appendix C. By definition, for some argument x, we have

$$J_0(x) = \frac{1}{2\pi}\int_{-\pi}^{\pi}\exp(jx\cos\theta)\,d\theta \tag{9.11}$$

We may therefore express the autocorrelation function of the complex signal $\tilde{x}_0(t)$ at the input of the moving receiver in the compact form

$$R_{\tilde{x}_0}(\tau) = P_0 J_0(2\pi\nu_{max}\tau) \tag{9.12}$$

The model described by the autocorrelation function of (9.12) is called the *Jakes model*. Figure 9.7a shows a plot of the autocorrelation $R_{\tilde{x}_0}(\tau)$ according to this model.

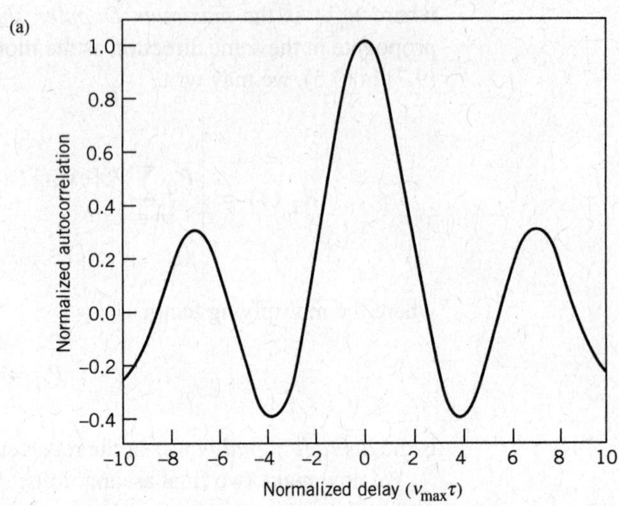

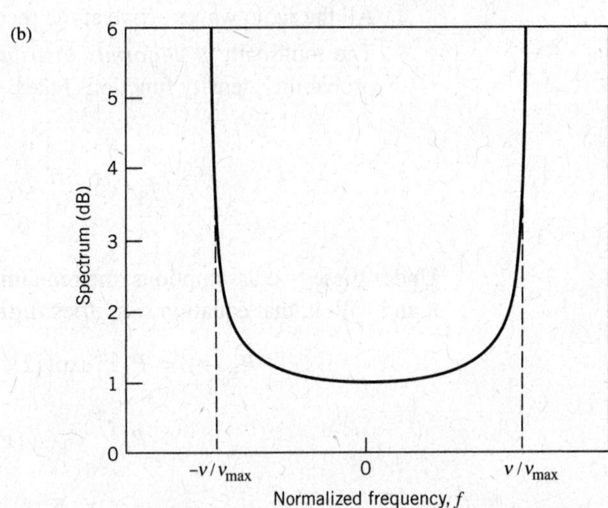

Figure 9.7
(a) Autocorrelation of the complex envelope of the received signal according to the Jakes model.
(b) Power spectrum of the fading process for the Jakes model.

According to the Wiener–Khintchine relations for a weakly (wide-sense) stationary process (discussed in Chapter 4), the autocorrelaton function and power spectrum form a Fourier-transform pair. Specifically, we may write

$$S_{\tilde{x}_0}(\nu) = \mathbb{F}[P_0 J_0(2\pi \nu_{max} \tau)] \tag{9.13}$$

At first sight, it might seem that a closed form solution of this transformation is mathematically intractable; in reality, however, the exact solution is given in (Jakes, 1974):

$$S_{\tilde{x}_0}(\nu) = \begin{cases} \dfrac{P_0}{\sqrt{1-(\nu/\nu_{max})^2}}, & \text{for } \nu < \nu_{max} \\ 0, & \text{for } \nu \ge \nu_{max} \end{cases} \tag{9.14}$$

and with it the model bears his name. Figure 9.7b plots the power spectrum in (9.14) versus the Doppler shift ν for $P_0 = 1$. This idealized graph has the shape of a "bathtub," exhibiting two symmetric integrable singularities at the end points $\nu = \pm\nu_{max}$.

EXAMPLE 1 Jakes Model Implemented as a FIR Filter

The objective of this example is to compute a FIR (TDL) filter that models the power spectrum of (9.14). To this end, we make use of the following relationships in light of material covered in Chapter 4 on stochastic processes:

1. The autocorrelation function and power spectrum of a weakly stationary process form a Fourier-transform pair, as already mentioned.
2. In terms of stochastic processes, the input–output behavior of a linear system, in the frequency domain, is described by

$$S_Y(f) = |H(f)|^2 S_X(f) \tag{9.15}$$

where $H(f)$ is the transfer function of the system, $S_X(f)$ is the power spectrum of the input process $X(t)$, and $S_Y(t)$ is the power spectrum of the output process $Y(t)$, both being weakly stationary.
3. If the input process $X(t)$ is Gaussian, then the output process $Y(t)$ is also Gaussian.
4. If the input $X(t)$ is uncorrelated, then the ouput $Y(t)$ will be correlated due to dispersive behavior of the system.

The issue at hand is to find the $H(f)$ required to produce the desired power spectrum of (9.14) using a white noise process of spectral density $N_0/2$ as the input process $X(t)$. Then, given the $S_Y(f)$ and setting the constant $K = N_0/2$, we may solve (9.15) for $H(f)$, obtaining

$$H(f) = \sqrt{\frac{S_Y(f)}{K}} \tag{9.16}$$

In other words, $H(f)$ is proportional to the square root of $S(f)$. (From a practical perspective, the constant K is determined by truncating the power-delay profile, an issue deferred to Section 9.14.)

In light of (9.14) and (9.16), we may now say that the $H(f)$ representing the desired Jakes FIR filter is given by (ignoring the constant K)

$$H(f) = \begin{cases} (1-f^2)^{-1/4}, & \text{for } -1 \le f \le 1 \\ 0, & \text{otherwise} \end{cases} \tag{9.17}$$

where $f = \nu/\nu_{max}$. Given this formula, we may then use inverse Fourier transformation to compute the corresponding impulse response of the Jakes FIR filter.

However, before proceeding further, an important aspect of using Jakes model to simulate a fading channel is to pay particular attention to the following point:

> The sampling rate of the input signal applied to the Jakes model and the sampled values of the fading process are highly different.

To be specific, the former is a multiple of the symbol rate and the latter is a multiple of the Doppler bandwidth, ν_{max}. In other words, the sampling rate is much larger than ν_{max}. It follows therefore that a *multiple sampling rate with interpolation* must be used in the simulation; the need for interpolation is to go from a discrete spectrum to its continuous version.

With this point in mind, a 512-point *inverse FFT algorithm* is applied to the transfer function of (9.17) for the following set of specifications:

maximum Doppler shift, $\nu_{max} = 100$ Hz

sampling frequency, $f_s = 16\,\nu_{max}$

We thus obtain the discrete-time version of the truncated impulse response h_n of the Jakes FIR filter plotted in Figure 9.8a.

Having computed h_n, we may go on to use the FFT algorithm to compute the corresponding transfer function $H(f)$ of the Jakes FIR filter; the result of this computation is plotted in Figure 9.8b, which has a bathtub-like shape of its own, as expected.

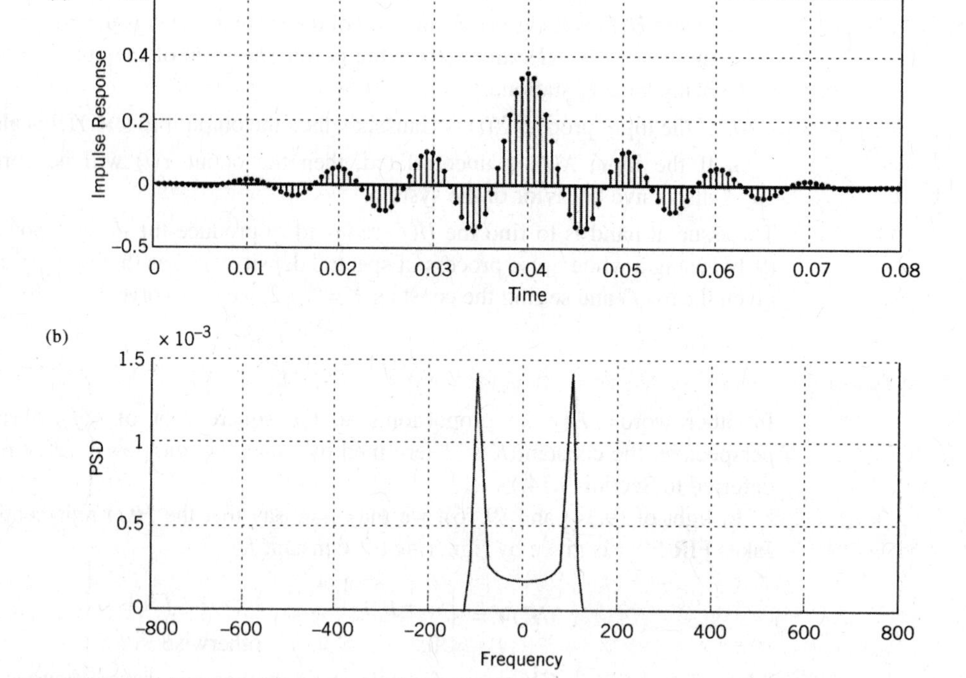

Figure 9.8 Jakes FIR filter. (a) Discrete impulse response. (b) Interpolated power spectral density (PSD).

EXAMPLE 2 Illustrative Generation of Fading Process Using the Jakes FIR Filter

To expand the practical utility of the Jakes FIR filter computed in Example 1 to simulate the fading process, the next thing we do is to pass a complex white noise process through the filter, with the noise having uncorrelated samples. Figure 9.9a displays the power spectrum of the resulting stochastic process at the filter output. Figure 9.9b shows the envelope of the output process, plotted on a logarithmic scale. This plot is typical of a fading correlated signal.

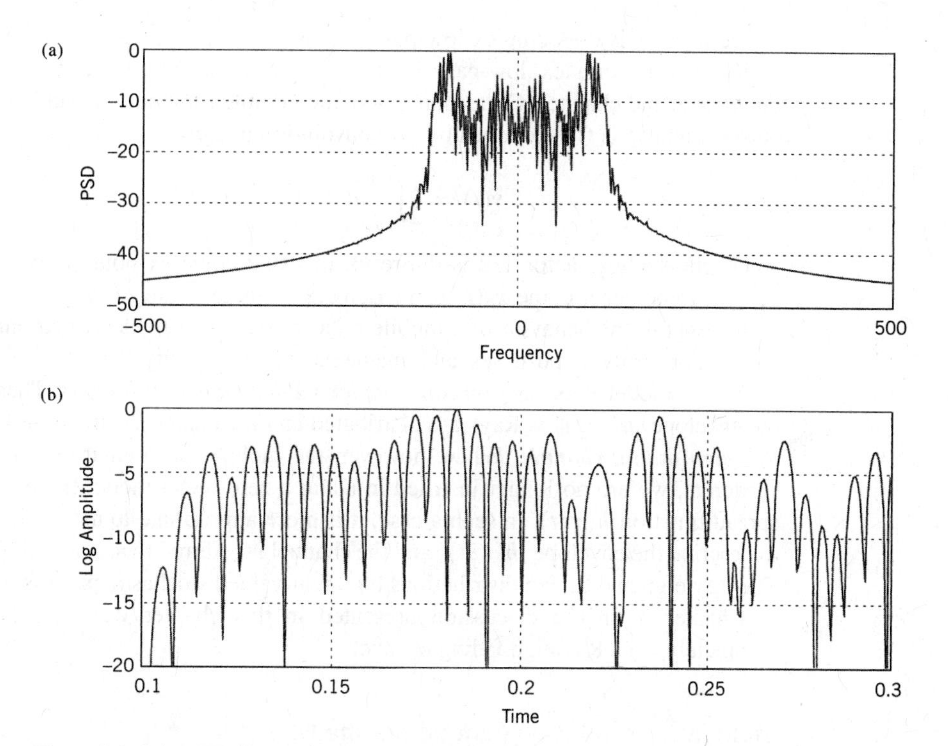

Figure 9.9 Jakes FIR filter driven by white Gaussian noise. (a) Output power spectrum. (b) Envelope of the output process.

9.4 Statistical Characterization of Wideband Wireless Channels

Physical characterization of the multipath environment described in Section 9.3 is appropriate for narrowband mobile radio transmissions where the signal bandwidth is small compared with the reciprocal of the spread in propagation path delays.

However, in real-life situations, we find that the signals radiated in a mobile radio environment occupy a *wide bandwidth*, such that statistical characterization of the wireless channel requires more detailed mathematical considerations, which is the objective of this section. To this end, we follow the complex notations described in Chapter 2 to simplify the analysis.

To be specific, we may express the transmitted band-pass signal as follows:

$$x(t) = \mathrm{Re}[\tilde{x}(t)\exp(\mathrm{j}2\pi f_c t)] \tag{9.18}$$

where $\tilde{x}(t)$ is the complex (low-pass) envelope of $x(t)$ and f_c is the carrier frequency. Since the channel is time varying due to multipath effects, the impulse response of the channel is delay dependent and, therefore, a time-varying function. Let the impulse response of the channel be expressed as

$$h(\tau;t) = \mathrm{Re}[\tilde{h}(\tau;t)\exp(j2\pi f_c t)] \tag{9.19}$$

where $\tilde{h}(\tau;t)$ is the complex low-pass impulse response of the channel and τ is a delay variable. The complex low-pass impulse response $\tilde{h}(\tau;t)$ is called the *delay-spread function* of the channel. Correspondingly, the complex low-pass envelope of the channel output, namely $\tilde{y}(t)$, is defined by the convolution integral

$$\tilde{y}(t) = \frac{1}{2}\int_{-\infty}^{\infty} \tilde{h}(\tau;t)\tilde{x}(t-\tau)\,d\tau \tag{9.20}$$

where the scaling factor $1/2$ is the result of using complex notation; see Chapter 2 for details. To be generic, the $\tilde{x}_0(t)$ in Section 9.2 has been changed to $\tilde{x}(t)$.

In general, the behavior of a mobile radio channel can be described only in statistical terms. For analytic purposes and mathematical tractability, the delay-spread function $\tilde{h}(\tau;t)$ is modeled as a *zero-mean complex-valued Gaussian process*. Then, at any time t the envelope $|\tilde{h}(\tau;t)|$ is Rayleigh distributed and the channel is therefore referred to as a *Rayleigh fading channel*. When, however, the mobile radio environment includes *fixed* scatterers, we are no longer justified in using a zero-mean model to describe the delay-spread function $\tilde{h}(\tau;t)$. In such a case, it is more appropriate to use a Rician distribution to describe the envelope $|\tilde{h}(\tau;t)|$ and the channel is referred to as a *Rician fading channel*. The Rayleigh and Rician distributions for a real-valued stochastic process were considered in Chapter 3. In the discussion presented in this chapter we focus largely, but not completely, on a Rayleigh fading channel.

Multipath Correlation Function of the Channel

The *time-varying transfer function* of the channel is defined as the Fourier transform of the delay-spread function $\tilde{h}(\tau;t)$ with respect to the delay variable τ, as shown by

$$\tilde{H}(f;t) = \int_{-\infty}^{\infty} \tilde{h}(\tau;t)\exp(-j2\pi f\tau)\,d\tau \tag{9.21}$$

where f denotes the frequency variable. The time-varying transfer function $\tilde{H}(f;t)$ may be viewed as a frequency transmission characteristic of the channel.

For a mathematically tractable statistical characterization of the channel, we make two assumptions motivated by physical considerations; hence the practical importance of the model resulting from these two assumptions.

ASSUMPTION 1 **Wide-Sense Stationarity**

With interest confined to fast fading in the short term, it is reasonable to assume that the complex impulse response $\tilde{h}(\tau;t)$ is wide-sense stationary.

As explained in Chapter 4, a stochastic process is said to be wide-sense (i.e., weakly) stationary if its mean is time independent and its autocorrelation function is dependent only on the difference between two time instants at which the process is observed. In what follows we use the "wide-sense stationary" terminology because of its common use in the wireless literature.

In the context of the discussion presented herein, this first assumption means that

- The expectation of $\tilde{h}(\tau;t)$ with respect to time t is dependent only on the delay τ.
- Insofar as time t is concerned, the expectation of the product $\tilde{h}^*(\tau_1;t_1) \times \tilde{h}(\tau_2, t_2)$ is dependent only on the time difference $\Delta t = t_2 - t_1$.

Because Fourier transformation is a *linear operation*, it follows that if the complex delay-spread function $\tilde{h}(\tau;t)$ is a zero-mean Gaussian wide-sense stationary process, then the complex time-varying transfer function $\tilde{H}(f;t)$ has similar statistics.

ASSUMPTION 2 **Uncorrelated Scattering**

The channel is said to be an uncorrelated scattering channel, *when contributions from two or more scatterers with different propagation delays are uncorrelated.*

In other words, the second-order expectation with respect to time t satisfies the requirement

$$\mathbb{E}[\tilde{h}^*(\tau_1;t_1)\tilde{h}(\tau_2;t_2)] = \mathbb{E}[\tilde{h}^*(\tau_1;t_1)\tilde{h}(\tau_1;t_2)]\delta(\tau_1 - \tau_2)$$

where $\delta(\tau_1 - \tau_2)$ is a Dirac-delta function defined in the delay domain. That is, the autocorrelation function of $\tilde{h}(\tau;t)$ is nonzero only when $\tau_2 \neq \tau_1$.

In the literature on statistical characterization of wireless channels, wide-sense stationarity is abbreviated as WSS and uncorrelated scattering is abbreviated as US. Thus, when both Assumptions 1 and 2 are satisfied simultaneously, the resulting channel model is said to be the *WSSUS model.*

Consider then the correlation function[3] of the delay-spread function $\tilde{h}(\tau;t)$. Since $\tilde{h}(\tau;t)$ is complex valued, we use the following definition for the correlation function:

$$R_{\tilde{h}}(\tau_1, t_1;\tau_2, t_2) = \mathbb{E}[\tilde{h}^*(\tau_1;t_1)\tilde{h}(\tau_2;t_2)] \tag{9.22}$$

where $\mathbb{E}$ is the statistical expectation operator, the asterisk denotes complex conjugation, τ_1 and τ_2 are propagation delays of the two paths involved in the calculation, and t_1 and t_2 are the times at which the outputs of the two paths are observed. Under the combined WSSUS channel model, we may reformulate the correlation function in (9.22) as shown by

$$\begin{aligned} R_{\tilde{h}}(\tau_1, \tau_2;\Delta t) &= \mathbb{E}[\tilde{h}^*(\tau_1;t)\tilde{h}(\tau_2;t+\Delta t)] \\ &= r_{\tilde{h}}(\tau_1;\Delta t)\delta(\tau_1 - \tau_2) \end{aligned} \tag{9.23}$$

where Δt is the difference between the observation times t_1 and t_2 and $\delta(\tau_1 - \tau_2)$ is the delta function in the τ-domain. Thus, using τ in place of τ_1 for mathematical convenience, the function in the second line of (9.23) is redefined as

$$r_{\tilde{h}}(\tau;\Delta t) = \mathbb{E}[\tilde{h}^*(\tau;t)\tilde{h}(\tau;t+\Delta t)] \tag{9.24}$$

The function $r_{\tilde{h}}(\tau;\Delta t)$ is called the *multipath correlation profile* of the channel. This new correlation function $r_{\tilde{h}}(\tau;\Delta t)$ provides a statistical measure of the extent to which the signal is *distorted in the time domain* as a result of transmission through the channel.

Spaced-Frequency, Spaced-Time Correlation Function of the Channel

Consider next statistical characterization of the channel in terms of the complex time-varying transfer function $\tilde{H}(f;t)$. Following a formulation similar to that described in (9.22), the correlation function of $\tilde{H}(f;t)$ is defined by

$$R_{\tilde{H}}(f_1, t_1; f_2, t_2) = \mathbb{E}[\tilde{H}^*(f_1;t_1)\tilde{H}(f_2;t_2)] \tag{9.25}$$

where f_1 and f_2 represent two frequencies in the spectrum of the transmitted signal. The correlation function $R_{\tilde{H}}(f_1, t_1; f_2, t_2)$ provides a statistical measure of the extent to which the signal is *distorted in the frequency-domain* by transmission through the channel. From (9.21), (9.22), and (9.25), it is apparent that the correlation functions $R_{\tilde{H}}(f_1, t_1; f_2, t_2)$ and $R_{\tilde{H}}(\tau_1, t_1; \tau_2, t_2)$ form a *two-dimensional Fourier-transform pair*, defined as follows:

$$R_{\tilde{H}}(f_1, t_1; f_2, t_2) \rightleftharpoons \int_{-\infty}^{\infty}\int_{-\infty}^{\infty} R_{\tilde{h}}(\tau_1, t_1; \tau_2, t_2) \exp[-j2\pi(f_1\tau_1 - f_2\tau_2)]\, d\tau_1\, d\tau_2 \tag{9.26}$$

Invoking wide-sense stationarity in the time domain, we may reformulate (9.25) as

$$R_{\tilde{H}}(f_1, f_2; \Delta t) = \mathbb{E}[\tilde{H}^*(f_1;t)\tilde{H}(f_2;t+\Delta t)] \tag{9.27}$$

Equation (9.27) suggests that the correlation function $R_{\tilde{H}}(f_1, f_2; \Delta t)$ may be measured by using pairs of spaced tones to carry out cross-correlation measurements on the resulting channel outputs. Such a measurement presumes stationarity in the time domain. If we also assume stationarity in the frequency domain, we may go one step further and write

$$\begin{aligned} R_{\tilde{H}}(f, f+\Delta f; \Delta t) &= r_{\tilde{H}}(\Delta f; \Delta t) \\ &= \mathbb{E}[\tilde{H}^*(f;t)\tilde{H}(f+\Delta f; t+\Delta t)] \end{aligned} \tag{9.28}$$

The new correlation function $r_{\tilde{H}}(\Delta f;\Delta t)$, introduced in the first line of (9.28), is in fact the Fourier transform of the multipath correlation profile $r_{\tilde{h}}(\tau;\Delta t)$ with respect to the delay-time variable τ, as shown by

$$r_{\tilde{H}}(\Delta f;\Delta t) = \int_{-\infty}^{\infty} r_{\tilde{h}}(\tau;\Delta t) \exp(-j2\pi\tau\Delta f)\, d\tau \tag{9.29}$$

The new function $r_{\tilde{H}}(\Delta f;\Delta t)$ is called the *spaced-frequency, spaced-time correlation function* of the channel, where the double use of "spaced" accounts for Δt and Δf.

Scattering Function of the Channel

Finally, we introduce another new function denoted by $S(\tau;\nu)$ that forms a Fourier-transform pair with the multipath correlation profile $r_{\tilde{h}}(\tau;\Delta t)$ with respect to the variable Δt; that is, by definition, we have

$$S(\tau;\nu) = \int_{-\infty}^{\infty} r_{\tilde{h}}(\tau;\Delta t) \exp(-j2\pi\nu\Delta t)\, d(\Delta t) \tag{9.30}$$

for the Fourier transform and

$$r_{\tilde{h}}(\tau;\Delta t) = \int_{-\infty}^{\infty} S(\tau;\nu) \exp(j2\pi\nu\Delta t)\, d\nu \tag{9.31}$$

for the inverse Fourier transform.

The function $S(\tau; \nu)$ may also be defined in terms of $r_{\tilde{H}}(\Delta f;\Delta t)$ by applying a form of *double Fourier transformation*:

> A Fourier transform with respect to the time variable Δt and an inverse Fourier transform with respect to the frequency variable Δf.

That is to say,

$$S(\tau;\nu) = \int_{-\infty}^{\infty}\int_{-\infty}^{\infty} r_{\tilde{H}}(\Delta f;\Delta t)\exp(-j2\pi\nu\Delta t)\exp(j2\pi\tau\Delta f)\,d(\Delta t)\,d(\Delta f) \tag{9.32}$$

Figure 9.10 displays the functional relationships between the three important functions: $r_{\tilde{h}}(\tau;\Delta t)$, $r_{\tilde{h}}(\Delta f;\Delta t)$, and $S(\tau;\nu)$ in terms of the Fourier transform and its inverse.

The function $S(\tau;\nu)$ is called the *scattering function* of the channel. For a physical interpretation of it, consider the transmission of a single tone of frequency f' relative to the carrier. The complex envelope of the resulting filter output is

$$\tilde{y}(t) = \exp(j2\pi f't)\tilde{H}(f';t) \tag{9.33}$$

The correlation function of $\tilde{y}(t)$ is given by

$$\begin{aligned}\mathbb{E}[\tilde{y}^*(t)\tilde{y}(t+\Delta t)] &= \exp(j2\pi f'\Delta t)\mathbb{E}[\tilde{H}^*(f';t)\tilde{H}^*(f';t+\Delta t)] \\ &= \exp(j2\pi f'\Delta t)r_{\tilde{H}}(0;\Delta t)\end{aligned} \tag{9.34}$$

where, in the last line, we made use of (9.28). Putting $\Delta f = 0$ in (9.29) and then using (9.31), we may write

$$\begin{aligned}r_{\tilde{H}}(0;\Delta t) &= \int_{-\infty}^{\infty} r_{\tilde{h}}(\tau;\Delta t)\,d\tau \\ &= \int_{-\infty}^{\infty}\left[\int_{-\infty}^{\infty} S(\tau;\nu)\,d\tau\right]\exp(j2\pi\nu\Delta t)\,d\nu\end{aligned} \tag{9.35}$$

Hence, we may view the integral inside the square brackets in (9.35), namely

$$\int_{-\infty}^{\infty} S(\tau;\nu)\,d\tau$$

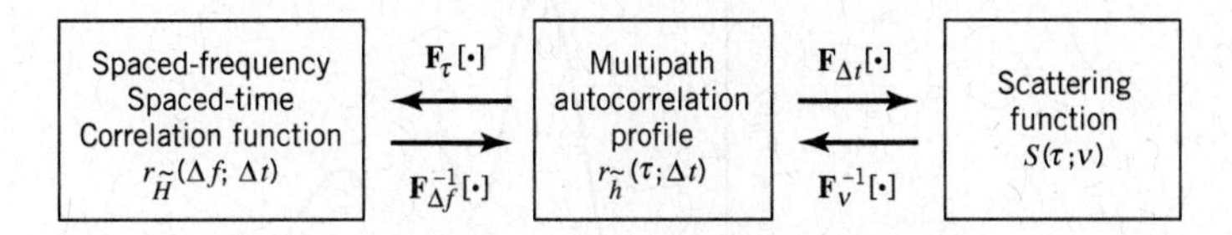

$\mathbf{F}_\tau[\cdot]$: Fourier transform with respect to delay τ

$\mathbf{F}_{\Delta f}^{-1}[\cdot]$: Inverse Fourier transform with respect to frequency increment Δf

$\mathbf{F}_{\Delta t}[\cdot]$: Fourier transform with respect to time increment Δt

$\mathbf{F}_\nu^{-1}[\cdot]$: Inverse Fourier transform with respect to Doppler shift ν

Figure 9.10 Functional relationships between the multipath correlation profile $r_{\tilde{h}}(\tau;\Delta t)$, the spaced-frequency spaced-time correlation function $r_{\tilde{H}}(\Delta f;\Delta t)$, and the scattering function $S(\tau;\nu)$.

as the *power spectral density* of the channel output relative to the frequency f' of the transmitted tone with the Doppler shift ν acting as the frequency variable. Generalizing this result, we may now make the statement:

> The scattering function $S(\tau; \nu)$ provides a statistical measure of the output power of the channel, expressed as a function of the time delay τ and the Doppler shift ν.

Power-Delay Profile

We continue statistical characterization of the wireless channel by putting $\Delta t = 0$ in (9.24) to obtain

$$\begin{aligned} P_{\tilde{h}}(\tau) &= r_{\tilde{h}}(\tau;0) \\ &= \mathbb{E}[|\tilde{h}(\tau;t)|^2] \end{aligned} \tag{9.36}$$

The function $P_{\tilde{h}}(\tau)$ describes the intensity (averaged over the fading fluctuations) of the scattering process at propagation delay τ for the WSSUS channel. Accordingly, $P_{\tilde{h}}(\tau)$ is called the *power-delay profile* of the channel. In any event, this profile provides an estimate of the average multipath power expressed as a function of the delay variable τ.

The power-delay profile may also be defined in terms of the scattering function $S(\tau;\nu)$ by averaging it over all potentially possible Doppler shifts. Specifically, setting $\Delta t = 0$ in (9.31) and then using the first line of (9.36), we obtain

$$P_{\tilde{h}}(\tau) = \int_{-\infty}^{\infty} S(\tau;\nu)\,d\nu \tag{9.37}$$

Figure 9.11 shows an example of the power-delay profile that depicts a typical plot of the power spectral density versus excess delay;[4] the excess delay is measured with respect to the time delay for the shortest echo path. The "threshold level" K included in Figure 9.11 defines the power level below which the receiver fails to operate satisfactorily.

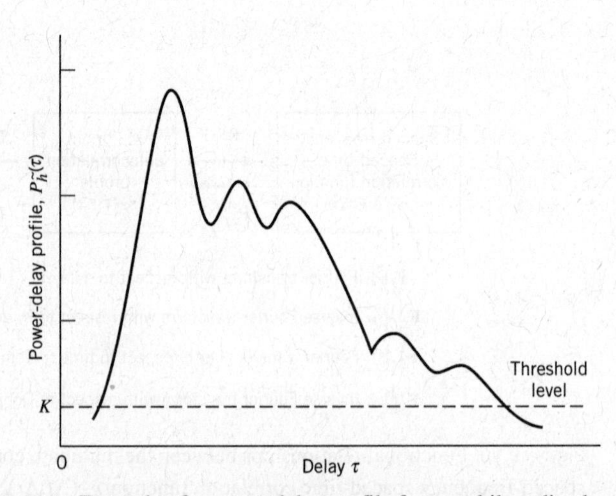

Figure 9.11 Example of a power-delay profile for a mobile radio channel.

Central Moments of $P_{\tilde{h}}(\tau)$

To characterize the power-delay profile of a WSSUS channel in statistical terms, we begin with the moment of order zero; that is, the integrated power averaged over the delay variable τ, as shown by

$$P_{\mathrm{av}} = \int_{-\infty}^{\infty} P_{\tilde{h}}(\tau)\,\mathrm{d}\tau \tag{9.38}$$

The *average delay*, normalized with respect to P_{av}, is defined in terms of the first-order moment by the formula

$$\tau_{\mathrm{av}} = \frac{1}{P_{\mathrm{av}}}\int_{-\infty}^{\infty} \tau P_{\tilde{h}}(\tau)\,\mathrm{d}\tau \tag{9.39}$$

Correspondingly, the second-order central moment, normalized with respect to P_{av}, is defined by the root-mean-square (RMS) formula

$$\sigma_{\tau} = \left[\frac{1}{P_{\mathrm{av}}}\int_{-\infty}^{\infty} (\tau - \tau_{\mathrm{av}})^2 P_{\tilde{h}}(\tau)\,\mathrm{d}\tau\right]^{1/2} \tag{9.40}$$

The new parameter σ_{τ} is called the *delay spread*, which has acquired a special stature among the parameters used to characterize the WSSUS channel.

From Chapter 2 on the representation of signals in a linear environment, we recall that the duration of a signal in the time domain is inversely related to the bandwidth of the signal in the frequency domain. Building on this time–frequency relationship, we may define the *coherence bandwidth* $B_{\mathrm{coherence}}$ of a WSSUS channel as follows:

$$B_{\mathrm{coherence}} = \frac{1}{\tau_{\mathrm{av}}} \tag{9.41}$$

In words:

> The coherence bandwidth of the WSSUS channel is that band of frequencies for which the frequency response of the channel is strongly correlated.

This statement is intuitively satisfying.

Doppler Power Spectrum

Consider next the issue of relating Doppler effects to time variations of the channel. In direct contrast to the power-delay profile, this time we set $\Delta f = 0$, which corresponds to the transmission of a single tone (of some appropriate frequency) over the channel. Under this condition, the spaced-frequency, spaced-time correlation function of the channel, described in (9.29), reduces to $r_{\tilde{H}}(0;\Delta t)$. Hence, evaluating the Fourier transform of this function with respect to the time variable Δt, we may write

$$S_{\tilde{H}}(\nu) = \int_{-\infty}^{\infty} r_{\tilde{H}}(0;\Delta t)\exp(-\mathrm{j}2\pi\nu\Delta t)\,\mathrm{d}(\Delta t) \tag{9.42}$$

The function $S_{\tilde{H}}(\nu)$ defines the power spectrum of the channel output expressed as a function of the Doppler shift ν; it is therefore called the *Doppler power spectrum* of the channel.

The Doppler-power spectrum of (9.42) may be interpreted in two insightful ways (Molisch, 2011):

1. The Doppler spectrum describes the frequency dispersion of a wireless channel, which results in the occurrence of transmission errors in narrowband mobile wireless communication systems.
2. The Doppler spectrum provides a measure of temporal variability of the channel, which, in mathematical terms, is described by the channel's correlation function $r_{\tilde{H}}(0;\Delta t)$ for $\Delta f = 0$.

As such, we may view the Doppler-power spectrum as another important statistical characterization of WSSUS channels.

The Doppler power spectrum may also be defined in terms of the scattering function by averaging it over all possible propagation delays, as shown by

$$S_{\tilde{H}}(\nu) = \int_{-\infty}^{\infty} S(\tau;\nu)\,d\tau \tag{9.43}$$

Typically, the Doppler shift ν assumes positive and negative values with almost equal likelihood. The mean Doppler shift is therefore effectively zero. The square root of the second moment of the Doppler spectrum is thus defined by

$$\sigma_{\nu} = \left(\frac{\int_{-\infty}^{\infty} \nu^2 S_{\tilde{H}}(\nu)\,d\nu}{\int_{-\infty}^{\infty} S_{\tilde{H}}(\nu)\,d\nu} \right)^{1/2} \tag{9.44}$$

The parameter σ_{ν} provides a measure of the width of the Doppler spectrum; therefore, it is called the *Doppler spread* of the channel.

Another useful parameter that is often used in radio propagation measurements is the *fade rate* of the channel. For a Rayleigh fading channel, the *average fade rate* is related to the Doppler spread σ_{ν} by the empirical rule:

$$f_{\text{fade rate}} = 1.475\,\sigma_{\nu} \text{ crossings per second} \tag{9.45}$$

As the name implies, the fade rate provides a measure of the rapidity of the channel fading phenomenon.

Some typical values encountered in a mobile radio environment are as follows:

- the delay spread σ_{τ} amounts to about 20 μs;
- the Doppler spread σ_{ν} due to the motion of a vehicle may typically occupy the range 40–100 Hz, but sometimes may well exceed 100 Hz.

One other parameter directly related to the Doppler spread is the *coherence time* of the channel. Here again, as with coherence bandwidth discussed previously, we may invoke the inverse time–frequency relationship to say that the coherence time of a multipath wireless channel is inversely proportional to the Doppler spread, as shown by

$$\begin{aligned} \tau_{\text{coherence}} &= \frac{1}{\sigma_{\nu}} \\ &\approx \frac{0.3}{2\nu_{\max}} \end{aligned} \tag{9.46}$$

where ν_{max} is the maximum Doppler shift due to motion of the mobile unit. In words:

> The coherence time of the channel is that duration for which the time response of the channel is strongly correlated.

Here again, this statement is intuitively satisfying.

Classification of Multipath Channels

The particular form of fading experienced by a multipath channel depends on whether the channel characterization is viewed in the frequency domain or the time domain:

1. When the channel is viewed in the frequency domain, the parameter of concern is the channel's *coherence bandwidth* $B_{\text{coherence}}$, which is a measure of the transmission bandwidth for which signal distortion across the channel becomes noticeable. A multipath channel is said to be *frequency selective* if the coherence bandwidth of the channel is small compared with the bandwidth of the transmitted signal. In such a situation, the channel has a filtering effect, in that two sinusoidal components with a frequency separation greater than the channel's coherence bandwidth are treated differently. If, however, the coherence bandwidth of the channel is large compared with the transmitted signal bandwidth, the fading is said to be *frequency nonselective*, or *frequency flat*.
2. When the channel is viewed in the time domain, the parameter of concern is the *coherence time* $\tau_{\text{coherence}}$, which provides a measure of the transmitted signal duration for which distortion across the channel becomes noticeable. The fading is said to be *time selective* if the coherence time of the channel is small compared with the duration of the received signal (i.e., the time for which the signal is in flight). For

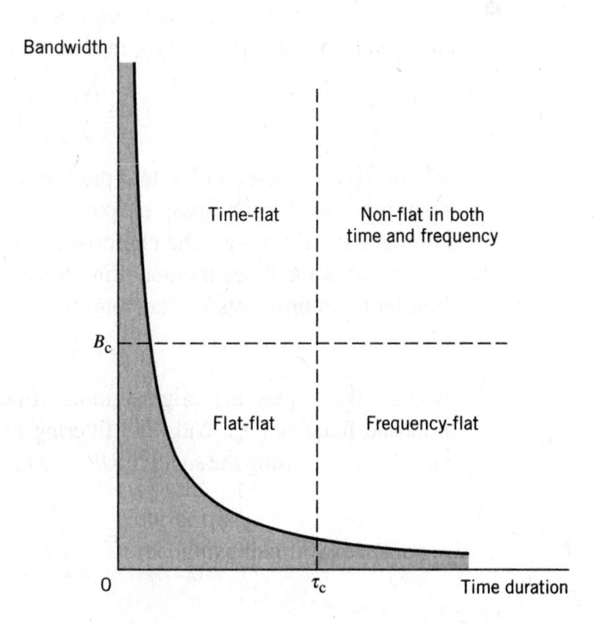

Figure 9.12
Illustrating the four classes of multipath channels: τ_c = coherence time, B_c = coherence bandwidth.

digital transmission, the received signal's duration is taken as the symbol duration plus the channel's delay spread. If, however, the channel's coherence time is large compared with the received signal duration, then the fading is said to be *time nonselective*, or *time flat*, in the sense that the channel appears to the transmitted signal as time invariant.

In light of this discussion, we may classify multipath channels as follows:

- *Flat-flat channel*, which is flat in both frequency and time.
- *Frequency-flat channel*, which is flat in frequency only.
- *Time-flat channel*, which is flat in time only.
- *Completely non-flat channel*, which is flat neither in frequency nor in time; such a channel is also referred to as a *doubly spread channel*.

The classification of multipath channels, based on this approach, is shown in Figure 9.12. The forbidden area, shown shaded in this figure, follows from the inverse relationship that exists between bandwidth and time duration.

9.5 FIR Modeling of Doubly Spread Channels

In Section 9.4, statistical analysis of the doubly spread channel was carried out by focusing on two complex low-pass entities, namely the impulse response $\tilde{h}(\tau;t)$ and the corresponding transfer function $\tilde{H}(f;t)$. Therein, mathematical simplification was accomplished by disposing of the midband frequency f_c of the actual band-pass character of the doubly spread channel. Despite this simplification, the analytic approach used in Section 9.4 is highly demanding in mathematical terms. In this section, we will take an "approximate" approach based on the use of a FIR filter to model the doubly spread channel.[5] From an engineering perspective. this new approach has a great deal of practical merit.

To begin, we use the *convolution integral* to describe the input–output relationship of the system, as shown in (9.20), reproduced here for convenience of presentation

$$\tilde{y}(t) = \frac{1}{2}\int_{-\infty}^{\infty} \tilde{h}(\tau;t)\tilde{x}(t-\tau)\,d\tau \tag{9.47}$$

where $\tilde{x}(t)$ is the complex low-pass input signal applied to the channel and $\tilde{y}(t)$ is the resulting complex low-pass output signal. Although this integral can be formulated in another equivalent way, the choice made in (9.47) befits modeling of a time-varying FIR system, as we will see momentarily. Speaking of the input signal $\tilde{x}(t)$, we assume that its Fourier transform satisfies the condition

$$\tilde{X}(f) = 0 \qquad \text{for } f > W \tag{9.48}$$

where $2W$ denotes the original input band-pass signal's bandwidth centered around the midband frequency f_c. With FIR filtering in mind, it is logical to expand the delayed input signal $\tilde{x}(t-\tau)$ using the *sampling theorem*, discussed in Chapter 6. Specifically, we write

$$\tilde{x}(t-\tau) = \sum_{n=-\infty}^{\infty} \tilde{x}(t-nT)\,\text{sinc}\left(\frac{\tau}{T_s}-n\right) \tag{9.49}$$

where T_s is the sampling period of the FIR filter chosen in accordance with the sampling theorem as follows:

$$\frac{1}{T_s} > 2W \tag{9.50}$$

The sinc function in (9.49) is defined by

$$\operatorname{sinc}\left(\frac{\tau}{T_s} - n\right) = \frac{\sin\left[\pi\left(\frac{\tau}{T_s} - n\right)\right]}{\pi\left(\frac{\tau}{T_s} - n\right)} \tag{9.51}$$

From the standpoint of the sampling theorem we could set $1/T_s = 2W$, but the choice made in (9.50) gives us more practical flexibility.

In (9.49) it is important to note that we have done the following:

- Dependence on the coordinate functions under the summation has been put on the delay variable τ in the sinc function.
- Dependence on the time-varying FIR coefficients has been put on time t.

This separation of variables is the key to the FIR modeling of a linear time-varying system. Note also that the sinc functions under the summation in (9.49) are orthogonal but not normalized.

Thus, substituting (9.49) into (9.47) and interchanging the order of integration and summation, which is permitted as we are dealing with a linear system, we get

$$\tilde{y}(t) = \sum_{n=-\infty}^{\infty} \tilde{x}\left(\frac{t}{T_s} - n\right)\left[\int_{-\infty}^{\infty} \tilde{h}(\tau;t)\operatorname{sinc}\left(\frac{t}{T_s} - n\right) d\tau\right] \tag{9.52}$$

To simplify matters, we now introduce the *complex tap-coefficients*[6] $\tilde{c}_n(t)$, defined in terms of the complex impulse response as follows:

$$\tilde{c}_n(t) = \frac{1}{2}\int_{-\infty}^{\infty} \tilde{h}(\tau;t)\operatorname{sinc}\left(\frac{t}{T_s} - n\right) d\tau \tag{9.53}$$

Accordingly, we may rewrite (9.52) in the much simplified summation form:

$$\tilde{y}(t) = \sum_{n=-\infty}^{\infty} \tilde{x}\left(\frac{t}{T_s} - n\right)\tilde{c}_n(t) \tag{9.54}$$

Examining (9.54) for insight, we may make our first observation:

> The uniformly sampled functions $\tilde{x}[(t/T) - n]$ are generated as tap-inputs by passing the complex low-pass input signal $\tilde{x}(t)$ through a TDL filter whose taps are spaced T seconds apart.

Turning next to (9.53) for insight, refer to Figure 9.13, where this equation is sketched for three different settings of the function $\operatorname{sinc}[(t/T_s) - n]$; the area shaded in the figure refers to the complex impulse response $\tilde{h}(\tau;t)$ that is assumed to be causal and occupying a finite duration. In light of the three different sketches shown in Figure 9.13, we may make our second observation.

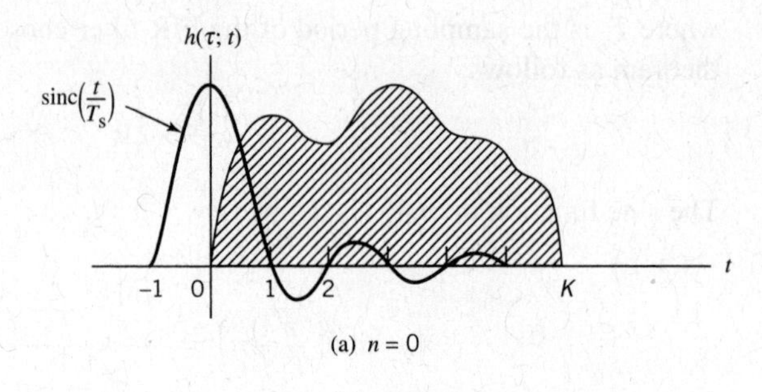

(a) $n = 0$

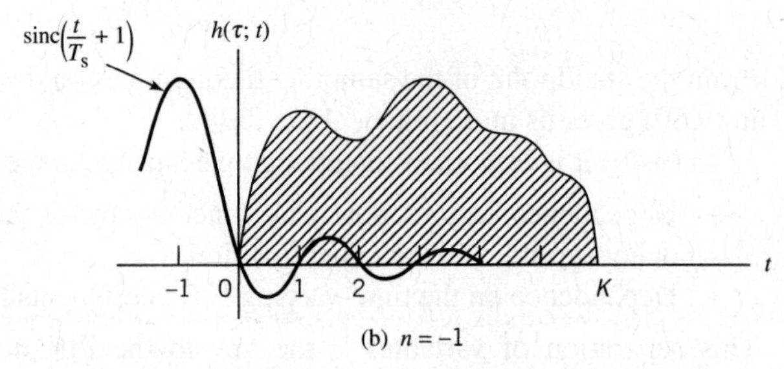

(b) $n = -1$

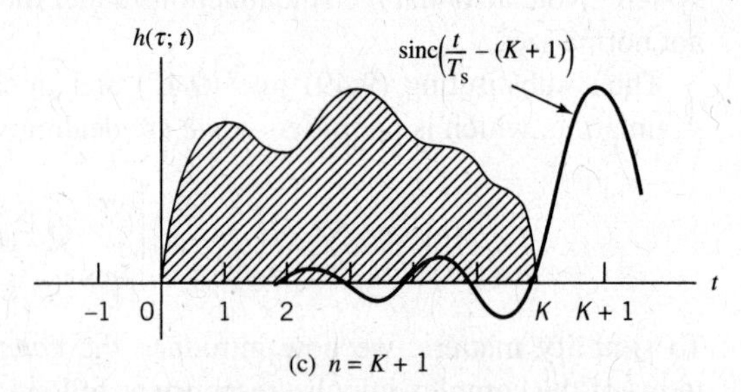

(c) $n = K + 1$

Figure 9.13 Illustrating the way in which location of the sinc weighting function shows up for varying n.

> Assuming that the integral in (9.53) is dominated by the mainlobe of the sinc function, the complex time-varying tap-coefficient $\tilde{c}_n(t)$ is essentially zero for negative values of discrete time n and all positive values of n greater than τ/T.

In accordance with these two observations, we may approximate (9.54) as follows:

$$\tilde{y}(t) \approx \sum_{n=0}^{K} \tilde{x}\left(\frac{t}{T_s} - n\right)\tilde{c}_n(t) \tag{9.55}$$

where K is the number of taps.

Equation (9.55) defines a *complex FIR model* for the representation of a complex low-pass time-varying system characterized by the complex impulse response $\tilde{h}(\tau;T)$. Figure 9.14 depicts a block diagram representation of this model, based on (9.55).

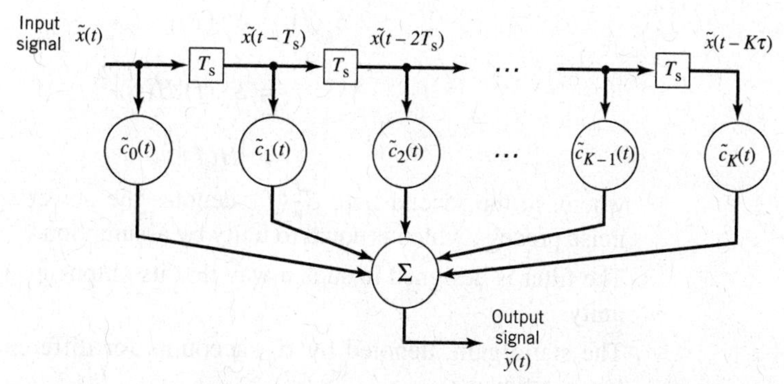

Figure 9.14 Complex FIR model of a complex low-pass time-varying channel.

Some Practical Matters

To model the doubly spread channel by means of a FIR filter in accordance with (9.55), we need to know the sampling rate $1/T_s$ and the number of taps K in this equation. To satisfy these two practical requirements, we offer the following empirical points:

1. The sampling rate of the FIR filter, $1/T_s$, is much higher than the maximum Doppler bandwidth of the channel, ν_{max}; typically, we find that $1/T_s$ is eight to sixteen times ν_{max}. Hence, knowing ν_{max}, we may determine a desirable value of the sampling rate $1/T_s$.
2. The number of taps K in (9.55) may be determined by truncating the power-delay profile $P_{\tilde{h}}(f)$ of the channel. Specifically, given a measurement of this profile, a suitable value of K is determined by choosing a threshold level below which the receiver fails to operate satisfactorily, as illustrated in Figure 9.11.

Generation of the Tap-Coefficients

To generate the tap-coefficients $\tilde{c}_n(t)$, we may use the scheme shown in Figure 9.15 that involves the following (Jeruchim *et al.*, 2000):

1. A complex white Gaussian process of zero mean and unit variance is used as the input.
2. A complex low-pass filter of transfer function $\tilde{H}(f)$ is chosen in such a way that it produces the desired Doppler power spectrum $S_{\tilde{H}}(f)$ where we have used f in place of the Doppler shift ν for convenience of presentation. In other words, we may set

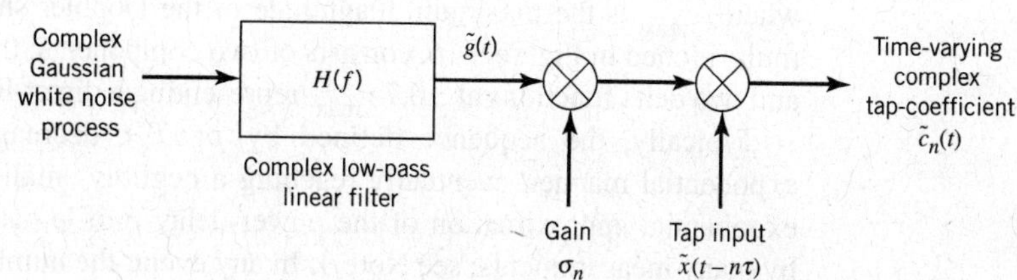

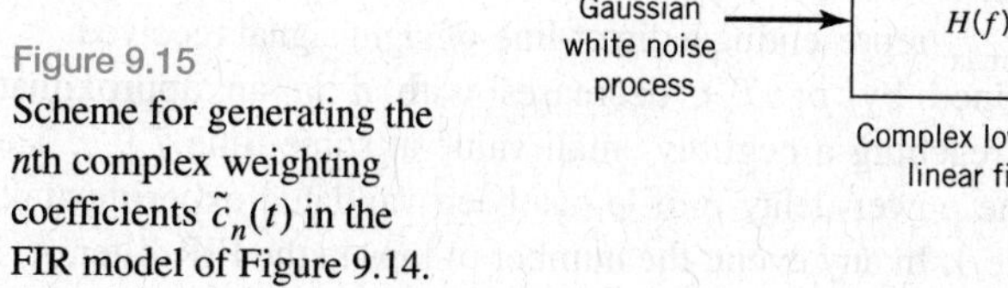

Figure 9.15
Scheme for generating the nth complex weighting coefficients $\tilde{c}_n(t)$ in the FIR model of Figure 9.14.

$$\begin{aligned} S_{\tilde{c}}(f) &= S_{\tilde{H}}(f) \\ &= S_{\tilde{w}}(f)|\tilde{H}(f)|^2 \\ &= |\tilde{H}(f)|^2 \end{aligned} \tag{9.56}$$

where, in the second line, $S_{\tilde{w}}(f)$ denotes the power spectral density of the white noise process, which is equal to unity by assumption.

3. The filter is designed in such a way that its output $\tilde{g}(t)$ has a normalized power of unity.
4. The static gain, denoted by σ_n, accounts for different variances of the different tap-coefficients.

EXAMPLE 3 **Rayleigh Processes**

For complex FIR modeling of a time-varying Rayleigh fading channel, we may use zero-mean complex Gaussian processes to represent the time-varying tap-coefficients $\tilde{c}_n(t)$, which, in turn, means that the complex impulse response of the channel $\tilde{h}(\tau;t)$ is also a zero-mean Gaussian process in the variable t.

Moreover, under the assumption of a *WSSUS channel*, the tap-coefficients $\tilde{c}_n(t)$ for varying n will be uncorrelated. The power spectral density of each tap-coefficient is specified by the Doppler spectrum. In particular, the variance σ_n^2 of the nth weight function is approximately given by

$$\mathbb{E}[|\tilde{c}_n(t)|^2] \approx T_s^2 p(n\tau) \tag{9.57}$$

where T_s is the sampling period of the FIR and $p(n\tau)$ is a discrete version of the *power-delay profile*, $P_{\tilde{h}}(\tau)$.

EXAMPLE 4 **Rician–Jakes Doppler Spectrum Model**

The Jakes model, discussed in Example 1, is well suited for describing the Doppler spectrum for a dense-scattering environment, exemplified by an urban area. However, in a *rural environment*, there is a high likelihood for the presence of one strong "direct line-of-sight" path, for which the FIR-based Rician model is an appropriate candidate. In such an environment, we may use the *Rician–Jakes Doppler spectrum* that has the following form (Tranter *et al.*, 2004):

$$S_{\tilde{c}}(f) = \frac{0.41}{\sqrt{1-(f/\nu_{max})^2}} + 0.91\,\delta(f \pm 0.7\,\nu_{max}) \tag{9.58}$$

where ν_{max} is the maximum magnitude of the Doppler shift. This partially empirical formula, plotted in Figure 9.16, consists of two components: the FIR Jakes filter of Example 1, and two delta functions at $\pm 0.7\,\nu_{max}$ representing a direct-line-of sight signal received.

Typically, the sequence defined by $p(nT_s)$ decreases with n in an approximate exponential manner, eventually reaching a neglibly small value at some time T_{max}. This exponential approximation of the power-delay profile has been validated experimentally by many measurements; see Note 4. In any event, the number of taps in the FIR filter, K, is

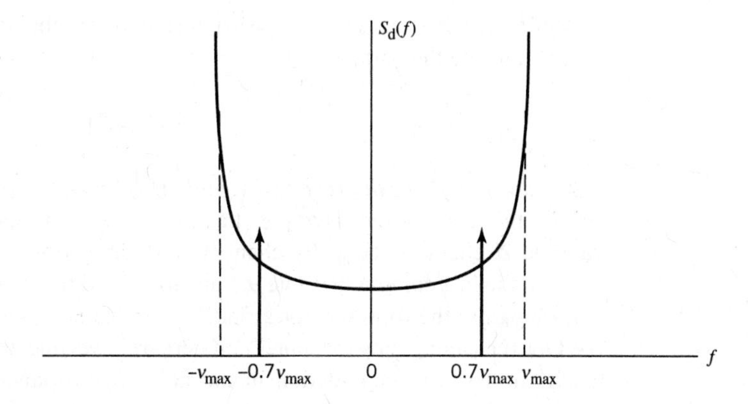

Figure 9.16
Illustrating the Rician–Jakes Doppler spectrum of (9.58).

approximately defined by the ratio T_{max}/T_s. The point made here on the number of taps K substantiates what has been made previously on Jakes model in Example 1 and in point 2 under Some Practical Matters in this section.

9.6 Comparison of Modulation Schemes: Effects of Flat Fading

We bring this first part of the chapter to an end by presenting the effects of flat fading on the behavior of different modulation schemes for wireless communications.

In Chapter 7 we studied the subject of signaling over AWGN channels using different modulation schemes and evaluated their performance under two different receiver conditions: coherence and noncoherence. For the purpose of comparison, we have reproduced the BER for a selected number of those modulation schemes in AWGN in Table 9.1.

Table 9.1 **Formulas for the BER of coherent and noncoherent digital receivers**

Signaling scheme	BER: AWGN channel	BER: Flat Rayleigh fading channel
(a) Binary PSK, QPSK, MSK using coherent detection	$Q\left(\sqrt{\frac{2E_b}{N_0}}\right)$	$\frac{1}{2}\left(1-\sqrt{\frac{\gamma_0}{1+\gamma_0}}\right)$
(b) Binary FSK using coherent detection	$Q\left(\sqrt{\frac{E_b}{N_0}}\right)$	$\frac{1}{2}\left(1-\sqrt{\frac{\gamma_0}{2+\gamma_0}}\right)$
(c) Binary DPSK	$\frac{1}{2}\exp\left(-\sqrt{\frac{E_b}{N_0}}\right)$	$\frac{1}{2(1+\gamma_0)}$
(d) Binary FSK using noncoherent detection	$\exp\left(-\sqrt{\frac{E_b}{2N_0}}\right)$	$\frac{1}{2+\gamma_0}$

E_b: transmitted energy per bit; $N_0/2$: power spectral density of channel noise;
γ_0: mean value of the received energy per bit-to-noise spectral density ratio.

Table 9.1 also includes the exact formulas for the BER for a flat Rayleigh fading channel, where the parameter

$$\gamma_0 = \frac{E_b}{N_0}\mathbb{E}[\alpha^2] \tag{9.59}$$

is the *mean value of the received signal energy per bit-to-noise spectral density ratio*. In (9.59), the expectation $\mathbb{E}[\alpha^2]$ is the mean value of the Rayleigh-distributed random variable α characterizing the channel. The derivations of the fading-channel formulas listed in the last column of Table 9.1 are addressed in Problems 9.1 and 9.2.

Comparing the formulas for a flat Rayleigh fading channel with the formulas for their AWGN (i.e., nonfading) channel counterparts, we find that the Rayleigh fading process results in a severe degradation in the noise performance of a wireless communication receiver with the degradation measured in terms of decibels of additional mean SNR spectral density ratio. In particular, the asymptotic decrease in the BER with γ_0 follows an *inverse law*. This form of asymptotic behavior is dramatically different from the case of a nonfading channel, for which the asymptotic decrease in the BER with γ_0 follows an *exponential law*.

In graphical terms, Figure 9.17 plots the formulas under part a of Table 9.1 compared with the BERs of binary PSK over the AWGN and Rayleigh fading channels. The figure also includes corresponding plots for the Rician fading channel with different values of the *Rice factor K*, discussed in Chapter 4. We see that as K increases from zero to infinity, the behavior of the receiver varies all the way from the Rayleigh channel to the AWGN

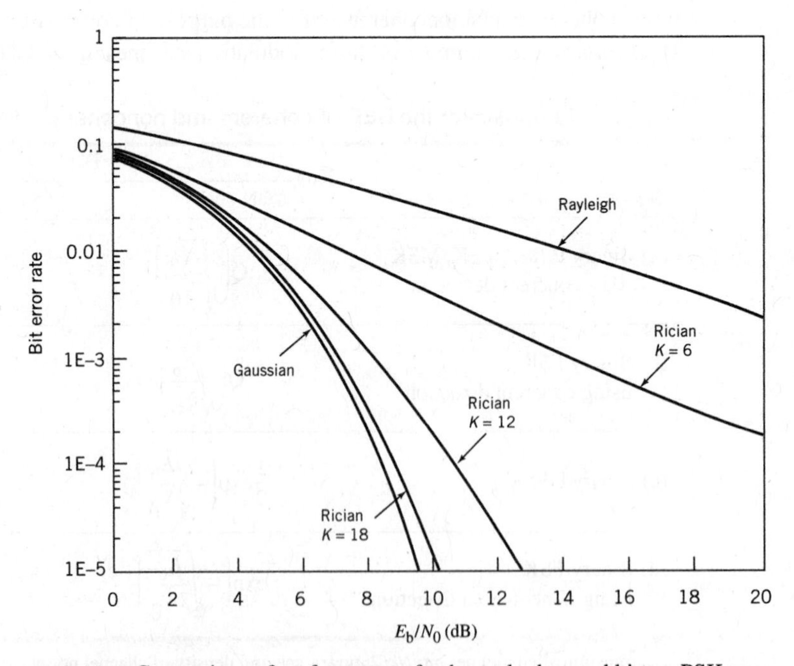

Figure 9.17 Comparison of performance of coherently detected binary PSK over different fading channels.

channel. The results plotted in Figure 9.17 for the Rician channel were obtained using simulations (Haykin and Moher, 2005). From Figure 9.17 we see that, as matters stand, we have a serious problem caused by channel fading. For example, at an SNR of 20 dB and the presence of Rayleigh fading, the use of binary PSK results in a BER of about 3×10^{-2}, which is not good enough for the transmission of speech or digital data over the wireless channel.

9.7 Diversity Techniques

Up to now, we have emphasized the multipath fading phenomenon as an inherent characteristic of a wireless channel, which indeed it is. Given this physical reality, how, then, do we make the communication process across the wireless channel into a *reliable* operation? The answer to this fundamental question lies in the use of *diversity*, which may be viewed as a form of *redundancy* in a spatial context. In particular, if several replicas of the information-bearing signal can be transmitted simultaneously over independently fading channels, then there is a good likelihood that at least one of the received signals will not be severely degraded by channel fading. There are several methods for making such a provision. In the context of the material covered in this book, we identify three approaches to diversity:

1. *Frequency diversity*, in which the information-bearing signal is transmitted using several carriers that are spaced sufficiently apart from each other to provide independently fading versions of the signal. This may be accomplished by choosing a frequency spacing equal to or larger than the coherence bandwidth of the channel.
2. *Time diversity*, in which the same information-bearing signal is transmitted in different time slots, with the interval between successive time slots being equal to or greater than the coherence time of the channel. We can still get some diversity if the interval is less than the coherence time of the channel, but at the expense of degraded performance. In any event, time diversity may be likened to the use of a repetition code for error-control coding.
3. *Space diversity*, in which multiple transmit or receive antennas, or both, are used with the spacing between adjacent antennas being chosen so as to ensure the independence of possible fading events occurring in the channel.

Among these three kinds of diversity, space diversity is the subject of interest in the second part of this chapter. Depending on which end of the wireless link is equipped with multiple antennas, we may identify three different forms of space diversity:

1. *Receive diversity,* which involves the use of a single transmit antenna and multiple receive antennas.
2. *Transmit diversity*, which involves the use of multiple transmit antennas and a single receive antenna.
3. *Diversity on both transmit and receive*, which combines the use of multiple antennas at both the transmitter and receiver.

Receive diversity is the oldest one of the three, with the other two being of more recent origin. In what follows, we will study these three different forms of diversity in this order.

9.8 "Space Diversity-on-Receive" Systems

In "space diversity on receive," multiple receiving antennas are used with the spacing between adjacent antennas being chosen so that their respective outputs are essentially independent of each other. This requirement may be satisfied by spacing the adjacent receiving antennas by as much as 10 to 20 radio wavelengths or less apart from each other. Typically, an elemental spacing of several radio wavelengths is deemed to be adequate for space diversity on receive. The much larger spacing is needed for elevated base stations, for which the angle spread of the incoming radio waves is small; note that the spatial coherence distance is inversely proportional to the angle spread. Through the use of diversity on receive as described here, we create a corresponding set of fading channels that are essentially independent. The issue then becomes that of combining the outputs of these statistically independent fading channels in accordance with a criterion that will provide improved receiver performance. In this section, we describe three different diversity-combining systems that do share a common feature: they all involve the use of linear receivers; hence the relative ease of their mathematical tractability.

Selection Combining

The block diagram of Figure 9.18 depicts a diversity-combining structure that consists of two functional blocks: N_r linear receivers and a logic circuit. This diversity system is said to be of a *selection combining* kind, in that given the N_r receiver outputs produced by a common transmitted signal, the logic circuit *selects* the particular receiver output with the *largest SNR* as the received signal. In conceptual terms, selection combining is the simplest form of space-diversity-on-receive system.

To describe the benefit of selection combining in statistical terms, we assume that the wireless communication channel is described by a *frequency-flat, slowly fading Rayleigh channel.* The implications of this assumption are threefold:

1. The frequency-flat assumption means that all the frequency components constituting the transmitted signal experience the same random attenuation and phase shift.
2. The slow-fading assumption means that fading remains essentially unchanged during the transmission of each symbol.
3. The fading phenomenon is described by the Rayleigh distribution.

Let $\tilde{s}(t)$ denote the complex envelope of the modulated signal transmitted during the symbol interval $0 \le t \le T$. Then, in light of the assumed channel, the complex envelope of the received signal of the kth diversity branch is defined by

$$\tilde{x}_k(t) = \alpha_k \exp(j\theta_k)\tilde{s}(t) + \tilde{w}_k(t), \qquad \begin{array}{l} 0 \le t \le T \\ k = 1, 2, \dots, N_r \end{array} \tag{9.60}$$

where, for the kth diversity branch, the fading is represented by the multiplicative term $\alpha_k \exp(j\theta_k)$ and the additive channel noise is denoted by $\tilde{w}_k(t)$. With the fading assumed to be slowly varying relative to the symbol duration T, we should be able to estimate and then remove the unknown phase shift θ_k at each diversity branch with sufficient accuracy, in which case (9.60) simplifies to

$$\tilde{x}_k(t) \approx \alpha_k \tilde{s}(t) + \tilde{w}_k(t), \qquad \begin{array}{l} 0 \le t \le T \\ k = 1, 2, \dots, N_r \end{array} \tag{9.61}$$

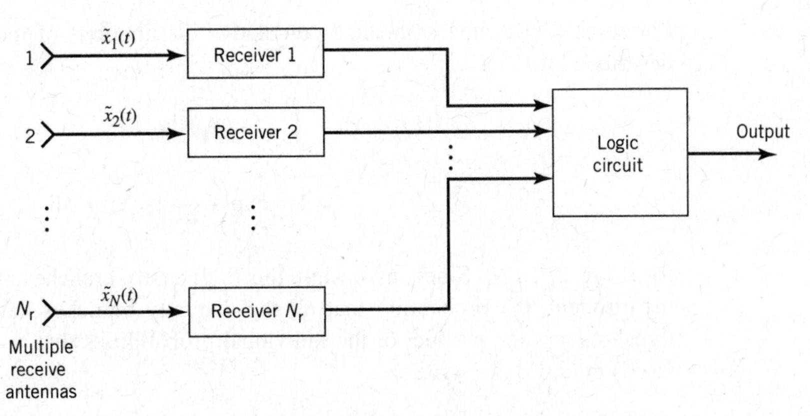

Figure 9.18 Block diagram of selection combiner, using N_r receive antennas.

The signal component of $\tilde{x}_k(t)$ is $\alpha_k \tilde{s}(t)$ and the noise component is $\tilde{w}_k(t)$. The average SNR at the output of the kth receiver is therefore

$$(\text{SNR})_k = \frac{\mathbb{E}[|\alpha_k \tilde{s}(t)|^2]}{\mathbb{E}[|\tilde{w}_k(t)|^2]}$$

$$= \left(\frac{\mathbb{E}[|\tilde{s}(t)|^2]}{\mathbb{E}[|\tilde{w}_k(t)|^2]}\right)\mathbb{E}[\alpha_k^2], \qquad k = 1, 2, \dots, N_r$$

Ordinarily, the mean-square value of $\tilde{w}_k(t)$ is the same for all k. Accordingly, we may express the $(\text{SNR})_k$ as

$$(\text{SNR})_k = \frac{E}{N_0}\mathbb{E}[\alpha_k^2], \qquad k = 1, 2, \dots, N_r \tag{9.62}$$

where E is the symbol energy and $N_0/2$ is the noise spectral density. For binary data, E equals the transmitted signal energy per bit E_b.

Let γ_k denote the *instantaneous* SNR measured at the output of the kth receiver during the transmission of a given symbol. Then, replacing the mean-square value $\mathbb{E}[|\alpha_k|^2]$ by the instantaneous value $|\alpha_k|^2$ in (9.62), we may write

$$\gamma_k = \frac{E}{N_0}\alpha_k^2, \qquad k = 1, 2, \dots, N_r \tag{9.63}$$

Under the assumption that the random amplitude α_k is Rayleigh distributed, the squared amplitude α_k^2 will be *exponentially distributed*[7] (i.e., chi-squared with two degrees of freedom, discussed in Appendix A). If we further assume that the average SNR over the short-term fading is the same, namely γ_{av}, for all the N_r diversity branches, then we may express the probability density functions of the random variables Γ_k pertaining to the individual branches as follows:

$$f_{\Gamma_k}(\gamma_k) = \frac{1}{\gamma_{av}}\exp\left(-\frac{\gamma_k}{\gamma_{av}}\right), \qquad \gamma_k \ge 0, \quad k = 1, 2, \dots, N_r \tag{9.64}$$

For some SNR γ, the associated cumulative distributions of the individual branches are described by

$$\begin{aligned}\mathbb{P}(\gamma_k \le \gamma) &= \int_{-\infty}^{\gamma} f_{\Gamma_k}(\gamma_k)\,\mathrm{d}\gamma_k \\ &= 1-\exp\left(-\frac{\gamma}{\gamma_{\mathrm{av}}}\right), \qquad \gamma \ge 0\end{aligned} \tag{9.65}$$

for $k = 1, 2,\ldots, N_r$. Since, by design, the N_r diversity branches are essentially statistically independent, the probability that all the diversity branches have an SNR less than the threshold γ is the product of the individual probabilities that $\gamma_k < \gamma$ for all k; thus, using (9.64) in (9.65), we write

$$\begin{aligned}\mathbb{P}(\gamma_k < \gamma) &= \prod_{k=1}^{N_r} \mathbb{P}(\gamma_k < \gamma) \\ &= \prod_{k=1}^{N_r}\left[1-\exp\left(-\frac{\gamma}{\gamma_{\mathrm{av}}}\right)\right] \\ &= \left[1-\exp\left(-\frac{\gamma}{\gamma_{\mathrm{av}}}\right)\right]^{N_r}, \qquad \gamma \ge 0\end{aligned} \tag{9.66}$$

for $k = 1, 2, \ldots, N_r$; note that the probability in (9.66) decreases with increasing N_r.

The cumulative distribution function of (9.66) is the same as the cumulative distribution function of the random variable Γ_{sc} described by the sample value

$$\gamma_{\mathrm{sc}} = \max\{\gamma_1, \gamma_2, \ldots, \gamma_{N_r}\} \tag{9.67}$$

which is less than the threshold γ if, and only if, the individual SNRs $\gamma_1, \gamma_2, \ldots, \gamma_{N_r}$ are all less than γ. Indeed, the cumulative distribution function of the selection combiner (i.e., the probability that all of the N_r diversity branches have an SNR less than γ) is given by

$$F_{\Gamma}(\gamma_{\mathrm{sc}}) = \left[1-\exp\left(-\frac{\gamma_{\mathrm{sc}}}{\gamma_{\mathrm{av}}}\right)\right]^{N_r}, \qquad \gamma_{\mathrm{sc}} \ge 0 \tag{9.68}$$

By definition, the probability density function $f_{\Gamma}(\gamma_{\mathrm{sc}})$ is the derivative of the cumulative distribution function $F_{\Gamma}(\gamma_{\mathrm{sc}})$ with respect to the argument γ_{sc}. Hence, differentiating (9.68) with respect to γ_{sc} yields

$$\begin{aligned}f_{\Gamma}(\gamma_{\mathrm{sc}}) &= \frac{\mathrm{d}}{\mathrm{d}\gamma_{\mathrm{sc}}}F_{\Gamma}(\gamma_{\mathrm{sc}}) \\ &= \frac{N_r}{\gamma_{\mathrm{av}}}\exp\left(-\frac{\gamma_{\mathrm{sc}}}{\gamma_{\mathrm{av}}}\right)\left[1-\exp\left(-\frac{\gamma_{\mathrm{sc}}}{\gamma_{\mathrm{av}}}\right)\right]^{N_r-1}, \qquad \gamma_{\mathrm{sc}} \ge 0\end{aligned} \tag{9.69}$$

For convenience of graphical presentation, we use the scaled probability density function

$$f_X(x) = \gamma_{\mathrm{av}} f_{\Gamma_{\mathrm{sc}}}(\gamma_{\mathrm{sc}})$$

where the sample value x of the normalized variable X is defined by

$$x = \gamma_{sc}/\gamma_{av}$$

Figure 9.19 plots $f_X(x)$ versus x for varying number of receive-diversity branches N_r under the assumption that the short-term SNRs for all the N_r branches share the common value γ_{av}. From this figure we make two observations:

1. As the number of diversity branches N_r is increased, the probability density function $f_X(x)$ of the normalized random variable $X = \Gamma/\gamma_{av}$ progressively moves to the right.
2. The probability density function $f_X(x)$ becomes more and more symmetrical and, therefore, Gaussian as N_r is increased.

Stated in another way, a frequency-flat, slowly fading Rayleigh channel is modified through the use of selection combining into a Gaussian channel provided that the number of diversity channels N_r is sufficiently large. Realizing that a Gaussian channel is a *digital communication theorist's dream*, we now see the practical benefit of using selection combining.

According to the theory described herein, the selection-combining procedure requires that we monitor the receiver outputs in a continuous manner and, at each instant of time, select the receiver with the strongest signal (i.e., the largest instantaneous SNR). From a practical perspective, such a selective procedure is rather cumbersome. We may overcome

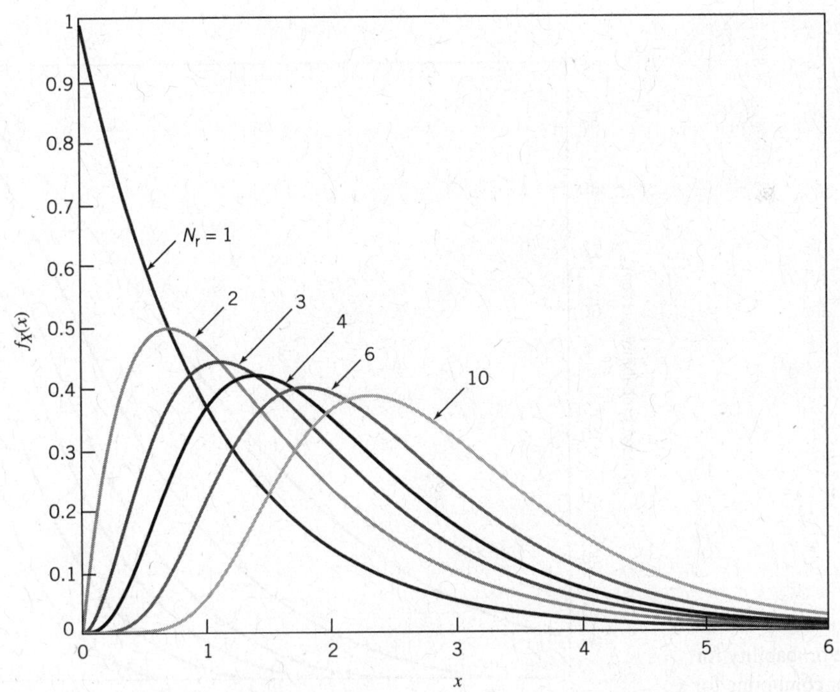

Figure 9.19 Normalized probability density function $f_X(x) = N_r \exp(-x)[1 - \exp(-x)]^{N_r - 1}$ for a varying number N_r of receive antennas.

this practical difficulty by adopting a *scanning* version of the selection-combining procedure:

- Start the procedure by selecting the receiver with the strongest output signal.
- Maintain using the output of this particular receiver as the combiner's output so long as its instantaneous SNR does not drop below a prescribed threshold.
- As soon as the instantaneous SNR of the combiner falls below the threshold, select a new receiver that offers the strongest output signal and continue the procedure.

This technique has a performance very similar to the nonscanning version of selective diversity.

EXAMPLE 5 Outage Probability of Selection Combiner

The *outage probability* of a diversity combiner is defined as the *percentage of time the instantaneous output SNR of the combiner is below some prescribed level for a specified number of branches*. Using the cumulative distribution function of (9.68), Figure 9.20 plots the outage curves for the selection combiner with N_r as the running parameter. The horizontal axis of the figure represents the instantaneous output SNR of the combiner relative to 0 dB (i.e., the 50-percentile point for $N_r = 1$) and the vertical axis represents the outage probability, expressed as a percentage. From the figure we observe the following:

> The fading depth introduced through the use of space diversity on receive diminishes rapidly with the increase in the number of diversity branches.

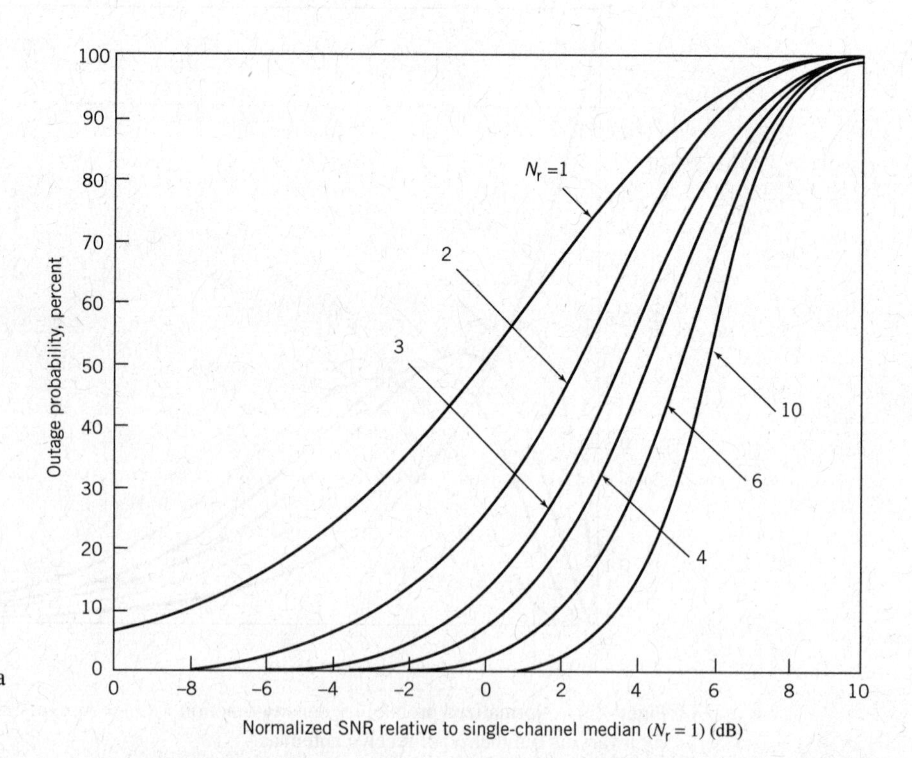

Figure 9.20 Outage probability for selector combining for a varying number N_r of receive antennas.

Maximal-Ratio Combining

The selection-combining technique just described is relatively straightforward to implement. However, from a performance point of view, it is not optimum, in that it ignores the information available from all the diversity branches except for the particular branch that produces the largest instantaneous power of its own demodulated signal.

This limitation of the selection combiner is mitigated by the *maximal-ratio combiner*,[8] the composition of which is described by the block diagram of Figure 9.21 that consists of N_r linear receivers followed by a linear combiner. Using the complex envelope of the received signal at the kth diversity branch given in (9.60), the corresponding complex envelope of the linear combiner output is defined by

$$\begin{aligned}\tilde{y}(t) &= \sum_{k=1}^{N_r} a_k \tilde{x}_k(t) \\ &= \sum_{k=1}^{N_r} a_k[\alpha_k \exp(j\theta_k)\tilde{s}(t) + \tilde{w}_k(t)] \\ &= \tilde{s}(t)\sum_{k=1}^{N_r} a_k\alpha_k \exp(j\theta_k) + \sum_{k=1}^{N_r} a_k\tilde{w}_k(t)\end{aligned} \tag{9.70}$$

where the a_k are *complex weighting parameters* that characterize the linear combiner. These parameters are changed from instant to instant in accordance with signal variations in the N_r diversity branches over the short-term fading process. The requirement is to design the linear combiner so as to maximize the output SNR of the combiner at each instant of time. From (9.70), we note the following two points:

1. The complex envelope of the output signal equals the first expression $\tilde{s}(t)\sum_{k=1}^{N_r} a_k\alpha_k \exp(j\theta_k)$.
2. The complex envelope of the output noise equals the second expression $\sum_{k=1}^{N_r} a_k\tilde{w}_k(t)$.

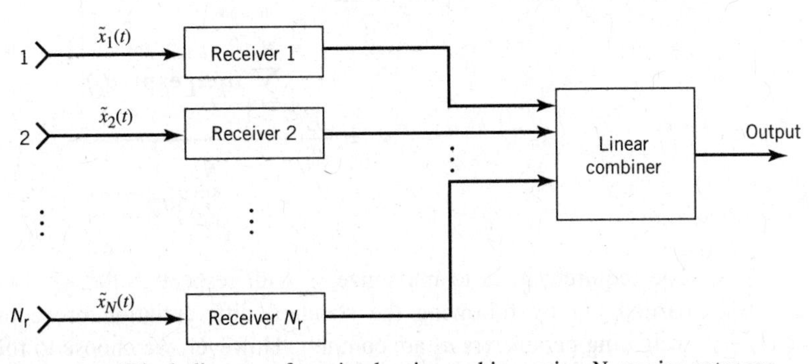

Figure 9.21 Block diagram of maximal-ratio combiner using N_r receive antennas.

Assuming that the $\tilde{w}_k(t)$ are mutually independent for $k = 1, 2, \ldots, N_r$, the output SNR of the linear combiner is therefore given by

$$
\begin{aligned}
(\text{SNR})_c &= \frac{\mathbb{E}\left[\left|\tilde{s}(t) \sum_{k=1}^{N_r} a_k \alpha_k \exp(j\theta_k)\right|^2\right]}{\mathbb{E}\left[\left|\sum_{k=1}^{N_r} a_k \tilde{w}_k(t)\right|^2\right]} \\
&= \frac{\mathbb{E}[|\tilde{s}(t)|^2]}{\mathbb{E}[|\tilde{w}_k(t)|^2]} \frac{\mathbb{E}\left[\left|\sum_{k=1}^{N_r} a_k \alpha_k \exp(j\theta_k)\right|^2\right]}{\mathbb{E}\left[\sum_{k=1}^{N_r} |a_k|^2\right]} \\
&= \frac{E}{N_0} \frac{\mathbb{E}\left[\left|\sum_{k=1}^{N_r} a_k \alpha_k \exp(j\theta_k)\right|^2\right]}{\mathbb{E}\left[\sum_{k=1}^{N_r} |a_k|^2\right]}
\end{aligned}
\tag{9.71}
$$

where E/N_0 is the *symbol energy-to-noise spectral density ratio.*

Let γ_c denote the *instantaneous output SNR* of the linear combiner. Then, using the two terms

$$
\left|\sum_{k=1}^{N_r} a_k \alpha_k \exp(j\theta_k)\right|^2 \quad \text{and} \quad \sum_{k=1}^{N_r} |a_k|^2
$$

as the instantaneous values of the expectations in the numerator and denominator of (9.71), respectively, we may write

$$
\gamma_c = \frac{E}{N_0} \frac{\left|\sum_{k=1}^{N_r} a_k \alpha_k \exp(j\theta_k)\right|^2}{\sum_{k=1}^{N_r} |a_k|^2} \tag{9.72}
$$

The requirement is to maximize γ_c with respect to the a_k. This maximization may be carried out by following the standard differentiation procedure, recognizing that the weighting parameters a_k are complex. However, we choose to follow a simpler procedure based on the Schwarz inequality, which was discussed in Chapter 7.

Let a_k and b_k denote any two complex numbers for $k = 1, 2, \ldots, N_r$. According to the *Schwarz inequality* for complex parameters, we have

$$\left|\sum_{k=1}^{N_r} a_k b_k\right|^2 \le \sum_{k=1}^{N_r} |a_k|^2 \sum_{k=1}^{N_r} |b_k|^2 \tag{9.73}$$

which holds with equality for $a_k = cb_k^*$, where c is some arbitrary complex constant and the asterisk denotes complex conjugation.

Thus, applying the Schwarz inequality to the instantaneous output SNR of (9.72), with a_k left intact and b_k set equal to $\alpha_k \exp(\mathrm{j}\theta_k)$, we obtain

$$\gamma_c \le \frac{E}{N_0} \frac{\sum_{k=1}^{N_r} |a_k|^2 \sum_{k=1}^{N_r} |\alpha_k \exp(\mathrm{j}\theta_k)|^2}{\sum_{k=1}^{N_r} |a_k|^2}$$

Canceling common terms in the numerator and denominator, we readily obtain

$$\gamma_c \le \frac{E}{N_0} \sum_{k=1}^{N_r} \alpha_k^2 \tag{9.74}$$

Equation (9.74) proves that, in general, γ_c cannot exceed $\sum_k \gamma_k$, where γ_k is as defined in (9.63). The equality in (9.74) holds for

$$\begin{aligned} a_k &= c[\alpha_k \exp(\mathrm{j}\theta_k)]^* \\ &= c\alpha_k^* \exp(-\mathrm{j}\theta_k), \qquad k = 1, 2, \ldots, N_r \end{aligned} \tag{9.75}$$

where c is some arbitrary complex constant.

Equation (9.75) defines the complex weighting parameters of the maximal-ratio combiner. Based on this equation, we may state that the optimal weighting factor a_k for the kth diversity branch has a magnitude proportional to the signal amplitude α_k and a phase that cancels the signal phase θ_k to within some value that is identical for all the N_r diversity branches. The phase alignment just described has an important implication: it permits the *fully coherent addition* of the N_r receiver outputs by the linear combiner.

Equation (9.74) with the equality sign defines the instantaneous output SNR of the maximal-ratio combiner, which is written as

$$\gamma_{\text{mrc}} = \frac{E}{N_0} \sum_{k=1}^{N_r} \alpha_k^2 \tag{9.76}$$

According to (9.62), $(E/N_0)\alpha_k^2$ is the *instantaneous output SNR* of the kth diversity branch. Hence, the maximal-ratio combiner produces an instantaneous output SNR that is the sum of the instantaneous SNRs of the individual branches; that is,

$$\gamma_{\text{mrc}} = \sum_{k=1}^{N_r} \gamma_k \tag{9.77}$$

The term "maximal-ratio combiner" has been coined to describe the combiner of Figure 9.21 that produces the optimum result given in (9.77). Indeed, we deduce from this result that the instantaneous output SNR of the maximal-ratio combiner can be large even when the SNRs of the individual branches are small. Since the instantaneous SNR produced by the selection combiner is simply the largest among the N_r terms of (9.77), it follows that:

> The selection combiner is clearly inferior in performance to the maximal-ratio combiner.

The maximal SNR γ_{mrc} is the sample value of a random variable denoted by Γ. According to (9.76), γ_{mrc} is equal to the sum of N_r exponentially distributed random variables for a frequency-flat, slowly fading Rayleigh channel. From Appendix A, the probability density function of such a sum is known to be *chi-square with* $2N_r$ *degrees of freedom*; that is,

$$f_\Gamma(\gamma_{mrc}) = \frac{1}{(N_r - 1)!}\frac{\gamma_{mrc}^{N_r-1}}{\gamma_{av}^{N_r}}\exp\left(-\frac{\gamma_{mrc}}{\gamma_{av}}\right) \tag{9.78}$$

Note that for $N_r = 1$, (9.69) and (9.78) assume the same value, which is to be expected.

Figure 9.22 plots the scaled probability density function, $f_X(x) = \gamma_{av} f_\Gamma(\gamma_{mrc})$, versus the normalized variable $x = \gamma_{mrc}/\gamma_{av}$ for varying N_r. Based on this figure, we may make

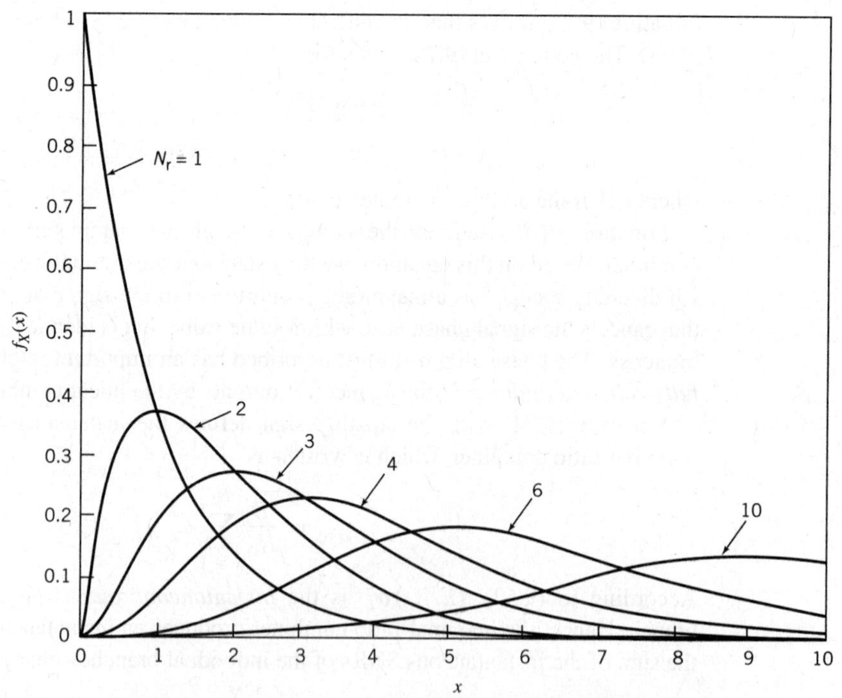

Figure 9.22 Normalized probability density function $f_X(x) = \frac{1}{(N_r - 1)}x^{N_r - 1}\exp(-x)$ for a varying number of N_r receive antennas.

observations similar to those for the selection combiner, except for the fact that for any N_r we find that the scaled probability density function for the maximal-ratio combiner is radically different from its counterpart for the selection combiner.

EXAMPLE 6 Outage Probability for Maximal-Ratio Combiner

The cumulative distribution function for the maximal-ratio combiner is defined by

$$\begin{aligned}\mathbb{P}(\gamma_{\text{mrc}} < x) &= \int_0^x f_\Gamma(\gamma_{\text{mrc}})\, d\gamma_{\text{mrc}} \\ &= 1 - \int_x^\infty f_\Gamma(\gamma_{\text{mrc}})\, d\gamma_{\text{mrc}}\end{aligned} \tag{9.79}$$

where the probability density function $f_\Gamma(\gamma_{\text{mrc}})$ is itself defined by (9.78). Using (9.79), Figure 9.23 plots the outage probability for the maximal-ratio combiner with N_r as a running parameter. Comparing this figure with that of Figure 9.20 for selection combining, we see that the outage-probability curves for these two diversity techniques are superficially similar. The *diversity gain*, defined as the E/N_0 saving at a given BER, provides a measure of the effectiveness of a diversity technique on an outage-probability basis.

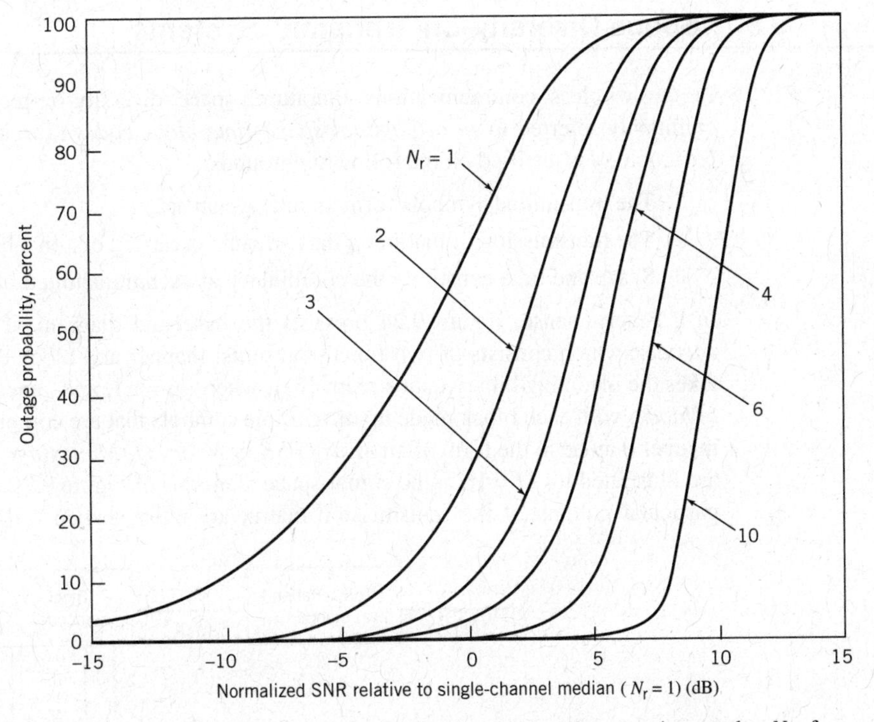

Figure 9.23 Outage probability of maximal-ratio combiner for a varying number N_r of receiver antennas.

Equal-Gain Combining

In a theoretical context, the maximal-ratio combiner is the *optimum* among linear diversity combining techniques, optimum in the sense that it produces the largest possible value of instantaneous output SNR. However, in practical terms, there are three important issues to keep in mind:[9]

1. Significant instrumentation is needed to adjust the complex weighting parameters of the maximal-ratio combiner to their exact values, in accordance with (9.75).
2. The additional improvement in output SNR gained by the maximal-ratio combiner over the selection combiner is not that large, and it is quite likely that the additional improvement in receiver performance is lost in not being able to achieve the exact setting of the maximal-ratio combiner.
3. So long as a linear combiner uses the diversity branch with the strongest signal, then other details of the combiner may result in a minor improvement in overall receiver performance.

Issue 3 points to formulation of the so-called *equal-gain combiner*, in which all the complex weighting parameters a_k have their phase angles set opposite to those of their respective multipath branches in accordance with (9.75). But, unlike the a_k in the maximal-ratio combiner, their magnitudes are set equal to some constant value, unity for convenience of use.

9.9 "Space Diversity-on-Transmit" Systems

In the wireless communications literature, space diversity-on-receive techniques are commonly referred to as *orthogonal space–time block codes* (Tarokh et al., 1999). This terminology is justified on the following grounds:

1. The transmitted symbols form an orthogonal set.
2. The transmission of incoming data streams is carried out on a block-by-block basis.
3. Space and time constitute the coordinates of each transmitted block of symbols.

In a generic sense, Figure 9.24 presents the baseband diagram of a *space–time block encoder*, which consists of two functional units: mapper and block encoder. The *mapper* takes the incoming binary data stream $\{b_k\}$, where $b_k = \pm 1$, and generates a new *sequence of blocks* with each block made up of multiple symbols that are complex. For example, the mapper may be in the form of an *M-ary PSK* or *M-ary QAM message constellation*, which are illustrated for $M = 16$ in the signal-space diagrams of Figure 9.25. All the symbols in a particular column of the transmission matrix are pulse-shaped (in accordance with the

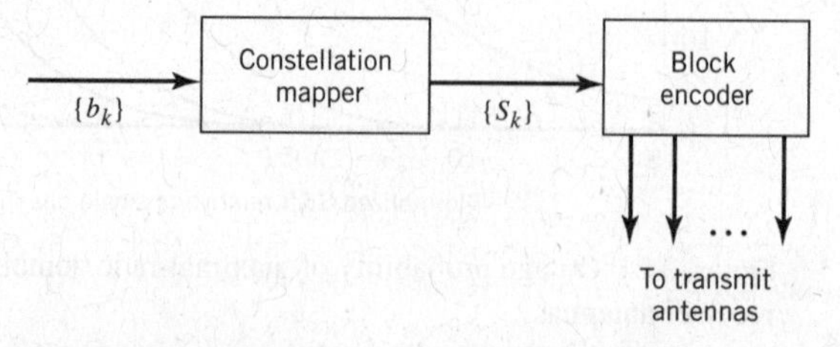

Figure 9.24 Block diagram of orthogonal space–time block encoder.

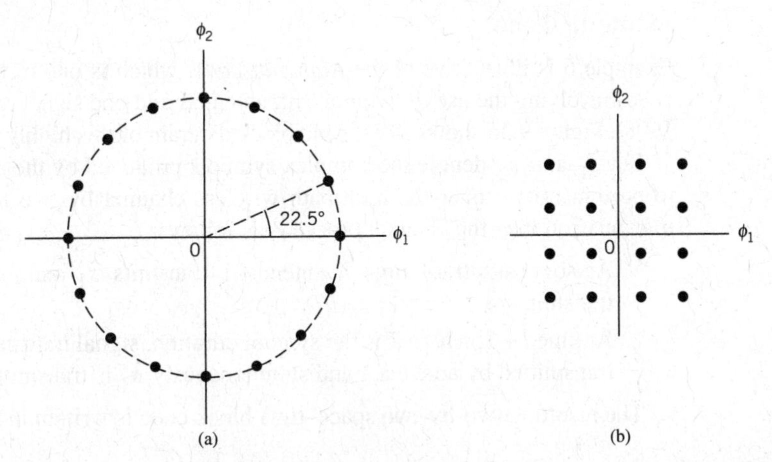

Figure 9.25 (a) Signal constellation of 16-PSK. (b) Signal constellation of 16-QAM.

criteria described in Chapter 8) and then modulated into a form suitable for simultaneous transmission over the channel by the transmit antennas. The pulse shaper and modulator are not shown in Figure 9.24 as the basic issue of interest is that of baseband data transmission with emphasis on the formulation of space–time block codes. The *block encoder* converts each block of complex symbols produced by the mapper into an l-by-N_t transmission matrix **S**, where l and N_t are respectively the *temporal* dimension and *spatial* dimension of the transmission matrix. The individual elements of the transmission matrix **S** are made up of linear combinations of $\tilde{s}_k$ and $\tilde{s}_k^*$, where the $\tilde{s}_k$ are complex symbols and the $\tilde{s}_k^*$ are their complex conjugates.

EXAMPLE 7

Quadriphase Shift Keying

As a simple example, consider the map portrayed by the QPSK, $M = 4$. This map is described in Table 9.2, where E is the transmitted signal energy per symbol.

The input dibits (pairs of binary bits) are *Gray encoded*, wherein only one bit is flipped as we move from one symbol to the next. (Gray encoding was discussed in Section 7.6 under "Quadriphase Shift Keying".) The mapped signal points lie on a circle of radius $\sqrt{E}$ centered at the origin of the signal-space diagram.

Table 9.2 Gray-encoded QPSK mapper

Dibit: $i = 1, 2, 3, 4$	Coordinates of mapped signal points: s_i, $i = 1, 2, 3, 4$
10	$\sqrt{E/2}(1, -1) = \sqrt{E}\exp(j7\pi/4)$
11	$\sqrt{E/2}(-1, -1) = \sqrt{E}\exp(j5\pi/4)$
01	$\sqrt{E/2}(-1, +1) = \sqrt{E}\exp(j3\pi/4)$
00	$\sqrt{E/2}(+1, +1) = \sqrt{E}\exp(j\pi/4)$

Alamouti Code

Example 6 is illustrative of the *Alamouti code*, which is one of the first space–time block codes involving the use of two transmit antennas and one signal receive antenna (Alamouti, 1998). Figure 9.26 shows a baseband block diagram of this highly popular spatial code.

Let $\tilde{s}_1$ and $\tilde{s}_2$ denote the complex symbols produced by the code's mapper, which are to be transmitted over the multipath wireless channel by two transmit antennas. Signal transmission over the channel proceeds as follows:

1. At some arbitrary time t, antenna 1 transmits $\tilde{s}_1$ and simultaneously antenna 2 transmits $\tilde{s}_2$.
2. At time $t + T$, where T is the symbol duration, signal transmission is switched to $-\tilde{s}_2^*$ transmitted by antenna 1 and simultaneously $\tilde{s}_1^*$ is transmitted by antenna 2.

The resulting two-by-two space–time block code is written in matrix form as follows:

$$\mathbf{S} = \begin{bmatrix} \tilde{s}_1 & \tilde{s}_2 \\ -\tilde{s}_2^* & \tilde{s}_1^* \end{bmatrix} \Rightarrow \text{Time} \tag{9.80}$$
$$\Downarrow \text{Space}$$

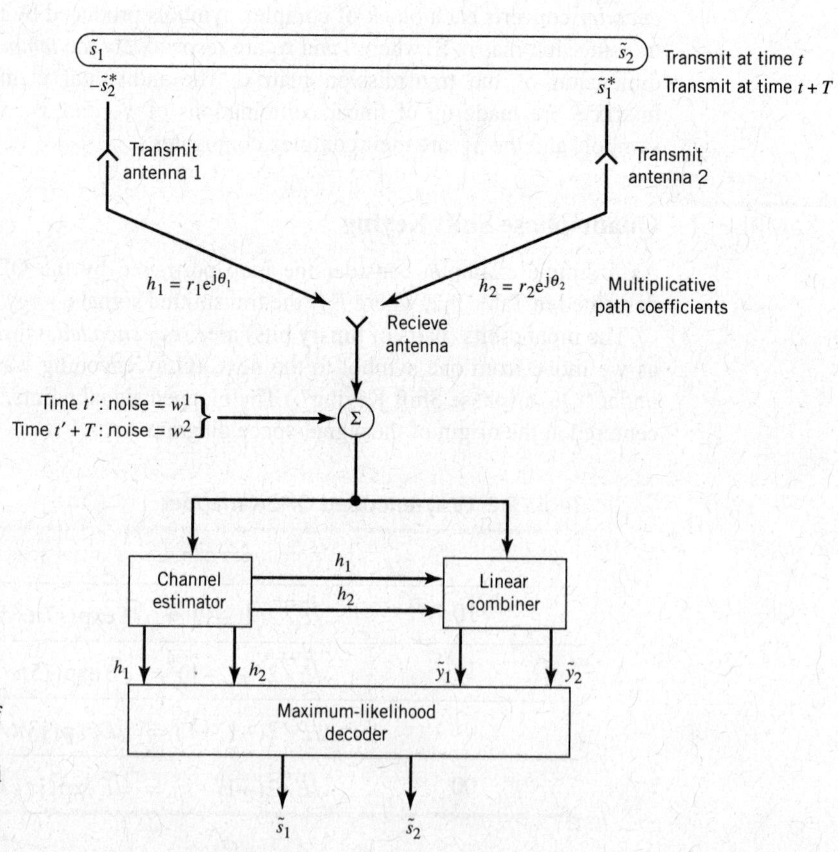

Figure 9.26 Block diagram of the transceiver (transmitter and receiver) for the Alamouti code. Note that $t' > t$ to allow for propagation delay.

This *transmission matrix* is a *complex-orthogonal matrix* (quaternion) in that it satisfies the condition for orthogonality in both the spatial and temporal senses. To demonstrate this important property of the Alamouti, let

$$\mathbf{S}^{\dagger} = \begin{bmatrix} \tilde{s}_1^* & -\tilde{s}_2 \\ \tilde{s}_2^* & \tilde{s}_1 \end{bmatrix} \Rightarrow \text{Space} \quad \Downarrow \text{Time} \tag{9.81}$$

denote the *Hermitian transpose* of **S**, which involves both transposition and complex conjugation. To demonstrate orthogonality in the spatial sense, we multiply the code matrix **S** by its Hermitian transpose $\mathbf{S}^{\dagger}$ on the right, obtaining

$$\begin{aligned}\mathbf{S}\mathbf{S}^{\dagger} &= \begin{bmatrix} \tilde{s}_1 & \tilde{s}_2 \\ -\tilde{s}_2^* & \tilde{s}_1^* \end{bmatrix}\begin{bmatrix} \tilde{s}_1^* & -\tilde{s}_2 \\ \tilde{s}_2^* & \tilde{s}_1 \end{bmatrix} \\ &= \begin{bmatrix} |\tilde{s}_1|^2 + |\tilde{s}_2|^2 & -\tilde{s}_1\tilde{s}_2 + \tilde{s}_2\tilde{s}_1 \\ -\tilde{s}_2^*\tilde{s}_1^* + \tilde{s}_1^*\tilde{s}_2^* & |\tilde{s}_2|^2 + |\tilde{s}_1|^2 \end{bmatrix} \\ &= (|\tilde{s}_1|^2 + |\tilde{s}_2|^2)\begin{bmatrix} 1 & 0 \\ 0 & 1 \end{bmatrix}\end{aligned} \tag{9.82}$$

Since the right-hand side of (9.81) is real valued, it follows that the alternative matrix product $\mathbf{S}^{\dagger}\mathbf{S}$, viewed in the temporal sense, yields exactly the same result. That is,

$$\mathbf{S}\mathbf{S}^{\dagger} = \mathbf{S}^{\dagger}\mathbf{S} = (|\tilde{s}_1|^2 + |\tilde{s}_2|^2)\mathbf{I} \tag{9.83}$$

where **I** is the two-by-two identity matrix.

In light of (9.80) and (9.83), we may now summarize three important properties of the Alamouti code:

PROPERTY 1 **Unitarity (Complex Orthogonality)**

The Alamouti code is an orthogonal space–time block code, in that its transmission matrix is a unitary matrix with the sum term $|\tilde{s}_1|^2 + |\tilde{s}_2|^2$ being merely a scaling factor.

As a consequence of this property, the Alamouti code achieves full diversity.

PROPERTY 2 **Full-Rate Complex Code**

The Alamouti code (with two transmit antennas) is the only complex space–time block code with a code rate of unity in existence.

Hence, for any signal constellation, full diversity of the code is achieved at the full transmission rate.

PROPERTY 3 **Linearity**

The Alamouti code is linear in the transmitted symbols.

We may therefore expand the transmission matrix $\mathbf{S}$ of the code as a linear combination of the transmitted symbols and their complex conjugates, as shown by

$$\mathbf{S} = \tilde{s}_1\mathbf{\Gamma}_{11} + \tilde{s}_1^*\mathbf{\Gamma}_{12} + \tilde{s}_2\mathbf{\Gamma}_{21} + \tilde{s}_2^*\mathbf{\Gamma}_{22} \tag{9.84}$$

where the four constituent matrices are themselves defined as follows:

$$\mathbf{\Gamma}_{11} = \begin{bmatrix} 1 & 0 \\ 0 & 0 \end{bmatrix}$$

$$\mathbf{\Gamma}_{12} = \begin{bmatrix} 0 & 0 \\ 0 & 1 \end{bmatrix}$$

$$\mathbf{\Gamma}_{21} = \begin{bmatrix} 0 & 1 \\ 0 & 0 \end{bmatrix}$$

$$\mathbf{\Gamma}_{22} = \begin{bmatrix} 0 & 0 \\ -1 & 0 \end{bmatrix}$$

In words, the Alamouti code is the only two-dimensional space–time code, the transmission matrix of which can be decomposed into the form described in (9.84).

Receiver Considerations of the Alamouti Code

The discussion presented thus far has focused on the Alamouti code viewed from the transmitter's perspective. We turn next to the design of the receiver for decoding the code.

To this end, we assume that the channel is frequency-flat and slowly time varying, such that the complex multiplicative distribution introduced by the channel at time t is essentially the same as that at time $t + T$, where T is the symbol duration. As before, the multiplicative distortion is denoted by $\alpha_k e^{j\theta_k}$ where we now have $k = 1, 2$, as indicated in Figure 9.25. Thus, with the symbols $\tilde{s}_1$ and $\tilde{s}_2$ transmitted simultaneously at time t, the complex received signal at some time $t' > t$, allowing for propagation delay, is described by

$$\tilde{x}_1 = \alpha_1 e^{j\theta_1}\tilde{s}_1 + \alpha_2 e^{j\theta_2}\tilde{s}_2 + \tilde{w}_1 \tag{9.85}$$

where $\tilde{w}_1$ is the complex channel noise at time t'. Next, with the symbols $-\tilde{s}_2^*$ and $\tilde{s}_1^*$ transmitted simultaneously at time $t + T$, the corresponding complex signal received at time $t' + T$ is

$$\tilde{x}_2 = -\alpha_1 e^{j\theta_1}\tilde{s}_2^* + \alpha_2 e^{j\theta_2}\tilde{s}_1^* + \tilde{w}_2 \tag{9.86}$$

where $\tilde{w}_2$ is the second complex channel noise at time $t' + T$. To be more precise, the noise terms $\tilde{w}_1$ and $\tilde{w}_2$ are circularly-symmetric complex-valued uncorrelated Gaussian random variables of zero mean and equal variance.

In the course of time from t' to $t' + T$, the channel estimator in the receiver has sufficient time to produce estimates of the multiplicative distortion represented by $\alpha_k e^{j\theta_k}$

for $k = 1, 2$. Hereafter, we assume that these two estimates are accurate enough for them to be treated as essentially exact; in other words, the receiver has knowledge of both $\alpha_1 e^{j\theta_1}$ and $\alpha_2 e^{j\theta_2}$. Accordingly, we may formulate the combination of two variables, $\tilde{x}_1$ in (9.85) and the complex conjugate of $\tilde{x}_2$ in (9.86), in matrix form as follows:

$$\begin{aligned}\tilde{\mathbf{x}} &= \begin{bmatrix} \tilde{x}_1 \\ \tilde{x}_2^* \end{bmatrix} \\ &= \begin{bmatrix} \alpha_1 e^{j\theta_1} & \alpha_2 e^{j\theta_2} \\ \alpha_2 e^{-j\theta_2} & -\alpha_1 e^{-j\theta_1} \end{bmatrix} \begin{bmatrix} \tilde{s}_1 \\ \tilde{s}_2 \end{bmatrix} + \begin{bmatrix} \tilde{w}_1 \\ \tilde{w}_2 \end{bmatrix}\end{aligned} \quad (9.87)$$

The nice thing about this equation is that the original complex signals s_1 and s_2 appear as the vector of two unknowns. It is with this goal in mind that $\tilde{x}_1$ and $\tilde{x}_2^*$ were used for the elements of the two-by-one received signal vector $\tilde{\mathbf{x}}$, in the manner shown on the right-hand side of (9.87).

According to (9.87), the channel matrix of the transmit diversity in Figure 9.25 is defined by

$$\begin{aligned}\mathbf{H} &= \begin{bmatrix} h_{11} & h_{12} \\ h_{21} & h_{22} \end{bmatrix} \\ &= \begin{bmatrix} \alpha_1 e^{j\theta_1} & \alpha_2 e^{j\theta_2} \\ \alpha_2 e^{-j\theta_2} & -\alpha_1 e^{-j\theta_1} \end{bmatrix}\end{aligned} \quad (9.88)$$

In a manner similar to the signal-transmission matrix $\tilde{\mathbf{S}}$, we find that the channel matrix $\mathbf{H}$ is also a *unitary matrix*, as shown by

$$\mathbf{H}^{\dagger}\mathbf{H} = (\alpha_1^2 + \alpha_2^2)\mathbf{I} \quad (9.89)$$

where, as before, $\mathbf{I}$ is the identity matrix and the sum term $\alpha_1^2 + \alpha_2^2$ is merely a scaling factor.

Using the definition of (9.88) for the channel matrix, we may rewrite (9.87) in the compact matrix form

$$\tilde{\mathbf{x}} = \mathbf{H}\tilde{\mathbf{s}} + \tilde{\mathbf{w}} \quad (9.90)$$

where

$$\tilde{\mathbf{s}} = \begin{bmatrix} \tilde{s}_1 \\ \tilde{s}_2 \end{bmatrix} \quad (9.91)$$

is the complex transmitted signal vector and

$$\tilde{\mathbf{w}} = \begin{bmatrix} \tilde{w}_1 \\ \tilde{w}_2 \end{bmatrix} \quad (9.92)$$

is the additive complex channel noise vector. Note that the column vector $\tilde{\mathbf{s}}$ in (9.91) is the same as the first row vector in the matrix $\tilde{\mathbf{S}}$ of (9.80).

We have now reached a point where we have to address the fundamental issue in designing the receiver:

How do we decode the Alamouti code, given the received signal vector $\tilde{\mathbf{x}}$?

To this end, we introduce a new complex two-by-one vector $\tilde{\mathbf{y}}$, defined as the matrix product of the received signal vector $\tilde{\mathbf{x}}$ and the Hermition transpose of the channel matrix $\mathbf{H}$ *normalized* with respect to the reciprocal sum term $\alpha_1^2 + \alpha_2^2$; that is,

$$\begin{aligned}\tilde{\mathbf{y}} &= \begin{bmatrix}\tilde{y}_1\\ \tilde{y}_2\end{bmatrix}\\ &= \left(\frac{1}{\alpha_1^2+\alpha_2^2}\right)\mathbf{H}^{\dagger}\tilde{\mathbf{x}}\end{aligned} \tag{9.93}$$

Substituting (9.90) into (9.93) and then making use of the unitarity property of the channel matrix described in (9.89), we obtain the mathematical basis for decoding of the Alamouti code:

$$\tilde{\mathbf{y}} = \tilde{\mathbf{s}} + \tilde{\mathbf{v}} \tag{9.94}$$

where $\tilde{\mathbf{v}}$ is a modified form of the complex channel noise $\tilde{\mathbf{w}}$, as shown by

$$\tilde{\mathbf{v}} = \left(\frac{1}{\alpha_1^2+\alpha_2^2}\right)\mathbf{H}^{\dagger}\tilde{\mathbf{w}} \tag{9.95}$$

Substituting (9.88) and (9.92) into (9.95), the expanded form of the complex noise vector $\tilde{\mathbf{v}}$ is defined as follows:

$$\begin{bmatrix}\tilde{v}_1\\ \tilde{v}_2\end{bmatrix} = \frac{1}{\alpha_1^2+\alpha_2^2}\begin{bmatrix}\alpha_1 e^{-j\theta_1}\tilde{w}_1 & + & \alpha_2 e^{j\theta_2}\tilde{w}_2^*\\ \alpha_2 e^{-j\theta_2}\tilde{w}_1 & - & \alpha_1 e^{j\theta_1}\tilde{w}_2^*\end{bmatrix} \tag{9.96}$$

Hence, we may go on to simply write

$$\tilde{y}_k = \tilde{s}_k + \tilde{v}_k, \qquad k = 1, 2 \tag{9.97}$$

Examination of (9.97) leads us to make the following statement insofar as the receiver is concerned:

The space–time channel is decoupled into a pair of scalar channels that are statistically independent of each other:

1. The complex symbol $\tilde{s}_k$ at the output of the kth space–time channel is identical to the complex symbol transmitted by the kth antenna for $k = 1, 2$; the decoupling shown clearly in (9.97) is attributed to complex orthogonality of the Alamouti code.
2. Assuming that the original channel noise $\tilde{w}_k$ is white Gaussian, then this statistical characterization is maintained in the modified noise $\tilde{v}_k$ appearing at the output of the kth space–time channel for $k = 1, 2$; this maintenance is attributed to the processing performed in the receiver.

This twofold statement hinges on the premise that the receiver has knowledge of the channel matrix **H**.

Moreover, with two transmit antennas and one receive antenna, the Alamouti code achieves the same level of diversity as a corresponding system with one transmit antenna and two receive antennas. It is in this sense that a wireless communication system based on the Alamouti code is said to enjoy a *two-level diversity gain.*

Maximum Likelihood Decoding

Figure 9.27 illustrates the signal-space diagram of an Alamouti-encoded system based on the QPSK constellation. The complex Gaussian noise clouds centered on the four signal points and with decreasing intensity illustrate the effects of complex noise term $\tilde{\mathbf{v}}$ on the linear combiner output $\tilde{\mathbf{y}}$.

In effect, the picture portrayed in Figure 9.27 is the graphical representation of (9.94) over two successive symbol transmissions at times t and $t + T$, repeated a large number of times.

Suppose that the two signal constellations in the top half of the signal-space diagram in Figure 9.27 represent the pair of symbols transmitted at time t, for which we write

$$\tilde{\mathbf{s}}_t = \begin{bmatrix} \tilde{s}_1 \\ \tilde{s}_2 \end{bmatrix}$$

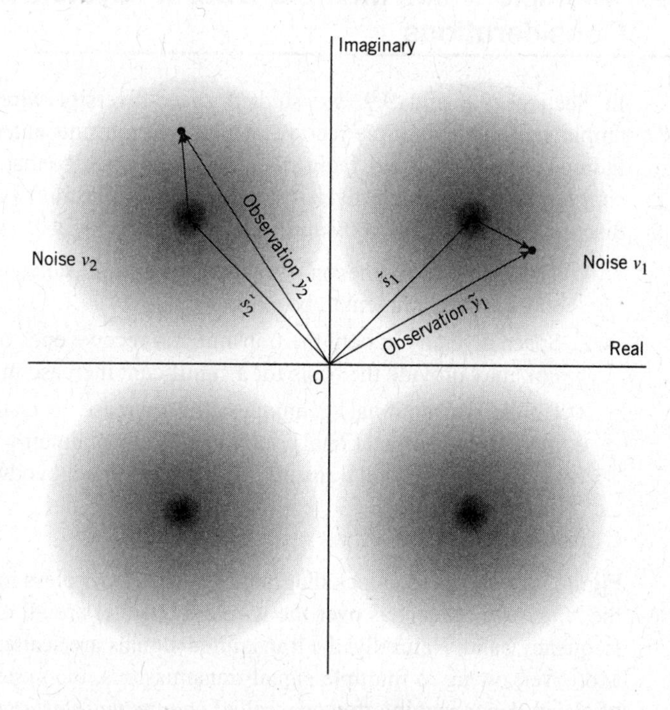

Figure 9.27 Signal-space diagram for Alamouti code, using the QPSK signal constellation. The signal points $\tilde{s}_1$ and $\tilde{s}_2$ and the corresponding linear normalized combiner outputs $\tilde{y}_1$ and $\tilde{y}_2$ are displayed in the top half of the figure.

Then, the remaining two signal constellations positioned in the right half of Figure 9.27 represent the other pair of symbols transmitted at $t + T$, for which we write

$$\tilde{\mathbf{s}}_{t+T} = \begin{bmatrix} -\tilde{s}_2^* \\ +\tilde{s}_1^* \end{bmatrix}$$

On this basis, we may now invoke the *maximum likelihood decoding rule,* discussed in Chapter 7, to make the three-fold statement:

1. Compute the composite *squared Euclidean distance metric*

$$\left\|\tilde{\mathbf{y}}_t - \tilde{\mathbf{s}}_t\right\|^2 + \left\|\tilde{\mathbf{y}}_{t+T} - \tilde{\mathbf{s}}_{t+T}\right\|^2$$

   produced by sending signal vectors $\tilde{\mathbf{s}}_t$ and $\tilde{\mathbf{s}}_{t+T}$, respectively.
2. Do this computation for all four possible signal pairs in the QPSK constellation.
3. Hence, the ML decoder selects the pair of signals for which the metric is the smallest.

The metric's component $\left\|\tilde{\mathbf{y}}_t - \tilde{\mathbf{s}}_t\right\|^2$ in part 1 of this statement is illustrated in Figure 9.27.

9.10 "Multiple-Input, Multiple-Output" Systems: Basic Considerations

In Sections 9.8 and 9.9, we studied space-diversity wireless communication systems employing either multiple receive or multiple transmit antennas to combat the multipath fading problem. In effect, fading was treated as a source that degrades performance, necessitating the use of space diversity on receive or transmit to mitigate it. In this section, we discuss *MIMO wireless communication*, which distinguishes itself in the following ways:[10]

1. The fading phenomenon is viewed not as a nuisance but rather as an environmental source of enrichment to be exploited.
2. Space diversity at both the transmit and receive ends of the wireless communication link may provide the basis for a significant increase in channel capacity.
3. Unlike conventional techniques, the increase in channel capacity is achieved by increasing computational complexity while maintaining the primary communication resources (i.e., total transmit power and channel bandwidth) fixed.

Coantenna Interference

Figure 9.28 shows the block diagram of a MIMO wireless link. The signals transmitted by the N_t transmit antennas over the wireless channel are all chosen to lie inside a common frequency band. Naturally, the transmitted signals are scattered differently by the channel. Moreover, owing to multiple signal transmissions, the system experiences a spatial form of signal-dependent interference, called *coantenna interference* (*CAI*).

Figure 9.29 illustrates the effect of CAI for one, two, and eight simultaneous transmissions and a single receive antenna (i.e., $N_t = 1, 2, 8$ and $N_r = 1$) using binary PSK; the transmitted binary PSK signals used in the simulation resulting in this figure were

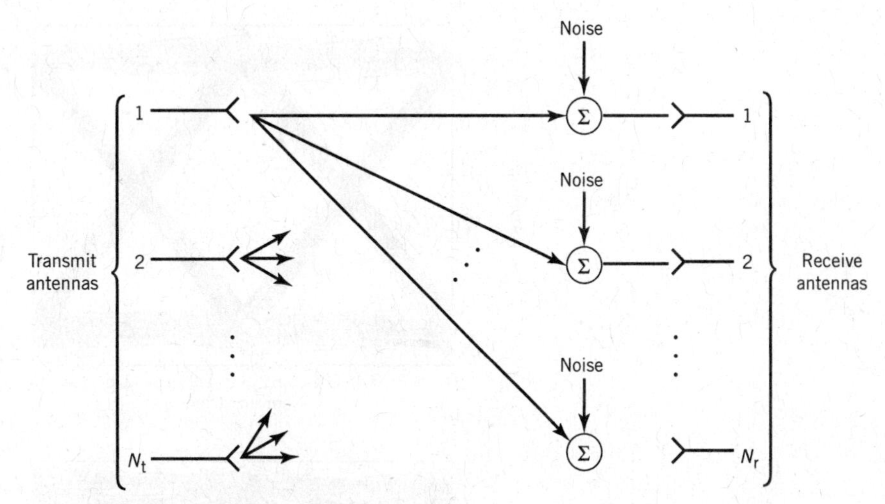

Figure 9.28 Block diagram of MIMO wireless link with N_t transmit antennas and N_r receive antennas.

different but they all had the same average power and occupied the same bandwidth. (Sellathurai and Haykin, 2008). Figure 9.29 clearly shows the difficulty that arises due to CAI when the number of transmit antennas N_t is large. In particular, with eight simultaneous signal transmissions, the *eye pattern* of the received signal is practically closed. The challenge for the receiver is how to mitigate the CAI problem and thereby make it possible to provide increased spectral efficiency.

In a theoretical context, the spectral efficiency of a communication system is intimately linked to the channel capacity of the system. To proceed with evaluation of the channel capacity of MIMO wireless communication, we begin by formulating a baseband channel model for the system as described next.

Basic Baseband Channel Model

Consider a MIMO narrowband wireless communication system built around a flat-fading channel, with N_t transmit antennas and N_r receive antennas. The antenna configuration is hereafter referred to as the pair (N_t, N_r). For a statistical analysis of the MIMO system in what follows, we use baseband representations of the transmitted and received signals as well as the channel. In particular, we introduce the following notation:

- The spatial parameter

$$N = \min\{N_t, N_r\} \tag{9.98}$$

  defines new degrees of freedom introduced into the wireless communication system by using a MIMO channel with N_t transmit antennas and N_r receive antennas.

- The N_t-by-1 vector

$$\tilde{\mathbf{s}}(n) = [\tilde{s}_1(n), \tilde{s}_2(n), \ldots, \tilde{s}_{N_t}(n)]^T \tag{9.99}$$

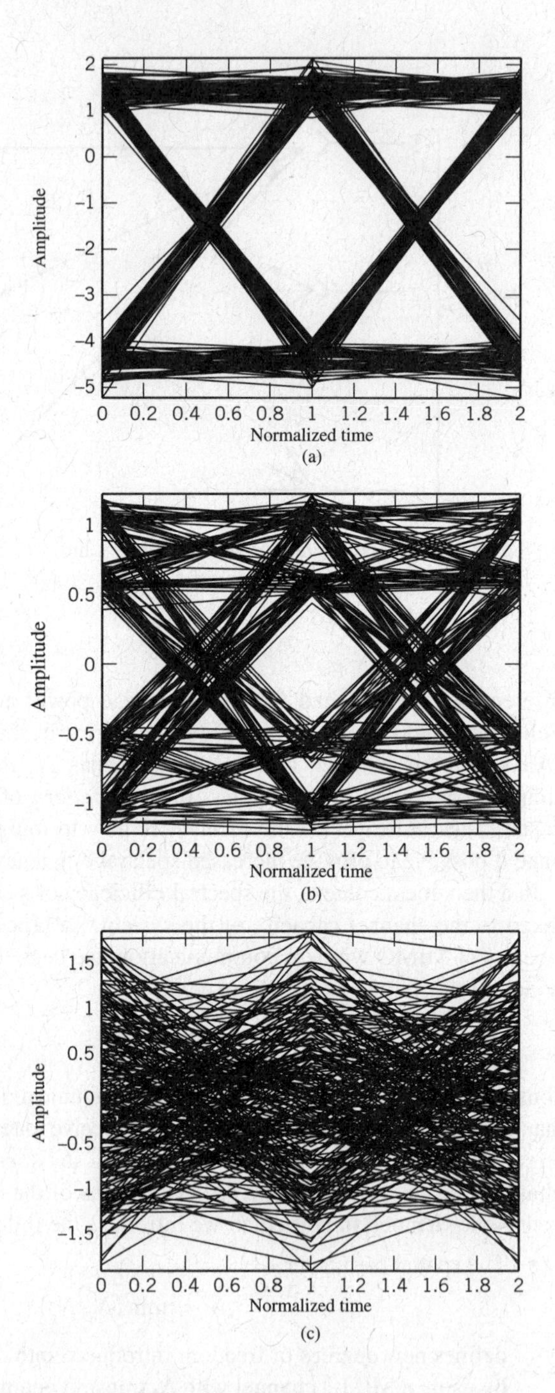

Figure 9.29 Effect of coantenna interference on the eye diagram for one receive antenna and different numbers of transmit antennas. (a) $N_t = 1$, (b) $N_t = 2$, (c) $N_t = 8$.

denotes the complex signal vector transmitted by the N_t antennas at discrete time n. The symbols constituting the vector $\tilde{\mathbf{s}}(n)$ are assumed to have zero mean and common variance σ_s^2. The total transmit power is *fixed* at the value

$$P = N_t \sigma_s^2 \tag{9.100}$$

For P to be maintained constant, the variance σ_s^2 (i.e., power radiated by each transmit antenna) must be inversely proportional to N_t.

- For a flat-fading Rayleigh distributing channel, we may use $\tilde{h}_{ik}(n)$ to denote the sampled complex gain of the channel coupling transmit antenna k to receive antenna i at discrete time n, where $i = 1, 2, \ldots, N_r$ and $k = 1, 2, \ldots, N_t$. We may thus express the N_r-by-N_t *complex channel matrix* as

$$\mathbf{H}(n) = \underbrace{\begin{bmatrix} \tilde{h}_{11}(n) & \tilde{h}_{12}(n) & \cdots & \tilde{h}_{1N_t}(n) \\ \tilde{h}_{21}(n) & \tilde{h}_{22}(n) & \cdots & \tilde{h}_{2N_t}(n) \\ \vdots & \vdots & & \vdots \\ \tilde{h}_{N_r1}(n) & \tilde{h}_{N_r2} & \cdots & h_{N_rN_t}(n) \end{bmatrix}}_{N_t \text{ transmit antennas}} \left.\vphantom{\begin{bmatrix} a \\ a \\ a \\ a \end{bmatrix}}\right\} \begin{matrix} N_r \\ \text{receive} \\ \text{antennas} \end{matrix} \tag{9.101}$$

- The system of equations

$$\tilde{x}_i(n) = \sum_{k=1}^{N_t} \tilde{h}_{ik}(n)\tilde{s}_k(n) + \tilde{w}_i(n) \qquad \begin{cases} i = 1, 2, \ldots, N_r \\ k = 1, 2, \ldots, N_t \end{cases} \tag{9.102}$$

defines the complex signal received at the ith antenna due to the transmitted symbol $\tilde{s}_k(n)$ radiated by the kth antenna. The term $\tilde{w}_i(n)$ denotes the additive complex channel noise perturbing $\tilde{x}_i(n)$. Let the N_r-by-1 vector

$$\tilde{\mathbf{x}}(n) = [\tilde{x}_1(n), \tilde{x}_2(n), \ldots, \tilde{x}_{N_r}(n)] \tag{9.103}$$

denote the complex received signal vector and the N_r-by-1 vector

$$\tilde{\mathbf{w}}(n) = [\tilde{w}_1(n), \tilde{w}_2(n), \ldots, \tilde{w}_{N_r}(n)]^T \tag{9.104}$$

denote the complex channel noise vector. We may then rewrite the system of equations (9.102) in the compact matrix form

$$\tilde{\mathbf{x}}(n) = \mathbf{H}(n)\tilde{\mathbf{s}}(n) + \tilde{\mathbf{w}}(n) \tag{9.105}$$

Equation (9.105) describes the *basic complex channel model for MIMO wireless communications*, assuming the use of a flat-fading channel. The equation describes the input–output behavior of the channel at discrete time n. To simplify the exposition, hereafter we suppress the dependence on time n by simply writing

$$\tilde{\mathbf{x}} = \mathbf{H}\tilde{\mathbf{s}} + \tilde{\mathbf{w}} \tag{9.106}$$

where it is understood that all four vector/matrix terms of the equation, **s**, **H**, **w**, and **x**, are in actual fact dependent on the discrete time n. Figure 9.30 shows the basic channel model of (9.106).

For mathematical tractability, we assume a *Gaussian model* made up of three elements:

1. N_t symbols, which constitute the transmitted signal vector $\tilde{\mathbf{s}}$ drawn from a *white complex Gaussian codebook*; that is, the symbols $\tilde{s}_1, \tilde{s}_2, \ldots, \tilde{s}_{N_t}$ are iid complex Gaussian random variables with zero mean and common variance σ_s^2. Hence, the correlation matrix of the transmitted signal vector **s** is defined by

$$\begin{aligned}\mathbf{R_s} &= \mathbb{E}[\mathbf{s}\mathbf{s}^\dagger] \\ &= \sigma_s^2 \mathbf{I}_{N_t}\end{aligned} \tag{9.107}$$

   where $\mathbf{I}_{N_t}$ is the N_t-by-N_t identity matrix.

2. $N_t \times N_r$ elements of the channel matrix **H**, which are also drawn from an ensemble of iid complex random variables with zero mean and unit variance, as shown by the complex distribution

$$h_{ik}: \quad \mathcal{N}(0, 1/\sqrt{2}) + \mathrm{j}\mathcal{N}(0, 1/\sqrt{2}) \qquad \begin{cases} i = 1, 2, \ldots, N_r \\ k = 1, 2, \ldots, N_t \end{cases} \tag{9.108}$$

   where $\mathcal{N}(\ldots)$ denotes a real Gaussian distribution. On this basis, we find that the amplitude component h_{ik} is *Rayleigh* distributed. It is in this sense that we sometimes speak of the MIMO channel as a *rich Rayleigh scattering environment*. By the same token, we also find that the squared amplitude component, namely $|h_{ik}|^2$, is a *chi-squared random variable* with the mean

$$\mathbb{E}[|h_{ik}|^2] = 1 \qquad \text{for all } i \text{ and } k \tag{9.109}$$

   (The chi-squared distribution is discussed in Appendix A.)

3. N_r elements of the channel noise vector **w**, which are iid complex Gaussian random variables with zero mean and common variance $\sigma_{\mathbf{w}}^2$; that is, the correlation matrix of the noise vector **w** is given by

$$\begin{aligned}\mathbf{R_w} &= \mathbb{E}[\mathbf{w}\mathbf{w}^\dagger] \\ &= \sigma_{\mathbf{w}}^2 \mathbf{I}_{N_r}\end{aligned} \tag{9.110}$$

   where $\mathbf{I}_{N_r}$ is the N_r-by-N_r identity matrix.

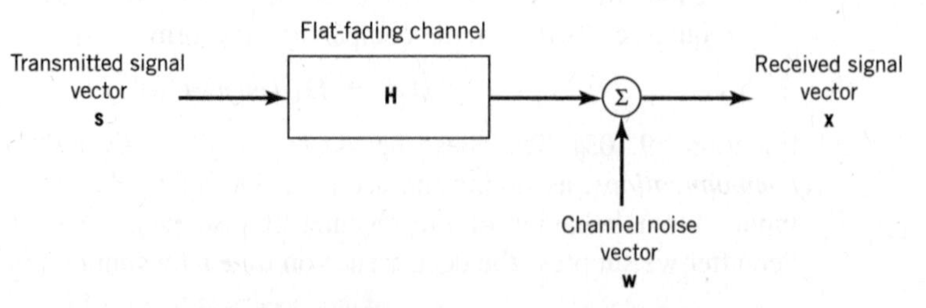

Figure 9.30 Depiction of the basic channel model of (9.106).

In light of (9.100) and the assumption that h_{ik} is a standard Gaussian random variable with zero mean and unit variance, the *average SNR* at each receiver input of the MIMO channel is given by

$$\rho = \frac{P}{\sigma_{\mathbf{w}}^2} = \frac{N_t \sigma_{\mathbf{s}}^2}{\sigma_{\mathbf{w}}^2} \quad (9.111)$$

which is, for a prescribed noise variance $\sigma_{\mathbf{w}}^2$, *fixed* once the total transmit power P is fixed. Note also that, first, all the N_t transmitted signals occupy a common channel bandwidth and, second, the average SNR ρ is independent of N_r.

The idealized Gaussian model just described of a MIMO wireless communication system is applicable to indoor local area networks and other wireless environments, where the extent of user-terminal mobilities is limited.[11]

9.11 MIMO Capacity for Channel Known at the Receiver

With the basic complex channel model of Figure 9.30 at our disposal, we are now ready to focus attention on the primary issue of interest: the channel capacity of a MIMO wireless link. In what follows, two special cases will be considered: the first case, entitled "ergodic capacity," assumes that the MIMO channel is weakly (wide-sense) stationary and, therefore, ergodic. The second case, entitled "outage capacity," considers a nonergodic MIMO channel under the assumption of quasi-stationarity from one burst of data transmission to the next.

Ergodic Capacity

According to *Shannon's information capacity* law discussed in Chapter 5, the capacity of a *real* AWGN channel, subject to the constraint of a fixed transmit power P, is defined by

$$C = B \log_2\left(1 + \frac{P}{\sigma_w^2}\right) \text{ bits/s} \quad (9.112)$$

where B is the channel bandwidth and σ_w^2 is the noise variance measured over the bandwidth B. Given a time-invariant channel, (9.112) defines the maximum data rate that can be transmitted over the channel with an arbitrarily small probability of error being incurred as a result of the transmission. With the channel used K times for the transmission of K symbols in T seconds, the transmission capacity per unit time is K/T times the formula for C given in (9.112). Recognizing that $K = 2BT$ in accordance with the sampling theorem discussed in Chapter 6, we may express the information capacity of the AWGN channel in the equivalent form

$$C = \frac{1}{2}\log_2\left(1 + \frac{P}{\sigma_w^2}\right) \text{ bits/(s Hz)} \quad (9.113)$$

Note that one bit per second per hertz corresponds to one bit per transmission.

With wireless communications as the medium of interest, consider next the case of a *complex* flat-fading channel with the receiver having perfect knowledge of the channel state. The capacity of such a channel is given by

$$C = \mathbb{E}\left[\log_2\left(1 + \frac{|h|^2 P}{\sigma_w^2}\right)\right] \text{ bits/(s Hz)} \tag{9.114}$$

where the expectation is taken over the gain of the channel $|h|^2$ and the channel is assumed to be stationary and ergodic. In recognition of this assumption, C is commonly referred to as the *ergodic capacity* of the flat-fading channel and the channel coding is applied across fading intervals (i.e., over an "ergodic" interval of channel variation with time).

It is important to note that the scaling factor of 1/2 is missing from the capacity formula of (9.114). The reason for this omission is that this equation refers to a complex baseband channel, whereas (9.113) refers to a real channel. The fading channel covered by (9.114) operates on a complex signal, namely a signal with in-phase and quadrature components. Therefore, such a complex channel is equivalent to two real channels with equal capacities and operating in parallel; hence the result presented in (9.114).

Equation (9.114) applies to the simple case of a *single-input, single-output (SISO) flat-fading channel*. Generalizing this formula to the case of a multiple-input, multiple-output MIMO flat-fading channel governed by the Gaussian model described in Figure 9.30, we find that the ergodic capacity of the MIMO channel is given by the following formula:[12]

$$C = \mathbb{E}\left[\log_2\left\{\frac{\det(\mathbf{R_w} + \mathbf{H}\mathbf{R_s}\mathbf{H}^\dagger)}{\det(\mathbf{R_w})}\right\}\right] \text{ bits/(s Hz)} \tag{9.115}$$

which is subject to the constraint

$$\max_{\mathbf{R_s}} \operatorname{tr}[\mathbf{R_s}] \le P$$

where P is the constant transmit power and tr[·] denotes the trace of the enclosed matrix. The expectation in (9.115) is over the random channel matrix $\mathbf{H}$, and the superscript dagger notes Hermitian transposition; $\mathbf{R_s}$ and $\mathbf{R_w}$ are respectively the correlation matrices of the transmitted signal vector $\mathbf{s}$ and channel noise vector $\mathbf{w}$. A detailed derivation of (9.115) is presented in Appendix E.

In general, it is difficult to evaluate (9.115) except for a Gaussian model. In particular, substituting (9.107) and (9.110) into (9.115) and simplifying yields

$$C = \mathbb{E}\left[\log_2\left\{\det\left(\mathbf{I}_{N_r} + \frac{\sigma_s^2}{\sigma_w^2}\mathbf{H}\mathbf{H}^\dagger\right)\right\}\right] \text{ bits/(s Hz)} \tag{9.116}$$

Next, invoking the definition of the average SNR ρ introduced in (9.111), we may rewrite (9.116) in the equivalent form

$$C = \mathbb{E}\left[\log_2\left\{\det\left(\mathbf{I}_{N_r} + \frac{\rho}{N_t}\mathbf{H}\mathbf{H}^\dagger\right)\right\}\right] \text{ bits/(s Hz)}, \quad \text{for } N_t \ge N_r \tag{9.117}$$

Equation (9.117), defining the ergodic capacity of a MIMO flat-fading channel, involves the determinant of an N_r-by-N_r sum matrix (inside the braces) followed by the logarithm to base 2. It is for this reason that this equation is referred to as the *log-det capacity formula* for a Gaussian MIMO channel.

As indicated in (9.117), the log-det capacity formula therein assumes that $N_t \geq N_r$ for the matrix product $\mathbf{H}\mathbf{H}^\dagger$ to be of full rank. The alternative case, $N_r \geq N_t$ makes the N_t-by-N_t matrix product $\mathbf{H}^\dagger\mathbf{H}$ to be of full rank, in which case the *log-det capacity formula* of the MIMO link takes the form

$$C = \mathbb{E}\left[\log_2\left\{\det\left(\mathbf{I}_{N_t} + \frac{\rho}{N_r}\mathbf{H}^\dagger\mathbf{H}\right)\right\}\right] \text{ bits/(s Hz)}, \qquad N_r \geq N_t \tag{9.118}$$

where, as before, the expectation is taken over the channel matrix $\mathbf{H}$.

Despite the apparent differences between (9.117) and (9.118), they are equivalent in that either one of them applies to all $\{N_r, N_t\}$ antenna configurations. The two formulas differentiate themselves only when the full-rank issue is of concern.

Clearly, the capacity formula of (9.114), pertaining to a complex, flat-fading link with a single antenna at both ends of the link, is a special case of the log-det capacity formula. Specifically, for $N_t = N_r = 1$ (i.e., no spatial diversity), $\rho = P/\sigma_w^2$, and $\mathbf{H} = h$ (with dependence on discrete-time n suppressed, (9.116) reduces to that of (9.114).

Another insightful result that follows from the log-det capacity formula is that if $N_t = N_r = N$, then, as N approaches infinity, the capacity C defined in (9.117) grows asymptotically (at least) linearly with N; that is,

$$\lim_{N \to \infty} \frac{C}{N} \geq \text{constant} \tag{9.119}$$

In words, the asymptotic formula of (9.119) may be stated as follows:

> The ergodic capacity of a MIMO flat-fading wireless link with an equal number of transmit and receive antennas N grows roughly proportionately with N.

What this statement teaches us is that, by increasing computational complexity resulting from the use of multiple antennas at both the transmit and receive ends of a wireless link, we are able to increase the *spectral efficiency* of the link in a far greater manner than is possible by conventional means (e.g., increasing the transmit SNR). The potential for this very sizable increase in the spectral efficiency of a MIMO wireless communication system is attributed to the key parameter

$$N = \min\{N_t, N_r\}$$

which defines the *number of degrees of freedom* provided by the system.

Two Other Special Cases of the Log-Det Formula: Capacities of Receive and Transmit Diversity Links

Naturally, the log-det capacity formula for the channel capacity of an N_t, N_r wireless link includes the channel capacities of receive and transmit diversity links as special cases:

1. *Diversity-on-receive channel.* The log-det capacity formula (9.118) applies to this case. Specifically, for $N_t = 1$, the channel matrix **H** reduces to a column vector and with it (9.118) reduces to

$$C = \mathbb{E}\left[\log_2\left\{\left(1 + \rho \sum_{i=1}^{N_r} |h_i|^2\right)\right\}\right] \text{ bits/(s Hz)} \tag{9.120}$$

Compared with the channel capacity of (9.114), for an SISO fading channel with $\rho = P/\sigma_w^2$, the squared channel gain $|h|^2$ is replaced by the sum of squared magnitudes $|h_i|^2$, $i = 1, 2, \ldots, N_r$. Equation (9.120) expresses the ergodic capacity due to the *linear combination* of the receive-antenna outputs, which is designed to maximize the information contained in the N_r received signals about the transmitted signal. This is simply a restatement of the maximal-ratio combining principle discussed in Section 9.8.

2. *Diversity-on-transmit channel.* The log-det capacity formula of (9.117) applies to this second case. Specifically, for $N_r = 1$, the channel matrix **H** reduces to a row vector, and with it (9.117) reduces to

$$C = \mathbb{E}\left[\log_2\left(1 + \frac{\rho}{N_t} \sum_{k=1}^{N_t} |h_k|^2\right)\right] \text{ bits/(s Hz)} \tag{9.121}$$

where the matrix product $\mathbf{H}\mathbf{H}^\dagger$ is replaced by the sum of squared magnitudes $|h_k|^2$, $k = 1, 2, \ldots, N_t$. Compared with case 1 on receive diversity, the capacity of the diversity-on-transmit channel is reduced because the total transmit power is being held constant, independent of the number of N_t transmit antennas.

Outage Capacity

To realize the log-det capacity formula of (9.117), the MIMO channel must be described by an ergodic process. In practice, however, the MIMO wireless channel is often nonergodic and the requirement is to operate the channel under *delay constraints*. The issue of interest is then summed up as follows:

> How much information can be transmitted across a nonergodic channel, particularly if the channel code is long enough to see just one random channel matrix?

In the situation described here, the rate of reliable information transmission (i.e., the strict Shannon-sense capacity) is zero, since for any positive rate there exists a nonzero probability that the channel would not support such a rate.

To get around this serious difficulty, the notion of *outage* is introduced into characterization of the MIMO link. (Outage was discussed previously in the context of diversity on receive in Section 9.8.) Specifically, we offer the following definition:

> The outage probability of a MIMO link is defined as the probability for which the link is in a state of outage (i.e., failure) for data transmitted across the link at a certain rate R, measured in bits per second per hertz.

To proceed on this probabilistic basis, it is customary to operate the MIMO link by transmitting data in the form of *bursts* or *frames* and invoke a *quasi-stationary model* governed by four points:

1. The burst is *long* enough to accommodate the transmission of a large number of symbols, which, in turn, permits the use of an idealized *infinite-time horizon* basic to information theory.
2. Yet, the burst is *short* enough to treat the wireless link as *quasi-stationary* during each burst; the slow variation is used to justify the assumption that the receiver has perfect knowledge of the channel state.
3. The channel matrix is permitted to change, from burst k to the next burst $k + 1$, thereby accounting for statistical variations of the link.
4. Different realizations of the transmitted signal vector **s** are drawn from a *white Gaussian codebook*; that is, the correlation matrix of **s** is defined by (9.107).

Points 1 and 4 pertain to signal transmission, whereas points 2 and 3 pertain to the MIMO channel itself.

To proceed with the evaluation of outage probability under this model, we first note that, in light of the log-det capacity formula (9.117), we may view the random variable

$$C_k = \log_2\left\{\det\left(\mathbf{I}_{N_r} + \frac{\rho}{N_t}\mathbf{H}_k\mathbf{H}_k^{\dagger}\right)\right\} \text{ bits/(s Hz) for burst } k \tag{9.122}$$

as the expression for a "sample realization" of the MIMO link. In other words, with the random-channel matrix $\mathbf{H}_k$ varying from one burst to the next, C_k will itself vary in a corresponding way. A consequence of this random behavior is that, occasionally, a sample drawn from the cumulative distribution function of the MIMO link results in a value for C_k that is inadequate to support reliable communication over the link. In this kind of situation the link is said to be in an *outage state*. Correspondingly, for a given transmission strategy, we define the *outage probability at rate R* as

$$P_{\text{outage}}(R) = \mathbb{P}\{C_k < R_k\} \text{ for some burst } k \tag{9.123}$$

Equivalently, we may write

$$P_{\text{outage}}(R) = \mathbb{P}\left\{\log_2\left\{\det\left(\mathbf{I}_{N_r} + \frac{\rho}{N_t}\mathbf{H}_k\mathbf{H}_k^{\dagger}\right)\right\} < R \text{ for some burst } k\right\} \tag{9.124}$$

On this basis, we may offer the following definition:

> The outage capacity of the MIMO link is the maximum bit rate that can be maintained across the link for all bursts of data transmissions (i.e., all possible channel states) for a prescribed outage probability.

By the very nature of it, the study of outage capacity can only be conducted using Monte Carlo simulation.

Channel Known at the Transmitter

The log-det capacity formula of (9.117) is based on the premise that the transmitter has *no* knowledge of the channel state. Knowledge of the channel state, however, can be made available to the transmitter by first estimating the channel matrix **H** at the receiver and then sending this estimate to the transmitter via a *feedback channel.* In such a scenario, the capacity is optimized over the correlation matrix of the transmitted signal vector **s**, subject to the power constraint; that is, the trace of this correlation matrix is less than or equal to the constant transmit power *P*. Naturally, formulation of the log-det capacity formula of a MIMO channel for which the channel is known in both the transmitter and receiver is more challenging than when it is only known to the receiver. For details of this formulation, the reader is referred to Appendix E.

9.12 Orthogonal Frequency Division Multiplexing

In Chapter 8 we introduced the DMT method as one discrete form of multichannel modulation for signaling over band-limited channels. *Orthogonal frequency division multiplexing* (OFDM)[13] is another clearly related form of multifrequency modulation.

OFDM is particularly well suited for high data-rate transmission over delay-dispersive channels. In its own way, OFDM solves the problem by following the engineering paradigm of "divide and conquer." Specifically, a large number of closely spaced *orthogonal subcarriers (tones)* is used to support the transmission. Correspondingly, the incoming data stream is divided into a number of low data-rate *substreams*, one for each carrier, with the *subchannels* so formed operating in parallel. For the modulation process, a modulation scheme such as QPSK is used.

What we have just briefly described here is essentially the same as the procedure used in DMT modulation. In other words, the underlying mathematical theory of DMT described in Chapter 8 applies equally well to OFDM, except for the fact that the signal constellation encoder does not include the use of loading for bit allocation. In addition, two other changes have to be made in the implementation of OFDM:

1. In the transmitter, an *upconverter* is included after the digital-to-analog converter to appropriately *translate* the transmitted frequency, so as to facilitate propogation of the transmitted signal over the radio channel.
2. In the receiver, a *downconverter* is included before the analog-to-digital converter to *undo* the frequency translation that was performed by the upconverter in the transmitter.

Figure 9.31 shows the block diagram of an OFDM system, the components of which are configured to accommodate the transmission of a binary data stream at 36 Mbit/s as an illustrative example. Parts a and b of the figure depict the transmitter and receiver of the system, respectively. Specifically, pertinent values of data carrier rates as well as sub-carrier frequencies at the various functional blocks are included in part a of the figure dealing with the transmitter. One last comment is in order: the front end of the transmitter and the back end of the receiver are allocated to forward error-correction encoding and decoding, respectively, for improved reliability of the system. (Error-control coding of the forward error-correction variety is discussed in Chapter 10.)

The Peak-to-Average Power Ratio Problem

A compelling practical importance of OFDM to wireless communications is attributed to the computational benefits brought about by the FFT algorithm that plays a key role in its

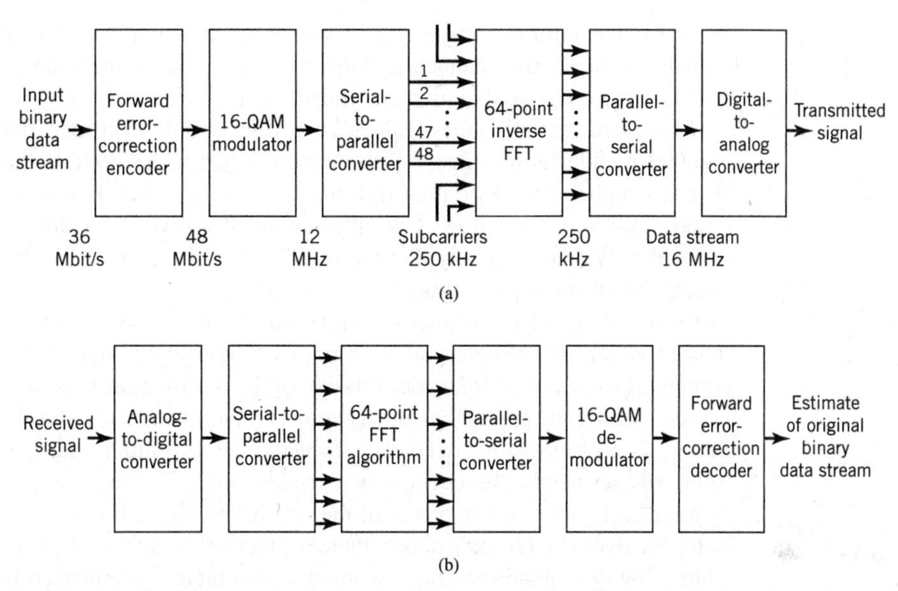

Figure 9.31 Block diagram of the typical implementation of an OFDM, illustrating the transmission of binary data at 36 Mbit/s.

implementation. However, OFDM suffers from the so-called PAPR problem. This problem arises due to the statistical probabilities of a large number of independent subchannels in the OFDM becoming superimposed on each other in some unknown fashion, thereby resulting in high peaks. For a detailed account of the PAPR problem and how to mitigate it, the reader is referred to Appendix G.

9.13 Spread Spectrum Signals

In previous sections of this chapter we described different methods for mitigating the effect of multipath interference in signaling over fading channels. In this section of the chapter, we describe another novel way of thinking about wireless communications, which is based on a class of signals called *spread spectrum signals*.[14]

A signal is said to belong to this class of signals if it satisfies the following two requirements:

1. *Spreading*. Given an information-bearing signal, spreading of the signal is accomplished in the transmitter by means of an independent *spreading signal*, such that the resulting spread spectrum signal occupies a bandwidth much larger than the bandwidth of the original information-bearing signal: the larger the better.
2. *Despreading*. Given a noisy version of the transmitted spread spectrum signal, *despreading* (i.e., recovering the original information-bearing signal) is achieved by *correlating* the received signal with a synchronized replica of the spreading signal in the receiver.

In effect, the information-bearing signal is spread (increased) in bandwidth *before* its transmission over the channel, and the received signal at the channel output is despread (i.e., decreased) in bandwidth by the same amount.

To explain the rationale of spread spectrum signals, consider, first, a scenario where there are no interfering signals at the channel output whatsoever. In this idealized scenario, an exact replica of the original information-bearing signal is reproduced at the receiver output; this recovery follows from the combined action of spreading and despreading, in that order. We may thus say that the receiver performance is *transparent* with respect to the combined spreading–despreading process.

Consider, next, a practical scenario where an additive narrowband interference is introduced at the receiver input. Since the interfering signal is introduced into the communication system *after* transmission of the information-bearing signal, its bandwidth is increased by the spreading signal in the receiver, with the result that its power spectral density is correspondingly reduced. Typically, at its output end, the receiver includes a filter whose bandwidth-occupancy matches that of the information-bearing signal. Consequently, the average power of the interfering signal is reduced, and the output SNR of the receiver is increased; hence, there is practical benefit in improved SNR to be gained from using the spread spectrum technique when there is an interfering signal (e.g., due to multipath) to deal with. Of course, this benefit is obtained at the expense of increased channel bandwidth.

Classification of Spread Spectrum Signals

Depending on how the use of spread spectrum signals is carried out, we may classify them as follows:

1. *Direct Sequence-Spread Spectrum*

   One method of spreading the bandwidth of an information-bearing signal is to use the so-called direct sequence-spread spectrum (DS-SS), wherein a *pseudo-noise* (PN) *sequence* is employed as the spreading sequence (signal). The PN sequence is a periodic binary sequence with noise-like properties, details of which are presented in Appendix J. The *baseband modulated signal*, representative of the DS-SS method, is obtained by multiplying the information-bearing signal by the PN sequence, whereby each information bit is chopped into a number of small time increments, called *chips*. The second stage of modulation is aimed at conversion of the baseband DS-SS signal into a form suitable for transmission over a wireless channel, which is accomplished by using M-ary PSK, discussed in Chapter 7. The family of spread spectrum systems so formed is referred to simply as *DS/MPSK systems,* a distinct characteristic of which is that spreading of the transmission bandwidth takes place *instantaneously*. Moreover, the signal-processing capability of these systems to combat the effect of interferers, commonly referred to as *jammers* be they friendly or unfriendly, is a function of the PN sequence length. Unfortunately, this capability is limited by physical considerations of the PN-sequence generator.

2. *Frequency Hop-Spread Spectrum*

   To overcome the physical limitations of DS/MPSK systems, we may resort to alternative methods. One such method is to force the jammer to occupy a wider spectrum by *randomly hopping* the input data-modulated carrier from one frequency to the next. In effect, the spectrum of the transmitter signal is *spread sequentially* rather than instantaneously; the term sequentially refers to the pseudo-randomly ordered sequence of frequency hops. This second type of spread spectrum in which the carrier hops randomly from one frequency to another is called *frequency hop-spread spectrum.* A commonly used modulation format used herein is that of M-ary FSK, which was also discussed in Chapter 7. The combination of the two modulation techniques, namely frequency hopping and M-ary FSK, is referred to simply as FH/MFSK. Since frequency-hopping does not cover over the entire spread spectrum instantaneously, we are led to consider the rate at which the hops occur. In this context, we may go on to identify two basic kinds of frequency hopping, which are the converse of each other, as summarized here:

- First, *slow-frequency hopping*, in which the symbol rate of the M-ary FSK signal, denoted by R_s, is an integer multiple of the hop rate, denoted by R_h; that is, several symbols of the input data sequence are transmitted for each frequency hop.
- Second, *fast-frequency hopping*, in which the hop rate R_h is an integer multiple of the M-ary FSK symbol rate R_s; that is, the carrier frequency will change (i.e., hop) several times during the transmission of one input-data symbol.

The spread spectrum technique of the FH variety is particularly attractive for military applications. But, compared with the alternative spread spectrum technique, DS/MPSK, the commercial use of FH/MFSK is insignificant, which is especially so in regard to fast frequency hopping. The limiting factor behind this statement is the expense involved in the employment of frequency synthesizers, which are basic to the implementation of FH/MFSK systems. Accordingly, the FH/MFSK will not be considered further.

Processing Gain of the DS/BPSK

Before closing this section on spread spectrum signals, it is informative to expand on the improvement in SNR gained at the receiver output, mentioned earlier on. To this end, consider the simple case of the DS/BPSK, in which the binary PSK, representing the second stage of modulation in the transmitter, is coherent; that is, the receiver is synchronized with the transmitter in all of its features. In Problem 9.34, it is shown that the processing gain of a spread spectrum signal compared to its unspread version is

$$\mathrm{PG} = \frac{T_b}{T_c} \tag{9.125}$$

where T_b is the bit duration and T_c is the *chip duration.* With PG expressed in decibels, in Problem 9.34 it is also shown that

$$10\log_{10}(\mathrm{SNR})_O = 10\log_{10}(\mathrm{SNR})_I + 10\log_{10}(\mathrm{PG})\ \mathrm{dB} \tag{9.126}$$

where $(\mathrm{SNR})_I$ and $(\mathrm{SNR})_O$ are the input SNR and output SNR, respectively. Furthermore, recognizing that the ratio T_b/T_c is equal to the number of chips contained in a single bit duration, it follows that the processing gain realized by the use of DS/BPSK increases with increasing length of a single period of the PN sequence, which was emphasized previously.

9.14 Code-Division Multiple Access

Modern wireless networks are commonly of a *multiuser* type, in that the multiple communication links within the network are shared among multiple users. Specifically, each individual user is permitted to share the available radio resources (i.e., time and frequency) with other users in the network and do so in an independent manner.

Stated in another way, a *multiple access technique* permits the radio resources to be shared among multiple users seeking to communicate with each other. In the context of time and frequency domains, we recall from Chapter 1 that frequency-division multiple access (FDMA) and time-division multiple access (TDMA) techniques allocate the radio resources of a wireless channel through the use of disjointedness (i.e., orthogonality) in frequency and time, respectively. On the other hand, the *code-division multiple access* (CDMA) technique, building on spread spectrum signals and benefiting from their attributes, provides an alternative to the traditional techniques of FDMA and TDMA; it does so by not requiring the bandwidth allocation of FDMA nor the time synchronization needed in TDMA. Rather, CDMA operates on the following principle:

> The users of a common wireless channel are permitted access to the channel through the assignment of a spreading code to each individual user under the umbrella of spread spectrum modulation.

This statement is testimony to what we said in the first paragraph of Section 9.13, namely that spread spectrum signals provide a novel way of thinking about wireless communications.

To elaborate on the way in which CDMA distinguishes itself from FDMA and TDMA in graphical terms, consider Figure 9.32. Parts a and b of the figure depict the ways in which the radio resources are distributed in FDMA and TDMA, respectively. To be specific:

- In FDMA, the channel bandwidth B is divided equally among a total number of K users, with each user being allotted a subband of width B/K and having the whole time resource T at its disposal.
- In TDMA, the time resource T is divided equally among the K users, with each user having total access to the frequency resource, namely the total channel bandwidth B, but for only T/K in each time frame.

In a way, we may therefore think of FDMA and TDMA as the dual of each other.

Turning next to Figure 9.32c, we see that CDMA operates in a manner entirely different from both FDMA and TDMA. Graphically, we see that each CDMA user has full access to the entire radio resources at every point in time from one frame to the next. Nevertheless, for the full utilization of radio resources to be achievable, it is necessary that the spreading codes assigned to all the K users form an *orthogonal set*.

In other words, orthogonality is a common requirement to the FDMA, TDMA, and CDMA, each in its own specific way. However, this requirement is easier to implement practically in FDMA and TDMA than it is in CDMA.

In an ideal CDMA system, to satisfy the orthogonality requirement, the cross-correlation between any two users of the system must be zero. Correspondingly, for this

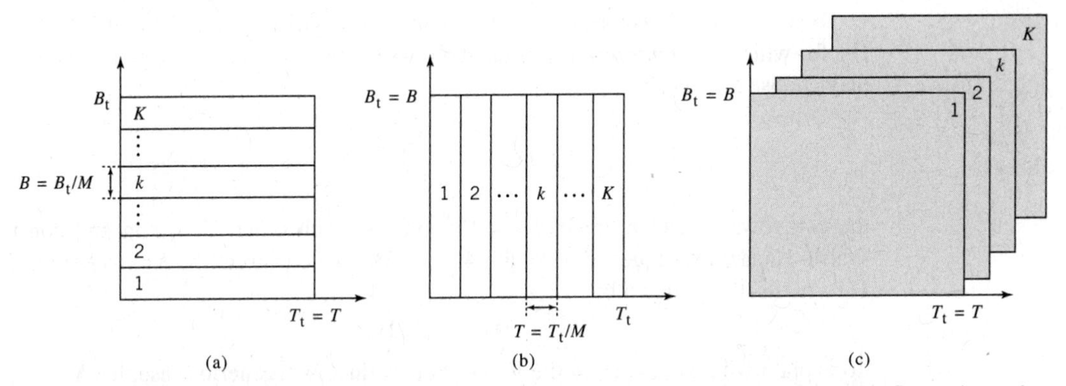

Figure 9.32 Resource distribution in (a) FDMA, (b) TDMA, and (c) CDMA. This figure shows the essence of multiple access as in Figure 1.2 with a difference: Figure 9.32 is quantitative in its description of multiple-access techniques.

ideal condition to be satisfied, we require that the cross-correlation function between the spreading sequences (codes) assigned to any two CDMA users of the system must be zero for all cyclic shifts in time. Unfortunately, ordinary PN sequences do not satisfy the orthogonality requirement because of their relatively poor cross-correlation properties.

Accordingly, we have to look to alternative spreading codes to satisfy the orthogonality requirements. Fortunately, such an endeavor is mathematically feasible, depending on whether synchrony of the CDMA receiver to its transmitter is required or not. In what follows, we describe the use of Walsh–Hadamard sequences for the synchronous case and Gold sequences for the asynchronous case.

Walsh–Hadamard Sequences

Consider the case of a CDMA system, for which synchronization among users of the system is *permissible*. Under this condition, perfect orthogonality of two spreading signals, $c_j(t)$ and $c_k(t)$, respectively assigned to users j and k for different time offsets, namely

$$R_{jk}(\tau) = \int_{-\infty}^{\infty} C_j(t)\, C_k^*(t-\tau)\, dt = 0 \qquad \text{for } j \neq k \tag{9.127}$$

reduces to

$$R_{jk}(0) = \int_{-\infty}^{\infty} C_j(t)\, C_k^*(t)\, dt = 0 \qquad \text{for } j \neq k \text{ and } \tau = 0 \tag{9.128}$$

where the asterisk denotes complex conjugation. It turns out that, for the special case described in (9.128), the orthogonality requirement can be satisfied exactly, and the resulting sequences are known as the *Walsh–Hadamard sequences (codes)*.[15]

To construct a Walsh–Hadamard sequence, we begin with a 2×2 matrix, denoted by $\mathbf{H}_2$, for which the *inner product* of its two rows (or two columns) is zero. For example, we may choose the matrix

$$\mathbf{H}_2 = \begin{bmatrix} +1 & +1 \\ +1 & -1 \end{bmatrix} \tag{9.129}$$

the two rows of which are indeed orthogonal to each other. To go on and construct a Walsh–Hadamard sequence of length 4 using $\mathbf{H}_2$, we construct the *Kronecker product* of $\mathbf{H}_2$ with itself, as shown by

$$\mathbf{H}_4 = \mathbf{H}_2 \otimes \mathbf{H}_2 \tag{9.130}$$

To explain what we mean by the Kronecker product in a generic sense, let $\mathbf{A} = \{a_{jk}\}$ and $\mathbf{B} = (b_{jk}\}$ denote $m \times m$ and $n \times n$ matrices, respectively.[16] Then, we may introduce the following rule:

> The Kronecker product of the two matrices **A** and **B** is made up of an $mn \times mn$ matrix, which is obtained from the matrix **A** by replacing its element a_{jk} in matrix **A** with the scaled matrix $a_{jk}\,\mathbf{B}$.

EXAMPLE 8 **Construction of Hadamard–Walsh $\mathbf{H}_4$ from $\mathbf{H}_2$**

For the example of (9.129) on matrix $\mathbf{H}_2$, applying the Kronecker product rule, we may express the $\mathbf{H}_4$ of (9.130) as follows:

$$\begin{aligned} \mathbf{H}_4 &= \begin{bmatrix} +1 \times \mathbf{H}_2 & +1 \times \mathbf{H}_2 \\ +1 \times \mathbf{H}_2 & -1 \times \mathbf{H}_2 \end{bmatrix} \\ &= \begin{bmatrix} +1 & +1 & +1 & +1 \\ +1 & -1 & +1 & -1 \\ +1 & +1 & -1 & -1 \\ +1 & -1 & -1 & +1 \end{bmatrix} \end{aligned} \tag{9.131}$$

The four rows (and columns) of $\mathbf{H}_4$ defined in (9.131) are indeed orthogonal to each other.

Carrying on in this manner, we may go on to construct the Hadamard–Walsh sequences $\mathbf{H}_6$, $\mathbf{H}_8$, and so on.

In practical terms, a synchronous CDMA system is achievable provided that a single transmitter (e.g., the base station of a cellular network) transmits individual data streams simultaneously, with each data stream being addressed to a specific CDMA user (e.g., mobile unit).

Gold Sequences

Whereas Walsh–Hadamard sequences are well suited for synchronous CDMA, *Gold sequences*, on the other hand, are well suited for applications in asynchronous CDMA;

therein, time- and phase-shifts between individual user signals, measured with respect to the base station in a cellular network, occur in a random manner; hence the adoption of asynchrony.

Gold sequences constitute a special class of maximal-length sequences, the generation of which is embodied in *Gold's theorem*, stated as follows:[17]

> Let $g_1(X)$ and $g_2(X)$ be a preferred pair of primitive polynomials of degree n whose corresponding linear feedback shift registers generate maximal-length sequences of period $2^n - 1$ and whose cross-correlation function has a magnitude less then or equal to
>
> $$2^{(n+1)/2} + 1 \quad \text{for } n \text{ odd} \tag{9.132}$$
>
> or
>
> $$2^{(n+1)/2} + 1 \quad \text{for } n \text{ even and } n \neq 0 \text{ mod } 4 \tag{9.133}$$
>
> Then, the linear feedback shift register corresponding to the product polynomial $g_1(X) \times g_2(X)$ will generage $2^n + 1$ different sequences, with each sequence having a period of $2^n = 1$ and the cross-correlation between any pair of such sequences satisfying the preceding condition.

To understand Gold's theorem, we need to define what we mean by a primitive polynomial. Consider a polynomial $g(X)$ defined over a *binary field* (i.e., a finite set of two elements, 0 and 1, which is governed by the rules of binary arithmetic). The polynomial $g(X)$ is said to be an *irreducible polynomial* if it cannot be factored using any polynomials from the binary field. An irreducible polynomial $g(X)$ of degree m is said to be a *primitive polynomial* if the smallest integer m for which the polynomial $g(X)$ divides the factor $X^n + 1$ is $n = 2^m - 1$. The topic of primitive polynomials is discussed in Chapter 10 on error-control coding.

EXAMPLE 9 Correlation Properties of Gold Codes

As an illustrative example, consider Gold sequences with period $2^7 - 1 = 127$. To generate such a sequence for $n = 7$ we need a preferred pair of PN sequences that satisfy (9.132) (n odd), as shown by

$$2^{(n+1)/2} + 1 = 2^4 + 1 = 17$$

This requirement is satisfied by the Gold-sequence generator shown in Figure 9.33 that involves the modulo-2 addition of these two sequences. According to Gold's theorem, there are a total of

$$2^n + 1 = 2^7 + 1 = 129$$

sequences that satisfy (9.132). The cross-correlation between any pair of such sequences is shown in Figure 9.34, which is indeed in full accord with Gold's theorem. In particular, the magnitude of the cross-correlation is less than or equal to 17.

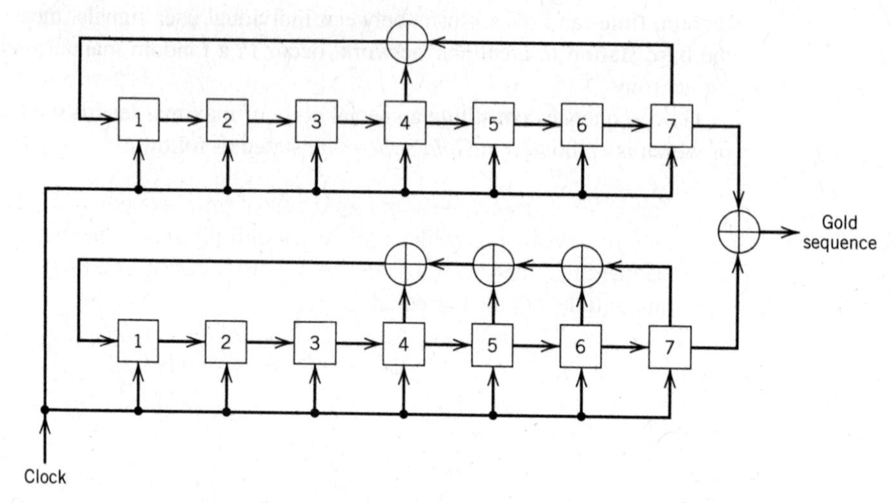

Figure 9.33 Generator for a Gold sequence of period $2^7 - 1 = 127$.

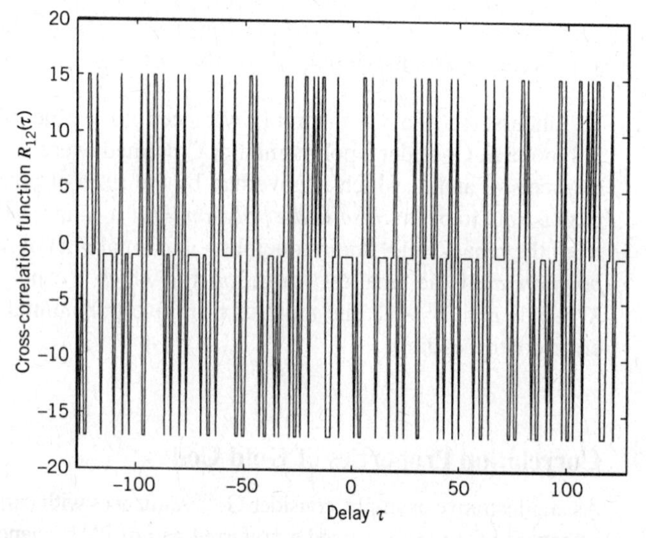

Figure 9.34 Cross-correlation function R_{12} of a pair of Gold sequences based on the two PN sequences [7,4] and [7,6,5,4].

9.15 The RAKE Receiver and Multipath Diversity

A discussion of wireless communications using CDMA would be incomplete without a description of the *RAKE receiver.*[18] The RAKE receiver was originally developed in the 1950s as a *diversity* receiver designed expressly to equalize the effect of multipath. First, and foremost, it is recognized that useful information about the transmitted signal is contained in the multipath component of the received signal. Thus, taking the viewpoint that multipath may be approximated as a linear combination of differently delayed echoes, as shown in the maximal ratio combiner of Figure 9.21, the RAKE receiver seeks to combat the effect of multipath by using a correlation method to detect the echo signals individually and then adding them algebraically. In this way, intersymbol interference due to multipath is dealt with by reinserting different delays into the detected echoes so that they perform a constructive rather than destructive role.

Figure 9.35 shows the basic idea behind the RAKE receiver. The receiver consists of a number of *correlators* connected in parallel and operating in a synchronous fashion with each other. Each correlator has two inputs: (1) a delayed version of the received signal and (2) a replica of the PN sequence used as the spreading code to generate the spread spectrum-modulated signal at the transmitter. In effect, the PN sequence acts as a *reference signal*. Let the nominal bandwidth of the PN sequence be denoted as $W = 1/T_c$, where T_c is the chip duration. From the discussion on PN sequences presented in Appendix J, we find that the autocorrelation function of a PN sequence has a single peak of width $1/W$, and it disappears toward zero elsewhere inside one period of the PN sequence (i.e., one symbol period). Thus, we need only make the bandwidth W of the PN sequence sufficiently large to identify the significant echoes in the received signal. To be sure that the correlator outputs all add constructively, two other operations are performed in the receiver by the functional blocks labeled "phase and gain adjustors":

1. An appropriate delay is introduced into each correlator output, so that the phase angles of the correlator outputs are in agreement with each other.
2. The correlator outputs are weighted so that the correlators responding to strong paths in the multipath environment have their contributions accentuated, while the correlators not synchronizing with any significant path are correspondingly suppressed.

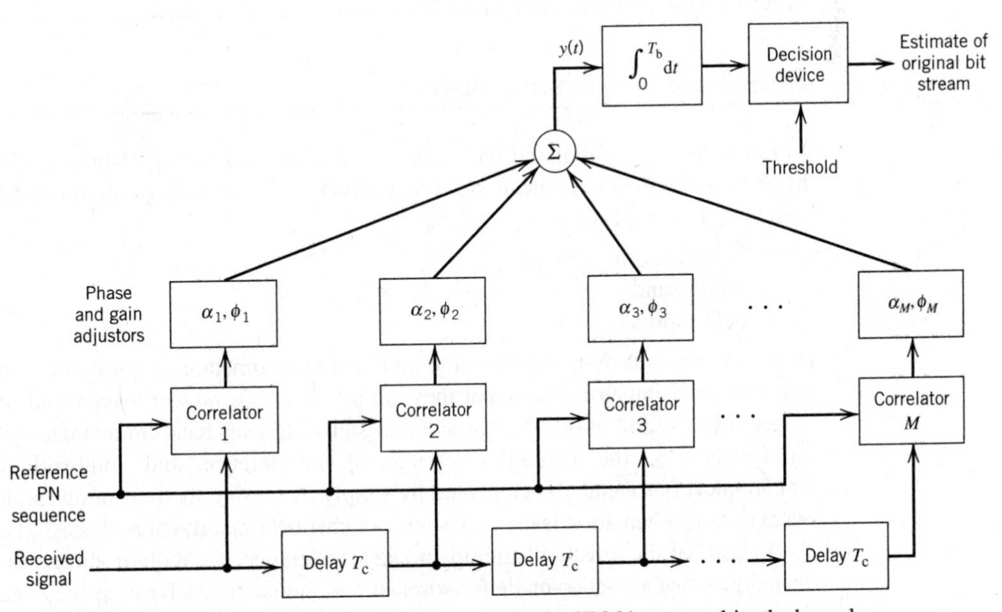

Figure 9.35 Block diagram of the RAKE receiver for CDMA over multipath channels.

The weighting coefficients a_k are computed in accordance with the *maximal ratio combining principle*, discussed in Section 9.8. Specifically, we recall that the SNR of a weighted sum, where each element of the sum consists of a signal plus additive noise of fixed power, is maximized when the amplitude weighting is performed in proportion to the pertinent signal strength. That is, the linear combiner output is

$$y(t) = \sum_{k=1}^{M} a_k z_k(t) \tag{9.134}$$

where $z_k(t)$ is the phase-compensated output of the kth correlator and M is the number of correlators in the receiver. Provided that we use enough correlators in the receiver to span a region of delays sufficiently wide to encompass all the significant echoes that are likely to occur in the multipath environment, the output $y(t)$ behaves essentially as though there was a single propagation path between the transmitter and receiver rather than a series of multiple paths spread in time.

To simplify the presentation, the receiver of Figure 9.35 assumes the use of binary PSK in performing spread spectrum modulation at the transmitter. Thus, the final operation performed in Figure 9.35 is that of integrating the linear combiner output $y(t)$ over the bit duration T_b and then determining whether binary symbol 1 or 0 was transmitted in that bit interval.

The RAKE receiver derives its name from the fact that the bank of parallel correlators has an appearance similar to the fingers of a rake; see Figure 9.36. Because spread spectrum modulation is basic to the operation of CDMA wireless communications, it is natural for the RAKE receiver to be central to the design of the receiver used in this type of multiuser radio communication.

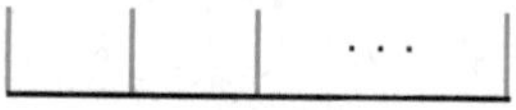

Figure 9.36
Picture of a rake, symbolizing the bank of correlators.

9.16 Summary and Discussion

In this chapter we discussed the topic of signaling over fading channels, which is at the heart of wireless communications. There are three major sources of signal degradation in wireless communications:

- co-channel interference,
- fading, and
- delay spread.

The latter two are by-products of the multipath phenomenon. A common characteristic of these channel impairments is that they are all *signal-dependent phenomena*. As it is with intersymbol interference that characterizes signaling over band-limited channels discussed in Chapter 8, the degrading effects of interference and multipath in wireless communications cannot be combated by simply increasing the transmitted signal, which is what is done when noise is the only source of channel impairment as discussed in Chapter 7.

To combat the effects of multipath and interference, we require the use of specialized techniques that are tailor-made for wireless communications. These specialized techniques include *space diversity*, which occupied much of the material presented in this chapter.

We discussed different forms of space diversity, the main idea behind which is that two or more propagation paths connecting the receiver to the transmitter are better than a single propagation path. In historical terms, the first form of space diversity used to mitigate the multipath fading problem was that of *receive diversity*, involving a single transmit antenna and multiple receive antennas. Under receive diversity, we discussed the selection combiner, maximal-ratio combiner, and equal-gain combiner:

- The selection combiner is the simplest form of receive diversity. It operates on the principle that it is possible to select, among N_r receive-diversity branches, a particular branch with the largest output SNR; the branch so selected defines the desired received signal.
- The maximal-ratio combiner is more powerful than the selection combiner by virtue of the fact that it exploits the full information content of all the N_r receive-diversity branches about the transmitted signal of interest; it is characterized by a set of N_r receive-complex weighting factors that are chosen to maximize the output SNR of the combiner.
- The equal-gain combiner is a simplified version of the maximal-ratio combiner.

We also discussed diversity-on-transmit techniques, which may be viewed as the dual of their respective diversity-on-receive techniques. Much of the discussion here focused on the Alamouti code, which is simple to design, yet powerful in performance, in that it realizes a two-level diversity gain: in other terms of performance, the Alamouti code is equivalent to a linear diversity-on-receive system with a single antenna and two receive antennas.

By far, the most powerful form of space diversity is the use of multiple antennas at both the transmit and receive ends of the wireless link. The resulting configuration is referred to as a MIMO wireless communication system, which includes the receive diversity and transmit diversity as special cases. The novel feature of the MIMO system is that, in a rich scattering environment, it can provide a high spectral efficiency, which may be simply explained as follows. The signals transmitted simultaneously by the transmit antennas arrive at the input of each receive antenna in an uncorrelated manner due to the rich scattering mechanism of the channel. The net result is a spectacular increase in the spectral efficiency of the wireless link. Most importantly, the spectral efficiency increases roughly linearly with the number of transmit or receive antennas, whichever is the smaller one of the two. This important result assumes that the receiver has knowledge of the channel state. The spectral efficiency of the MIMO system can be further enhanced by including a feedback channel from the transmitter to the receiver, whereby the channel state is also made available to the transmitter and with it the transmitter is enabled to exercise control over the transmitted signal.

Multiple Access Considerations

An issue of paramount practical importance in wireless communications is that of multiple access to the wireless channel, in the context of which the following two approaches are considered to be the dominant ones:

1. *Orthogonal frequency division multiple access* (*OFDMA*), which is the multi-user version of OFDM that was discussed in Section 9.12. In OFDMA multiple access is accomplished through the assignment of subchannels (subcarriers) to individual users.

Naturally, OFDMA inherits the distinctive features of OFDM. In particular, OFDMA is well suited for high data-rate transmissions over delay-dispersive channels, realized by exploiting the principle of "divide and conquer." Accordingly, OFDMA is computationally efficient in using the FFT algorithm. Moreover, OFDMA lends itself to the combined use of MIMO, hence the ability to improve spectral efficiency and take advantage of channel flexibility.

2. *Code-division multiple access* (*CDMA*), which distinguishes itself by exploiting the underlying principle of spread spectrum signals, discussed in Section 9.13. To be specific, through the combined process of spectrum spreading in the transmitter and corresponding spectrum despreading in the receiver, a certain amount of processing gain is obtained, hence the ability of CDMA users to occupy the same channel bandwidth. Moreover, CDMA provides a flexible procedure for the allocation of resources (i.e., PN codes) among a multiplicity of active users. Last but by no means least, in using the RAKE, viewed as an adaptive TDL filter, CDMA is enabled to match the receiver input to the channel output by adjusting tap delays as well as tap weights, thereby enhancing receiver performance in the presence of multipath.

To conclude, OFDMA and CDMA provide two different approaches for the multiple access of active users to wireless channels, each one of which builds on its own distinctive features.

Problems

Effect of Flat Fading on the BER of Digital Communications Receivers

9.1 Derive the BER formulas listed in the right-hand side of Table 9.2 for the following signaling schemes over flat fading channels:

a. Binary PSK using coherent detection
b. Binary FSK using coherent detection
c. Binary DPSK
d. Binary FSK using noncoherent detection

9.2 Using the formulas derived in Problem 9.1, plot the BER charts for the schemes described therein.

Selective Channels

9.3 Consider a time-selective channel, for which the modulated received signal is defined by

$$x(t) = \sum_{n=1}^{N} \alpha_n(t) m(t) \cos(2\pi f_c t + \phi(t) + \sigma_n(t))$$

where $m(t)$ is the message signal, $\phi(t)$ is the result of angle modulation; the amplitude $\alpha_n(t)$ and phase $\sigma_n(t)$ are contributed by the nth path, where $n = 1, 2, \ldots, N$.

a. Using complex notation, show that the received signal is described as follows:

$$\tilde{x}(t) = \tilde{\alpha}(t)\tilde{s}(t)$$

where

$$\tilde{\alpha}(t) = \sum_{n=1}^{N} \tilde{\alpha}_n(t)$$

What is the formula for $\tilde{s}(t)$?

b. Show that the delay-spread function of the multipath channel is described by

$$\tilde{h}(\tau;t) = \tilde{\alpha}(t)\delta(\tau)$$

where $\delta(\tau)$ is the Dirac delta function in the τ-domain. Hence, justify the statement that the channel described in this problem is a *time-selective channel.*

c. Let $S_{\tilde{\alpha}}(f)$ and $S_{\tilde{s}}(f)$ denote the Fourier transforms of $\tilde{\alpha}(t)$ and $\tilde{s}(t)$, respectively. What then is the Fourier transform of $\tilde{x}(t)$?

d. Using the result of part c, justify the statement that the multipath channel described herein can be approximately frequency-flat. What is the condition that would satisfy this description?

9.4 In this problem, we consider a multipath channel embodying large-scale effects. Specifically, using complex notation, the received signal at the channel output is described by

$$\tilde{x}(t) = \sum_{l=1}^{L} \tilde{\alpha}_l \tilde{s}(t-\tau_l)$$

where $\tilde{\alpha}_l$ and τ_l denote the amplitude and time delay associated with the lth path in the channel for $l = 1, 2, \ldots, L$. Note that $\tilde{\alpha}_l$ is assumed to be constant for all l.

a. Show that the delay-spread function of the channel is described by

$$\tilde{h}(\tau;t) = \sum_{l=1}^{L} \tilde{\alpha}_l \delta(\tau-\tau_l)$$

where $\delta(\tau)$ is the Dirac delta function expressed in the τ-domain.

b. This channel is said to be time-nonselective. Why?

c. The channel does exhibit a frequency-dependent behavior. To illustrate this behavior, consider the following delay-spread function:

$$\tilde{h}(\tau;t) = \delta(\tau) + \tilde{\alpha}_2 \delta(\tau-\tau_2)$$

where τ_2 is the time delay produced by the second path in the channel. Plot the magnitude (amplitude) response of the channel for the following specifications:

i. $\tilde{\alpha}_2 = 0.5$

ii. $\tilde{\alpha}_2 = j/2$

iii. $\tilde{\alpha}_2 = -j$

where $j = \sqrt{-1}$. Comment on your results.

9.5 Expanding on the multipath channel considered in Problem 9.4, a more interesting case is characterized by the scenario in which the received signal at the channel output is described as follows:

$$\tilde{x}(t) = \sum_{l=1}^{L} \tilde{\alpha}_l(t)\tilde{s}(t-\tau_l(t))$$

where the amplitude $\tilde{\alpha}_l(t)$ and time delay $\tau_l(t)$ for the lth path are both time dependent for $l = 1, 2, \ldots, L$.

a. Show that the delay-spread function of the multipath channel described herein is given by

$$\tilde{h}(\tau;t) = \sum_{l=1}^{L} \tilde{\alpha}_l(t)\delta(\tau-\tau_l(t))$$

where $\delta(\tau)$ is the Dirac delta function in the τ-domain. This channel is said to exhibit both large- and small-scale effects. Why?

b. The channel is also said to be both time selective and frequency selective. Why?

c. To illustrate the point made under b, consider the following channel description:

$$\tilde{h}(\tau;t) = \tilde{\alpha}_1(t)\delta(\tau) + \tilde{\alpha}_2(t)\delta(t-\tau_2)$$

where $\tilde{\alpha}_1(t)$ and $\tilde{\alpha}_2(t)$ are both Rayleigh processes.

For selected $\tilde{\alpha}_1(t)$, $\tilde{\alpha}_2(t)$ and τ_2, do the following:

i. At each time $t = 0$, compute the Fourier transform of $\tilde{h}(\tau;t)$.

ii. Hence, plot the magnitude spectrum of the channel, that is, $|\tilde{H}(f;t)|$, expressed as a function of both time t and frequency f.

Comment on the results so obtained.

9.6 Consider a multipath channel where the delay-spread function is described by

$$\tilde{h}(\tau;t) = \sum_{l=1}^{L} \tilde{\alpha}_l(t)\delta(\tau-\tau_l)$$

where the scattering processes attributed to the time-varying amplitude $\tilde{\alpha}_l(t)$ and fixed delay τ_l are uncorrelated for $l = 1, 2, \ldots, L$.

a. Determine the correlation function of the channel, namely $R_{\tilde{h}}(\tau_1, t_1; \tau_2, t_2)$.

b. With a Jakes model for the scattering process described in (9.12), find the corresponding formula for the correlation function of the channel under part a of the problem.

c. Hence, justify the statement that the multipath channel described in this problem fits a WSSUS model.

9.7 Revisit the Jakes model for a fast fading channel described in (9.12). Let the coherence time be defined as that range of values Δt over which the correlation function defined in (9.12) is greater than 0.5.

For some prescribed maximum Doppler shift ν_{max}, find the coherence time of the channel.

9.8 Consider a multipath channel for which the delay-spread function is given by

$$\tilde{h}(\tau;t) = \sum_{l=1}^{L} \tilde{\alpha}_l(t)\delta(t-\tau_l)$$

where the amplitude $\tilde{\alpha}_l(t)$ is time varying but the time delay τ_l is fixed. As in Problem 9.4, the scattering processes are described by the Jakes model in (9.12). Determine the power-delay profile of the channel, $P_{\tilde{h}}(\tau)$.

9.9 In real-life situations, the wireless channel is *nonstationary* due to the presence of moving objects of different kinds and other physical elements that can significantly affect radio propagation. Naturally, different types of wireless channels have different degrees of nonstationarity.

Even though many wireless communication channels are indeed highly nonstationary, the WSSUS model described in Section 9.4 still provides a reasonably accurate account of the statistical characteristics of the channel. Elaborate on this statement.

"Space Diversity-on-Receive" Systems

9.10 Following the material presented on Rayleigh fading in Chapter 4, derive the probability density function of (9.64).

9.11 A receive-diversity system uses a selection combiner with two diversity paths. The outage occurs when the instantaneous SNR γ drops below $0.25\gamma_{av}$, where γ_{av} is the average SNR.

Determine the probability of outage experienced by the receiver.

9.12 The average SNR in a selection combiner is 20 dB. Compute the probability that the instantaneous SNR of the selection combiner drops below $\gamma = 10$ dB for the following number of receive antennas:

a. $N_r = 1$

b. $N_r = 2$

c. $N_r = 3$

d. $N_r = 4$.

Comment on your results.

9.13 Repeat Problem 9.12 for $\gamma = 15$ dB.

9.14 In Section 9.8 we derived the optimum values of (9.75) for complex weighting factors of the maximal-ratio combiner using the Cauchy–Schwartz inequality.

This problem addresses the same issue, but this time we use the standard maximization procedure. To simplify matters, the number of diversity paths N_r is restricted to two, with the complex weighting parameters denoted by a_1 and a_2. Let

$$a_k = x_k + \mathrm{j}y_k, \qquad k = 1, 2$$

The complex derivative with respect to a_k is defined by

$$\frac{\partial}{\partial a_k^*} = \frac{1}{2}\left(\frac{\partial}{\partial x_k} + \mathrm{j}\frac{\partial}{\partial y_k}\right), \qquad k = 1, 2$$

Applying this formula to the combiner's output SNR γ_c of (9.71), derive the optimum γ_{mrc} in (9.75).

9.15 As discussed in Section 9.8, an *equal-gain combiner* is a special form of the maximal-ratio combiner for which the weighting factors are all equal. For convenience of presentation, the weighting parameters are set to unity.

Assuming that the instantaneous SNR γ is small compared with the average SNR γ_{av}, derive an approximate formula for the probability density function of the random variable Γ represented by the sample γ.

9.16 Compare the performances of the following linear "diversity-on-receive" techniques:

a. Selection combiner.

b. Maximal-ratio combiner.

c. Equal-gain combiner.

Base the comparison on signal-to-noise improvement, expressed in decibels for the following number of diversity branches: $N_r = 2, 3, 4, 5, 6$.

9.17 Show that the maximum-likelihood decision rule for the maximal-ratio combiner may be formulated in the following two equivalent forms:

a. If

$$[(\alpha_1^2 + \alpha_2^2)|s_i|^2 - y_1 s_i^* - y_1^* s_i] < [(\alpha_1^2 + \alpha_2^2)|s_k|^2 - y_1 s_k^* - y_1 s_k], \qquad k \neq i$$

then choose symbol s_i over s_k.

b. If, by the same token,

$$[(\alpha_1^2 + \alpha_2^2 - 1)|s_i|^2 + d^2(y_1, s_i)] < [(\alpha_1^2 + \alpha_2^2 - 1)|s_k|^2 + d^2(y_1, s_k)], \qquad k \neq i$$

then choose symbol s_i over s_k. Here, $d^2(y_1,s_i)$ denotes the squared Euclidean distance between the signal points y_1 and s_i.

9.18 It may be argued that, in a rather loose sense, transmit-diversity and receive-diversity antenna configurations are the dual of each other, as illustrated in Figure P9.18.

a. Taking a general viewpoint, justify the mathematical basis for this duality.

b. However, we may cite the example of frequency-division diplexing (FDD) for which, in a strict sense, we find that the duality depicted in Figure P9.18 is violated. How is it possible for the violation to arise in this example?

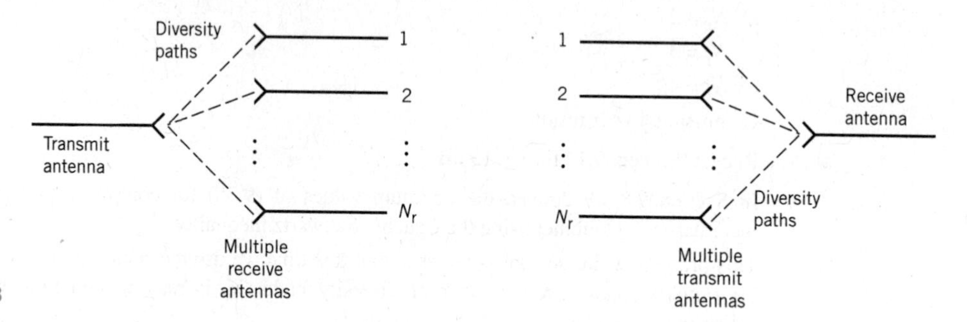

Figure P9.18

"Space Diversity-on-Transmit" Systems

9.19 Show that the two-by-two channel matrix in (9.88), defined in terms of the multiplicative fading factors $\alpha_1 e^{j\theta_1}$ and $\alpha_2 e^{j\theta_2}$, is a unitary matrix, as shown by

$$\begin{bmatrix} \alpha_1 e^{j\theta_1} & \alpha_2 e^{j\theta_2} \\ \alpha_2 e^{-j\theta_2} & -\alpha_1 e^{-j\theta_1} \end{bmatrix}^{\dagger} \begin{bmatrix} \alpha_1 e^{j\theta_1} & \alpha_2 e^{j\theta_2} \\ \alpha_2 e^{-j\theta_2} & -\alpha_1 e^{-j\theta_1} \end{bmatrix} = (\alpha_1^2 + \alpha_2^2) \begin{bmatrix} 1 & 0 \\ 0 & 1 \end{bmatrix}$$

9.20 Derive the formula for the average probability of symbol error incurred by the Alamouti code.

9.21 Figure P9.22 shows the extension of orthogonal space–time codes to the Alamouti code, using two antennas on both transmit and receive. The sequence of signal encoding and transmissions is identical to that of the single-receiver case of Figure 9.18. Part a of the table below defines the channels between the transmit and receive antennas. Part b of the table defines the outputs of the receive antennas at times t' and $t' + T$, where T is the symbol duration.

a. Derive expressions for the received signals $\tilde{x}_1, \tilde{x}_2, \tilde{x}_3$, and $\tilde{x}_4$, including the respective additive noise components expressed in terms of the transmitted symbols.

b. Derive expressions for the line of combined outputs in terms of the received signals.

c. Derive the maximum-likelihood decision rule for the estimates $\tilde{s}_1$ and $\tilde{s}_2$.

		Receive antenna 1	Receive antenna 2
a.	Transmit antenna 1	h_1	h_3
	Transmit antenna 2	h_2	h_4
b.	Time t'	$\tilde{x}_1$	$\tilde{x}_3$
	Time $t' + T$	$\tilde{x}_2$	$\tilde{x}_4$

9.22 This problem explores a new interpretation of the Alamouti code. Let

$$\tilde{s}_i = s_i^{(1)} + js_i^{(2)}, \qquad i = 1, 2$$

Figure P9.22

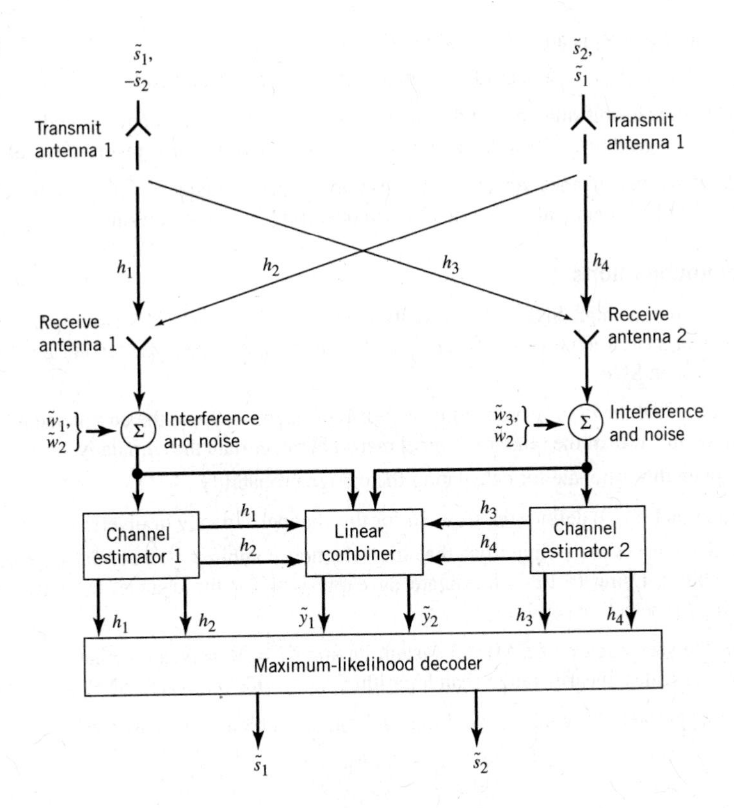

where $s_i^{(1)}$ and $s_i^{(2)}$ are both real numbers. The complex entry $\tilde{s}_i$ in the 2-by-2 Alamouti code is represented by the 2-by-2 real orthogonal matrix

$$\begin{bmatrix} s_i^{(1)} & s_i^{(2)} \\ -s_i^{(2)} & s_i^{(1)} \end{bmatrix}, \qquad i = 1, 2$$

Likewise, the complex-conjugated entry $\tilde{s}_i^*$ is represented by the 2-by-2 real orthogonal matrix

$$\begin{bmatrix} s_i^{(1)} & -s_i^{(1)} \\ s_i^{(2)} & s_i^{(2)} \end{bmatrix}, \qquad i = 1, 2$$

a. Show that the 2-by-2 complex Alamouti code $\mathbf{S}$ is equivalent to the 4-by-4 *real* transmission matrix

$$\mathbf{S}_4 = \left[\begin{array}{cc|cc} s_1^{(1)} & s_1^{(2)} & s_2^{(1)} & s_2^{(2)} \\ -s_1^{(2)} & s_1^{(1)} & -s_2^{(2)} & s_2^{(1)} \\ \hline -s_2^{(1)} & s_2^{(2)} & s_1^{(1)} & -s_1^{(2)} \\ -s_2^{(2)} & -s_2^{(1)} & s_1^{(2)} & s_1^{(1)} \end{array}\right]$$

b. Show that $\mathbf{S}_4$ is an orthogonal matrix.

c. What is the advantage of the complex code $\mathbf{S}$ over the real code $\mathbf{S}_4$?

9.23 For two transmit antennas and simple receive antenna, the Alamouti code is said to be the only optimal space–time block. Using the log-det formula of (9.117), justify this statement.

9.24 Show that the channel capacity of the Alamouti code is equal to the sum of the channel capacities of two SISO systems with each one of them operating at half the original bit rate.

MIMO Wireless Communications

9.25 Show that, at high SNRs, the capacity gain of a MIMO wireless communication system with the channel state known to the receiver is $N = \min\{N_t, N_r\}$ bits per second per hertz for every 3 dB increase in SNR.

9.26 To calculate the outage probability of MIMO systems, we use the complementary cumulative distribution function of the random channel matrix $\mathbf{H}$ rather than the cumulative probability function itself. Explain this rationale for calculating the outage probability.

9.27 Equation (9.120) defines the formula for the channel capacity of diversity-on-receive channel. In Section 9.8 we pointed out that the selection combiner is a special case of the maximal-ratio combiner. Using (9.120), formulate an expression for the channel capacity of wireless diversity using the selection combiner.

9.28 For the special case of a MIMO system having $N_t = N_r = N$, show that the ergodic capacity of the system scales linearly, rather than logarithmically, with increasing SNR as N approaches infinity.

9.29 In this problem we continue with the solution to Problem 9.28, namely

$$C \rightarrow \left(\frac{\lambda_{av}}{\log_e 2}\right)\rho \qquad \text{as } N \rightarrow \infty$$

where $N_t = N_r = N$ and λ_{av} is the average eigenvalue of the matrix produced $\mathbf{H}\mathbf{H}^\dagger = \mathbf{H}^\dagger\mathbf{H}$. What is the value of the constant?

a. Justify the asymptotic result given in (9.119); that is,

$$\frac{C}{N} \geq \text{constant}$$

b. What conclusion can you draw from this asymptotic result?

9.30 Suppose that an additive, temporally stationary, Gaussian interference $\mathbf{v}(t)$ corrupts the basic complex channel model of (9.105). The interference $\mathbf{v}(t)$ has zero mean and correlation matrix $\mathbf{R}_v$. Evaluate the effect of the interference $\mathbf{v}(t)$ on the ergodic capacity of the MIMO link.

9.31 Consider a MIMO link for which the channel may be considered to be essentially "constant for k users of the channel."

a. Starting with the basic channel model of (9.105), formulate the input–output relationship of this link with the input being described by the N_r-by-k matrix

$$\mathbf{S} = [\mathbf{s}_1, \mathbf{s}_2, \ldots, \mathbf{s}_k]$$

b. How is the log-det capacity formula of the link correspondingly modified?

9.32 In a MIMO channel, the ability to exploit space-division multiple-access techniques for spectrally efficient wireless communications is determined by the rank of the complex channel matrix $\mathbf{H}$. (The rank of a matrix is defined by the number of independent columns in the matrix.) For a given (N_t, N_r) antenna configuration, it is desirable that the rank of $\mathbf{H}$ equal the minimum one of N_t transmit and N_r receive antennas, for it is only then that we are able to exploit the full potential of the MIMO antenna configuration. Under special conditions, however, the rank of the channel matrix **H** is reduced to unity, in which case the scattering (fading) energy flow across the MIMO link is effectively confined to a very narrow pipe, and with it, the channel capacity is severely degraded.

Under the special conditions just described, a physical phenomenon known as the *keyhole channel* or *pinhole channel* is known to arise. Using a propagation layout of the MIMO link, describe how this phenomenon can be explained.

OFDMA and CDMA

9.33 Parts a and b of Figure 9.31 show the block diagrams of the transmitter and receiver of an OFDM system, formulated on the basis of digital signal processing. It is informative to construct an analog interpretation of the OFDM system, which is the objective of this problem.

a. Construct the analog interpretations of parts a and b in Figure 9.31.

b. With this construction at hand, compare the advantages and disadvantages of the digital and analog implementations of OFDM.

9.34 Figure P9.34 depicts the *model* of a DS/BPSK system, where the order of spectrum spreading and BPSK in the actual system has been interchanged; this is feasible because both operations are linear. For system analysis, we build on signal-space theoretic ideas of Chapter 7, using this model and assuming the presence of a jammer at the receiver input. Thus, whereas signal-space representation of the transmitted signal, $x(t)$, is one-dimensional, that of the jammer, $j(t)$, is two-dimensional.

a. Derive the processing gain formula of (9.125).

b. Next, ignoring the benefit gained from coherent detection, derive the SNR formula of (9.126).

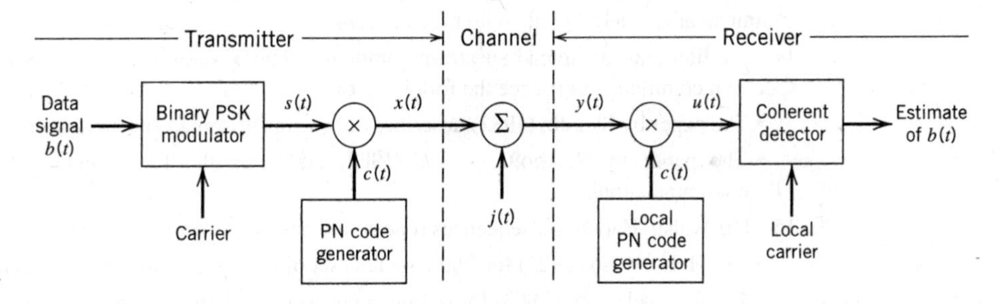

Figure P9.34

Notes

1. Local propagation effects are discussed in Chapter 1 of the classic book by Jakes (1974). For a comprehensive treatment of this subject, see the books by Parsons (2000) and Molisch (2011).

2. Bessel functions are discussed in Appendix C.

3. To be precise, we should use the terminology "autocorrelation" function rather then "correlation" function as we did in Section 9.3. However, to be consistent with the literature, hereafter we use the terminology "correlation function" for the sake of simplicity.

4. On the basis of many measurements, the power-delay profile may be approximated by the one-sided exponential functions (Molisch, 2011):

$$P_{\tilde{h}}(\tau)\begin{cases} = \exp(-\tau/\sigma_{\tau}), & \text{for } \tau \geq 0 \\ = 0, & \text{otherwise} \end{cases}$$

For a more generic model, the power-delay profile is viewed as the sum of several one-sided exponential functions representing multiple clusters of interacting objects, as shown by

$$P_{\tilde{h}}(\tau) = \sum_i \left(\frac{P_i}{\sigma_{\tau,i}}\right) P_{\tilde{h}}(\tau - \tau_{0,i})$$

where P_i, $\tau_{0,i}$, and $\sigma_{\tau,i}$ are respectively the power, delay, and delay spread of the ith cluster.

5. The approximate approach described in Section 9.5 follows Van Trees (1971).

6. The complex *tap-coefficient* $\tilde{c}_n(t)$ is also referred to as the tap-gain or tap-weight.

7. The chi-squared distribution with two degrees of freedom is described in Appendix A.

8. The term "maximal-ratio combiner" was coined in a classic paper on linear diversity combining techniques by Brennan (1959).

9. The three-point exposition presented in this section on maximal-ratio combining follows the chapter by Stein in Schwartz *et al.* (1966: 653–654).

10. The idea of MIMO for wireless communications was first described in the literature by Foschini (1996). In the same year, Teletar (1996) derived the capacity of multi-antenna Gaussian channels in a technical report.

11. As a result of experimental measurements, the model is known to be decidedly non-Gaussian owing to the *impulsive nature* of human-made electromagnetic interference and natural noise.

12. Detailed derivation of the ergodic capacity in (9.115) is presented in Appendix E.

13. The idea of OFDM has a long history, dating back to Chang (1966). Then, Weinstein and Ebert (1971) used the FFT algorithm and guard intervals for the first digital implementation of OFDM. The first use of OFDM for mobile communications is credited to Cemini (1985).

In the meantime, OFDM has developed into an indispensable tool for broadband wireless communications and digital audio broadcasting.

14. The literature on spread spectrum communications is enormous. For classic papers on spread spectrum communications, see the following two:

- The paper by Scholtz (1982) describes the origins of spread spectrum communications.
- The paper by Pickholtz, *et al.* (1982) addresses the fundamentals of spread spectrum communications.

15. The Walsh–Hadamard sequences (codes) are named in honor of two pioneering contributions:

- Joseph L. Walsh (1923) for finding a new set of orthogonal functions with entries ± 1.
- Jacques Hadamard (1893) for finding a new set of square matrices also with entries ± 1, which had all their rows (and columns) orthogonal.

For more detailed treatments of these two papers, see Harmuth (1970), and Seberry and Yamada (1992), respectively.

16. To be rigorous mathematically, we should speak of the matrices **A** and **B** to be over the Galois field, GF(2). To explain, for any *prime p*, there exists a *finite field* of p elements, denoted by GF(P). For any positive integer b, we may expand the finite field GF(p) to a field of p^b elements, which is called an *extension field* of GF(p) and denoted by GF(p^b). Finite fields are also called *Galois fields* in honor of their discoverer.

Thus, for the example of (9.129), we have a Galois field of $p = 2$ and thus write GF(2). Correspondingly, for the $\mathbf{H}_4$ in (9.130) we have the Galois field GF(2^2) = GF(4)

17. The original papers on Gold sequences are Gold (1967, 1968). A detailed discussion of Gold sequences is presented in Holmes (1982).

18. The classic paper on the RAKE receiver is due to Price and Green (1958). For a good treatment of the RAKE receiver, more detailed than that presented in Section 9.15, see Chapter 5 in the book by Haykin and Mohr (2005). For application of the RAKE receiver in CDMA, see the book by Viterbi (1995).

CHAPTER
10

Error-Control Coding

10.1 Introduction

In the previous three chapters we studied the important issue of data transmission over communication channels under three different channel-impairment scenarios:

- In Chapter 7 the focus of attention was on the kind of channels where AWGN is the main source of channel impairment. An example of this first scenario is a satellite-communication channel.
- In Chapter 8 the focus of attention was intersymbol interference as the main source of channel impairment. An example of this second scenario is the telephone channel.
- Then, in Chapter 9, we focused on multipath as a source of channel impairment. An example for this third scenario is the wireless channel.

Although, indeed, these three scenarios are naturally quite different from each other, they do share a common practical shortcoming: *reliability.* This is where the need for error-control coding, the topic of this chapter, assumes paramount importance.

Given these physical realities, the task facing the designer of a digital communication system is that of providing a cost-effective facility for transmitting information from one end of the system at a rate and level of reliability and quality that are acceptable to a user at the other end.

From a communication theoretic perspective, the key system parameters available for achieving these practical requirements are limited to two:

- transmitted signal power, and
- channel bandwidth.

These two parameters, together with the power spectral density of receiver noise, determine the signal energy per bit-to-noise power spectral density ratio, E_b/N_0. In Chapter 7 we showed that this ratio uniquely determines the BER produced by a particular modulation scheme operating over a Gaussian noise channel. Practical considerations usually place a limit on the value that we can assign to E_b/N_0. To be specific, in practice, we often arrive at a modulation scheme and find that it is not possible to provide acceptable data quality (i.e., low enough error performance). For a fixed E_b/N_0, the only practical option available for changing data quality from problematic to acceptable is to use *error-control coding*, which is the focus of attention in this chapter. In simple terms, by incorporating a fixed number of redundant bits into the structure of a codeword at the transmitter, it is feasible to provide reliable communication over a noisy channel, provided that Shannon's code theorem, discussed in Chapter 5, is satisfied. In effect, channel bandwidth is traded off for reliable communication.

Another practical motivation for the use of coding is to reduce the required E_b/N_0 for a fixed BER. This reduction in E_b/N_0 may, in turn, be exploited to reduce the required transmitted power or reduce the hardware costs by requiring a smaller antenna size in the case of radio communications.

10.2 Error Control Using Forward Error Correction

Error control for data integrity may be exercised by means of *forward error correction* (FEC).[1] Figure 10.1a shows the model of a digital communication system using such an approach. The discrete source generates information in the form of binary symbols. The *channel encoder* in the transmitter accepts message bits and adds *redundancy* according to a prescribed rule, thereby producing an encoded data stream at a higher bit rate. The *channel decoder* in the receiver exploits the redundancy to decide which message bits in the original data stream, given a noisy version of the encoded data stream, were actually transmitted. The combined goal of the channel encoder and decoder is to minimize the effect of channel noise. That is, the number of errors between the channel encoder input (derived from the source) and the channel decoder output (delivered to the user) is minimized.

For a fixed modulation scheme, the addition of redundancy in the coded messages implies the need for *increased transmission bandwidth.* Moreover, the use of error-control coding adds *complexity* to the system. Thus, the design trade-offs in the use of error-control coding to achieve acceptable error performance include considerations of bandwidth and system complexity.

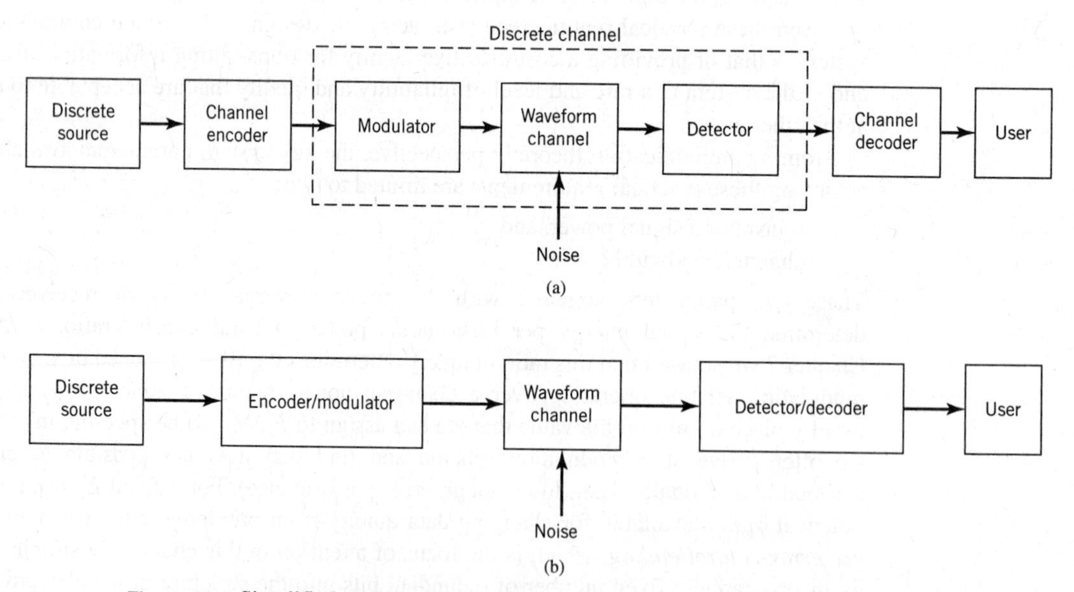

Figure 10.1 Simplified models of a digital communication system. (a) Coding and modulation performed separately. (b) Coding and modulation combined.

There are many different error-correcting codes (with roots in diverse mathematical disciplines) that we can use. Historically, these codes have been classified into *block codes* and *convolutional codes*. The distinguishing feature for this particular classification is the presence or absence of *memory* in the encoders for the two codes.

To generate an (n, k) block code, the channel encoder accepts information in successive k-bit *blocks*; for each block, it adds $n - k$ redundant bits that are algebraically related to the k message bits, thereby producing an overall encoded block of n bits, where $n > k$. The n-bit block is called a *codeword*, and n is called the *block length* of the code. The channel encoder produces bits at the rate $R_0 = (n/k)R_s$, where R_s is the bit rate of the information source. The dimensionless ratio $r = k/n$ is called the *code rate*, where $0 < r < 1$. The bit rate R_0, coming out of the encoder, is called the *channel data rate*. Thus, the code rate is a dimensionless ratio, whereas the data rate produced by the source and the channel data rate produced by the encoder are both measured in bits per second.

In a convolutional code, the encoding operation may be viewed as the *discrete-time convolution* of the input sequence with the impulse response of the encoder. The duration of the impulse response equals the memory of the encoder. Accordingly, the encoder for a convolutional code operates on the incoming message sequence, using a "sliding window" equal in duration to its own memory. This, in turn, means that in a convolutional code, unlike in a block code, the channel encoder accepts message bits as a continuous sequence and thereby generates a continuous sequence of encoded bits at a higher rate.

In the model depicted in Figure 10.1a, the operations of channel coding and modulation are performed separately in the transmitter; and likewise for the operations of detection and decoding in the receiver. When, however, bandwidth efficiency is of major concern, the most effective method of implementing forward error-control correction coding is to combine it with modulation as a single function, as shown in Figure 10.1b. In this second approach, coding is redefined as a process of imposing certain patterns on the transmitted signal and the resulting code is called a *trellis code*.

Block codes, convolutional codes, and trellis codes represent the *classical family of codes* that follow traditional approaches rooted in algebraic mathematics in one form or another. In addition to these classical codes, we now have a "new" generation of coding techniques exemplified by *turbo codes* and *low-density parity-check (LDPC) codes*. These new codes are not only fundamentally different, but they have also already taken over the legacy coding schemes very quickly in many practical systems. Simply put, turbo codes and LDPC codes are structured in such a way that decoding can be split into a number of manageable steps, thereby making it possible to *construct powerful codes in a computationally feasible manner*, which is not attainable with the legacy codes. Turbo codes and LDPC codes are discussed in the latter part of the chapter.

10.3 Discrete Memoryless Channels

Returning to the model of Figure 10.1a, the waveform channel is said to be *memoryless* if in a given interval the detector output depends only on the signal transmitted in that interval and not on any previous transmission. Under this condition, we may model the combination of the modulator, the waveform channel, and the demodulator (detector) as a *discrete memoryless channel*. Such a channel is completely described by the set of *transition probabilities* denoted by $p(j|i)$, where i denotes a modulator input symbol, j

denotes a demodulator output symbol, and $p(j|i)$ is the probability of receiving symbol j given that symbol i was sent. (Discrete memoryless channels were described previously at some length in Chapter 5 on information theory.)

The simplest discrete memoryless channel results from the use of binary input and binary output symbols. When binary coding is used, the modulator has only the binary symbols 0 and 1 as inputs. Likewise, the decoder has only binary inputs if binary quantization of the demodulator output is used; that is, a *hard decision* is made on the demodulator output as to which binary symbol was actually transmitted. In this situation, we have a *binary symmetric channel* with a *transition probability diagram* as shown in Figure 10.2. From Chapter 5, we recall that the binary symmetric channel, assuming a channel noise modeled as AWGN, is completely described by the *transition probability.* Hard-decision decoding takes advantage of the special algebraic structure that is built into the design of channel codes; the decoding is therefore relatively easy to perform.

However, the use of hard decisions prior to decoding causes an irreversible loss of valuable information in the receiver. To reduce this loss, *soft-decision* coding can be used. This is achieved by including a multilevel quantizer at the demodulator output, as illustrated in Figure 10.3a for the case of binary PSK signals. The input–output characteristic of the quantizer is shown in Figure 10.3b. The modulator has only binary symbols 0 and 1 as inputs, but the demodulator output now has an alphabet with Q symbols. Assuming the use of the three-level quantizer described in Figure 10.3b, we have $Q = 8$. Such a channel is called a *binary input, Q-ary output discrete memoryless channel.* The corresponding channel transition probability diagram is shown in Figure 10.3c. The form of this distribution, and consequently the decoder performance, depends on the location of the representation levels of the quantizer, which, in turn, depends on the signal level and noise variance. Accordingly, the demodulator must incorporate automatic gain control if an effective multilevel quantizer is to be realized. Moreover, the use of soft decisions complicates the implementation of the decoder. Nevertheless, soft-decision decoding offers significant improvement in performance over hard-decision decoding by taking a probabilistic rather than an algebraic approach. It is for this reason that soft-decision decoders are also referred to as *probabilistic decoders*.

Channel Coding Theorem Revisited

In Chapter 5 on information theory we established the concept of *channel capacity*, which, for a discrete memoryless channel, represents the maximum amount of information that

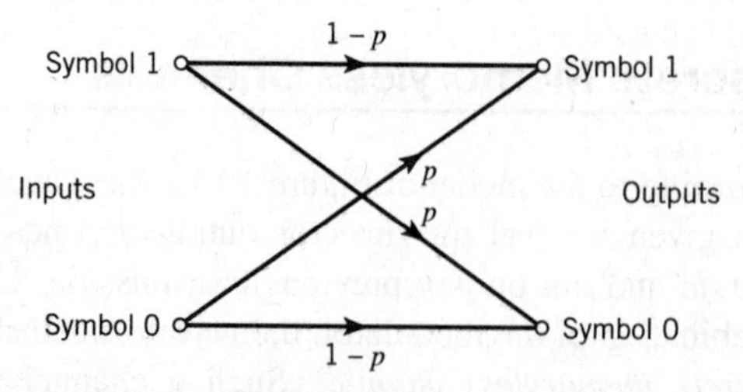

Figure 10.2 Transition probability diagram of binary symmetric channel.

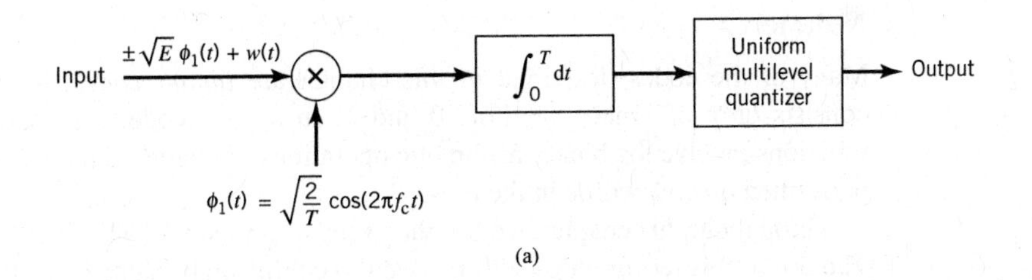

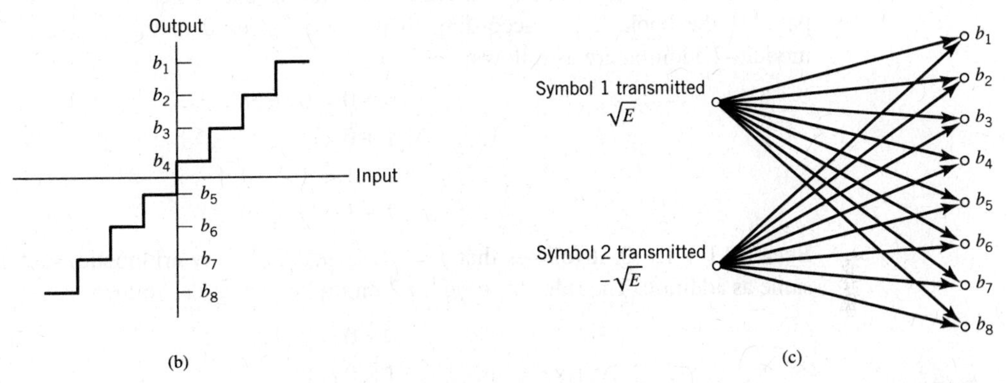

Figure 10.3 Binary input, Q-ary output discrete memoryless channel. (a) Receiver for binary PSK. (b) Transfer characteristic of a multilevel quantizer. (c) Channel transition probability diagram. Parts (b) and (c) are illustrated for eight levels of quantization.

can be transmitted per channel use in a reliable manner. The *channel coding theorem* states:

> If a discrete memoryless channel has capacity C and a source generates information at a rate less than C, then there exists a coding technique such that the output of the source may be transmitted over the channel with an arbitrarily low probability of symbol error.

For the special case of a binary symmetric channel, the theorem teaches us that if the code rate r is less than the channel capacity C, then it is possible to find a code that achieves error-free transmission over the channel. Conversely, it is not possible to find such a code if the code rate r is greater than the channel capacity C. Thus, the channel coding theorem specifies the channel capacity C as a *fundamental limit* on the rate at which the transmission of reliable (error-free) messages can take place over a discrete memoryless channel. The issue that matters here is not the SNR, so long as it is large enough, but how the channel input is encoded.

The most unsatisfactory feature of the channel coding theorem, however, is its nonconstructive nature. The theorem asserts the existence of good codes but does not tell us how to find them. By *good codes* we mean families of channel codes that are capable of providing reliable transmission of information (i.e., at arbitrarily small probability of symbol error) over a noisy channel of interest at bit rates up to a maximum value less than the capacity of that channel. The error-control coding techniques described in this chapter provide different methods of designing good codes.

Notation

Many of the codes described in this chapter are *binary codes*, for which the alphabet consists only of binary symbols 0 and 1. In such a code, the encoding and decoding functions involve the binary arithmetic operations of *modulo-2 addition and multiplication* performed on codewords in the code.

Throughout this chapter, we use the ordinary plus sign (+) to denote modulo-2 addition. The use of this terminology will not lead to confusion because the whole chapter relies on binary arithmetic. In so doing, we avoid use of the special symbol $\oplus$, as we did in previous parts of the book. Thus, according to the notation used in this chapter, the rules for modulo-2 addition are as follows:

$$0 + 0 = 0$$
$$1 + 0 = 1$$
$$0 + 1 = 1$$
$$1 + 1 = 0$$

Because $1 + 1 = 0$, it follows that $1 = -1$. Hence, in binary arithmetic, subtraction is the same as addition. The rules for modulo-2 multiplication are as follows:

$$0 \times 0 = 0$$
$$1 \times 0 = 0$$
$$0 \times 1 = 0$$
$$1 \times 1 = 1$$

Division is trivial, in that we have

$$1 \div 1 = 1$$
$$0 \div 1 = 0$$

and division by 0 is not permitted. Modulo-2 addition is the EXCLUSIVE-OR operation in logic and modulo-2 multiplication is the AND operation.

10.4 Linear Block Codes

By definition:

> A code is said to be linear if any two codewords in the code can be added in modulo-2 arithmetic to produce a third codeword in the code.

Consider, then, an (n,k) linear block code, in which k bits of the n code bits are always identical to the message sequence to be transmitted. The $(n - k)$ bits in the remaining portion are computed from the message bits in accordance with a prescribed encoding rule that determines the mathematical structure of the code. Accordingly, these $(n - k)$ bits are referred to as *parity-check bits*. Block codes in which the message bits are transmitted in unaltered form are called *systematic codes*. For applications requiring *both* error detection and error correction, the use of systematic block codes simplifies implementation of the decoder.

Let $m_0, m_1, \ldots, m_{k-1}$ constitute a block of k arbitrary message bits. Thus, we have 2^k distinct message blocks. Let this sequence of message bits be applied to a linear block encoder, producing an n-bit codeword whose elements are denoted by $c_0, c_1, \ldots, c_{n-1}$. Let $b_0, b_1, \ldots, b_{n-k-1}$ denote the $(n-k)$ *parity-check bits* in the codeword. For the code to possess a systematic structure, a codeword is divided into two parts, one of which is occupied by the message bits and the other by the parity-check bits. Clearly, we have the option of sending the message bits of a codeword before the parity-check bits, or vice versa. The former option is illustrated in Figure 10.4, and its use is assumed in the following.

According to the representation of Figure 10.4, the $(n-k)$ leftmost bits of a codeword are identical to the corresponding parity-check bits and the k rightmost bits of the codeword are identical to the corresponding message bits. We may therefore write

$$c_i = \begin{cases} b_i, & i = 0, 1, \ldots, n-k-1 \\ m_{i+k-n}, & i = n-k, n-k+1, \ldots, n-1 \end{cases} \tag{10.1}$$

The $(n-k)$ parity-check bits are *linear sums* of the k message bits, as shown by the generalized relation

$$b_i = p_{0i}m_0 + p_{1i}m_1 + \ldots + p_{k-1,i}m_{k-1} \tag{10.2}$$

where the coefficients are defined as follows:

$$p_{ij} = \begin{cases} 1 & \text{if } b_i \text{ depends on } m_j \\ 0 & \text{otherwise} \end{cases} \tag{10.3}$$

The coefficients p_{ij} are chosen in such a way that the rows of the generator matrix are linearly independent and the parity-check equations are *unique*. The p_{ij} used here should not be confused with the $p(j|i)$ introduced in Section 10.3.

The system of (10.1) and (10.2) defines the mathematical structure of the (n,k) linear block code. This system of equations may be rewritten in a compact form using matrix notation. To proceed with this reformulation, we respectively define the 1-by-k *message vector* $\mathbf{m}$, the 1-by-$(n-k)$ parity-check vector $\mathbf{b}$, and the 1-by-n code vector $\mathbf{c}$ as follows:

$$\mathbf{m} = [m_0, m_1, \ldots, m_{k-1}] \tag{10.4}$$

$$\mathbf{b} = [b_0, b_1, \ldots, b_{n-k-1}] \tag{10.5}$$

$$\mathbf{c} = [c_0, c_1, \ldots, c_{n-1}] \tag{10.6}$$

Note that all three vectors are *row vectors*. The use of row vectors is adopted in this chapter for the sake of being consistent with the notation commonly used in the coding literature. We may thus rewrite the set of simultaneous equations defining the parity check-bits in the compact matrix form

$$\mathbf{b} = \mathbf{mP} \tag{10.7}$$

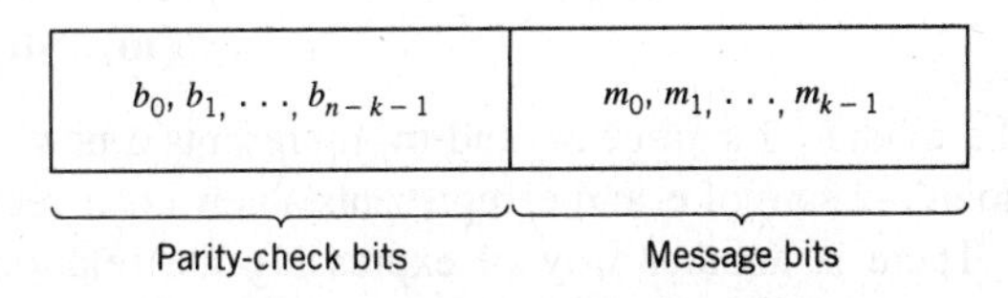

Figure 10.4 Structure of systematic codeword.

The $\mathbf{P}$ in (10.7) is the k-by-$(n-k)$ *coefficient matrix* defined by

$$\mathbf{P} = \begin{bmatrix} p_{00} & p_{01} & \cdots & p_{0,n-k-1} \\ p_{10} & p_{11} & \cdots & p_{1,n-k-1} \\ \vdots & \vdots & & \vdots \\ p_{k-1,0} & p_{k-1,1} & \cdots & p_{k-1,n-k-1} \end{bmatrix} \tag{10.8}$$

where the element p_{ij} is 0 or 1.

From the definitions given in (10.4)–(10.6), we see that $\mathbf{c}$ may be expressed as a partitioned row vector in terms of the vectors $\mathbf{m}$ and $\mathbf{b}$ as follows:

$$\mathbf{c} = \begin{bmatrix} \mathbf{b} & \vdots & \mathbf{m} \end{bmatrix} \tag{10.9}$$

Hence, substituting (10.7) into (10.9) and factoring out the common message vector $\mathbf{m}$, we get

$$\mathbf{c} = \mathbf{m}\begin{bmatrix} \mathbf{P} & \vdots & \mathbf{I}_k \end{bmatrix} \tag{10.10}$$

where $\mathbf{I}_k$ is the k-by-k *identity matrix*:

$$\mathbf{I}_k = \begin{bmatrix} 1 & 0 & \cdots & 0 \\ 0 & 1 & \cdots & 0 \\ \vdots & \vdots & \ddots & \vdots \\ 0 & 0 & \cdots & 1 \end{bmatrix} \tag{10.11}$$

Define the k-by-n *generator matrix*

$$\mathbf{G} = \begin{bmatrix} \mathbf{P} & \vdots & \mathbf{I}_k \end{bmatrix} \tag{10.12}$$

The generator matrix $\mathbf{G}$ of (10.12) is said to be in the *canonical form*, in that its k rows are linearly independent; that is, it is not possible to express any row of the matrix $\mathbf{G}$ as a linear combination of the remaining rows. Using the definition of the generator matrix $\mathbf{G}$, we may simplify (10.10) as

$$\mathbf{c} = \mathbf{mG} \tag{10.13}$$

The full set of codewords, referred to simply as *the code*, is generated in accordance with (10.13) by passing the message vector $\mathbf{m}$ range through the set of all 2^k binary k-tuples (1-by-k vectors). Moreover, the sum of any two codewords in the code is another codeword. This basic property of linear block codes is called *closure*. To prove its validity, consider a pair of code vectors $\mathbf{c}_i$ and $\mathbf{c}_j$ corresponding to a pair of message vectors $\mathbf{m}_i$ and $\mathbf{m}_j$, respectively. Using (10.13), we may express the sum of $\mathbf{c}_i$ and $\mathbf{c}_j$ as

$$\begin{aligned} \mathbf{c}_i + \mathbf{c}_j &= \mathbf{m}_i\mathbf{G} + \mathbf{m}_j\mathbf{G} \\ &= (\mathbf{m}_i + \mathbf{m}_j)\mathbf{G} \end{aligned}$$

The modulo-2 sum of $\mathbf{m}_i$ and $\mathbf{m}_j$ represents a new message vector. Correspondingly, the modulo-2 sum of $\mathbf{c}_i$ and $\mathbf{c}_j$ represents a new code vector.

There is another way of expressing the relationship between the message bits and parity-check bits of a linear block code. Let $\mathbf{H}$ denote an $(n-k)$-by-n matrix, defined as

$$\mathbf{H} = \left[\mathbf{I}_{n-k} \;\vdots\; \mathbf{P}^{\mathrm{T}}\right] \tag{10.14}$$

where $\mathbf{P}^{\mathrm{T}}$ is an $(n-k)$-by-k matrix, representing the transpose of the coefficient matrix $\mathbf{P}$, and $\mathbf{I}_{n-k}$ is the $(n-k)$-by-$(n-k)$ identity matrix. Accordingly, we may perform the following multiplication of partitioned matrices:

$$\mathbf{H}\mathbf{G}^{\mathrm{T}} = \left[\mathbf{I}_{n-k} \;\vdots\; \mathbf{P}^{\mathrm{T}}\right]\begin{bmatrix}\mathbf{P}^{\mathrm{T}} \\ \text{----} \\ \mathbf{I}_k\end{bmatrix}$$

$$= \mathbf{P}^{\mathrm{T}} + \mathbf{P}^{\mathrm{T}}$$

where we have used the fact that multiplication of a rectangular matrix by an identity matrix of compatible dimensions leaves the matrix unchanged. In modulo-2 arithmetic, the matrix sum $\mathbf{P}^{\mathrm{T}} + \mathbf{P}^{\mathrm{T}}$ is $\mathbf{0}$. We therefore have

$$\mathbf{H}\mathbf{G}^{\mathrm{T}} = \mathbf{0} \tag{10.15}$$

Equivalently, we have $\mathbf{G}\mathbf{H}^{\mathrm{T}} = \mathbf{0}$, where $\mathbf{0}$ is a new null matrix. Postmultiplying both sides of (10.13) by $\mathbf{H}^{\mathrm{T}}$, the transpose of $\mathbf{H}$, and then using (10.15), we get the inner product

$$\begin{aligned}\mathbf{c}\mathbf{H}^{\mathrm{T}} &= \mathbf{m}\mathbf{G}\mathbf{H}^{\mathrm{T}} \\ &= \mathbf{0}\end{aligned} \tag{10.16}$$

The matrix $\mathbf{H}$ is called the *parity-check matrix* of the code and the equations specified by (10.16) are called *parity-check equations*.

The generator equation (10.13) and the parity-check detector equation (10.16) are basic to the description and operation of a linear block code. These two equations are depicted in the form of block diagrams in Figure 10.5a and b, respectively.

Syndrome: Definition and Properties

The generator matrix $\mathbf{G}$ is used in the encoding operation at the transmitter. On the other hand, the parity-check matrix $\mathbf{H}$ is used in the decoding operation at the receiver. In the context of the latter operation, let $\mathbf{r}$ denote the 1-by-n *received vector* that results from

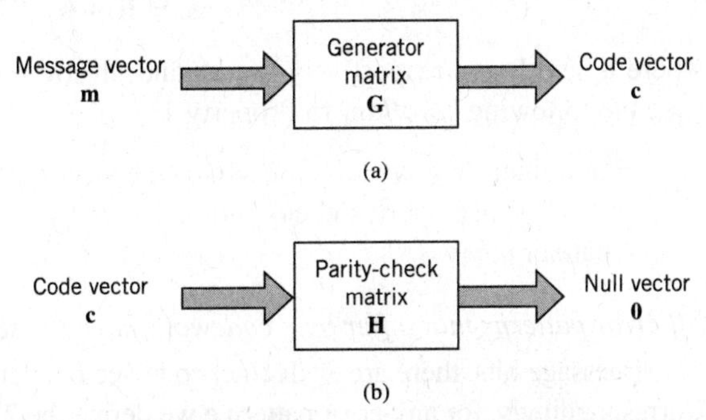

Figure 10.5 Block diagram representations of the generator equation (10.13) and the parity-check equation (10.16).

sending the code vector $\mathbf{c}$ over a noisy binary channel. We express the vector $\mathbf{r}$ as the sum of the original code vector $\mathbf{c}$ and a new vector $\mathbf{e}$, as shown by

$$\mathbf{r} = \mathbf{c} + \mathbf{e} \tag{10.17}$$

The vector $\mathbf{e}$ is called the *error vector* or *error pattern*. The ith element of $\mathbf{e}$ equals 0 if the corresponding element of $\mathbf{r}$ is the same as that of $\mathbf{c}$. On the other hand, the ith element of $\mathbf{e}$ equals 1 if the corresponding element of $\mathbf{r}$ is different from that of $\mathbf{c}$, in which case an *error* is said to have occurred in the ith location. That is, for $i = 1, 2, \ldots, n$, we have

$$e_i = \begin{cases} 1 & \text{if an error has occurred in the } i\text{th location} \\ 0 & \text{otherwise} \end{cases} \tag{10.18}$$

The receiver has the task of decoding the code vector $\mathbf{c}$ from the received vector $\mathbf{r}$. The algorithm commonly used to perform this decoding operation starts with the computation of a 1-by-$(n - k)$ vector called the *error-syndrome vector* or simply the *syndrome*.[2] The importance of the syndrome lies in the fact that it depends only upon the error pattern.

Given a 1-by-n received vector $\mathbf{r}$, the corresponding syndrome is formally defined as

$$\mathbf{s} = \mathbf{r}\mathbf{H}^{\mathrm{T}} \tag{10.19}$$

Accordingly, the syndrome has the following important properties.

PROPERTY 1 *The syndrome depends only on the error pattern and not on the transmitted codeword.*

To prove this property, we first use (10.17) and (10.19), and then (10.16) to write

$$\begin{aligned} \mathbf{s} &= (\mathbf{c} + \mathbf{e})\mathbf{H}^{\mathrm{T}} \\ &= \mathbf{c}\mathbf{H}^{\mathrm{T}} + \mathbf{e}\mathbf{H}^{\mathrm{T}} \\ &= \mathbf{e}\mathbf{H}^{\mathrm{T}} \end{aligned} \tag{10.20}$$

Hence, the parity-check matrix $\mathbf{H}$ of a code permits us to compute the syndrome $\mathbf{s}$, which depends only upon the error pattern $\mathbf{e}$.

To expand on Property 1, suppose that the error pattern $\mathbf{e}$ contains a pair of errors in locations i and j caused by the additive channel noise, as shown by

$$\mathbf{e} = [0\ldots01_i0\ldots01_j0\ldots0]$$

Then, substituting this error pattern into (10.20) yields the syndrome

$$\mathbf{s} = \mathbf{h}_i + \mathbf{h}_j$$

where $\mathbf{h}_i$ and $\mathbf{h}_j$ are respectively the ith and jth rows of the matrix $\mathbf{H}^{\mathrm{T}}$. In words, we may state the following *corollary* to Property 1:

> For a linear block code, the syndrome $\mathbf{s}$ is equal to the sum of those rows of the transposed parity-check matrix $\mathbf{H}^{\mathrm{T}}$ where errors have occurred due to channel noise.

PROPERTY 2 *All error patterns that differ by a codeword have the same syndrome.*

For k message bits, there are 2^k distinct code vectors denoted as $\mathbf{c}_i$, where $i = 0, 1, \ldots, 2^k - 1$. Correspondingly, for any error pattern $\mathbf{e}$ we define the 2^k distinct vectors $\mathbf{e}_i$ as follows

$$\mathbf{e}_i = \mathbf{e} + \mathbf{c}_i \qquad \text{for } i = 0, 1, \ldots, 2^k - 1 \tag{10.21}$$

The set of vectors ($\mathbf{e}_i$, $i = 0, 1, \ldots, 2^k - 1$) defined in (10.21) is called a *coset* of the code. In other words, a coset has exactly 2^k elements that differ at most by a code vector. Thus, an (n,k) linear block code has 2^{n-k} possible cosets. In any event, multiplying both sides of (10.21) by the matrix $\mathbf{H}^{\mathrm{T}}$ and again using (10.16), we get

$$\begin{aligned} \mathbf{e}_i\mathbf{H}^{\mathrm{T}} &= \mathbf{e}\mathbf{H}^{\mathrm{T}} + \mathbf{c}_i\mathbf{H}^{\mathrm{T}} \\ &= \mathbf{e}\mathbf{H}^{\mathrm{T}} \end{aligned} \tag{10.22}$$

which is independent of the index i. Accordingly, we may say:

Each coset of the code is characterized by a unique syndrome.

We may put Properties 1 and 2 in perspective by expanding (10.20). Specifically, with the matrix $\mathbf{H}$ having the systematic form given in (10.14), where the matrix $\mathbf{P}$ is itself defined by (10.8), we find from (10.20) that the $(n - k)$ elements of the syndrome $\mathbf{s}$ are linear combinations of the n elements of the error pattern $\mathbf{e}$, as shown by

$$\begin{aligned} s_0 &= e_0 + e_{n-k}p_{00} + e_{n-k+1}p_{10} + \cdots + e_{n-k}p_{k-1,0} \\ s_1 &= e_1 + e_{n-k}p_{01} + e_{n-k+1}p_{11} + \cdots + e_{n-k}p_{k-1,1} \\ &\vdots \\ s_{n-k-1} &= e_{n-k-1} + e_{n-k}p_{0,n-k-1} + \cdots + e_{n-1}p_{(k-1,n-k+1)} \end{aligned} \tag{10.23}$$

This set of $(n - k)$ linear equations clearly shows that the syndrome contains information about the error pattern and may, therefore, be used for error detection. However, it should be noted that the set of equations (10.23) is *underdetermined*, in that we have more unknowns than equations. Accordingly, there is *no* unique solution for the error pattern. Rather, there are 2^n error patterns that satisfy (10.23) and, therefore, result in the same syndrome, in accordance with Property 2 and (10.22). In particular, with 2^{n-k} possible syndrome vectors, the information contained in the syndrome $\mathbf{s}$ about the error pattern $\mathbf{e}$ is *not* enough for the decoder to compute the exact value of the transmitted code vector. Nevertheless, knowledge of the syndrome $\mathbf{s}$ reduces the search for the true error pattern $\mathbf{e}$ from 2^n to 2^{n-k} possibilities. Given these possibilities, the decoder has the task of making the best selection from the cosets corresponding to $\mathbf{s}$.

Minimum Distance Considerations

Consider a pair of code vectors $\mathbf{c}_1$ and $\mathbf{c}_2$ that have the same number of elements. The *Hamming distance*, denoted by $d(\mathbf{c}_1,\mathbf{c}_2)$, between such a pair of code vectors is defined as the number of locations in which their respective elements differ.

The *Hamming weight* $w(\mathbf{c})$ of a code vector $\mathbf{c}$ is defined as the number of nonzero elements in the code vector. Equivalently, we may state that the Hamming weight of a code vector is the distance between the code vector and the all-zero code vector. In a corresponding way, we may introduce a new parameter called the *minimum distance* $d_{\min}$, for which we make the statement:

The minimum distance $d_{\min}$ of a linear block code is the smallest Hamming distance between any pair of codewords.

That is, the minimum distance is the same as the smallest Hamming weight of the difference between any pair of code vectors. From the closure property of linear block codes, the sum (or difference) of two code vectors is another code vector. Accordingly, we may also state:

> The minimum distance of a linear block code is the smallest Hamming weight of the nonzero code vectors in the code.

The minimum distance d_{min} is related to the structure of the parity-check matrix $\mathbf{H}$ of the code in a fundamental way. From (10.16) we know that a linear block code is defined by the set of all code vectors for which $\mathbf{cH}^T = \mathbf{0}$, where $\mathbf{H}^T$ is the transpose of the parity-check matrix $\mathbf{H}$. Let the matrix $\mathbf{H}$ be expressed in terms of its columns as shown by

$$\mathbf{H} = [\mathbf{h}_1, \mathbf{h}_2, \ldots, \mathbf{h}_n] \tag{10.24}$$

Then, for a code vector $\mathbf{c}$ to satisfy the condition $\mathbf{cH}^T = \mathbf{0}$, the vector $\mathbf{c}$ must have ones in such positions that the corresponding rows of $\mathbf{H}^T$ sum to the zero vector $\mathbf{0}$. However, by definition, the number of ones in a code vector is the Hamming weight of the code vector. Moreover, the smallest Hamming weight of the nonzero code vectors in a linear block code equals the minimum distance of the code. Hence, we have another useful result stated as follows:

> The minimum distance of a linear block code is defined by the minimum number of rows of the matrix $\mathbf{H}^T$ whose sum is equal to the zero vector.

From this discussion, it is apparent that the minimum distance d_{min} of a linear block code is an important parameter of the code. Specifically, d_{min} determines the *error-correcting capability of the code*. Suppose an (n,k) linear block code is required to detect and correct all error patterns over a binary symmetric channel, and whose Hamming weight is less than or equal to t. That is, if a code vector $\mathbf{c}_i$ in the code is transmitted and the received vector is $\mathbf{r} = \mathbf{c}_i + \mathbf{e}$, we require that the decoder output $\hat{\mathbf{c}} = \mathbf{c}_i$ whenever the error pattern $\mathbf{e}$ has a Hamming weight

$$w(\mathbf{e}) \le t$$

We assume that the 2^k code vectors in the code are transmitted with equal probability. The best strategy for the decoder then is to pick the code vector closest to the received vector $\mathbf{r}$; that is, the one for which the Hamming distance $d(\mathbf{c}_i,\mathbf{r})$ is the smallest. With such a strategy, the decoder will be able to detect and correct all error patterns of Hamming weight $w(\mathbf{e})$, provided that the minimum distance of the code is equal to or greater than $2t + 1$. We may demonstrate the validity of this requirement by adopting a geometric interpretation of the problem. In particular, the transmitted 1-by-n code vector and the 1-by-n received vector are represented as points in an n-dimensional space. Suppose that we construct two spheres, each of radius t, around the points that represent code vectors $\mathbf{c}_i$ and $\mathbf{c}_j$ under two different conditions:

1. Let these two spheres be disjoint, as depicted in Figure 10.6a. For this condition to be satisfied, we require that $d(\mathbf{c}_i,\mathbf{c}_j) \ge 2t + 1$. If, then, the code vector $\mathbf{c}_i$ is transmitted and the Hamming distance $d(\mathbf{c}_i,\mathbf{r}) \le t$, it is clear that the decoder will pick $\mathbf{c}_i$, as it is the code vector closest to the received vector $\mathbf{r}$.
2. If, on the other hand, the Hamming distance $d(\mathbf{c}_i,\mathbf{c}_j) \le 2t$, the two spheres around $\mathbf{c}_i$ and $\mathbf{c}_j$ intersect, as depicted in Figure 10.6b. In this second situation, we see that if $\mathbf{c}_i$

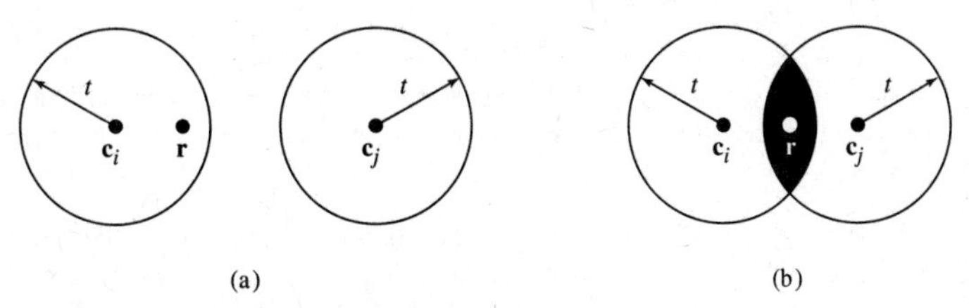

Figure 10.6 (a) Hamming distance $d(\mathbf{c}_i,\mathbf{c}_j) \geq 2t + 1$. (b) Hamming distance $d(\mathbf{c}_i,\mathbf{c}_j) < 2t$. The received vector is denoted by $\mathbf{r}$.

is transmitted, there exists a received vector $\mathbf{r}$ such that the Hamming distance $d(\mathbf{c}_i,\mathbf{r}) \leq t$, yet $\mathbf{r}$ is as close to $\mathbf{c}_j$ as it is to $\mathbf{c}_i$. Clearly, there is now the possibility of the decoder picking the vector $\mathbf{c}_j$, which is wrong.

We thus conclude the ideas presented thus far by saying:

> An (n,k) linear block code has the power to correct all error patterns of weight t or less if, and only if, $d(\mathbf{c}_i,\mathbf{c}_j) \geq 2t + 1$, *for all* $\mathbf{c}_i$ *and* $\mathbf{c}_j$.

By definition, however, the smallest distance between any pair of code vectors in a code is the minimum distance $d_{\min}$ of the code. We may, therefore, go on to state:

> An (n,k) linear block code of minimum distance $d_{\min}$ can correct up to *t errors if, and only if,*
>
> $$t \leq \left\lfloor \frac{1}{2}(d_{\min} - 1) \right\rfloor \tag{10.25}$$
>
> where $\lfloor\ \rfloor$ *denotes the largest integer less than or equal to the enclosed quantity.*

The condition described in (10.25) is important because it gives the error-correcting capability of a linear block code a quantitative meaning.

Syndrome Decoding

We are now ready to describe a syndrome-based decoding scheme for linear block codes. Let $\mathbf{c}_1, \mathbf{c}_2, \ldots, \mathbf{c}_{2^k}$ denote the 2^k code vectors of an (n, k) linear block code. Let $\mathbf{r}$ denote the received vector, which may have one of 2^n possible values. The receiver has the task of partitioning the 2^n possible received vectors into 2^k disjoint subsets $D_1, D_2, \ldots, D_{2^k}$ in such a way that the ith subset D_i corresponds to code vector $\mathbf{c}_i$ for $1 \leq i \leq 2^k$. The received vector $\mathbf{r}$ is decoded into $\mathbf{c}_i$ if it is in the ith subset. For the decoding to be correct, $\mathbf{r}$ must be in the subset that belongs to the code vector $\mathbf{c}_i$ that was actually sent.

The 2^k subsets described herein constitute a *standard array* of the linear block code. To construct it, we exploit the linear structure of the code by proceeding as follows:

1. The 2^k code vectors are placed in a row with the all-zero code vector $\mathbf{c}_1$ as the leftmost element.
2. An error pattern $\mathbf{e}_2$ is picked and placed under $\mathbf{c}_1$, and a second row is formed by adding $\mathbf{e}_2$ to each of the remaining code vectors in the first row; it is important that

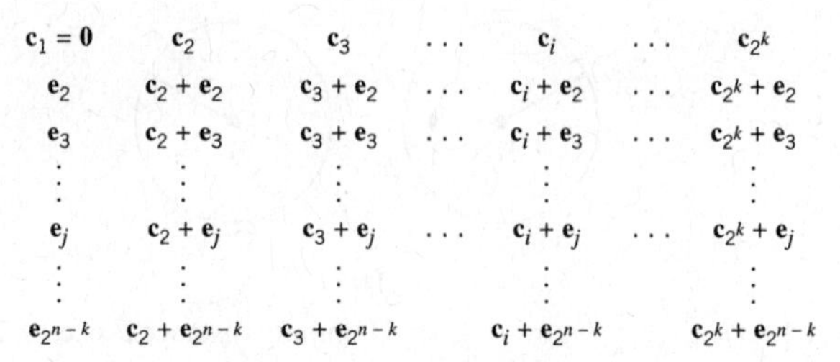

Figure 10.7 Standard array for an (n,k) block code.

the error pattern chosen as the first element in a row has not previously appeared in the standard array.

3. Step 2 is repeated until all the possible error patterns have been accounted for.

Figure 10.7 illustrates the structure of the standard array so constructed. The 2^k columns of this array represent the disjoint subsets $D_1, D_2, \ldots, D_{2^k}$. The 2^{n-k} rows of the array represent the cosets of the code, and their first elements $\mathbf{e}_2, \ldots, \mathbf{e}_{2^{n-k}}$ are called *coset leaders*.

For a given channel, the probability of decoding error is minimized when the most likely error patterns (i.e., those with the largest probability of occurrence) are chosen as the coset leaders. In the case of a binary symmetric channel, the smaller we make the Hamming weight of an error pattern, the more likely it is for an error to occur. Accordingly, the standard array should be constructed with each coset leader having the minimum Hamming weight in its coset.

We are now ready to describe a decoding procedure for linear block codes:

1. For the received vector $\mathbf{r}$, compute the syndrome $\mathbf{s} = \mathbf{r}\mathbf{H}^T$.
2. Within the coset characterized by the syndrome $\mathbf{s}$, identify the coset leader (i.e., the error pattern with the largest probability of occurrence); call it $\mathbf{e}_0$.
3. Compute the code vector

$$\mathbf{c} = \mathbf{r} + \mathbf{e}_0 \tag{10.26}$$

as the decoded version of the received vector $\mathbf{r}$.

This procedure is called *syndrome decoding*.

EXAMPLE 1 **Hamming Codes**

For any positive integer $m \geq 3$, there exists a linear block code with the following parameters:

code length	$n = 2^m - 1$
number of message bits	$k = 2^m - m - 1$
number of parity-check bits	$n - k = m$

Such a linear block code for which the error-correcting capability $t = 1$ is called a Hamming code.[3] To be specific, consider the example of $m = 3$, yielding the (7,4) Hamming code with $n = 7$ and $k = 4$. The generator of this code is defined by

$$\mathbf{G} = \underbrace{\begin{bmatrix} 1\,1\,0 \\ 0\,1\,1 \\ 1\,1\,1 \\ 1\,0\,1 \end{bmatrix}}_{\mathbf{P}} \begin{matrix} \vdots \\ \vdots \\ \vdots \\ \vdots \end{matrix} \underbrace{\begin{bmatrix} 1\,0\,0\,0 \\ 0\,1\,0\,0 \\ 0\,0\,1\,0 \\ 0\,0\,0\,1 \end{bmatrix}}_{\mathbf{I}_k}$$

which conforms to the systematic structure of (10.12).

The corresponding parity-check matrix is given by

$$\mathbf{H} = \underbrace{\begin{bmatrix} 1\,0\,0 \\ 0\,1\,0 \\ 0\,0\,1 \end{bmatrix}}_{\mathbf{I}_{n-k}} \begin{matrix} \vdots \\ \vdots \\ \vdots \end{matrix} \underbrace{\begin{bmatrix} 1\,0\,1\,1 \\ 1\,1\,1\,0 \\ 0\,1\,1\,1 \end{bmatrix}}_{\mathbf{P}^{\mathrm{T}}}$$

The operative property embodied in this equation is that the columns of the parity-check matrix **P** consist of all the nonzero m-tuples, where $m = 3$.

With $k = 4$, there are $2^k = 16$ distinct message words, which are listed in Table 10.1. For a given message word, the corresponding codeword is obtained by using (10.13). Thus, the application of this equation results in the 16 codewords listed in Table 10.1.

In Table 10.1, we have also listed the Hamming weights of the individual codewords in the (7,4) Hamming code. Since the smallest of the Hamming weights for the nonzero codewords is 3, it follows that the minimum distance of the code is 3, which is what it should be by definition. Indeed, all Hamming codes have the property that the minimum distance $d_{\min} = 3$, independent of the value assigned to the number of parity bits m.

To illustrate the relation between the minimum distance $d_{\min}$ and the structure of the parity-check matrix **H**, consider the codeword 0110100. In matrix multiplication, defined

Table 10.1 **Codewords of a (7,4) Hamming code**

Message word	Codeword	Weight of codeword	Message word	Codeword	Weight of codeword
0000	0000000	0	1000	1101000	3
0001	1010001	3	1001	0111001	4
0010	1110010	4	1010	0011010	3
0011	0100011	3	1011	1001011	3
0100	0110100	3	1100	1011100	4
0101	1100101	4	1101	0001101	3
0110	1000110	3	1110	0101110	4
0111	0010111	4	1111	1111111	7

by (10.16), the nonzero elements of this codeword “sift” out the second, third, and fifth columns of the matrix **H**, yielding

$$\begin{bmatrix}0\\1\\0\end{bmatrix} + \begin{bmatrix}0\\0\\1\end{bmatrix} + \begin{bmatrix}0\\1\\1\end{bmatrix} = \begin{bmatrix}0\\0\\0\end{bmatrix}$$

We may perform similar calculations for the remaining 14 nonzero codewords. We thus find that the smallest number of columns in **H** that sums to zero is 3, reconfirming the defining condition $d_{\min} = 3$.

An important property of binary Hamming codes is that they satisfy the condition of (10.25) with the equality sign, assuming that $t = 1$. Thus, assuming single-error patterns, we may formulate the error patterns listed in the right-hand column of Table 10.2. The corresponding eight syndromes, listed in the left-hand column, are calculated in accordance with (10.20). The zero syndrome signifies no transmission errors.

Suppose, for example, the code vector [1110010] is sent and the received vector is [1100010] with an error in the third bit. Using (10.19), the syndrome is calculated to be

$$\mathbf{s} = [1100010]\begin{bmatrix}1 & 0 & 0\\0 & 1 & 0\\0 & 0 & 1\\1 & 1 & 0\\0 & 1 & 1\\1 & 1 & 1\\1 & 0 & 1\end{bmatrix}$$

$$= \begin{bmatrix}0 & 0 & 1\end{bmatrix}$$

From Table 10.2 the corresponding coset leader (i.e., error pattern with the highest probability of occurrence) is found to be [0010000], indicating correctly that the third bit of the received vector is erroneous. Thus, adding this error pattern to the received vector, in accordance with (10.26), yields the correct code vector actually sent.

Table 10.2 **Decoding table for the (7,4) Hamming code defined in Table 10.1**

Syndrome	Error pattern
000	0000000
100	1000000
010	0100000
001	0010000
110	0001000
011	0000100
111	0000010
101	0000001

10.5 Cyclic Codes

Cyclic codes form a subclass of linear block codes. Indeed, many of the important linear block codes discovered to date are either cyclic codes or closely related to cyclic codes. An advantage of cyclic codes over most other types of codes is that they are easy to encode. Furthermore, cyclic codes possess a well-defined mathematical structure, which has led to the development of very efficient decoding schemes for them.

A binary code is said to be a *cyclic code* if it exhibits two fundamental properties:

PROPERTY 1 **Linearity Property**

The sum of any two codewords in the code is also a codeword.

PROPERTY 2 **Cyclic Property**

Any cyclic shift of a codeword in the code is also a codeword.

Property 1 restates the fact that a cyclic code is a linear block code (i.e., it can be described as a parity-check code). To restate Property 2 in mathematical terms, let the n-tuple $c_0, c_1, \ldots, c_{n-1}$ denote a codeword of an (n, k) linear block code. The code is a cyclic code if the n-tuples

$$(c_{n-1}, c_0, \ldots, c_{n-2})$$
$$(c_{n-2}, c_{n-1}, \ldots, c_{n-3})$$
$$\vdots$$
$$(c_1, c_2, \ldots, c_{n-1}, c_0)$$

are all codewords in the code.

To develop the algebraic properties of cyclic codes, we use the elements $c_0, c_1, \ldots, c_{n-1}$ of a codeword to define the code polynomial

$$\mathbf{c}(X) = c_0 + c_1X + c_2X^2 + \cdots + c_{n-1}X^{n-1} \tag{10.27}$$

where X is an indeterminate. Naturally, for binary codes, the coefficients are 1s and 0s. Each power of X in the polynomial $\mathbf{c}(X)$ represents a one-bit *shift* in time. Hence, multiplication of the polynomial $\mathbf{c}(X)$ by X may be viewed as a shift to the right. The key question is: How do we make such a shift *cyclic*? The answer to this question is addressed next.

Let the code polynomial $\mathbf{c}(X)$ in (10.27) be multiplied by X^i, yielding

$$X^i\mathbf{c}(X) = c_0X^i + c_1X^{i+1} + \cdots + c_{n-i-1}X^{n-1} + \cdots + c_{n-1}X^{n+i-1}$$

Recognizing, for example, that $c_{n-i} + c_{n-i} = 0$ in modulo-2 addition, we may manipulate the preceding equation into the following compact form:

$$X^i\mathbf{c}(X) = \mathbf{q}(X)(X^n + 1) + \mathbf{c}^{(i)}(X) \tag{10.28}$$

where the polynomial $\mathbf{q}(X)$ is defined by

$$\mathbf{q}(X) = c_{n-i} + c_{n-i+1}X + \cdots + c_{n-1}X^{i-1} \tag{10.29}$$

As for the polynomial $\mathbf{c}^{(i)}X$ in (10.28), it is recognized as the code polynomial of the codeword $(c_{n-i}, \ldots, c_{n-1}, c_0, c_1, \ldots, c_{n-i-1})$ obtained by applying i cyclic shifts to the codeword $c_0, c_1, \ldots, c_{n-i-1}, c_{n-i}, \ldots, c_{n-1}$. Moreover, from (10.28) we readily see that $\mathbf{c}^{(i)}(X)$ is the remainder that results from dividing $X^i\mathbf{c}(X)$ by $(X^n + 1)$. We may thus formally state the cyclic property in polynomial notation as follows:

If $\mathbf{c}(X)$ is a code polynomial, then the polynomial

$$\mathbf{c}^{(i)}(X) = X^i\mathbf{c}(X) \ \mathrm{mod}(X^n + 1) \tag{10.30}$$

is also a code polynomial for any cyclic shift i; the term mod is the abbreviation for modulo.

The special form of polynomial multiplication described in (10.30) is referred to as *multiplication modulo* $X^n + 1$. In effect, the multiplication is subject to the constraint $X^n = 1$, the application of which restores the polynomial $X^i\mathbf{c}(X)$ to order $n-1$ for all $i < n$. Note that, in modulo-2 arithmetic, $X^n + 1$ has the same value as $X^n - 1$.

Generator Polynomial

The polynomial $X^n + 1$ and its factors play a major role in the generation of cyclic codes. Let $\mathbf{g}(X)$ be a polynomial of degree $n-k$ that is a factor of $X^n + 1$; as such, $\mathbf{g}(X)$ is the polynomial of least degree in the code. In general, $\mathbf{g}(X)$ may be expanded as follows:

$$\mathbf{g}(X) = 1 + \sum_{i=1}^{n-k-1} g_i X^i + X^{n-k} \tag{10.31}$$

where the coefficient g_i is equal to 0 or 1 for $i = 1, \ldots, n-k-1$. According to this expansion, the polynomial $\mathbf{g}(X)$ has two terms with coefficient 1 separated by $n-k-1$ terms. The polynomial $\mathbf{g}(X)$ is called the *generator polynomial* of a cyclic code. A cyclic code is uniquely determined by the generator polynomial $\mathbf{g}(X)$ in that each code polynomial in the code can be expressed in the form of a polynomial product as follows:

$$\mathbf{c}(X) = \mathbf{a}(X)\mathbf{g}(X) \tag{10.32}$$

where $\mathbf{a}(X)$ is a polynomial in X with degree $k-1$. The $\mathbf{c}(X)$ so formed satisfies the condition of (10.30) since $\mathbf{g}(X)$ is a factor of $X^n + 1$.

Suppose we are given the generator polynomial $\mathbf{g}(X)$ and the requirement is to encode the message sequence $(m_0, m_1, \ldots, m_{k-1})$ into an (n, k) *systematic* cyclic code. That is, the message bits are transmitted in unaltered form, as shown by the following structure for a codeword (see Figure 10.4):

$$(\underbrace{b_0, b_1, \ldots, b_{n-k-1}}_{n-k \text{ parity-check bits}}, \underbrace{m_0, m_1, \ldots, m_{k-1}}_{k \text{ message bits}})$$

Let the *message polynomial* be defined by

$$\mathbf{m}(X) = m_0 + m_1X + \cdots + m_{k-1}X^{k-1} \tag{10.33}$$

and let

$$\mathbf{b}(X) = b_0 + b_1X + \cdots + b_{n-k-1}X^{n-k-1} \tag{10.34}$$

Then, according to (10.1), we want the code polynomial to be in the form

$$\mathbf{c}(X) = \mathbf{b}(X) + X^{n-k}\mathbf{m}(X) \tag{10.35}$$

To this end, the use of (10.32) and (10.35) yields

$$\mathbf{a}(X)\mathbf{g}(X) = \mathbf{b}(X) + X^{n-k}\mathbf{m}(X)$$

Equivalently, invoking modulo-2 addition, we may also write

$$\frac{X^{n-k}\mathbf{m}(X)}{\mathbf{g}(X)} = \mathbf{a}(X) + \frac{\mathbf{b}(X)}{\mathbf{g}(x)} \tag{10.36}$$

Equation (10.36) states that the polynomial $\mathbf{b}(X)$ is the remainder left over after dividing $X^{n-k}\mathbf{m}(X)$ by $\mathbf{g}(X)$.

We may now summarize the steps involved in the encoding procedure for an (n,k) cyclic code, assured of a systematic structure. Specifically, we proceed as follows:

Step 1: Premultiply the message polynomial $\mathbf{m}(X)$ by X^{n-k}.

Step 2: Divide $X^{n-k}\mathbf{m}(X)$ by the generator polynomial $\mathbf{g}(X)$, obtaining the remainder $\mathbf{b}(X)$.

Step 3: Add $\mathbf{b}(X)$ to $X^{n-k}\mathbf{m}(X)$, obtaining the code polynomial $\mathbf{c}(X)$.

Parity-Check Polynomial

An (n,k) cyclic code is uniquely specified by its generator polynomial $\mathbf{g}(X)$ of order $(n-k)$. Such a code is also uniquely specified by another polynomial of degree k, which is called the *parity-check polynomial*, defined by

$$\mathbf{h}(X) = 1 + \sum_{i=1}^{k-1} h_i X^i + X^k \tag{10.37}$$

where the coefficients h_i are 0 or 1. The parity-check polynomial $\mathbf{h}(X)$ has a form similar to the generator polynomial, in that there are two terms with coefficient 1, but separated by $k-1$ terms.

The generator polynomial $\mathbf{g}(X)$ is equivalent to the generator matrix $\mathbf{G}$ as a description of the code. Correspondingly, the parity-check polynomial $\mathbf{h}(X)$ is an equivalent representation of the parity-check matrix $\mathbf{H}$. We thus find that the matrix relation $\mathbf{HG}^{\mathrm{T}} = \mathbf{0}$ presented in (10.15) for linear block codes corresponds to the relationship

$$\mathbf{g}(X)\mathbf{h}(X) \bmod (X^n + 1) = \mathbf{0} \tag{10.38}$$

Accordingly, we may make the statement:

> The generator polynomial $\mathbf{g}(X)$ and the parity-check polynomial $\mathbf{h}(X)$ are factors of the polynomial $X^n + 1$, as shown by
>
> $$\mathbf{g}(X)\mathbf{h}(X) = X^n + 1 \tag{10.39}$$

This statement provides the basis for selecting the generator or parity-check polynomial of a cyclic code. In particular, if $\mathbf{g}(X)$ is a polynomial of degree $(n-k)$ and it is also a factor of $X^n + 1$, then $\mathbf{g}(X)$ is the generator polynomial of an (n,k) cyclic code. Equivalently, if

$\mathbf{h}(X)$ is a polynomial of degree k and it is also a factor of $X^n + 1$, then $\mathbf{h}(X)$ is the parity-check polynomial of an (n,k) cyclic code.

A final comment is in order. Any factor of $X^n + 1$ with degree $(n - k)$ can be used as a generator polynomial. The fact of the matter is that, for large values of n, the polynomial $X^n + 1$ may have many factors of degree $n - k$. Some of these polynomial factors generate good cyclic codes, whereas some of them generate bad cyclic codes. The issue of how to select generator polynomials that produce good cyclic codes is very difficult to resolve. Indeed, coding theorists have expended much effort in the search for good cyclic codes.

Generator and Parity-Check Matrices

Given the generator polynomial $\mathbf{g}(X)$ of an (n,k) cyclic code, we may construct the generator matrix $\mathbf{G}$ of the code by noting that the k polynomials $\mathbf{g}(X)$, $X\mathbf{g}(X)$, ..., $X^{k-1}\mathbf{g}(X)$ span the code. Hence, the n-tuples corresponding to these polynomials may be used as rows of the k-by-n generator matrix $\mathbf{G}$.

However, the construction of the parity-check matrix $\mathbf{H}$ of the cyclic code from the parity-check polynomial $\mathbf{h}(X)$ requires special attention, as described here. Multiplying (10.39) by $\mathbf{a}(x)$ and then using (10.32), we obtain

$$\mathbf{c}(X)\mathbf{h}(X) = \mathbf{a}(X) + X^n\mathbf{a}(X) \tag{10.40}$$

The polynomials $\mathbf{c}(X)$ and $\mathbf{h}(X)$ are themselves defined by (10.27) and (10.37) respectively, which means that their product on the left-hand side of (10.40) contains terms with powers extending up to $n + k - 1$. On the other hand, the polynomial $\mathbf{a}(X)$ has degree $k - 1$ or less, the implication of which is that the powers of $X^k, X^{k+1}, \ldots, X^{n-1}$ do *not* appear in the polynomial on the right-hand side of (10.40). Thus, setting the coefficients of $X^k, X^{k-1}, \ldots, X^{n-1}$ in the expansion of the product polynomial $\mathbf{c}(X)\mathbf{h}(X)$ equal to zero, we obtain the following set of $n - k$ equations:

$$\sum_{i=j}^{j+k} c_i h_{k+j-i} = 0 \qquad \text{for } 0 \le j \le n-k-1 \tag{10.41}$$

Comparing (10.41) with the corresponding relation (10.16), we may make the following important observation:

> The coefficients of the parity-check polynomial $\mathbf{h}(X)$ involved in the polynomial multiplication described in (10.41) are arranged in reversed order with respect to the coefficients of the parity-check matrix $\mathbf{H}$ involved in forming the inner product of vectors described in (10.16).

This observation suggests that we define the *reciprocal of the parity-check polynomial* as follows:

$$\begin{aligned} X^k\mathbf{h}(X^{-1}) &= X^k\left(1 + \sum_{i=1}^{k-1} h_i X^{-i} + X^{-k}\right) \\ &= 1 + \sum_{i=1}^{k-1} h_{k-1} X^i + X^k \end{aligned} \tag{10.42}$$

which is also a factor of $X^n + 1$. The n-tuples pertaining to the $(n-k)$ polynomials $X^k h(X^{-1}), X^{k+1} h(X^{-1}), \ldots, X^{n-1} h(X^{-1})$ may now be used in rows of the $(n-k)$-by-n parity-check matrix **H**.

In general, the generator matrix **G** and the parity-check matrix **H** constructed in the manner described here are not in their systematic forms. They can be put into their systematic forms by performing simple operations on their respective rows, as illustrated in Example 1.

Encoding of Cyclic Codes

Earlier we showed that the encoding procedure for an (n,k) cyclic code in systematic form involves three steps:

- multiplication of the message polynomial $\mathbf{m}(X)$ by X^{n-k},
- division of $X^{n-k}\mathbf{m}(X)$ by the generator polynomial $\mathbf{g}(X)$ to obtain the remainder $\mathbf{b}(X)$, and
- addition of $\mathbf{b}(X)$ to $X^{n-k}\mathbf{m}(X)$ to form the desired code polynomial.

These three steps can be implemented by means of the encoder shown in Figure 10.8, consisting of a *linear feedback shift register* with $(n-k)$ stages.

The boxes in Figure 10.8 represent *flip-flops*, or *unit-delay elements*. The flip-flop is a device that resides in one of two possible states denoted by 0 and 1. An *external clock* (not shown in Figure 10.8) controls the operation of all the flip-flops. Every time the clock ticks, the contents of the flip-flops (initially set to the state 0) are shifted out in the direction of the arrows. In addition to the flip-flops, the encoder of Figure 10.8 includes a second set of logic elements, namely *adders*, which compute the modulo-2 sums of their respective inputs. Finally, the *multipliers* multiply their respective inputs by the associated coefficients. In particular, if the coefficient $g_i = 1$, the multiplier is just a direct "connection." If, on the other hand, the coefficient $g_i = 0$, the multiplier is "no connection."

The operation of the encoder shown in Figure 10.8 proceeds as follows:

1. The gate is switched on. Hence, the k message bits are shifted into the channel. As soon as the k message bits have entered the shift register, the resulting $(n-k)$ bits in the register form the parity-check bits. (Recall that the parity-check bits are the same as the coefficients of the remainder $\mathbf{b}(X)$.)

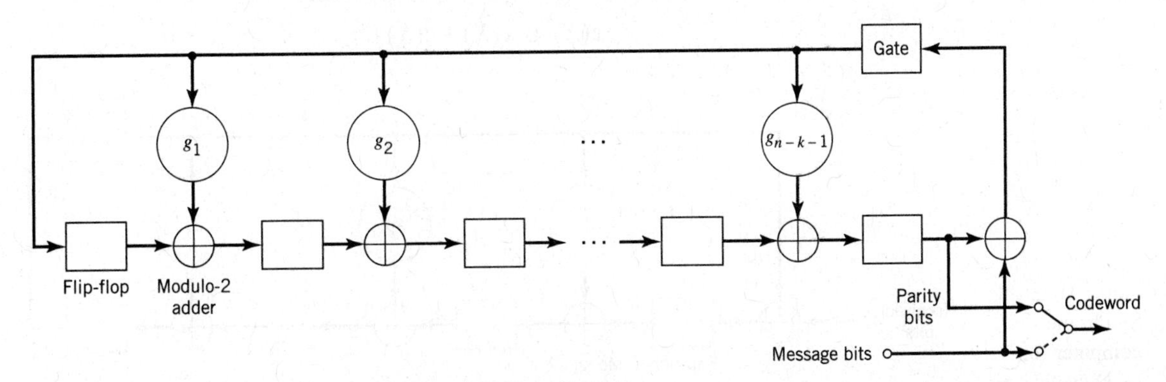

Figure 10.8 Encoder for an (n,k) cyclic code.

2. The gate is switched off, thereby breaking the feedback connections.
3. The contents of the shift register are read out into the channel.

Calculation of the Syndrome

Suppose the codeword $(c_0, c_1, \ldots, c_{n-1})$ is transmitted over a noisy channel, resulting in the received word $r_0, r_1, \ldots, r_{n-1}$. From Section 10.3, we recall that the first step in the decoding of a linear block code is to calculate the syndrome for the received word. If the syndrome is zero, there are no transmission errors in the received word. If, on the other hand, the syndrome is nonzero, the received word contains transmission errors that require correction.

In the case of a cyclic code in systematic form, the syndrome can be calculated easily. Let the received vector be represented by a polynomial of degree $n-1$ or less, as shown by

$$\mathbf{r}(X) = r_0 + r_1 X + \cdots + r_{n-1} X^{n-1}$$

Let $\mathbf{q}(X)$ denote the quotient and $\mathbf{s}(X)$ denote the remainder, which are the results of dividing $\mathbf{r}(X)$ by the generator polynomial $\mathbf{g}(X)$. We may therefore express $\mathbf{r}(X)$ as follows:

$$\mathbf{r}(X) = \mathbf{q}(X)\mathbf{g}(X) + \mathbf{s}(X) \qquad (10.43)$$

The remainder $\mathbf{s}(X)$ is a polynomial of degree $n-k-1$ or less, which is the result of interest. It is called the *syndrome polynomial* because its coefficients make up the $(n-k)$-by-1 syndrome $\mathbf{s}$.

Figure 10.9 shows a *syndrome calculator* that is identical to the encoder of Figure 10.8 except for the fact that the received bits are fed into the $n-k$ stages of the feedback shift register from the left. As soon as all the received bits have been shifted into the shift register, its contents define the syndrome $\mathbf{s}$.

The syndrome polynomial $\mathbf{s}(X)$ has the following useful properties that follow from the definition given in (10.43).

PROPERTY 1 *The syndrome of a received word polynomial is also the syndrome of the corresponding error polynomial.*

Given that a cyclic code with polynomial $\mathbf{c}(X)$ is sent over a noisy channel, the received word polynomial is defined by

$$\mathbf{r}(X) = \mathbf{c}(X) + \mathbf{e}(X)$$

where $\mathbf{e}(X)$ is the error polynomial. Equivalently, we may write

$$\mathbf{e}(X) = \mathbf{r}(X) + \mathbf{c}(X)$$

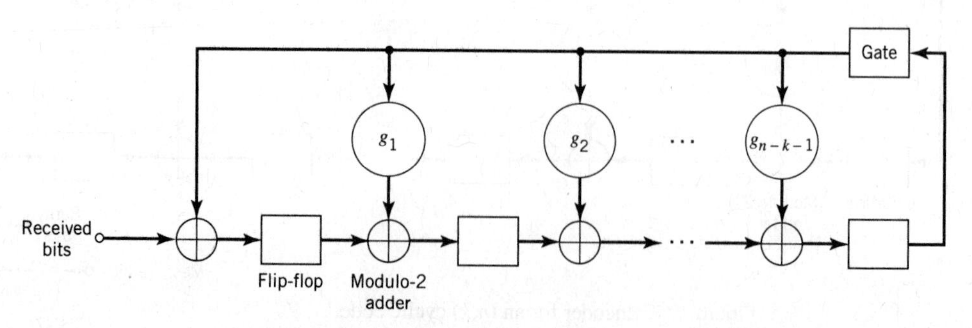

Figure 10.9 Syndrome computer for (n, k) cyclic code.

Hence, substituting (10.32) and (10.43) into the preceding equation, we get

$$\mathbf{e}(X) = \mathbf{u}(X)\mathbf{g}(X) + \mathbf{s}(X) \tag{10.44}$$

where the quotient is $\mathbf{u}(X) = \mathbf{a}(X) + \mathbf{q}(X)$. Equation (10.44) shows that $\mathbf{s}(X)$ is also the syndrome of the error polynomial $\mathbf{e}(X)$. The implication of this property is that when the syndrome polynomial $\mathbf{s}(X)$ is nonzero, the presence of transmission errors in the received vector is detected.

PROPERTY 2 *Let $\mathbf{s}(X)$ be the syndrome of a received word polynomial $\mathbf{r}(X)$. Then, the syndrome of $X\mathbf{r}(X)$, representing a cyclic shift of $\mathbf{r}(X)$, is $X\mathbf{s}(X)$.*

Applying a cyclic shift to both sides of (10.43), we get

$$X\mathbf{r}(X) = X\mathbf{q}(X)\mathbf{g}(X) + X\mathbf{s}(X) \tag{10.45}$$

from which we readily see that $X\mathbf{s}(X)$ is the remainder of the division of $X\mathbf{r}(X)$ by $\mathbf{g}(X)$. Hence, the syndrome of $X\mathbf{r}(X)$ is $X\mathbf{s}(X)$ as stated. We may generalize this result by stating that if $\mathbf{s}(X)$ is the syndrome of $\mathbf{r}(X)$, then $X^i\mathbf{s}(X)$ is the syndrome of $X^i\mathbf{r}(X)$.

PROPERTY 3 *The syndrome polynomial $\mathbf{s}(X)$ is identical to the error polynomial $\mathbf{e}(X)$, assuming that the errors are confined to the $(n-k)$ parity-check bits of the received word polynomial $\mathbf{r}(X)$.*

The assumption made here is another way of saying that the degree of the error polynomial $\mathbf{e}(X)$ is less than or equal to $(n-k-1)$. Since the generator polynomial $\mathbf{g}(X)$ is of degree $(n-k)$, by definition, it follows that (10.44) can only be satisfied if the quotient $\mathbf{u}(X)$ is zero. In other words, the error polynomial $\mathbf{e}(X)$ and the syndrome polynomial $\mathbf{s}(X)$ are one and the same. The implication of Property 3 is that, under the aforementioned conditions, error correction can be accomplished simply by adding the syndrome polynomial $\mathbf{s}(X)$ to the received vector $\mathbf{r}(X)$.

EXAMPLE 2 **Hamming Codes Revisited**

To illustrate the issues relating to the polynomial representation of cyclic codes, we consider the generation of a (7,4) cyclic code. With the block length $n = 7$, we start by factorizing $X^7 + 1$ into three *irreducible polynomials*:

$$X^7 + 1 = (1 + X)(1 + X^2 + X^3)(1 + X + X^3)$$

By an "irreducible polynomial" we mean a polynomial that cannot be factored using only polynomials with coefficients from the binary field. An irreducible polynomial of degree m is said to be *primitive* if the smallest positive integer n for which the polynomial divides $X^n + 1$ is $n = 2^m - 1$. For the example at hand, the two polynomials $(1 + X^2 + X^3)$ and $(1 + X + X^3)$ are primitive. Let us take

$$\mathbf{g}(X) = 1 + X + X^3$$

as the generator polynomial, whose degree equals the number of parity-check bits. This means that the parity-check polynomial is given by

$$\begin{aligned}\mathbf{h}(X) &= (1 + X)(1 + X^2 + X^3) \\ &= 1 + X + X^2 + X^4\end{aligned}$$

whose degree equals the number of message bits $k = 4$.

Next, we illustrate the procedure for the construction of a codeword by using this generator polynomial to encode the message sequence 1001. The corresponding message vector is given by

$$\mathbf{m}(X) = 1 + X^3$$

Hence, multiplying $\mathbf{m}(X)$ by $X^{n-k} = X^3$, we get

$$X^{n-k}\mathbf{m}(X) = X^3 + X^6$$

The second step is to divide $X^{n-k}\mathbf{m}(X)$ by $\mathbf{g}(X)$, the details of which (for the example at hand) are given below:

$$\begin{array}{rlllll}
 & X^3 & & +X & & \\
\cline{2-6}
X^3+X+1 \big) & X^6 & & +X^3 & & \\
 & X^6 & +X^4 & +X^3 & & \\
\cline{2-6}
 & & X^4 & & & \\
 & & X^4 & & +X^2 & +X \\
\cline{3-6}
 & & & & X^2 & +X
\end{array}$$

Note that in this long division we have treated subtraction the same as addition since we are operating in modulo-2 arithmetic. We may thus write

$$\frac{X^3 + X^6}{1 + X + X^3} = X + X^3 + \frac{X + X^2}{1 + X + X^3}$$

That is, the quotient $\mathbf{a}(X)$ and remainder $\mathbf{b}(X)$ are as follows, respectively:

$$\mathbf{a}(X) = X + X^3$$

$$\mathbf{b}(X) = X + X^2$$

Hence, from (10.35) we find that the desired code vector is

$$\begin{aligned}
\mathbf{c}(X) &= \mathbf{b}(X) + X^{n-k}\mathbf{m}(X) \\
&= X + X^2 + X^3 + X^6
\end{aligned}$$

The codeword is therefore 0111001. The four rightmost bits, 1001, are the specified message bits. The three leftmost bits, 011, are the parity-check bits. The codeword thus generated is exactly the same as the corresponding one shown in Table 10.1 for a (7,4) Hamming code.

We may generalize this result by stating that:

> Any cyclic code generated by a primitive polynomial is a Hamming code of minimum distance 3.

We next show that the generator polynomial $\mathbf{g}(X)$ and the parity-check polynomial $\mathbf{h}(X)$ uniquely specify the generator matrix $\mathbf{G}$ and the parity-check matrix $\mathbf{H}$, respectively.

To construct the 4-by-7 generator matrix $\mathbf{G}$, we start with four vectors represented by $\mathbf{g}(X)$ and three cyclic-shifted versions of it, as shown by

$$\begin{aligned} \mathbf{g}(X) &= 1 + X + X^3 \\ X\mathbf{g}X &= X + X^2 + X^4 \\ X^2\mathbf{g}(X) &= X^2 + X^3 + X^5 \\ X^3\mathbf{g}(X) &= X^3 + X^4 + X^6 \end{aligned}$$

The vectors **g**(X), X**g**(X), X^2**g**(X), and X^3**g**(X) represent code polynomials in the (7,4) Hamming code. If the coefficients of these polynomials are used as the elements of the rows of a 4-by-7 matrix, we get the following generator matrix:

$$\mathbf{G}' = \begin{bmatrix} 1 & 1 & 0 & 1 & 0 & 0 & 0 \\ 0 & 1 & 1 & 0 & 1 & 0 & 0 \\ 0 & 0 & 1 & 1 & 0 & 1 & 0 \\ 0 & 0 & 0 & 1 & 1 & 0 & 1 \end{bmatrix}$$

Clearly, the generator matrix $\mathbf{G}'$ so constructed is not in systematic form. We can put it into a systematic form by adding the first row to the third row, and adding the sum of the first two rows to the fourth row. These manipulations result in the desired generator matrix:

$$\mathbf{G} = \begin{bmatrix} 1 & 1 & 0 & 1 & 0 & 0 & 0 \\ 0 & 1 & 1 & 0 & 1 & 0 & 0 \\ 1 & 1 & 1 & 0 & 0 & 1 & 0 \\ 1 & 0 & 1 & 0 & 0 & 0 & 1 \end{bmatrix}$$

which is exactly the same as that in Example 1.

We next show how to construct the 3-by-7 parity-check matrix **H** from the parity-check polynomial **h**(X). To do this, we first take the reciprocal of **h**(X), namely $X^4\mathbf{h}(X^{-1})$. For the problem at hand, we form three vectors represented by $X^4\mathbf{h}(X^{-1})$ and two shifted versions of it, as shown by

$$\begin{aligned} X^4\mathbf{h}(X^{-1}) &= 1 + X^2 + X^3 + X^4 \\ X^5\mathbf{h}(X^{-1}) &= X + X^3 + X^4 + X^5 \\ X^6\mathbf{h}(X^{-1}) &= X^2 + X^4 + X^5 + X^6 \end{aligned}$$

Using the coefficients of these three vectors as the elements of the rows of the 3-by-7 parity-check matrix, we get

$$\mathbf{H}' = \begin{bmatrix} 1 & 0 & 1 & 1 & 1 & 0 & 0 \\ 0 & 1 & 0 & 1 & 1 & 1 & 0 \\ 0 & 0 & 1 & 0 & 1 & 1 & 1 \end{bmatrix}$$

Here again we see that the matrix $\mathbf{H}'$ is not in systematic form. To put it into a systematic form, we add the third row to the first row to obtain

$$\mathbf{H} = \begin{bmatrix} 1 & 0 & 0 & 1 & 0 & 1 & 1 \\ 0 & 1 & 0 & 1 & 1 & 1 & 0 \\ 0 & 0 & 1 & 0 & 1 & 1 & 1 \end{bmatrix}$$

which is exactly the same as that of Example 1.

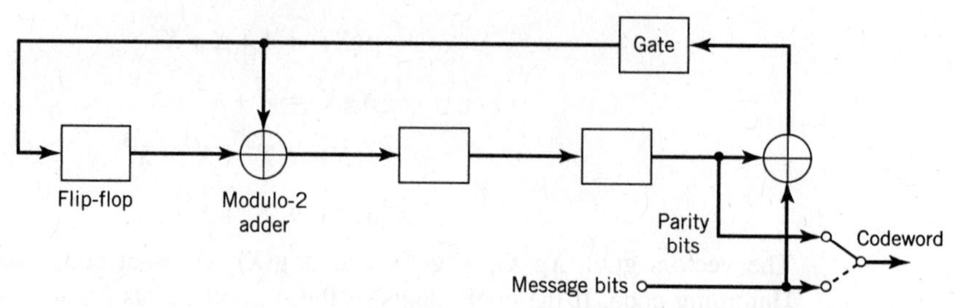

Figure 10.10 Encoder for the (7,4) cyclic code generated by $\mathbf{g}(X) = 1 + X + X^3$.

Figure 10.10 shows the encoder for the (7,4) cyclic Hamming code generated by the polynomial $\mathbf{g}(X) = 1 + X + X^3$. To illustrate the operation of this encoder, consider the message sequence (1001). The contents of the shift register are modified by the incoming message bits as in Table 10.3. After four shifts, the contents of the shift register, and therefore the parity-check bits, are (011). Accordingly, appending these parity-check bits to the message bits (1001), we get the codeword (0111001); this result is exactly the same as that determined earlier in Example 1.

Table 10.3 Contents of the shift register in the encoder of Figure 10.10 for message sequence (1001)

Shift	Input bit	Contents of shift register
		000 (initial state)
1	1	110
2	0	011
3	0	111
4	1	011

Figure 10.11 shows the corresponding syndrome calculator for the (7,4) Hamming code. Let the transmitted codeword be (0111001) and the received word be (0110001); that is, the middle bit is in error. As the received bits are fed into the shift register, initially set to zero, its contents are modified as in Table 10.4. At the end of the seventh shift, the syndrome is identified from the contents of the shift register as 110. Since the syndrome is nonzero, the received word is in error. Moreover, from Table 10.2, we see that the error pattern corresponding to this syndrome is 0001000. This indicates that the error is in the middle bit of the received words, which is indeed the case.

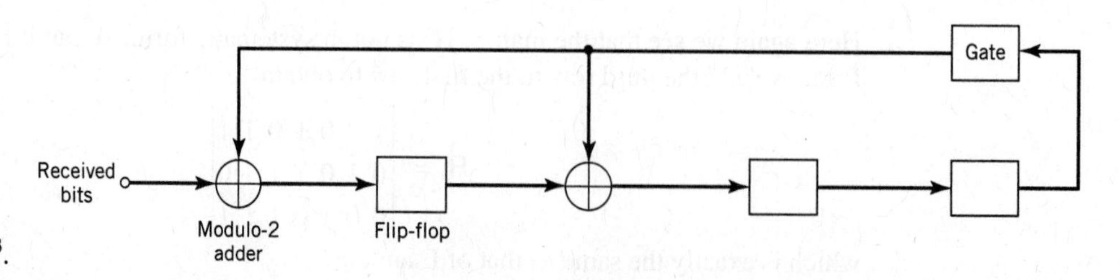

Figure 10.11 Syndrome calculator for the (7,4) cyclic code generated by the polynomial $\mathbf{g}(X) = 1 + X + X^3$.

Table 10.4 **Contents of the syndrome calculator in Figure 10.11 for the received word (0110001)**

Shift	Input bit	Contents of shift register
		000 (initial state)
1	1	100
2	0	010
3	0	001
4	0	110
5	1	111
6	1	001
7	0	110

EXAMPLE 3 **Maximal-Length Codes**

For any positive integer $m \geq 3$, there exists a *maximal-length code*[4] with the following parameters:

block length: $n = 2^m - 1$

number of message bits: $k = m$

minimum distance: $d_{\min} = 2^{m-1}$

Maximal-length codes are generated by vectors of the form

$$\mathbf{g}(X) = \frac{1 + X^n}{\mathbf{h}(X)} \tag{10.46}$$

where $\mathbf{h}(X)$ is any primitive polynomial of degree m. Earlier we stated that any cyclic code generated by a primitive polynomial is a Hamming code of minimum distance 3 (see Example 2). It follows, therefore, that maximal-length codes are the *dual* of Hamming codes.

The polynomial $\mathbf{h}(X)$ defines the feedback connections of the encoder. The generator polynomial $\mathbf{g}(X)$ defines one period of the maximal-length code, assuming that the encoder is in the initial state 00 … 01. To illustrate this, consider the example of a (7,3) maximal-length code, which is the dual of the (7,4) Hamming code described in Example 2. Thus, choosing

$$\mathbf{h}(X) = 1 + X + X^3$$

we find that the generator polynomial of the (7,3) maximal-length code is

$$\mathbf{g}(X) = 1 + X + X^2 + X^4$$

Figure 10.12 shows the encoder for the (7,3) maximal-length code. The period of the code is $n = 7$. Thus, assuming that the encoder is in the initial state 001, as indicated in Figure 10.12, we find the output sequence is described by

$$\underbrace{1\,0\,0}_{\text{initial state}} \quad \underbrace{1\,1\,1\,0\,1\,0\,0}_{\mathbf{g}(X) = 1 + X + X^2 + X^4}$$

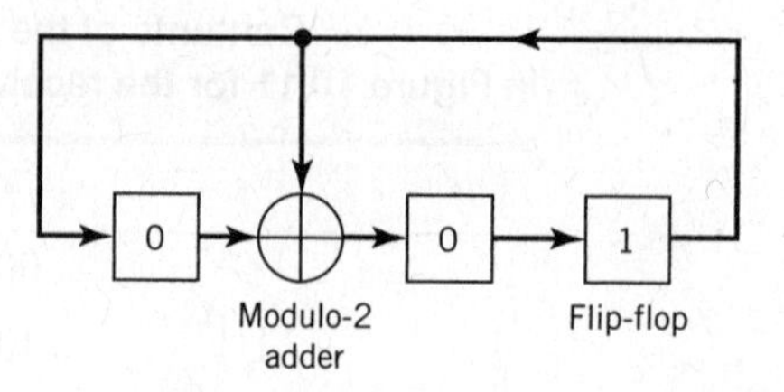

Figure 10.12 Encoder for the (7,3) maximal-length code; the initial state of the encoder is shown in the figure.

This result is readily validated by cycling through the encoder of Figure 10.12.

Note that if we were to choose the other primitive polynomial

$$\mathbf{h}(X) = 1 + X^2 + X^3$$

for the (7,3) maximal-length code, we would simply get the "image" of the code described above, and the output sequence would be "reversed" in time.

Reed–Solomon Codes

A study of cyclic codes for error control would be incomplete without a discussion of *Reed–Solomon codes,*[5] albeit briefly.

Unlike the cyclic codes considered in this section, Reed–Solomon codes are *nonbinary codes*. A cyclic code is said to be nonbinary in that given the code vector

$$\mathbf{c} = (c_0, c_1, \ldots, c_{n-1})$$

the coefficients $\{c_i\}_{i=0}^{n-1}$ are not binary 0 or 1. Rather, the c_i are themselves made up of sequences of 0s and 1s, with each sequence being of length k. A Reed–Solomon code is therefore said to be a q-ary code, which means that the size of the alphabet used in construction of the code is $q = 2^k$. To be specific, a Reed–Solomon (n,k) code is used to encode m-bit symbols into blocks consisting of $n = 2^m - 1$ symbols; that is, $m(2^m - 1)$ bits, where $m \geq 1$. Thus, the encoding algorithm expands a block of k symbols to n symbols by adding $n - k$ redundant symbols. When m is an integer power of 2, the m-bit symbols are called *bytes*. A popular value of m is 8; indeed, 8-bit Reed–Solomon codes are extremely powerful.

A t-error-correcting Reed–Solomon code has the following parameters:

block length	$n = 2^m - 1$ symbols
message size	k symbols
parity-check size	$n - k = 2t$ symbols
minimum distance	$d_{\min} = 2t + 1$ symbols

The block length of the Reed–Solomon code is one less than the size of a code symbol, and the minimum distance is one greater than the number of parity-check symbols. Reed–Solomon codes make highly efficient use of redundancy; block lengths and symbol sizes can be adjusted readily to accommodate a wide range of message sizes. Moreover, Reed–Solomon codes provide a wide range of code rates that can be chosen to optimize performance, and efficient techniques are available for their use in certain practical applications. In particular, a distinctive feature of Reed–Solomon codes is their ability to correct *bursts of errors*, hence their application in wireless communications to combat the fading phenomenon.

10.6 Convolutional Codes

In block coding, the encoder accepts a k-bit message block and generates an n-bit codeword, which contains $n - k$ parity-check bits. Thus, codewords are produced on a block-by-block basis. Clearly, provision must be made in the encoder to buffer an entire message block before generating the associated codeword. There are applications, however, where the message bits come in *serially* rather than in large blocks, in which case the use of a buffer may be undesirable. In such situations, the use of *convolutional coding* may be the preferred method. A convolutional coder generates redundant bits by using *modulo-2 convolutions*; hence the name *convolutional codes*[6].

The encoder of a binary convolutional code with rate $1/n$, measured in bits per symbol, may be viewed as a *finite-state machine* that consists of an M-stage shift register with prescribed connections to n modulo-2 adders and a multiplexer that serializes the outputs of the adders. A sequence of message bits produces a coded output sequence of length $n(L + M)$ bits, where L is the length of the message sequence. The *code rate* is therefore given by

$$\begin{aligned} r &= \frac{L}{n(L+M)} \\ &= \frac{1}{n(1+M/L)} \quad \text{bits/symbol} \end{aligned} \tag{10.47}$$

Typically, we have $L \gg M$, in which case the code rate is approximately defined by

$$r \approx \frac{1}{n} \quad \text{bits/symbol} \tag{10.48}$$

An important characteristic of a convolutional code is its constraint length, which we define as follows:

> The constraint length of a convolutional code, expressed in terms of message bits, is the number of shifts over which a single incoming message bit can influence the encoder output.

In an encoder with an M-stage shift register, the *memory* of the encoder equals M message bits. Correspondingly, the constraint length, denoted by ν, equals $M + 1$ shifts that are required for a message bit to enter the shift register and finally come out.

Figure 10.13 shows a convolutional encoder with the number of message bits $n = 2$ and constraint length $\nu = 3$. In this example, the code rate of the encoder is 1/2. The encoder operates on the incoming message sequence, one bit at a time, through a convolution process; it is therefore said to be a *nonsystematic* code.

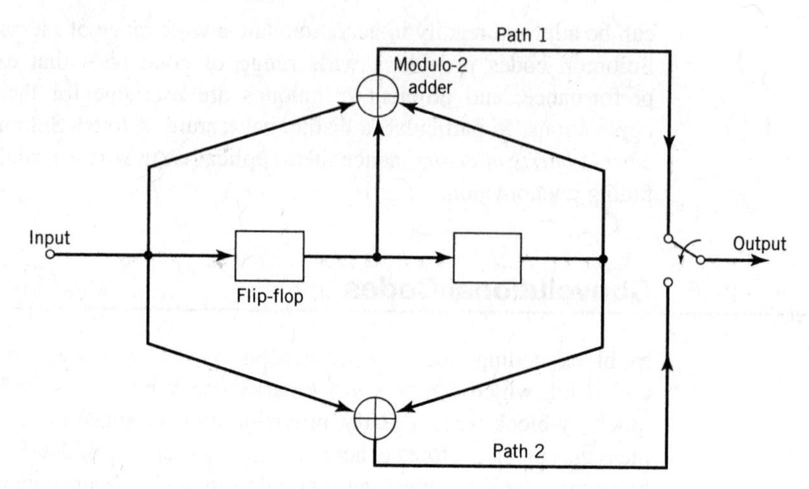

Figure 10.13
Constraint length-3, rate -1/2 convolutional encoder.

Each path connecting the output to the input of a convolutional encoder may be characterized in terms of its *impulse response*, defined as follows:

> The impulse response of a particular path in the convolutional encoder is the response of that path in the encoder to symbol 1 applied to its input, with each flip-flop in the encoder set initially to the zero state.

Equivalently, we may characterize each path in terms of a generator polynomial, defined as the *unit-delay transform* of the impulse response. To be specific, let the *generator sequence* $\left(g_0^{(i)}, g_1^{(i)}, g_2^{(i)}, \ldots, g_M^{(i)}\right)$ denote the impulse response of the ith path, where the coefficients $g_0^{(i)}, g_1^{(i)}, g_2^{(i)}, \ldots, g_M^{(i)}$ equal symbol 0 or 1. Correspondingly, the *generator polynomial* of the ith path is defined by

$$g^{(i)}(D) = g_0^{(i)} + g_1^{(i)}D + g_2^{(i)}D^2 + \cdots + g_M^{(i)}D^M \tag{10.49}$$

where D denotes the *unit-delay variable*. The complete convolutional encoder is described by the set of generator polynomials $\{g^{(i)}(D)\}_{i=1}^{M}$.

EXAMPLE 4

Convolutional Encoder

Consider again the convolutional encoder of Figure 10.13, which has two paths numbered 1 and 2 for convenience of reference. The impulse response of path 1 (i.e., upper path) is (1, 1, 1). Hence, the generator polynomial of this path is

$$g^{(1)}(D) = 1 + D + D^2$$

The impulse response of path 2 (i.e., lower path) is (1, 0, 1). The generator polynomial of this second path is

$$g^{(2)}(D) = 1 + D^2$$

For an incoming message sequence given by (10011), for example, we have the polynomial representation

$$m(D) = 1 + D^3 + D^4$$

As with Fourier transformation, convolution in the time domain is transformed into multiplication in the D-domain. Hence, the output polynomial of path 1 is given by

$$\begin{aligned} c^{(1)}(D) &= g^{(1)}(D)m(D) \\ &= (1+D+D^2)(1+D^3+D^4) \\ &= 1+D+D^2+D^3+D^6 \end{aligned}$$

where it is noted that the sums $D^4 + D^4$ and $D^5 + D^5$ are both zero in accordance with the rules of binary arithmetic. We therefore immediately deduce that the output sequence of path 1 is (1111001). Similarly, the output polynomial of path 2 is given by

$$\begin{aligned} c^{(2)}(D) &= g^{(2)}(D)m(D) \\ &= (1+D^2)(1+D^3+D^4) \\ &= 1+D^2+D^3+D^4+D^5+D^6 \end{aligned}$$

The output sequence of path 2 is therefore (1011111). Finally, *multiplexing* the two output sequences of paths 1 and 2, we get the encoded sequence

$$\mathbf{c} = (11, 10, 11, 11, 01, 01, 11)$$

Note that the message sequence of length $L = 5$ bits produces an encoded sequence of length $n(L + \nu - 1) = 14$ bits. Note also that for the shift register to be restored to its initial all-zero state, a terminating sequence of $\nu - 1 = 2$ zeros is appended to the last input bit of the message sequence. The terminating sequence of $\nu - 1$ zeros is called the *tail of the message*.

Code Tree, Trellis Graph, and State Graph

Traditionally, the structural properties of a convolutional encoder are portrayed in graphical form by using any one of three equivalent graphs: code tree, trellis graph, and state graph.

Although, indeed, these three graphical representations of a convolutional encoder look different, their compositions follow the same underlying rule:

> A code branch produced by input bit 0 is drawn as a solid line, whereas a code branch produced by input bit 1 is a dashed line.

Hereafter, we refer to this convention as the *graphical rule* of a convolutional encoder.

We will use the convolutional encoder of Figure 10.13 as a running example to illustrate the insights that each one of these three diagrams provides.

Code Tree

We begin the graphical representation of a convolutional encoder with the *code tree* of Figure 10.14. Each branch of the tree represents an input bit, with the corresponding pair of output bits indicated on the branch. The convention used to distinguish the input bits 0 and 1 follows the graphical rule described above. Thus, a specific *path* in the tree is traced from left to right in accordance with the message sequence. The corresponding coded bits on the branches of that path constitute the message sequence (10011) applied to the input of the encoder of Figure 10.13. Following the procedure just described, we find that the corresponding encoded sequence is (11, 10, 11, 11, 01), which agrees with the first five pairs of bits in the encoded sequence $\{c_i\}$ that was derived in Example 4.

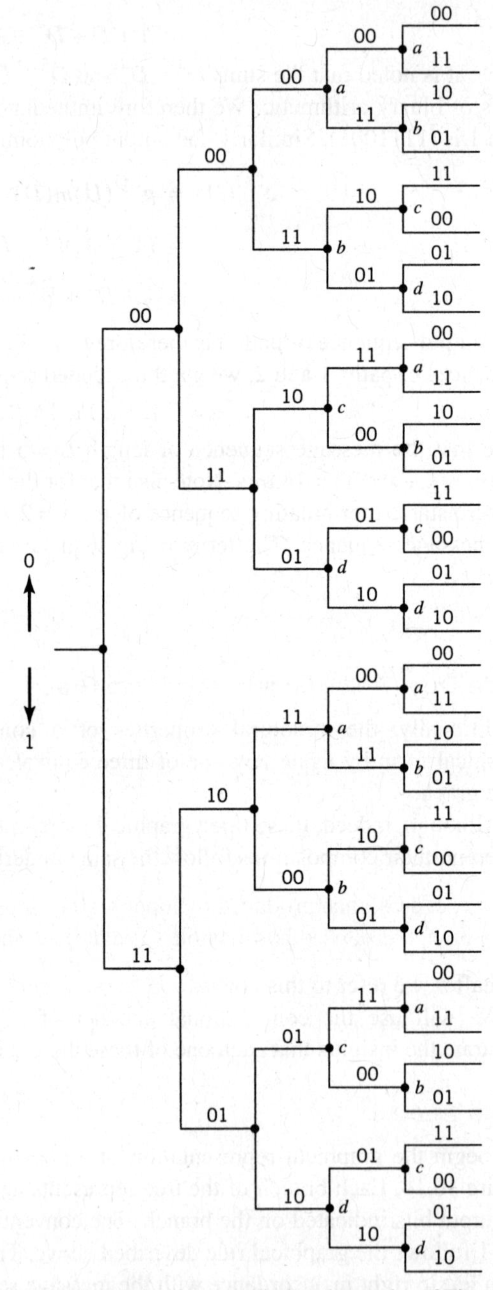

Figure 10.14 **Code tree for the convolutional encoder of Figure 10.13.**

Trellis Graph

From Figure 10.14, we observe that the tree becomes *repetitive* after the first three branches. Indeed, beyond the third branch, the two nodes labeled a are identical and so are all the other node pairs that are identically labeled. We may establish this repetitive property of the tree by examining the associated encoder of Figure 10.13. The encoder has memory $M = \nu - 1 = 2$ message bits. We therefore find that, when the third message bit enters the encoder, the first message bit is shifted out of the register. Consequently, after the third branch, the message sequences (100 m_3m_4...) and (000 m_3m_4...) generate the same code symbols, and the pair of nodes labeled a may be joined together. The same reasoning applies to the other nodes in the code tree. Accordingly, we may collapse the code tree of Figure 10.14 into the new form shown in Figure 10.15, which is called a *trellis*. It is so called since a trellis is a treelike structure with re-emerging branches. The convention used in Figure 10.15 to distinguish between input symbols 0 and 1 is as follows:

> A code branch in a trellis produced by input binary symbol 0 is drawn as a solid line, whereas a code branch produced by an input 1 is drawn as a dashed line.

As before, each message sequence corresponds to a specific path through the trellis. For example, we readily see from Figure 10.15 that the message sequence (10011) produces the encoded output sequence (11, 10, 11, 11, 01), which agrees with our previous result.

The Notion of State

In conceptual terms, a trellis is more instructive than a tree. We say so because it brings out explicitly the fact that the associated convolutional encoder is in actual fact a *finite-state machine*. Basically, such a machine consists of a tapped shift register and, therefore, has a finite state; hence the name of the machine. Thus, we may conveniently say the following:

> The state of a rate $1/n$ convolutional encoder is determined by the smallest number of message bits stored in memory (i.e., the shift register).

For example, the convolutional encoder of Figure 10.13 has a shift register made up of two memory cells. With the message bit stored in each memory cell being 0 or 1, it follows that this encoder can assume any one of $2^2 = 4$ possible states, as described in Table 10.5.

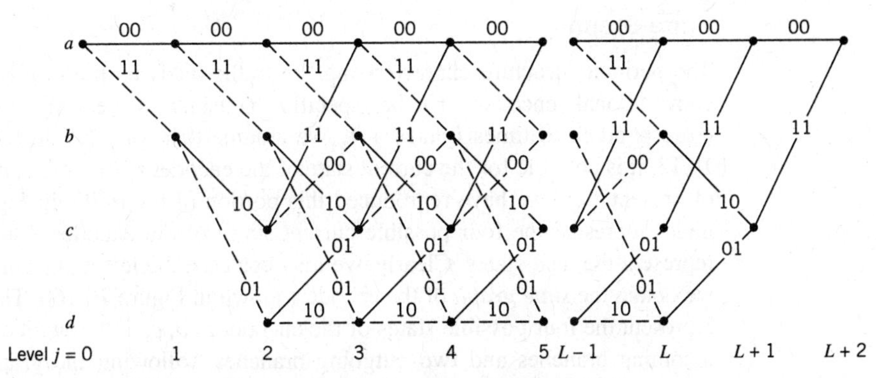

Figure 10.15 Trellis for the convolutional encoder of Figure 10.13.

Table 10.5 **State table for the convolutional encoder of Figure 10.13**

State	Binary description
a	00
b	10
c	01
d	11

In describing a convolutional encoder, the notion of state is important in the following sense:

> Given the current message bit and the state of the encoder, the codeword produced at the output of the encoder is completely determined.

To illustrate this statement, consider the general case of a rate $1/n$ convolutional encoder of constraint length ν. Let the state of the encoder at time-unit j be denoted by

$$S = (m_{j-1}, m_{j-2}, \ldots, m_{j-\nu+1})$$

The jth codeword c_j is completely determined by the state S together with the current message bit m_j.

Now that we understand the notion of state, the trellis graph of the simple convolutional encoder of Figure 10.13 for $\nu = 3$ is presented in Figure 10.15. From this latter figure, we now clearly see a unique characteristic of the trellis diagram:

> The trellis depicts the evolution of the convolutional encoder's state across time.

To be more specific, the first $\nu - 1 = 2$ time-steps correspond to the encoder's departure from the initial zero state and the last $\nu - 1 = 2$ time-steps correspond to the encoder's return to the initial zero state. Naturally, not all the states of the encoder can be reached in these two particular portions of the trellis. However, in the central portion of the trellis, for which time-unit j lies in the range $\nu - 1 \le j \le L$, where L is the length of the incoming message sequence, we do see that all the four possible states of the encoder are reachable. Note also that the central portion of the trellis exhibits a *fixed periodic structure*, as illustrated in Figure 10.16a.

State Graph

The periodic structure characterizing the trellis leads us next to the state diagram of a convolutional encoder. To be specific, consider a central portion of the trellis corresponding to times j and $j + 1$. We assume that for $j \ge 2$ in the example of Figure 10.13, it is possible for the current state of the encoder to be a, b, c, or d. For convenience of presentation, we have reproduced this portion of the trellis in Figure 10.16a. The left nodes represent the four possible current states of the encoder, whereas the right nodes represent the next states. Clearly, we may coalesce the left and right nodes. By so doing, we obtain the *state graph* of the encoder, shown in Figure 10.16b. The nodes of the figure represent the four possible states of the encoder a, b, c, and d, with each node having two incoming branches and two outgoing branches, following the graphical rule described previously.

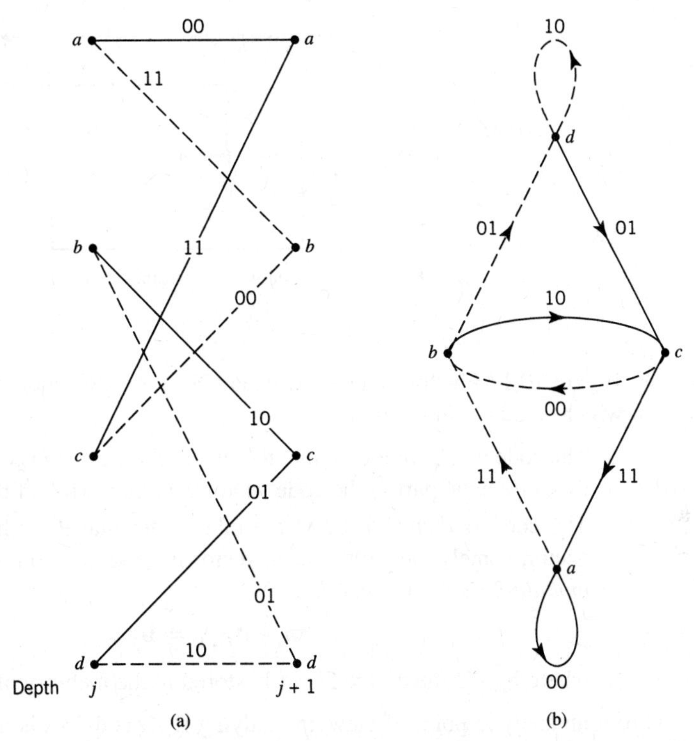

Figure 10.16 (a) A portion of the central part of the trellis for the encoder of Figure 10.13. (b) State graph of the convolutional encoder of Figure 10.13.

The binary label on each branch represents the encoder's output as it moves from one state to another. Suppose, for example, the current state of the encoder is (01), which is represented by node c. The application of input symbol 1 to the encoder of Figure 10.13 results in the state (10) and the encoded output (00). Accordingly, with the help of this state diagram, we may readily determine the output of the encoder of Figure 10.13 for any incoming message sequence. We simply start at state a, the *initial all-zero state*, and walk through the state graph in accordance with the message sequence. We follow a solid branch if the input is bit 0 and a dashed branch if it is bit 1. As each branch is traversed, we output the corresponding binary label on the branch. Consider, for example, the message sequence (10011). For this input, we follow the path *abcabd*, and therefore output the sequence (11, 10, 11, 11, 01), which agrees exactly with our previous result. Thus, the input–output relation of a convolutional encoder is also completely described by its state graph.

Recursive Systematic Convolutional Codes

The convolutional codes described thus far in this section have been feedforward structures of the nonsystematic variety. There is another type of linear convolutional codes that are the exact opposite, being recursive as well as systematic; they are called *recursive systematic convolutional (RSC) codes*.

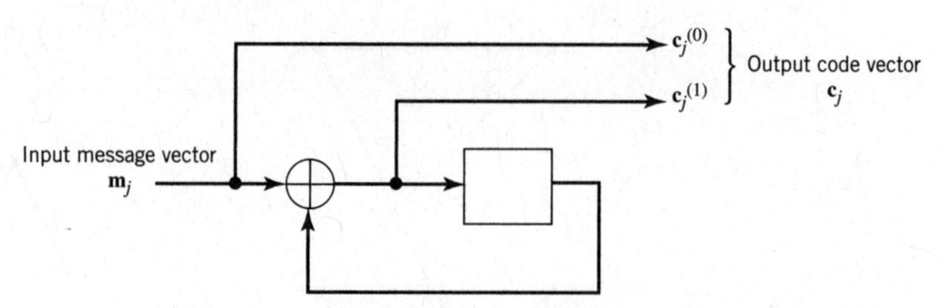

Figure 10.17 Example of a recursive systematic convolutional (RSC) encoder.

Figure 10.17 illustrates a simple example of an RSC code, two distinguishing features of which stand out in the figure:

1. The code is *systematic*, in that the incoming message vector $\mathbf{m}_j$ at time-unit j defines the systematic part of the code vector $\mathbf{c}_j$ at the output of the encoder.
2. The code is *recursive* by virtue of the fact that the other constituent of the code vector, namely the parity-check vector $\mathbf{b}_j$, is related to the message vector $\mathbf{m}_j$ by the *modulo-2 recursive equation*

$$\mathbf{m}_j + \mathbf{b}_{j-1} = \mathbf{b}_j \tag{10.50}$$

where $\mathbf{b}_{j-1}$ is the past value of $\mathbf{b}_j$ stored in the memory of the encoder.

From an analytic point of view, in studying RSC codes, it is more convenient to work in the transform D-domain than the time domain. By definition, we have

$$\mathbf{b}_{j-1} = D[\mathbf{b}_j] \tag{10.51}$$

and therefore rewrite (10.50) in the equivalent form:

$$\mathbf{b}_j = \frac{1}{1+D}[\mathbf{m}_j] \tag{10.52}$$

where the transfer function $1/(1+D)$ operates on $\mathbf{m}_j$ to produce $\mathbf{b}_j$. With the code vector $\mathbf{c}_j$ consisting of the message vector $\mathbf{m}_j$ followed by the parity-check vector $\mathbf{b}_j$, we may express the code vector $\mathbf{c}_j$ produced in response to the message vector $\mathbf{m}_j$ as follows:

$$\begin{aligned}\mathbf{c}_j &= (\mathbf{m}_j, \mathbf{b}_j) \\ &= \left(1, \frac{1}{1+D}\right)\mathbf{m}_j\end{aligned} \tag{10.53}$$

It follows, therefore, that the *code generator* for the RSC code of Figure 10.17 is given by the matrix

$$\mathbf{G}(D) = \left(1, \frac{1}{1+D}\right) \tag{10.54}$$

Generalizing, we may now make the statement:

> For recursive systematic convolutional codes, the transform-domain matrix $\mathbf{G}(D)$ is easier to use as the code generator than the corresponding time-domain matrix $\mathbf{G}$ whose entries contain sequences of infinite length.

The same statement applies equally well to the parity-check generator $\mathbf{H}(D)$ compared with its time-domain counterpart $\mathbf{H}$.

The rationale behind making convolutional codes recursive is to feed one or more of the tap-outputs in the shift register back to the encoder input, which, in turn, makes the internal state of the shift register depend on past outputs. This modification, compared with a feedforward convolutional code, affects the behavior of error patterns in a profound way, which is emphasized in the following statement:

> A single error in the systematic bits of an RSC code produces an infinite number of parity-check errors due to the use of feedback in the encoder.

This property of recursive convolutional codes turns out to be one of the key factors behind the outstanding performance achieved by the class of turbo codes, to be discussed in Section 10.12. Therein, we shall see that feedback plays a key role not only in the encoder of turbo codes but also the decoder. For reasons that will become apparent later, further work on turbo codes will be deferred to Section 10.12.

10.7 Optimum Decoding of Convolutional Codes

In the meantime, we resume the discussion on convolutional codes whose encoders are of the feedforward variety, aimed at the development of two different decoding algorithms, each of which is optimum according to a criterion of its own.

The first algorithm is the *maximum likelihood (ML) decoding algorithm*; the decoder is itself referred to as the *maximum likelihood decoder* (maximum likelihood estimation was discussed in Chapter 3). A distinctive feature of this decoder is that it produces a codeword as output, the conditional probability of which is always maximized on the assumption that each codeword in the code is equiprobable. From Chapter 3 on probability theory, we recall that the conditional probability density function of a random variable X given a quantity θ can be rethought as the likelihood function of θ with that function being dependent on X, given a parameter θ. We may therefore make the statement:

> In the maximum likelihood decoding of a convolutional code, the metric to be maximized is the likelihood function of a codeword, expressed as a function of the noisy channel output.

The second algorithm is the *maximum a posteriori (MAP) probability decoding algorithm*; the decoder is correspondingly referred to as a *MAP decoder*. In light of this second algorithm's name, we may make the statement:

> In MAP decoding of a convolutional code, the metric to be maximized is the posterior of a codeword, expressed as the product of the likelihood function of a given bit and the a priori probability of that bit.

These two decoding algorithms, *optimal* in accordance with their own respective criteria, are distinguished from each other as follows:

> The ML decoding algorithm produces the most likely codeword as its output. On the other hand, the MAP decoding algorithm operates on the received sequence on a bit-by-bit basis to produce the most likely symbol as output.

Stated in another way, we may say:

> The ML decoder minimizes the probability of selecting the wrong codeword, whereas the MAP decoder minimizes the decoded BER.

Typically, the ML decoder is simpler to implement; hence its popular use in practice. However, the MAP decoding algorithm is preferred over the ML decoding algorithm in the following two situations:

1. The information bits are not equally likely.
2. Iterative decoding is used in the receiver, in which case the a priori probabilities of the message bits change from one iteration to the next; such a situation arises in turbo decoding, which is discussed in Section 10.12.

Applications of the Two Decoding Algorithms

The ML decoding algorithm is applied to convolutional codes in Section 10.8; in so doing, we are, in effect, opting for a simple approach to decode convolutional codes. This simple approach is also applicable to another class of codes, called trellis-coded modulation, which is discussed in Section 10.15.

Then, in Section 10.9 we move on to study the MAP decoding algorithm; the length of that section and the illustrative example in Section 10.10 are testimony to the complexity of this second approach to decoding convolutional codes. Equipped with the MAP algorithm and its modified forms, Section 10.12 and 10.13 discuss their application to turbo codes. It is in the material covered in those two sections that we find the practical benefits of feedback in decoding turbo codes.

10.8 Maximum Likelihood Decoding of Convolutional Codes

We begin the discussion of decoding convolutional codes by first describing the underlying theory of maximum likelihood decoding. The description is best understood by focusing on a trellis that represents each time step in the decoding process with a separate state graph.

Let $\mathbf{m}$ denote a *message vector* and $\mathbf{c}$ denote the corresponding *code vector* applied by the encoder to the input of a discrete memoryless channel. Let $\mathbf{r}$ denote the *received vector*, which, in practice, will invariably differ from the transmitted code vector $\mathbf{c}$ due to additive channel noise. Given the received vector $\mathbf{r}$, the decoder is required to make an *estimate* $\hat{\mathbf{m}}$ of the message vector $\mathbf{m}$. Since there is a one-to-one correspondence between the message vector $\mathbf{m}$ and the code vector $\mathbf{c}$, the decoder may equivalently produce an estimate $\hat{\mathbf{c}}$ of the code vector. We may then put

$$\hat{\mathbf{m}} = \mathbf{m} \text{ if and only if } \hat{\mathbf{c}} = \mathbf{c}$$

Otherwise, a *decoding error* is committed in the receiver. The *decoding rule* for choosing the estimate $\hat{\mathbf{c}}$, given the received vector $\mathbf{r}$, is said to be optimum when the *probability of decoding error* is minimized. In light of the material presented on signaling over AWGN channel in Chapter 7, we may state:

For equiprobable messages, the probability of decoding error is minimized if the estimate $\hat{\mathbf{c}}$ is chosen to maximize the log-likelihood function.

Let $\mathbb{P}(\mathbf{r}|\mathbf{c})$ denote the conditional probability of receiving $\mathbf{r}$, given that $\mathbf{c}$ was sent. The log-likelihood function equals $\ln\mathbb{P}(\mathbf{r}|\mathbf{c})$, where ln denotes the natural logarithm. The *maximum likelihood decoder* for decision making is described as follows:

Choose the estimate $\hat{\mathbf{c}}$ for which the log-likelihood function $\ln\mathbb{P}(\mathbf{r}|\mathbf{c})$ is maximum.

Consider next the special case of a binary symmetric channel. In this case, both the transmitted code vector $\mathbf{c}$ and the received vector $\mathbf{r}$ represent binary sequences of some length N. Naturally, these two sequences may differ from each other in some locations because of errors due to channel noise. Let c_i and r_i denote the ith elements of $\mathbf{c}$ and $\mathbf{r}$, respectively. We then have

$$\mathbb{P}(\mathbf{r}|\mathbf{c}) = \prod_{i=1}^{N} p(r_i|c_i)$$

Correspondingly, the log-likelihood function is

$$\ln\mathbb{P}(\mathbf{r}|\mathbf{c}) = \sum_{i=1}^{N} \ln p(r_i|c_i) \tag{10.55}$$

The term $p(r_i|c_i)$ in (10.55) denotes a *transition probability*, which is defined by

$$p(r_i|c_i) = \begin{cases} p, & \text{if } r_i \neq c_i \\ 1-p, & \text{if } r_i = c_i \end{cases} \tag{10.56}$$

Suppose also that the received vector $\mathbf{r}$ differs from the transmitted code vector $\mathbf{c}$ in exactly d places in the codeword, By definition, the number d is the *Hamming distance* between the vectors $\mathbf{r}$ and $\mathbf{c}$. Hence, we may rewrite the log-likelihood function in (10.55) as follows:

$$\begin{aligned} \ln p(\mathbf{r}|\mathbf{c}) &= d\ln p + (N-d)\ln(1-p) \\ &= d\ln\left(\frac{p}{1-p}\right) + N\ln(1-p) \end{aligned} \tag{10.57}$$

In general, the probability of an error occurring is low enough for us to assume $p < 1/2$. We also recognize that $N\ln(1-p)$ is a constant for all $\mathbf{c}$. Accordingly, we may restate the maximum-likelihood decoding rule for the binary symmetric channel as follows:

Choose the estimate $\hat{\mathbf{c}}$ that minimizes the Hamming distance between the received vector $\mathbf{r}$ and the transmitted vector $\mathbf{c}$.

That is, for the binary symmetric channel, the maximum-likelihood decoder for a convolutional code reduces to a *minimum distance decoder.* In such a decoder, the received vector $\mathbf{r}$ is compared with each possible transmitted code vector $\mathbf{c}$, and the particular one closest to $\mathbf{r}$ is chosen as the correct transmitted code vector. The term "closest" is used in the sense of minimum number of differing binary symbols (i.e., Hamming distance) between the code and received vectors under investigation.

The Viterbi Algorithm

The equivalence between maximum likelihood decoding and minimum distance decoding for the binary symmetric channel implies that we may decode a convolutional code by choosing a path in the code tree whose coded sequence differs from the received sequence in the fewest number of places. Since a code tree is equivalent to a trellis, we may equally limit our choice to the possible paths in the trellis representation of the code. The reason for preferring the trellis over the tree is that the number of nodes at each time instant does not continue to grow as the number of incoming message bits increases; rather, it remains constant at $2^{\nu-1}$, where ν is the constraint length of the code.

Consider, for example, the trellis diagram of Figure 10.15 for a convolutional code with rate $r = 1/2$ and constraint length $\nu = 3$. We observe that, at time-unit $j = 3$, there are two paths entering any of the four nodes in the trellis. Moreover, these two paths will be identical onward from that point. Clearly, a minimum distance decoder may make a decision at that point as to which of those two paths to retain, without any loss of performance. A similar decision may be made at time-unit $j = 4$, and so on. This sequence of decisions is exactly what the *Viterbi algorithm*[7] does as it walks through the trellis. The algorithm operates by computing a *metric* (i.e., discrepancy) for every possible path in the trellis; hence the following statement:

> The metric for a particular path is defined as the Hamming distance between the coded sequence represented by that path and the received sequence.

Thus, for each node (state) in the trellis of Figure 10.15 the algorithm compares the two paths entering the node. The path with the lower metric is retained and the other path is discarded. This computation is repeated for every time-unit j of the trellis in the range $M \le j \le L$, where $M = \nu - 1$ is the encoder's memory and L is the length of the incoming message sequence. The paths that are retained by the algorithm are called *survivor* or *active paths*. For a convolutional code of constraint length $\nu = 3$, for example, no more than $2^{\nu-1} = 4$ survivors and their metrics will ever be stored. The list of $2^{\nu-1}$ paths computed in the manner just described is always guaranteed to contain the maximum-likelihood choice.

A difficulty that may arise in the application of the Viterbi algorithm is the possibility that when the paths entering a state are compared, their metrics are found to be identical. In such a situation, we simply make the choice by flipping a fair coin (i.e., simply make a random guess).

To sum up:

> The Viterbi algorithm is a maximum-likelihood decoder, which is optimum for an AWGN channel as well as a binary symmetric channel.

The algorithm proceeds in a step-by-step fashion, as summarized in Table 10.6.

Table 10.6 **Summary of the Viterbi algorithm**

The Viterbi algorithm is a maximum likelihood decoder, which is optimal for any discrete memoryless channel. It proceeds in three basic steps. In computational terms, the so-called *add–compare–select (ACS) operation* in Step 2 is at the heart of the Viterbi algorithm.

Initialization

Set the all-zero state of the trellis to zero.

Computation Step 1: time-unit j

Start the computation at some time-unit j and determine the metric for the path that enters each state of the trellis. Hence, identify the survivor and store the metric for each one of the states.

Computation Step 2: time-unit $j + 1$

For the next time-unit $j + 1$, determine the metrics for all $2^{\nu-1}$ paths that enter a state where ν is the constraint length of the convolutional encoder; hence do the following:

a. *Add* the metrics entering the state to the metric of the survivor at the preceding time-unit j;
b. *Compare* the metrics of all 2^{ν} paths entering the state;
c. *Select* the survivor with the largest metric, store it along with its metric, and discard all other paths in the trellis.

Computation Step 3: continuation of the search to convergence

Repeat Step 2 for time-unit $j < L + L'$, where L is the length of the message sequence and L' is the length of the termination sequence.

Stop the computation once the time-unit $j = L + L'$ is reached.

EXAMPLE 5

Correct Decoding of Received All-Zero Sequence

Suppose that the encoder of Figure 10.13 generates an all-zero sequence that is sent over a binary symmetric channel and that the received sequence is (0100010000 ...). There are two errors in the received sequence due to noise in the channel: one in the second bit and the other in the sixth bit. We wish to show that this double-error pattern is correctable through the application of the Viterbi decoding algorithm.

In Figure 10.18 we show the results of applying the algorithm for time-unit $j = 1, 2, 3, 4, 5$. We see that for $j = 2$ there are (for the first time) four paths, one for each of the four states of the encoder. The figure also includes the metric of each path for each level in the computation.

In the left side of Figure 10.18, for time-unit $j = 3$ we show the paths entering each of the states, together with their individual metrics. In the right side of the figure we show the four survivors that result from application of the algorithm for time-unit $j = 3, 4, 5$. Examining the four survivors in the figure for $j = 5$, we see that the all-zero path has the smallest metric and will remain the path of smallest metric from this point forward. This clearly shows that the all-zero sequence is indeed the maximum likelihood choice of the Viterbi decoding algorithm, which agrees exactly with the transmitted sequence.

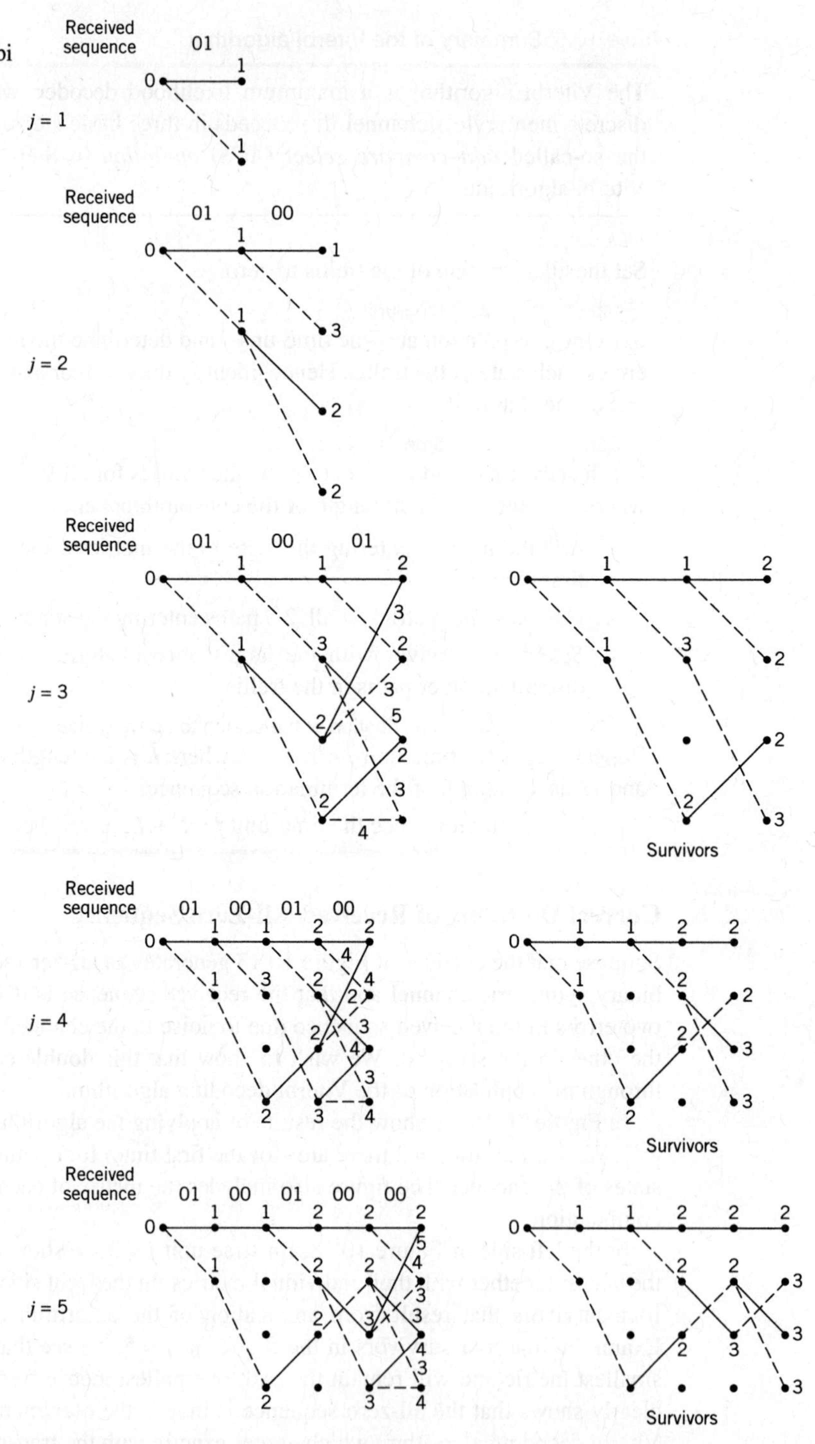

Figure 10.18
Illustrating steps in the Viterbi algorithm for Example 5.

EXAMPLE 6

Incorrect Decoding of Received All-Zero Sequence

Suppose next that the received sequence is (1100010000 …), which contains three errors compared with the transmitted all-zero sequence; two of the errors are adjacent to each other and the third is some distance away.

In Figure 10.19, we show the results of applying the Viterbi decoding algorithm for levels $j = 1, 2, 3, 4$. We see that in this second example on Viterbi decoding the correct path has been eliminated by time-unit $j = 3$. Clearly, a triple-error pattern is uncorrectable by the Viterbi algorithm when applied to a convolutional code of rate 1/2 and constraint length $\nu = 3$. The exception to this algorithm is a triple-error pattern spread over a time span longer than one constraint length, in which case it is likely to be correctable.

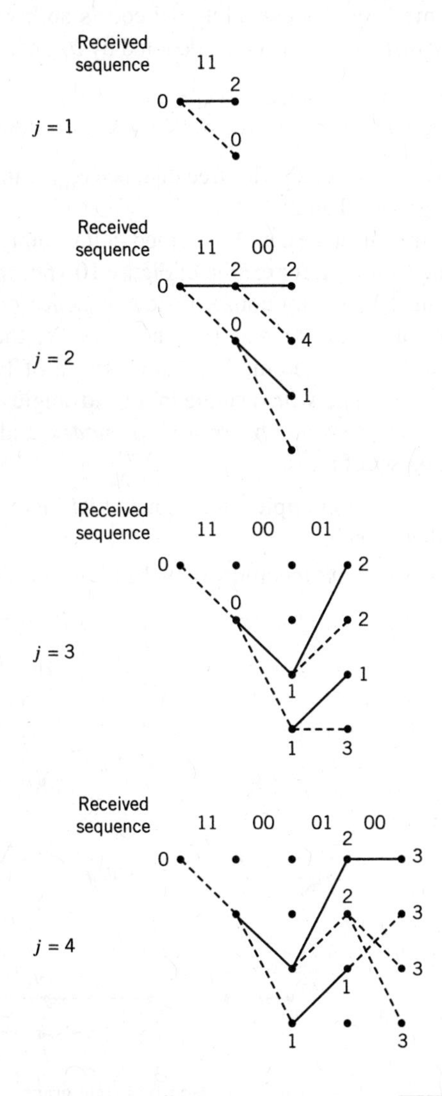

Figure 10.19 Illustrating breakdown of the Viterbi algorithm in Example 6.

What Have We Learned from Examples 5 and 6?

In Example 5 there were two errors in the received sequence, whereas in Example 6 there were three errors, two of which were in adjacent symbols and the third one was some distance away. In both examples the encoder used to generate the transmitted sequence was the same. The difference between the two examples was attributed to the fact that the number of errors in Example 6 was beyond the error-correcting capability of the maximum likelihood decoding algorithm, which is the next topic for discussion.

Free Distance of a Convolutional Code

The performance of a convolutional code depends not only on the decoding algorithm used but also on the distance properties of the code. In this context, the most important single measure of a convolutional code's ability to combat errors due to channel noise is the *free distance* of the code, denoted by d_{free}; it is defined as follows:

> The free distance of a convolutional code is given by the minimum Hamming distance between any two codewords in the code.

A convolutional code with free distance d_{free} can, therefore, correct t errors if, and only if, d_{free} is greater than $2t$.

The free distance can be obtained quite simply from the state graph of the convolutional encoder. Consider, for example, Figure 10.16b, which shows the state graph of the encoder of Figure 10.13. Any nonzero code sequence corresponds to a complete path beginning and ending at the 00 state (i.e., node a). We thus find it useful to split this node in the manner shown in the modified state graph of Figure 10.20, which may be viewed as a *signal-flow graph* with a single input and single output.

A signal-flow graph consists of *nodes* and directed *branches*; it operates by the following set of rules:

1. A branch multiplies the signal at its input node by the *transmittance* characterizing that branch.
2. A node with incoming branches *sums* the signals produced by all of those branches.

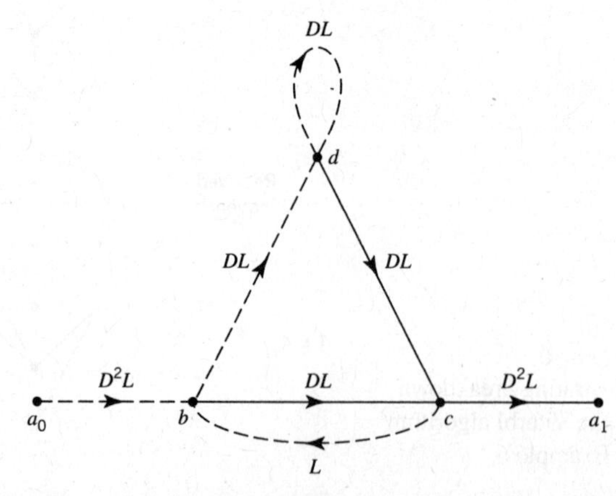

Figure 10.20 Modified state graph of convolutional encoder.

3. The signal at a node is applied equally to all the branches outgoing from that node.
4. The *transfer function* of the graph is the ratio of the output signal to the input signal.

Returning to the signal-flow graph of Figure 10.20, the exponent of D on a branch in this graph describes the Hamming weight of the encoder output corresponding to that branch; the symbol D used here should not be confused with the unit-delay variable in Section 10.6 and the symbol L used herein should not be confused with the length of the message sequence. The exponent of L is always equal to one, since the length of each branch is one. Let $T(D,L)$ denote the *transfer function of the signal-flow graph*, with D and L playing the role of dummy variables. For the example of Figure 10.20, we may readily use rules 1, 2, and 3 to obtain the following input-output relations:

$$\left.\begin{aligned} b &= D^2 L a_0 + Lc \\ c &= DLb + DLd \\ d &= DLb + DLd \\ a_1 &= D^2 Lc \end{aligned}\right\} \tag{10.58}$$

where a_0, b, c, d, and a_1 denote the node signals of the graph. Solving the system of four equations in (10.58) for the ratio a_1/a_0, we obtain the transfer function

$$T(D, L) = \frac{D^5 L^3}{1 - DL(1 + L)} \tag{10.59}$$

Using the binomial expansion, we may equivalently express $T(D,L)$ as follows:

$$\begin{aligned} T(D, L) &= D^5 L^3 (1 - DL(1 + L))^{-1} \\ &= D^5 L^3 \sum_{i=0}^{\infty} (DL(1 + L))^i \end{aligned}$$

Setting $L = 1$ in this formula, we thus get the *distance transfer function* expressed in the form of a power series as follows:

$$T(D, 1) = D^5 + 2D^6 + 4D^7 + \cdots \tag{10.60}$$

Since the free distance is the minimum Hamming distance between any two codewords in the code and the distance transfer function $T(D,1)$ enumerates the number of codewords that are a given distance apart, it follows that the exponent of the first term in the expansion of $T(D,1)$ in (10.60) defines the free distance. Thus, on the basis of this equation, the convolutional code of Figure 10.13 has the free distance $d_{\text{free}} = 5$.

This result indicates that up to two errors in the received sequence are correctable, as two or fewer transmission errors will cause the received sequence to be at most at a Hamming distance of 2 from the transmitted sequence but at least at a Hamming distance of 3 from any other code sequence in the code. In other words, in spite of the presence of any pair of transmission errors, the received sequence remains closer to the transmitted sequence than any other possible code sequence. However, this statement is no longer true if there are three or more *closely spaced* transmission errors in the received sequence. The observations made here reconfirm the results reported earlier in Examples 5 and 6.

Asymptotic Coding Gain

The transfer function of the encoder's state graph, modified in a manner similar to that illustrated in Figure 10.20, may be used to evaluate a *bound on the BER* for a given decoding scheme; details of this evaluation are, however, beyond the scope of our present discussion.[8] Here, we simply summarize the results for two special channels, namely the binary symmetric channel and the binary-input AWGN channel, assuming the use of binary PSK with coherent detection.

1. *Binary symmetric channel.*

   The binary symmetric channel may be modeled as an AWGN channel with binary PSK as the modulation in the transmitter followed by hard-decision demodulation in the receiver. The transition probability p of the binary symmetric channel is then equal to the BER for the uncoded binary PSK system. From Chapter 7 we recall that for large values of E_b/N_0, denoting the ratio of signal energy per bit-to-noise power spectral density, the BER for binary PSK without coding is dominated by the exponential factor $\exp(-E_b/N_0)$. On the other hand, the BER for the same modulation scheme with convolutional coding is dominated by the exponential factor $\exp(-d_{\text{free}} r E_b/2N_0)$, where r is the code rate and d_{free} is the free distance of the convolutional code. Therefore, as a *figure of merit* for measuring the improvement in error performance made by the use of coding with hard-decision decoding, we may set aside the E_b/N_0 to use the remaining exponent to define the *asymptotic coding gain* (in decibels) as follows:

$$G_a = 10\log_{10}\left(\frac{d_{\text{free}} r}{2}\right) \text{ dB} \tag{10.61}$$

2. *Binary-input AWGN channel.*

   Consider next the case of a memoryless binary-input AWGN channel with no output quantization (i.e., the output amplitude lies in the interval $(-\infty, \infty)$). For this channel, theory shows that for large values of E_b/N_0 the BER for binary PSK with convolutional coding is dominated by the exponential factor $\exp(-d_{\text{free}} r E_b/N_0)$, where the parameters are as previously defined. Accordingly, in this second case, we find that the asymptotic coding gain is defined by

$$G_a = 10\log_{10}(d_{\text{free}} r) \text{ dB} \tag{10.62}$$

Comparing (10.61) and (10.62) for cases 1 and 2, respectively, we see that the asymptotic coding gain for the binary-input AWGN channel is greater than that for the binary symmetric channel by 3 dB. In other words, for large E_b/N_0, the transmitter for a binary symmetric channel must generate an additional 3 dB of signal energy (or power) over that for a binary-input AWGN channel if we are to achieve the same error performance. Clearly, there is an advantage to be gained by using an unquantized demodulator output in place of making hard decisions. This improvement in performance, however, is attained at the cost of increased decoder complexity due to the requirement for accepting analog inputs.

It turns out that the asymptotic coding gain for a binary-input AWGN channel is approximated to within about 0.25 dB by a binary input Q-ary output discrete memoryless channel with the number of representation levels $Q = 8$. This means that, for practical purposes, we may avoid the need for an analog decoder by using a *soft-decision decoder* that performs finite output quantization (typically, $Q = 8$), and yet realize a performance close to the optimum.

Practical Limitations of the Viterbi Algorithm

When the received sequence is very long, the storage requirement of the Viterbi algorithm becomes too high, in which case some compromises must be made. The approach usually taken in practice is to "truncate" the path memory of the decoder as follows:

> A decoding window of length l is specified and the algorithm operates on a corresponding frame of the received sequence, always stopping after l steps. A decision is then made on the "best" path and the symbol associated with the first branch on that path is released to the user. The symbol associated with the last branch of the path is dropped. Next, the decoding window is moved forward one time interval. A decision on the next code frame is made, and the process is repeated.

Naturally, decoding decisions made in the way just described are no longer truly maximum likelihood, but they can be made almost as good provided that the decoding window is chosen long enough. Experience and analysis have shown that satisfactory results are obtained if the decoding window length l is on the order of five times the constraint length ν of the convolutional code or more.

10.9 Maximum a Posteriori Probability Decoding of Convolutional Codes

Summarizing the discussion on convolutional decoding presented in Section 10.8, we may say that, given a received vector $\mathbf{r}$ that is the noisy version of a convolutionally encoded vector $\mathbf{c}$, the Viterbi algorithm computes the code vector $\hat{\mathbf{c}}$ for which the log-likelihood function is maximum; for a binary symmetric channel, the code vector $\hat{\mathbf{c}}$ minimizes the Hamming distance between the received vector $\mathbf{r}$ and the transmitted vector $\mathbf{c}$. For the more general case of an AWGN channel, this result is equivalent to finding the vector $\hat{\mathbf{c}}$ that is the closest to the received vector $\mathbf{r}$ in Euclidean distance. Simply put then: given the vector $\mathbf{r}$, the Viterbi algorithm finds the most likely vector $\hat{\mathbf{c}}$ that minimizes the conditional probability $\mathbb{P}(\hat{\mathbf{c}} \neq \mathbf{c}|\mathbf{r})$, which is the *sequence error* or the *word error rate*.

In practice, however, we are often interested in the BER, defined as the conditional probability $\mathbb{P}(\hat{m}_i \neq m_i|\mathbf{r})$, where m_i is an estimate of the ith bit of message vector $\hat{\mathbf{m}}$. Recognizing the fact that the BER can indeed assume a value different from the sequence error, we need a probabilistic decoding algorithm that minimizes the BER.

Bahl, Cocke, Jelinek, and Raviv (1974) are credited for deriving an algorithm that maximizes the a posteriori probabilities of the states in the decoding model as well as the transition probability from one state to another. In the course of time, this decoding algorithm has become known as the *BCJR algorithm* in honor of its four inventors. The BCJR algorithm is applicable to any linear code, be it of a block or convolutional kind. However, as we may well expect, computational complexity of the BCJR algorithm is greater than that of the Viterbi algorithm. But, when the message bits in the received vector **r** are equally likely, the Viterbi algorithm is preferred over the BCJR algorithm. When, however, the message bits are not equally likely, then the BCJR algorithm provides a better decoding performance than the Viterbi algorithm. Moreover, in iterative decoding exemplified by turbo decoding (to be discussed in Section 10.12), the a priori probabilities of the message bits may change from one iteration to the next; in such a scenario, the BCJR algorithm provides the best performance.

Henceforth, the two terminologies, BCJR algorithm and maximum a posteriori probability (MAP) decoding algorithm, are used interchangeably.

The MAP Decoding Algorithm

The function of the MAP decoder is to compute the values of *log-a-posteriori ratios*, on the basis of which estimates of the original message bits are computed in the receiver. In what follows, we derive the *MAP decoding algorithm* for the case of rate $= 1/n$ convolutional codes applied to a binary input–continuous output AWGN channel.[9]

Henceforth, in this section, we use the mapping of bits 0 and 1 as follows:

$$\begin{aligned} &\text{bit } 0 \rightarrow \text{level } -1 \\ &\text{bit } 1 \rightarrow \text{level } +1 \end{aligned}$$

Thus, given a message sequence of block length L, we express the message vector **m** as follows:

$$\mathbf{m} = (m_0, m_1, \ldots, m_{L-1})$$

where

$$m_j = \pm 1 \qquad \text{for } j = 0, 1, \ldots, L-1$$

The individual elements in the message vector **m** are referred to as *message bits*. In any event, the vector **m** is encoded into the codeword **c**, which, in turn, produces the noisy received signal vector **r** at the channel output. Note, however, the elements of the vector **r** can assume positive as well as negative values, which, in theory, can be infinitely large due to the analog nature of the additive channel noise.

Before proceeding further, there are two natural logarithmic concepts, namely log-likelihood ratios, that will occupy our attention in deriving the MAP decoding algorithm:

1. *A priori L-values*, denoted by $L_a(m_j)$, which define the natural logarithmic ratio of a priori probabilities of message bits, $m_j = -1$ and $m_j = +1$, generated by a source at the encoder input in the transmitter.
2. *A posteriori L-values*, denoted by $L_p(m_j)$, which define the log-likelihood ratio of the conditional a posteriori probabilities of the message bits $m_j = -1$ and $m_j = +1$, given the channel output at the decoder input in the receiver.

In what follows, we will focus on $L_p(m_j)$ first, deferring the discussion of $L_a(m_j)$ until later in this section.

With the message $m_j = \pm 1$, there are two conditional probabilities to be considered: $\mathbb{P}(m_j = +1|\mathbf{r})$ and $\mathbb{P}(m_j = -1|\mathbf{r})$. These two probabilities are called the *a posteriori probabilities (APPs)*. In terms of these two APPs, the *log-a-posteriori L-value* is defined by

$$L_p(m_j) = \ln\left(\frac{\mathbb{P}(m_j = +1|\mathbf{r})}{\mathbb{P}(m_j = -1|\mathbf{r})}\right) \tag{10.63}$$

Hereafter, for the sake of brevity, we refer to the $L_{\mathrm{p}}(m_j)$ simply as the *a posteriori L-value* of message bit m_j at time-unit j. Having computed a set of L_{p}-values, the decoder makes a *hard decision* by applying the two-part formula:

$$\hat{m}_j = \begin{cases} +1 \text{ if } L_{\mathrm{p}}(m_j) > 0, \\ -1 \text{ if } L_{\mathrm{p}}(m_j) < 0, \end{cases} \qquad j = 0, 1, \ldots, L-1 \tag{10.64}$$

where L is the length of the message sequence; L must not be confused with the two L-values, $L_{\mathrm{a}}(m_j)$ and $L_{\mathrm{p}}(m_j)$.

Given the received vector $\mathbf{r}$, the conditional probability $\mathbb{P}(m_j = +1|\mathbf{r})$ is expressed in terms of the joint probability density function $f(m_j = +1, \mathbf{r})$ as follows:

$$\mathbb{P}(m_j = +1|\mathbf{r}) = \frac{f(m_j = +1|\mathbf{r})}{f(\mathbf{r})}$$

where $f(\mathbf{r})$ is the probability density function of the received vector $\mathbf{r}$; this formula follows from the definition of joint probability.

Similarly, we may express the second conditional probability $\mathbb{P}(m_j = -1|\mathbf{r})$ as follows:

$$\mathbb{P}(m_j = -1|\mathbf{r}) = \frac{f(m_j = -1|\mathbf{r})}{f(\mathbf{r})}$$

Accordingly, using these two conditional properties and canceling the common term $f(\mathbf{r})$, we may reformulate the a posteriori L-values of (10.63) in the equivalent form

$$L_{\mathrm{p}}(m_j) = \ln\left(\frac{f(m_j = +1|\mathbf{r})}{f(m_j = -1|\mathbf{r})}\right) \tag{10.65}$$

which sets the stage for deriving the MAP decoding algorithm.

Lattice-based Framework for the Derivation

With computational complexity being at a premium, we propose to exploit the lattice structure of the convolutional code as the basis for deriving the MAP decoding algorithm. To this end, let Σ_j^+ denote the set of all *state-pairs* for which the states $s_j = s'$ and $s_{j+1} = s$ correspond to message bit $m_j = +1$. We may then express the conditional probability density function $f(m_j = +1|\mathbf{r})$ in the expanded form:

$$f(m_j = +1|\mathbf{r}) \propto \sum_{(s', s) \in \Sigma_j^+} f(s_j = s', s_{j+1} = s, \mathbf{r}) \tag{10.66}$$

where the symbol $\propto$ stands for proportionality. In a similar way, we may reformulate the other conditional probability density function as follows:

$$f(m_j = -1|\mathbf{r}) \propto \sum_{(s', s) \in \Sigma_j^-} f(s_j = +1, s_j = -1, \mathbf{r}) \tag{10.67}$$

where Σ_j^- is the set of all state-pairs for which the state-pair $s_j = s'$ and $s_j + 1 = s$ corresponds to the message bit $m_j = -1$. Hence, substituting (10.66) and (10.67) into (10.65) and recognizing that the proportionality factor is common to both (10.66) and

(10.67), thereby canceling out, the a posteriori L_p-value of message bit m_j at time-unit j takes the following equivalent form:

$$L_p(m_j) = \ln\left(\frac{\sum\limits_{(s', s) \in \Sigma_j^+} f(s_j = s', s_{j+1} = s, \mathbf{r})}{\sum\limits_{(s', s) \in \Sigma_j^-} f(s_j = s', s_{j+1} = s, \mathbf{r})}\right) \tag{10.68}$$

Equation (10.68) provides the mathematical basis for forward–backward computation of the MAP decoding algorithm. In this context, it is important to note the following point in (10.68):

> Every branch in the trellis, connecting a state at time-unit j to a state at the next time-unit $j + 1$, is always in one of the two summation terms in (10.68).

Forward–Backward Recursions: Background Terminology and Assumptions

Our next task is to show how the pair of joint probability density functions in (10.68) can be computed *recursively*, using forward and backward recursions.

With this important point in mind, we introduce some new and relevant terminology. First, we express the received vector $\mathbf{r}$ as the *triplet*

$$\mathbf{r} = (\mathbf{r}_{t>j}, \mathbf{r}_j, \mathbf{r}_{t<j})$$

where the two new terms $\mathbf{r}_{t<j}$ and $\mathbf{r}_{t>j}$ denote those portions of the received vector $\mathbf{r}$ that appear *before* and *after* time-unit j, respectively. Moreover, we simplify the notation by using s' and s in place of $s_j = s'$ and $s_{j+1} = s$, respectively, recognizing that the time-unit j is implicitly contained in the $L_p(m_j)$.

In particular, the joint probability density function common to the numerator and denominator in (10.68) is now rewritten as

$$f(s_j = s', s_{j+1} = s, \mathbf{r}) = f(s', s, \mathbf{r}_{t>j}, \mathbf{r}_j, \mathbf{r}_{t<j}) \tag{10.69}$$

Moreover, before proceeding further, we find it instructive to introduce two assumptions that are basic to derivation of the MAP decoding algorithm:

1. *Markovian Assumption*

   > In a convolutional code represented by a trellis, the present state of the encoder depends only on two entities: the immediate past state and the input message bit.

   Under this assumption, convolutional encoding of the message vector performed in the transmitter is said to be a *Markov chain*.

2. *Memoryless Assumption*

   > The channel connecting the receiver to the transmitter is memoryless.

   In other words, the channel has no knowledge of the past.

Resuming the discussion on the log a posteriori L-value, $L_{\mathrm{p}}(m_j)$ in (10.68), we use the definition of joint probability density function to express the right-hand side of (10.69) as follows:

$$f(s', s, \mathbf{r}_{t>j}, \mathbf{r}_j, \mathbf{r}_{t<j}) = f(\mathbf{r}_{t>j}|s', s, \mathbf{r}_j, \mathbf{r}_{t<j})f(s', s, \mathbf{r}_j, \mathbf{r}_{t<j})$$

Focusing on the conditional probability density function on the right-hand side of this equality, we invoke the Markovian assumption to recognize that the vector $\mathbf{r}_{t>j}$ representing the received vector $\mathbf{r}$ after time-unit j subsumes knowledge of the following three entities:

- the state $s' = s_j$,
- the vector $\mathbf{r}_j$ at time-unit j, and
- the vector $\mathbf{r}_{j<t}$ received before time-unit j.

Accordingly, we may simplify matters by writing

$$f(\mathbf{r}_{t>j}|s', s, \mathbf{r}_j, \mathbf{r}_{t<j}) = f(\mathbf{r}_{t>j}|s) \tag{10.70}$$

where s denotes the state s_{j+1}.

Next, we again use the definition of joint probability density function to write

$$f(s', s, \mathbf{r}_j, \mathbf{r}_{t<j}) = f(s, \mathbf{r}_j|s', \mathbf{r}_{t<j})\ f(s', \mathbf{r}_{t<j})$$

Focusing on the second conditional probability density function $f(s, \mathbf{r}_j|s', \mathbf{r}_{t<j})$ and invoking the Markovian assumption one more time, we recognize that the received vector $\mathbf{r}_j$ at time-unit j subsumes knowledge of the past vector $\mathbf{r}_{t<j}$. Hence, we may further simplify matters by writing

$$f(s, \mathbf{r}_j|s', \mathbf{r}_{t<j}) = f(s, \mathbf{r}_j|s') \tag{10.71}$$

where the states $s = s_{j+1}$ and $s' = s_j$.

Collecting the results obtained in (10.70) and (10.71), we are finally ready to express the probability density function common to the numerator and denominator of (10.68) as follows:

$$f(s', s, \mathbf{r}) = f(\mathbf{r}_{t>j}|s)f(s, \mathbf{r}_j|s')f(s', \mathbf{r}_{t<j}) \tag{10.72}$$

which provides the mathematical basis for recursive implementation of the MAP decoding algorithm.

Three New Algorithmic Metrics

To simplify the computational steps involved in deriving the algorithm, we now introduce the following three *algorithmic metrics:*

$$\alpha_j(s') = f(s', \mathbf{r}_{t<j}) \tag{10.73}$$

$$\gamma_j(s', s) = f(s, \mathbf{r}_j|s') \tag{10.74}$$

$$\beta_{j+1}(s) = f(\mathbf{r}_{t>j}|s) \tag{10.75}$$

Using these three metrics, we may finally express the probability density function common to the numerator and denominator of (10.68) in the simplified form:

$$f(s', s, \mathbf{r}) = \beta_{j+1}(s)\gamma_j(s', s)\alpha_j(s') \tag{10.76}$$

in light of which, hereafter, the three metrics are referred to as follows:

forward metric	$\alpha_j(s')$
branch metric	$\gamma_j(s', s)$
backward metric	$\beta_{j+1}(s)$

As the names would imply, the forward and backward metrics play key roles in the forward and backward recursions of the MAP decoding algorithm, respectively. As for the branch metric, its role is to couple these two recursions to work together in a harmonious manner.

Forward Recursion

Updating the forward metric has the effect of moving from state s' at time-unit j to state s at time-unit $j + 1$; hence, we write

$$\begin{aligned}\alpha_{j+1}(s) &= f(s, \mathbf{r}_{t<j+1}) \\ &= \sum_{s' \in \sigma_j} f(s', s, \mathbf{r}_{t<j+1})\end{aligned}$$

where σ_j is the set of all the states at time-unit j. Using the definition of a joint probability density function, we write

$$\begin{aligned}\alpha_{j+1}(s) &= \sum_{s' \in \sigma_j} f(s, \mathbf{r}_j|s', \mathbf{r}_{t<j})f(s', \mathbf{r}_{t<j}) \\ &= \sum_{s' \in \sigma_j} f(s, \mathbf{r}_j|s')f(s', \mathbf{r}_{t<j})\end{aligned}$$

where, in the second line, we used the Markovian assumption for $\mathbf{r}_j$ subsuming $\mathbf{r}_{t<j}$. Hence, using the defining equations for the branch and forward metrics in (10.74) and (10.73), respectively, we simplify matters by writing

$$\alpha_{j+1}(s) = \sum_{s' \in \sigma_j} \gamma_j(s', s)\alpha_j(s') \tag{10.77}$$

For obvious reasons, (10.77) is called the *forward recursion*; this recursion is illustrated graphically in Figure 10.21a.

Backward Recursion

To formulate the recursion for the backward metric, we move from state s at time-unit j + 1 back to state s' at time-unit j. Adapting the use of (10.75) to the scenario just described, we write

$$\beta_j(s') = f(\mathbf{r}_{t>j-1}|s')$$

The portion of received vector denoted by $\mathbf{r}_{t>j-1}$ may be equivalently expressed as follows:

$$\begin{aligned}\mathbf{r}_{t>j-1} &= \mathbf{r}_{t+1>j} \\ &= (\mathbf{r}_j, \mathbf{r}_{t>j})\end{aligned}$$

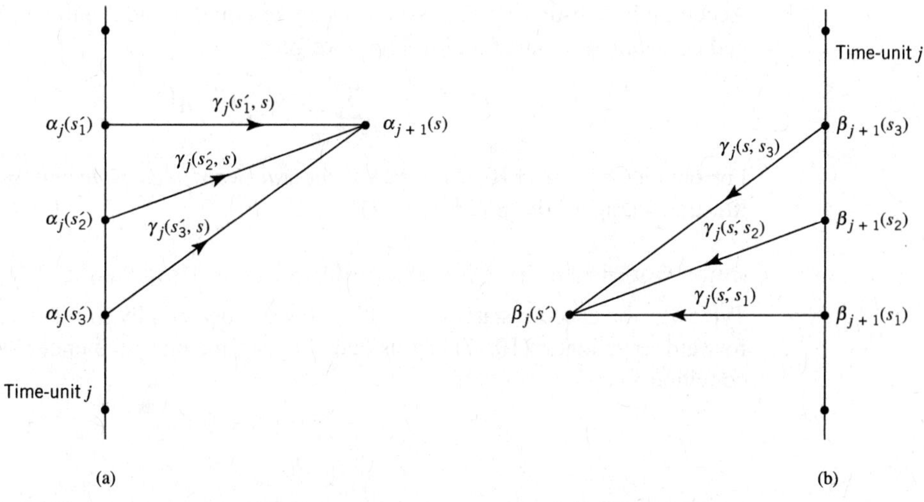

Figure 10.21 Illustrating the computation of forward-metric and backward-metric recursions.

Correspondingly, the backward metric $\beta_j(s')$ is reformulated as shown by

$$\begin{aligned}\beta_j(s') &= f(\mathbf{r}_j, \mathbf{r}_{t>j}|s') \\ &= \sum_{s \in \sigma_{j+1}} f(s, \mathbf{r}_j, \mathbf{r}_{t>j}|s')\end{aligned}$$

where σ_{j+1} is the set of all states at time-unit j. Here again, using the definition of joint probability density function, we write

$$\begin{aligned}\beta_j(s') &= \sum_{s \in \sigma_{j+1}} f(s, \mathbf{r}_j, \mathbf{r}_{t>j}|s') \\ &= \sum_{s \in \sigma_{j+1}} \frac{1}{\mathbb{P}(s')} f(s', s, \mathbf{r}_j, \mathbf{r}_{t>j}) \\ &= \sum_{s \in \sigma_{j+1}} \frac{1}{\mathbb{P}(s')} f(\mathbf{r}_{t>j}|s', s, \mathbf{r}_j) f(s', s, \mathbf{r}_j)\end{aligned}$$

To simplify matters, we note the following two points:

1. Under the memoryless assumption, the received vector $\mathbf{r}_{t>j}$ at the channel output depends only on the state in which the encoder was residing at $j - 1$, namely s. We may, therefore, write

$$\begin{aligned}f(\mathbf{r}_{t>j}|s', s, \mathbf{r}_j) &= f(\mathbf{r}_{t>j}|s) \\ &= \beta_{j+1}(s)\end{aligned}$$

2. Invoking the definition of joint probability density function one more time, we have

$$\begin{aligned}f(s', s, \mathbf{r}_j) &= f(s, \mathbf{r}_j|s')\mathbb{P}(s') \\ &= \gamma_j(s', s)\mathbb{P}(s')\end{aligned}$$

Accordingly, substituting the two results under points 1 and 2 into the formula for $\beta_j(s')$ and canceling the common term $\mathbb{P}(s')$ we get

$$\beta_j(s') = \sum_{s \in \sigma_{j+1}} \gamma_j(s', s)\beta_{j+1}(s) \tag{10.78}$$

For obvious reasons, (10.78) is called the *backward recursion*; this second recursion is illustrated graphically in Figure 10.21b.

Initial Conditions for Forward and Backward Recursions

Typically, the encoder starts in the all-zero state, denoted by $s_0 = \mathbf{0}$. Correspondingly, the forward recursion of (10.77) begins operating at time-unit $j = 0$ under the following initial condition:

$$\alpha_0(s) = \begin{cases} 1, & s = \mathbf{0} \\ 0, & s \neq \mathbf{0} \end{cases} \tag{10.79}$$

which follows from the fact that the convolutional encoder starts in the all-zero state. Thus, $\alpha_{j+1}(s)$ is recursively computed forward in time at $j = 0, 1, \ldots, K-1$, where the overall length of the input data stream is

$$K = L + L'$$

in which L and L' denote the lengths of the message and termination sequences.

Similarly, the backward recursion of (10.78) begins at time-unit $j = K$ under the following initial condition:

$$\beta_K(s) = \begin{cases} 1, & s = \mathbf{0} \\ 0, & s \neq \mathbf{0} \end{cases} \tag{10.80}$$

Since the encoder ends in the all-zero state, we recursively compute $\beta_j(s')$ backward in time at $j = K-1, K-2, \ldots, 0$.

Branch Metric Evaluation for the AWGN Channel

Thus far, we have accounted for all the issues important to the MAP decoder except for the discrete-input, continuous-output AWGN channel, which naturally comes into play in evaluating the branch metric: a necessary requirement. This issue was discussed in Example 10 in Chapter 5. For this evaluation, we first rewrite the defining equation (10.74) as follows:

$$\begin{aligned} \gamma_j(s', s) &= f(s, \mathbf{r}_j | s') \\ &= \frac{1}{\mathbb{P}(s')} f(s, s', \mathbf{r}_j) \\ &= \left(\frac{\mathbb{P}(s', s)}{\mathbb{P}(s')}\right) \cdot \left(\frac{f(s', s, \mathbf{r}_j)}{\mathbb{P}(s', s)}\right) \\ &= \mathbb{P}(s|s') f(\mathbf{r}_j | s', s) \end{aligned}$$

which may be transformed into a more desirable form that involves the message bit m_j and the corresponding code vector $\mathbf{c}_j$, as shown by

$$\gamma_j(s', s) = \mathbb{P}(m_j) f(\mathbf{r}_j | \mathbf{c}_j) \tag{10.81}$$

Justification for this transformation may be explained as follows:

1. The transition from the state $s' = s_j$ to the new state $s = s_{j+1}$ is attributed to the message bit inputing the convolutional encoder at time-unit j; hence, we may substitute the probability $\mathbb{P}(m_j)$ for the conditional probability $\mathbb{P}(s|s')$.
2. The state transition (s, s') may be viewed as another way of referring to the code vector $\mathbf{c}_j$; hence, we may substitute the conditional probability $\mathbb{P}(\mathbf{r}_j|\mathbf{c}_j)$ for $f(\mathbf{r}_j|s, s')$.

In (10.81), m_j is the message bit at the encoder's input and $\mathbf{c}_j$ is the code vector defining the encoded bits pertaining to the state transition $s' \to s$ at time-unit j. When this state transition is a valid one, the conditional probability density function $f(\mathbf{r}_j \mid \mathbf{c}_j)$, defining the input–output statistical behavior of the channel, assumes the following form:

$$f(\mathbf{r}_j|\mathbf{c}_j) = \left(\sqrt{\frac{E_s}{\pi N_0}}\right)^n \exp\left(-\frac{E_s}{N_0}\|\mathbf{r}_j - \mathbf{c}_j\|^2\right) \tag{10.82}$$

where E_s is transmitted energy per symbol, n is the number of bits in each codeword, $N_0/2$ is the power spectral density of the additive white Gaussian channel noise, and $\|\mathbf{r}_j - \mathbf{c}_j\|^2$ is the squared Euclidean distance between the transmitted vector $\mathbf{c}_j$ at the channel input and the received vector $\mathbf{r}_j$ at the channel output at time-unit j. Thus, substituting (10.82) into (10.81) yields

$$\gamma_j(s', s) = \mathbb{P}(m_j)\left(\sqrt{\frac{E_s}{\pi N_0}}\right)^n \exp\left(-\frac{E_s}{N_0}\|\mathbf{r}_j - \mathbf{c}_j\|^2\right) \tag{10.83}$$

This equation holds if, and only if, the state transition $s' \to s$ at time-unit j is a valid one; otherwise, the state-transition probability $p(s', s)$ is zero, in which case the branch metric $\gamma_j(s', s)$ is also zero.

A priori L-value, $L_a(m_j)$

At this point in the discussion, we are ready to revisit the a priori L-value $L_a(m_j)$, introduced previously on page 624. Specifically, with the message bit m_j taking the value +1 or −1, we may follow the format of (10.63) to define the *a priori L-value* of m_j as follows:

$$\begin{aligned} L_a(m_j) &= \ln\frac{\mathbb{P}(m_j = +1)}{\mathbb{P}(m_j = -1)} \\ &= \ln\left(\frac{\mathbb{P}(m_j = +1)}{1 - \mathbb{P}(m_j = +1)}\right) \end{aligned} \tag{10.84}$$

where, in the second line, we used the following axiom from probability theory:

$$\mathbb{P}(m_j = +1) + \mathbb{P}(m_j = -1) = 1$$

or, equivalently,

$$\mathbb{P}(m_j = -1) = 1 - \mathbb{P}(m_j = +1)$$

Solving the second line of (10.84) for $\mathbb{P}(m_j = +1)$ in terms of the *a priori L-value* $L_a(m_j)$, we get

$$\mathbb{P}(m_j = +1) = \frac{1}{1 + \exp(-L_a(m_j))}$$

Correspondingly,

$$\mathbb{P}(m_j = -1) = \frac{\exp(-L_a(m_j))}{1 + \exp(-L_a(m_j))}$$

This latter pair of equations for the two probabilities of $m_j = -1$ and $m_j = +1$ may be combined into a single equation, as shown by

$$\mathbb{P}(m_j) = \left(\frac{\exp(-L_a(m_j)/2)}{1 + \exp(-L_a(m_j))}\right)\exp\left(\frac{1}{2}m_j L_a(m_j)\right) \tag{10.85}$$

where $m_j = \pm 1$. The important point to note in (10.85) is that the first term on the right-hand side of the equation turns out to be independent of $m_j = \pm 1$; hence, this term may be treated as a constant.

Turning next to the exponential term in (10.83), we may express the exponent of the second term as follows:

$$\begin{aligned} -\frac{E_s}{N_0}\|\mathbf{r}_j - \mathbf{c}_j\|^2 &= -\frac{E_s}{N_0}\left[\sum_{l=1}^{n}(r_{jl} - c_{jl})^2\right] \\ &= -\frac{E_s}{N_0}\left[\sum_{l=1}^{n}(r_{jl}^2 - 2r_{jl}c_{jl} + c_{jl}^2)\right] \\ &= -\frac{E_s}{N_0}(\|\mathbf{r}_j\|^2 - 2\mathbf{r}_j^T\mathbf{c}_j + \|\mathbf{c}_j\|^2) \end{aligned} \tag{10.86}$$

where E_s is the transmitted symbol energy, and the terms inside the parentheses are

$$\|\mathbf{r}_j\|^2 = \sum_{l=1}^{n}(r_{jl})^2 \tag{10.87}$$

$$\mathbf{r}_j^T\mathbf{c}_j = \sum_{l=1}^{n} r_{jl}c_{jl} \tag{10.88}$$

$$\|\mathbf{c}_j\|^2 = \sum_{l=1}^{n}(c_{jl})^2 = n \tag{10.89}$$

The terms r_{jl} and c_{jl} denote the individual bits in the received vector $\mathbf{r}_j$ and code vector $\mathbf{c}_j$ at time-unit j, and n denotes the number of bits in each of $\mathbf{r}_j$ and $\mathbf{c}_j$. Note also that in (10.88) the term $\mathbf{r}_j^T\mathbf{c}_j$ denotes the inner product of the vectors $\mathbf{r}_j$ and $\mathbf{c}_j$.

In light of (10.87) to (10.89), we make three observations:

1. The term $(E_s/N_0)\|\mathbf{r}_j\|^2$ depends only on the channel SNR and the squared magnitude of the received vector $\mathbf{r}_j$.
2. The third product term $(E_s/N_0)\|\mathbf{c}_j\|^2$ depends only on the channel SNR and the squared magnitude of the transmitted code vector $\mathbf{c}_j$.
3. The remaining product term $2(E_s/N_0)\mathbf{r}_j^T\mathbf{c}_j$ is the only one that contains useful information for detection in the receiver by virtue of the inner product $\mathbf{r}_j^T\mathbf{c}_j$ that *correlates* the received vector $\mathbf{r}$ with the transmitted code vector $\mathbf{c}$, as shown in (10.88).

In light of these observations and the observation made previously that the bracketed fractional term in (10.85) does not depend on whether the symbol m_j is +1 or −1, we may simplify the formula for the transition metric $\gamma_j(s', s)$ in (10.83) as follows:

$$\gamma_j(s', s) = A_j B_j \exp\left(\frac{1}{2}m_j L_a(m_j)\right)\exp\left(\frac{1}{2}L_c(\mathbf{r}_j^T\mathbf{c}_j)\right), \quad j = 0, 1, \ldots, L-1 \tag{10.90}$$

where L_c denotes the *channel reliability factor*, defined by

$$L_c = \frac{4E_s}{N_0} \tag{10.91}$$

As for the two multiplying factors A_j and B_j, they are respectively defined by

$$A_j = \frac{\frac{1}{2}\exp(-L_a(m_j))}{1+\exp(-L_a(m_j))}, \quad j = 0, 1, \ldots, L-1 \tag{10.92}$$

and

$$B_j = \left(\sqrt{\frac{E_s}{\pi N_0}}\right)^n \exp\left[-\frac{E_s}{N_0}(\|\mathbf{r}_j\|^2 + n)\right], \quad j = 0, 1, \ldots, L-1 \tag{10.93}$$

where, as before, n is the number of bits in each transmitted codeword.

Equations (10.90), (10.92), and (10.93) apply to the message bits of length L. However, for the *termination bits* we have

$$\mathbb{P}(m_j) = 1 \text{ and } L_a(m_j) = \pm\infty, \quad j = L, L+1, \ldots, K-1 \tag{10.94}$$

for each valid state transition; the K in (10.94) denotes the combined length of the message and termination bits. Accordingly, (10.90) for the termination bits simplifies to

$$\gamma_j(s', s) = B_j \exp\left(\frac{1}{2}L_c(\mathbf{r}_j^T\mathbf{c}_j)\right), \quad j = L, L+1, \ldots, K-1 \tag{10.95}$$

Examining (10.92), we find that the factor A_j is independent of the algebraic sign of message bit m_j; it is therefore a constant. Moreover, from (10.76) and the follow-up formulas of (10.77) and (10.78) for updating recursive computations of the forward and backward metrics, we find that the joint probability density function $f(s', s, \mathbf{r})$ contains the factors

$$\prod_{j=0}^{L-1} A_j \text{ and } \prod_{j=0}^{K-1} B_j$$

With these factors being common to every term in the numerator and denominator of (10.68), they both cancel out and may, therefore, be ignored. Thus, we may simplify (10.90) and (10.95) into the following two-part formula:

$$\gamma_j(s', s) = \begin{cases} \exp\left(\frac{1}{2}m_j L_a(m_j)\right)\exp\left(\frac{1}{2}L_c(\mathbf{r}_j^T \mathbf{c}_j)\right), & j = 0, 1, \dots, L-1 \text{ for message bits} \\ \exp\left(\frac{1}{2}L_c(\mathbf{r}_j^T \mathbf{c}_j)\right), & j = L, L+1, \dots, K-1 \text{ for termination bits} \end{cases} \tag{10.96}$$

One last comment is in order. When the original message bits are *equally likely*, we have

$$\mathbb{P}(m_j) = \frac{1}{2} \text{ and } L_a(m_j) = 0 \quad \text{for all } j \tag{10.97}$$

Under these two conditions, we have a simple expression for the transition metric for the entire stream of bits, as shown by

$$\gamma_j(s', s) = \exp\left(\frac{1}{2}L_c(\mathbf{r}_j^T \mathbf{c}_j)\right), \qquad j = 0, 1, \dots, K-1 \tag{10.98}$$

The a Posteriori *L*-Value Finalized

With the forward and backward recursions as well as the branch metric that ties them together all now at hand, we are equipped to finalize the formula for computing the a posteriori *L*-value $L_p(m_j)$ defined way back in (10.68). Specifically, using (10.69) and (10.76), we may now write

$$\begin{aligned} L_p(m_j) &= \ln\left(\frac{\sum_{(s', s) \in \Sigma_j^+} f(s_j = s', s_{j+1} = s, \mathbf{r})}{\sum_{(s', s) \in \Sigma_j^-} f(s_j = s', s_{j+1} = s, \mathbf{r})}\right) \\ &= \ln\left(\frac{\sum_{(s', s) \in \Sigma_j^+} f(s', s, \mathbf{r})}{\sum_{(s', s) \in \Sigma_j^-} f(s', s, \mathbf{r})}\right) \\ &= \ln\left(\frac{\sum_{(s', s) \in \Sigma_j^+} \beta_{j+1}(s)\gamma_j(s', s)\alpha_j(s')}{\sum_{(s', s) \in \Sigma_j^-} \beta_{j+1}(s)\gamma_j(s', s)\alpha_j(s')}\right) \end{aligned} \tag{10.99}$$

It is the a posterior *L*-value $L_p(m_j)$ defined in the last line of (10.99), which is delivered by the MAP decoding algorithm given the received vector **r**.

Summary of the MAP Algorithm

Starting with given AWGN channel values, namely E_s/N_0 and received vector $\mathbf{r}_j$ at time-unit j, the *computational flow diagram* of Figure 10.22 provides a visual summary of the key recursions involved in using the MAP decoding algorithm. Specifically, the functional blocks pertaining to the forward metric $\alpha_{j+1}(s)$, the backward metric $\beta_j(s')$, the branch metric $\gamma_j(s', s)$, and the a posteriori L-value $L_p(m_j)$ are all identified together with their respective equation numbers.

Modifications of the MAP Decoding Algorithm

The MAP algorithm, credited to Bahl *et al.* (1974), is roughly three times as computationally complex as the Viterbi algorithm. It was on account of this high computational complexity that the MAP algorithm was largely ignored in the literature for almost two decades. However, its pioneering application in turbo codes by Berrou *et al.* (1993) re-ignited interest in the MAP algorithm, which, in turn, led to the formulation of procedures for significant reductions in computational complexity.

Specifically, we may mention the following two modifications of the MAP algorithm, the first one being exact and the second one being approximate:

1. *Log-MAP Algorithm*

   Examination of the forward and backward metrics of the MAP algorithm for continuous-output AWGN channels reveals that they are sums of exponential terms, one for each valid state transition in the trellis. This finding, in turn, leads to the idea of simplifying the MAP computations by making use of the following identity (Robertson *et al.*, 1995):

$$\ln(e^x + e^y) = \max(x, y) + \ln(1 + e^{-|x-y|}) \tag{10.100}$$

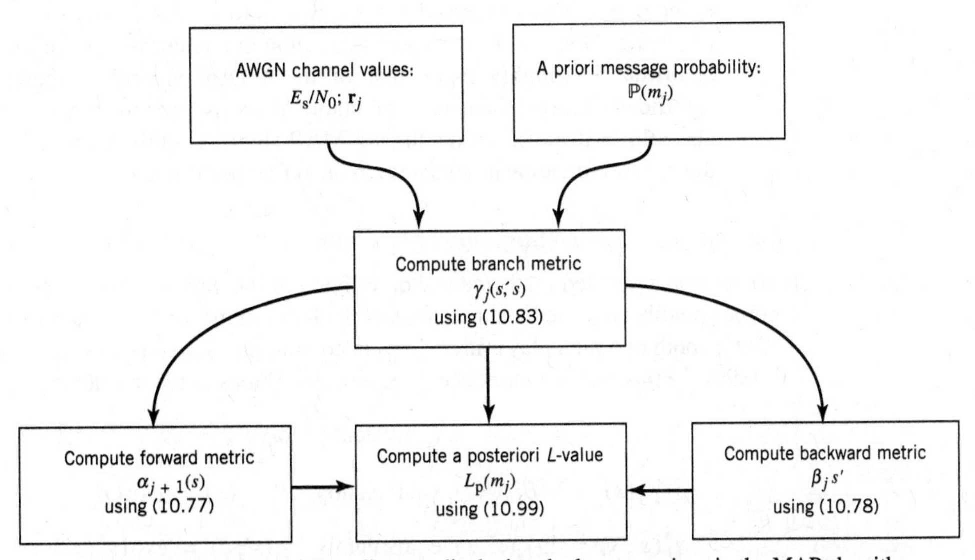

Figure 10.22 Computational flow diagram displaying the key recursions in the MAP algorithm.

where the computationally difficult operation $\ln(e^x + e^y)$ is replaced by the sum of two simpler computations:

a. the *max function,* $\max(x,y)$, equals x or y, depending on which is larger;
b. the *correction term,* $\ln(1 + e^{-|x-y|})$, may be evaluated using a look-up table.

The resulting algorithm, called the *log-MAP algorithm*, is considerably simpler in implementation and provides greater numerical stability than the original MAP algorithm. We say so because its formulation is based on two relatively simple entities: a max function and a look-up table. Note, however, that in developing the log-MAP algorithm, no approximations whatsoever are made.[10]

2. *Max-log-MAP Algorithm*

We may simplify the computational complexity of the MAP decoding algorithm even further by ignoring the correction term $\ln(1 + e^{-|x-y|})$ altogether. In effect, we simply use the approximation

$$\ln(e^x + e^y) \approx \max(x, y) \tag{10.101}$$

The correction term, ignored in this approximate formula, is bounded by

$$0 < \ln(1 + e^{-|x-y|}) \le \ln(2) = 0.693$$

The approximate formula of (10.101) yields reasonably good results whenever the condition

$$|\max(x, y)| \ge 7$$

holds. The decoding algorithm that uses the max function $\max(x,y)$ in place of $\ln(e^x, e^y)$ is called the *max-log-MAP algorithm*. In this simplified algorithm, the max function plays a role similar to the ACS described previously in the Viterbi algorithm; we therefore find that the forward recursion in the max-log-MAP algorithm is equivalent to a forward Viterbi algorithm, and the backward recursion in the max-log-MAP algorithm is equivalent to a Viterbi algorithm performed in the backward direction. In other words, computational complexity of the max-log-MAP algorithm is roughly twice that of the Viterbi algorithm, thereby providing a significant improvement in computational terms over the original MAP decoding algorithm. However, unlike the log-MAP algorithm, this improvement is attained at the expense of some degradation in decoding performance.

Details of the Max-Log-MAP Algorithm

To develop a detailed mathematical description of the max-log-MAP algorithm, we have to come up with simplified computations of the forward metric $\alpha_{j+1}(s)$ and backward metric $\beta_j(s')$, both of which play critical roles in computing the log-a-posteriori L-value $L(m_j)$ in (10.99). To this end, we introduce three new definitions in the log-domain:

$$\alpha_j^*(s') = \ln\alpha_j(s'), \text{ equivalently, } \alpha_j(s') = \exp(\alpha_j^*(s')) \tag{10.102}$$

$$\beta_{j+1}^*(s) = \ln\beta_{j+1}(s), \text{ equivalently, } \beta_{j+1}(s) = \exp(\beta_{j+1}^*(s)) \tag{10.103}$$

$$\gamma_j^*(s', s) = \ln\gamma_j(s', s), \text{ equivalently, } \gamma_j(s', s) = \exp(\gamma_j^*(s', s)) \tag{10.104}$$

where the asterisk for all three metrics is intended to signify the use of natural logarithm and must, therefore, not be confused with complex conjugation.

The motivation for these new definitions is to exploit the physical presence of exponentials in the forward and backward metrics so as to facilitate applying the approximate formula of (10.101). Thus, substituting the recursion of (10.104) into (10.77), we get

$$\begin{aligned}\alpha^*_{j+1}(s) &= \ln\left(\sum_{s'\in\sigma_j}\gamma_j(s',s)\alpha_j(s')\right)\\ &= \ln\left(\sum_{s'\in\sigma_j}\exp(\gamma^*_j(s',s)+\alpha^*_j(s'))\right)\end{aligned} \tag{10.105}$$

where σ_j is a subset of Σ_j. Hence, application of the approximate formula of (10.101) yields

$$\alpha^*_{j+1}(s) \approx \max_{s'\in\sigma_j}(\gamma^*_j(s',s)+\alpha^*_j(s')),\qquad j = 0, 1, \ldots, K-1 \tag{10.106}$$

Equation (10.106) indicates that, for each path in the trellis from the old state s' at time-unit j to the updated state s at time-unit $j+1$, the max-log MAP algorithm adds the branch metric $\gamma^*_j(s',s)$ to the old value $\alpha^*_j(s')$ to produce the updated value $\alpha^*_{j+1}(s)$; this update is the "maximum" of all the α^* values of the previous paths terminating on the state $s_{j+1} = s$, that is, $j = 1, 0, \ldots, K-1$. The process just described may be thought of as that of selecting the one particular path viewed as the "survivor" with all the other paths in the trellis reaching the state s being discarded. We may, therefore, view (10.106) as a mathematical basis for describing the forward recursion in the max-log-MAP algorithm in exactly the same way as the forward recursion in the Viterbi algorithm.

Proceeding in a manner similar to that for the forward recursion, we may write

$$\begin{aligned}\beta^*_j(s') &= \ln\left(\sum_{s\in\sigma_{j+1}}\gamma_j(s',s)\beta_{j+1}(s)\right)\\ &= \ln\left(\sum_{s\in\sigma_{j+1}}\exp(\gamma^*_j(s',s)+\beta^*_{j+1}(s))\right)\end{aligned} \tag{10.107}$$

whose approximate form is given by

$$\beta^*_j(s') \approx \max_{s\in\sigma_{j+1}}(\gamma^*_j(s',s)+\beta^*_{j+1}(s)),\qquad j = K-1, \ldots, 1, 0 \tag{10.108}$$

Next, proceeding onto the branch metric, we may similarly write the two-part formula

$$\gamma^*_j(s',s) = \begin{cases}\frac{1}{2}m_jL_{\mathrm{a}}(m_j)+\frac{1}{2}L_{\mathrm{c}}\mathbf{r}_j^{\mathrm{T}}\mathbf{c}_j, & j = 0, 1, \ldots, L-1 \text{ for message bits}\\ \frac{1}{2}L_{\mathrm{c}}\mathbf{r}_j^{\mathrm{T}}\mathbf{c}_j, & j = L, L+1, \ldots, K-1 \text{ for termination bits}\end{cases} \tag{10.109}$$

where, for the message bits in the first line of the equation, the additive term $\frac{1}{2}m_jL(m_j)$ accounts for a priori information.

At long last, using (10.105), (10.107), and (10.109) in (10.99), we may finally express the a posteriori L-value for the log-MAP algorithm as follows:

$$L_{\mathrm{p}}(m_j) = \ln\left(\frac{\sum_{(s',s)\in\Sigma_j^+} \beta_{j+1}(s)\gamma_j(s',s)\alpha_j(s')}{\sum_{(s',s)\in\Sigma_j^-} \beta_{j+1}(s)\gamma_j(s',s)\alpha_j(s')}\right)$$
$$= \ln \sum_{(s',s)\in\Sigma_j^+} \exp(\beta_{j+1}^*(s) + \gamma_j^*(s',s) + \alpha_j^*(s')) - \ln \sum_{(s',s)\in\Sigma_j^-} \exp(\beta_{j+1}^*(s) + \gamma_j^*(s',s) + \alpha_j^*(s')) \quad (10.110)$$

A couple of reminders:

- Σ_j^+ is the set of all state pairs $s_j = s'$ and $s_{j+1} = s$ that correspond to the original message bit $m_j = +1$ at time-unit j.
- Σ_j^- is the set of all other state pairs $s_j = s'$ and $s_{j+1} = s$ that correspond to the original message bit $m_j = -1$ at time-unit j.

Correspondingly, the approximate form of the $L_{\mathrm{p}}(m_j)$ in the max-log-MAP algorithm is defined by

$$L_{\mathrm{p}}(m_j) \approx \max_{(s',s)\in\Sigma_j^+}(\beta_{j+1}^*(s) + \gamma_j(s',s) + \alpha_j^*(s')) - \max_{(s',s)\in\Sigma_j^-}(\beta_{j+1}^*(s) + \gamma_j^*(s',s) + \alpha_j^*(s')) \quad (10.111)$$

10.10 Illustrative Procedure for Map Decoding in the Log-Domain

In the preceding section we described three different algorithms for decoding a convolutional code, as summarized here:

1. *The BCJR algorithm*, which distinguishes itself from the Viterbi algorithm in that it performs MAP decoding on a bit-by-bit basis. However, a shortcoming of this algorithm is its computational complexity, which, as mentioned previously, is roughly three times that of the Viterbi algorithm for the same convolutional code.
2. *The log-MAP-algorithm*, which simplifies the BCJR algorithm by replacing the computationally difficult logarithmic operation, namely $\ln(e^x + e^y)$, with the so-called *max function* plus a *look-up table* for evaluating $\ln(1 + e^{-|x-y|})$ in accordance with (10.100). The attractive feature of this second algorithm is twofold:
   - transformation of the BCJR algorithm into the log-MAP algorithm is exact;
   - its computational complexity is twice that of the Viterbi algorithm, thereby providing a significant reduction in complexity compared to the BCJR algorithm.

3. *The max-log-MAP algorithm*, which simplifies computational complexity even further by doing away with the look-up table; this simplification may result in some degradation in decoding performance depending on the application of interest.

In this section, we illustrate how the simpler of the latter two algorithms, namely the max-log-MAP algorithm, is used to decode an RSC code by way of an example.

EXAMPLE 7 Max-Log-MAP Decoding of Rate 3/8 Recursive Systematic Convolutional Code over AWGN Channel

In this example, we revisit the simple RSC code discussed previously at the tail end of Section 10.6 on convolutional codes.

For convenience of presentation, the two-state RSC encoder of Figure 10.17 is reproduced in Figure 10.23a. The message vector applied to the encoder is denoted by

$$\mathbf{m} = \{m_j\}_{j=0}^{3}$$

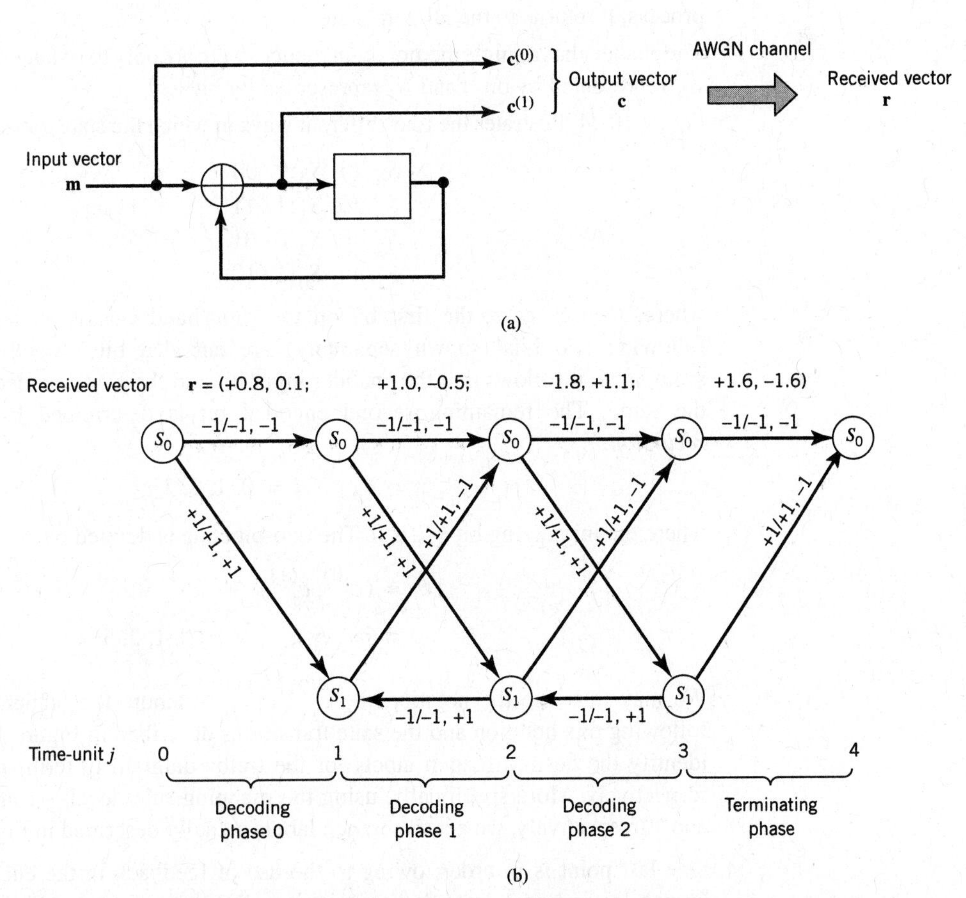

Figure 10.23 (a) Block diagram of rate-3/8, two-state recursive systematic convolutional (RSC) encoder. (b) Trellis graph of the encoder.

which produces the encoded output vector

$$\mathbf{c} = \left\{c_j^{(0)}, c_j^{(1)}\right\}_{j=0}^{3}$$

Correspondingly, the received vector at the channel output is denoted by

$$\mathbf{r} = \left\{r_j^{(0)}, r_j^{(1)}\right\}_{j=0}^{3}$$

The first three elements of the message vector $\mathbf{m}$, namely m_0, m_1, and m_2, are *message bits*. The last element, m_3, is a *termination bit*. With the encoded output vector, $\mathbf{c}$, consisting of eight bits, it follows that the code rate $r = 3/8$.

Figure 10.23b shows the trellis diagram of the RSC encoder. The underlying points covering the ways in which the branches of the trellis diagram have been labeled should be carefully noted:

1. The encoder is initialized to the *all-zero state* and, on termination of the encoding process, it returns to the all-zero state.
2. The encoder has a single memory unit; hence, there are only two states denoted by: S_0 represented by bit 0 and S_1 represented by bit 1.
3. Figure 10.24 illustrates the four different ways in which the state transitions take place:

$$\begin{aligned} &S_0 \text{ to } S_0{:}0/00 \\ &S_0 \text{ to } S_1{:}1/11 \\ &S_1 \text{ to } S_1{:}0/01 \\ &S_1 \text{ to } S_0{:}1/10 \end{aligned}$$

   where, in each case, the first bit on the right-hand side is an input bit and the following two bits (shown separately) are encoded bits. Since the encoder is systematic, it follows that the encoder input bit and the first encoded bit are exactly the same. The remaining second encoded bit is determined by the *modulo-2 recursion*:

$$m_j + b_{j-1} = b_j, \qquad j = 0, 1, 2, 3 \tag{10.112}$$

   where the initializing bit b_{-1} is 0. The two-bit code is defined by

$$\begin{aligned} \mathbf{c}_j &= (c_j^{(0)}, c_j^{(1)}) \\ &= (m_j, b_j), \qquad j = 0, 1, 2, 3 \end{aligned}$$

   We may thus use the notation $m_j/c_j^{(0)}, c_j^{(1)}$ to denote the branch labels. Hence, following this notation and the state transitions described in Figure 10.23b, we may identify the desired branch labels for the trellis diagram in terms of bits 0 and 1, respectively. More specifically, using the mapping rule: levels -1 and $+1$ for bits 0 and 1, respectively, we get the branch labels actually described in Figure 10.23b.
4. One last point is in order: owing to the use of feedback in the encoder, the lower branch leaving each state does not necessarily correspond to a bit 1 (level $+1$) and the upper branch does not necessarily correspond to a bit 0 (level -1).

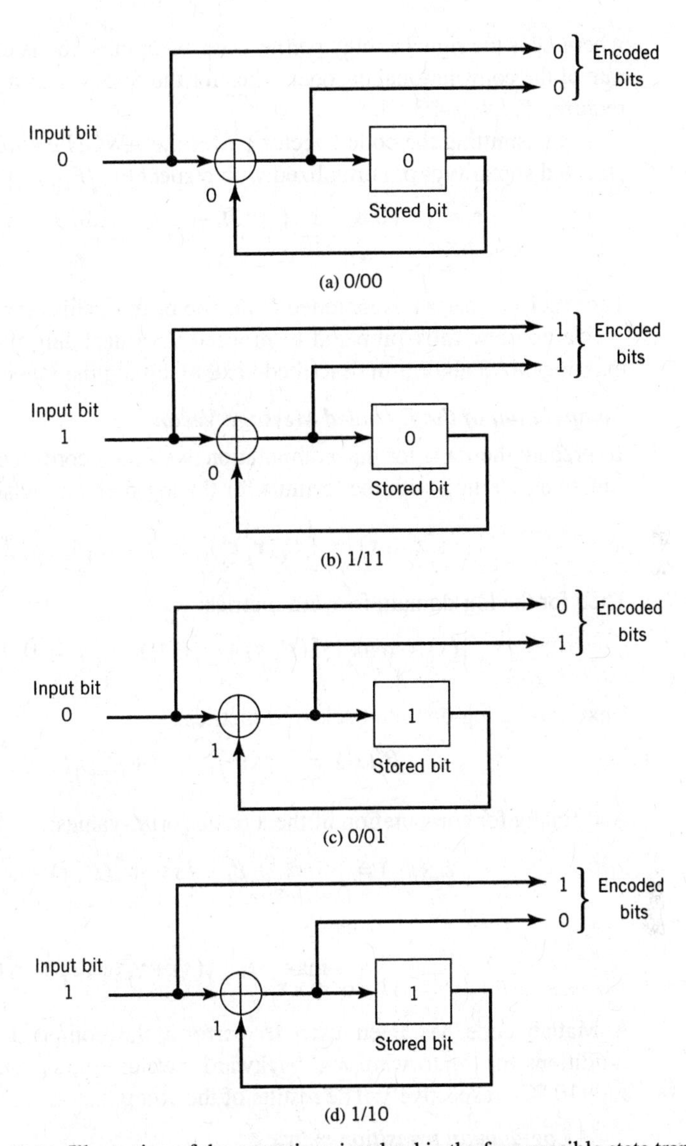

Figure 10.24 Illustration of the operations involved in the four possible state transitions.

To continue the background material for the example, we need to bring in a *mapper* that transforms the encoded signal into a form suitable for transmission over the AWGN channel. To this end, consider the simple example of binary PSK as the mapper. We may then express the SNR at the channel output (i.e., receiver input) as follows (see Problem 10.35):

$$\begin{aligned}(\mathrm{SNR})_{\text{channel output}} &= \frac{E_s}{N_0} \\ &= r\left(\frac{E_b}{N_0}\right)\end{aligned} \tag{10.113}$$

where E_b is the signal energy per message bit applied to the encoder input, and r is the code rate of the convolutional encoder. Thus for the SNR = 1/2, that is –3.01 dB and $r = 3/8$, the required E_b/N_0 is 4/3.

In transmitting the coded vector **c** over the AWGN environment, it is assumed that the received signal vector, normalized with respect to $\sqrt{E_s}$, is given by

$$\mathbf{r} = (\underbrace{+0.8, 0.1}_{\mathbf{r}_0}\ ;\ \underbrace{+1.0, -0.5}_{\mathbf{r}_1}\ ;\ \underbrace{-1.8, 1.1}_{\mathbf{r}_2}\ ;\ \underbrace{+1.6, -1.6}_{\mathbf{r}_3})$$

The received vector **r** is included at the top of the trellis diagram in Figure 10.23b.

We are now fully prepared to proceed with decoding the received vector **r** using the max-log-MAP algorithm described next, assuming the message bits are equally likely.

Computation of the Decoded Message Vector

To prepare the stage for this computation, we find it convenient to reproduce the following equations, starting with the formula for the log-domain transition metrics:

$$\gamma_j^*(s', s) = \frac{1}{2}L_c(\mathbf{r}_j^T \mathbf{c}_j), \qquad j = 0, 1, \ldots, K-1 \tag{10.114}$$

Then for the log-domain forward metrics:

$$\alpha_{j+1}^*(s) \approx \max_{s' \in \sigma_j^+} (\gamma_j^*(s', s) + \alpha_j^*(s')), \qquad j = 0, 1, \ldots, K-1 \tag{10.115}$$

Next, for the log-domain backward metrics:

$$\beta_j^*(s') = \max_{s \in \sigma_{j+1}} (\gamma_j^*(s', s) + \beta_{j+1}^*(s)) \tag{10.116}$$

And finally for computation of the a posteriori L-values:

$$\begin{aligned} L_p(m_j) = & \max_{(s, s') \in \Sigma_j^+} \beta_{j+1}^*(s) + \gamma_j^*(s', s) + \alpha_j^*(s') \\ & -\max_{(s, s') \in \Sigma_j^-} \beta_{j+1}^*(s) + \gamma_j^*(s', s) + \alpha_j^*(s') \end{aligned} \tag{10.117}$$

A Matlab code has been used to perform the computation, starting with the initial conditions for the forward and backward metrics, $\alpha_0(s)$ and $\beta_K(s')$, defined in (10.79) and (10.80), respectively. The results of the computation are summarized as follows:

1. *Log-domain transition metrics*

$$\text{Gamma } 0: \begin{cases} \gamma_0^*(S_0, S_0) = -0.9 \\ \gamma_0^*(S_0, S_1) = 0.9 \end{cases}$$

$$\text{Gamma } 1: \begin{cases} \gamma_1^*(S_0, S_1) = -0.5 \\ \gamma_1^*(S_1, S_0) = 1.5 \\ \gamma_1^*(S_0, S_1) = 0.5 \\ \gamma_1^*(S_1, S_1) = -1.5 \end{cases}$$

$$\text{Gamma 2}: \begin{cases} \gamma_2^*(S_0, S_0) = 0.7 \\ \gamma_2^*(S_1, S_0) = -2.9 \\ \gamma_2^*(S_0, S_1) = -0.7 \\ \gamma_2^*(S_1, S_1) = 2.9 \end{cases}$$

$$\text{Gamma 3}: \begin{cases} \gamma_3^*(S_0, S_0) = 0 \\ \gamma_3^*(S_1, S_0) = 3.2 \end{cases}$$

2. *Log-domain forward metrics*

$$\text{Alpha 0}: \begin{cases} \alpha_0^*(S_0) = 0 \\ \alpha_0^*(S_1) = 0 \end{cases}$$

$$\text{Alpha 1}: \begin{cases} \alpha_1^*(S_0) = -0.9 \\ \alpha_1^*(S_1) = 0.9 \end{cases}$$

$$\text{Alpha 2}: \begin{cases} \alpha_2^*(S_0) = 2.4 \\ \alpha_2^*(S_1) = -0.4 \end{cases}$$

3. *Log-domain backward metrics*

$$\beta_K: \begin{cases} \beta_K^*(S_0) = 0 \\ \beta_K^*(S_1) = 0 \end{cases}$$

$$\text{Beta 3}: \begin{cases} \beta_3^*(S_0) = 0 \\ \beta_3^*(S_1) = 3.2 \end{cases}$$

$$\text{Beta 2}: \begin{cases} \beta_2^*(S_0) = 2.5 \\ \beta_2^*(S_1) = 6.1 \end{cases}$$

$$\text{Beta 1}: \begin{cases} \beta_1^*(S_0) = 6.6 \\ \beta_1^*(S_1) = 4.6 \end{cases}$$

4. A posteriori L-values

$$\left.\begin{aligned} L_{\mathrm{p}}(m_0) &= -0.2 \\ L_{\mathrm{p}}(m_1) &= 0.2 \\ L_{\mathrm{p}}(m_2) &= -0.8 \end{aligned}\right\} \tag{10.118}$$

5. *Final decision*

Decoded version of the original message vector

$$\hat{\mathbf{m}} = [-1, 1, -1] \tag{10.119}$$

In binary form, we may equivalently write

$$\hat{\mathbf{m}} = [0, 1, 0]$$

Two Final Remarks on Example 7

1. In arriving at the decoded output of (10.119) we made use of the termination bit, m_3. Although m_3 is *not* a message bit, the same procedure was used to calculate its a posteriori L-value. Lin and Costello (2004) showed that this kind of calculation is a necessary requirement in the iterative decoding of turbo codes. Specifically, with the turbo decoder consisting of two stages, "soft-output" a posteriori L-values are passed as a priori inputs to a second decoder.
2. In Example 7, we focused attention on the application of the max-log-MAP algorithm to decode the rate-3/8 RSC code produced by the two-state encoder of Figure 10.23a. The procedure described herein, embodying six steps, applies equally well to the log-MAP algorithm with no approximations. In Problem 10.34 at the end of the chapter, the objective is to show that the corresponding decoded output is (+1, +1, −1), which is different from that of Example 7. Naturally, in arriving at this new result, the calculations are somewhat more demanding but more accurate in the final decision-making.

10.11 New Generation of Probabilistic Compound Codes

Traditionally, the design of good codes has been tackled by constructing codes with a great deal of algebraic structure, for which there are feasible decoding schemes. Such an approach is exemplified by the linear block codes, cyclic codes, and convolutional codes discussed in preceding sections of this chapter. The difficulty with these traditional codes is that, in an effort to approach the theoretical limit for Shannon's channel capacity, we need to increase the codeword length of a linear block code or the constraint length of a convolutional code, which, in turn, causes the computational complexity of a maximum likelihood or maximum a posteriori decoder to increase exponentially. Ultimately, we reach a point where complexity of the decoder is so high that it becomes physically impractical.

Ironically enough, in his 1948 paper, Shannon showed that the "average" performance of a randomly chosen ensemble of codes results in an exponentially decreasing decoding error with increasing block length. Unfortunately, as it was with his coding theorem, Shannon did not provide guidance on how to construct randomly chosen codes.

The Turbo Revolution Followed by LDPC Rediscovery

Interest in the use of randomly chosen codes was essentially dormant for a long time until the new idea of *turbo coding* was described by Berrou *et al.* (1993); that idea was based on two design initiatives:

1. The design of a good code, the construction of which is characterized by random-like properties.
2. The iterative design of a decoder that makes use of soft-output values by exploiting the maximum a posteriori decoding algorithm due to Bahl *et al.* (1974).

By exploiting these two ideas, it was experimentally demonstrated that turbo coding can approach the Shannon limit at a computational cost that would have been infeasible with traditional algebraic codes. Therefore, it can be said that the invention of turbo coding deserves to be ranked among the major technical achievements in the design of communication systems in the 20th century.

What is also remarkable is the fact that the discovery of turbo coding and iterative decoding flamed theoretical as well as practical interest in some prior work by Gallager (1962, 1963) on *LDPC codes*. These codes also possess the information-processing power to approach the Shannon limit in their own individual ways. The important point to note here is the fact that both turbo codes and LDPC codes are capable of approaching the Shannon limit at a similar level of computational complexity, provided that they both have a sufficiently long codeword. Specifically, turbo codes require a long turbo interleaver, whereas LDPC codes require a longer codeword at a given code rate (Hanzo, 2012).

We thus have two basic classes of *probabilistic compound coding techniques*: turbo codes and LDPC codes, which complement each other in the following sense:

> Turbo encoders are simple to design but the decoding algorithm can be demanding. In contrast, LDPC encoders are relatively complex but they are simple to decode.

With these introductory remarks, the stage is set for the study of turbo codes in Section 10.12, followed by LDPC codes in Section 10.14.

10.12 Turbo Codes

Turbo Encoder

As mentioned in the preceding section, the use of a good code with random-like properties is basic to turbo coding. In the first successful implementation of turbo codes[11], Berrou *et al.* achieved this design objective by using *concatenated codes*. The original idea of concatenated codes was conceived by Forney (1966). To be more specific, concatenated codes can be of two types: *parallel* or *serial*. The type of concatenated codes used by Berrou *et al.* was of the parallel type, which is discussed in this section. Discussion of the serial type of concatenated codes will be taken up in Section 10.16.

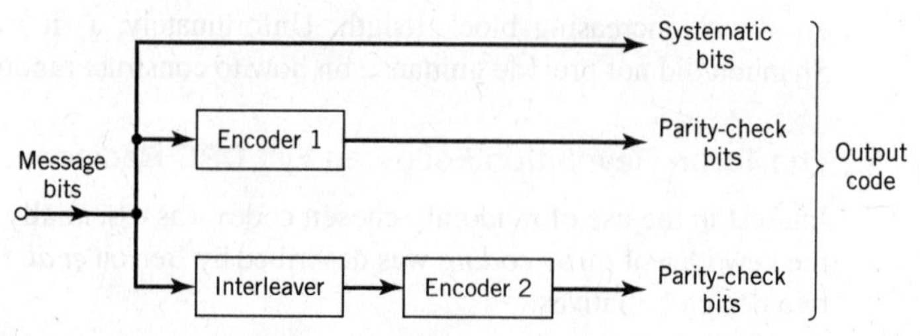

Figure 10.25 Block diagram of turbo encoder of the parallel type.

Figure 10.25 depicts the most basic form of a turbo code generator that consists of two constituent systematic encoders, which are concatenated by means of an interleaver.

The *interleaver* is an input–output mapping device that permutes the ordering of a sequence of symbols from a fixed alphabet in a completely deterministic manner; that is, it takes the symbols at the input and produces identical symbols at the output but in a different temporal order. Turbo codes use a *pseudo-random interleaver*, which operates only on the systematic (i.e., message) bits. (Interleavers are discussed in Appendix F.) The size of the interleaver used in turbo codes is typically very large, on the order of several thousand bits.

There are two reasons for the use of an interleaver in a turbo code:

1. The interleaver ties together errors that are easily made in one half of the turbo code to errors that are exceptionally unlikely to occur in the other half; this is indeed one reason why the turbo code performs better than a traditional code.
2. The interleaver provides robust performance with respect to mismatched decoding, a problem that arises when the channel statistics are not known or have been incorrectly specified.

Ordinarily, but not necessarily, the same code is used for both constituent encoders in Figure 10.25. The constituent codes recommended for turbo codes are *short constraint-length RSC codes*. The reason for making the convolutional codes recursive (i.e., feeding one or more of the tap outputs in the shift register back to the input) is to make the internal state of the shift register depend on past outputs. This affects the behavior of the error patterns, with the result that a better performance of the overall coding strategy is attained.

EXAMPLE 8 Two-State Turbo Encoder

Figure 10.26 shows the block diagram of a specific turbo encoder using an identical pair of two-state RSC constituent encoders. The generator matrix of each constituent encoder is given by

$$\mathbf{G}(D) = \left(1, \frac{1}{1+D}\right)$$

The input sequence of bits has length $K = 4$, made up of three message bits and one termination bit. (This RSC encoder was discussed previously in Section 10.9.) The input vector is given by

$$\mathbf{m} = (m_0, m_1, m_2, m_3)$$

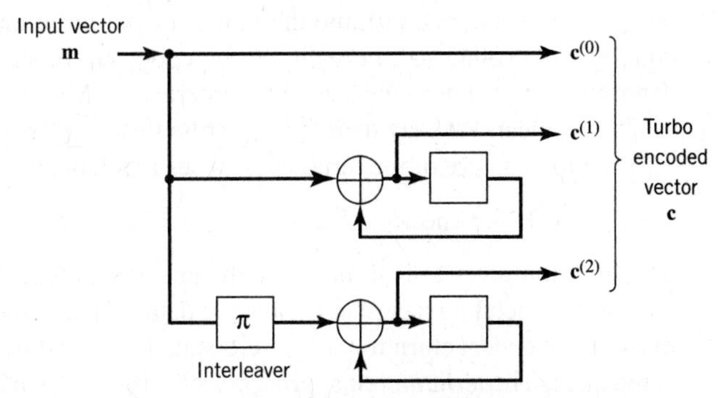

Figure 10.26 Two-state turbo encoder for Example 8.

The parity-check vector produced by the first constituent encoder is given by

$$\mathbf{b}^{(1)} = (b_0^{(1)}, b_1^{(1)}, b_2^{(1)}, b_3^{(1)})$$

Similarly, the parity-check vector produced by the second constituent encoder is given by

$$\mathbf{b}^{(2)} = (b_0^{(2)}, b_1^{(2)}, b_2^{(2)}, b_3^{(2)})$$

The transmitted code vector is therefore defined by

$$\mathbf{c} = (\mathbf{c}^{(0)}, \mathbf{c}^{(1)}, \mathbf{c}^{(2)})$$

With the convolutional code being systematic, we thus have

$$\mathbf{c}^{(0)} = \mathbf{m}$$

As for the remaining two sub-vectors constituting the code vector **c**, they are defined by

$$\mathbf{c}^{(1)} = \mathbf{b}^{(1)}$$

and

$$\mathbf{c}^{(2)} = \mathbf{b}^{(2)}$$

The transmitted code vector **c** is therefore made up of 12 bits. However, recalling that the termination bit m_3 is not a message bit, it follows that the code rate of the turbo code described in Figure 10.26 is

$$r = \frac{3}{12} = \frac{1}{4}$$

One last point is in order: with each RSC encoder having two states, the interleaver has a two-by-two (row–column) structure. Note also that the interleaver in Figure 10.26 is denoted by the symbol π, which is a common usage; this practice is adopted throughout the book.

In Figure 10.25, the input data stream is applied directly to encoder 1 and the pseudo-randomly reordered version of the same data stream is applied to encoder 2. The systematic bits (i.e., original message bits) and the two sets of parity-check bits generated by the two encoders constitute the output of the turbo encoder. Although the constituent codes are convolutional, in reality, turbo codes are block codes with the block size being determined by the periodic size of the interleaver. Moreover, both RSC encoders in Figure 10.25 are linear. We may therefore describe turbo codes generally as *linear block codes.*

The block nature of the turbo code raises a practical issue:

How do we know the beginning and the end of a codeword?

The common practice is to initialize the encoder to the *all-zero state* and then encode the data. After encoding a certain number of data bits, a number of tail bits are added so as to make the encoder return to the all-zero state at the end of each block; thereafter, the cycle is repeated. The *termination* approaches of turbo codes include the following:

- A simple approach is to terminate the first RSC code in the encoder and leave the second one undetermined. A drawback of this approach is that the bits at the end of the block due to the second RSC code are more vulnerable to noise than the other bits. Experimental work has shown that turbo codes exhibit a leveling off in performance as the SNR increases. This behavior is not like an error floor; rather, it has the appearance of an error floor compared with the steep drop in error performance at low SNR. The *error floor* is affected by a number of factors, the dominant one of which is the choice of interleaver.
- A more refined approach is to terminate both constituent codes in the encoder in a symmetric manner. Through the combined use of a good interleaver and dual termination, the error floor can be reduced by an order of magnitude compared to the simple termination approach.

In the original version of the turbo encoder described in Berrou *et al.* (1993), the parity-check bits generated by the two encoders in Figure 10.25 were punctured prior to data transmission over the channel to maintain the rate at 1/2. A *punctured code* is constructed by deleting certain parity-check bits, thereby increasing the data rate; the message bits in the puncturing process are of course unaffected. Basically, puncturing is the inverse of extending a code. It should, however, be emphasized that the use of a puncture map is not a necessary requirement for the generation of turbo codes.

As mentioned previously, the encoding scheme of Figure 10.25 is of the *parallel concatenation* type, the novelty of which is twofold:

- the use of *RSC codes* and
- the insertion of a pseudo-random interleaver between the two encoders.

The net result of parallel concatenation is a turbo code that appears essentially *random* to the channel by virtue of the pseudo-random interleaver, yet it possesses sufficient structure for the decoding to be physically realizable. Coding theory asserts that a code chosen at random is capable of approaching Shannon's channel capacity, provided that the block size is sufficiently large. This is indeed the reason behind the impressive performance of turbo codes, as discussed next.

Performance of Turbo Codes

Figure 10.27 shows the error performance of a 1/2-rate turbo code with a large block size for binary data transmission over an AWGN channel.[12] The code uses an interleaver of size 65,536 bits and a MAP decoder.

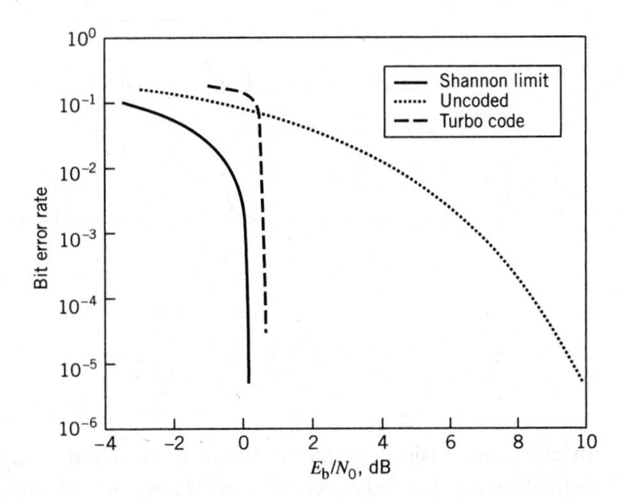

Figure 10.27 Noise performance of 1/2 rate, turbo code and uncoded transmission for AWGN channel; the figure also includes Shannon's theoretical limit on channel capacity for code rate $r = 1/2$.

For the purpose of comparison, Figure 10.27 also includes two other curves for the same AWGN channel:

- uncoded transmission (i.e., code rate $r = 1$);
- Shannon's theoretical limit for code rate 1/2, which follows from Figure 5.18b.

From Figure 10.27, we may draw two important conclusions:

1. Although the BER for the turbo-coded transmission is significantly higher than that for uncoded transmission at low E_b/N_0, the BER for the turbo-coded transmission drops very rapidly once a critical value of E_b/N_0 has been reached.
2. At a BER of 10^{-5}, the turbo code is less than 0.5 dB from Shannon's theoretical limit.

Note, however, attaining this highly impressive performance requires that the size of the interleaver or, equivalently, the block length of the turbo code be large. Also, the large number of iterations needed to improve performance increases the decoder latency. This drawback is due to the fact that the digital processing of information does not lend itself readily to the application of feedback, which is a distinctive feature of the turbo decoder.

Extrinsic Information

Before proceeding to describe the operation of the turbo decoder, we find it desirable to introduce the notion of extrinsic information. The most convenient representation for this new concept is in terms of the log-likelihood ratio, in which case extrinsic information is computed as the difference between two a posteriori L-values as depicted in Figure 10.28. Formally, *extrinsic information*, generated by a decoding stage for a set of systematic (message) bits, is defined as follows:

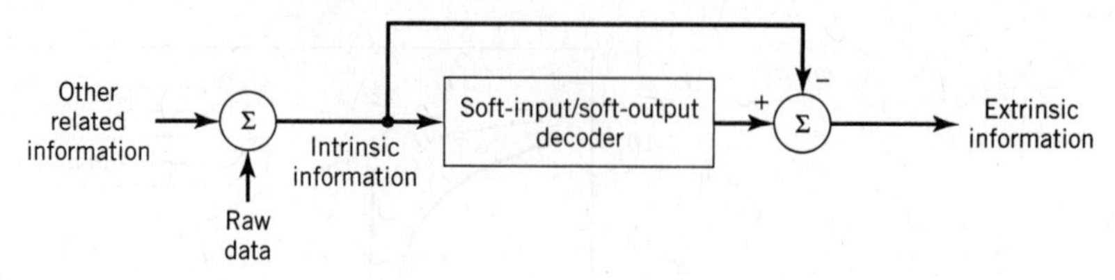

Figure 10.28 Block diagram for illustrating the concept of extrinsic information.

> Extrinsic information is the difference between the log-likelihood ratio computed at the output of a decoding stage and the intrinsic information represented by the log-likelihood ratio applied to the input of that decoding stage.

In effect, extrinsic information is the *incremental information* gained by exploiting the dependencies that exist between a message bit of interest and incoming raw data bits processed by the decoder. Extrinsic information plays a key role in the iterative decoding process, as discussed next.

Turbo Decoder

Figure 10.29a shows the block diagram of the two-stage *turbo decoder*. Using a MAP decoding algorithm discussed in Section 10.9, the decoder operates on noisy versions of the systematic bits and the two sets of parity-check bits in two decoding stages to produce an estimate of the original message bits.

A distinctive feature of the turbo decoder that is immediately apparent from the block diagram of Figure 10.29a is the use of *feedback*, manifesting itself in producing extrinsic information from one decoder to the next in an iterative manner. In a way, this decoding process is analogous to the feedback of exhaust gases experienced in a turbo-charged engine; indeed, turbo codes derive their name from this analogy. In other words, the term "turbo" in turbo codes has more to do with the decoding rather than the encoding process.

In operational terms, the turbo encoder in Figure 10.29a operates on *noisy* versions of the following inputs, obtained by demultiplexing the channel output, $\mathbf{r}_j$,

- systematic (i.e., message) bits, denoted by $\mathbf{r}_j^{(0)}$;
- parity-check bits corresponding to encoder 1 in Figure 10.25, denoted by $\mathbf{r}_j^{(1)}$;
- parity-check bits corresponding to encoder 2 in Figure 10.25, denoted by $\mathbf{r}_j^{(2)}$.

The net result of the decoding algorithm, given the received vector $\mathbf{r}_j$, is an estimate of the original message vector, namely $\hat{\mathbf{m}}$, which is delivered at the decoder output to the user.

Another important point to note in the turbo decoder of Figure 10.29a is the way in which the interleaver and de-interleaver are positioned inside the feedback loop. Bearing in mind the fact that the definition of extrinsic information requires the use of intrinsic information, we see that decoder 1 operates on three inputs:

- the noisy systematic (i.e., original message) bits,
- the noisy parity-check bits due to encoder 1, and
- de-interleaved extrinsic information computed by decoder 2.

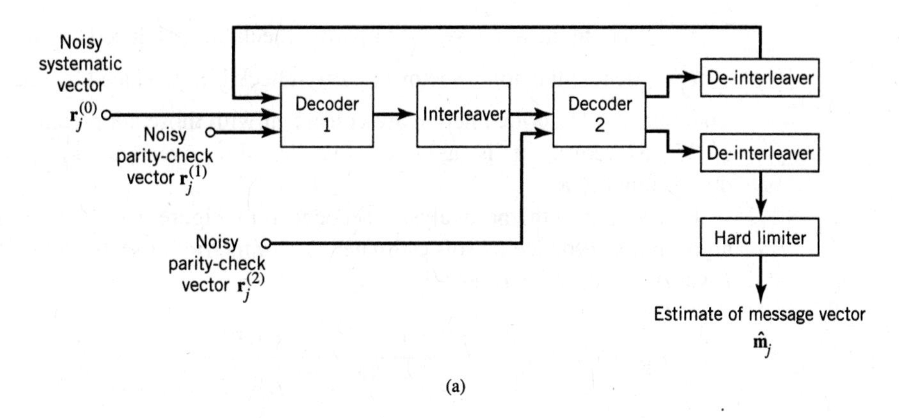

(a)

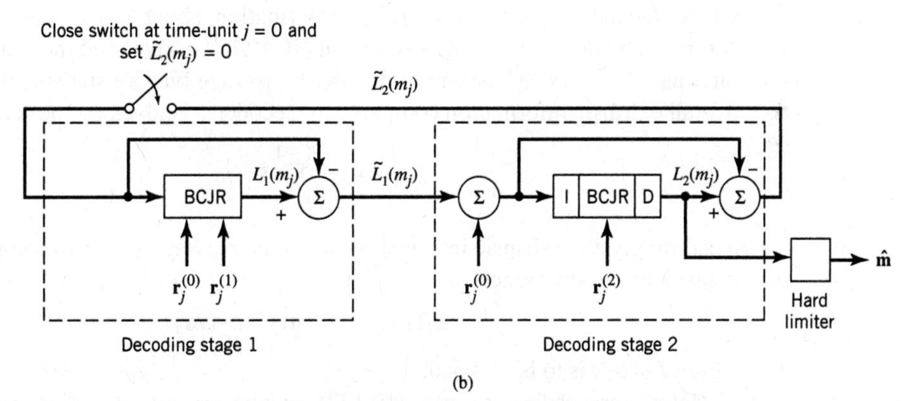

(b)

Figure 10.29 (a) Block diagram of turbo decoder. (b) Extrinsic form of turbo decoder, where I stands for interleaver, D for deinterleaver, and BCJR for BCJR for BCJR algorithm for log-MAP decoding.

In a complementary manner, decoder 2 operates on two inputs of its own:

- the noisy parity-check bits due to encoder 2 and
- the interleaved version of the extrinsic information computed by decoder 1.

For this iterative exchange of information between the two decoders inside the feedback loop to *continuously reinforce each other*, the de-interleaver and interleaver would have to separate the two decoders in the manner depicted in Figure 10.29a. Moreover, the structure of the decoder in the receiver is configured to be consistent with the structure of the encoder in the transmitter.

Mathematical Feedback Analysis

To put the two-state turbo decoding process just described on a mathematical basis, we structure the flow of information around the feedback loop as depicted in Figure 10.29a. For the sake of simplicity without loss of generality, we assume the use of a code rate $r = 1/3$ parallel concatenated convolutional code without puncturing. At time-unit j, let

$\mathbf{r}_j^{(0)}$ denote the noisy vector of systematic bits,

$\mathbf{r}_j^{(1)}$ denote the noisy vector of parity-check bits produced by encoder 1, and

$\mathbf{r}_j^{(2)}$ denote the noisy vector of parity-check bits produced by encoder 2.

The notations described herein are consistent with those adopted in the encoder of Figure 10.25. Moreover, it is assumed that all three vectors, $\mathbf{r}_j^{(0)}$, $\mathbf{r}_j^{(1)}$, $\mathbf{r}_j^{(2)}$, are of dimensionality K.

Proceeding with the analysis, decoder 1 in Figure 10.29b uses the BCJR decoding algorithm to produce a "soft estimate" of symmetric bit m_j by computing the a posteriori L-values for decoder 1, namely

$$L_1(m_j) = \ln\left(\frac{\mathbb{P}(m_j = +1 \mid \mathbf{r}_j^{(0)}, \mathbf{r}_j^{(1)}, \tilde{L}_2(\mathbf{m}))}{\mathbb{P}(m_j = -1 \mid \mathbf{r}_j^{(0)}, \mathbf{r}_j^{(1)}, \tilde{L}_2(\mathbf{m}))}\right), \qquad j = 0, 1, \ldots, K-1 \tag{10.120}$$

where $\tilde{L}_2(\mathbf{m})$ denotes the extrinsic information about the message vector $\mathbf{m}$ that is computed by decoder 2. Note also that in (10.120) we have used the usual mapping: +1 for bit 1 and –1 for bit 0. Assuming that the L message bits are statistically independent, the overall extrinsic information computed by decoder 1 is given by the summation:

$$L_1(\mathbf{m}) = \sum_{j=0}^{K-1} L_1(m_j) \tag{10.121}$$

Accordingly, the extrinsic information about the message vector $\mathbf{m}$ computed by decoder 1 is given by the difference

$$\tilde{L}_1(\mathbf{m}) = L_1(\mathbf{m}) - \tilde{L}_2(\mathbf{m}) \tag{10.122}$$

where $\tilde{L}_2(\mathbf{m})$ is to be defined.

Before proceeding to use (10.122) in the second decoding stage, the extrinsic information $\tilde{L}_1(\mathbf{m})$ is reordered (i.e., de-interleaved) to compensate for the pseudo-random interleaving introduced originally in the turbo encoder in the manner indicated in both Figure 10.29b. In addition to $\tilde{L}_1(\mathbf{m})$, the input applied to decoder 2 also includes the vector of noisy parity-check bits $\mathbf{r}^{(2)}$. Accordingly, by using the BCJR algorithm, decoder 2 produces a more refined soft estimate of the message vector $\mathbf{m}$. Next, as indicated in Figure 10.29b, this refined estimate of the message vector is re-interleaved to compute the a posteriori L-values for decoder 2, namely

$$\tilde{L}_2(\mathbf{m}) = \sum_{j=0}^{K-1} L_2(m_j) \tag{10.123}$$

where

$$\tilde{L}_2(m_j) = \ln\left(\frac{\mathbb{P}(m_j = +1 \mid \mathbf{r}_j^{(2)}, \tilde{\mathbf{L}}_1(\mathbf{m}))}{\mathbb{P}(m_j = -1 \mid \mathbf{r}_j^{(2)}, \tilde{\mathbf{L}}_1(\mathbf{m}))}\right), \qquad j = 0, 1, \ldots, K-1 \tag{10.124}$$

Accordingly, the extrinsic information $\tilde{L}_2(\mathbf{m})$ fed back to the input of decoder 1 is given by

$$\tilde{L}_2(\mathbf{m}) = L_2(\mathbf{m}) - \tilde{L}_1(\mathbf{m}) \tag{10.125}$$

and with it the feedback loop, embodying constituent decoders 1 and 2, is closed.

As indicated in Figure 10.29b, the decoding process is *initiated* by setting the a posteriori extrinsic L-value

$$\tilde{L}_2(m_j) = 0, \qquad \text{for } j = 0 \tag{10.126}$$

The decoding process is *stopped* when it reaches a point at which no further improvement in performance is attainable. At this point, an estimate of the message vector $\mathbf{m}$ is computed by hard-limiting the a priori L-value at the output of decoder 2, yielding

$$\hat{\mathbf{m}} = \text{sgn}(L_2(\mathbf{m})) \tag{10.127}$$

To conclude the discussion on turbo decoding, two other points are noteworthy:

1. Although the noisy vector of systematic bits $\mathbf{r}^{(0)}$ is applied only to decoder 1, its influence on decoder 2 manifests itself *indirectly* through the a posteriori extrinsic L-value, $\tilde{L}_1(\mathbf{m})$, computed by decoder 1.
2. Equations (10.122) and (10.125) assume that the a posteriori extrinsic L-values, $\tilde{L}_1(\mathbf{m})$ and $\tilde{L}_2(\mathbf{m})$, passed between decoders 1 and 2, are statistically independent of the message vector $\mathbf{m}$. In reality, however, this condition applies only to the first iteration of the decoding process. Thereafter, the extrinsic information becomes less helpful in realizing successively more reliable estimates of the message vector $\mathbf{m}$.

EXAMPLE 9 **UMTS Codec Using Binary PSK Modulation**[13]

In this example, we study the *Universal Mobile Telecommunications Systems* (UMTS) standard's codec. To simplify the study, binary PSK modulation is used for data transmission over an AWGN channel. The basic RSC encoder of the UMTS turbo codes is as follows:

code-rate	$r = 1/3$
constraint length	$\nu = 4$
memory length	$m = 3$

The UMTS Turbo Encoder

Figure 10.30a shows the block diagram of the UMTS turbo encoder, which consists of two identical concatenated RSC encoders, operating in parallel with an interleaver separating them. To be specific:

- Each encoder is made up of a *linear feedback shift register* (LFSR) whose number of flip-flops $m = 3$; in each LFSR, therefore, we have a finite-state machine with

$$2^m = 2^3 = 8 \text{ states}$$

- The encoding process is initialized by setting each LFSR to the all-zero state.
- To activate the encoding process, the two switches in Figure 10.30a are closed, thereby applying the message vector $\mathbf{m}$ to the top RSC encoder and applying the interleaved version of $\mathbf{m}$, namely $\mathbf{n}$, to the bottom RSC encoder. The length of $\mathbf{m}$ is denoted by K.
- Each RSC constituent encoder produces a sequence of parity-check bits, the length of which is $K + m$.
- Once the encoding process is completed, a set of m bits is appended to each block of encoded bits, so as to force each LFSR back to the initial all-zero state.

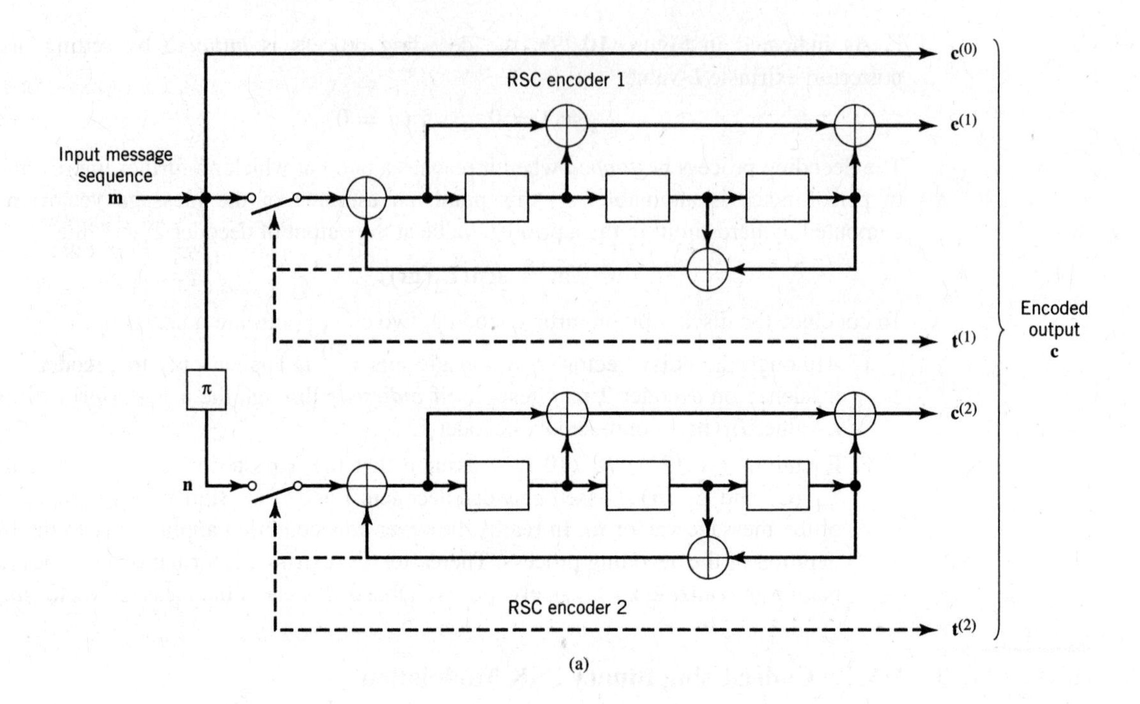

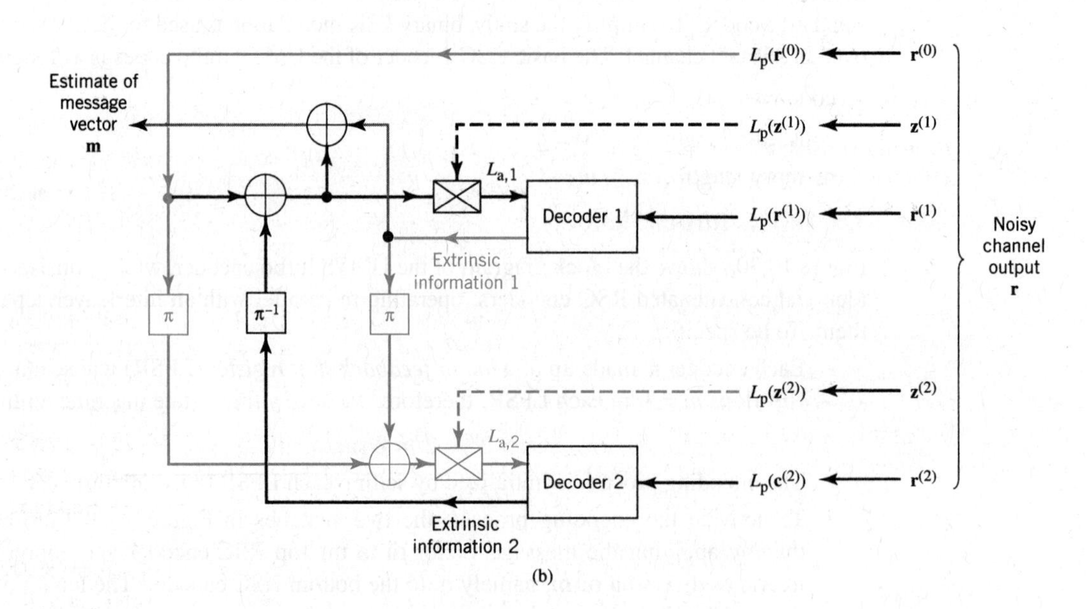

Figure 10.30 Block diagram of UMTS codec. (a) Encoder, (b) Decoder. *Notes:* 1. The received vectors $\{\mathbf{r}^{(0)}, \mathbf{z}^{(1)}, \mathbf{r}^{(1)}, \mathbf{z}^{(2)}, \mathbf{r}^{(2)}\}$ correspond to the transmitted vectors $\{\mathbf{c}^{(0)}, \mathbf{t}^{(0)}, \mathbf{c}^{(1)}, \mathbf{t}^{(2)}, \mathbf{c}^{(2)}\}$. 2. The block labeled π : interleaver. The block labeled π^{-1} : de-interleaver.

From this description, it is apparent that the *overall code-rate* of the turbo code is lower than the UMTS code-rate, namely 1/3, as shown by

$$r_{\text{overall}} = \frac{K}{3K + 4m}$$

Note that if we set the memory length $m = 0$, the code rate r_{overall} is increased again to 1/3.

On the basis of this description, each block of the *multiplexed output* of the turbo encoder is composed as follows:

- $\mathbf{c}^{(0)}$ vector of systematic bits (i.e., message bits), followed by
- $\mathbf{c}^{(1)}$ and $\mathbf{c}^{(2)}$ pair of vectors, representing the parity-check bits produced by the top and bottom RSC encoders, respectively, then followed by
- $\mathbf{t}^{(1)}$ and $\mathbf{t}^{(2)}$ pair of vectors, representing encoder termination-tail bits for forcing the top and bottom RSC constituent encoders back to all-zero state, respectively.

In the UMTS standard, the block length of the turbo code lies in the range [40, 5114].

The UMTS Turbo Decoder

Figure 10.30b shows a block diagram of the UMTS decoder. Specifically, proceeding from top to bottom on the right-hand side of the figure, we have five sequences of a posteriori L-values computed in the receiver, namely $L_p(\mathbf{c}^{(0)})$, $L_p(\mathbf{t}^{(1)})$, $L_p(\mathbf{c}^{(1)})$, $L_p(\mathbf{t}^{(2)})$, and $L_p(\mathbf{c}^{(2)})$; these L-values correspond to the encoded sequences $\mathbf{c}^{(0)}$, $\mathbf{t}^{(1)}$, $\mathbf{c}^{(1)}$, $\mathbf{t}^{(2)}$, and $\mathbf{c}^{(2)}$, respectively.

Considering, first, how decoder 1 operates in the receiver, we find from Figure 10.30b that it receives two input sequences of L-values, the first one of which, namely the a posteriori L-value $L_p(\mathbf{c}^{(1)})$, comes directly from the channel. The other input, the a priori L-value denoted by $L_{a,1}$, is made up of three components:

1. The a posteriori L-value, $L_p(\mathbf{c}^{(0)})$, which accounts for the received systematic bits, $\mathbf{c}^{(0)}$.
2. The reordered version of the extrinsic information produced by decoder 2, resulting from the de-interleaver π^{-1}.
3. The a posteriori L-value, $L_p(\mathbf{t}^{(1)})$ attributed to the systematic vector of termination bits, $\mathbf{t}^{(1)}$, which is appended to the sum of components 1 and 2 to complete $L_{a,1}$.

In a corresponding but slightly different way, decoder 2 receives two input sequences of L-values, the first one of which, namely the a posteriori L-value $L_p(\mathbf{c}^{(2)})$, comes directly from the channel. The other input, a priori L-value $L_{a,2}$, is also made up of three components:

1. The reordered version of a posteriori L-value, $L_p(\mathbf{c}^{(0)})$ is due to the received vector of systematic bits, $\mathbf{c}^{(0)}$, where the reordering is produced by the interleaver to the left of the de-interleaver π^{-1}.
2. The reordered version of extrinsic information is produced by decoder 1, where the reordering is performed by the second interleaver, π, to the right of the de-interleaver π^{-1}.
3. The a posteriori L-value, $L_p(\mathbf{t}^{(2)})$ is attributed to the systematic vector of termination bits, $\mathbf{t}^{(2)}$; this time, however, $L_p(\mathbf{t}^{(2)})$ is removed before it is interleaved and passed to decoder 2.

Simulation Results

In Figure 10.31, we have plotted the BER chart for the iterative decoding process, using the turbo codec of Figure 10.30. The results were obtained for the case of 5000 systematic bits, as follows:

- For a prescribed E_b/N_0 ratio, the bit errors were averaged over 50 Monte Carlo runs.
- Each point in the BER chart was the result of 100 bits per point-count in the decoding process.
- The computations were repeated for different values of E_b/N_0.

The remarkable points to observe from Figure 10.31 are summarized here:

1. In the course of just four iterations, the BER of the UMTS decoder drops to 10^{-14} at an SNR = 3 dB, which, for all practical purposes, is zero.
2. The steepness of the BER plot on iteration 4 is showing signs of the turbo cliff, but is not there yet. Unfortunately, to get there would require a great deal more computation.[14] (The turbo cliff is illustrated in Figure 10.32 in the next section).

Rudimentary Comparison of Viterbi and MAP Algorithms

For a rather rudimentary but plausible approach, to address the issue of computational complexity in a fair-minded way, consider a convolutional code that has $m = 6$ states and, therefore, requires

$$2^6 = 64$$

ACS operations for Viterbi decoding.

To match this computational complexity, using the turbo decoder of Figure 10.29b with 16 ACS operations, we need the following number of decoding iterations:

$$\frac{64}{16} = 4$$

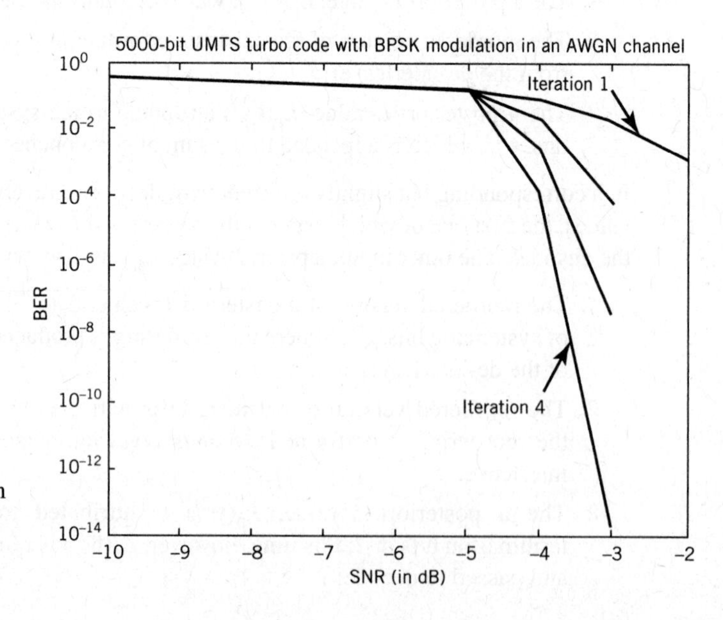

Figure 10.31
The bit error rate (BER) diagram for the UMTS-turbo decoder, using 5000 systematic bits and –3 dB SNR.

Correspondingly, Figure 10.31 plots the BER chart for the turbo decoder for the sequences of decoding iterations: 1, 2, 3, and 4. Of special interest is the BER chart for four iterations, for which we find that the turbo decoder with BER $\approx 10^{-14}$ outperforms the Viterbi decoder significantly for the same computational complexity, namely a total of 64 ACS operations.

10.13 EXIT Charts

In an idealized BER chart exemplified by that in Figure 10.32, we may identify three distinct regions, described as follows:

a. *Low BER region*, for which the E_b/N_0 ratio is correspondingly low.
b. *Waterfall region*, also referred to as the *turbo cliff* in the turbo coding literature, which is characterized by a persistent reduction in BER over the span of a small fraction of dB in SNR.
c. *BER floor region*, where a rather small improvement in decoding performance is achieved for medium to large values of SNRs.

As informative as the BER chart of Figure 10.32 is, from a practical perspective it has a serious drawback. Simply put, the BER chart lacks insight into the underlying dynamics (i.e., convergence behavior) of iterative decoding algorithms, particularly around the turbo-cliff region. Furthermore, since the BER occurs at low BERs, excessive simulation runs are required.

The question is: how do we overcome this serious drawback of the BER chart? The answer lies in using the *extrinsic information chart*, or EXIT chart for short, which was formally introduced by ten Brink (2001).

The EXIT chart is insightful because it provides a graphical procedure for visualizing the underlying dynamics of the turbo decoding process for a prescribed E_b/N_0. Moreover,

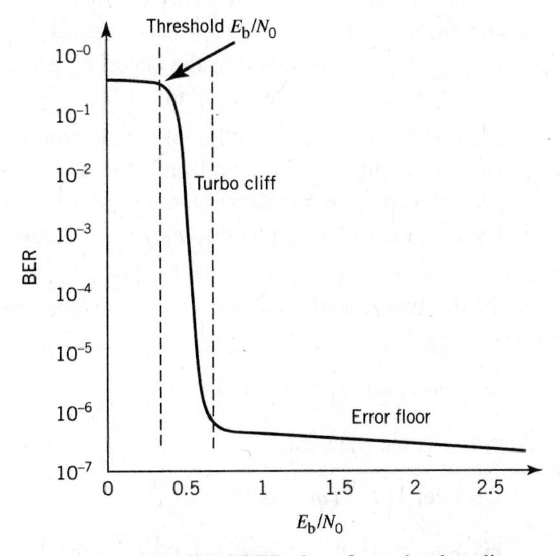

Figure 10.32 Idealized BER chart for turbo decoding.

the procedure provides a tool for the design of turbo codes characterized by good performance in the turbo-cliff region. In any event, development of the EXIT chart exploits the idea of mutual information in Shannon's information theory, which was discussed previously in Chapter 5.

Development of the EXIT Chart

Consider a constituent decoder in the turbo decoder of Figure 10.29b, which is, for convenience of presentation, labeled decoder 1; the other constituent decoder is labeled decoder 2. Let $I_1(m_j; L_a(m_j))$ denote the mutual information between a transmitted message bit m_j and the a priori L-value $L_a(m_j)$ for a prescribed E_b/N_0. Correspondingly, let $L_p(m_j)$ denote the a posteriori L-value of message bit m_j and let $I_2(m_j; L_p(m_j))$ denote the mutual information between m_j and $L_p(m_j)$ for the same E_b/N_0. Then, with $I_2(m_j; L_p(m_j))$ viewed as a function f $I_1(m_j; L_a(m_j))$, we may express the *extrinsic information transfer characteristic* of constituent decoder 1 for some operator $T(\cdot)$ and the prescribed E_b/N_0 as follows:

$$I_2(m_j; L_p(m_j)) = T(I_1(m_j; L_a(m_j))) \qquad (10.128)$$

In the continuum, it is shown that both mutual informations, I_1 and I_2, lie within the range [0,1]. Thus, a plot of $I_2(m_j; L_p(m_j))$ versus $I_1(m_j; L_a(m_j))$, depicted in Figure 10.33a, displays graphically the extrinsic information transfer characteristic of the constituent decoder 1.

Since the two constituent decoders are similar and they are connected together sequentially inside a closed feedback loop, it follows that the extrinsic information transfer characteristic of constituent decoder 2 is the *mirror image* of the curve in Figure 10.33a with respect to the straight line $I_1 = I_2$, as shown in Figure 10.33b. With this relationship in mind, we may go on to put the transfer characteristic curves of the two constituent decoders side by side, but keeping the same horizontal and vertical axes of Figure 10.33a. We thus get the composite picture depicted in Figure 10.33c. In effect, this latter figure represents the input–output extrinsic transfer characteristic of the two constituent decoders working together in a turbo-decoding algorithm for the prescribed E_b/N_0.

To elaborate on the practical utility of Figure 10.33a, suppose that the iterative turbo-decoding algorithm begins with $I_1^{(1)} = 0$, representing the initial condition of constituent decoder 1 for the first iteration in the decoding process. Then, in proceeding forward, we keep the following two points in mind:

- First, the a posteriori L-value of constituent decoder 1 becomes the a priori L-value of constituent decoder 2, and similarly when these two decoders are interchanged, as we proceed from one iteration to the next.
- Second, the message bits $m_1, m_2, m_3, \ldots$ occur on consecutive iterations.

Hence, we will experience the following sequence of extrinsic information transfers between the two constituent decoders from one message bit to the next one for some prescribed E_b/N_0:

Initial condition: $I_1^{(1)}(m_1) = 0$.

Iteration 1: message bit, m_1

Decoder 1: $I_1^{(1)}(m_1)$ defines $I_2^{(1)}(m_1)$.

Decoder 2: $I_2^{(1)}(m_1)$ initiates $I_1^{(2)}(m_2)$ for iteration 2.

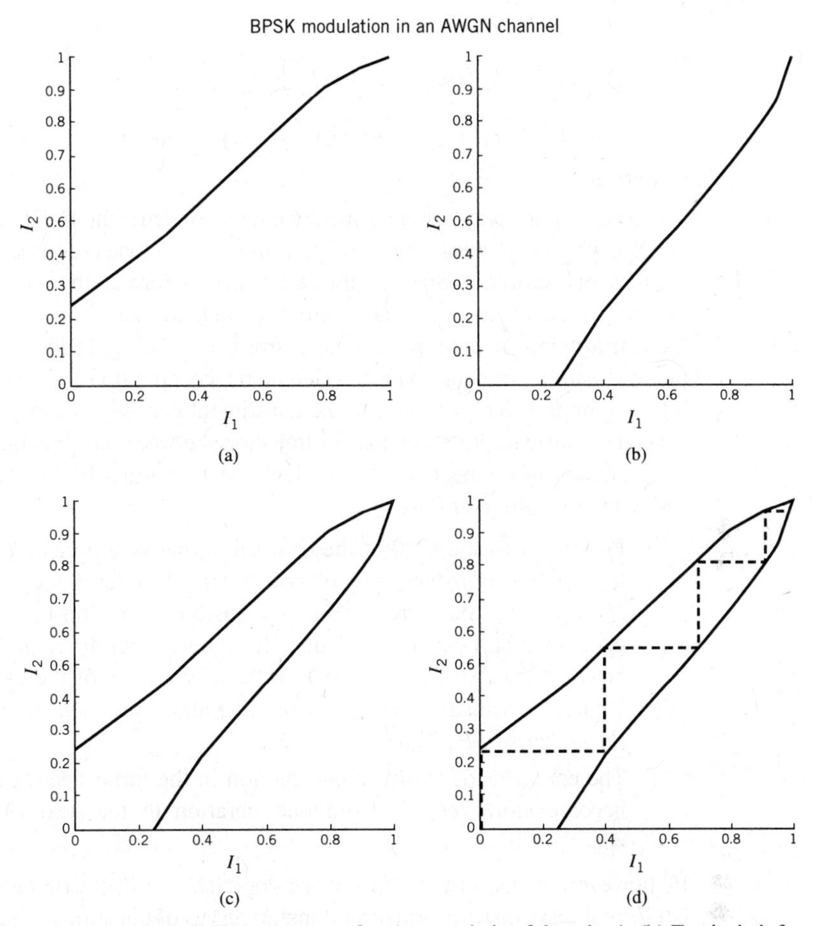

Figure 10.33 (a) Extrinsic information transfer characteristic of decoder 1; (b) Extrinsic information transfer characteristic of decoder 2; (c) Input-output extrinsic transfer characteristic of the two constituent decoders working together; (d) EXIT chart, including the staircase (shown dashed) embracing the extrinsic information transfer characteristics of both constituent decoders.

Iteration 2: message bit, m_2

Decoder 1: $I_1^{(2)}(m_2)$ defines $I_2^{(2)}(m_2)$.

Decoder 2: $I_2^{(2)}(m_2)$ initiates $I_1^{(3)}(m_3)$ for iteration 3.

Iteration 3: message bit, m_3

Decoder 1: $I_1^{(3)}(m_3)$ defines $I_2^{(3)}(m_3)$.

Decoder 2: $I_2^{(3)}(m_3)$ initiates $I_1^{(4)}(m_4)$ for iteration 4.

Iteration 4: message bit, m_4

Decoder 1: $I_1^{(4)}(m_4)$ defines $I_2^{(4)}(m_4)$.

Decoder 2: $I_2^{(4)}(m_4)$ initiates $I_1^{(4)}(m_5)$ for iteration 5.

and so on.

Proceeding in the way just described, we may construct the EXIT chart illustrated in Figure 10.33d, which embodies a trajectory that moves from one constituent decoder to the other in the form of a *staircase*. Specifically, the extrinsic information transfer curve from constituent decoder 1 to constituent decoder 2 proceeds in a horizontal manner and, by the same token, the extrinsic information transfer curve from constituent decoder 2 to constituent decoder 1 proceeds in a vertical manner. Hereafter, construction of the sequence of extrinsic information transfer curves from one constituent decoder to another is called the staircase-shaped extrinsic information transfer trajectory between constituent decoders 1 and 2.

Examination of the EXIT chart depicted in Figure 10.33d prompts us to make the following two observations:

a. Provided that the SNR at the channel output is sufficiently high, then the extrinsic information transfer curve of constituent decoder 1 stays above the straight line $I_1 = I_2$, while the corresponding extrinsic information transfer curve of constituent decoder 2 stays below this line. It follows, therefore, that an open *tunnel* exists between the extrinsic information transfer curves of the two constituent decoders. Under this scenario, the turbo-decoding algorithm *converges to a stable solution for the prescribed* E_b/N_0.

b. The estimates of extrinsic information in the turbo-decoding algorithm continually become *more reliable* from one iteration to the next as the stable solution is approached.

If, however, in contrast to the picture depicted in Figure 10.33d, *no* open tunnel exists between the extrinsic information transfer curves of constituent decoders 1 and 2 when the prescribed E_b/N_0 is relatively low, then the turbo-decoding algorithm fails to converge (i.e., the turbo-decoding algorithm is unstable). This behavior is illustrated in the EXIT chart of Figure 10.34 where the SNR has been reduced compared to that in Figure 10.33.

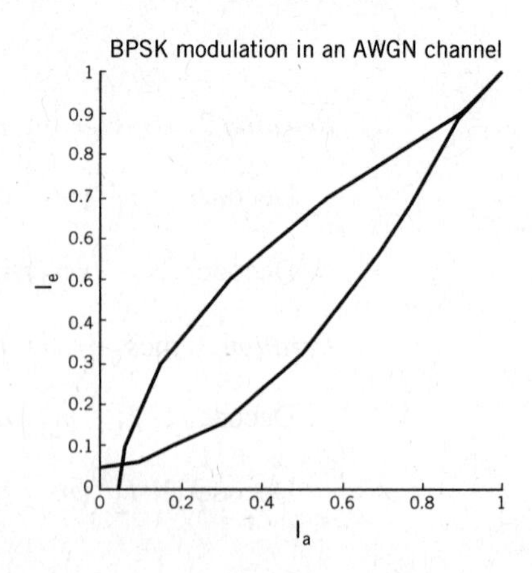

Figure 10.34
EXIT chart demonstrating nonconvergent behavior of the turbo decoder when the E_b/N_0 is reduced compared to that in Figure 10.33d.

The stage is now set for us to introduce the following statement:

> The E_b/N_0 threshold of a turbo-decoding algorithm is that smallest value of E_b/N_0 for which an open tunnel exists in the EXIT chart.

It is the graphical simplicity of this important statement that makes the EXIT chart such a useful practical tool in the design of iterative decoding algorithms.

Moreover, if it turns out that the EXIT curves of the two constituent decoders do not intersect before the (1, 1)-point of perfect convergence and the staircase-shaped decoding trajectory also succeeds in reaching this critical point, then a vanishingly low BER is expected (Hanzo, 2012).

Approximate Gaussian Model

For an approximate model needed to display the underlying dynamics of iterative decoding algorithms, the first step is to assume that the a priori L-values for message bit m_j, namely $L_a(m_j)$, constitute independent Gaussian random variables. With $m_j = \pm 1$, the $L_a(m_j)$ assumes a variance σ_a^2 and a mean value $(\sigma_a^2/2)\,m_j$. Equivalently, we may express the statistical dependence of L_a on m_j as follows:

$$L_a(m_j) = \left(\frac{\sigma_a^2}{2}\right) m_j + n_a \tag{10.129}$$

where n_a is the sample value of a zero-mean Gaussian random variable with variance σ_a^2.

The rationale for the approximate Gaussian model just described is motivated by the following two points (Lin and Costello, 2003):

a. For an AWGN channel with soft (i.e., unquantized) output, the log-likelihood ratio, L-value, denoted by $L_a(m_j|r_j^{(0)})$ of a transmitted message bit m_j given the receiver signal $r_j^{(0)}$, may be modeled as follows (see Problem 10.36):

$$L_a(m_j|r_j^{(0)}) = L_c r_j^{(0)} + L_a(m_j) \tag{10.130}$$

where $L_c = 4(E_s/N_0)$ is the channel reliability factor defined in (10.91) and $L_a(m_j)$ is the a priori L-value of message bit m_j. The point to note here is that the product terms $L_c r_j^{(0)}$ for varying j are independent Gaussian random variables with variance $2L_c$ and mean $\pm L_c$.

b. Extensive Monte Carlo simulations of the a posteriori extrinsic L-values, $L_e(m_j)$, for a constituent decoder with large block length appear to support the Gaussian-model assumption of (10.129); see Wiberg *et al.* (1999).

Accordingly, using the Gaussian approximation of (10.129), we may express the conditional probability density function of the a priori L-value as follows:

$$f_{L_a}(\xi|m_j) = \frac{1}{\sqrt{2\pi}\,\sigma_a}\exp\left[-\frac{(\xi - m_j\sigma_a^2/2)^2}{2\sigma_a^2}\right] \tag{10.131}$$

where ξ is a dummy variable, representing a sample value of $L_a(m_j)$. Note also that ξ is continuous whereas, of course, m_j is discrete. It follows that in formulating the mutual information between the message bit $m_j = +1$ and a priori L-value $L_a(m_j)$ we have a binary input AWGN channel to deal with; such a channel was discussed previously in Example 5 of Chapter 5 on information theory. Building on the results of that example, we may express the first desired mutual information, denoted by $I_1(m_j; L_a)$, as follows:

$$I_1(m_j;L_a) = \frac{1}{2} \sum_{m_j = -1, +1} \int_{-\infty}^{\infty} f_{L_a}(\xi|m_j) \log_2\left(\frac{2f_{L_a}(\xi|m_j)}{f_{L_a}(\xi|m_j = -1) + f_{L_a}(\xi|m_j = +1)}\right) d\xi \quad (10.132)$$

where the summation accounts for the binary nature of the information bit m_j and the integral accounts for the continuous nature of L_a. Using (10.131) and (10.132) and manipulating the results, we get (ten Brink, 2001):

$$I_1(m_j;L_a) = 1 - \int_{-\infty}^{\infty} \frac{\exp\left[-\dfrac{(\xi - m_j\sigma_a^2/2)^2}{2\sigma_a^2}\right]}{\sqrt{2\pi}\,\sigma_a} \log_2[1 + \exp(-\xi)]\, d\xi \quad (10.133)$$

which, as expected, depends solely on the variance σ_a^2. To emphasize this fact, let the new function

$$\mathcal{J}(\sigma_a) := I_1(m_j;L_a) \quad (10.134)$$

with the following two limiting values:

$$\lim_{\sigma_a \to 0} \mathcal{J}(\sigma_a) = 0$$

and

$$\lim_{\sigma_a \to \infty} \mathcal{J}(\sigma_a) = 1$$

In other words, we have

$$0 \le I_1(m_j;L_a) \le 1 \quad (10.135)$$

Moreover, $\mathcal{J}(\sigma_a)$ increases monotonically with increasing σ_a, which means that if the value of the mutual information $I_1(m_j;L_a)$ is given, then the corresponding value of σ_a is uniquely determined by the inverse formula:

$$\sigma_a = \mathcal{J}^{-1}(I_1) \quad (10.136)$$

and with it, the corresponding Gaussian random variable $L_a(m_j)$ defined in (10.129) is obtained.

Referring back to (10.128), we note that for us to construct the EXIT chart we also need to know the second mutual information between the message bit m_j and the a posteriori extrinsic L-value $L_p(m_j)$. To this end, we may build on the formula of (10.132) to write

$$I_2(m_j;L_p) = \frac{1}{2} \sum_{m_j = -1, +1} \int_{-\infty}^{\infty} f_{L_p}(\xi|m_j) \log_2\left(\frac{2f_{L_p}(\xi|m_j)}{f_{L_p}(\xi|m_j = -1) + f_{L_p}(\xi|m_j = +1)}\right) d\xi \quad (10.137)$$

where, in a manner similar to the a priori mutual information $I_1(m_j;L_a(m_j))$, we also have

$$0 \le I_2(m_j;L_p) \le 1 \quad (10.138)$$

Accordingly, with the two mutual informations $I_1(m_j;L_a)$ and $I_2(m_j;L_p)$ at hand, we may go on to compute the EXIT chart for an iterative decoding algorithm by merely focusing on a single constituent decoder in the turbo decoding algorithm.

The next issue to be considered is how to perform this computation, which we now address.

Histogram Method for Computing the EXIT Chart

For turbo codes having long interleavers, the approximate Gaussian model of (10.129) is good enough for practical purposes. Hence, we may use this model to formulate the traditional *histogram method*, described in ten Brink (2001) to compute the EXIT chart. Specifically, (10.137) is used to compute the mutual information $I_2(m_j;L_p)$ for a prescribed E_b/N_0. To this end, *Monte Carlo simulation* (i.e., histogram measurements) is used to compute the required probability density function, $f_{L_p}(\xi|L_p(m_j))$, on which *no* Gaussian assumption can be imposed for obvious reasons. Computaton of this probability density function is central to the EXIT chart, which may proceed in a step-by-step manner for a prescribed E_b/N_0, as follows:

Step 1: Apply the independent Gaussian random variable defined in (10.129) to constituent decoder 1 in the turbo decoder. The corresponding value of the mutual information $I_1(m_j;L_p)$ is obtained by choosing the variance σ_a^2 in accordance with (10.129).

Step 2: Using Monte Carlo simulation, compute the probability density function $f_{L_p}(\xi|L_p)$. Hence, compute the second mutual information $I_2(m_j;L_p)$, and with it a certain point for the extrinsic information transfer curve of constituent decoder 1 is determined.

Step 3: Continue Steps 1 and 2 until we have sufficient points to construct the extrinsic information transfer curve of constituent decoder 1.

Step 4: Construct the extrinsic information transfer curve of constituent decoder 2 as the mirror image of the curve for constituent decoder 1 computed in Step 3, respecting the straight line $I_1 = I_2$.

Step 5: Construct the EXIT chart for the turbo decoder by combining the extrinsic information transfer curves of constituent decoders 1 and 2.

Step 6: Starting with some prescribed initial condition, for example $I_1(m_1) = 0$ for message bit m_1, construct the staircase information transfer trajectory between constituent decoders 1 and 2.

A desirable feature of the histogram method for computing the EXIT chart is the fact that, except for the approximate Gaussian model of (10.129), there are *no* other assumptions needed for the computations involved in Steps 1 through 6.

Averaging Method for Computing EXIT Charts

For another method to compute EXIT charts, we may use the so-called *averaging method*, which represents an alternative approach to the histogram method.

As a reminder, the basic issue in computing an EXIT chart is to measure the mutual information between the information bits, m_j, at the turbo encoder input in the transmitter and the corresponding L-values produced at the output of the corresponding BCJR decoder in the receiver. Due to the inherently nonlinear input–output characteristic of the BCJR decoder, the underlying distribution of the L-values is not only unknown, but also highly likely to be non-Gaussian as well, thereby complicating the measurement. To get around this difficulty, we may invoke the *ergodic theorem*, which was discussed previously in Chapter 4 on stochastic processes. As explained therein, under certain conditions, it is feasible to replace the operation of ensemble averaging (i.e., expectation) with time averaging. Thus, in proceeding along this ergodic path, we have a new nonlinear transformation, where the *time average* of a large set of L-value samples, available at the output of the BCJR decoder, provides an estimate of the desired mutual information; moreover, it does so without requiring knowledge of the original data (i.e., the m_j). It is for this reason that the second method of computing EXIT charts is called the averaging method.[15]

Just as the use of a single constituent decoder suffices for computing an EXIT chart in the histogram method, the same decoding scenario applies equally well to the averaging method. It is with this point in mind that the underlying scheme for the averaging method is as depicted in Figure 10.35. Most importantly, this scheme is designed in such a way that the following requirements are satisfied:

1. Implementations of channel estimation, carrier receiver, modulation, and demodulation are all perfect.
2. The turbo decoder is perfectly synchronized with the turbo encoder.
3. The BCJR algorithm or exact equivalent (i.e., the log-MAP algorithm) is used to optimize the turbo decoder.

Moreover, the following analytic correspondences between the constituent encoder 1 at the top of Figure 10.35 and the turbo decoder 2 at the bottom of the figure are carefully noted: the code vectors $\mathbf{c}^{(0)}$, $\mathbf{c}^{(1)}$, and termination vector $\mathbf{t}^{(1)}$ in the encoder map onto the a posteriori L-values $L_p(\mathbf{r}^{(0)})$, $L_p(\mathbf{r}^{(1)})$, and $L_p(\mathbf{z}^{(1)})$ in the decoder, respectively.

It can therefore be justifiably argued that in light of these rigorous requirements, the underlying algorithm for the averaging method is well designed and therefore trustworthy in the following sense: in the course of computing the EXIT chart, the algorithm trusts what the computed L-values actually say; that is, they do not under or over represent their confidence in the message bits. This important characteristic of the averaging method is to be contrasted against the histogram method. Indeed, it is for this reason that the histogram method compares the L-values against the true values of the message bits, hence requiring knowledge of them.

In summary, we may say that trustworthy L-values are those L-values that satisfy the *consistency condition.* A simple way of testing this condition is to do the following: use the averaging and histogram methods to compute two respective sets of L-values. If, then, both methods yield the same value for the mutual information, then the consistency condition is satisfied (Maunder, 2012).

Procedure for Measuring the EXIT Chart

Referring back to the scheme of Figure 10.35, the demultiplexer outputs denoted by $\mathbf{r}^{(0)}$, $\mathbf{r}^{(1)}$and $\mathbf{z}^{(1)}$ represent the L-values corresponding to the encoder outputs $\mathbf{c}^{(0)}$, $\mathbf{c}^{(1)}$ and $\mathbf{t}^{(1)}$, respectively. Thus, following the way in which the turbo decoder of Figure 10.33b was described, the internally generated input applied to BCJR decoder 1 assumes exactly the

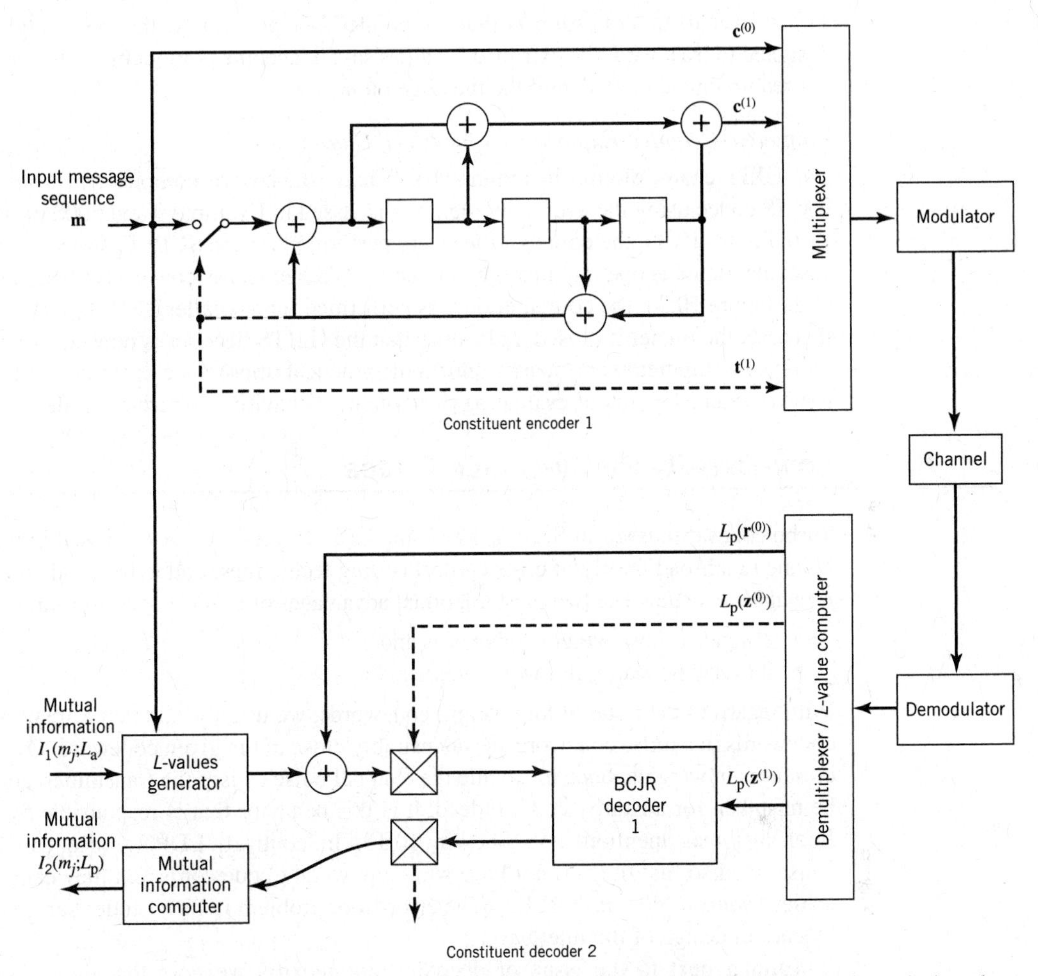

Figure 10.35 Schematic diagram for computing the EXIT chart for the UMTS-turbo code, based on the averaging method.

same value as that produced in computing the BER. With the objective being that of constructing an EXIT chart, we need to provide values for the mutual information that was previously denoted by $I_1(m_j;\, L_a)$, where m_j is the jth message bit and L_a is the corresponding a priori L-value. As indicated, the mutual information is the externally supplied input applied to the block labeled *L-value generator*. We may therefore assign to $I_1(m_j;\, L_a)$ any values that we like. However, recognizing that $0 \le I_1(m_j;\, L_a) \le 1$, a sensible choice of values for $I_1(m_j;\, L_a)$ would be the set $\{0.0, 0.1, 0.2, 0.3, 0.4, 0.5, 0.6, 0.7, 0.8, 0.9, 1.0\}$. Such a choice provides inputs for eleven different experiments based on the averaging method.

For each one of these inputs, the constituent decoder 1 produces a corresponding value for the a posteriori extrinsic L-value, L_p, which is applied to the block labeled *mutual-information computer* in Figure 10.35. The resulting output of this second computation is the second desired mutual information, namely $I_2(m_j;\, L_e)$. At this point, a question that begs itself is: how can this computation be performed in the absence of the message m_j?

The answer to this question is that, as pointed out previously, the averaging method is designed to trust what the extrinsic L-values say; hence, the computation of $I_2(m_j; L_e)$ does not require any knowledge of the message bit m_j.

Computer-Oriented Experiments for EXIT Charts

The EXIT charts plotted in Figures 10.33 and 10.34 were computed in Matlab for the UMTS codec using the averaging method discussed in Example 9 and 5000 message bits.

In Figure 10.33, the computations were performed for the SNR: $E_b/N_0 = -4$ dB. In this case, the tunnel is open, indicating that the UMTS decoder is convergent (stable).

In Figure 10.34, the computation was performed for a smaller SNR: $E_b/N_0 = -6$ dB. In this case, the tunnel is closed, indicating that the UMTS decoder is nonconvergent.

These computer experiments confirm the practical importance of EXIT charts when the issue of interest is that of evaluating the dynamic behavior of a turbo decoder.

10.14 Low-Density Parity-Check Codes

Turbo codes discussed in Section 10.12 and LDPC codes[16] to be discussed in this section belong to a broad family of error-control coding techniques, collectively called *compound probabilistic codes*. The two most important advantages of LDPC codes over turbo codes are:

- absence of low-weight codewords and
- iterative decoding of lower complexity.

With regard to the issue of low-weight codewords, we usually find that a small number of codewords in a turbo codeword are undesirably close to the given codeword. Owing to this closeness in weights, once in a while the channel noise causes the transmitted codeword to be mistaken for a nearby code. Indeed, it is this behavior that is responsible for the error floor that was mentioned in Section 10.13. In contrast, LDPC codes can be easily constructed so that they do not have such low-weight codewords and they can, therefore, achieve *vanishingly small* BERs. (The error-floor problem in turbo codes can be alleviated by careful design of the interleaver.)

Turning next to the issue of decoding complexity, we note that the computational complexity of a turbo decoder is dominated by the MAP algorithm, which operates on the trellis for representing the convolutional code used in the encoder. The number of computations in each recursion of the MAP algorithm scales linearly with the number of states in the trellis. Commonly used turbo codes employ trellises with 16 states or more. In contrast, LDPC codes use a simple parity-check trellis that has just two states. Consequently, the decoders for LDPC codes are significantly simpler to design than those for turbo decoders. However, a practical objection to the use of LDPC codes is that, for large block lengths, their encoding complexity is high compared with turbo codes.

It can be argued that LDPC codes and turbo codes complement each other, giving the designer more flexibility in selecting the right code for extraordinary decoding performance.

Construction of LDPC Codes

LDPC codes are specified by a parity-check matrix denoted by **A**, which is purposely chosen to be *sparse*; that is, the code consists mainly of 0s and a small number of 1s. In particular, we speak of (n, t_c, t_r) LDPC codes, where n denotes the block length, t_c denotes the weight (i.e., number of 1s) in each column of the matrix **A**, and t_r denotes the weight of each row with $t_r > t_c$. The rate of such an LDPC code is defined by

$$r = 1 - \frac{t_c}{t_r}$$

whose validity may be justified as follows. Let ρ denote the *density* of 1s in the parity-check matrix **A**. Then, following the terminology introduced in Section 10.4, we may set

$$t_c = \rho(n-k)$$

and

$$t_r = \rho n$$

where $(n - k)$ is the number of rows in **A** and n is the number of columns (i.e., the block length). Therefore, dividing t_c by t_r, we get

$$\frac{t_c}{t_r} = 1 - \frac{k}{n} \tag{10.139}$$

By definition, the code rate of a block code is k/n; hence, the result of (10.139) follows. For this result to hold, however, the rows of **A** must be *linearly independent.*

The structure of LDPC codes is well portrayed by bipartite graphs, which were introduced by Tanner (1981) and, therefore, are known as *Tanner graphs.*[17] Figure 10.36 shows such a graph for the example code of $n = 10$, $t_c = 3$, and $t_r = 5$. The left-hand nodes in the graph are *variable (symbol) nodes*, which correspond to elements of the codeword. The right-hand nodes of the graph are *check nodes*, which correspond to the set of parity-check constraints satisfied by codewords in the code. LDPC codes of the type exemplified by the graph of Figure 10.36 are said to be *regular*, in that all the nodes of a similar kind have exactly the same degree. In Figure 10.36, the degree of the variable nodes is $t_c = 3$ and the degree of the check nodes is $t_r = 5$. As the block length n approaches infinity, each check node is connected to a vanishingly small fraction of variable nodes; hence the term "low density."

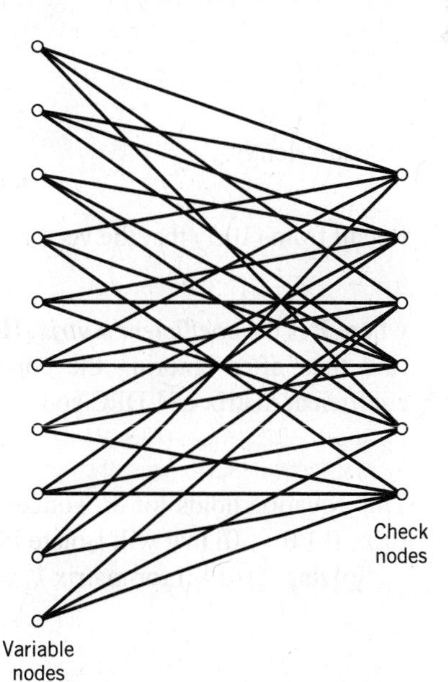

Figure 10.36
Bipartite graph of the (10, 3, 5) LDPC code.

The matrix $\mathbf{A}$ is constructed by putting 1s in $\mathbf{A}$ at *random*, subject to the *regularity constraints*:

- each column of matrix $\mathbf{A}$ contains a small fixed number t_c of 1s;
- each row of the matrix contains a small fixed number t_r of 1s.

In practice, these regularity constraints are often violated slightly in order to avoid having linearly dependent rows in the parity-check matrix $\mathbf{A}$.

Unlike the linear block codes discussed in Section 10.4, the parity-check matrix $\mathbf{A}$ of LDPC codes is *not* systematic (i.e., it does not have the parity-check bits appearing in diagonal form); hence the use of a symbol different from that used in Section 10.4. Nevertheless, for coding purposes, we may derive a generator matrix $\mathbf{G}$ for LDPC codes by means of Gaussian elimination performed in modulo-2 arithmetic; this procedure is illustrated later in Example 10. Following the terminology introduced in Section 10.4, the 1-by-n code vector $\mathbf{c}$ is first partitioned as shown by

$$\mathbf{c} = [\mathbf{b} \vdots \mathbf{m}]$$

where $\mathbf{m}$ is the k-by-1 message vector and $\mathbf{b}$ is the $(n-k)$-by-1 parity-check vector; see (10.9). Correspondingly, the parity-check matrix $\mathbf{A}$ is partitioned as

$$\mathbf{A}^T = \begin{bmatrix} \mathbf{A}_1 \\ \text{----} \\ \mathbf{A}_2 \end{bmatrix} \tag{10.140}$$

where $\mathbf{A}_1$ is a square matrix of dimensions $(n-k) \times (n-k)$ and $\mathbf{A}_2$ is a rectangular matrix of dimensions $k \times (n-k)$; transposition symbolized by the superscript T is used in the partitioning of matrix $\mathbf{A}$ for convenience of presentation. Imposing a constraint on the LDPC code similar to that of (10.16) we may write

$$[\mathbf{b} \vdots \mathbf{m}] \begin{bmatrix} \mathbf{A}_1 \\ \text{----} \\ \mathbf{A}_2 \end{bmatrix} = \mathbf{0}$$

or, equivalently,

$$\mathbf{bA}_1 + \mathbf{mA}_2 = \mathbf{0} \tag{10.141}$$

Recall from (10.7) that the vectors $\mathbf{m}$ and $\mathbf{b}$ are related by

$$\mathbf{b} = \mathbf{mP}$$

where $\mathbf{P}$ is the *coefficient matrix*. Hence, substituting this relation into (10.141), we readily find that, after ignoring the common factor $\mathbf{m}$ for any nonzero message vector, the coefficient matrix of LDPC codes satisfies the condition

$$\mathbf{PA}_1 + \mathbf{A}_2 = \mathbf{0} \tag{10.142}$$

This equation holds for *all* nonzero message vectors and, in particular, for $\mathbf{m}$ in the form $[0 \ldots 0\ 1\ 0 \ldots 0]$ that will isolate individual rows of the generator matrix.

Solving (10.142) for matrix $\mathbf{P}$, we get

$$\mathbf{P} = \mathbf{A}_2\mathbf{A}_1^{-1} \tag{10.143}$$

where $\mathbf{A}_1^{-1}$ is the *inverse* of matrix $\mathbf{A}_1$, which is naturally defined in modulo-2 arithmetic. Finally, building on (10.12), the generator matrix of LDPC codes is defined by

$$\begin{aligned}\mathbf{G} &= [\mathbf{P} \;\vdots\; \mathbf{I}_k] \\ &= [\mathbf{A}_2\mathbf{A}_1^{-1} \;\vdots\; \mathbf{I}_k]\end{aligned} \tag{10.144}$$

where $\mathbf{I}_k$ is the k-by-k identity matrix.

It is important to note that if we take the parity-check matrix $\mathbf{A}$ for some arbitrary LDPC code and just pick $(n-k)$ columns of $\mathbf{A}$ at random to form a square matrix $\mathbf{A}_1$, there is *no* guarantee that $\mathbf{A}_1$ will be *nonsingular* (i.e., the inverse $\mathbf{A}_1^{-1}$ will exist), even if the rows of $\mathbf{A}$ are linearly independent. In fact, for a typical LDPC code with large block length n, such a randomly selected $\mathbf{A}_1$ is highly unlikely to be nonsingular because it is very likely that at least one row of $\mathbf{A}_1$ will be all 0s. Of course, when the rows of $\mathbf{A}$ are linearly independent, there will be *some* set of $(n-k)$ columns of $\mathbf{A}$ that will result in a nonsingular $\mathbf{A}_1$, to be illustrated in Example 10. For some construction methods for LDPC codes, the first $(n-k)$ columns of $\mathbf{A}$ may be guaranteed to produce a nonsingular $\mathbf{A}_1$, or at least do so with a high probability, but that is *not* true in general.

EXAMPLE 10 (10, 3, 5) LDPC Code

Consider the Tanner graph of Figure 10.34 pertaining to a (10, 3, 5) LDPC code. The parity-check matrix of the code is defined by

$$\mathbf{A} = \underbrace{\begin{bmatrix} 1&1&0&1&0&1 \\ 0&1&1&0&1&0 \\ 1&0&0&0&1&1 \\ 0&1&1&1&0&1 \\ 1&0&1&0&1&0 \\ 0&0&0&1&0&0 \end{bmatrix}}_{\mathbf{A}_1^T}\!\!\!\!\begin{matrix}\vdots\\ \vdots\\ \vdots\\ \vdots\\ \vdots\\ \vdots\end{matrix}\!\!\!\!\underbrace{\begin{bmatrix} 0&0&1&0 \\ 1&1&0&0 \\ 0&0&1&1 \\ 1&0&0&0 \\ 0&1&0&1 \\ 1&1&1&1 \end{bmatrix}}_{\mathbf{A}_2^T}$$

which appears to be random, while maintaining the regularity constraints: $t_c = 3$ and $t_r = 5$. Partitioning the matrix $\mathbf{A}$ in the manner just described, we write

$$\mathbf{A}_1 = \begin{bmatrix} 1&0&1&0&1&0 \\ 1&1&0&1&0&0 \\ 0&1&0&1&1&0 \\ 1&0&0&1&0&1 \\ 0&1&1&0&1&0 \\ 1&0&1&1&0&0 \end{bmatrix}$$

$$\mathbf{A}_2 = \begin{bmatrix} 0&1&0&1&0&1 \\ 0&1&0&0&1&1 \\ 1&0&1&0&0&1 \\ 0&0&1&0&1&1 \end{bmatrix}$$

To derive the inverse of matrix $\mathbf{A}_1$, we first use (10.140) to write

$$\underbrace{[\ b_0, b_1, b_2, b_3, b_4, b_5\ ]}_{\mathbf{b}} \underbrace{\begin{bmatrix} 1&0&1&0&1&0 \\ 1&1&0&1&0&0 \\ 0&1&0&1&1&0 \\ 1&0&0&1&0&1 \\ 0&1&1&0&1&0 \\ 1&0&1&1&0&0 \end{bmatrix}}_{\mathbf{A}_1} = \underbrace{[\ u_0, u_1, u_2, u_3, u_4, u_5\ ]}_{\mathbf{u}\,=\,\mathbf{m}\mathbf{A}_2}$$

where we have introduced the vector $\mathbf{u}$ to denote the matrix product $\mathbf{m}\mathbf{A}_2$. By using *Gaussian elimination, modulo-2*, the matrix $\mathbf{A}_1$ is transformed into lower diagonal form (i.e., all the elements above the main diagonal are zero), as shown by

$$\mathbf{A}_1 \rightarrow \begin{bmatrix} 1&0&0&0&0&0 \\ 1&1&0&0&0&0 \\ 0&1&1&0&0&0 \\ 1&0&1&1&0&0 \\ 0&1&0&1&1&0 \\ 1&0&0&1&0&1 \end{bmatrix}$$

This transformation is achieved by the following modulo-2 additions performed on the columns of square matrix $\mathbf{A}_1$:

- columns 1 and 2 are added to column 3;
- column 2 is added to column 4;
- columns 1 and 4 are added to column 5;
- columns 1, 2, and 5 are added to column 6.

Correspondingly, the vector $\mathbf{u}$ is transformed as shown by

$$\mathbf{u} \rightarrow [u_0, u_1, u_0 + u_1 + u_2, u_1 + u_3, u_0 + u_3 + u_4, u_0 + u_1 + u_4 + u_5]$$

Accordingly, premultiplying the transformed matrix $\mathbf{A}_1$ by the parity vector $\mathbf{b}$, using successive eliminations in modulo-2 arithmetic working backwards and putting the solutions for the elements of the parity vector $\mathbf{b}$ in terms of the elements of the vector $\mathbf{u}$ in matrix form, we get

$$\underbrace{[\ u_0, u_1, u_2, u_3, u_4, u_5\ ]}_{\mathbf{u}} \underbrace{\begin{bmatrix} 0&0&1&0&1&1 \\ 1&0&1&0&0&1 \\ 1&1&1&0&0&0 \\ 1&1&0&0&1&0 \\ 0&1&0&0&1&1 \\ 1&1&1&1&0&1 \end{bmatrix}}_{\mathbf{A}_1^{-1}} = \underbrace{[\ b_0, b_1, b_2, b_3, b_4, b_5\ ]}_{\mathbf{b}}$$

The inverse of matrix $\mathbf{A}_1$ is therefore

$$\mathbf{A}_1^{-1} = \begin{bmatrix} 0&0&1&0&1&1 \\ 1&0&1&0&0&1 \\ 1&1&1&0&0&0 \\ 1&1&0&0&1&0 \\ 0&1&0&0&1&1 \\ 0&1&1&1&0&1 \end{bmatrix}$$

Using the given value of $\mathbf{A}_2$ and the value of $\mathbf{A}_1^{-1}$ just found, the matrix product $\mathbf{A}_2\mathbf{A}_1^{-1}$ is given by

$$\mathbf{A}_2\mathbf{A}_1^{-1} = \begin{bmatrix} 1&0&0&1&1&0 \\ 0&0&0&1&1&1 \\ 0&0&1&1&1&0 \\ 0&1&0&1&1&0 \end{bmatrix}$$

Finally, using (10.144), the generator of the (10, 3, 5) LDPC code is defined by

$$\mathbf{G} = \left[\begin{array}{cccccc:cccc} 1&0&0&1&1&0&1&0&0&0 \\ 0&0&0&1&1&1&0&1&0&0 \\ 0&0&1&1&1&0&0&0&1&0 \\ 0&1&0&1&1&0&0&0&0&1 \end{array}\right]$$
$$\quad\underbrace{\qquad\qquad\qquad}_{\mathbf{A}_2\mathbf{A}_1^{-1}}\quad\underbrace{\qquad\quad}_{\mathbf{I}_k}$$

It is important to recognize that the LDPC code described in this example is intended only for the purpose of illustrating the procedure involved in the generation of such a code. In practice, the block length n is orders of magnitude larger than that considered in this example. Moreover, in constructing the matrix **A**, we may constrain all pairs of columns to have a *matrix overlap* (i.e., inner product of any two columns in matrix **A**) not to exceed one; such a constraint, over and above the regularity constraints, is expected to improve the performance of LDPC codes. Unfortunately, with a small block length as that considered in this example, it is difficult to satisfy this additional requirement.

Minimum Distance of LDPC Codes

In practice, the block length of an LDPC code is large, ranging from 10^3 to 10^6, which means that the number of codewords in a particular code is correspondingly large. Consequently, the algebraic analysis of LDPC codes is rather difficult. As such, it is much more productive to perform a *statistical analysis* on an ensemble of LDPC codes. Such an analysis permits us to make statistical statements about certain properties of member codes in the ensemble. An LDPC code with these properties can be found with high probability by a *random selection from the ensemble*, hence the inherent probabilistic structure of the code.

Among these properties, the minimum distance of the member codes is of particular interest. From Section 10.4, we recall that the minimum distance of a linear block code is the smallest Hamming distance between any pair of code vectors in the code. In contrast, we say:

> Over an ensemble of LDPC codes, the minimum distance of a member code in the ensemble is naturally a random variable.

Elsewhere,[18] it is shown that as the block length n increases for fixed $t_c \geq 3$ and $t_r > t_c$, the probability distribution of the minimum distance can be overbounded by a function that approaches a unit step function at a fixed fraction $\Delta_{t_c t_r}$ of the block length n. Thus, for large n, practically all the LDPC codes in the ensemble have a minimum distance of at least $n\Delta_{t_c t_r}$. Table 10.7 presents the rate and $\Delta_{t_c t_r}$ of LDPC codes for different values of the weight-pair (t_c, t_r). From this table we see that for $t_c = 3$ and $t_r = 6$, the code rate r attains its highest value of 1/2 and the fraction $\Delta_{t_c t_r}$ attains its smallest value; hence the preferred choice of $t_c = 3$ and $t_r = 6$ in constructing the LDPC code.

Probabilistic Decoding of LDPC Codes

At the transmitter, a message vector $\mathbf{m}$ is encoded into a code vector $\mathbf{c} = \mathbf{mG}$, where $\mathbf{G}$ is the generator matrix for a specified weight-pair (t_c, t_r) and, therefore, minimum distance $d_{\min}$. The vector $\mathbf{c}$ is transmitted over a noisy channel to produce the received vector

$$\mathbf{r} = \mathbf{c} + \mathbf{e}$$

where $\mathbf{e}$ is the error vector due to channel noise; see (10.17). By construction, the matrix $\mathbf{A}$ is the parity-check matrix of the LDPC code; that is, $\mathbf{AG}^T = \mathbf{0}$. Given the received vector $\mathbf{r}$, the bit-by-bit decoding problem is to find the most probable vector $\hat{\mathbf{c}}$ that satisfies the condition $\hat{\mathbf{c}}\mathbf{A}^T = \mathbf{0}$ in accordance with the constraint imposed on matrix $\mathbf{A}$ in (10.140).

In what follows, a bit refers to an element of the received vector $\mathbf{r}$ and a check refers to a row of matrix $\mathbf{A}$. Let $\mathcal{J}(i)$ denote the set of bits that participate in check i. Let $\mathcal{I}(j)$ denote the set of checks in which bit j participates. A set of $\mathcal{J}(i)$ that excludes bit j is denoted by $\mathcal{J}(i)\backslash j$. Likewise, a set of $\mathcal{I}(j)$ that excludes check i is denoted by $\mathcal{I}(j)\backslash i$.

Table 10.7 **The rate and fractional term of LDPC codes for varying weight-pairs***

t_c	t_r	Rate r	$\Delta_{t_c t_r}$
5	6	0.167	0.255
4	5	0.2	0.210
3	4	0.25	0.122
4	6	0.333	0.129
3	5	0.4	0.044
3	6	0.5	0.023

The decoding algorithm has two alternating steps: a *horizontal step* and a *vertical step*, which run along the rows and columns of matrix **A**, respectively. In the course of decoding, two probabilistic quantities associated with nonzero elements of matrix **A** are alternately updated. One quantity, denoted by P_{ij}^{x}, defines the probability that bit j is symbol x (i.e., symbol 0 or 1), given the information derived via checks performed in the horizontal step except for check i. The second quantity, denoted by Q_{ij}^{x}, defines the probability that check i is satisfied given that bit j is fixed at the value x and the other bits have the probabilities P_{ij} where we have $j' \in \mathcal{J}(i)\backslash j$.

The LDPC decoding algorithm then proceeds as follows.[19]

Initialization

The variables $P_{ij}^{(0)}$ and $P_{ij}^{(1)}$ are set equal to the a priori probabilities $p_j^{(0)}$ and $p_j^{(1)}$ of symbols 0 and 1, respectively, with $p_j^{(0)} + p_j^{(1)} = 1$ for all j.

Horizontal Step

In the horizontal step of the algorithm, we run through the checks i. To this end, define

$$\Delta P_{ij} = P_{ij}^{(0)} - P_{ij}^{(1)}$$

For each weight-pair (i, j), compute

$$\Delta Q_{ij} = \prod_{j' \in \mathcal{J}(i)\backslash j} \Delta P_{ij'}$$

Hence, set

$$Q_{ij}^{(0)} = \frac{1}{2}(1 + \Delta Q_{ij})$$

$$Q_{ij}^{(1)} = \frac{1}{2}(1 - \Delta Q_{ij})$$

Vertical Step

In the vertical step of the algorithm, values of the probabilities $P_{ij}^{(0)}$ and $P_{ij}^{(1)}$ are updated using the quantities computed in the horizontal step. In particular, for each bit j, compute

$$P_{ij}^{(0)} = \alpha_{ij} p_j^{(0)} \prod_{i' \in \mathcal{J}(i)\backslash j} \Delta Q_{i'j}^{(0)}$$

$$P_{ij}^{(1)} = \alpha_{ij} p_j^{(1)} \prod_{i' \in \mathcal{J}(i)\backslash j} \Delta Q_{i'j}^{(1)}$$

where the scaling factor α_{ij} is chosen so as to satisfy the condition

$$P_{ij}^{(0)} + P_{ij}^{(1)} = 1 \qquad \text{for all } ij$$

In the vertical step, we may also update the *pseudo-posterior probabilities*:

$$P_j^{(0)} = \alpha_j p_j^{(0)} \prod_{i \in \mathcal{J}(j)} Q_{ij}^{(0)}$$

$$P_j^{(1)} = \alpha_j p_j^{(1)} \prod_{i \in \mathcal{J}(j)} Q_{ij}^{(1)}$$

where α_j is chosen so as to make

$$P_j^{(0)} + P_j^{(1)} = 1 \qquad \text{for all } j$$

The quantities obtained in the vertical step are used to compute a tentative estimate $\hat{\mathbf{c}}$. If the condition $\hat{\mathbf{c}}\mathbf{A}^T = \mathbf{0}$ is satisfied, the decoding algorithm is terminated. Otherwise, the algorithm goes back to the horizontal step. If after some maximum number of iterations (e.g., 100 or 200) there is no valid decoding, then a decoding failure is declared. The decoding procedure described herein is a special case of the general low-complexity sum–product algorithm.

Simply stated, the *sum–product algorithm*[20] passes probabilistic quantities between the check nodes and variable nodes of the Tanner graph. On account of the fact that each parity-check constraint can be represented by a simple convolutional coder with one bit of memory, we find that LDPC decoders are simpler to implement than turbo decoders, as stated earlier on in the section.

In terms of performance, however, we may say the following in light of experimental results reported in the literature:

> Regular LDPC codes do not appear to come as close to Shannon's limit as do their turbo code counterparts.

Irregular LDPC Codes

Thus far in this section, we have focused attention on regular LDPC codes, which distinguish themselves in the following way: referring to the Tanner (bipartite) graph in Figure 10.36, all variable nodes on the left-hand side of the graph have the same degree and likewise for the check nodes on the right-hand side of the graph.

To go beyond the performance attainable with regular LDPC codes and thereby come increasingly closer to the Shannon limit, we look to irregular LDPC codes, in the context of which we introduce the following definition:

> An LDPC code is said to be irregular if the variable nodes in the code's Tanner graph have multiple degrees, and so do the check nodes in the graph.

To be specific, an irregular LDPC code distinguishes itself from its regular counterpart in that its Tanner graph involves the following two degree distributions:

a. The degree distribution of the variable nodes in the Tanner graph of an irregular LDPC code is described by:

$$\lambda(X) = \sum_{d=1}^{d_N} \lambda_d X^{d-1} \tag{10.145}$$

where X denotes a node variable in the code's Tanner graph, λ_d denotes the fraction of variable nodes with degree d in the graph, and d_N denotes the maximum degree of a variable node in the graph.

b. Correspondingly, the degree distribution of the check nodes in the irregular code's Tanner graph is described by

$$\rho(X) = \sum_{d=1}^{d_c} \rho_d X^{d-1} \tag{10.146}$$

where X denotes the check node in the code's Tanner graph, ρ_d denotes the fraction of check nodes with degree d in the graph, and d_c denotes the maximum degree of a check node in the graph.

The irregular LDPC code embodies the regular LDPC code as a special case. Specifically, (10.145) and (10.146) simplify as follows for the variable and check nodes of a regular LDPC code, respectively:

$$\lambda(X) = X^{\omega_N - 1} \quad \text{for } \lambda_d = 1 \text{ and } d_N = \omega_N \tag{10.147}$$

and

$$\rho(X) = X^{\omega_c - 1} \quad \text{for } \rho_d = 1 \text{ and } d_c = \omega_c \tag{10.148}$$

By exploiting the two degree distributions of (10.145) and (10.146) for the variable and check nodes, respectively, irregular LDPC codes are commonly constructed on the basis of their Tanner graphs. Such an approach is exemplified by the irregular LDPC codes reported in Richardson *et al.* (2001), and Richardson and Urbanke (2001).[21]

10.15 Trellis-Coded Modulation

In the different approaches to channel coding described up to this point in the chapter, the one common feature that describes them all may be summarized as follows:

> Encoding is performed separately from modulation in the transmitter, and likewise for decoding and detection in the receiver.

Moreover, error control is provided by transmitting additional redundant bits in the code, which has the effect of lowering the information bit rate per channel bandwidth. That is, bandwidth efficiency is traded for increased power efficiency.

To attain a more effective utilization of available resources, namely bandwidth and power, coding and modulation would have to be treated as a combined (single) entity. We may deal with this new paradigm by invoking the statement

> Coding is redefined as *the process of imposing certain patterns on the modulated signal in the transmitter.*

Indeed, this definition includes the traditional idea of parity-check coding.

Trellis codes for band-limited channels result from the treatment of modulation and coding as a *combined* entity rather than as two separate operations. The combination itself is referred to as *trellis-coded modulation* (TCM).[22] This form of signaling has three basic requirements:

1. The number of signal points in the constellation used is larger than what is required for the modulation format of interest with the same data rate; the additional signal points allow redundancy for forward error-control coding without sacrificing bandwidth.
2. Convolutional coding is used to introduce a certain dependency between successive signal points, such that only certain *patterns* or *sequences of signal points* are permitted for transmission.
3. *Soft-decision decoding* is performed in the receiver, in which the permissible sequence of signals is modeled as a trellis structure; hence the name *trellis codes*.

Requirement 3 is the result of using an enlarged signal constellation. By increasing the size of the constellation, the probability of symbol error increases for a fixed SNR. Hence, with hard-decision demodulation, we would face a performance loss before we begin. Performing soft-decision decoding on the combined code and modulation trellis ameliorates this problem.

In an AWGN channel, we look to the following approach:

> Maximum likelihood decoding of trellis codes consists of finding that particular path through the trellis with *minimum squared Euclidean distance* to the received sequence.

Thus, in the design of trellis codes, the emphasis is on maximizing the Euclidean distance between code vectors (equivalently codewords) rather than maximizing the Hamming distance of an error-correcting code. The reason for this approach is that, except for conventional coding with binary PSK and QPSK, maximizing the Hamming distance is not the same as maximizing the squared Euclidean distance. Accordingly, in what follows, the Euclidean distance between code vectors is adopted as the distance measure of interest. Moreover, while a more general treatment is possible, the discussion is (by choice) confined to the case of *two-dimensional constellations of signal points*. The implication of such a choice is to restrict the development of trellis codes to multilevel amplitude and/or phase modulation schemes such as *M*-ary PSK and *M*-ary QAM.

EXAMPLE 11 Two-level Partitioning of 8-PSK Constellation

The approach used to design this restricted type of trellis codes involves partitioning an *M*-ary constellation of interest successively into 2, 4, 8, ... subsets with size $M/2$, $M/4$, $M/8$, ..., and having progressively larger increasing minimum Euclidean distance between their respective signal points. Such a design approach by *set-partitioning* represents the key idea in the construction of efficient coded modulation techniques for band-limited channels.

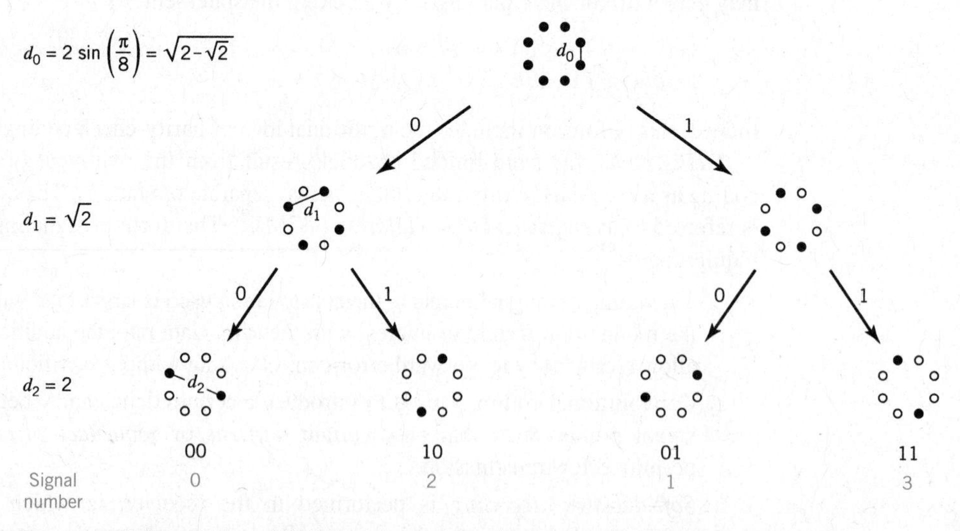

Figure 10.37 Partitioning of 8-PSK constellation, which shows that $d_0 < d_1 < d_2$.

In Figure 10.37, we illustrate the partitioning procedure by considering a circular constellation that corresponds to 8-PSK. The figure depicts the constellation itself and the two and four subsets resulting from two levels of partitioning. These subsets share the common property that the minimum Euclidean distances between their individual points follow an increasing pattern, namely:

$$d_0 < d_1 < d_2$$

EXAMPLE 12 **Three-level Partitioning of QAM Constellation**

For a different two-dimensional example, Figure 10.38 illustrates the partitioning of a rectangular constellation corresponding to 16-QAM. Here again, we see that the subsets have increasing within-subset Euclidean distances, as shown by

$$d_0 < d_1 < d_2 < d_3$$

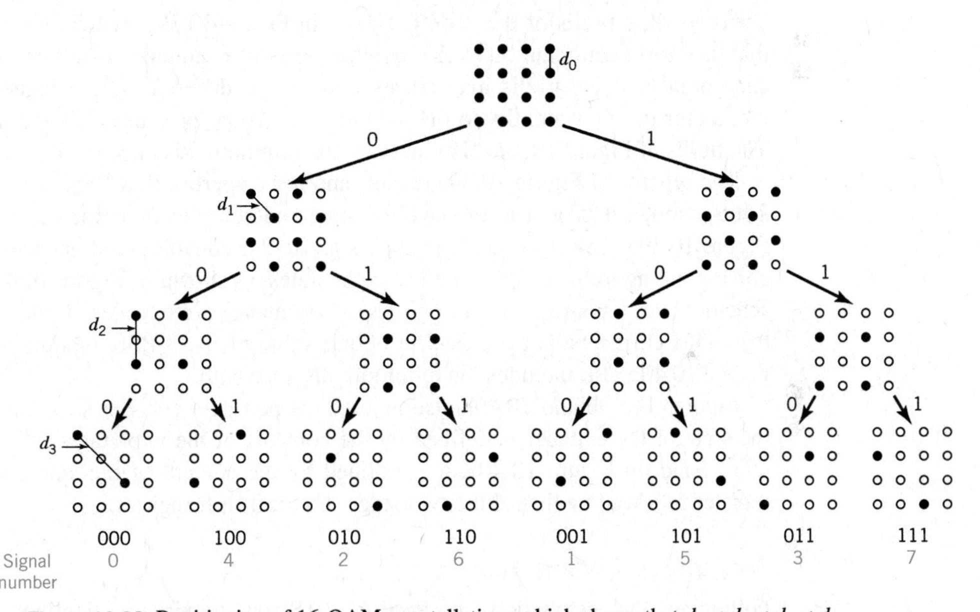

Figure 10.38 Partitioning of 16-QAM constellation, which shows that $d_0 < d_1 < d_2 < d_3$.

Based on the subsets resulting from successive partitioning of a two-dimensional constellation, illustrated in Examples 11 and 12, we may devise relatively simple, yet highly effective coding schemes. Specifically, to send n bits/symbol with *quadrature modulation* (i.e., one that has in-phase and quadrature components), we start with a two-dimensional constellation of 2^{n+1} signal points appropriate for the modulation format of interest; a circular grid is used for M-ary PSK and a rectangular one for M-ary QAM. In any event, the constellation is partitioned into four or eight subsets. One or two incoming message bits per symbol enter a rate-1/2 or rate-2/3 binary convolutional encoder, respectively; the resulting two or three coded bits per symbol determine the selection of a particular subset. The remaining uncoded messege bits determine which particular signal point from the selected subset is to be signaled over the AWGN channel. This class of trellis codes is known as *Ungerboeck codes* in recognition of their originator.

Since the modulator has memory, we may use the *Viterbi algorithm* (discussed in Section 10.8) to perform maximum likelihood sequence estimation at the receiver. Each branch in the trellis of the Ungerboeck code corresponds to a subset rather than an individual signal point. The first step in the detection is to determine the signal point within each subset that is closest to the received signal point in the Euclidean sense. The signal point so determined and its metric (i.e., the squared Euclidean distance between it and the received point) may be used thereafter for the branch in question, and the Viterbi algorithm may then proceed in the usual manner.

Ungerboeck Codes for 8-PSK

The scheme of Figure 10.39a depicts the simplest Ungerboeck 8-PSK code for the transmission of 2 bits/symbol. The scheme uses a rate-1/2 convolutional encoder; the corresponding trellis of the code is shown in Figure 10.39b, which has four states. Note that the most significant bit of the incoming message sequence is left uncoded. Therefore, each branch of the trellis may correspond to two different output values of the 8-PSK modulator or, equivalently, to one of the four two-point subsets shown in Figure 10.37. The trellis of Figure 10.39b also includes the minimum distance path.

The scheme of Figure 10.40a depicts another Ungerboeck 8-PSK code for transmitting 2 bits/sample; it is next in the level of increased complexity, compared to the scheme of Figure 10.39a. This second scheme uses a rate-2/3 convolutional encoder. Therefore, the corresponding trellis of the code has eight states, as shown in Figure 10.40b. In this latter scheme, both bits of the incoming message sequence are encoded. Hence, each branch of the trellis corresponds to a specific output value of the 8-PSK modulator. The trellis of Figure 10.40b also includes the minimum distance path.

Figures 10.39b and 10.40b also include the pertinent encoder states. In Figure 10.39a, the state of the encoder is defined by the contents of the two-stage shift register. On the other hand, in Figure 10.40a, it is defined by the content of the single-stage (top) shift register followed by that of the two-stage (bottom) shift register.

Asymptotic Coding Gain

Following the discussion in Section 10.8 on maximum likelihood decoding of convolutional codes, we define the asymptotic coding gain of Ungerboeck codes as follows:

$$G_{\mathrm{a}} = 10 \log_{10}\left(\frac{d_{\mathrm{free}}^2}{d_{\mathrm{ref}}^2}\right) \tag{10.149}$$

where d_{free} is the *free Euclidean distance* of the code and d_{ref} is the minimum Euclidean distance of an uncoded modulation scheme operating with the same signal energy per bit. For example, by using the Ungerboeck 8-PSK code of Figure 10.39a, the signal constellation has eight message points and we send two message bits per signal point. Hence, uncoded transmission requires a signal constellation with four message points. We

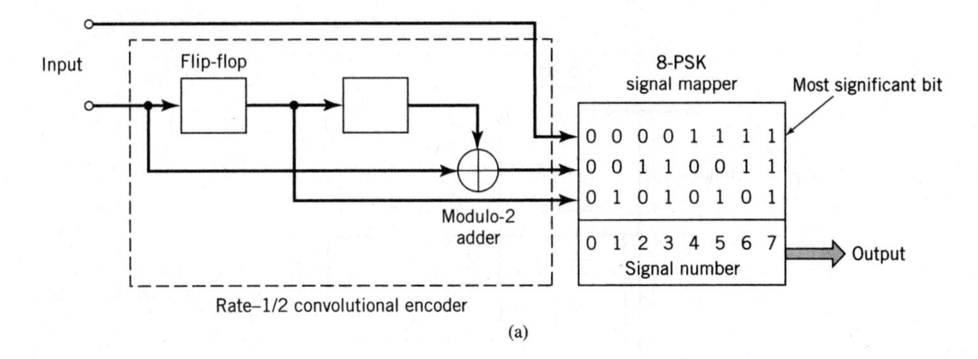

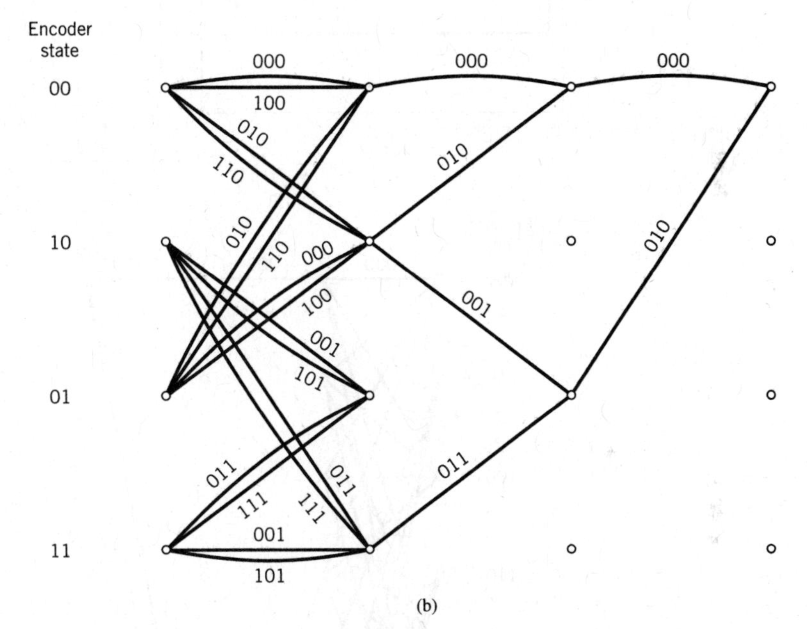

Figure 10.39 (a) Four-state Ungerboeck code for 8-PSK; the mapper follows Figure 10.37. (b) Trellis of the code.

may therefore regard uncoded 4-PSK as the frame of reference for the Ungerboeck 8-PSK code of Figure 10.39a.

The Ungerboeck 8-PSK code of Figure 10.39a achieves an asymptotic coding gain of 3 dB, which is calculated as follows:

1. Each branch of the trellis in Figure 10.39b corresponds to a subset of two antipodal signal points. Hence, the free Euclidean distance d_{free} of the code can be no larger than the Euclidean distance d_2 between the antipodal signal points of such a subset. We may therefore write

$$d_{\text{free}} = d_2 = 2$$

where the distance d_2 is defined in Figure 10.41a.

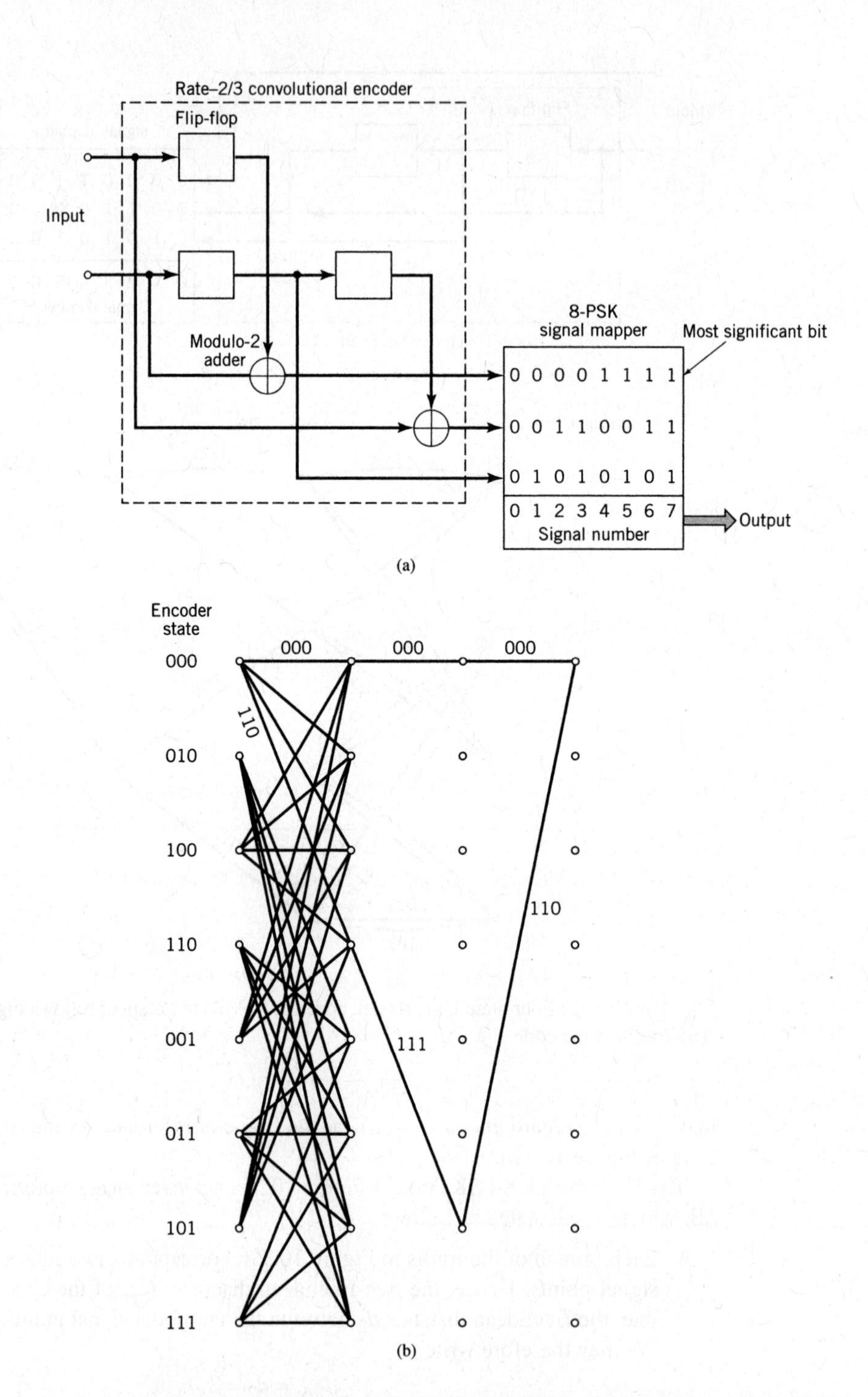

Figure 10.40 (a) Eight-state Ungerboeck code for 8-PSK; the mapper follows Figure 10.37. (b) Trellis of the code with only some of the branches shown.

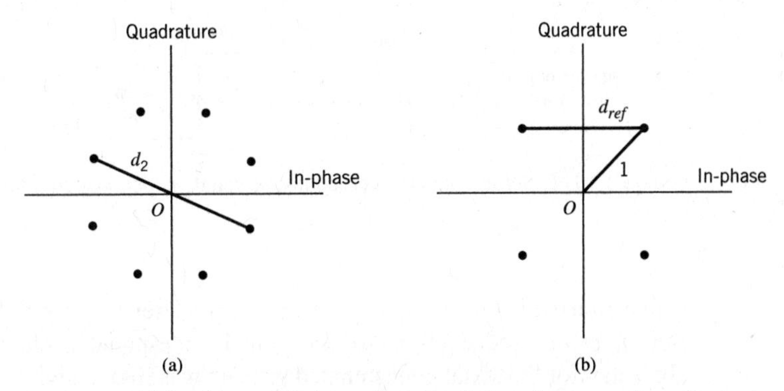

Figure 10.41 Signal-space diagrams for calculation of asymptotic coding gain of Ungerboeck 8-PSK code: (a) definition of distance d_2; (b) definition of reference distance d_{ref}.

2. From Figure 10.41b, we see that the minimum Euclidean distance of an uncoded QPSK, viewed as the frame of reference operating with the same signal energy per bit, assumes the following value:

$$d_{\text{ref}} = \sqrt{2}$$

Hence, as previously stated, the use of (10.149) yields an asymptotic coding gain of

$$10 \log_{10} 2 = 3 \text{ dB}.$$

The asymptotic coding gain achievable with Ungerboeck codes increases with the number of states in the convolutional encoder. Table 10.8 presents the asymptotic coding gain (in dB) for Ungerboeck 8-PSK codes for increasing number of states, expressed with respect to uncoded 4-PSK. Note that improvements on the order of 6 dB require codes with a very large number of states.

Table 10.8 **Asymptotic coding gain of Ungerboeck 8-PSK codes, with respect to uncoded 4-PSK**

Number of states	4	8	16	32	64	128	256	512
Coding gain (dB)	3	3.6	4.1	4.6	4.8	5	5.4	5.7

10.16 Turbo Decoding of Serial Concatenated Codes

In Section 10.12 we pointed out that there are two types of concatenated codes: parallel and serial. The original turbo coding scheme involved a *parallel concatenated code*, since the two encoders operate in parallel on the same set of message bits. We now turn our attention in this section to a *serial concatenation* scheme as depicted in Figure 10.42, comprised of an "outer" encoder whose output feeds an "inner" encoder. Whereas the serial concatenation idea can be traced to as early as Shannon's seminal work, the

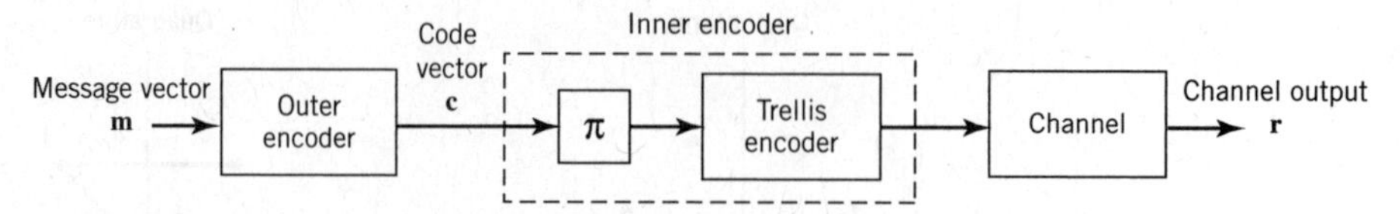

Figure 10.42 Serial concatenated codes; as usual, π denotes an interleaver.

connection with turbo coding occurred only after the parallel concatenated scheme of Berrou *et al.* (see Section 10.12) gained widespread acclaim. The iterative decoding algorithm for the serial concatenated scheme was first analyzed in detail by Benedetto and coworkers (Benedetto and Montorsi, 1996; Benedetto *et al.*, 1998); the algorithm follows a similar logic to the parallel concatenated scheme, in the form of information exchange between the two decoders as in Figure 10.43. This iterative information exchange is observed to significantly improve the overall error-correction abilities of the decoder, just as in the conventional turbo decoder. We shall review the basics of the iterative decoding algorithm in what follows in order to emphasize the common points with the iterative algorithm described in Section 10.12.

The particular interest in the serial concatenated scheme, however, becomes apparent once we recognize that the inner encoder–decoder pair need not be a conventional error-correction code, but in fact may assume more general forms that are often encountered in communication systems. A few examples may be highlighted as follows:

1. The inner encoder may in fact be a TCM stage, as studied in Section 10.15. The iterative decoding algorithm connecting the trellis-coded demodulator with the outer error-correction code leads to *turbo TCM*.[23]
2. The inner encoder may be the communication channel itself, which is of interest when the channel induces ISI. The output symbols of the channel may then be expressed as a convolution between the input symbol sequence and the channel impulse response, and the decoder operation corresponds to channel equalization (Chang and Hancock, 1966). Combining the equalizer with the outer channel decoder gives rise to *turbo equalization*.[24]

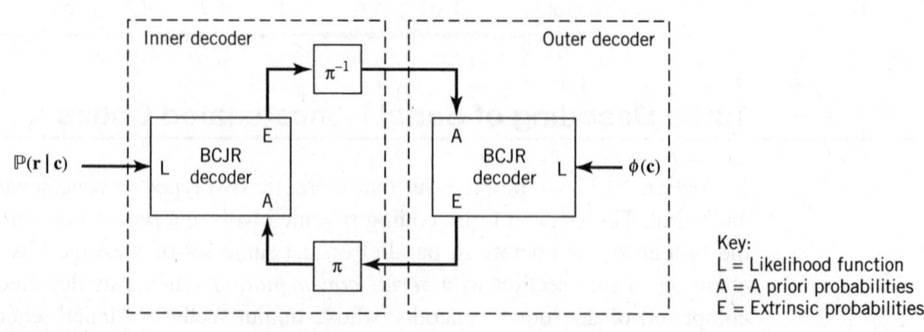

Figure 10.43 Iterative decoder structure.

3. In multi-user communication systems, the inner encoder may represent a single user's access point to a shared channel through, say, direct-sequence CDMA, in which users sharing a channel are distinguished by an assigned repetition code. The inner decoder is a multiuser detector, which aims to separate the multiple users into distinct symbol streams; when combined with the outer decoder using information exchange, a *turbo CDMA* system results.[25]

The above list is by no means exhaustive, but merely represents some of the more commonly studied variants of iterative receiver design. We will focus here on the basic iterative decoding scheme using an error-correction code for the inner encoder–decoder pair, and then briefly illustrate its applications to turbo equalization.

Serial Turbo Codes

Consider first the case in which both encoders in the serial concatenation of Figure 10.42 implement forward error-correction coding. For efficiency reasons, we assume that the outer encoder implements a systematic code, so that the codeword **c** it produces appears as follows:

$$\mathbf{c} = [\,\mathbf{b} \mid \mathbf{m}\,] \tag{10.150}$$

in which **m** contains the k message bits and **b** contains the $n - k$ parity-check bits. By choosing a recursive systematic encoder, the corresponding decoding operation can exploit the BCJR algorithm discussed in Section 10.9.

The second, or "inner," encoder is also based on a trellis code (although not necessarily systematic) so that it, too, will admit an efficient decoder using the MAP decoding algorithm. As illustrated in Figure 10.42, the inner encoder also integrates an interleaver, denoted by π, which permutes the order of the bits in the code vector **c** prior to the second encoding operation. Without this interleaver, the serial concatenation of two trellis codes would merely give a larger-dimension trellis code having limited error-correction capabilities. The inclusion of the interleaver alters markedly the minimum distance properties of the code, and constitutes an essential ingredient in obtaining a good error-correction code.

Probabilistic Considerations

The output from the inner encoder is sent across the channel, which may be a binary symmetric channel or an AWGN channel, to produce the received vector **r**. The simplest way to decode the received signal is to cascade the corresponding inner and outer decoders. A refined approach is to allow information exchange between the two decoders, to trigger the turbo effect; this idea is illustrated in Figure 10.43, and the manner of information exchange is developed in what follows.

To begin, the inner decoder aims to obtain the bitwise a posteriori probability ratios

$$\frac{\mathbb{P}(c_i = +1 \mid \mathbf{r})}{\mathbb{P}(c_i = -1 \mid \mathbf{r})}, \qquad i = 1, 2, \ldots, n \tag{10.151}$$

As $\mathbb{P}(c_i|\mathbf{r})$ is a marginal probability calculated from the conditional probability $\mathbb{P}(\mathbf{c}|\mathbf{r})$, the bit-wise a posteriori probability ratio may be developed into the new form

$$\frac{\mathbb{P}(c_i=+1|\mathbf{r})}{\mathbb{P}(c_i=-1|\mathbf{r})} = \frac{\sum_{\mathbf{c}:c_i=+1} \mathbb{P}(\mathbf{c}|\mathbf{r})}{\sum_{\mathbf{c}:c_i=-1} \mathbb{P}(\mathbf{c}|\mathbf{r})}$$

$$= \frac{\sum_{\mathbf{c}:c_i=+1} \mathbb{P}(\mathbf{r}|\mathbf{c})\,\mathbb{P}(\mathbf{c})}{\sum_{\mathbf{c}:c_i=-1} \mathbb{P}(\mathbf{r}|\mathbf{c})\,\mathbb{P}(\mathbf{c})}, \qquad i = 1, 2, \ldots, n \tag{10.152}$$

In the second line of (10.152), we used Bayes' rule

$$\mathbb{P}(\mathbf{c}|\mathbf{r}) = \mathbb{P}(\mathbf{r}|\mathbf{c})\mathbb{P}(\mathbf{c})/\mathbb{P}(\mathbf{r})$$

to expose the likelihood function $\mathbb{P}(\mathbf{r}|\mathbf{c})$ as well as the *a priori* probability function $\mathbb{P}(\mathbf{c})$; the term $\mathbb{P}(\mathbf{r})$ is common to the numerator and denominator, and so cancels in the ratio.

We now make the assumption that the a priori probability $\mathbb{P}(\mathbf{c})$ factors into the product of its marginals; that is,

$$\mathbb{P}(\mathbf{c}) = \mathbb{P}(c_1)\mathbb{P}(c_2)\ldots\mathbb{P}(c_n) \tag{10.153}$$

Strictly speaking, this is incorrect, since $\mathbf{c}$ contains both the message bits $\mathbf{m}$ and parity-check bits $\mathbf{b}$ from the outer encoder, and we know that the message bits determine the parity-check bits once the outer encoder is specified. The reason for invoking this assumption is to facilitate decoding via the BCJR algorithm. In particular, inserting this factorization of the *a priori* probability function into the *a posteriori* probability ratio, we may continue our development as shown by

$$\frac{\mathbb{P}(c_i=+1|\mathbf{r})}{\mathbb{P}(c_i=-1|\mathbf{r})} = \frac{\sum_{\mathbf{c}:c_i=+1} \mathbb{P}(\mathbf{r}|\mathbf{c})\prod_{j=1}^{n}\mathbb{P}(c_j)}{\sum_{\mathbf{c}:c_i=-1} \mathbb{P}(\mathbf{r}|\mathbf{c})\prod_{j=1}^{n}\mathbb{P}(c_j)}$$

$$= \underbrace{\frac{\mathbb{P}(c_i=+1)}{\mathbb{P}(c_i=-1)}}_{\text{prior ratio}} \times \underbrace{\frac{\sum_{\mathbf{c}:c_i=+1} \mathbb{P}(\mathbf{r}|\mathbf{c})\prod_{j=1, j\neq i}^{n}\mathbb{P}(c_j)}{\sum_{\mathbf{c}:c_i=-1} \mathbb{P}(\mathbf{r}|\mathbf{c})\prod_{j=1, j\neq i}^{n}\mathbb{P}(c_j)}}_{\text{extrinsic information ratio}}, \qquad i = 1, 2, \ldots, n \tag{10.154}$$

We obtained the second line in (10.154) by noting that each term in the numerator contains the factor $\mathbb{P}(c_i = +1)$ and, similarly, each term in the denominator contains the factor $\mathbb{P}(c_i = -1)$, hence the reason for the prior ratio factoring out of the expression. The remaining term is the extrinsic information ratio for bit c_i from the inner decoder.

To facilitate passing information to the outer decoder, we may interpret each extrinsic information ratio from the inner decoder as the probability ratio of an auxiliary probability mass function $T(\mathbf{c})$ that fulfills two properties:

- The probability mass function $T(\mathbf{c})$ factors into its bitwise marginal functions according to

$$T(\mathbf{c}) = T_1(c_1)T_2(c_2)\ \dots\ T_n(c_n) \tag{10.155}$$

- The bitwise marginal evaluations, each, sum to one, $T_i(+1) + T_i(-1) = 1$, and are chosen such that their ratios match the extrinsic information ratios from the inner decoder:

$$\frac{T_i(c_i = +1)}{T_i(c_i = -1)} = \frac{\displaystyle\sum_{\mathbf{c}:c_i = +1} \mathbb{P}(\mathbf{r}|\mathbf{c}) \prod_{\substack{j=1 \\ j \neq i}}^{n} \mathbb{P}(c_j)}{\displaystyle\sum_{\mathbf{c}:c_i = -1} \mathbb{P}(\mathbf{r}|\mathbf{c}) \prod_{\substack{j=1 \\ j \neq i}}^{n} \mathbb{P}(c_j)}, \qquad i = 1, 2, \dots, n \tag{10.156}$$

Now, we note that by natural taking logarithms, the *log extrinsic ratio* becomes

$$\ln[T_i(+1)/T_i(-1)] = L_{\mathrm{p}}(c_i) - L_{\mathrm{a}}(c_i) \tag{10.157}$$

where $L_{\mathrm{p}}(c_i)$ is the *log posterior ratio* and $L_{\mathrm{a}}(c_i)$ is the *log prior ratio*.[26]

Next, we note that the outer decoder does not have the usual channel likelihood evaluations available, but must instead take information from the inner decoder. While many possibilities in this direction may be envisaged, a successful iterative decoding algorithm results by replacing the *a posteriori* probability according to

$$\mathbb{P}(\mathbf{c}|\mathbf{r}) \leftarrow \phi(\mathbf{c})T(\mathbf{c}) \tag{10.158}$$

in which $\phi(\mathbf{c})$ is the indicator function for the outer code, that is

$$\phi(\mathbf{c}) = \begin{cases} 1, & \text{if } \mathbf{c} \text{ is a code vector} \\ 0, & \text{otherwise} \end{cases} \tag{10.159}$$

We may think of the function $\phi(\mathbf{c})$ as replacing the conventional channel likelihood function $\mathbb{P}(\mathbf{r}|\mathbf{c})$, since it vanishes whenever $\mathbf{c}$ is not a code vector, and $T(\mathbf{c}) = T_1(c_1)\ \dots\ T_n(c_n)$ as replacing the *a priori* probability on each bit, since it factors into the product of its marginals. The conventional posterior probability ratio for the outer decoder is thus replaced with

$$\frac{\mathbb{P}(c_i = +1|\mathbf{r})}{\mathbb{P}(c_i = -1|\mathbf{r})} = \frac{\displaystyle\sum_{\mathbf{c}:c_i = +1} \phi(\mathbf{c}) \prod_{j=1}^{n} T_j(c_j)}{\displaystyle\sum_{\mathbf{c}:c_i = -1} \phi(\mathbf{c}) \prod_{j=1}^{n} T_j(c_j)}$$

$$= \underbrace{\frac{T_i(c_i = +1)}{T_i(c_i = -1)}}_{\text{prior ratio}} \underbrace{\frac{\displaystyle\sum_{\mathbf{c}:c_i = +1} \phi(\mathbf{c}) \prod_{\substack{j=1 \\ j \neq i}}^{n} T_j(c_j)}{\displaystyle\sum_{\mathbf{c}:c_i = -1} \phi(\mathbf{c}) \prod_{\substack{j=1 \\ j \neq i}}^{n} T_j(c_j)}}_{\text{extrinsic information ratio}}, \quad i = 1, 2, \dots, n \tag{10.160}$$

In (10.160) the second line is obtained upon noting that each term in the numerator (denominator) contains a factor $T_i(c_i = +1)$ $(T_i(c_i = -1))$. This separates the "pseudo-prior" ratio from the outer decoder's extrinsic information ratio.

Now, to couple the information from the outer decoder back to the inner decoder, we map the outer decoder's extrinsic information values to a probability mass function $U(\mathbf{c})$ which, akin to $T(\mathbf{c})$ introduced above, fulfills two properties:

1. The probability mass function $U(\mathbf{c})$ factors into the product of its bitwise marginal functions according to

$$U(\mathbf{c}) = U_1(c_1)\, U_2(c_2) \dots U_n(c_n) \tag{10.161}$$

2. The bitwise marginal evaluations each sum to one, $U_i(+1) + U_i(-1) = 1$, and are chosen such that their ratios match the extrinsic information ratios for the outer decoder:

$$\frac{U_i(c_i = +1)}{U_i(c_i = -1)} = \frac{\displaystyle\sum_{\mathbf{c}:c_i = +1} \phi(\mathbf{c}) \prod_{\substack{j=1 \\ j \neq i}}^{n} T_j(c_j)}{\displaystyle\sum_{\mathbf{c}:c_i = -1} \phi(\mathbf{c}) \prod_{\substack{j=1 \\ j \neq i}}^{n} T_j(c_j)}, \quad i = 1, 2, \dots, n \tag{10.162}$$

The marginal probability functions $U_i(c_i)$ then replace the *a priori* probability values $\mathbb{P}(c_i)$ in the inner decoder, and the procedure iterates, thus defining the turbo decoder. In this fashion, we say the following:

> The turbo decoder for serially concatenated codes follows the same logic as for parallel concatenated codes, in that the extrinsic information values furnished from one decoder replace the a priori probability values required of the other.

Turbo Equalization

In high-speed communication systems, further signal degradation can come from multipath artifacts in wireless environments, or reflection artifacts due to impedance mismatches in wireline systems. When such degradations have a temporal duration commensurate with the symbol period, the signal impinging on the receiver at a given sample instant is the composite influence of successive transmitted symbols, giving rise to ISI. Its severity worsens as the symbol period diminishes relative to the "delay spread" of the system, meaning that higher rate data systems must contend with ISI as a significant channel distortion mechanism.

In such cases, the received symbol r_i at sample instant i is a weighted combination of a set of successive transmitted symbols, according to

$$r_i = \sum_{k=0}^{L-1} h_k s_{i-k} + \zeta_i \tag{10.163}$$

Here, ζ_i is the additive background noise, $\{s_i\}$ is the transmitted symbol sequence obtained by interleaving the bit sequence $\{c_i\}$ (and possibly followed by symbol mapping if using TCM), and $\{h_0, h_1, ..., h_{L-1}\}$ is the channel impulse response of length L.

If we consider a simple case in which each s_i is antipodal ($s_i = \pm 1$) and $k = 0, 1, 2$, we see that the noise-free channel outputs can be obtained through the trellis graph of Figure 10.44: The transitions are determined by whether the input symbol is $s_i = +1$ or $s_i = -1$, while the noise-free outputs are drawn from a finite set comprised of sums and differences of the channel impulse response coefficients. Thus, a convolutional channel which induces ISI may itself be viewed as a trellis code, and the BCJR algorithm may be applied directly to estimate the a posteriori probabilities of the transmitted symbols, and thus of the codeword bits $\{c_i\}$. The new result is that we have the traditional MAP equalizer.

A turbo equalizer results upon noting that the convolutional channel and its MAP equalizer may be viewed as the inner encoder–decoder pair of a serial cascade scheme albeit one dictated by the communication channel and thus beyond the designer's control. The outer encoder may again be chosen as a recursive systematic trellis code, whose decoder is coupled with the MAP equalizer in precisely the same manner: the extrinsic probabilities from one decoder are used in place of the a priori probabilities of the other, resulting in an iterative decoding and equalization scheme.

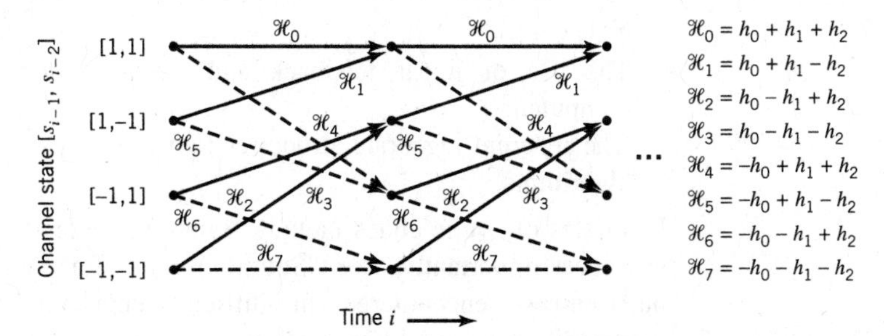

Figure 10.44 Trellis graph for a three-tap channel model, with transition branches listing the noise-free channel outputs. Solid transitions occur when the channel input is $s_i = +1$; dashed transitions occur when $s_i = -1$.

10.17 Summary and Discussion

In this rather long chapter we studied error-control coding techniques that have established themselves as indispensable tools for reliable digital communication over noisy channels. The effect of errors occurring during transmission is reduced by adding redundancy to the data prior to transmission in a controlled manner. The redundancy is used to enable a decoder in the receiver to detect and correct errors.

Regardless of how they are designed, error-control coding schemes rely on Shannon's 1948 landmark paper, particularly the celebrated *coding theorem*, which asserts the following statement:

> Given the proper strategy for encoding the incoming message bits, unavoidable errors produced by a noisy channel can be corrected without having to sacrifice the rate of data transmission.

The coding theorem was discussed in Chapter 5 on information theory. Restating it here, for the last time in the final chapter of the book, is intended to emphasize the importance of the theorem, which will last forever.

In a historical context, error-control coding schemes may be divided into two broadly defined families:

1. *Legacy Codes*

   As the name would imply, the family of legacy codes embodies several kinds of *linear codes* that originated in 1950 and, in the course of three decades or so, broadened its scope in depth as well as breadth. A distinctive feature of legacy codes is that of exploiting *abstract algebraic structures* built into their design in different ways and increasing mathematical abstraction.

   Specifically, legacy codes cover the following four schemes:

   a. *Linear block codes*, the first kind of which were described independently by Golay in 1949 and Hamming in 1950. Hamming codes are simple to construct and just as easy to decode using a look-up table based on the notion of syndrome. It is because of their computational simplicity and the ability to operate at high data rates, that we find that Hamming codes are widely used in digital communications.

   b. *Cyclic codes*, which form an importance subclass of linear block codes. Indeed, many of the block codes used in practice are cyclic codes for two compelling reasons:

      - The use of linear feedback shift registers for encoding and syndrome computation.
      - The inherent algebraic structure used to develop various practical decoding algorithms.

      Examples of cyclic codes include Hamming codes for digital communications, and most importantly, Reed–Solomon codes for combatting both random and burst errors encountered in difficult environments such as deep-space communications and compact discs.

   c. *Convolutional codes*, which distinguish themselves from linear block codes in the use of *memory* in the form of a finite-state shift register for implementing the encoder. For decoding convolutional codes, the Viterbi algorithm (based on maximum likelihood decoding) is commonly used; this algorithm is designed to minimize the symbol-error rate on a symbol-by-symbol basis.

d. *Trellis coded modulation*, which distinguishes itself from linear convolutional codes in the combined use of encoding and modulation in a single entity. The next result of so doing is the achievement of significant coding gains over conventional uncoded multilevel modulation schemes without having to sacrifice bandwidth efficiency in decoding.

2. *Probabilistic Compound Codes*

This second family of error-control coding schemes is exemplified by turbo codes and LDPC codes, which, as different as they are from each other, share a common property:

Random encoding of a linear block kind.

To be more specific, in their own individual ways, they are both revolutionary:

In practical terms, turbo codes and LDPC codes have made it possible to achieve coding gains on the order of 10 dB, thereby approaching the Shannon limit not attainable by the legacy codes.

Moreover, in some specialized cases, very long rate-1/2 irregular LPDC codes have approached the Shannon limit to within 0.0045 dB for AWGN channels, which is truly remarkable (Chung *et al.*, 2001).

These impressive coding gains have been exploited to dramatically extend the range of digital communication receivers, substantially increase the bit rates of digital communication systems, or significantly decrease the transmitted signal energy per symbol. The benefits have significant implications for the design of wireless communications and deep-space communications, just to mention two important applications of digital communications. Indeed, turbo codes have already been standardized for use on both of these applications.

One last comment is in order: Turbo codes have not only impacted digital communications in the different ways just described, but the turbo decoding paradigm has also impacted applications outside the traditional scope of error-control coding. One such example is that of *turbo equalization*, briefly described in Section 10.16. Indeed, we may justifiably say the following as the last statement of the chapter:

The turbo-decoding paradigm, by virtue of its broadly defined scope of applications, stands out as one of the ground-breaking achievements in modern telecommunications.

Problems

Soft-Decision Coding

10.1 Consider a binary input Q-ary output discrete memoryless channel. The channel is said to be symmetric if the channel transition probability $p(j|i)$ satisfies the condition

$$p(j|0) = p(Q-1-j|1), \qquad j = 0, 1, \ldots, Q-1$$

Suppose that the channel input bits 0 and 1 are equally likely. Show that the channel output symbols are also equally likely; that is,

$$p(j) = \frac{1}{Q}, \qquad j = 0, 1, \ldots, Q-1$$

10.2 Consider the quantized demodulator for binary PSK signals shown in Figure 10.3a. The quantizer is a four-level quantizer, normalized as in Figure P10.2. Evaluate the transition probabilities of the binary input–quarternary output discrete memoryless channel so characterized. Hence, show that it is a symmetric channel. Assume that the transmitted signal energy per bit is E_b and the AWGN has zero mean and power spectral density $N_0/2$.

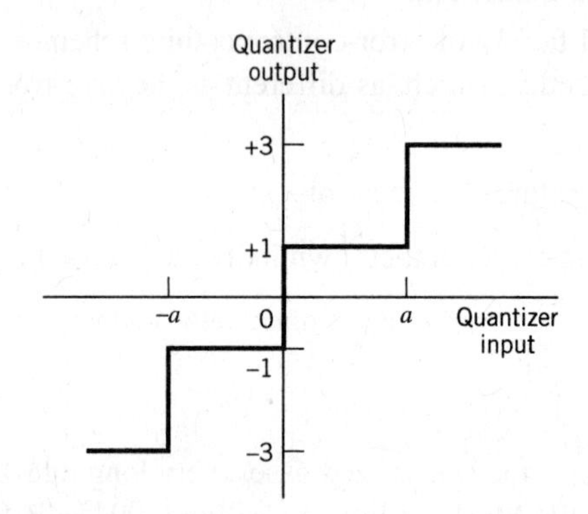

Figure P10.2

10.3 Consider a binary input AWGN channel, in which the bits 1 and 0 are equally likely. The bits are transmitted over the channel by means of phase-shift keying. The code symbol energy is E and the AWGN has zero mean and power spectral density $N_0/2$. Show that the channel transition probability is given by

$$p(y|0) = \frac{1}{\sqrt{2\pi}}\exp\left[-\frac{1}{2}\left(y + \sqrt{\frac{2E}{N_0}}\right)^2\right], \qquad -\infty < y < \infty$$

Linear Block Codes

10.4 Hamming codes are said to be perfect single-error correcting codes. Justify the fact that Hamming codes are *perfect*.

10.5 Consider the following statement:

An (n, k) code is often said to be a good code.

Explain the conditions under which this statement is justified.

10.6 In a *repetition code*, a single message bit is encoded into a block of identical bits to produce an (n, 1). Considering the (5, 1) repetition code, evaluate the syndrome for:

a. All five possible single-error patterns.

b. All ten possible double-error patterns.

10.7 In a *single-parity-check code*, a single parity bit is appended to a block of k message bits (m_0, m_1, …, m_{k-1}). The single parity bit b_0 is chosen so that the codeword satisfies the *even parity rule*:

$$m_0 + m_1 + \ldots + m_{k-1} + b_{k-1} = 0, \qquad \text{mod } 2$$

For $K = 3$, set up the 2^k possible codewords in the code defined by this rule.

10.8 Compare the parity-check matrix of the (7,4) Hamming code considered in Example 1 with that of a (4,1) repetition code.

10.9 Consider the (7,4) Hamming code of Example 1. The generator matrix **G** and the parity-check matrix **H** of the code are described in that example. Show that these two matrices satisfy the condition

$$\mathbf{H}\mathbf{G}^{\mathrm{T}} = \mathbf{0}$$

10.10 a. For the (7,4) Hamming code described in Example 1, construct the eight codewords in Hamming's dual code.

b. Find the minimum distance of the dual code determined in part a.

Linear Cyclic Codes

10.11 For an application that requires error detection *only*, we may use a *nonsystematic* code. In this problem, we explore the generation of such a cyclic code. Let $\mathbf{g}(X)$ denote the generator polynomial, and $\mathbf{m}(X)$ denote the message polynomial. We define the code polynomial $\mathbf{c}(X)$ simply as

$$\mathbf{c}(X) = \mathbf{m}(X)\mathbf{g}(X)$$

Hence, for a given generator polynomial, we may readily determine the codewords in the code. To illustrate this procedure, consider the generator polynomial for a (7,4) Hamming code:

$$\mathbf{g}(X) = 1 + X + X^3$$

Determine the 16 codewords in the code, and confirm the nonsystematic nature of the code.

10.12 The polynomial $1 + X^7$ has $1 + X + X^3$ and $1 + X^2 + X^3$ as primitive factors. In Example 10.2, we used $1 + X + X^3$ as the generator polynomial for a (7,4) Hamming code. In this problem, we consider the adoption of $1 + X^2 + X^3$ as the generator polynomial. This should lead to a (7,4) Hamming code that is different from the code analyzed in Example 2. Develop the encoder and syndrome calculator for the generator polynomial:

$$\mathbf{g}(X) = 1 + X^2 + X^3$$

Compare your results with those in Example 2.

10.13 Consider the (7,4) Hamming code defined by the generator polynomial

$$\mathbf{g}(X) = 1 + X + X^3$$

The codeword 0111001 is sent over a noisy channel, producing the received word 0101001 that has a single error. Determine the syndrome polynomial $\mathbf{s}(X)$ for this received word, and show that it is identical to the error polynomial $\mathbf{e}(X)$.

10.14 The generator polynomial of a (15,11) Hamming code is defined by

$$\mathbf{g}(X) = 1 + X + X^4$$

Develop the encoder and syndrome calculator for this code, using a systematic form for the code.

10.15 Consider the (15,4) maximal-length code that is the dual of the (15,11) Hamming code of Problem 10.14.

Find the generator polynomial $\mathbf{g}(X)$; hence, determine the output sequence assuming the initial state 0001. Confirm the validity of your result by cycling the initial state through the encoder.

10.16 Consider the (31,15) Reed–Solomon code.

a. How many bits are there in a symbol of the code?

b. What is the block length in bits?

c. What is the minimum distance of the code?

d. How many symbols in error can the code correct?

Convolutional Codes

10.17 A convolutional encoder has a single-shift register with two states, (i.e., constraint length $\nu = 3$), three modulo-2 adders, and an output multiplexer. The generator sequences of the encoder are as follows:

$$g^{(1)} = (1, 0, 1)$$
$$g^{(2)} = (1, 1, 0)$$
$$g^{(3)} = (1, 1, 1)$$

Draw the block diagram of the encoder.

10.18 Consider the rate $r = 1/2$, constraint length $\nu = 2$ convolutional encoder of Figure P10.18. The code is systematic. Find the encoder output produced by the message sequence 10111...

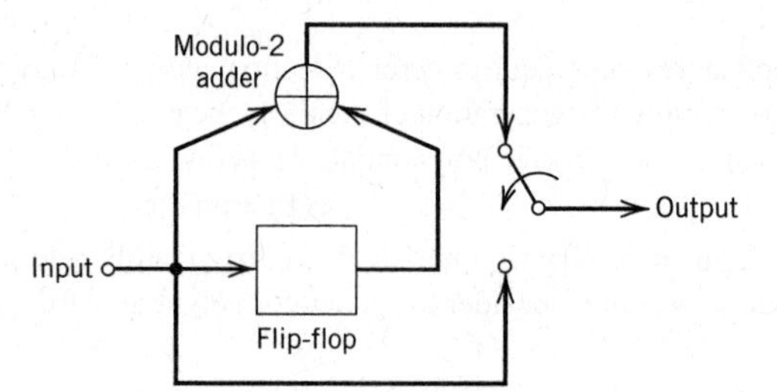

Figure P10.18

10.19 Figure P10.19 shows the encoder for a rate $r = 1/2$, constraint length $\nu = 4$ convolutional code. Determine the encoder output produced by the message sequence 10111....

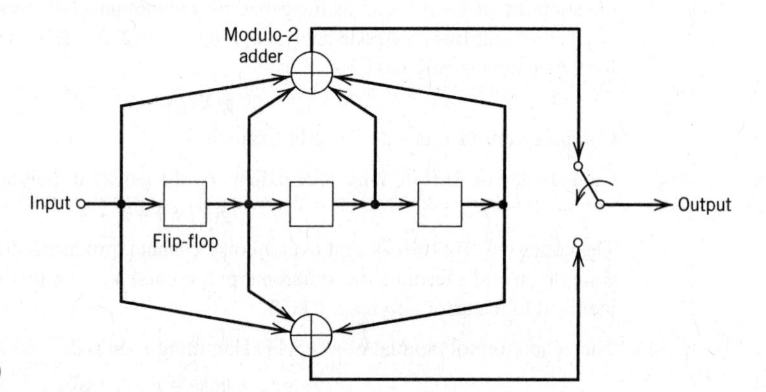

Figure P10.19

10.20 Consider the encoder of Figure P10.20 for a rate $r = 2/3$, constraint length $\nu = 2$ convolutional code. Determine the code sequence produced by the message sequence 10111....

10.21 Construct the code tree for the convolutional encoder of Figure P10.19. Trace the path through the tree that corresponds to the message sequence 10111..., and compare the encoder output with that determined in Figure P10.19.

10.22 Construct the trellis graph for the encoder of Figure P10.19, assuming a message sequence of length 5. Trace the path through the trellis corresponding to the message sequence 10111.... Compare the resulting encoder output with that found in Problem 10.19.

10.23 Construct the state graph for the encoder of Figure P10.19. Starting with the all-zero state, trace the path that corresponds to the message sequence 10111... and compare the resulting code sequence with that determined in Problem 10.19.

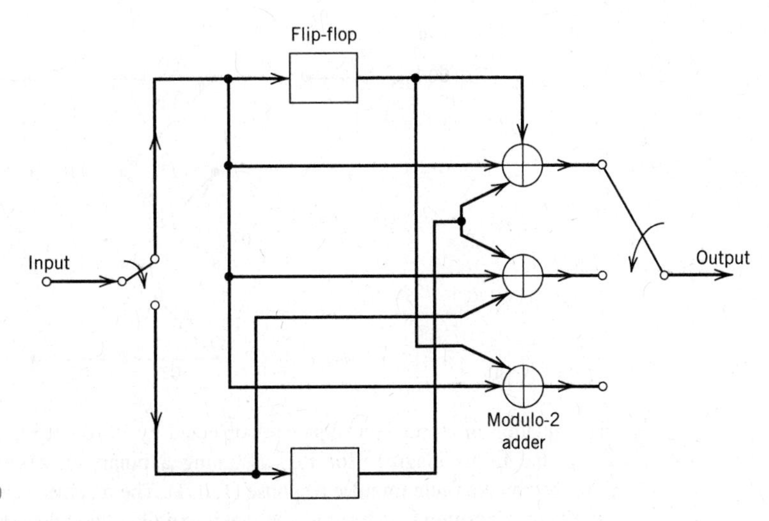

Figure P10.20

10.24 Consider the encoder of Figure 10.13.

a. Construct the state graph for this encoder.

b. Starting from the all-zero state, trace the path that corresponds to the message sequence 10111... Compare the resulting sequence with that determined in Problem 10.19.

10.25 By viewing the minimum shift keying (MSK) scheme as a finite-state machine, construct the trellis diagram for the MSK. (A description of the MSK is presented in Chapter 7).

10.26 Consider a rate-1/2, constraint length-7 convolutional code with free distance $d_{\text{free}} = 10$. Calculate the asymptotic coding gain for the following two channels:

a. Binary symmetric channel.

b. Binary input AWGN channel.

10.27 The transform-domain generator matrix $\mathbf{G}(D)$ of an RSC encoder includes ratios of polynomials in the delay variable D, whereas, in the case of a nonrecursive convolutional encoder $\mathbf{G}(D)$ is simply a polynomial in D. Justify the $\mathbf{G}(D)$ for these two cases.

10.28 Consider an eight-state RSC encoder, the generator matrix of which is given by

$$\mathbf{g}(D) = \left[1, \frac{1 + D + D^2 + D^3}{1 + D + D^2}\right]$$

where D is the delay variable.

a. Construct the block diagram of this encoder.

b. Formulate the parity-check equation that embodies all the message as well as parity-check bits in the time domain.

10.29 Describe the similarities and differences between traditional encoders and RSC encoders.

The Viterbi Algorithm

10.30 The trellis diagram of a rate-1/2, constraint length-3 convolutional code is shown in Figure P10.30. The all-zero sequence is transmitted and the received sequence is 1000100000.... Using the Viterbi decoding algorithm, compute the decoded sequence.

10.31 In Section 10.8, we described the Viterbi algorithm for maximum likelihood decoding of a convolutional code. Another application of the Viterbi algorithm is for maximum likelihood

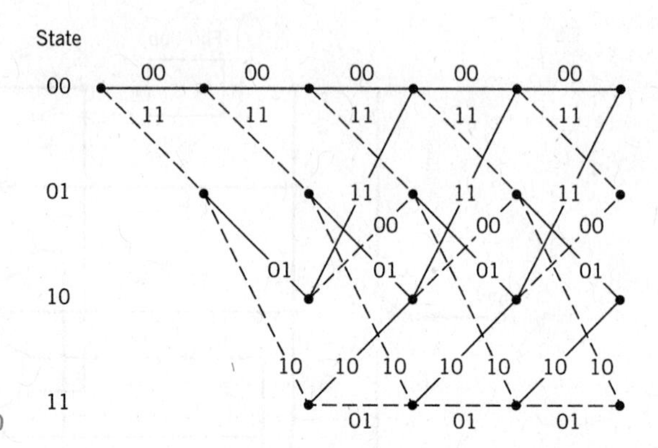

Figure P10.30

demodulation of a received sequence corrupted by ISI due to a dispersive channel. Figure P10.31 shows the trellis diagram for ISI, assuming a binary data sequence. The channel is discrete, described by the finite impulse response (1, 0, 1). The received sequence is (1.0, –0.3, –0.7, …). Use the Viterbi algorithm to determine the maximum likelihood decoded version of the sequence.

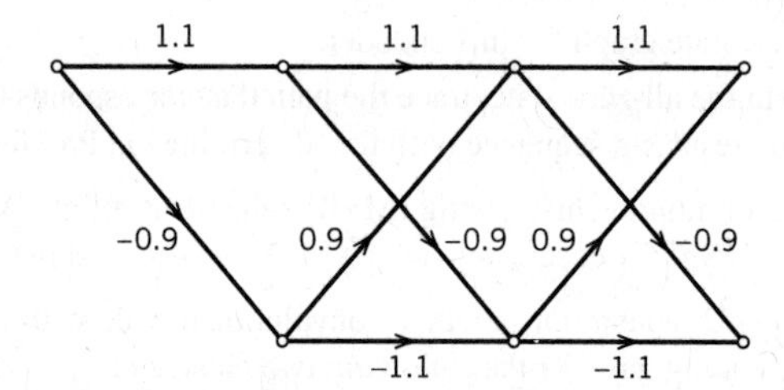

Figure P10.31

10.32 In dealing with channel equalization, a primary objective is to undo the convolution performed by a linear communication channel on the source signal. This task is well suited for the Viterbi equalizer functioning as a channel equalizer.

a. What is the underlying idea in the Viterbi algorithm that ties channel equalization and convolutional decoding together?

b. Suppose that the channel has memory defined by 2^l, where l is an integer.

What is the required length of the window for the Viterbi equalizer? Justify your answers for both parts a and b of the question.

The MAP Algorithm

10.33 Refer back to (10.92), where

$$A_j = \left(\frac{\frac{1}{2}\exp L_a(-m_j)}{1 + \exp(L_a(-m_j))} \right), \qquad j = 0, 1, 2, \ldots$$

Verify that the factor A_j is a constant, regardless of whether the message bit m_j is –1 or +1.

10.34 In Example 7, we used the max-log-MAP algorithm to decode the three message bits at the output of the RSC encoder depicted in Figure 10.23. The computations were obtained using a Matlab code. Parts a and b of the figure pertain to the block diagram of the encoder and its trellis, respectively. The five computational steps described therein apply equally well to the log-MAP algorithm.

a. Repeat Example 7, but this time develop a Matlab code for using the log-MAP algorithm to compute the decoded binary output of the encoder in Figure 10.23a.

b. Confirm the decoded binary output produced in part a by performing the five tasks involved in the log-MAP algorithm, doing all the computations in the traditional way.

c. Compare the decoded output product using the log-MAP algorithm with that reported in Example 7.

Comment on your results.

10.35 Figure P10.35 depicts two processing stages involved in the MAP decoding algorithm. The first stage is a convolutional encoder of rate, $r = k/n$, producing the code vector $\mathbf{c}$ in response to the message vector $\mathbf{m}$. The second stage is a mapper, represented by binary PSK. The signal energy per message bit at the encoder input is denoted by E_b; the noise spectral density of the AWGN channel is $N_0/2$. Let E_s denote the signal energy per symbol transmitted by the binary PSK mapper. Show that the SNR measured at the channel output is given by

$$(\text{SNR})_{\text{out}} = \frac{E_s}{N_0}$$
$$= r\left(\frac{E_b}{N_0}\right)$$

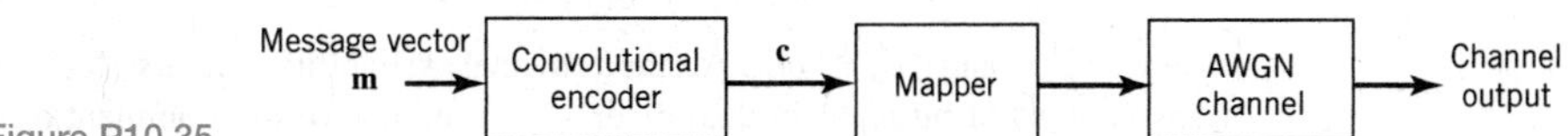

Figure P10.35

10.36 Consider an AWGN channel with unquantized output, assuming the binary code maps $0 \rightarrow -1$ and $1 \rightarrow +1$. Given a received signal $r_j^{(0)}$ at the channel output in response to a transmitted message bit m_j before decoding, the a posteriori L-value is defined by

$$L(m_j|r_j^{(0)}) = \ln\left(\frac{\mathbb{P}(m_j = +1|r_j^{(0)})}{\mathbb{P}(m_j = -1|r_j^{(0)})}\right)$$

a. Show that

$$L(m_j|r_j^{(0)}) = -\frac{E_s}{N_0}[(r_j^{(0)}-1)^2 - (r_j^{(0)}+1)^2]$$
$$+\ln\left(\frac{\mathbb{P}(m_j = +1)}{\mathbb{P}(m_j = -1)}\right)$$

where E_s is the transmitted signal energy per encoded symbol.

b. The channel reliability factor is defined by the following formula, assuming that both m_j and $r_j^{(0)}$ are normalized by the factor $\sqrt{E_s}$ by

$$L_c = 4E_s/N_0$$

where E_s/N_0 is the channel output SNR. Hence, show that

$$L(m_j|r_j^{(0)}) = L_c r_j^{(0)} + L_a(m_j)$$

where $L_a(m_j)$ is the a priori probability of message bit m_j.

10.37 In this problem, we expand on Problem 10.36 by considering a binary fading wireless channel, where the channel noise is additive, white, and Gaussian. As in Problem 10.36, start with the log-likelihood ratio of a transmitted message bit m_j conditioned on the corresponding matching filtered output r_j at time-unit j:

$$L(m_j|r_j) = \ln\left(\frac{\mathbb{P}(m_j = +1|r_j)}{\mathbb{P}(m_j = -1|r_j)}\right)$$

Let a denote the fading amplitude, which distinguishes this problem from Problem 10.36.

a. Show that

$$L(m_j|r_j) = L_c r_j + L(m_j)$$

where

$$L(m_j) = \ln\left(\frac{\mathbb{P}(m_j = +1)}{\mathbb{P}(m_j = -1)}\right)$$

and

$$L_c = \frac{4aE_s}{N_0}$$

is the modified channel reliability factor.

b. For statistically independent transmissions as in dual diversity, show that the log-likelihood ratio takes the expanded form:

$$L(m_j|r_j^{(1)}, r_j^{(2)}) = L_c^{(1)} + L_c^{(2)} + L(m_j)$$

where $L_c^{(1)}$ and $L_c^{(2)}$ denote the channel reliability factors for the two simultaneous transmissions of bit m_j as in dual diversity. Given this result, comment on the benefit gained by the use of diversity.

Turbo Codes

10.38 Let $r_c^{(1)} = p/q_1$ and $r_c^{(2)} = p/q_2$ be the code rates of RSC encoders 1 and 2 in the turbo encoder of Figure 10.26. Find the code rate of the turbo code.

10.39 The feedback nature of the constituent codes in the turbo encoder of Figure 10.26 has the following implication: a single bit error corresponds to an infinite sequence of channel errors. Illustrate this using a message sequence consisting of symbol 1 followed by an infinite number of symbols 0.

10.40 Consider the following generator matrices for rate-1/2 turbo codes:

$$\text{4-state encoder: } \mathbf{g}(D) = \left[1, \frac{1 + D + D^2}{1 + D^2}\right]$$

$$\text{8-state encoder: } \mathbf{g}(D) = \left[1, \frac{1 + D^2 + D^3}{1 + D + D^2 + D^3}\right]$$

$$\text{16-state encoder: } \mathbf{g}(D) = \left[1, \frac{1 + D^4}{1 + D + D^2 + D^3 + D^4}\right]$$

a. Construct the block diagram for each one of these RSC encoders.

b. Set up the parity-check equation associated with each encoder.

10.41 Turbo decoding relies on the feedback of extrinsic information. The fundamental principle adhered to in the turbo decoder is to avoid feeding a decoding state information that stems from the constituent decoder itself. Explain the justification for this principle in conceptual terms.

10.42 Suppose a communication receiver consists of two components: a demodulator and a decoder. The demodulator is based on a Markov model of the combined modulator and channel, and the decoder is based on a Markov model of a forward error-correction code. Discuss how the turbo principle may be applied to construct a joint demodulator–decoder for this system.

10.43 Summarize the properties/attributes of turbo codes by expanding on the following six issues:

a. Structural composition of the turbo encoder and decoder.
b. Improvement in the speed of decoding attributed to the two constituent decoders at the expense of increased computational complexity.
c. Similarity of turbo decoding to the use of feedback in nonlinear control theory.
d. Feeding extrinsic information from constituent decoder 1 to constituent decoder 2, back and forth, thereby maintaining statistical independence between the bits from one iteration to the next.
e. Typical termination of the turbo decoding process after a relatively small number of iterations, somewhere in the range of 10 to 20.
f. Relatively small degradation in decoding performance of the Max-log-MAP algorithm in the order of 0.5 dB, compared with the MAP algorithm.

10.44 Present a comparative evaluation of convolutional codes and turbo codes in terms of the encoding and decoding strategies as well as other matters that pertain to signaling over wireless communications. Specifically, address the following issues in the comparative evaluation:

a. Encoding
b. Decoding
c. Fading wireless channels
d. Latency (i.e., delay incurred in transmission over the channel).

10.45 Referring back to the eight-state Ungerboeck 8-PSK of Figure 10.40, show that the asymptotic coding gain of this code is 3.5; see Table 10.8.

LDPC Codes

10.46 The generator polynomial of the (7,8) cyclic maximal-length code is given by

$$\mathbf{g}(X) = 1 + X + X^2 + X^4$$

Show that this code is an LDPC code by constructing its Tanner graph.

10.47 Consider the (7,4) cyclic Hamming code, whose generator polynomial is given by

$$\mathbf{g}(X) = 1 + X + X^3$$

Construct the Tanner graph of this code, demonstrating that it is another example of an LDPC code.

10.48 The expanded version of the cyclic Hamming code is obtained as follows. If $\mathbf{H}$ is parity-check matrix of the cyclic Hamming code, then the parity-check matrix of its extended version is defined by

$$\mathbf{H}' = \left[\begin{array}{ccc:c} 1 & 1 & 1 \cdots & 1 \\ \hdashline & & & 0 \\ & \mathbf{H} & & \vdots \\ & & & 0 \end{array}\right]$$

$$= \left[\begin{array}{ccccccc:c} 1&1&1&1&1&1&1&1 \\ \hdashline 1&0&0&1&0&1&1&0 \\ 0&1&0&1&1&1&0&0 \\ 0&0&1&0&1&1&1&0 \end{array}\right]$$

whereby the distance between every pair of codewords in the extended code is now even.

Construct the Tanner graph of the extended cyclic Hamming code (8, 4).

10.49 In light of the linear cyclic codes considered in Problems 10.46 to 10.48, comment on the relationship between this class of codes and LDPC codes.

10.50 In Note 20, we introduced the idea of rateless codes, emphasizing the relationship that exists between the new class of codes and LDPC codes. Which features distinguish rateless codes from LDPC codes?

10.51 Develop a list comparing LDPC codes with turbo codes.

Notes

1. Feedforward error correction (FEC) relies on the controlled use of redundancy in the transmitted codeword for both the *detection and correction* of errors incurred during the course of transmission over a noisy channel. Irrespective of whether the decoding of the received codeword is successful, no further processing is performed at the receiver. Accordingly, channel coding techniques suitable for FEC require only a *one-way link* between the transmitter and receiver.

There is another approach known as *automatic-repeat request* (ARQ) for solving the error-control problem. The underlying philosophy of ARQ is quite different from that of FEC. Specifically, ARQ uses redundancy merely for the purpose of *error detection*. Upon the detection of an error in a transmitted codeword, the receiver requests a repeat transmission of the corrupted codeword, which necessitates the use of a *return path* (i.e., a feedback channel from the receiver to the transmitter).

For a comprehensive treatment of error-control coding, see Lin and Costello (2004) and Moon (2005).

2. In medicine, the term *syndrome* is used to describe a pattern of symptoms that aids in the diagnosis of a disease. In coding, the error pattern plays the role of the disease and parity-check failure that of a symptom. This use of syndrome was coined by Hagelbarger (1959).

3. The first error-correcting codes, known as Hamming codes, were invented by Hamming at about the same time as the conception of information theory by Shannon; for details, see the classic paper by Hamming (1950).

4. Maximal-length codes, also referred to as *m*-sequences, are discussed further in Appendix J; they provide the basis for pseudo-noise (PN) sequences, which play a key role in the study of spread spectrum signals in Chapter 9.

5. Reed–Solomon codes are so named in honor of their originators; see their classic paper (Reed and Solomon, 1960).

The book edited by Wicker and Bhargava (1994) contains an introductory chapter on Reed–Solomon codes; a historical overview of the codes written by Reed and Solomon themselves; and chapters on the applications of Reed–Solomon codes to exploration of the solar system, the compact disc, automatic repeat-request protocols, and spread-spectrum multiple-access communications.

In a historical context, Reed–Solomon codes are a subclass of the Bose–Chaudhuri and Hocquenghem (BCH) codes that represent a large class of powerful random error-correcting cyclic codes. However, it is important to recognize that the Reed–Solomon codes were discovered independently of the pioneering works by Hocquenghem (1959) and Bose and Ray-Chaudhuri (1960).

For detailed mathematical treatments of binary BCH codes and nonbinary BCH codes with emphasis on Reed–Solomon codes, see Chapters 6 and 7 of the book by Li and Costello (2004), respectively.

6. Convolutional codes were invented by Elias (1955) as an alternative to linear block codes. The aim of that classic paper was to formulate a new class of codes with as much structure as practically feasible without loss of performance in using them over binary symmetric and AWGN channels.

7. In a classic paper, Viterbi (1967) proposed a decoding algorithm for convolutional codes that has become known as the Viterbi algorithm. The algorithm was recognized by Forney (1972, 1973) to be a maximum likelihood decoder. Readable accounts of the *Viterbi algorithm* are presented in the book by Lin and Costello (2004).

The discussion presented in this chapter is confined to the classical Viterbi algorithm involving hard decisions. For iterative decoding applications with soft outputs, Hagenauer and Hoeher (1989) described the so-called *soft-output Viterbi algorithm (SOVA)*. For detailed discussion of both versions of the Viterbi algorithm, the reader is referred to Lin and Costello (2004).

8. For details of the evaluation of asymptotic coding gain for binary symmetric and binary-input AWGN channels, see Lin and Costello (2004).

9. At first sight, derivation of the MAP decoding algorithm appears to be complicated. In reality, however, the derivation is straight forward, given knowledge of probability theory. The derivation presented herein follows the book by Lin and Costello (2004).

10. For detailed mathematical description of the log-MAP algorithm, the reader is referred to the book (Lin and Costello, 2004).

11. Costello and Forney (2007) surveyed the evolution of coding on the road to channel capacity for AWGN channels over the course of 50 years, going back to the classic paper of Claude Shannon (1948). Proceeding in a stage-by-stage manner through the history of codes over band-limited channels, they came to the paper written by Berrou, *et al.*, (1993) on turbo codes, which was presented at the IEEE International Communications Conference (ICC) in Geneva, Switzerland; therein, the achievement of a performance near the Shannon limit with modest decoding complexity was claimed by its three co-authors. Listening to this claim, the coding research community at the conference were stunned, with comments being whispered to the effect: "It cannot be true; they must have made a 3 dB error." However, in the course of a year, the claims reported by Berrou were confirmed by various laboratories. And, with it, the turbo revolution was launched.

12. The plots presented in Figure 10.26 follow those in the book by Frey (1998).

13. Example 9 is based on the Ph.D. thesis by Li (2011), with useful comments by Maunder (2012).

14. For the case when the interleaver's length is high, as in the simulation results plotted in the BER chart of Figure 10.31, finding the floor region can be extremely time consuming. Indeed, it is for this reason that the number of iterations in Figure 10.31 was limited to four.

15. The averaging method emanated from the Ph.D. thesis of Land (2005); this method is also described in Land *et al.* (2004). The first reference to the averaging method was made under "private communication" in Hagenauer (2004).

16. The LDPC codes, introduced by Gallager (1960, 1963), were dormant for more than three decades. Lack of interest in these codes in the 1960s and 1970s may well have been attributed to the fact that the computers of those days were not powerful enough to cope with LDPC codes of long block lengths. But, reflecting back over the 1980s, it is surprising to find that lack of interest in LDPC codes by the coding community persisted for all those years except for a single paper: Tanner (1981) proposed a graphical representation for studying the structure of Gallager's LDPC codes (as well as other codes) for the purpose of iterative decoding; such graphs are now called the *Tanner graphs*. In any event, it was not until the introduction of turbo codes and iterative decoding by Berrou *et al.* (1993) that interest in LDPC codes was rekindled. Two factors were responsible for this rekindled interest (Hanzo, 2012):

- the protection of turbo codes by a patent and unwillingness of industry to pay royalties, and
- the rediscovery of LDPC codes by MacKay and Neal (1996; MacKay, 1999).

And with it, the LDPC rediscovery was launched.

17. In a historical context, Tanner's classic paper was also forgotten for well over a decade, until its rediscovery by Wiberg (1996) in his seminal thesis.

18. For a detailed treatment of the statement that the probability distribution of the minimum distance of an LDPC code approaches a unit step function of the block length for certain values of weight-pair (t_c, t_r), see Gallager (1962, 1963).

19. The decoding algorithm of LDPC codes described in Section 10.14 follows MacKay and Neal (1996, 1997).

20. The sum–product algorithm (SPA) is a computationally efficient, soft-in soft-out (SISO), iterative deciding algorithm based on belief propagation. The notion of belief propagation was originally described in Pearl (1988), wherein it was used to study statistical inference in Bayesian networks. For a detailed exposition of SPA for the iterative decoding of LDPC codes, see MacKay (1999).

In a related context, a relationship exists between LDPC codes and a new class of erasure codes known as rateless codes, pioneered by Luby (2002). An erasure code is said to be rateless if, ideally, it satisfies two requirements:

- First, encoding symbols are generated in the transmitter from an incoming data stream in on-line manner, such that their number is potentially limitless.
- Second, a decoder in the receiver recovers a replica of the data from an aggregate set of the generated encoding symbols, which is only slightly longer than the original data stream.

Rateless codes are designed for channels without feedback and whose statistics are not known a priori. One such channel is the Internet packet switching, where the probability of packet erasure is unknown. In any event, rateless codes are basically low-density generator-matrix codes, which are decoded using the SPA used to decode LDPC codes; hence the relationship between them. This relationship is discussed in detail in Bonello, Chen, and Hanzo (2011).

21. In a historical context, the discovery of irregular LDPC codes was originally spearheaded by Luby *et al.* (1997, 2001), resulting from the substantial efforts that were invested in the development of LDPC codes after the onset of the turbo revolution.

In terms of performance attainable by irregular LDPC codes, Chung *et al.* (2001) were the first to demonstrate that several very long rate-1/2 irregular LDPC codes for AWGN channels could be designed to approach the Shannon limit within 0.0045 dB, which is truly remarkable.

22. Trellis-coded modulation was invented by Ungerboeck (1982); its historical evolution is described in Ungerboeck (1987). Table 10.8 is adapted from this latter paper.

Trellis-coded modulation may be viewed as a form of *signal-space coding*—a viewpoint discussed at an introductory level in Chapter 14 of the book by Lee and Messerschmitt (1994). For an extensive treatment of trellis-coded modulation, see the books by Schlegel (1997) and Lin and Costello (2004: 875–880).

23. A concatenated coding scheme using trellis-coded modulation first appeared in Robertson and Wörz (1998), appropriately dubbed *Turbo TCM* using a parallel concatenation scheme, and has met with further refinements in Hanzo *et al.* (2003), Koca and Levy (2004), and Sun *et al.* (2004).

The serial concatenation scheme can likewise apply, in which the outer encoder is still a recursive systematic encoder, while the inner encoder implements an Ungerboeck code for modulating the symbols to be sent over the communication channel. As the Ungerboeck code imposes a trellis structure, the inner decoder may be implemented with the MAP algorithm to obtain the bitwise *a posteriori* probabilities; the extrinsic information extraction from this inner decoder follows the same steps as in Section 10.16, and the coupling of decoders as per Figure 10.43 carries over immediately.

24. For turbo equalization and related issues, see Douillard *et al.* (1995); Supnithi *et al.* (2003); Jiang *et al.* (2004); Kötter *et al.* (2004); Rad and Moon (2005); Lopes and Barry (2006); Regalia (2010).

25. For turbo CDMA, see the papers by Alexander *et al.* (1999) and Wang and Poor (1999).

The topic of DS-CDMA was discussed in Chapter 9.

26. In formulating (10.157), we have introduced *log posterior ratio* and *log prior ratio* so as to avoid confusion with the traditional log likelihood ratio, particularly so when the ratio of interest in this section is not always between likelihood function evaluations.

APPENDIX

 Advanced Probabilistic Models

In the study of digital communications presented in preceding chapters, the Gaussian, Rayleigh, and Rician distributions featured in the formulation of probabilistic models in varying degrees. In this appendix we describe three relatively advanced distributions:

- the chi distribution;
- the log-normal distribution;
- the Nakagami distribution.

The chi distribution is featured in the study of diversity-on-receive techniques in Chapter 9 on signaling across fading channels. Just as importantly, the log-normal distribution was mentioned in passing in the context of shadowing in wireless communications, also in Chapter 9. The Nakagami distribution is the most advanced of all the three:

- it includes the Rayleigh distribution as a special case;
- its shape is similar to the Rician distribution;
- it is flexible in its applicability.

A.1 The Chi-Square Distribution

A *chi-square* χ^2 *distributed random variable* is produced, for example, when a Gaussian random variable is passed through a squaring device. Viewed in this manner, there are two kinds of χ^2 distributions:

1. *Central* χ^2 *distribution*, which is produced when the Gaussian random variable has zero mean.
2. *Noncentral* χ^2 *distribution*, which is produced when the Gaussian random variable has a nonzero mean.

In this appendix, we will discuss only the central form of the distribution.

Consider, then, a standard Gaussian random variable X, which has zero mean and unit variance, as shown by

$$f_X(x) = \frac{1}{\sqrt{2\pi}}\exp\left(-\frac{x^2}{2}\right), \qquad -\infty < x < \infty \tag{A.1}$$

Let the variable X be applied to a *square-law device*, producing a new random variable Y, whose sample value is defined by

$$y = x^2 \tag{A.2}$$

or, equivalently,

$$x = \pm\sqrt{y} \tag{A.3}$$

The cumulative distribution function of the random variable Y produced at the output of the square-law device is therefore defined by

$$F_Y(y) = \int_{-\sqrt{y}}^{\sqrt{y}} f_X(x)\,dx \tag{A.4}$$

Differentiating $F_Y(y)$ with respect to y yields the probability density function (pdf):

$$\begin{aligned} f_Y(y) &= \frac{\partial}{\partial y}\left(\sqrt{y} - f_X(x)\right)\bigg|_{x=-\sqrt{y}}\left(\frac{\partial}{\partial y}(-\sqrt{y})\right) \\ &= \frac{f_X(\sqrt{y}) + f_X(-\sqrt{y})}{2\sqrt{y}} \end{aligned} \tag{A.5}$$

Substituting (A.1) into (A.5), we get

$$\begin{aligned} f_Y(y) &= \frac{1}{2\sqrt{y}}\left[\frac{1}{\sqrt{2\pi}}\exp\left(-\frac{y}{2}\right) + \frac{1}{\sqrt{2\pi}}\exp\left(-\frac{y}{2}\right)\right] \\ &= \frac{1}{\sqrt{2\pi y}}\exp\left(-\frac{y}{2}\right), \qquad 0 \le y < \infty \end{aligned} \tag{A.6}$$

The distribution described in (A.6) is called the *chi-square* (χ^2) *distribution with one degree of freedom.*

The first two moments of Y are given by

$$\mathbb{E}[Y] = 1$$

$$\mathbb{E}[Y^2] = 3$$

and its variance is

$$\text{var}[Y] = 2$$

Note, however, that these values are based on the standard Gaussian distribution with zero mean and unit variance. For the general case of an ordinary Gaussian distribution with zero mean and variance σ^2, the mean, mean-square value, and variance of the X^2 random variable Y are, respectively, as follows:

$$\mathbb{E}[Y] = \sigma^2$$

$$\mathbb{E}[Y^2] = 3\sigma^4$$

$$\text{var}[Y] = 2\sigma^4$$

In its most general setting, derivation of the chi-square distribution follows from a set of iid random variables denoted by $\{X_i^2\}_{i=1}^{n}$, on the basis of which a new random variable is defined as follows:

$$Y = \sum_{i=1}^{n} X_i^2 \tag{A.7}$$

On this basis, the pdf of the random variable Y is defined by

$$f_Y(y) = \begin{cases} \dfrac{y^{(n/2)-1}}{\left(2^{n/2}\Gamma\left(\frac{n}{2}\right)\sigma^n\right)}\exp\left(-\dfrac{y}{2\sigma^2}\right), & y > 0 \\ 0, & \text{otherwise} \end{cases} \tag{A.8}$$

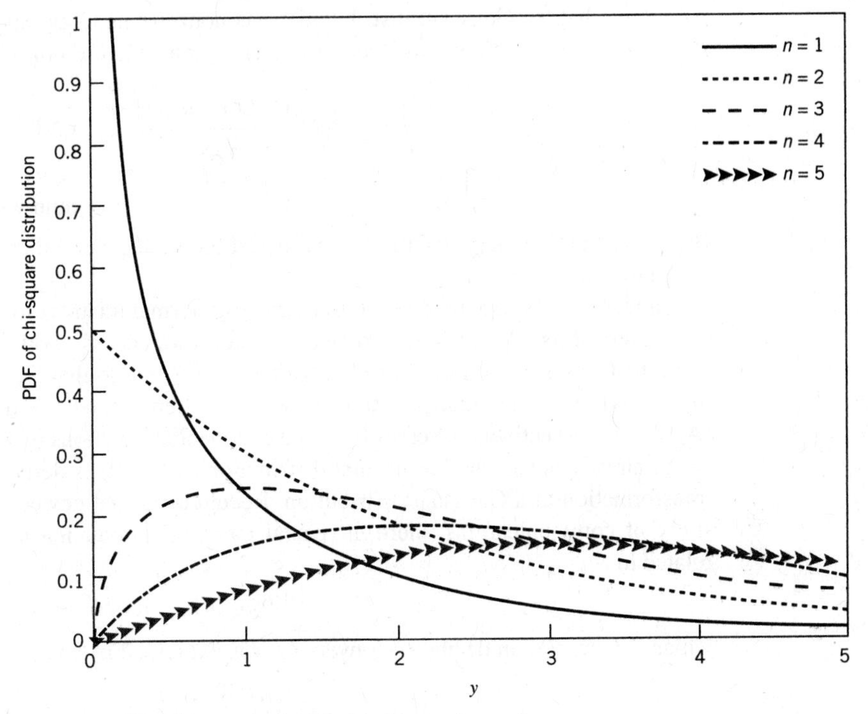

Figure A.1 The chi-square distribution for varying order n.

where $\Gamma(\lambda)$ is Euler's *gamma function*, defined by (Abramowitz and Stegun, 1965)

$$\Gamma(\lambda) = \int_0^{\infty} t^{\lambda-1} \exp(-t)\, dt \tag{A.9}$$

As such, the random variable Y is said to have the *chi-square distribution of order n*. When $n = 1$, $\Gamma(1/2) = \sqrt{\pi}$, and we get the special case described in (A.6); this special case of the χ^2 distribution is also referred to as the *one-sided exponential distribution*. Figure A.1 plots the χ^2 distribution for varying orders: $n = 1, 2, 3, 4, 5$.

A.2 The Log-Normal Distribution

To proceed next with the log-normal distribution, let X and Y be two random variables that are related to each other through the logarithmic transformation

$$Y = \ln(X) \tag{A.10}$$

where ln is the natural logarithm. Conversely, we have

$$X = \exp(Y) \tag{A.11}$$

In light of this logarithmic transformation, the random variable X is said to be *log-normally distributed* if the other random variable Y is *normally* (i.e., Gaussian) distributed.

Assuming that the Gaussian-distributed Y has nonzero mean μ_Y and variance σ_Y^2, then a straightforward transformation based on (A.11) yields the *log-normal distribution*:

$$f_X(x) = \begin{cases} \dfrac{1}{\sqrt{2\pi}\sigma_Y x}\exp\left[-\dfrac{(\ln(x)-\mu_Y)^2}{2\sigma_Y^2}\right], & x \geq 0 \\ 0, & \text{otherwise} \end{cases} \tag{A.12}$$

By the same token, a probability model based on the log-normal distribution of (A.12) is called the *log-normal model.*

Unlike the chi-square distribution, the log-normal distribution has two adjustable parameters of its own, the nonzero mean μ_Y and variance σ_Y^2, both of which are inherited from the Gaussian distributed random variable Y. Note also that the mean and variance of the log-normally distributed random variable X, represented by the sample value x in (A.12), are respectively different from the exponential functions of μ_Y and σ_Y^2.

As already noted, the log-normal distribution of (A.12) is derived via the logarithmic transformation of a Gaussian-distribution. Recognizing that power plays a key role in the study of communications, there is special merit in introducing a new random variable related to X:

$$z = 10\log_{10}(x) \tag{A.13}$$

which is measured in decibels. Conversely, x is expressed in terms of z as follows:

$$x = 10^{z/10} \tag{A.14}$$

Hence, using (A.14) in (A.13), we get

$$Y = cZ \tag{A.15}$$

where the constant is

$$c = \frac{\ln(10)}{10} \tag{A.16}$$

Equation (A.15) shows that both Y and Z are Gaussian distributed, differing by the scaling factor c.

Accordingly, the mean and variance of the Gaussian-distributed random variable Z are respectively defined by

$$\mu_Z = \frac{1}{c}\mu_Y, \qquad \sigma_Z^2 = \frac{1}{c^2}\sigma_Y^2 \tag{A.17}$$

Equivalently, we may write

$$\mu_Y = c\mu_Z, \qquad \sigma_Y^2 = c^2\sigma_Z^2 \tag{A.18}$$

To visualize the log-normal distribution defined in (A.12), we propose to proceed as follows:[1]

1. The mean μ_Y is maintained at the constant value, $\mu_Y = 0$ dB.
2. The standard deviation σ_Y (that is, the square root of the variance σ_Y^2) is assigned three different values: $\sigma_Y = 1, 5, 10$ dB.

With decibel as the logarithmic measure of interest, the new variable x in the log-normal distribution of (A.12) is also measured in decibels. Thus, using the assigned values of μ_Y and σ_Y under points (1) and (2) in (A.12), we get the plots displayed in Figure A.2.

Examining Figure A.2, we make two observations that are of particular interest:

1. The log-normal distribution exhibits long tails for $\sigma_Z \geq 6$ dB; hence its appropriateness as a model for the *shadow-fading phenomenon* in wireless communications. From a practical perspective, a standard deviation lying in the range $6 \leq \sigma_Z \leq 8$ dB is typical for shadowing, in which case we see that the distribution of Figure A.2 is quite *asymmetric with a small "modal" value*. In other words, $6 \leq \sigma_Z \leq 8$ dB is the *mode* or the *most likely range* of shadowing.
2. When the standard deviation σ_Z is reduced below this range, the log-normal distribution tends to become more symmetric and, therefore, Gaussian, centered roughly around $x = 1$ dB.

Useful Properties of the Log-Normal Distribution

Over and above having the characteristic of long tails, the log-normal distribution has two other useful properties:[2]

PROPERTY 1 *The product (or quotient) of log-normal variables is log-normal.*

This property follows from the fact that the exponents of the random variable Y or Z add (or subtract). Since the exponents are Gaussian distributed, they remain Gaussian after the addition (or subtraction); hence the validity of Property 1.

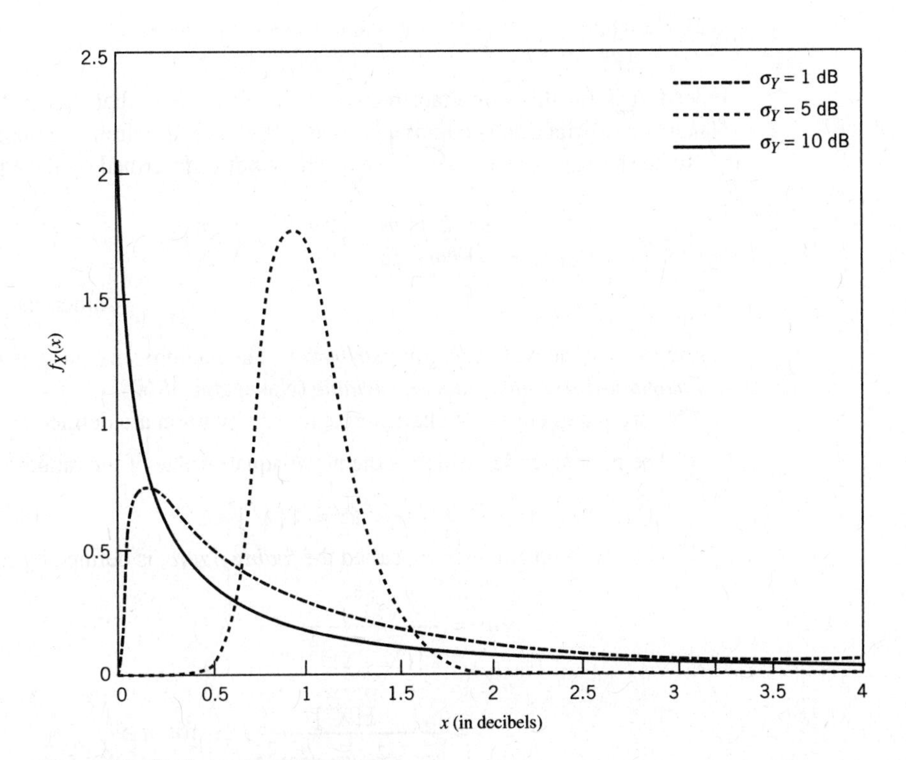

Figure A.2 The log-normal distribution.

As a corollary to this property, we may also state:

> The amplitude and power of a log-normal random variable are both log-normal.

PROPERTY 2 *The product of a large number of iid random variables is asymptotically log-normal.*

This property is the counterpart of the *central limit theorem*, involving the addition of a large number of iid random variables. The reason for this second property is rather obvious for two reasons:

- First, the example of the random variables involved in forming the product add.
- Second, applying the central limit theorem to the addition of the example, the result asymptotically converges to a Gaussian distribution; hence the validity of Property 2.

A.3 The Nakagami Distribution

As different as the distributions covered until this point are, namely the Rayleigh and Rician distributions derived in Chapter 4, as well as the chi-square and log-normal distributions derived in this appendix, all four of them share a common factor:

> They are derived from the Gaussian distribution through respective transformations.

In the last part of this appendix we describe another distribution, namely the Nakagami distribution, which is different from all the others in the following sense:

> Through the use of simulation, the Nakagami distribution can be fitted directly to real-life data.

Indeed, it is for this important reason (and a few others that will be discussed) that the Nakagami distribution is commonly used as a model for wireless communications.

To be specific, a random variable X whose pdf is described by the equation

$$f_X(x) = \begin{cases} \dfrac{2}{\Gamma(m)}\left(\dfrac{m}{\Omega}\right)^m x^{2m-1}\exp\left(-\dfrac{m}{\Omega}x^2\right), & x \geq 0 \\ 0, & \text{otherwise} \end{cases} \tag{A.19}$$

is said to have the *Nakagami-m distribution*. The random variable X is itself referred to as a *Nakagami-distributed random variable* (Nakagami, 1960).

The two parameters that characterize this distribution are defined as follows:

1. The parameter Ω, which is the mean-square value of the random variable X; that is,

$$\Omega = \mathbb{E}[X^2] \tag{A.20}$$

2. The second parameter, m, called the *fading figure*, is defined by the ratio:

$$\begin{aligned} m &= \frac{\Omega^2}{\mathbb{E}[(X^2-\Omega)^2]} \\ &= \frac{\mathbb{E}[X^2]}{\mathbb{E}[(X^2-\mathbb{E}[X^2])^2]}, \qquad \text{for } m \geq \frac{1}{2} \end{aligned} \tag{A.21}$$

Note the restriction that is placed on m for (A.21) to hold. Close examination of the definitions embodied in (A.20) and (A.21) reveals that the statistical characterization of the fading figure m involves two moments:

- the *mean-square value* of the random variable X in the numerator and
- the *variance* of the squared random variable X^2 in the denominator.

It follows, therefore, that the fading figure m is dimensionless.

For visualization, the Nakagami-m distribution is plotted in Figure A.3 for varying values of m. Two observations from these plots are noteworthy:

1. For $m = 1/2$, the Nakagami-m distribution reduces to the Rayleigh distribution; in other words:

   The Rayleigh distribution is a special case of the Nakagami distribution.

2. The Nakagami and Rician distributions have a similar shape.

To elaborate on point 2, for $m > 1$ we find that the fading figure m can be computed from the dimensionless *Rice factor* K (discussed in Chapter 4), as shown in (Stüber, 1996):

$$m = \frac{(K+1)^2}{2K+1} \tag{A.22}$$

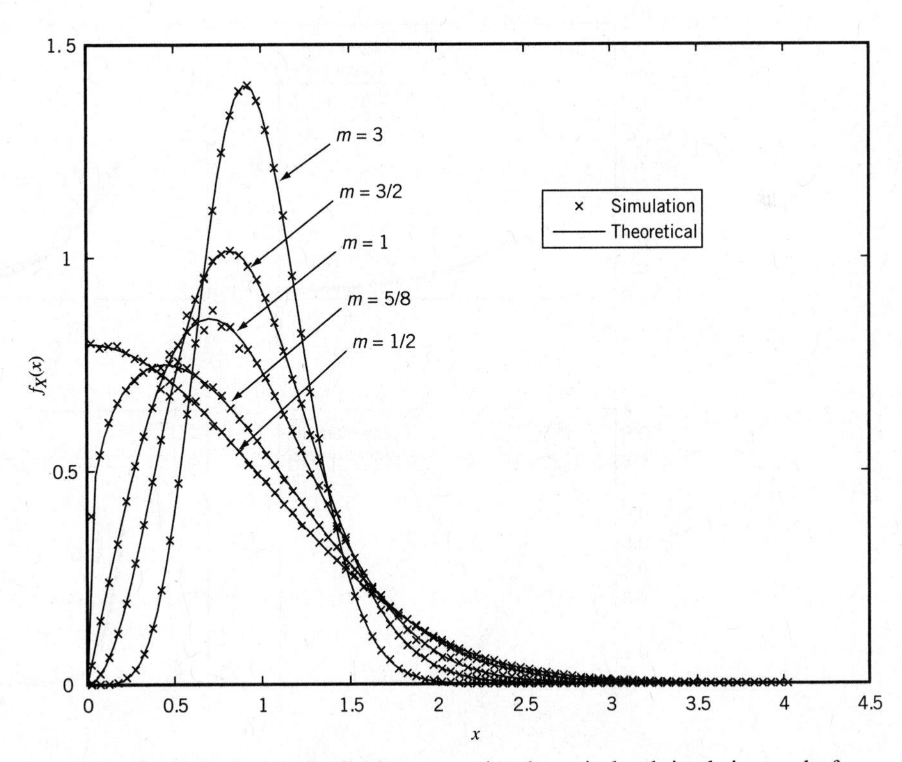

Figure A.3 The Nakagami-m distribution, presenting theoretical and simulation results for varying fading figure m.

Conversely,

$$K = \frac{(m^2 - m)^{1/2}}{m - (m^2 - m)^{1/2}} \tag{A.23}$$

A cautionary note is in order, however. Although the Nakagami-*m* and Rician distributions appear to have good agreement insofar as their shapes are concerned, they have different slopes at the origin, $x = 0$; this difference has a significant impact on the achievable diversity, with the advantage residing in the Nakagami distribution (Molisch, 2011).

From a practical perspective, the Nakagami-*m* distribution has the following attributes, in accordance with (A.20) and (A.21):

> The two parameters, Ω and m, lend themselves to computation from experimentally measured data in a relatively straightforward manner.

This succinct statement re-emphasizes the point we made at the beginning of this subsection:

> Through the use of simulation, real-life data can be fitted into the Nakagami distribution.

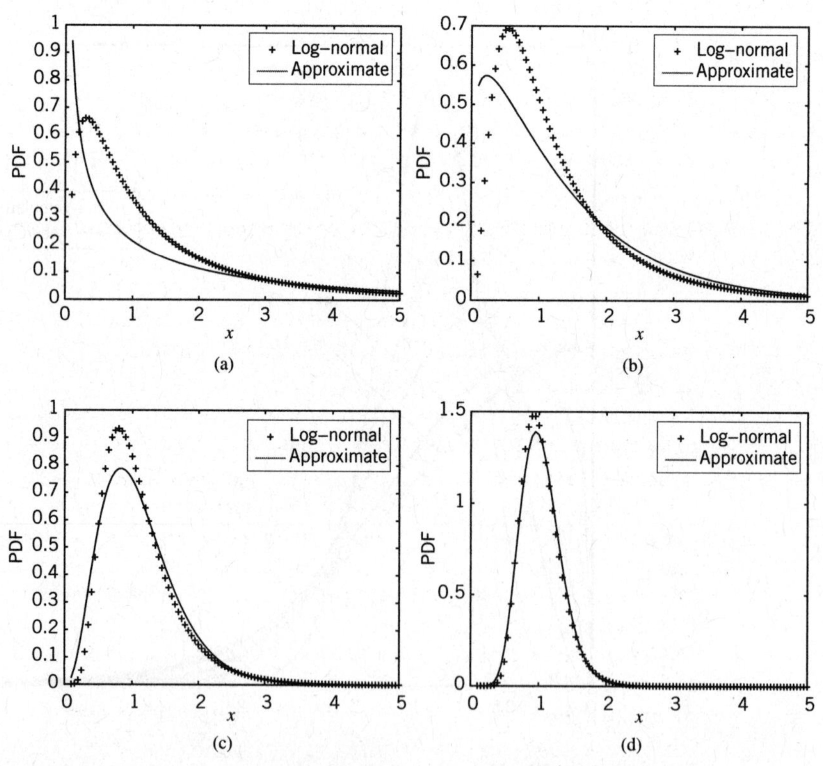

Figure A.4 A set of sample functions of log-normal distribution and its approximation with the Nakagami distribution as the fading figure *m* is increased.

Indeed, with this important point in mind, the plots presented in Figure A.3 actually include points (denoted by crosses) that pertain to an arbitrarily selected wireless data.[3]

Figure A.4 provides further demonstration of the inherent flexibility of the Nakagami-m distribution in approximating the log-normal distribution. It is clearly shown that the approximation gets gradually better as the fading figure m is increased.

It is not surprising, therefore, to find that the Nakagami-m distribution outperforms the Rayleigh and Rician distributions, particularly so in urban wireless communication environments.[4]

Notes

1. The visualization procedure described herein for the log-normal distribution follows Cavers (2000).

Two other procedures for visualizing the log-normal distribution are described in the literature, as summarized here:

- In Proakis and Salehi (2008), the standard deviation $\sigma_Y = 1$ and the mean μ_Y are varied, with both μ_Y and σ_Y measured in volts.
- In Goldstein (2005), a new random variable Ψ defined as the ratio of transmit-to-receive power, is used in place of x, and a new formula for the log-normal distribution is derived. In so doing, the use of power measured in decibels plays a prominent role in a new formulation of the log-normal distribution. However, this new formulation takes values for $0 \leq \psi < \infty$, which raises a physically unacceptable scenario; specifically, for $\psi < 1$, the receive-power assumes a value greater than the transmit-power.
- Fortunately, the probability of this unacceptable scenario arising is very small, provided that the mean μ_ψ, expressed in decibels, is positive and large. It is thus claimed that the log-normal model based on the random variable ψ captures the underlying physical model very accurately when the mean μ_ψ is very large compared to 0 dB.

2. The properties of the log-normal distribution described herein follow Cavers (2000).

3. The procedure used to compute the simulated points in the plots presented in Figure A.3 follows Matthaiou and Laurenson (2007).

4. This note provides additional noteworthy material on the Nakagami-m distribution. In Turin *et al.* (1972) and Suzuki (1977), it is demonstrated that the Nakagami-m distribution provides the best statistical fit to measured data in urban wireless environments.

Two other papers of interest are Braun and Dersch (1991), in which a physical interpretation of the Nakagami-m distribution is presented, and Abdi *et al.*, (2000), in which the statistical characteristics of the Nakagami and Rician distributions are summarized.

Moreover, there are three other papers on the Nakagami distribution that deserve attention. Given a set of real-life fading-channel data, various papers have been published on how to estimate the parameter m in the Nakagami model. In Zhang (2002), numerical results are presented to show that none of the previously published results exceed the classical one by Greenwood and Durand (1960). The correlated Rayleigh fading lends itself readily to simulate a fading channel by virtue of its relationship to a complex Gaussian process. Unfortunately, this is not so with the Nakagami distribution. In Zhang (2000), a decomposition technique is described for the efficient generation of a correlated Nakagami fading channel.

In Zhang (2003), a generic correlated Nakagami-m model is described using a multiple joint characteristic function, which allows for an *arbitrary* covariance matrix and *distinct real* fading parameters.

APPENDIX

B Bounds on the *Q*-Function

Following Chapter 3, we define the Q-function as

$$Q(x) = \frac{1}{\sqrt{2\pi}} \int_x^{\infty} \exp\left(-\frac{1}{2}t^2\right) dt \tag{B.1}$$

which represents *the area under the tail of the standard Gaussian distribution.* In this appendix, we derive some useful bounds on the Q-function for large positive values of x.

To this end, we change the variable of integration in (B.1) by setting

$$z = x - t \tag{B.2}$$

and then recast (B.1) in the form

$$Q(x) = \frac{1}{\sqrt{2\pi}} \exp\left(-\frac{x^2}{2}\right) \int_{-\infty}^{0} \exp(xz) \exp\left(-\frac{1}{2}z^2\right) dz \tag{B.3}$$

For any real z, the value of $\exp(-1/2z^2)$ lies between the successive partial sums of the power series:

$$1 - \frac{z^2/2}{1!} + \frac{(z^2/2)^2}{2!} - \frac{(z^2/2)^3}{3!} + \cdots$$

Therefore, for $x > 0$ we find that, on using $(n + 1)$ terms of this series, the Q-function lies between the values taken by the integral

$$\frac{1}{\sqrt{2\pi}} \exp\left(-\frac{x^2}{2}\right) \int_{-\infty}^{0} \left[1 - \frac{(z^2/2)}{1!} + \frac{(z^2/2)^2}{2!} - \cdots \pm \frac{(z^2/2)^n}{n!}\right] \exp(xz)\, dz$$

for even n and odd n. We now make another change in the integration variable by setting

$$v = -xz \tag{B.4}$$

and also use the definite integral

$$\int_0^{\infty} v^n \exp(-v)\, dv = n! \tag{B.5}$$

Doing so, we obtain the following asymptotic expansion for the Q-function, assuming that $x > 0$:

$$Q(x) \approx \frac{\exp(-x^2/2)}{\sqrt{2\pi}x} \left[1 - \frac{1}{x^2} + \frac{1 \times 3}{x^4} - \cdots \pm \frac{1 \times 3 \times 5 \cdots (2n-1)}{x^{2n}}\right] \tag{B.6}$$

For large positive values of x, the successive terms of the series on the right-hand side of (B.6) decrease very rapidly. We thus deduce two simple bounds on the Q-function, one lower and the other upper, as shown by

$$\frac{\exp(-x^2/2)}{\sqrt{2\pi}x}\left(1 - \frac{1}{x^2}\right) < Q(x) < \frac{\exp(-x^2/2)}{\sqrt{2\pi}x} \tag{B.7}$$

For large positive x, a second bound on the Q-function is obtained by simply ignoring the multiplying factor $1/x$ in the upper bound of (B.7), in which case we write

$$Q(x) < \frac{1}{\sqrt{2\pi}} \exp\left(-\frac{x^2}{2}\right) \tag{B.8}$$

Figure B.1 contains plots of the following quantities:

- tabulated values of the Q-function presented in Table 3.1;
- the lower and upper bounds of (B.7);
- the upper bound of (B.8).

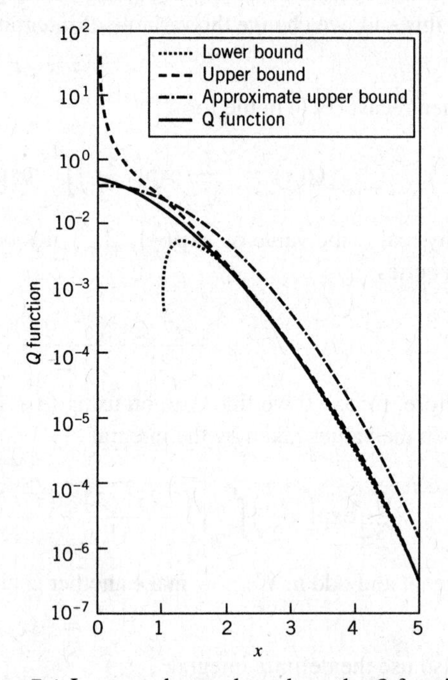

Figure B.1 Lower and upper bounds on the Q-function.

APPENDIX

C Bessel Functions

C.1 Series Solution of Bessel's Equation

In a certain class of differential or difference equations encountered in many branches of science and engineering, Bessel functions and their modified versions feature commonly in their solutions, just as cosines and sines feature commonly in trigonometry.

For example, in spectral analysis of analog frequency-modulated (FM) signals (discussed briefly in Chapter 2), the analysis involves the use of Bessel functions of infinite order; see Haykin (2001) for details of this analysis. For yet another example, in studying the Jakes FIR model in Chapter 9 on signaling over fading channels, we found that the Bessel functions of zero order featured in the autocorrelation function at the input of the mobile receiver. Then, in Chapter 7 on signaling over AWGN channels, the modified Bessel function of zero order featured in arriving at the nondata-aided recursive algorithm for symbol timing in the receiver.

These motivating examples prompt us to devote this appendix to mathematical analysis of Bessel functions and their modified versions.

In its most basic form, *Bessel's equation of order n* is written as

$$x^2\frac{d^2y}{dx^2} + x\frac{dy}{dx} + (x^2 - n^2)y = 0 \tag{C.1}$$

which is one of the most important of all variable-coefficient differential equations. For each n, a solution of this equation is defined by the power series

$$J_n(x) = \sum_{m=0}^{\infty} \frac{(-1)^m(1/2)^{n+2m}}{m!(n+m)!} \tag{C.2}$$

The function $J_n(x)$ is called a *Bessel function of the first kind of order n*. Equation (C.1) has two coefficient functions to deal with: $1/x$ and $(x - n^2/x^2)$. Hence, it has no finite singular points except for the origin. It follows, therefore, that the series expansion of (C.2) converges for all $x > 0$. This equation may thus be used to numerically calculate $J_n(x)$ for $n = 0, 1, 2, \ldots$. Table C.1 gives values of $J_n(x)$ for different order n and varying x.

The function $J_n(x)$ may also be expressed in the form of an integral as

$$J_n(x) = \frac{1}{\pi}\int_0^{\pi} \cos(x\sin\theta - n\theta)\, d\theta \tag{C.3}$$

or, equivalently,

$$J_n(x) = \frac{1}{2\pi}\int_0^{\pi} \exp(jx\sin\theta - jn\theta)\, d\theta \tag{C.4}$$

C.2 Properties of the Bessel Function

The Bessel function $J_n(x)$ has the following properties:

1.
$$J_n(x) = (-1)^n J_{-n}(x) \tag{C.5}$$

To prove this relation, we replace θ by $(\pi - \theta)$ in (C.3). Then, noting that $\sin(\pi - \theta)$ is equal to $\sin\theta$, we get

$$\begin{aligned} J_n(x) &= \frac{1}{\pi}\int_0^\pi \cos(x\sin\theta + n\theta - n\pi)\,\mathrm{d}\theta \\ &= \frac{1}{\pi}\int_0^\pi [\cos(n\pi)\cos(x\sin\theta + n\theta) + \sin(n\pi)\sin(x\sin\theta + n\theta)]\,\mathrm{d}\theta \end{aligned}$$

For integer values of n, we have

$$\begin{aligned} \cos(n\pi) &= (-1)^n \\ \sin(n\pi) &= 0 \end{aligned}$$

Therefore,

$$J_n(x) = \frac{(-1)^n}{\pi}\int_0^\pi \cos(x\sin\theta + n\theta)\,\mathrm{d}\theta \tag{C.6}$$

From (C.3), we also find that by replacing n with $-n$:

$$J_{-n}(x) = \frac{1}{\pi}\int_0^\pi \cos(x\sin\theta + n\theta)\,\mathrm{d}\theta \tag{C.7}$$

The desired result follows immediately from (C.6) and (C.7).

2.
$$J_n(x) = (-1)^n J_n(-x) \tag{C.8}$$

This relation is obtained by replacing x with $-x$ in (C.3), and then using (C.6).

3.
$$J_{n-1}(x) + J_{n+1}(x) = \frac{2n}{x}J_n(x) \tag{C.9}$$

This *recurrence formula* is useful in constructing tables of Bessel coefficients; its derivation follows from the power series of (C.2).

4. For small values of x, we have

$$J_n(x) \approx \frac{x^n}{2^n n!} \tag{C.10}$$

This relation is obtained simply by retaining the first term in the power series of (C.2) and ignoring the higher order terms. Thus, when x is small, we have

$$J_0(x) \approx 1$$

$$J_0(x) \approx \frac{x}{2}$$

$$J_n(x) \approx 0 \qquad \text{for } n > 1 \tag{C.11}$$

5. For large values of x, we have

$$J_n(x) \approx \sqrt{\frac{2}{\pi x}} \cos\left(x - \frac{\pi}{4} - \frac{n\pi}{2}\right) \tag{C.12}$$

This property shows that, for large values of x, the Bessel function $J_n(x)$ behaves like a sine wave with progressively decreasing amplitude.

6. With x real and fixed, $J_n(x)$ approaches zero as the order n goes to infinity.

7.
$$\sum_{n=-\infty}^{\infty} J_n(x)\exp(jn\phi) = \exp(jx\sin\phi) \tag{C.13}$$

To prove this property, consider the sum $\Sigma_{n=-\infty}^{\infty} J_n(x)\exp(jn\phi)$ and use (C.4) for $J_n(x)$ to obtain

$$\sum_{n=-\infty}^{\infty} J_n(x)\exp(jn\phi) = \frac{1}{2\pi} \sum_{n=-\infty}^{\infty} \exp(jn\phi) \int_{-\pi}^{\pi} \exp(jx\sin\theta - jn\theta)\, d\theta$$

Interchanging the order of integration and summation:

$$\sum_{n=-\infty}^{\infty} J_n(x)\exp(jn\phi) = \frac{1}{2\pi} \int_{-\pi}^{\pi} d\theta \exp(jx\sin\theta) \sum_{n=-\infty}^{\infty} \exp[jn(\phi - \theta)] \tag{C.14}$$

We now invoke the following relation from Fourier transform theory:

$$\delta(\phi) = \frac{1}{2\pi} \sum_{n=-\infty}^{\infty} \exp[jn(\phi)], \qquad -\pi \le \phi \le \pi \tag{C.15}$$

where $\delta(\phi)$ is the delta function. Therefore, using (C.15) in (C.14) and then applying the sifting property of the delta function, we get

$$\begin{aligned} \sum_{n=-\infty}^{\infty} J_n(x)\exp(jn\phi) &= \int_{-\pi}^{\pi} \exp(jx\sin\theta)\delta(\phi - \theta)\, d\theta \\ &= \exp(jx\sin\phi) \end{aligned}$$

which is the desired result.

8.
$$\sum_{n=-\infty}^{\infty} J_n^2(x) = 1 \qquad \text{for all } x \tag{C.16}$$

To prove this property, we may proceed as follows. We observe that $J_n(x)$ is real; hence, multiplying (C.4) by its own complex conjugate and summing over all possible values of n, we get

$$\sum_{n=-\infty}^{\infty} J_n^2(x) = \frac{1}{(2\pi)^2} \sum_{n=-\infty}^{\infty} \int_{-\pi}^{\pi}\int_{-\pi}^{\pi} \exp(jx\sin\theta - jn\theta - jx\sin\phi + jn\phi)\, d\theta\, d\phi$$

Interchanging the order of double integration and summation:

$$\sum_{n=-\infty}^{\infty} J_n^2(x) = \frac{1}{(2\pi)^2}\int_{-\pi}^{\pi}\int_{-\pi}^{\pi} \mathrm{d}\theta\,\mathrm{d}\phi\,\exp[\mathrm{j}x(\sin\theta - \sin\phi)]\sum_{n=-\infty}^{\infty}\exp[\mathrm{j}n(\phi-\theta)] \tag{C.17}$$

Using (C.15) in (C.17) and then applying the sifting property of the delta function, we finally get

$$\sum_{n=-\infty}^{\infty} J_n^2(x) = \frac{1}{2\pi}\int_{-\pi}^{\pi} \mathrm{d}\theta = 1$$

which is the desired result.

C.3 Modified Bessel Function

Consider the *modified Bessel equation*:

$$x^2\frac{\mathrm{d}^2y}{\mathrm{d}x^2} + x\frac{\mathrm{d}y}{\mathrm{d}x} - (x^2 + n^2)y = 0 \tag{C.18}$$

With $\mathrm{j}^2 = -1$, we may rewrite this equation as

$$x^2\frac{\mathrm{d}^2y}{\mathrm{d}x^2} + x\frac{\mathrm{d}y}{\mathrm{d}x} + (\mathrm{j}^2x^2 - n^2)y = 0$$

from which it is therefore evident that (C.18) is nothing but Bessel's equation, namely (C.1), rewritten with x replaced by jx. Thus, replacing x by jx in (C.2) and again noting that $-1 = \mathrm{j}^2$, we get

$$J_n(\mathrm{j}x) = \sum_{m=0}^{\infty}\frac{(-1)^m(\mathrm{j}x/2)^{n+2m}}{m!(n+m)!}$$

$$= \mathrm{j}^n\sum_{m=0}^{\infty}\frac{x/2^{n+2m}}{m!(n+m)!}$$

Next we note that $J_n(\mathrm{j}x)$ multiplied by a constant will still be a solution of Bessel's equation. Accordingly, we multiply $J_n(\mathrm{j}x)$ by the constant j^{-n}, obtaining

$$\mathrm{j}^{-n}J_n(\mathrm{j}x) = \sum_{m=0}^{\infty}\frac{(1/2x)^{n+2m}}{m!(n+m)!}$$

This new function is called the *modified Bessel function of the first kind of order n*, denoted by $I_n(x)$. We may thus formally express a solution of the modified Bessel equation (C.18) as

$$I_n(x) = \mathrm{j}^{-n}J_n(\mathrm{j}x)$$

$$= \sum_{m=0}^{\infty}\frac{(1/2x)^{n+2m}}{m!(n+m)!} \tag{C.19}$$

The modified Bessel function $I_n(x)$ is a monotonically increasing real function of the argument $x \geq 0$ for all n, as shown in Figure C.1 for $n = 0, 1$.

The modified Bessel function $I_n(x)$ is identical to the original Bessel function $J_n(x)$ except for an important difference:

> The terms in the series expansion of (C.19) are all positive, whereas they alternate in sign in the series expansion of (C.2).

The relationship between $J_n(x)$ and $I_n(x)$ is analogous to the way in which the trigonometric functions cos x and sin x are related to the hyperbolic functions cosh x and sinh x, respectively.

An interesting property of the modified Bessel function $I_n(x)$ is derived from (C.13). Specifically by replacing x by jx and the angle ϕ by $\theta - \pi/2$ in this equation and then invoking the definition of $I_n(x)$ in the first line of (C.19), we obtain

$$\sum_{n=-\infty}^{\infty} I_n(x)\exp(jn\theta) = \exp(x\cos\theta) \tag{C.20}$$

From this relation it follows that

$$I_n(x) = \frac{1}{2\pi}\int_{-\pi}^{\pi} \exp(x\cos\theta)\cos(n\theta)\,d\theta \tag{C.21}$$

This integral formula for $I_n(x)$ may, of course, also be derived from (C.4) by making the appropriate changes.

When the argument x is small, we obtain the following asymptotic estimates directly from the series representation of (C.19):

$$I_0(x) \to 1 \qquad \text{for } x \to 0 \tag{C.22}$$

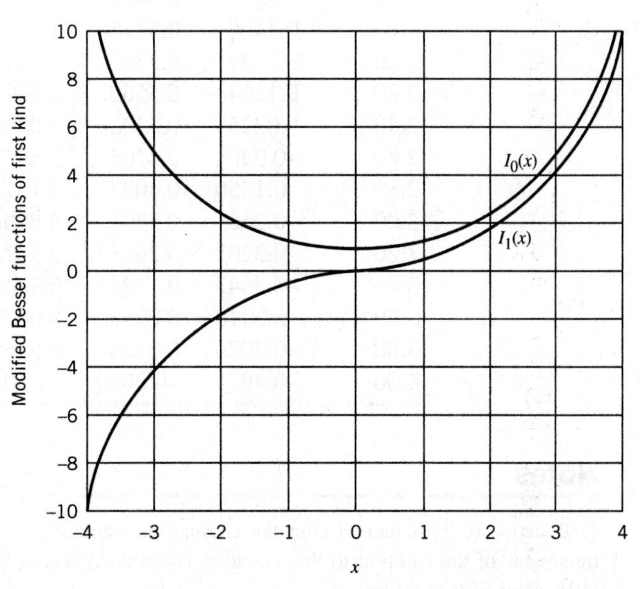

Figure C.1 Plots of modified Bessel functions of the first kind $I_0(x)$ and $I_1(x)$.

and

$$I_n(x) \to 0 \qquad \text{for } n \geq 1 \text{ and } x \to 0 \tag{C.23}$$

For large values of x we have the following asymptotic estimate for $I_n(x)$, which is valid for all integers $n \geq 0$:

$$I_n(x) \approx \frac{\exp(x)}{\sqrt{2\pi x}} \qquad \text{for } x \to \infty \tag{C.24}$$

Note that this asymptotic behavior of $I_n(x)$ is independent of the order n for large values of x.

Tables

In numerical terms, Table C.1 provides a limited set of values of the Bessel function $J(x)$ and modified Bessel function $I(x)$. More extensive tables of these two functions are given in Abramowitz and Stegun (1965).

Table C.1 **Values of Bessel functions and modified Bessel functions of the first kind**

x	$J_0(x)$	$J_1(x)$	$I_0(x)$	$I_1(x)$
0.00	1.0000	0.0000	1.0000	0.0000
0.20	0.9900	0.0995	1.0100	0.1005
0.40	0.9604	0.1960	1.0404	0.2040
0.60	0.9120	0.2867	1.0920	0.3137
0.80	0.8463	0.3688	1.1665	0.4329
1.00	0.7652	0.4401	1.2661	0.5652
1.20	0.6711	0.4983	1.3937	0.7147
1.40	0.5669	0.5419	1.5534	0.8861
1.60	0.4554	0.5699	1.7500	1.0848
1.80	0.3400	0.5815	1.9896	1.3172
2.00	0.2239	0.5767	1.1796	1.5906
2.20	0.1104	0.5560	2.6291	1.9141
2.40	0.0025	0.5202	3.0493	2.2981
2.60	−0.0968	0.4708	3.5533	2.7554
2.80	−0.1850	0.4097	4.1573	3.3011
3.00	−0.2601	0.3391	4.8808	3.9534
3.20	−0.3202	0.2613	5.7472	4.7343
3.40	−0.3643	0.1792	6.7848	5.6701
3.60	−0.3918	0.0955	8.0277	6.7927
3.80	−0.4026	0.0128	9.5169	8.1404
4.00	−0.3971	−0.0660	11.3019	9.7595

Notes

1. Equation (C.1) is named after the German mathematician and astronomer Bessel. For detailed treatments of the solution to this equation and related issues, see the books by Wylie and Barrett (1982) and Watson (1966).

APPENDIX

D Method of Lagrange Multipliers

D.1 Optimization Involving a Single Equality Constraint

Consider the minimization of a real-valued function $f(\mathbf{w})$ that is a quadratic function of a parameter vector $\mathbf{w}$, subject to the *constraint*

$$\mathbf{w}^{\dagger}\mathbf{s} = g \tag{D.1}$$

where $\mathbf{s}$ is a prescribed vector and g is a complex constant; the superscript $\dagger$ denotes Hermitian transposition. We may redefine the constraint by introducing a new function $c(\mathbf{w})$ that is linear in $\mathbf{w}$, as shown by

$$\begin{aligned} c(\mathbf{w}) &= \mathbf{w}^{\dagger}\mathbf{s} - g \\ &= 0 + \mathrm{j}0 \end{aligned} \tag{D.2}$$

In general, the vectors $\mathbf{w}$ and $\mathbf{s}$ and the function $c(\mathbf{w})$ are all *complex*. For example, in a beamforming application, the vector $\mathbf{w}$ represents a set of complex weights applied to the individual sensor outputs and $\mathbf{s}$ represents a steering vector whose elements are defined by a prescribed "look" direction; the function $f(\mathbf{w})$ to be minimized represents the mean-square value of the overall beamformer output. In a harmonic retrieval application, for another example, $\mathbf{w}$ represents the tap-weight vector of an FIR filter and $\mathbf{s}$ represents a sinusoidal vector whose elements are determined by the angular frequency of a complex sinusoid contained in the filter input; the function $f(\mathbf{w})$ represents the mean-square value of the filter output. In any event, assuming that the issue is one of minimization, we may state the constrained optimization problem as follows:

$$\text{Minimize a real-valued function } f(\mathbf{w})\text{, subject to the constraint } c(\mathbf{w}) = 0 + \mathrm{j}0 \tag{D.3}$$

The *method of Lagrange multipliers* converts the problem of constrained minimization just described into one of unconstrained minimization by the introduction of *Lagrange multipliers*. First, we use the real function $f(\mathbf{w})$ and the complex constraint function $c(\mathbf{w})$ to define a new real-valued function

$$h(\mathbf{w}) = f(\mathbf{w}) + \lambda_1 \mathrm{Re}[c(\mathbf{w})] + \lambda_2 \mathrm{Im}[c(\mathbf{w})] \tag{D.4}$$

where λ_1 and λ_2 are *real Lagrange multipliers* and

$$c(\mathbf{w}) = \mathrm{Re}[c(\mathbf{w})] + \mathrm{j}\,\mathrm{Im}[c(\mathbf{w})] \tag{D.5}$$

Now we define a complex *Lagrange multiplier*:

$$\lambda = \lambda_1 + \mathrm{j}\lambda_2 \tag{D.6}$$

The $\mathrm{Re}[\cdot]$ and $\mathrm{Im}[\cdot]$ in (D.4) and (D.5) denote real and imaginary operators, respectively. We may then rewrite (D.4) in the form

$$h(\mathbf{w}) = f(\mathbf{w}) + \mathrm{Re}[\lambda^* c(\mathbf{w})] \tag{D.7}$$

where the asterisk denotes complex conjugation.

Next, we minimize the function $h(\mathbf{w})$ with respect to the vector $\mathbf{w}$. To do this, we set the conjugate derivative $\partial h/(\partial \mathbf{w}^*)$ equal to the null vector:

$$\frac{\partial f}{\partial \mathbf{w}^*} + \frac{\partial}{\partial \mathbf{w}^*}(\mathrm{Re}[\lambda^* c(\mathbf{w})]) = \mathbf{0} \tag{D.8}$$

The system of simultaneous equations consisting of (D.8) and the original constraint given in (D.2) defines the optimum solutions for the vector $\mathbf{w}$ and the Lagrange multiplier λ. We call (D.8) the *adjoint equation* and (D.2) the *primal equation* (Dorny, 1975).

APPENDIX

E Information Capacity of MIMO Channels

The topic of multiple-input multiple-output (MIMO) links for wireless communications was discussed in Chapter 9 on signaling over fading channels. To get a measure of the transmission efficiency of MIMO links therein, we resorted to the notion of outage capacity, which is naturally of practical interest. However, in light of its mathematical sophistication, we deferred discussion of the information capacity of MIMO links rooted in Shannon's information theory to this appendix.

To be specific, in this appendix we discuss two different aspects of information capacity:

1. The channel state is known to the receiver but not the transmitter;
2. The channel state is known to both the receiver and the transmitter.

The discussion will proceed in this order.

E.1 Log-Det Capacity Formula of MIMO Channels

Consider a communication channel with multiple antennas.[1] Let the N_t-by-1 vector **s** denote the transmitted signal vector and the N_r-by-1 vector **x** denote the received signal vector. These two vectors are related by the *input–output relation of the channel*:

$$\mathbf{x} = \mathbf{H}\mathbf{s} + \mathbf{w} \tag{E.1}$$

where **H** is the *channel matrix* of the link and **w** is the additive channel noise vector. The vectors **s**, **w**, and **x** are realizations of the random vectors **S**, **W**, and **X**, respectively.

In what follows in this appendix, the following assumptions are made:

1. The channel is stationary and ergodic.
2. The channel matrix **H** is made up of iid Gaussian elements.
3. The transmitted signal vector **s** has zero mean and correlation matrix $\mathbf{R_s}$.
4. The additive channel noise vector **w** has zero mean and correlation matrix $\mathbf{R_w}$.
5. Both **s** and **w** are governed by Gaussian distributions.

In this section, we also assume that the channel state **H** is known to the receiver but not the transmitter. With both **H** and **x** unknown to the transmitter, the primary issue of interest is to determine $I(\mathbf{s};\mathbf{x},\mathbf{H})$, which denotes the mutual information between the transmitted signal vector **s** and both the received signal vector **x** and the channel matrix **H**. Extending the definition of mutual information introduced in Chapter 5 to the problem at hand, we write

$$I(\mathbf{S};\mathbf{X},\mathbf{H}) = \iiint_{\mathcal{H}\,\mathcal{X}\,\mathcal{S}} f_{\mathbf{S},\mathbf{X},\mathbf{H}}(\mathbf{s},\mathbf{x},\mathbf{H})\log_2\left(\frac{f_{\mathbf{S}|\mathbf{X},\mathbf{H}}(\mathbf{s}|\mathbf{x},\mathbf{H})}{f_{\mathbf{X},\mathbf{H}}(\mathbf{x},\mathbf{H})}\right) d\mathbf{s}\,d\mathbf{x}\,d\mathbf{H} \tag{E.2}$$

where $\mathcal{S}$, $\mathcal{X}$, and $\mathcal{H}$ are the respective spaces pertaining to the random vectors **S** and **X** and matrix **H**.

Using the definition of a joint probability density function (pdf) as the product of a conditional pdf and an ordinary pdf, we write

$$f_{\mathbf{S},\mathbf{X},\mathbf{H}}(\mathbf{s},\mathbf{x},\mathbf{H}) = f_{\mathbf{S},\mathbf{X}|\mathbf{H}}(\mathbf{s},\mathbf{x}|\mathbf{H})f_{\mathbf{H}}(\mathbf{H})$$

We may therefore rewrite (E.2) in the equivalent form

$$\begin{aligned}
I(\mathbf{S};\mathbf{X},\mathbf{H}) &= \int_{\mathscr{H}} f_{\mathbf{H}}(\mathbf{H})\left[\iint_{\mathscr{X}\mathscr{S}} f_{\mathbf{S},\mathbf{X}|\mathbf{H}}(\mathbf{s},\mathbf{x}|\mathbf{H})\log_2\left(\frac{f_{\mathbf{S}|\mathbf{X},\mathbf{H}}(\mathbf{s}|\mathbf{x},\mathbf{H})}{f_{\mathbf{X},\mathbf{H}}(\mathbf{x},\mathbf{H})}\right)d\mathbf{s}\,d\mathbf{x}\right]d\mathbf{H} \\
&= \mathbb{E}_{\mathbf{H}}\left[\iint_{\mathscr{X}\mathscr{S}} f_{\mathbf{S},\mathbf{X}|\mathbf{H}}(\mathbf{s},\mathbf{x}|\mathbf{H})\log_2\left(\frac{f_{\mathbf{S}|\mathbf{X},\mathbf{H}}(\mathbf{s}|\mathbf{x},\mathbf{H})}{f_{\mathbf{X},\mathbf{H}}(\mathbf{x},\mathbf{H})}\right)d\mathbf{s}\,d\mathbf{x}\right] \\
&= \mathbb{E}_{\mathbf{H}}[I(\mathbf{s};\mathbf{x}|\mathbf{H})]
\end{aligned} \tag{E.3}$$

where the expectation is with respect to the channel matrix **H** and

$$I(\mathbf{s};\mathbf{x}|\mathbf{H}) = \iint_{\mathscr{X}\mathscr{S}} f_{\mathbf{S},\mathbf{X}|\mathbf{H}}(\mathbf{s},\mathbf{x}|\mathbf{H})\log_2\left(\frac{f_{\mathbf{S}|\mathbf{X},\mathbf{H}}(\mathbf{s}|\mathbf{x},\mathbf{H})}{f_{\mathbf{X},\mathbf{H}}(\mathbf{x},\mathbf{H})}\right)d\mathbf{s}\,d\mathbf{x}$$

is the conditional mutual information between the transmitted signal vector **s** and received signal vector **x**, given the channel matrix **H**. However, by assumption, the channel state is unknown to the transmitter. Therefore, it follows that, insofar as the receiver is concerned, $I(\mathbf{s};\mathbf{x}|\mathbf{H})$ is a random vector; hence the expectation with respect to **H** in (E.3). The quantity resulting from this expectation is therefore deterministic, defining the mutual information jointly between the transmitted signal vector **s** and both the received signal vector **x** and channel matrix **H**. The result so obtained is indeed consistent with what we know about the notion of joint mutual information.

Next, applying the vector form of the first line in (5.81) to the mutual information $I(\mathbf{s};\mathbf{x}|\mathbf{H})$, we have

$$I(\mathbf{s};\mathbf{x}|\mathbf{H}) = h(\mathbf{x}|\mathbf{H}) - h(\mathbf{x}|\mathbf{s},\mathbf{H}) \tag{E.4}$$

where $h(\mathbf{x}|\mathbf{H})$ is the conditional differential entropy of the channel output **x** given **H**, and $h(\mathbf{x}|\mathbf{s},\mathbf{H})$ is the conditional differential entropy of **x**, given both **s** and **H**. Both of these entropies are random quantities, because they both depend on **H**.

To proceed further, we now invoke the assumed Gaussianity of both **s** and **H**, in which case **x** also assumes a Gaussian description. Under these circumstances, we may use the result of Problem 5.32 to express the entropy of the received signal **x** of dimension N_r, given **H**, as

$$h(\mathbf{x}|\mathbf{H}) = N_r + N_r\log_2(2\pi) + \log_2(\det(\mathbf{R}_{\mathbf{x}}))\ \text{bits} \tag{E.5}$$

where $\mathbf{R}_{\mathbf{x}}$ is the correlation matrix of **x** and $\det(\mathbf{R}_{\mathbf{x}})$ is its determinant. Recognizing that the transmitted signal vector **s** and channel noise vector **w** are independent of each other, we find from (E.1) that the correlation matrix of the received signal vector **x** is given by

$$\begin{aligned}\mathbf{R_x} &= [\mathbf{xx}^\dagger] \\ &= \mathbb{E}[(\mathbf{Hs}+\mathbf{w})(\mathbf{Hs}+\mathbf{w})^\dagger] \\ &= \mathbb{E}[(\mathbf{Hs}+\mathbf{w})(\mathbf{s}^\dagger\mathbf{H}^\dagger+\mathbf{w}^\dagger)] \\ &= \mathbb{E}[\mathbf{Hss}^\dagger\mathbf{H}^\dagger]+\mathbb{E}[\mathbf{ww}^\dagger], \qquad (\mathbb{E}[\mathbf{sw}^\dagger]=\mathbf{0}) \\ &= \mathbf{H}\mathbb{E}[\mathbf{ss}^\dagger]\mathbf{H}^\dagger+\mathbf{R_w} \\ &= \mathbf{HR_sH}^\dagger+\mathbf{R_w}\end{aligned} \tag{E.6}$$

where † denotes Hermitian transposition,

$$\mathbf{R_s} = \mathbb{E}[\mathbf{ss}^\dagger] \tag{E.7}$$

is the correlation matrix of the transmitted signal vector **s**, and

$$\mathbf{R_w} = \mathbb{E}[\mathbf{ww}^\dagger] \tag{E.8}$$

is the correlation matrix of the channel noise vector **w**. Hence, using (E.6) in (E.5), we get

$$h(\mathbf{x}|\mathbf{H}) = N_r + N_r\log_2(2\pi) + \log_2\{\det(\mathbf{R_w}+\mathbf{HR_sH}^\dagger)\} \text{ bits} \tag{E.9}$$

where N_r is the number of elements in the receiving antenna. Next, we note that since the vectors **s** and **w** are independent and the sum of **w** plus **Hs** equals **x** as indicated in (E.1), then the conditional differential entropy of **x**, given both **s** and **H**, is simply equal to the differential entropy of the additive channel noise vector **w**; that is,

$$h(\mathbf{x}|\mathbf{s},\mathbf{H}) = h(\mathbf{w}) \tag{E.10}$$

The entropy $h(\mathbf{w})$ is given by (see Problem 5.32)

$$h(\mathbf{w}) = N_r + N_r\log_2(2\pi) + \log_2\{\det(\mathbf{R_w})\} \text{ bits} \tag{E.11}$$

Thus, using (E.9), (E.10), and (E.11) in (E.4), we get

$$\begin{aligned}I(\mathbf{s};\mathbf{x}|\mathbf{H}) &= \log_2\{\det(\mathbf{R_w}+\mathbf{HR_sH}^\dagger)\} - \log_2\{\det(\mathbf{R_w})\} \\ &= \log_2\left\{\frac{\{\det(\mathbf{R_w}+\mathbf{HR_sH}^\dagger)\}}{\{\det(\mathbf{R_w})\}}\right\}\end{aligned} \tag{E.12}$$

As remarked previously, the conditional mutual information $I(\mathbf{s};\mathbf{x}|\mathbf{H})$ is a random variable. Hence, using (E.12) in (E.3), we finally formulate the *ergodic capacity* of the MIMO link as the expectation

$$C = \mathbb{E}_{\mathbf{H}}\left[\log_2\left\{\frac{\{\det(\mathbf{R_w}+\mathbf{HR_sH}^\dagger)\}}{\{\det(\mathbf{R_w})\}}\right\}\right] \text{ bits/(sHz)} \tag{E.13}$$

which is subject to the constraint

$$\max_{\mathbf{R_s}} \operatorname{tr}[\mathbf{R_s}] \le P$$

where P is constant transmit power and tr[.] denotes the trace operator, which extracts the sum of the diagonal elements of the enclosed matrix.

Equation (E.13) is the desired *log-det formula* for the ergodic capacity of the MIMO link. This formula is of general applicability, in that correlations among the elements of the transmitted signal vector **s** and among those of the channel noise vector **w** are permitted. However, the assumptions made in its derivation involve the Gaussianity of **s**, **H**, and **w**.

E.2 MIMO Capacity for Channel Known at the Transmitter

The log-det formula of (E.13) for the ergodic capacity of a MIMO flat-fading channel assumes that the channel state is only known at the receiver. What if the channel state is also known perfectly at the transmitter? Then the channel state becomes known to the entire system, which means that we may treat the channel matrix **H** as a constant. Hence, unlike the partially known case treated in Section E.1, there is no longer the need for invoking the expectation operator in formulating the log-det capacity. Rather, the problem becomes one of constructing the optimal $\mathbf{R_s}$ (i.e., the correlation matrix of the transmitted signal vector **s**) that maximizes the ergodic capacity. To simplify the construction procedure, we consider a MIMO channel for which the number of elements in the receiving antenna N_r and the number of elements in the transmitting antenna N_t have a common value, denoted by N.

Accordingly, using the assumption of additive white Gaussian noise with variance $\sigma_{\mathbf{w}}^2$ in the log-det capacity formula of (E.13), we get

$$C = \log_2\left\{\det\left(I_N + \frac{1}{\sigma_{\mathbf{w}}^2}\mathbf{H}\mathbf{R_s}\mathbf{H}^{\dagger}\right)\right\} \text{ bits/(s Hz)} \tag{E.14}$$

We can now formally postulate the optimization problem at hand as follows:

Maximize the ergodic capacity C of (E.14) with respect to the correlation matrix $\mathbf{R_s}$, subject to two constraints, expressed as

1. Nonnegative definite **R**s, which is a necessary requirement for a correlation matrix.
2. Global power constraint

$$\text{tr}[\mathbf{R_s}] = P \tag{E.15}$$

where P is the total transmit power.

To proceed with construction of the optimal $\mathbf{R_s}$, we first use the *determinant identity*:

$$\det(\mathbf{I} + \mathbf{AB}) = \det(\mathbf{I} + \mathbf{BA}) \tag{E.16}$$

Application of this identity to (E.14) yields

$$C = \log_2\left\{\det\left(\mathbf{I}_N + \frac{1}{\sigma_{\mathbf{w}}^2}\mathbf{R_s}\mathbf{H}^{\dagger}\mathbf{H}\right)\right\} \text{ bits/(s Hz)} \tag{E.17}$$

Diagonalizing the matrix product $\mathbf{H}^{\dagger}\mathbf{H}$ by invoking the *eigendecomposition* of a Hermitian matrix, we write

$$\mathbf{U}^{\dagger}(\mathbf{H}^{\dagger}\mathbf{H})\mathbf{U} = \mathbf{\Lambda} \tag{E.18}$$

where $\mathbf{\Lambda}$ is a diagonal matrix made up of the eigenvalues of $\mathbf{H}^{\dagger}\mathbf{H}$, and $\mathbf{U}$ is a unitary matrix whose columns are the associated eigenvectors.[2] We may therefore rewrite (E.18) in the equivalent form

$$\mathbf{H}^{\dagger}\mathbf{H} = \mathbf{U}\mathbf{\Lambda}\mathbf{U}^{\dagger} \tag{E.19}$$

where by definition we have used the fact that the matrix product $\mathbf{U}\mathbf{U}^{\dagger}$ is equal to the identity matrix. Substituting (E.18) into (E.17), we get

$$C = \log_2\left\{\det\left(\mathbf{I}_N + \frac{1}{\sigma_{\mathbf{w}}^2}\mathbf{R}_{\mathbf{s}}\mathbf{U}\mathbf{\Lambda}\mathbf{U}^{\dagger}\right)\right\} \tag{E.20}$$

Next, applying the determinant identity of (E.16) to the formula, we get

$$\begin{aligned} C &= \log_2\left\{\det\left(\mathbf{I}_N + \frac{1}{\sigma_{\mathbf{w}}^2}\mathbf{\Lambda}\mathbf{U}^{\dagger}\mathbf{R}_{\mathbf{s}}\mathbf{U}\right)\right\} \\ &= \log_2\left\{\det\left(\mathbf{I}_N + \frac{1}{\sigma_{\mathbf{w}}^2}\mathbf{\Lambda}\bar{\mathbf{R}}_{\mathbf{s}}\right)\right\} \text{ bits/(s Hz)} \end{aligned} \tag{E.21}$$

where

$$\bar{\mathbf{R}}_{\mathbf{s}} = \mathbf{U}^{\dagger}\mathbf{R}_{\mathbf{s}}\mathbf{U} \tag{E.22}$$

Note that the transformed correlation matrix $\bar{\mathbf{R}}_{\mathbf{s}}$ is nonnegative definite. Since $\mathbf{U}\mathbf{U}^{\dagger} = \mathbf{I}$, we also have

$$\begin{aligned} \text{tr}[\bar{\mathbf{R}}_{\mathbf{s}}] &= \text{tr}[\mathbf{U}^{\dagger}\mathbf{R}_{\mathbf{s}}\mathbf{U}] \\ &= \text{tr}[\mathbf{U}\mathbf{U}^{\dagger}\mathbf{R}_{\mathbf{s}}] \\ &= \text{tr}[\mathbf{R}_{\mathbf{s}}] \end{aligned} \tag{E.23}$$

where, in the second line, we used the equality $\text{tr}[\mathbf{AB}] = \text{tr}[\mathbf{BA}]$. It follows, therefore, that maximization of the ergodic capacity of (E.21) can be carried out equally well over the transformed correlation matrix $\bar{\mathbf{R}}_{\mathbf{s}}$.

One other important point to note is that any nonnegative definite matrix $\mathbf{A}$ satisfies the *Hadamard inequality*

$$\det(\mathbf{A}) \le \prod_k a_{kk} \tag{E.24}$$

where the a_{kk} are the diagonal elements of matrix $\mathbf{A}$. Hence, applying this inequality to the determinent term in (E.21), we may write

$$\det\left(\mathbf{I}_N + \frac{1}{\sigma_{\mathbf{w}}^2}\mathbf{\Lambda}\bar{\mathbf{R}}_{\mathbf{s}}\right) \le \prod_{k=1}^{N}\left(1 + \frac{1}{\sigma_{\mathbf{w}}^2}\lambda_k \bar{r}_{s,kk}\right) \tag{E.25}$$

where λ_k is the kth eigenvalue of the matrix product $\mathbf{H}\mathbf{H}^\dagger$ and $\bar{r}_{s,kk}$ is the kth diagonal element of the transformed matrix $\bar{\mathbf{R}}_\mathbf{s}$. Equation (E.25) holds only when $\bar{\mathbf{R}}_\mathbf{s}$ is a diagonal matrix, which is the very condition that maximizes the ergodic capacity C.

To proceed further, we now use (E.21) and (E.25) with the equality sign to express the ergodic capacity as

$$\begin{aligned} C &= \log_2\left\{\prod_{k=1}^{N}\left(1+\frac{1}{\sigma_\mathbf{w}^2}\lambda_k\bar{r}_{s,kk}\right)\right\} \\ &= \sum_{k=1}^{N}\log_2\left(1+\frac{1}{\sigma_\mathbf{w}^2}\lambda_k\bar{r}_{s,kk}\right) \\ &= \sum_{k=1}^{N}\log_2\left\{\lambda_k\left(\lambda_k^{-1}+\frac{1}{\sigma_\mathbf{w}^2}\bar{r}_{s,kk}\right)\right\} \\ &= \sum_{k=1}^{N}\log_2\lambda_k+\sum_{k=1}^{N}\log_2\left(\lambda_k^{-1}+\frac{1}{\sigma_\mathbf{w}^2}\bar{r}_{s,kk}\right) \end{aligned} \tag{E.26}$$

where only the second sum term is clearly adjustable through $\bar{r}_{s,kk}$. We may therefore reformulate the optimization problem at hand as follows:

> Given the set of eigenvalues $\{\lambda_k\}_{k=1}^{N}$ pertaining to the matrix product $\mathbf{H}\mathbf{H}^\dagger$, determine the optimal set of autocorrelations $\{\bar{r}_{s,kk}\}_{k=1}^{N}$ that maximizes the summation
>
> $$\sum_{k=1}^{N}\left(\frac{1}{\lambda_k}+\frac{1}{\sigma_\mathbf{w}^2}\bar{r}_{s,kk}\right)$$
>
> subject to the constraint
>
> $$\sum_{k=1}^{N}\bar{r}_{s,kk} = P \tag{E.27}$$

The global power constraint of (E.27) follows from (E.23) and the trace definition of a trace:

$$\mathrm{tr}[\bar{\mathbf{R}}_\mathbf{s}] = \sum_{k=1}^{N}\bar{r}_{skk} \tag{E.28}$$

Water-Filling Interpretation of (E.26)

The solution to the reformulated optimization problem that was initiated after (E.14) may be determined through the discrete spatial version of the *water-filling procedure*, which is described in Chapter 5. Effectively, the solution to the water-filling problem says that, in a multiple-channel scenario, we transmit more signal power in the better channels and less signal power in the poorer channels. To be specific, imagine a vessel whose bottom is defined by the set of N dimensionless discrete levels

$$\left\{\frac{\mu-(\sigma_{\mathrm{w}}^2/\lambda_k)}{\lambda_k}\right\}_{k=1}^{N}$$

and pour "water" into the vessel in an amount corresponding to the total transmit power P. The power P is optimally divided among the N eigenmodes of the MIMO link in accordance with their corresponding "water levels" in the vessel, as illustrated in Figure E.1 for a MIMO link with $N = 6$. The "water-fill level," denoted by the dimensionless parameter μ and indicated by the dashed line in the figure, is chosen to satisfy the constraint of (E.27). On the basis of the spatially discrete water-filling picture portrayed in Figure E.1, we may now finally postulate the optimal $\bar{r}_{skk}$ to be

$$\bar{r}_{skk} = \left(\mu-\frac{\sigma_{\mathrm{w}}^2}{\lambda_k}\right)^+, \qquad k = 1, 2, \ldots, N \tag{E.29}$$

The superscript "+" applied to the right parenthesis in (E.29) signifies retaining only those terms in the right-hand side of the equation that are positive (i.e., the terms that pertain to those eigenmodes of the MIMO link for which the water levels lie below the constant μ).

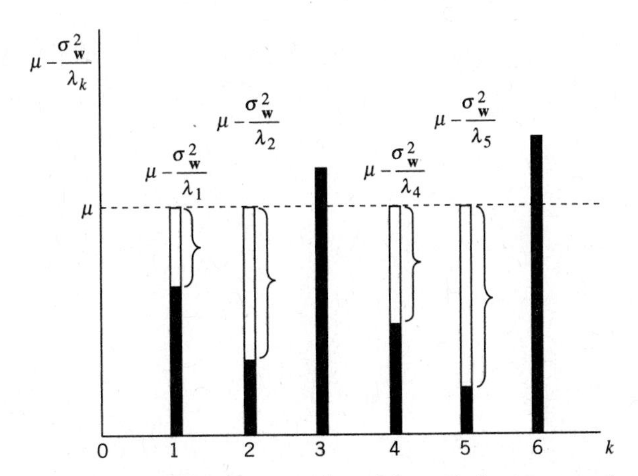

Figure E.1 Water-filling interpretation of the optimization procedure.

We may thus finally state that if the channel matrix $\mathbf{H}$ is known to both the transmitter and the receiver of a MIMO link with $N_r = N_t = N$, then the maximum value of the capacity of the MIMO link is defined by

$$\begin{aligned} C &= \sum_{k=1}^{N} \log_2\left(1 + \frac{1}{\sigma_{\mathbf{w}}^2}\lambda_k \bar{r}_{skk}\right) \\ &= \sum_{k=1}^{N} \log_2\left\{1 + \frac{1}{\sigma_{\mathbf{w}}^2}\lambda_k\left(\mu - \frac{\sigma_{\mathbf{w}}^2}{\lambda_k}\right)^+\right\} \\ &= \sum_{k=1}^{N} \log_2\left(\frac{\mu\lambda_k}{\sigma_{\mathbf{w}}^2}\right)^+ \end{aligned} \tag{E.30}$$

where, as stated previously, the constant μ is chosen to satisfy the global power constraint of (E.27).

Notes

1. The first detailed derivation of the log-det capacity formula for a stationary MIMO channel was presented by Telatar in an AT&T technical memorandum published in 1995 and republished as a journal paper (Telatar, 1999).
2. Given a complex-valued matrix $\mathbf{A}$, the eigendecomposition of $\mathbf{A}$ is defined by $\mathbf{U}^{\dagger}\mathbf{A}\mathbf{U} = \Lambda$.

APPENDIX

F Interleaving

Previous chapters of the book, going back to Chapter 5, have shown us how a digital wireless communication system can be separated by function into source-coding and channel-coding applications on the transmitting side and the corresponding inverse functions on the receiving side. In Chapter 6, we also learned how analog signals can be converted into a digital format. The motivation behind these techniques is to minimize the amount of information that has to be transmitted over a wireless channel. Such minimization has potential benefits in the allocation of two primary resources, namely transmit power and channel bandwidth, available to wireless communications:

1. *Reducing the amount of data that must be transmitted, which usually means that less power has to be consumed*; power consumption is always a serious concern for mobile units that are typically battery operated.
2. *Reducing the spectral (or radio-frequency) resources, which are required for satisfactory performance*; this reduction enables us to increase the number of users who can share the same but limited channel bandwidth.

Moreover, insofar as channel coding is concerned, forward error-correction (FEC) coding, discussed in Chapter 10, provides a powerful technique for transmitting information-bearing data reliably from a source to a sink across the wireless channel.

However, to obtain the maximum benefit from FEC coding in wireless communications, we require an additional technique known as *interleaving*.[1] The need for this new technique is justified on the grounds that, in light of the material presented in Chapter 9, we know that wireless channels have *memory* due to multipath fading that results from the arrival of signals at the receiver via multiple propagation paths of different lengths. Of particular concern is *fast* fading, which arises out of reflections from objects in the local vicinity of the transmitter, the receiver, or both. The term *fast* refers to the speed of fluctuations in the received signal due to these reflections, relative to the speeds of other propagation phenomena. Compared with transmit data rates, even fast fading can be relatively slow. That is, fast fading can be approximately constant over a number of transmission symbols, depending upon the data transmission speed and the mobile unit's velocity. Consequently, fast fading may be viewed as a time-correlated form of channel impairment, the presence of which results in statistical dependence among continuous (sets of) symbol transmissions. That is, instead of being isolated events, transmission errors due to fast fading tend to occur in *bursts*.

Now, most FEC channel codes are designed to deal with a limited number of bit errors, assumed to be randomly distributed and statistically independent from one bit to the next. To be specific, in Section 10.8 on convolutional decoding, we indicated that the Viterbi algorithm, as powerful as it is, will fail if there are $d_{\text{free}}/2$ closely spaced bit errors in the received signal, where d_{free} is the free distance of the convolutional code. Accordingly, in the design of a *reliable* wireless communication system, we are confronted with *two conflicting phenomena:*

- a wireless channel that produces bursts of correlated bit errors;
- a convolutional decoder that cannot handle error bursts.

Interleaving is an indispensable technique for resolving these two conflicting phenomena. First and foremost, however, it is important to note that for interleaving we do *not* need the exact statistical characterization of the wireless channel. Rather, we only require knowledge of the *coherence time* for fast fading, which is approximately given by (see (9.46))

$$\tau_{\text{coherence}} \approx \frac{0.3}{2\nu_{\text{max}}} \tag{F.1}$$

where ν_{max} is the maximum Doppler shift. Consequently, we would expect an error burst to occupy typically a time duration equal to $\tau_{\text{coherence}}$. To deal with bad situations of this kind in wireless communications, we do two things:

- An *interleaver* (i.e., a device that performs interleaving) is used to randomize the order of encoded bits *after* the channel encoder in the transmitter.
- A *de-interleaver* (i.e., a device that performs de-interleaving) is used to undo the randomization *before* the data reach the channel decoder in the receiver.

Interleaving has the net effect of breaking up any error bursts that may occur during the course of data transmission over the wireless channel and spreading them over the duration of operation of the interleaver. In so doing, the likelihood of a correctable received sequence is significantly improved. In the transmitter, the interleaver is placed after the channel encoder; in the receiver, the de-interleaver is placed before the channel decoder.

Three types of interleaving are commonly used in practice, and are discussed next.

F.1 Block Interleaving

In basic terms, a *classical block interleaver* acts as a memory buffer, as shown in Figure F.1. Data are written into this $N \times L$ rectangular array from the channel encoder in column fashion. Once the array is filled, it is read out in row fashion and its contents are sent to the transmitter. At the receiver, the inverse operation is performed: the contents of the array in the receiver are written row-wise with data; once the array is filled, it is read out column-wise into the decoder. Note that the (N,L) interleaver and de-interleaver described herein are both *periodic* with the fundamental period $T = NL$.

Suppose the correlation time or error-burst-length time corresponds to L received bits. Then, at the receiver, we expect that the effect of an error burst would corrupt the equivalent of one row of the de-interleaver block. However, since the de-interleaver block is read columnwise, all of these "bad" bits would be separated by $N - 1$ "good" bits when the burst is read into the decoder. If N is greater than the constraint length of the convolutional code being employed, then the Viterbi decoder will correct all of the errors in the error burst.

In practice, owing to the frequency of error bursts and the presence of other errors caused by channel noise, the interleaver should ideally be made as large as possible. However, *an interleaver introduces delay* into the transmission of the message signal, in that we must fill the $N \times L$ array before it can be transmitted. This is an issue of particular concern in real-time applications such as voice, because it limits the usable block size of the interleaver and necessitates a compromise solution.

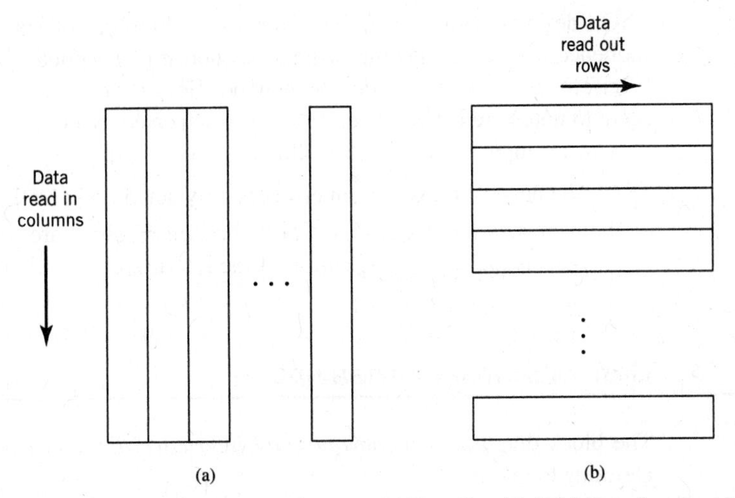

Figure F.1 Block interleaver structure. (a) Data "read in." (b) Data "read out."

EXAMPLE 1 Interleaving

Figure F.2a depicts an original sequence of encoded words, with each word consisting of five symbols. Figure F.2b depicts the interleaved version of the encoded sequence, with the symbols shown in reordered positions. An error burst occupying five symbols, caused by channel impairment, is also shown alongside Figure F.2b. Note that the manner in which the encoded symbols are reordered by the interleaver is the same from one word to the next.

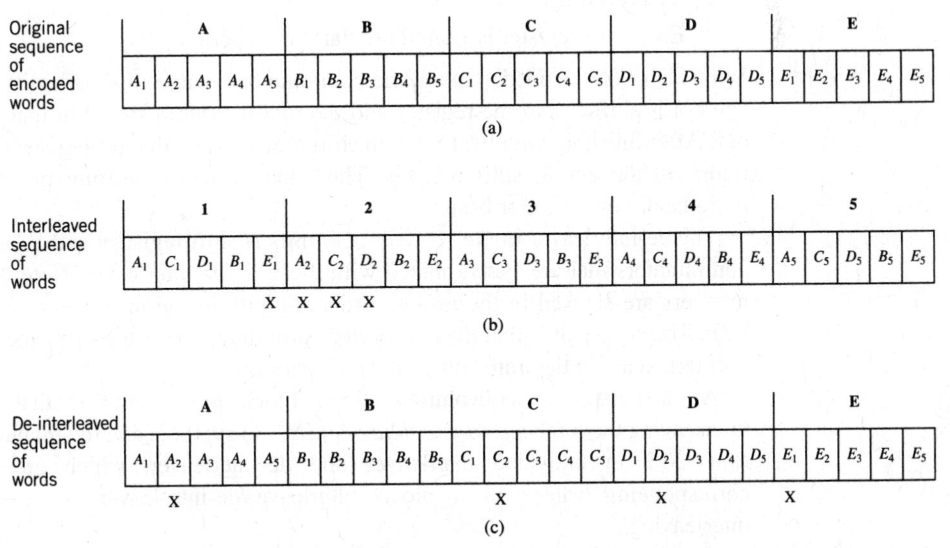

Figure F.2 Interleaving example. (a) Original sequence. (b) Interleaved sequence. (c) De-interleaved sequence.

On de-interleaving in the receiver, the scrambling of symbols is undone, yielding a sequence that resembles the original sequence of encoded symbols, as shown in Figure F.2c. This figure also includes the new positions of the transmission errors. The important point to note here is that the error burst is dispersed as a result of de-interleaving.

This example teaches us the following:

1. The burst of transmission errors is only acted upon by the de-interleaver.
2. Insofar as the encoded symbols that are received are error free, the de-interleaver cancels the scrambling action of the interleaver.

F.2 Convolutional Interleaving

The block diagram of a *convolutional interleaver/de-interleaver* is shown in Figure F.3. Defining the period

$$T = LN$$

the interleaver is referred to as an ($L \times N$) *convolutional interleaver*, which has properties similar to those of the ($L \times N$) block interleaver.

The sequence of encoded bits to be interleaved in the transmitter is arranged in blocks of L bits. For each block, the encoded bits are sequentially shifted into and out of a bank of N registers by means of two synchronized input and output *commutators*. The interleaver, depicted in Figure F.3a, is structured as follows:

1. The zeroth shift register provides no storage; that is, the incoming encoded symbol is transmitted immediately.
2. Each successive shift register provides a storage capacity of L symbols more than the preceding shift register.
3. Each shift register is visited regularly on a periodic basis.

With each new encoded symbol, the commutators switch to a new shift register. The new symbol is shifted into the register and the oldest symbol stored in that register is shifted out. After finishing with the ($N - 1$)th shift register (i.e., the last register), the commutators return to the zeroth shift register. Thus, the switching/shifting procedure is repeated periodically on a regular basis.

The de-interleaver in the receiver also uses N shift registers and a pair of input/output commutators that are synchronized with those in the interleaver. Note, however, the shift registers are stacked in the *reverse* order to those in the interleaver, as shown in Figure F.3b. The net result is that the de-interleaver in the receiver performs the inverse operation to interleaving in the transmitter, and so it should.

An advantage of convolutional over block interleaving is that in convolutional interleaving the total *end-to-end delay* is $L(N - 1)$ symbols and the memory requirement is $L(N - 1)/2$ in both the interleaver and de-interleaver, which are one-half of the corresponding values in a block interleaver/de-interleaver for a similar level of interleaving.

The description of the convolutional interleaver/de-interleaver in Figure F.3b is presented in terms of shift registers. The actual implementation of the system can also be

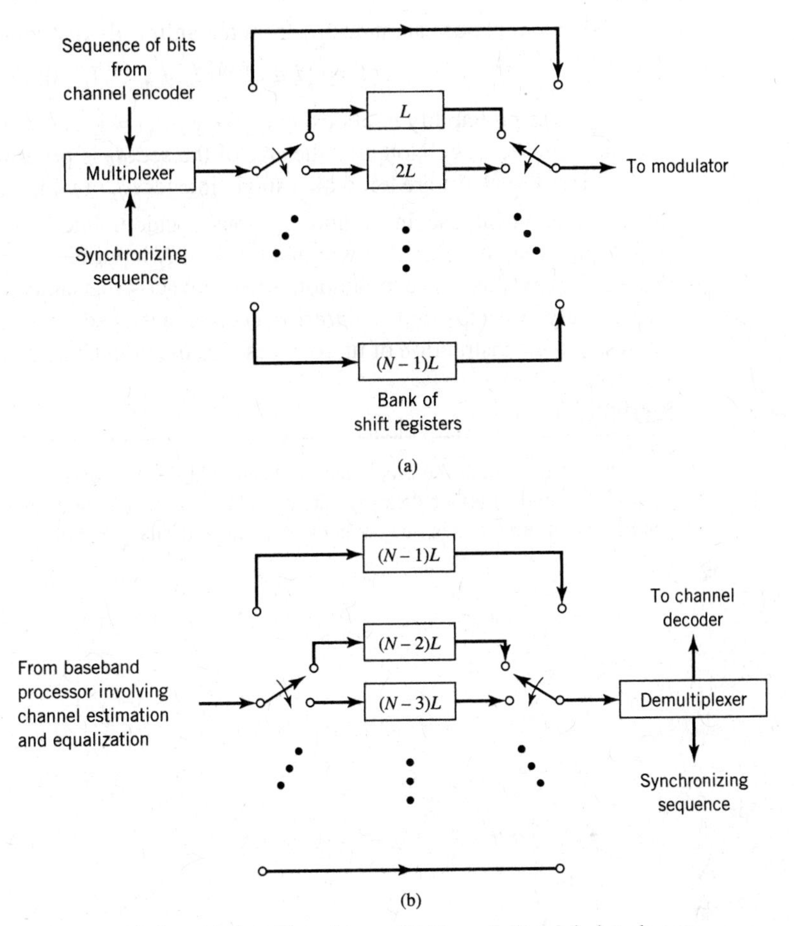

Figure F.3 (a) Convolutional interleaver. (b) Convolutional de-interleaver.

accomplished with a *random access memory* unit in place of shift registers. This alternative implementation simply requires that access to the memory units be appropriately controlled.

F.3 Random Interleaving

In a *random interleaver*, a block of N input bits is written into the interleaver in the order in which they are received, but they are read out in a random manner. Typically, the permutation of the input bits is defined by a *uniform distribution*. Let $\pi(i)$ denote the permuter location of the ith input bit, where $i = 1, 2, \ldots, N$. The set of integers denoted by $\{\pi(i)\}_{i=1}^{N}$, defining the order in which the stored input bits are read out of the interleaver, is generated according to the following two-step algorithm:

1. Choose an integer i_1 from the uniformly distributed set $\mathcal{A} = \{1, 2, \ldots, N\}$, with the probability of choosing i_1 being $p(i_1) = 1/N$. The chosen integer i_1 is set to be $\pi(i)$.

2. For $k > 1$, choose an integer i_k from the uniformly distributed set

$$\mathcal{A}_k = \{i \in \mathcal{A}, i \neq i_1, i_2, \ldots, i_{k-1}\}$$

with the probability of choosing i_k being $p(i_k) = 1/(N - k + 1)$. The chosen integer i_k is set to be $\pi(k)$. Note that the size of the set $\mathcal{A}_k$ is progressively reduced for $k > 1$. When $k = N$, we are left with a single integer i_N, in which case i_N is set to be $\pi(N)$.

To be of practical use in communications, random interleavers are configured to be *pseudo-random*, meaning that within a block of N input bits the permutation is random as described above, but the permutation order is exactly the same from one block to the next. Accordingly, *pseudo-random interleavers* are designed *off-line*; they are of particular interest in the construction of turbo codes, discussed in Chapter 10.

Notes

1. Interleaving of both the block and convolutional types is discussed in some detail in Clark and Cain (1981) and in lesser detail in Sklar (2001). For a treatment of interleaving viewed from the perspective of turbo codes, see the book (Vucetic and Yuan, 2000).

APPENDIX

G The Peak-Power Reduction Problem in OFDM

In Section 9.11 we discussed the multicarrier transmission technique, namely orthogonal frequency-division multiplexing (OFDM), which is of particular importance to wireless communications due to the computational benefits offered by the fast Fourier transform (FFT) algorithm. However, envelope variations are a frequently cited drawback of OFDM because of the *peak-power limited problem*. This problem arises due to the statistical possibility of a large number of independent subchannels in the OFDM becoming constructively superimposed, thereby resulting in high peaks. In the literature, the practical issue of envelope variations is described in terms of the *peak-to-average power ratio*, commonly abbreviated as PAPR.[1]

In this section, we discuss the PAPR problem in wireless communications and how it can be reduced.

G.1 PAPR Properties of OFDM Signals

Consider a single modulation interval, that is, a single symbol of OFDM, the duration of which is denoted by T_s. In its most basic form, the transmitted OFDM signal is described by

$$s(t) = \sum_{n=0}^{N-1} s_n \exp(j2\pi n \Delta f t), \qquad 0 \le t \le T_s \tag{G.1}$$

where the term Δf denotes the frequency separation between any two adjacent subchannels in the OFDM. By definition, the frequency separation Δf and symbol duration T_s are related by the time–bandwidth product:

$$T_s \Delta f = 1 \tag{G.2}$$

This condition is required to satisfy the orthogonality requirement among the N subchannels of the OFDM.

Typically, the coefficients in OFDM, denoted by s_n in (G.1) are taken from a fixed modulation constellation, exemplified by M-ary phase-shift-keying (PSK) or M-ary quadrature amplitude modulation (QAM) techniques, which were discussed in Chapter 7. With $s(t)$, in its baseband form, being a complex-valued signal with an amplitude and phase that characterize it, we may express the *time-averaged power* of an individual symbol of the OFDM signal in (G.1) as follows:

$$\begin{aligned} \bar{P} &= \frac{1}{T_s}\int_0^{T_s} |s(t)|^2\, dt \\ &= \sum_{n=0}^{N-1} |s_n|^2 \end{aligned} \tag{G.3}$$

where the summation in the second line of the equation follows for *Parseval's theorem*, discussed in Chapter 2. With the OFDM coefficient s_n being a random variable, which it is in a wireless environment, it follows that the time-averaged power $\bar{P}$ is itself a random variable. It follows therefore that the *ensemble-averaged power* of the OFDM signal is given by the expectation

$$\begin{aligned} P_{av} &= \mathbb{E}[\bar{P}] \\ &= \mathbb{E}[|s(t)|^2] \quad \text{for} \quad 0 \le t \le T_s \end{aligned} \tag{G.4}$$

In an OFDM signal based on M-ary PSK, for example, we have $|s_n| = 1$ for all n. In this special case, (G.4) yields

$$P_{av} = N \tag{G.5}$$

As pointed out previously, the metric of interest commonly used in the literature for assessing the issue of statistical peak-power variations in the use of OFDM for wireless communications is the peak-to-average power ratio (PAPR), for which we offer the following definition:

$$\begin{aligned} \xi &= \frac{\max_{0 \le t \le T_s} |s(t)|^2}{P_{av}} \\ &= \frac{\max_{0 \le t \le T_s} |s(t)|^2}{\mathbb{E}[|s(t)|^2]} \end{aligned} \tag{G.6}$$

where, in words, the term in the numerator denotes the maximum value of the instantaneous power (i.e., peak power) of the OFDM signal measured across the symbol interval, $0 \le t \le T_s$, and the denominator denotes average power, hence PAPR. The formula used in (G.6) refers to the baseband formulation of the PAPR problem.[2]

Recognizing that PAPR is, in reality, a random variable distributed across each OFDM symbol, a statistical interpretation of it is useful. To this end, we may express the probability of the event that an OFDM symbol, denoted by $s(t)$ as defined in (G.1), exceeds the *peak value* ξ_p with probability P_c as follows:

$$\mathbb{P}[\xi > \xi_p] = P_c \tag{G.7}$$

To expand on this definition, we say that the PAPR is less than some prescribed value ξ_p for $100(1 - P_c)$ of the OFDM symbols, in which case we may refer to $100(1 - P_c)$ as a percentile PAPR.

G.2 Maximum PAPR in OFDM Using *M*-ary PSK

Consider an OFDM system based on M-ary PSK for its modulation scheme. For this special application of OFDM, the PAPR is always less than or equal to N, where N is the number of subchannels. To justify this statement, we first note that for M-ary PSK,

$$|s_n| = 1 \quad \text{for} \quad 1 \le n \le N$$

Hence, PAPR is lower banded as follows:

$$\xi > 1 \tag{G.8}$$

For the upper band on the PAPR under M-ary PSK, we may write:

$$\begin{aligned}\xi &= \max_{0 \le t \le T_s} \left| \frac{1}{\sqrt{N}} \sum_{n=0}^{N-1} s_n \exp(\mathrm{j}2\pi n \Delta f t) \right|^2 \\ &= \frac{1}{N} \max_{0 \le t \le T_s} |s(t)|^2 \\ &\le \frac{1}{N} \left(\sum_{n=0}^{N-1} |s_n \exp(\mathrm{j}2\pi n \Delta f t)| \right)^2 \\ &\le N\end{aligned} \tag{G.9}$$

We may therefore go on to say:

In an OFDM system using M-ary PSK for modulation, the PAPR is bounded as follows:

$$1 < \xi \le N \tag{G.10}$$

where N is the number of subchannels in the system.

EXAMPLE 1 **PAPR for OFDM Using *M*-ary PSK**

Consider the example of an OFDM system using M-ary PSK for which $M = 8$; that is, the number of subchannels is

$$N = 2^8 = 256$$

For such an OFDM system, the upper bound on the PAPR, expressed in decibels, can be as high as the value

$$\begin{aligned}10 \log_{10}(256) &= 10 \times 8 \times \log_{10}(2) \\ &= 10 \times 8 \times 3.01 \\ &= 24.08 \text{ dB}\end{aligned}$$

The possibility that the PAPR attains such an upper band is inversely proportional to 2^N, where N is the number of subchannels. It follows that, fortunately in practice, the probability that the upper bound in (G.10) is attained is negligibly small when N is large (Tellambura and Friese, 2006).

G.3 Clipping-Filtering: A Technique for PAPR Reduction

From the discussion just presented, we clearly see the need for reducing the PAPR for commercial viability of OFDM in wireless communications.[3]

Considering the nature of envelope variations in the OFDM signal $s(t)$ (these are responsible for the PAPR problem), an obvious approach for addressing this problem is to do the following:

- First, clip $s(t)$, such that its envelope is limited to a certain desired maximum value.
- Second, use a linear filter so as to reduce the distortion produced by the clipping.

A system configuration for this PAPR-reduction scheme may proceed as follows: the OFDM modulator constitutes the first functional block of the system, followed by the envelope-peak clipper, then a linear filter, and finally an up-converter for translating the complex baseband signal into a real-valued RF signal ready for transmission over the wireless channel.

For the clipping, we may consider two types of nonlinear devices: complex baseband hard clipper and high-power transistor amplifier. Now, when a modulated signal is passed through a nonlinear device, two forms of distortion arise, namely[4]

- amplitude modulation-to-phase modulation (AM/PM) conversion and
- amplitude modulation-to-amplitude modulation (AM/AM) conversion.

The above-mentioned nonlinear devices are of practical interest because the AM/PM conversion can be eliminated almost completely through the use of a suitable pre-distorter. However, the AM/PM conversion remains to be an issue of concern. Specifically, the process of AM/AM conversion results in the production of two kinds of distortion:

- out-of-band (OOB) distortion and
- in-band (IB) distortion,

which are related; in any event, they both can be viewed as another source of noise. The IB noise cannot be reduced by filtering and, therefore, results in a degradation of error performance. The OOB noise can be reduced by the filter but also causes the "regrowth" of some original peaks. To reduce the overall regrowth of signal peaks, we may repeat the operation of clipping followed by filtering.

As mentioned previously, high peak values are extremely rare; in particular, a PAPR greater than 14 dB is almost impossible. Consequently, in typical wireless applications, we find that the use of clipping-filtering techniques can reduce the PAPR down to about 10 dB and yet maintain OOB noise at acceptable levels.[5]

Notes

1. The discussion on the PAPR problem presented herein closely follows the chapter article in Tellambura and Friese (2006). Another review paper of interest is Han and Lee (2005).

2. In the context of (G.6), strictly speaking, $|s(t)|^2$ is the envelope but not the transmitted signal; as such, (G.5) embodies the peak-to-mean envelope power ratio (PMEPR). Nevertheless, the PAPR is the term commonly used in the literature.

3. In a way, this same statement also applies to the use of discrete multitone modulation (DMT) for digital subscriber lines (DSLs) in baseband data transmission, which was discussed in Chapter 8.

4. The AM/PM and AM/PM conversions in power amplifiers are considered in Appendix H.

5. Further reduction in PAPR can be accomplished through the use of sophisticated modulation and coding techniques; for a discussion of these and other PAPR-reduction techniques, see Tellambura and Friese (2006). Unfortunately, there is no single "best" technique for solving the PAPR-reduction problem.

APPENDIX

H Nonlinear Solid-State Power Amplifiers

One of the most critical constraints imposed on the design of hand-held devices (terminals) in mobile radio communications is that of *limited battery power*. These devices are designed for the purpose of a certain battery life or time taken for recharging the battery; the corresponding electronic circuitry must therefore respect the underlying power budget. Moreover, a significant consumer of power in mobile radio is the *transmit power amplifier*. Attention must therefore be paid to solid-state power amplifiers in mobile radio, hence this appendix.

Another point to keep in mind is that power amplifiers are inherently *nonlinear*, regardless of where they are used in the design of communication systems. In this context we may classify nonlinearities into one of two types:

- low-pass or band-pass;
- memoryless or with memory.

In this appendix, we focus attention on *band-pass nonlinearities*.

H.1 Power Amplifier Nonlinearities

There are many amplifier designs, and they have been traditionally categorized in the electronics literature as Class A, Class B, Class AB, Class C, Class D, and so on, typically increasingly nonlinear. Although Class A is considered to be a linear amplifier, no amplifier is truly linear; what linearity means in this context is that the operating point is chosen such that the amplifier behaves linearly over the signal range. The drawback of the Class A amplifier is that it is power inefficient. Typically, 25% or less of the input power is actually converted to radio-frequency (RF) power; the power that is left is converted to heat and, therefore, wasted. The remaining amplifier classes are designed to provide increasingly improved power efficiency, but at the expense of making the amplifier increasingly more nonlinear.

Figure H.1 shows the measured gain characteristic of a solid-state power amplifier at two different frequencies: 1626 GHz and 1643 GHz. The curves show that the amplifier gain is approximately constant; that is, the amplifier is linear over a wide range of inputs. However, as the input level increases, the gain decreases, indicating that the amplifier is saturating. It can also be seen that there is a significant difference in amplifier performance at different frequencies. If this amplifier is operated at an average input level of –10 dBm with an amplitude swing of ±2 dB, then the amplifier would be considered linear. If, however, the input signal has an amplitude swing of ±10 dB, the amplifier would be considered nonlinear. The fact that the gain is not constant over all input levels means that the amplifier introduces *amplitude distortion* in the form of amplitude modulation (AM). Since the amplitude distortion depends upon the input level, it is typically referred to as *AM-to-AM conversion*.

An ideal amplifier does not affect the phase of an input signal, except possibly for a constant phase rotation. Unfortunately, a practical amplifier behaves quite differently, as illustrated in Figure H.2, which shows the phase characteristic of the same power amplifier

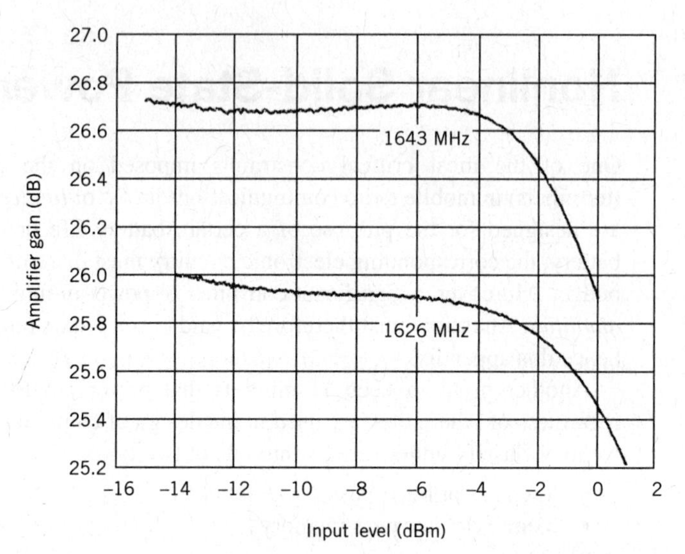

Figure H.1 Gain characteristic of a solid-state amplifier at two different operating frequencies: 1626 MHz and 1643 MHz.

considered in Figure H.1. The fact that the phase characteristic is not constant over all input levels means that the amplifier introduces *phase distortion* in the form of phase modulation (PM). Since the phase distortion depends upon the input level, this second form of distortion is typically called *AM-to-PM conversion*.

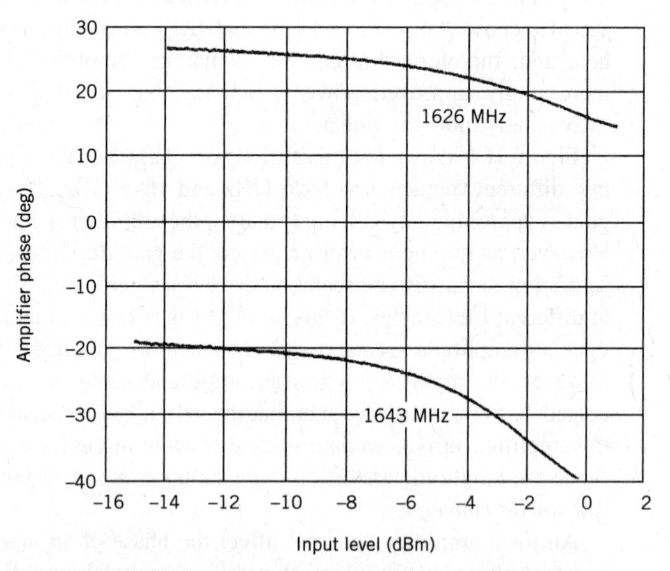

Figure H.2 Phase characteristic of a nonlinear amplifier at two different operating frequencies: 1626 MHz and 1643 MHz.

An amplifier with "ideal" nonlinearity acts linearly up to a given point, whereafter it sets a hard limit on the input signal. This can sometimes be achieved by placing appropriate compensation around a nonideal amplifier. With this ideal nonlinearity, the phase distortion is assumed to be zero. In reality, however, we have amplitude distortion as well as phase distortion, as illustrated in Figure H.3. The operating point of the amplifier is often specified as the *input back-off*, defined as the root-mean-square (rms) input signal level $V_{\text{in, rms}}$ relative to the saturation input level $V_{\text{in, sat}}$ in decibels. That is, we define

$$\text{Input back-off} = 10\log_{10}\left(\frac{V_{\text{in, rms}}}{V_{\text{in, sat}}}\right)^2 \tag{H.1}$$

Alternatively, the operating point can be expressed in terms of the *output back-off*, defined as

$$\text{Output back-off} = 10\log_{10}\left(\frac{V_{\text{out, rms}}}{V_{\text{out, sat}}}\right)^2 \tag{H.2}$$

where $V_{\text{out, rms}}$ is the rms output signal and $V_{\text{out, sat}}$ is the saturation output level. In both (H.1) and (H.2), the closeness to saturation determines the amount of distortion introduced by the amplifier.

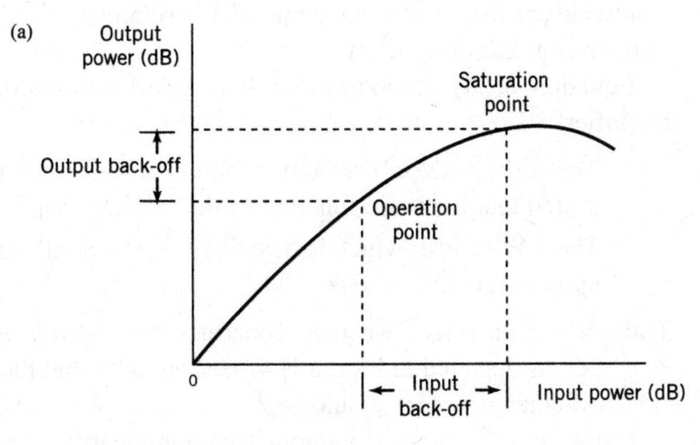

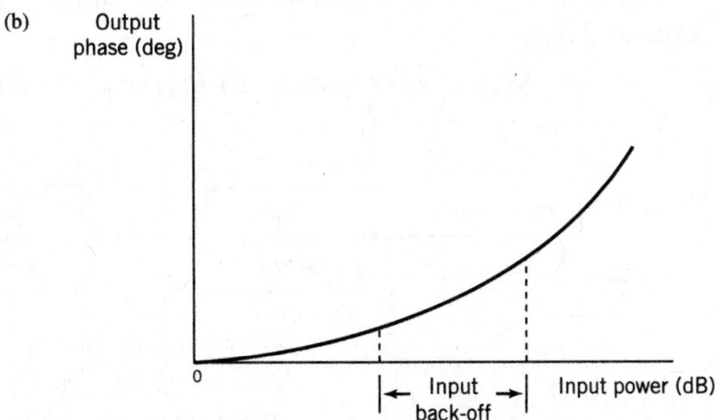

Figure H.3 Characterization of post-amplifier nonlinearity. (a) AM–AM conversion. (b) AM–PM conversion.

Thus, the *operating point* of the amplifier can be expressed in terms of the input back-off (IBO), defined as the input power measured relative to the saturation input level, both in decibels. Alternatively, it is expressed in terms of the *output back-off (OBO)*, defined as the output power measured relative to the saturation output level, again both in decibels.

H.2 Nonlinear Modeling of Band-Pass Power Amplifiers

Consider a band-pass power amplifier, producing measurable output in response to band-pass inputs. In practice, we typically find that characterization of the amplifier is achieved by performing measurements on it, and then using the measurements to formulate an empirically based model.

With this empirical approach to the nonlinear modeling of the power amplifier in mind, let the hybrid modulated signal

$$x(t) = a(t)\cos(2\pi f_c t + \theta(t)) \tag{H.3}$$

be applied to the input of the amplifier, producing the output

$$y(t) = g(a(t))\cos[2\pi f_c t + \theta(t) + \phi(a(t))] \tag{H.4}$$

where $g(\cdot)$ and $\phi(\cdot)$ are nonlinear functions of their respective arguments. This input–output relationship characterization of the amplifier is justifiable provided that the bandwidth of the modulated signal $x(t)$ is relatively small, compared with the bandwidth of the power amplifier itself.

Equation (H.4) embodies the two basic conversion characteristics of the power amplifier:

1. The AM-to-AM conversion, which is described by the nonlinear amplitude function $g(a(t))$ that is an odd function of the original amplitude $a(t)$.
2. The AM-to-PM, which is described by the nonlinear phase function $\phi(a(t))$ that is an even function of $a(t)$.

Thus, based on (H.4), we may construct the *cascade nonlinear model* of a band-pass amplifier, as depicted in Figure H.4. Herein, note that the AM/PM converter precedes the AM/AM converter, as it should be.

Using a well-known trigonometric nonlinearity, we may reformulate (H.4) in the expanded form

$$y(t) = y_I(t)\cos(2\pi f_c t + \theta(t)) - y_Q(t)\sin(2\pi f_c t + \theta(t)) \tag{H.5}$$

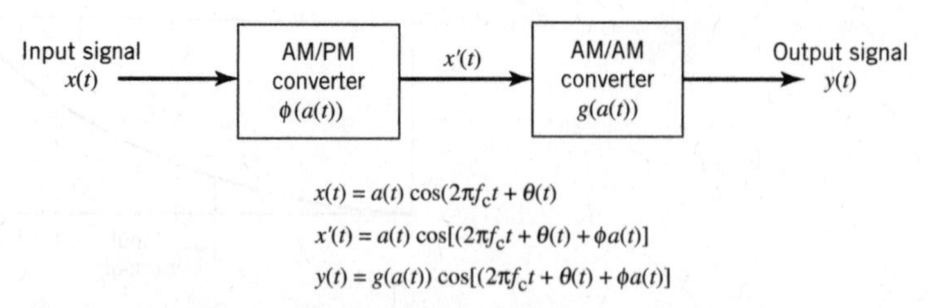

Figure H.4 Cascade nonlinear model of a band-pass power amplifier, driven by a hybrid-modulated input signal.

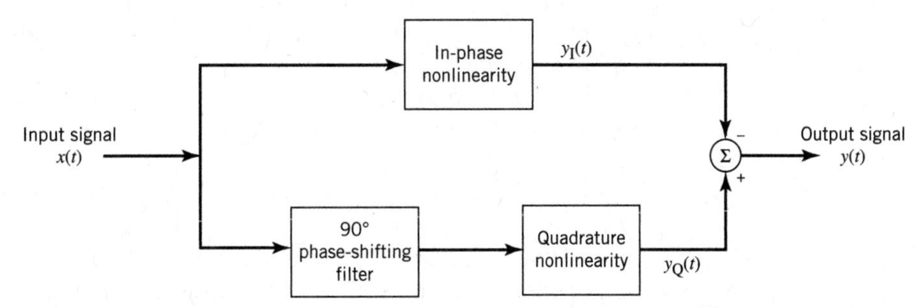

Figure H.5 Quadrature nonlinear model of a band-pass power amplifier driven by a hybrid-modulated signal.

For the *in-phase component* of the power amplifier output we have

$$y_1(t) = g(a(t)) \cos(\theta(t)) \tag{H.6}$$

and for its *quadrature component* we have

$$y_Q(t) = g(a(t)) \sin(\theta(t)) \tag{H.7}$$

Based on this second characterization of the power amplifier given in (H.7), we may construct the *quadrature nonlinear model* of the amplifier, depicted in Figure H.5. With the availability of such a model, the road is paved for Monte Carlo simulations to study the nonlinear behavior of solid-state power amplifiers that are of the band-pass variety.[2]

Notes

1. A model described in (Saleh, 1981) is well-suited for studying the in-phase and quadrature components of the output produced by a nonlinear power amplifier.

2. For detailed discussion of band-pass nonlinearity in power amplifiers, the reader is referred to the book (Tranter *et al.* 2004).

APPENDIX

I Monte Carlo Integration

In a generic sense, *Monte Carlo simulation*[1] is an invaluable experimental tool for tackling difficult problems that are mathematically intractable; but the tool is imprecise in that it provides statistical estimates. Nevertheless, provided that the Monte Carlo simulation is conducted properly, valuable insight into a problem of interest is obtained, which would be difficult otherwise.

In this appendix, we focus on *Monte Carlo integration*, which is a special form of Monte Carlo simulation. Specifically, we address the difficult integration problem encountered in Chapter 5 dealing with computation of the differential entropy $h(Y)$, based on the conditional probability density function of (5.102) in Chapter 5.

To elaborate, we may say:

> Monte Carlo integration is a computational tool, which is used to integrate a given function defined over a prescribed area of interest that is not easy to sample in a random and uniform manner.

Let W denote the difficult area over which random sampling of the differential entropy $h(Y)$ is to be performed. To get around this difficulty, let V denote an area so configured that it incudes the area W and is easy to randomly sample. Desirably, the selected area V enclosed W as closely as possible for the simple reason that samples picked outside of W are of no practical interest.

Suppose now we pick a total of N samples in the area V, randomly and uniformly. Then according to Press, *et al.* (1998), the basic *Monte Carlo integration theorem* states that a computed "estimate" of the integral defining the differential entropy $h(Y)$ is given by

$$h(Y) \approx V \times \langle h \rangle \pm V \times \left(\frac{1}{N}(\langle h^2 \rangle - \langle h \rangle^2)\right)^{\frac{1}{2}} \tag{I.1}$$

where the average value (i.e., mean)

$$\langle h \rangle = \frac{1}{N}\sum_{i=1}^{N} h(y_i) \tag{I.2}$$

and the mean-square value

$$\langle h^2 \rangle = \frac{1}{N}\sum_{i=1}^{N} h^2(y_i) \tag{I.3}$$

The y_i in (I.2) and (I.3) is the ith sample of the random variable Y picked from the area V. The "plus or minus" sign in the approximate formula of (I.1) should not be viewed as a rigorous bound. Rather, it represents a "one standard-deviation error" that results from the use of Monte Carlo integration.

Clearly, the larger we make the number of samples N, the smaller this error will be, resulting in a more accurate integration. However, this improvement is attained at the cost of increased computational complexity.

Notes

1. Monte Carlo simulation derives its name from the city, Monte Carlo, Monaco, which is widely known for its casino gambling: a "game of chance."
The term "Monte Carlo" was introduced into the technical literature by von Neumann and Ulam during World War II. Its adoption was intended as a codeword for the secret work that was going on the time in Los Alamos, New Mexico, USA.

APPENDIX

J Maximal-Length Sequences

Basically, maximal-length sequences, also referred to in the literature as *m-sequences*, are *linear cyclic codes*, the generation of which is realized by using a linear feedback-shift register (LFSR) as discussed in Chapter 10 on error-control coding; Figure J.1 is an illustrative example of LFSR. However, from a practical perspective insofar as this book is concerned, it is the pseudo-noise (PN) characteristic that befits their use in producing spread-spectrum signals, an issue that was discussed in Section 9.13 of Chapter 9. In short, a maximal-length sequence viewed as a "carrier" may be used to spread the spectrum of an incoming message sequence in the transmitter and despread the received signal so as to recover the original message signal at the receiver output.

It is therefore apropos that we begin the discussion of maximal-length sequences in this appendix by discussion their basic properties, illustrated by the LFSR as the sequence generator.

J.1 Properties of Maximal-Length Sequences

Maximal-length sequences[1] have many of the properties possessed by a truly *random binary sequence*. A random binary sequence is a sequence in which the presence of binary symbol 1 or 0 is equally probable. Maximal-length sequences have the following properties.

PROPERTY 1 **Balance Property**

In each period of a maximal-length sequence, the number of 1s is always one more than the number of 0s.

PROPERTY 2 **Run Property**

Among the runs of 1s and of 0s in each period of a maximal-length sequence, one-half the runs of each kind are of length one, one-fourth are of length two, one-eighth are of length three, and so on as long as these fractions represent meaningful numbers of runs.

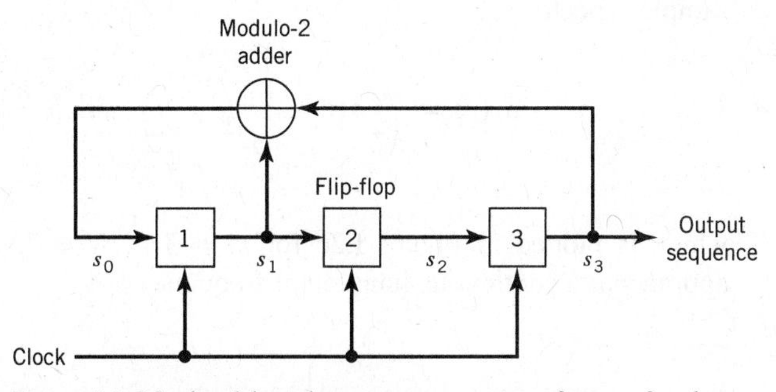

Figure J.1 Maximal-length sequence generator for $m = 3$, where m is the number of flip-flops in the generator.

By a "run" we mean a subsequence of identical symbols (1s and 0s) within one period of the sequence. The length of this subsequence is the length of the run. For a maximal-length sequence generated by a linear feedback shift register (LFSR) of length m, the total number of runs is $(N + 1)/2$, where $N = 2^m - 1$.

PROPERTY 3 Correlation Property

The autocorrelation function of a maximal-length sequence is periodic and binary valued.

As mentioned previously, the period of a maximum-length sequence is defined by

$$N = 2^m - 1 \tag{J.1}$$

where m is the length of the LFSR. Let binary symbols 0 and 1 of the sequence be denoted by the levels -1 and $+1$, respectively. Let $c(t)$ denote the resulting waveform of the maximal-length sequence, as illustrated in Figure J.2a for $N = 7$. Henceforth, the period of the waveform $c(t)$ is

$$T_b = NT_c \tag{J.2}$$

where T_c is the duration assigned to binary symbol 1 or 0 in the maximal-length sequence. Let $c(t)$ denote the maximal-length sequence, the autocorrelation function of which is defined by

$$R_c(\tau) = \frac{1}{T_b}\int_{-T_b/2}^{T_b/2} c(t)c(t-\tau)\,dt \tag{J.3}$$

where the lag τ lies in the interval $(-T_b/2, T_b/2)$. Applying this formula to $c(t)$, we get

$$R_c(\tau) = \begin{cases} 1 - \dfrac{N+1}{NT_c}|\tau|, & |\tau| \le T_c \\ -\dfrac{1}{N}, & \text{for the remainder of the period} \end{cases} \tag{J.4}$$

This result is plotted in Figure J.2b for the case of $m = 3$ or $N = 7$.

From Fourier transform theory, covered in Chapter 2, we know that periodicity in the time domain is transformed into uniform sampling in the frequency domain. This interplay between the time and frequency domains is borne out by the power spectral density of the maximal-length wave $c(t)$. Specifically, taking the Fourier transform of (J.4), we get the sampled spectrum

$$S_c(f) = \frac{1}{N^2}\delta(f) + \frac{1+N}{N^2}\sum_{\substack{n=-\infty \\ n \ne 0}}^{\infty} \text{sinc}^2\left(\frac{n}{N}\right)\delta\left(f - \frac{n}{NT_c}\right) \tag{J.5}$$

which is plotted in Figure J.2c for $m = 3$ or $N = 7$. As N approaches infinity, $S_c(f)$ approaches a continuous function of frequency f.

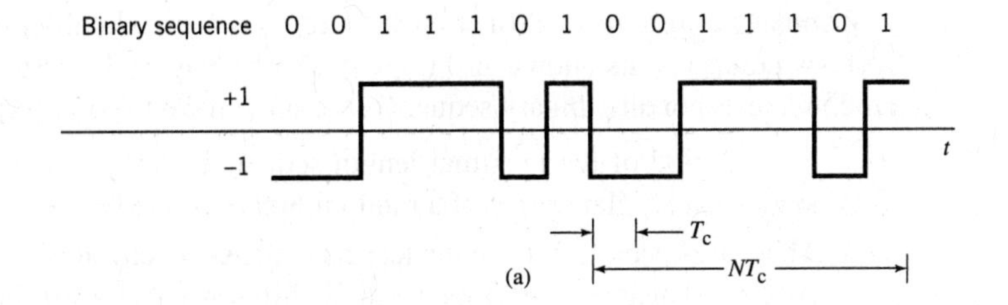

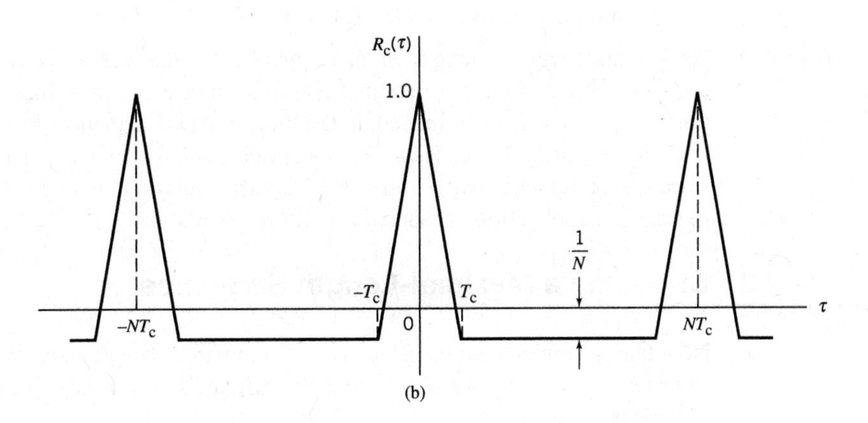

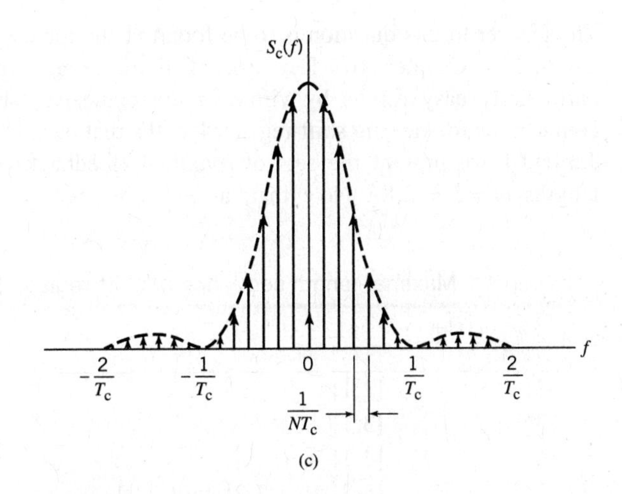

Figure J.2 (a) Waveform of maximal-length sequence for length $m = 3$ or period $N = 7$. (b) Autocorrelation function. (c) Power spectral density. All three parts refer to the output of the feedback shift register of Figure J.1.

Comparing the results of Figure J.2c for a maximal-length sequence with the corresponding results shown in Figure 4.12 of Chapter 4 on stochastic processes, for a random corresponding binary sequence, we may make two observations:

1. For a period of the maximal-length sequence, the autocorrelation function $R_c(\tau)$ is somewhat similar to that of a random binary sequence.
2. The waveforms of both sequences have the same envelope, $\text{sinc}^2(fT)$, for their power spectral densities. The fundamental difference between them is that whereas the random binary sequence has a continuous spectral density characteristic, the corresponding characteristic of a maximal-length sequence is discrete, consisting of delta functions spaced $(1/NT_c)$ Hz apart.

As the shift-register length m or, equivalently, the period N of the maximal-length sequence is increased, the maximal-length sequence becomes increasingly similar to the random binary sequence. Indeed, in the limit, the two sequences become identical when N is made infinitely large. However, the price paid for making N large is an increasing storage requirement, which imposes a practical limit on how large N can actually be made in practical applications of spread spectrum modulation.

J.2 Choosing a Maximal-Length Sequence

Now that we understand the properties of a maximal-length sequence and the fact that we can generate it using a linear feedback shift register, the key question that we need to address is:

> How do we find the feedback logic for a desired period N?

The answer to this question is to be found in the theory of error-control codes, which is covered in Chapter 10. The task of finding the required feedback logic is made particularly easy for us by virtue of the extensive tables of the necessary feedback connections for varying shift-register lengths that have been compiled in the literature. In Table J.1 we present the sets of maximal (feedback) taps pertaining to shift-register lengths $m = 2, 3, ..., 8$.[2] Note that, as m increases, the number of alternative schemes

Table J.1 Maximal-length sequence of shift-register lengths 2–8

Shift-register length, m	Feedback taps
2*	[2,1]
3*	[3,1]
4	[4,1]
5*	[5,2], [5,4,3,2], [5,4,2,1]
6	[6,1], [6,5,2,1], [6,5,3,2]
7*	[7,1], [7,3], [7,3,2,1], [7,4,3,2], [7,6,4,2], [7,6,3,1], [7,6,5,2], [7,6,5,4,2,1], [7,5,4,3,2,1]
8	[8,4,3,2], [8,6,5,3], [8,6,5,2], [8,5,3,1], [8,6,5,1], [8,7,6,1], [8,7,6,5,2,1], [8,6,4,3,2,1]

(codes) is enlarged. Also, for every set of feedback connections shown in this table, there is an "image" set that generates an identical maximal-length code, reversed in time sequence. Note also that the particular sets, identified with an asterisk in Table J.1, correspond to *Mersenne prime length sequences*, for which the period N is a prime number.

EXAMPLE 1 Maximal-Length Code Generation

Consider a maximal-length sequence requiring the use of a linear feedback-shift register of length $m = 5$. For feedback taps, we select the set [5,2] from Table J.1. The corresponding configuration of the code generator is shown in Figure J.3a. Assuming that the initial state is 10000, the evolution of one period of the maximal-length sequence generated by this scheme is shown in Table J.2, where we see that the generator returns to the initial 10000 after 31 iterations; that is, the period is 31, which agrees with the value obtained from (J.2).

Suppose, next, we select another set of feedback taps from Table J.1, namely [5,4,2,1]. The corresponding code generator is as shown in Figure J.3b. For the initial state 10000, we now find that the evolution of the maximal-length sequence is as shown in Table J.3. Here again, the generator returns to the initial state 10000 after 31 iterations, and so it should. But the maximal-length sequence generated is different from that shown in Table J.2.

Clearly, the code generator of Figure J.3a has an advantage over that of Figure J.3b, as it requires fewer feedback connections.

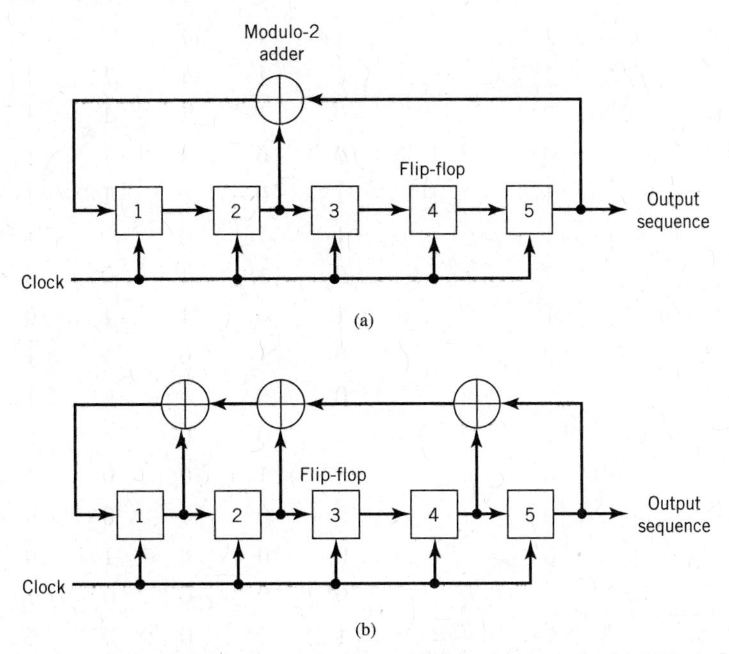

Figure J.3 Two different configurations of feedback shift register of length m = 5. (a) Feedback connections [5,2]. (b) Feedback connections [5,4,2,1].

Table J.2 **Evolution of the maximal-length sequence generated by the feedback-shift register of Figure J.3a**

Feedback Symbol	State of shift register					Output symbol
	1	0	0	0	0	
0	0	1	0	0	0	0
1	1	0	1	0	0	0
0	0	1	0	1	0	0
1	1	0	1	0	1	0
1	1	1	0	1	0	1
1	1	1	1	0	1	0
0	0	1	1	1	0	1
1	1	0	1	1	1	0
1	1	1	0	1	1	1
0	0	1	1	0	1	1
0	0	0	1	1	0	1
0	0	0	0	1	1	0
1	1	0	0	0	1	1
1	1	1	0	0	0	1
1	1	1	1	0	0	0
1	1	1	1	1	0	0
1	1	1	1	1	1	0
0	0	1	1	1	1	1
0	0	0	1	1	1	1
1	1	0	0	1	1	1
1	1	1	0	0	1	1
0	0	1	1	0	0	1
1	1	0	1	1	0	0
0	0	1	0	1	1	0
0	0	0	1	0	1	1
1	1	0	0	1	0	1
0	0	1	0	0	1	0
0	0	0	1	0	0	1
0	0	0	0	1	0	0
0	0	0	0	0	1	0
1	1	0	0	0	0	1

Code generated: 0000101011101100011111001101001.

Table J.3 Evolution of the maximal-length sequence generated by the feedback-shift register of Figure J.3b

Feedback symbol	State of Shift Register					Output symbol
	1	0	0	0	0	
1	1	1	0	0	0	0
0	0	1	1	0	0	0
1	1	0	1	1	0	0
0	0	1	0	1	1	0
1	1	0	1	0	1	1
0	0	1	0	1	0	1
0	0	0	1	0	1	0
1	1	0	0	1	0	1
0	0	1	0	0	1	0
0	0	0	1	0	0	1
0	0	0	0	1	0	0
1	1	0	0	0	1	0
0	0	1	0	0	0	1
1	1	0	1	0	0	0
1	1	1	0	1	0	0
1	1	1	1	0	1	0
1	1	1	1	1	0	1
1	1	1	1	1	1	0
0	0	1	1	1	1	1
1	1	0	1	1	1	1
1	1	1	0	1	1	1
0	0	1	1	0	1	1
0	0	0	1	1	0	1
1	1	0	0	1	1	0
1	1	1	0	0	1	1
1	1	1	1	0	0	1
0	0	1	1	1	0	0
0	0	0	1	1	1	0
0	0	0	0	1	1	1
0	0	0	0	0	1	1
1	1	0	0	0	0	1

Code generated: 0000110101001000101111101100111.

Notes

1. For further details on maximal-length sequences, see Golomb (1964: 1–32), Simon, *et al.* (1985: 283–295), and Peterson and Weldon (1972). The last reference includes an extensive list of polynomials for generating maximal-length sequences. For a tutorial paper on PN sequences, see Sarwate and Pursley (1980).

2. Table J.1 is extracted from the book by Dixon (1984: 81–83), where feedback connections of maximal-length sequences are tabulated for shift-register length m extending up to 89.

APPENDIX

K Mathematical Tables

Table K.1 Trigonometric identities

$$\exp(\pm j\theta) = \cos\theta \pm j\sin\theta$$

$$\cos\theta = \frac{1}{2}[\exp(j\theta) + \exp(-j\theta)]$$

$$\sin\theta = \frac{1}{2j}[\exp(j\theta) - \exp(-j\theta)]$$

$$\sin^2\theta + \cos^2\theta = 1$$

$$\cos^2\theta - \sin^2\theta = \cos(2\theta)$$

$$\cos^2\theta = \frac{1}{2}[1 + \cos(2\theta)]$$

$$\sin^2\theta = \frac{1}{2}[1 - \cos(2\theta)]$$

$$2\sin\theta\cos\theta = \sin(2\theta)$$

$$\sin(\alpha \pm \beta) = \sin\alpha\cos\beta \pm \cos\alpha\sin\beta$$

$$\cos(\alpha \pm \beta) = \cos\alpha\cos\beta \mp \sin\alpha\sin\beta$$

$$\tan(\alpha \pm \beta) = \frac{\tan\alpha \pm \tan\beta}{1 \mp \tan\alpha\tan\beta}$$

$$\sin\alpha\sin\beta = \frac{1}{2}[\cos(\alpha - \beta) - \cos(\alpha + \beta)]$$

$$\cos\alpha\cos\beta = \frac{1}{2}[\cos(\alpha - \beta) + \cos(\alpha + \beta)]$$

$$\sin\alpha\cos\beta = \frac{1}{2}[\sin(\alpha - \beta) + \sin(\alpha + \beta)]$$

Table K.2 **Series expansions**

Taylor series

$$f(x) = f(a) + \frac{f'(a)}{1!}(x-a) + \frac{f''(a)}{2!}(x-a)^2 + \cdots + \frac{f^{(n)}(a)}{n!}(x-a)^n + \cdots$$

where

$$f^{(n)}(a) = \left.\frac{\mathrm{d}^n f(x)}{\mathrm{d}x^n}\right|_{x=a}$$

MacLaurin series

$$f(x) = f(0) + \frac{f'(0)}{1!}x + \frac{f''(0)}{2!}x^2 + \cdots + \frac{f^{(n)}(0)}{n!}x^n + \cdots$$

where

$$f^{(n)}(0) = \left.\frac{\mathrm{d}^n f(x)}{\mathrm{d}x^n}\right|_{x=0}$$

Binomial series

$$(1+x)^n = 1 + nx + \frac{n(n-1)}{2!}x^2 + \cdots, \qquad |n| < 1$$

Exponential series

$$\exp x = 1 + x + \frac{1}{2!}x^2 + \cdots$$

Logarithmic series

$$\log(1+x) = x - \frac{1}{2}x^2 + \frac{1}{3}x^3 - \cdots$$

Trigonometric series

$$\sin x = x - \frac{1}{3!}x^3 + \frac{1}{5!}x^5 - \cdots$$

$$\cos x = 1 - \frac{1}{2!}x^2 + \frac{1}{4!}x^4 - \cdots$$

$$\tan x = x + \frac{1}{3}x^3 + \frac{2}{15}x^5 + \cdots$$

$$\sin^{-1} x = x + \frac{1}{6}x^3 + \frac{3}{40}x^5 + \cdots$$

$$\tan^{-1} x = x - \frac{1}{3}x^3 + \frac{1}{5}x^5 - \cdots, \qquad |x| < 1$$

$$\operatorname{sinc} x = 1 - \frac{1}{3!}(\pi x)^2 + \frac{1}{5!}(\pi x)^4 - \cdots$$

Table K.3 **Integrals**

Indefinite integrals

$$\int x\sin(ax)\,dx = \frac{1}{a^2}[\sin(ax) - ax\cos(ax)]$$

$$\int x\cos(ax)\,dx = \frac{1}{a^2}[\cos(ax) + ax\sin(ax)]$$

$$\int x\exp(ax)\,dx = \frac{1}{a^2}\exp(ax)(ax - 1)$$

$$\int x\exp(ax^2)\,dx = \frac{1}{2a}\exp(ax^2)$$

$$\int \exp(ax)\sin(bx)\,dx = \frac{1}{a^2 + b^2}\exp(ax)[a\sin(bx) - b\cos(bx)]$$

$$\int \exp(ax)\cos(bx)\,dx = \frac{1}{a^2 + b^2}\exp(ax)[a\cos(bx) + b\sin(bx)]$$

$$\int \frac{dx}{a^2 + b^2x^2} = \frac{1}{ab}\tan^{-1}\left(\frac{bx}{a}\right)$$

$$\int \frac{x^2\,dx}{a^2 + b^2x^2} = \frac{x}{b^2} - \frac{a}{b^3}\tan^{-1}\left(\frac{bx}{a}\right)$$

Definite integrals

$$\int_0^\infty \frac{x\sin(ax)}{b^2 + x^2}\,dx = \frac{\pi}{2}\exp(-ab), \qquad a > 0,\ b > 0$$

$$\int_0^\infty \frac{\cos(ax)}{b^2 + x^2}\,dx = \frac{\pi}{2b}\exp(-ab), \qquad a > 0,\ b > 0$$

$$\int_0^\infty \frac{\cos(ax)}{(b^2 - x^2)}\,dx = \frac{\pi}{4b^3}[\sin(ab) - ab\cos(ab)], \qquad a > 0,\ b > 0$$

$$\int_0^\infty \operatorname{sinc}(x)\,dx = \int_0^\infty \operatorname{sinc}^2(x)\,dx = \frac{1}{2}$$

$$\int_0^\infty \exp(-ax^2)\,dx = \frac{1}{2}\sqrt{\frac{\pi}{a}}, \qquad a > 0$$

$$\int_0^\infty x^2\exp(-ax^2)\,dx = \frac{1}{4a}\sqrt{\frac{\pi}{a}}, \qquad a > 0$$

Table K.4 **Useful constants**

Physical constants	
Boltzmann's constant	$k = 1.38 \times 10^{-23}$ J/K
Planck's constant	$h = 6.626 \times 10^{-34}$ J s
Electron (fundamental charge)	$q = 1.602 \times 10^{-19}$ C
Speed of light in vacuum	$c = 2.998 \times 10^{8}$ m/s
Standard (absolute) temperature	$T_0 = 273$ K
Thermal voltage	$V_T = 0.026$ V at room temperature
Thermal energy kT at standard temperature	$kT_0 = 3.77 \times 10^{-21}$ J
1 Hz = 1 cycle/s; 1 cycle = 2π radians	
1 W = 1 J/s	
Mathematical constants	
Base of natural logarithm	e = 2.7182818
Logarithm of *e* to base 2	$\log_2 e = 1.442695$
Logarithm of 2 to base e	$\log_2 e = 0.693147$
Logarithm of 2 to base 10	$\log_{10} 2 = 0.30103$
Pi	$\pi = 3.1415927$

Table K.5 **Recommended unit prefixes**

Multiples and submultiples	Prefixes	Symbols
10^{12}	tera	T
10^{9}	giga	G
10^{6}	mega	M
10^{3}	kilo	k
10^{-3}	milli	m
10^{-6}	micro	m
10^{-9}	nano	n
10^{-12}	pico	p

Glossary

Conventions and Notations

1. The symbol | | means the absolute value or magnitude of the complex quantity contained within.
2. The symbol arg( ) means the phase angle of the complex quantity contained within the brackets.
3. The symbol Re[ ] means the "real part of" and Im[ ] means the "imaginary part of" the complex quantity contained within the brackets.
4. The natural logarithm is denoted by ln.
5. Logarithms to bases 2 and 10 are denoted by $\log_2$ and $\log_{10}$, respectively.
6. The use of an asterisk as superscript denotes complex conjugate; e.g., x^* is the complex conjugate of x.
7. The symbol $\rightleftharpoons$ indicates a Fourier-transform pair, e.g., $g(t) \rightleftharpoons G(f)$, where a lowercase letter denotes the time function and a corresponding uppercase letter denotes the frequency function.
8. The symbol **F**[ ] indicates the Fourier-transform operation on a time function enclosed within the brackets, e.g., $\mathbf{F}[g(t)] = G(f)$.

   The symbol $\mathbf{F}^{-1}$[ ] indicates the inverse Fourier-transform operation of a frequency function enclosed within the brackets, e.g., $\mathbf{F}^{-1}[G(f)] = g(t)$.
9. The symbol ★ denotes convolution, e.g.,

$$x(t) \star (t) = \int_{-\infty}^{\infty} x(\tau)h(t-\tau)\, d\tau$$

10. In Chapter 10 on error-control coding, the symbol $\oplus$ is used in the figures, but when it comes to binary arithmetic, the modulo-2 addition is denoted by an ordinary plus sign throughout that chapter; the same statement applies to Appendix J on maximal-length codes.
11. The use of subscript T_0 indicates that the pertinent function $g_{T_0}(t)$, say, is a periodic function of time t with period T_0.
12. The use of a hat over a function indicates one of two things:
    a. The Hilbert transform of a function; e.g., the function $\hat{g}(t)$ is the Hilbert transform of $g(t)$.
    b. The estimate of an unknown parameter, e.g., the quantity $\hat{\alpha}(\mathbf{x})$ is an estimate of the unknown parameter α, based on the observation vector $\mathbf{x}$.
13. The impulse response of a linear time-invariant system is denoted by $h(t)$, and its transfer function is denoted by $H(f)$; the two of them, $h(t)$ and $H(f)$, form a Fourier-transform pair.
14. The use of a tilde over a function indicates the complex envelope of a narrowband signal; e.g., the function $\tilde{g}(t)$ is the complex envelope of the narrowband signal $g(t)$. The exception to this convention is in Section 10.12, where, in the description of turbo decoding, the tilde in

$\tilde{L}_i(m_j)$, is used to signify extrinsic information and thereby distinguish it from log-likelihood ratio.

15. The use of subscript + indicates the pre-envelope of a signal; e.g., the function $g_+(t)$ is the pre-envelope of the signal $g(t)$. We may thus write $g_+(t) = g(t) + j\hat{g}(t)$, where $\hat{g}(t)$ is the Hilbert transform of $g(t)$. The use of subscript – indicates that $g_-(t) = g(t) - j\hat{g}(t) = g_+^*(t)$.

16. The use of subscripts I and Q indicates the in-phase and quadrature components of a narrowband signal, a narrowband random process, or the impulse response of a narrowband filter, with respect to the carrier $\cos(2\pi f_c t)$.

17. For a low-pass message signal, the highest frequency component or message bandwidth is denoted by W. The spectrum of this signal occupies the frequency interval $-W \leq f \leq W$ and is zero elsewhere. For a band-pass signal with carrier frequency f_c, the spectrum occupies the frequency intervals $f_c - W \leq f \leq f_c + W$ and $-f_c - W \leq f \leq -f_c + W$, so $2W$ denotes the bandwidth of the signal. The (low-pass) complex envelope of this band-pass signal has a spectrum that occupies the frequency interval $-W \leq f \leq W$.

    For a low-pass filter, the bandwidth is denoted by B. A common definition of filter bandwidth is the frequency at which the magnitude response of the filter drops by 3 dB below the zero-frequency value. For a band-pass filter with mid-band frequency f_c the bandwidth is denoted by $2B$, centered on f_c. The complex low-pass equivalent of this band-pass filter has a bandwidth equal to B.

    The transmission bandwidth of a communication channel, required to transmit a modulated signal, is denoted by B_T.

18. Random variables or random vectors are uppercase (e.g., X or $\mathbf{X}$) and their sample values are lowercase (e.g., x or $\mathbf{x}$). The symbol $\mathbb{P}[\ ]$ signifies the probability of an event enclosed within the brackets; for example, $\mathbb{P}[X \leq x]$ signifies the probability that the occurence of random variable X assumes a value equal to or less than the sample value x.

19. A vertical bar in an expression means "given that" or "conditional on"; e.g., $f_X(x|H_0)$ is the probability density function of the random variable X given that hypothesis H_0 is true.

20. The symbol $\mathbb{E}[\ ]$ means the expected value of the random variable enclosed within; the $\mathbb{E}$ acts as an operator.

21. The symbol var[ ] means the variance of the random variable enclosed within.

22. The symbol cov[ ] means the covariance of the two random variables enclosed within.

23. The average probability of symbol error is denoted by P_e.

    In the case of binary signaling techniques, p_{10} denotes the conditional probability of error given that symbol 0 was transmitted, and p_{01} denotes the conditional probability of error given that symbol 1 was transmitted. The a priori probabilities of symbols 0 and 1 are denoted by p_0 and p_1, respectively.

24. The symbol $\langle\ \rangle$ denotes the time average of the sample function enclosed within.

25. Boldface letter denotes a vector or matrix. The inverse of a square matrix $\mathbf{R}$ is denoted by $\mathbf{R}^{-1}$. The transpose of a vector $\mathbf{w}$ is denoted by $\mathbf{w}^T$. The Hermitian transpose of a complex-valued vector $\mathbf{x}$ is denoted by $\mathbf{x}^\dagger$; Hermitian transposition involves both transposition and complex conjugation.

26. The length of a vector $\mathbf{x}$ is denoted by $\| \mathbf{x} \|$. The Euclidean distance between the vectors $\mathbf{x}_i$ and $\mathbf{x}_j$ is denoted by $d_{ij} = \| \mathbf{x}_i - \mathbf{x}_j \|$.

27. The inner product of two real-valued vectors **x** and **y** is denoted by $\mathbf{x}^T\mathbf{y}$; their outer product is denoted by $\mathbf{x}\mathbf{y}^T$. If the vectors **x** and **y** are complex valued, their inner product is $\mathbf{x}^\dagger\mathbf{y}$, and their outer product is $\mathbf{x}\mathbf{y}^\dagger$.

28. In set theory, the symbols $\cup$ and $\cap$ stand for the union and intersection, respectively, of two random variables A and B, for example.

    The symbol A^c stands for the complement of random variable A.

29. In stochastic processes theory, $M_{XX}(t_1, t_2)$ stands for the autocorrelation of a stochastic process $X(t)$ sampled at times t_1 and t_2 when no conditions are imposed on $X(t)$. In the special case of a weakly (wide-sense) stationary process $X(t)$, the autocorrelation function is denoted by $R_{XX}(\tau)$ for some time shift τ, and sometimes this symbol is simplified to $R_X(\tau)$; the time shift τ is also referred to as delay. Similar notations are used for cross-correlation, namely $M_{XX}(t_1, t_2)$ for a pair of generic stochastic processes $X(t)$ and $Y(t)$, and $R_{XY}(\tau)$ for the special case of two weakly (wide-sense) processes.

30. In information theory, the symbol $H(S)$ denotes the entropy of a discrete event S. For a continuous random variable denoted by X, the symbol $h(X)$ is used to denote its differential entropy.

    Given a pair of continuous random variables X and Y, their mutual information is denoted by $I(X; Y)$.

    Channel capacity is denoted by C.

31. In error-control coding, the code rate is denoted by r.

    The syndrome in decoding of linear block codes is denoted by S.

    In convolutional codes, the symbol $L(\mathbf{m}_j|\mathbf{r}_j)$ is used to denote the log-likelihood ratio of the message vector $\mathbf{m}_j$ given the received vector $\mathbf{r}_j$ at time-step j.

    For MAP (maximum a posteriori) decoding, the following symbols are used:

    - The L-value denotes a log-likelihood ratio of two conditional probabilities, the numerator pertaining to binary symbol 1 and the denominator pertaining to binary symbol 0.
    - $L_a(m_j)$ denotes the a priori L-value at time-step j of the decoding algorithm for message bit m_j.
    - $L_p(m_j)$ denotes the a posteriori L-value at time-step j of the decoding algorithm for message bit m_j.
    - L_c denotes the transmission reliability factor.
    - The symbols $\alpha_j(s)$, $\gamma_j(s, s')$, and $\beta_{j+1}(s')$ denote the forward metric for state S at time-step j, the transition metric for going from state s' to s at time-step j, and the backward metric for state s' at time-step $j+1$, respectively.

32. Lastly and rather importantly: to avoid confusion in the use of italics throughout the book, d is used to denote a differential and j is used to denote the square root of –1.

Functions

1. Rectangular functions:

$$\mathrm{rect}(t) = \begin{cases} 1, & -\frac{1}{2} < t < \frac{1}{2} \\ 0, & |t| > \frac{1}{2} \end{cases}$$

2. Unit-step function:

$$u(t) = \begin{cases} 1, & t > 0 \\ 0, & t < 0 \end{cases}$$

3. Signum function:

$$\mathrm{sgn}(t) = \begin{cases} 1, & t > 0 \\ 0, & t = 0 \\ -1, & t < 0 \end{cases}$$

4. (Dirac) delta function:

$$\delta(t) = 0, \quad t \neq 0$$

$$\int_{-\infty}^{\infty} \delta(t)\, dt = 1$$

or, equivalently,

$$\int_{-\infty}^{\infty} g(t)\delta(t - t_0)\, dt = g(t_0)$$

5. Sinc function:

$$\mathrm{sinc}(x) = \frac{\sin(\pi x)}{\pi x}$$

6. Sine integral:

$$\mathrm{Si}(u) = \int_0^u \frac{\sin x}{x} dx$$

7. Q-function:

$$Q(u) = \frac{1}{\sqrt{\pi}} \int_0^{\infty} \exp\left(-\frac{1}{2}t^2\right) dt$$

8. Binomial coefficient:

$$\binom{n}{k} = \frac{n!}{(n-k)!k!}$$

9. Bessel function of the first kind of order n:

$$J_n(x) = \frac{1}{2\pi} \int_{-\pi}^{\pi} \exp(\mathrm{j}x \sin\theta - \mathrm{j}n\theta)\, d\theta$$

10. Modified Bessel function of the first kind of zero order:

$$I_0(x) = \frac{1}{2\pi} \int_{-\pi}^{\pi} \exp(x \cos\theta)\, d\theta$$

Abbreviations

ADC	analog-to-digital converter
ADM	adaptive delta modulation
ADPCM	adaptive differential pulse-code modulation
ADSL	asymmetric digital subscriber line
AM	amplitude modulation
APP	a posteriori probability
ASK	amplitude-shift keying

AWGN	additive white Gaussian noise
BCJR	Bahl, Cocke, Jelinek, and Raviv (algorithm)
BER	bit error rate (chart)
BPF	band-pass filter
BSC	binary symmetric channel
cdf	cumulative distribution function
CDM	code-division multiplexing
CDMA	code-division multiple access
codec	coder/decoder
CPFSK	continuous-phase frequency-shift keying
CW	continuous wave
DAC	digital-to-analog converter
dB	decibel
dBW	decibel referenced to 1 W
dBmW	decibel reference to 1 mW
DC	direct current
DEM	demodulator
DFT	discrete Fourier transform
DM	delta modulation
DMT	discrete multitone
DPCM	differential pulse-code modulation
DPSK	differential phase-shift keying
DSB-SC	double sideband-suppressed carrier
DS/BPSK	direct sequence/binary phase-shift keying (for spread spectrum signals)
DSL	digital subscriber line
DTV	digital television
exp	exponential, e.g., e^x is written as $\exp(x)$; both are used interchangeably
FFT	fast Fourier transform (algorithm)
FIR	finite-duration impulse-response (filter)
FM	frequency modulation
FSK	frequency-shift keying
GMSK	Gaussian filtered MSK
Hz	hertz
IDFT	inverse discrete Fourier transform
IF	intermediate frequency
IFFT	inverse fast Fourier transform (algorithm)
IIR	infinite-duration impulse response (filter)
I/O	input/output
ISI	intersymbol interference
LDM	linear delta modulation
LFSR	linear finite-shift register
LMS	least-mean-square (algorithm)
ln	natural logarithm
$\log_2$	logarithm to base 2
$\log_{10}$	logarithm to base 10
LPC	linear predictive coding (model)

LPF	low-pass filter
MAP	maximum a posteriori (probability)
ML	maximum likelihood
mmse	minimum mean-square error
modem	modulator–demodulator
ms	millisecond
μs	microsecond
nm	nanometer
NRZ	nonreturn-to-zero
OFDM	orthogonal frequency-division multiplexing
OFDMA	orthogonal frequency-division multiple access
OOK	on–off keying
PAM	pulse-amplitude modulation
PAPR	peak-to-average power ratio
PCM	pulse-code modulation
pdf	probability distribution function
PG	processing gain
PSK	phase-shift keying
QAM	quadrature amplitude modulation
QPSK	quadriphase-shift keying
RC	raised cosine (spectrum)
RF	radio frequency
rms	root mean-square
RS	Reed–Solomon (code)
RSC	recursive systematic convolutional (code)
RZ	return-to-zero
s	second
SIR	signal-to-interference ratio
SNR	signal-to-noise ratio
SRRC	square-root raised cosine (spectrum)
TCM	trellis-coded modulation
TDL	tapped-delay line (filter)
TV	television
UHF	ultrahigh frequency
UMTS	Universal Mobile Telecommunication System
V	volt
W	watt
$\phi_X(x)$	characteristic function of random variable X with sample value x
π	interleaver
π^{-1}	de-interleaver

Bibliography

References

Abdi, A., and M. Kaveh. (2000) Performance comparison of three different estimators for the Nakagami m parameter using Monte Carlo Simulation. *IEEE Communication Letters*, 4, 119–121.

Abramowitz, M. and I.A. Stegun. (1965) *Handbook of Mathematical Functions with Formulas, Graphs, and Mathematical Tables,* New York, NY: Dover Publications.

Alamouti, A. (1998) A simple transmitter diversity scheme for wireless communications. *IEEE Journal on Selected Areas in Communications*, 16, 1451–1458.

Alexander, P.D., A.J. Grant, and M.C. Reed. (1998) Performance analysis of an iterative decoder for code-division multiple access. *European Transactions on Telecommunications*, 9, 419–426.

Anderson, J.B. (2005) *Digital Transmission Engineering,* 2nd ed. Piscataway, NJ: IEEE Press.

Anderson, R.R., and J. Salz. (1965) Spectra of digital FM. *Bell System Technical Journal*, 44, 1165–1189.

Arens, R. (1957) Complex processes for envelopes of normal noise. *IRE Transactions on Information Theory*, IT-3, 204–207.

Arthurs, E., and H. Dym. (1962) On the optimum detection of digital signals in the presence of white Gaussian noise—A geometric interpretation and a study of three basic data transmission systems. *IRE Transactions on Communication Systems*, CS-10, 336–372.

Bahl, L.R., J. Cocke, F. Jelinek, and J. Ravis. (1974) Optimal decoding of linear codes for minimizing symbol error rate. *IEEE Transactions on Information Theory*, IT-20, 284–287.

Bellamy, J.C. (1991) *Digital Telephony*, 2nd ed., New York: Wiley.

Bello, P.A. (1963) Characterization of randomly time-variant linear channels. *IEEE Transactions on Communication Systems*, CS-11, 360–393.

Benedetto, S., D. Divsalar, G. Montorsi, and F. Pollara. (1998) Analysis, design, and iterative decoding of double serial concatenated codes with interleavers. *IEEE Journal on Selected Areas in Communications,* 16: 231–244.

Benedetto, S., G. Montorsi. (1996) Iterative decoding of serially concatenated convolutional codes. *Electronics Letters*, 32: 1186–1188.

Benedetto, S., G. Montorsi. (1996) Unveiling turbo codes: some results on parallel concatenated coding schemes. *IEEE Transactions on Information Theory*, 42, 409–428.

Bennett, W.R. (1948) Spectra of quantized signals. *Bell System Technical Journal*, 27, 446–472.

Bennett, W.R., (1970) *Introduction to Signal Transmission*, New York: McGraw-Hill.

Berger, T. (1971) *Rate Distortion Theory: A Mathematical Basis for Data Compression,* Englewood Cliffs, NJ: Prentice-Hall.

Bernardo, J.M. and A.F.M. Smith (1998) *Bayesian Theory,* New York: Wiley.

Berrou, C., A. Glavieux, and P. Thitmajshima. (1993) Near Shannon limit error-correction coding and decoding: turbo codes. *International Conference on Communications*, 1064–1090, Geneva, Switzerland, May.

Bertsekas, P.D., and J.N. Tsitsiklis. (2008) *Introduction to Probability Theory,* 2nd ed. Nashua, NH: Athena Scientific.

Blahut, R.E. (1987) *Principles and Practice of Information Theory,* Reading, MA: Addison-Wesley.

Bornello, N., S. Chen, and L.L. Hanzo. (2011) Low-density parity-check codes and their rateless relatives. *Communications Surveys and Tutorials, IEEE,* 13, 3–25.

Bose, R.C., and D.K. Ray-Chaudhuri. (1960) On a class of error correcting binary group codes. *Information and Control*, 3, 68–79.

Bracewell, R.N. (1986) *The Fourier Transform and Its Applications,* 2nd ed. New York: McGraw-Hill.

Braun, W., and U. Dersch. (1991) Physical mobile radio channel model. *IEEE Transactions on Vehicular Technology,* 40, 472–482.

Brennan, D.G. (1959) Linear diversity combining techniques. *Proceedings of the IRE*, 47, 1075–1102.

Brigham, E.O. (1988) *The Fast Fourier Transform and its Applications*, Englewood Cliffs, NJ: Prentice-Hall.

Cavers, J.K. (2000) *Mobile Channel Characteristics*, Norwell, MA: Kluwer Academic Publishers.

Chaing, R.W., and J.C. Hancock. (1966) On receiver structures for channels having memory. *IEEE Transactions on Information Theory,* 12, 463–468.

Chang, R.W. (1966) Synthesis of band-limited orthogonal signals for multichannel data transmission. *Bell System Technical Journal,* 45, 1775–1796.

Chennakshu, S., and G.J. Saulnier. (1993) Differential detection of π/4 –shifted DQPSK for digital cellular radio. *IEEE Transactions on Vehicular Technology*, 42, 46–57.

Chung, S.Y., G.D. Forney, Jr., T.J. Richardson, and R. Urbanke. (2001) On the design of low-density parity-check codes within 0.0045 dB from the Shannon limit. *IEEE Communication Letters*, 5, 58–60.

Cimini, L.J., Jr., and Y. Li. (1999) Orthogonal frequency division multiplexing for wireless communications. *Tutorial Notes, TU 18, International Conference on Communications,* Vancouver, BC, Canada.

Cioffi, J.M., V. Oksman, J.-J Werner, T. Pollet, P.M.P. Spruyt, J.S. Chow, and K.S. Jacobsen. (1999) Very-high-speed digital subscriber lines. *IEEE Communications Magazine*, 37, 72–79.

Clark, G.C., Jr., and J.B. Cain. (1981) *Error-Correction Coding for Digital Communications,* New York, NY: Plenum Publishers.

Clarke, R. (1968) A statistical theory of mobile radio reception. *Bell System Technical Journal*, 47, 957–1000.

Costello, D.J., Jr. and G.D. Forney, Jr. (2007) Channel coding: the road to channel capacity. *Proceedings of the IEEE*, 95, 1146–1177.

Cover, T.M., and J.A. Thomas. (2006) *Elements of Information Theory*, 2nd, ed., Hoboken, NJ: Wiley.

Cramér, H., and M.R. Leadbetter. (1967) *Stationary and Related Stochastic Processes: Sample Function Properties and Their Applications*, New York: Wiley

deBuda, R. (1972) Coherent demodulation of frequency-shift keying with low deviation ratio. *IEEE Transactions on Communications*, COM-20, 429–435.

Dixon, R.C. (1984) *Spread Spectrum Systems*, 2nd ed., New York: Wiley.

Doelz, M.I., and E.H. Heald. (1961) Minimum shift data communication system, U.S. Patent 2977417, March 1961.

Doob, L.J. (1953) *Stochastic Processes,* New York: Wiley.

Dorny, C.N. (1975) *A Vector Space Approach to Models and Optimization*, New York: Wiley.

Douillard, C., M. Jezequel, C. Berrou. (1995) Iterative correction of intersymbol interference: turbo equalization. *European Transactions on Telecommunications*, 6: 507–511.

Dungundji, J. (1958) Envelopes and pre-envelopes of real waveforms. *IRE Transactions on Information Theory,* IT-4, 53–57.

Elias, P. (1955) Coding for noisy channels. *IRE Convention Record,* part 4, pp. 37–46.

Fine, T.L. (2006) *Probability and Probabilistic Reasoning for Electrical Engineering*, Upper Saddle River, NJ: Prentice-Hall.

Fisher, R.A. (1912) On the absolute criteria for frequency curves. *Messenger of Mathematics*, 41, 155–160.

Fisher, R.A. (1922) On the mathematical foundation of theoretical statistics. *Philosophical Transactions of the Royal Society of London*, Series A, 222, 309–368.

Forney, G.D., Jr. (1966) *Concatenated Codes*, Cambridge, MA: MIT Press.

Forney, G.D., Jr. (1972) Maximum likelihood sequence estimation of digital sequences in the presence of intersymbol interference. *IEEE Transactions on Information Theory,* IT-18, 363–378.

Forney, G.D., Jr. (1973) The Viterbi algorithm. *Proceedings of the IEEE*, 61, 268–278.

Foschini, G.J. (1996) Layered space-time architecture for wireless communication in a fading environment when using multi-element antennas. *Bell Labs Technical Journal*, Autumn, 41–59.

Frey, R.S. (1998) *Graphical Models for Machine Learning and Digital Communications*, Cambridge, MA: MIT Press.

Gabor, D. (1946) Theory of communications. *Journal of IEE (London)*, 93, Part III, 429–457.

Gallager, R.G. (1960) *Low-density Parity Check Codes, Sc.D. Thesis, MIT.*

Gallager, R.G. (1962) Low-density parity-check codes. *IRE Transactions on Information Theory,* IT-8, 21–28.

Gallager, R.G. (1963) *Low-density Parity-check Codes*, Cambridge, MA: MIT Press.

Gallager, R.G. (1968) *Information Theory and Reliable Communication*, New York: Wiley.

Gardner, W.A. (1987) Introduction to Einstein's contribution to time-series analysis. *IEEE ASSP Magazine,* 4, 4–5.

Gauss, C.F. (1809) *Theory of the Motion of Heavenly Bodies Moving About the Sun in Conic Sections,* a translation of Gauss's "Theoria Motus", with an appendix.

Gersho, A., and R.M. Gray (1992) *Vector Quantization and Signal Compression,* Boston, MA: Kluwer Academic Publishers.

Gibby, R.A., and J.W. Smith (1965). Some extensions of Nyquist's telegraph transmission theory. *Bell System Technical Journal*, 44, 1487–1510.

Gitlin, R.D., and E.Y. Ho (1975) The performance of staggered quadrature amplitude modulation in the presence of phase jitter. *IEEE Transactions on Communications*, COM-23, 348–352.

Gold, R. (1967) Optimal binary sequences for spread spectrum multiplexing. *IEEE Transactions on Information Theory,* IT-13, 619–621.

Gold, R. (1968) Maximal recursive sequences with 3-valued recursive cross correlation functions. *IEEE Transactions on Information Theory,* IT-14: 154–156.

Goldstein, A. (2005) *Wireless Communications*, New York: Cambridge University Press.

Golomb, S.W. (1967) *Shift Register Sequences,* San Francisco, CA: Holden-Day.

Greenwood, J.A. and D. Durant (1960) Aids for fitting the Gamma distribution by maximum likelihood. *Technometrics*, 2, 55–65.

Hadamard, J., (1893) Resolution d'une question relative aux determinants. *Bull. des Sci. Math.*, 17, 240–246.

Hagelbarger, D.W. (1959) Recurrent codes: easily mechanized, burst-correcting binary codes. *Bell System Technical Journal,* 38, 969–984.

Hagenauer, J. (2004) The EXIT chart-introduction to extrinsic information transfer in interactive processing, *EUSIPCO*, Vienna, Austria.

Hagenauer, J., and P. Hoher. (1989) A Viterbi decoding algorithm with soft-decision outputs and its applications, *Proceedings of the IEEE Globecom*, 47.11–47.17, Dallas, Texas.

Hamming, R.W. (1950) Error detecting and error correcting codes. *Bell System Technical Journal*, 29, 147–160.

Hamming, R.W. (1973) *Numerical Methods for Scientists and Engineers*, 2nd ed. New York, NY: McGraw-Hill.

Hanzo, L. (2012), Private Communication.

Hanzo, L., and T. Keller (2006) *OFDM and MC-CDMA: A primer*, Chichester, UK: Wiley.

Hanzo, L., L.–L Yang, E. Kuan, and K. Yen. (2003) *Single- and Multi-Carrier DS-CDMA: Multi-User Detection, Space-Time Spreading Synchronisation and Standards,* Chichester, UK: Wiley.

Hanzo, L., T.H. Liew, and B.L. Yeap (2002) *Turbo Coding, Turbo Equalization and Space-time Coding for Transmission over Fading Channels*, Chichester, UK: Wiley.

Harmuth, H.F. (1970) Transmission of Information by Orthogonal Functions, New York: Springer Verlag.

Hartley, T.V.L. (1928) Transmission of information. *Bell System Technical Journal*, 7, 535–563.

Haykin, S. (2001) *Communication Systems,* 4th ed. New York, NY: Wiley.

Haykin, S. (2002) *Adaptive Filter Theory*, 4th ed. Englewood Cliffs, NJ: Prentice-Hall.

Haykin, S. and B. Van Veen. (2002) *Signals and Systems*, 2nd ed., New York: Wiley.

Haykin, S., and M. Moher. (2005) *Modern Wireless Communications*, Upper Saddle River, NJ: Prentice Hall.

Hocquenghem, A. (1959) Codes coreteurs d'erreurs. *Chiffres*, 2, 147–156.

Holmes, J.K. (1982) *Coherent Spread Spectrum Systems*, New York: Wiley.

Huffman, D.A. (1952) A method for the construction of minimum redundancy codes. *Proceedings of the IRE*, 40, 1098–1101.

Ishizuka, M., and K. Hirade. (1980) Optimum Gaussian filler and deviation-frequency locking scheme for coherent detection of MSK. *IEEE Transactions on Communications*, COM-28, 850–857.

Jakes, W.C., Jr., ed. (1974) *Microwave Mobile Communications,* New York: Wiley.

Jayant, N.S. and P. Noll. (1984) *Digital Coding of Waveforms: Principles and Applications to Speech and Video,* Englewood Cliffs, NJ: Prentice-Hall.

Jeruchim, M.C., P. Balaban, and K.S. Shanmugan. (2000) *Simulation of Communication Systems*, 2nd ed., New York: Kluwer Academic Publishers.

Jiang, S., L. Ping, H. Sun, and C.S. Leung. (2004) Modified LMMSE turbo equalization. *IEEE Communications Letters*, 8, 174–176.

Johnson, J.B. (1928) Thermal agitation of electricity in conductors. *Physical Review, Second Series*, 32, 97–109.

Kammler, D.W. (2000) *A First Course in Fourier Analysis*, Upper Saddle River, NJ: Prentice-Hall.

Khintchin, A. (1957) *Mathematical Foundations of Information Theory,* New York: Dover Publications.

Khintchine, A.I. (1934) Korrelationstheorie der stationären stochastischen Prozesse. *Mathematische Annalen*, 1, 109, 415–458.

Koca, M., and B.C. Levy. (2004) *T*urbo space–time equalization of TCM for broadband wireless channels. *IEEE Transactions on Wireless Communications*, 2, 50–59.

Koetter, R., A.C. Singer, and M. Tüchler. (2004) Turbo equalization. *IEEE Signal Processing Magazine*, 21, 67–80.

Kotel'nikov, V.A. (1960) *The Theory of Optimum Noise Immunity,* New York: McGraw-Hill.

Kullback, S. (1968) *Information Theory and Statistics*, New York: Dover Publications.

Land, I. (2005) Reliability information in channel decoding-practical methods and information theoretical bounds, *Ph.D. Dissertation, University of Kiel, Germany.*

Land, I., P.A. Hoeher, and S. Gilgorevic. (2004) Computation of symbol-wise mutual information in transmission systems with log APP decoders and applications to EXIT charts, *International ITG conference on source and channel coding*, Erlargon, Germany, 195–202

Lechleider, J.W. (1989) Line codes for digital subscriber lines. *IEEE Communications Magazine,* 27, 25–32.

Lee, E.A., and D.G. Messerschmilt. (1994) *Digital Communications*, Boston, MA: Kluwer Academic.

Li, L., (2012) Energy-efficient implementation of turbo codes for wireless sensor networking channel coding systems, *Ph.D. Thesis*, University of Southampton, England.

Lighthill, M.J. (1958) *Introduction to Fourier Analysis and Generalized Functions*, New York: Cambridge University Press.

Lim, D.-W., S.-J. Heo, and J.-S. No. (2009) An overview of peak-to-average power ratio reduction schemes for OFDM signals. *Journal of Communications and Networks*, 11, 229–239.

Lin, S. and D.J. Costello, Jr. (2004) *Error Control Coding: Fundamentals and Applications*, 2nd ed. Englewood Cliffs, NJ: Prentice-Hall.

Lin, S., and D.J. Costello, Jr. (1983) *Error-control Coding: Fundamentals and Applications*, Englewood Cliffs, NJ: Prentice-Hall.

Lindsey, W.C. (1972) Synchronization Systems in Communication and Control, Englewood Cliffs, NJ: Prentice-Hall.

Lindsey, W.C. and M.K. Simon. (1973) *Telecommunication System Engineering*, Englewood Cliffs, NJ: Prentice-Hall.

Lloyd, S.P. (1957) Least squares quantization in PCM, Unpublished Bell Laboratories Technical Note. This Report was reprinted in *IEEE Transactions on Information Theory,* IT-28, 129–137, 1982.

Loéve, M. (1963) *Probability Theory,* Princeton, NJ: Van Nostrand.

Lopes, R.R., and J.R. Barry. (2006) *The soft-feedback equalizer for turbo equalization of highly dispersive channels. IEEE Transactions on Communications*, 54, 783–788.

Luby, M., M. Mitzenmacher, M.A. Shokrollahi, and D.A. Spielman. *IEEE Transactions on Information Theory,* 47, 585–598.

Luby, M., M. Mitzenmacher, M.A. Shokrollahi, and V. Stermann. (1997) Practical loss resilient codes, *Proceeding of the 29th Symposium: System Theory and Computing*, 150–159.

Luby, M., Mitzenmacher, M.A. Shokrollahi, and D.A. Spielman. (2001) Efficient erasure-correcting codes. *IEEE Transactions on Information Theory,* 47, 569–584.

Lucky, R.W. (1989) *Silicon Dreams: Information, Man, and Machine*, New York: St. Martin's Press.

MacKay, D.J.C. (1999) Good error-correcting codes based on very sparse matrices. *IEEE Transactions on Information Theory,* 45, 399–431.

MacKay, D.J.C. (2003) *Information Theory, Inference, and Learning Algorithms*, Cambridge, UK: Cambridge University Press.

MacKay, D.J.C., R.M. Neal. (1997) Near Shannon limit performance of low-density parity-check codes. *Electronics Letters*, 33, 457–458, and (1996), 32, 1645–1646.

Matthaiou, M., and D.I. Laurensen. (2007) Rejection method for generating Nagakami-m independent deviates. *Electronics Letters*, 43.

Maunder, R.G. (2012) Private communication.

Max, J. (1960) Quantizing for minimum distortion. *IRE Transactions on Information Theory*, IT-6, 7–12.

McDonough, R.M. and A.D. Whalen. (1995) *Detection of Signals in Noise*, 2nd ed., New York: Academic Press.

McEliece, R.J. (2004) *The Theory of Information and Coding*, Cambridge, UK: Cambridge University Press.

McEliece, R.J., D.J.C. MacKay, and J.F. Cheng. (1998) Turbo coding as an instance of Pearl's propogation algorithm. *IEEE Journal on Selected Areas of Communication*, 16, 140–152.

Mengali, A., and N. D'Andrea. (1997) *Synchronization Techniques for Digital Receivers*, New York: Plennum.

Meyer, H., and G. Ascheid. (1998) *Synchronization in Digital Communications*, vol. 1, New York: Wiley.

Molisch, A.F. (2011) *Wireless Communications*, Second Edition, Chichester, UK: Wiley.

Moon, T.K. (2005) *Error Correction Coding: Mathematical Methods and Algorithms*, Hoboken, NJ: Wiley.

Murota, K., and K. Hirade. (1981) GMSK modulation for digital mobile radio telephone. *IEEE Transactions on Communications*, COM-29, 1044–1050.

Nakagami, M. (1960) The m-distribution – a general formula of intensity distribution of rapid fading, in W.C. Hoffman, editor, *Statistical Methods of Radio Propagation*, 3–36, New York: Pergamon Press.

Nyquist, H. (1924) Certain factors affecting telegraph speed. *Bell System Technical Journal*, 3, 324–346.

Nyquist, H. (1928a) Thermal agitation of electric charge in conductors. *Physical Review*, second series, 32, 110–113.

Nyquist, H. (1928b) Certain topics in telegraph transmission theory. *Transactions of the American Institute of Electrical Engineers*, 47, 617–644.

Paley, R.E.A.C., and N. Wiener. (1934) Fourier transforms in the complex domain. *American Mathematical Society Colloquium Publication*, 19, 16–17.

Parsons, J.D. (2000) *The Mobile Radio Propagation Channel*, 2nd ed., New York: Wiley.

Pasupathy, S. (1974) Nyquist's Third criterion. *Proceedings of the IEEE*, 62, 860–861.

Pasupathy, S. (1979) Minimum shift keying—A spectrally efficient modulation. *IEEE Communications Magazine*, 17, no. 4, 14–22.

Pearl, J., (1988) *Probabilistic Reasoning in Intelligence Systems: Networks of Plausible Inference*, San Mateo, CA: Morgan Kaufman.

Peterson, W.W., and E.J. Weldon, Jr. (1972) Error Correcting Codes, 2nd ed., Cambridge, MA: MIT Press.

Pickholtz, R.L., D.L. Schilling, and L.B. Milstein. (1982) Theory of spread-spectrum communications – A tutorial. *IEEE Transactions on Communications*, COM-30, 855–884.

Press, W.H., B.P. Flannery, S.A. Teukolsky, and W.T. Vetterling. (1988) *Numerical Recipes in C: The Art of Scientific Computing*, Cambridge, UK: Cambridge University Press.

Price, R. and P.E. Green, Jr. (1958) A communication technique for multipath channels. *Proceedings of the IRE*, 46, 555–570.

Priestley M.B. (1981) *Spectral Analysis and Time Series, vol. 1: Univeriate Series*, London, UK: Academic Press.

Proakis, J.G. and M. Salehi. (2008) *Digital Communications*, 5th ed., New York: McGraw-Hill.

Rad, F.R., and J. Moon. (2005) *Turbo equalization utilizing soft decision feedback. IEEE Transactions on Magnetics*, 41, 2998–3000.

Reed, I.S., and G. Solomon. (1960) Polynomial codes over certain finite fields. *Journal of Society for Industrial and Applied Mathematics*, 8, 300–304.

Regalia, P.A. (2010) Turbo equalization, in *Adaptive Signal Processing: Next Generation Solutions,* edited by T. Adali and S. Haykin, Hoboken, NJ: Wiley, 143–210.

Rice, S.O. (1945) Mathematical analysis of random noise. *Bell System Technical Journal,* 24, 46–156.

Rice, S.O. (1948) Statistical properties of a sine-wave plus random noise. *Bell System Technical Journal,* 27, 109–157.

Rice, S.O. (1982) Envelopes of narrow-band signals. *Proceedings of the IEEE*, 70, 692–699.

Richardson, T.J., A. Shokrolliahi and R. Urbank. (2001) Design of capacity-approaching irregular low-density parity-check codes. *IEEE Transactions on Information Theory*, 47, 619–637.

Richardson, T.J., and R. Urbanke. (2001) The capacity of low-density parity-check codes under message-passing decoding. *IEEE Transactions on Information Theory*, 47, 599–618.

Rief, F. (1965) *Fundamentals of Statistical and Thermal Physics*, New York: McGraw-Hill.

Robert, C.R. (2001) *The Bayesian Choice*, New York: Springer.

Robertson, P. and T. Wörz. (1998) Bandwidth-efficient turbo trellis-coded modulation using punctured component codes. *IEEE Journal on Selected Areas in Communications*, 16, 206–218.

Robertson, P., (1994) Illuminating the structure of code and decoder of parallel concatenated recursive systematic (Turbo) codes, *Proceedings of the IEEE GLOBECOM*, 1298–1303, San Francisco, CA.

Saleh, A.A.M. (1981) Frequency-independent and frequency-dependent nonlinear models of TWT amplifiers. *IEEE Transactions on Communications*, 29, 1715–1720.

Sayar, B. and S. Pasupathy. (1987), Nyquist 3 pulse shaping in continuous phase modulation. *IEEE Transactions on Communications*, 35, 57–67.

Schlegel, C. (1997) *Trellis Coding*, 2nd ed., Piscataway, NJ: IEEE Press.

Scholtz, R.A. (1982) The origins of spread-spectrum communications. *IEEE Transactions on Communications*, COM-30, 822–854.

Schwartz, M., W.R. Bennett, and S. Stein. (1966) *Communication Systems and Techniques,* New York: McGraw-Hill.

Seberry, J. and M. Yamada. (1992) Hadamard matrices, sequences, and block designs, *Contemporary Design Theory: A Collection of Surveys,* edited by J.H. Diniz and D.R. Stinson, 431–469, New York: Wiley.

Shannon, C.E. (1948) A mathematical theory of communications. *Bell System Technical Journal,* 27, 379–423, 623–656.

Simon, M.K. and D. Divsalar. (1992) On the implementation and performance of angle and double differential detection

schemes. *IEEE Transactions on Communications*, 40, 278–291.
Simon, M.K., J.K. Omura, R.A. Scholtz, and B. Levitt. (1985) *Spread Spectrum Communications, vols I, II, and III*, New York: Computer Science Press.
Sklar, B. (2001) *Digital Communications: Fundamentals and Applications*, 2nd ed., Upper Saddle River, NJ: Prentice-Hall.
Slepian, D. (1974) *Key Papers in the Development of Information Theory*, Piscataway, NJ: IEEE Press.
Sloane, N.J.A., and A.D. Wyner. (1993) *Claude Shannon: Collected Papers*, Piscatataway, NJ:, IEEE Press.
Snyder, D.L. (1975) *Random Point Processes*, New York: Wiley.
Starr, T., J.M. Cioffi, and P.J. Silverman. (1999) *Understanding Digital Subscriber Line Technology* Englewood Cliffs, NJ: Prentice-Hall.
Stein, S. (1964) Unified analysis of certain coherent and noncoherent binary communication systems, *IEEE Transactions on Information Theory*, IT-10, 43–51.
Stüber, G.L. (1996) *Principles of Mobile Communication*, Boston: Kluwer Academic Publishers.
Sun, T.W., R.D. Wesel, M.R. Shane, and K. Jarett. (2004) Superposition turbo TCM for multirate broadcast. *IEEE Transactions on Communications*, 52, 368–371.
Sunde, R.D. (1969) *Communications Systems Engineering*, Eaglewood Cliffs, NJ: Prentice-Hall.
Supnithi, P., R. Lopes, and S.W. McLaughlin. (2003) Reduced-complexity turbo equalization for high-density magnetic recording systems. *IEEE Transactions on Magnetics*, 39, 2585–2587.
Suzuki, H. (1977) A statistical model for urban multipath propogation. *IEEE Transactions on Communications*, 25, 673–680.
Tanenbaum, A.S. (2003) *Communication Networks*, 3rd ed., Eaglewood Cliffs, NJ: Prentice-Hall.
Tanner, R.M. (1981) A recursive approach to low complexity codes. *IEEE Transactions on Information Theory*, IT-27, 533–547.
Tarokh, V., H. Jafarkhami, and A.R. Calderbank. (1999) Space-time block coding for wireless communications: Performance results. *IEEE Journal on Selected Areas in Communications*, 17, 451–460.
Telatar, L.E. (1995) Capacity of multi-antenna Gaussian Channels. *ATT-Bell Labs*, *Murray Hill*, Tech. Memo.
Tellambura, C. and M. Friese. (2006) Peak Power Reduction Techniques, in Y. Li, G. Stuber, and K. Wilson, editors, *OFDM for Wireless Communications*, New York: Springer.
ten Brink, S. (2001) Convergence behaviour of iteratively decoded parallel concatenated codes. *IEEE Transactions on Communications*, 49, 1727–1737.
Titchmarsh, E.C. (1950) *Introduction to the Theory of Fourier Integrals*, London: Oxford University Press.
Tranter, W.M., K.S. Shanmugan, T.S. Rappaport, and K.L, Kosbar. (2004) *Principles of Communication Systems Simulation with Wireless Applications*, Upper Saddle River, NJ: Prentice-Hall.
Ungerboeck, G. (1982) Channel coding with multilevel/phase signals. *IEEE Transactions on Information Theory*, IT-28, 55–67.
Ungerboeck, G. (1987) Trellis-coded modulation with redundant signal sets. Parts 1 and 2, *IEEE Communications Magazine*, 25 (2), 5–21.
Van Trees, H.L. (1968) *Detection, Estimation, and Modulation Theory, Part I.* New York: Wiley.
Van Trees, H.L. (1971) *Detection, Estimation, and Modulation Theory, Part III*, New York: Wiley.
Viterbi, A.J. (1967) Error bounds for convolutional codes and an asymptotically optimum decoding algorithm. *IEEE Transactions on Information Theory*, IT-13, 160–269.
Viterbi, A.J. (1995) *CDMA: Principles of Spread Spectrum Communication*, Englewood Cliffs: NJ, Prentice-Hall.
Walsh, J.L. (1923) A closed set of orthogonal functions. *American Journal of Mathematics*, 45, 5–24.
Wang, X. and H.V. Poor. (1999) Iterative (turbo) soft interference cancellation and decoding for coded CDMA. *IEEE Transactions on Communications*, 47, 1046–1050.
Watson, G.N. (1966) *A Treatise on the Theory of Bessel Functions*, New York: Cambridge University Press.
Weinstein, S.B., and P.M. Ebert. (1971) Data transmission by frequency-division multiplexing using the discrete Fourier transform. *IEEE Transactions on Communications*, COM-19, 628–634.
Wiberg, N., (1996) Codes and decoding on general graphs, *Ph.D. Dissertation*, Linkoping University, Linkoping, Sweden.
Wiberg, N., H.-A. Loeliger, and R. Kotter. (1995) Codes and iterative decoding on general graphs. *European Transactions of Telecommunications*, 6, 513–525
Wicker, S.B. and V.K. Bhargava, eds. (1994) *Reed–Solomon Codes.* Piscataway, NJ: IEEE Press.
Widrow, B. and I. Kollar. (2008) *Quantization Noise: Roundoff Error in Digital Computation, Signal Processing, Control, and Communications*, Cambridge, UK: Cambridge University Press.
Wiener, N. (1930) Generalized Harmonic Analysis. *Acta Mathmatica*, 55, 117–258.
Wozencraft, J.M. and I.M. Jacobs. (1965) *Principles of Communication Engineering.* New York: Wiley.
Wylie, C.R., and L.C. Barrett. (1999) *Advanced Engineering Mathematics*, New York: McGraw-Hill
Yaglom, A.M. (1962) *An Introduction to the Theory of Stationary Random Functions*, translated and edited by R.A. Silverman, Englewood Cliffs, NJ: Prentice-Hall.
Zhang, Q.T. (2000) Decomposition technique for efficient generation of correlated Nakagami fading channels. *IEEE Journal on Selected Areas in Communications*, 18, 2385–2392.
Zhang, Q.T. (2002) A note on the estimation of Nakagami-m fading parameter. *IEEE Communications Letters*, 6, 237–238.
Zhang, Q.T. (2003) A generic correlated Nakagami model for wireless communications, *IEEE Transactions on communications,* 51, 1745–1748.
Ziv, J. and A. Lempel. (1977) A universal algorithm for sequential data compression. *IEEE Transactions on Information Theory*, IT-23, 337–343.
Ziv, J. and A. Lempel. (1978) Compression of individual sequences via variable-rate coding. *IEEE Transactions on Information Theory*, IT-24, 530–536.

Further Reading

Books

Benedetto, S. and E. Biglieri. (1999) *Principles of Digital Transmission with Wireless Applications,* New York: Kluwer Academic Publishers.

Biglieri, E., D. Divsalar, P.J. McLane, and M.K. Simon. (1991) *Introduction to Trellis-Coded Modulation with Applications,* New York: Macmillan.

Blahut, R.E. (1990) *Digital Transmission of Information* Reading, MA: Addison-Wesley.

Feller, W. (1968) *An Introduction to Probability Theory and its Application*, Vol. 1, 3rd ed. New York: Wiley.

Kolmogorov, A.N. (1956) *Foundations of the Theory of Probability,* New York, NY: Chelsea Publishing.

Leon-Garcia, A. (1994) *Probability and Random Processes for Electrical Engineering*, 2nd ed. Reading, MA: Addison-Wesley.

Lucky, R.W., J. Salz, and E.J. Weldon, Jr. (1968) *Principles of Data Communication,* New York: McGraw-Hill.

Marks II, R.J. (1991) *Introduction to Shannon Sampling and Interpolation Theory,* New York: Springer-Verlag.

Shannon, C.E., and W. Weaver. (1949) *The Mathematical Theory of Communication,* Urbana, IL: University of Illinois Press.

Viterbi, A.J. and J.K. Omura. (1979) *Principles of Digital Communication and Coding,* New York: McGraw-Hill.

Wax, N. ed. (1954) *Selected Papers on Noise and Stochastic Processes,* New York: Dover Publications.

Wiener, N. (1949) *The Extrapolation, Interpolation, and Smoothing of Stationary Time Series, with Engineering Applications,* New York: Wiley.

Papers, Reports, Patents

Amoroso, F. (1980) The bandwidth of digital data signals. *IEEE Communications Magazine*, 18, no. 6, 13–24.

Berrou, C., and A. Glavieux. (1996) Near optimum error correcting coding and decoding: turbo codes. *IEEE Transactions on Communications*, 44, 1261–1271.

Berrou, C., and A. Glavieux. (1998) Reflections on the Prize Paper: Near optimum error-correcting coding and decoding turbo codes. *IEEE Information Theory Society Newsletter*, 48 (2) 1, 24–31.

Cutler, C.C. (1952) Differential quantization of communication signals. *US Patent 1-505-361.*

DeJager, F.E. (1952) Delta modulation, a method of PCM transmission using the 1-unit code. *Phillips Research Reports*, 7, 442–46.

Hagenauer, J. (1997) The turbo principle: tutorial introduction and state of the art. *Proceedings of 1st International Symposium on Turbo Codes,* Brest, France, 1–11.

Hagenauer, J., E. Offer, and L. Papke. (1996) Iterative decoding of binary block and convolutional codes. *IEEE Transactions on Information Theory*, 42, 429–445.

Hill, Jr., F.S. (1974) On time-domain representations for vestigial sideband signals. *Proceedings of the IEEE*, 62, 1032–1033.

Lodge, J., R. Young, P. Hoeher, and J. Hagenauer. (1993) Separable MAP 'filters' for the decoding of product and concatenated codes. *Proceedings of the IEEE International Conference on Communications*, Geneva, Switzerland, May, 1740–1745.

North, D.O. (1963) An analysis of the factors which determine signal/noise discrimination in pulsed carrier systems. *Proceedings of the IEEE*, 51, 1016–1027 (reprint of a classified RCA Report published in 1943).

Oliver, H. B.M., J.R. Pierce, and C.E. Shannon. (1948) The philosophy of PCM. *Proceedings of the IRE,* 36, 1324–1331.

Reeves, A.H. (1975) The past, present and future of PCM. *IEEE Spectrum*, 12 (5), 58–63.

Schouten J.S., F. DeJager, and J.A. Greefkes. (1952) Delta modulation, a new modulation system for telecommunication. *Phillips Technical Review*, 13, 237–245.

Shannon, C.E. (1976) Communication in the presence of noise. *Proceedings of the IRE*, 37, 10–21, 1949.

Slepian, S. (1976) On bandwidth. *Proceedings of the IEEE*, 64, 292–300.

Sundberg, C.E. (1986) Continuous phase modulation. *IEEE Communications Magazine*, 24 (4), 25–38,

Turin, G.L. (1960) An introduction to matched filters, *IRE Transactions on Information Theory*, IT-6, 311–329.

Wei, L.-F. (1987) Trellis-coded modulation with multidimensional constellations. *IEEE Transactions on Information Theory*, IT-33, 483–501.

Werner, J.J. (1992) Tutorial on carrierless AM/PM–Part I: Fundamentals and digital CAP transmitter. *AT&T Bell Laboratories Report*, June 23, 1992.

Werner, J.J. (1993) Tutorial on carrierless AM/PM–Part II: Performance of bandwidth-efficient line codes. *AT&T Bell Laboratories Report*, February 6, 1993.

Wyner, A.D. (1981) Fundamental limits in information theory. *Proceedings of the IEEE*, 69, 239–251.

Abbreviations:

IEE: Institution of Electrical Engineers, UK
IEEE: Institute of Electrical and Electronics Engineers, USA
IRE: Institute of Radio Engineers, USA
EUSIPCO: European Signal Processing Conference, Europe

Index

G

Q

R

S

T

Credits

Figures 2.1, 2.4, 2.13, 2.28, 2.30, 2.32, 2.33, 2.34, 2.35, 4.19, 4.20, 4.21, 4.24, P4.5, P4.16, P4.20, P4.21, P4.24, P4.28, 7.36, 7.37, 10.8, 10.9, 10.10, 10.11, 10.12: Haykin, Simon, *Communication Systems, 3rd Edition* © 1994 John Wiley & Sons, Inc. This material is reproduced with permission of John Wiley & Sons, Inc.

Figures 2.18, 2.19, 2.20, 2.22, 2.27, 2.31, 4.1, 4.2, 4.3, 4.4, 4.5, 4.6, 4.7, 4.8, 4.9, 4.11, 4.12, 4.14, 4.15, 4.16, 4.17, 4.18, 4.22, 4.23, Table 4.1, 5.1, 5.2, 5.3, 5.4, 5.5, 5.6, 5.7, 5.8, 5.10, 5.11, 5.12, 5.14, 5.15, 5.16, 5.17, 5.18, 5.19, 5.20, 5.21, P5.5, P5.15, P5.24, P5.25, P5.41, 6.1, 6.2, 6.3, 6.4, 6.5, 6.6, 6.7, 6.8, 6.9, 6.11, 6.14, 6.15, 6.21, 6.22, 6.23, 6.24, 6.25, 7.1, 7.2, 7.3, 7.4, 7.5, 7.6, 7.7, 7.8, 7.9, 7.10, 7.11, 7.12, 7.13, 7.14, 7.15, 7.16, 7.17, 7.18, 7.19, 7.20, 7.22, 7.24, 7.25, 7.26, 7.27, 7.28, 7.29, 7.30, 7.31, 7.32, 7.34, 7.35, 7.38, 7.39, 7.40, 7.41, 7.42, 7.43, 7.44, 7.45, 7.46, 7.47, 7.48, 7.49, 7.50, P7.3, P7.4, P7.5, P7.12, P7.13, P7.17, P7.41, 8.1, 8.2, 8.3, 8.4, 8.5, 8.6, 8.17, 8.18, 8.19, 8.20, 8.21, 8.22, 8.23, 8.24, 8.25, 8.26, 8.27, 8.28, 8.29, 8.30, P8.1, P8.14, 10.1 (a)(b), 10.2, 10.3 (a)(b)(c), 10.4, 10.5, 10.6, 10.7, 10.13, 10.14, 10.15, 10.16 (a)(b), 10.18, 10.19, 10.20, 10.25, 10.27, 10.28, 10.29 (a)(b), 10.34, 10.37, 10.38, 10.39, 10.40, 10.41, J.1, J.2, J.3: Haykin, Simon, *Communication Systems, 4th Edition* © 2001 John Wiley & Sons, Inc. This material is reproduced with permission of John Wiley & Sons, Inc.

Figures 2.2, 2.6, 2.7, 2.11, 2.14: Haykin, Simon, *Introduction to Analog and Digital Communications, 1st Edition* © 1989 John Wiley & Sons, Inc. This material is reproduced with permission of John Wiley & Sons, Inc.

Table 3.1: Reproduced from the book *Introduction to Probability, 2nd edition*, by D.P. Bertsekas and J.N. Tsitsiklis, with permission.

Figures 3.4, 3.7, 3.10, 3.11, 8.12, 8.13, 8.14, 8.15, 8.16: Haykin, Simon, *Introduction to Analog and Digital Communications, 2nd Edition* © 2007 John Wiley & Sons, Inc. This material is reproduced with permission of John Wiley & Sons, Inc.

Figure 6.1: Adapted from Bennett, 1948, with permission of AT&T.

Figures 7.33, 8.9, 8.10, 9.7a, 9.7b, 9.17, 9.18, 9.19, 9.20, 9.21, 9.22, 9.23, 9.24, 9.25, 9.26, 9.27, 9.28, 9.29, 9.30, 9.31, P9.18, P9.21, E.1, F.1, F.2, F.3, H.1, H.2: Haykin, Simon O.; Moher, Michael, *Modern Wireless Communications, 1st Edition*, © 2005. Reprinted by permission of Pearson Education, Inc., Upper Saddle River, NJ.

Figure 7.34: © 1981 IEEE. Reprinted, with permission, from Murota, K.; Hirade, K.; "GMSK Modulation for Digital Mobile Radio Telephony," Communications, IEEE Transactions on , vol.29, no.7, pp. 1044–1050, Jul 1981.

Figures 9.1, 9.4, 9.5: From Parsons, J.D., *The Mobile Radio Propagation Channel, 1st Edition*, 1992 John Wiley & Sons, Inc. This material is reproduced with permission of John Wiley & Sons, Inc.

Figure 9.11: Parsons, J.D., *The Mobile Radio Propagation Channel, 2nd Edition* © 2000 John Wiley & Sons, Inc. This material is reproduced with permission of John Wiley & Sons, Inc.

Figure 9.13: Van Trees, Harry, *Detection, Estimation, and Modulation Theory, Part III: Radar-Sonar Signal Processing and Gaussian Signals in Noise* © 2001 John Wiley & Sons, Inc. This material is reproduced with permission of John Wiley & Sons, Inc.